COMMON IONS AND THEIR CHARGES

Name	Symbol and charge
Aluminum	Al^{3+}
Ammonium	NH_4^+
Antimony(III) or antimonous	Sb^{3+}
Arsenic(III) or arsenious	As^{3+}
Barium	Ba^{2+}
Bismuth	Bi^{3+}
Calcium	Ca^{2+}
Cadmium	Cd^{2+}
Cerium(IV) or ceric	Ce^{4+}
Cerium(III) or cerous	Ce^{3+}
Cesium	Cs^+
Chromium(III) or chromic	Cr^{3+}
Chromium(II) or chromous	Cr^{2+}
Cobalt(III) or cobaltic	Co^{3+}
Cobalt(II) or cobaltous	Co^{2+}
Copper(II) or cupric	Cu^{2+}
Copper(I) or cuprous	Cu^+
Gallium(III)	Ga^{3+}
Gold(III) or auric	Au^{3+}
Gold(I) or aurous	Au^+
Hydrogen	H^+
Hydronium	H_3O^+
Iron(III) or ferric	Fe^{3+}
Iron(II) or ferrous	Fe^{2+}
Lead(II) or plumbous	Pb^{2+}
Lithium	Li^+
Magnesium	Mg^{2+}
Mercury(II) or mercuric	Hg^{2+}
Mercury(I) or mercurous	Hg_2^{2+}
Nickel(II)	Ni^{2+}
Potassium	K^+
Rubidium	Rb^+
Silver	Ag^+
Sodium	Na^+
Strontium	Sr^{2+}
Thallium(III) or thallic	Tl^{3+}
Thallium(I) or thallous	Tl^+
Tin(IV) or stannic	Sn^{4+}
Tin(II) or stannous	Sn^{2+}
Titanium(IV) or titanic	Ti^{4+}
Titanium(III) or titanous	Ti^{3+}
Vanadium	V^{3+}
Zinc	Zn^{2+}

Name	
Acetate	$C_2H_3O_2^-$
Arsenate	AsO_4^{3-}
Arsenite	AsO_3^{3-}
Benzoate	$C_7H_6O_2^{2-}$
Borate	BO_3^{3-}
Bromate	BrO_3^-
Bromide	Br^-
Carbonate	CO_3^{2-}
Chlorate	ClO_3^-
Chloride	Cl^-
Chlorite	ClO_2^-
Chromate	CrO_4^{2-}
Cyanate	CNO^-
Cyanide	CN^-
Dihydrogen phosphate	$H_2PO_4^-$
Fluoride	F^-
Hydride	H^-
Hexacyanoferrate(III) or ferricyanide	$Fe(CN)_6^{3-}$
Hexacyanoferrate(II) or ferrocyanide	$Fe(CN)_6^{4-}$
Hydrogen carbonate or bicarbonate	HCO_3^-
Hydrogen oxalate or bioxalate	$HC_2O_4^-$
Hydrogen phthalate or biphthalate	$HC_8H_4O_4^-$
Hydrogen sulfate or bisulfate	HSO_4^-
Hydrogen sulfide or bisulfide	HS^-
Hydrogen sulfite or bisulfite	HSO_3^-
Hydroxide	OH^-
Hypochlorite	ClO^-
Iodate	IO_3^-
Iodide	I^-
Monohydrogen phosphate	HPO_4^{2-}
Nitrate	NO_3^-
Nitrite	NO_2^-
Orthosilicate	SiO_4^{4-}
Oxalate	$C_2O_4^{2-}$
Oxide	O^{2-}
Perchlorate	ClO_4^-
Periodate	IO_4^-
Permanganate	MnO_4^-
Peroxide	O_2^{2-}
Phosphate	PO_4^{3-}
Pyrophosphate	$P_2O_7^{4-}$
Sulfate	SO_4^{2-}
Sulfite	SO_3^{2-}
Thiocyanate	SCN^-
Thiosulfate	$S_2O_3^{2-}$

The Holt Chemistry Program

Foundations of Chemistry *Toon and Ellis* *Second Edition*

Laboratory Experiments for Foundations of Chemistry
Teachers Edition for Foundations of Chemistry
Teachers Edition for Laboratory Experiments for
Foundations of Chemistry
Tests for Foundations of Chemistry

Modern Chemistry *Metcalfe, Williams, and Castka*
Chemistry Problems *Castka*
Radioactivity: Fundamentals and Experiments *Hermias and Joecile*
Scientific Experiments in Chemistry *Manufacturing Chemists Association*
Semimicro Chemistry *Debruyne, Kirk, and Beers*
Scientific Experiments in Environmental Pollution *Weaver (MCA)*

Holt Library of Science

Life and the Physical Sciences *Morowitz*
A Tracer Experiment *Kamen*
Viruses, Cells, and Hosts *Sigel and Beasley*
Photosynthesis *Rosenberg*
Chemistry in the Space Age *Gardner*

Some of John Dalton's chemical symbols

Ernest R. Toon
George L. Ellis

Foundations of Chemistry

CONSULTANT RUSSELL C. BOVIE, Science Department Chairman, Arcadia High School, Arcadia, California.

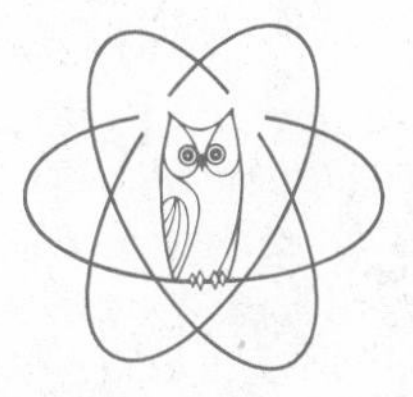

Holt, Rinehart and Winston, Inc.
New York • Toronto • London • Sydney

ERNEST R. TOON, Professor of Chemistry, Los Angeles Valley College, Van Nuys, California.

GEORGE L. ELLIS, Los Angeles City Schools, former Chemistry Instructor, North Hollywood High School, North Hollywood, California.

Acknowledgment of source appears with each photograph.

Cover: Growth trigons on natural diamond magnified 125 times. The crystal was photographed through a Nikon Françon-Yamamoto interference microscope by the distinguished photomicrographer Julius Weber of Mamaroneck, New York.

Printed in the United States of America
ISBN 0-03-088484-5
5 6 7 0 7 1 9 8 7 6 5 4 3

Preface

Foundations of Chemistry and **Laboratory Experiments for Foundations of Chemistry** together provide a modern, enriched, chemistry program which is designed to help the student see how chemical principles and concepts are developed from experimental observations and data, and how these principles can be used to explain phenomena which he encounters in many of his daily activities as well as in the laboratory. Special attention is given to identification of problems associated with modern living and the development of attitudes, understandings, and skills which provide the student with a background that enables him to analyze carefully and act wisely on issues that confront him as a citizen of the nuclear-space age.

Wherever possible and practical, chemical principles and theories are developed on the basis of experimental data. In some instances the student collects and interprets his own data. Whenever this is not practical, the data and method of obtaining it are described. The historical evolution of an important idea is traced wherever it will not disrupt the continuity of ideas and where it will help to give the student an insight into the way in which a scientist develops theories and unifying principles from experimental data. The observations which the student makes in the laboratory and the discussions in the laboratory manual give added meaning and depth to the concepts discussed in the text. The laboratory manual is closely integrated with the text and contains specific experiments which are designed for each chapter in the text. A number of experiments are designed so that the student collects data which can be used as a basis for developing a concept.

Many of these concepts which are now regarded as essential in an up-to-date text are, by nature, abstract and difficult for students to grasp. In this text, these concepts are introduced after the necessary background has been developed and only after a need for them has been established. Concepts are first introduced qualitatively in the simplest terms and then slowly developed in a sequence of steps to a depth that gives real meaning and significance to them. They are illustrated with examples and then applied at every opportunity. The quantitative aspects of a concept are introduced later for those who wish to pursue it to a greater depth.

The text is large, not because it is a dictionary of chemical facts, but because principles and concepts have been considered in greater detail and depth than is customary. This type of developmental writing requires more space than merely setting down definitions, followed by only cursory discussions of the topic in question. Because emphasis is placed on continuity and a smooth flow of ideas, some chapters are long; however, subdivision of such chapters helps compensate for their length. In this book the more quantitative or abstract sections are located in the latter half of the chapters, where they can be omitted without disrupting the continuity of the subject, thus allowing the instructor much latitude in his

Preface

selection of topics and depth of coverage. The text is flexible and suitable for use with an honors class, and either a science-oriented or liberal-arts-oriented college-preparatory class. Because of its scope and in-depth development of concepts, the text is also ideally suited for introductory college courses.

The Foundations of Chemistry program contains many features which are designed to make the study of chemistry interesting, rational, understandable, and relevant. Some of these features are

1. Chapter *previews* which provide a smooth transition from one chapter to another. These previews given an overview of the ideas and sequence of topics developed within the chapter, and set the stage for the introduction of the material which follows.

2. *Looking ahead* sections at the end of each chapter summarize the highlights of a chapter and establish the need and reason for including the subject matter about to be covered. Both *Looking Ahead* and *Previews* are "bridges" between ideas, and tend to unify the text.

3. Chapters are divided into closely related subchapters, and subdivided further into specifically labeled, closely related subtopics. This arrangement helps the student see the logical sequence of ideas, and makes short digestible assignments possible.

4. Biographies of important scientists give the student an historical perspective and reveal some aspects of the personalities of these men, some of the problems they faced, and, in addition, the significance and impact of some of their discoveries. Interesting thought- and discussion-provoking quotations by a well-known scientist or personality are provided for each chapter.

5. Special features at the end of some of the chapters provide interesting discussions of applications or phenomena which are related to the principles treated in these chapters, and constitute essentially enough material from which a chemistry appreciation course could be designed.

6. A large number of two-color illustrations and line drawings, augmented with extensive captions, supplement the text discussion.

7. Margin notes are used to emphasize an important point, to extend a discussion, to define a previously unused term, or to supply incidental relevant information that does not fit into the mainstream of the discussion.

Preface

8. Important principles, laws, concepts, and key statements are printed in italics and/or bold-face type.

9. A large number of well-placed solved problems followed immediately by Follow-up Problems for the student which are similar to the example helps the student to develop proper problem-solving technique.

10. An exceptionally large number of graded end-of-chapter questions are designed so that all students, regardless of aptitude, can answer enough of them to feel some degree of success and gain confidence in their ability to master the material.

11. A laboratory manual of experiments specifically designed for each chapter in the text helps to tie experiment to theory and emphasizes the importance and significance of experimentation.

12. A completely revised testing program for the whole course enables student and instructor to evaluate progress and achievement.

In the second edition, we have eliminated a few of the more abstract, difficult, and nonessential sections that appeared in the first edition. These include nonaqueous solvents, the quantitative aspects of entropy and free energy, equivalent mass, normality, and certain concepts from physics. We have introduced, briefly discussed, and applied the concepts of equivalent mass and normality in one of the experiments in the laboratory manual for those who may still feel that these concepts are an essential part of the program. At the suggestion of those instructors who are using the first edition we have made several changes in the organization of the second edition. The principal changes are

1. The chapter dealing with chemical bonding in the first edition has been divided into two chapters and has been completely revised. The principles of covalent bonding are developed first in Chapter 8. The ionic bond is then developed in Chapter 9 as a logical extension of the polar covalent bond. The remainder of Chapter 9 is devoted to a discussion of intermolecular bonds, metallic bonds, and the relationship between structure and the properties of aggregates.

2. Chapter 4 in the first edition dealing with the mole concept and the quantitative relationships in chemical reactions has been divided into two chapters and completely revised. The mole concept is now developed in Chapter 2 following a discussion of Gay-Lussac's observations and Avogadro's hypothesis. Quantitative

Preface

aspects of reactions are then introduced in Chapter 3.

3. Chapter 1 has been expanded to include the introduction of the conversion factor-unit cancellation method of problem solving. Ample opportunity is provided for the student to apply this technique to quantitative problems in later chapters.

4. The chapter dealing with the characteristics of liquids and solids has been moved so that it now follows the chapters dealing with bonding and properties of aggregates. This permits the use of bonding concepts to explain the observed behavior of these condensed phases of matter.

5. The chapter dealing with chemical equilibrium has been divided into two chapters and completely revised, with the material related to solubility product being covered in a separate chapter.

6. The chapter dealing with acid-base equilibria has been completely rewritten. The sections covering nonaqueous solvents and normality have been deleted, and the concepts of pH and K_w have been moved forward to the chapter dealing with properties of solutions.

7. The number of solved problems, follow-up problems, and end-of-chapter problems has been greatly expanded to provide more practice for the student and greater selectivity for the instructor.

8. The Appendix has been modified to include a section dealing with the use of the slide rule, and a four-place table of logarithms.

9. A Periodic Table has been added to the inside back cover of the text.

We wish to express our sincere appreciation to our consultant, Mr. Russell C. Bovie, science department chairman, Arcadia High School, Arcadia, California. He prepared a detailed critique of the first edition and made many valuable suggestions for improving the text and laboratory manual. We also gratefully acknowledge the suggestions from our editor, Elbert C. Weaver, former chairman of chemistry at Phillips Academy, Andover, Massachusetts, and from William F. Knaack, professor of chemistry at Los Angeles Valley College. A special thanks is also given to Dorothy Szalay who did an outstanding job of typing the final draft of the manuscript for both the text and laboratory manual. Credit for the distinctive and attractive artwork goes to Mr. George Mass of Versatron Corporation.

CONTENTS

Contents

Contents

Contents

Foundations of Chemistry

1

A Perspective

Preview

You are about to begin the study of chemistry, a fascinating science. Chemistry is like a majestic skyscraper. The concrete secure foundation of chemistry consists of countless experimentally observed facts. The theories, principles, and laws developed from these observations are like an elevator which runs from the bottom to the top of the edifice. To reach the top you must go from floor to floor. At each level, new observations are made and new ideas, essential to continued progress, are proposed.

In this text, the 21 chapters are the individual floors of the building. The previews to the chapters are rather like windows. They give you an overall picture with little detail. They are designed to furnish insight into what lies ahead and to provide a reason or motivation for future study.

In this introductory chapter, we shall first review some of mankind's general problems, the solution to which chemistry will play an important role. In addition, we shall discuss the general organization and goals of chemistry as well as the qualities of scientists and the methods they use to achieve their goals. Finally, we shall introduce the commonly used system of measurement which enables us to describe the properties and behavior of matter quantitatively.

"I do not know what I may appear to the world; but to myself I seem to have been only like a boy playing on the sea-shore, and diverting myself in now and then finding a smoother pebble or a prettier shell than ordinary, whilst the great ocean of truth lay all undiscovered before me."

Sir Isaac Newton (1642–1727)

MAN AND THE ENVIRONMENT

1-1 Changes in the Relationship Between Man and His Environment Calculations based on astronomical and geological data suggest that the universe may have been in existence from 5 to 10 billion years. Fossil evidence indicates that man has existed on the earth from 1 to 2 million years. Archeological discoveries suggest that civilized societies have existed for approximately 10 000 years. During his brief existence, man has made remarkable progress toward adjusting to his environment and becoming its master rather than its slave. From the time of the cave man to the present, there has been a constant change in the relationship between man

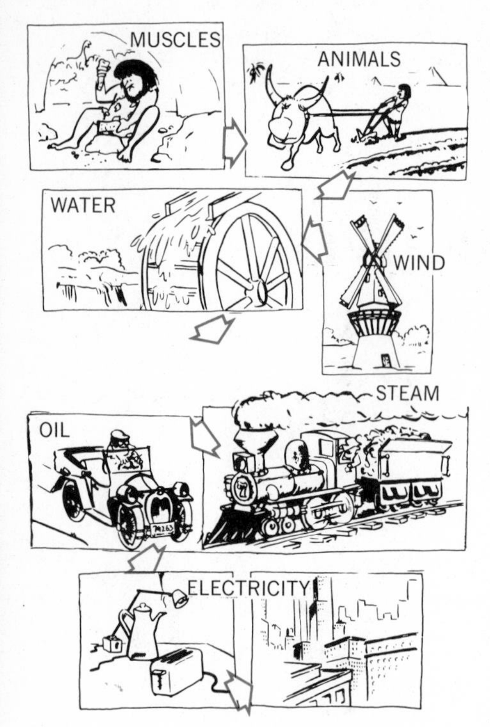

1-1 From cave to penthouse. Energy from various sources, and power devices which convert energy into useful work are used by man to advance his civilization.

and his environment. Whereas the cave man was engaged in a life-and-death struggle with the hazards of his prehistoric environment, modern man in highly civilized societies enjoys luxuries and a standard of living far beyond those enjoyed by the royalty of a few centuries ago.

There are, of course, many reasons for the change in the relationship between man and his environment. Among the most important factors are the use of natural resources and of energy other than that supplied by the muscles of men and animals. Primitive man was capable of fashioning tools and weapons from stone, but no amount of chipping and grinding altered the basic characteristics of the stone. He merely used the resources of nature as he found them. Modern man applies his scientific knowledge and uses sources of energy to change nature's raw materials into products which he uses to advance his civilization and raise his standard of living. **The progress of civilization has paralleled man's progress in understanding the uses of the resources of nature.**

1-2 Problems Faced by Modern Man Unfortunately, the resources of nature are not limitless. The development of our highly industrialized society has provided many benefits for mankind but it has also caused new problems which must be solved. Two of these are (1) *depletion of the world's natural resources,* and (2) *pollution of the environment.* Underlying and contributing to these problems is the *alarming rate of population growth.* A few statistics will illustrate the gravity of these problems.

From 6000 B.C. until 1650 A.D. the human population of the earth is estimated to have increased from about five million to 500 million. The "doubling time," or time required for the population to double during this period was about one thousand years. From 1650 until 1850 the world population again doubled which means that the doubling time had diminished to about 200 years. By 1930 the population was estimated to be about 2 billion, thus reducing the doubling time to 80 years. At present, the doubling time is about 37 years and getting shorter. The present world population is about 3.3 billion. By the year 2000, if present trends continue, the world population will be in excess of 7 billion with a doubling time of fewer than 20 years. It is apparent that twice as many people will require twice as much food, water, shelter, energy, manufactured products, hospitals, schools, transportation facilities, and other items relevant to living in our modern world. It is questionable whether even a highly advanced country such as the United States, with all its assets and capabilities, can meet the demands which will be imposed upon it. These problems must not only be solved, but solved without further degradation of our environment.

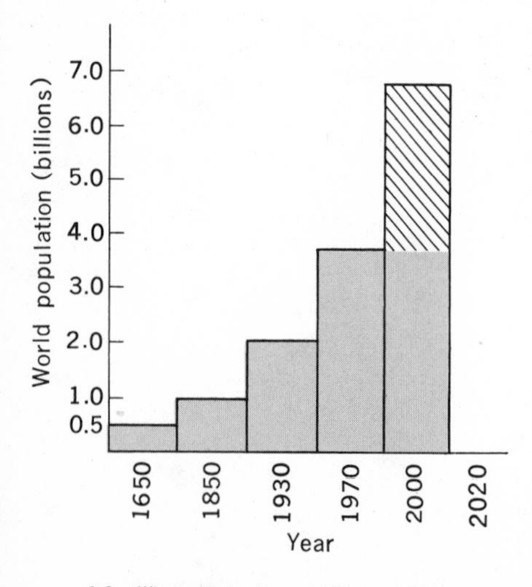

1-2 What will be the world's population in the year 2020 assuming that the "doubling" time in the year 2000 is 20 years?

Man is now aware that he lives on a planet which must be viewed as a kind of space ship. The occupants of this planet, as in a space ship, require a life-support system. For millions of years nature has worked out a delicately balanced relationship between the plants, animals, energy, and chemicals of this space capsule. The plants and animals through the centuries have adapted to living in their specialized ways in the streams, lakes, land masses, and oceans of this planet. Living organisms produce waste materials which are assimilated into the system and used by other organisms. These natural processes which produce our life-giving food, oxygen, and other materials necessary for living constitute our life-support system for this planet. If man continues to disturb the balance of these natural processes by polluting the environment and over-populating the earth, he will be faced with more famine, disease, war, and other perils than ever before. Even though many of the problems which we face seem to be social and economic in nature, their solution will definitely require a more advanced technology than that we have today. Chemistry will continue to play a major role in the solution to problems of famine, population control, conservation of natural resources, restoration of a healthful environment, and in maintenance of an adequate life-support system for our space ship.

Insects alone consume nearly one-third of all the crops planted by man. Control of insects is a problem faced by modern scientists.

FOLLOW-UP QUESTIONS

1. (a) What are some specific ways in which the earth's environment is being polluted? (b) Give specific examples of the effects of environmental pollution on animal, plant, and human life. (c) Discuss ways in which the chemical industry can help solve the pollution problem. (d) What can you as an individual citizen do to help?
2. Discuss the moral, political, and economic aspects of Dr. Borgstrom's statement in the margin.

1-3 The Physical Universe Space-ship earth is a small planet revolving about an average star, the sun, which is located in the Milky Way galaxy. The Milky Way is a galaxy of average size which contains about 100 billion stars. There are tens of billions of galaxies, and the universe contains at least 1 000 000 000 000 000 000 000 000 (10^{24}) stars. All stars emit light, a form of radiant energy. The analysis of light emitted by stars enables scientists to determine their composition. Studies of the types of substances present at various stages in the life of a star indicate that all objects in the universe are composed of the same fundamental materials as the earth. Scientists classify materials as *matter*. Thus, they broadly interpret the entire physical universe in terms of two major concepts: *matter* and *energy*. The science of chemistry, the study of which you are about to begin, is concerned with the properties, transformation, and uses of materials and energy.

"The United States with six percent of the world's population, is consuming more than one-third of the world's production of raw materials."

Dr. Greg Borgstrom

THE SCOPE OF CHEMISTRY

1-4 Branches of Chemistry **Chemistry, physics, biology, and earth science are the four main branches of natural science.** There is a great deal of overlapping in these four areas, as is implied by the specialized fields of *biophysics, biochemistry, physical chemistry, geochemistry,* and *geophysics.*

The scope of chemistry is far too great to be mastered by any one person. For convenience, chemistry may be subdivided into several branches (Fig. 1-3). Keep in mind that the branches are not separate and independent but overlap.

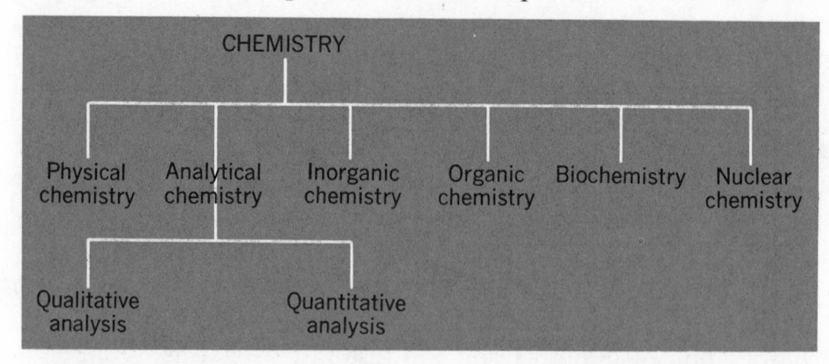

1-3 Branches of chemistry.

Physical chemistry deals with the investigation of the laws and theories of all branches of chemistry. Among its primary goals is the investigation of the *structure and transformation of matter* and the *interrelationships of energy and matter.* The subject matter, the experimental techniques, and the instruments used are common to both chemistry and physics.

Analytical chemistry is concerned with the separation, identification, and composition of all kinds of matter. Analytical procedures are used in every branch of chemistry. Two broad classifications of analytical chemistry are *qualitative analysis* and *quantitative analysis.*

Qualitative analysis involves the separation and identification of the individual components of materials. It answers the question, "What is present?" ***Quantitative analysis*** is, in addition, concerned with *how much* of each component is present.

Organic chemistry is the study of carbon-containing materials that are compounds. The term *organic* was derived from the original belief that these types of compounds were found only in living organisms. This branch of chemistry now deals with commonly used synthetic substances such as plastics, drugs, dyes, explosives, and detergents. There are many times more organic compounds known than those of all the other kinds combined.

Inorganic chemistry covers the chemistry of all of the elements and their compounds with the exception of carbon and its compounds. Thus, this area of chemistry comprises the investigation of those substances which are *not organic* such as nonliving matter and minerals found in the earth's crust.

Biochemistry is that branch of chemistry which includes the study of the materials and processes that occur in living organisms. These materials are largely organic (carbon) compounds.

Nuclear chemistry deals with the changes in the nuclei of atoms and the uses of these changes, especially in the study of how substances react. Radioactive nuclei, both natural and man-made, decompose spontaneously.

In this book we shall draw from all branches of chemistry, but emphasis is placed on elementary physical chemistry. For the quantitative application of principles we shall frequently turn to the subject matter of quantitative analysis. We shall attempt to achieve a balance between the descriptive and theoretical aspects of the science. Our main objective will be to develop and/or present, in a clear and logical fashion, the fundamental laws, principles, concepts, facts, and methods which are essential to understanding the science of chemistry and the nature of the physical world in which you live.

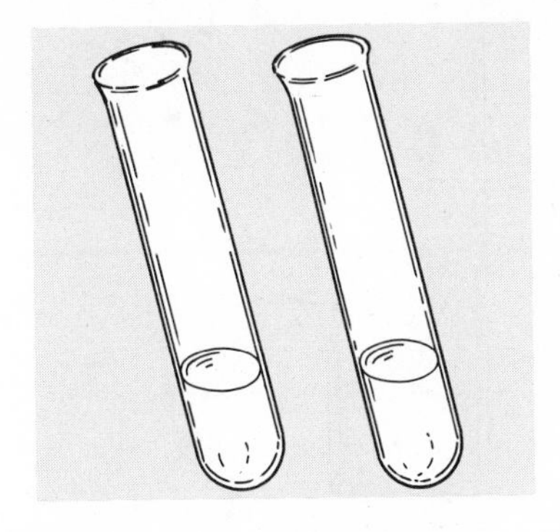

1-4 The initial state of a chemical system.

FOLLOW-UP QUESTIONS

1. Which branch of chemistry would be primarily concerned with these investigations? (a) The effect of birth-control pills on human metabolism. (b) The synthesis of a new dye. (c) An efficient method for separating a metal from its ore. (d) The role of particles in the atomic nucleus. (e) A fuel cell for converting the chemical energy in a gaseous fuel directly into electric energy. (f) The search for non-polluting fuels. (g) The desalinization of sea water. (h) How electric impulses are transmitted in a nerve.

SYSTEMS

1-5 Description of a Chemical System and Its Surroundings **All branches of chemistry involve the study of chemical systems and the changes they undergo.** We shall tentatively define a system as that part of the environment and its materials which has been isolated for study and experimentation. Everything else in the environment is referred to as the ***surroundings.*** For example, if we wish to investigate the behavior of ether, we can isolate the ether in a beaker. The contents of the beaker constitute the system while the beaker is part of the surroundings. Observation reveals that when the ether evaporates, the beaker becomes colder. We interpret this observation by saying that the system has absorbed heat from the surroundings.

Much of your laboratory work will involve chemical reactions. It is important that you be able to describe systems before and after a reaction has occurred. Let us briefly examine a reaction encountered by all beginning chemistry students. The reaction involves mixing a solution of ordinary salt (sodium chloride and water) with a solution of silver nitrate (silver nitrate and water). Our initial observation shows that the components of the system in its ***initial state*** are two transparent, colorless solutions (Fig. 1-4). An observation made immediately after mixing shows that the components of the system in its present state (Fig. 1-5) consist of a rather curdy, white solid and a liquid in which the solid is suspended.

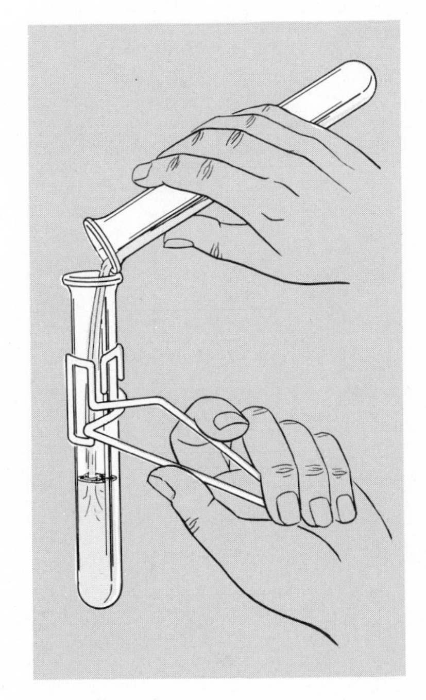

1-5 The solid formed when two solutions are mixed is called a precipitate.

The solid may be separated from the liquid by pouring the components of the system through a funnel fitted with a filter paper (Fig. 1-6). The solid which remains on the filter paper is known

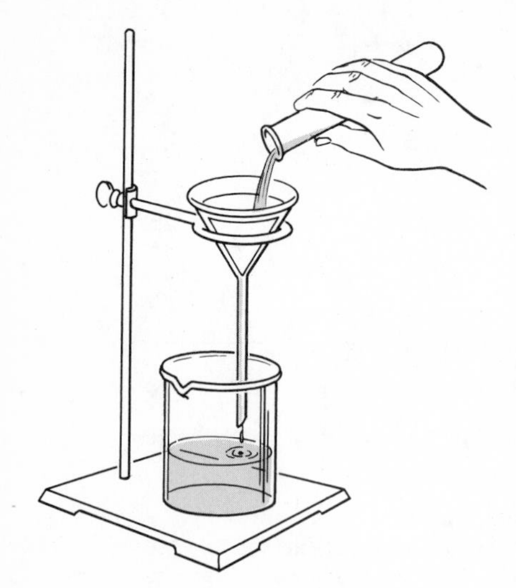

1-6 Filtration is the process used to separate an undissolved solid from a liquid.

as the ***residue,*** and the liquid which passes through the filter paper is called the ***filtrate.*** Thus, the final, rather superficial observation shows that the components of the system in its ***final state*** are a white, solid residue and a colorless, transparent, liquid filtrate (Fig. 1-7). Further examination and analysis would reveal more detailed characteristics of the system in its final state.

Laboratory experiments are the cornerstone of chemical knowledge. The principles developed in all branches of chemistry are based on laboratory experiments. You will find that laboratory work is the most enjoyable part of chemistry. In introductory laboratory experiments you have an opportunity to observe several chemical systems. Your laboratory experiences should be interesting, stimulating, and informative. At this point, you are not expected to understand all of the chemistry involved in an experiment. Introductory experiments are primarily designed to acquaint you with laboratory equipment, materials, and procedures and to stimulate your interest in the subject.

Experimental data and observations also lead to the development of chemical concepts, theories, and models which help us to understand our observations. For example, the first modern concept of an atom and the atomic theory were developed by John Dalton. He carried on *chemical reactions* from which he made experimental observations. His theory explained the reactions to him. Let us briefly examine the origin and postulates of Dalton's atomic theory and model.

Reaction **A chemical change in which a new substance is formed.**

Element **A simple form of matter that has not been decomposed by chemical means.**

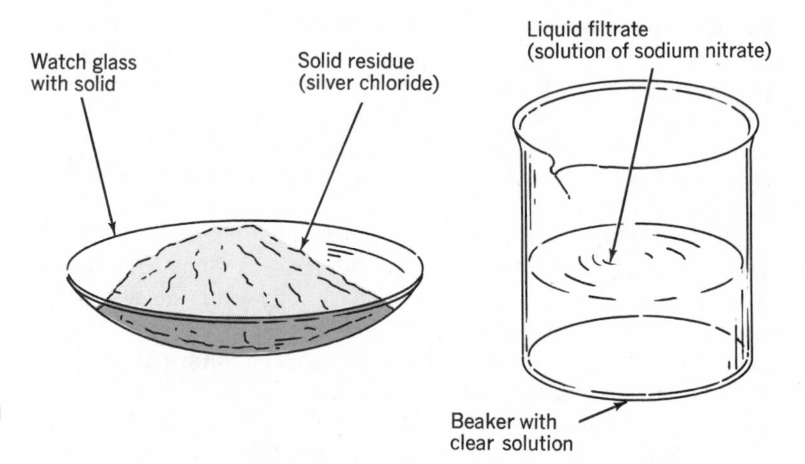

1-7 The components of the system in its final state are a solid residue and a liquid filtrate.

DALTON'S CONCEPT OF ATOMS

1-6 Experimental Basis and Postulates of Dalton's Atomic Theory During the 18th century, Antoine Lavoisier studied the mass relationships between the reactants and products involved in a chemical reaction. As a result of his observations, he suggested that mass was conserved when elements reacted and formed compounds. Additional observations by Joseph Proust (1754–1826), in 1797, indicated that when elements combined and formed compounds, they did so in definite proportions by mass. In 1803, Dalton proposed an atomic theory to explain the mass relationships existing among the substances which take part in chemical reactions. According to Dalton, all substances are composed of small, hard (dense), indivisible particles of matter that resemble tiny billiard balls. He called these particles atoms. Dalton believed that each element consisted of a particular kind of atom, and he attributed the varying properties of the elements to the differences in their atoms. He further proposed that the most important physical difference in the atoms of the various elements was a difference in mass. Accordingly, he assigned separate mass values (*atomic masses*) to each of the known elements. These were relative masses based on an arbitrarily chosen atomic mass of 1 for hydrogen, the lightest element.

Products are substances formed by a chemical reaction.

Compound A substance composed of two or more elements in definite composition by mass.

Nucleus Central, positively charged part of an atom.

Atom The structural building block of an element.

To account for Proust's observations, Dalton proposed that during the formation of chemical compounds, the atoms of elements unite in a definite numerical ratio. Thus, the composition by mass of a given compound is always the same. Dalton further postulated that the total number of atoms of each kind does not change as a result of a reaction. No atoms are gained or lost in a chemical change. This idea explains the conservation of matter noted by Lavoisier.

1-8 The shadows of atoms observed with the field emission microscope are produced using the same principle used in making the magnified shadow of the boy's hand.

The group of postulates or assumptions which were advanced by Dalton to explain the nature and behavior of chemical systems is known as Dalton's Atomic Theory. The postulates of this theory are summarized in Table 1-1.

TABLE 1-1
SUMMARY OF DALTON'S ATOMIC THEORY

1. All substances are composed of small, dense, indestructible particles called atoms.
2. Atoms of a given substance are identical in mass, size, and shape.
3. An atom is the smallest part of an element that enters into a chemical change.
4. Molecules of a compound are produced by the combination of the atoms of two or more different elements.

Some of the points of Dalton's original theory have been modified in accordance with more recent discoveries, but the particle nature of matter and the existence of atoms are now accepted by all scientists. The *field-ion microscope,* developed in 1955 by Erwin Müller (1899–) of the Pennsylvania State University, furnishes the best direct evidence for the *atomic nature of matter.* This instrument, which magnifies by a factor of 4 million, reproduces on a fluorescent screen an image of each atom in a metallic crystal.

1-9 A photograph showing the atomic lattice structure of a tungsten surface taken using a field ion microscope (magnified 2 million times). The dots are images of the individual atoms in the tungsten crystal.

Atoms are far too small to be observed directly. The best we can do is to develop a tentative mental picture of the concept. These mental pictures, called ***models,*** help scientists to understand and

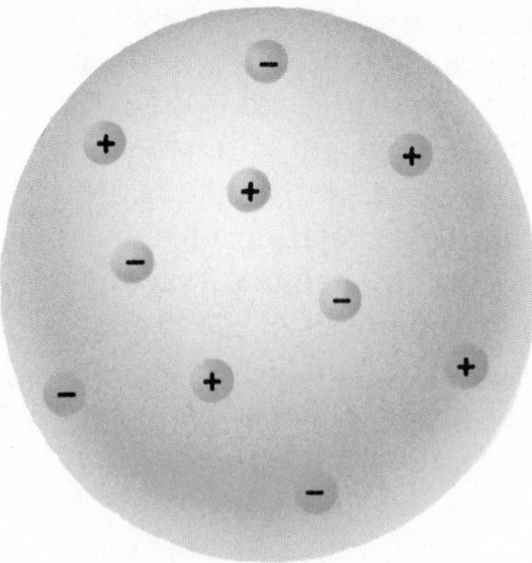

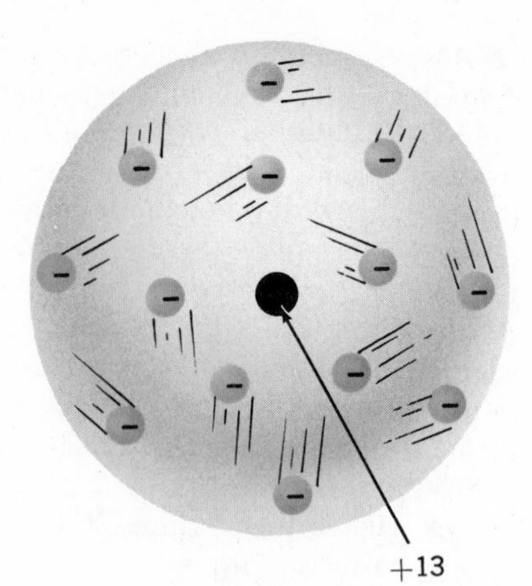

1-10 Dalton's concept of the atom (left), with Thomson's concept of atoms (center). Unlike Dalton's individual spheres, Thomson's model accounted for the electric nature of matter. Rutherford visualized the atom as a tiny, positively charged nucleus surrounded by rapidly moving electrons (right).

explain abstract concepts. Models serve as useful guides to man's thinking in his search for nature's secrets. Thus, Dalton's billiard-ball model of atoms helped to clarify the mass relationships that had been observed to exist among the elements in chemical compounds. Although this model is extremely useful both then and now, it is important to avoid taking models too literally. They all have limitations and fall short of reality. The model of atoms has been modified many times since Dalton's time as a result of the work and discoveries of many scientists. Let us briefly examine the evolution of the atomic model from 1800 to the present. We shall investigate several of these concepts in more detail later.

EVOLUTION OF AN ATOMIC MODEL

1-7 Shortcomings of Dalton's Atomic Model Dalton's indivisible "billiard-ball" atom, conceived in 1803, served as a satisfactory model for the explanation of mass relationships in chemical reactions but did not explain how or why atoms combined in certain ratios. It did not account for the attractive forces existing between particles of matter nor hint of a possible relationship between electricity and matter. During the 19th century, a number of experiments involving electric decomposition of solutions and passage of electricity through gases in sealed tubes suggested that atoms are divisible, and contain electrically charged particles. The negatively charged particles were named *electrons* and the positively charged particles were called *protons.*

1-8 Thomson's Concept of Atoms As a result of his experiments designed to investigate the nature and characteristics of these charged particles, J. J. Thomson (1856–1940), a British scientist, proposed an improved model. He suggested that atoms consist of a solid bulk of positive charge with electrons dispersed throughout them. Thomson's concept became known as the "plum pudding" model of atoms.

1-9 Rutherford's Concept of Atoms Further experimentation by Ernest Rutherford (1871–1937) and his co-workers, in 1911, revealed that the positive charge (protons) and mass of atoms were concentrated in the center (***nucleus***) of atoms. In this model atoms are pictured as a tiny, dense, positively charged nucleus surrounded by electrons moving at inconceivably fast speeds at relatively great distances from the nucleus but still within an atom. X-rays, discovered by Wilhelm Roentgen (1845–1923) in 1895, were used by H. G. J. Moseley (1889–1915) to determine the positive charge on the nucleus of atoms. In a neutral atom, the nuclear charge, known as the ***atomic number,*** also represents the number of electrons outside the nucleus. The ***neutron,*** a nuclear particle which contributes mass but no charge to the nucleus, was discovered in 1932 by James Chadwick (1891-   ), a British scientist.

1-10 Bohr's Concept of Atoms In 1913, Niels Bohr (1885–1962) used Rutherford's concept of the nucleus, concepts from Max Planck's (1858–1947) quantum theory, and other experimental data to develop his well-known "satellite" or "solar system" model of atoms. In Bohr's model, the electrons are arranged in *definite energy levels* (shells) and follow a prescribed orbit around the nucleus.

1-11 The Modern Concept of Atoms During the 1920's, the discovery of the wavelike properties of electrons led to the wave-mechanical model of atoms in which atoms are conceived to be a positively charged nucleus surrounded by pulsating electron waves. In this model, the electrons are associated with definite energy levels but they do not follow a prescribed trajectory. Instead, they are described in terms of the probability of being found in certain regions of space about the nucleus. These regions of space are called ***orbitals.*** The wave-mechanical model of atoms is widely used at present to explain the behavior of atomic and molecular systems.

The evolution of the atomic model from Dalton's simple "billiard ball atoms" to the highly mathematical, abstract, and sophisticated wave-mechanical model illustrates the importance of experimental investigation. As new evidence accumulates, theories and models must be modified accordingly. It should be noted however that no

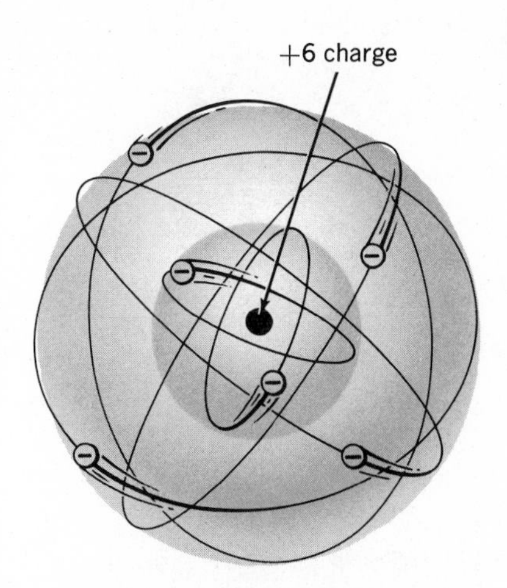

1-11 The electrons in Bohr's atomic model were confined to definite orbits. A neutral carbon atom shown here has a 6+ nuclear charge (6 protons). The 6+ charge on the nucleus is balanced by the negative charge of the 6 electrons revolving about the nucleus. Two of the electrons are located in the first orbit and 4 of them are found in the second. The nucleus, formed by the merging of 6 protons and 6 neutrons comprises essentially the entire mass of the atom.

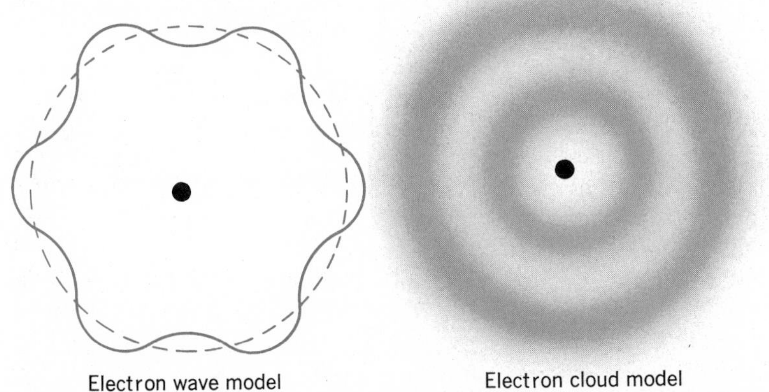

1-12 In the wave-mechanical model of atoms, electrons are not confined to orbits of fixed radius. Rather, they may be thought of as occupying regions of space represented by an electron charge cloud.

matter how refined a model of atoms becomes, it can never depict a true atomic system.

One of the most important objectives of science is to *relate the properties of large samples of matter (macroscopic samples) to the nature of the individual atoms or molecules in the sample.* The atomic theory will help us achieve this objective. The atomic model is one of the fundamental concepts on which modern chemistry is built. We shall use it and the ***Kinetic Molecular Theory,*** developed in Chapter 4, to help us explain and interpret the behavior of the systems we investigate throughout this book. The value and satisfaction you derive from your investigation of chemical systems will, in large part, depend on your attitude and approach. In the next section, we shall discuss additional goals, methods, and attitudes of scientists which, if applied, will make your study of chemistry easier, more enjoyable, and more beneficial.

ANTOINE LAURENT LAVOISIER
1743–1794

Lavoisier was one of the first scientists to stress the use of the chemical balance as an essential tool for studying mass relationships in chemical changes. He used the best balance in France and performed many quantitative experiments which clearly showed that ordinary combustion is a combination with oxygen. Because of his accurate and scientific explanation of the process of combustion, he is sometimes called the father of modern chemistry. As a result of his patient weighings of products and reactants in various chemical systems, he became convinced that matter is conserved in any chemical reaction. He was the co-discoverer of oxygen with Priestley of England and Scheele of Sweden but never tried to take credit for its discovery.

Lavoisier was very much interested in the general welfare. He devised a system for lighting the streets of Paris, proposed government insurance for poor people, suggested a number of other public projects including the cleaning and fumigating of the prison cells, and better treatment of the prisoners. During the 1788 famine, he loaned his own money without interest to the towns of Blois and Romorantin for the purchase of barley for the people. In spite of his humanitarian activities, the leaders of the French Revolution could not forgive him for being a nobleman and having a government position. Charges were brought against him for "watering the tobacco of soldiers" and appropriating revenue that belonged to the state. Lavoisier was guillotined. Considering the enormous contributions made by Lavoisier during his relatively short life time, it is logical to speculate that he might have made even more discoveries. As Joseph Lagrange, the great French mathematician said, "It required only a moment to sever that head, and perhaps a century will not be sufficient to produce another like it."

GOALS, METHODS, AND CHARACTERISTICS OF SCIENTISTS

1-12 Goals of Scientists ***Chemistry is an experimental science concerned primarily with the search for and the acquisition of knowledge about the fundamental nature and characteristics of the materials of which the universe is composed.*** Learning all the facts related to the substances which make up the universe is impossible. Fortunately, scientists have discovered a few unifying principles which correlate a large number of facts and concepts. These principles provide a basis for the systematic organization of a large body of knowledge.

Discovery of these broad principles and generalizations is the goal of much scientific research. These principles enable us to analyze, to discriminate, to associate, and to predict behavior in new situations. For example, one of the most important unifying principles developed by chemists is that the *properties of elements are periodic in nature.* This means that when elements are arranged sequentially in order of a fundamental or identifying characteristic (number of protons in the nucleus of their atoms), elements with similar or related physical and chemical properties recur at regular intervals. The grouping of elements in the Periodic Table enables us to observe relationships among elements. Elements having similar properties are grouped together in the same column (Group) of the Periodic Table. The behavior and properties of a given element in the chart may be predicted by noting its position in the Table and the properties of some of its neighbors. For example, sodium (element number 11), is a rather soft metal that reacts violently with water yielding hydrogen gas and a strongly alkaline solution. With this knowledge, we might predict that potassium (element number 19) would be a soft metal that reacts in about the same manner. Experiments confirm our prediction. We shall

PERIODIC TABLE OF THE ELEMENTS

IA	IIA											IIIA	IVA	VA	VIA	VIIA	0
H 1																	He 2
Li 3	Be 4											B 5	C 6	N 7	O 8	F 9	Ne 10
Na 11	Mg 12											Al 13	Si 14	P 15	S 16	Cl 17	Ar 18
K 19	Ca 20	Sc 21	Ti 22	V 23	Cr 24	Mn 25	Fe 26	Co 27	Ni 28	Cu 29	Zn 30	Ga 31	Ge 32	As 33	Se 34	Br 35	Kr 36
Rb 37	Sr 38	Y 39	Zr 40	Nb 41	Mo 42	Tc 43	Ru 44	Rh 45	Pd 46	Ag 47	Cd 48	In 49	Sn 50	Sb 51	Te 52	I 53	Xe 54
Cs 55	Ba 56	*	Hf 72	Ta 73	W 74	Re 75	Os 76	Ir 77	Pt 78	Au 79	Hg 80	Tl 81	Pb 82	Bi 83	Po 84	At 85	Rn 86
Fr 87	Ra 88	†	Ku 104	Ha 105													

* Lanthanides	La 57	Ce 58	Pr 59	Nd 60	Pm 61	Sm 62	Eu 63	Gd 64	Tb 65	Dy 66	Ho 67	Er 68	Tm 69	Yb 70	Lu 71
† Actinides	Ac 89	Th 90	Pa 91	U 92	Np 93	Pu 94	Am 95	Cm 96	Bk 97	Cf 98	Es 99	Fm 100	Md 101	No 102	Lr 103

1-13 A Periodic Table of the elements. The elements are arranged sequentially in order of their atomic numbers. The atomic number represents the nuclear charge and the total number of electrons in a neutral atom of the element. The Roman numerals at the top of the columns indicate the number of electrons in the outermost shell of an atom. Elements in the same columns display marked similarities in their properties.

make extensive use of the Periodic Table and the principles underlying it throughout the entire book. In general, emphasis will be placed on the understanding and application of unifying principles and generalizations rather than on the memorizing of isolated facts.

1-13 Methods of Scientists The scientific approach to problem solving and discovery requires experiments, careful observation, and accurate measurement. The importance of these activities is illustrated by comparing the work of two scientists who proposed conflicting theories about ordinary burning (combustion).

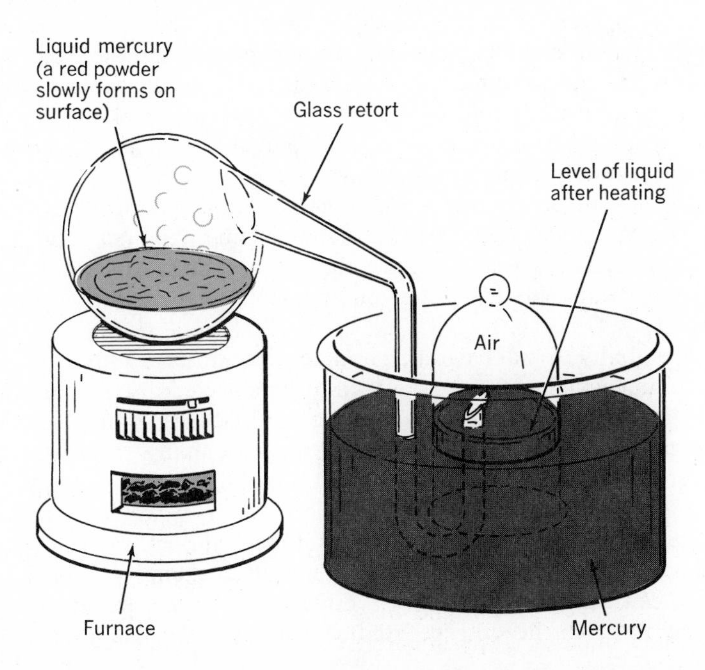

1-14 When oxygen of the enclosed air reacts with mercury in the retort, the volume of enclosed air decreases by one fifth and the water level rises. The increase in mass of the substance in the retort equals the decrease in mass of the gas in the jar.

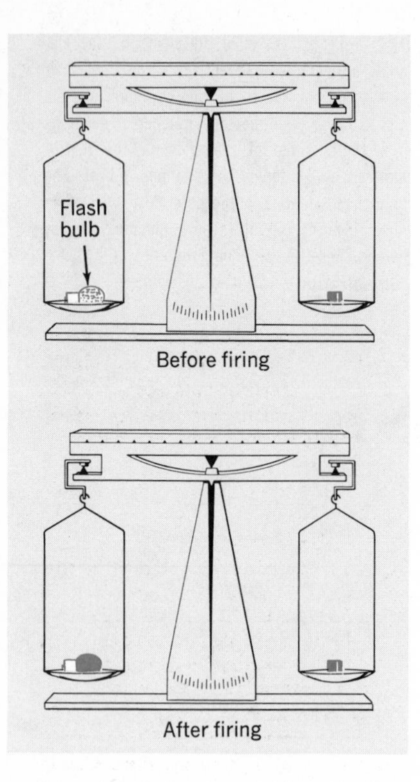

1-15 The mass of the magnesium and oxygen in the flashbulb before firing is equal to the mass of the new substance formed (magnesium oxide) after firing. This is parallel to the observation made by Lavoisier when he reacted tin and oxygen and formed an oxide of tin.

You all have seen a piece of wood or paper burn. The flames shoot upward, and the wood or paper is reduced to ashes which weigh less than the original material. In the 17th century, these observations led Georg Ernst Stahl (1660–1734) to propose that all combustible materials possess a "fire substance" called *phlogiston.* He believed that wood was composed of ash and phlogiston. When wood burned, the phlogiston escaped and the ash remained. To explain the apparent increase in mass observed when metals burned, it was necessary to say that phlogiston might sometimes have negative mass. Thus, phlogiston was a substance whose properties varied with the source.

In the 18th century, Antoine Lavoisier noted that when metals are heated in air, an apparent increase in their mass occurs. His explanation of the phenomenon was based on quantitative measurements. He first carefully weighed a closed container containing the metal and filled with air. He then caused the metal to burn and reweighed the container and its contents. The experiment is like weighing a flashbulb before and after ignition. Lavoisier's experimental data showed that the container and its contents neither gained nor lost mass during the combustion. Additional experiments showed that the increase in the mass of the solid was equal to the decrease in the mass of the air in the container. This led him to propose that combustion is a combination of a metal with a gas from the air. Lavoisier's experiments emphasize the importance of accurate measurement. His work led to the development of the very important ***Law of Conservation of Mass*** which states that matter cannot be created or destroyed. This is a fundamental law of nature and one to which we shall refer many times.

It is apparent from the preceding discussion that the success of experimental work, the development of concepts, and the discovery of new principles often depend upon the ability of the experimenter to make accurate, detailed observations. You can check your alertness to detail, your ingenuity, and your ability to make careful, scientific observations by observing a burning candle and preparing a list of your observations. Your teacher may wish to read a list of observations made by a professional chemist. You may find that a beginner fails to note many details. Our calling your attention to these details should enable you to make more penetrating observations.

"Chance favors the prepared mind."

Louis Pasteur

The ability to make detailed and accurate observations, coupled with intellectual curiosity, can lead to unexpected discoveries. The relevant statement by Louis Pasteur (1822–1895), printed in the margin, is worth noting and is the basis of many scientific discoveries that were made by accident. One of the most interesting accidental discoveries was that made in 1856 by William Perkin (1838–1907), a 17-year-old chemistry student at the City of London School. Perkin, who had become interested in chemistry at age 13, had been assigned the task of synthesizing quinine. In the course of his research, he observed that a black tar-like residue, which he was trying to remove from a dirty test tube, dissolved in alcohol and produced a beautiful purple solution. Rather than discarding the mixture and forgetting the event, young Perkin followed up his chance discovery with carefully planned experiments designed to determine the origin of the color. He immediately obtained a patent which founded the synthetic dye industry which today produces over 1000 shades of color. Other interesting examples of chance discoveries are listed in the margin.

1-14 Characteristics of Scientists It is possible that you may some day make an important contribution to scientific knowledge. In your study of chemistry, you will have the opportunity to find answers to problems by means of your own experiments. This experience should give you a great deal of personal satisfaction. In a way, you may feel the same excitement of discovery experienced by every scientist when he finds answers to his problems. You should try to develop the characteristics, attitudes, and techniques of skillful scientists, some of which are listed below.

1. An inquiring mind.
2. Accurate and critical observations.
3. Recording of data in a thorough, neat, and organized fashion.
4. Alertness to recognize the unexpected.
5. Willingness to reject old ideas and to accept new ones when sufficient data warrant it.
6. Resistance to the tendency to make generalizations on the basis of insufficient data.

The greatest assets in your search for knowledge are your intellectual curiosity and your ability to reason and understand. You are certain to find satisfaction in understanding more about your environment through the study of chemistry. The basic knowledge you will obtain is essential to many professions, including medicine, pharmacy, home economics, and engineering. You will also discover that scientific attitudes and methods can be applied to everyday problems. The study of chemistry should broaden your view and enrich your life. The need for such enrichment becomes apparent when you consider the vast number of words that have been added to our daily vocabulary as a result of the discoveries of the "space age." This text is designed to give you a sound and working knowledge of chemistry that will enable you to understand better the nature of the world in which you live.

Do not expect answers and explanation for all of the questions that may be raised in your mind. There are *fundamental properties of nature* for which there are no explanations at present. You will also find that there are exceptions to almost all general trends and rules. This should not cause you to lose faith in established principles. Remember that even identical twins are unique individuals and react differently under the same conditions, so that it is not possible to predict with complete accuracy the manner in which

Chance Discoveries

Vulcanization	Charles Goodyear	1840
Photography	Louis Daguerre	1838
Dynamite	Alfred Nobel	1867
X rays	Wilhelm Roentgen	1895
Insulin	Frederick Banting	1920
Penicillin	Alexander Fleming	1929

TEST YOUR AWARENESS

First read the sentence enclosed in the box below.

FINISHED FILES ARE THE RESULT OF YEARS OF SCIENTIFIC STUDY COMBINED WITH THE EXPERIENCE OF MANY YEARS.

Now count the F's in the sentence. Count them only once and do not go back and count them again. See discussion in margin on p. 14.

"Nothing great was ever achieved without enthusiasm."

Ralph Waldo Emerson (1803–1882)

either twin will react in a given situation. Such is the case with the substances about which you will study in chemistry. In general, you will learn how to make fairly accurate predictions. There may be, however, exceptions for which no apparent explanation is available. The best way to check the validity of predictions is through laboratory experiments. Scientific theories must agree with, and be supported by, experimental evidence if they are valid.

"When you can measure what you are speaking about and express it in numbers, you know something about it, and when you cannot measure it, when you cannot express it in numbers, your knowledge is of a meager and unsatisfactory kind."

Lord Kelvin

Chemistry is a quantitative science. Chemistry is quantitative in the sense that a chemist is able to reproduce experiments and obtain results which agree to the extent that he is able to make precise measurements. The development of precision in chemistry would have been impossible without the use of measurements. The statement by Lord Kelvin (1824–1907), printed in the margin, suggests that the ability to express and interpret observations and ideas quantitatively is a needed background for a sound knowledge and understanding of chemical concepts and principles. Let us briefly review some of the techniques used by scientists to represent quantitative measurements. A more detailed discussion is provided in Appendix 1.

Answer to alertness test
There are six F's. Because the F in "of" sounds like a *V*, it seems to disappear, and most people count only three F's. It is really remarkable how frequently we fail to perceive things as they really are.

SCIENTIFIC NOTATION

1-15 Uncertainty in Measurements All quantitative laboratory measurements involve a degree of uncertainty. Suppose we use a ruler to measure the length of the piece of metal shown in Fig. 1-16. You would agree that the metal is slightly more than 2.5 centimeters (cm) long. All would not agree on the value of the third digit. Some might read 2.51 cm while others would say that the length is 2.52 or 2.53 cm. In other words, the third digit is uncertain but useful in that it conveys the information that the metal is slightly longer than 2.5 cm. The certain digits plus one uncertain digit are known as ***significant figures.*** Thus, the measurement 2.53 cm represents three significant figures (two certain and one uncertain). The number of significant figures conveys information related to the ***precision*** or ***reproducibility*** of the measurement. All but the last figure are considered precise or reproducible.

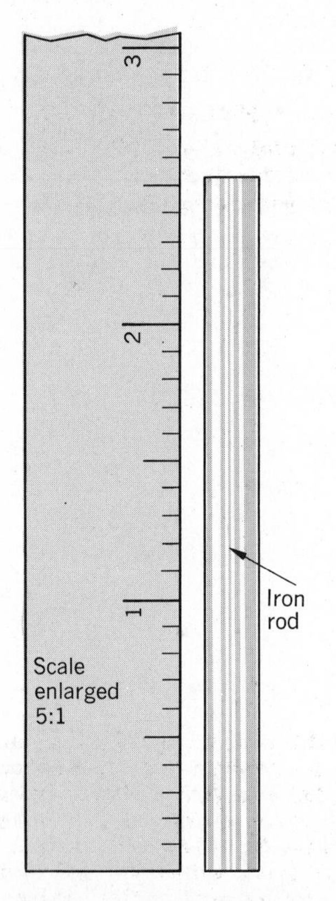

1-16 Assume that the measuring device shown above is graduated in centimeters and determine the length of the rod beside it to three significant figures. How does your measurement compare with that of your classmates?

If the extent of the uncertainty in the last figure is known, it is frequently indicated by a plus-or-minus number following the measured value. If our measuring instrument permits us to measure the length of the metal to the nearest 0.03 cm, then the uncertainty is represented by the figure, 2.53 ± 0.03 cm. Occasionally, it is desirable to represent the uncertainty in terms of a percentage. The fraction 0.03/2.53 equals 1.2%. Therefore, an uncertainty of 0.03 cm in a measurement of 2.53 cm may be expressed as $2.53 \pm 1.2\%$. It is apparent that for definite uncertainty the degree of uncertainty in the measurement decreases as the magnitude of the measurement increases.

FOLLOW-UP QUESTIONS

1. Melting points are often used to help identify unknown substances. Most thermometers used in school laboratories are accurate to $\pm 0.5\,^{\circ}C$. What is the percentage uncertainty in a temperature reading recorded as $20.0\,^{\circ}C$? **Ans. $\pm 2.5\%$ or $\pm 3\%$.**

2. If the temperature in part 1 above was recorded as $100.0\,^{\circ}C$, what is the percentage uncertainty? **Ans. $\pm 0.5\%$.**

1-16 Precision and Accuracy The terms *precision* and *accuracy* are widely used in any scientific work where quantitative measurements are made. ***Precision*** is a measure of the degree to which results of a given experiment "check." A high degree of precision is obtained when several results for the same experiment agree closely. The ***accuracy*** of a result is the degree to which the experimental value agrees with the true or accepted value. It is possible to have a high degree of precision with poor accuracy. This occurs if the same error is involved in repeated trials of the experiment.

The percentage error in a result may be expressed as

$$\frac{\textit{difference between experimental and true value}}{\textit{true value}} \times (100)$$

FOLLOW-UP QUESTION

In an experiment designed to determine the percentage of water in a substance, a student obtained these results

Trial 1	14.95%
Trial 2	14.94%
Trial 3	14.94%
Trial 4	14.93%
Average value	14.94%

The correct value is known to be 14.74%. (a) In general, what can you say about the degree of precision? (b) What is the percentage error in the average value?

Ans. (a) High degree of precision, (b) 1.4%.

1-17 Exponential Notation When making *chemical calculations,* students, as well as scientists, are constantly required to use very large and very small numbers. The simple arithmetic operations of multiplication, division, and extraction of roots would be very cumbersome if these numbers were not expressed in *exponential form.* In exponential, sometimes called ***scientific notation,*** a number is written as a product of two numbers. The first is called the *digit term.* It is one or greater than one but less than ten. The second number is called the *exponential term* and is written as 10 with an exponent. For example, the speed of light, 30,000,000,000 cm/sec, is expressed in scientific notation as 3×10^{10} cm/sec. This value indicates that the speed of light is not exactly 3×10^{10} cm/sec. That is, there is some uncertainty in the figure 3. To show a greater degree of certainty, we might write 3.0×10^{10} cm/sec. This number indicates that there is no uncertainty in the figure 3 but that the zero which follows the decimal point is uncertain. We can be sure however that the speed of light is not 3.1×10^{10} cm/sec or 2.9×10^{10} cm/sec. Handbooks give the accepted speed of light as 2.997925×10^{10} cm/sec. This value indicates an uncertainty in the figure 5. The rules for using significant figures in mathematical operations are given in Appendix 1B.

In most cases three significant figures will be used throughout this text. Following the rules of significant figures, however, often leads to mechanically awkward mathematical setups. For example, we would have to write 1.00×10^2 ml every time we wished to indicate one hundred ml to three significant figures. To avoid this cumbersome notation, we shall treat numbers such as 100 and 150 as though they were expressed to three significant figures.

At this point it would be highly desirable for you to become acquainted with the operation and use of the slide rule. You will find that mathematical operations which are usually tedious and time-consuming can be performed rapidly and accurately with this device. A detailed discussion of the slide rule is provided in Appendix 1C. You may wish to refer to this discussion before attempting the problems listed below.

FOLLOW-UP PROBLEMS

1. Indicate the number of significant figures in each: (a) 0.003, (b) 0.1030, (c) 46.01, (d) 284.0, (e) 4.7×10^3, (f) 2×10^{27}, (g) 1.006, (h) 6.02×10^{23}, (i) 0.23×10^{10}, (j) 271 000, (k) 12.40×10^{-3}, (l) $0.000\,3 \times 10^{17}$, (m) 14 000, (n) 14 000.0.
Ans. (a) 1, (b) 4, (e) 2, (j) 3, (m) 2, (n) 6.

2. Perform these operations and express the answers to the proper number of significant figures. Use scientific notation for multiplication and division.
(a) 4.02 + 6.083, (b) \$12 + 47¢, (c) 33.64 + 12.78, (d) 140 + 21, (e) 447 − 21.3, (f) 12.03 − 1.1, (g) \$8.87 − \$1, (h) 3.36 + 12.28 − 4.499, (i) 3.47×12.06, (j) $2.01 \div 4.1$, (k) $12\,000 \div 4.44$, (l) $\dfrac{0.09 \times 4.73 \times 12.80}{41 \times 0.0063}$, (m) $0.002\,00 \times 581$, (n) (0.000 018)(95).
Ans. (a) 10.10, (b) \$12, (k) 2.7×10^3, (l) 20, (m) 1.16, (n) 1.7×10^{-3}

3. Complete these exercises, expressing the answers in scientific notation. Express the digit term to the proper number of significant figures.
(a) $4.71 \times 10^{-5} \times 6.60 \times 10^7$
(b) $3.7 \times 10^7 \times 4.59 \times 10^{21}$
(c) $\dfrac{8.47 \times 10^7}{3.3 \times 10^3}$
(d) $\dfrac{4.49 \times 10^3 \times 8.3 \times 10^{12}}{6.68 \times 10^7}$
(e) $\dfrac{12.41 \times 10^{-5} \times 3.6 \times 10^4 \times 4.773 \times 10^8}{0.049 \times 10^4 \times 8.772 \times 10^3}$
Ans. (a) 3.11×10^3, (b) 1.7×10^{29}, (c) 2.6×10^4, (d) 5.6×10^8, (e) 4.8×10^2.

MEASUREMENT

1-18 The Metric System In order to describe the properties and behavior of matter quantitatively, scientists have developed several systems of measurement. Communication between scientific laboratories throughout the world requires that there be uniform and standard systems of measurement for mass, distance, time, temperature, energy, pressure, and other measurable quantities. The *metric system of measurement* is used in most scientific laboratories. It is a decimal system in which the basic units may be converted into larger or smaller units by moving the decimal point. The larger and smaller units are named by attaching a prefix to the basic unit. The commonly used prefixes are listed in Table 1-2.

TABLE 1-2
COMMONLY USED PREFIXES

micro	one-millionth of the basic unit	1×10^{-6}
milli	one-thousandth of the basic unit	1×10^{-3}
centi	one-hundredth of the basic unit	1×10^{-2}
deci	one-tenth of the basic unit	1×10^{-1}
kilo	one-thousand times the basic unit	1×10^3

The metric system units commonly used in chemistry are listed in Table 1-3. You may wish to refer to Appendix 2 for a further discussion of the metric system.

The metric system of measurement used in chemistry is known as the centimeter-gram-second (cgs) system.

TABLE 1-3
METRIC SYSTEM UNITS

mass	gram (g)
length	centimeter (cm)
volume	cubic centimeter (cm^3 or cc), milliliter (ml)

At this point it is worth noting that the terms *mass* and *weight* are not synonymous. **Mass is the quantity of matter present in a material body.** The *mass of a body is constant and is independent of location.* Mass can be measured by its resistance to change in velocity. The chemist uses weight as a measure of the mass of an object. ***Weight*** is a *force.* It is a measure of the gravitational attraction between the earth and the object being weighed. Weight depends on the mass of the object, the mass of the earth, and the distance of the object from the center of the earth. Whereas *mass is a fixed property of an object and remains constant, the weight of an object may vary.* For example, an object having a weight of 600 pounds on the earth would weigh 100 pounds on the moon and be weightless in outer space. Its mass would be the same on the moon, on earth, and in outer space. Since the weight of an object remains essentially constant everywhere on the earth's surface, the terms *mass and weight are commonly used interchangeably.* For the sake of accuracy, scientists prefer to use the term *mass* where it applies. In order to insure accuracy in experimental data, an equal-arm balance is used to compare the mass of a quantity of

Velocity is the rate of motion in a given direction.

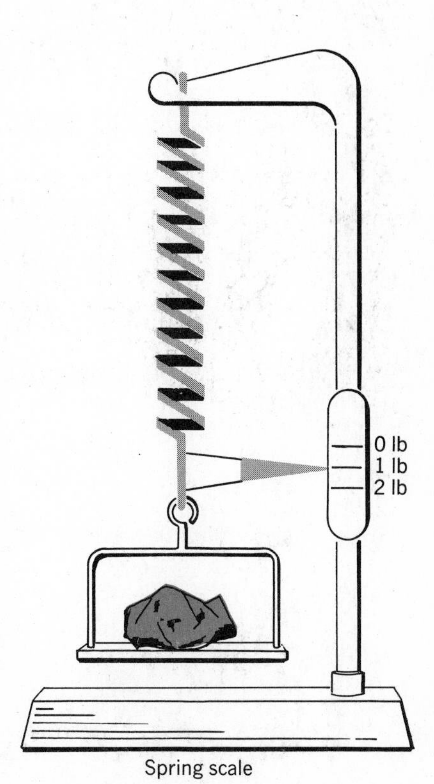

1-17 The deflection of the pointer on this spring scale varies with the magnitude of the gravitational force. What would be the approximate weight of this rock on the moon?

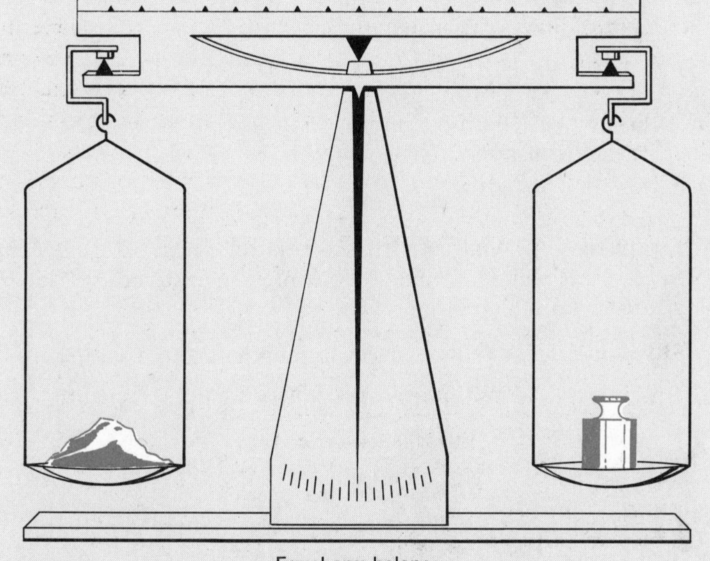

1-18 The mass of an object may be determined by placing it on the left pan of an equal-arm balance. The object is counterbalanced with standard masses on the right pan. When balanced, the mass of the object equals the total standard mass.

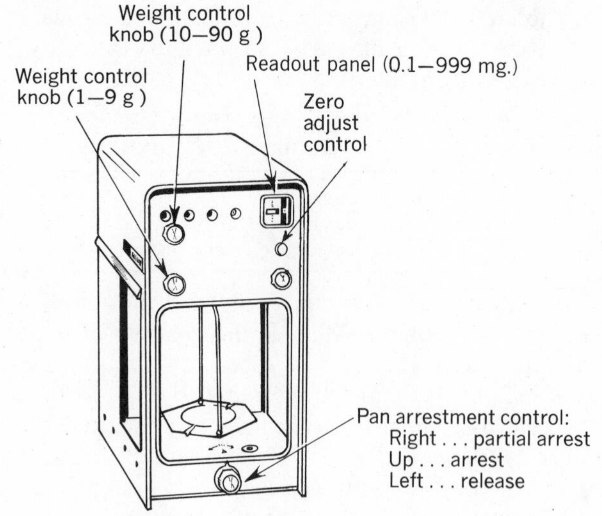

1-19 The single pan balance is replacing the equal-arm balance in many school laboratories. After zero adjusting the balance, the substance to be weighed is placed on the pan. Standard masses are added or removed inside the case by turning dials until equilibrium is indicated on the lighted readout panel. The total mass of the sample is obtained by reading the dial positions and the image of an engraved scale projected on a ground glass plate.

matter to a known standard mass. Metric-system mass units and conversion factors are listed in Table 1-4. Linear units and conversions are listed in Table 1-5.

FOLLOW-UP QUESTION

Which of these statements is open to criticism?
(a) An object has a mass of 5.0 grams.
(b) An object weighs 10.0 grams.
(c) The weight of an object is 20.0 grams.

1-19 Derived Units All units used in chemistry can be derived from the fundamental quantities of *mass* (m), *length* (l), *temperature* (T), *time* (t), and the *mole* (described in Section 2-20). For example, the dimension of area may be derived by squaring l, the fundamental dimension of length. That is,

$$\textbf{area} = l \times l = l^2$$

Since the unit of length commonly used in chemistry is the centimeter, the unit of area is

$$\textbf{cm} \times \textbf{cm} = \textbf{cm}^2$$

The space-filling characteristics of matter is called its *volume.* Volume is three dimensional. It is expressed in a unit of length taken to the third dimension. That is,

$$\textbf{volume} = l \times l \times l = l^3 \text{ (if all lengths are the same)}$$

The commonly used basic unit of volume in the metric system is the *cubic centimeter* (cm^3 or cc).

$$\textbf{cm} \times \textbf{cm} \times \textbf{cm} = \textbf{cm}^3$$

One cubic centimeter is the volume of a cube one centimeter on an edge.

TABLE 1-4
METRIC SYSTEM MASS UNITS AND CONVERSIONS

1 kilogram (kg)	= 1000 gram (g)
1 gram (g)	= 1000 milligrams (mg)
1 milligram	= 1000 micrograms (μg)
1 kilogram	= 2.2 pounds (lb)
454 grams	= 1 pound

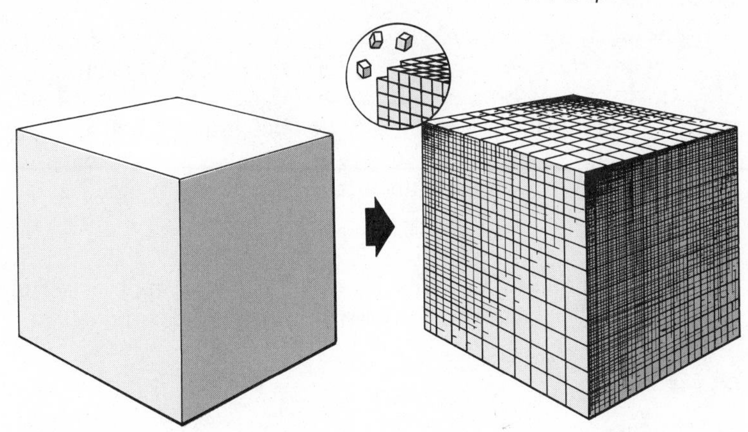

1-20 A cubic decimeter. The large cube is 10.0 centimeters on an edge. Its volume is 1000 cubic centimeters or 1.0 liter. The small cube is 1.0 cm on an edge and has a volume of 1.0 cm^3. What is the volume of the small cube in mm^3 and how many small cubes are there in the large cube?

The ***liter,*** a frequently used larger volume unit, is now defined as the volume equal to the cubic decimeter. A cubic decimeter is the volume of a cube whose edges are 1 decimeter (10 cm) in length. Thus, one liter is equal to 1000 cm^3. From these definitions it may be seen that *1 milliliter (ml) is equal to 1 cm^3*. Some practical equivalent volumes used in the field of chemistry are listed in Table 1-6. Note the relationship between the volume and mass of water at 4°C.

TABLE 1-5
LINEAR UNITS AND CONVERSIONS

1 meter (m)	=	10 decimeters (dm)
1 meter	=	100 centimeters (cm)
1 meter	=	1000 millimeters (mm)
1 meter	=	10^6 microns (μ)
1 meter	=	10^9 millimicrons (mμ)
1 meter	=	10^{10} Å (Angstrom units)
1 centimeter	=	10^8 Å
2.54 cm.	=	1 inch*

*This is an exact relationship and is accepted by the United States Bureau of Standards as the only exact relationship between the metric and British systems of measurement.

TABLE 1-6
VOLUME UNITS AND CONVERSIONS

1 cubic centimeter (cm^3)	=	1 milliliter (ml)
1000 cubic centimeters	=	1 liter (ℓ)
1 liter	=	1 cubic decimeter (dm^3)
1 liter of water at 4°C	=	1 kilogram
1 milliliter of water at 4°C	=	1 gram (g)
1 liter	=	1.06 quarts (qt)

Density is a derived quantity which is widely used to describe different kinds of matter. When the mass of a sample of matter is determined at a given temperature, its volume (V) is definite. Doubling the mass of the sample doubles its volume. This observation may be stated in mathematical terms by saying that mass is directly proportional to volume. The relationship is expressed symbolically as

$$\mathbf{m} \propto \mathbf{V}$$

Density (d) is the proportionality constant that relates the mass of a substance to its volume. That is,

$$\mathbf{m} = \mathbf{dV} \qquad 1\text{-}1$$

The units and significance of density may be seen by solving Equation 1-1 for density.

$$\mathbf{d} = \mathbf{m}/\mathbf{V} \qquad 1\text{-}2$$

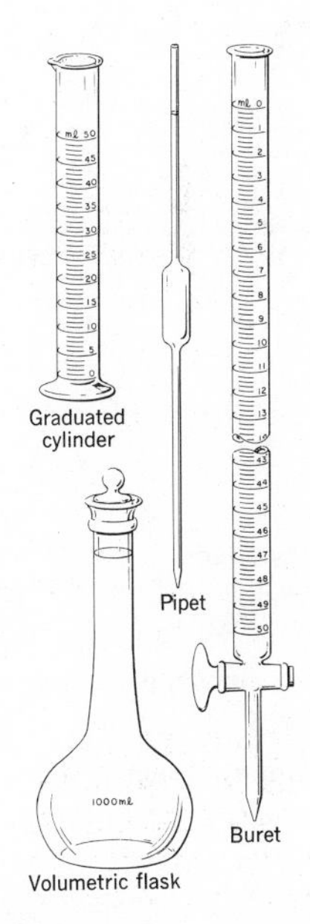

1-21 Graduated cylinders, volumetric flasks, pipets, and burets are used for measuring the volume of liquids. Which device do you think involves the greater degree of uncertainty, a graduated cylinder or a buret?

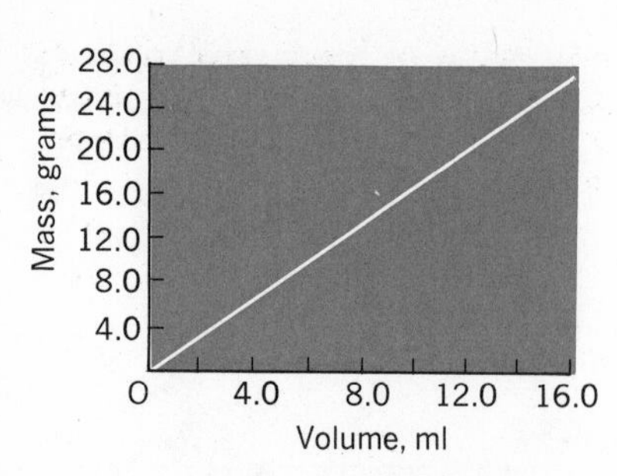

Mass of magnesium vs volume of magnesium

1-22 A graph showing the relationship between the mass of magnesium and the volume of magnesium.

If one plots the mass of various samples of a substance on the y axis against their volume on the x axis, the points fall on a straight line which passes through the origin. A plot of mass vs volume (temperature constant) for samples of magnesium metal is shown left. This graph is characteristic of *two variables which are directly proportional to each other.* This relationship is represented as

$$m \propto V$$

The proportionality may be converted to an equality by inserting a proportionality constant, density in this case, which relates mass to volume. Thus,

$$m = dV$$

The constant of proportionality, d, is the *slope* or *inclination* of the straight line. Thus, the slope of the line is the ratio of mass to volume ($d = m/V$). Graphs are often used to show or determine relationships between plotted numbers.

FOLLOW-UP PROBLEM

(a) Use several points on the above graph to determine the density of magnesium.
(b) Is the slope constant at the different points?
(c) What is the significance of the point (0,0)?

Interpretation of Equation 1-2 shows that *density is the mass per unit volume of a substance.* When mass is expressed in grams and volume in milliliters, the unit of density is *g/ml.*

We can use the concept of density to show quantitatively the difference between the three physical states of nitrogen. Nitrogen gas, a component of the atmosphere, has a density of 0.001 25g/ml at standard conditions (0°C and 760 torr, STP), liquid nitrogen has a density of 0.808 g/ml, and solid nitrogen has a density of 1.026 g/ml. These data can be used to show that one gram of solid nitrogen occupies 0.97 ml, one g of liquid nitrogen occupies 1.24 ml, and one g of nitrogen gas occupies approximately 900 ml at room temperature and pressure (about 25°C and 750 torr). Notice that when the solid is converted into a gas, the volume increases by almost 1000-fold. If one gram of nitrogen contains the same number and kind of molecules, regardless of the physical state or phase, then the molecules must be further apart in the gas than in the liquid or solid. In other words, solid nitrogen is the most *compact* and *dense* form of nitrogen because the molecules are packed closer together than in the liquid or solid.

Gas Measurement
Standard conditions are 0°C and 1 atm or 760 torr.

STP means temperature 0°C and pressure 760 torr.

1 torr = 1 mm of mercury

FOLLOW-UP QUESTION

Account for the variation in volume of one g of nitrogen in terms of the spacing of the particles which make up the solid, liquid, and gaseous samples.

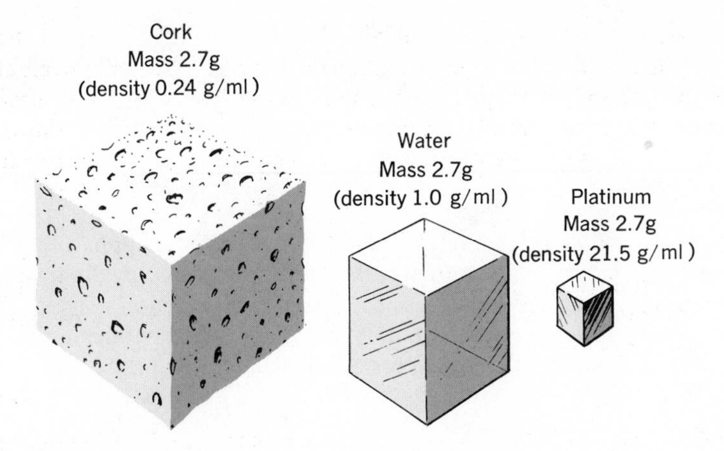

1-23 Each of the volumes shown contains the same mass of material. It is apparent that platinum is a more compact form of matter than either cork or water.

The densities of some common substances are listed in Table 1-7 below.

TABLE 1-7
DENSITIES OF COMMON SUBSTANCES

Solids

Name		*Density (g/ml at 20°C)*
Aluminum	Al	2.7
Carbon (d)	C	3.5
Copper	Cu	8.92
Gold	Au	19.3
Iodine	I	4.93
Iron	Fe	7.86
Lead	Pb	11.3
Magnesium	Mg	1.74
Nickel	Ni	8.9
Platinum	Pt	21.4
Potassium	K	0.86
Sodium chloride	NaCl	2.16
Silver	Ag	10.5
Sodium	Na	0.97
Sugar	$C_{12}H_{22}O_{11}$	1.6
Sulfur	S	2.0
Tin	Sn	7.3
Titanium	Ti	4.5

Liquids

Name		*Density (g/ml at 20°C)*
Acetone	$(CH_3)_2CO$	0.791
Benzene	C_6H_6	0.879
Bromine	Br_2	3.12
Carbon tetrachloride	CCl_4	1.594
Chloroform	$CHCl_3$	1.489
Ether, diethyl	$(C_2H_5)_2O$	0.704
Ethyl alcohol	C_2H_5OH	0.789
Glycerol	$C_3H_5(OH)_3$	1.261
Mercury	Hg	13.555
Water, 4°C	H_2O	1.000

Gases

Name		*Density (g/ℓ at STP)*
Air	(a mixture)	1.29
Carbon dioxide	CO_2	1.96
Carbon monoxide	CO	1.25
Chlorine	Cl_2	3.17
Hydrogen	H_2	0.090
Methane	CH_4	0.714
Nitrogen	N_2	1.25
Oxygen	O_2	1.43

1-20 Conversion Factors and Unit Cancellation In this book you will encounter many problems which deal with quantitative aspects of chemistry. It should be emphasized that the only way to master the problem-solving techniques which are essential to your understanding and application of the principles of chemistry, is to practice by working as many problems as possible. When solving problems, all quantities should be labeled with the proper units. The same mathematical operations should be applied to the units as to the numbers. There are many methods available for solving problems. We shall use calculations involved in converting one linear unit into another to illustrate the use of scientific notation and a technique often used for solving chemical problems. The method involves the use of *conversion factors* and *unit cancellation,* and it is called the *factor-label* method of problem solving. The techniques can be readily applied to any problem that involves changing one unit to another.

Example 1-1

Light travels at a speed of 6.70×10^8 miles per hour. Express this speed in centimeters per second.

Solution

We must multiply the given quantity, 6.70×10^8 mi/hr by conversion factors which will change miles in the numerator to centimeters, and hours in the denominator to seconds. Known relationships are

1 mile = 5.28×10^3 feet **1 hour = 6.00×10^1 minutes**
1 foot = 1.20×10^1 inches **1 min = 6.00×10^1 seconds**
1 inch = 2.54 centimeters

Each of these relationships may be expressed in terms of a ratio (conversion factor). The ratios must be constructed so that all but the desired units of cm/sec cancel.
We can determine the form of the ratio by first expressing the problem in terms of units.

Desired units		Given units		Factor converts miles to feet		Factor converts feet to inches		Factor converts inches to centimeters		Factor converts hours to minutes		Factor converts minutes to seconds
$\frac{cm}{sec}$	=	$\frac{miles}{hr}$	×	$\frac{feet}{mile}$	×	$\frac{in.}{foot}$	×	$\frac{cm}{in.}$	×	$\frac{hr}{min}$	×	$\frac{min}{sec}$

Using the known relationships we can now complete each ratio.

$$\frac{cm}{sec} = \frac{6.70 \times 10^8 \text{ mi}}{\text{hr}} \times \frac{5.28 \times 10^3 \text{ ft}}{\text{mi}} \times \frac{1.20 \times 10^1 \text{ in.}}{\text{ft}} \times \frac{2.54 \text{ cm}}{\text{in.}} \times \frac{1 \text{ hr}}{6.00 \times 10^1 \text{ min}} \times \frac{1 \text{ min}}{6.00 \times 10^1 \text{ sec}}$$

$$= 3.00 \times 10^{10} \text{ cm/sec.}$$

FOLLOW-UP PROBLEMS

1. A clad metal quarter has a mass of about 6.25 g. Express this mass in milligrams. **Ans. 6.25×10^3 mg.**

2. A nickel weighs exactly 5.000×10^3 mg. Express this mass in grams. **Ans. 5.000 g.**

3. Calculate the number of grams of sugar in a 10.0-pound bag of sugar. **Ans. 4.54×10^3.**

4. A sprinter runs a 100-meter dash. How many yards does he run? **Ans. 109 yards.**

5. A sprinter runs a 100-yard dash. How many meters does he run? **Ans. 91.4 m.**

6. An electron is traveling at the speed of 2.5×10^{10} cm/sec. Use the relationships given in Example 1-1 to express this speed in miles per hour. **Ans. 5.6×10^8 mi/hr.**

7. The wavelength of a uniform wave is the distance between corresponding points on adjacent waves (Fig. 5-21 on p. 163). A spectral line in the spectrum of sodium has a wavelength of 5.890×10^{-5} cm. Express this wavelength in Angstroms. **Ans. 5.890×10^3 Å.**

8. How many liters in 1.50×10^3 ml? **Ans 1.50 liters.**

9. Calculate the volume of a cylindrical solid having a diameter of 2.00 cm and a height of 10.0 cm. V (cylinder) = πr^2h. **Ans. 31.4 cm^3.**

10. A graduated cylinder contains 45.0 ml of water. When an iron rod is submerged in the water, the liquid level rises to the 50.0 ml mark. What is the volume of the iron rod? **Ans. 5.0 ml.**

11. A cube of iron is 4.0 cm on an edge. Calculate (a) the volume of the cube, (b) the total area of all the faces. **Ans. (a) 64 cm^3, (b) 96 cm^2.**

12. A rectangular piece of brass is 5.0 cm long, 4.0 cm wide, and 3.0 cm deep. Calculate (a) the volume of the brass, (b) the total area of all the faces of the rectangular solid. **Ans. (a) 60 cm^3, (b) 94 cm^2.**

13. There are 2.54 cm in 1.00 in. Calculate the number of cm^3 in 1.00 $in.^3$ **Ans. 16.4 cm^3.**

14. Calculate the number of cm^3 in 1.00 m^3. **Ans. 1.00×10^6 cm^3.**

15. The inside dimensions of a rectangular tank are length, 30 cm, width, 20 cm, depth, 10 cm. (a) How many grams of water does the tank hold at 4°C? (b) Express this mass in kg. **Ans. (a) 6×10^3 g, (b) 6 kg.**

16. The inside dimensions of a cylindrical vessel are diameter, 10 cm, height, 50 cm. (a) How many grams of water can the vessel hold at 4°C? (b) Express this mass in kilograms. **Ans. (a) 4×10^3 g, (b) 4 kg.**

17. One liter of hydrogen gas weighs 0.090 g at STP. A balloon when filled with hydrogen at STP has a diameter of 100 cm. (a) What mass of hydrogen does the balloon hold? (The formula for the volume of a sphere is $\frac{4}{3}\pi r^3$.) (b) How many grams of air will the balloon hold if under the same conditions, one liter of air weighs 1.29 g? **Ans. (a) 47 g, (b) 670 g.**

18. The lifting ability of a balloon is the difference in its mass when filled with air and when filled with a gas less dense than air. If the balloon in Problem 17 has a mass of 100 g, what payload will it lift when filled with hydrogen gas? **Ans. 520 g.**

19. An aluminum cylinder weighing 50.0 g is placed in a graduated vessel containing 40.0 ml of water. The new water level is 58.5 ml. What is the density of the aluminum? **Ans. 2.70 g/ml.**

20. The density of mercury is 13.6 g/ml. How many ml of mercury occupy a 1-lb (454-g) bottle? **Ans. 33.4 ml.**

21. A regular block of gold with dimensions 2.0 cm × 2.0 cm × 6.0 cm weighs 463 g. What is the density of gold? **Ans. 19 g/ml.**

1-21 Percentage The concept of percentage is widely used in chemical calculations and, therefore, warrants a brief discussion. The objective of many quantitative analytical procedures is to determine the percentage of a component in a mixture. The percentage of a component in a mixture simply represents the parts of component per 100 parts of mixture. In terms of mass percentage, the relationship is

$$\textbf{percentage of component} = \frac{\textbf{mass of component}}{\textbf{mass of mixture}}(\textbf{100}) \quad 1\text{-}3$$

Example 1-2
A silver medal is an alloy of copper and silver and has a mass of 2.50 grams. Analysis of the medal reveals that it contains 2.25 g of silver. (a) What is the percentage of silver in the medal? (b) How many grams of the alloy are required to yield 200 g of pure silver? (c) How many grams of silver are needed to prepare 500 g of alloy?

Solution
(a) Substitute the data in Equation 1-3.

$$\text{percentage silver} = \frac{2.25 \text{ g silver}}{2.50 \text{ g alloy}}(100) = 90.0 \text{ percent.}$$

(b) The answer to (a) reveals that the mass ratio of alloy/silver is 100/90. In other words, every 100 g of alloy yields only 90 g of silver. Multiply the desired quantity (200 g silver) by a factor which contains silver in the denominator and alloy in the numerator. After setting up the dimensions so that the desired unit is obtained, insert the numerical values.

$$200 \text{ g silver} \times \frac{100 \text{ g alloy}}{90 \text{ g silver}} = 222 \text{ g alloy.}$$

(c) Multiply the desired quantity (500 g alloy) by the factor which converts grams of alloy to grams of silver.

$$500 \text{ g alloy} \times \frac{90 \text{ g silver}}{100 \text{ g alloy}} = 450 \text{ g silver.}$$

FOLLOW-UP PROBLEM

Analysis of a 5.000-g sample of brass reveals that it contains 3.50 g of copper. (a) What is the percentage of copper in the sample? (b) How many grams of brass yield 140 g of pure copper? (c) The only other component of this alloy is zinc. How many grams of zinc are needed to prepare 250 g of brass?

Ans. (a) 70.0% Cu, (b) 200 g, (c) 75.0 g Zn.

LOOKING AHEAD

Now that we are familiar with some of the goals, methods, and tools of chemistry, we can use them to help us investigate the types and nature of the chemical materials and chemical changes which constitute our environment. The first step in such an investigation is to obtain and study the behavior and characteristics of the simplest kinds of matter. In Chapter 2 we shall introduce methods used by chemists for separating naturally occurring materials into simpler substances. Another goal will be to investigate the properties of the different kinds of matter and describe their characteristics in terms of general laws. One of our ultimate objectives will be to explain the behavior of the macroscopic samples we investigate in terms of their fundamental particles. We shall base our explanations and develop our models on the evidence presented. This means that our models and explanations will be modified and become more detailed and sophisticated as new evidence accumulates from one chapter to the next.

The special features which appear at the end of some chapters contain interesting discussions of material related to the chapter. It is not practical to include them in the mainstream of the discussion.

Special Feature: Communication Of Scientific Information

Perhaps the most amazing feature of this century is the staggering rate at which scientific knowledge is accumulating and technological advances are being made. Persons eighty years old, such as your grandparents, have witnessed changes in the area of transportation from horse to bicycle to motorcycle to automobile to propeller aircraft to jet aircraft to rocket. The same rapid evolution of ideas and applications is characteristic of all areas of applied and basic science. Underlying the rapid development observed in all areas of human scientific endeavor is the communication of scientific information. It is the communication and interplay of ideas that lead to progress. Communication of information is equally as important as the research which produces the information.

"If I have seen farther than others, it is because I have stood on the shoulders of giants."—Isaac Newton (1642–1727)

There are countless examples of the fruitfulness of communication between the men who contributed to the rapid development of chemical knowledge.

As methods and avenues of communications improved, scientists published their ideas and scientific progress mushroomed accordingly. The ideas and discoveries of one man served as a stepping stone for the work of another. In Chapter 1, we superficially traced the evolution of the atomic model. One of the most familiar models is the "satellite" model developed by Neils Bohr in 1913. As we shall show later the revolutionary quantum theory proposed in 1900 by Max Planck and the concept of the positively charged nucleus developed by Ernest Rutherford in 1911 served as the stepping stones which led to Bohr's contribution.

Credit for a scientific discovery almost always goes to the person who first persuasively and clearly communicates his discovery to the world. In some instances, failure of a scientist to publish his results promptly, and in widely-read publication, resulted in credit being given to another person. For example, Dmitri Mendeleyev (1834–1907) of Russia and Lothar Meyer (1830–1895) of Germany both discovered periodicity in the properties of the elements about the same time. Credit is generally given to Mendeleyev because he communicated his findings to the world. Although John Dalton is usually credited with proposing the first modern atomic theory, there is evidence that William Higgins (1763–1825) proposed an atomic theory earlier but failed to communicate his ideas. A third example of interest involves Joseph Henry (1797–1878), the science advisor to President Lincoln, who proposed a number of electric principles and invented electric devices (transformer, dynamo, and others) generally credited to Michael Faraday (1791–1867) of England. Henry's work was little known nor appreciated by most of his contemporaries even though his accomplishments were probably greater than those of Faraday. The reason is that Henry's work was published long after his research was completed and it was published in the *American Journal of Science,* a paper read by few people, usually not by those in the centers of learning in Europe.

There are probably more scientists living and working today than in all the past ages of man combined. With the development of complex scientific instruments, the acquisition of knowledge and data has been constantly accelerating during the last century. This "explosion" of information has produced a communication problem for students and workers in the fields of science. Organization and interpretation of such data would require many lifetimes. Fortunately, the resourceful mind of man has developed the computer.

1-24 Computers serve many functions. They are used to index information, to analyze data, to solve complex mathematical equations, and to monitor and control operations of industrial plants and space vehicles. The photograph shows the control (nerve) center of a large oil refinery. Process control computers minimize operational errors, increase product yield and quality, lower costs, enhance plant safety, and provide processing flexibility.

Without the aid of computers, scientists would be hopelessly trapped in a jungle of data accumulating from thousands of research laboratories all over the world. With the introduction of computers, many of the problems of processing and storing data have been largely solved. Computers, however, do not eliminate the need for intensive scientific training as is believed by some uninformed people. The use of most scientific instruments, including the computer, demands intelligent and highly trained individuals.

All citizens should play a role in the decisions made by their government representatives. Many decisions are related to scientific and technological developments. It is apparent that all citizens should be informed as to the latest developments taking place in the world of science. As chemistry students, you should be aware of sources of such information. There are at least three excellent sources of current information for the beginning chemistry student. These are *Chemistry,* and *Chemical and Engineering News,* both published by the American Chemical Society; and *Scientific American,* a popular publication that covers all areas of science. In addition to these publications, every chemistry student should be acquainted with the *Handbook of Chemistry and Physics* published by the Chemical Rubber Publishing Company, or similar handbooks. These handbooks provide a wealth of data related to the chemical and physical properties of elements, compounds, crystals, solutions, and various mixtures.

The important role played by publications as a source of ideas for the present and future generations was noted by René Descartes (1596–1650) in the following extract: ". . . to induce intelligent man to try to advance farther by contributing each, according to his inclination and ability, to the necessary experiments and also by publishing their findings. Thus, the last would start where their predecessors had stopped, and by joining the lives and works of many people, we would proceed much farther together than each would have done by himself."

Special Feature: Water Pollution

Man has always depended upon a source of fresh water for his survival. Early Egyptian, Babylonian, and Greek cultures were located near rivers or in river valleys. The river served not only as a source of fresh water but also as a means of disposing of waste products. By the eighteenth century in England and in Europe, rivers near large population centers were polluted. The human population explosion and the growth of industry over the past 150 years has now caused a serious world-wide water-pollution problem. At the present time, most of the rivers and lakes of the world are already polluted or at least threatened with human waste products.

A human could theoretically exist on as little as one-half gallon of water per day although at the present time each man, woman, and child in the United States, directly or indirectly, "uses" 1900 gallons of water per day. The indirect use is illustrated by the data in Table 1-8.

TABLE 1-8
WATER CONSUMPTION IN INDUSTRIAL PRODUCTION

Product	*Gallons of Water Required for Production*
1 loaf of bread	250
1 lb of meat (beef)	1000
1 lb of vegetables or fruit	2000
1 ton of paper	100 000
1 ton of steel (amount in one small automobile)	100 000
$\frac{1}{2}$ ton of gasoline (about 130 gal)	1 000 000

Over 97 percent of the world's water is sea water and an additional 2 percent is tied up in polar ice caps. This means that all of the rivers, streams, lakes, and reservoirs in the world constitute only one percent of the earth's water. As population increases and industry grows, our supply of fresh water per person rapidly diminishes. Trouble ahead or at hand is obvious.

The water used by agriculture and industry is returned to streams, lakes, and the ocean often with various pollutants present. These pollutants may contain metallic salts of mercury, arsenic, copper, and lead, as well as nutrients such as phosphates and nitrates. Organic pesticides such as DDT are also discharged into our water systems. In rivers and streams which are still capable of supporting aquatic life, food chains sometimes tend to concentrate certain toxic materials such as mercury and pesticide poisons. This presents a real danger to humans who, at the end of the food chain, consume the fish. Many species of birds and other animals are similarly threatened.

Many rivers and lakes are becoming so polluted with these contaminants that they are considered to be "dead." In cases like Lake Erie, the world's thirteenth largest body of fresh water, even decades of strict enforcement of antipollution policies may not return the water to its natural state. Lake Erie has been termed an ecological disaster since virtually all aquatic life in the lake is in trouble.

A condition known as eutrophication occurs in water where excessive nutrients have been discharged. Manufacturing plants often dump excessive waste products such as vegetable oils or milk products into a river. Slaughter houses may discharge blood into the same river. When these and other organic materials are mixed with a combination of nitrates from agricultural fertilizers and phosphates from human wastes, detergents and other sources, a rich nutrient solution results. It may seem that this would tend to promote the growth of aquatic life. If the concentration of these pollutants were small enough, this could be true. In many rivers, streams, and lakes, however, the concentration of these nutrients is so great that conditions known as "algae blooms" occur. These are very rapid and dense growths of algae and other plants. This rapid growth of plants is even further accelerated in bodies of water if the temperature is increased. Thus, the thermal pollution or heating of the water by industrial as well as natural processes also contributes to our environmental pollution. Algae are, of course, a green, oxygen-producing plant and most of the oxygen produced goes into the air. When the dense algae growths and other plants die, the resulting decay of the organic matter uses up the oxygen dissolved in the water. The decaying process is represented by the expression

$$\textbf{organic plant material} + O_2 \xrightarrow{\text{bacteria}} CO_2 + H_2O$$

The loss of oxygen dissolved in the water results in the death of fish and other animals present. The subsequent decay of the fish lowers the oxygen concentration even further so that the body of water becomes "dead."

The solution to the problem of eutrophication of lakes and streams is not a simple one. One method of restoring normal biological cycles to these lifeless bodies of water is to aerate them mechanically and return the oxygen supply. A better method, of course, is to stop discharging nutrient effluents which caused the problem. This is ultimately the solution we must find because all other methods of restoring life to our streams and lakes will fail if we continue to pollute them heavily. This solution will also entail great expense because it requires many industries and towns to clean their wastes.

Since algae require at least fifteen elements for growth, a method of controlling them would be to limit one of their essential requirements. At present it appears that phosphates are probably the easiest factor of control. Phosphates have been used extensively in detergents as well as in a variety of agricultural fertilizers. Detergent manufacturers are now making products which they claim to be free of phosphates. They claim that many of the products they manufacture are biodegradable. This means they contain materials which fit into nature's cycles rather than interfere with them. An example of a biodegradable material is a detergent which is readily broken down by bacterial action.

Solutions to the water-pollution problem are being devised although it is possible that several decades will be required before many of these solutions can be extensively implemented. It has been estimated that over 100 billion dollars could be required to clean up all of the domestic water pollution in the United States alone. It is essential to our existence that this environmental problem be solved even if this high cost estimate is accurate.

QUESTIONS

1. Give examples of each (a) nature's raw materials which are processed by the chemical industry, (b) consumer goods produced by the chemical industry from raw materials, (c) other industries using materials supplied by the chemical industry, (d) reactions or applications where the chemist is more interested in energy considerations than in the material substances produced.

2. (a) What are some specific natural resources that are in danger of being depleted? (b) What steps can be taken to help conserve our planet's resources?

3. How has chemical technology contributed to each? (a) environmental beauty, (b) population problems, (c) food production.

4. Under what conditions should a model be modified?

5. (a) What are the reasons that the mass of iron rust produced when a piece of iron is burned is greater than that of the original iron, but the mass of the ash remaining after a piece of wood is burned is less than that of the original wood? (b) Prove that matter was conserved in both cases?

6. The gravitational force on the moon is only about $\frac{1}{6}$ of that on the earth. A beaker of sand is just counterbalanced on a platform balance with a one-kg mass. If this whole apparatus were transported to the moon, would they still be balanced? The same beaker is weighed on a bathroom spring scale on earth and found to weigh 2.2 pounds. What mass would the bathroom scale indicate for this same beaker of sand on the moon?

7. Are chemical facts more often obtained by interpretation of chemical theory or from experiments? Explain.

8. Why should original experimental data be recorded in ink in permanently bound record books and dated?

9. What branch of science would be most directly concerned with (a) the temperature changes that occur when a gas expands, (b) the composition of a rock, (c) the classification of trees, (d) the digestion of food in the stomach, (e) how burning occurs, (f) identifying contaminants in food, (g) the intensity of light emitted by a firefly, (h) the size of a drop of water, (i) how laundry bleach works, (j) how to make a new plastic?

10. Give examples showing the relationship of chemistry to geology, zoology, physics, agriculture, astronomy, bacteriology, meteorology, psychology, botany, and anthropology.

11. Refer to biology and physics as well as your chemistry text and give examples of concepts or applications from each of these fields of science that are used in each of the other two: (a) chemistry, (b) physics, (c) biology.

12. Classify these statements as to whether they are observations or explanations. (a) The moon goes around the earth. (b) The moon moves through the sky. (c) A flame is hot. (d) A fire uses oxygen. (e) Steel rusts when exposed to moist air. (f) Rust weighs more than the iron from which it was formed. (g) Iron combines with oxygen when it rusts.

PROBLEMS

A. Scientific notation

1. Express each of these numbers as a number between 1 and 10 multiplied by 10 raised to a power. (a) 22 400, (b) 454, (c) 0.454, (d) 300, (e) 0.000 01, (f) 0.000 100 5, (g) 186 000, (h) 30 000 000 000, (i) 12, (j) 0.082.

B. Significant figures

2. How many significant figures in each of these numbers? (a) 275, (b) 0.0565, (c) 4025, (d) 2250, (e) 1.18×10^3, (f) 5 hundredths, (g) 15 thousand, (h) 5 thousandths.

3. Round off to 3 significant figures. Express as a simple number times a power of 10 if the situation warrants. (a) 3.578, (b) 5.322, (c) 4.505, (d) 75.45, (e) 0.003 064, (f) 3481.

4. Express the answers to these problems using the correct number of significant figures.
(a) $2.345 + 0.0761 + 13.51 + 0.0025 =$ **Ans. 15.93.**
(b) $0.0653 + 44.21 + 7.898 + 0.0048 =$
(c) $856.20 - 0.678 =$ **Ans. 855.52.**
(d) $245.61 - 0.672 =$

(e) $\frac{42.7 \times 2.11 \times 444}{156} =$ **Ans. 256.**

(f) $\frac{25.2 \times 3.45 \times 333}{221} =$

(g) $\frac{85.10 \times 2.656 \times 10^{-5} \times 4.666 \times 10^{-3}}{6.443 \times 10^{-25} \times 2.051 \times 10^{23}} =$ **Ans. 7.981×10^{-5}.**

(h) $\frac{48.20 \times 7.544 \times 10^{-6} \times 5.233 \times 10^{-4}}{3.771 \times 10^{-26} \times 3.072 \times 10^{24}} =$

C. Slide rule

5. Use a slide rule to solve these problems. Approximate your answers by using mental arithmetic.
(a) 1.5×3.0 **Ans. 4.5.**
(b) 3.0×6.0
(c) $4.0 \times 37 \times 2.1$ **Ans. 3.1×10^2.**
(d) $\frac{15}{3.0}$
(e) $\frac{2.5 \times 3.0}{1.5}$ **Ans. 5.0.**
(f) $\frac{25.6 \times 273 \times 7.40 \times 10^2}{3.00 \times 10^2 \times 7.60 \times 10^2}$
(g) $\frac{6.6 \times 10^{-27} \times 3.0 \times 10^{10}}{6.0 \times 10^{-5}}$ **Ans. 3.3×10^{-12}.**
(h) $\frac{2.45 \times 0.895 \times 31.7}{145.6 \times 0.0206}$
(i) $\sqrt{380}$ **Ans. 19.5.**
(j) $\sqrt{14.70}$
(k) $\sqrt{3.50}$ **Ans. 1.87**

D. Mathematical operations

6. Carry out these algebraic operations. Refer to Appendix 1E or an algebra book as needed.

(a) $\frac{5.00}{X} = \frac{2.0}{3.0}$ **Ans. 7.5.**

(b) $X = \frac{(5.00 \times 10^{-4})^3}{(5.00 \times 10^{-2})^2}$

(c) $\frac{7.20 \times 10^2}{273} = P\left(\frac{2.00}{3.00 \times 10^2}\right)$ **Ans. $P = 3.96 \times 10^2$.**

(d) $\frac{139\,X}{74.5} + \frac{139(10.0 - X)}{119} = 14.4$

(e) $50.0 = \frac{X^2}{(2.00 - X)(2.00 - X)}$

(f) $50.0 = \frac{X^2}{(1.0 - X)(2.0 - X)}$ **Ans. 2.1 and 0.98.**

(g) $X = \log 3600$

(h) $X = -\log 3.6 \times 10^{-3}$ **Ans. 2.4.**

(i) $\log X = 2.5$

(j) $\log X = 3.500$ **Ans. 3162.**

(k) $X = \log 10^2 + \log 10^{-4}$

E. Conversion factors and unit-cancellation

7. Solve these problems in two steps. First set up conversion factors using dimensions only. Then put in the proper numbers. (a) A sprinter runs the 100-yard dash in 11.0 seconds. Express his speed in miles per hour. (b) A jet airplane flies at the speed of 700 miles per hour. Express this speed in centimeters per second. **Ans. 3.13×10^4 cm/sec.**
(c) At 25°C, hydrogen molecules travel at an average speed of 1770 meters per second. Express this speed in kilometers per hour. **Ans. 6360 km/hr.**
(d) Light travels at a velocity of 186 000 miles per second. A light-year is the distance that light travels in one year. How many miles is it to the nearest star which is approximately 4.50 light-years away? **Ans. 2.64×10^{13} mi.**
(e) A drop of water contains approximately 1.70×10^{21} molecules. If the water were evaporating at the rate of one million molecules per second, how many years would it take for the drop of liquid water to disappear completely? **Ans. 5.39×10^7 years.**
(f) A gold atom has a mass of 3.30×10^{-22} g. Calculate the mass of 6.02×10^{23} gold atoms. (g) One querk contains 2.0 warts; 3.0 querks make 1.0 *zags;* 5.0 zags compose 6.0 nerfs, and 4.0 nerfs join to make 5.0 wigs. How many warts are contained in 13 wigs? **Ans. 52 warts.**

F. Measurement

8. Convert these quantities into grams (a) 5270 mg, (b) 2.2 kg, (c) 2.50 lb, (d) 16.0 oz, (e) 75 ml of water at 4°C.

9. Express these quantities in terms of liters (a) 120 ml, (b) 250 cm^3, (c) 22400 ml, (d) 300 g of water at 4°C, (e) 1.5 ml.

10. Express these quantities in centimeters (a) 15 mm, (b) 760 mm, (c) 4861 Å, (d) 5500 microns, (e) 330 millimicrons, (f) 76 meters, (g) 0.10 kilometers, (h) 8.0 inches, (i) 2.5 yards.

11. Calculate the number of cm^2 in 1.00 ft^2.

12. Calculate the volume of each of these solids. (a) A rectangular solid 5.0 cm long, 2.0 cm wide, and 1.5 cm deep. (b) A cube 10 cm on an edge. (c) A cylinder 10 cm high and 2.0 cm in diameter. (d) A sphere having a diameter of 4.00 cm.

13. A special flea can hop 5.60 mm, skip 8.90 cm, and jump 0.820 m. What is the total distance in meters he (she or it?) can travel in 20 hops, 10 skips, and 2 jumps? **Ans. 2.54 m.**

14. If a 3.220-kg cat eats a 213.6-g rat which has just popped a 4.120-mg piece of cheese into its mouth, the cat will then weigh in at how many kilograms? (Total = cat + rat + cheese.) **Ans. 3.434 kg.**

G. Density and Specific Gravity

15. Gallium is a metal which melts at 27°C. Its density at this temperature is 5.88 g/ml. How many grams of liquid gallium fills a 25.0-ml vessel? **Ans. 147 g.**

16. The density of benzene at 25°C is 0.879 g/ml. What volume does 500 g of benzene occupy? **Ans. 569 ml.**

17. When 17.6 g of a metal is placed in a graduated vessel containing 10.00 ml of water, the water level rises to the 12.20-ml mark. What is the density of the metal? **Ans. 8.00 g/ml.**

18. A 50-ml graduated cylinder weighs 57.60 g. When filled with an unknown liquid to the 50.00-ml mark, the cylinder and contents have a total mass of 97.6 g. What is the density of the liquid? **Ans. 0.800 g/ml.**

19. A volume of 15.0 ml of concentrated sulfuric acid has a mass of 27.9 g. (a) What is the density of the acid? (b) What is the mass of 100 ml of the acid? **Ans. (a) 1.86 g/ml, (b) 186 g.**

20. A mass of 940 g of sugar is dissolved in 400 g of water, yielding 1000 ml of sugar solution. (a) Calculate the density of the solution in g/ml, (b) Calculate the percentage by mass of sugar in the solution, (c) How many milliliters of this solution contain 235 g of sugar? **Ans. (a) 1.34 g/ml, (b) 70.0%, (c) 25.0 × 10 ml.**

21. A glass bulb with a stopcock has a mass of 33.20 g when completely evacuated at 20°C. The bulb has a mass of 33.33 g when filled with an unknown gas at 20°C. The bulb holds 25.0 ml of water. What is the density in g/l of the unknown gas at 20°C? **Ans. 5.20 g/l.**

22. Two identical graduated cylinders are balanced on a platform balance. When 150 ml of alcohol having a density of 0.80 g/ml is placed in one of the cylinders, how many ml of carbon tetrachloride having a density of 1.60 g/ml must be placed in the second cylinder to restore the balance? **Ans. 75 ml.**

23. Two identical graduated cylinders are balanced on a platform balance. When 50.0 ml of carbon tetrachloride is placed in one of the cylinders, 91 ml of benzene poured into the other restores balance. What is the density of the benzene? **Ans. 0.88 g/ml.**

H. Percentage Problems

24. Calculate the percentage of *X*, *Y*, and *Z* in a mixture containing 4.00 g *X*, 1.00 g *Y*, and 5.00 g *Z*. **Ans. *X* = 40.0%, *Y* = 10.0%, *Z* = 50.0%.**

25. A sample of brass weighing 4.50 g contains 3.15 g of copper. (a) What is the percentage by mass of copper in the brass? (b) How many grams of copper are required to produce 500 g of brass? **Ans. (a) 70.0%, (b) 350 g.**

26. How many pounds of materials *X* and of *Y* must be added to 200 pounds of material *Z* so that the final mixture contains 0.500 lb of *X* and 0.600 lb of *Y* per 25.0 lb of mixture? **Ans. 9.21 lb.**

2

Fundamental Concepts of Matter

Preview

One of our objectives in studying chemistry is to investigate the fundamental nature of the basic materials which compose our environment. We are surrounded by a vast number of different kinds of materials. If we are to achieve our objective, it is necessary that we start our study with the simplest type of materials.

In this chapter we shall first describe methods available for separating and resolving materials, as found in nature, into the simplest pure substances. Quantitative measurements provide data related to the formation and composition of these substances. Explanation of these data in terms of the nature and behavior of the particles which make up the substances constitutes a major goal of science and is one of our objectives.

"The chemists are a strange class of mortals, impelled by an almost insane impulse to seek their pleasure among smoke and vapor, soot and flame, poisons and poverty, yet among all these evils I seem to live so sweetly, that may I die if I would change places with the Persian King."

Attributed to Johann Joachim Becher (1635–1682)

The discovery that pure substances react only in definite atomic and molecular ratios gives rise to a need for a method of measuring a specific number of molecules of a given substance. In other words, the chemist needs to know how much of a given chemical is needed to produce a required amount of product in a specific reaction. He bases these calculations on the masses and the number of molecules or other kinds of particles taking part in a reaction. This problem leads us to trace the development of the atomic mass concept, the establishment of a scale of relative atomic masses, and the definition of the ***mole*** as the chemical unit used by the chemist to "count" atoms and molecules.

In this chapter, the concept of relative atomic masses and the mole are used to determine experimentally the formula of a compound. In the next chapter we shall use the mole concept to develop quantitative relationships between substances which participate in a chemical reaction.

In the last major section of this chapter, we shall present evidence which suggests that atoms in some compounds behave as though they carry an electric charge. We shall use these charges to develop a convenient method for writing formulas.

Finally, we shall investigate the characteristics and naming of compounds which you will encounter throughout your chemistry course.

MIXTURES

2-1 Types of Mixtures Many samples of matter have a *nonuniform composition.* Such samples are called ***heterogeneous mixtures.*** Consider a piece of granite. Physical examination reveals that it consists of nonuniformly distributed particles which a chemist or geologist identifies as quartz, mica, and feldspar. Visible physical boundaries separate regions (***phases***) having different compositions. Each region or phase in a heterogeneous sample has a *uniform composition* and is said to be *homogeneous.* Small crystals of mica are homogeneous pieces of matter and may be removed from the heterogeneous granite.

2-1 Granite is a heterogeneous mixture composed of minerals representing four mineral groups (B. M. Shaub).

Other materials such as the air we breathe or beverages we drink appear to have a uniform composition and contain only one kind of matter. However, experiments designed to separate these materials into simpler components reveal that they are ***homogeneous mixtures.*** In a practical sense, homogeneous refers to how the sample appears under an ordinary microscope. Of course, on the atomic or molecular level, all samples of matter are heterogeneous. The solutions which you will use in almost every laboratory experiment are homogeneous mixtures. Because of their importance in your activities, we shall briefly examine their components and some of their characteristics at this time. A more detailed investigation will be made later.

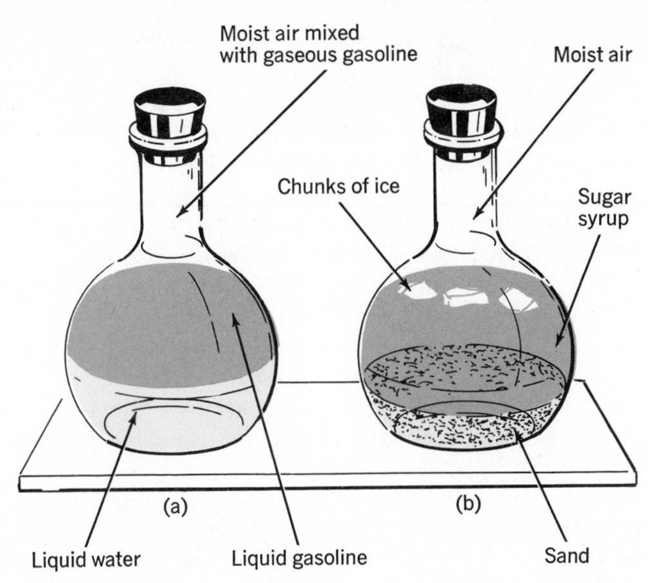

2-2 The system in flask *a* consists of three phases, while the system in flask *b* consists of four phases.

2-2 True Solutions Consider a solution prepared by adding a small quantity of table salt to water. When the salt has dissolved, the resulting solution is clear. The transparency of true solutions leads to the conclusion that the dissolved particles are too small to reflect light. The dissolved substance is known as the ***solute.*** Water, the dissolving medium, is called the ***solvent.*** In general, the component present in the greater quantity is considered to be the solvent. The one exception is a solid-liquid solution in which the liquid is always considered to be the solvent.

A sample taken from one part of a solution has the same properties and composition as a sample taken from any other part. This observation leads to the conclusion that *true solutions are single-phased, homogeneous systems consisting of two or more components.*

The composition of a true solution can be varied by varying the ratio of solute to solvent. Using simple apparatus you can readily show that properties of solutions such as boiling points, freezing points, density, and conductivity of electricity depend on the ratio of dissolved solute to solvent.

When a true solution is allowed to stand, the solute particles do not settle out on the bottom of the container. The solute particles are of molecular dimension and they pass unchanged through ordinary filter paper.

Substances in each of the three phases of matter may act as either solutes or solvents. In general, we shall be concerned with solutions in which the solvent is a liquid. The solute may be a solid, liquid, or gas. Salt dissolved in water, water dissolved in alcohol, and carbon dioxide (gas) dissolved in water are familiar examples of each of the three types of solutes dissolved in a liquid that form a true solution. The extent to which a solute dissolves in a solvent is known as its ***solubility.***

Gaseous solutes may be separated from liquids by heating the liquid or by reducing the pressure of the gas on the surface of the liquid. You have all seen the fine bubbles of dissolved air which rise to the surface and escape when you heat a pan of water. Also, you may have noticed the flat taste of soda pop which has lost its dissolved carbon dioxide by being allowed to stand at room temperature. These observations suggest that the *solubility of a gas in a liquid decreases when the temperature increases.*

The effect of reducing the pressure on a liquid containing a dissolved gas may be observed by removing the cap from a bottle of soda pop. The sudden reduction of pressure permits the dissolved carbon dioxide to escape rapidly (*effervesce*) into the atmosphere. Quantitative measurements lead to the conclusion that the *solubility of a gas in a liquid is proportional to the pressure of the gas on the surface of the liquid.*

In order to obtain the simplest kinds of matter required for our investigation, it is necessary to separate mixtures into pure substances. Pure substances can be recognized by their characteristic properties. Modern analytical techniques enable the chemist to separate and determine the composition of even the most complex mixtures. Let us examine some of the methods devised by chemists for the *physical* separation of mixtures. Physical methods of separation do not alter the composition of the components.

SEPARATION OF MIXTURES

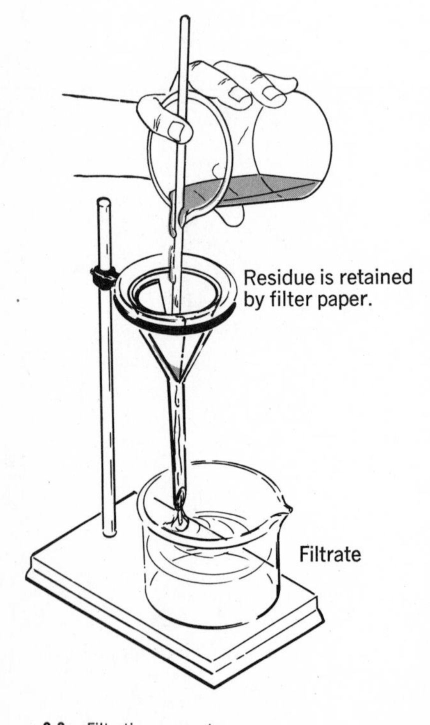

2-3 Filtration apparatus.

2-3 Filtration Consider a heterogeneous mixture of water and clay or sand. Such a mixture is a multi-phase system known as a ***suspension.*** Many suspensions can be separated into their components by *filtration.* When the mixture is allowed to stand, the particles of clay or sand slowly settle to the bottom of the container. The solid particles may be separated from the liquid by pouring the mixture through a filtration apparatus similar to that shown in Fig. 2-3. The large solid particles are retained by the filter paper, and the liquid passes through the paper into the receiving vessel. It should be noted that the solid particles in a suspension are not called the solute.

In general, a mixture of two solids can be separated by filtration provided there is a solvent available in which one of the solids is highly soluble and the other solid is not soluble. For example, salt and sand can be separated using water as a solvent for the salt. Similarly, carbon disulfide can be used to dissolve the sulfur in a mixture of sulfur and powdered iron. In both examples, the undissolved component can be separated by filtration.

2-4 Distillation In ordinary laboratory work, as well as in large-scale industrial operations, it is often necessary to separate solutes from solvents. One of the processes in the refining of crude oil consists of separating it into various grades of oils such as

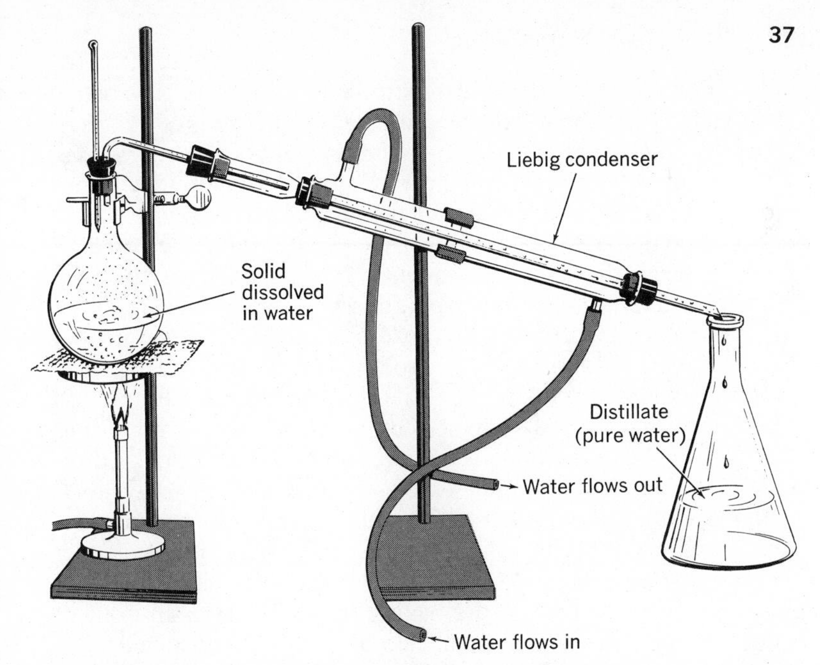

2-4 Distillation apparatus.

gasoline, kerosene, and a variety of other products. Physical methods of separating solutes from solvents are based on differences in their physical properties such as boiling points or solubilities.

In general, solids have much higher boiling points than liquids. If the solid and solvent are stable with respect to heat (do not decompose when heated) and nonflammable, and if one desires to recover only the solid solute, then ordinary evaporation may be employed to drive off the liquid. This technique must be avoided if the solid is readily decomposed upon heating.

If both the solid solute and liquid solvent are to be recovered, then ***simple distillation*** can be used to separate the components of a true solution. During the distillation process the liquid is *vaporized,* and the vapor is passed through a cooling tube where it is *condensed* to a liquid (*distillate*). The solid remains behind as a *residue* in the boiler.

Liquids which dissolve in one another in all proportions are said to be ***miscible.*** Miscible liquids may be separated by a process known as ***fractional distillation.*** The separation of such a mixture is based on the difference in boiling points of the liquids composing the solution. When a mixture of two miscible liquids such as alcohol and water is heated, alcohol, which has the lower boiling point of the two, vaporizes more rapidly than the water. Therefore, at the beginning of the distillation, the vapor and the distillate formed by condensing the vapor contain a greater percentage of alcohol than water. As the distillation proceeds, the distillate becomes richer in water. When the first fraction, the distillate from the first distillation, is redistilled, the second fraction has a higher percentage of alcohol. Repeated distillation of each fraction leads to mixtures that have different concentrations of the original components. Thus, the application of fractional distillation to such raw materials as petroleum yields fractions which are processed further to give us drugs, cosmetics, fuels, and a variety of other chemicals. Fractional distillation not only serves as the heart of the petroleum industry but more importantly, as the heart of the chemical industry.

2-5 Fractional Crystallization (Recrystallization) A common problem faced by chemistry students, as well as by the chemical industry, is the synthesis or preparation of pure substances. Often the nature of the process leads to a solid product contaminated by a relatively small quantity of a solid impurity used in the preparation of the pure product. For example, the commercial process for synthesizing ordinary baking soda (sodium bicarbonate) leaves the baking soda contaminated with ammonium chloride. Fortunately, baking soda is quite soluble in hot water but only moderately soluble in cold water. On the other hand, ammonium chloride is quite soluble in cold water and extremely soluble in hot water. These observations suggest that when a hot solution containing the mixture is cooled, the baking soda will crystallize out of solution while the ammonium chloride remains dissolved. The process is called ***fractional crystallization (recrystallization)***. In general, the impure mixture is dissolved in a minimum amount of hot solvent and allowed to cool. The purified product that crystallizes on cooling is then filtered to separate it from the liquid. Successive recrystallizations are sometimes necessary to obtain a pure product. The main problem faced by the experimenter is to find a solvent in which the solubility of the desired product is much greater at high temperatures than at low ones.

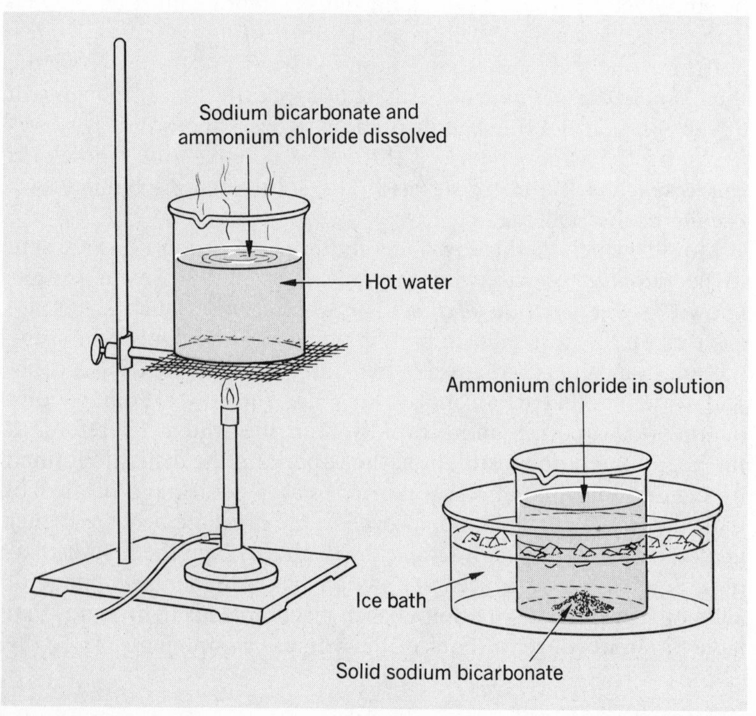

2-5 Separation and purification by recrystallization. At 60°C, 58 g of ammonium chloride and 17 g of sodium bicarbonate dissolve in 100 g of water. At 0°C, 30 g of ammonium chloride and 7 g of sodium bicarbonate dissolve in 100 g of water. If a mixture containing 17 g of sodium bicarbonate and 20 g of ammonium chloride is added to 100 g of water, heated to 60°C and then cooled to 0°C, how many grams of pure sodium bicarbonate can be filtered out of the solution? **Ans. 10 g.**

2-6 Extraction A significant difference between the solubility of a solute in two different solvents can be used as the basis for a physical separation provided the two solvents are ***immiscible.*** That is, the two solvents must separate into two distinct layers after they are mixed well and allowed to stand. For example, water and carbon tetrachloride are two immiscible liquids. The data in solubility tables reveal that iodine is 85 times more soluble in carbon tetrachloride than in water, whereas salt is soluble in water but not in carbon tetrachloride. These observations suggest that we might be able to separate the solutes in an *aqueous* (water) solution of iodine and salt by using the extraction technique. You can verify

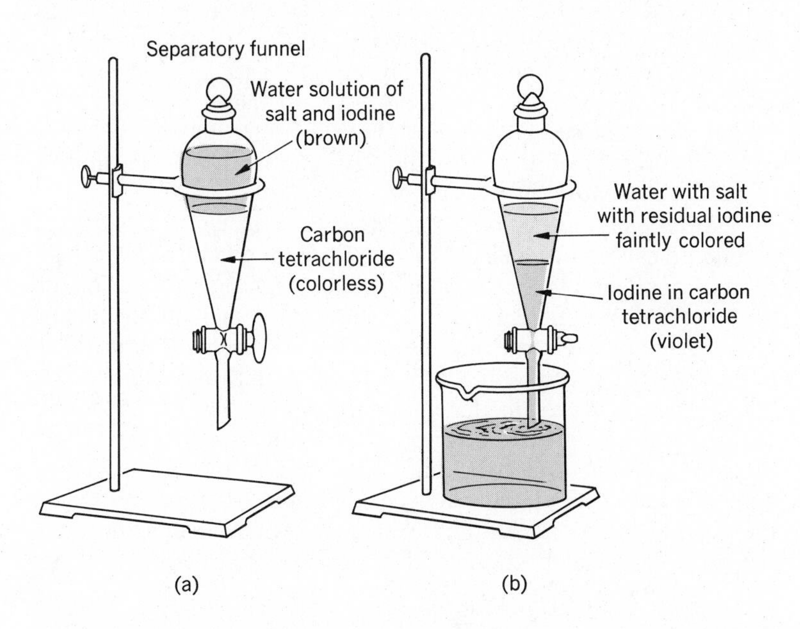

2-6 Iodine may be extracted from salt water by shaking in a separatory funnel with carbon tetrachloride. The carbon tetrachloride is denser than water and collects at the bottom of the funnel where it may be drained into a beaker. The salt does not dissolve in carbon tetrachloride and may be recovered from the water layer.

this suggestion by shaking a water solution of iodine and salt with some carbon tetrachloride in a *separatory funnel.* You will discover that the carbon tetrachloride layer turns violet as the iodine is removed from the water layer. Repeated extractions with fresh carbon tetrachloride removes virtually all of the iodine and leaves the salt in the water layer.

2-7 Recent Developments in Separation and Purification Techniques During the last 20 to 30 years a number of techniques have been developed which enable scientists to separate complex mixtures rapidly and efficiently into their components. Some of these newer methods and the principles underlying their use are given.

1. ***Chromatography,*** a process based on differences in the rate at which the components of a mixture move through a stationary phase under the influence of a moving phase. The stationary phase may be a solid resin packed in a column or a solution supported by a coarse paper or inactive solid. The components distribute

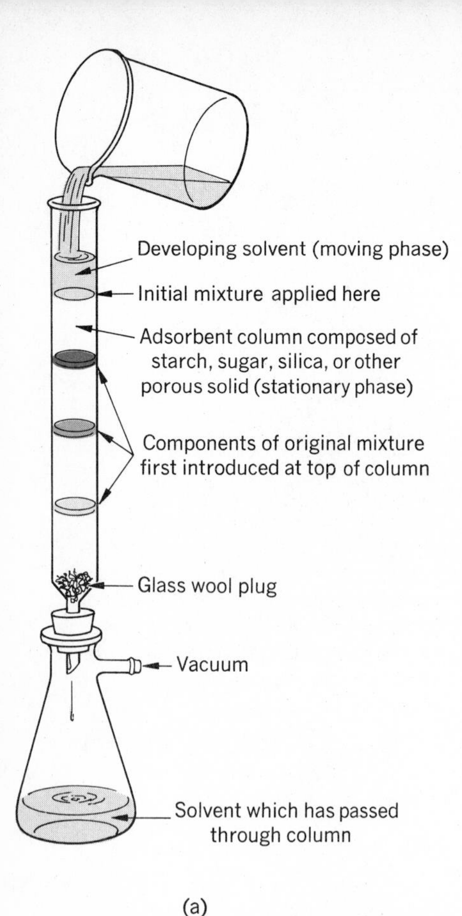

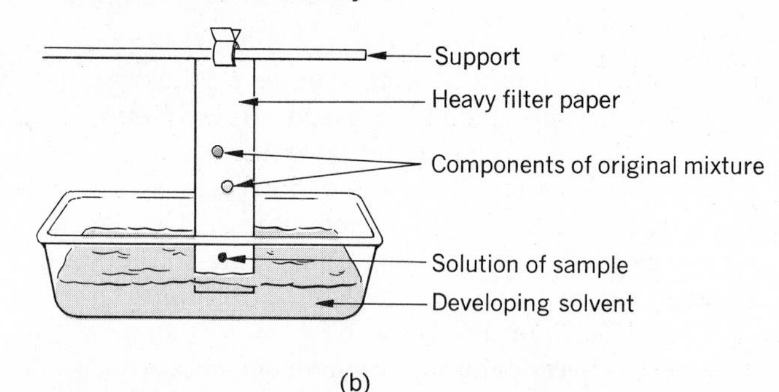

2-7 Separation of a mixture by chromatographic methods. (a) Column chromatography. The components of the mixture migrate at different rates through the column under the influence of a mobile phase (developing solvent). As the various bands (components) pass out the end of the column, they can be collected in separate containers. (b) Paper chromatography. Capillary action causes the solvent to move up the paper. The components of the mixture migrate at different rates and appear as spots at different points on the paper. The components may be recovered by cutting the paper.

themselves on the stationary phase as the moving phase goes over it. You can separate the components of a washable ink such as black, Quink by ***paper chromatography.*** The set-up is illustrated in Fig. 2-7(b). A ringstand and clamp may be used to hold the rod and the paper. A simple developing solvent may be prepared by adding one volume of methyl alcohol (CH_3OH), to nine volumes of water. Because of its importance, chromatography is the subject of a Special Feature at the end of this chapter.

2. ***Electrophoresis*** refers to the movement of charged particles in a liquid caused by an electromotive force (derived from a voltage source) applied to electrodes in contact with the suspension or solution. This process is often used to separate the components of blood serum or a mixture of proteins. One characteristic of proteins is that they can carry either a positive or a negative charge. The net charge on a protein depends partially on the acidity of the solution. For each degree of acidity, each protein has its own characteristic pattern of electric charge and rate of movement. Thus, when a mixture is placed between oppositely charged terminals, the component proteins migrate at their own characteristic rates toward the oppositely charged electrode.

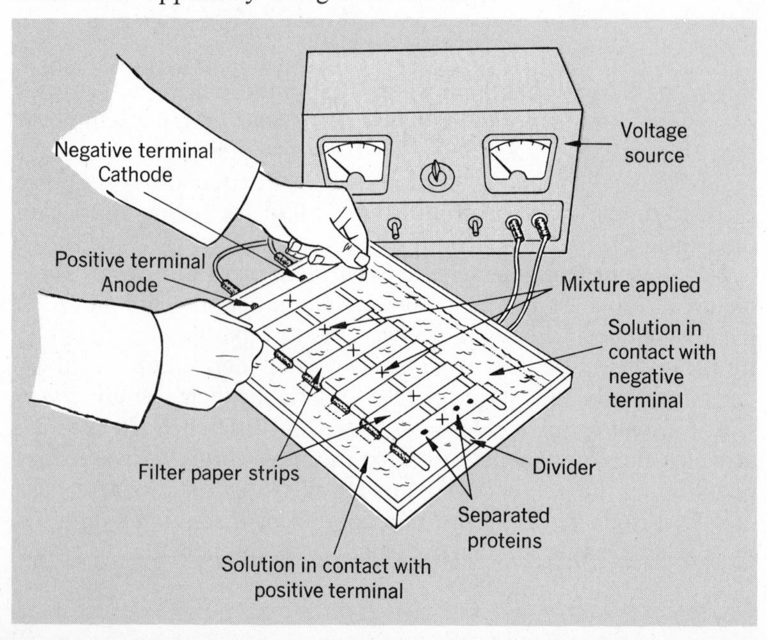

2-8 Electrophoresis apparatus. A drop of solution containing a mixture of proteins is placed in the center of paper strips. The strips are then clamped into a tray with one end of the paper immersed in a solution which contacts the positive terminal. When the voltage is turned on, each charged protein moves along the paper strip at its own characteristic rate toward the oppositely charged terminal. When the process is completed and the paper is sprayed with the proper indicator, the separated proteins appear as colored spots.

3. ***Ultracentrifugation,*** a process based on differences in the rate of sedimentation under the influence of a centrifugal force which may involve a gravitation force hundreds of thousands of times greater than that of the earth. This large centrifugal force is developed as a result of the tremendous speed reached by the revolving ultracentrifuge. These devices can spin at more than 5×10^4 revolutions per minute. The ultracentrifuge is often used to determine the purity of a protein solution and the relative masses of protein molecules.

PURE SUBSTANCES

2-8 Purity of Components We have now explored physical methods used by scientists to separate mixtures into simpler components. We can check the purity of the components by measuring the melting points of the solids or the boiling points of the liquids. *A pure solid has a sharp and constant melting point. A pure liquid* has a constant boiling point at a known pressure. If the substance is not pure, the temperature steadily changes as the substance melts or boils. Consider the distillation of a sodium chloride (NaCl) solution. The solution is resolved into two pure components: solid sodium chloride (salt) crystals and liquid water. Experimental measurements show that sodium chloride melts sharply at 801°C and water has a constant boiling point of 100°C at one atmosphere of pressure. We also find that it is impossible to resolve either sodium chloride or water into simpler substances by any physical method. These observations confirm that sodium chloride and water are pure substances.

2-9 Resolution of Pure Substances Resolution of pure substances into simpler pure substances whose properties differ from the original involves a ***chemical change.*** Electric energy is frequently used to bring about a chemical change. Both water and melted sodium chloride may be decomposed into simpler pure substances by an electric current. The process is called ***electrolysis.*** A simplified schematic diagram illustrating the electrolysis of melted sodium chloride is shown in Fig. 2-9. The apparatus used to electrolyze water is shown in Fig. 2-10.

An ordinary automobile battery is connected to electrodes which are located in the melted salt, and in the water which has been slightly acidified to increase its conductivity. As the electric current passes through the acidified water, a gas is produced at each electrode. The gases may be identified by their physical and chemical properties.

Physical properties may be observed without changing the identity or the composition of the substance. Physical properties commonly used to help identify substances are *phase, color, odor, melting point, boiling point, heat conductivity, electric conductivity in the solid and liquid phases, brittleness,* and *density.*

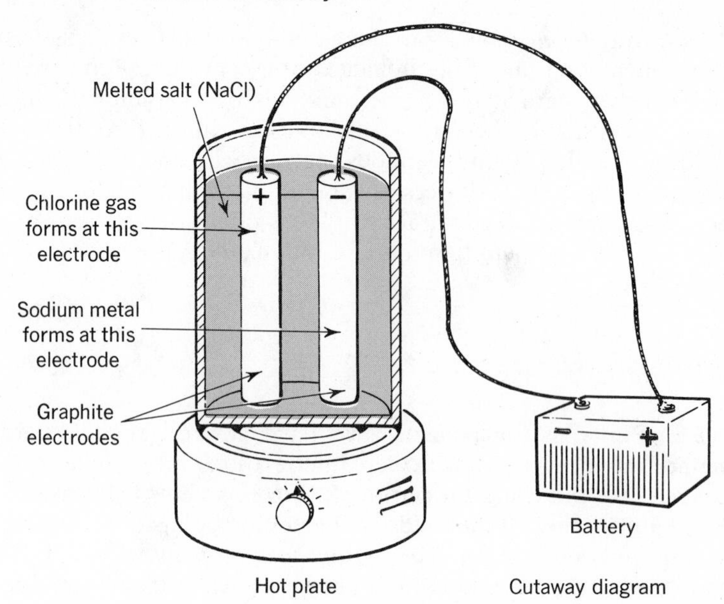

2-9 Electrolysis of melted salt (sodium chloride). The battery serves as an electron pump and as a source of energy. The electric energy supplied by the battery causes a chemical change to occur. The electric energy is converted into chemical energy of the elements being formed. Therefore, the sum of the energy of the elements formed is greater than that of the original compound (sodium chloride).

The ***chemical properties*** of a substance describe the manner in which the substance reacts in the presence of another substance.

In the electrolysis of water, the gas collected at the negative electrode is odorless, colorless, and has a density of 0.089 g/l at 0°C and 760 torr. It is combustible, does not support combustion, and explodes when it is mixed with air and ignited. These properties can be used to help identify the gas as hydrogen. The gas collected at the positive electrode is odorless, colorless, and has a density of 1.43 g/l at 0°C and 760 torr. It does not burn, but supports combustion and reacts with almost all other elements. These properties are characteristic of oxygen gas. Neither hydrogen nor oxygen can be resolved into simpler pure substances. Thus, pure substances may be grouped into two categories. Those that cannot be decomposed by chemical methods into simpler pure substances are called ***elements;*** those that can be resolved into simpler pure substances are called ***compounds.*** *Elements may be considered the chemical building blocks of matter.*

Electrolysis of melted sodium chloride results in the production of a silvery metal (sodium) at the negative electrode and a yellow-green gas (chlorine) at the positive electrode. The properties of the three substances are itemized in Table 2-1. From the data in Table 2-1 note that the electrolysis process altered the composition of the compound and produced new substances having properties and compositions entirely different from the original materials. Changes in which new substances are formed are called ***chemical changes.*** Chemical changes also involve physical changes. On the basis of the changes observed during the electrolysis of water and of sodium chloride we may conclude that the *properties of a compound do not resemble those of the elements in the compound.*

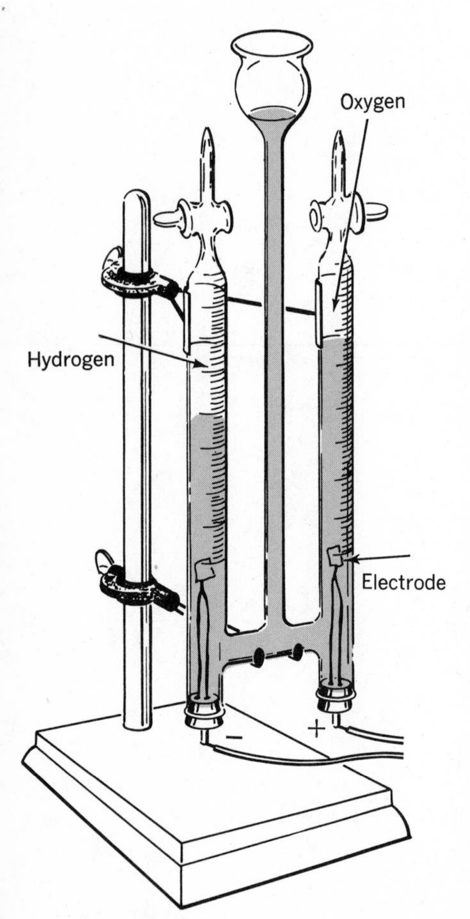

2-10 Electrolysis of water. The markings on the collection tubes reveal that the volume ratio of hydrogen gas to oxygen gas is 2 to 1. Which has the greater energy content, the water or the two gases formed by its decomposition?

TABLE 2-1
PROPERTIES OF SODIUM, CHLORINE, AND SODIUM CHLORIDE

Name	Type of Substance	Phase	Color	Toxicity	Relative Chemical Reactivity	Density at 25°C (g/ml)	Melting Point (°C)	Boiling Point (°C)	Electric Conductivity
Sodium	element	solid	silvery	toxic	high	0.97	97.8	892	excellent in solid and liquid phase
Chlorine	element	gas	yellow-green	toxic	high	0.0032	−100.98	−34.6	nonconductor
Sodium chloride	compound	solid	colorless (white)	non-toxic	low	2.165	801	1413	good in liquid phase; non-conductor in solid phase

2-10 Elements Instrumental analysis of naturally occurring or synthetically produced matter on earth or on any of the stars in the universe reveals that at present there is a total of 105 different elements. As far as is known at present, these elements are the chemical building blocks of the totality of matter in the physical universe. The occurrence of the elements in nature varies widely. Their relative abundance is shown in Table 2-2. These data reveal that two elements, oxygen and silicon, constitute about three-fourths of the earth's crust. Six other elements, aluminum, iron, calcium, sodium, potassium, and magnesium, make up almost all of the remaining one-fourth. Although hydrogen constitutes only 0.9 percent of the earth's crust, it is by far the most abundant element in the universe. The drawing, Fig. 2-11, shows that about 70 percent of the elements were discovered during the 18th and 19th centuries. Only a few were discovered in nature in the 20th century, but 16 were synthesized in scientific laboratories.

There is no standard system for the naming of the elements. Some were named for famous scientists; others were named after localities. A few of the elements are grouped in Table 2-3 according to the systems used for naming them.

The naming of the elements varies, of course, according to language. In contrast, the name of each element is represented by a *symbol* that is uniformly used throughout the world. Most of the symbols are derived from the first one or two letters of the element's name in English. For example, the symbol for aluminum is Al and that for bromine is Br. The symbols for elements such as lead (Pb) and copper (Cu) are derived from their Latin names which are, respectively, *plumbum* and *cuprum.* Common elements whose symbols are related to their early names are listed in Table 2-4.

The system of letter symbols was originated by the Swedish chemist, Jöns Jakob Berzelius (1779–1848). When a symbol contains two letters, the first is capitalized and the second is lower case (small letter). A list of the elements with their symbols is found on the inside back cover of this book. It is not necessary to memorize all of these names and symbols, but it is valuable for you to know

TABLE 2-2
RELATIVE ABUNDANCE OF ELEMENTS BY MASS PERCENTAGE

Element	Earth*	Atmos-phere	Sea Water	Dry Soil	Vege-tation Dry	Human Body	Universe (Approx.)
Oxygen	49.52	23.2	85.79	47.3	42.9	65.0	
Silicon	25.75			27.7	3.0		
Aluminum	7.51			7.8			
Iron	4.70		2×10^{-6}	4.9	0.04	0.004	
Calcium	3.39		0.05	3.47	0.62	1.6	
Sodium	2.64		1.14	3.0	0.43	0.3	
Potassium	2.40		0.04	2.46	1.68	0.4	
Magnesium	1.94		0.14	2.24	0.38	0.05	
Hydrogen	0.88	0.007	10.67	0.22	6.1	10.2	99
Titanium	0.58			0.50			
Chlorine	0.188		2.07	0.06	0.22	0.3	
Phosphorus	0.12			0.12	0.56	0.9	
Carbon	0.087	0.01	0.002	0.19	44.3	17.5	
Manganese	0.08			0.08			
Sulfur	0.048		0.09	0.12	0.37	0.2	
Barium	0.047						
Chromium	0.033						
Nitrogen	0.030	75.5			1.62	2.4	
Fluorine	0.027		1×10^{-4}	0.10			
Zirconium	0.026						
Nickel	0.018						
Strontium	0.017		0.001				
Vanadium	0.016						
Copper	0.010		1×10^{-6}				
Uranium	0.008		1×10^{-7}				
Tungsten	0.005						
Zinc	0.008		5×10^{-7}				
Lead	0.002		5×10^{-7}				
Cobalt	0.001						
Boron	0.001		4×10^{-4}				
Tin	1×10^{-4}						
Bromine	1×10^{-4}		0.008				
Mercury	1×10^{-5}						
Iodine	1×10^{-5}		4×10^{-8}				
Silver	1×10^{-6}		3×10^{-8}				
Platinum	1×10^{-7}						
Gold	1×10^{-7}		6×10^{-10}				
Radium	1×10^{-10}		5×10^{-16}				
Argon		1.3					
Neon		0.0013					
Helium		7×10^{-1}					1
Krypton		3×10^{-4}					
Xenon		4×10^{-5}					

*Average composition 10-mile crust, hydrosphere, and atmosphere.

TABLE 2-3
SYSTEMS FOR NAMING ELEMENTS
Elements Named After

Famous Scientists	*Localities*	*Properties*	*Heavenly Bodies*	*Mythological Characters*
Curium	berkelium	bromine	cerium	iridium
Einsteinium	californium	chlorine	helium	mercury
Lawrencium	germanium	gold	plutonium	thorium
Fermium	polonium	hydrogen	uranium	vanadium

TABLE 2-4
FORMER NAMES OF COMMON ELEMENTS

Element	*Origin of Symbol*	*Symbol*
Antimony	Stibium	Sb
Copper	Cuprum	Cu
Gold	Aurum	Au
Iron	Ferrum	Fe
Lead	Plumbum	Pb
Mercury	Hydrargyrum	Hg
Potassium	Kalium	K
Silver	Argentum	Ag
Sodium	Natrium	Na
Tin	Stannum	Sn

TABLE 2-5
COMMON ELEMENTS

Metals		*Nonmetals*	
Aluminum	(Al)	Boron	(B)
Barium	(Ba)	Bromine	(Br)
Calcium	(Ca)	Carbon	(C)
Copper	(Cu)	Chlorine	(Cl)
Gold	(Au)	Hydrogen	(H)
Iron	(Fe)	Iodine	(I)
Lead	(Pb)	Nitrogen	(N)
Magnesium	(Mg)	Oxygen	(O)
Mercury	(Hg)	Phosphorus	(P)
Platinum	(Pt)	Silicon	(Si)
Potassium	(K)	Sulfur	(S)
Silver	(Ag)		
Sodium	(Na)		
Tin	(Sn)		
Uranium	(U)		
Zinc	(Zn)		

the names and symbols of the more common ones, some of which are listed in Tables 2-4 and 2-5. The *symbol of an element may represent,* among other quantities, *one atom of an element.* Atoms may be thought of as the structural building blocks of the elements. We shall find that *atoms are the smallest identifiable unit of an element that can take part in a chemical change characteristic of that element.*

Examination of the properties of the known elements reveals that they may be grouped on the basis of their common properties into three categories: *metals, nonmetals,* and *metalloids.* The properties of individual metals vary to a great extent. Lithium is soft and light while platinum is hard and dense. Mercury is a liquid, but tungsten is extremely difficult to melt. Regardless of the individual variation in their properties, metals do have a number of characteristics in common. The general properties of metals are listed in Table 2-6. It can be seen that nonmetals generally have characteristics opposite to those of metals. The properties in Table 2-6 are those that can either be observed directly or measured. The description of a substance in terms of observable or measurable properties constitute what is known as an ***operational definition.***

TABLE 2-6
PROPERTIES OF METALS AND NONMETALS

Metals	*Nonmetals*
Have a luster	Have low luster
Good conductors of heat	Poor conductors of heat
Good conductors of electricity	Poor conductors of electricity
Malleable and ductile	Neither malleable nor ductile

Approximately 80 percent of the elements are classified as metals. In general, metals (except mercury) are solids at room temperature and relatively more dense than nonmetals. The chemical properties of metals are quite different from those of nonmetals.

Sometimes it is difficult to classify an element as either a metal or a nonmetal. Elements such as silicon, germanium, and arsenic are generally grouped with the nonmetals but have both metallic and nonmetallic characteristics. Elements with intermediate characteristics are generally called ***metalloids.*** These elements are characterized by a slight electric conductivity which increases with a temperature rise. They are known as ***semiconductors.*** This property is responsible for their use in the electronics industry as transistor material. The existence of metalloids is evidence that there is no sharp dividing line between the metals and the nonmetals but rather, a gradual transition from metallic to nonmetallic properties.

2-11 Organization of the Elements One of the great milestones in the progress of chemistry was the discovery by Dmitri Mendeleyev that the elements could be systematically arranged in such a way that elements having similar properties would be grouped together. The ***Periodic Table*** shown on the inside back cover of your text is the result of this discovery. In this table, the metallic elements are located on the left side of the heavy line which runs diagonally across the table.

In the Periodic Table, the columns of elements are called ***groups.*** A Roman numeral and a capital letter are used to identify each group. In general, elements in the same group have similar properties. They are rather like the members of your family. Individual members differ from one another in their mass, size, and characteristics. However, they usually resemble each other more than they do their neighbors.

Each square in the table contains as basic information the *symbol, relative atomic mass,* and identifying number (*atomic number*) of a specific element. The elements in several of the groups have family names. These are itemized below.

1. Group IA elements are known as the ***alkali family.*** The members of this family are all highly reactive metals with relatively low densities. They are easily corroded by air and all react vigorously with water and form hydrogen gas.

2. Group IIA elements are known as the ***alkaline-earth family.*** These are all silvery metals which react readily with oxygen forming oxides and with acids releasing hydrogen gas.

3. Group VIIA elements are known as the ***halogen family.*** These elements are all relatively reactive nonmetals which are never found free in nature. At room temperatures, fluorine (F) and chlorine (Cl) are gases, bromine (Br) is a liquid, and iodine (I) and astatine (At) are solids.

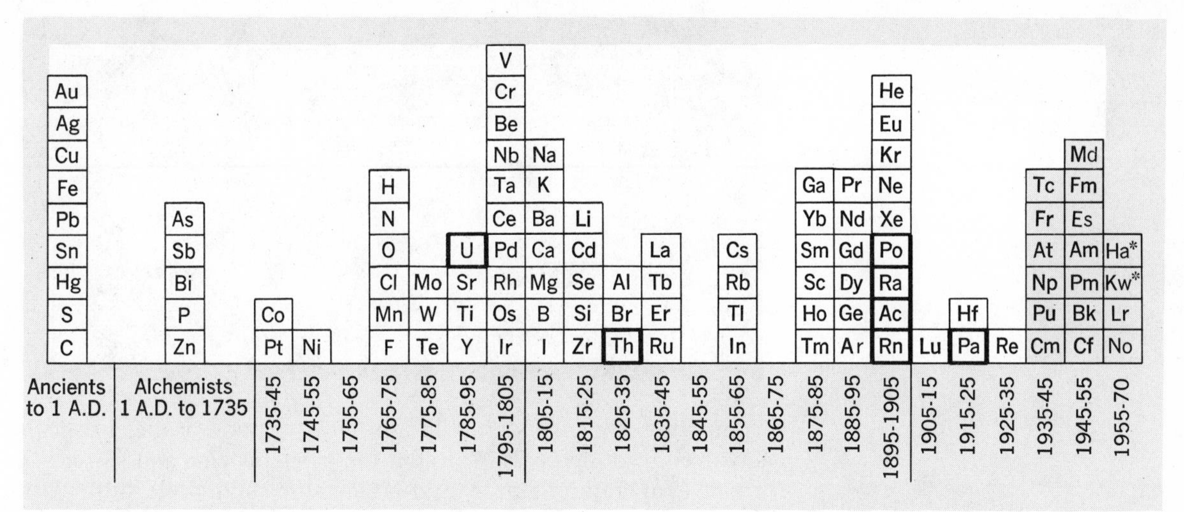

*Kurchatavium and hahnium are tentative names for 104 and 105.

2-11 Discovery of the elements. The heavy squares show the natural radioactive elements; shaded squares are synthetic elements.

4. Group O elements are known as the ***noble gases.*** These elements are all nonmetals and are gases at room temperatures. With rare exceptions, they do not react with other elements.

5. Collectively, the elements in the *B* groups are known as the ***transition elements.*** These elements are all metallic and include many common metals such as iron (Fe), copper (Cu), and silver (Ag).

6. The elements in the two long rows beneath the table are referred to as the ***inner transition elements.*** They are all metallic. With the exception of uranium (U), most of these elements are unimportant to students who are beginning chemistry.

You will find it helpful to become acquainted with the organization of the Periodic Table and the information it conveys. There will be many occasions for you to refer to this table throughout your chemistry course. We shall make a detailed study of its significance and theoretical basis in subsequent chapters.

The relationships between the different kinds of matter discussed in this chapter are shown in Fig. 2-12. Classifying the different kinds of matter into various categories and arranging the elements systematically in the Periodic Table helps to systematize and simplify the study of chemistry. For example, knowing the location and the properties of one element of a given family will enable you, in a general way, to predict the properties of the other elements in that family. With the exception of a few noble gases, all elements react and form compounds. Let us now investigate the formation and composition of compounds.

2-12 Compounds First we shall examine the quantitative relationship that exists between elements which combine and form a

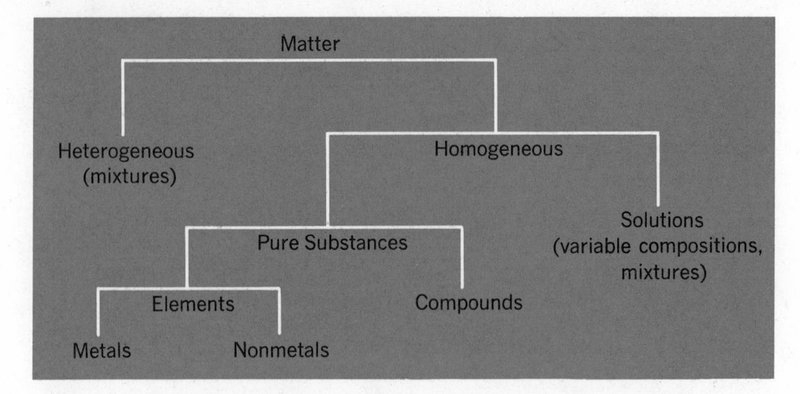

2-12 Classification of matter.

compound. We shall start with a brief description of the chemicals involved in the process. Consider the elements zinc and sulfur. At room temperature, zinc is a grey, metallic solid while sulfur is a yellow, nonmetallic solid. When zinc and sulfur react chemically, a white solid remains. It possesses neither the metallic properties of zinc nor the nonmetallic properties of sulfur. The white solid is a compound called *zinc sulfide.*

The names of compounds are derived from the names of the elements they contain. Compounds composed of only two kinds of elements are called ***binary compounds.*** *The name of a binary compound containing a metal and a nonmetal consists of the name of the metal followed by the stem of the name of the nonmetal to which is added the suffix -ide. The names of compounds are represented by formulas* just as the names of elements are represented by symbols. The formulas of compounds are derived by combining the symbols of the elements involved in a definite numerical ratio. Thus, *formulas indicate the different kinds of elements in a compound and the ratio in which the atoms of these elements are present.* The ratio is determined experimentally by methods described later in this chapter. In the case of our compound, experimental data reveal that the formula for zinc sulfide is *ZnS.*

FOLLOW-UP QUESTION

1. Name these compounds (a) the solid formed by the reaction of zinc with oxygen whose formula is ZnO, (b) the gas formed by the reaction of hydrogen with chlorine whose formula is HCl (assume hydrogen behaves as a metal in this reaction), (c) the solid formed by the reaction of calcium with bromine whose formula is $CaBr_2$, (d) the solid formed by the reaction of aluminum with carbon whose formula is Al_4C_3, (e) the solid formed by the reaction of potassium with iodine whose formula is KI, (f) the solid formed by the reaction of magnesium with nitrogen whose formula is Mg_3N_2, (g) the solid formed by the reaction of silver with sulfur whose formula is Ag_2S.

The process in which the chemical composition of any substance is formed is called a ***chemical reaction.*** Chemical reactions are represented by ***chemical equations.*** Both a *word equation* (1) and a *formula equation* (2) for the reaction between zinc and sulfur are shown on p. 49.

$$\text{(1)}\quad \underset{\text{(element)}}{\text{zinc}} + \underset{\text{(element)}}{\text{sulfur}} \longrightarrow \underset{\text{(compound)}}{\text{zinc sulfide}}$$

$$\text{(2)}\quad \underset{\text{reactants}}{Zn + S} \longrightarrow \underset{\text{product}}{ZnS}$$

These equations may be read: zinc plus sulfur yields zinc sulfide, or zinc reacts with sulfur and yields (produces) zinc sulfide.

The zinc and the sulfur are known as ***reactants,*** and the zinc sulfide is called the ***product.*** Reactants always appear on the left side of the equation; products appear on the right side. It may be observed experimentally that 65.4 g of zinc dust reacts with 32.1 g of sulfur and forms 97.5 g of zinc sulfide. Since no zinc or sulfur is left over as a result of this reaction, no matter is lost during the course of the reaction. The above data illustrate the *principle of conservation of matter* which was introduced by Lavoisier. The modern version is known as the ***Law of Conservation of Mass*** and states *in an ordinary chemical reaction the mass of the system remains constant.* All chemical reactions are described by this law.

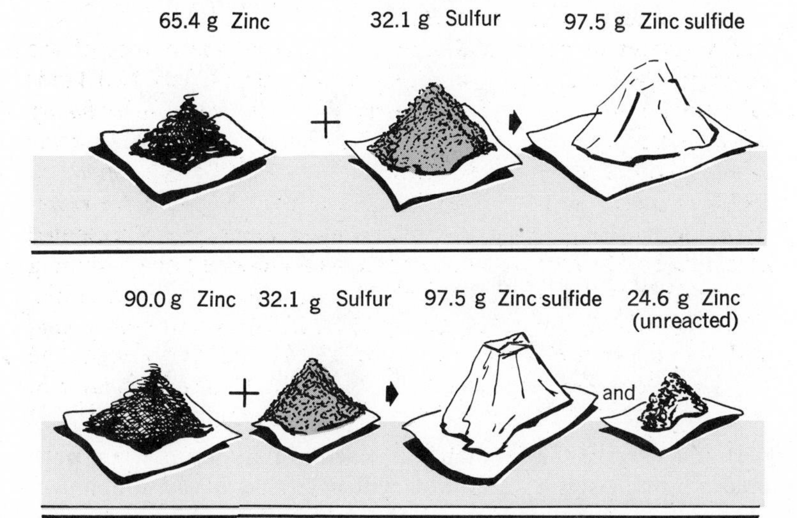

2-13 Zinc and sulfur always react in the same mass ratio shown in the illustration. If they are not present in this ratio, one of the reactants remains in excess.

If zinc and sulfur were present in any mass ratio other than 65.4 to 32.1, an excess of one or the other would remain upon completion of the reaction and could be separated from the product by an appropriate physical process. When zinc sulfide from any source is quantitatively analyzed, it is found to contain zinc and sulfur in the same fixed ratio of 65.4 parts zinc to 32.1 parts sulfur by mass. The definite composition of this compound is an illustration of the ***Law of Definite Composition.*** This law, which describes most chemical compounds, was discovered by a Frenchman, Joseph Louis Proust (1754–1826), in 1797. It may be stated as follows: *when elements combine and form a specific compound, they do so in definite proportions by mass.* If the mass composition of a compound is known, then application of this law allows us to calculate the quantity of an element which may be obtained by decomposing a given amount of its compound. The Example 2-1 illustrates the application of this principle.

Example 2-1

Iron oxide has a definite composition of 112.0 parts iron to 48.0 parts oxygen by mass. How many grams of iron can be obtained from 125 g of iron oxide?

Solution

The composition of the iron oxide indicates that 112.0 g of iron is combined with 48.0 g of oxygen in 160.0 g of iron oxide. By mass, the fraction of iron in the compound is $\frac{112.0}{160.0}$. This means that $\frac{112.0}{160.0}$ of the 125 g of iron oxide is iron. Using unit cancellation it can be seen that multiplying grams of iron per gram of iron oxide by grams of iron oxide yields grams of iron.

$$\frac{112.0 \text{ g iron}}{160.0 \text{ g iron oxide}} \times 125 \text{ g iron oxide} = 87.5 \text{ g iron}$$

Note that we are actually multiplying the given quantity (125 g iron oxide) by a conversion factor that converts grams of iron oxide to grams of iron.

FOLLOW-UP PROBLEM

Analysis of water reveals that it is composed of 2.0 parts of hydrogen to 16.0 parts of oxygen by mass. (a) How many grams of oxygen gas can be obtained by decomposing 549 g of water? (b) What is the mass percentage of oxygen in water? **Ans. (a) 488 g, (b) 88.9%.**

If you observe the spontaneous reaction that occurs when powered zinc and sulfur react, you will note that a vast amount of heat energy is liberated. When 65.4 g of zinc and 32.1 g of sulfur react, enough heat is liberated to increase the temperature of a pint of water from room temperature to its boiling point. We may use the following equation to show that heat energy is involved in this reaction.

$$\mathbf{Zn + S \longrightarrow ZnS + q}$$

Here, q represents the quantity of heat involved. The quantity of heat absorbed or liberated during a reaction is usually expressed in ***calories* (*cal*)** or ***kilocalories* (*kcal*).** *The calorie is the amount of heat required to raise the temperature of 1 gram of water 1 Celsius degree. One kilocalorie equals 1000 cal.* It is of interest to note that the so-called calorie values assigned to foods are actually kilocalorie values. In the reaction described above, 48.5 kcal is liberated. This observation indicates that there is more energy stored in the two reactants than in the product. This stored chemical energy is a type of *potential energy* of particular interest to chemists. It is sometimes called the *heat content of a chemical substance.* In general, *potential energy* is the stored energy that matter possesses because of its position, condition, composition, or electric charge.

When *heat energy is liberated* during the course of a reaction, the reaction is *exothermic,* the products have *less* heat content than the reactants, and the *energy of the chemical system decreases.* This means that heat is evolved, so that the energy content of the surroundings increases. ***The Law of Conservation of Energy*** requires that the *heat lost by the system be equal to that gained by the surroundings.* When *heat energy is absorbed* during the course of a reaction, the reaction is *endothermic,* the products have a *greater* heat content than the reactants, and the *energy of the system increases.* Since heat is absorbed from the surroundings, the energy content of the surroundings decreases. The quantity of heat absorbed in an endothermic reaction may be represented by placing the heat term on the reactant side of the equation. For example, the formation of nitric oxide gas from nitrogen gas and oxygen gas is an endothermic reaction. The *thermochemical equation* for the reaction may be written

$$\mathbf{q + N_2(g) + O_2(g) \longrightarrow 2NO(g)}$$

It should be noted that the heat term may be algebraically transposed to the opposite side of the equation. That is q is subtracted from both sides. Of course, the sign of q becomes changed. The above equation then reads

$$\mathbf{N_2(g) + O_2(g) \longrightarrow 2NO(g) - q}$$

FOLLOW-UP PROBLEM

When 16 g of oxygen reacts with 2.0 g of hydrogen and forms 18 g of water vapor, 59.6 kcal is liberated. The equation for the reaction is

$$2H_2(g) + O_2(g) \longrightarrow 2H_2O(g)$$

(a) Which has the greater heat content, the reactants or the products? (b) Is the reaction exothermic or endothermic? (c) Convert the equation to a thermochemical equation by adding a heat term. (d) How much energy is liberated when 8.0 g of oxygen reacts with 1.0 g of hydrogen? **Ans. (d) 29.8 kcal.**

In future chapters we shall extend and modify our model of atoms so that we can explain observed energy relationships in terms of structure and behavior. In this chapter, we are mainly interested in observing and explaining, in terms of the atomic theory, the mass relationships observed to exist between reacting substances. Let us now see how the laws of chemical combination are consistent with Dalton's simple atomic model.

The Law of Conservation of Mass led Dalton to postulate that atoms are indestructible. Thus, the atoms of the reactants in a chemical reaction simply undergo a rearrangement as they combine

C + O O → O C O

1 carbon atom and 2 oxygen atoms rearranged form 1 molecule of carbon dioxide

2-14 Dalton's atomic theory explains the conservation of matter in a chemical reaction. The atoms involved in a reaction undergo rearrangement but their number and mass remain constant. Hence the products of a reaction have the same mass as that of the starting materials.

and form new products. This means that the total number and mass of the atoms which react have to be identical to the number and mass of the atoms in the products.

The observation that a given compound always has a constant mass composition was the basis of his postulate that atoms of elements unite in a definite numerical ratio. The postulate that each element is composed of a particular kind of atom was based on his observation that properties of different elements were not the same. Dalton's billiard-ball atoms are used in Fig. 2-14 to represent the formation of carbon dioxide.

The validity and the value of the atomic theory of matter was established when it was successfully used to predict the ***Law of Multiple Proportions.*** This law is based on the mass relationships observed when the same two elements form more than one compound. Consider nitrogen and oxygen. These two elements are present in a number of different compounds known generally as the *oxides of nitrogen.* Experimental evidence reveals that in one of the compounds, 14 g of nitrogen has combined with 16 g of oxygen, while in another compound, 14 g of nitrogen has combined with 32 g of oxygen. These data are summarized in Table 2-7.

TABLE 2-7
NITROGEN COMPOUNDS WITH OXYGEN

	Nitrogen (fixed mass)	*Oxygen*
Compound 1	14 g	16 g
Compound 2	14 g	32 g

Observe that the mass of oxygen which combines with a fixed mass (14 g) of nitrogen in each compound is in the ratio of 16 to 32, or 1 to 2. The ratio of the mass of oxygen in the two compounds is 1 to 2, so the ratio of the number of oxygen atoms in the two compounds must also be 1 to 2. These and similar data obtained from the analyses of other compounds led to the formulation of the ***Law of Multiple Proportions.*** This law, illustrated in Fig. 2-15, states that when *two elements combine and form more than one compound, the masses of one element that combine with a fixed mass of the other are in the ratio of small whole numbers.* This law and the data on which it is based show that the same elements may combine and form different compounds according to a definite pattern rather than in a random or a haphazard fashion. We shall see later that these patterns depend upon individual atomic structures.

FOLLOW-UP QUESTION

A certain nonmetal reacts with oxygen and forms an oxide in which the mass percentage of the nonmetal is 42.8%. What is the mass percentage of the same nonmetal in a different oxide?

Ans. 27.3% (Other answers possible).

The ability to determine correct molecular formulas of pure substances is vital to the progress of chemical science. In turn, the determination of formulas requires an accurate scale of relative atomic masses.

ATOMIC AND MOLECULAR MASSES

2-13 Dalton's Scale The determination of relative atomic masses and molecular formulas was a formidable problem to the chemists of the early 19th century. By chemical analysis they could determine the *relative combining (reacting) masses* of elements. However, they were unable to determine molecular formulas or other relevant factors. Therefore, they sometimes confused the combining mass with the atomic mass of an element. For example, when Dalton prepared the first table of relative atomic masses, he assigned hydrogen, the lightest element, an atomic mass of 1 and oxygen a relative atomic mass of 8. The value for oxygen was based on the analysis of water in which 8 g of oxygen is combined with 1 g of hydrogen. Dalton had no way to determine the molecular formula of water. He assumed, incorrectly, that the elements combined in the simplest (1 to 1) atomic ratio and, therefore, identified the combining mass as the atomic mass. With a correct formula, the problem of determining relative atomic masses from experimentally determined reacting masses is relatively simple.

FOLLOW-UP PROBLEM

Use Dalton's standard of H = 1.0 atomic mass unit (amu), the experimentally derived percentage composition of water (88.8 percent oxygen, 11.2 percent hydrogen by mass) and the correct formula for water (H_2O) to calculate the relative atomic mass of oxygen.

Ans. 15.9.

In addition to reacting masses, a standard and other data such as molecular formulas are needed to establish a scale of relative atomic masses. Let us briefly examine a number of experimental observations and imaginative proposals made between 1800 and 1860 which led to an important milestone in the evolution of chemical science: the establishment of a complete scale of relative atomic masses.

It was primarily the observation and study of the physical characteristics and chemical behavior of gases that led to an understanding of the atomic and molecular theories of matter.

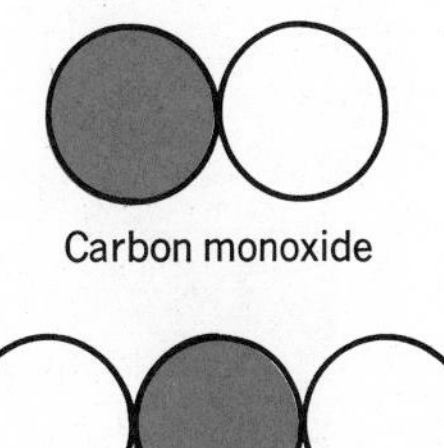

2-15 Dalton's atomic model was useful in explaining the Law of Multiple Proportions. In carbon monoxide, 1.33 g of oxygen is combined with 1.00 g of carbon. How many grams of oxygen are combined with 1.00 g of carbon in carbon dioxide? Explain the relative masses in terms of the Law of Multiple Proportions.

2-14 Gay-Lussac's Law of Combining Volumes In 1809, Joseph Louis Gay-Lussac (1778–1850) studied the relative volumes of gases involved in chemical reactions. Some of his observations are listed below.

1. One volume of hydrogen gas reacts with 1 volume of chlorine gas and gives 2 volumes of hydrogen chloride gas.

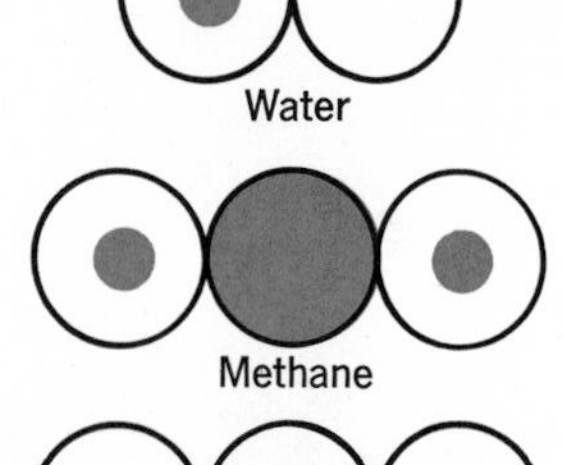

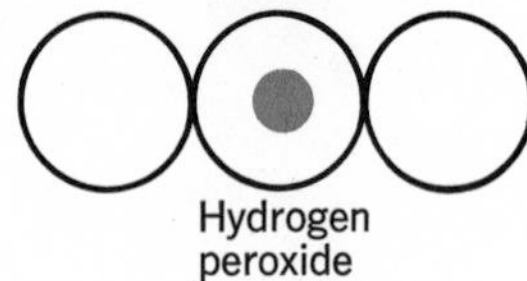

2-16 Dalton invented pictorial symbols for the elements and combined these symbols to represent compounds. Note that he believed that the formula for water was HO. Because of this belief, he mistakenly assigned oxygen an atomic mass of 8 relative to his standard of H = 1.

2. Three volumes of hydrogen gas react with 1 volume of nitrogen gas and give 2 volumes of ammonia gas.
3. One volume of oxygen gas reacts with 2 volumes of hydrogen gas and gives 2 volumes of gaseous water.

These observations led Gay-Lussac to propose the ***Law of Combining Volumes:*** *gases react chemically in a volume ratio of small whole numbers.* These integer relationships observed by Gay-Lussac suggest that there is some relationship between gaseous volumes and the number of gaseous particles. These observations could not be explained by Dalton's theory which assumed the formula for water was HO. Rather than modify his theory, Dalton questioned the accuracy of Gay-Lussac's data even though they had been verified by several prominent chemists of that day.

2-15 Avogadro's Hypothesis and Concept of Molecules In 1811, Amadeo Avogadro (1776–1856) made a proposal that explained Gay-Lussac's observations without violating Dalton's concept of indivisible atoms. He proposed that *equal volumes of gases at the same conditions of temperature and pressure contain equal numbers of molecules.* He reasoned that elemental gases such as hydrogen, oxygen, nitrogen, and chlorine, were composed of *molecules containing more than one atom.* To the scientist of that day, this idea seemed to contradict Dalton's experimentally based atomic theory which stated that elements were composed of atoms. Scientists of that day were unable to distinguish between atoms and molecules. For this reason, Avogadro's proposal lay dormant until several years after his death in 1856. In 1860, Avogadro's pupil, Stanislao Cannizzaro (1826–1910), gave such a clear explanation of the molecular hypothesis that practically all scientists accepted it. Lothar Meyer (1830–1895), one of the leading German chemists, wrote concerning Cannizzaro's presentation, "It was as if the scales fell from my eyes."

Let us see how Avogadro's reasoning is consistent with Gay-Lussac's observations. Consider the reaction between hydrogen gas and chlorine gas. It is observed that 1 volume of hydrogen gas reacts with 1 volume of chlorine gas and produces 2 volumes of hydrogen chloride gas. According to Avogadro, the smallest possible volume must have at least 1 particle. Therefore, 2 volumes of hydrogen chloride must have a minimum of 2 particles. Since analysis shows that hydrogen chloride contains hydrogen and chlorine atoms, then each of the particles in the 2 volumes of product must have at least 1 hydrogen atom and 1 chlorine atom. That is, the simplest formula of a hydrogen chloride molecule is HCl. This means that the minimum 2 atoms of hydrogen in the product were derived from 1 particle in the reactant. Thus, that particle which Avogadro called a molecule must have at least 2 hydrogen atoms. Therefore, the simplest molecular formula for elemental hydrogen is H_2. The same reasoning holds for chlorine. The reaction is represented diagrammatically in Fig. 2-17. We can use the formulas of the participants in this reaction to distinguish between two kinds of molecules. Molecules of elemental substances such as oxygen, hydrogen, sulfur, and chlorine are composed of only one kind of atom whereas molecules of compounds are composed of two or more different kinds of atoms.

It follows that the mass of a hydrogen molecule (H_2) is twice that of a hydrogen atom. Therefore, if the relative molecular mass of H_2 is known, then the relative atomic mass can be calculated.

2-16 Molecular Masses Let us use Avogadro's "equal volumes - equal numbers of molecules" hypothesis to devise a method for determining relative atomic masses. If equal volumes of different gases contain the same number of molecules, then the relative masses of a given volume of two gases must represent the relative masses of their molecules. For example, if 1 liter of oxygen gas has a mass of 1.430 g and 1 liter of methane gas has a mass of 0.715 g at the same conditions, then the ratio of the mass of a methane molecule to that of an oxygen molecule is 0.715 g to 1.430 g or 1 to 2. In 1906, a molecular mass of 32.000 for oxygen was chosen as the standard for molecular masses. Since a methane molecule is only half as massive as the oxygen molecule, the relative molecular mass of methane is

$$\frac{1}{2} \times 32.000 = 16.000$$

This means that 16 grams of methane gas occupies the same volume and contains the same number of molecules as 32 grams of oxygen gas at the same conditions of temperature and pressure. The ratio of the gas densities

$$\frac{\text{methane}}{\text{oxygen}} = \frac{0.715 \text{ g}/\ell}{1.430 \text{ g}/\ell}$$

is equal to the ratio of the molecular masses

$$\frac{16.00}{32.00}$$

2-17 To explain the volume ratios noted below, Avogadro proposed that elemental gases were composed of molecules containing more than one atom. Application of Avogadro's Principle enables us to use volumes of gases at the same conditions of temperature and pressure to count or measure relative numbers of molecules.

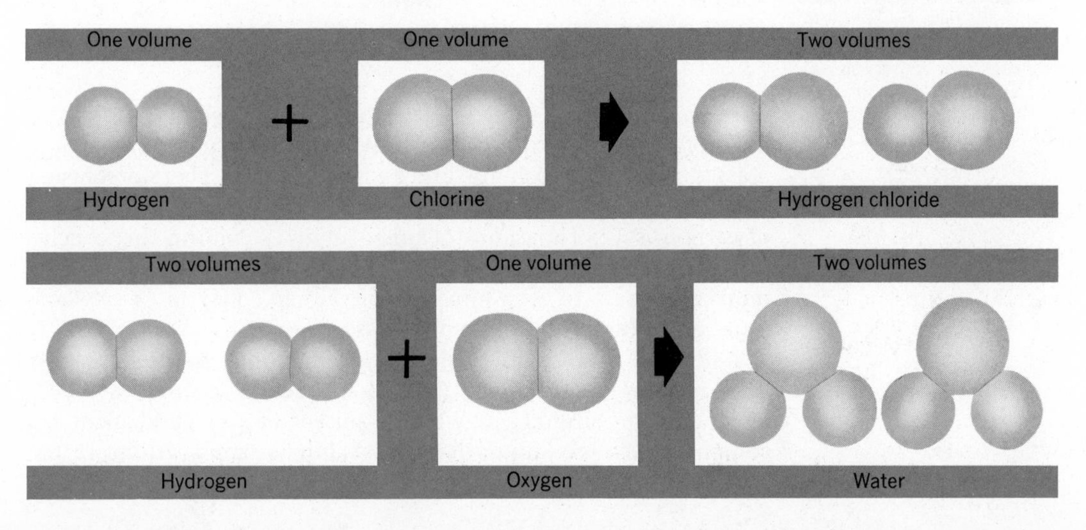

JONS JAKOB BERZELIUS
1779–1848

Berzelius was a medical doctor and chemist from a small village near Stockholm, Sweden. His work embraced almost every phase of chemistry. He was the first person to use the term *catalyst* in a modern sense, although the word had previously appeared in alchemical writings. Berzelius recognized that these substances enhanced chemical reactions but were themselves unchanged or regenerated in the process. He wrote a book on the effect of these mysterious materials. By his careful analysis and observation, Berzelius recognized that two compounds could have identical chemical formulas but different structures. For this phenomenon, he coined the word *isomerism*. In 1841, he studied different forms of the same element which he called *allotropes*.

This versatile scientist was the first person to determine accurately the atomic weights of a large number of elements. He was also the first scientist to use letters for the symbols of elements rather than the odd assortment of circles, lines, and dots used by Dalton. He discovered the elements selenium, thorium, and cerium and aided in the discovery of vanadium. He was the first to employ his letter symbols as formulas for minerals and other compounds similar to the way in which we use them today. Beginning in 1821, Berzelius published a paper called "Annual Reports on the Progress of the Physical Sciences." This publication greatly stimulated the progress of chemistry throughout Europe.

Before his death, he became a great celebrity. Both the King and Crown Prince of Sweden were his pupils, and he was visited in his laboratory by the Czar and Prince of Russia. He was made a baron in 1835 and elected to the upper chamber of the Swedish Diet. Berzelius never retired from his laboratory but kept on working and publishing papers as long as he lived.

Thus, the *relative molecular masses of gases may be determined by measuring their densities.* When one of the gases is assigned a standard molecular mass, then Avogadro's principle may be applied to determine the relative molecular mass of the other gas.

2-17 Atomic Masses from Molecular Masses It was Cannizzaro who used experimentally determined molecular masses to determine the relative atomic masses of elements. Cannizzaro reasoned that *in a large number of compounds containing a given element, at least one of them will be composed of molecules that contain a single atom of the element.* His approach involved the following procedure:

1. Establish an arbitrary standard and use Avogadro's principle to determine the relative molecular masses of a large number of compounds containing a given element.
2. Determine by chemical analysis the composition of each compound.
3. Use the molecular mass and percentage composition of the compounds to determine the relative mass of the given element in one molecular mass of each compound.
4. Select the smallest mass of the given element found in a molecular mass of any of its compounds as its atomic mass.

We can illustrate Cannizzaro's proposal with these data:

Name of Compound (all contain carbon)	*Molecular Mass*	*Mass Percent of Carbon*	*Relative Mass Carbon per Molecular Mass of Compound*
Butane	58.0	82.8	48.0
Carbon monoxide	28.0	42.9	12.0
Ethyl alcohol	46.0	52.3	24.0
Carbon tetrachloride	154.0	7.79	12.0
Benzene	78.0	92.3	72.1
Propane	44.0	81.7	36.0

It is apparent that 12 is the smallest mass of carbon in one molecular mass of any of the compounds listed above. Furthermore, all the masses in the last column are divisible by 12. Therefore, using Cannizzaro's idea, we may conclude that the probable atomic mass of carbon is 12. Thousands of other analyses confirm this conclusion. We may also conclude that the mass of an element in one molecular mass of its compounds is either equal to or a whole number multiple of the atomic mass.

2-18 The Carbon-12 Scale Relative atomic masses are now very accurately determined using instrumental techniques. These techniques reveal that, in general, samples of a given element are composed of a mixture of atoms that differ in mass. Atoms of the same element having different masses are called *isotopes.* The percentage of each isotope in a mixture of naturally-occurring isotopes is essentially constant. Both the isotopic composition and relative isotopic mass may be determined by means of instrumental analysis. The atomic mass of an element may be calculated from its known isotopic composition. For example, if 80.0 percent of the atoms in a sample of element *X* have an atomic mass of 20.0 amu and 20.0 percent have an atomic mass of 18.0 amu, then the weighted-average atomic mass of the element is

$$(0.800)(20.0 \text{ amu}) + (0.200)(18.0 \text{ amu}) = 19.6 \text{ amu}$$

The most recent atomic masses are average relative masses based on the carbon-12 isotope whose mass is exactly 12.000 amu. For example, the atomic mass of magnesium, based on the carbon-12 standard, is 24.312 amu. This number indicates that the average relative mass of the isotopes of magnesium is 24.312/12.0000, or approximately twice that of a carbon-12 atom. The latest atomic masses of the elements are tabulated in the table on the inside back cover of this text.

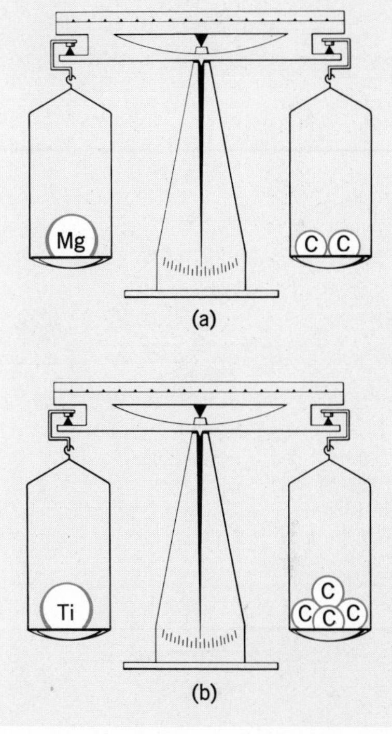

2-18 A magnesium atom has approximately twice the mass of a carbon-12 atom. Therefore, magnesium has an atomic mass of 24. What is the approximate atomic mass of a titanium atom suggested by the analogy in (b)?

2-19 Molecular Masses from Atomic Masses We can use tabulated atomic masses to calculate molecular masses of substances. The molecular mass of a substance is the sum of the masses of the atoms which compose the molecule. This number expresses how massive a molecule is relative to the mass of the carbon-12 atom. For example, the molecular mass of carbon dioxide (CO_2) is

$$1 \text{ (atomic mass of C)} + 2 \text{ (atomic mass of O)}$$
$$1 \text{ (12 amu)} + 2 \text{ (16 amu)} = 44 \text{ amu}$$

FOLLOW-UP PROBLEM

Calculate the molecular masses of (a) SO_2, (b) P_4O_{10}, (c) UF_6, (d) NH_3, (e) CCl_4.

Ans. (a) 64.0, (b) 284, (c) 352, (d) 17.0, (e) 154.

Individual atoms have actual masses in the magnitude of 1×10^{-23} grams, an inconceivably small number. The masses of atoms are, in actuality, so small that they have little value for us in terms of ordinary laboratory work where countless atoms are involved. Since the atomic mass unit (amu) represents only $\frac{1}{12}$ the mass of a carbon-12 atom, it is a practical unit only on the atomic level.

One atomic mass unit (amu) has an actual mass of 1.7×10^{-24} grams.

When we carry out a reaction, we cannot count or weigh out individual atoms or molecules because of their negligible sizes and the countless number involved. For practical laboratory work we need *a chemical unit based on mass in which the basic unit always contains the same number of particles.* If we can find the ratio in which reacting particles combine, we could then weigh out masses of the reacting substances which contain the proper ratio of particles. Let us see how such a practical unit could be devised.

THE MOLE CONCEPT

2-20 Gram-Molecular Mass Consider Gay-Lussac's observation that 1 volume of hydrogen gas reacts with 1 volume of chlorine gas and yields 2 volumes of hydrogen chloride gas. According to Avogadro's Principle, this means that 1 molecule of hydrogen gas reacts with 1 molecule of chlorine gas and yields 2 molecules of hydrogen chloride. The reaction may be represented as

$$\textbf{1 volume } H_2(g) + \textbf{1 volume } Cl_2(g) \longrightarrow \textbf{2 volumes } HCl(g)$$

or

$$\textbf{1 molecule } H_2 + \textbf{1 molecule } Cl_2 \longrightarrow \textbf{2 molecules } HCl$$

We may express the same reaction in terms of molecular masses as

$$\textbf{2 amu } H_2 + \textbf{71 amu } Cl_2 \longrightarrow \textbf{73 amu } HCl$$

Now since amu are relative, we can use any practical mass units in place of atomic mass units. Using grams we obtain

$$\textbf{2 g } H_2 + \textbf{71 g } Cl_2 \longrightarrow \textbf{73 g } HCl$$

In the above equation the molecular masses expressed in grams are called ***gram-molecular masses.*** Our line of reasoning suggests that 2 g (1 gram-molecular mass) of H_2 occupies the same volume as 71 g (1 gram-molecular mass) of Cl_2 at the same conditions of temperature and pressure. According to Avogadro's principle 2 g H_2 and 71 g Cl_2 contain the same number of molecules because their volumes are identical. Thus, if we weigh out 71 g of Cl_2 and 2 g of H_2, we shall have the proper ratio of H_2 molecules to Cl_2 molecules (1 to 1) for the reaction. We can extend our reasoning to show that *one gram-molecular mass of all molecular substances contain equal numbers of molecules.* Thus, the gram-molecular mass provides the basis for a chemical unit, based on mass, which always contains the same number of particles of a pure substance.

The chemical unit that always contains the same number of particles is called the ***mole.*** It is possible, using experimental methods described later, to determine the number of particles in a mole as well as the volume occupied by a mole of gaseous substance at a given temperature and pressure. It turns out that there are 6.02×10^{23} particles in a mole. This inconceivably large number is known as ***Avogadro's number*** and is symbolized as N.

The mole is a convenient, practical unit for measuring small particles. It is rather analogous to calling 12 large objects, such as 12 eggs, a dozen. The unit of dozen or gross would not be practical for measuring atoms, molecules, or other small particles. There is no balance that could weigh a dozen or a gross of atoms. However, we can use laboratory balance to weigh one mole of hydrogen (6.02×10^{23} molecules or 2.0 grams) or 1 mole of chlorine. Furthermore, we can express the reaction between hydrogen and chlorine in terms of moles.

$$\textbf{1 mole } H_2 + \textbf{1 mole } Cl_2 \longrightarrow \textbf{2 moles HCl}$$

The standard for the mole is 12.0000 g of carbon-12. One mole of a substance is the mass of substance that contains the same number of units as exactly 12.0000 g of carbon-12. Since 12.0000 g (1 mole) of carbon-12 contains 6.02×10^{23} carbon-12 atoms, one mole of any chemical species or entity contains this number of units. Thus, a mole refers to a specific number of particles and a definite mass of atoms or molecules. When using the term mole, it is important to identify the kind of particle involved. For example, a mole of oxygen atoms, often referred to as a gram-atom, involves 6.02×10^{23} oxygen atoms and has a mass of 16 grams. A mole of oxygen molecules, however, involves 6.02×10^{23} O_2 molecules which has a mass of 32 g.

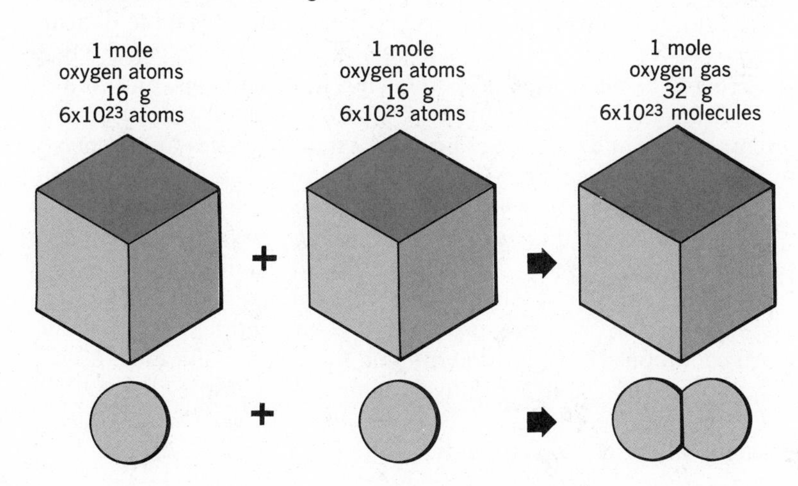

2-19 A mole of diatomic oxygen molecules has twice the mass of a mole of oxygen atoms.

FOLLOW-UP PROBLEM

(a) How many moles of oxygen in 8.0 g of O_2? (b) How many molecules of O_2 are there in 8.0 g of O_2 gas? (c) How many oxygen atoms are combined when 8.0 g of oxygen gas forms? **Ans. (a) 0.25 moles, (b) 1.5×10^{23} molecules, (c) 3.0×10^{23} atoms.**

2-21 Molar Volume When dealing with gaseous systems it is often desirable and convenient to measure volumes rather than masses. Small quantities of gases are difficult to weigh accurately because of their relatively low densities. This situation suggests that it would be desirable to express the mole in terms of volumes. This would enable us to "count" out a desired number of gaseous molecules by measuring the volume of the gas. Let us see how this can be accomplished.

In our development of the mole concept, it was stated that 2 g (1 mole) of hydrogen gas occupies the same volume as 71 g (1 mole)

2-20 (a) At the right, volumes and masses of one mole of selected solid elements. Note that the masses vary and the volumes vary. Calculate and cite a reason for the relative densities. (b) Below, volumes and masses of one mole of selected liquids. Calculate and cite a reason for the relative densities. (c) At the bottom of the page volumes and masses of one mole of selected gases. Note that the equal volumes equal number of molecules principle applies only to gases.

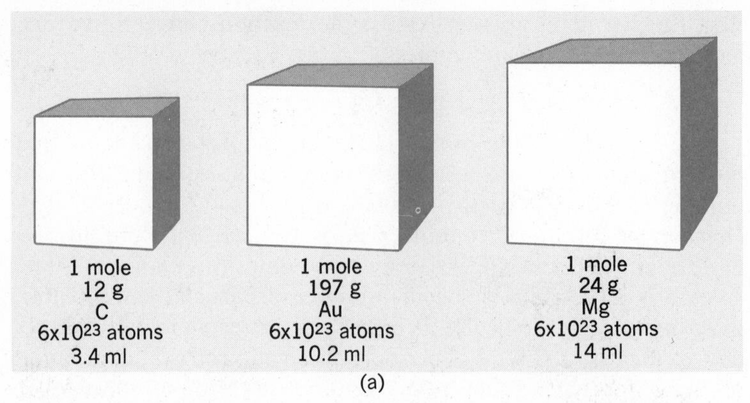

(a)

or chlorine gas at the same temperature and pressure. The same line of reasoning applied to other gaseous systems indicates that 1 mole of any gas should occupy the same volume under identical conditions. You will note that we have been very careful to indicate that gas volumes are related to temperature and pressure. There are many familiar examples involving tires, balloons, and pumps which reveal that the volume of a given mass of gas varies with temperature and pressure. This means that it is necessary to specify a temperature and pressure when referring to a definite volume of gas. *Scientists conventionally use the average atmospheric (air) pressure at sea level as a standard and refer to it as 1 atmosphere (atm) pressure.* The *melting point of ice at 1 atm pressure is zero degrees Celsius (0° C).* This value is conventionally used as a *standard temperature.* We shall use 1 atm pressure and 0° C as the standard conditions (STP) for determining the volume of a mole of gas.

Experimentally, it is relatively simple to measure the mass of a given volume of a gas (determine its density). This is done in school laboratory experiments. Multiplying the mass of a mole of gas by the experimentally determined ratio $\frac{\text{volume } (\ell)}{\text{mass (g)}}$ yields the volume of a mole. That is,

$$\frac{\textbf{grams}}{\textbf{mole}} \times \frac{\textbf{liters}}{\textbf{grams}} = \frac{\textbf{liters}}{\textbf{mole}}$$

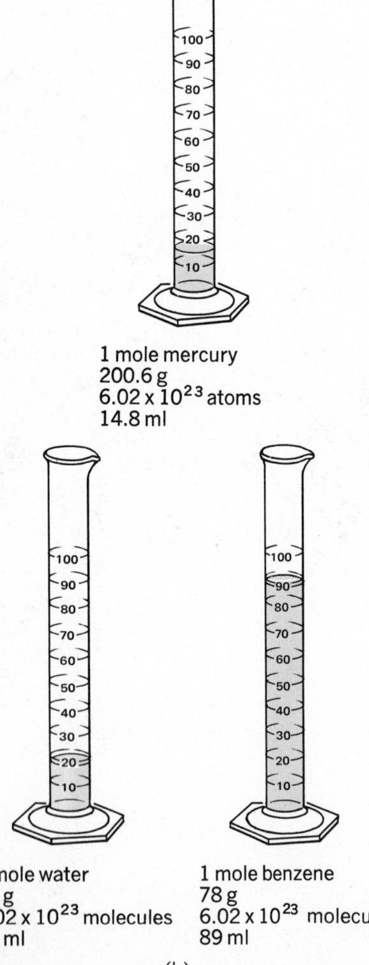

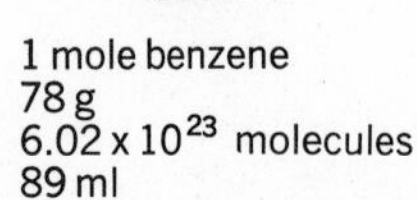

(b)

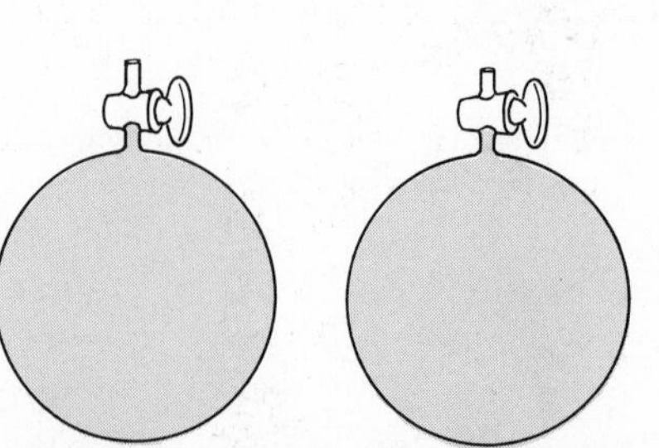

(c)

The density of oxygen gas at 0°C and 1 atm pressure (STP) is 1.43 g/l. Therefore, the volume of 1 mole of O_2 at STP is

$$\frac{32.0 \text{ g}}{1 \text{ mole}} \times \frac{1 \ \ell}{1.43 \text{ g}} = 22.4 \ \ell/\text{mole}$$

FOLLOW-UP QUESTION

Hydrogen gas (H_2) has a density of 0.089 g/ℓ at STP. What is the volume of 1 mole of H_2 at STP? **Ans. 22.4 ℓ.**

The above answers are consistent with our expectations that a mole of any gas should occupy the same volume. The volume of 22.4 ℓ is known as the ***gram-molecular volume*** or ***molar volume*** of a gas at STP. We have now developed the concept of the mole to the point where we can interpret it in terms of the following quantities: (a) number of grams, (b) number of liters of gas at STP, and (c) number of particles. Applied to the case of oxygen gas (O_2), one mole can be expressed in these terms.

1. A gram-molecular mass of 32.0 g/mole.
2. A gram-molecular volume of 22.4 ℓ/mole, or 22 400 ml/mole at STP.
3. A specific number of particles, 6.02×10^{23} molecules/mole.

2-22 Quantitative Interpretation of a Formula The concept of the mole is one of the most widely used and important concepts in chemistry. We shall have many occasions to use the relationships developed above. For example, we can use conversion factors based on the above relationships along with the unit cancellation technique to calculate for a given substance, either the grams in a specific number of moles or the moles in a specific number of grams.

The relationships between numbers of atoms, molecules, moles, grams, and liters of gas at STP are represented in Fig. 2-21 and illustrated in Example 2-2.

Example 2-2

These problems refer to ethane (C_2H_6) a gas at STP.

(a) Calculate the mass of 1.00 mole of C_2H_6.

The formula, C_2H_6, reveals that 2 moles of carbon atoms are combined with 6 moles of hydrogen atoms in 1 mole of C_2H_6.

mass of one mole

$$= 2 \text{ moles C} \times \frac{12.0 \text{ g C}}{1 \text{ mole C}} + 6 \text{ moles H} \times \frac{1.0 \text{ g H}}{1 \text{ mole H}}$$

$$= 24.0 \text{ g C} + 6.0 \text{ g H}$$
$$= 30.0 \text{ g } C_2H_6$$

(b) Calculate mass in grams of 0.25 mole C_2H_6.

Multiply the given quantity, 0.25 mole C_2H_6 by a factor that converts moles to grams.

$$0.25 \text{ mole } C_2H_6 \times \frac{30.0 \text{ g } C_2H_6}{1 \text{ mole } C_2H_6} = 7.5 \text{ g } C_2H_6$$

(c) Calculate the moles of hydrogen atoms in 1.50 g of compound.

Multiply the given quantity, 1.50 g C_2H_6, by a conversion factor which converts g C_2H_6 into moles of H atoms.

$$1.50 \text{ g } C_2H_6 \times \frac{6 \text{ moles H atoms}}{30.0 \text{ g } C_2H_6} = 0.30 \text{ mole H atoms}$$

(d) Calculate the mass of compound which contains 6.0 g of carbon.

$$6.0 \text{ g C} \times \frac{30.0 \text{ g } C_2H_6}{24.0 \text{ g C}} = 7.5 \text{ g } C_2H_6$$

(e) Calculate the number of molecules in 10.0 g of compound.

$$10.0 \text{ g } C_2H_6 \times \frac{1 \text{ mole } C_2H_6}{30.0 \text{ g } C_2H_6} \times \frac{6.02 \times 10^{23} \text{ molecules } C_2H_6}{1 \text{ mole } C_2H_6} = 2.01 \times 10^{23} \text{ molecules}$$

(f) Calculate the grams of carbon combined with 1.0 mole of hydrogen atoms.

$$1.0 \text{ mole H atoms} \times \frac{24 \text{ g C}}{6 \text{ mole H atoms}} = 4.0 \text{ g C}$$

Note that the ratio of the atoms in the formula C_2H_6 is also the ratio of the moles of atoms.

(g) Calculate the moles of carbon atoms combined with 1.0 g of hydrogen.

$$1.0 \text{ g H} \times \frac{2.0 \text{ mole C atoms}}{6.0 \text{ g H}} = 0.33 \text{ mole C atoms}$$

(h) Calculate the volume in liters occupied by 60.0 g of compound at STP.

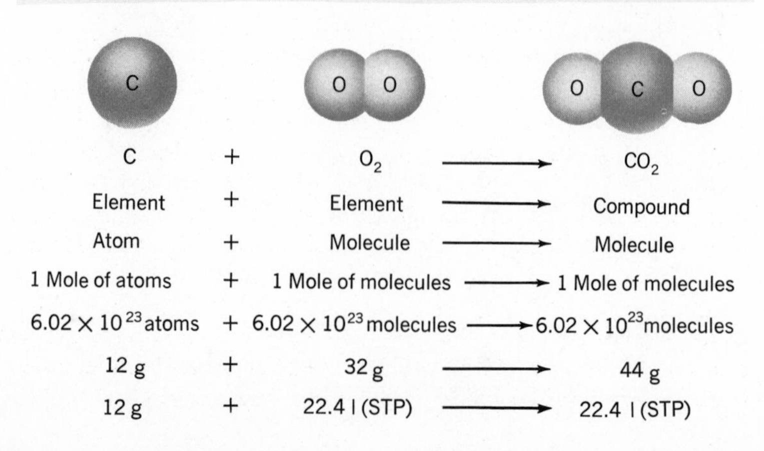

2-21 Relationships between atoms, molecules, moles, and grams.

$$60.0 \text{ g } C_2H_6 \times \frac{1 \text{ mole } C_2H_6}{30.0 \text{ g } C_2H_6} \times \frac{22.4 \ \ell \ C_2H_6}{\text{mole } C_2H_6} = 44.8 \ \ell \ C_2H_6$$

(i) Calculate the mass of gas in 5.6 liters of gas measured at STP.

$$5.6 \ \ell \ C_2H_6 \times \frac{1 \text{ mole } C_2H_6}{22.4 \ \ell \ C_2H_6} \times \frac{30.0 \text{ g } C_2H_6}{\text{mole } C_2H_6} = 7.5 \text{ g } C_2H_6$$

(j) Calculate the grams of carbon in 180.0 g of C_2H_6.

$$180.0 \text{ g } C_2H_6 \times \frac{24.0 \text{ g C}}{30.0 \text{ g } C_2H_6} = 144 \text{ g C}$$

(k) Calculate the mass percent of carbon in C_2H_6.

$$\text{mass percentage of carbon} = \frac{\text{mass of carbon in one mole of compound}}{\text{mass of one mole of compound}}(100)$$

$$\text{Percent C} = \frac{24.0 \text{ g carbon}}{30.0 \text{ g } C_2H_6} \times 100 = 80.0\%$$

FOLLOW-UP PROBLEMS

These problems refer to dinitrogen oxide (nitrous oxide, N_2O) a gas at STP. (a) Calculate the mass of 1.00 mole of gas. (b) Calculate the mass in grams of 1.50 moles of N_2O. (c) Calculate the number of moles of nitrogen atoms in 2.20 g of compound. (d) Calculate the mass of N_2O which contains 4.00 g of oxygen. (e) Calculate the number of molecules in 4.40 g N_2O. (f) Calculate the mass in grams of oxygen combined with 0.50 mole of nitrogen atoms. (g) Calculate the number of moles of N atoms combined with 1.6 g of O. (h) Calculate the number of gram-atoms of N combined with 0.10 gram-atom of O. (i) Calculate the volume in liters occupied by 22.0 g N_2O at STP. (j) Calculate the mass of gas in 16.8 l of N_2O at STP. (k) Calculate the grams of N in 110.0 g of N_2O. (l) Calculate the mass percentage of O in the compound.

Ans. (a) 44.0 g, (b) 66.0 g, (c) 0.100 mole N atoms, (d) 11.0 g, (e) 6.02×10^{22} molecules, (f) 4.0 g, (g) 0.20 mole N atoms, (h) 0.20 g atoms O, (i) 11.2 ℓ N_2O, (j) 33.0 g, (k) 70.0 (g), (l) 36.4 percent.

It is not usually convenient nor practical to go into the laboratory every time we wish to determine the percentage composition of a compound. As long as we know its formula, we can use readily available atomic masses to calculate the formula or molecular masses and percentage composition of compounds. In turn, the composition of a compound can be used to calculate the actual mass of a given element that can be obtained by decomposing a given mass of compound. Conversely, given the mass of element needed, we can calculate the mass of compound which must be decomposed. In addition, application of the same principles enables us to determine the percentage of an element in an impure sample of an ore containing the element. These calculations are illustrated in the following examples.

Example 2-3

Calculate the mass percentage of oxygen in water and the number of grams of oxygen that could be obtained by decomposing 50.0 g of water.

Solution

1. Calculate the gram-molecular mass of water.

$$2(1.0 \text{ g}) + 1(16.0 \text{ g}) = 18.0 \text{ g}$$

2. Divide the total mass of the element (oxygen) in one mole of compound by the gram-molecular mass of the compound and multiply by 100.

$$\frac{16.0 \text{ g oxygen}}{18.0 \text{ g water}} \times 100 = 88.9 \text{ percent oxygen}$$

3. Multiply the given mass of water (50.0 g) by the percentage of oxygen.

$$\frac{88.9 \text{ parts oxygen}}{100 \text{ parts water}} \times 50.0 \text{ g water} = 44.4 \text{ g oxygen}$$

Alternatively, multiply the given mass of water by the mass ratio of oxygen to water.

$$50.0 \text{ g water} \times \frac{16.0 \text{ g oxygen}}{18.0 \text{ g water}} = 44.4 \text{ g oxygen}$$

Example 2-4

Calculate the mass of magnesium chloride ($MgCl_2$) which must be electrolyzed (decomposed by electricity) in order to obtain 120 g of magnesium.

Solution

1. The mass of one mole of $MgCl_2$ is

$$1(24.0 \text{ g}) + 2(35.5 \text{ g}) = 95.0 \text{ g}$$

2. The mass ratio of magnesium chloride to magnesium in the compound is

$$\frac{95.0 \text{ g } MgCl_2}{24.0 \text{ g Mg}}$$

3. Multiplying the mass ratio by the required quantity of magnesium yields

$$\frac{95.0 \text{ g } MgCl_2}{24.0 \text{ g Mg}} \times 120 \text{ g Mg} = 475 \text{ g } MgCl_2$$

The set-up for a one-step solution is

$$120 \text{ g Mg} \times \frac{\text{mole Mg}}{24.0 \text{ g Mg}} \times \frac{\text{mole } MgCl_2}{\text{mole Mg}} \times \frac{95.0 \text{ g } MgCl_2}{\text{mole } MgCl_2} = 475 \text{ g } MgCl_2$$

Example 2-5

A sample of iron ore weighing 2.80 g is to be analyzed for the percentage of iron. All of the iron in the sample is converted into iron oxide (Fe_2O_3) which has a mass of 1.00 g. What is the percentage of iron in the ore?

Solution

1. The mass of one mole of Fe_2O_3 is

$$2(56.0 \text{ g}) + 3(16.0 \text{ g}) = 112.0 + 48.0 = 160.0 \text{ g}$$

2. The mass ratio of iron to iron oxide is

$$\frac{112.0 \text{ g iron}}{160.0 \text{ g iron oxide}}$$

3. Multiplying the ratio of iron oxide by the number of grams of Fe_2O_3 gives

$$\frac{112.0 \text{ g Fe}}{160.0 \text{ g } Fe_2O_3} \times 1.00 \text{ g } Fe_2O_3 = 0.700 \text{ g iron}$$

4. The percentage of iron in the ore is

$$\frac{\text{g iron}}{\text{g ore}} \times 100 = \frac{0.700 \text{ g Fe}}{2.80 \text{ g ore}} \times 100 = 25.0 \text{ percent}$$

The set-up for a one-step solution is

$$\frac{112.0 \text{ g Fe}}{160.0 \text{ g } Fe_2O_3} \times \frac{1.00 \text{ g } Fe_2O_3}{2.80 \text{ g ore}} \times 100 = 25.0 \text{ percent}$$

FOLLOW-UP PROBLEMS

1. Calculate (a) the mass percentage of sulfur in sulfur dioxide (SO_2) and (b) the number of grams of sulfur that could be obtained by decomposing 2000 grams of the oxide. **Ans. (a) 50.0 percent sulfur, (b) 1000 g sulfur.**
2. Calculate the mass of water (H_2O) which must be electrolyzed in order to obtain 80.0 g of oxygen. **Ans. 90.0 g.**
3. An ore sample weighing 8.32 g is analyzed for the percentage of chromium. All of the chromium is converted to Cr_2O_3, which has a mass of 1.52 g. What is the percentage of chromium in the ore? **Ans. 12.5 percent.**

2-23 Experimental Determination of a Chemical Formula One problem faced by the chemist is the determination of the composition and structure of new compounds. Thousands of new substances are synthesized each year. The majority of these are organic compounds. That is, they contain, among other elements, carbon and hydrogen. One step in the identification of the compound involves the determination of its simplest formula. This may be accomplished by means of a *carbon-hydrogen analysis.*

In this process, a weighed sample of the substance is completely burned in pure oxygen. If the compound contains only carbon and hydrogen the products of combustion are carbon dioxide (CO_2) and water (H_2O). These gases are absorbed in preweighed tubes that contain appropriate absorbing material and which are included in the

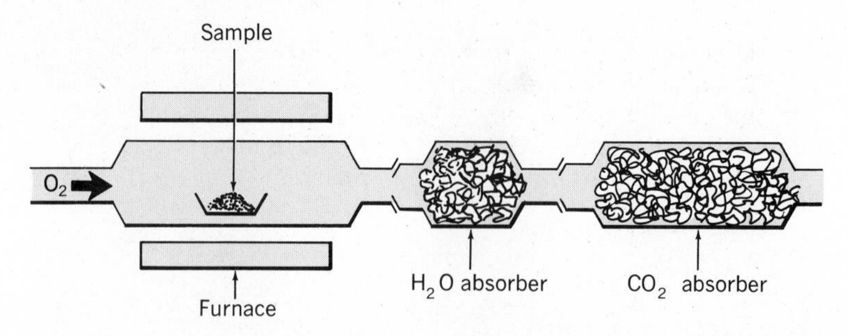

2-22 Simplified schematic diagram of apparatus for determining the percent of hydrogen, carbon, and oxygen in a compound by combustion analysis. The gas absorption tubes are weighed before and after the combustion.

combustion apparatus shown in Fig. 2-22. The masses of carbon and hydrogen in the original sample may be determined from the mass of the carbon dioxide and water absorbed. If the original compound also contained oxygen, its mass is equal to the difference between the mass of the original sample and the combined masses of the carbon and hydrogen. Thus, an analysis of the products of combustion enables us to determine the *mass* or the percentage of carbon, hydrogen, and oxygen in our compound.

The *simplest formula* of a compound represents the *mole ratio* in which the atoms of the different elements are present. The calculations involve converting the experimentally derived mass or percentage ratio into a simple (whole number) mole ratio.

Example 2-6
When a 15.00-mg sample of an organic compound containing only carbon, hydrogen, and oxygen is completely burned, 29.30 mg of CO_2 and 15.00 mg of H_2O are produced. The oxygen is obtained by difference. What is the simplest formula of the compound? Atomic and molecular masses: C = 12.00, O = 16.00, H = 1.00, CO_2 = 44.00, H_2O = 18.00.

Solution
1. Calculate the mg of carbon in 29.30 mg of CO_2. This is the quantity of C in the original sample.

$$29.30 \text{ mg } CO_2 \times \frac{12.00 \text{ mg C}}{44.00 \text{ mg } CO_2} = 8.00 \text{ mg C}$$

2. Calculate the mg of H in 15.00 mg of H_2O. This is the quantity of H in the original sample.

$$15.00 \text{ mg } H_2O \times \frac{2.00 \text{ mg H}}{18.00 \text{ mg } H_2O} = 1.67 \text{ mg H}$$

3. Calculate the mg of O in the sample by difference.

$$\text{total C + H in original sample} = 8.00 \text{ mg} + 1.67 \text{ mg} = 9.67 \text{ mg}$$

$$\text{mg O} = 15.00 \text{ mg sample} - 9.67 \text{ mg (C + H)} = 5.33 \text{ mg O}$$

4. Convert the mg of each element to millimoles (mmole).

Carbon

$$8.00 \text{ mg} \times \frac{1 \text{ m mole C}}{12.00 \text{ mg C}} = 0.67 \text{ mmole C}$$

Hydrogen

$$1.67 \text{ mg H} = \frac{1 \text{ m mole H}}{1.00 \text{ mg H}} = 1.67 \text{ mmole H}$$

Oxygen

$$5.33 \text{ mg O} \times \frac{1.00 \text{ mmole O}}{16.00 \text{ mg O}} = 0.33 \text{ mmole O}$$

5. Simplify the millimole ratio so that the element that is present in the smallest mole quantity (oxygen) is represented by 1 mmole. Divide the number of mmoles of all three elements by 0.33.

$$0.33 \text{ mmole O}/0.33 = 1 \text{ mmole O}$$
$$0.67 \text{ mmole C}/0.33 = 2 \text{ mmole C}$$
$$1.67 \text{ mmole H}/0.33 = 5 \text{ mmole H}$$

6. Since the ratio in which the atoms combine is numerically equal to the mmole ratio, the simplest formula is C_2H_5O.

FOLLOW-UP PROBLEM

At 10.12-mg sample of a compound containing only carbon, hydrogen, and oxygen is completely burned and 14.08 mg of carbon dioxide and 8.64 mg of water are produced. What is the simplest formula of the compound?

Ans. CH_3O.

2-24 Molecular Formulas Chemical analysis provides only the mass, and hence, the mole ratio in which elementary substances are present in a compound. The mole ratio, which can be converted into an atomic ratio, does not necessarily give the correct number of atoms in a molecule (the molecular formula). For example, the simplest formula (ratio of atoms) of a compound containing 92.2 percent carbon and 7.8 percent hydrogen by mass is CH. Compounds with the formulas C_2H_2, C_3H_3, C_4H_4, C_5H_5, and C_6H_6 all contain 92.8 percent carbon and 7.8 percent hydrogen. The molecular formula may be obtained by multiplying the number of atoms in the simple formula by an integer *n*. In other words, *n* represents the ratio of the number of atoms in the molecular formula to the number of atoms in the simplest formula. The relationship may be expressed as

$$\mathbf{n} = \frac{\textbf{number of atoms in molecular formula}}{\textbf{number of atoms in simplest formula}}$$

This means that *n* is also the ratio of the molecular mass represented by the molecular formula to the molecular mass represented by the simplest formula.

To determine the value of n and the molecular formula, it is necessary to obtain an *experimentally determined gram-molecular mass* for the compound. The value of n can be determined by substituting the known gram molecular mass in the formula

$$\mathbf{n} = \frac{\textbf{experimentally derived molecular mass}}{\textbf{calculated simplest molecular mass}}$$

In our example, the simplest molecular mass derived from the simplest formula (CH) is 13. If the experimentally determined mass is approximately 78, then the ratio is $\mathbf{n} = \frac{78}{13} = \mathbf{6}$. This means there are six times as many atoms in the molecular formula as in the simple one. Thus, the molecular formula is $(CH)_6$ or C_6H_6, a compound known as benzene. Because the value of n must be an integer, it is often necessary to "round off" the value of n to the nearer whole number.

The experimental molecular masses of compounds may be determined in a number of ways. Some of these will be described later. We have already indicated how Avogadro's Principle may be applied in the laboratory to determine the molecular mass of a gas.

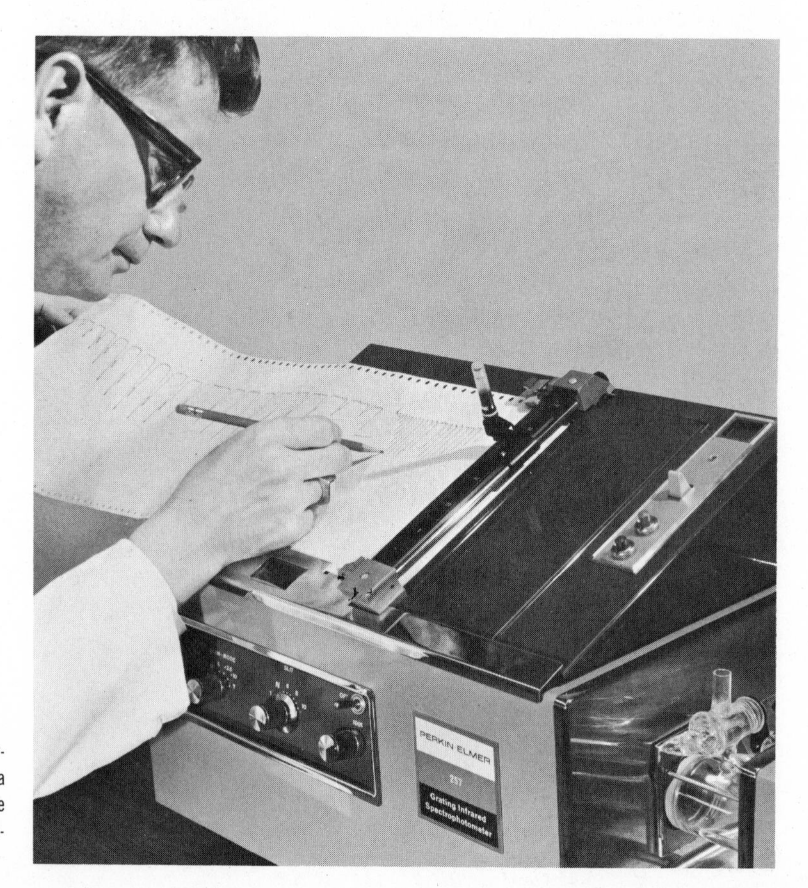

2-23 An infrared spectrophotometer. This important tool of the analytical chemist furnishes data (infrared spectrogram) which helps him determine the arrangement of atoms in molecules. (Perkin-Elmer Corp.)

FOLLOW-UP PROBLEM

The simplest formula of a certain compound is C_2H_5. A given volume of this gas has a mass of 0.47 g. The same volume of oxygen at the same temperature and pressure weighs 0.26 g. What is the molecular formula of this compound? **Ans. C_4H_{10} (butane).**

Molecular formulas reveal only the composition of molecules. They convey no information related to the structure, properties, or how the compound was made. Structural formulas and space-filling models are often used to show other important molecular characteristics. Data required to construct these models are usually obtained by means of instrumental analysis. Modern instruments such as infrared, ultraviolet, nuclear magnetic resonance, and mass spectrometers furnish data which enable scientists to determine such characteristics of molecules as composition, mass, interatomic distances, interatomic angles, and interatomic forces of attraction. For example, instrumental analysis reveals this information about a water molecule.

Interatomic forces of attraction are referred to as chemical bonds and are measured in terms of the energy required to separate the bonded atoms.

1. It is composed of two hydrogen atoms chemically bonded to a single oxygen atom.
2. The molecular mass of a water molecule is 18.015 atomic mass units.
3. The bond length (distance between the nucleus of a hydrogen atom and the nucleus of an oxygen atom) is 0.000 000 009 58 cm (9.58×10^{-9} cm).
4. The bond angle (angle formed by lines joining the nuclei of the two H atoms to the nucleus of the O atom) is 104.45°.
5. The bond strength (energy needed to separate a hydrogen atom from an oxygen atom) is 1.98×10^{-19} calories.

A structural formula and space-filling model of a water molecule is shown in Fig. 2-24. In future chapters we shall seek an explanation for these characteristics and use the characteristics to help explain the observed properties of substances. In the meantime, let us examine a convenient method often used for predicting and writing formulas of compounds. The method involves using the actual or apparent charges exhibited by atoms in compounds.

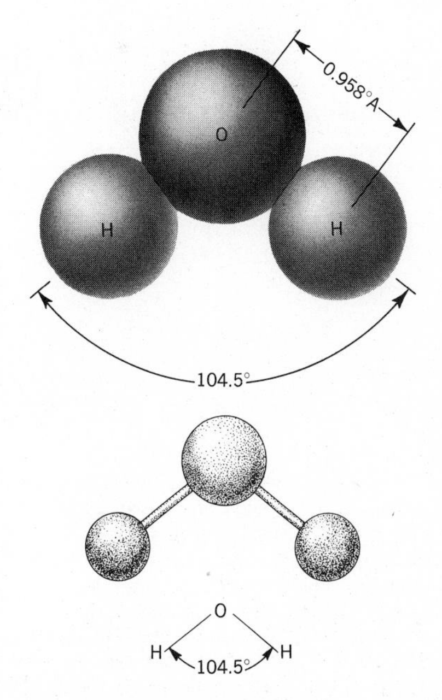

2-24 Space-filling model, ball-and-stick model, and structural formula for water.

FORMULAS FROM IONIC CHARGES

2-25 Evidence for the Existence of Ions There is considerable experimental evidence which indicates that certain types of compounds are composed of atoms that carry a net electric charge. For example, the electrolysis of melted sodium chloride (NaCl), described earlier, provides evidence that the liquid is composed of oppositely charged particles present in such a ratio that the liquid is electrically neutral. The observed flow of an electric current

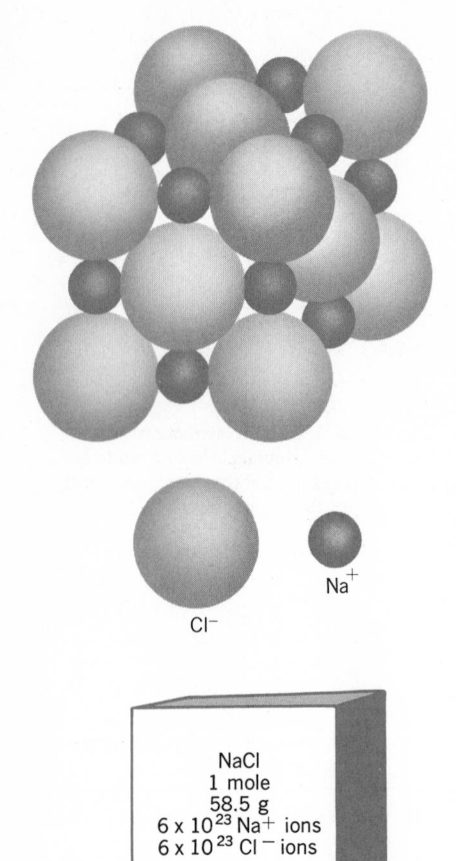

2-25 Sodium chloride crystal (top) contain equal numbers of positive sodium ions (Na^+) and negative chloride ions (Cl^-) so that electric neutrality is maintained. Not all ions in the crystals are shown.

means that an electric charge is being transported between the electrodes placed in the melted salt. The production of metallic sodium at the negative electrode suggests that sodium atoms in the compound carry a positive charge which is neutralized when they reach the negative electrode. The resulting neutral sodium atoms have entirely different properties than the charged sodium atoms. The production of electrically neutral chlorine gas at the positive electrode suggests that chlorine atoms in NaCl carry a negative charge which is neutralized when they reach the positive electrode. Instrumental analysis (X-ray diffraction) of solid sodium chloride provides additional evidence that the solid is an aggregate composed of oppositely charged particles called ions. Such a solid is called an *ionic crystal.*

Atoms or groups of chemically combined atoms which carry a charge are called ***ions.*** The charge carried by an ion determines how many oppositely charged ions are combined with it in a compound. Ions carrying a positive charge are called ***cations,*** whereas those carrying a negative charge are called ***anions.*** When an ion is composed of more than one kind of atom, it is called a ***poly-atomic ion.*** Thus, a sodium ion (Na^+) is a cation, a chloride ion (Cl^-) is an anion, and a sulfate ion (SO_4^{2-}) is a polyatomic anion. The charge carried by an ion determines the ratio in which it is found in combination with other ions. The charge on a simple ion such as Na^+ or Cl^- is often called its ***oxidation state.*** A list of common positive and negative ions is provided in Tables 2-8 and 2-9. Note that some elements such as iron can exhibit more than one oxidation state.

You may use the electric charges carried by ions to predict the simplest formulas of compounds that contain any pair of positive and negative ions. A few suggestions which may help you to write formulas correctly are listed and illustrated below.

Technically, masses assigned to the formulas of ionic substances are known as *formula masses.* Formula masses are calculated in the same way as molecular masses. For convenience and simplicity, we shall use the term molecular mass regardless of the nature of the substance.

2-26 Writing Formulas

1. Write the symbol and charge of the positive ion first, followed by those of the negative ion. For example, to write the formula for calcium hydroxide, first write

$$Ca^{2+}\ OH^-$$

2. The total positive charge in a formula of a substance must be equal to the total negative charge. The algebraic sum of the ionic charges in a formula must be *zero*.

3. The total positive *or* negative charge is the lowest common multiple of the charges on the two ions; that is, the lowest number that is divisible by the absolute charge on one of the ions. A calcium ion has a charge of 2+. A hydroxide ion has a charge of 1−. The lowest number divisible by both 2 and 1 is 2. Thus, 2 is the lowest common multiple. Dividing the charge on each ion into the lowest common multiple yields the number of each kind of ion in the formula. This step indicates that 1 calcium ion and 2 hydroxide ions are needed to write a formula for an electrically neutral compound.

TABLE 2-8
PARTIAL LIST OF COMMON POSITIVE IONS

Formula	1+ Name	Formula	2+ Name	Formula	3+ Name	Formula	4+ Name
NH_4^+	ammonium	Ba^{2+}	barium	Al^{3+}	aluminum	Ce^{4+}	cerium(IV)
Cs^+	cesium	Cd^{2+}	cadmium	Sb^{3+}	antimony(III)		or ceric
Cu^+	copper(I)	Ca^{2+}	calcium		or antimonous	Sn^{4+}	tin(IV)
H^+	hydrogen	Cr^{2+}	chromium(II)	As^{3+}	arsenic(III)		or stannic
Li^+	lithium		or chromous		or arsenious	Ti^{4+}	titanium(IV)
Hg_2^{2+}	mercury(I)	Co^{2+}	cobalt(II)	Bi^{3+}	bismuth		or titanic
	or mercurous		or cobaltous	Ce^{3+}	cerium(III)		
K^+	potassium	Cu^{2+}	copper(II)		or cerous		
Rb^+	rubidium		or cupric	Cr^{3+}	chromium(III)		
Na^+	sodium	Fe^{2+}	iron(II)		or chromic		
Ag^+	silver		or ferrous	Co^{3+}	cobalt(III)		
Tl^+	thallium(I)	Pb^{2+}	lead(II)		or cobaltic		
	or thallous		or plumbous	Fe^{3+}	iron(III)		
Au^+	gold(I) or	Mg^{2+}	magnesium		or ferric		
	aurous	Hg^{2+}	mercury(II)	Ga^{3+}	gallium		
			or mercuric	Tl^{3+}	thallium(III)		
		Ni^{2+}	nickel(II)		or thallic		
		Sr^{2+}	strontium	Ti^{3+}	titanium(III)		
		Sn^{2+}	tin(II)		or titanous		
			or stannous	V^{3+}	vanadium		
		Zn^{2+}	zinc	Au^{3+}	gold(III)		
					or auric		

TABLE 2-9
PARTIAL LIST OF SOME COMMON NEGATIVE IONS

Formula	1− Name	Formula	1− Name	Formula	2− Name	Formula	3− Name
$C_2H_3O_2^-$	acetate	IO_3^-	iodate	$C_7H_6O_2^{2-}$	benzoate	AsO_4^{3-}	arsenate
HCO_3^-	bicarbonate or	I^-	iodide	CO_3^{2-}	carbonate	AsO_3^{3-}	arsenite
	hydrogen carbonate	NO_3^-	nitrate	CrO_4^{2-}	chromate	BO_3^{3-}	borate
$HC_2O_4^-$	bioxalate or	NO_2^-	nitrite	$Cr_2O_7^{2-}$	dichromate	PO_4^{3-}	phosphate
	hydrogen oxalate	ClO_4^-	perchlorate	HPO_4^{2-}	mono-	$Fe(CN)_6^{3-}$	ferricyanide
$HC_8H_4O_4^-$	biphthalate or	MnO_4^-	permanganate		hydrogen	*Formula*	*4− Name*
	hydrogen phthalate	SCN^-	thiocyanate		phosphate	$Fe(CN)_6^{4-}$	ferrocyanide
HSO_4^-	bisulfate or			$C_2O_4^{2-}$	oxalate	$P_2O_7^{4-}$	pyrophosphate
	hydrogen sulfate			O^{2-}	oxide	SiO_4^{4-}	orthosilicate
HS^-	bisulfide or			SO_4^{2-}	sulfate		
	hydrogen sulfide			S^{2-}	sulfide		
BrO_3^-	bromate			SO_3^{2-}	sulfite		
Br^-	bromide			$S_2O_3^{2-}$	thiosulfate		
ClO_3^-	chlorate						
Cl^-	chloride						
ClO_2^-	chlorite						
CN^-	cyanide						
$H_2PO_4^-$	dihydrogen phosphate						
F^-	fluoride						
OH^-	hydroxide						

4. When more than one positive or negative ion is needed to complete the formula, indicate the number required with a subscript. If a polyatomic ion is involved, enclose the ion in parentheses and place the subscript outside the parentheses. Following these directions leads to $Ca(OH)_2$ as the formula for calcium hydroxide. The formulas $CaOH_2$ or CaO_2H_2 are incorrect for calcium hydroxide. Neither of these formulas indicates the presence of a hydroxide ion (OH^-).

FOLLOW-UP PROBLEMS

1. Write formulas for these compounds. Refer to Tables 2-8 and 2-9 as needed. The Roman numerals in parenthesis indicate the positive charge on the metallic ion. This is the convention used in the Stock system of nomenclature discussed in the Special Feature at the end of this chapter. a) silver phosphate, b) copper(I) iodide, c) tin(IV) bromide, d) mercury(I) chloride, e) iron(II) oxalate, f) cadmium cyanide, g) cobalt(II) permanganate, h) aluminum acetate, i) sodium thiosulfate, j) lead(IV) oxide, k) iron(III) chromate, l) barium iodate, m) ammonium thiocyanate, n) nickel(II) chlorate, o) potassium perchlorate, p) ammonium sulfide, q) calcium bicarbonate, r) antimony(III) nitrate, s) zinc arsenate, t) sodium sulfite, u) potassium nitrate, v) arsenic(III) sulfate, w) mercury(II) bromide, x) potassium monohydrogen phosphate, y) chromium(III) oxide, z) ammonium bisulfate.

TYPES AND NAMES OF COMPOUNDS

2-27 Dissociation of Ionic Solids in Aqueous Solution Most solids which are composed of positive metallic and negative nonmetallic ions are referred to as *ionic solids* or *salts*. In general, ionic solids dissolve in water and form solutions which conduct an electric current. This observation can be explained by assuming that the ions that compose ionic solids (salts) *dissociate* when they are added to water. We can represent the dissociation of NaCl by the following equation.

Solutes which form ionic solutions are called *electrolytes*.

$$Na^+Cl^-(s) \longrightarrow Na^+(aq) + Cl^-(aq)$$

The designation (s) for solid, (g) for gas, (l) for liquid, and (aq) for solutes in water solution are placed to the right of each formula in an equation.

In general, we can say that the solute in aqueous solutions of ionic solids exists entirely as ions. Note that a salt such as magnesium chloride ($MgCl_2$) dissociates and forms twice as many chloride (Cl^-) ions as magnesium ions (Mg^{2+}). *Coefficients* are placed in front of the formulas and symbols to indicate the relative numbers of ions formed by the dissociation of a substance. Thus, the equation for the dissociation of magnesium chloride is

$$Mg^{2+}(Cl^-)_2(s) \longrightarrow Mg^{2+}(aq) + 2Cl^-(aq)$$

Also, remember that many polyatomic ions behave as a unit. This means that a salt such as sodium sulfate (Na_2SO_4) dissociates into sodium (Na^+) and sulfate (SO_4^{2-}) ions.

$$(Na^+)_2SO_4^{2-}(s) \longrightarrow 2\,Na^+(aq) + SO_4^{2-}(aq)$$

There is no evidence that sulfate ions dissociate into other ions or atoms in an aqueous solution. Later in the text we shall find that some polyatomic ions which contain hydrogen do dissociate. We shall omit these ions at this point.

2-28 Acids and Bases Experimental data show that solutions of many compounds whose formulas begin with the symbol for hydrogen (except water) all have a number of common properties. These solutions all turn blue litmus (a chemical indicator) red, conduct electricity, react with compounds called carbonates and produce carbon dioxide gas, and neutralize solutions of compounds containing hydroxide ions. These properties constitute an operational definition of an *acid.*

We may assume that the electric conductivity of acid solutions is connected with the presence of ions. We therefore might assume that the properties common to all acid solutions are caused by the presence of hydrogen ions (H^+). It will be shown later that in aqueous solution hydrogen ions always combine with water molecules and form *hydronium ions* (H_3O^+). *Therefore, H_3O^+ ions are responsible for the characteristics of acids.*

When tested, the solutions of compounds containing hydroxide ions (OH^-), all feel slippery, turn red litmus blue, conduct electricity, and neutralize acid solutions. These observable properties constitute the *operational definition* of a *basic* solution. On the basis of these observations, we may assume that hydroxide ions are common to all basic solutions and that they are responsible for basic characteristics.

Special attention should be called to an aqueous solution of ammonia (NH_3), a common household cleansing agent. The formula (NH_3) shows no hydroxide ions, yet a solution formed by passing NH_3 gas into water is basic. *Ammonia water* contains more hydroxide ions than hydrogen (hydronium) ions. This suggests that NH_3 gas reacts with water and forms hydroxide ions. Tests also confirm the presence of ammonium ions (NH_4^+) and NH_3 molecules. We may represent this information by the following equation.

$$NH_3(g) + H_2O(l) \rightleftharpoons NH_4^+(aq) + OH^-(aq)$$

Laboratory solutions of ammonia are frequently labeled and referred to as *ammonium hydroxide.* Note that a double arrow is used in the above equation to indicate, among other features, that all four species represented in the equation (NH_3, H_2O, NH_4^+, OH^-) are present in the solution.

When hydrochloric acid reacts with sodium hydroxide it can form a solution which has neither acid nor basic properties. In this reaction, sometimes called a ***neutralization reaction,*** hydronium ions are removed from the acid solution and hydroxide ions are removed from the basic solution. The equation for the reaction is

$$H_3O^+ + OH^- \longrightarrow 2H_2O$$

The resulting solution is said to be neutral if equal numbers of H_3O^+ and OH^- ions react. *Neutral solutions contain no excess of hydronium or hydroxide ions. Basic solutions have more hydroxide than hydronium ions, and acid solutions have more hydronium than hydroxide ions.*

At this point we may use our observations to define an *acid* as *any substance which contains hydrogen and yields an aqueous solution which contains more hydronium ions than hydroxide ions.* Similarly, a *base* may be tentatively defined as *any substance which furnishes a greater quantity of hydroxide than hydronium ions in water solution.*

We have already noted that, in general, salts may be described as ionic solids. Ordinary salt (NaCl) crystals can be prepared by evaporating the solution formed by neutralizing hydrochloric acid (HCl) with sodium hydroxide (NaOH). The reaction in terms of word equation is expressed as

$$\textbf{acid} + \textbf{base} \longrightarrow \textbf{salt} + \textbf{water}$$

and in terms of a formula equation as

$$HCl + NaOH \longrightarrow NaCl + H_2O$$

The reaction described in the above word equation is typical of aqueous acid-base reactions. That is, *acids and bases react chemically and form water.* A salt remains in the solution. It should be noted that our first definition of acids, bases, and neutralization reactions are, like our first model of the atom, based on limited observations and, therefore, tentative. We shall modify, extend, and expand our definitions whenever old ones fail to explain new observations.

Because you will be using a number of different acids throughout your chemistry course, it is essential that you recognize their formulas and be able to name them. We shall, therefore, briefly consider the naming of acids. Additional information related to the naming of chemical compounds is found later in this chapter.

2-29 Naming Acids Binary acids contain hydrogen and a second nonmetallic element. It is customary to write the symbol for hydrogen first in the formula for an acid. The names of all binary acids begin with the prefix hydro- and end with the suffic -ic. The central part of the name is derived from the name of the nonmetal. A water solution of HCl is called hydro-chlor-ic acid. It is important to note that a water solution of HCl contains very few molecules of HCl. Other common binary acids are listed in Table 2-10.

TABLE 2-10
COMMON BINARY ACIDS

Formula	Acid Name	Compound Name
HF	hydrofluoric acid	hydrogen fluoride
HCl	hydrochloric acid	hydrogen chloride
HBr	hydrobromic acid	hydrogen bromide
HI	hydriodic acid	hydrogen iodide
H_2S	hydrosulfuric acid	hydrogen sulfide

In the pure state, HCl and the other compounds listed in Table 2-10 are named as binary hydrogen compounds. The word *acid* denotes an aqueous solution of the pure compound. A water solution of hydrogen cyanide (HCN) is called hydrocyanic acid even though HCN is not a binary compound. This is an exception to the general rule of naming.

Acids containing hydrogen, oxygen, and a third element are called ***oxyacids.*** The names for oxyacids end with the suffix -ic or the suffix -ous depending on the number of oxygen atoms present. Many elements form several oxyacids. For example, chlorine forms oxyacids having these formulas and names.

Formula	*Name*
HClO	**hypochlorous acid**
$HClO_2$	**chlorous acid**
$HClO_3$	**chloric acid**
$HClO_4$	**perchloric acid**

Generally, oxyacids of a given element have different numbers of oxygen atoms in their formulas. If the element forms only two oxyacids, the suffix -ic is used to designate the acid with the higher oxygen content. The suffix -ous is used to designate the acid with the lower oxygen content. If an element forms more than two oxyacids the prefix per- is used to denote the acid having a higher oxygen content than the regular -ic acid. The prefix hypo- is used to indicate a lower oxygen content than the regular -ous acid. A list of common oxyacids is provided in Table 2-11. Rather than list additional rules for naming acids, it seems more desirable to list the common acids and their names and introduce others along with their names as the situation demands.

TABLE 2-11
COMMON OXYACIDS

Formula	*Acid Name*
HNO_2	nitrous acid
HNO_3	nitric acid
H_2SO_3	sulfurous acid
H_2SO_4	sulfuric acid
H_3PO_3	phosphorous acid
H_3PO_4	phosphoric acid
H_2CO_3	carbonic acid
$HC_2H_3O_2$	acetic acid

LOOKING AHEAD

Most of our laboratory work involves observing and interpreting chemical reactions. We now need a concise method of representing qualitatively and quantitatively chemical reactions which we observe in the laboratory and discuss in the text. In this chapter, we have acquired the necessary concepts and tools for this task. The formulas which we learned to write and name furnish the "raw materials" for qualitatively describing a chemical reaction. The mole concept and relative atomic and molecular mass concepts allow us to discuss the quantitative aspect of reactions. The laws and principles developed in this chapter furnish the guide lines for our calculations.

With this arsenal of knowledge at our disposal, we shall turn our attention in the next chapter to writing, balancing, and interpreting equations for chemical reactions.

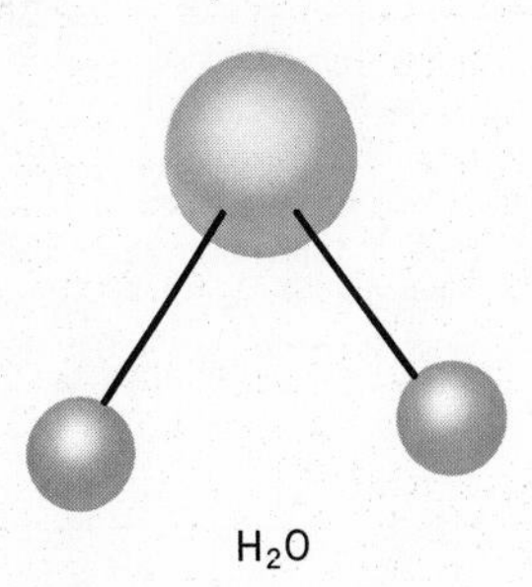

2-26 Common or classical names are used for many familiar substances.

Special Feature: Nomenclature

Names of compounds are uniquely associated with specific formulas. The two systems used by chemists today are (a) ***the classical system*** and (b) ***the Stock system.*** In the Stock system, named after Alfred Stock (1876–1946), a brilliant 20th-century inorganic chemist, Roman numerals are used to indicate the positive charge on a metallic ion in a compound. For example, Fe_2O_3 is called *iron(III) oxide.*

JOHN DALTON
1766–1844

One of the most important milestones in the progress of science was the introduction of the atomic theory by John Dalton. All science is fundamentally based on the concept of the atomic nature of matter. Although the idea that matter was composed of atoms had been proposed before 1800, it was Dalton who converted it into a real and usable working hypothesis. He assigned symbols to the elements and combined symbols to represent compounds. In addition, he prepared a list of atomic masses for 14 elements. Although his symbols were later improved upon by Berzelius, they were a great step ahead of the hieroglyphics of the alchemists. In alchemical writings, there were over 200 names and symbols for the element mercury.

Dalton published his theory in "A New System of Chemical Philosophy." He theorized that atoms of a given element were all alike and that they all had the same mass. He chose hydrogen, the lightest element, as a standard for atomic mass. He assigned a unit mass to the hydrogen atom and based the masses of other atoms on the mass of the hydrogen atom.

It is interesting to note that Dalton was able to make so many scientific observations and contributions in spite of the fact that he was afflicted with red-green color blindness. When he was presented to the King, protocol required that he wear a red robe. Since he was a Quaker, it was considered improper for him to wear bright clothing. When questioned about the red robe, Dalton replied that to him the robe was a simple green color, the same as the leaves on the trees.

In the classical system, different suffixes are used to indicate different charges on a metallic ion. The suffix -ous is used to denote the lower charge, and the suffix -ic is used to denote the higher charge. Thus, FeO is named *ferrous oxide,* and Fe_2O_3 is named *ferric oxide.* In this text we shall use the Stock system exclusively for identifying the charges on metallic ions. A few rules are given and illustrated below. Others will be introduced as they are needed.

Compounds containing only two elements are known as ***binary compounds.*** The names of binary compounds are derived from the names of the two elements involved. Binary compounds may contain a metal combined with a nonmetal or two nonmetals chemically combined. *The name of a binary compound containing a metal and nonmetal consists of the name of the metal followed by the stem of the name of the nonmetal to which is added the suffix -ide.* The charge (oxidation state) of a metal having multiple oxidation states is designated by a Roman numeral enclosed in parentheses immediately following the name of the metal. The Stock system and classical system names are given for a number of these compounds in Table 2-12. It should be noted that mercury(I) ions are somewhat unusual. They are not simple ions like most metallic ions, but diatomic ions composed of two charged mercury atoms bonded together. The ion $(Hg_2)^{2+}$ carries a 2+ charge which is equivalent to each of the bonded atoms having an oxidation state of 1+.

Binary compounds composed of two nonmetals are molecular rather than ionic. That is, macroscopic samples consist of an aggregate of neutral molecules rather than an aggregate of ions. These compounds may be named by using either the classical or the Stock system. To use the Stock system, it is necessary to determine the oxidation state (apparent charge) of the nonmetal whose symbol

TABLE 2-12
NAMES AND FORMULAS OF BINARY COMPOUNDS OF METALS AND NONMETALS

Formula	*Stock Name*	*Classical Name*
KCl	potassium chloride	potassium chloride
CaO	calcium oxide	calcium oxide
$SnCl_4$	tin(IV) chloride	stannic chloride
$SnCl_2$	tin(II) chloride	stannous chloride
Al_4C_3	aluminum carbide	aluminum carbide
CdS	cadmium sulfide	cadmium sulfide
Ca_3P_2	calcium phosphide	calcium phosphide
Mg_3N_2	magnesium nitride	magnesium nitride
Hg_2Cl_2	mercury(I) chloride	mercurous chloride
$HgCl_2$	mercury(II) chloride	mercuric chloride
MnO_2	manganese(IV) oxide	manganese dioxide
Mn_2O_5	manganese(V) oxide	manganese pentoxide
Mn_2O_7	manganese(VII) oxide	manganese heptoxide
CuCl	copper(I) chloride	cuprous chloride
$CuCl_2$	copper(II) chloride	cupric chloride

is written first in the formula. The oxidation state of this element is written as a Roman numeral enclosed in parentheses following the symbol. For example, N_2O is named nitrogen(I) oxide. The oxidation state (apparent charge) of an atom in a molecule is determined by assigning a fixed common oxidation state to the other atom(s) in the formula. Oxygen exhibits an oxidation state of 2− in most of its compounds. If we assign oxygen an oxidation state of 2− in N_2O, then the nitrogen atom must exhibit a 1+ (I) oxidation state in order to make the molecule electrically neutral. In general, the symbol of the most electropositive element is written first. In the classical system, prefixes are used to designate the number of atoms of each element in the formula. Commonly used prefixes are listed with their meanings in Table 2-13. Examples of nonmetallic binary compounds with their Stock and classical names are listed in Table 2-14.

Compounds containing metallic ions and polyatomic negative ions are named by stating the name of the metallic ion followed by the name of the polyatomic ion. Roman numerals are not needed unless the metallic ion has a variable charge. The formulas and names of some of these compounds are listed in Table 2-15.

TABLE 2-13
COMMONLY USED PREFIXES

Prefix	*Meaning*
mono-	1
di	2
tri-	3
tetra-	4
penta-	5
hexa-	6
hepta-	7
octa-	8

TABLE 2-14
NAMES AND FORMULAS OF BINARY COMPOUNDS (Two Nonmetals)

Formula	*Stock Name*	*Classical Name*
SO_2	sulfur(IV) oxide	sulfur dioxide
N_2O	nitrogen(I) oxide	nitrous oxide or dinitrogen monoxide
NO	nitrogen(II) oxide	nitric oxide or nitrogen monoxide
N_2O_3	nitrogen(III) oxide	dinitrogen trioxide
N_2O_4	nitrogen(IV) oxide	dinitrogen tetroxide
N_2O_5	nitrogen(V) oxide	dinitrogen pentoxide
SF_6	sulfur(VI) fluoride	sulfur hexafluoride
Cl_2O_7	chlorine(VII) oxide	dichlorine heptoxide
S_2Cl_2	sulfur(I) chloride	disulfur dichloride
CO	carbon(II) oxide	carbon monoxide
CO_2	carbon(IV) oxide	carbon dioxide

TABLE 2-15
NAMES AND FORMULAS OF COMPOUNDS CONTAINING A METALLIC AND A POLYATOMIC NEGATIVE ION

Formula	*Name*
$Al_2(SO_4)_3$	aluminum sulfate
$Sn(NO_3)_4$	tin(IV) nitrate
$NaHCO_3$	sodium hydrogen carbonate or sodium bicarbonate
Na_3PO_4	trisodium phosphate
Na_2HPO_4	disodium monohydrogen phosphate
NaH_2PO_4	sodium dihydrogen phosphate
$Ca(C_2H_3O_2)_2$	calcium acetate
$BaCrO_4$	barium chromate
$K_2Cr_2O_7$	potassium dichromate
$Ca(ClO)_2$	calcium hypochlorite
K_2SO_3	potassium sulfite
$KHC_8H_4O_4$	potassium hydrogen phthalate or potassium biphthalate
$Na_2C_2O_4$	sodium oxalate
$NaC_7H_5O_2$	sodium benzoate
$KMnO_4$	potassium permanganate

FOLLOW-UP PROBLEM

Name these compounds: (a) Ag_3PO_4, (b) $NaIO_3$, (c) $SbCl_3$, (d) Hg_2SO_4, (e) I_2O_5, (f) $MgCO_3$, (g) $Mn(OH)_2$, (h) CrF_2, (i) I_2O_7, (j) SiO_2, (k) $Zn(HCO_3)_2$, (l) $CaCrO_4$.

Special Feature: Chromatography

The 19th century physiologist, Claude Bernard (1813–1878), said that progress in science frequently depends upon the development of good methods. Chromatography is an excellent example of a method developed for separating mixtures. A Russian botanist, Mikhail Tsvett (1872–1920), in 1906, discovered that the colored chlorophyll pigments in green leaves could be separated by passing an ether solution of these pigments through a tube of solid powdered calcium carbonate. The name of the process, chromatography, literally means "color writing."

Chromatography is based upon the fact that different kinds of molecules are *adsorbed* to a different extent by different kinds of substances. Thus when a mixture of pigments is passed through a column containing an adsorbent which could be alumina, (Al_2O_3), starch, charcoal, or even powdered sugar, the components tend to be adsorbed to different degrees. When an appropriate solvent is poured through the column, the components of the mixture separate into bands as they move down the column at their own specific rate. The components of the mixture are like the athletes in a 440-yard run. They start together but are strung out at the end of the race. Column chromatography is illustrated in Fig. 2-7.

The field of chromatography was relatively dormant for the first twenty-five years after Tsvett first separated the components of chlorophyll. Then slowly new applications were found until today there are many thousands of applications. Chromatography is used to make many different types of separations. Mixtures of gases, liquids, or dissolved solids can be fractionated into their component parts by chromatography. The method is particularly effective for separating and identifying fragile or elusive molecules which would be destroyed or chemically modified by other means.

Complex mixtures of amino acids obtained by hydrolyzing (breaking down) proteins can be separated using a process known as *paper chromatography*. This method employs an absorbent sheet of paper in place of the chromatographic column. The paper consists of cellulose containing adsorbed water. The adsorbed water is the stationary phase and an organic solvent is the moving phase. The original sample is placed on the corner of a piece of the paper which is suspended vertically. The solvent is passed through the sample so that the components are separated downward in a vertical line from the origin of the sample. The different amino acids travel at different rates depending on the extent to which they are retained by the stationary water phase. The paper can then be turned at right angles so that this line of components is at the top. A new solvent is then passed over these components so that further separation can take place in case any of the original spots contain two or more components. The result is called a *two-dimensional paper chromatogram* where the components are separated out over the whole sheet of paper. The method of paper chromatography has been standardized to the degree that it can be used to identify different compounds by their final position on a paper chromatogram. A normal sample of decomposed hemoglobin gives a particular pattern on a paper chromatogram while an abnormal sample gives quite a different and distinctive pattern. Colorless components can be detected by spraying the paper with a solution that forms a pigment with the component. If desired, the amino acids can be dissolved and then analyzed separately. The researchers who discovered the mechanism of the photosynthetic process, used paper chromatography as one of their most important tools. Paper chromatography is illustrated in Fig. 2-7. In this illustration, the solvent moves up the strip by capillary flow.

Gas chromatography is an effective method of separating complex mixtures of gases. In a typical gas chromatographic separation and analysis, a sample which contains a mixture of gases is injected into a stream of carrier gas at the inlet of the chromatographic column. The column is packed with an inactive, pulverized, solid material coated with a layer of a nonvolatile liquid known as a partitioner. As the carrier gas propels the sample through the column, the components of the mixture arrive at the outlet at different times (Fig. 2-27). This separation is caused by the interaction of

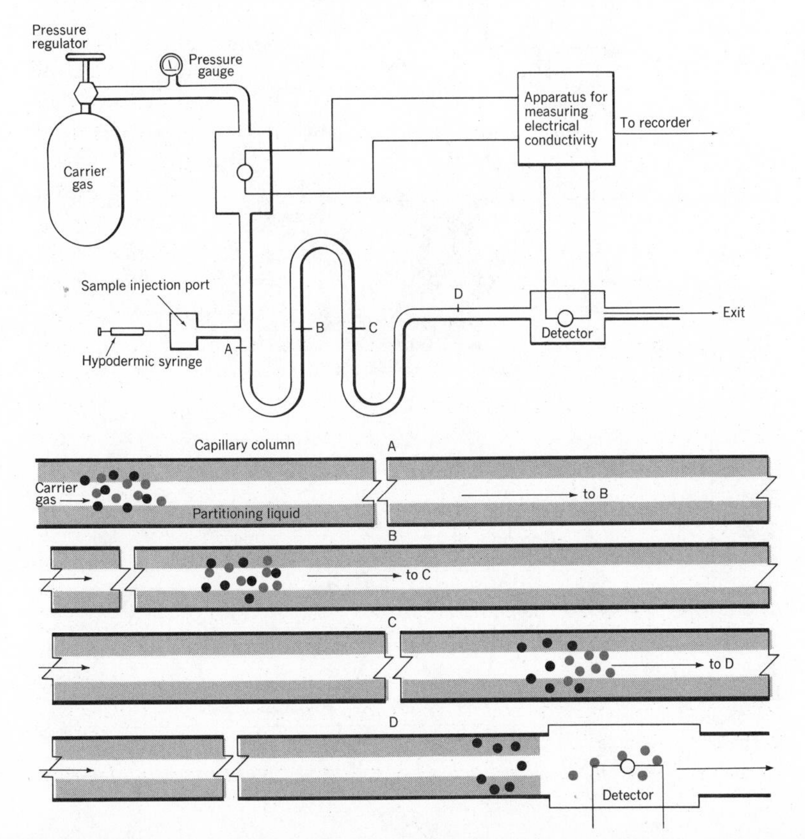

2-27 Schematic representation of a gas chromatograph. The dots represent molecules of the gas mixture being analyzed. Molecules of the carrier gas are not shown.

the components with the partitioner liquid which selectively interferes with the progress of each component. The rate at which the components pass through the column depends, in part, on how they partition themselves between the liquid and vapor phases. The less soluble, more volatile gases are swept ahead and separated from the less volatile ones. At the outlet, the gases pass into a detector. The detector may use a thermal-conductivity cell or other device which signals the change in the composition of the gas. The signals may be fed to a strip recorder which produces a chart known as a gas chromatogram. Analysis of the chromatogram reveals qualitative and quantitative information. Using modern gas chromatographs, samples with mass of a few thousandths of a gram can be analyzed in a few seconds. Components with mass as little as a trillionth of a gram can be detected. This means a gas chromatograph can identify insecticides used on vegetables and fruits even after thorough washing. It can be used to detect and measure adulterants in food, water, and other supposedly pure substances as well as some of the components in smog, cigarette smoke, and other agents hazardous to health. The potential of this versatile method is almost unlimited in its application.

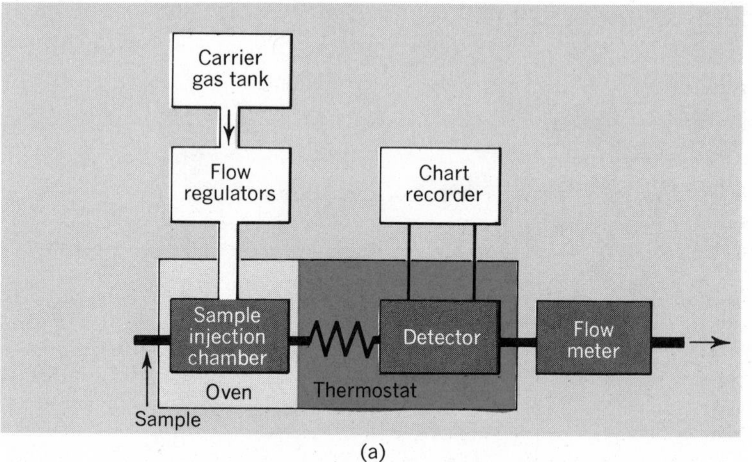

(a)

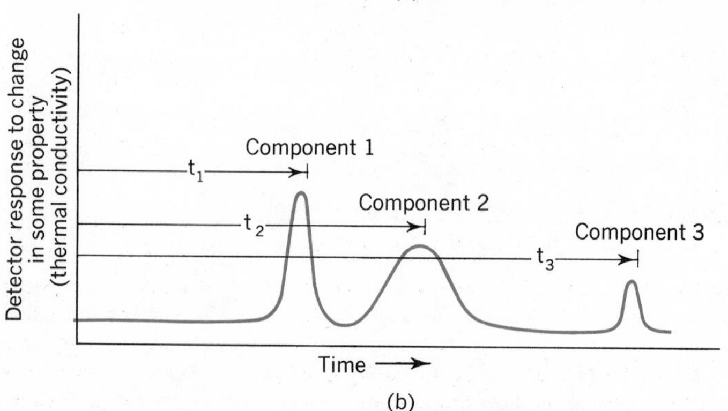

(b)

2-28 (a) Block diagram of apparatus used for gas chromatographic analysis. Helium, hydrogen, nitrogen, and carbon dioxide are the most commonly used carrier gases. (b) A gas chromatogram for the analysis of a three-component system such as a mixture of propane, butane, and pentane. t represents the time for the solute (component) peak to reach the detector. The components may be qualitatively identified by the time required for the peak to reach the detector. The relative amount of each component is related to the area under its respective peak.

Diagram taken from Fundamentals of Quantitative Analysis—Skoog and West, Holt, Rinehart, and Winston.

QUESTIONS

1. What is the difference between a homogeneous mixture and a heterogeneous mixture? Give examples of each.

2. Explain how a homogeneous mixture differs from a pure substance in terms of (a) composition, (b) method of separating components, and (c) identity of components.

3. What are the different components of a solution called?

4. How many phases are present in each of these: (a) a salt solution, (b) a salt solution and small stones, (c) a 70 percent solution of ethyl alcohol, (d) an oil, water, and sand mixture, (e) the atmosphere, (f) sterling silver (an alloy of silver and copper)?

5. You are given two liquids to identify as either solutions or pure substances. Explain how to identify them.

6. Describe fractional distillation.

7. Which of these pairs of liquids are immiscible: (a) gasoline and kerosene, (b) salad oil and water, (c) water and alcohol, (d) paint thinner and linseed oil, (e) mercury and water?

8. What two ways can be used to (a) increase the quantity of oxygen dissolved in water, (b) decrease the nitrogen dissolved in water?

9. Is the Law of Definite Proportions evidence of the validity of Dalton's Atomic Theory? Explain.

10. (a) What is an atom? (b) What is a molecule? (c) Is it possible to isolate an atom of a compound? Explain. (d) Is it possible to isolate a molecule of an element? Justify your answer with examples.

11. Name five gases whose molecules are diatomic at room temperature.

12. Elements A and B unite and form two different compounds.

0.3 g of A + 0.4 g of B $\longrightarrow$ 0.7 g of compound X
18 g of A + 48 g of B $\longrightarrow$ 66 g of compound Y

Show how the Law of Multiple Proportions is illustrated by these data.

13. Using the oxides of nitrogen (N_2O, NO, N_2O_3, NO_2), demonstrate the Law of Multiple Proportions.

14. Write a symbol for these elements. You may use the Periodic Table but do not refer to a list of elements with their symbols.

(a) Sodium	(e) Phosphorus	(i) Mercury
(b) Calcium	(f) Sulfur	(j) Iron
(c) Aluminum	(g) Iodine	(k) Cobalt
(d) Carbon	(h) Neon	(l) Antimony

15. Write the names of the elements whose symbols are listed. Do not refer to a list of elements.

(a) K	(e) Si	(i) Pb
(b) Mg	(f) As	(j) Ag
(c) B	(g) Br	(k) Cu
(d) Li	(h) Sn	(l) Mn

16. Name these binary compounds containing metallic and nonmetallic elements. (a) KCl, (b) Al_2O_3, (c) NaI, (d) ZnS, (e) $LiBr$, (f) Ba_3N_2, (g) CaF_2, (h) Mg_3P_2, (i) CaC_2.

17. How many atoms of each element are represented in these formulas?

(a) Na_2O	(e) $Zn(C_2H_3O_2)_2$	(h) $Zn_3(PO_4)_2$
(b) $MgCl_2$	(f) $CuSO_4 \cdot 5H_2O$	(i) $Fe_4[Fe(CN)_6]_3$
(c) H_3PO_4	(g) $NaHCO_3$	(j) $C_{12}H_{22}O_{11}$
(d) KH_2PO_4		

18. Give the origin of some chemical symbols for three elements that do not correspond to the English names.

19. Element A combines with element B and forms compound AB. Heat is given off by the reaction. (a) Write an equation to express this reaction. (b) Is this reaction endothermic or exothermic? (c) Do the reactants or products have the greater energy content? (d) Discuss the conservation laws as they relate to this system.

20. Identify these changes as exothermic or endothermic: (a) burning coal in oxygen, (b) using food as fuel in our bodies, (c) green plants making starch out of carbon dioxide and water in sunlight, (d) changing water to water vapor, (e) shooting off a flashbulb, (f) melting ice.

21. A piece of calcium is placed in an unreactive sealed quartz tube with pure oxygen. The entire system weighs 40.0 g. After heating, a white powder is noted in the tube. Explain, using word equations, what probably took place between the calcium and oxygen in the tube. What does the whole system weigh now? What principle did you use to answer this question?

22. Which of these are exothermic reactions?

(a) $2HgO + 43.3 \text{ kcal} \longrightarrow 2Hg + O_2$
(b) $Fe_2O_3 + 3CO \longrightarrow 2Fe + 3CO_2 + 6.6 \text{ kcal}$
(c) $N_2 + O_2 \longrightarrow 2NO - 43.2 \text{ kcal}$

23. Rewrite each of the above equations so that the coefficient in front of the first formula to the right of the arrow is 1. Show the heat change associated with one mole of this substance.

24. Name these compounds: (a) NH_4F, (b) $Pb(ClO_3)_2$, (c) K_3PO_4, (d) $FeSO_4$, (e) $BaCrO_4$, (f) $Cu(OH)_2$, (g) $Al_2(SO_4)_3$, (h) ICl, (i) OF_2, (j) $CuSO_3$, (k) $Ca(NO_2)_2$, (l) $Sn(NO_3)_4$.

25. Write formulas for (a) iron(III) sulfate, (b) sodium sulfide, (c) barium sulfite, (d) potassium bisulfite, (e) ammonium bisulfide, (f) calcium bisulfate, (g) magnesium acetate, (h) barium monohydrogen phosphate (i) mercury(I) nitrate, (j) mercury(II) bromide, (k) potassium oxalate, (l) copper(II) arsenate, (m) tin(IV) oxide, (n) zinc phosphide, (o) silver thiocyanate, (p) cadmium cyanide, (q) antimony(III) sulfide, (r) cobalt(II) bromate, (s) thallium(III) hydroxide, (t) titanium(IV) chloride, (u) calcium bicarbonate, (v) chromium(II) iodide, (w) sodium dichromate, (x) nickel(II) chlorate, (y) aluminum perchlorate, (z) manganese carbonate.

26. Show by writing equations the dissociation of each of these ionic substances in water solutions: (a) KCl, (b) $Ca(C_2H_3O_2)_2$, (c) NH_4Br, (d) $NaNO_3$, (e) Li_2CO_3, (f) $MgSO_4$.

PROBLEMS

1. (a) The mass of 2.5×10^4 grapes is 50 kg and that of an equal number of oranges is 1.2×10^3. What is the mass ratio of a single grape to a single orange? (b) A mole of carbon atoms has a mass of 12 g, and a mole of magnesium atoms 24 g. What is the mass ratio of a single carbon atom to a single magnesium atom?

2. A reaction involves 23.0 g of NO_2. (a) How many grams of SO_2 react with the NO_2 assuming an equal number of SO_2 molecules is required? (b) What volume of SO_2 is needed at STP?

3. Assume that the molecular mass of O_2 gas were 48 instead of 32. (a) What would be the value of Avogadro's number in the new system? (b) How many atoms of oxygen would there be in a molecule of oxygen? (c) How would the mass in grams of a single oxygen molecule compare with its mass in the $O_2 = 32$ system? (d) What would be the atomic mass of nitrogen?

4. Lime is formed when calcium and oxygen react in a mass ratio of 2.5 g Ca to 1.0 g oxygen. (a) How many grams of oxygen are needed to combine with 40 g of calcium? (b) What is the mass of the product?

Ans. (a) 16 g, (b) 56 g.

5. A mass of 100.3 g of Hg combines with 8.00 g of oxygen and forms an oxide of mercury. How many grams of oxygen could be obtained by the complete decomposition of 50.0 grams of the oxide? **Ans. 3.70 g.**

6. Aluminum and oxygen combine in a mass ratio of 9.00 to 8.00. If a flashbulb contains 5.4×10^{-3} g of aluminum, what mass of oxygen must be present for complete combustion of the aluminum?

Ans. 4.8×10^{-3} g.

7. The elements A and Z form a series of binary compounds with each other. One member of the series contains 84.6% Z by mass. Find the percent of Z in another member of the series.

8. Determine the molecular mass of these compounds: (a) methane (CH_4), (b) phosphorus trichloride (PCl_3), (c) ammonia (NH_3), (d) sulfuric acid (H_2SO_4), (e) silicon dioxide (SiO_2), (f) carbon tetrachloride (CCl_4), (g) nitrogen(IV) oxide (NO_2), (h) nitrogen(V) oxide (N_2O_5), (i) glucose ($C_6H_{12}O_6$).

9. How many moles in each quantity? (a) 1000 g water, (b) 28.0 g of carbon dioxide (CO_2), (c) 24.0 g of O_2, (d) 92.0 g of ethyl alcohol (C_2H_5OH).

10. How many molecules in the masses of each substance listed in Problem 9?

11. What is the mass of (a) 0.25 mole of sodium hydroxide (NaOH), (b) 1.50 moles of H_2SO_4, (c) 1.0×10^{-3} mole of $Ba(OH)_2$, (d) 3.0×10^{23} molecules of water?

Ans. (c) 0.17 g.

12. (a) What is the average mass of a water molecule? (b) How many water molecules are there in 1 drop (0.05 g) of water? (c) How many years will it take for the drop to evaporate if one billion water molecules leave the liquid each second? **Ans. (c) 5.3×10^6 yr.**

13. These problems refer to H_2S, a gaseous compound. (a) Calculate the grams of 1.00 mole. (b) Calculate the moles of hydrogen atoms in 0.100 mole. (c) Calculate the grams of compound which contains 2.00 moles of sulfur atoms. (d) Calculate the number of moles which contain 4.8 g of sulfur. (e) Calculate the mass which contains 6.4 grams of hydrogen. (f) How many molecules in 10.3 g? (g) How many atoms in 10.3 g? (h) Calculate the atoms of sulfur combined with 20 atoms of hydrogen. **Ans. (c) 68 g, (e) 109 g, (h) 10 atoms S.**

14. Calculate the mass percentage composition of each compound. (a) $MgCl_2$, (b) Na_2SO_4, (c) Fe_2O_3, (d) $C_7H_5N_3O_6$, (e) $AlBr_3 \cdot 6H_2O$.

15. How many grams of bromine can be obtained by the decomposition of 92.0 g of $MgBr_2$? **Ans. 80.0 g.**

16. How many grams of oxygen can be obtained by the decomposition of 245 g of $KClO_3$?

17. An ore of nickel contains 3.00 moles of NiS per kg of ore. How many grams of nickel could be obtained from 1.0 kg of this ore?

18. How many pounds of phosphorus can be obtained by the decomposition of 10.0 pounds of calcium phosphate [$Ca_3(PO_4)_2$]? **Ans. 2.00 lb.**

19. An iron ore contains 50.0% Fe_2O_3. How many grams of pure iron could be obtained from 200 g of this ore?

20. A sample of iron, mass 50.0 g, is completely converted into Fe_2O_3. What mass of Fe_2O_3 is obtained? **Ans. 71.5 g.**

21. The chromium in a chromium ore is completely converted into Cr_2O_3. If 0.450 g of Cr_2O_3 is obtained from a 2.00-g sample of the ore, what is the percentage of chromium in the ore?

22. Cinnabar is an ore containing mercury sulfide (HgS). A sample of impure cinnabar contains 85.0% HgS. How many grams of mercury can be obtained from 150 g of the ore?

23. Pyrolusite is an ore containing manganese dioxide (MnO_2). The ore may be analyzed by isolating the manganese as $Mn_2P_2O_7$. If 1.00 g of $Mn_2P_2O_7$ is obtained from a 0.750-g sample of the ore, what is the percentage of Mn in the ore? **Ans. (a) 51.7 percent.**

24. A 12.00-g sample of a mixture of $KClO_3$ and KCl is heated until all the oxygen has been removed from the sample. The product, entirely KCl, has a total mass of 9.00 g. What is the percentage of $KClO_3$ in the original mixture?

25. A compound of carbon and hydrogen is found to contain 75.0% carbon and 25.0% hydrogen. (a) How many moles of C are contained in a 100-g sample? (b) How many moles of H are contained in a 200-g sample? (c) How many moles of hydrogen atoms (gram-atoms of H) are combined with 1 mole of carbon atoms (1 gram-atom of C)? (d) What is the simplest formula for this compound?

26. Calculate the simplest formula for each compound: (a) 92.3% C, 7.7% H, (b) 75.7% As, 24.3% O, (c) 31.9% K, 28.9% Cl, 39.2% O, (d) 29.1% Na, 40.5% S, 30.4% O. **Ans. (d) $Na_2S_2O_3$.**

27. A titanium (Ti) chloride is analyzed by converting all the titanium into 1.20 g of TiO_2 and all the chloride into 6.45 g of AgCl. What is the simplest formula of the original compound?

28. When a 3.00-g sample of a compound containing only C, H, and O was completely burned, 1.17 g of H_2O and 2.87 g of CO_2 were formed. What is the simplest formula of the compound?

29. A compound contains 82.7% carbon and 17.3% hydrogen. The density of its vapor at STP is 2.59 g/ℓ. Assume molecular mass of O_2 is 32.0. What is the molecular formula of the compound? **Ans. C_4H_{10}.**

30. The composition of nicotine is 74.0% C, 8.7% H, and 17.3% N. The molecular mass of nicotine is 162. What is its molecular formula?

31. A certain volume of arsenic (g) has a mass of 4.56 g. The same volume of O_2 at the same conditions has a mass of 0.477 g. (a) What is the molecular mass of arsenic (g)? Assume $O_2 = 32.0$. (b) What is the molecular formula of arsenic vapor? (c) How many atoms are there per molecule of arsenic?

32. A compound whose formula is XCl_2 contains 34.05% chlorine by mass. Chlorine has an atomic mass of 35.45. What is the atomic mass of X?

33. A compound whose formula is Y_2O_5 contains 44.0 percent O by mass. What is the atomic mass of Y?

34. Indium reacts with sulfur and forms a compound having the formula In_2S_3. Analysis reveals that 3.26 g of compound contains 2.30 g of indium. Assume the atomic mass of sulfur is 32.1 and calculate the atomic mass of indium. **Ans. 115.**

3

Qualitative and Quantitative Interpretation of Chemical Reactions

Preview

"We must lay it down as an incontestable axiom, that in all the workings of art and nature, nothing is made from nothing; an identical quantity of matter exists both before and after the experiment . . . upon this principle, the whole art of performing chemical experiments depends.

Antoine Lavoisier,
Elements of Chemistry (1789)

Chemistry is a quantitative science, and, consequently, a chemist deals with measured amounts of substances. He must be able to determine how much of a given substance is needed to produce a required amount of product in a specific reaction. He bases these calculations on the masses and numbers of molecules or other kinds of particles participating in a reaction.

The clue to the relative numbers of each kind of molecule (or other particle) that participates in a reaction is obtained from a balanced chemical equation. These equations show the relative number of moles of all participants in a reaction. Because the quantitative interpretation of a chemical reaction is based on a balanced equation, we shall first learn how to balance a skeleton equation which shows only the formulas of the reactants and products.

We shall then use balanced equations, the mole concept, and relative atomic and molecular masses to calculate the quantities of products.

Many of the reactions we encounter in the laboratory are carried out in aqueous solutions. Furthermore, some of the most important processes and reactions that take place in solution involve ions. Because of the importance of reactions between ions, we shall examine in some detail one of the simplest types of ionic reactions, a *precipitation reaction*. In our discussion we shall draw on the background in Chapter 2 and the data from related experiments.

We shall also use the mole concept to develop a widely-used method for expressing quantitatively the amount of solute in a given solution. The application of this concentration unit, called *molarity*, will be illustrated when we solve quantitative problems involving solutions. The calculations are not difficult, but considerable practice is required to develop skill in handling them.

THE CHEMICAL EQUATION

3-1 Representation of Laboratory Observations One of the most efficient, useful devices for communicating information related to chemical changes is a chemical equation. When properly constructed, an equation conveys not only qualitative but also quantitative information related to the nature and quantity of the participants involved in the change. It may also include the energy change associated with the reaction. As you would expect, the concept of the mole plays a key role in the quantitative interpretation of reactions as represented by chemical equations.

An equation for a chemical reaction represents the changes that occur during a reaction. Equations, therefore, are based on laboratory experimentation and observation. The correct formulas of the participants in a reaction must be known or determined in the laboratory by methods described earlier. In addition, the masses or the volumes of the reactants and products must be determined. We shall use the electrolysis of water to demonstrate how an equation can be developed from laboratory data.

Qualitatively, the products can be identified as hydrogen and oxygen gas by their properties. We have already used Gay-Lussac's observations and Cannizzaro's and Avogadro's Principles to deduce that the formulas for these gases are H_2 and O_2. Various methods of analysis also reveal that the formula for water is H_2O.

Experimental data show that when 36.0 g of water is decomposed, 32.0 g of oxygen gas and 4.0 g of hydrogen gas are produced. The moles of each participant are calculated:

$$\mathbf{36.0\ g\ H_2O \times \frac{1\ mole\ H_2O}{18.0\ g\ H_2O} = 2.0\ moles\ H_2O}$$

$$\mathbf{32.0\ g\ O_2 \times \frac{1\ mole\ O_2}{32.0\ g\ O_2} = 1.0\ mole\ O_2}$$

$$\mathbf{4.0\ g\ H_2 \times \frac{1\ mole\ H_2}{2.0\ g\ H_2} = 2.0\ moles\ H_2}$$

The result of this experiment may be expressed by the word equation

2 moles water ⟶ 1 mole oxygen gas + 2 moles hydrogen gas

The same information may be conveyed by the formula equation

$$\mathbf{2H_2O(l) \longrightarrow O_2(g) + 2H_2(g)} \qquad \text{3-1}$$

Since the phase in which a pure substance exists varies with temperature and pressure, it is important to specify the state at the conditions of the experiment. The observations (s) for solid, (g) for gas, (l) for liquid, and (aq) for solutes in water solution are placed in parentheses to the right of each formula.

Equation 3-1 is called a ***balanced equation.*** Let us examine its features. First of all, it shows that atoms and therefore mass are both conserved during the reaction. There are four hydrogen atoms and two oxygen atoms in the water molecules shown on the left side of the equation and also four hydrogen and two oxygen atoms in the molecules of product shown on the right side of the equation. In accordance with the ***Law of Conservation of Mass,*** *the number of atoms in the system undergoing change must be constant.* Note that the atoms have rearranged themselves so as to form three molecules of product from two molecules of reactant. The total number of atoms on each side of the equation must be equal but the total number of molecules or moles is not necessarily equal.

The numbers in front of the formulas are called ***coefficients.*** Since Equation 3-1 was derived from experimental data, it is apparent that the *coefficients in a balanced equation stand for the relative number of moles of substances involved in the reaction.* Equation 3-1 shows that when water is decomposed, hydrogen gas and oxygen gas are always formed in the ratio of two moles of hydrogen to one of oxygen. A balanced equation is necessary to enable us to predict the quantities of reactants needed to produce required amounts of specified products. It is not practical to go into the laboratory to deduce formulas and an equation for every reaction. As the course unfolds, you will be able to write formulas and equations. In the meantime, you can learn to balance and to interpret equations for which the formulas are given for the reactants and products.

3-2 Balancing Equations Some equations may be balanced by following rules. To balance equations given in the exercises that follow, you will use the inspection method. It consists of first equalizing the symbol for the atom that is obviously unbalanced. The number of atoms represented can be balanced on the left and right sides by placing the required coefficients in front of the appropriate formulas. These coefficients may upset the balance of some atom in another formula. The atom whose balance has been disturbed is balanced by placing the proper coefficient in front of the formula of the substance containing the unbalanced atom. This process is continued until each of the different atoms is balanced. Hint: balance coefficients only. Do not change correct formulas of compounds.

Example 3-1

Oxygen gas is frequently obtained in the laboratory by decomposing potassium chlorate. The skeleton equation for this reaction is

$$KClO_3(s) + \text{heat} \longrightarrow KCl(s) + O_2(g)$$

Our objective is to balance this equation.

Solution

1. Balance the oxygen atoms. Put a coefficient 2 in front

of $KClO_3$ and a coefficient 3 in front of O_2, giving

$$2KClO_3(s) + \text{heat} \longrightarrow KCl(s) + 3O_2(g)$$

2. Place a coefficient 2 in front of KCl to give

$$2KClO_3(s) + \text{heat} \longrightarrow 2KCl(s) + 3O_2(g)$$

3. Count the atoms on both the left and right sides of the equation. If the equation is correctly balanced, an equal number of each kind of atom is represented on both sides of the equation.

Left	*Right*
K = 2	K = 2
Cl = 2	Cl = 2
O = 6	O = 6

FOLLOW-UP PROBLEM

Balance these expressions.

(a) $HgO(s) \longrightarrow Hg(l) + O_2(g)$
(b) $K(s) + Br_2(l) \longrightarrow KBr(s)$
(c) $CaCO_3(s) \longrightarrow CaO(s) + CO_2(g)$
(d) $Al_4C_3(s) + H_2O(l) \longrightarrow CH_4(g) + Al(OH)_3(s)$
(e) $Mg_3N_2(s) + H_2O(l) \longrightarrow NH_3(g) + Mg(OH)_2(s)$
(f) The thermal decomposition of calcium chlorate [$Ca(ClO_3)_2$]. Hint: see Example 3-1 for a clue to the formulas of the products.

A type of reaction frequently encountered involves the combustion or burning of a compound containing only carbon and hydrogen. These compounds, often used as fuels, are called ***hydrocarbons.*** Natural gas and gasoline are examples of fuels containing hydrocarbons. The *complete combustion* of these fuels always yields carbon dioxide (CO_2) and water as products. Incomplete combustion

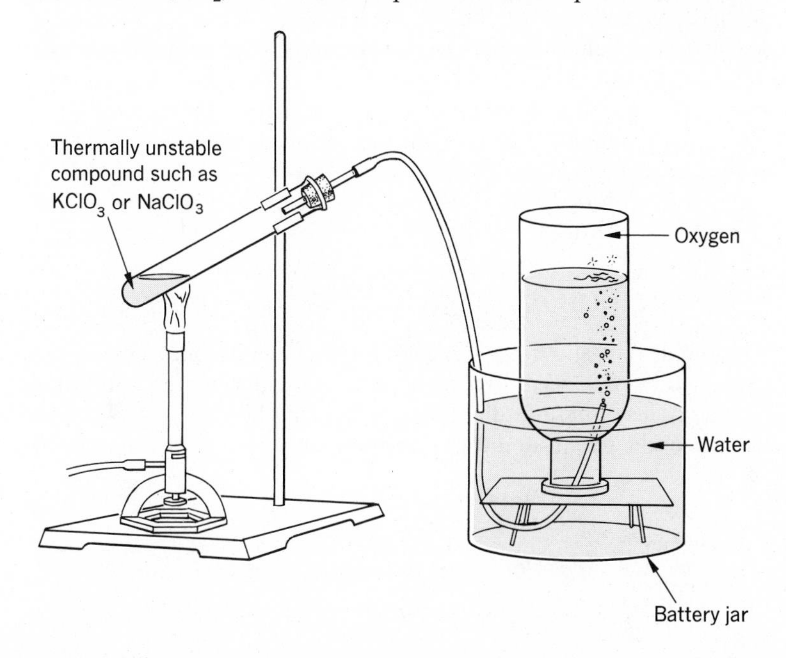

3-1 Decomposition of a relatively unstable compound by heating. This apparatus is commonly used in the laboratory preparation of oxygen gas. Oxygen gas, which is only slightly soluble in water, may be collected by displacement of water. Chlorates should not be allowed to contact the rubber stopper nor be contaminated with any substance such as sulfur or carbon which readily combines with oxygen.

yields carbon monoxide (CO) or carbon (C) as products. We shall use the complete combustion of butane (C_4H_{10}) to illustrate the technique for balancing equations for this type of reaction. The skeleton equation is

$$C_4H_{10} + O_2 \longrightarrow CO_2 + H_2O$$

The atoms in this type of equation should be balanced in the sequence: carbon, hydrogen, and oxygen. A fractional coefficient may be necessary to bring the oxygen atoms into balance. If whole-number coefficients are desired, then all coefficients may be multiplied by the multiplier needed to convert the fractions into an integer. Following these suggestions, the first coefficient to be written in the above equation should be a 4 in front of CO_2. The second coefficient should be a 5 in front of H_2O. At this point count 13 oxygen atoms on the right side of the equation. Thus, a coefficient of 13/2 in front of O_2 brings the equation into balance.

$$C_4H_{10} + 13/2O_2 \longrightarrow 4CO_2 + 5H_2O$$

This equation may be cleared of fractional coefficients by multiplying each term by 2. This operation yields

$$2C_4H_{10} + 13O_2 \longrightarrow 8CO_2 + 10H_2O$$

FOLLOW-UP PROBLEM

Complete and balance equations for (a) The complete combustion of propane (C_3H_8), (b) The complete combustion of ethane (C_2H_6), (c) The reaction of carbon dioxide with carbon forming carbon monoxide.

The ability to balance and interpret equations should enable us to make calculations involving the relative masses of substances involved in chemical reactions, if we understand mole, mass, and volume relationships. These are known as ***stoichiometric calculations.*** The word *stoichiometry* is derived from the Greek word *stoicheion* (element). Let us now examine an equation and interpret it in terms of molecules, moles, and grams.

3-3 Quantitative Interpretation of Chemical Reactions Consider the combustion of ethyl alcohol (C_2H_5OH). The equation for the reaction is

$$C_2H_5OH(l) + 3O_2(g) \longrightarrow 2CO_2(g) + 3H_2O(g) \qquad 3\text{-}2$$

The coefficients show that for each mole of alcohol molecules used, three moles of oxygen molecules are needed, with the formation of two moles of carbon dioxide and three moles of water. We shall use these coefficients and conversion factors to show the *proper mole ratios.*

Example 3-2
How many moles of carbon dioxide are produced by burning 1.50 moles C_2H_5OH?

Solution

From Equation 3-2, we obtain the mole relationship between C_2H_5OH and CO_2 and construct a conversion factor

$$\frac{2 \text{ moles } CO_2}{1 \text{ mole } C_2H_5OH}$$

Multiplying the mole ratio (conversion factor) by the given number of moles of C_2H_5OH yields

$$1.50 \text{ moles } C_2H_5OH \times \frac{2 \text{ moles } CO_2}{1 \text{ mole } C_2H_5OH} = 3.00 \text{ moles } CO_2$$

Example 3-3

How many grams of CO_2 are produced when 1.50 moles of C_2H_5OH is burned?

Solution

The answer to Example 3-2 is 3.00 moles of CO_2, produced from 1.50 moles C_2H_5OH. We can multiply 3.00 moles of CO_2 by a factor that changes moles CO_2 into grams of CO_2. The factor is the gram-molecular mass in units of g/mole derived from the formula CO_2 and the atomic masses of carbon and oxygen.

$$3.00 \text{ moles } CO_2 \times \frac{44.0 \text{ g } CO_2}{\text{mole } CO_2} = 132 \text{ g } CO_2$$

The complete setup is

$$1.50 \text{ moles } C_2H_5OH \times \frac{2 \text{ moles } CO_2}{\text{mole } C_2H_5OH} \times \frac{44.0 \text{ g } CO_2}{\text{mole } CO_2} = 132 \text{ g } CO_2$$

Example 3-4

How many grams of CO_2 are produced when 23 g C_2H_5OH is burned?

Solution

1. Convert grams C_2H_5OH into moles C_2H_5OH.

$$23 \text{ g } C_2H_5OH \times \frac{\text{mole } C_2H_5OH}{46 \text{ g } C_2H_5OH} = 0.50 \text{ moles } C_2H_5OH$$

2. Convert moles C_2H_5OH into moles CO_2.

$$0.50 \text{ moles } C_2H_5OH \times \frac{2 \text{ moles } CO_2}{\text{mole } C_2H_5OH} = 1.0 \text{ mole } CO_2$$

3. Convert moles CO_2 into grams of CO_2.

$$1.0 \text{ mole } CO_2 \times \frac{44 \text{ g } CO_2}{\text{mole } CO_2} = 44 \text{ g } CO_2$$

The complete setup is

$$23\text{ g } C_2H_5OH \times \frac{\text{mole } C_2H_5OH}{46\text{ g } C_2H_5OH} \times \frac{2\text{ moles } CO_2}{\text{mole } C_2H_5OH} \times \frac{44\text{ g } CO_2}{\text{mole } CO_2} = 44\text{ g } CO_2$$

Example 3-5

How many liters of CO_2 are produced at STP when 23.0 g C_2H_5OH is burned?

Solution

Since 23 g of C_2H_5OH was burned in Example 3-4, we may use the mass of CO_2 obtained as the starting point in this problem. In this case, we must first convert 44.0 g CO_2 into moles and then convert moles into liters at STP. The conversion factor for the second step is $\frac{22.4\text{ l}}{\text{mole}}$.

$$44.0\text{ g } CO_2 \times \frac{1\text{ mole } CO_2}{44.0\text{ g } CO_2} = 1.00\text{ mole } CO_2$$

$$1.00\text{ mole } CO_2 \times \frac{22.4\ \ell\ CO_2}{\text{mole } CO_2} = 22.4\ \ell\ CO_2$$

The complete setup is

$$23.0\text{ g } C_2H_5OH \times \frac{\text{mole } C_2H_5OH}{46.0\text{ g } C_2H_5OH} \times \frac{2.00\text{ moles } CO_2}{\text{moles } C_2H_5OH} \times \frac{22.4\ \ell\ CO_2}{\text{mole } CO_2} = 22.4\ \ell\ CO_2$$

Example 3-6

How many grams of CO_2 are formed when 32 g O_2 reacts with 23 g C_2H_5OH?

Solution

Unless the reactants are present in the mole ratio indicated by the balanced equation, one of them is in excess and is not completely consumed. The amount of product is determined by the reactant that is not in excess. This means that we must take a preliminary calculation to determine which reactant is in excess.

1. Calculate the moles of each reactant.

$$32\text{ g } O_2 \times \frac{\text{mole } O_2}{32\text{ g } O_2} = 1.0\text{ mole } O_2$$

$$23\text{ g } C_2H_5OH \times \frac{\text{mole } C_2H_5OH}{46\text{ g } C_2H_5OH} = 0.50\text{ mole } C_2H_5OH$$

If it is not obvious which is in excess, we must calculate the moles of one reactant needed to react with the given quantity of the second reactant.

2. Calculate the moles of O_2 required to react with 0.50 mole C_2H_5OH.

$$0.50 \text{ mole } C_2H_5OH \times \frac{3 \text{ moles } O_2}{\text{mole } C_2H_5OH} = 1.5 \text{ moles } O_2$$

(from equation)

Since 1.5 moles of O_2 is required and only 1.0 mole is available, the C_2H_5OH must be in excess and the amount of CO_2 produced will be determined by the O_2.

3. Calculate the moles and then grams of CO_2 produced.

$$1.0 \text{ mole } O_2 \times \frac{2.0 \text{ moles } CO_2}{3.0 \text{ moles } O_2} \times \frac{44.0 \text{ g } CO_2}{\text{mole } CO_2} = 29 \text{ g } CO_2$$

FOLLOW-UP QUESTIONS

These questions refer to the equation

$$3Ag(s) + 4HNO_3(aq) \longrightarrow 3AgNO_3(aq) + NO(g) + 2H_2O(l)$$

1. How many moles of NO are produced when 1.5 mole of Ag reacts with excess HNO_3? **Ans. 0.50.**

2. How many grams of NO are produced when 1.5 mole Ag reacts with excess HNO_3? **Ans. 15 g.**

3. How many liters of NO are produced at STP when 162 g Ag reacts with excess HNO_3? **Ans. 11.2 l.**

4. How many grams of $AgNO_3$ are produced when 154 g of Ag reacts with 189 g HNO_3? **Ans. 242 g.**

REACTIONS INVOLVING IONS IN AQUEOUS SOLUTION

3-4 Precipitation Reactions Many industrial, metabolic, and other reactions which we encounter in ordinary laboratory practice take place between ions in aqueous solution. A precise description of these reactions requires an equation which includes ions as participants. These equations, known as ***ionic equations,*** show the state and condition of all reactants and products. *Net ionic equations* are those from which chemical species which do not actually participate in the reaction are omitted. These nonparticipating ions, which are still a part of the system, are chemically unreactive ions and are sometimes called "spectator" ions. Let us illustrate the technique by writing and balancing a net ionic equation for a ***precipitation reaction,*** a simple type of ionic reaction. In this type of reaction a slightly soluble salt called a ***precipitate*** forms as the result of the very strong attractive forces between oppositely charged ions. Slightly soluble precipitates are sometimes erroneously referred to as *insoluble.* What is really implied is that the precipitate dissolves to a negligible extent in water.

Later we shall develop principles that enable us to predict whether or not a precipitate forms when oppositely charged ions are mixed in solution. In the meantime, we shall identify precipitates by referring to general solubility rules or tables of solubility, both of which are derived from experimental data. For convenience, a list of general solubility rules is provided in Table 3-1. Recall from Chapter 2 that soluble ionic salts as well as many acids and bases form ionic aqueous solutions. Let us examine the reaction that may occur when solutions of soluble ionic solids are mixed.

TABLE 3-1
GENERAL SOLUBILITY RULES

1. Salts of sodium, ammonium, and potassium are soluble ionic solids.
2. Nitrates and acetates are soluble ionic solids. $AgC_2H_3O_2$ is a moderately soluble ionic solid.
3. Chlorides, bromides, and iodides are soluble except the silver, mercury (I), and lead compounds ($PbCl_2$ is moderately soluble in hot water).
4. Sulfates are soluble except $BaSO_4$, $SrSO_4$, and $PbSO_4$; $CaSO_4$ and Hg_2SO_4 are sparingly soluble.
5. Carbonates and phosphates are slightly soluble except Na, K, NH_4.
6. Hydroxides are slightly soluble except NaOH, KOH, and $Ba(OH)_2$. $Ca(OH)_2$ is sparingly soluble.
7. Sulfides are only slightly soluble except Na, K, NH_4. Mg, Ca, Ba, and Al sulfides decompose in water.
8. Silver salts are slightly soluble except $AgNO_3$ and $AgNO_2$. $AgClO_4$, $AgC_2H_3O_2$, and Ag_2SO_4 are moderately soluble.

Consider the reaction that occurs when a solution of lead nitrate [$Pb(NO_3)_2$] is added to a solution of sodium iodide (NaI). General solubility rules or various experimental procedures can be used to show that $Pb(NO_3)_2$ and NaI are both soluble ionic compounds that contain ions which react, forming a precipitate. When a colorless lead nitrate solution [$Pb(NO_3)_2$] is added to a colorless sodium iodide solution (NaI), a yellow solid forms and settles to the bottom of the vessel. The slightly soluble yellow solid is the ***precipitate.*** Let us first identify the precipitate and then write an equation for the reaction. It is reasonable to assume that the yellow precipitate resulted from the reaction of a positive ion from one solution with a negative ion from the other. A lead nitrate solution contains Pb^{2+} and NO_3^- ions, while a sodium iodide solution contains Na^+ and I^- ions. Therefore, we might conclude that the yellow precipitate is either sodium nitrate or lead iodide. Referring to the general solubility rules, we note that sodium and nitrate compounds are soluble. Thus, sodium nitrate ($NaNO_3$) is soluble, and the yellow precipitate is probably lead iodide (PbI_2). We can verify this prediction experimentally by filtering out the lead iodide and subjecting it to further tests. Also other soluble lead compounds react with different soluble iodides and form the same precipitate. In addition, the *filtrate* (solution passing through the filter paper) when tested, shows the presence of Na^+ and NO_3^- ions.

The above reaction can be expressed in terms of either a *formula equation* or an *ionic equation.* In a formula equation, all substances are represented by electrically neutral formulas. We have identified the reactants and products of the reaction. On the basis of our tests, we can write

lead nitrate(aq) + sodium iodide(aq) $\longrightarrow$
lead iodide(s) + sodium nitrate(aq) 3-3

The word equation can be converted into a skeleton formula equation by substituting formulas for words. Formulas for ionic substances represent the ratios in which the ions are present just as molecular formulas represent the combining ratios of atoms. Positive and negative ions combine in ratios that yield neutral compounds. Since lead ion has a 2^+ charge and iodide ions have a charge of 1^-, two iodide ions are required to neutralize the charge on one lead ion. The number of each ion required is indicated by a subscript. Thus, the formula for lead iodide is PbI_2. If more than one polyatomic ion is involved, as in the case of lead nitrate, then enclose the ion in parenthesis and place the subscript outside the parenthesis. Following these procedures yields

$$\mathbf{Pb(NO_3)_2(aq) + NaI(aq) \longrightarrow PbI_2(s) + NaNO_3(aq)}$$

as the skeleton formula equation for the precipitation reaction discussed above. The expression may then be balanced by the inspection method. These procedures yield

$$\mathbf{Pb(NO_3)_2(aq) + 2NaI(aq) \longrightarrow PbI_2(s) + 2NaNO_3(aq)}$$

The disadvantage of the formula equation is that it does not convey precise information as to the nature of the actual reactants

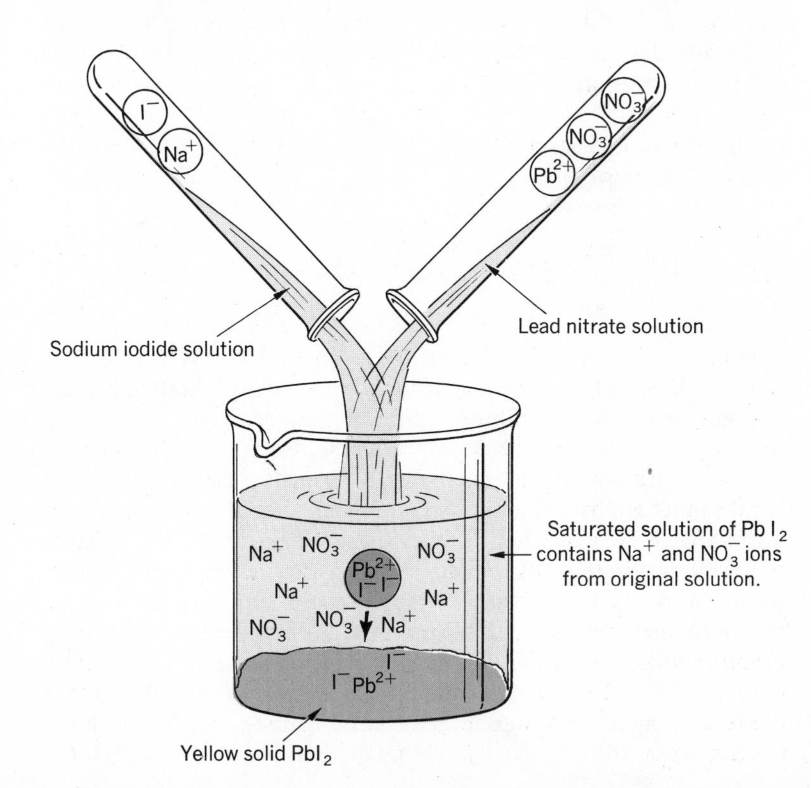

3-2 A precipitation reaction. An instantaneous reaction occurs when lead ions from a soluble lead compound are mixed with an aqueous solution containing iodide ions from a soluble iodide compound. The saturated solution contains an extremely small concentration of Pb^{2+} and I^- ions.

and products. To write an ionic equation for the reaction, write the formulas of any soluble salt in dissociated (ionic) form. General solubility rules (Table 3-1) can be used to help identify soluble and slightly soluble substances. The formulas of slightly soluble ionic salts should be written in the form of neutral formulas and identified by placing (*s*) after the formula.

Following these directions, Equation 3-3 becomes

$$Pb^{2+} + 2NO_3^- + 2Na^+ + 2I^- \longrightarrow PbI_2(s) + 2Na^+ + 2NO_3^-$$

The ionic equation may be converted to a *net ionic equation* by eliminating those ions that are found in the solution both before and after the reaction. These are the unreactive ions. The Na^+ and NO_3^- ions are not included in the equation even though they are a part of the system. Their omission yields

$$Pb^{2+}(aq) + 2I^-(aq) \longrightarrow PbI_2(s)$$

The equation now implies that PbI_2 precipitates in a solution containing appreciable concentrations of Pb^{2+} and I^- ions, regardless of the source of the ions. Net-ionic equations are described by the Law of Conservation of Matter and the Law of Conservation of Charge. That is, the total atoms (or ions) of each type on the reactant side must equal that on the product side. In addition, the total ionic charge on the left side of the equation must equal that on the right. In the above equation, the net ionic charge on both the left and right sides of the equation is zero. Note again that it is often desirable to bypass the formula equation and write the net ionic equation directly.

FOLLOW-UP PROBLEM

Write word, formula, and net-ionic equations for the precipitation reactions which occur when solutions of these ionic compounds in solutions are mixed. If no reaction occurs, write *no reaction*. (a) iron(III) nitrate + calcium hydroxide, (b) silver nitrate + sodium sulfide, (c) potassium sulfate + lead nitrate, (d) barium iodide + ammonium carbonate, (e) sodium chromate + silver nitrate, (f) potassium nitrate + ammonium sulfate.

Up to this point we have considered only the qualitative aspects of ionic reactions in solution. We are now ready to discuss the quantitative aspects. This requires an exact method of expressing the concentration of a solution. You will recall that stoichiometric calculations are based on mole relationships in balanced equations. Therefore, it is desirable to express the concentration of a solution in a way that permits us to measure out a desired number of moles of solute by measuring the volume of a solution. In order to compare volumes of different solutions required to react with each other, chemists have devised a concentration unit based on the mole concept. This unit, known as the *molar concentration,* identifies the number of moles of solute in a specified volume of solution. Most stock solutions are labeled in terms of their molar concentration.

CONCENTRATION OF SOLUTIONS

3-5 Molarity The ***molar concentration*** of a solution is defined as the *moles of solute per liter of solution.* Molar concentration or *molarity* is symbolized by M and expressed in formula form by

$$M = \frac{\text{moles of solute}}{\text{liters of solution}} \quad 3\text{-}4$$

A parallel definition in terms of millimoles and milliliters (ml) is

$$M = \frac{\text{millimoles of solute}}{\text{milliliters of solution}} \quad 3\text{-}5$$

Note that the volume of a solution varies with temperature so that the molar concentration (unlike molal concentration) also varies with temperature.

A one molar ($1M$ or M) solution of a pure substance may be prepared by adding enough water to one mole of substance to make a total of one liter of solution. A calibrated vessel known as a volumetric flask (Fig. 3-3) is often used to prepare a solution with a precisely known molarity. On the basis of the definition we can calculate the molarity of solutions from various data. The calculations are illustrated by examples.

Concentrations of substances such as ordinary salt, which do not exist as molecules in solution, are frequently expressed in terms of *formality,* defined as the number of formula masses of solute per liter of solution. *Formality* or *formal concentration* is symbolized by *F*. For convenience, we shall represent the concentration of all solutions in terms of molarity regardless of the nature of the solute.

Example 3-7

What is the molarity of a solution prepared by dissolving 17.1 g of ordinary sugar ($C_{12}H_{22}O_{11}$) in enough water to make 0.500 ℓ of solution?

Solution

Calculate the moles of solute and liters of solution and then substitute in Equation 3-4.

moles solute:

$$17.1 \text{ g sugar} \times \frac{\text{mole sugar}}{342 \text{ g sugar}} = 0.0500 \text{ moles sugar}$$

molarity:

$$\frac{0.0500 \text{ moles sugar}}{0.500\ \ell \text{ solution}} = \frac{0.100 \text{ moles}}{\ell} = 0.100\ M$$

Example 3-8

Concentrated nitric acid has a density of 1.42 g/ml and is 72.0 percent HNO_3 by mass. What is its molar concentration?

Solution

1. Calculate the mass of 1 liter of solution.

$$\frac{1000 \text{ ml}}{\text{liter}} \times \frac{1.42 \text{ g}}{\text{ml}} = \frac{1.42 \times 10^3 \text{ g solution}}{\ell \text{ of solution}}$$

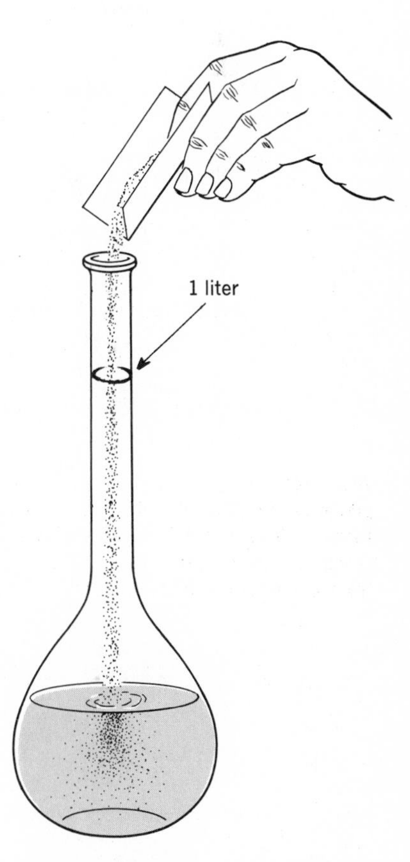

3-3 A 1-M sugar ($C_{12}H_{22}O_{11}$) solution is prepared by adding 342 g of sugar to some water in a volumetric flask. After the solute dissolves, water is added to fill the flask to the 1-liter mark. The solution contains 6.02×10^{23} sugar molecules.

2. Calculate the mass of solute (HNO_3) in one liter of solution.

$$\frac{1.42 \times 10^3 \text{ g solution}}{\ell \text{ solution}} \times \frac{72.0 \text{ g } HNO_3}{100 \text{ g solution}} = \frac{1.02 \times 10^3 \text{ g } HNO_3}{\ell \text{ solution}}$$

3. Calculate the moles of HNO_3 in one liter of solution.

$$\frac{1.02 \times 10^3 \text{ g } HNO_3}{\ell \text{ solution}} \times \frac{\text{mole } HNO_3}{63.0 \text{ g } HNO_3} = \frac{16.2 \text{ moles } HNO_3}{\ell \text{ solution}}$$

Therefore, the solution is 16.2 *M*. The setup for a one-step solution is

$$\frac{1.00 \times 10^3 \text{ ml solution}}{\ell \text{ solution}} \times \frac{1.42 \text{ g solution}}{\text{ml solution}} \times \frac{72.0 \text{ g } HNO_3}{100 \text{ g solution}} \times \frac{1 \text{ mole } HNO_3}{63.0 \text{ g } HNO_3} = \frac{16.2 \text{ moles } HNO_3}{\ell \text{ solution}}$$

FOLLOW-UP PROBLEMS

1. Commercial sulfuric acid is a water solution that has a density of 1.84 g/ml and is 98.0 percent H_2SO_4 (m.m. 98.0) by mass. What is the molarity of the solution? **Ans. 18.4 *M*.**

2. What is the molarity of a solution made by dissolving 10.0 g of NaOH (m.m. 40.0) in 0.200 ℓ of solution? **Ans. 1.25 *M*.**

3. Calculate the molarity of a solution containing 5.0 mg of $BaCl_2$ per ml of solution. **Ans. 0.024 *M*.**

For quantitative chemical reactions involving solutions, we usually need to calculate the quantity (moles, grams) of solute. The molar concentration of a solution enables us to determine the number of moles and hence, the number of grams of solute in any volume of the solution. For example, 100 ml (0.100 ℓ) of a 0.0500-*M* solution of sugar ($C_{12}H_{22}O_{11}$) contains

$$\frac{\textbf{0.0500 moles sugar}}{\textbf{liter of solution}} \times \textbf{0.100 liter solution} = \textbf{0.00500 moles sugar}$$

and

$$\textbf{0.00500 moles sugar} \times \frac{\textbf{342 g sugar}}{\textbf{mole sugar}} = \textbf{1.71 g sugar}$$

These relationships are expressed in general terms below. In any solution,

$$\textbf{moles of solute} = MV\textbf{(liters)} \qquad 3\text{-}6$$

or

$$\textbf{millimoles (mmole) of solute} = MV\textbf{(ml)} \qquad 3\text{-}7$$

The mass of solute can be calculated from the number of moles using the relationships

$$\textbf{g of solute} = \textbf{moles of solute} \times \textbf{molecular mass} \qquad 3\text{-}8$$

or

$$\textbf{mg of solute} = \textbf{millimoles of solute} \times \textbf{millimolecular mass} \qquad 3\text{-}9$$

The absolute numerical values for molecular mass and millimolecular mass are identical; only the units differ. *The unit for millimolecular mass is mg/millimole.*

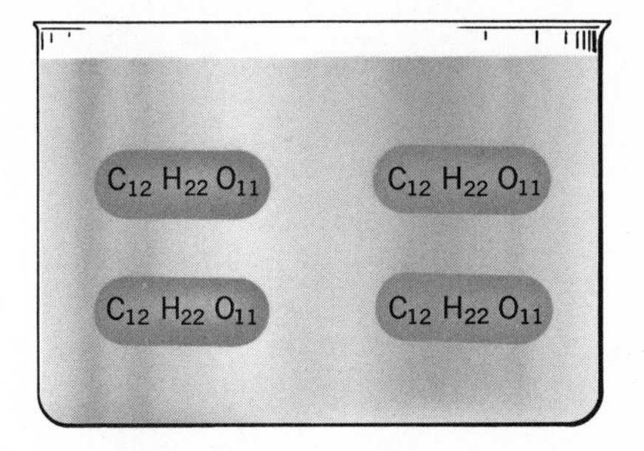

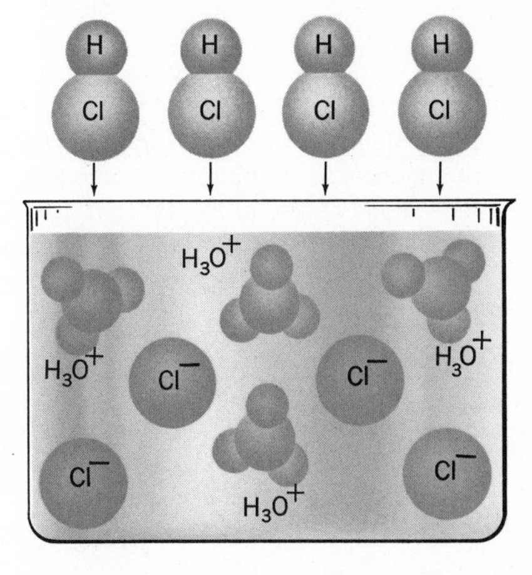

3-4 Four sugar molecules yield four particles in solution. Four HCl molecules by reaction with water yield eight particles, four hydronium ions and four chloride ions.

Example 3-9
Calculate the grams of silver nitrate needed to prepare 0.500 ℓ of a 0.100-*M* solution.

Solution
1. Calculate the moles of solute needed.

$$\text{moles solute} = 0.100\ \text{moles}/\ell \times 0.500\ \ell = 0.0500\ \text{moles}$$

2. Calculate the grams of solute needed.

$$\text{grams } AgNO_3 = 0.0500 \text{ moles } AgNO_3 \times \frac{170\text{ g } AgNO_3}{\text{mole } AgNO_3} = 8.50\text{ g } AgNO_3$$

The problem could be solved in one step.

$$\frac{0.100\text{ moles}}{\ell} \times 0.500\ \ell \times \frac{170\text{ g } AgNO_3}{\text{mole } AgNO_3} = 8.50\text{ g } AgNO_3$$

FOLLOW-UP PROBLEMS

1. Calculate the grams of $BaCl_2$ needed to prepare 200 ml of a 0.500-*M* solution. **Ans. 20.8 g.**

2. Concentrated hydrochloric acid has a density of 1.2 g/ml and is 36 percent HCl by mass. What is its molar concentration? **Ans. 12*M*.**

3. The following questions refer to a 0.20-*M* solution of $BaCl_2$. Formula mass of $BaCl_2$ is 208. (a) Calculate the moles of $BaCl_2$ in 200 ml of solution. (b) Calculate the mg and grams of $BaCl_2$ in the solution. (c) Calculate the volume in ml that contains 5.0 mmoles of $BaCl_2$. (d) Calculate the volume in ml that contains 41.6 mg of $BaCl_2$.
Ans. (a) 4.0×10 millimoles, (b) 8.3×10^3 mg and 8.3 g, (c) 25 ml, (d) 1.0 ml.

3-6 Ion Concentration One mole of a nonelectrolyte such as sugar furnishes just one mole of solute particles (neutral molecules). One mole of an electrolyte, however, furnishes more than one mole of solute particles. If the electrolyte is completely dissociated, then the concentration of each ion may be determined from the molarity of the solute and the equation that shows the dissociation. Consider a 0.10-*M* solution of magnesium chloride, $MgCl_2$. The equation for the dissociation is

$$MgCl_2(s) \longrightarrow Mg^{2+}(aq) + 2Cl^-(aq)$$

This equation reveals that when dissociation is complete, one mole of magnesium ions and two moles of chloride ions are in solution for each mole of magnesium chloride dissolved. Therefore, a 0.10-*M* solution $MgCl_2$ prepared by dissolving 105 g of the salt in one liter of solution has a Mg^{2+} ion concentration of 0.10 *M*, a Cl^- concentration of 0.20 *M* and contains no undissociated magnesium chloride.

FOLLOW-UP PROBLEMS

1. Three hundred ml of 0.20-*M* HNO_3 solution is added to 100 ml of 0.15-*M* $NaNO_3$ solution. Assume that both compounds are completely dissociated. Find the concentrations of the (a) H^+, (b) Na^+, (c) NO_3^- ions in the resulting solution.
Ans. (a) 0.15 *M*, (b) 0.038 *M*, (c) 0.19 *M*.

2. How many mg of $BaCl_2$ are needed to prepare 300 ml of a solution containing 1.0 mg of Ba^{2+} ions per ml of solution?
Ans. 455 mg.

3-7 Dilution of Solutions Frequently, it is necessary to prepare dilute solutions of specified concentrations from more concentrated stock solutions. The process consists of measuring out the volume of concentrated solution that contains the number of moles of solute needed to prepare the dilute solution and adding enough water to give the final desired volume. Note that adding water to a concentrated solution does not change the number of moles of solute. The number of moles of solute in either the dilute or concentrated solution will be equal to $MV_{(\ell)}$. Equating the moles of solute in a concentrated solution to moles of solute in the dilute solution yields the dilution equation

$$\underset{\text{concentrated solution}}{M_cV_c} = \underset{\text{dilute solution}}{M_dV_d}$$

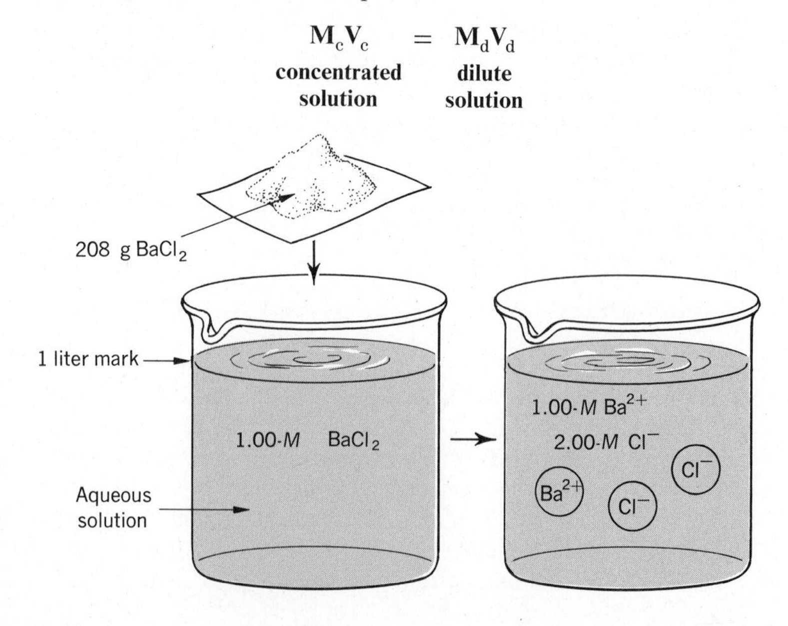

3-5 The molar concentration of the ions in solutions of electrolytes differ from the molar concentration of the original components. 208 g of $BaCl_2$ dissolved in one liter of solution yields a 1.0-*M* solution of $BaCl_2$. In this solution, the concentration of the barium ions (Ba^{2+}) is 1.0 *M* and that of the chloride ions (Cl^-) is 2.0 *M*. There are no barium chloride molecules.

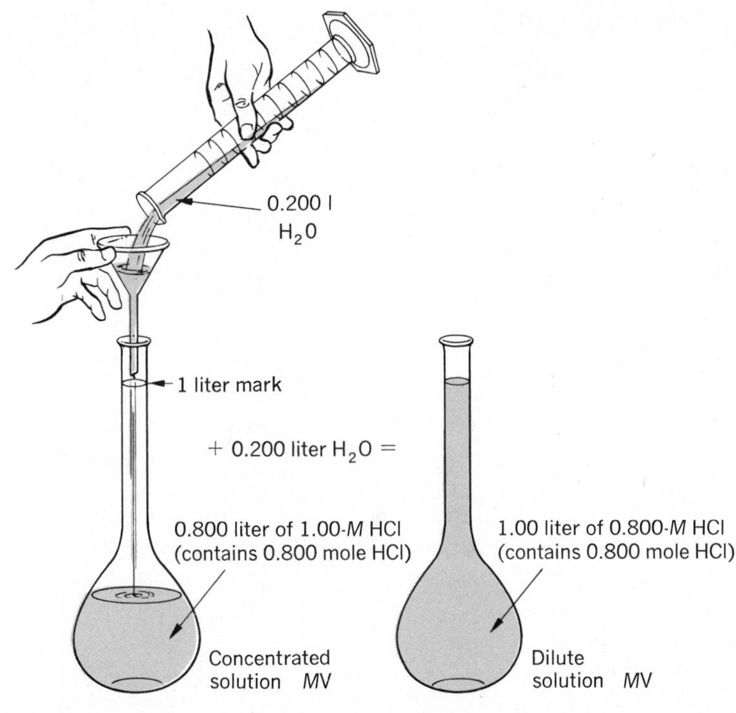

3-6 The number of moles of solute particles in the dilute solution is identical to the number in the concentrated solution on the left. If 10.0 ml of the dilute solution were removed, its concentration would be 0.800-*M*.

The volume (V) may be expressed in any unit, but the same unit must be used on both sides of the equation.

Example 3-10
Calculate the volume of 0.400-*M* $BaCl_2$ solution that must be measured out to prepare 500 ml of a 0.100-*M* solution.

Solution 1
1. Determine the number of moles of $BaCl_2$ required to prepare the dilute solution.

$$MV = 0.100 \text{ mole}/\ell \times 0.500\ \ell = 0.0500 \text{ mole}$$

2. Determine the volume of 0.400-*M* $BaCl_2$ that contains 0.0500 mole of solute.

$$V = \text{moles}/M$$

$$V = 0.0500 \text{ mole} \times \frac{1 \text{ liter}}{0.400 \text{ mole}} = 0.125\ \ell = 125 \text{ ml}$$

Solution 2
1. Itemize the data, substitute in the dilution equation, $M_cV_c = M_dV_d$, and solve for V_c.

$$M_c = 0.400 \qquad V_c = \text{unknown}$$
$$M_d = 0.100 \qquad V_d = 500 \text{ ml}$$

$$V_c = \frac{M_dV_d}{M_c} = 0.100 \times 500/0.400 = 125 \text{ ml}$$

FOLLOW-UP PROBLEMS

1. Calculate (a) the volume of 2.0-*M* $CaCl_2$ that must be used to prepare 1200 ml of 0.80-*M* $CaCl_2$, (b) the volume of water that must be added.
Ans. (a) 480 ml, (b) 720 ml.

2. What volume in liters of 12.0-*M* HCl should be added to 3.00 liters of 1.00-*M* HCl to give 6.00 liters of 6.00-*M* HCl on dilution with water? **Ans. 2.75 liters.**

Let us now illustrate how molarity can be used to calculate the quantity of product formed in a chemical reaction.

3-8 Application of Molarity to Reactions in Solutions Calculations involving solutions must take into account the number of moles of solute. Earlier we learned that the moles of solute can be determined from a knowledge of the molar concentration and the volume of solution used. That is,

$$\textbf{moles of solute} = \textbf{molarity}(M) \times \textbf{volume}(V)\textbf{ in liters}$$

Once the moles of reactant solute has been determined, the quantity of a product may be calculated. The application of molarity to the stoichiometry of solutions is illustrated in this example.

Example 3-11
Calculate the grams of PbI_2 (m.m. 461.0) formed when 0.500 liter of 0.100-*M* $Pb(NO_3)_2$ reacts with an excess of KI solution. The equation is

$$Pb(NO_3)_2(aq) + 2KI(aq) \longrightarrow PbI_2(s) + 2KNO_3(aq)$$

Solution
1. Calculate the number of moles of $Pb(NO_3)_2$ in 0.500 ℓ of 0.100-*M* solution.

$$\text{moles } Pb(NO_3)_2 = MV = 0.500\ \ell \times \frac{0.100 \text{ mole}}{\ell}$$

$$= 0.0500 \text{ mole } Pb(NO_3)_2$$

2. Use the coefficients in the balanced equation to determine the number of moles of PbI_2 that correspond to 1 mole of $Pb(NO_3)_2$.

$$\frac{1 \text{ mole } PbI_2}{\text{mole } Pb(NO_3)_2}$$

3. Determine the moles of PbI_2 that correspond to 0.0500 mole of $Pb(NO_3)_2$.

$$\frac{1 \text{ mole } PbI_2}{\text{mole } Pb(NO_3)_2} \times 0.0500 \text{ mole } Pb(NO_3)_2 = 0.0500 \text{ mole } PbI_2$$

4. Convert moles of PbI_2 to grams of PbI_2.

$$0.0500 \text{ mole } PbI_2 \times \frac{461.0 \text{ g } PbI_2}{\text{mole } PbI_2} = 23.0 \text{ g } PbI_2$$

Example 3-12
Calculate the number of ml of 2.00-*M* HNO_3 solution required to react with 216 g of Ag (at.m. 108.0) according to the equation

$$3Ag(s) + 4HNO_3(aq) \rightarrow 3AgNO_3(aq) + NO(g) + 2H_2O(\ell)$$

Solution

1. Calculate the number of moles of Ag in 216.0 g of Ag.

$$216.0 \text{ g Ag} \times \frac{1 \text{ mole Ag}}{108.0 \text{ g Ag}} = 2.00 \text{ moles Ag}$$

2. Calculate the number of moles of HNO_3 required to react with 2.00 moles of Ag.

$$\frac{4 \text{ moles } HNO_3}{3 \text{ moles Ag}} \times 2.00 \text{ moles Ag} = 2.67 \text{ moles } HNO_3$$

3. Calculate the number of liters of 2.00-*M* HNO_3 solution that contains 2.67 moles HNO_3. Remember, a 2.00-*M* solution contains 2.00 moles of solute per liter.

$$2.67 \text{ moles } HNO_3 \times \frac{1.00 \ \ell \text{ sol'n}}{2.00 \text{ moles } HNO_3} = 1.34 \ \ell \text{ sol'n.}$$

4. Convert liters of HNO_3 solution to ml of HNO_3 solution.

$$1.34 \ \ell \text{ sol'n} \times \frac{1000 \text{ ml sol'n}}{\ell \text{ sol'n}} = 1340 \text{ ml sol'n}$$

The set-up for a one-step solution is

$$216.0 \text{ g Ag} \times \frac{\text{mole Ag}}{108 \text{ g Ag}} \times \frac{4 \text{ moles } HNO_3}{3 \text{ moles Ag}} \times \frac{1.00 \ \ell \text{ sol'n}}{2.00 \text{ moles } HNO_3} \times \frac{1000 \text{ ml sol'n}}{\ell \text{ sol'n}} = 1340 \text{ ml sol'n.}$$

FOLLOW-UP PROBLEMS

1. Calculate in ml the volume of 0.500-*M* NaOH required to react with 3.0 g of acetic acid (m.m. 60.0). The equation is

$$NaOH(aq) + HC_2H_3O_2(aq) \longrightarrow NaC_2H_3O_2(aq) + H_2O(l)$$

Ans. 100 ml.

2. Calculate the number of grams of AgCl (m.m. 143.5) formed when 0.200 ℓ of 0.200-*M* $AgNO_3$ reacts with an excess of $CaCl_2$. The equation is

$$2AgNO_3(aq) + CaCl_2(aq) \longrightarrow 2AgCl(s) + Ca(NO_3)_2(aq)$$

Ans. 5.74 g.

3. Calculate (a) the least volume of NaCl solution required to precipitate all the Ag^+ ions as AgCl, and (b) the mass of AgCl formed when an excess of 0.100-*M* solution of NaCl is added to 0.100 ℓ of 0.200-*M* $AgNO_3$.

Ans. (a) 0.200 ℓ, (b) 2.87 g.

4. Calculate (a) the mass of $BaSO_4$ (m.m. 233) formed when excess 0.200-*M* Na_2SO_4 solution is added to 0.500 ℓ of 0.500-*M* $BaCl_2$ solution, and (b) the minimum volume of the Na_2SO_4 solution needed to precipitate the Ba^{2+} ions from the $BaCl_2$ solution.

Ans. (a) 58.2 g, (b) 1.25 ℓ.

An important application of molar concentration and ionic reactions is found in the processes of analytical chemistry. Let us briefly examine this important branch of chemistry.

ANALYTICAL CHEMISTRY

3-9 Quantitative Analysis The importance of analytical chemistry and analytical techniques cannot be overemphasized. They are a vital part of virtually all branches of chemistry. In addition, the physicist, biologist, geologist, and investigators in almost all fields of pure and applied science use analytical data and techniques in their work. In the applied sciences, physicians, pharmacists, engineers, and others use analytical processes. It is safe to say that modern industry, technology, and society as we know it could not exist without the contributions of analytical chemistry and analytical techniques. Millions of quantitative analytical tests are performed each year in monitoring industrial processes and in testing the quality, safety, purity, and characteristics of virtually all consumer goods. The list of products involved is almost limitless but includes food products, drugs, fuels, paints, fertilizers, automobiles, construction materials, and countless household items.

The laboratory work associated with quantitative analytical chemistry affords one an opportunity to develop traits and abilities which can be applied and be of value in almost any profession. These traits include these abilities:

1. Planning and organizing experimental procedures.
2. Carrying out experimental procedures carefully and with approved laboratory techniques.
3. Making accurate observations.
4. Recording and interpreting observations.
5. Understanding and communicating the results of an experiment.

(a)

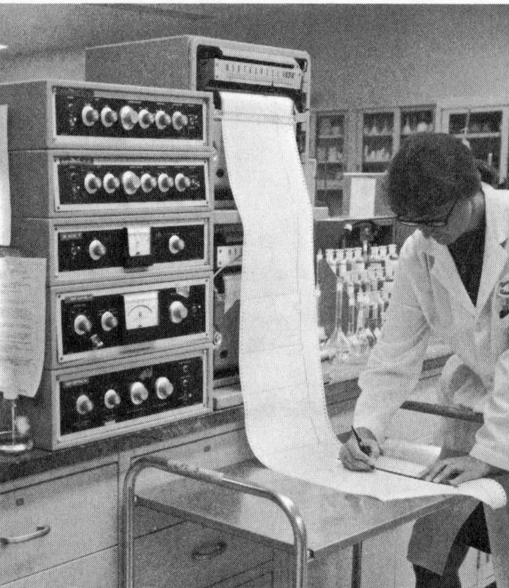

(b)

3-7 A modern analytical laboratory. (a) A portion of the Food and Drug Administration pesticide laboratory showing separatory funnels, chromatographic columns, and evaporative-concentrators used in the extraction and clean-up of pesticide samples. (b) A gas chromatograph used in the quantitative analysis of the pesticide residue in the sample extract. Gas chromatography is discussed on p. 79. Photographs by Donald D. Dechert, FDA Laboratories, Los Angeles, Calif.

There are two main divisions of analytical chemistry: ***qualitative analysis*** and ***quantitative analysis.*** The methods of qualitative analysis are designed and used to determine which constituents are present in a sample. The amounts or percentage of constituents are determined by methods of quantitative analysis. The choice of method depends upon the nature of the constituents. Therefore, if the constituents are not known, it is necessary to perform a qualitative analysis prior to designing a method for a quantitative analysis. The classical chemical methods used in quantitative analysis may be grouped under two general classifications: ***volumetric analysis*** and ***gravimetric analysis.*** Both volumetric and gravimetric analysis require some type of calculation based on mole, mass, and/or volume relationships shown by formulas and equations. At this point we shall consider only the simplest type of gravimetric analysis, but later in the course you will have an opportunity to perform the more complex operations and calculations of volumetric analysis.

3-10 Gravimetric Analysis A gravimetric analysis is based upon the precise measurement of the mass of a sample of unknown composition and the mass of a known compound isolated from the sample. For example, if we wish to determine the percentage of silver in an alloy, we can dissolve the alloy, add sodium chloride solution to precipitate the Ag^+ ions as AgCl, and finally dry and weigh the precipitate. From the known mass of precipitate, mass of sample, and composition of AgCl, we can calculate the percentage of silver in the alloy. You might also use this technique to determine the percentage of an element in a mixture of compounds. Suppose we wish to determine the percentage of the element chlorine in a sample known to contain sodium chloride and sodium nitrate. One method would be to isolate the chloride in the form of pure silver chloride by precipitation with a silver nitrate solution. We would first dissolve the mixture in water and then add a solution of silver nitrate as the precipitating agent. The chloride ions in the mixture is the only species that reacts with the silver ions; thus the amount of silver chloride formed depends only on the quantity of sodium chloride in the sample. The mass and percentage of NaCl in the sample can be calculated from the mass of the AgCl isolated from the original sample. Typical calculations are illustrated in these examples.

Example 3-13

A sample of impure sodium chloride weighing 1.00 g is dissolved in water and completely reacted with silver nitrate solution. The dried precipitate of AgCl has a mass of 1.48 g. Calculate the percentage of NaCl in the original impure sample.

$$AgNO_3(aq) + NaCl(aq) \longrightarrow AgCl(s) + NaNO_3(aq)$$

Net ionic equation: $Ag^+(aq) + Cl^-(aq) \longrightarrow AgCl(s)$

Solution

1. Convert grams of AgCl to moles AgCl

$$1.48 \text{ g AgCl} \times \frac{1 \text{ mole AgCl}}{143 \text{ g AgCl}} = 0.0103 \text{ mole AgCl}$$

2. Convert moles AgCl to moles NaCl using the coefficients in the balanced equation to obtain the mole ratio.

$$0.0103 \text{ mole AgCl} \times \frac{1 \text{ mole NaCl}}{\text{mole AgCl}} = 0.0103 \text{ mole NaCl}$$

3. Convert moles NaCl to grams NaCl

$$0.0103 \text{ mole NaCl} \times \frac{58.5 \text{ g NaCl}}{\text{mole NaCl}} = 0.6020 \text{ g NaCl}$$

4. Calculate percentage of NaCl on sample

$$\frac{0.6020 \text{ g NaCl}}{1.000 \text{ g sample}}(100) = 60.2 \text{ percent NaCl}$$

FOLLOW-UP PROBLEMS

1. A sample of potassium chloride mixed with potassium nitrate weighs 0.500 g. The sample is dissolved and an excess of $AgNO_3$ is added to the solution. The resulting precipitate of AgCl when dried weighs 0.750 g. What is the percentage of KCl in the original mixture?
Ans. 78.0 percent.

2. A silver-copper alloy having a mass of 0.500 g is dissolved in HNO_3 and the Ag^+ ions are precipitated as AgCl(s). What is the percentage of silver in the alloy if the dried precipitate has a mass of 0.598 g?
Ans. 89.9%.

Example 3-14

A sample of sodium chloride is known to contain a maximum of 97.5 percent NaCl. (a) How many ml of 0.100-*M* $AgNO_3$ are required to precipitate the chloride ions from a 1.00-g sample? (b) What is the maximum mass of AgCl that could be formed? The net ionic equation is

$$Ag^+(aq) + Cl^-(aq) \longrightarrow AgCl(s)$$

Solution

(a) Calculate the maximum number of grams and moles of NaCl and of Cl^- ion in the sample. Convert the moles of Cl^- to moles of Ag^+ ion and to moles of $AgNO_3$. Determine the volume of solution required to contain the required moles of $AgNO_3$.

$$1.00 \text{ g sample} \times \frac{97.5 \text{ g NaCl}}{100 \text{ g sample}} \times \frac{1 \text{ mole NaCl}}{58.5 \text{ g NaCl}} \times \frac{1 \text{ mole Cl}^-}{\text{mole NaCl}}$$

$$\times \frac{1 \text{ mole Ag}^+}{\text{mole Cl}^-} \times \frac{1 \text{ mole AgNO}_3}{\text{mole Ag}^+} \times \frac{1 \text{ } \ell \text{ AgNO}_3 \text{ sol'n}}{0.100 \text{ mole AgNO}_3}$$

$$\times \frac{1000 \text{ ml}}{\ell}$$

$$\text{(b) } 1.00 \text{ g sample} \times \frac{97.5 \text{ g NaCl}}{100 \text{ g sample}} \times \frac{1 \text{ mole NaCl}}{58.5 \text{ g NaCl}}$$

$$\times \frac{1 \text{ mole AgCl}}{\text{mole NaCl}} \times \frac{143 \text{ g AgCl}}{\text{mole AgCl}} = 2.39 \text{ g AgCl}$$

LOOKING AHEAD

In Chapter 2 we introduced the type of data that led to the formulation of four laws of chemical combination: Law of Conservation of Mass, Law of Definite Composition, Law of Multiple Proportions, and Law of Combining Volumes. We found that these laws could be interpreted in terms of Dalton's atomic theory modified to include Avogadro's concept of molecules. Application of these laws, the mole concept, and use of a scale of relative atomic masses, first introduced by Dalton, provide the basis for the calculations discussed in this chapter.

For continued progress it is necessary that we expand our concept of atoms to include their electric nature, and a description of their motion and behavior in the systems we investigate. In the next chapter, we shall consider the translational motion and behavior of atoms and molecules in gases. The electric nature of atoms will then be investigated in Chapter 5.

Of the three phases of matter, the gaseous phase is by far the simplest and easiest to study. Many of the observations that led Dalton to propose the particle nature of matter involved the behavior of gaseous systems. We now need tools which enable us to describe quantitatively the behavior of gases at various conditions of temperature and pressure. The physical laws of gas behavior are these tools.

In the next chapter we shall examine the behavior of gases, use experimental data to develop the laws describing their behavior, and construct a model that will help explain the laws and observations made on macroscopic samples in terms of the behavoir of the individual particles in the sample.

Special Feature: The Solid Earth-Geochemistry

One of the serious problems facing mankind in the 21st century is the demand that will be made upon the natural resources of spaceship earth. Up to now a relatively small percentage of the world's population has been drawing heavily upon earth's nonrenewable resources. The United States consumes more than one-third of the world's production of raw materials. It is apparent that the population explosion coupled with the increasing technological capabilities of many other nations will place an enormous strain on our planet's mineral, energy, and other resources.

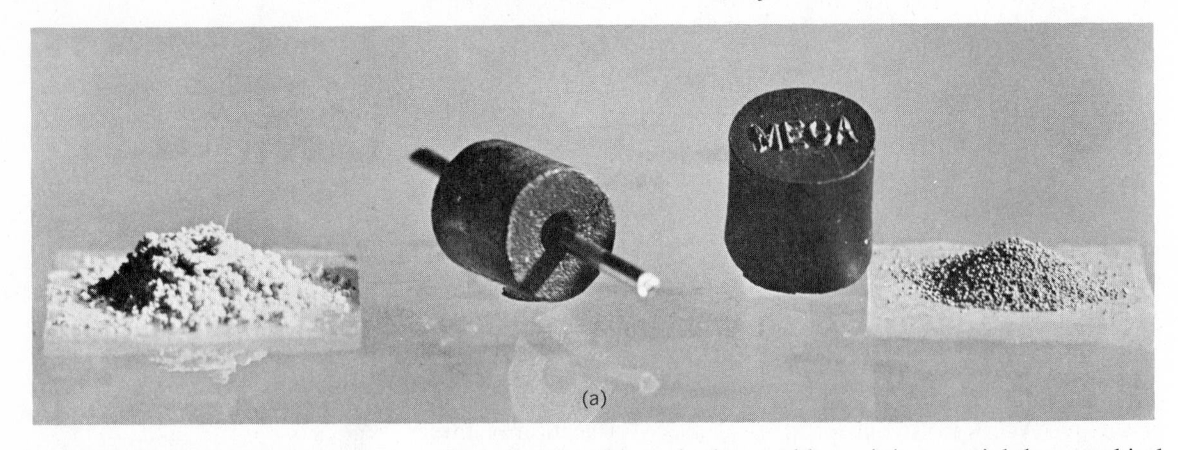

3-8 (a) Industrial diamonds formed by moulding diamond dust (left and right side of photograph) at 1 000 000 pounds per square inch and 4000 degrees Fahrenheit in a special high-pressure machine invented by Dr. H. Tracy Hall. The diamond with MEGA moulded on it weighs 20 carats and the diamond with the hole weighs 12 carats. These black diamonds are used primarily in industrial tools where extreme hardness and wear resistance are necessary. (b) High-pressure temperature machine used to mould diamond dust. This machine has six hydraulic rams (like a hydraulic truck jack). Each ram has a thrust of 1200 tons and could lift 500 automobiles off the ground at the same time.

Photographs and data courtesy Dr. H. Tracy Hall and McCartney Manufacturing Co., Baxter Springs, Kansas.

To help solve this and other problems, it is essential that mankind explore the composition and the nature of the processes occurring in the solid earth. Information about the inner earth is obtained from research on the earth's magnetic and gravitational fields. Laboratory simulations of the physical conditions in the inner earth have given scientists much new information about how minerals react to the great pressures, the high temperatures, and the chemical environments believed to exist in the inner earth. Geologists are experimenting with high-temperature and pressure reaction cells that can subject minerals to pressures of over 50 000 atmospheres. For short periods of time by using explosive charges, they have produced pressures of up to 5 million atmospheres. This exceeds the pressure at the center of the earth which is believed to be about 3.6 million atmospheres. Minerals when placed under these conditions undergo many changes in structure. Geochemists are busy injecting a large variety of materials into these high-pressure reaction cells. Man-made diamonds are produced from diamond dust in cells of this type. Diamond is composed of pure carbon just as graphite but has a slightly higher density and a different crystalline structure. As geologists improve these high-pressure reaction cells, their ability to produce diamonds of greater size also improves. A 20-carat diamond produced by this method is shown in Fig. 3-8. In nature, diamonds are believed to be produced about 250 miles down in the mantle (Fig. 3-10) and brought to the surface by volcanic action. The most productive diamond mines have been in the "pipes" or cores of extinct volcanoes.

Better understanding of the earth's interior can lead to greater use of geothermal energy from hot springs and steam wells. Energy from these sources may be converted to electric energy. These non-polluting energy sources can offer an alternative to burning fossil fuels in the decades ahead while nuclear power plants and other energy sources are being developed.

Nuclear reactors provide many benefits but also can cause a problem of radioactive waste materials. An understanding of the earth's interior structure is needed to enable industry to dispose of radioactive as well as nonradioactive waste products intelligently. Many of these materials have been disposed of in the sea which can be a dangerous and very undesirable practice because the oceans are rapidly becoming contaminated. Waste materials deposited in abandoned mines have contaminated water supplies. In one case it is believed that waste products pumped underground in Colorado caused several hundred minor earthquakes.

Knowledge of the structure of the earth's crust helps geologists to find new mineral deposits. Prospectors have long known that by examining and analyzing minerals at the surface you could obtain some clues as to what minerals might be found below. Methods of prospecting depending on surface deposits of minerals are severely limited however as movements in the earth's crust have often obscured mineral deposits so that the surface may give no clue as to what lies below. By applying knowledge from all phases of earth science, new deposits of much needed gold, silver, platinum, tin, mercury, petroleum, natural gas, sulfur, and other chemical raw materials are being found in the earth's crust. New techniques developed by geologists and geochemists will allow us to find these deposits and remove them in the most efficient manner. Efficient methods developed by chemists for processing these materials will make them less expensive and available to more people.

3-9 High quality synthetic diamonds. These man-made diamonds weigh about one carat (200 mg) each and are usually flawless. It is interesting to note that most natural diamonds are not flawless. This suggests that natural diamonds can be distinguished from the synthetic ones by the flaws they contain. The diamonds are synthesized from the black graphite shown at the top of the photograph.

Courtesy H. M. Strong, General Electric Research Laboratories.

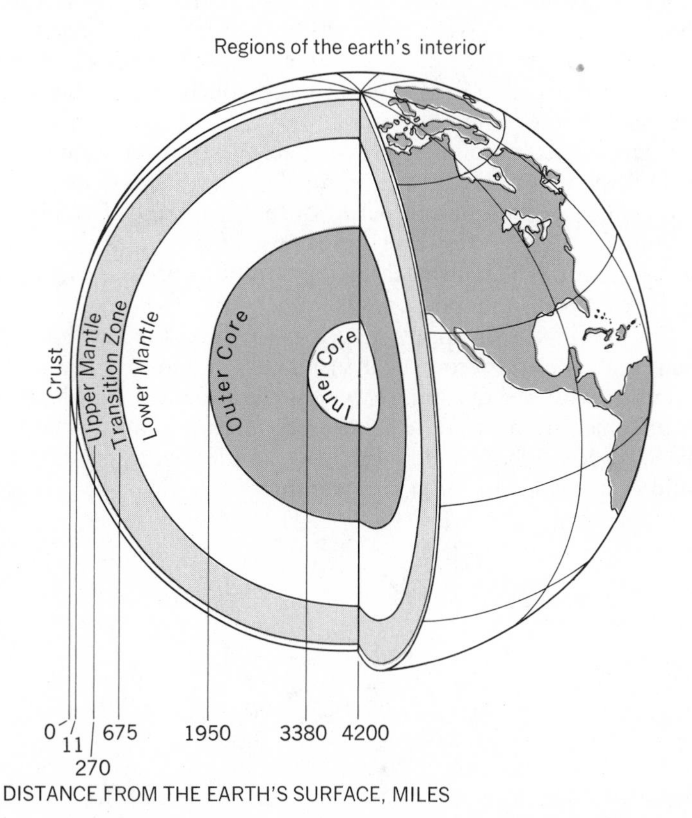

3-10 This model of the interior of the earth is the result of combining many types of geological data. Probably the most important source of information about the interior of the earth is derived from the seismic waves which are initiated either by earthquakes or by explosive charges.

An interesting case involves the rare-earth element, europium. In 1950, there was no particular use for this element whose cost was $1,600 per pound. However, it was soon discovered that europium oxide was the best material to use as a red phosphor on the screens of color-television sets. With the discovery of new deposits, the price has dropped to $600 per pound. The average color television set contains about one gram of europium which costs $1.40.

It is interesting to note that the primary clues to the earth's interior structure are obtained from seismic waves produced either by earthquakes or by man-made explosions. Although there are only about 20 to 30 major earthquakes in an average year, there are about a million minor tremors per year or one about every two minutes. Seismic waves from these quakes give geologists a continuous source of information for analyzing the earth's structure.

Using his knowledge of the earth's gravitational fields, the English scientist, Henry Cavendish (1731–1810) in 1798 "weighed" the earth and found it to have a mass of 6.6×10^{22} tons. Recent research has indicated that his data were remarkably accurate. From measurements of its size and mass the average density of the earth was found to be about 5.52 g/ml. Sea water has a density of just slightly more than one gram/ml while surface rocks in the earth's crust average about 2.8 g/ml. Because the average density of the earth is much greater than that of the surface materials, it is obvious that the core must have a high density. This information, combined with knowledge of the earth's magnetic field and the composition of meteorites has led most scientists to conclude that the core of the earth is predominantly iron.

Many aspects of solid-earth research are purely theoretical. These include theories related to the origin of the earth and moon, theories explaining the "wandering poles" of the earth, and theories of the drifting continents. However, research in these areas may lead to many practical benefits such as the ability to predict earthquakes and location of mineral deposits.

Considering how rapidly man is populating the earth and consuming its natural resources it seems reasonable to agree with the scientists who say that man's very survival depends on his ability to understand and use the solid earth in an intelligent manner in the next few decades. Perhaps your limitless curiosity about the solid earth will direct you into the interesting fields of geology or geochemistry.

QUESTIONS

1. (a) List all the information that can be obtained from a chemical equation. (b) Is it essential in an equation for the total number of molecules of products to equal exactly the number of molecules of the reactants? Explain.

2. Balance these skeleton equations.
(a) $Al(s) + O_2(g) \longrightarrow Al_2O_3(s)$
(b) $Al(s) + Fe_2O_3(s) \longrightarrow Al_2O_3(s) + Fe(s)$
(c) $KClO_4(s) \longrightarrow KCl(s) + O_2(g)$
(d) $Ca(OH)_2(aq) + HNO_3(aq) \longrightarrow Ca(NO_3)_2(aq) + H_2O(l)$

(e) $Cr_2(SO_4)_3(aq) + NaOH(aq) \longrightarrow Cr(OH)_3(s) + Na_2SO_4(aq)$
(f) $Cu(s) + AgNO_3(aq) \longrightarrow Cu(NO_3)_2(aq) + Ag(s)$
(g) $CH_4(g) + O_2(g) \longrightarrow CO_2(g) + H_2O(g)$
(h) $C_2H_6(g) + O_2(g) \longrightarrow CO_2(g) + H_2O(l)$
(i) $SiO_2(s) + HF(aq) \longrightarrow SiF_4(g) + H_2O(l)$
(j) $MgO(s) + H_3PO_4(aq) \longrightarrow Mg_3(PO_4)_2(s) + H_2O(l)$
(k) $PbO_2(s) \longrightarrow PbO(s) + O_2(g)$
(l) $NaBr(aq) + Cl_2(g) \longrightarrow Br_2(l) + NaCl(aq)$
(m) $Sb_2S_3(s) + HCl(aq) \longrightarrow H_3SbCl_6(aq) + H_2S(g)$

3. Balance these equations. It is not necessary to write the number *1* as a coefficient.
(a) $Na(s) + Cl_2(g) \longrightarrow NaCl(s)$
(b) $Ca(s) + Br_2(l) \longrightarrow CaBr_2(s)$
(c) $Al(s) + O_2(g) \longrightarrow Al_2O_3(s)$
(d) $C(s) + O_2(g) \longrightarrow CO_2(g)$
(e) $P_4(s) + Cl_2(g) \longrightarrow PCl_5(s)$
(f) $SO_2(g) + O_2(g) \longrightarrow SO_3(g)$
(g) $Sn(s) + O_2(g) \longrightarrow SnO(s)$
(h) $Al_4C_3(s) + H_2O(l) \longrightarrow Al(OH)_3(s) + CH_4(g)$
(i) $Fe_3O_4(s) + H_2(g) \longrightarrow Fe(s) + H_2O(g)$
(j) $KClO_3(s) \longrightarrow KCl(s) + O_2(g)$
(k) $HgO(s) \longrightarrow Hg(l) + O_2(g)$

4. What are three criteria that must be satisfied in a balanced net-ionic equation?

5. Explain what is meant by an unreactive or "spectator" ion.

6. Write net-ionic equations for these precipitation reactions when solutions containing these solutes are mixed. If no reaction occurs, write *no reaction*.

(a) potassium hydroxide + aluminum chloride
(b) silver nitrate + calcium iodide
(c) sodium nitrate + ammonium chloride
(d) barium chloride + sodium phosphate
(e) antimony(III) chloride + sodium sulfide
(f) ammonium oxalate + calcium nitrate
(g) lead nitrate + potassium chromate
(h) barium nitrate + ammonium carbonate
(i) bismuth(III) nitrate + potassium sulfide
(j) iron(III) sulfate + calcium hydroxide

7. Does the molarity of a solution vary with temperature? Explain.

8. Does one liter of a 1.0-*M* NaOH solution completely neutralize one liter of a 1.0-*M* H_2SO_4 solution? Explain. The skeleton equation for the reaction is

$$NaOH(aq) + H_2SO_4(aq) \longrightarrow Na_2SO_4(aq) + H_2O(l)$$

9. Lead chloride ($PbCl_2$) is moderately soluble, especially in warm water. Explain why the precipitation reaction

$$Pb^{2+} + 2Cl^- \longrightarrow PbCl_2(s)$$

should not be used as the basis for a quantitative gravimetric analysis for either Pb^{2+} or Cl^- ions.

10. In which of these equal volumes of 1-*M* aqueous solutions is there the least number of solute particles? (a) NaCl, (b) $CaCl_2$, (c) sulfuric acid (H_2SO_4). Assume complete dissociation.

PROBLEMS

1. One mole of N_2 combines with one mole of O_2 according to the equation:

$$N_2(g) + O_2(g) \longrightarrow 2NO(g)$$

How many (a) atoms are there in each molecule of N_2? (b) moles of oxygen combine with 0.5 mole of N_2? (c) moles of NO are formed when the reaction is 5 moles of N_2 reacting with excess O_2? (d) moles of oxygen atoms could be derived from 2 moles of NO?

2. Questions (a) through (f) refer to the equation:

$$3Cu(s) + 8HNO_3(aq) \longrightarrow 3Cu(NO_3)_2(aq) + 2NO(g) + 4H_2O(l)$$

(a) How many moles of NO are produced by the reaction of 4.0 moles of copper with excess HNO_3? (b) How many moles of HNO_3 are required to react completely with 5.0 moles of copper? (c) How many moles of $Cu(NO_3)_2$ are produced when 1.0 mole Cu reacts with 2.0 mole HNO_3? (d) How many moles of NO are produced by the reaction of 6.35 g of Cu with excess HNO_3? (e) What mass of NO is produced by the reaction of 6.35 g of Cu with excess HNO_3? (f) What mass of H_2O is produced when 6.35 g of Cu reacts with 12.6 g HNO_3?

3. Ammonia is produced synthetically by the reaction:

$$N_2(g) + 3H_2(g) \longrightarrow 2NH_3(g) + 22 \text{ kcal}$$

Assume the reaction is complete and answer these questions:
(a) Is this an exothermic or endothermic process? (b) How many moles of NH_3 are formed when one mole of N_2 reacts with excess hydrogen? (c) If 18×10^{23} molecules of H_2 react with sufficient nitrogen, how many

moles of NH_3 are produced? (d) Show that both nitrogen and hydrogen atoms are conserved in this reaction. (e) When 0.1 mole of N_2 combines with 0.3 mole of H_2, how many moles of NH_3 are produced?

4. Calculate the mass of silver that could be obtained by the reaction of a large excess of copper metal with 5.10 g of $AgNO_3$ in aqueous solution. The equation for the reaction is

$$Cu(s) + 2AgNO_3(aq) \longrightarrow 2Ag(s) + Cu(NO_3)_2(aq)$$

Ans. 3.24 g Ag.

5. (a) Which element is in excess when 3.00 g of Mg is ignited in 2.20 g of pure oxygen? (b) What mass is in excess? (c) What mass of MgO is formed?

Ans. (c) 5.00 g.

6. Given: $XeF_6(\ell) + 3H_2O(\ell) \longrightarrow XeO_3(s) + 6HF(g)$. How many moles of (a) water are needed to produce 1 mole of XeO_3? (b) XeO_3 does 59.8 g represent? (c) water are needed to produce 59.8 g of XeO_3?

7. Given: $3Fe_2O_3 + CO \longrightarrow 2Fe_3O_4 + CO_2$. How many grams of Fe_2O_3 can be converted to Fe_3O_4 by 14.0 g of CO?

8. How many grams of Al_2S_3 are formed when 5.00 g of Al is heated with 10.0 g S? **Ans. 13.9 g.**

9. Calcium carbonate undergoes thermal decomposition according to the equation:

$$CaCO_3(s) \longrightarrow CaO(s) + CO_2(g)$$

(a) How many grams of CaO and of CO_2 are formed when 50.0 g of $CaCO_3$ is completely decomposed? (b) When a 100-g sample of impure $CaCO_3$ (containing unreactive silica) was completely decomposed, the residue has a mass of 60.0 g. What was the percentage of $CaCO_3$ in the original sample? **Ans. (b) 91.0 percent.**

10. How many grams of CO_2 gas are released when a 50.0-g sample of Na_2CO_3 that is 60.0 percent pure reacts completely with excess HCl?

$$Na_2CO_3(aq) + 2HCl(aq) \longrightarrow CO_2(g) + H_2O(l) + 2NaCl(aq)$$

11. How many grams of 97.0 percent pure NaCl are required to produce 100.0 g of pure HCl when reacted with H_2SO_4?

$$NaCl(s) + H_2SO_4(l) \longrightarrow NaHSO_4(s) + HCl(g)$$

12. What mass of iron is required to convert 500.0 lbs of PbS into Pb? The reaction is

$$PbS(s) + Fe(s) \longrightarrow Pb(g) + FeS(s)$$

13. When MoO_3 and Zn are heated together they react

$$3Zn(s) + 2MoO_3(s) \longrightarrow Mo_2O_3(s) + 3ZnO(s)$$

What mass of ZnO is formed when 20.0 g of MoO_3 is reacted with 10.0 g of Zn?

14. How many liters of pure O_2 (STP) are required to react completely with 2.43 g of magnesium?

15. The oxidation of NH_3 is an important reaction in the preparation of nitric acid. The equation is

$$4NH_3(g) + 5O_2(g) \longrightarrow 6H_2O(g) + 4NO(g)$$

(a) How many liters of O_2 (STP) are needed to react with 5.00 kg of NH_3? (b) How many liters of air are required assuming air is 20.0 percent oxygen by volume?

Ans. (b) 41,200 ℓ.

16. (a) How many liters of CO_2 at STP is consumed by a plant in the production of 454 g of a simple sugar?

$$6CO_2(g) + 6H_2O(l) \longrightarrow C_6H_{12}O_6(s) + 6O_2(g)$$

(b) How many liters of air are needed to supply the required CO_2 in part (a) assuming air is 0.040 percent CO_2 by volume? **Ans. (b) 8.45×10^5 ℓ.**

17. When 11.10 g of compound XCl_2 is reacted with excess $AgNO_3$, 28.66 g of AgCl is produced.

$$XCl_2(aq) + 2AgNO_3(aq) \longrightarrow 2AgCl(s) + X(NO_3)_2(aq)$$

What is the atomic mass of X?

18. Impure nitrates may be analyzed by measuring the volume of NO generated during the reaction:

$$2NaNO_3(s) + 4H_2SO_4(\ell) + 3Hg(\ell) \longrightarrow 3HgSO_4(aq) + Na_2SO_4(aq) + 4H_2O(\ell) + 2NO(g)$$

An impure sample of $NaNO_3$ weighing 2.50 g generated 400 ml of NO measured at STP. Calculate the percentage of $NaNO_3$ in the original sample.

Ans. 60.8 percent.

19. What is the molar concentration of each of these solutions? (a) 3.65 g of HCl dissolved in 200 ml of solution, (b) 4.0 g of NaOH dissolved in 150 ml of solution, (c) 5.1 g of NH_3 dissolved in 100 ml of solution, (d) 1.59 g Na_2CO_3 dissolved in 500 ml of solution, (e) 26.8 g $H_2C_2O_4 \cdot 2H_2O$ dissolved in 0.400 ℓ of solution. (Oxalic acid dihydrate contains two moles of water per mole of oxalic acid in the crystal.)

20. Calculate the molarity of these solutions: (a) 1.00 ℓ containing 119 g KCl, (b) 2.00 liters containing 223.5 g $CuBr_2$, (c) 0.250 liters containing 13.35 g aluminum chloride, (d) 0.150 ℓ containing 13.0 g cobalt(II)

chloride, (e) 0.250 ℓ containing 4.33 g chromium(III) iodide. **Ans. (c) 0.400 *M*, (d) 0.667 *M*.**

21. Concentrated H_2SO_4 has a density of 1.84 g/ml and is 98.0 percent H_2SO_4. What volume of concentrated acid is needed to prepare 0.500 l of a 6.00-*M* solution?

22. The table below shows the density of some 100 percent pure reagents used in the laboratory. Calculate the molarity of each.

Name	*Formula*	*Density (g/ml)*
Acetic acid	$HC_2H_3O_2$	1.049
Sulfuric acid	H_2SO_4	1.841
Phosphoric acid	H_3PO_4	1.834
Ethanol	C_2H_5OH	0.789

Ans. Acetic acid 17.5 *M*.

23. A solution of sulfuric acid is labeled 6.0 *M*. How many grams of H_2SO_4 are contained in 0.100 ℓ of solution?

24. How many moles of NaOH are contained in (a) 1.0 liter of a 1.0-*M* solution? (b) 0.50 ℓ of a 1.0-*M* solution? (c) 0.50 ℓ of a 0.50-*M* solution? (d) 0.125 ℓ of a 0.50-*M* solution? (e) 5.0×10^{-2} ℓ of a 5.0×10^{-3}-*M* solution?

25. Calculate the mass of solute needed to make these solutions: (a) 1.0 ℓ of 1.0-*M* sodium hydroxide, (b) 0.500 ℓ of 2.00-*M* calcium nitrate, (c) 2.0 ℓ of 0.50-*M* potassium bromide, (d) 0.200 ℓ of 0.75-*M* zinc chloride, (e) 0.125 ℓ of 2.40-*M* ammonium chloride. **Ans. (b) 164 g, (d) 20 g.**

26. What mass of solute is required to prepare the volume of each of these solutions? (a) 500 ml of 3-*M* $CaCl_2$? (b) 750 ml of 2.0-*M* H_2SO_4? (c) 5.0 ℓ of 0.50-*M* $H_2C_2O_4$ (oxalic acid)? (d) 0.125 ℓ of 8.00-*M* $HC_2H_3O_2$? (e) 5.0×10^4 ml of 0.10-*M* KBr? (f) 1.0×10^{-3} ℓ of 1.0×10^{-3}-*M* NH_4Cl? (g) 150 ml of 0.50-*M* $ZnCl_2$?

27. Stock solutions of acids and bases for general use in the laboratory may be 6-*M* solutions. How many milliliters of 6.0-*M* stock solution are needed to prepare each of these solutions? (a) 0.50 ℓ of 1.0-*M* HCl, (b) 124 ml of 0.25-*M* H_2SO_4, (c) 250 ml of 0.50-*M* NH_3(aq), (d) 0.200 ℓ of 2.00-*M* NaOH, (e) 75 ml of 0.10-*M* H_3PO_4. **Ans. (a) 83 ml, (b) 5.2 ml, (e) 1.3 ml.**

28. A 0.5-*M* solution of SrI_2 is totally dissociated into Sr^{2+} ions and I^- ions. What is the molar concentration of each ion?

29. Twenty milliliters of 18-*M* H_2SO_4 is diluted to a total volume of 1.0 ℓ. Find the molar concentration of this solution. **Ans. 0.36 *M*.**

30. Tell how to prepare a solution of HNO_3 which contains 0.1 m mole per ml from a 6-*M* stock solution of HNO_3.

31. Fifty ml of 0.50-*M* HCl is mixed with 30 ml of 0.28-*M* HCl. Calculate the molarity of the resulting mixture. **Ans. 0.42 *M*.**

32. One hundred ml of 0.50-*M* HNO_3 is mixed with 400 ml of 0.25-*M* $Ca(NO_3)_2$. Assume both compounds are 100 percent dissociated. Find the concentration of these ions in the mixtures: (a) $H^+(H_3O^+)$, (b) Ca^{2+}, (c) NO_3^-.

33. (a) Two liters of 6.0-*M* HNO_3 is added to 1.0 ℓ of 2.0-*M* HNO_3. What is the concentration of HNO_3 in the mixture? Assume volumes are additive. (b) Calculate the volume of 12-*M* HCl that must be added to 2.5 ℓ of 1.0-*M* HCl in order to obtain 15 ℓ of 1.0-*M* HCl. **Ans. (a) 4.67 *M*.**

34. What is the molar concentration of the solutes in these solutions? (a) A silver-nitrate solution containing 10 mg Ag^+ per ml of solution. (b) A cadmium-chloride solution containing 100 mg Cd^{2+} ions per ml of solution. (c) A magnesium-sulfate solution containing 1.2 mg of $MgSO_4$ per ml of solution. **Ans. (b) 0.89-*M*.**

35. How many mg of each cation in 1.0 ml of these solutions? (a) 0.20-*M* $LiNO_3$, (b) 1.50-*M* $CuSO_4$, (c) 1.0-*M* $Al(NO_3)_3$.

36. A sugar solution is prepared by dissolving 772 g of sucrose ($C_{12}H_{22}O_{11}$) in 515 g of H_2O. The volume of the solution is 1000 ml. Calculate the (a) density of the solution, (b) mass percentage of sugar, (c) molarity of the solution, (d) mg of sugar per ml of solution, (e) volume of solution which contains 193 g of sugar, (f) ml of water needed to dilute 1000 ml of the solution to a 0.50-*M* concentration.

37. Barium nitrate and potassium sulfate solutions react and form a precipitate. What is the precipitate? How many ml of 0.40-*M* $Ba(NO_3)_2$ solution are required to precipitate completely the sulfate ions in 25 ml of 0.80-*M* K_2SO_4 solution? **Ans. 50 ml.**

38. What volume of 0.25-*M* KCl solution is required to precipitate completely the silver ions from a 60-ml sample of 0.60-*M* $AgNO_3$ solution?

39. What mass of silver chloride can be precipitated from a silver-nitrate solution by 200 ml of a solution of 0.50-*M* $CaCl_2$? **Ans. 28.7 g.**

40. An impure sample of Na_2SO_4 has a mass of 1.56 g.

This sample is dissolved and allowed to react with $BaCl_2$ solution. The precipitate has a mass of 2.50 g. Calculate the percentage of Na_2SO_4 in the original sample.

Ans. 97.6 percent.

41. An impure, 0.500-g sample of NaCl was dissolved in 20.0 ml of water. The chloride ions were precipitated completely by addition of a $AgNO_3$ solution. The dried AgCl precipitate has a mass of 1.15 g. (a) How many moles of AgCl formed? (b) How many moles of NaCl were in the sample? (c) How many grams of NaCl were in the sample? (d) What was the percentage of NaCl in the impure sample?

42. An impure sample of Na_2SO_4 has a mass of 1.65 g and is dissolved in water. Addition of $BaCl_2$ solution produced a precipitate of barium sulfate with mass 2.32 g. What is the percentage of Na_2SO_4 in the impure sample?

43. A sample known to contain only NaCl and KCl has a mass of 1.00 g. The sample is dissolved and treated with $AgNO_3$ until precipitation is complete. The precipitate of AgCl has a mass of 2.32 g. What is the percentage of NaCl in the mixture?

(1) $NaCl + AgNO_3 \longrightarrow AgCl(s) + NaNO_3$ or
$Ag^+(aq) + Cl^-(aq) \longrightarrow AgCl(s)$

(2) $KCl + AgNO_3 \longrightarrow AgCl(s) + KNO_3$ or
$Ag^+(aq) + Cl^-(aq) \longrightarrow AgCl(s)$

Ans. 76.1 percent NaCl.

44. A mixture of Na_2SO_4 and K_2SO_4 having a total mass 0.500 g, was dissolved in water. Barium chloride was added as a precipitating agent. The dried $BaSO_4$ resulting from the reaction had a mass of 0.715 g. What is the percentage of each component in the original mixture?

4

The Behavior of Gases

Preview

We are surrounded by an atmosphere of gases which, on a clear, calm day, cannot be seen, smelled, felt, tasted, or heard. We are familiar with such things as spray cans, footballs, automobile tires, balloons, air hammers, air brakes, and other devices that depend on gases for their operation. In addition, many chemical reactions involve substances in the gaseous phase. Perhaps the reaction most familiar to you is the burning of a gas to cook your food or to heat your home.

"We have shown that the strengths required to compress air are in reciprocal proportions, or thereabouts, to the spaces comprehending the same portion of the air."

Robert Boyle (1627–1691)

Historically, observations made on gaseous systems played an important role in the development of the atomic theory and other important scientific concepts. Because the gaseous phase of matter is the simplest of the three phases, it is the easiest to study. The essentially independent action of each molecule in a sample of gas makes it possible to relate the properties of a macroscopic sample of gas to the behavior of the individual molecules in the sample. All elements can exist as individual molecules in the gaseous phase. The temperature at which different elements change into gases at one atmosphere pressure varies over a wide range. Helium, with a boiling point of −269°C, and tungsten, with a boiling point of 5900°C, represent the extremes. In this chapter we shall concern ourselves with those substances that exist as gases at ordinary temperatures and pressures.

We shall first look for characteristics common to all substances in the gaseous phase and then use these observations as a basis for introducing the *Kinetic Molecular Theory,* one of most successful of all scientific theories.

At this point we shall describe experiments designed to determine the relationship between the volume, pressure, and temperature of a gaseous system. The data from these experiments will then be used to develop quantitative laws which describe the physical behavior of gases. We shall illustrate the laws and interpret them in terms of the *Kinetic Molecular Theory.*

Also, we shall consider reactions which yield gaseous products. Application of stoichiometric principles developed in Chapter 3 and the gas laws developed in this chapter will enable us to calculate the volumes of gases produced in chemical reactions at any conditions of temperature and pressure.

GENERAL CHARACTERISTICS OF GASES

4-1 Molecular Motion Suppose we remove all of the air from a stoppered flask by means of a vacuum pump and introduce a colored gas into the evacuated vessel. Immediately the colored material is observed to be uniformly distributed throughout the flask. This observation suggests that matter in the gaseous phase is in constant motion and shows that a gas fills all the space available to it. If the gas is then transferred to another vessel having a different volume and shape, the gas will occupy the second vessel completely. Thus, we may conclude that *gases do not have definite shapes or volumes.* Observation over long periods of time shows that the gas does not settle out. This may be interpreted to mean that the *gas molecules remain in motion.* Apparently there is no net loss of energy during collisions between molecules and *negligible attractive forces* between molecules. Their ability to maintain movement and to fill any available space suggests that gas molecules have a great deal of energy. The energy of motion is called ***kinetic energy.*** Experiments show that the kinetic energy possessed by a moving particle depends on the mass of the particle and its velocity. The quantitative relationship is

Inertia is that property possessed by a body in motion that tends to keep it in motion with the same direction and same velocity.

1 calorie = 4.18 × 10⁷ ergs
The butter that you eat with your toast in the morning (2 pats) liberates about 4 × 10¹² (1 trillion) ergs when it is completely used by your body. This is approximately the amount of kinetic energy possessed by a 10-pound ball traveling through space at 500 mi/hr.

$$\text{K.E.} = \tfrac{1}{2}\text{mv}^2 \qquad 4\text{-}1$$

where m is the mass expressed in grams and v is the velocity usually expressed in cm/sec. Substituting grams for m and cm/sec for v in Equation 4-1 the unit of energy in the cgs system has the dimensions of g cm^2/sec^2. This unit is called an ***erg.***

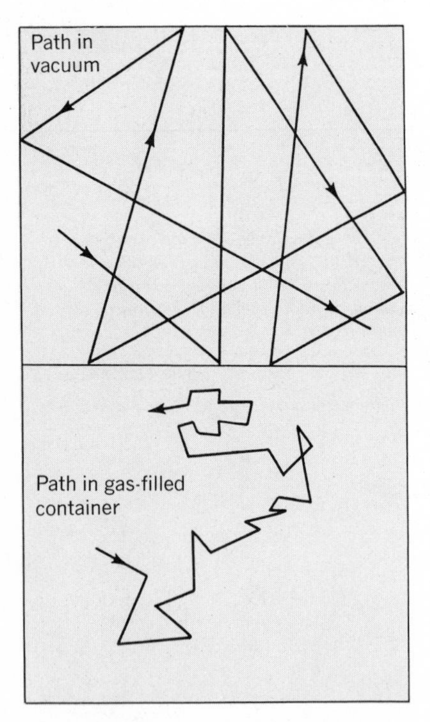

4-1 Mean free paths of molecules. A single molecule in a vacuum would follow a path similar to that shown above. In a gas-filled container, a molecule makes countless collisions. Thus, its mean free path is shorter in air than it is in a vacuum.

The effect of molecular motion can be seen by examining smoke from a burning match under a microscope. The tiny, solid quivering smoke particles are observed to move about haphazardly in a zigzag path. The particles do not show any appreciable tendency to settle to the bottom of the container. If the chamber is warmed, the particles are observed to move faster. This constant motion of very small, but visible, particles suggests that they are suspended in a medium of even smaller but faster-moving particles. Thus, as the tiny smoke particles float in the air, they are struck irregularly on all sides by the rapidly moving molecules of gas which compose the air. Since the collisions are random, the particles follow an erratic, zigzag path. It may be shown that at room temperature molecules of gases in the air travel at an average speed of about 1300 ft/sec, the speed of an ordinary rifle bullet. At higher temperatures, the average velocity and kinetic energy are greater.

4-2 Mean Free Path The rapid motion of gas molecules might lead us to predict that odors should be propagated instantly across a room and that a localized change in temperature, caused by the opening of a window, should be felt immediately at the opposite side of the room. Experience shows that this is not true. Gas molecules in a room are somewhat similar to a large number of billiard balls rolling around on a table. Molecules travel only a short distance before they collide with other particles and are deflected from their original paths. The average length of the free path between two successive collisions is known as the ***mean free path.*** The shorter the path, the more slowly any condition depending on molecular motion is propagated.

The mean free path depends on both the *concentration of the molecules* and on their *size.* The greater the number of molecules per unit volume and the greater their size, the shorter is the mean free path.

4-3 Gas Pressure and Compressibility Another characteristic of gases may be observed by blowing up a balloon. As the gas is introduced, the balloon expands, becomes firmer, and finally bursts. This observation may be explained by assuming that *gases exert pressure* on the inner walls of their containers. How can this be explained in terms of our concept of the molecular motion of gases? Sometimes it is possible to relate the behavior of a submicroscopic system to that of a well-understood macroscopic system. Thus, the pressure of a gas on the walls of its container may be explained in terms of the behavior of *ideal* billiard balls on an *ideal* table.

The motion of a billiard ball can be described by assuming that an ideal table has perfectly elastic cushions so that the speed of the ball is not decreased when it rebounds from a cushion. That is, when a ball pushes on a cushion and rebounds with no loss of energy, the collision is said to be a *perfectly elastic collision.* If there were no friction with the table top or air resistance, the ball, once struck, would travel indefinitely in a zigzag path. If several balls were placed on the table and set in motion, they would behave similarly. If the balls were *perfectly elastic,* there would be *no loss of energy* when two of them collided. Their directions would change but *inertia* would keep the balls in motion forever. An increase in the number of balls results in a greater number of collisions and more frequent pushes on the cushion. The principles and mathematics of the billiard-ball system mentioned above are well understood. Since the behavior of gas molecules is similar to that of the ideal billiard-ball system, the same mathematics may be used to describe the pressure exerted by gas molecules. In this manner, the ideal billiard-ball system serves as a model for our understanding of the behavior of gases.

Scientists visualize a gas as a collection of molecules moving around in a container, making perfectly elastic collisions with the walls. Even the smallest sample of gas contains countless molecules. Therefore, a vast number of collisions and "pushes" on the walls of the container occur each second. The sum of all the "pushes" on a given area accounts for the *pressure* of the gas. For example, blowing up a balloon increases the number of gas molecules within the balloon and, consequently, the number of collisions. An increased number of collisions results in more outward "pushes" on the wall, causing the expansion of the balloon.

Experiments such as that illustrated in Fig. 4-3 show that an

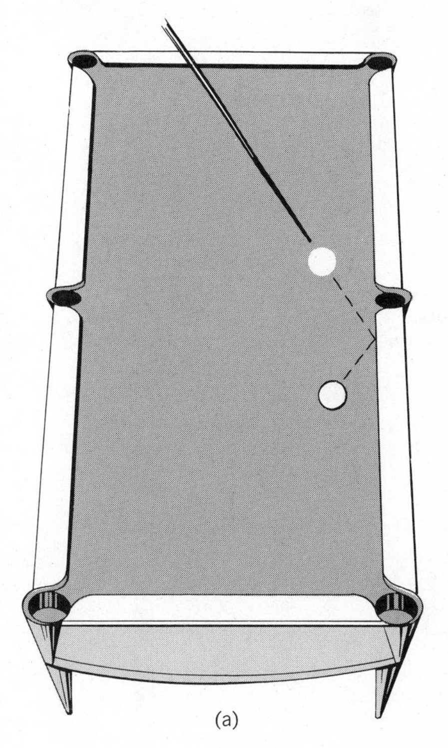

(a)

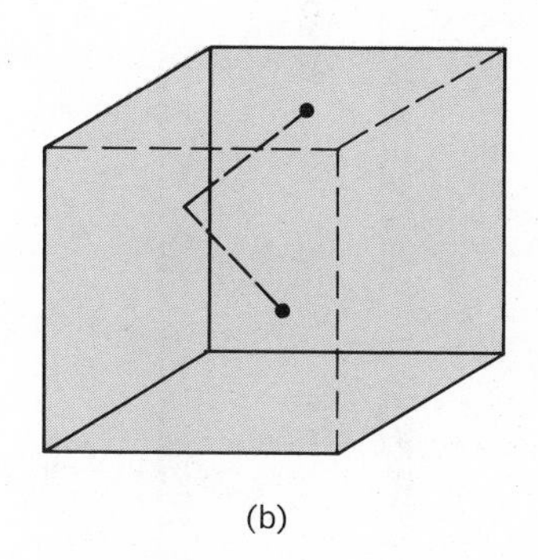

(b)

4-2 The behavior of ideal gas molecules in (b) is similar to that of ideal billiard balls on an ideal billiard table in (a). The laws of motion which describe the billiard-ball system can be applied to the motion of ideal gas molecules. By applying mathematical principles and the laws of motion to the gaseous systems, scientists have developed general laws expressed in mathematical terms which accurately describe the behavior of ideal gases.

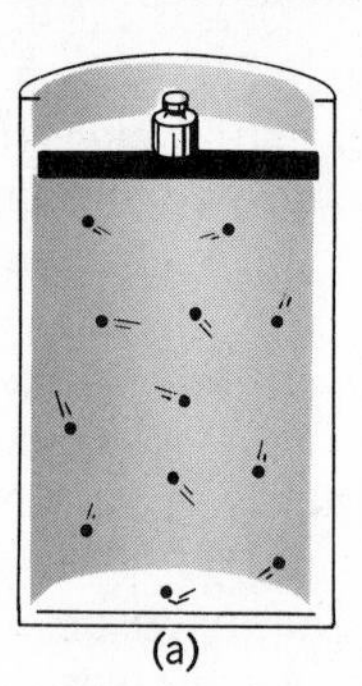

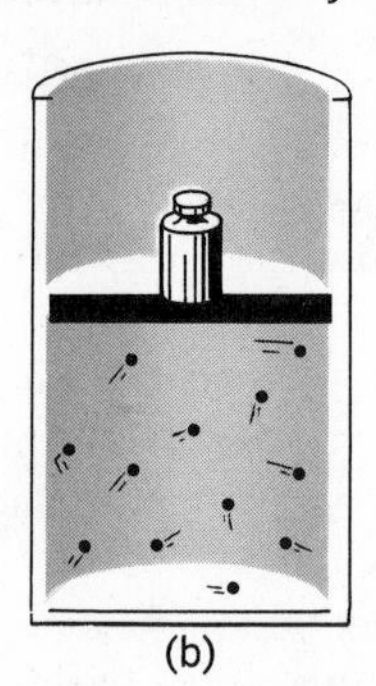

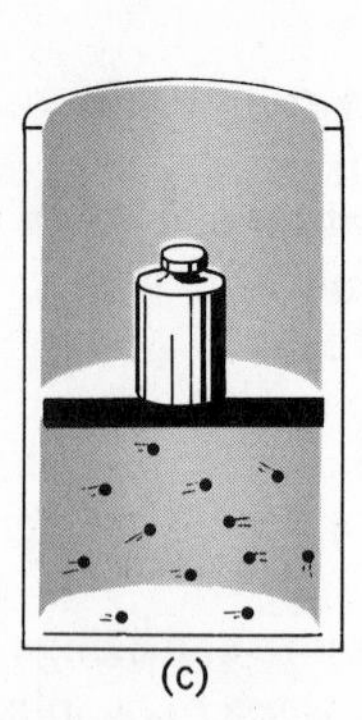

4-3 As mass is added to the piston, the pressure on the gas increases, and its volume correspondingly decreases.

increase in the external pressure exerted on a gas produces a large decrease in volume. In other words, *gases are highly compressible.* If you have ever owned a bicycle and have used a hand-pump to fill a tire with air, you yourself have demonstrated the compressibility of gases. When you pushed down on the pump, you pushed the molecules together (compressed the gas). Compressing a gas decreases the distances between molecules. It may be shown at room temperature and pressure, that a confined gas is more than 99 percent empty space. Thus, the molecules themselves occupy relatively small space. This vast distance between molecules accounts for the relative ease of compression and the extent to which gases may be compressed. When a gas is compressed, the number of molecules striking a unit area of wall per second is greater than that before compression. The result is greater pressure. Let us now define the concept of pressure more rigorously and identify pressure units commonly used in chemistry.

Pressure is *the force per unit area* which (in this case) gas molecules exert on the walls of their container. Gas pressure is usually measured in chemistry laboratories by means of the Torricellian-type barometer, shown in Fig. 4-4. This type of apparatus was first constructed by Evangelista Torricelli (1608–1647) around the year 1640. To reproduce Torricelli's experiment, take a glass tube of about $\frac{1}{4}$ in. in diameter and about 3 ft. long, sealed at one end, and fill it with mercury. Place your finger over the open end, invert, and submerge the closed end in a dish containing mercury. Remove your finger and note that the column of mercury drops to a height of about 760 mm above the mercury in the dish.

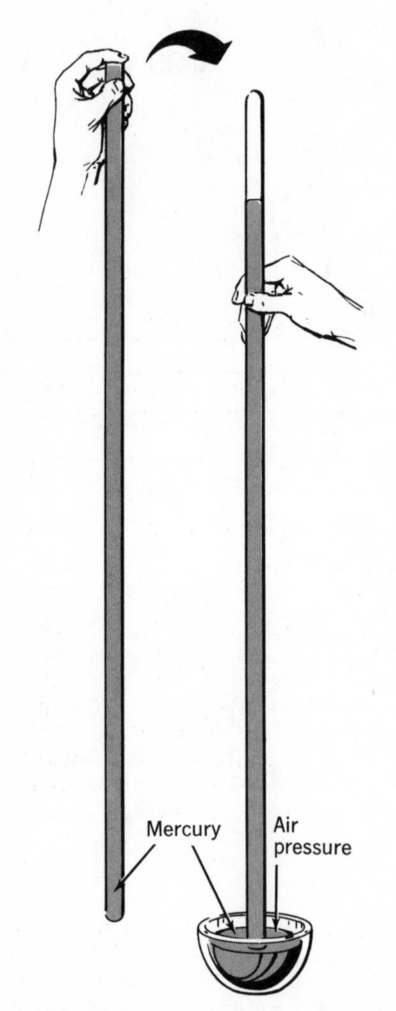

4-4 The downward pressure of the atmosphere on the surface of the mercury is transmitted through the mercury reservoir and exerts an upward force in the barometer tube which is just balanced by the downward force of the mass of the mercury in the tube.

The empty space above the mercury column contains a little mercury vapor but is essentially free of other molecules. The height of the mercury column depends on the force per unit area exerted by the atmosphere on the mercury in the dish. At sea level, the force per unit area exerted by a column of air extending to the "top" of the atmosphere is approximately equal to the force per unit area exerted by a column of mercury 760 mm in height. This quantity is known as ***one standard atmosphere of pressure*** when the mercury is at 0°C.

In chemistry, gas pressures have traditionally been expressed in terms of *mm of mercury* or in *atmospheres* (*atm*). Another pressure unit, the ***torr,*** named in honor of Torricelli, is also used. One ***torr*** is a pressure unit equal to the pressure exerted by a column of mercury 1 mm high at 0°C. In other words, one atmosphere is equal to 760 torr.

FOLLOW-UP PROBLEM

Express a pressure of 745 torr in terms of atmospheres. **Ans. 0.980 atm.**

4-4 Molecular Velocities and Energies In a given sample of gas, it may be shown experimentally that there is a wide range in speed of the individual molecules. A molecule in a gas is much like a ping-pong ball during an active game. At times the ball is lobbed slowly over the net, and at other times it is driven vigorously across the table. Occasionally, the ball may drop to the floor and may roll nearly to a stop. This variation in direction and speed is also characteristic of a molecule in a sample of gas. If a single molecule exhibits this variation, then it is easy to imagine the possible great range of velocities for a large number of molecules in a sample of gas at any given instant.

One method of determining relative velocities of molecules in a gas is shown in Fig. 4-6. The experimentally determined distribution of molecular velocities of nitrogen gas molecules at 0°C and 100°C is shown in Fig. 4-7. The velocities and masses of atoms can be used to determine their kinetic energies. It can be seen that at 100°C there are fewer molecules with low velocities than at 0°C but more with high velocities, even though the range is much the same for both temperatures.

If we were able to take a room full of air at ordinary temperatures and separate out a thimbleful of the fastest-moving particles (air is a mixture of molecules and atoms), the air in our sample would be at the same temperature as the interior of a raging furnace. At the other extreme, if we were able to separate out a thimbleful of the slowest-moving particles in this same room, the sample would consist of liquid air much colder than the air of the Arctic during the winter. If the air in this room were heated to a higher temperature, there would still be a great range in molecular velocities. The warm room would still have some very slow-moving particles as well as some fast ones, but the average velocity and energy of motion (*kinetic energy*) would be higher.

Fig. 4-7 reveals that *warming a sample of gas merely shifts the distribution of energy in the direction of greater average kinetic energy.* There is no assurance that warming a sample of gas will cause any single molecule to increase in speed. Probability tends to favor an increase in speed for any single molecule, but it is possible that, as the temperature increases, the speed of a given molecule might decrease.

The energy distribution curves shown in Fig. 4-7 were worked out separately by two scientists in the 19th century. Two of the

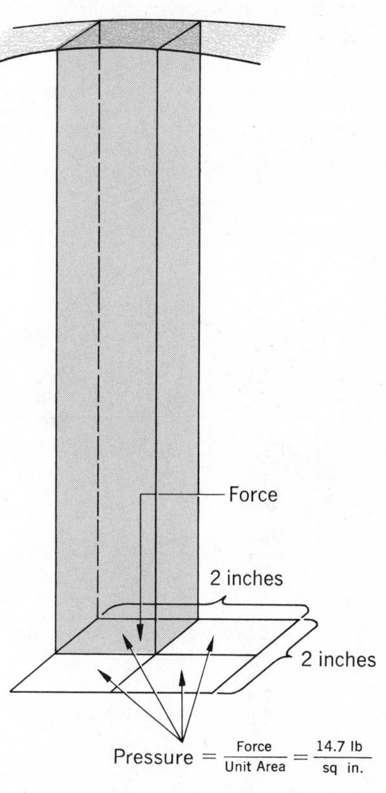

Total force exerted on 2.00 in. x 2.00 in. surface

$$= 4.00 \text{ sq in.} \times \frac{14.7 \text{ lb}}{1 \text{ sq in.}} = 58.8 \text{ lb}$$

4-5 The weight of the column of air is proportional to the area of the column. The pressure, which is the ratio of the weight (a force) to the area, does not change. The pressure of the atmosphere varies with humidity and location. "One atmosphere" is defined as the pressure exerted by a column of mercury exactly 760 millimeters high measured at 0°C at sea level at 45° latitude.

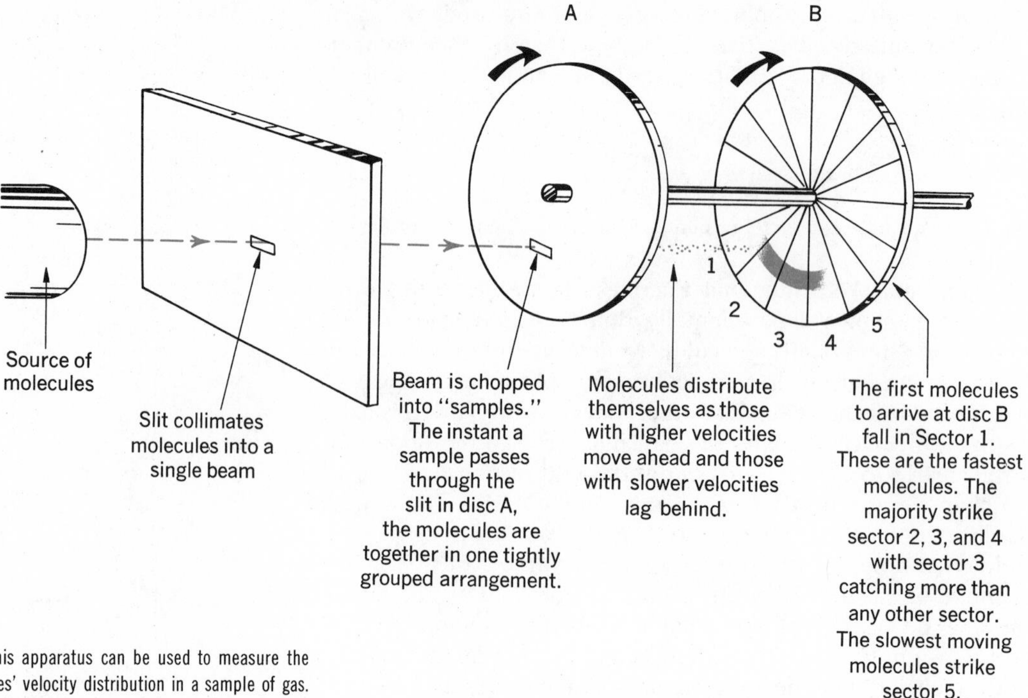

4-6 This apparatus can be used to measure the molecules' velocity distribution in a sample of gas. Discs A and B are rapidly rotated on a common axle in the direction shown. The slower moving molecules take longer to reach disc B than the faster ones. The relative numbers of molecules striking each section of disc B may be determined. These data can be related to the relative velocities and plotted to give a graph similar to that in Fig. 4-7.

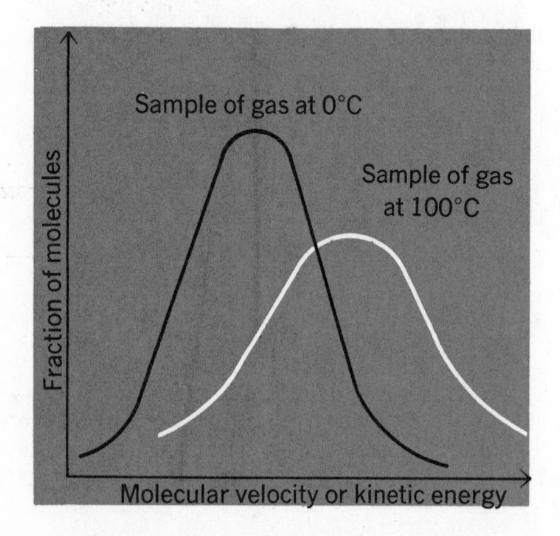

4-7 The range of molecular velocity and kinetic energy in a warm and cold gas are about the same. At the higher temperature, a greater percentage of the molecules have a higher velocity and a higher kinetic energy of motion.

world's greatest theoretical physicists, James Clerk Maxwell (1831–1879) and Ludwig Boltzmann (1844–1906), both predicted these curves from theoretical calculations. The curves are known as *Maxwell-Boltzmann distributions* of kinetic energies. They are based on statistical theory and the theory of probability. Statistical methods such as these are used whenever scientists must deal with situations involving a large number of random events.

Observations such as those described above led to a model for gaseous systems which is described by the *Kinetic Molecular Theory of Gases.* This theory extends the atomic theory so that it includes the factors of motion and energy.

KINETIC MOLECULAR THEORY OF GASES

4-5 Kinetic Molecular Theory of Gases From observations and measurements of the behavior of real gases, scientists have developed a theory and constructed a model of an ideal gas that helps us understand the nature of a gas at the submicroscopic level. The main assumptions of the ***Kinetic Molecular Theory*** are summarized below.

1. Ideal gases consist of small particles (molecules or atoms) that are far apart in comparison to their own size. These particles are considered to be dimensionless

points which occupy zero volume. The volume of *real* gas molecules is assumed to be negligible for most purposes.

2. Molecules are in rapid and random, straight-line motion. This motion can be described by well-defined and established laws of motion.

3. The collisions of molecules with the walls of a container or with other molecules are perfectly elastic; that is, no loss of energy occurs.

4. There are no attractive forces between molecules or between molecules and the walls of the container with which they collide.

5. At any particular instant, the molecules in a given sample of gas do not all possess the same amount of energy. The average kinetic energy of all the molecules is proportional to the temperature.

It should be noted that the assumptions above are made only for a "perfect" or ideal gas. They cannot be rigorously applied (mathematically) to real gases but can be used to explain their observed behavior qualitatively.

We are now ready to investigate the quantitative aspects of gas behavior. We shall have an opportunity to test the Kinetic Molecular Theory by using it to test the results of our experiments.

In order to describe a gaseous system, it is necessary to identify certain measurable quantities. These are: *number of moles, volume, temperature, and pressure.* We shall find as a result of our experiments that these factors are interdependent. Our objective is to discover the relationships.

LAWS DESCRIBING THE BEHAVIOR OF GASES

4-6 Variation of Gas Volume with Pressure Changes The relationship between the pressure and the volume of a sample of gas at a constant temperature may be determined by using the apparatus shown in Fig. 4-8. This apparatus consists of a manometer and a pump composed of a piston in a graduated glass tube. When the mercury levels in both arms of the manometer are equal, the pressure of the gas in the graduated tube is equal to the atmospheric pressure exerted on the mercury in the right arm of the manometer.

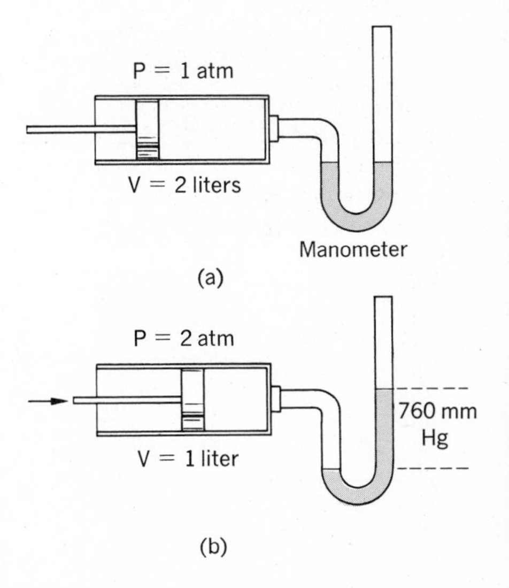

4-8 A version of a Boyle's law apparatus.

When the sample of air in the tube is compressed by the pump the mercury drops in the left side of the manometer and rises in the right side. In Fig. 4-8(b), the mercury in the right side of the manometer is 760 mm higher than in the left side. The pressure on the gas in the graduated tube is now atmospheric pressure plus 760 mm of mercury or 760 torr. It can be seen that the volume of gas is smaller at the higher pressure. By setting the piston at

TABLE 4-1
PRESSURE-VOLUME DATA
(Number of Moles of Gas and Temperature Constant)

Volume	Pressure	Pressure X Volume
600 ml	1 atm	
300 ml	2 atm	
200 ml	3 atm	
150 ml	4 atm	
120 ml	5 atm	
100 ml	6 atm	

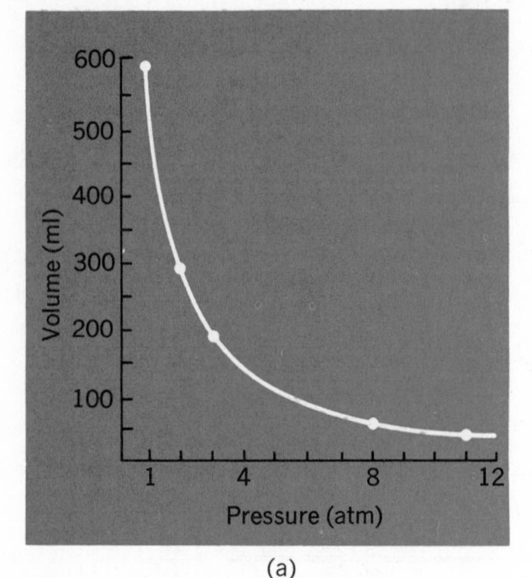

(a)

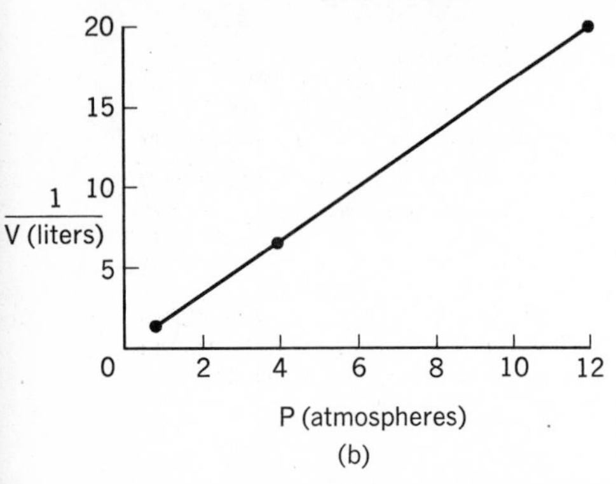

(b)

4-9 (a) A Pressure-Volume graph. (b) A plot of P_{atm} versus $1/V$. Chemists attempt where possible to arrange data so that straight lines can be obtained. In this way, fewer experimental points are necessary to establish the mathematical relationship between the variables. The graph in (b) shows that P and $1/V$ are *directly* proportional; therefore it follows that P is inversely proportional to V.

several different positions, a series of pressure and corresponding volume measurements may be obtained. Sample data are listed in Table 4-1. These data are usually plotted as shown in Fig. 4-9. The form of the graph in Fig. 4-9(a) (a rectangular hyperbola) reveals an inverse relationship between the two variables. The straight-line plot of P vs 1/V shown in Fig. 4-9(b) confirms this relationship.

Simple mathematical relationships between two variables such as pressure and volume may often be determined by testing the data. When a series of simultaneous values for two factors are inversely proportional to each other, their product is constant. If you check the data in Table 4-1, you will find that it satisfies this criterion. Therefore, the mathematical relationship between the pressure and volume of a gas at constant temperature may be expressed as

$$\mathbf{PV = k} \text{ or } \mathbf{P = k \cdot \frac{1}{V}}$$

where k is a constant, the value of which is determined by the mass of the gas sample and the temperature. Thus, we may conclude that **the volume of a fixed number of moles of gas varies inversely with the pressure at constant temperature.** This relationship is known as ***Boyle's Law*** in honor of Robert Boyle (1627–1691) who discovered it in 1660. Using the data in Table 4-1, we can develop an expression for Boyle's Law which can readily be used to solve pressure-volume problems. These data show that each set of PV values is equal to a constant. Therefore, they are equal to each other. This may be expressed as

$$\mathbf{P_1V_1 = P_2V_2} \qquad 4\text{-}2$$

where V_1 is the initial volume, V_2 the final volume, P_1 the initial pressure, and P_2 the final pressure.

Let us apply Boyle's Law to a problem we are likely to encounter in the laboratory.

Example 4-1
A student collects 25 ml of gas at 720 torr. What volume would this gas occupy at 780 torr? There is no change in temperature or mass.

Solution
It is usually advisable to itemize the data first.

$V_1 = 25$ ml $\qquad P_1 = 720$ torr
$V_2 =$ unknown $\qquad P_2 = 780$ torr

The thought processes involved in applying the cortect principle to the solution of the problem are summarized below.
1. How does the pressure change?

Ans. The pressure increases.

2. What happens to the volume when the pressure increases?

Ans. Since pressure and volume are inversely proportional, an increase in pressure causes a decrease in volume.

3. Must the original volume be multiplied by a number larger or smaller than 1?

Ans. Since the volume decreases, the original volume must be multiplied by a pressure ratio that is less than 1. This ratio is 720 torr/780 torr.

The final volume, V_2, is equal to the original volume multiplied by the pressure ratio.

$$\text{final volume} = \text{original volume} \times \text{pressure ratio}$$

$$V_2 = 25 \text{ ml} \times \frac{720 \text{ torr}}{780 \text{ torr}} = 23 \text{ ml}$$

Alternatively, the data could be substituted in Equation 4-2 which could then be solved for V_2.

FOLLOW-UP PROBLEM

A gas measuring 525 ml is collected at 785 torr. What volume does this gas occupy at 745 torr? **Ans. 553 ml.**

We can use the Kinetic Molecular Theory to explain Boyle's Law qualitatively. At a constant temperature, the molecules in a cylinder such as that shown in Fig. 4-10 are moving at a constant average velocity. When these moving molecules strike the walls of the cylinder and the piston, they exert pressure. The pressure is proportional to the number of molecules that strike a unit area in unit time. When the force on the piston is increased, it moves down until the internal pressure exerted by the gas molecules equals that exerted on the top of the piston provided the mass of the piston is neglected. When the external pressure is doubled, the volume decreases to one-half of its original value. Since only half as much space is available to the molecules, there will be twice as many collisions per unit area per second. Thus, when the volume is halved, the pressure is doubled.

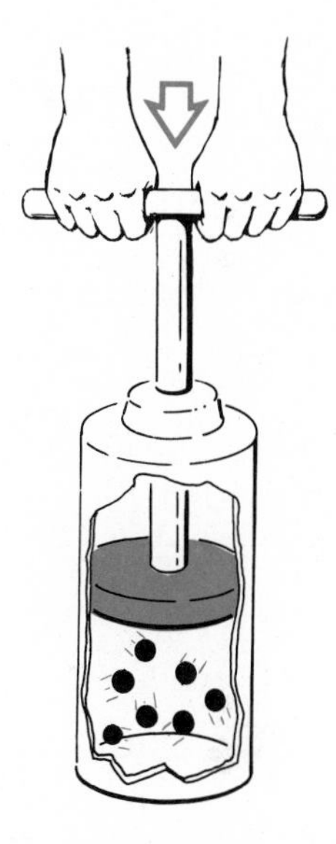

4-10 When the plunger is pushed downward, the molecules come together increasing the number of collisions per unit time on the walls of the container.

The doubling of pressure is also related to the average decrease in time between successive collisions per molecule.

4-7 Deviations of Real Gases from Ideal Behavior Strictly speaking, Boyle's Law describes the behavior of an "Ideal" gas. Studies made on real gases reveal deviations from ideal behavior. In 1873, the Dutch physicist, Johannes van der Waals (1837–1923), proposed that deviations from ideal are observed because molecules of real gases occupy volume and *exert attractive forces on each other.* The attractive forces are known as ***van der Waals forces.*** Experimental evidence shows that *real gases approach the behavior of an ideal gas at low pressures and high temperatures.*

The degree of interaction between molecules varies with their

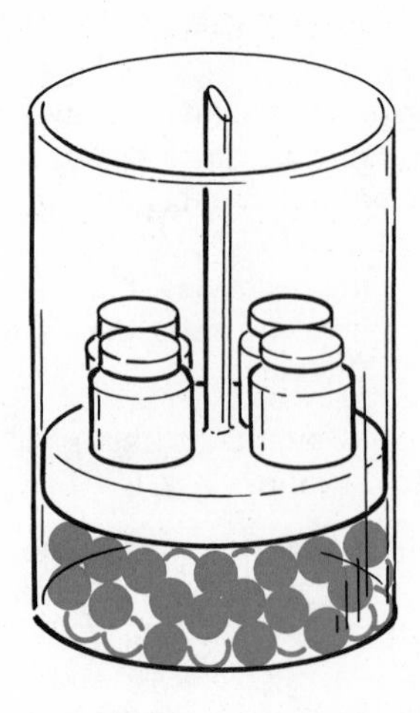

4-11 Under extreme pressure, the volume occupied by the molecules themselves become a significant fraction of the volume of the sample of gas.

composition and shapes. We shall find later that large, irregularly-shaped molecules such as sulfur dioxide behave as little magnets and have more surface area than small, monatomic, symmetrical molecules like helium. We would predict, therefore, that helium would behave more closely to an ideal manner than sulfur dioxide. Experimental evidence verifies our prediction. Helium remains a gas until it is cooled to $-269°C$. Sulfur dioxide liquefies when it is cooled to only $-10°C$. The molecules of an ideal gas are considered to be "point masses," occupying no volume. In theory, the sample of ideal gas would continue to contract in volume with increased pressure and reduced temperature until a volume of zero was reached.

4-8 Temperature and Its Measurement Before investigating the relationship between the volume and temperature of a gas it will be instructive to discuss the concepts of temperature and temperature scales. In general, we associate high temperatures with hot objects and low temperatures with cold objects. Whenever a hot object is placed in contact with a cold object, the cold one always becomes warmer. No one has ever observed a hot object becoming hotter when placed in contact with a colder object. We interpret these observations by saying that *heat (thermal energy) "flows" from regions of high temperature to regions of low temperature. Accordingly, temperature may be thought of as a number which indicates the direction that heat (thermal energy) will flow between two different systems.* For example, when you sip coffee having a temperature of 90°C, you can be sure that heat (thermal energy) will be transferred to your mouth (36°C) and that the coffee will feel hot. The magnitude of the temperature may be thought of as a measure of the degree of "hotness" or the "intensity of heat." If you wish to increase the temperature of your coffee, you place it over a flame. Thermal energy (heat) flows into the coffee, the kinetic energy of its molecules increases and the temperature rises. Thus, the measurable temperature is a reflection of the kinetic energy of the molecules of the system. When two systems reach the same temperature, there is no net flow of thermal energy between them and they are in ***thermal equilibrium.*** This means that when two systems are at the same temperature, the average kinetic energy of their molecules is equal, regardless of whether the systems consist of solid, liquid, or gaseous phases, or all three.

All samples of matter are influenced by changes in temperature. The slight expansion of mercury with increasing temperature makes it a useful liquid in thermometers. A mercury thermometer consists of a sealed capillary tube connected to a bulb filled with mercury. The fixed points of 0°C and 100°C represent, respectively, (a) the height of the mercury column when the bulb is immersed in an ice-water mixture and (b) the height when surrounded by the vapor of water boiling at 1 atm pressure. The distance between the fixed points is divided into 100 equal spaces, each space representing 1C°. A change from 8°C to 9°C does not represent quite the same change in temperature as the change from 85°C to 86°C. This is because the length of the mercury column does not change linearly with temperature. In fact, no two liquids expand uniformly with temperature. For example, a temperature reading of 20°C on a mercury thermometer does not represent exactly the same degree of "hotness" as a reading of 20°C on an alcohol thermometer. Ordinarily, this small amount of error is ignored in making temperature measurements (Fig. 4-12).

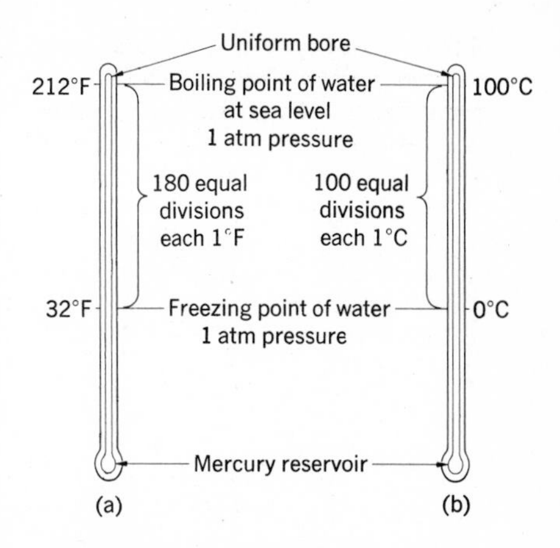

4-12 Relationship between the Celsius and Fahrenheit temperature scales. Perhaps you can develop a formula showing the relationship between the two scales.

Now that we have defined a temperature scale, let us determine whether volumes of gases vary linearly with temperature changes.

4-9 Variation of Gas Volume with Temperature Changes The relationship between gas volumes and temperatures may be determined experimentally by using a gas thermometer. In such a thermometer, the change in volume of a gas sample can be related to a change in temperature. With this apparatus, Fig. 4-13, the volume occupied by the gas at various Celsius temperatures can be determined. Typical data are in Table 4-2. The Kelvin temperature (T) is obtained by adding 273 to the Celsius temperature (t).

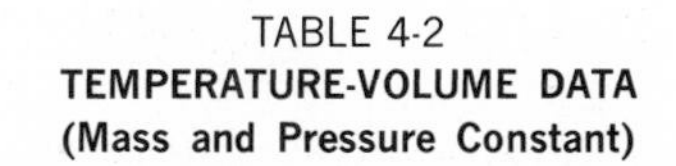
TABLE 4-2
TEMPERATURE-VOLUME DATA
(Mass and Pressure Constant)

Volume (V)	*Celsius Temperature (t)*	*Kelvin Temperature (T)*	*V/T*
1.55 liters	150.0°C	423.0°K	
1.37 liters	100.0°C	373.0°K	
1.18 liters	50.0°C	323.0°K	
1.00 liters	0.00°C	273.0°K	
0.815 liters	− 50.0°C	223.0°K	
0.633 liters	− 100.0°C	173.0°K	

In general, the expansion of the mercury is caused by the increase in atomic motion and the corresponding increase in interatomic distance.

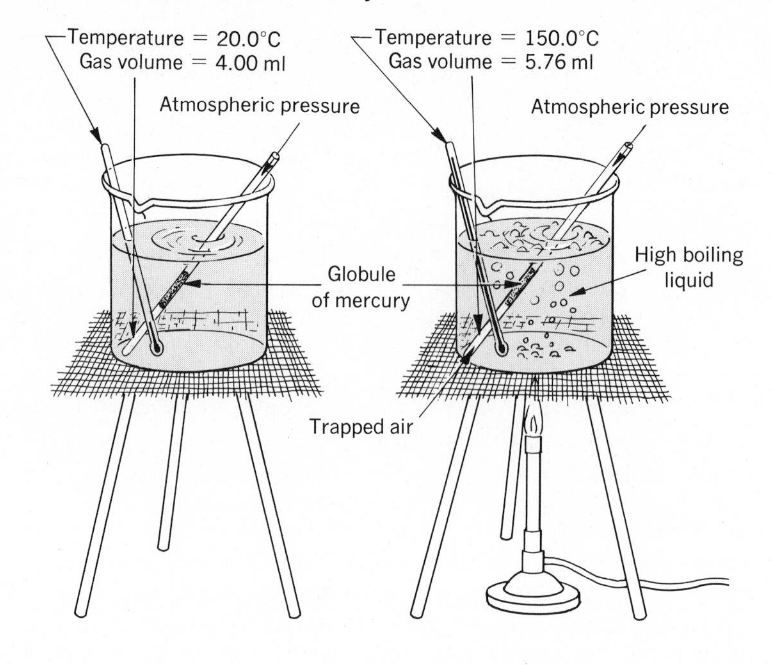

4-13 A simple Charles' law apparatus for student experimentation. A constant mass of gas is trapped under the globule of mercury. When the system is heated, the liquid mercury moves so as to keep a constant pressure. For each temperature, the corresponding volume is indicated by the marks on the tube. That is, the ratio of the lengths of the gas column at the two temperatures is equal to the ratio of the volumes. Explain why this is a valid assumption.

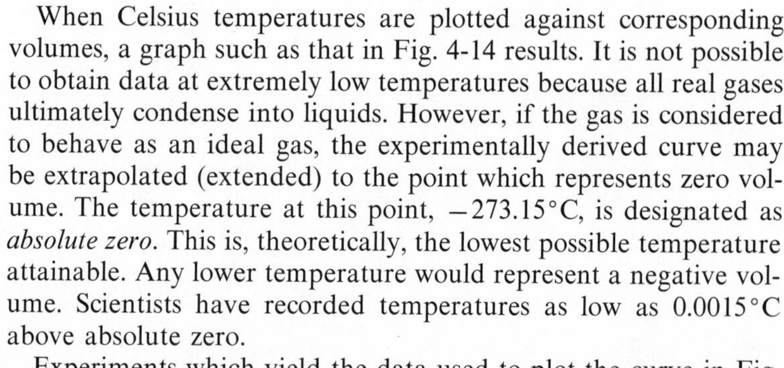

When Celsius temperatures are plotted against corresponding volumes, a graph such as that in Fig. 4-14 results. It is not possible to obtain data at extremely low temperatures because all real gases ultimately condense into liquids. However, if the gas is considered to behave as an ideal gas, the experimentally derived curve may be extrapolated (extended) to the point which represents zero volume. The temperature at this point, −273.15°C, is designated as *absolute zero*. This is, theoretically, the lowest possible temperature attainable. Any lower temperature would represent a negative volume. Scientists have recorded temperatures as low as 0.0015°C above absolute zero.

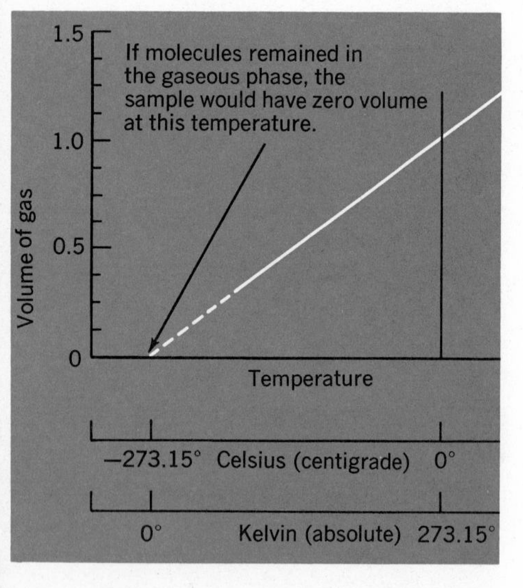

4-14 The straight line graph indicates that the variables V and T are directly proportional to each other. That is $V/T = k$. Why is it not possible to collect data at the extreme left of this graph?

Experiments which yield the data used to plot the curve in Fig. 4-14 reveal that when the temperature of a given mass of any gas at constant pressure is changed from 0°C to 1°C, the volume increases by $\frac{1}{273}$ or 0.37% of its original value. That is,

$$\mathbf{V_1 = V_0 + \frac{1}{273}V_0} \qquad \mathbf{V_1 = V_0 + \frac{t_1}{273}V}$$

$$\mathbf{V_1 = V_0\left(1 + \frac{t_1}{273}\right)} \qquad \mathbf{V_1 = V_0\left(\frac{273 + t_1}{273}\right)} \tag{4-3}$$

where V_0 is the original volume at 0°C, V_1 is the new volume at t_1, the new Celsius temperature. When the gas is heated from 0°C to 2°C, the volume is observed to increase by $\frac{2}{273}$ of its original value. That is,

$$\mathbf{V_2 = V_0 + \frac{2}{273}V_0} \qquad \mathbf{V_2 = V_0 + \frac{t_2}{273}V_0}$$

$$V_2 = V_0\left(1 + \frac{t_2}{273}\right) \qquad V_2 = V_0\left(\frac{273 + t_2}{273}\right) \qquad 4\text{-}4$$

The ratio of the volumes of a gas at two different Celsius temperatures may be obtained by dividing Equation 4-3 by Equation 4-4. The result is

$$\frac{V_1}{V_2} = \frac{V_0\left(\frac{273 + t_1}{273}\right)}{V_0\left(\frac{273 + t_2}{273}\right)} = \frac{273 + t_1}{273 + t_2} \qquad 4\text{-}5$$

Equation 4-5 defines a temperature scale in which all Celsius temperatures are increased by 273°. This *absolute temperature* scale is known as the Kelvin scale in honor of Lord Kelvin (1824–1907), an English physicist. Degrees on the scale are called *degrees Kelvin* °K, and are expressed as

$$T = 273° + t°C$$

Substituting T_1 for $273 + t_1$ and T_2 for $273 + t_2$ in Equation 4-5 yields

$$\frac{V_1}{V_2} = \frac{T_1}{T_2} \qquad 4\text{-}6$$

Equation 4-6 shows that for a fixed amount of any ideal gas at constant pressure the ratio of two temperatures on the Kelvin scale is equal to the ratio of the volumes of the gas at those temperatures. This principle is the basis for the gas thermometer, which can be used to define the temperature scale and to calibrate Celsius thermometers. The straight-line graph which intersects the axis at −273°C indicates that the volume and Kelvin temperature are directly proportional to each other. When a series of simultaneous values for two variables are directly proportional to each other, their quotient is a constant. You may wish to complete Table 4-2 and verify this relationship. This relationship, discovered in 1787, is known as the ***Law of Charles (1746–1823) and Gay-Lussac,*** after two French scientists. It may be expressed mathematically as $V/T =$ *a constant.* This constant is not the same as the PV constant. On the basis of our analysis, we may conclude that **the volume of a given mass of gas at constant pressure is directly proportional to its Kelvin temperature.**

It is important to note that the mathematical relationships which represent gas laws and other fundamental laws of nature are simplified by using the Kelvin scale. In all gas problems involving a temperature factor, the Celsius temperature should be converted to Kelvin temperature by adding 273. We shall use 273°K as the standard temperature. Thus, *standard conditions of temperature and pressure (STP) are 273°K and 1 atm* (760 mm of Hg or 760 torr) respectively (Fig. 4-15).

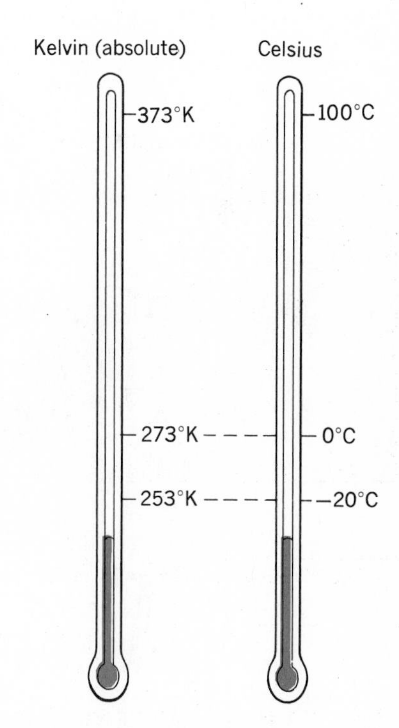

4-15 Relationship between the Celsius and the Kelvin temperature scales. Can you determine absolute zero on the Fahrenheit scale?

Example 4-2
A gas occupies 50.0 ml at standard temperature. What volume will it occupy at 335°C, pressure unchanged?

Solution
1. The data should first be itemized. Be sure to convert the Celsius temperature to Kelvin.

$V_1 = 50.0$ ml $\qquad T_1 = 273°K$
$V_2 =$ unknown $\qquad T_2 = 335 + 273 = 608°K$

2. The data show that there is an increase in temperature from the initial to the final state of the system. This means that the initial volume will increase.
3. The temperature factor must be a ratio greater than 1. Use the initial and final temperatures to set up a ratio greater than 1. The actual ratio is

$$608°K/273°K$$

4. Multiply the original volume by the temperature factor.

$$V_2 = 50.0 \text{ ml} \times \frac{608°K}{273°K} = 111 \text{ ml}$$

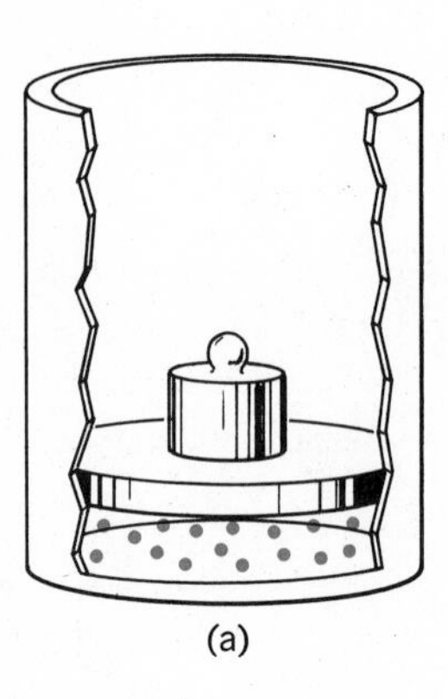

(a)

4-16 The effect of temperature change on the volume of a gas; at (a) above, when the temperature is lowered, and at (b) below, when the temperature is raised.

FOLLOW-UP PROBLEMS

1. Convert these temperatures to the Kelvin scale: (a) −40°C, (b) 37°C, (c) 100°C, (d) 273°C.

2. Convert these temperatures to the Celsius scale: (a) 40°K, (b) 273°K.

3. A gas occupies 515 ml at 27°C. What volume will it occupy at standard temperature? **Ans. 468 ml.**

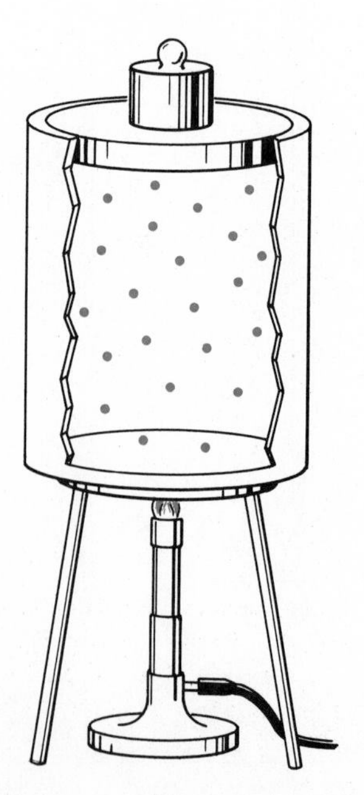

(b)

Let us interpret the preceding problem in terms of the Kinetic Molecular Theory. Imagine that a fixed number of molecules is confined inside a cylinder with a freely-moving piston whose mass is negligible, as shown in Fig. 4-16. At the lower temperature, the molecules move relatively slowly and exert a certain pressure on the inside walls of the cylinder. As the cylinder is warmed, the molecules increase in speed and strike the walls more frequently and more vigorously. This forces the movable piston upward so that the volume inside the cylinder is increased. The piston continues to rise until the pressure exerted by the molecules on the inside again balances the atmospheric pressure.

The cylinder at the higher temperature is shown in Fig. 4-16(b). The relationships between several factors in Fig. 4-16 are compared in Table 4-3. In both illustrations, the piston is free to move. Therefore the pressure both inside and outside the cylinder is always 1 atm. The greater velocity of the molecules in Fig. 4-16(b) is just offset by the greater intermolecular distance. Thus, at the higher temperature, there are fewer collisions per unit area.

4-10 Variation of Gas Pressure with Temperature Changes If the volume of the container were fixed, as shown in Fig. 4-17(a),

TABLE 4-3

RELATIVE MAGNITUDE OF FACTORS USED TO DESCRIBE BEHAVIOR OF GASES AT TWO DIFFERENT TEMPERATURES

Factor	*Fig. 4-16(a)*	*Fig. 4-16(b)*
Pressure	equal	equal
Number of molecules	same	same
Temperature	low	high
Velocity of molecules	low	high
Density of gas	high	low
Distance between molecules	small	great
Total volume and wall area	small	great
Collisions per unit area	great	small

then an increase in temperature would result in an increase in pressure. At the higher temperature, the molecules not only strike the walls more vigorously but also more frequently. Laboratory experiments show that *the pressure of a gas at constant volume is directly proportional to the Kelvin temperature.* That is, doubling the Kelvin temperature of a given gas at constant volume doubles the pressure. The relationship may be expressed mathematically as

$$\frac{P_1}{T_1} = \frac{P_2}{T_2} \quad \text{or} \quad \frac{P}{T} = \text{a constant Fig. 4-17(b).}$$

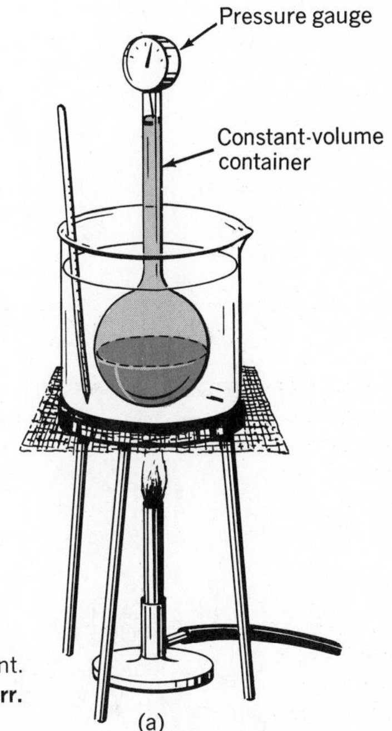

FOLLOW-UP PROBLEM

A gas at 27.0°C exerts a pressure of 745 torr. What pressure is exerted by the same gas at 0°C? Volume and mass are held constant.
Ans. 678 torr.

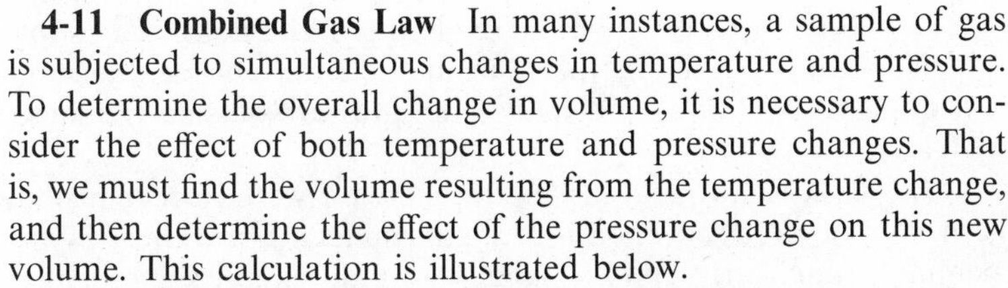

4-11 Combined Gas Law In many instances, a sample of gas is subjected to simultaneous changes in temperature and pressure. To determine the overall change in volume, it is necessary to consider the effect of both temperature and pressure changes. That is, we must find the volume resulting from the temperature change, and then determine the effect of the pressure change on this new volume. This calculation is illustrated below.

Suppose we have a 68.0-ml sample of carbon dioxide gas at 30.0°C and 725 torr. Let us find the volume of this sample at STP. We shall first calculate what volume this sample occupies at 273°K and 725 torr. Since a decrease in temperature decreases the volume, the new volume is

$$68 \text{ ml} \times \frac{273°\text{K}}{303°\text{K}} = 61.3 \text{ ml}$$

This 61.3-ml sample is at 273°K and 725 torr. Next we calculate the effect of increasing the pressure to 760 torr. Since an increase in pressure decreases the volume, the final volume is

$$61.3 \text{ ml} \times \frac{725 \text{ torr}}{760 \text{ torr}} = 58.5 \text{ ml}$$

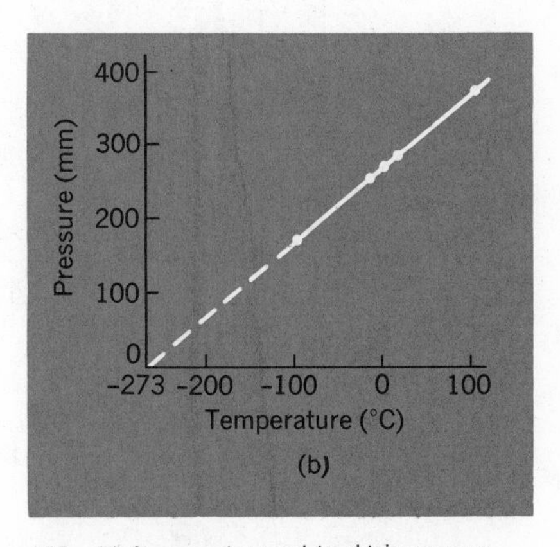

4-17 (a) An apparatus used to obtain pressure-temperature data. (b) The pressure-temperature graph for an ideal gas. What does the graph reveal about the relationship between the variables *P* and *T*?

The one-step solution is

$$\mathbf{68.0\ ml} \times \frac{\mathbf{273°K}}{\mathbf{303°K}} \times \frac{\mathbf{725\ torr}}{\mathbf{760\ torr}} = \mathbf{58.5\ ml}$$

In solving this problem, we use a combination of Boyle's and Charles' Laws. It can be shown experimentally that for a constant mass of gas, PV/T = a constant. The value of the constant depends on the mass or the number of moles of gas. This so-called Combined Gas Law is sometimes mathematically expressed as

$$\frac{P_1V_1}{T_1} = \frac{P_2V_2}{T_2} \qquad 4\text{-}7$$

Thus, the answer to the preceding problem could have been obtained by substituting the itemized data in Equation 4-7.

FOLLOW-UP PROBLEM

A gas occupies 68.0 ml at standard temperature and pressure. What volume will it occupy at 30.0°C and 725 torr? **Ans. 79.0 ml.**

4-12 Variation in Moles of Gas with Changes in Pressure and Temperature Earlier we remarked that the value of the PV constant depends on the temperature and the number of molecules of gas. The number of molecules may be expressed in terms of the *mass of gas* or the *moles of gas.* Let us examine a system with a fixed volume in which the number of molecules (mass or moles) of gas may vary with changes in temperature and pressure.

Consider a 5.00-liter flask equipped with a rubber stopper containing a fine capillary tube. At STP this flask holds 7.14 g of oxygen. What changes take place in the mass of oxygen when we heat this flask to 120°C and reduce the pressure outside the flask to 700 torr?

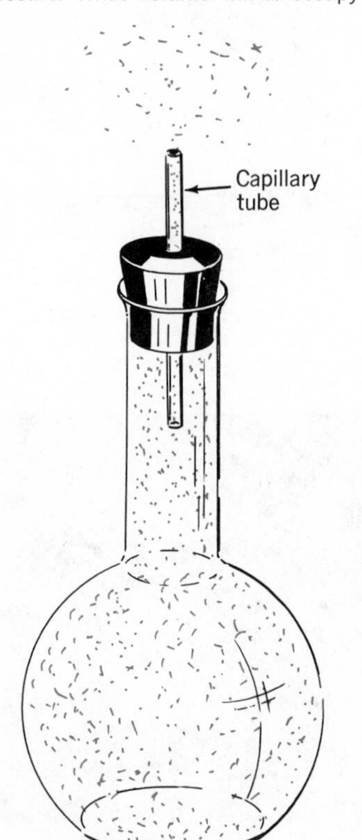

4-18 As the temperature is increased and the outside pressure reduced, molecules escape through the capillary tube until the pressure within the flask equals that outside the flask.

As the flask is warmed, a pressure increase would occur inside the flask if it were not open to the atmosphere. Since there is an opening through the capillary tube, we would expect some of the molecules to escape. In addition, the decrease in external pressure from 760 torr to 700 torr would also cause molecules to escape through the capillary tube, since the pressure inside and outside the flask tend to equalize (Fig. 4-18).

Experiments show that the original mass of 7.10 g of oxygen would be diminished to 4.50 g as a result of the temperature and pressure changes. The mass of oxygen present in the final state of the system can be calculated thus:

$$\mathbf{7.14\ g} \times \frac{\mathbf{273°K}}{\mathbf{393°K}} \times \frac{\mathbf{700\ torr}}{\mathbf{760\ torr}} = \mathbf{4.57\ g}$$

If the mass had been expressed in moles, the same principles and procedures would apply. They indicate that **at constant volume and**

temperature, the pressure is directly proportional to the number of moles (*n*) of gas. This may be expressed mathematically as

$$\mathbf{P = n \times Constant}$$

Where the constant depends on the volume and temperature. Similarly, the experiment above shows an *inverse relationship between Kelvin temperature and number of moles at constant volume* and may be expressed as

$$\mathbf{T = C/n \quad or \quad nT = C}$$

where C is a constant that depends on pressure and volume.

FOLLOW-UP PROBLEM

An apparatus similar to that described above contains 2.25 g of methane (CH_4) at 120°C and 700 torr. (a) What mass of this gas does the flask hold at STP? (b) How many moles of methane does the flask hold at STP? **Ans. (a) 3.51 g, (b) 0.219 mole.**

4-13 The Equation of State for Gases The equation of state for gases shows how the variables which determine the state of the system are related. The four variables, pressure (P), volume (V), Kelvin temperature (T), and number of moles (n) are related by a general equation which may be derived from the individual gas laws or from the Kinetic Molecular Theory.

We shall give a simplified derivation based on the individual gas laws. The following is a summary of relationships derived from experimental observations.

1. At constant temperature, the volume of a fixed mass (number of moles) of gas is inversely proportional to the pressure. This relationship may be expressed mathematically as

$$\mathbf{V \propto \frac{1}{P}}$$

where $\propto$ is a symbol for proportionality.

2. At constant pressure, the volume of a fixed mass of gas is directly proportional to its Kelvin temperature, T. That is,

$$\mathbf{V \propto T}$$

3. From Avogadro's Principle that equal volumes of gases under the same conditions of temperature and pressure contain the same number of molecules (or moles), it follows that at constant temperature and pressure, the volume of a gas is directly proportional to the number of molecules. We can express this relationship as

$$\mathbf{V \propto n}$$

where n is the number of moles of gas.

Mathematically, it is true that a quantity which is proportional to each of three separate quantities is proportional to their product. Therefore

$$\mathbf{V \propto \frac{1}{P} \times T \times n}$$

It is also mathematically true that a proportion may be changed to an equality by inserting a constant. It follows that

$$\mathbf{V = a\ constant \times \frac{1}{P} \times T \times n}$$

which can be rearranged to give

$$\mathbf{PV = n \times constant \times T}$$

This constant, known as the gas constant, is symbolized by the letter R which has the same value for all gases whose behavior approaches that of an ideal gas. Substituting R in the preceding equation yields

$$\mathbf{PV = nRT} \qquad \text{4-8}$$

Equation 4-8 is known as the ***Ideal Gas Equation.*** In this equation, R is called the ***universal gas constant.*** The value of R depends on the dimensions used for pressure, volume, and temperature. To evaluate R, it is necessary to have available simultaneous values for n, P, T, and V. A convenient set of values are standard conditions of temperature and pressure, 1 mole of gas, and the volume of a mole of gas at STP. The molar volume of the ideal gas is obtained from measurements made on real gases at high temperatures and low pressures, since under these conditions they behave closely to an ideal manner. The calculation of 22.414 liters/mole is then made by extrapolating the volume to STP by using Boyle's and Charles' Laws.

Solving Equation 4-8 for R and substituting standard condition values of P, V, and T for 1 mole of gas yields

$$\mathbf{R = \frac{22.4\ \ell \times 1\ atm}{1\ mole \times 273^\circ K} = 0.082\ \ell\ atm/mole\ ^\circ K}$$

The value of 0.082 ℓ atm/mole °K may be used to solve problems with the Ideal Gas Law provided you express volume in liters, pressure in atm, n in moles, and temperature in °K.

Note that the individual gas laws and relationships between pairs of variables can be deduced from the Ideal Gas Equation. For example, if we wish to determine the pressure-volume relationship (Boyle's Law), we observe that n and T are constant. Since R is also constant, then PV = constant and the two factors (P, V) are inversely proportional to each other. Use the Ideal Gas Equation and derive a law mathematically (a) when P is constant, and (b) when V is constant.

TABLE 4-4
MOLAR VOLUMES OF COMMON GASES

Gas	*Molecular Mass*	*Molar Volume (liters, S.T.P.)*
Hydrogen	2.016	22.430
Helium	4.003	22.426
Ideal gas	—	22.414
Oxygen	32.00	22.392
Methane	16.043	22.360
Chlorine	70.914	22.063
Sulfur dioxide	64.066	21.888

Example 4-3
Sixteen grams of oxygen gas is introduced into an evacuated 10.0-l flask at 77.0°C. What is the pressure in atmospheres in the container?

Solution

1. In this problem, only one state of the system is given. The Ideal Gas Equation, PV = nRT, may therefore be used. However, in order to use the value of 0.0820 for *R*, it is first necessary to express the mass in moles and the temperature in °*K*.

$$n = 16.0 \text{ g} \times 1 \text{ mole}/32.0 \text{ g} = 0.500 \text{ mole}$$
$$T = 77.0°C + 273 = 350°K$$
$$V = 10.0 \ \ell$$

2. Substituting these values in PV = nRT and solving for P yields

$$P = \frac{0.500 \text{ mole} \times 0.082 \ \ell \text{ atm} \times 350°K}{10.0 \ \ell \text{ mole } °K} = 1.43 \text{ atm.}$$

Pressure measurements in many laboratories are expressed in torr rather than in atmospheres. Thus, it is often convenient to express the general gas constant as

$$R = \frac{62.3 \text{ l torr}}{°K \text{ mole}}$$

or

$$R = \frac{6.23 \times 10^4 \text{ ml torr}}{°K \text{ mole}}$$

FOLLOW-UP PROBLEM

Fifty-six grams of nitrogen is introduced into an evacuated 20.0-liter flask at −73.0°C. What is the pressure (a) in atmospheres and (b) in torr in the container?

Ans. (a) 1.64 atm, (b) 1.25×10^3 torr.

Equation 4-8 can be modified to show a relationship unique to gases between density and molecular mass. The number of moles of substance may be expressed as

$$\mathbf{n = g/molecular\ mass\ (m.m.)}$$

Substituting this value in Equation 4-8 gives

$$\mathbf{PV = gRT/m.m.}$$

Rearranging gives

$$\mathbf{P = \frac{g}{V}\frac{RT}{m.m.}}$$

Substituting density, *d*, for *g*/*V*, we obtain

$$\mathbf{P = dRT/m.m.} \qquad 4\text{-}9$$

Thus, the molecular mass of a gas may be determined experimentally by measuring its density. Equation 4-9 reveals that density and pressure are directly proportional where as density and Kelvin temperature are inversely proportional.

Example 4-4

The density of a certain gas at 27.0°C and 740 torr is 2.53 g/l. Assuming that the gas behaves ideally, calculate its molecular mass.

Solution

This problem may be solved by using Equation 4-9 or by correcting the given density to standard conditions and multiplying by the standard gram-molecular volume of an ideal gas. Note that when pressure is expressed in torr and volume in liters, R has a value of 62.3 l torr/mole °K.

Rearranging Equation 4-9, gives $\text{m.m.} = \frac{dRT}{P}$

Thus, $\text{m.m.} = \frac{2.53\text{ g}}{\text{l}} \times \frac{62.3\text{ l torr}}{\text{mole °K}} \times \frac{300\text{°K}}{740\text{ torr}} = 64.0\text{ g/mole.}$

FOLLOW-UP PROBLEM

A certain gas is found to have a density of 0.16 g/ℓ at 25°C and 745 torr. What is the molecular mass of the gas?

Ans. 4.0 g/mole.

4-14 Dalton's Law of Partial Pressures Many samples of gases we encounter are mixtures. The atmosphere is a familiar example. It was John Dalton who, during his study of atmospheric gases, first observed a relationship between the total pressure of a mixture of gases and the individual pressures exerted by the separate components. Dalton found that when several nonreacting gases were mixed, the pressure of the mixture was equal to the sum of the individual pressures exerted by each gas alone at the given temperature. For example, at 0°C, 44 g of carbon dioxide and 32 g of oxygen each exert a pressure of 1.0 atm if they are placed in separate 22.4-ℓ containers. When they are both placed in the same 22.4-ℓ container at 0°C, the pressure exerted by the mixture is 2.0 atm. The pressure exerted by each component in a mixture is called the ***partial pressure*** of that component. If we assume all components in a mixture are at the same temperature and occupy the same volume, then the differences in the partial pressures of the components are related to the differences in the number of moles of each component (Fig. 4-19).

The Ideal Gas Equation (4-8) can be applied to determine the pressure of each gas in the mixture. The total pressure of the mixture can be determined by adding the partial pressures of all of the gases. If just the total pressure of the mixture is required, then it can be determined directly by using the Ideal Gas Equation,

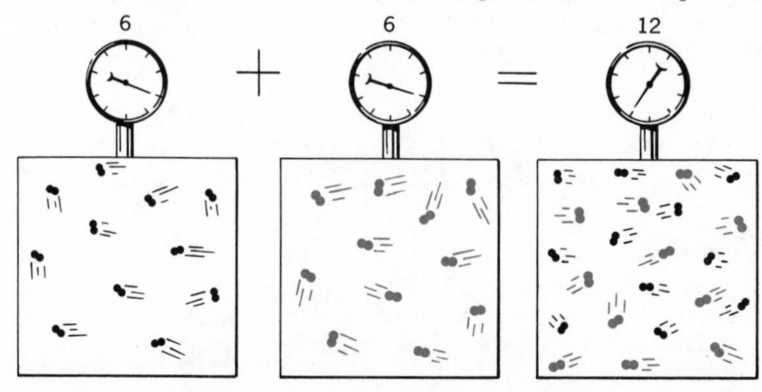

4-19 The pressure of a mixture of ideal gases is the sum of the pressures that each gas would exert if it were the only gas present in the container.

letting n equal the total moles of all gases present. Dalton's Law may be expressed as

$$\mathbf{P_T = P_a + P_b + P_c + \cdots}$$

when P_T is the total pressure of the mixture and P_a, P_b and P_c are the partial pressures of the components. In other words, the law states that the *total pressure of a mixture of gases is the sum of the partial pressures of the components.*

Example 4-5
Determine (a) the total pressure and (b) the partial pressure of hydrogen and oxygen in a gaseous mixture composed of 1.00 g of H_2 and 32.0 g of O_2. The mixture is contained in a 22.4-liter container at 27.0°C.

Solution
(a) 1. Determine the number of moles of each gas.

$$H_2 = 1.00\text{ g} \times 1.00\text{ mole}/2.00\text{ g} = 0.500\text{ mole}$$
$$O_2 = 32.0\text{ g} \times 1.00\text{ mole}/32.0\text{ g} = 1.00\text{ mole}$$

2. Determine the total moles of all gases by adding the moles of each component.

$$0.500\text{ mole } H_2 + 1.00\text{ mole } O_2 = 1.50\text{ mole}$$

3. Substitute n = 1.50 mole, V = 22.4 l, and T = 300°K in the Ideal Gas Equation and solve for P_T (Total Pressure).

$$P_T = \frac{1.50\text{ mole} \times 0.0820\ \ell\text{ atm/mole °K} \times 300\text{°K}}{22.4\ \ell}$$

$$= 1.65\text{ atm}$$

(b) 1. Repeat the calculation using n = 0.500 mole.

$$P_{H_2} = \frac{0.500\text{ mole} \times 0.0820\ \ell\text{ atm/mole °K} \times 300\text{°K}}{22.4\ \ell}$$

$$= 0.550\text{ atm}$$

$$P_{O_2} = P_T - P_{H_2} = 1.65\text{ atm} - 0.550\text{ atm} = 1.10\text{ atm.}$$

The fraction of the total pressure exerted by each gas is equal to the fraction of the molecules of that gas in the sample. In the preceding problem, there is a total of 1.5 mole. The fraction of O_2 molecules in the mixture is

$$\frac{\textbf{1.0 mole}}{\textbf{1.5 mole}} = \frac{2}{3}$$

and the fraction of H_2 molecules is

$$\frac{\textbf{0.5 mole}}{\textbf{1.5 mole}} = \frac{1}{3}$$

This means that $\frac{2}{3}$ of the total pressure is caused by O_2, and $\frac{1}{3}$ by

ROBERT BOYLE
1627–1691

Robert Boyle was the seventh son of Richard Boyle, the Earl of Cork, Ireland. Robert showed an early interest in natural philosophy, religion, and languages. He learned Latin and French before entering Eton at eight years of age. When he was eleven years old, Boyle, like many boys from wealthy families of England, took a "grand tour" of Europe. This tour lasted for eight years. While in Switzerland, France, and Italy, he studied at various universities. In 1654, after returning from Europe, he moved to Oxford where he became a member of the "Invisible College," a discussion group devoted to the study of the natural sciences. This group was later to become the Royal Society. Boyle served as president of this society from 1680 until his death in 1691.

In Boyle's most famous work, "The Sceptical Chymist," he gives a very modern definition of an element and distinguishes between an element and a compound. Boyle was an experimental chemist who introduced the word *analysis* to describe an investigative experiment. He used vegetable juices as *indicators* and carried out fractional distillation of various solutions. As a result of his experiments, he discovered methyl alcohol and acetone.

Boyle defined the function of chemistry as an "investigation of properties and substances through experimental means." He often admitted his skepticism of his own theories as well as those of others. He said, "The chemists view their task as the preparation of medicines and the extraction and transmutation of metals. I have tried to deal with chemistry from a quite different viewpoint, not as a physician or an alchemist, but as a philosopher. If men had the progress of true science more at heart than their own interests, one could easily persuade them that they would render the greatest possible service to the world by devoting all their effort to setting up experiments and making observations and not by proclaiming any theories without having tested the relevant phenomena."

H_2. These fractions are known as *mole fractions,* and they have no units. Thus, the partial pressure of a gas in a mixture may be expressed as

$$\textbf{partial pressure} = \textbf{mole fraction} \times \textbf{total pressure}$$

The mole fraction of a given component may be expressed as

$$\textbf{mole fraction of component} = \frac{\textbf{moles of component}}{\textbf{total moles of all components}}$$

FOLLOW-UP PROBLEM

A gaseous mixture consists of 11.0 g of CO_2 and 48.0 g of O_2. The volume of the container is 22.4 liters, and the temperature is 273°C. Calculate (a) the moles of each gas, (b) the total moles of all gases, (c) the mole fraction of each gas, (d) the partial pressure of each gas, (e) the total pressure of the mixture.
Ans. (a) CO_2 = 0.250 mole, O_2 = 1.50 mole, (b) 1.75 mole, (c) CO_2 = 0.14, O_2 = 0.86, (d) CO_2 = 0.50 atm, O_2 = 3.0 atm, (e) 3.5 atm.

Dalton's Law is often applied in experiments involving the measurement of gases produced in a chemical reaction. Frequently, the gas is collected by bubbling the gas through water into a flask or gas-collection tube filled with water as shown in Fig. 4-20.

Collecting a gas by displacing water means that the sample of gas is "wet," or saturated with water vapor. Thus, the total pressure of the "wet" sample which can be measured will be the sum of the partial pressure of the dry sample plus the partial pressure of the water vapor. To find the pressure of the dry gas alone, subtract the partial pressure (vapor pressure) of the water vapor from the total pressure. Water-vapor pressures vary with temperature, so it is necessary to consult a table, such as Table 4-4, to find the data needed. These data can then be used as shown in Example 4-6 to determine the amount of gas collected.

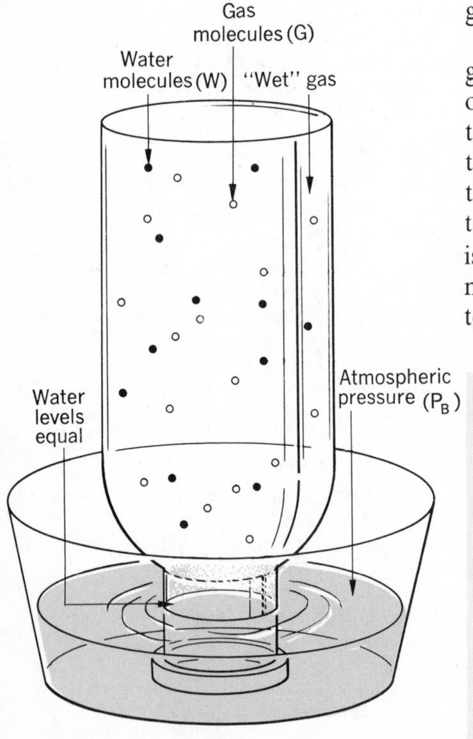

4-20 The total pressure inside the gas collection bottle is the sum of the partial pressures of the gas collected and the water vapor.

Example 4-6
When 81.4 ml of hydrogen is collected by displacement of water, the water levels inside and outside the gas-collection vessel are equal. This means the pressure of the mixture of gas in the vessel is equal to barometric pressure. The barometric pressure is 740.0 torr and the temperature is 23.0°C. Find the number of moles of hydrogen gas in this sample.

Solution
1. The pressure inside the tube is the sum of the partial pressures of the two components, hydrogen gas and water vapor. Table 4-5 shows that the partial pressure of the water vapor at 23.0°C is 21.1 torr. The partial pressure of the hydrogen is

$$740.0 \text{ torr} - 21.1 \text{ torr} = 718.9 \text{ torr}$$

2. Solving Equation 4-8 for n and substituting the data yield

$$n = \frac{PV}{RT} = \frac{719 \text{ torr} \times 81.4 \text{ ml}}{6.23 \times 10^4 \frac{\text{ml torr}}{°\text{K mole}} \times 296°\text{K}}$$

$$n = 3.17 \times 10^{-3} \text{ mole.}$$

FOLLOW-UP PROBLEM

A student collects 41.4 ml of oxygen gas at 745 torr and 26.0°C by displacement of water. The water levels inside and outside the collection tube are equal. (a) How many moles of oxygen are present in this sample? (b) How many molecules are present?
Ans. (a) 0.0016 mole, (b) 9.6 × 10^{20} molecules.

Let us now consider an interesting application of the kinetic theory that helped solve an extremely important problem related to the production of fissionable material in the early days of the atomic era. The problem was how to separate mixtures of gases which have the same chemical properties but different masses. The practical solution to the problem was based on differences in density.

4-15 Graham's Law of Diffusion The fundamental principle underlying the separation noted above is that *at a given temperature the molecules of any ideal gas have the same average kinetic energy* ($\frac{1}{2}mv^2$). For two gases, A and B, at a given temperature the relationship may be expressed mathematically as

$$\tfrac{1}{2}\mathbf{m}_A\mathbf{v}_A^2 = \tfrac{1}{2}\mathbf{m}_B\mathbf{v}_B^2$$

$$\frac{\mathbf{v}_A^2}{\mathbf{v}_B^2} = \frac{\mathbf{m}_B}{\mathbf{m}_A}$$

$$\frac{\mathbf{v}_A}{\mathbf{v}_B} = \sqrt{\frac{\mathbf{m}_B}{\mathbf{m}_A}} \qquad \text{4-10}$$

where m_A and m_B represent the molecular masses of the two gases. Equation 4-10 is a mathematical expression of the principle proposed in 1829 by the Scottish chemist, Thomas Graham (1805–1869). The principle, known as ***Graham's Law of Diffusion,*** states that *under identical conditions, the rates at which different gases diffuse are inversely proportional to the square root of their densities* (or *molecular masses*).

We can determine the relative average velocities of hydrogen and oxygen molecules by substituting their molecular masses in Equation 4-10.

$$\frac{\mathbf{v}_{H_2}}{\mathbf{v}_{O_2}} = \sqrt{\frac{\mathbf{32.0}}{\mathbf{2.0}}} = \sqrt{\mathbf{16.0}} = \mathbf{4.0}$$

$$\mathbf{v}_{H_2} = \mathbf{4\ v}_{O_2}$$

This value indicates that the molecular velocity of H_2 is, on the average, 4 times that of O_2 (Fig. 4-21).

TABLE 4-5
EQUILIBRIUM WATER-VAPOR PRESSURES AT VARIOUS TEMPERATURES

Temperature (°C)	*Pressure* (torr)
5	6.5
10	9.2
15	12.8
16	13.6
17	14.5
18	15.5
19	16.5
20	17.5
21	18.7
22	19.8
23	21.1
24	22.4
25	23.8
26	25.2
27	26.7
28	28.3
29	30.0
30	31.8
40	55.3
50	92.5
60	149.4
70	233.7
75	289.1
78	327.3
80	355.1
85	433.6
90	525.8
95	633.9
100	760

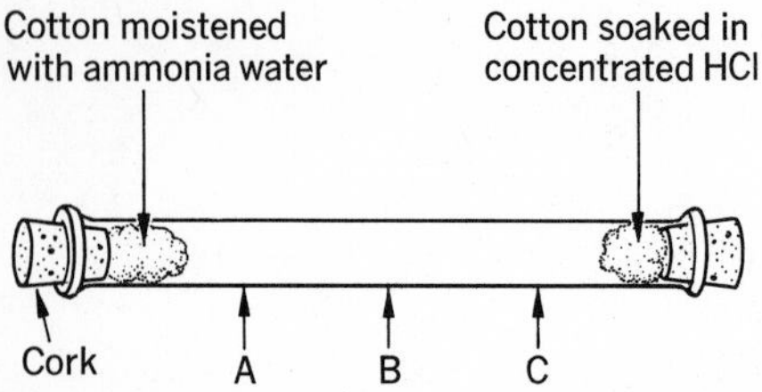

Cotton moistened with ammonia water

Cotton soaked in concentrated HCl

Cork A B C

4-21 NH_3 gas from a concentrated ammonia-water solution reacts with HCl gas from a concentrated hydrochloric acid solution and produces a white cloud of solid NH_4Cl. Does the cloud form near A, B, or C?

One of the properties of all gases is that they rapidly spread out to fill their containers. This tendency to move about rapidly and to intermingle readily as a result of molecular motion is known as ***diffusion.*** As indicated by the comparison of relative velocities of the hydrogen and oxygen molecules, gases diffuse at different rates. At a given temperature, a gas consisting of very light molecules, such as hydrogen, diffuses much more rapidly than a gas consisting of more massive molecules. A simple analogy will show qualitatively the logic of this statement. Imagine ten men who have identical times for running a mile. Now require five of them to carry a five-pound ball. At the end of the mile how would the runners be distributed? It is likely that the ones carrying the extra mass would lag.

This was the principle applied by scientists when they were faced with the problem of separating the isotopes of uranium in 1942. One isotope, uranium-238, has 3 more neutrons per atom than uranium-235. This means that the former is approximately 3 atomic-mass units heavier than the latter. In nature there is only one uranium-235 atom for every 140 uranium-238 atoms. When a mixture of molecules of gaseous UF_6 containing these atoms is passed through a series of porous barriers, the more massive molecules carrying the extra neutrons move more slowly and lag behind the lighter ones. Thus, the lighter molecules tend to emerge first. After many repetitions of this process, the lighter molecules become separated from the heavier ones. The gaseous diffusion method turned out to be the most efficient method of separating large amounts of the desired fissionable uranium-235 from uranium-238 during World War II. It is interesting to note that the gaseous diffusion plant at Oak Ridge, Tennessee, consumed more electric energy per year during the war years than all of France.

Gases that diffuse rapidly also *effuse* rapidly. ***Effusion*** refers to the rate at which a gas passes through a small hole called an orifice. Gaseous effusion may be used for determining the molecular masses of gases. In this instance the rate of effusion of a known gas is used as a reference. Gaseous effusion is illustrated in Example 4-7.

Example 4-7

A small bicycle pump is filled with helium (He) gas. With constant pressure, the gas is forced out through a small aperture in two seconds. The same pump is filled with hydrogen bromide (HBr) gas. Using the same pressure, how long will it take to force out this gas?

Solution

1. According to Graham's Law, $\frac{v_1}{v_2} = \sqrt{\frac{m_2}{m_1}}$, the molecular mass of He gas $= 4.0 = m_1$; the molecular mass of HBr gas $= 81 = m_2$.

2. Substituting molecular masses, we obtain the ratio

$$\frac{v_1}{v_2} = \sqrt{\frac{81}{4.0}} = \frac{9}{2} = 4.5$$

This indicates that He diffuses (or effuses) 4.5 times as fast as HBr. We may also say that HBr will take 4.5 times as long to effuse. The time required for the HBr to be pumped out would be 4.5 × 2.0 seconds or 9.0 seconds.

FOLLOW-UP PROBLEM

1. Calculate the relative rates of diffusion of $H_2(g)$ and $Br_2(g)$ at the same conditions.

Ans. H_2 diffuses 8.9 times as rapidly.

REACTIONS IN GASEOUS SYSTEMS

4-16 Application of the Ideal Gas Law to Reactions Involving Gases One reason for studying the physical behavior of gases is to gain an understanding of the nature and behavior of individual atoms and molecules in a macroscopic sample of matter. The Kinetic Molecular Theory, developed earlier, enabled us to interpret volume, pressure, and temperature relationships in terms of the behavior of individual molecules.

A second reason is to develop principles and generalizations which we can apply to chemical reactions involving gases. The Ideal Gas Equation can be used to help us predict the volume of gas which is evolved in a specific reaction under a given set of conditions. All we need to know is the temperature, the pressure, and the number of moles of gas evolved.

Example 4-8
Calculate the volume of CO_2 produced when 24.0 g of C is completely burned. All measurements are made at 745 torr and 25.0°C. The equation for the reaction is

$$C(s) + O_2(g) \longrightarrow CO_2(g)$$

Solution
1. Calculate the moles in 24.0 g C.

$$24.0 \text{ g C} \times \frac{1 \text{ mole C}}{12.0 \text{ g C}} = 2.00 \text{ moles C}$$

2. Calculate the moles of CO_2 produced. Use a conversion factor derived from the coefficients in the equation.

$$\frac{1 \text{ mole } CO_2}{\text{mole C}} \times 2.00 \text{ moles C} = 2.00 \text{ moles } CO_2$$

3. Change the temperature to °K. 25.0 + 273 = 298°K

4. Substitute $n = 2.00$ moles, $P = 745$ torr, and $T = 298°K$, and $R = 62.3$ ℓ torr/mole °K into the Ideal Gas Law equation and solve for the volume.

$$PV = nRT$$

$$V = \frac{nRT}{P} = \frac{(2.00 \text{ moles})\left(\frac{62.3 \text{ ℓ torr}}{\text{moles°K}}\right)(298°K)}{745 \text{ torr}}$$

$$= 49.8 \text{ liters}$$

Example 4-9

An 11.2-liter container is filled with hydrogen gas at STP. Then 40.0 g of liquid bromine (Br_2) is introduced and the mixture heated to 101°C. At this temperature the reaction

$$H_2(g) + Br_2(g) \longrightarrow 2HBr(g)$$

goes to completion. Calculate the final pressure in the container at 101°C.

Solution

1. Calculate the moles of each reactant.

$$1.00 \text{ g } H_2 \times \frac{1 \text{ mole } H_2}{2.00 \text{ g } H_2} = 0.500 \text{ mole } H_2$$

$$40.0 \text{ g } Br_2 \times \frac{1 \text{ mole } Br_2}{160 \text{ g } Br_2} = 0.250 \text{ mole } Br_2$$

2. Determine which reagent is in excess and the quantity in excess. The balanced equation shows that H_2 and Br_2 react in a 1:1 mole ratio. Therefore H_2 is in excess.

0.500 moles (orig) – 0.250 moles (used) = 0.250 moles H_2 in excess

3. Determine moles of product

$$0.250 \text{ moles } Br_2 \times \frac{2 \text{ moles HBr}}{1 \text{ mole } Br_2} = 0.500 \text{ moles HBr}$$

Therefore, the total moles of gas in the container at the end of the reaction is

$$0.250 \text{ moles } H_2 + 0.500 \text{ moles HBr} = 0.750 \text{ moles gas}$$

4. Substitute $n = 0.750$, $T = 374°K$, $V = 11.2$ ℓ and $R = 62.3$ ℓ torr/mole °K in $PV = nRT$ and solve for P.

$$P = \frac{nRT}{V} = \frac{(0.75 \text{ mole})\left(\frac{62.3 \text{ ℓ torr}}{\text{mole°K}}\right)(374°K)}{11.2 \text{ ℓ}}$$

$$= 1.55 \times 10^3 \text{ torr.}$$

Example 4-10

What volume of oxygen can be collected by displacement

of water at 745 torr and 23.0°C by the complete decomposition of 5.00 g of $KClO_3$? See Table 4-5 for water-vapor pressure.

$$2KClO_3(s) + \text{heat} \longrightarrow 2KCl(s) + 3O_2(g)$$

Solution

Calculate the moles of O_2 produced, substitute the data in $PV = nRT$, and solve for V.

$$5.00 \text{ g } KClO_3 \times \frac{1.00 \text{ mole } KClO_3}{122.6 \text{ g } KClO_3} \times \frac{3 \text{ moles } O_2}{2 \text{ moles } KClO_3} = 6.12 \times 10^{-2} \text{ moles } O_2$$

$$V = \frac{nRT}{P} = \frac{6.12 \times 10^{-2} \text{ moles}}{724 \text{ torr}} \times \frac{62.3 \text{ } \ell \text{ torr}}{\text{mole}^\circ K} \times 296^\circ K = 1.56 \text{ } \ell$$

FOLLOW-UP PROBLEMS

1. How many grams of magnesium are required to yield 1.12 ℓ of H_2 collected by displacement of water at 26.0°C and 748 torr?

$$Mg(s) + 2HCl(aq) \longrightarrow MgCl_2(aq) + H_2(g)$$

Ans. 1.06 g.

2. Calculate the volume of H_2 produced at 27°C and 740 torr when 48 g of Mg reacts with excess HCl. Equation is in Problem 1, above. **Ans. 50.5 ℓ.**

3. Calculate the final pressure in a 5.00-liter evacuated tank when 25.5 g of NH_3 gas is pumped in and allowed to react completely with 36.5 g of HCl at 85.0°C. Note that the product is a solid which does not increase the total pressure significantly.

$$NH_3(g) + HCl(g) \longrightarrow NH_4Cl(s)$$

Ans. 2.94 atm.

LOOKING AHEAD

In the first four chapters we have developed and applied the laws of chemical combination and the Kinetic Molecular Theory. These laws and theories were explained in terms of a simple model of atoms and molecules. The explanations did not require a knowledge of the composition of atoms. We have, however, observed many phenomena which cannot be explained in terms of Dalton's simple atomic model. The model did not explain the reason that atoms combine in certain ratios nor did it account for the forces of attraction existing between particles of matter. Also, there was no hint in Dalton's proposal of a possible relationship between electricity and matter. Furthermore it completely fails to explain the properties of substances and the observed similarities and regularities in the behavior of specific groups of elements.

Perhaps some of you are wondering why some substances such as sodium chloride (NaCl) are ionic solids while other substances such as sugar ($C_{12}H_{22}O_{11}$) are solids composed of neutral molecules. Others may be curious as to why some substances dissolve readily in a given solvent and others do not, or why some acids form highly ionic solutions while others form solutions containing relatively few ions.

We are now ready to seek explanations of these observations and phenomena in terms of the composition of the atoms and molecules which make up macroscopic samples of matter. In the next chapter we shall describe some of the experiments which revealed the general composition of atoms and led to the development of the modern concepts of atomic structure and behavior of electrons.

Special Feature: The Atmosphere

Certainly, one of the most important gaseous systems to you as an inhabitant of spaceship earth is the layer of air called the atmosphere. Let us briefly consider the evolution of our atmosphere, its composition, natural balances, and chemical processes which result in its pollution and a variation in its composition that poses a threat to animal and plant life.

The earth's early geological history is somewhat uncertain although most geologists agree that it was formed about four and one-half billion years ago and was originally in a liquid condition. After the planet solidified, great volcanic activity took place for hundreds of millions of years. Decomposition of minerals in the earth's interior formed vast quantities of gases which were spewed into the atmosphere. Similar processes are still taking place today. The gigantic eruptions at Krakatoa in 1883 and on Bali in 1963 blew many thousands of tons of gas into the atmosphere. The analysis of the gases from a modern Hawaiian volcano is shown in Table 4-6. No doubt, these are some of the gases which made up the solid earth's primitive atmosphere. It is also possible that high-temperature reactions involving carbon, hydrogen, and nitrogen in the minerals beneath the surface added ammonia (NH_3) and methane (CH_4) to the early atmosphere.

TABLE 4-6
ANALYSIS OF GASES FROM A HAWAIIAN VOLCANO

H_2O (steam)	79%
CO_2	12%
SO_2	6.5%
N_2	1.5%
Traces of H_2, CO, Cl_2, and Ar	

You will note that the gases given off from volcanos include no oxygen, the gas upon which all life depends. Most of our atmospheric oxygen was probably produced by photosynthesis in simple plants which evolved in the ocean. It was only here deep in the ocean that the sun's penetrating and lethal ultraviolet rays were sufficiently filtered out so that a living organism could survive. The production of oxygen in the atmosphere was extremely slow because most of that produced by the first organisms deep in the ocean was dissolved by the ocean water. Calculations indicate that when atmospheric oxygen had reached about one percent of its present level, sufficient ozone was produced in the upper atmosphere to filter out enough ultraviolet rays to permit oxygen-producing organisms to exist in the upper layers of the ocean. Ozone is a triatomic form of oxygen (a molecular allotrope) that is produced when ultraviolet rays from the sun strike ordinary oxygen gas. The equation for the production of ozone is

$$3O_2 \xrightarrow[\text{light}]{\text{ultraviolet}} 2O_3$$

Molecular allotropes **are composed of only a single kind of atom but differ in the structural arrangement of the atoms. Hence the allotropic forms of an element have different properties.**

Ozone has a particular ability to absorb ultraviolet rays from the sun. The present concentration of ozone in the upper atmosphere is just the proper amount to permit an intensity of ultraviolet light to reach the earth's surface which is optimal for both plants and animals. When the first ozone appeared in the upper atmosphere, photosynthesis began to take place on a relatively large scale. The earth's early CO_2-rich atmosphere was no doubt very conducive to plant growth. The photosynthesis reaction

$$nCO_2(g) + nH_2O(l) + \text{energy} \longrightarrow (CH_2O)_n(s) + nO_2(g)$$

tends to reduce the concentration of CO_2 in the atmosphere as well as increase the concentration of O_2.

By the time man appeared on earth, the oxygen concentration was about what it is today, approximately 21 percent. The respiratory action of animals tends to decrease the oxygen concentration in the atmosphere. The process can be represented as the reverse of photosynthesis.

$$\text{glucose} + \text{oxygen} \longrightarrow \text{water} + \text{carbon dioxide} + \text{energy}$$

Thus the photosynthetic process of plants and the respiratory process of animals both act to keep the composition of the atmosphere in equilibrium. The composition of our lower atmosphere is shown in Table 4-7.

TABLE 4-7
COMPOSITION OF THE ATMOSPHERE

Gases	*Percentage by Volume*
Nitrogen	78.0900
Oxygen	20.9400
Argon	0.9300
Carbon dioxide	0.0318
Other noble gases	0.002428
Methane	0.00015
Hydrogen	0.00005
Nitrogen oxides	0.0000251
Carbon monoxide	0.00001
Ozone	0.000002
Sulfur dioxide	0.00000002

In addition to the components listed in Table 4-7, air also contains water vapor, in variable amounts. Water vapor plays an important role in atmospheric circulation. The sun, of course, is the primary driving force behind the motion of the atmosphere. The motion is continual and the mixing of the gases in the atmosphere is thorough. The sun evaporates large amounts of water near the equator and the air moves in large cyclonic patterns from the equator toward both poles. The condensation of water vapor releases energy which continues driving these great air masses. The mixing and recycling of the atmosphere is so thorough that the next breath of air you inhale may contain molecules that were once inhaled by Christopher Columbus or George Washington. Someone has calculated that each breath you inhale actually contains over 2 billion (2×10^9) atoms that were once breathed by the great artist, Leonardo da Vinci who died in 1519. One thing is certain, each breath of air you inhale also contains traces of radioactive waste products such as iodine-131 and strontium-90, as well as oxides of nitrogen and other chemical pollutants.

Your inspection of Table 4-7 probably led you to wonder why the atmospheric concentration of nitrogen gas is so high (79 percent) when volcanic emissions contain such a small percentage of

"Nuclear reactions in atomic power plants do not generate carbon dioxide but pose other problems related to the disposal of radioactive wastes and the thermal heating of water in natural bodies of water."

this gas. The answer seems to be that nitrogen is a relatively chemically inactive gas which remains in the air to a far greater extent than either oxygen or carbon dioxide. About two billion years ago, nitrogen-fixing bacteria developed. These bacteria enable atmospheric nitrogen to combine chemically with other elements. The fixing or chemical combining of nitrogen with other elements is accomplished in nature by bacteria in nodules of certain plant roots, by blue-green algae, and by lightning in electric storms. In an electric storm, nitrogen combines with oxygen and forms oxides of nitrogen which are further chemically changed and absorbed into the soil. Plants use these nitrogen compounds in the soil and produce proteins. Plant and animal waste and dead organisms return these fixed-nitrogen compounds to the soil. These materials decompose with the aid of denitrifying bacteria, returning free nitrogen to the atmosphere. This cycle has been taking place since the first animals appeared on the earth. Man has already seriously upset the nitrogen cycle by manufacturing large quantities of nitrogen fertilizers. At present, it is unlikely that denitrifying processes are keeping up with man's fixation processes. No one can accurately predict what all of the effects of disturbing this balance will have on the environment.

Another balance that man must consider is the cycle that maintains a 0.13 per cent concentration of carbon dioxide in the atmosphere. Natural events that add carbon dioxide to the atmosphere are animal respiration, combustion processes such as forest fires,

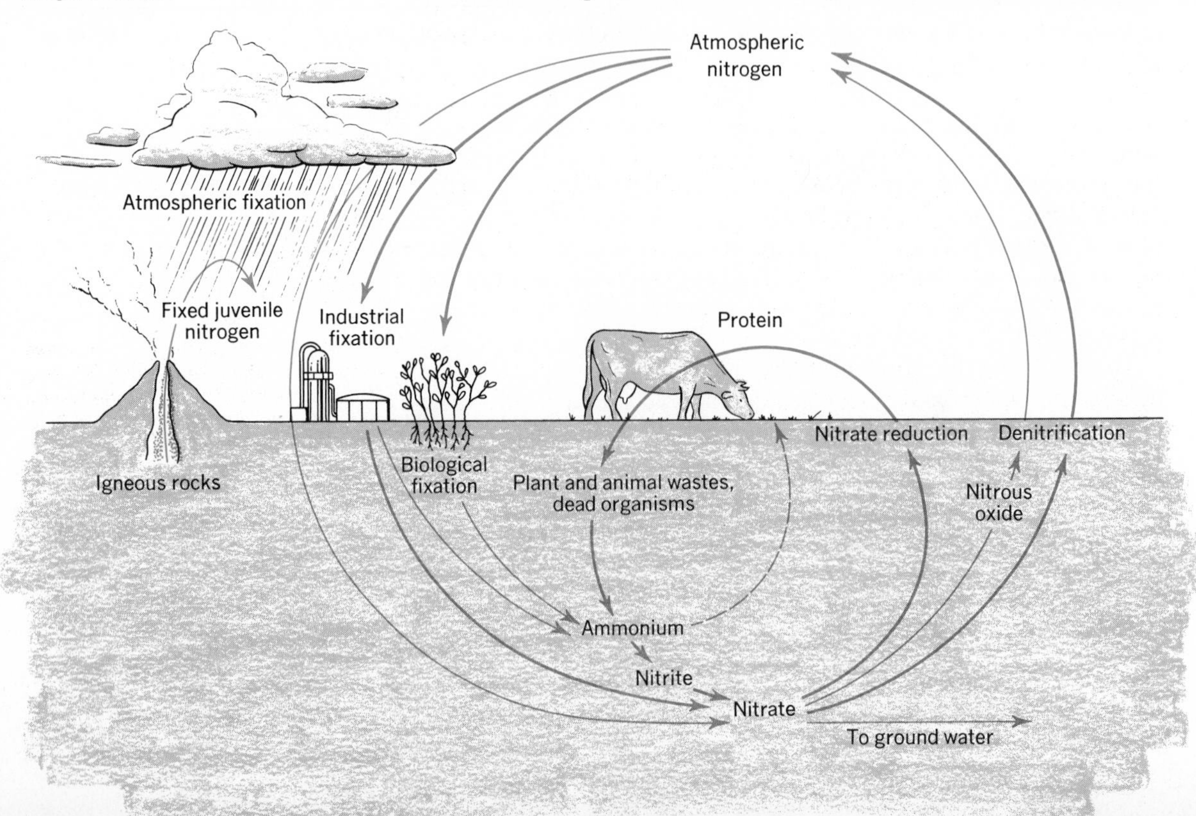

4-22 The nitrogen cycle. Industrial fixation of nitrogen is removing nitrogen from the atmosphere faster than the natural denitrifying processes return nitrogen to the air.

The disruption of nature's balanced carbon dioxide cycle by human population.

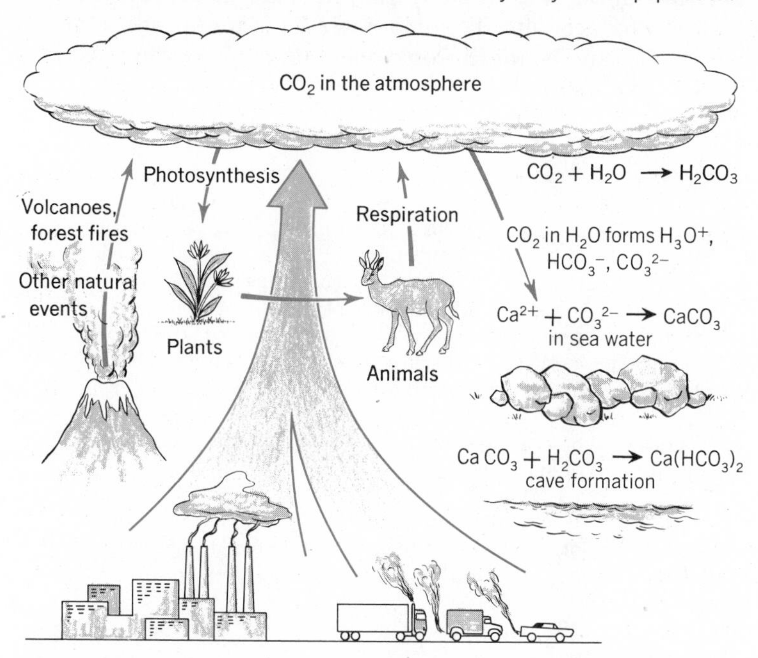

4-23 The carbon dioxide balance in nature may be upset by man's infusion of excessive amounts of carbon dioxide into the atmosphere.

and volcanic emissions. Carbon dioxide is removed from the atmosphere by photosynthesis reactions, the weathering of some rocks, and dissolution in water of the atmosphere, lakes, and oceans. It is possible that the burning of fossil fuels such as coal, oil, and natural gas may add carbon dioxide to the atmosphere faster than it can be removed by natural processes. It is estimated that by the year 2000 there will be over 0.14 per cent carbon dioxide in the atmosphere. It has been predicted that this extra carbon dioxide will have a "greenhouse" effect and result in the warming of the atmosphere. This is the result of carbon dioxide's ability to absorb infrared radiation. This heating of the atmosphere is called "thermal pollution." If we are causing the temperature of the atmosphere to increase, then the long-range effects could involve a partial melting of polar icecaps and a rise in the level of the oceans.

The Ca^{2+} and Mg^{2+} ions in sea water have an enormous ability to absorb CO_2 and form carbonates. That is, CO_2 reacts with water and forms carbonate ions (CO_3^{2-}), which in turn react with Ca^{2+} and Mg^{2+} and form slightly soluble $CaCO_3$ and $MgCO_3$."

In contrast to the warming effect produced by excess carbon dioxide, a cooling effect is caused by the presence of particles such as soot, smoke, dust, and droplets of water in the upper atmosphere. These particles can scatter light and thus reduce the amount of infrared radiation from the sun that reaches the earth's surface. At present, very effective methods are available for reducing particle pollution.

There are, of course, other serious atmospheric pollutants. Photochemical smog and other pollutants are discussed later. It is apparent that man is probably having a profound and unpredictable effect upon his own environment by polluting the atmosphere. As inhabitants of spaceship earth, we cannot help being concerned about the uncontrolled use of fossil fuels and the careless emission of waste products into the atmosphere. It is to our advantage to learn all we can about the problem and to do all we can as citizens to help solve the problem.

QUESTIONS

1. Explain how the kinetic-molecular model accounts for (a) the compressibility of gases, (b) gas pressure, (c) miscibility of gases, (d) diffusion of gases.

2. Use the Kinetic Molecular Theory to explain how an odor moves across a room. If molecules travel at velocities of rifle bullets, why do odors travel so slowly?

3. Explain: It is impossible to construct a physical model of a gas, using particles like billiard balls, which accurately depicts the behavior of a real gas.

4. (a) If two beginning ping-pong players were playing with ball A and two experts were playing with ball B, which ball is likely to have the greater average kinetic energy? (b) Is it possible to make an accurate prediction of the speed of either of these ping-pong balls at any given instant? (c) Assume that a large number of ping-pong players in a large gymnasium are all playing simultaneously. Compare the range of speeds (kinetic energies) of all these ping-pong balls with the range of kinetic energy of the molecules of air in the room.

5. Assume a sample of neon gas whose molecules all have exactly the same amount of kinetic energy. After a short period of time in a container together, would all maintain equal kinetic energies?

6. Which would be the wiser policy, to buy oxygen gas by the pound or by the cubic foot? Explain.

7. Container A holds a 10-g sample of hydrogen, and container B holds a 10-g sample of oxygen, both gases at 25°C and 1 atm. Compare the (a) volumes of the two samples, (b) average kinetic energies of the individual molecules in the two samples, (c) number of moles of gas in the two samples, (d) average speed of a hydrogen molecule with the average speed of an oxygen molecule. Express answers in terms of $>$, $=$, $<$.

8. What is meant by the Maxwell-Boltzmann distribution of kinetic energy?

9. According to the curves in Fig. 4-7, what is the effect of an increase in temperature on the kinetic energy of gas molecules?

10. Explain why a gas gets warmer when it is compressed.

11. What is meant by the "mean free path" of a molecule?

12. A molecule at sea level travels only about 1×10^{-6} m between collisions. If this molecule has a velocity of 430 m/sec, how many collisions does it have in one second?

13. What are the standard conditions for gas measurements?

14. Weather balloons are partially inflated and then released; yet, as they ascend, the balloons expand. Explain.

PROBLEMS

1. A mass of gas occupies 1 liter at 1 atm. At what pressure does the gas occupy (a) 2 liters, (b) 0.5 liter?
Ans. (a) 0.5 atm.

2. A 10-g sample of an ideal gas occupies a volume of 8 liters at STP. (a) What volume does this sample occupy at 273°K and 1200 torr? (b) What volume does it occupy at standard temperature and 2 atm pressure?
Ans. (a) 5 ℓ, (b) 4 ℓ.

3. Calculate the final volume of each of these samples of gases. Assume pressure is constant. (a) 0.200 ℓ of a gas at 600°K is cooled to 300°K, (b) 0.200 ℓ of a gas at 327°C is cooled to 27.0°K, (c) 22.4 ℓ of a gas at 0.00°C is warmed to 32.0°C, (d) 0.250 ℓ of a gas at 25.0°C is cooled to 0.00°C.
Ans. (a) 0.100 ℓ, (d) 0.229 ℓ.

4. A certain mass of oxygen gas measures 0.300 ℓ at 15.0°C. At what Celsius temperature does this mass of oxygen measure 0.250 ℓ keeping pressure constant?
Ans. −33.0°C.

5. A 0.300 ℓ sample of oxygen at STP is brought to 27.0°C and 800.0 torr. Calculate the new volume. **Ans. 0.314 ℓ.**

6. A sample of gas occupies 275 ml at 52.0°C and 720.0 torr. Change its volume to STP. **Ans. 219 ml.**

7. A sample of gas measuring 0.500 ℓ at 720 torr and 87.0°C is heated to such a temperature that the volume becomes 1.00 ℓ. If this temperature is 127°C, what is the new pressure? **Ans. 400 torr.**

8. Calculate the Kelvin temperature at which 0.400 ℓ of gas initially at 0.00°C and 740 torr expands to 0.500 ℓ at 700 torr.

9. What change in pressure permits 25.0 ml of neon at 735 torr to expand to a volume of 40.0 ml?

10. A 0.500-liter container filled with nitrogen at STP is heated to 177°C. What is the pressure of the nitrogen at the new temperature? **Ans. 1.25×10^3 torr.**

11. A sample of carbon dioxide gas occupies 50.0 ℓ at 25°C and 725 torr. What will the volume be at 77°C and 788 torr?

12. A certain gas measures 500 ml and at STP has a mass of 0.450 g. What mass of this same gas is present in a 1.00-liter container measured at 27.0°C and 745 torr? **Ans. 0.802 g.**

13. What temperature change in Celsius degrees is needed to cause the pressure of a gas to change from 10.0 atm at 600°C to 25.0 atm if the volume remains unchanged?

14. What volume is occupied by 0.250 g of O_2 measured at 25.0°C and 755 torr?

15. What is the molecular mass of a gas if 2.82 g of the gas occupies 3.16 liters at STP?

16. What is the density of the gas in Problem 15 at 27.0°C and 745 torr? **Ans. 0.795 g/ℓ.**

17. What is the molecular mass of a gas if 0.560 ℓ of it weighs 1.55 g at 0°C and 800 torr? **Ans. 58.8.**

18. A mixture of gases contains 4.35 percent oxygen molecules by volume. How many oxygen molecules are in 1.00 liter of the mixture at STP?

19. The density of a gas at 273°K and 760 torr is 4.00 g/ℓ. What is its molecular mass in g/mole?

20. Nitrogen is collected by displacement of water giving a total pressure of 745 torr. The pressure of the water vapor is 25.0 torr. What is the partial pressure of the nitrogen?

21. Calculate the density of these gases at STP (a) methane (CH_4), (b) ethane (C_2H_6), (c) propane (C_3H_8). **Ans. (b) 1.34 g/ℓ.**

22. The pressure attained by a laboratory vacuum pump is 1×10^{-9} torr. How many molecules are present in one ml of such a vacuum? **Ans. 3.5×10^7 molecules/ml.**

23. A mixture of 2.0 g He, 7.0 g N_2, 16 g O_2, and 10.0 g argon occupies a container. The total internal pressure is 900 torr. Find the partial pressure exerted by each gas. **Ans. P_{N_2} = 150 torr.**

24. What is the volume of a mixture of 0.0400 mole of nitrogen and 0.100 mole of oxygen (a) at STP, (b) at 27°C and 760 torr. **Ans. (a) 1.12 ℓ, (b) 1.23 ℓ.**

25. A gas is collected in a tube by displacing water. The levels of water inside and outside the tube are equal and the volume is 248 ml. The temperature of the gas and water is 29°C. A barometer in the laboratory reads 780 torr. (a) What is the pressure caused by water vapor? (b) What is the partial pressure of the gas alone? (c) What is the volume of this gas at STP? (d) How many moles of gas are in the sample? (e) How many molecules are there in the sample? **Ans. (d) 0.0098 moles.**

26. A rigid 5.0 ℓ tank at STP contains nitrogen gas (N_2). (a) How many moles of N_2 are present? (b) How many grams are present? (c) How many molecules are present?
Ans. (a) 0.22 mole, (b) 6.2 g, (c) 1.3×10^{23} molecules.

27. A tank contains B ml of gas at a pressure P_2. A ml of a gas at pressure P_1 is added. Express the resulting pressure in terms of A, B, P_1, and P_2.

28. How many grams of potassium chlorate ($KClO_3$) are required to produce 30.0 ℓ of O_2 measured at 27.0°C and 745 torr?

29. How many liters of H_2 can be collected by displacement of water at 22.0°C and 752 torr when 0.755 g of Al reacts with excess HCl?

$$2Al(s) + 6HCl(aq) \longrightarrow 2AlCl_3(aq) + 3H_2(g)$$

30. A bicycle tire pump is filled with helium. Under constant pressure the helium is expelled in 5 seconds. How long will it take to expel sulfur dioxide (SO_2) from

this same pump under the same constant pressure?
Ans. 20 sec.

31. An unknown gas effuses through a capillary in 60 seconds. The same volume of hydrogen escapes in 10 seconds. Calculate the molecular mass of the unknown gas.

32. Calculate the relative rates of diffusion of the uranium hexafluorides, $^{238}UF_6$ and $^{235}UF_6$. The superscripts represent atomic mass units.

33. How many grams of copper are needed to produce 224 ml of NO collected by displacement of water at 29.0°C and 748 torr?

$$3Cu(s) + 8HNO_3(aq) \longrightarrow 3Cu(NO_3)_2(aq) + 2NO(g) + 4H_2O(\ell)$$

34. How many liters of oxygen measured at 30.0°C and 755 torr can be obtained from 10.0 g of $KClO_3$?

35. When 12.0 grams of sodium react with water, how many liters of hydrogen are produced at 20.0°C and 745 torr?

$$2Na(s) + 2H_2O(l) \longrightarrow H_2(g) + 2NaOH(aq)$$

36. A reaction occurs in a mixture of 2 liters of $H_2(g)$ and 3 liters of $Cl_2(g)$ (a) How many liters of HCl(g) are produced? (b) Which reactant is in excess and how many liters?

$$H_2(g) + Cl_2(g) \longrightarrow 2HCl(g).$$

Ans. (a) 4 liters, (b) Cl_2, 1 liter.

37. A mass of 130.8 g of Zn reacts with excess hydrochloric acid.

$$Zn(s) + 2HCl(aq) \longrightarrow ZnCl_2(aq) + H_2(g)$$

Calculate (a) the liters of H_2 evolved at STP, (b) the liters of H_2 evolved at 27.0°C and 740 torr.

38. A rocket contains 3.00 kg of C_2H_6 and sufficient liquid oxygen for complete combustion. The gases emitted from the burning rocket go through a temperature change from 00.0°C to 1200°C. The combustion in the rocket engine is

$$C_2H_6(g) + 3\tfrac{1}{2}O_2(g) \longrightarrow 2CO_2 + 3H_2O(g)$$

(a) How many kg of liquid oxygen must be carried in the rocket? (b) To what volume does a liter of any gas at 0°C expand when heated to 1200°C? (c) Is the work of lifting the rocket largely accomplished by increased numbers of gas molecules caused by combustion or by thermal expansion of the gases? (d) How many moles (*a*) of CO_2 and (*b*) of H_2O are produced in this total reaction?
Ans. (a) 11.2 kg, (b) 5.4 ℓ.

39. If 0.335 g of iron reacts with dilute sulfuric acid, what volume does the evolved hydrogen occupy at 47.0°C and 750 torr?

$$Fe(s) + H_2SO_4(aq) \longrightarrow FeSO_4(aq) + H_2(g)$$

40. A 2.00-g sample of ammonium nitrate (NH_4NO_3) exploded. What is the total volume of gas produced at 527°C and 745 torr? The reaction is

$$NH_4NO_3(s) \longrightarrow 2H_2O(g) + N_2(g) + \tfrac{1}{2}O_2(g)$$

Ans. 5.85 ℓ.

41. A 114-g sample of carbon disulfide is completely burned in 105 ml of oxygen at STP. What is the total volume of the final mixture of gases at STP?

$$CS_2(l) + 3O_2(g) \longrightarrow CO_2(g) + 2SO_2(g)$$

42. A balloon filled with hydrogen gas has a lifting power of 9.00 kg. The atmosphere has a density of 1.29 g per liter. When the balloon was filled with an unknown gas, its lifting power was found to be 3.00 kg. Calculate the molecular mass of the unknown gas. Neglect the mass of the balloon. Refer to Problem 18, p. 23.
Ans. 20.0.

43. A container is filled with a mixture of CO_2 and N_2 at 1.00 atm. The mixture is passed over an absorbent which removes the CO_2 and is then returned to the container. The pressure in the container drops to 540 torr. What is the percentage by mass of N_2 in the original mixture if the temperature remains constant?
Ans. 61 percent.

44. What mass can be lifted by a balloon containing 200 kg of CH_4 at 30.0°C and 750.0 torr? Assume air is 80.0 percent N_2 and 20.0 percent O_2 by volume.
Ans. 160 kg.

45. Calculate or graph the absolute zero of temperature on the X scale from the following data made on a sample of CO_2 having a mass of 25.8 g.

Pressure units	*Volume units*	*Temperature °X*
550	40.0	−80.0°X
1000	24.0	25.0°X

Ans. −1235°X

5

Radiant Energy and the Composition of Atoms

Preview

We are now ready to seek an explanation for the observed macroscopic properties of substances in terms of the composition and behavior of the particles which compose the aggregates. This objective will require several chapters. The first step is to acquire an understanding of the composition and structure of atoms, the simplest, identifiable component of an aggregate involved in chemical reactions. Because modern atomic theory is abstract and difficult to visualize, we shall start with the simple aspects in this chapter and proceed by steps to the more complex aspects later. Thus, we shall trace the evolution of the model of an atom from its simple beginning at the turn of the 19th century to its highly sophisticated form used by scientists in the last half of the 20th century.

In this chapter, we shall first review some of the observations which revealed the general composition of atoms. Most of the knowledge that scientists have accumulated concerning the internal structures of atoms and molecules has been obtained by analyzing the radiant energy (light) emitted or absorbed by these particles. In order to see the relationship between the structure of atoms and the radiant energy they absorb, or emit, you must understand some of the characteristics of radiant energy. We shall, therefore, devote a section of this chapter to the investigation of the *nature and characteristics of radiant energy*.

Finally, we shall introduce the *Bohr theory of atomic structure*. This theory will help us to visualize a useful model of atoms and to introduce important concepts which will assist us in the development of more modern theories later.

"Now the most startling result of Faraday's laws is perhaps this: if we accept the hypothesis that the elementary substances are composed of atoms, we cannot avoid concluding that electricity also, positive as well as negative, is divided into elementary portions which behave like atoms of electricity."

Hermann Ludwig Ferdinand Von Helmholtz (1821–1894)

THE ELECTRIC NATURE OF MATTER AND COMPOSITION OF ATOMS

5-1 Fundamental Concepts of Electric Charge The electric nature of matter was observed over 2000 years ago by the Greeks. They discovered that objects could be electrified by rubbing. When two pieces of hard-rubber rod are rubbed with fur, they acquire an electric charge and repel one another. When two pieces of glass

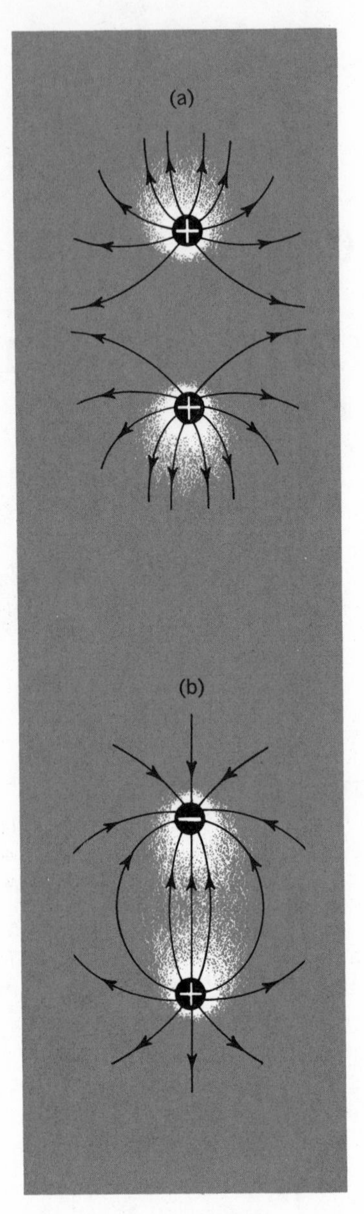

5-1 (a) The electric fields around similarly charged particles interact to repel the particles. (b) The electric fields around oppositely charged particles interact to attract the particles.

rod are rubbed with a piece of silk, they also acquire an electric charge and repel one another. When one of the charged rubber rods is brought near one of the charged glass rods, the two rods are attracted to one another. Since the rods exert a force on one another when separated, it is apparent that the region surrounding the charged rods is changed in some way. This region of influence that exerts a force on any charged object within the region is called an electric field. An electric field has no mass, can exist in a vacuum, and is always found in the space surrounding a charged body.

The experiments above provide evidence that there are two types of electric charge: negative and positive. They also illustrate the fundamental relationships describing charged bodies: like charges repel one another; unlike charges attract one another. The rubber rod is arbitrarily assigned a negative charge. That acquired by the glass rod is said to be positive. The force operating between the charged rods is called an electrostatic or coulombic force. The magnitude of the force between two charged objects is shown by the following relationship which was experimentally derived by Charles Coulomb (1736–1806), a French scientist.

$$\mathbf{F} \propto \mathbf{q_1 q_2 / r^2}$$

or

$$\mathbf{F_{vacuum} = k q_1 q_2 / r^2} \qquad 5\text{-}1$$

where F is the force, q_1 and q_2 are the charges on the two objects, r is the distance between the charges, and k is a constant of proportionality whose value depends on the units used for the other quantities. This relationship, ***Coulomb's Law,*** includes an example of the ***Inverse Square Principle,*** and states that **the electrostatic force between two charged objects is directly proportional to the product of the charges and inversely proportional to the square of the distance between them.** In other words, when the distance between two oppositely charged particles is doubled, the force of attraction between them is reduced to one-fourth the original value.

5-2 Electric Potential Energy In chemistry, one of the most important kinds of potential energy is that related to the forces which act between charged bodies. This is called *electric potential energy.* Changes in electric potential energy occur when the distance between charged particles increases or decreases. They are somewhat analogous to the potential energy changes which accompany changes in the height of an object above the ground. Energy is required to lift an object from a low shelf to a high shelf where the object has greater potential energy. Likewise, energy is required to separate a positively charged object from a negatively charged one or a north pole of a magnet from a south pole. The work that is done or the energy that is required to separate the charged particles results in the system's having a higher energy content. Consequently, the oppositely charged particles have more potential energy when they are far apart.

Be sure you understand the following facts:

1. When oppositely charged objects approach one another the attractive forces operate to decrease the potential energy of the system.

2. When similarly charged objects approach one another the repulsive forces operate to increase the potential energy of the system.

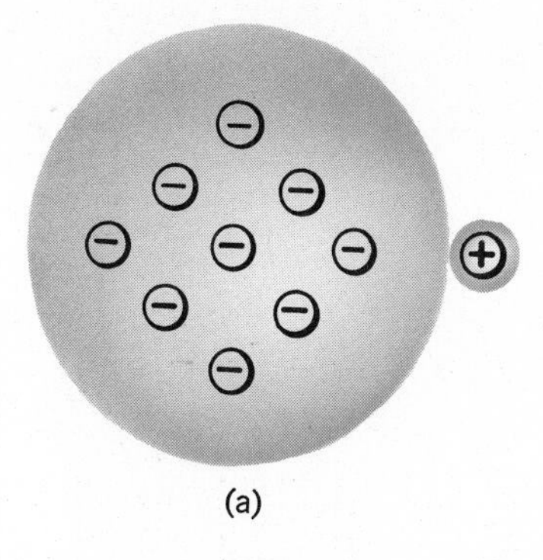
(a)

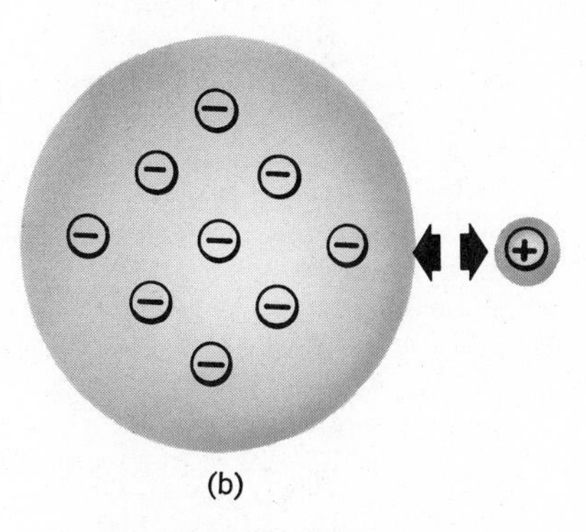
(b)

5-2 A force must be exerted through a distance in order to separate the small positive particle from the large negative particle. Which system, (a) or (b), contains more potential energy?

5-3 Transfer of Electric Charge The electric charge associated with the charged rod mentioned above is often referred to as static electricity. Electric charge may also be transferred through a conductor from one point to another. When you turn on a flashlight, an electric charge moves through a conductor (the filament in the light bulb) from the negative terminal of the dry cell to the positive pole. The movement of charge through a conductor is known as an ***electric current.*** Some of the first evidence for the electric nature of atoms and the particulate nature of electricity was obtained by studying the effect of an electric current on chemical compounds. Let us briefly review the nature and significance of these experiments.

5-4 The Particle Nature of Electricity In 1834, Michael Faraday (1791–1867) studied the effect of an electric current on aqueous (water) solutions of compounds such as *acids, bases,* and *salts.* He gave the name ***electrolyte*** to those substances whose water solution conduct an electric current, and defined ***electrolysis*** as decomposition by an electric current. His source of current was a battery which he connected to electrodes immersed in the solution. The elemental products formed at the electrodes led Faraday to suggest that under the influence of an electric current, certain compounds separate into electrically charged atoms.

Faraday introduced the word *ion* to describe an atom which bears a net electric charge. To distinguish between the ions associated with the two different electrodes, Faraday coined the words *anode, cathode, anion,* and *cation.* The anode is defined as the *electrode toward which the anions (negative ions) migrate,* and the cathode is defined as the *electrode toward which the cations (positive ions) migrate.* Faraday mistakenly believed that ions were made at the electrodes during electrolysis. We now know that certain types of solid crystals (salts) are composed of ions which are merely separated in the dissolving process by the action of water. Other types of substances may react with water and form ions. As a result of his experiments, Faraday suggested that the *chemical combination of atoms is a result of electric attractions between them.*

Faraday's quantitative studies of electrolysis led him to propose that a definite amount of a substance was evolved at an electrode by a fixed quantity of electricity. For example, the quantity of electricity that produces 8 g ($\frac{1}{4}$ mole) of oxygen (O_2) also produces 35.5 g ($\frac{1}{2}$ mole) of chlorine (Cl_2), 127 g ($\frac{1}{2}$ mole) of iodine (I_2), 59.3 g ($\frac{1}{2}$ mole) of tin, and 108 g (1 mole) of silver. Faraday called these numbers the electrochemical equivalents of the elements. We may refer to these quantities as *chemical equivalents* or *gram-equivalent masses* of the elements. It can be seen that the gram-equivalent mass of an element is equal to or a fraction of the gram-molecular mass (the mole). This means that the gram-equivalent mass and gram-molecular mass (mole) are related by a *small whole number.* In other words, the gram-equivalent mass may be obtained by dividing the gram-molecular mass by some integer. Because one gram-molecular mass of a substance contains a definite number of particles (6.02×10^{23}), one gram-equivalent mass of a substance also contains a definite number of particles. The number depends on the integer relationship between the gram-molecular and gram-equivalent mass. In the case of tin, one mole has a mass of 118 g while one gram-equivalent mass has a mass of 59 g. Therefore, one gram-equivalent mass of tin contains only

$$\frac{1}{2} \times \mathbf{6.02 \times 10^{23} = 3.01 \times 10^{23}\ atoms}$$

The quantity of electricity associated with the evolution of one gram-equivalent mass of a substance is called the *faraday.*

The fact that a fixed quantity of electricity (1 faraday) is always associated with a definite quantity of charged atoms (1 gram-equivalent mass) suggests that electricity is particulate in nature. This interpretation of electricity, based on Faraday's experiments, was given in 1874 by C. J. Stoney (1826–1911), an English physicist. In 1891, Stoney named the unit particle of electricity, the *electron.* An electron may be symbolized by e^-. It turns out the *one faraday of electricity is equivalent to the electric charge carried by 6.02×10^{23} electrons* (1 mole of electrons). Since one faraday deposits 3.01×10^{23} atoms of tin, it is apparent that two electrons are associated with the deposition of each atom of tin. The relationships are summarized in Table 5-1. We shall investigate electric units, Faraday's Laws, and the concept of the chemical equivalent in later chapters. In this section our primary concern is to trace historically the development of ideas related to atomic structure.

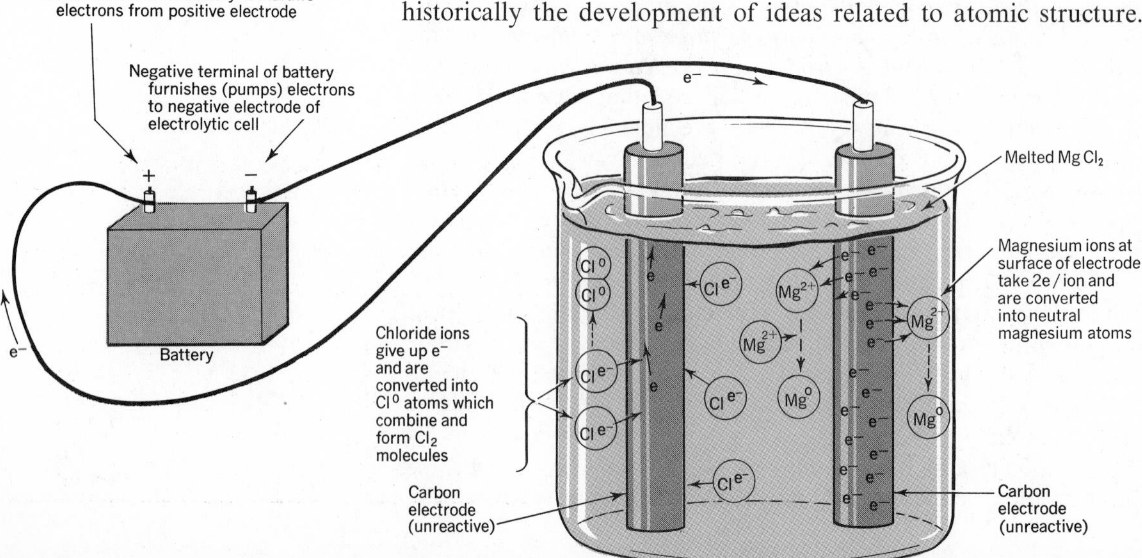

5-3 Electrolysis of magnesium chloride. Two electrons are required to convert 1 doubly charged magnesium ion to 1 neutral magnesium atom. Therefore, 1 mole of electrons (1 faraday of electricity) yields one-half mole (12 g) of magnesium. Chlorine molecules contain 2 chlorine atoms. Since the loss of 1 electron is required to convert 1 chloride ion to a chlorine atom, the loss of 1 mole of electrons (1 faraday) will yield 1 mole of chlorine atoms or $\frac{1}{2}$ mole of chlorine molecules (35.5 g). The ratio 12 g Mg to 35.5 g Cl is the mass ratio of Mg to Cl in the compound $MgCl_2$. These masses are known as the combining or gram-equivalent masses of the elements. The combining or equivalent masses of elements can be obtained by dividing the atomic mass by the number of electrons transferred per atom.

Example 5-1
Use the relationships $1F = 6.02 \times 10^{23}\ e^-$, $6.02 \times 10^{23}\ e^- = 1$ mole electrons, $1F = 1$ electric equivalent $= 1$ chemical equivalent to solve these problems. (a) How many electrons in 0.500 faraday of electricity?

Solution

$$0.500\ F \times \frac{6.02 \times 10^{23}\ e^-}{F} = 3.01 \times 10^{23}\ e^-.$$

(b) How many electrons in 3.00 moles of electrons?

Solution

$$3.00 \text{ moles } e^- \times \frac{6.02 \times 10^{23}\ e^-}{1 \text{ mole } e^-} = 1.81 = 10^{24}\ e^-.$$

(c) How many moles of electrons in 6.23×10^{18} electrons?

Solution

$$6.23 \times 10^{18}\ e^- \times \frac{1 \text{ mole } e^-}{6.02 \times 10^{23}\ e^-} = 1.03 \times 10^{-5} \text{ moles } e^-.$$

(d) How many electric and/or chemical equivalents correspond to 0.10 mole of electrons?

Solution

$$0.10 \text{ mole } e^- \times \frac{1 \text{ chem eq}}{\text{mole } e^-} = 0.10 \text{ chem equivalent.}$$

(e) One chemical equivalent mass of calcium has a mass of 20.0 g. How many grams of calcium are produced in an electrolysis experiment if 0.100 mole of electrons (0.100 F) is passed through a cell?

Solution

$$0.100 \text{ mole } e^- \times \frac{1 \text{ chem eq}}{\text{mole } e^-} \times \frac{20.0 \text{ g Ca}}{\text{chem eq}} = 2.00 \text{ g Ca}$$

TABLE 5-1
RELATIONSHIPS BETWEEN FACTORS RELATED TO ELECTROLYSIS OF CHEMICAL SUBSTANCES

Substance	*No. of e^- Used (1 mole)*	*No. of Faradays and/or Chemical Equivalents*	*No. of Grams Discharged*	*Mass (g) of 1 Chemical Equivalent (g-equiv. mass)*	*Mass (g) of 1 Mole (g-molecular mass)*	*No. of Chemical Equiv. Per Mole*	*No. of e^- to Discharge One Atom (a) or Molecule (m)*
O_2	6.02×10^{23}	1	8.00	8.00	32.0	4	4 per (m)
Cl_2	6.02×10^{23}	1	35.5	35.5	71.0	2	2 per (m)
I_2	6.02×10^{23}	1	127	127	254	2	2 per (m)
Sn	6.02×10^{23}	1	59.3	59.3	118.6	2	2 per (a)
Ag	6.02×10^{23}	1	108	108	108	1	1 per (a)
Al	6.02×10^{23}	1	8.99	8.99	26.98	3	3 per (a)
Ca	6.02×10^{23}	1	20.0	20.0	40.0	2	2 per (a)

FOLLOW-UP PROBLEM

(a) How many electrons in 2.00×10^{-2} faradays of electricity? (b) How many electrons in 0.250 moles of electrons? (c) How many moles of electrons in 1.20×10^{15} electrons? (d) How many electrons must be passed through a cell to produce 0.300 chemical equivalents of O_2 gas? (e) How many grams of O_2 gas can be liberated by passing 0.200 moles of electrons through a solution? The gram-equivalent mass of oxygen is 8.00 g. **Ans. (a) 1.20×10^{22}, (b) 1.50×10^{23}, (c) 1.99×10^{-9}, (d) 1.81×10^{23}, (e) 1.60 g.**

The nature of the electron was investigated by a number of scientists in the late 19th century. In general, the experiments consisted of studying the passage of electricity through gases in partially evacuated tubes known as gas-discharge or Crookes tubes. Some of the apparatus, experiments, and results are described below.

5-5 The Nature of Cathode Rays In 1879, Sir William Crookes (1832–1919) devised an apparatus consisting of a glass tube containing two metal plates, called *electrodes,* which are connected to a source of electricity. A similar apparatus is shown in Fig. 5-4. The plate connected to the negative terminal of the source is called the *cathode.* The plate connected to the positive terminal is called the *anode* and consists of a movable metal cross which may be placed in either a horizontal or a vertical position. Most of the gas particles have been removed from the tube (the tube has been partially evacuated) so that they will not interfere with the passage of electricity.

When the electric current is flowing and the anode is in a horizontal position, the glass opposite the cathode glows with a greenish light. This phenomenon is known as *fluorescence.* When the metal cross is placed in a vertical position, the rays traveling in a straight line from the cathode strike the cross but do not penetrate it. As a result, the shadow of the cross appears on the glass. This experiment shows that some kind of ray coming from the negative terminal travels in a straight line, transmits energy, and does not penetrate a metal sheet. Such a ray is called a *cathode ray.*

To determine more about the nature of cathode rays, a tube

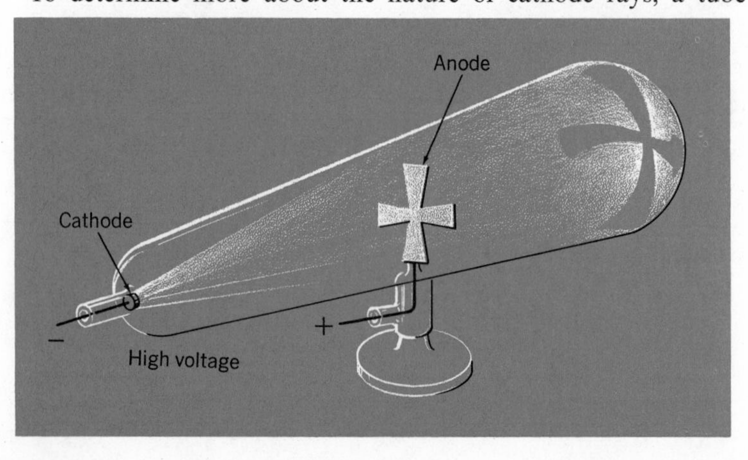

5-4 Cathode rays transmit energy.

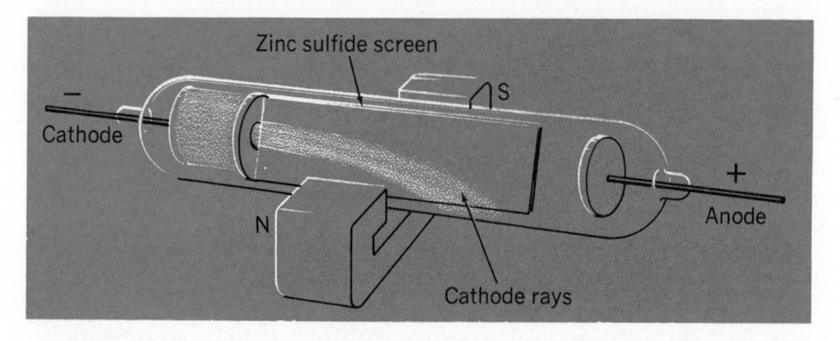

5-5 Cathode rays consist of negatively charged particles.

similar to that illustrated in Fig. 5-5 may be used. This tube has a metal screen which is coated with zinc sulfide and placed parallel to the path of the rays. Zinc sulfide is one of a number of substances which fluoresce when bombarded by cathode rays, thus permitting the detection of these rays. When the electrodes are connected to a source of high-voltage electricity, the path of the rays is seen as a green line on the zinc-sulfide screen. A magnet placed outside the tube, as illustrated, causes the beam to be deflected downward. When the poles of the magnet are reversed the beam is deflected upward. Interpretation of the effect of the *magnetic lines of force* (called a *magnetic field*) indicates that the *cathode rays are composed of negatively charged particles* (electrons).

5-6 The Charge-to-Mass Ratio of an Electron In 1897, J. J. Thomson (1856–1940), a British physicist, deflected a beam of cathode rays by applying an electric field of known strength. He then varied a magnetic field which was perpendicular to the electric field until the beam returned to its original undeflected position (Fig. 5-6). Using the data from his experiments, Thomson was able to determine the ratio of the charge to mass (e/m) of an electron and thus show that cathode rays consist of discrete particles of matter. By using different voltages, various metals for electrodes, and various gases in the tube, Thomson was able to show that the cathode rays always had the same properties. He concluded that *atoms of all substances contain the same kind of negative particles.* In 1906 he was awarded the Nobel Prize.

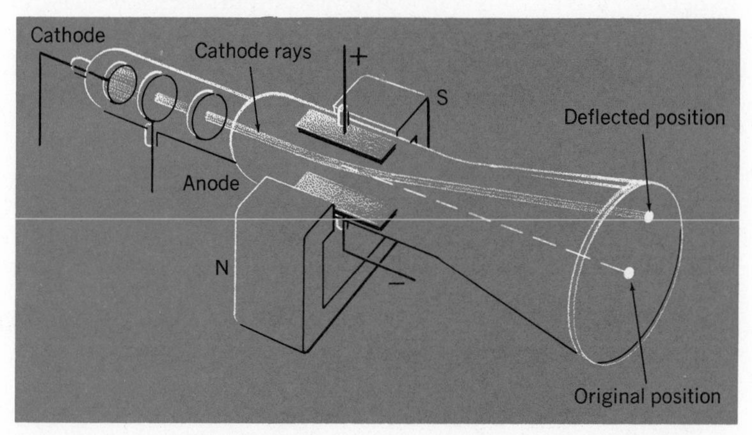

5-6 J. J. Thomson used an apparatus such as this to determine the e/m ratio for the electron.

5-7 The Charge on an Electron In 1909, Robert A. Millikan (1868–1953), an American physicist, accurately determined the charge of an electron by his famous "oil drop" experiment. He measured the time required for electrically charged oil droplets to rise a measured distance when subjected to an electric field. He also measured the time required for the droplets to fall a specified distance in the absence of an electric field. He repeated these measurements with droplets containing various amounts of charge. Calculation of the charges on the droplets showed that they were always a multiple of the same small number. The apparatus he used is schematically represented in Fig. 5-7.

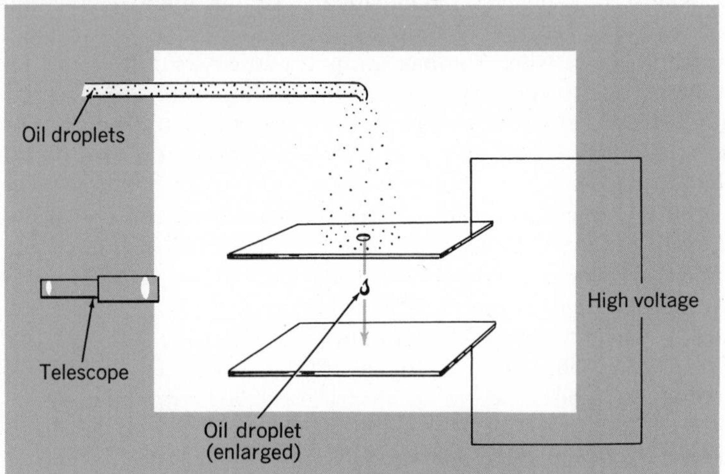

5-7 Robert A. Millikan determined the charge on a single electron by using an apparatus similar to that shown in the schematic diagram.

Since the values for the different charges on the droplets were all multiples of a fundamental charge, Millikan concluded that the fundamental charge was the absolute charge of a single electron. Knowledge of the charge determined by Millikan and the e/m ratio determined by Thomson enabled scientists to calculate the mass of an electron. The mass is approximately 9.1×10^{-28} g or 5.5×10^{-4} amu. This is about $\frac{1}{1837}$ of the mass of a hydrogen atom, the lightest of all atoms. For our practical purposes, the mass of an electron is negligible when compared with the total mass of an atom. For this work, Millikan was awarded the Nobel Prize in 1932.

An electron has a negative charge of 4.8×10^{-10} electrostatic units (esu). In general, we shall not be concerned with the absolute

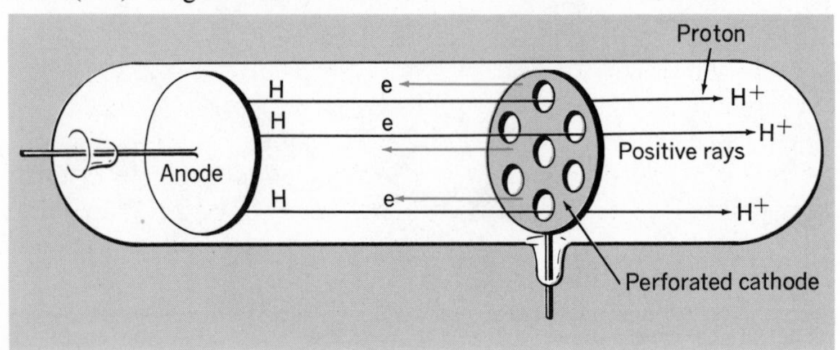

5-8 Rapidly moving electrons from the cathode collide with and remove electrons from hydrogen atoms. By their behavior, positive rays which pass through the perforated cathode may be identified as protons.

charge of an electron. We are primarily interested in relative charges. Since an electron carries the smallest unit of negative electricity, it has a relative charge of 1−. An electron is designated by the symbol e^-, and the superscript indicates that it bears a unit negative charge (1−). The charges on all other atomic particles are based on the 1− charge of an electron.

5-8 Protons Since atoms are electrically neutral particles, the identification of electrons as negative components of all atoms naturally led to the search for a positive particle. A beam consisting of positive particles was discovered in 1885 by E. Goldstein (1850–1931). The characteristics of the particles in this beam were investigated by J. J. Thomson, who used a cathode-ray tube similar to that used in the investigation of electrons. The tube, pictured in Fig. 5-8, contains a perforated cathode which permits the passage of positive rays away from the anode.

The least massive positive particles were found when hydrogen gas was used in the tube. When an electron is removed from a hydrogen atom, a single, positively charged particle remains. This particle is called a proton. It has a charge equal and opposite to that of an electron. A proton has a mass of 1.6725×10^{-24} g, or 1.007 276 63 amu. This is 1837 times the mass of an electron. A proton is one of the fundamental particles of all atoms and carries a unit positive charge (1+).

5-9 The Nucleus A more detailed investigation of an atom requires a probe smaller than an atom itself. A series of discoveries around the turn of the 20th century provided the tools for probing experiments.

In 1895, Wilhelm Roentgen (1845–1923), a German scientist, was operating a cathode-ray tube. He noticed that a fluorescent chemical in a bottle on a nearby shelf was glowing. It continued to glow even when a piece of cardboard was placed between it and the cathode-ray tube. Very penetrating rays were obviously coming from the tube. Like light rays, these unknown rays were not deflected by a magnetic field. Because of their mysterious nature, Roentgen named them X rays. The discovery of the X rays turned out to be one of the most important events of the century. X rays are used in basic research by scientists in their investigations of the structure and interrelationships of matter and energy and also in applied technology by doctors, dentists, and others. See Fig. 5-10 at the right.

One discovery frequently leads quickly to another. While studying the nature of X rays in 1896, Henri Becquerel (1852–1908), a French scientist, discovered that some chemical substances spontaneously decompose and give off very penetrating rays. This phenomenon is called ***radioactivity.*** The three common types of radiation emitted by radioactive substances are called *alpha particles, beta particles,* and *gamma rays.* The nature and significance of each type of radiation are considered in detail later.

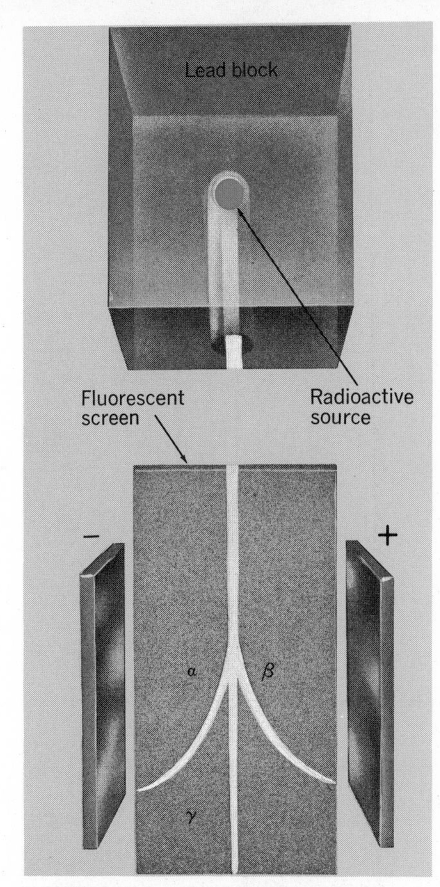

5-9 When a beam from a radioactive source passes between oppositely charged plates, alpha particles are bent toward the negative plate, beta particles are bent toward the positive plate, and gamma radiation is undeflected.

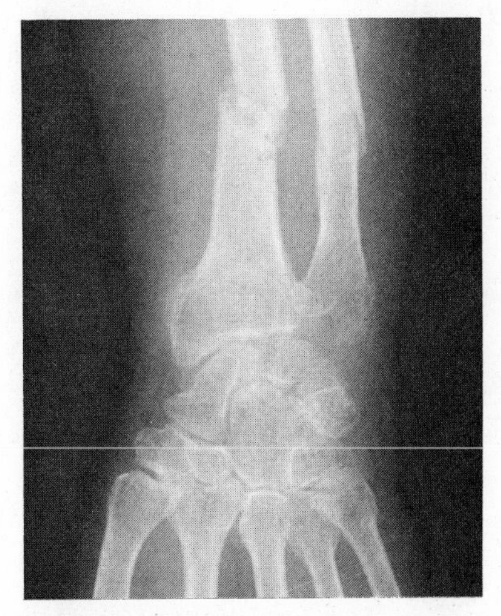

5-10 X ray photograph of a fracture in each of the bones of the lower part of the arm.

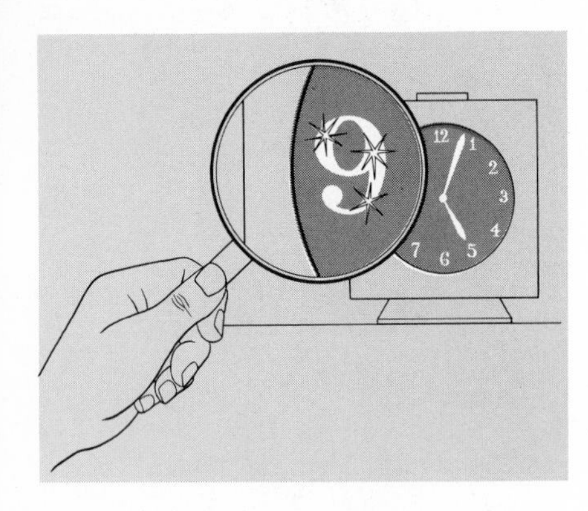

5-11 Scintillations from the face of a clock.

It was learned through experimentation that an alpha particle is a fragment of a helium atom which carries two units of positive charge (2+) and has a mass of 4 amu. As such, it was sufficiently small to serve as a probe to be used in the study of the interior structure of atoms. The presence of alpha particles can be detected by the glow produced when they strike a fluorescent chemical. You can observe this effect by focusing a strong magnifying glass on the luminous dial of a watch or clock in a dark room. Rather than seeing an even glow, you will notice individual sparks called *scintillations.* They occur when an alpha particle from an atom of radium traveling at the rate of 10 000 mi/sec, strikes a crystal of zinc sulfide.

Around 1911, Ernest Rutherford (1871–1937) and his co-workers used alpha particles emitted from a sample of radioactive substance as subatomic "bullets" for probing into atoms. His targets were the atoms in a piece of very thin gold foil, and his detectors were fluorescent screens. The arrangement in Fig. 5-12 shows the apparatus that might be used.

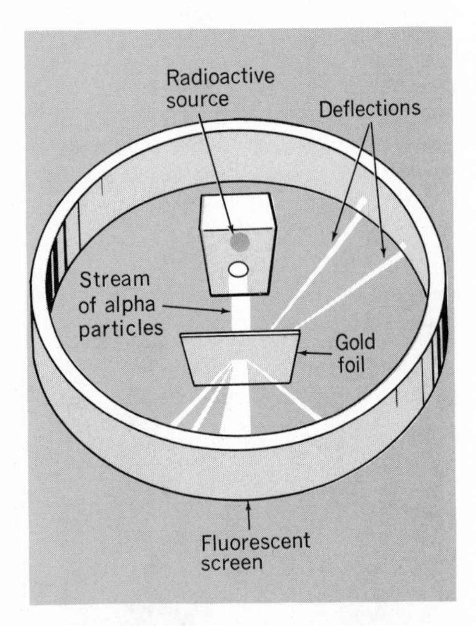

5-12 Rutherford used the newly discovered alpha particles as a probe to study the interior of atoms. From the results of scattering experiments, the model of the nuclear atom was developed.

By placing the screens at different positions around the source, Rutherford was able to determine the path of the alpha-particle bullets. To his astonishment, most of the alpha particles passed through the gold foil without being deflected. This observation helped to refute Dalton's idea that atoms were solid spheres. Rutherford found that single flashes occasionally occurred far off to the side. Most astonishing of all were the occasional flashes which he observed on the screen almost directly behind the source. This meant that several of the relatively massive alpha particles were bouncing almost straight back to the source. Knowing the high velocity and relatively large mass of an alpha particle, Rutherford remarked, "It was about as believable as if you had fired a 15-inch shell at a piece of tissue paper, and it came back and hit you." It was not possible that an electron with its negligible mass could be responsible for the deflection of a massive alpha particle. Such an occurrence would be similar to the deflection of a steamroller by a marble. From the experimental data, Rutherford reasoned that the atoms of gold must be nearly empty space. Each atom, however, must contain an extremely tiny, dense core bearing a positive charge. He named this dense, central portion of the atom the ***nucleus.*** See Fig. 5-13 on p. 157.

Calculations based on the data collected by Rutherford and others reveal that almost the entire mass of an atom is concentrated in the nucleus and that the diameter of the entire atom is over 100 000 times that of the nucleus. In other words, atoms are mostly empty space. They consist of a tiny, dense positively-charged nucleus surrounded by electrons moving at inconceivable speeds at great distances from the nucleus. The following analogy will give you an idea of the vast emptiness of an atom. If the nucleus of a hydrogen atom were enlarged to the size of a tennis ball, the outside of the atom, represented by the average distance of its single electron from the nucleus, would be about two miles away.

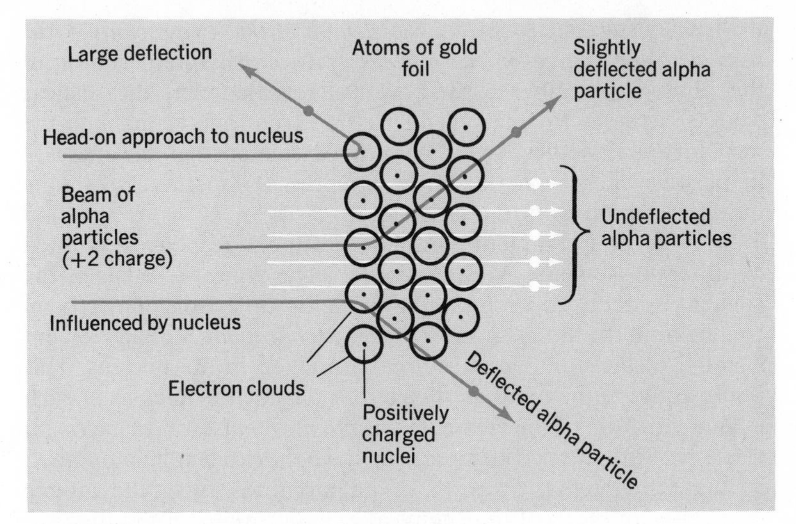

5-13 Alpha particles shot at gold foil which is only a few atoms thick. The nearer the alpha particle comes to the nucleus, the greater is the angle of deflection.

You can see why Rutherford was prompted to make the 15-inch shell and tissue paper comparison. See Fig. 5-15 on p. 158.

Although an atom is mostly space, an electron moves around the nucleus so rapidly that it effectively occupies all of the space around the nucleus. This situation may be visualized by drawing an analogy between an electron and an airplane propeller. The area swept by the propeller blade is mostly empty space as is the region surrounding the nucleus of an atom. Although the blade *actually occupies* only a small part of the total area, it *effectively occupies* the entire area when it is spinning at the rate of 100 revolutions per second. Placing your finger in the "empty space" within the radius of the spinning blade for more than a small fraction of a second would result in the loss of your finger. The electron, which never stops moving, does not actually occupy a significant amount of space about a nucleus but would appear to fill effectively all of the space because of its rapid motion.

The fact that the mass of an atom is concentrated in its nucleus is illustrated by the following example. Imagine a piece of matter composed only of atoms whose electrons had been stripped away. This sample of matter would consist of closely packed nuclei. A single drop of this matter approximately the same size as a drop of water would weigh millions of tons. Astronomers have identified dwarf stars having apparent densities of several tons per cubic inch. The atoms of which these stars are composed contain, on the average, much less space than the atoms that make up the earth.

5-14 An analogy of Rutherford's astonishing results.

5-10 Atomic Number Rutherford postulated that the nucleus carries a positive charge, but it was another Englishman, H. G. J. Moseley (1887–1915), who actually determined the charges on the nuclei of most of the atoms. Moseley observed that when he placed a variety of elements on the anode of an X-ray tube and bombarded each with electrons (cathode rays), each element produced X rays with a wavelength (Sec. 5-14) different from that produced by each of the other elements.

5-15 The nucleus of a hydrogen atom is 10^{-13} cm in diameter; that of the entire atom is 10^{-8} cm. If the nucleus were the size of a tennis ball, its electron would most likely be found at a distance of 2 miles from the nucleus (8 times the height of the Empire State Building).

Moseley suggested that the *wavelength of the X rays emitted was related to the charge on the nucleus of an atom.* Interpretation of the photographically-recorded X rays revealed that the nuclear charge increases by one unit in passing from one element to the next in the Periodic Table. Moseley postulated that the increase in positive charge on the nucleus represents an increase in the number of protons in the nucleus.

In 1914, after completing his study of the X-ray beams emitted by different elements, Moseley wrote, "The atomic number of the element is identified with the number of positive units of electricity contained in the atomic nucleus." The ***atomic number*** of an element is equal to the number of protons contained in the nucleus. This is also equal to the number of electrons outside the nucleus of each uncharged atom of the element. *The atomic number of each element is unique.* Since no two elements can have the same atomic number, it is a fundamental property of an atom. As such, the atomic number is an index of the chemical and also many of the physical properties of atoms.

Rutherford showed that *the mass of an atom is concentrated in the nucleus.* He could not, however, account for the total known mass of the atom in terms of the number of protons in the nucleus. For example, a helium nucleus (alpha particle) has a charge of 2+ furnished by the 2 protons, which also contribute approximately 2 units of mass. Since helium atoms have a mass of 4 amu, there are still 2 units of mass to be accounted for. Rutherford postulated the existence of a neutral particle to account for the missing mass. The existence of this particle was not proved for 20 years because its lack of charge made its detection difficult.

5-11 Neutrons In 1932, James Chadwick (1891–), a British physicist, bombarded the element beryllium (atomic number 4) with alpha particles. He found that a beam of rays with very highly penetrating power was emitted from the beryllium. These rays were not affected by a magnet. This meant that they were neutral and, in this respect, similar to X rays. Chadwick also noticed that the new rays traveled at about $\frac{1}{10}$ the speed of light. Since X rays travel at the speed of light, the new beam could not be pure radiant energy. When other atoms were bombarded by these new rays, the nature of the collisions indicated that the new beam must be composed of particles. On the basis of his experiments, Chadwick was able to show that the new nuclear particles had a mass of approximately 1 amu and were electrically neutral. Accordingly, the particle was called a ***neutron.*** The mass of a neutron is now known to be 1.008 665 4 amu or 1.6748×10^{-24} g. As you will learn later, neutrons play the leading role in the drama of the development of nuclear energy. They serve as the key which unlocks the energy of the nucleus and produces the radioisotopes which are used in medicine, industry, agriculture, and research.

The discovery of neutrons completed the roster of particles which are the three major building blocks of an atom. These particles and their characteristics are summarized in Table 5-2.

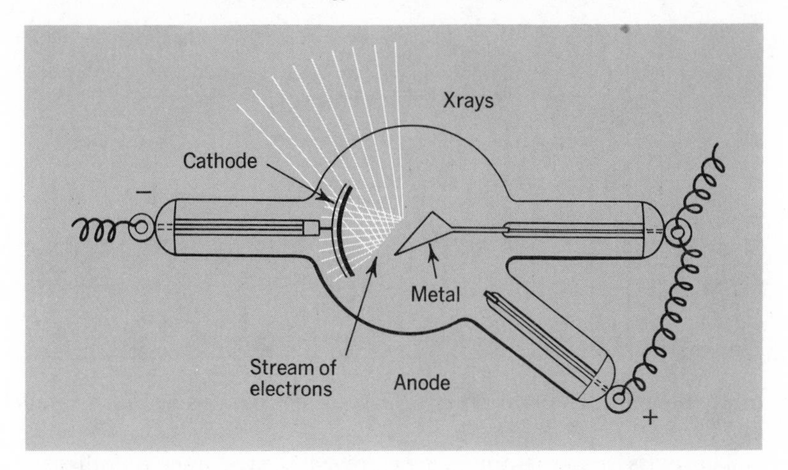

5-16 Moseley made X-ray tubes with different metals as cathodes.

TABLE 5-2
FUNDAMENTAL PARTICLES OF ATOMS

Particle	Charge	Mass (amu)	Mass (g)
Electron	1−	0.000 548 597	9.1091×10^{-28}
Proton	1+	1.007 276 63	1.6725×10^{-24}
Neutron	0	1.008 665 41	1.6748×10^{-24}

You have, no doubt, noticed that the relative masses of the proton and neutron are given to nine significant figures in Table 5-2. Perhaps you wonder how the relative masses of such small particles can be determined to such a high degree of accuracy. To satisfy your curiosity we shall examine briefly the method used by scientists to obtain these amazing data.

5-12 Isotopes and Their Masses When a neutral atom loses or gains an electron, it becomes a charged particle called an ***ion.*** There are a number of ways in which substances may be ionized. One method is the bombardment of atoms of a gas with high-energy electrons. When neon is bombarded with high-energy electrons, the neon atoms lose one or more electrons and become neon ions. For all practical purposes, the mass of an ion is the same as the mass of its parent atom. We are justified in making this assumption because the mass of an electron is negligible. This means that the relative masses of atoms can be determined by measuring the relative masses of their ions.

The relative masses of ions may be very accurately determined with a *mass spectrograph.* This instrument was developed by F. W. Aston (1877–1945) in 1919. A schematic diagram of a mass spectrograph is shown in Fig. 5-18. This instrument changes (ionizes) atoms of a specific element into positive ions by bombarding them with electrons. The ions are accelerated by an electric field and passed through a system of slits. They are then deflected by a magnetic field. Ions having the smaller charge-to-mass ratio are deflected the least. All ions having the same charge-to-mass ratio may be focused on the same line on a photographic plate. The intensity of the line on the plate is an indication of the relative number of ions that strike at that particular point. The developed plate is known as a ***mass spectrogram.*** An alternate method involves focusing the ion beam on a collector plate. The magnitude of the current produced by the beam is an indication of the relative number of ions that strike at a given position. A recorder attached to the collector plate may also be used to trace a mass spectrogram

(a)

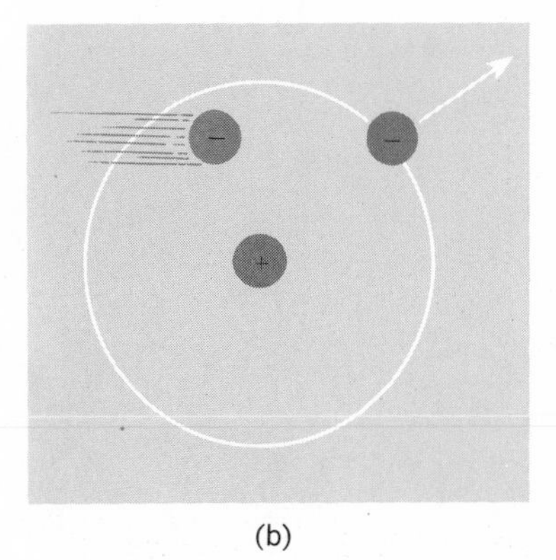

(b)

5-17 Analogy of bombardment. A high speed marble would knock out marbles from a ring just as a high speed electron would knock out electrons from atoms.

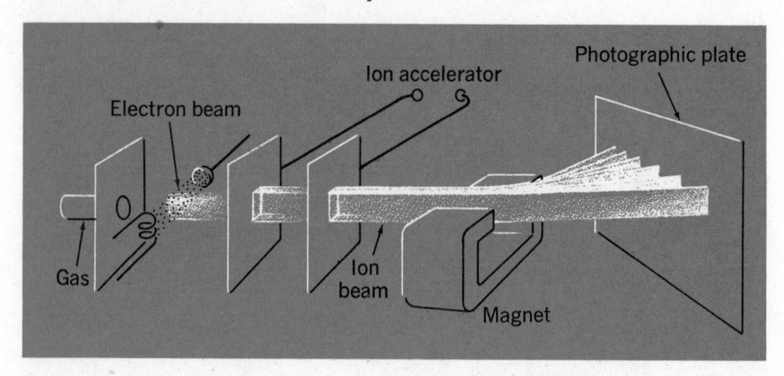

5-18 In the mass spectrograph, vaporized atoms enter at the gas inlet, are ionized by a beam of electrons, and accelerated through slits. The beam of ions is then bent by a magnetic field and allowed to strike a photographic plate.

Acceleration can be defined as a rate of change in the velocity of a particle.

of an element. The two types of mass spectrograms are illustrated in Fig. 5-20.

The mass spectrogram of neon reveals three lines of different intensity. This is interpreted to mean that ordinary neon contains three types of atoms that differ in mass, although the nuclear charges and the number of electrons are identical for all neon atoms. Atoms of the same element which differ only in mass are known as ***isotopes.*** They have the *same atomic number but different atomic masses.* It has been observed experimentally that the chemical properties of the isotopes of a given element are the same. Because of the similarity in chemical properties, the separation of isotopes is based on differences in physical properties. Since isotopes have similar properties, the number of electrons and the nuclear charge must be important factors in determining the chemical properties of a substance.

The composition of isotopes may be deduced from the total charge on the nucleus (atomic number) and its mass number. The total charge on the nucleus (Z) is equal to the number of protons that go to make up the nucleus and the total mass (A) is approximately equal to the sum of the masses of its neutrons and protons. The number of neutrons (N) is given by the expression

$$\mathbf{N = A - Z}$$

The number of neutrons in the oxygen-16 isotope (atomic number 8) is given by the expression

$$\mathbf{N = 16 - 8 = 8}$$

The neutral atom also has 8 electrons.

FOLLOW-UP PROBLEM

How many neutrons and electrons in the Cd-116 isotope (atomic number 48)? **Ans. 68 neutrons, 48 electrons.**

The fractional atomic masses of the elements listed in the table on the back inside cover result in part from the fact that elements occurring in nature are mixtures of isotopes. The atomic mass of an element may be calculated from experimentally determined isotopic masses and composition. The technique is illustrated in the following example.

Example 5-2

Interpretation of the mass spectrogram of neon yields this composition for ordinary neon: neon-20, 90.92 percent; neon-21, 0.26 percent; neon-22, 8.82 percent. Calculate the average atomic mass of neon.

Solution

The analysis indicates that in 10,000 neon atoms, 9092 atoms have a mass of 20.00 amu, 26 atoms have a mass of 21.00 amu, and 882 atoms have a mass of 22.00 amu. The weighted average is

$$\frac{(9092)(20.00 \text{ amu}) + (26)(21.00 \text{ amu}) + (882)(22.00 \text{ amu})}{10\,000} = 20.18 \text{ amu}$$

or

$$(0.0882)(22.00 \text{ amu}) + (0.9092)(20.00 \text{ amu}) + (0.0026)(21.00 \text{ amu}) = 20.18 \text{ amu.}$$

The actual masses have been determined to a much higher degree of accuracy than that indicated by the data given above. The mass of neon-20 is actually 19.998 798 amu.

5-19 Isotopes of hydrogen: hydrogen, deuterium, and tritium. What is the atomic mass of each isotope?

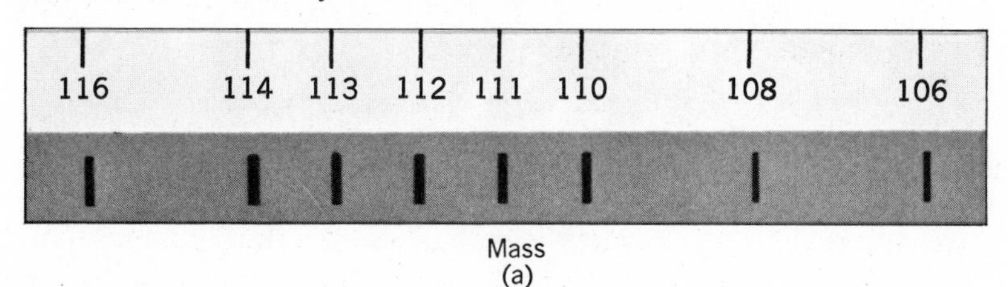

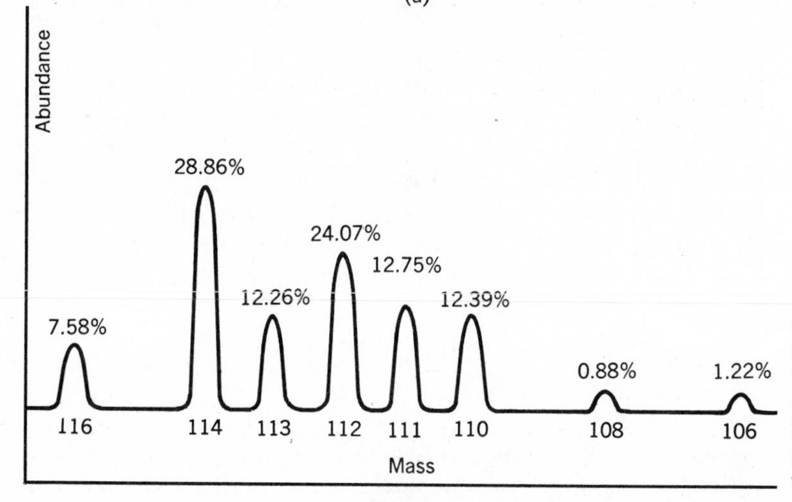

5-20 (a) The mass spectrogram of cadmium recorded photographically. The numbers represent the masses of the eight natural isotopes of this element. (b) The mass spectrogram of cadmium as traced by a recorder attached to the collector. The percentage of each isotope in the sample is related to the height of the peaks.

FOLLOW-UP PROBLEM

Ordinary chlorine contains 75.40 percent chlorine-35 and 24.60 percent chlorine-37. Calculate the average atomic mass of chlorine from the isotopic composition given above. Assume that the isotopic masses are: chlorine-35 = 35.00; chlorine-37 = 37.00.

Ans. 35.46 amu.

Example 5-3
Calculate the mass in grams of 1 amu.

Solution
The relevant relationships are:
(1) 1 amu = $\frac{1}{12}$ the mass of the carbon-12 atom.
(2) 1 mole (gram-atom) of carbon-12 contains 6.02×10^{23} atoms and has a mass of 12.0 g.
From (2) we can construct a ratio (conversion factor) which relates grams to number of atoms. The mass of a single C-12 atom is 1 carbon atom times this factor.

$$1 \text{ carbon-12 atom} \times \frac{12.0 \text{ g}}{6.02 \times 10^{23} \text{ carbon atoms}} = 1.99 \times 10^{-23} \text{ g}$$

The mass in grams of 1 amu is

$$\tfrac{1}{12} \times 1.99 \times 10^{-23} \text{ g} = 1.66 \times 10^{-24} \text{ g}$$

FOLLOW-UP PROBLEM

The first chemical tests made on artificially produced elements involved microgram quantities. On September 22, 1942, the first weighable sample of a man-made element was isolated. This historical sample of pure plutonium oxide had a mass of 2.77 micrograms. Assume that the average atomic mass of plutonium is 242 and calculate the number of atoms of plutonium in the sample. Assume the oxide to be PuO_2.

Ans. 6.06×10^{15} atoms.

Rutherford's concept that an atom consists of a tiny, positively charged nucleus surrounded by a cloud of rapidly moving electrons is an improvement on Dalton's indivisible hard-sphere model. It does not, however, satisfactorily explain the arrangement and behavior of electrons about the nucleus. The electron configuration in atoms may be determined by analyzing the light absorbed or emitted by atoms. In order to relate the electronic structure of an atom to the light absorbed or emitted by it, we must first examine the nature and characteristics of radiant energy.

THE NATURE AND CHARACTERISTICS OF RADIANT ENERGY

5-13 The Wave Nature of Radiant Energy You are familiar with many types of radiant energy. Visible light, infrared and ultraviolet light, X rays, and gamma radiation from radioactive materials are forms of this type of energy.

In 1856, James Clerk Maxwell (1831–1879) wrote a famous theoretical paper in which he attempted to describe the nature of radiant energy. He proposed the existence of waves which were related to both electricity and magnetism and called them ***electromagnetic waves.*** His observations resulted in a set of four mathematical expressions, now referred to as *Maxwell's equations,* which generalize the behavior of this type of wave. Since light and all the other types of radiant energy described above may be explained in terms of these equations, they may be classified as electromagnetic waves.

It is easier to illustrate the behavior of waves than to attempt to explain what a wave is. In general, we think of waves as disturbances which originate from vibrating sources. For example, a vibrating object held in a tub of water acts as a center of disturbance and generates waves which radiate outward from the center of disturbance to the edge of the tub. These waves transmit energy, as may be shown by placing a cork in the water. The cork gains the kinetic energy of motion transmitted to it by the wave, but it does not move toward the edge of the tub. Although the wave itself travels from point to point, nothing in the medium (individual water molecules) or on it (the cork) moves along with the wave toward the edge of the tub. Once the wave has passed, the cork is left at its original place. It is the disturbance (wave) that travels, not the medium. These examples are used to illustrate two basic facts pertaining to all waves: *they originate from a center of disturbance and they transmit energy from the center of disturbance to some distant point.*

Some waves require a physical medium for their propagation (transmission). Water and sound waves are examples of this type. There can be no water waves without water and no sound waves without air or some other form of matter. The ringing of a bell inside an evacuated jar cannot be heard because there is no material medium (air) available for the propagation of the wave. *Electromagnetic waves however, require no medium for transmission.* Light and radio waves from the sun and distant stars travel through space and transmit energy to the earth. Maxwell demonstrated that these electromagnetic waves could be generated by an accelerating electrically charged particle, much in the same way that a boat generates waves as it moves along the surface of a still lake.

A ***uniform wave*** may be thought of as a continuous sequence of alternating crests and troughs, as shown in Fig. 5-21. On the basis of these crests and troughs, all waves may be described in terms of three fundamental characteristics: (a) wavelength, (b) frequency, and (c) velocity.

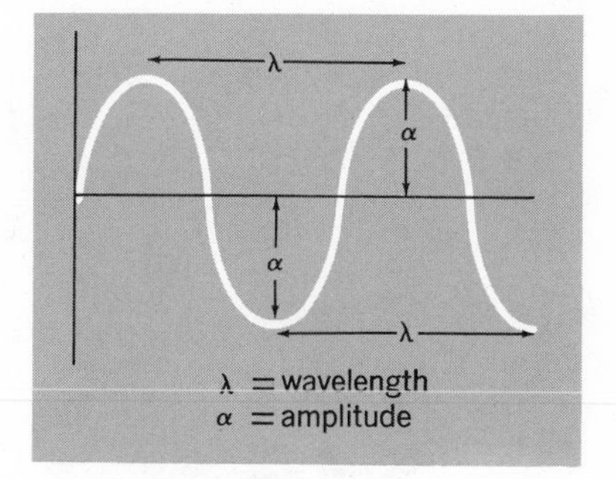

5-21 The wavelength of a uniform wave is a measure of the distance crest to crest. This distance is designated by λ. The field strength or height of the wave is called the amplitude, sometimes designated by α.

5-14 Wavelength Wavelength is the distance between corresponding points on adjacent waves. Consider the waves in Fig. 5-21. The distance between adjacent crests or troughs is called the ***wavelength.*** It is symbolized by the Greek letter lambda (λ). Wavelengths

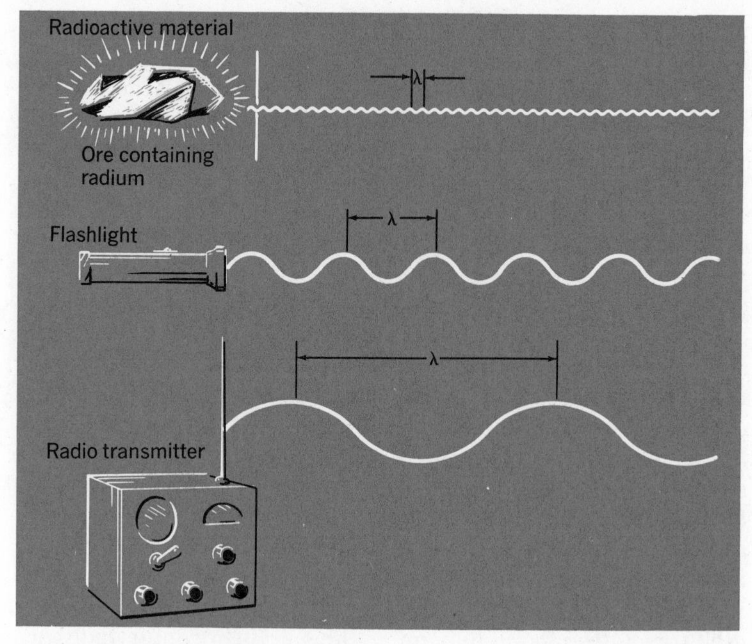

5-22 The wavelength and the amplitude of an electromagnetic wave depend upon the nature of the system generating the wave. Gamma rays have a wavelength of a few Ångstroms; visible light waves, a few thousand Ångstroms; radio waves, many meters.

of electromagnetic waves may be expressed in any linear unit such as meters or centimeters. We shall be concerned mostly with visible light which involves very short wavelengths, usually expressed in terms of an ***Ångstrom,*** symbolized as Å. There are 100 million (10^8) Ångstroms in 1 centimeter. Centimeters can be converted to Ångstrom units by multiplying the number of centimeters by 10^8 Å/cm. For example, the value of 3.50×10^{-5} cm expressed in Ångstroms is 3.50×10^{-5} cm $\times$ 10^8 Å/cm $= 3.5 \times 10^3$ Å.

FOLLOW-UP PROBLEM

Light with a wavelength of 3.00×10^3 Å lies in the ultraviolet region of the spectrum. Calculate the wavelength of this radiation in centimeters.

Ans. 3.00×10^{-5} cm.

5-15 Frequency Frequency is the number of waves which pass a given point in a unit of time. In Fig. 5-23, the distance between two adjacent soldiers is analogous to one wavelength (1λ), and the number of rows of soliders passing a particular point each minute is analogous to the frequency of vibration of a wave. Frequency is usually designated by the greek letter nu, (ν), although it is sometimes designated by *f*. The unit of frequency is cycles per second or, more simply, reciprocal seconds, sec^{-1}. When we use the term cycle in this sense, we are referring to a complete cycle of change. This involves the departure of a wave crest, the passing of the trough, and the arrival of the next crest. It is apparent from the figure that *as the wavelength decreases, frequency increases.*

5-16 The Velocity of a Wave The velocity with which a wave travels is equal to the product of the wavelength and the frequency.

This relationship, which applies to all waves, may be expressed as

$$\mathbf{V} = \nu\lambda \qquad 5\text{-}2$$

in which V is the velocity, ν the frequency, and λ the wavelength.

The velocity of a wave depends on the medium through which the wave travels. The ultimate velocity is assumed to be the velocity of propagation of electromagnetic waves through a vacuum. This velocity has been experimentally determined to be 3.0×10^{10} cm/sec (186 000 mi/sec). The latter figure is what is generally referred to as the speed of light. The constant velocity of light in a vacuum is symbolized by c. The speed of any wave in any other medium is less than this value. There is, however, only a negligible difference between the value of c measured in a vacuum and the value measured in air. Since we shall be primarily concerned with electromagnetic radiation traveling in air, we may substitute c for V in Equation 5-2. In addition, we shall usually be interested in calculating the frequency of electromagnetic radiation. In view of these facts, a more useful form of this equation is

$$\nu = c/\lambda \qquad 5\text{-}3$$

Example 5-4

The wavelength of the radiation which produces the yellow line in the spectrum of sodium is approximately 6.0×10^3 Å. What is the frequency of the radiation that produces this line?

Solution

1. Since the velocity of light is given in cm/sec, it is first necessary to convert Ångstroms to centimeters.

$$6.0 \times 10^3 \text{ Å} \times 10^{-8} \text{ cm/Å} = 6.0 \times 10^{-5} \text{ cm}$$

2. Substituting in Equation 5-3 yields

$$\nu = \frac{c}{\lambda} = \frac{3.0 \times 10^{10} \text{ cm sec}^{-1}}{6.0 \times 10^{-5} \text{ cm}} = 5.0 \times 10^{14} \text{ sec}^{-1}$$

5-17 The Electromagnetic Spectrum The arrangement of all types of electromagnetic radiation in order of increasing wavelength is known as the electromagnetic spectrum. The different types of electromagnetic radiation differ from one another only in their wavelengths and frequencies. The wavelengths of different types of radiation vary from a few millionths of a centimeter to many miles. Each wavelength is determined by the source of the radiation. The visible region of the spectrum ranges from the *shorter wavelength violet light* to the *longer wavelength red light.*

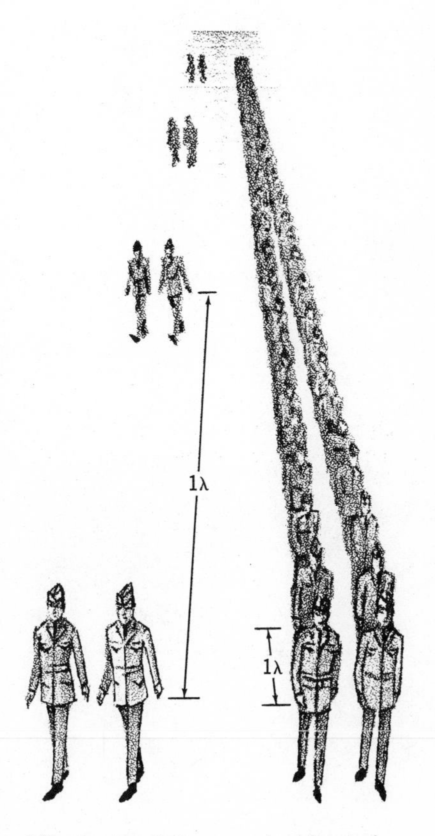

5-23 Assuming both groups of soldiers to be moving at the same velocity, which group has the greater wavelength? Which has the greater frequency?

5-24 Visible light which includes wavelengths from 4000 Å to 7600 Å represents only a small portion of the electromagnetic spectrum.

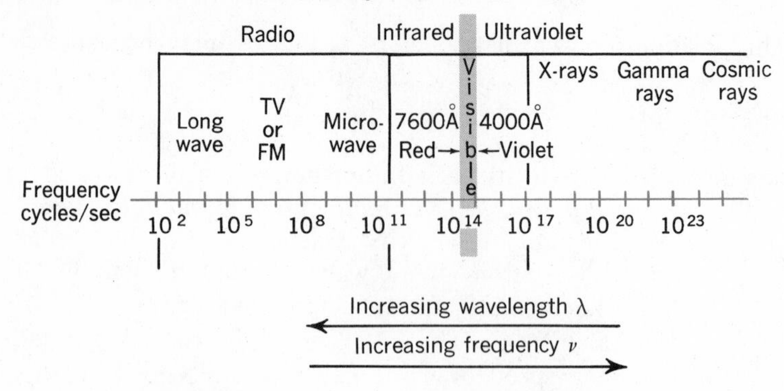

"I do not know what I may appear to the world; but to myself I seem to have been only like a boy playing on the sea-shore, and diverting myself in now and then finding a smoother pebble or a prettier shell than ordinary, whilst the great ocean of truth lay all undiscovered before me."

Sir Isaac Newton (1642–1727)

This means that the human eye can detect electromagnetic waves that range in wavelength from 3800 Å to 7600 Å (0.000 038 cm to 0.000 076 cm). Fig. 5-24 shows that the visible wavelengths constitute only a minute portion of the entire electromagnetic spectrum.

The region just above this visible spectrum includes the *infrared frequencies.* Infrared energy may be produced by vibrations of molecules. It can also set molecules in motion and, therefore, according to the Kinetic Molecular Theory, may increase the temperature of a sample of matter. The terms *infrared energy* and *heat* are often used interchangeably. Just above the infrared region are the *microwave frequencies.* Microwaves have longer wavelengths than infrared waves and are used in telephone transmission and for determining a number of important characteristics of molecules. Beyond the microwaves in the spectrum are the short and long radio waves.

Just below the violet end of the visible spectrum we find the *ultraviolet region* of the spectrum. These waves of relatively short wavelength are components of the sun's rays and cause suntan or sunburn. Chemists have developed chemicals that are used in suntan oils and lotions to reflect or filter out this short-wavelength radiation. Below the ultraviolet, we move into the *X-ray region* and next, in order of decreasing wavelength, are the *gamma rays* that emanate from the nuclei of radioactive atoms. *Cosmic rays* have the shortest wavelength and are the most energetic of all radiations. This radiation originates in outer space and rains down on the earth night and day. In our study of chemistry, we shall be concerned mainly with the visible and the near-visible portions of the spectrum.

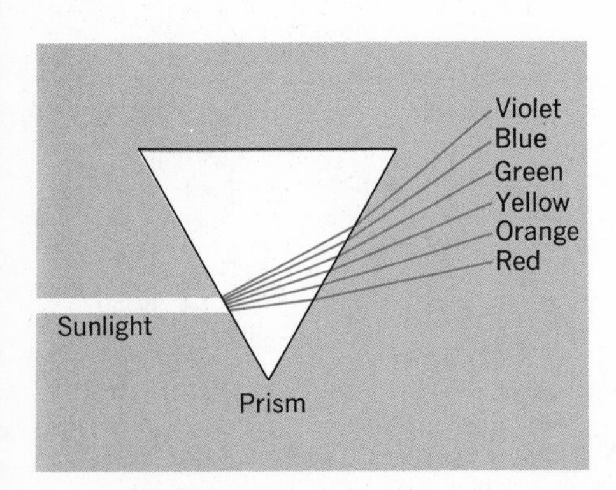

5-25 When sunlight is passed through a prism, it is dispersed into its component colors. Red light travels faster through the medium of the prism (glass) and is bent the least.

5-18 The Dispersion of Light Light may be dispersed by a glass prism into its component colors. When sunlight (white light) is passed through a prism, the light is dispersed, resulting in a rainbow of colors. This rainbow or series of color bands constitutes a ***continuous solar spectrum*** of all the visible wavelengths. A discontinuous spectrum differs in that certain wavelengths are missing. This dispersion of light by a prism was explained by Isaac Newton (1642–1727). From his experiments, Newton came to the following conclusions.

1. White light is a mixture of many colors (wavelengths).
2. A prism does not make colors; it merely separates colors that already exist.
3. Each component of the white light is bent (refracted) through a different angle by a prism, resulting in the spectrum.

Violet light with the shortest wavelength is refracted to the greatest extent, and the longer-wavelength red light is refracted the least. Refraction occurs when light passes from one medium to another; for example, from the air into the glass of a prism and then back into the air. This is explained by the fact that the speed of light varies in different media, and speed depends on two factors: the nature of the medium and the wavelength of the light. Thus, white light is dispersed by a prism because each of the component wavelengths travels with a different velocity through the glass. Red light travels the fastest and is refracted the least. It appears on the side of the spectrum toward the apex of the prism. Violet light travels the slowest and is refracted to the greatest extent. Consequently, it appears on the side of the visible spectrum toward the base of the prism (Fig. 5-25).

In 1859, Robert Bunsen (1811–1899) and Gustav Kirchhoff (1824–1887) developed the ***spectroscope,*** an instrument which can be used to study the spectra of luminous sources. A spectroscope contains a prism or a diffraction grating for dispersing the light and a telescope to enable the observer to examine the spectra. In some spectroscopes, the telescope is replaced by a photographic film on which the spectrum is reproduced. This instrument is called a spectrograph, and the photograph of the spectrum is called a spectogram. A spectogram enables a scientist to measure the wavelengths of the components of various light samples. A simple spectroscope and a cutaway view of a spectroscope are shown in Fig. 5-26. A schematic view is given in Fig. 5-27.

5-19 Atomic-Emission Spectra When the light emitted by energized atoms, such as the mercury atoms in a mercury lamp or the sodium atoms in a sodium lamp, is examined with a spectroscope, a series of colored lines separated by dark spaces is observed. Each line corresponds to a specific wavelength. A spectrum containing only specific wavelengths is called a ***discontinuous*** or ***line***

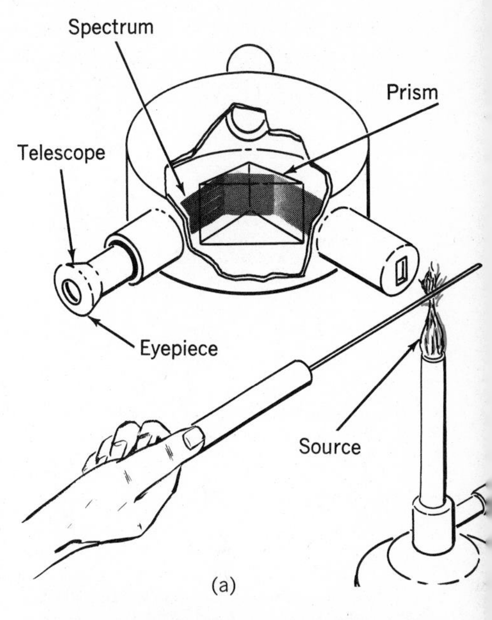

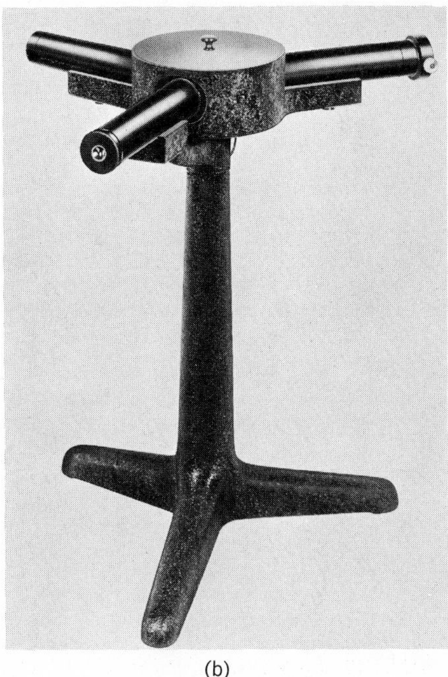

5-26 A simple prism spectroscope. Light from the source is separated into its component colors by the prism. Analysis of the distribution of the colors provides data related to the structure and composition of the source. (Bausch & Lomb)

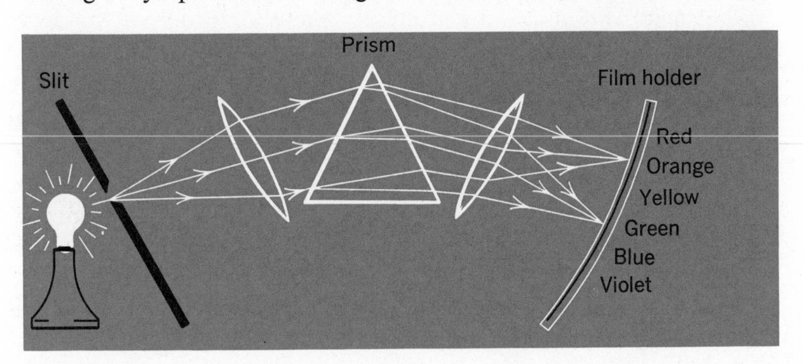

5-27 A prism spectroscope separates light into its component wavelengths.

MAX PLANCK
1858–1947

Max Planck, a German theoretical physicist, is best known for his revolutionary suggestion that radiation emitted by matter consists of packets of energy, or quanta. This was the first suggestion that the classical laws of physics might not always apply to atomic systems. This quantum theory was used by Einstein to describe the photoelectric effect, by Bohr to develop his concept of quantized energy levels, and by many other scientists who played leading roles in the development of modern physical theories.

Planck was a versatile man with many interests. When he entered the University of Munich, he seriously considered a career in music rather than in physics. He chose physics in spite of the suggestion from his adviser that little remained to be discovered in the field of physics. In 1918, at the age of 60, he was awarded the Nobel Prize in physics. Planck balanced his intellectual activities with many hobbies, two of which were music and mountain climbing.

Planck was a man of unflinching courage, loyal to his friends and dedicated to what he believed was true. In 1934, while at the Kaiser Wilhelm Institute in Berlin, Planck showed his courage by attending a memorial service for Fritz Haber even though the Reichministry of Education warned him against going to the service because of Haber's non-Aryan background. He was held in disfavor by the Nazis because of his support of Jewish scientists. One of his own sons was executed by the Nazis for participating in a plot to assassinate Hitler.

spectrum. The line or ***atomic-emission spectrum*** of an element is unique; that is, no two elements have the same spectrum. Spectroscopic analysis, therefore, may be used to identify an element.

In 1884, Johann Balmer (1825–1898) of Switzerland energized atoms of hydrogen gas and examined the visible radiation with a spectroscope. He found four prominent colored lines, illustrated in Color Plate I (opposite p. 274) in the hydrogen spectrum:

1. A red line with wavelength 6.563×10^{-5} cm
2. A blue-green line with wavelength 4.861×10^{-5} cm
3. A blue line with wavelength 4.340×10^{-5} cm
4. A violet line with wavelength 4.102×10^{-5} cm

It was not until 1913 that the origin of the spectral lines was satisfactorily explained by Neils Bohr (1885–1962), the great Danish physicist. A revolutionary proposal by Max Planck in 1901 provided the clue which enabled Bohr to explain the line spectra of atoms. It also set the stage for the "satellite" or "solar system" model of atoms developed in 1913 by Bohr.

THE QUANTUM THEORY

5-20 The Particulate Nature of Radiant Energy As a result of studying the relative energy of the light of different frequencies radiated from incandescent bodies (those which glow because of intense heat), Planck proposed that radiant energy, such as light and heat, is not emitted continuously but in little packets called ***quanta.*** He suggested that the *quantum of energy* depends upon the frequency of the radiation. Assuming that the source of the light from the hot bodies was vibrating atoms, he proposed that vibrating atoms could have only certain allowed energy values (*quantized energy*). In other words, *energy could only be emitted or absorbed in discrete units* by vibrating atoms. This is rather analogous to a tuning fork which remains motionless until another fork in the room is vibrating with its own specific frequency.

To explain his theory quantitatively, Planck developed an equation that *related the allowed energy of a vibrating atom to its frequency.* This equation is

$$\mathbf{E} = \mathbf{h}\nu \qquad 5\text{-}4$$

where E is the energy in ergs, ν is the frequency in reciprocal seconds (sec^{-1}) and h is a fundamental constant of nature, Planck's constant, which is equal to 6.62×10^{-27} erg-sec. According to Planck's hypothesis, atoms would not start vibrating until energy corresponding to a certain allowed value was absorbed. The atoms would then start vibrating with an amplitude corresponding to this quantized value of energy. Planck further postulated that when more energy was added, the amplitude would not gradually increase but would change suddenly when energy corresponding to twice the original quantum had been supplied.

Equation 5-4 can be used to show that high-frequency violet light has more energy than lower-frequency red light. If you have had experience with sun lamps which emit ultraviolet light and heat lamps which emit infrared energy, you might conclude that heat lamps give off more energy because of the warmth you feel. According to Equation 5-4, this is not the case. The infrared energy from a heat lamp increases the motion of atoms and molecules, resulting in an increase in temperature. The rays from the ultraviolet lamp in contrast have sufficient energy to produce chemical changes in your skin. X rays and gamma rays have even higher frequencies and are more energetic. The calculation of the energy associated with light of a specific color is illustrated by the following example.

Example 5-5
A certain violet line in a spectrum has a wavelength of 4100 Å. What is the energy of the photons (quanta) which give rise to this line?

Solution
1. Converting 4100 Å to centimeters gives

$$4100\text{ Å} \times 1\text{ cm}/10^8\text{ Å} = 4.1 \times 10^{-5}\text{ cm}$$

2. Substituting in Equation 5-2 gives

$$\nu = \frac{3.0 \times 10^{10}\text{ cm/sec}}{4.1 \times 10^{-5}\text{ cm}} = 7.3 \times 10^{14}\text{ sec}^{-1}$$

3. Substituting ν in Equation 5-3 gives

$$\begin{aligned} E &= 6.6 \times 10^{-27}\text{ erg sec} \times 7.3 \times 10^{14}\text{ sec}^{-1} \\ &= 4.8 \times 10^{-12}\text{ ergs} \end{aligned}$$

FOLLOW-UP PROBLEM

A certain red line in the spectrum of an element has a wavelength of 6500 Å. What is the energy of the photons (quanta) which give rise to this line?

Ans. 3.0×10^{-12} **ergs.**

Albert Einstein (1879–1955) expanded Planck's idea of the particle nature of radiant energy to include light or free radiant energy dissociated from all matter. Einstein assumed that the energy of electromagnetic radiation was not distributed over the entire wave front but was concentrated in packets of energy called ***photons.*** Photons are not material bodies as are atoms. They may be thought of as "massless" bundles of energy. Although it is rather difficult to visualize something that has both particle and wave characteristics, Einstein did not deny the possibility of this coexistence.

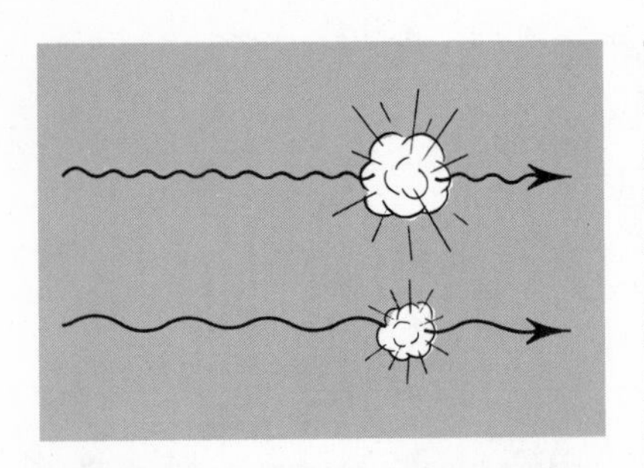

5-28 The energy of a short wavelength photon is greater than that of a long wavelength photon. We may visualize these bundles of energy as being guided by the waves associated with them.

The best we can do is say that a wave of frequency, ν, carries its energy in bundles (photons) of size $h\nu$. In other words, a wave guides the flow of energy which is transported in bundles of definite size. As a result of these discoveries, *light is said to have a dual nature.* Under certain conditions it exhibits particle characteristics. Under other conditions, it behaves as a wave. For example, light is exhibiting wave characteristics when it undergoes *diffraction* and particle characteristics when it dislodges electrons from certain metals (exhibits the photoelectric effect). Thus, *energy is transferred either by means of waves or by means of particles.* Since a stream of particles cannot be diffracted like waves, diffraction can be used to determine whether an energy transfer involves waves or particles. Let us briefly examine the phenomena of diffraction and the photoelectric effect.

5-21 Diffraction ***Diffraction*** refers to the ability of a wave to bend around the edges of an obstacle or an opening. It can best be explained by the following comparison. A stream of particles moving in a path perpendicular to a barrier passes through a narrow slit in the barrier without being deflected into the shadow region behind the barrier. When, however, a wave strikes a barrier with a narrow slit, the wave that emerges on the other side spreads out beyond the boundary of the slit into the shadow region of the barrier. The water waves in Fig. 5-29 illustrate diffraction. You can observe the diffraction of light by looking through a pinhole in a piece of cardboard at a bright automobile headlight which is 10 to 12 feet away. You can observe a diffraction pattern that consists of a series of dark and bright rings surrounding a central bright spot. Alternatively, you can observe a diffraction pattern by looking at a source of light through a linen handkerchief held close to your eye. The linen has many slits, formed by the woven threads, which serve as openings through which light waves pass. The waves passing through the slits bend around the edges of the slits and produce a diffraction pattern that can be observed as a visible spectrum. Diffraction gratings, which produce an atomic emission spectrum, are often used in spectroscopes to resolve light from excited atoms into component wavelengths. The diffraction grating is usually a piece of glass or plastic on which thousands of parallel lines have been ruled or scratched.

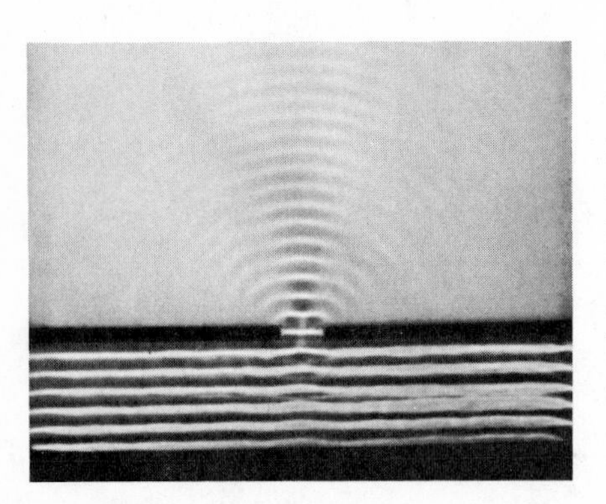

5-29 When water waves strike a barrier having a small slit, they are diffracted as they pass through the slit. The circular waves which are propagated beyond the barrier spread out into the shadow region of the barrier. (Physics, 2nd Edition, D. C. Heath.)

Since a stream of particles cannot be diffracted like waves, diffraction can be used to identify wave phenomena. Some of the most precise and valuable measurements in science are made with the use of instruments based on the principle of diffraction. The diffraction of X rays is used to determine ionic radii, crystal structure, and other characteristics of material particles.

5-22 The Photoelectric Effect Maxwell thought that electromagnetic radiation was continuous and developed the *wave model of light.* This model is successful in explaining the phenomenon of diffraction but cannot explain the ***photoelectric effect.*** The photoelectric effect refers to the ability of light to eject electrons from a piece of metal (Fig. 5-30). It was first noted by J. J. Thomson. Upon irradiating metals in a vacuum, he observed that the particles that left the surface of the metal were identical in charge and mass to cathode-ray particles. In fact, these particles were electrons.

According to classical wave theory, the energy of light depends on the *intensity* (intensity is related to the number of photons). It can be shown, however, that light of very great intensity but low frequency cannot produce the photoelectric effect. To explain the photoelectric effect, it is necessary to use the *"particle" theory of light,* in which the energy of the photon depends on the frequency of its wave. When light having a certain frequency strikes an electron in a metal, the photon imparts its energy to the electron and disappears. Part of the energy is used to break the electron away from the atom in the metal surface, and the rest is used to give the electron a certain velocity. Increasing the frequency increases the velocity of the photoelectron. Increasing the intensity (number of photons per unit time per unit area) does not change the velocity. These observations led to the acceptance of the particulate nature of light.

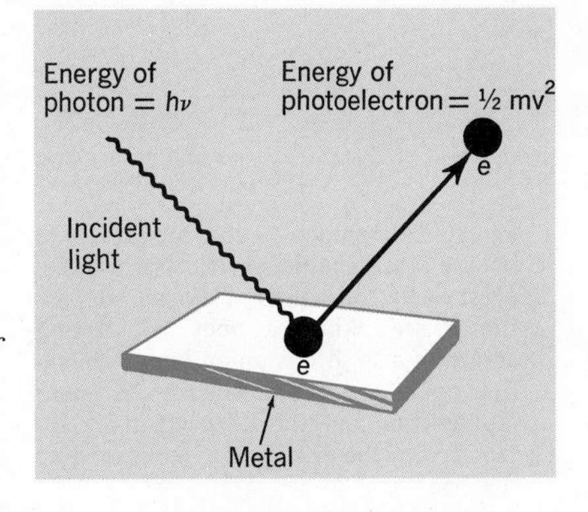

5-30 In the photoelectric effect, a photon of energy, $h\nu$, ejects a photoelectron from a metal. The velocity of the photoelectron depends on the frequency of the incident photon and not on the total number of photons.

Any experiment designed to show the wave nature of light is successful. Likewise, any experiment which attempts to prove the particulate nature of light does so. As a result, electromagnetic energy is said to be dualistic in nature. Scientists use parts of both models, depending on their experiments. Let us now investigate the role played by the quantum theory in the development of Bohr's atomic model.

THE BOHR MODEL OF ATOMIC STRUCTURE

5-23 Energy States of a Hydrogen Atom In 1913, Niels Bohr, using Rutherford's concept of a nuclear atom, Planck's quantum theory, and great imagination, proposed a new model of atomic structure. This new model explained two important observations that could not be explained by either the Dalton or the Rutherford model. These were: the *stability of atoms* and the *wavelength of the lines in the hydrogen spectrum.*

According to the classical, electrodynamic theory developed by Maxwell in 1856, the electrons whirling around the nucleus in Rutherford's model of atoms should be attracted to and spiral into the nucleus. In other words, the atom should be unstable and should collapse readily. The observation that matter exists is evidence that Rutherford's atomic model is incorrect and incomplete. In addition to the instability of atoms, the classical theory predicts that a continuous range of energies would be emitted by an atom, so that a continuous atomic-emission spectrum should be observed.

Bohr proposed a theory of atomic structure based on Planck's quantum theory. Bohr's revolutionary model was based on three postulates.

1. *Electrons can occupy only certain specific energy levels which are sometimes referred to as orbits or shells.* An electron does not emit energy when it is in one of these permitted energy levels. This postulate explains the stability of atoms because it does not permit an electron to lose energy continuously and spiral into the nucleus.

2. *Energy is radiated only when an electron falls from a higher to a lower energy level.* Atoms can emit or absorb energy only in specific amounts (quanta). The amount of energy absorbed or liberated is equal to the difference in energy possessed by the electron in the two different energy levels. This postulate, which incorporates Planck's quantum concept, explains the discontinuous atomic-emission spectra of atoms. That is, it permits an atom to emit radiation of a specific wavelength only.

3. *The angular momentum of an electron revolving about the nucleus is quantized.* That is, Bohr said the angular momentum can take only specific values which are equal to

$$\frac{nh}{2\pi}$$

where h is Planck's constant and n is an integer that is restricted to any whole-number value. It turns out that n identifies the energy level (shell) in which the electron is located.

Linear momentum is the product of a body's mass and linear velocity. It is expressed by the equation, linear momentum = mv. When a body is rotating about an axis, it has an angular momentum defined as the product of its linear momentum and the perpendicular distance from the axis. This is expressed by the equation

$$\text{angular momentum} = mvr$$

where r is the distance of the body from the axis of rotation.

Bohr's model of the atom may be visualized as a miniature solar system with the electrons analogous to planets. In the case of stable atoms, Bohr assumed that an electron is kept in its orbit by an inward centripetal force which was caused by and equal to the electrostatic force of attraction exerted by the nucleus. By expressing these relationships in quantitative terms and then introducing his quantized angular momentum requirement, Bohr was able to derive an expression which enabled him to calculate the radii of shells which represent the different energy levels within a hydrogen atom. This expression is

$$r = n^2 \frac{h^2}{4\pi^2 mZe^2} \qquad 5\text{-}5$$

where Z equals the nuclear charge, e equals the charge on an electron, and m is the mass of the electron. For hydrogen, $Z = 1$. The radius of the first energy level (first Bohr orbit) may be calculated using the expression

$$r = (1)^2 \frac{(6.62 \times 10^{-27})^2}{4(3.14)^2(9.11 \times 10^{-28})(4.80 \times 10^{-10})^2(1)}$$
$$= 5.29 \times 10^{-9}\ \text{cm}$$
$$= 0.529\ \text{Å}$$

Since all factors in this equation except n are constants, the radius of any Bohr orbit may be calculated using the expression

$$r = (n)^2(0.529\ \text{Ångstroms})$$

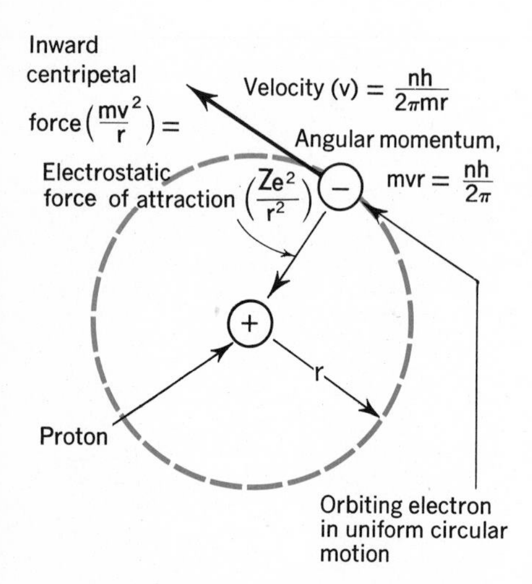

5-31 To account for the stability of a hydrogen atom, Bohr equated the inward centripetal force (mv^2/r) to the inward electrostatic force (Ze^2/r^2)

$$mv^2/r = Ze^2/r^2$$

When the velocity (v) obtained from the quantized angular momentum requirement is substituted in this equation, an equation for the radius of the Bohr orbits (Equation 5-5) is obtained.

The total energy of an electron is equal to the sum of its kinetic and potential energy. It can be shown that the total energy is

$$E = -Ze^2/2r \tag{5-6}$$

Substituting the value of *r* from Equation 5-5 in Equation 5-6 yields

$$E = \frac{-2\pi^2 Z^2 e^4 m}{n^2 h^2} \tag{5-7}$$

where *E* represents the energy associated with an electron when it occupies energy levels having the different allowed values of *n*.

If we wish to describe the energy state of a single atom, then a convenient unit of energy to use is one electron volt (ev). One electron volt is the amount of energy a single electron absorbs when it passes through a potential difference of one volt. One electron volt is an extremely small energy unit. It represents about the same amount of energy that the smallest imaginable particle of chalk dust would lose if it were to fall over a cliff as high as the thickness of this page. Two useful conversion factors which relate electron volts to ergs, calories, and kilocalories are given below.

$$\begin{aligned} 1 \text{ electron volt} &= 1.60 \times 10^{-12} \text{ ergs} \\ 1 \text{ electron volt} &= 3.83 \times 10^{-20} \text{ cal} \\ &= 3.83 \times 10^{-23} \text{ kcal} \end{aligned}$$

The energy associated with a hydrogen atom when its electron is in the first level is given by the expression

$$\begin{aligned} E &= \frac{-2(3.14)^2(1)^2(4.80 \times 10^{-10})^4(9.11 \times 10^{-28})}{(1)^2(6.62 \times 10^{-27})^2} \\ &= -2.18 \times 10^{-11} \text{ erg/atom} \end{aligned}$$

$$E_{ev} = \frac{-2.18 \times 10^{-11} \text{ erg}}{\text{atom}} \times \frac{1 \text{ ev}}{1.6 \times 10^{-12} \text{ erg}} = -13.6 \text{ ev}$$

At this point, it will be instructive to interpret the significance of this negative number. In Section 6-7, we shall describe an experiment in which the energy to remove an electron completely from an atom is measured. Applied to a neutral hydrogen atom with an electron in its lowest energy level, the experiment reveals that 13.6 ev is required to overcome the attractive force between the electron and its nucleus. In other words, energy is added to the atomic system to remove an electron. This means *that the atomic system has a greater potential energy when an electron is far from the nucleus than when close to it.* If we arbitrarily assign the atomic system zero potential energy when the electron is completely free from the nuclear attraction, then the value of the potential energy decreases and has a negative sign as the electron approaches the nucleus. Applied to a hydrogen atom, we can say that the atomic system has 13.6 ev less energy when the electron is in the first energy level than when it is at infinity. Since the potential energy

HENRY GWYN JEFFREYS MOSELEY
1887–1915

Occasionally a man will accomplish an amount of work in his youth that few could do in an entire lifetime. Such was the case with Henry Moseley. Although he was only four years old when his father died, his mother prepared him for school so well that he entered Eton College at nine years of age with a King's Scholarship. He was so fascinated with mathematics that he taught himself algebra at the age of nine.

After graduating from Trinity College, Oxford, he went to work for the celebrated scientist Ernest Rutherford at Manchester.

Moseley read and followed closely the experiments of Max von Laue and William and Henry Bragg, who were using X rays to study crystal structure. It was his interest in this work that led him to his greatest scientific contribution. Moseley's experiments consisted of bombarding targets made out of different metals with a beam of electrons. He noted that each metal emitted X rays of a characteristic frequency. He repeated this experiment for numerous elements and found that the heavier the element, the shorter was the wavelength and the more penetrating were the X rays. When he organized the observed X rays in order of increasing frequency, Moseley found an occasional gap which indicated to him that an element was missing. He interpreted the regular increase in the frequency of the X rays in terms of a regular increase in positive charge on the nuclei of atoms. Moseley's atomic number concept solved the riddle of the rare earth (inner transition) elements which had perplexed Mendeleyev and other scientists for many years and explained the apparent irregularities in the location in the Periodic Table of such elements as potassium and argon. In addition, his work stimulated further search for new elements, just as Mendeleyev's work had done 50 years before.

Moseley volunteered for military service during World War I and became a signal officer. He was killed in action in Gallipoli in 1915 at age 27.

with the electron at maximum distance from the nucleus is zero, it must have a negative value at all other distances. Hence, the energy value of the atomic system in its lowest energy state (electron closest to the nucleus) is −13.6 ev. Interpretation of the atomic-emission spectrum of hydrogen shows that 13.6 ev is evolved when an electron which has been removed from the attraction of the nucleus returns to the first, or lowest possible energy level of a hydrogen atom.

Note that all of the factors in Equation 5-7 except n are constants. Thus, the permitted energy states of a hydrogen atom may be calculated, using the equation

$$\mathbf{E_n = \frac{-13.6}{n^2}\ ev/atom} \qquad 5\text{-}8$$

when n represents the energy level containing the electron.

FOLLOW-UP PROBLEM

What is the energy state of a hydrogen atom expressed in ev/atom when its electron is in the second energy level of the atom?

Ans. −3.4 ev.

The potential energy of a hydrogen atom is at a minimum when its electron is in the energy level nearest the nucleus; thus, the total energy of a hydrogen atom is at a minimum. For all other levels, the potential energy is greater and has a more positive value. Thus, the energy associated with an electron occupying the second energy level is −3.4 ev, which is more positive than −13.6 ev. The allowed radii and energies associated with each energy level may be calculated and represented schematically by the solar-system diagram, Fig. 5-32, or the energy-level diagram, Fig. 5-33.

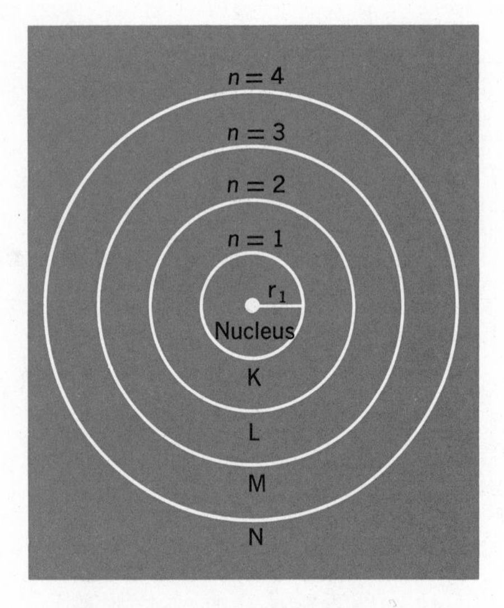

5-32 A solar system model of the hydrogen atom. By using Bohr's equation, the distance from the nucleus to the lowest energy electron shell (r_1) is calculated to be 0.529 Å.

The energy associated with a mole of hydrogen atoms is usually expressed in terms of kcal/mole. Using the conversion factor

$$\mathbf{1\ kilocalorie = 4.18 \times 10^{10}\ ergs}$$

we may express the energy associated with a mole of hydrogen atoms as

$$\mathbf{\frac{-2.18 \times 10^{-11}\ erg}{atom} \times \frac{1\ kcal}{4.18 \times 10^{10}\ erg} \times \frac{6.02 \times 10^{23}\ atoms}{1\ mole\ atoms} = -313.6\ kcal/mole}$$

This value indicates that 313.6 kcal is evolved when a mole of electrons free from the nuclear attraction drops into the first energy levels of a mole of hydrogen atoms. Conversely, it means that 313.6 kcal is required to remove a mole of electrons completely from the first energy levels of a mole of hydrogen atoms completely away from the attraction of the nuclei. Thus, the energy states of a hydrogen atom may be calculated, using the equation

$$E_n = \frac{-313.6}{n^2} \text{ kcal/mole of electrons}$$

Electron volts per atom may be converted directly into kilocalories per mole of atoms by using the relationship

$$\frac{\text{ev}}{\text{atom}} \times 23.1 = \frac{\text{kcal}}{\text{mole of atoms}}$$

Perhaps you can derive the "magic" number 23.1.

FOLLOW-UP PROBLEM

What is the energy state of a hydrogen atom expressed in kcal/mole of electrons when its electron is in the second energy level of the atom?

Ans. −78.4 kcal/mole.

5-24 Origin of Atomic-Emission Spectral Lines The energy levels in a hydrogen atom may be identified in terms of a principal quantum number known as n. Thus, the energy level closest to the nucleus is designated as either the $n = 1$ level or the first level. Succeeding levels are designated as $n = 2$, $n = 3$, $n = 4$, $n = 5$, $n = 6$, and $n = 7$. When an electron is completely independent of the nuclear attraction, $n = \infty$. When an electron is in the first energy level, the atom has minimum potential energy and is said to be in the ***ground state.*** When electric energy is applied to hydrogen gas in a discharge tube, energy is absorbed by the atoms as electrons are raised to higher energy levels. That is, work is done in separating an electron from the nucleus which attracts it. Thus, the potential energy of an atom is greater when the electron is in the second level than when it is in the first level. When an electron is in a level other than the first level, the atom is said to be in an *excited state.* Since the excited state is unstable, the electrons return almost immediately (10^{-8} sec) to their more stable lower energy levels. The difference in the energy possessed by an electron in the two different energy levels is emitted by the atom as a discrete amount of radiant energy. For example, when an electron drops from the second to the first energy level, the energy evolved is equal to $E_2 - E_1$. That is, e

$$\textbf{energy evolved} = -3.4 \text{ ev} - (-13.6 \text{ ev}) = 10.2 \text{ ev.}$$

This value may be converted to either ergs/atom or kcal/mole of atoms by using the proper conversion factors.

Since photon units are involved in this transfer of energy, we may represent the emission of radiation by an excited atom as

$$E_2 - E_1 = h\nu$$

where E_2 is the energy of the atom in the higher energy state, E_1 is the energy of the atom in the lower energy state, and $h\nu$ is the energy of the photon. The frequency of the wave associated with

Energy level	Potential energy (kcal/mole)	(ev)
$n = \infty$	0	0
$n = 6$	−8.7	−0.4
$n = 5$	−12.5	−0.5
$n = 4$	−19.6	−0.9
$n = 3$	−34.8	−1.5
$n = 2$	−78.4	−3.4
$n = 1$	−313.6	−13.6

5-33 An energy level diagram for the hydrogen atom. Negative values indicate the energy evolved when a mole of electrons drop from $n = \infty$ to a lower level. Positive values indicate the energy absorbed in removing a mole of electrons from a given level to $n = \infty$.

a photon may be expressed as

$$\nu = \frac{E_2 - E_1}{h}$$

Frequency is related to wavelength by the equation

$$\lambda = c/\nu$$

The "line," as it is called, is merely the image of a vertical slit through which the light travels. The light that one sees comes from the contributions of many photons of the same energy being emitted simultaneously by identical electron transitions in a large number of atoms.

Thus, each electron transition from a higher to a lower level produces a photon of specific wavelength which contributes to the "line" in the spectrum of the element. The greater the distance between the levels, the greater is the energy and the shorter is the wavelength of the emitted radiation.

The following analogy may help you to understand the energy relationship that exists between the excited and unexcited states of an atom. Let us consider the steps on a staircase as being analogous to the energy levels in an atom (Fig. 5-34). A ball resting on the floor represents a position of minimum potential energy analogous to the ground state in a hydrogen atom. When resting on any step of the staircase, the ball is in an unstable, higher potential-energy position. The ball has maximum potential energy on the top step. This is analogous to a free electron which is completely away from the influence of the nucleus. When the ball falls from the top step, it bounces down the stairs to a position of minimum potential energy on the floor below. Similarly, an electron in an atom is permitted to make only definite jumps. The electron cannot stop between the energy levels any more than the ball can hang in midair between the steps.

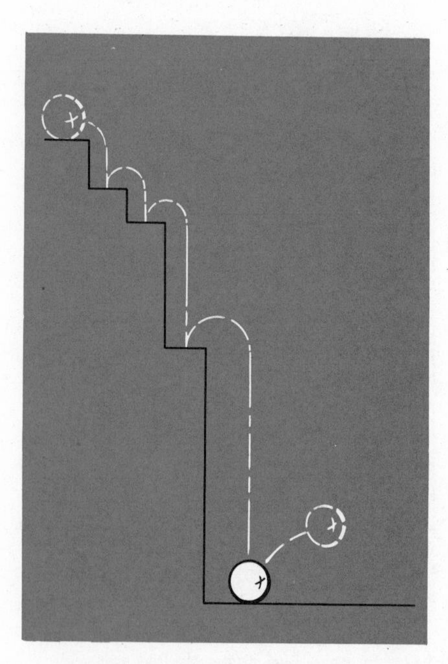

5-34 A ball rolling down a staircase is analogous to an electron dropping from a higher to a lower energy level within an atom. The bottom of the staircase is analogous to the lowest energy level within the atom.

The electrons, in making transitions from outer to inner levels, release energy, the emission of which is indicated by the production of spectral lines. In a sample of hydrogen gas, there are countless numbers of atoms. Individual atoms in the sample absorb and liberate different amounts of energy. For example, an electron in some of the hydrogen atoms may drop from the fourth to the second major energy level. These particular atoms emit energy which corresponds to the difference in energy between the fourth and second energy levels. This transition contributes to the blue-green line observed in the hydrogen spectrum. In other hydrogen atoms, an electron may drop from the third level to the second level. These transitions represent a smaller energy gap and, hence, these atoms emit radiation of longer wavelength which is observed as the red line in the visible spectrum of hydrogen. The visible lines in the hydrogen spectrum constitute what is known as the ***Balmer series.*** It can be shown that these lines are produced when an electron jumps from the 3rd, 4th, 5th, or 6th level to the 2nd energy level (Fig. 5-35). The intensity of the color of a particular line depends on the number of electrons simultaneously making the transition represented by the line.

Bohr was able to calculate theoretically the wavelength of the known visible lines in the hydrogen spectrum and also to predict

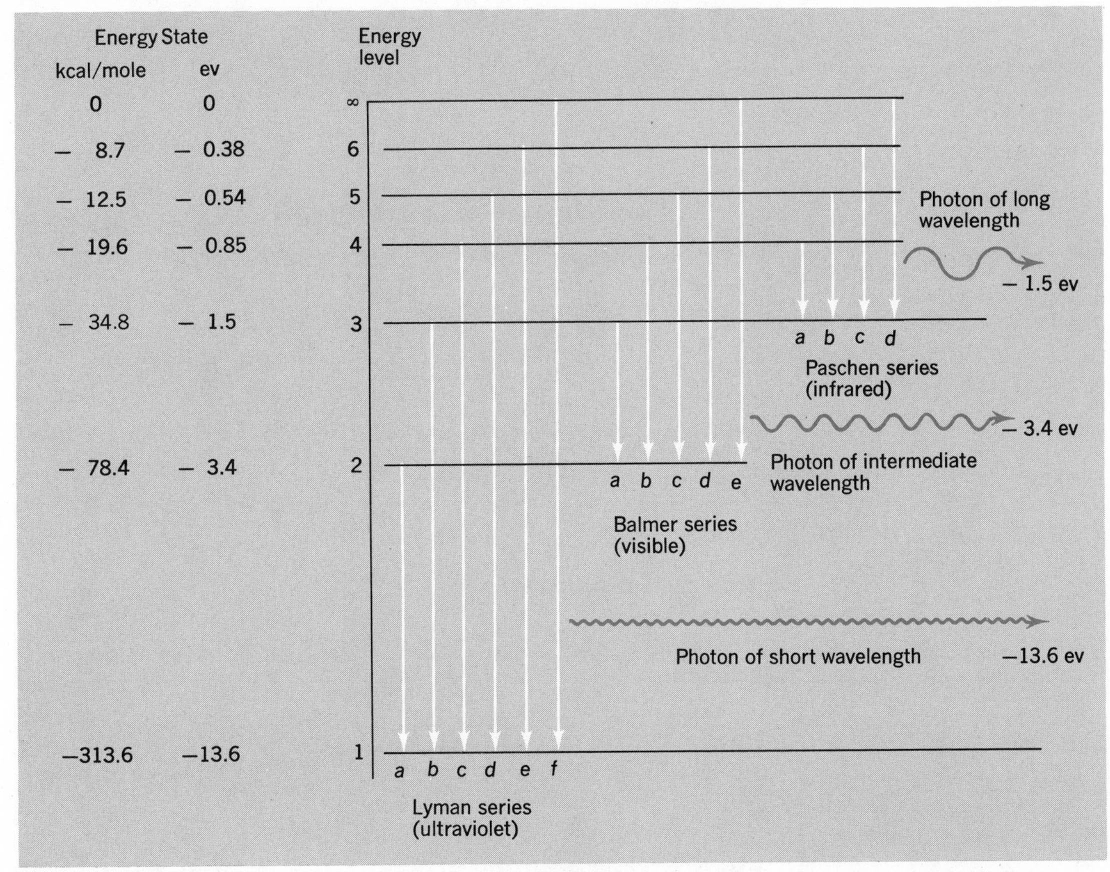

5-35 When an electron falls from a higher to a lower energy level, a photon is emitted. The greater the difference in energy levels, the greater will be the energy of the emitted photon.

the existence of an undiscovered series of lines in the ultraviolet region of the hydrogen spectrum. From his own theoretically derived equation, he predicted the wavelengths of these unknown lines. The subsequent discovery of the ***Lyman series*** of lines with wavelengths similar to those predicted by Bohr gave added support to Bohr's original theory. This series originates when electrons fall from the 6th, 5th, 4th, 3rd, and 2nd levels to the first energy level. Perhaps you can use the energy level diagram in Fig. 5-35 to explain why the Lyman series is in the ultraviolet rather than the visible region of the spectrum.

The next example illustrates how the wavelength of a spectral line may be calculated using the different energy states of a hydrogen atom.

Example 5-6
Show that the transition of an electron from n = 4 to n = 2 in a hydrogen atom gives rise to the blue-green spectral line with a wavelength of 4861 Å.

Solution
1. The energy evolved when a mole of electrons falls from

energy level 4 to energy level 2 may be determined by referring to the energy-level diagram shown in Fig. 5-35.

$$E_{evolved} = E_4 - E_2 = -0.85 \text{ ev} - (-3.40 \text{ ev})$$
$$= 2.55 \text{ ev}$$

2. Substitute known values for the factors in the equation, $(E_4 - E_2) = \frac{hc}{\lambda}$ and solve for λ. Using electron volts, the setup is

$$-0.85 \text{ ev} - (-3.40 \text{ ev}) = \frac{(6.62 \times 10^{-27} \text{ erg sec})(3.0 \times 10^{10} \text{ cm/sec})}{\lambda}$$

$$\lambda = \frac{(6.62 \times 10^{-27} \text{ erg sec})(3.0 \times 10^{10} \text{ cm/sec})}{(2.55 \text{ ev})(1.6 \times 10^{-12} \text{ erg/ev})}$$
$$= 4.87 \times 10^{-5} \text{ cm}$$

FOLLOW-UP PROBLEM

Show that the transition of an electron from $n = 5$ to $n = 2$ in a hydrogen atom gives rise to the blue line with a wavelength of 4340 Å.

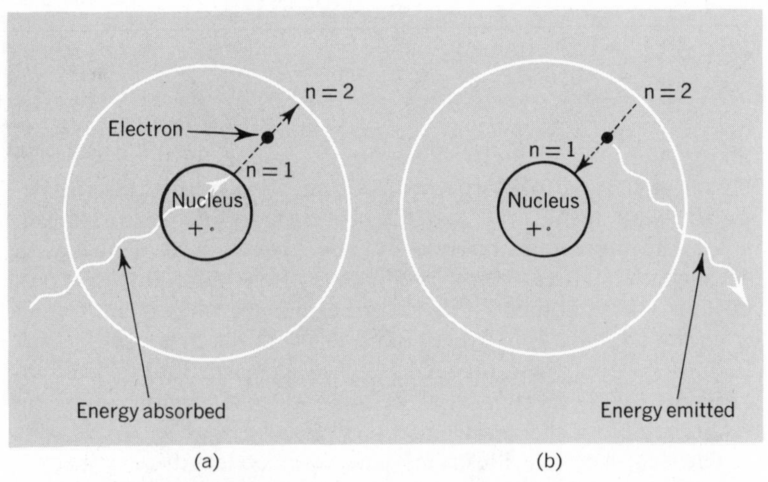

5-36 When the electron in the first energy level of the hydrogen atom absorbs 10.2 ev of energy, it makes a *quantum jump* to the second level. When it returns to the first level, it gives up 10.2 ev of energy. This energy is associated with ultraviolet radiation.

When excited, each element emits radiant energy with a unique set of wavelengths. In other words, no two elements have the same spectrum. Therefore, the spectrum of an element is like a set of fingerprints and can be used to identify the element. See Color Plate I between pp. 274 and 275. Helium in the sun's atmosphere was first "discovered" by means of spectral analysis. It was later found on earth and again identified by its spectrum. The composition of a star billions of miles away can be determined by spectroscopic analysis of its emitted light.

5-25 Quantum Jumps The following demonstration can be used to show that atoms can absorb or emit energy of only specific values. In this demonstration two different sources of ultraviolet light are used. Try to predict which lamp uses mercury atoms as a source of light.

In a darkened room, mercury vapor is allowed to pass between a source of ultraviolet light and a fluorescent screen (filter paper coated with anthracene), as shown in Fig. 5-37). Although mercury vapor is invisible, dark clouds of "smoke" are seen drifting across the screen. (Do not try this demonstration yourself. Mercury vapor is extremely poisonous.) The "smoke" is the shadow of mercury atoms which have absorbed the ultraviolet light that strikes them. In the case of the short-wavelength ultraviolet light, the photons contain the exact amount of energy needed to effect a *quantum jump;* that is, to promote an electron in a mercury atom from its unexcited (ground) state to a higher energy level. Consequently, this energy is absorbed by the atom and does not reach the screen as light energy. As a result, the atom casts a shadow. If ultraviolet light of longer wavelength is used, a shadow is not produced. Since the longer-wavelength photon does not possess sufficient energy to promote an electron from one specific level to a higher level, the energy of the photon is not absorbed. As you may have predicted, the lamp emitting the longer-wavelength light does not use mercury as the source of the radiation.

5-26 The Electron Population of Energy Levels The electron population of each energy level in atoms having more than one electron can be determined by spectroscopic analysis and explained

NIELS BOHR
1885–1962

Niels Bohr, one of the greatest and most famous atomic physicists of the 20th century, proposed the familiar solar-system picture of the atom. The influence exerted by Bohr and his many brilliant students on the course of modern atomic theory cannot be overestimated.

In 1913, he used the concept of the nuclear atom proposed by Ernest Rutherford and the quantum theory proposed by Max Planck to develop the solar-system model of the atom. This model explained the spectral lines of hydrogen observed by Johann Balmer, the Swiss spectroscopist. Bohr was awarded the 1922 Nobel Prize in physics for his work on atomic structure.

His studies and theories on the nature of the atomic nucleus enabled him to play a leading role in the development of nuclear energy. Bohr, collaborating with Otto Frisch and Lise Meitner in 1939, interpreted the atomic nucleus as a "drop of water" which might be capable of rupturing under certain circumstances. It was he who brought the information that it was possible to split the uranium nucleus to the United States.

When World War II ended, Bohr returned to Denmark to the Copenhagen Institute for Theoretical Physics. For his intense interest and work on the problem of the peaceful use of atomic energy, he received the first Atoms for Peace Award in 1957.

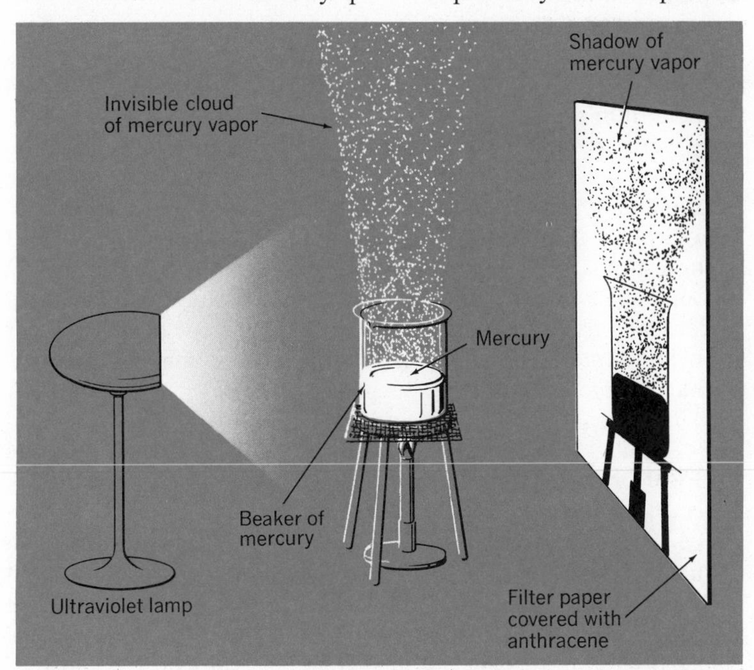

5-37 The invisible cloud of mercury atoms absorbs the ultraviolet light from the lamp and casts a shadow on the fluorescent screen. Since only light of the specific wavelength, 2537 Å is absorbed, an ultraviolet lamp emitting this wavelength must be used.

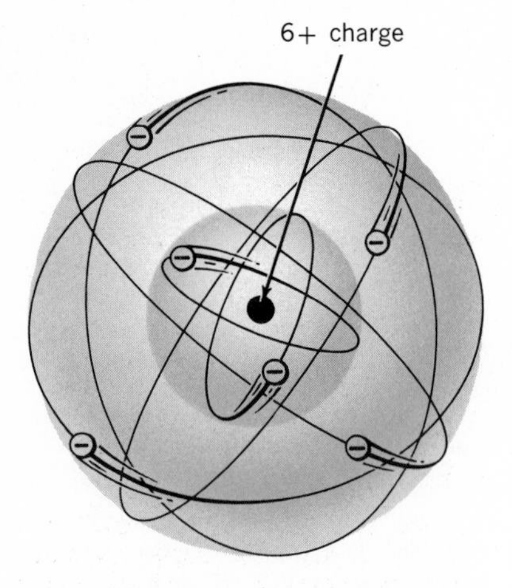

5-38 Bohr model of a carbon atom. Two electrons occupy and fill the first energy level and four electrons occupy but do not fill the second energy level. How many more electrons may be accommodated in the second energy level?

in terms of the more advanced atomic theory discussed in the next chapter. It will be shown that the maximum number of electrons allowed in a given principal energy level is $2n^2$ where n is the number of the principal energy level. Thus, the maximum number of electrons in a given energy level is limited by the value of n.

Using the $2n^2$ formula, we can show that the first energy level can contain a maximum of 2 electrons and that the second level can contain a maximum of 8 electrons. Isolated atoms in the ground state have minimum potential energy. This means that the electrons occupy the lowest available energy levels. Thus, an isolated chlorine atom (17 electrons) has 2 electrons in the first level, 8 electrons in the second level, and 7 electrons in the third level. Beyond calcium, element number 20, we need additional information regarding the distribution of electrons in atoms. This information is provided in Chapter 6.

LOOKING AHEAD

Bohr's theory satisfactorily explained the stability and atomic emission spectra of a hydrogen atom. It also contributed the important concept of fixed energy levels or states, and enabled scientists to construct a model of an atom that is easy to visualize.

Unfortunately, this model does not satisfactorily explain the electron configuration, nature, and behavior of atoms containing many electrons. First of all, Bohr's model does not explain the spectra of multi-electron atoms nor does it account for the observed energy differences between the electrons which occupy the same principal energy level. In addition, it fails to explain the shapes and characteristics of molecules. Furthermore, some of the factors found in Bohr's equations were not a natural outcome of his theory. They were arbitrarily introduced by Bohr because they fixed the positions of electron orbits, enabling him to predict correctly the experimentally observed wavelengths of the hydrogen spectrum.

To explain the electron configuration and behavior of multi-electron atoms, it is necessary to develop a more complex theory and model of atomic structure. The need for a more adequate model leads us to investigate the modern concept of an atom in the next chapter. The model is known as the quantum or wave-mechanical model.

Special Feature: The Laser

The discoveries made, apparatus used, and concepts developed by Maxwell (electromagnetic radiation theory), Crookes (cathode ray tube), Roentgen (X rays), Thomson (electron characteristics), Planck (quantum theory), Einstein (photoelectric effect), Bohr (energy states [levels] of atoms, quantum jumps) revolutionized science and provided the tools which led to the development of many of our amazing technological devices. The operation of a number of these devices can be explained in terms of the behavior of electrons, the nature of electromagnetic radiation, and the interaction between the two. This chapter's two special features are designed to acquaint you with technological applications of fundamental concepts and principles. In these cases, basic research supplied the underlying principles and applied research produced the application. In this feature we shall examine the amazing laser and in the following feature we shall briefly explain color television.

"It is not enough that you should understand about applied science in order that your work may increase man's blessing. Concern for man himself and his fate must always form the chief interest of all technical endeavors, concern for the great unsolved problems of the organization of labor and the distribution of goods, in order that the creations of our mind shall be a blessing and not a curse to mankind. Never forget this in the midst of your diagrams and equations."

Albert Einstein,
from an address given at
California Institute of Technology, 1931

The laser is a device which produces a high-intensity beam of coherent light capable of performing amazing feats. The intensity is so great that it can instantly burn needle-like holes through such high-melting materials as diamond and hard steel alloys. Although lasers are still in the research stage, many varied and practical uses have been found for them. They have been used to "weld detached retinas in place in the eye, melt the interior of a tooth without harming the surroundings, weld high-melting metals, cut out garments from fabric, and in communication setups. Their potential in the area of communication is staggering. It has been calculated that every person in the United States could be talking in pairs on the telephone and all of their voices could be transmitted at the same time on a single laser beam using only four-fifths of the beam's capacity. Let us examine the principle underlying these unusual devices.

The proposal of the German physicist, Max Planck, that light consisted of energy packets called photons or quanta, supplies the theoretical basis for the understanding of the laser. Historically, the laser was an outgrowth of a device known as a maser. The term maser is an acronym derived from the words *m*icrowave *a*mplification by *s*timulated *e*mission of *r*adiation. Microwaves consist of photons of electromagnetic energy like light or infrared radiation but are of a much longer wavelength. Professor Charles H. Townes (1915–   ) formerly of Columbia University, received a Nobel Prize for his work on the development of the maser which was the forerunner of the laser. The first maser was built and operated by Townes in 1954. This device consists of an oven which emits a beam of energized ammonia molecules which are separated so that the energized (spinning) molecules are caught in the cavity of a receiving device. A beam of microwaves is then directed into this cavity where it is absorbed by the activated ammonia molecules. The excited molecules then emit a new beam of microwave photons which are extremely uniform in wavelength. This is known as a *maser beam*. A beam of microwave or ordinary light photons which are very uniform in wavelength is said to be coherent. Beams of coherent and incoherent photons are compared in Fig. 5-39.

Coherent radiation consists of photons of nearly the same wavelength. These "in-step" photons travel in straight beams with a minimum of scattering.

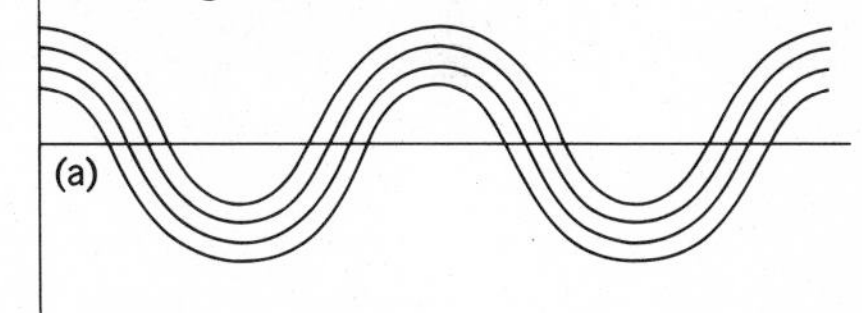

Incoherent radiation consists of photons of varying wavelengths. Beams of these "out-of-step" photons tend to scatter.

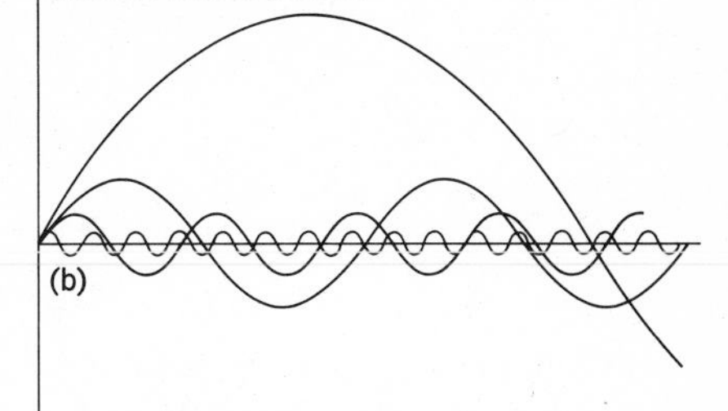

5-39 A comparison of coherent and incoherent photon waves is shown in this diagram of wave models.

Researchers on masers predicted that a similar type of coherent radiation involving light rather than microwaves might be possible.

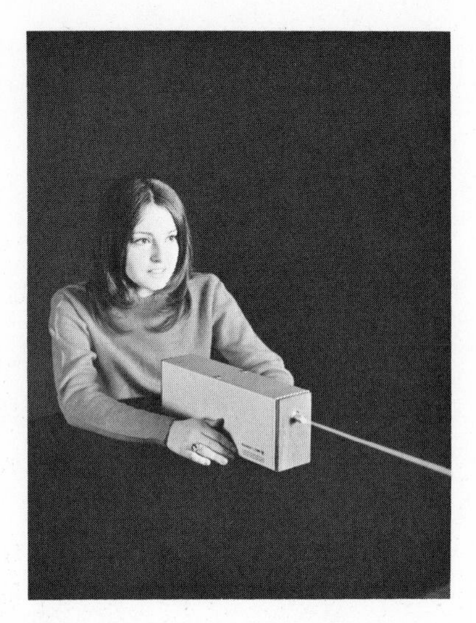

5-40 A laser emitting a beam of coherent photons.

The first laser was successfully put into operation by Theodore H. Maiman of the Hughes Research Laboratory in Malibu, California in 1960. He caused an intense flash of red light to be emitted from a pink synthetic ruby rod when light from a powerful strobe was "pumped" into the side of the rod. The laser, whose name is derived from the acronym, *l*ight *a*mplification by *s*timulated *e*mission of *r*adiation, has been a subject of intense research since its invention.

Let us now examine the mechanism of this device. We know that when an electron within an atom drops from a higher to a lower potential energy level, a photon is emitted. In this chapter we have seen how an invisible cloud of mercury vapor may absorb photons and cast a shadow on a fluorescent screen. An electron within the mercury atom has, in this case, absorbed a photon and moved to a higher energy level within the atom. Absorption of a photon by an electron is depicted in Fig. 5-41(A). Fluorescence, which is essentially the opposite of absorption, is depicted in Fig. 5-41(B). You have probably seen minerals which have been exposed to an ultraviolet lamp (sometimes called a black light) glow in the dark. In this case an electron, which has been excited from a lower to a higher energy level, spontaneously drops back to its original level, with a photon being emitted. Laser action, which is depicted in Fig. 5-41(C), involves the drop of an electron and the subsequent emission of a photon when stimulated by a photon of the same wavelength. The original photon which stimulated the emission is there also because no absorption has taken place.

The ruby rod in the laser experiment performed by Maiman in 1960 was a synthetic ruby with both ends polished and coated with

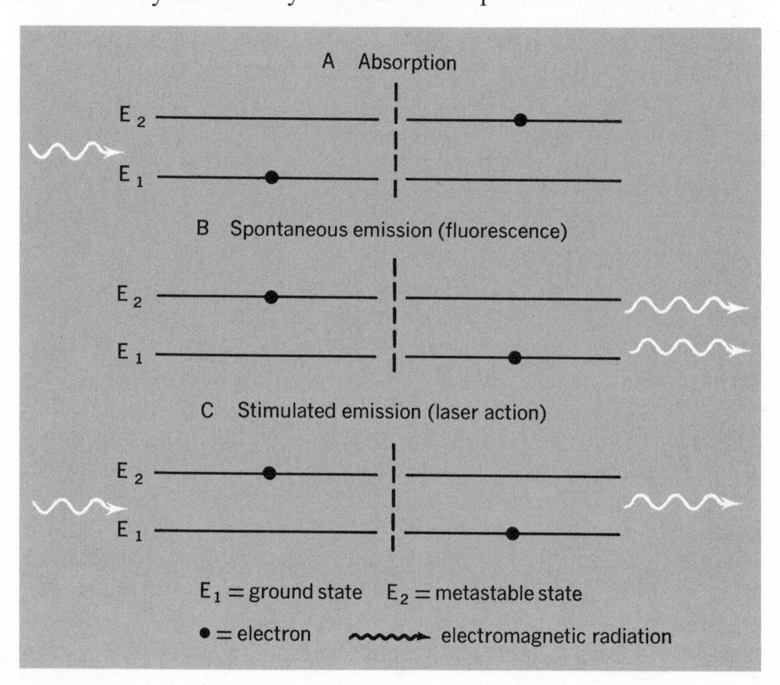

5-41 (A) Absorption of light by an electron raises it to an excited state. (B) Spontaneous emission of light is called fluorescence. (C) Stimulated emission of light may occur when electromagnetic radiation is applied to an electron in an excited (metastable) state.

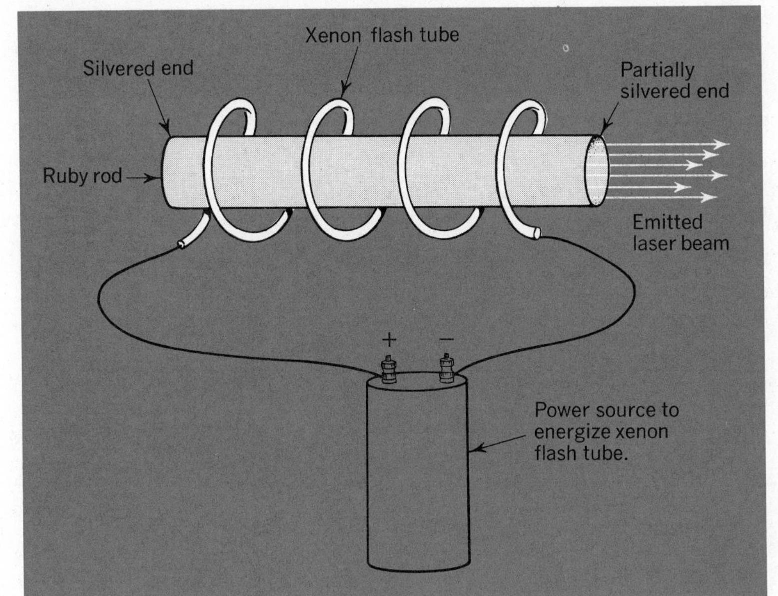

5-42 Schematic diagram of the essentials of a ruby laser.

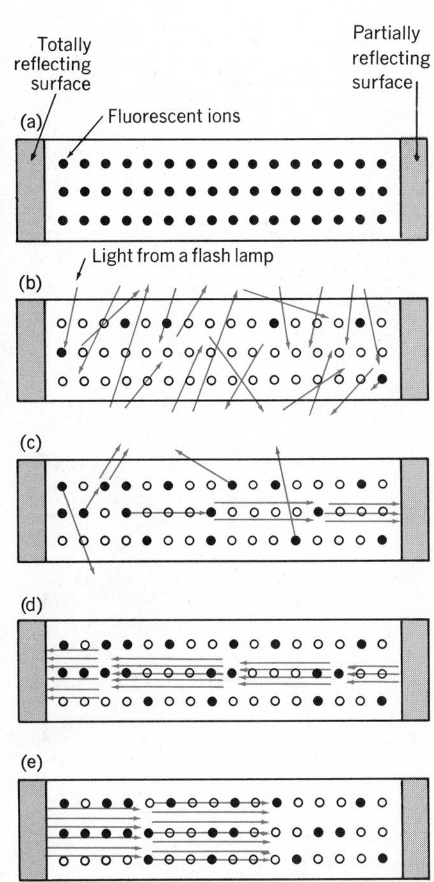

5-43 Sequence of laser action. (a) A laser rod containing fluorescent ions. (b) Light from a flash lamp exciting some ions. (c) Excited ions emitting light spontaneously, some of which is lost through the surface of the rod. (d) The radiation emitted parallel to the long axis of the rod initiates laser action and begins to bounce from end to end. (e) Radiations continue to bounce. (f) Radiation is finally emitted as a laser beam through the partially transmitting end of the rod.

silvery reflecting material. One end was coated with a totally reflecting surface and the other with a partially reflecting surface as shown in Fig. 5-42. The ruby rod consists of an aluminum oxide medium with a dispersion of 0.05 percent chromium atoms. The chromium atoms which impart the red color to the ruby, also are responsible for the laser action. The xenon flash tube surrounding the ruby rod, as shown in Fig. 5-42, "pumps" photons into the rod. Although some of the photons are lost out the side of the rod, many are propagated along the long axis of the rod. These cause excited atoms to emit other photons as shown in Fig. 5-42. The radiation continues to build up more photons of the same wavelength and bounce back and forth between the silvered ends of the rod until the beam is emitted through the partially silvered end of the rod. The sequence of the laser action is shown in Fig. 5-43.

The intensity of the energy emitted by the laser staggers the imagination. When sunlight is focused with the best lenses, a concentration of photons of about 500 watts per square centimeter can be obtained. This is sufficient to burn almost any flammable objects and to melt most other materials. A laser beam can focus over 100 million watts per square centimeter. The energy output of a ruby laser is such that if it were maintained for one second, enough energy could be produced to lift a 150-pound man over 500 feet in this single second. The ruby laser flash actually takes place in less than $\frac{1}{2000}$ of a second so these large energy outputs have really never been attained. The coherent light of a laser beam has such a small tendency to scatter that it can be focused into a spot the size of a small fraction of the point of a needle. A laser burst that can instantly burn needle-like holes through a diamond or hard steel alloys may consist of about 2×10^{20} photons. These photons

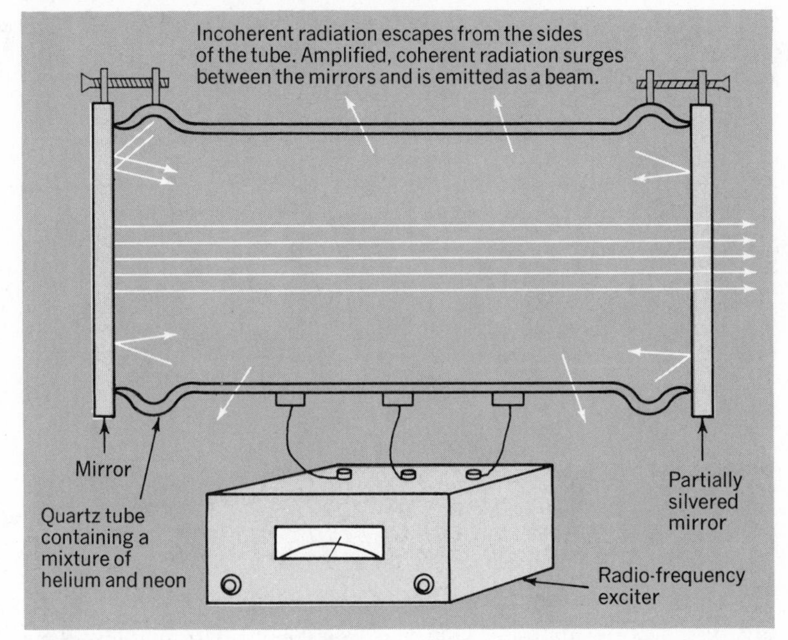

5-44 A schematic diagram of a gas laser.

could be emitted from a ruby rod containing about $\frac{1}{3000}$ of a mole or 0.014 grams of chromium. The coherent light from a laser can travel in such straight lines that a beam projected to the moon is scattered to a circle on the moon of only about two miles in diameter. Ordinary incoherent light from a straight beam would have spread to several moon diameters before reaching the moon.

Research with crystals other than ruby have produced a host of types of lasers. Gas lasers have been developed by substituting for the ruby rod a quartz tube filled with helium and neon. Gas lasers emit a steady stream of coherent infrared light rather than sudden bursts of energy. Fig. 5-44 shows the essentials of a gas laser. It is now possible to purchase a gas laser instrument for aligning an airplane wing for about $400.

Scientists are conducting research on the possibility of using gas lasers for the detection of minor earth tremors. It is possible that this research could lead to a system of predicting earthquakes. Gas lasers have been used to determine the velocity of light to a degree of precision never before attained using incoherent light. No one can tell what wonderful uses the laser will have, but research indicates that there are many ways in which this tool can serve man.

5-45 Electrons at work. The motion and energy of electrons are utilized in radios, record players, tape recorders, television sets, countless types of machinery and a host of other devices which lighten man's work load and make his leisure hours more enjoyable.

Special Feature: Television

The color television camera has three independent systems for each of the three primary colors: red, blue, and green. As the light from a scene enters the lens of the camera the light is separated by a system of mirrors into these three primary colors. Fig. 5-48 shows a schematic representation of how the scene is transmitted

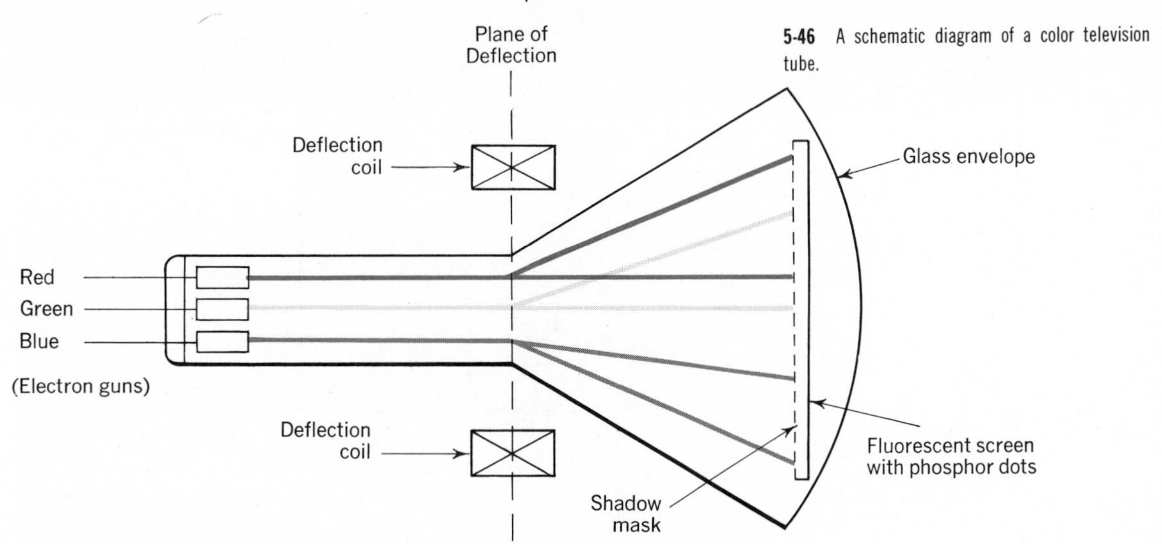

5-46 A schematic diagram of a color television tube.

through the camera, to the transmitting antenna. The camera tube has a "light-sensitive" screen where the image from the lens is projected. Electrons are knocked loose from this screen by the photoelectric effect. This means that a photon of light can cause an electron to be ejected from a sample of matter. The electron that is removed from this photosensitive screen then strikes the target. Thus, only a few electrons impinge on the target which correspond to the darker areas of a scene. An electron gun shoots a scanning beam of electrons onto the target as shown in Fig. 5-47. This beam sweeps across the target in repeating horizontal lines so that 525 sweeps are made from the top to the bottom of the screen. Thirty complete cycles are made by the scanning beam each second. This means that each horizontal sweep is made in about $\frac{1}{16000}$ of a second. You can see that any rapid action that takes place in front of the camera will be instantly recorded by this scanning beam. The electrons from the electron gun bounce back off the target and form a returning beam. This beam is amplified by a photomultiplier tube and sent to the transmitting antenna. The electric pulses in the antenna are converted into a video signal which is an electromagnetic wave. This wave travels at the speed of light to the antennas of television receivers. The theory of the television camera and transmitting antenna was first suggested by James Clerk Maxwell. Maxwell's theory of electromagnetic radiation stated that electric oscillations in a wire would send out electromagnetic radiation. The frequency of the oscillation in the wire determines the frequency of the wave that is produced.

The electromagnetic oscillations sent out by the television transmitting antenna in turn produce electronic oscillations in the receiving antennas of television sets. The signals from the antenna are transmitted to three electron guns in the back of the picture tube of the television set. Each electron gun corresponds to the three primary colors picked up by the camera. See Fig. 5-46.

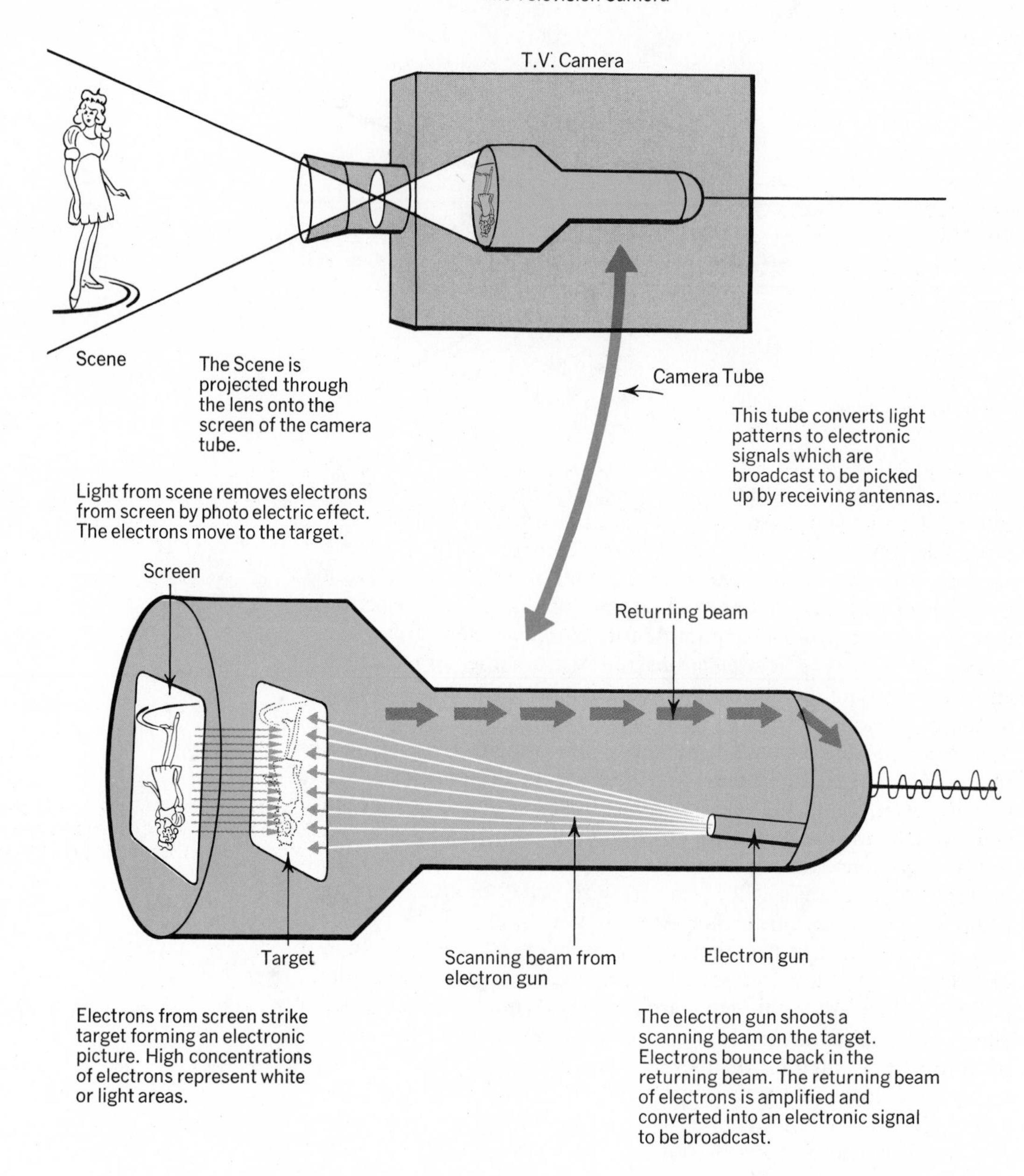

5-47 A schematic representation of a television camera tube picking up and recording an image for transmission.

The screen is composed of an array of phosphor dots which glow red, blue, or green respectively when a high-energy electron strikes the dot. When red and green are mixed the result is yellow. If the red, green, and blue are mixed with the proper intensities the result on the television screen is white. Mixing colors on a television screen is altogether different from mixing the pigments of paint. The image on the screen seems to show all colors but is really just a mixture of the three primary colors. The phosphor dots are clustered on the screen in groups of three, one for each color. Each dot is less than one-half millimeter in diameter. The phosphors on this screen are composed of rare-earth oxides. When a high-energy electron from the electron gun strikes the phosphor dot, the kinetic energy of the electron is converted to visible light as the phosphor dot glows. Electrons within the phosphor are promoted to higher energy levels by the high-energy electron from the electron gun. As these electrons drop back to their original energy level, they emit a photon of a specific wavelength. We therefore observe that the dot glows with a specific color. The electron guns in the picture tube scan the screen in the same way that the electron gun of the camera scanned the target. The screen therefore is scanned with 525 horizontal lines with 30 complete scanning cycles in each second. In effect, the scanning beam traces the picture you see dot by dot. The beam moves so fast and the dots are so close together that the human eye sees a complete picture.

The shadow mask which is located between the electron guns and the screen contains over 200 000 tiny holes which correspond to the phosphor dots on the screen. The shadow mask is aligned in such a manner that the red, blue, and green electron guns respectively can impinge their electrons only on the red, blue, and green phosphor dots. The color hues and intensity of color are controlled by the electronic circuitry of the television set. Increasing the voltage increases the kinetic energy of the electrons and brightness of the picture. Fig. 5-48 on pages 188 and 189 shows the composite of the entire television system from the scene to the viewer. Fig. 4-47 on the opposite page shows a schematic representation of a scene being transmitted through the camera to the antenna.

It should be noted that the high-energy electrons in color television tubes, like those in Roentgen's cathode ray tube, may produce X rays.

Television has been under study for nearly 50 years but color television is only one of the many examples of how man has effectively put electrons to work for his benefit.

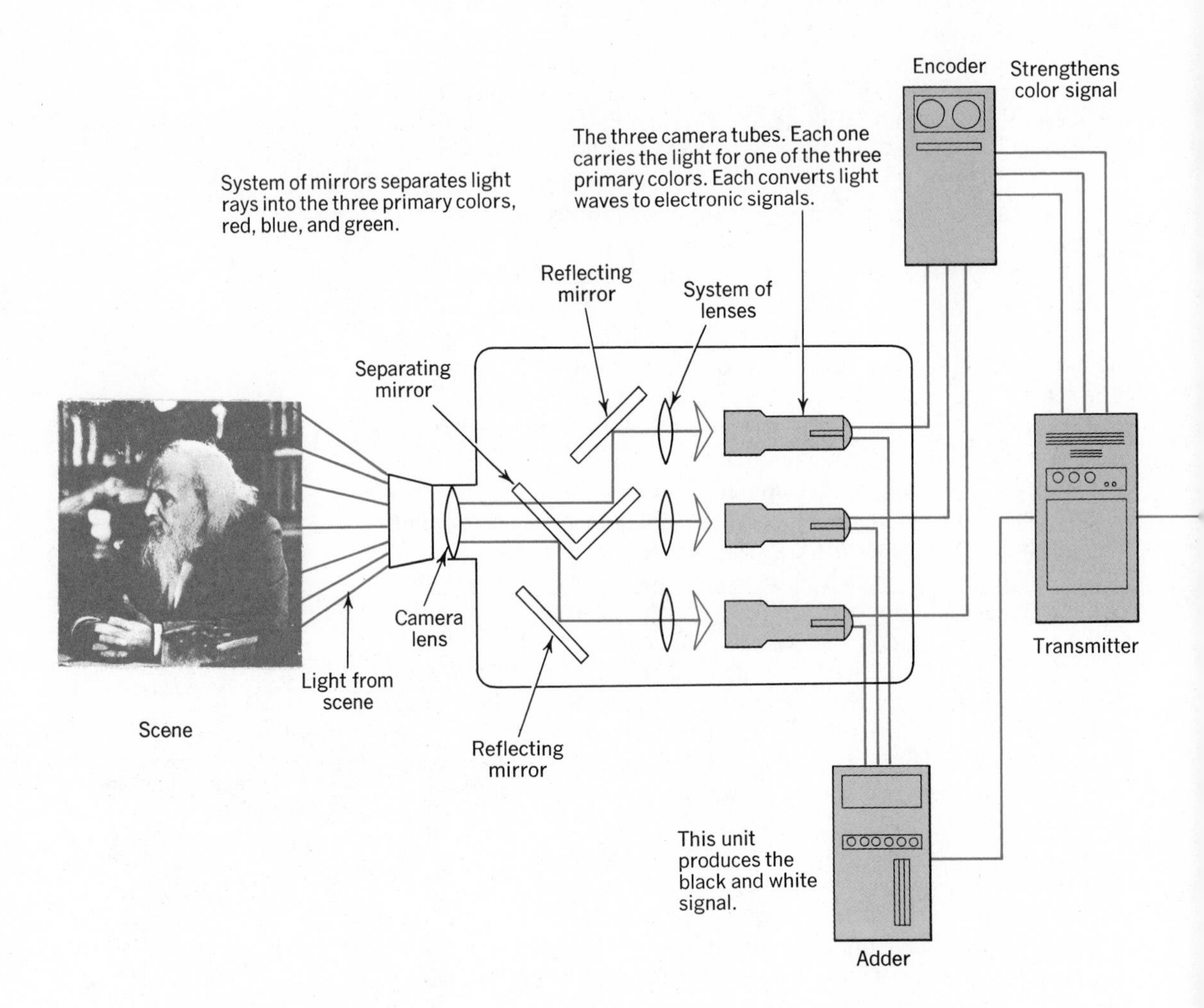

5-48 A schematic representation of the stages from the reception of an image by the television camera to the transmission and formation of the image on the television receiver screen on the home set.

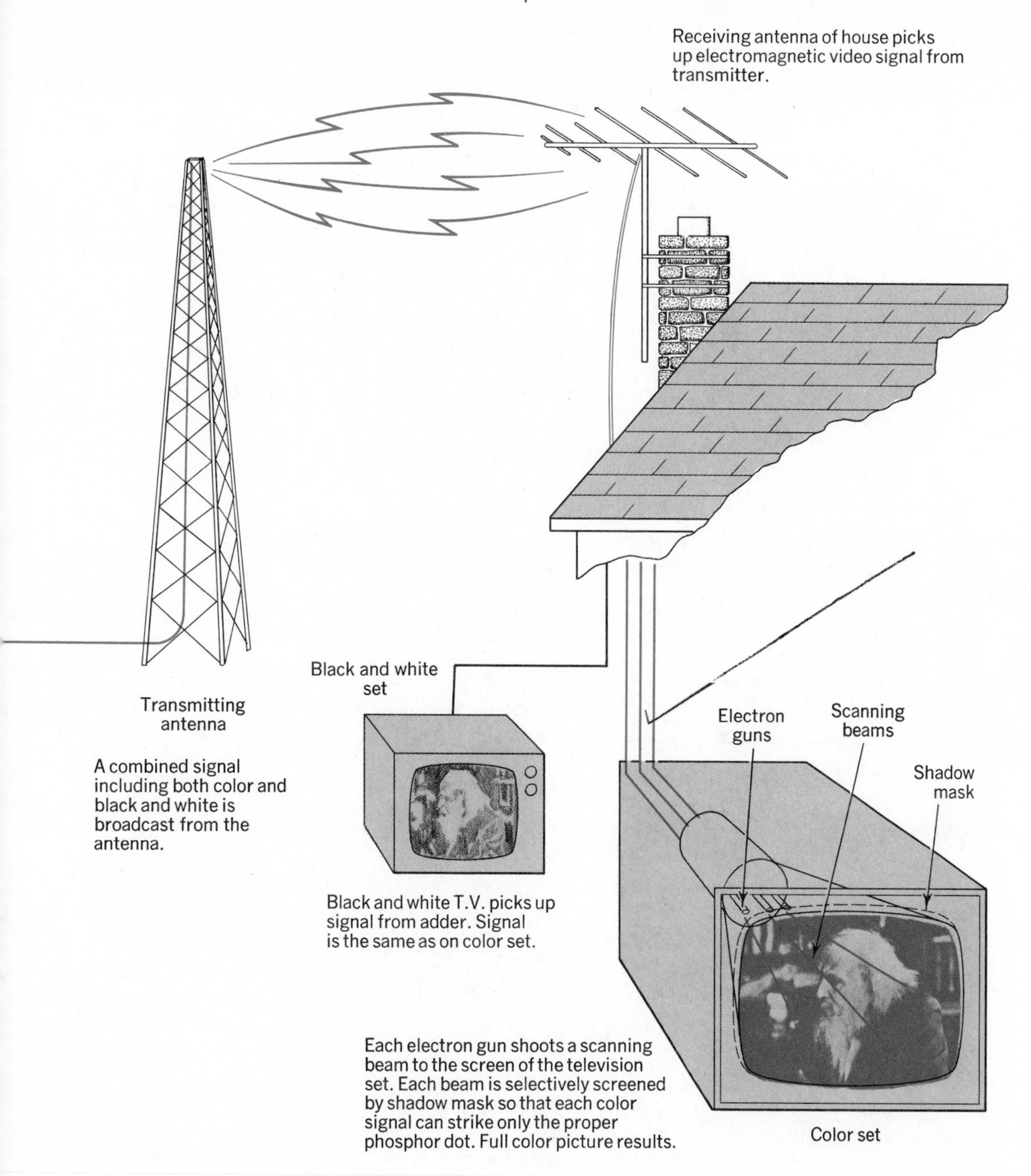

Color Television Receiver

QUESTIONS

1. What is the effect on the force of attraction between two oppositely charged particles when the distance between them is tripled?

2. What is the effect on the force of attraction between bodies if (a) the electric charge on one body is doubled, (b) the charge on both bodies is doubled?

3. Two spheres, A and B, separated by a distance, d, each has a charge of 1+. The force of repulsion is F. How does the value of F change if (a) the charge on A becomes 3+, (b) the charge on both A and B becomes 2+, (c) the charge on A becomes zero, (d) the charges remain as 1+ on each sphere, but the spheres are separated by a distance of 4d?

4. What does the atomic number of an atom reveal about its composition?

5. Which property of an atom is more closely related to its chemical behavior, atomic mass or atomic number? Explain.

6. Copy the table and fill in the blanks. *Do not write in this book.*

Element	Atomic Number	Mass Number	Number of Protons	Number of Neutrons	Number of Electrons
	13	27			
	19			20	
Nickel				30	
Xenon				78	
		238	92		

7. Calculate the maximum number of electrons that may be contained in each of the first six energy (quantum) levels of an atom.

8. The atomic masses given in the Periodic Table are not whole numbers. Explain.

9. How do radiant-energy waves differ from sound or water waves?

10. (a) State the three fundamental characteristics of all waves. (b) How are they related?

11. The wavelengths of X rays are much shorter than those of ultraviolet or visible light. Show qualitatively, using the relationships $E = h\nu$ and $c = \lambda\nu$, why continued exposure to X rays is more damaging than exposure to sunlight.

12. Describe Max Planck's contribution to the understanding of the nature of light.

13. (a) What is the photoelectric effect? (b) How did Einstein use the quantum theory to explain the photoelectric effect?

14. What experimental evidence requires that the emission of energy by an atom be quantized?

15. (a) Explain why scientists continue to view electromagnetic radiation as having a dual nature? (b) Why do they not decide upon using exclusively either the wave model or the particle model?

16. Write the number of electrons in each principal energy level of these atoms in the ground state. (a) He, (b) F, (c) Na, (d) P, (e) Ar.

PROBLEMS

1. If Thomson determined e/m to be 1.8×10^{8} coul/g and Millikan found e^{-} to be 1.6×10^{-19} coul/electron, calculate the mass of an electron.

Ans. 8.9×10^{-28} g/e⁻.

2. Radio dials are calibrated in frequency. A station broacasts on a frequency of 600 kilocycles (600 on the dial). Calculate the wavelength of the signal.

3. Some earthquake waves travel at a velocity of 5 km/sec. What is the wavelength of such a wave if the earth tremors are 10 per second? **Ans. 0.5 km.**

4. Some experimental lights use salts of lithium because excited lithium atoms emit a photon which is in the red ($\lambda = 7.0 \times 10^{3}$ Å) region of the spectrum. Calculate the number of electron volts emitted from an excited lithium atom as it returns to its ground state. There are 1.6×10^{-12} ergs/ev. **Ans. 1.8 ev.**

5. The functioning part of a photoelectric cell (electric eye) is coated on one side with cesium metal or a compound of cesium. A voltage is applied across the cell, and when light strikes the cesium, an electron is removed, completing the circuit. In order to remove an

electron from a cesium atom, 3.89 ev is required. (a) Calculate the wavelength of light that would be necessary to remove the electron from a neutral cesium atom and activate the cell. Use the conversion factor 1 ev $= 1.6 \times 10^{-12}$ ergs. (b) Would photons of visible light have enough energy? **Ans. (a) 3200 Å, (b) No.**

6. If the energy needed to remove an electron from one of the energy levels of sodium is 8.2×10^{-12} ergs per electron, calculate (a) the energy in ergs required to ionize a mole of atoms, (b) the energy in cal/mole (there are 4.18×10^7 ergs/cal), (c) the energy in kcal/mole, (d) the wavelength of a photon which is just capable of dislodging this electron.

Ans. (a) 4.9×10^{12} ergs/mole, (d) 2400 Å.

7. Hydrogen atoms are ionized by radiation of 912 Å. Calculate the ionization energy in (a) ergs per electron, (b) cal per mole of electrons, (c) kcal per mole of electrons, (d) ev per atom (23.1 kcal of energy is needed to remove 1 mole of electrons, each requiring 1 ev).

Ans. (b) 3.1×10^5 cal/mole, (d) 14 ev.

8. Approximately 14 ev of energy is required to remove an electron from the first energy level of a hydrogen atom. It can be shown that approximately 10 ev is required to cause the electron to go from the 1st to the 2nd level. What is the energy in ev required to remove the electron completely from the atom when it is in the 2nd energy level? **Ans. 4 ev.**

9. The characteristic color imparted to a bunsen-burner flame when ordinary salt (NaCl) is heated is the result of an electronic transition involving 2.1 ev. What is the wavelength of this radiation and what color is imparted to the flame? The approximate wavelengths for different colors are red—6500 Å, yellow—6000 Å, blue—4500 Å, violet—4000 Å. **Ans. 5.9×10^3 Å.**

10. In an experiment to determine the ionization energy of hydrogen, a high-velocity electron transfers its energy to a hydrogen atom causing a hydrogen electron to jump from energy level 1 to energy level 5. The hydrogen electron then cascades down to energy level 2, and then to energy level 1. Calculate the energy of the photons emitted by that hydrogen atom. Express your answer in (a) ev per atom, (b) ergs per atom, and (c) kcal per mole of atoms. Refer to Table 5-32 for energy states of a hydrogen atom. Alternatively, you may use Equation 5-8 to calculate the energy states.

Ans. (a) 2.85 ev/atom and 10.2 ev/atom.

11. Calculate the energy in ev required to cause an electronic transition in a hydrogen atom from (a) n = 1 to n = 2, and (b) from n = 3 to n = 4.

Ans. (b) 0.66 ev/atom.

12. (a) Calculate the energy state of a hydrogen atom when the principal quantum number of the electron is 7. (b) What is the significance of the minus sign in your answer?

13. Millikan's oil-drop experiment involved determining charges on thousands of oil drops. Each oil drop contained an unknown number of unit charges, yet Millikan was able to calculate the charge on a single electron. Assume that you are weighing samples of uniformly large oranges on a balance with a sensitivity of ± 1 oz. Each sample being weighed contains an unknown number of oranges. (a) Explain how you would determine the mass of a single orange with ± 1 oz. (b) Would one or two weighings be adequate?

14. In an oil-drop experiment designed to determine the charge on an electron, the charges found on different oil droplets are

6.72×10^{-9} esu	3.84×10^{-9} esu
5.28×10^{-9} esu	2.88×10^{-9} esu
4.32×10^{-9} esu	1.44×10^{-9} esu

Use these data to calculate (a) the charge on an electron in esu, and (b) the number of electrons attached to an oil drop with a charge of 8.19×10^{-9} esu.

Ans. (a) 4.8×10^{-10} esu, (b) 17 electrons.

15. Some dwarf stars are believed to be composed of almost pure nuclear material. If a dime were composed of pure nuclear material, calculate its mass in tons. Assume that a dime occupies 1 cm^3 and that nuclear particles have a density of about 5×10^{14} g/cm^3.

Ans. 6×10^8 tons.

16. What volume does an atom of osmium occupy if the density of osmium is 22.5 g/cm^3 and 1 mole has a mass of 190 g? **Ans. 1.4×10^{-23} cm^3/atom.**

17. Assuming each osmium atom is a sphere enclosed in a cube whose volume has been calculated in Problem 16, what is the radius of the atom?

Ans. 1.2×10^{-8} cm.

18. Indium ions have a charge of 3^+. (a) How many electrons are required to convert one indium ion into an indium atom? (b) How many electrons are required to convert one mole of indium ions into one mole of indium atoms? (c) How many faradays of electricity are required to convert one mole of indium ions into one mole of indium atoms? (d) How many chemical (electric) equivalents are there per mole of indium ions?

Ans. (a) 3, (b) $3 \times 6.02 \times 10^{23}$, (c) 3, (d) 3.

19. In the electrolysis of calcium chloride, 2 faradays of electricity is required to convert one mole of calcium

ions into one mole of calcium atoms. (a) How many electrons are required to convert one mole of calcium ions into one mole of calcium atoms? (b) How many electrons are required to convert one calcium ion into one calcium atom? (c) How many chemical and electric equivalents are there per mole of calcium atoms?

Ans. (a) $2 \times 6.02 \times 10^{23}$, (b) 2, (c) 2.

20. Calculate the average atomic mass for each element whose isotopic composition is shown below.

Argon – 36 = 0.34%	Potassium – 39 = 93.10%
Argon – 38 = 0.06%	Potassium – 40 = 0.01%
Argon – 40 = 99.60%	Potassium – 41 = 6.89%

21. Complete this table. Use the equations and conversion factors developed in this chapter as needed.

Electron Transition (higher to lower level)	*Energy Evolved (kcal/mole)*	*Energy Evolved (ev/atom)*	*Energy Evolved (erg/atom)*	*Frequency (sec^{-1})*	*Wavelength (cm)*	*Wavelength (Ångstroms)*	*Series (Lyman, Balmer, Paschen)*	*Region of Spectrum (visible, IR or UV)*
$\infty \longrightarrow 3$	34.8	1.51	2.41×10^{-12}	3.64×10^{14}	8.25×10^{-5}	8250	Paschen	Infrared
$6 \longrightarrow 3$	26.1							
$5 \longrightarrow 3$	22.3							
$4 \longrightarrow 3$	15.2							
$\infty \longrightarrow 2$	78.4							
$6 \longrightarrow 2$	69.7							
$5 \longrightarrow 2$	65.9							
$4 \longrightarrow 2$	58.8							
$3 \longrightarrow 2$	43.6							
$\infty \longrightarrow 1$	313.6							
$6 \longrightarrow 1$	304.9							
$5 \longrightarrow 1$	301.1							
$4 \longrightarrow 1$	294.0							
$3 \longrightarrow 1$	278.8							
$2 \longrightarrow 1$	235.2							

6

The Electron Configuration of Atoms

Preview

In order to explain the behavior of multi-electron atoms, we must acquire a knowledge of their electron configurations. The Bohr atom model provided the important concept of specific (quantized) energy states or levels and an easy-to-visualize picture of an atom. Modern quantum or wave-mechanical theory, developed during the 1920's, retains the concept of quantized energy states but abandons (invalidates) the easy-to-visualize satellite model with its electrons in fixed trajectories (orbits).

Modern theory views the atom as a positively charged nucleus surrounded by pulsating electron waves. Although we shall continue to refer to an electron as a particle, it is important to remember that scientists use the solutions to a complicated wave equation to describe mathematically the behavior of an electron in terms of its wave characteristics.

The transition from the easy-to-visualize satellite model is made easier by drawing analogies to it. In the Bohr model we visualized electrons as traveling around the nucleus in orbits of fixed radius. For the wave-mechanical model we translate the solutions to the wave equations into particle language by saying that the electrons are moving about in such a way as to occupy effectively designated regions of space around the nucleus called *orbitals*. We shall attach recognizable shapes and spatial orientations to electron orbitals.

"It is a truth very certain that when it is not in our power to determine what is true we ought to follow what is most probable."

René Descartes (1596–1650)
from *Discourse on Method*

Orbitals may be viewed as *electron charge clouds* produced by the rapid motion of negatively charged electrons. In other words, a charge cloud represents the distribution of electron charge that would be obtained by observing the motion of an electron over a period of time. Modern theory does not attempt to define a trajectory for an electron. Rather, it describes the electron locations in terms of probability. Probability information tells us in which regions of space about the nucleus an electron spends most of its time.

The models and illustrations of electron orbitals which you will observe and learn to recognize, represent the boundaries of the volume within which an electron with a given energy might be found a given percentage (90%) of the time. The shapes and spatial orientations of the orbitals are important because they will help us to understand and predict the way in which atoms react (form and break chemical bonds).

As you might expect, the wave-mechanical theory is abstract and mathematical. Therefore, in this chapter we shall first present a simplified version of electronic configuration in terms of wave-mechanical concepts. In this overview, you will learn to express the electron configuration of atoms in terms of wave-mechanical language. In our discussion, we shall relate the electron configuration of an atom to its location in the Periodic Table. In other words, we shall explain the form of the Periodic Table and the observed periodicity in the properties of the elements in terms of the electron configuration of the atoms.

The essential features of electronic configuration needed as minimum background for understanding material in subsequent chapters are presented in Sec. 6-1 through 6-8. After completing these sections, you should be able to use energy-level diagrams and quantum (wave)-mechanical principles to designate the electron population of the orbitals and principal energy levels.

In the last half of the chapter, we shall discuss the origin and theoretical basis of the wave-mechanical model. The content of these sections is, by nature, rather difficult. It is designed and included to give you an idea of how scientists view an atom, but is not a prerequisite or essential to your progress.

THE ELECTRON CONFIGURATION OF MULTI-ELECTRON ATOMS AND IONS

6-1 Energy Sublevels Experimental data show that all of the electrons which occupy a given principal energy level of a multi-electron atom do not have the same energy. A clue to the energies of the electrons in a principal energy level is found in the energy required to remove them from the attractive force of the nucleus. Consider an isolated magnesium atom (atomic number 12) which has 12 electrons. Using the $2n^2$ rule noted in the last chapter, we find that the lowest energy level (n = 1) has two electrons, the second energy level ($n = 2$) has eight electrons, and the third energy level ($n = 3$) has two electrons. Spectroscopic analysis can be used to determine the energy that is required to remove completely any one of the electrons from a magnesium atom. The results of such an analysis applied to 20 different atoms are in Table 6-1. Let us see how the data in Table 6-1 provide evidence for energy sublevels. Consider a magnesium atom. Removal of either of the two electrons in the third energy level requires 8 electron volts. Apparently these two electrons are located in the same sublevel. It should be noted that after one of the electrons in the third level is removed, the inter-electron repulsion decreases so that the remaining electron is closer to the nucleus and held more firmly than before. Thus, 15 ev is required to remove the other electron. Of the eight electrons located in the second energy level, any one of a group of six can be removed by 52 ev while either of the remaining pair can be removed by 85 ev. These data suggest that the electrons in the second energy level are arranged in two energy sublevels which have different energy values. Either of the two electrons in the first energy level may be removed by 1200 ev. This suggests that there is only one energy sublevel in the first principal energy level. An extension of Table 6-1 would show that **the number**

TABLE 6-1
ENERGIES (ELECTRON VOLTS) REQUIRED TO REMOVE ELECTRONS FROM ATOMS

Total Electrons	1	2	3	4	5	6	7	8	9	10	11	12	13	14	15	16	17	18	19	20
Principal Energy Level	1		2								3								4	
Orbital	1s		2s		2p						3s		3p						4s	
Hydrogen	14																			
Helium	25	25																		
Lithium	65	65	5																	
Beryllium	120	120	9	9																
Boron	200	200	13	13	8															
Carbon	300	300	17	17	11	11														
Nitrogen	450	450	20	20	15	15	15													
Oxygen	550	550	28	28	14	14	14	14												
Fluorine	700	700	38	38	17	17	17	17	17											
Neon	900	900	50	50	22	22	22	22	22	22										
Sodium	1000	1000	60	60	40	40	40	40	40	40	5									
Magnesium	1200	1200	85	85	52	52	52	52	52	52	8	8								
Aluminum	1600	1600	110	110	75	75	75	75	75	75	10	10	6							
Silicon	1800	1800	140	140	100	100	100	100	100	100	13	13	8	8						
Phosphorus	2200	2200	180	180	130	130	130	130	130	130	17	17	11	11	11					
Sulfur	2400	2400	220	220	160	160	160	160	160	160	20	20	10	10	10	10				
Chlorine	2800	2800	260	260	190	190	190	190	190	190	24	24	13	13	13	13	13			
Argon	3000	3000	320	320	225	225	225	225	225	225	29	29	16	16	16	16	16	16		
Potassium	3400	3400	380	380	280	280	280	280	280	280	35	35	26	26	26	26	26	26	4	
Calcium	3600	3600	430	430	320	320	320	320	320	320	43	43	28	28	28	28	28	28	6	6

of different kinds of energy sublevels in a principal level is equal to the principal quantum number, *n*.

Within a given energy level, the energy sublevels are designated in order of increasing energy as *s*, *p*, *d*, and *f*. Thus, the first principal energy level contains only an *s* sublevel; the second contains *s* and *p* sublevels; the third contains *s*, *p*, and *d* sublevels; and the fourth contains *s*, *p*, *d*, and *f* sublevels. We can represent the relative energies of the sublevels and their relationship to the principal energy levels as shown at the right.

Principal Energy Level (n)	Number of Energy Sublevels = n	Designation and Relative Energies of Energy Sublevels
1	1	s
2	2	$s<p$
3	3	$s<p<d$
4	4	$s<p<d<f$

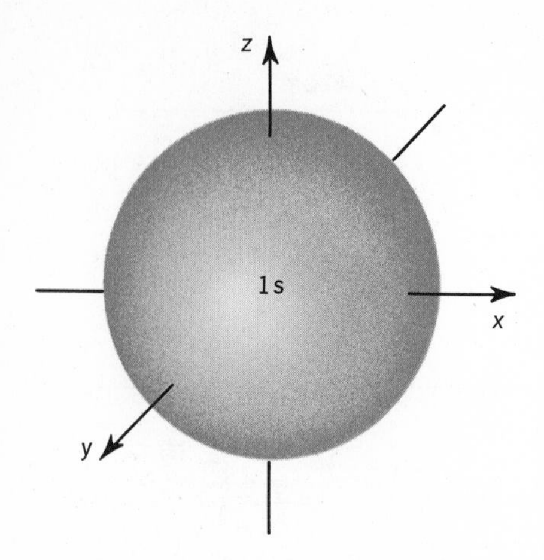

6-1 The 1s atomic orbital. The nucleus is at the intersection of the x, y, and z axes. The ls electron, the lowest energy electron in an isolated atom, is most likely to be found within the boundary of the spherical region of space shown above.

Assuming that the electrons are found in the lowest energy sublevels first, we can say that in a gaseous neon atom, two electrons occupy the *s* sublevel of the first energy level, two electrons occupy the *s* sublevel of the second energy level, and six electrons can occupy the *p* sublevel.

6-2 Orbitals It can be shown that under the influence of a magnetic field the electrons in the energy sublevels are most apt to be found in certain preferred regions of space about the nucleus. These regions of space are called ***orbitals.*** The spatial orientation of the orbitals is illustrated in Fig. 6-1, Fig. 6-2, and Fig. 6-3. In order to help us visualize orbitals, boundaries are established which define their shapes. It can be seen in Fig. 6-1 that *s* orbitals are spherical regions of space and in Fig. 6-2 that *p* orbitals, which have two major lobes, are oriented with their axes along one of the three axes in space. Furthermore, the three *p* orbitals are perpendicular to each other. They are identified as p_x, p_y, p_z, where the subscripts refer to the orientation of the orbital axis with respect to the space axes (Cartesian coordinates). The *d* orbitals are shown in Fig. 6-3.

Quantum-mechanical calculations show that the first principal energy level in an atom has only an *s* orbital, the second energy level has one *s* and three *p* orbitals, the third energy level has one *s*, three *p*, and five *d* orbitals, and the fourth energy level has one *s*, three *p*, five *d*, and seven *f* orbitals. A little reflection shows that **the total number of orbitals available in a given energy level is equal to n^2.** More detailed analysis and calculations reveal that **each orbital can contain a maximum of two electrons.** This means that the **maximum number of electrons in a given energy level is $2n^2$.** The relationships discussed above are tabulated in Table 6-2 and schematically summarized in the space above Table 6-2.

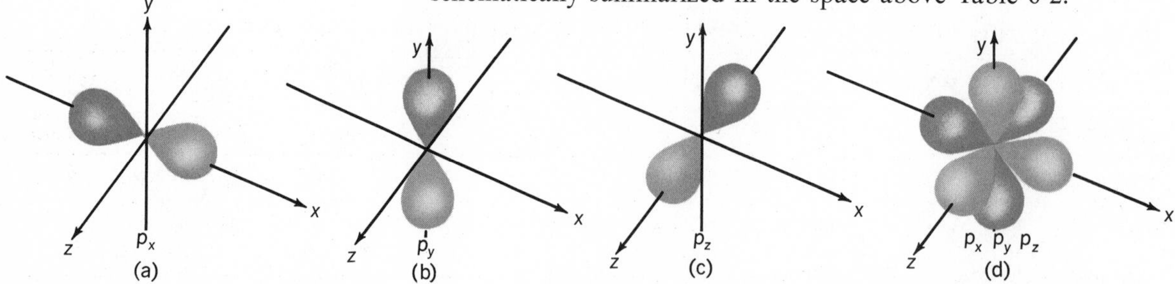

6-2 *p* orbitals. Unlike s orbitals, *p* orbitals have directional characteristics. *p* electrons are most likely to be found in the regions of space directed along the *x*, *y*, and *z* axes. In an atom, the three 2*p* orbitals are oriented as shown in (d).

6-3 Distribution of Electrons in Orbitals Up to this point we have identified the orbital composition and maximum electron population of the principal energy levels. We found that each orbital, each energy sublevel, and each energy level can accommodate only a specific number of electrons. We are now ready to explain how the electrons are distributed in the orbitals of specific gaseous atoms in their lowest energy (ground) states. This means we must indicate the number of electrons in each orbital of each energy level.

To acquaint you with the electron configuration of atoms we shall employ a useful but *imaginary* process. In this process we begin with a hydrogen atom which contains one proton in its nucleus and one electron in the 1*s* orbital. We then "add" particles to construct an atom with any atomic number we wish. Each time we "add" a proton to the nucleus, we "add" an electron to an available orbital. In other words, we imagine that electrons "enter and finally fill" the system of empty orbitals shown in Fig. 6-4. The relative energies of the orbitals in multi-electron atoms are schematically depicted in Fig. 6-4 on the next page.

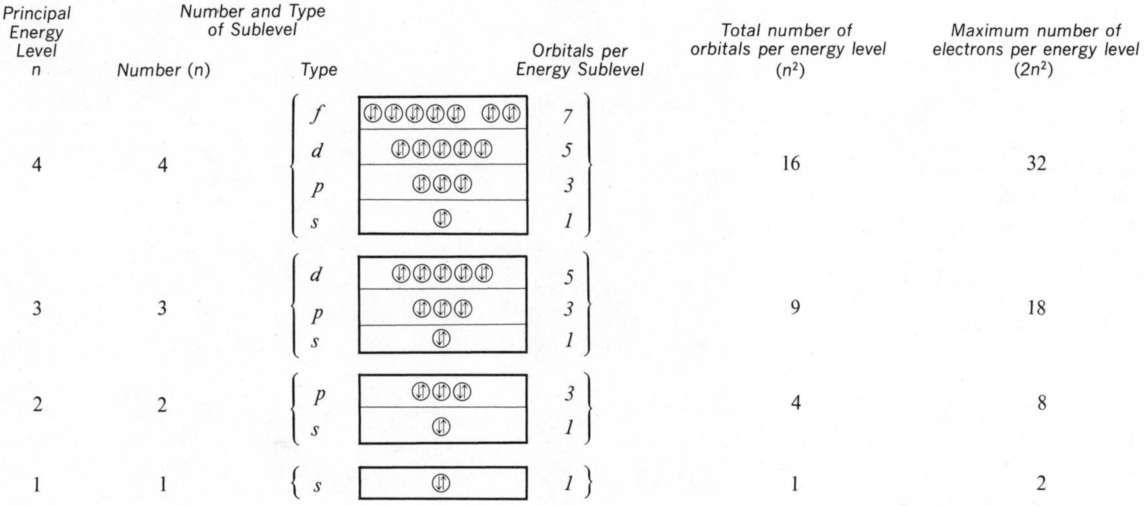

Principal Energy Level n	Number and Type of Sublevel: Number (n)	Type	Orbitals	Orbitals per Energy Sublevel	Total number of orbitals per energy level (n^2)	Maximum number of electrons per energy level ($2n^2$)
4	4	*f*	⇅⇅⇅⇅⇅ ⇅⇅	7	16	32
		d	⇅⇅⇅⇅⇅	5		
		p	⇅⇅⇅	3		
		s	⇅	1		
3	3	*d*	⇅⇅⇅⇅⇅	5	9	18
		p	⇅⇅⇅	3		
		s	⇅	1		
2	2	*p*	⇅⇅⇅	3	4	8
		s	⇅	1		
1	1	*s*	⇅	1	1	2

TABLE 6-2
PRINCIPAL ENERGY LEVEL, ENERGY SUBLEVEL, AND ORBITAL DISTRIBUTION OF ELECTRONS

Shell	Energy Level	Principal Quantum Number n	Energy Sub-levels	Orbitals per Sub-level	Orbital Designation	Maximum Electrons per Orbital	Maximum Electrons per Energy Sublevel	Maximum Electrons per Principal Energy Level
K	1	1	s	1	1*s*	2	2	2
L	2	2	s	1	2*s*	2	2	8
			p	3	$2p_xp_yp_z$	2	6	
M	3	3	s	1	3*s*	2	2	18
			p	3	$3p_{xy}p_yp_z$	2	6	
			d	5	$3d_{xy}d_{xz}d_{yz}$ $d_{x^2-y^2}d_{z^2}$	2	10	
N	4	4	s	1	4*s*	2	2	32
			p	3	$4p_xp_yp_z$	2	6	
			d	5	$4d_{xy}d_{xz}d_{yz}$ $d_{x^2-y^2}d_{z^2}$	2	10	
			f	7	unimportant for our purpose	2	14	

6-3 The spatial orientations and shapes of d orbitals. Each orbital can accommodate a maximum of 2 electrons. One electron in a d_{yz} orbital has an equal probability of being found in any one of the four lobes.

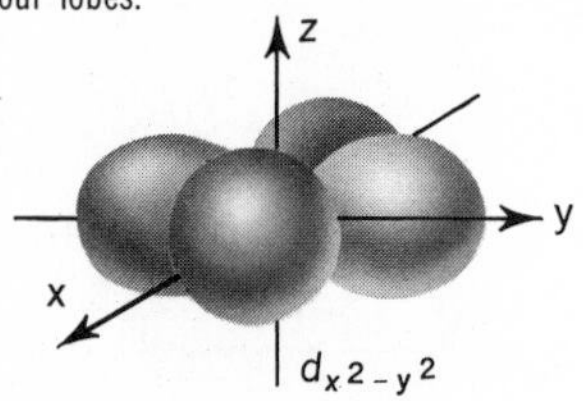

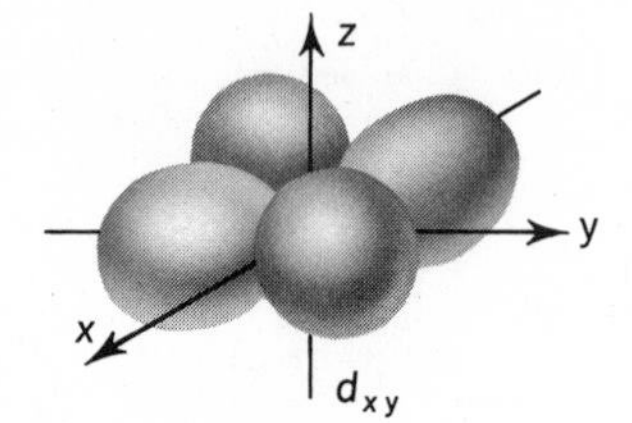

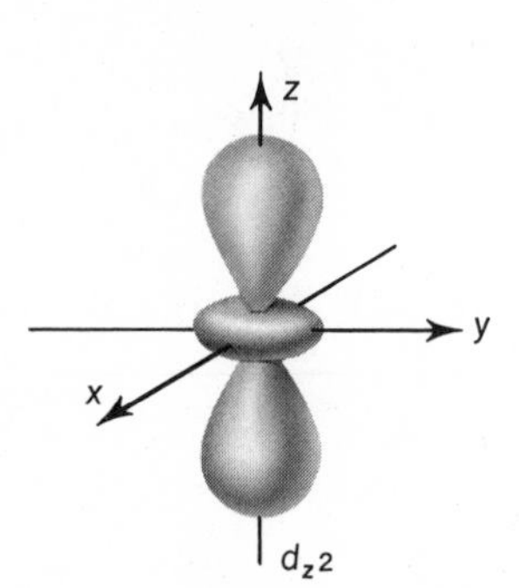

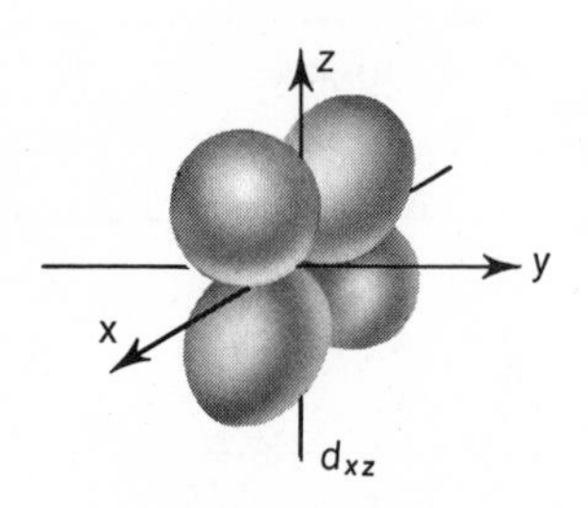

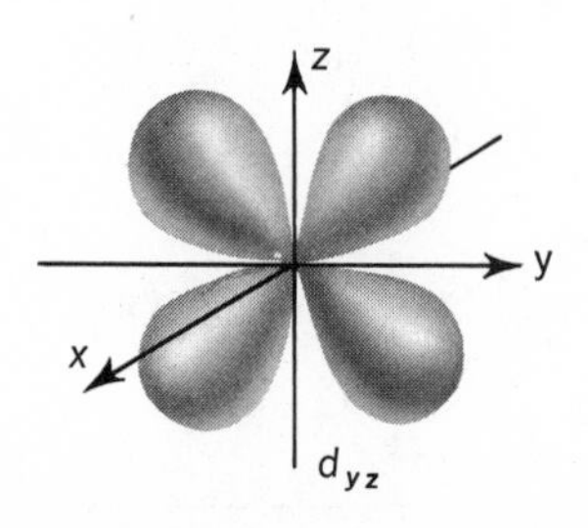

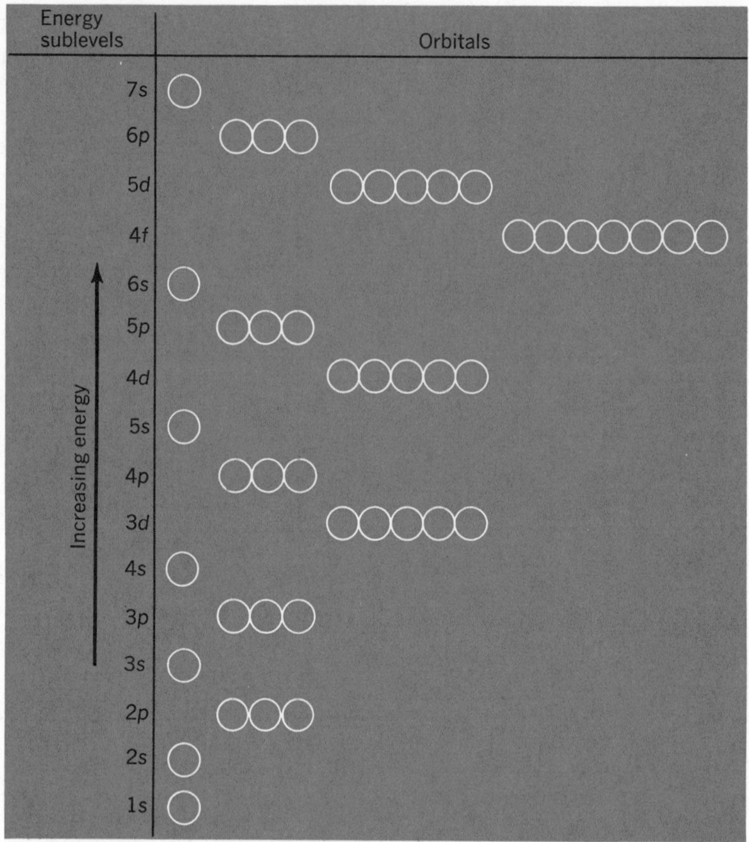

6-4 Relative energies of orbitals in isolated multi-electron atoms. Each circle represents an orbital capable of holding two electrons. The lowest-energy available orbitals are filled first.

In reality, elements are not formed in this way. No magical process takes place in which protons are "added" to a nucleus and electrons are systematically "added" to empty orbitals one at a time. Thus, the term "entering and filling of orbitals" is actually misleading and erroneous if you interpret it literally. A discussion of the possible formation, or origin, of the elements is given in the special feature at the end of Chapter 7.

Let us now examine rules and devices which will enable us to represent properly the electron configuration of atoms in the ground state. The energy level diagram shown in Fig. 6-4 and the rules listed below may be used to write the electron configuration of gaseous atoms in the ground state. If an energy-level diagram is not available, the mnemonic diagram shown in Fig. 6-5 will help you remember the sequence. Both show that, in order of increasing energy, the orbital sequence is 1*s*, 2*s*, 2*p*, 3*s*, 3*p*, 4*s*, 3*d*, 4*p*, 5*s*, 4*d*, 5*p*, 6*s*, 4*f*, 5*d*, 6*p*, 7*s*, 5*f*.

Remember that this is an "order of filling" diagram and not a sequence of energy levels which is correct for the ground state of *all* atoms. Because of the close spacing of some of the higher energy-sublevels, experiments indicate some overlapping and "criss-crossing" of these sublevels. This makes it impossible to construct an "all-inclusive" energy-level diagram. You will, however, find that the sequence above is consistent with the arrangement of atoms in the Periodic Table. Hence, we find it to be useful for correlating and predicting.

Rules for Writing Electron Configurations of Gaseous Atoms in the Ground State

1. Electrons occupy the lowest energy orbital of the lowest energy level first.
2. No electron pairing takes place in *p*, *d*, or *f* orbitals until each orbital of the given set contains one electron. This is known as ***Hund's Rule.***
3. No orbital can contain more than two electrons.

When writing electron configurations, each orbital must be identified with a principal quantum number. For example, a *p* orbital in the 4th energy level is called a 4*p* orbital. The number of electrons in each orbital is indicated by a superscript. A single electron in a 4*p* orbital would be represented as $4p^1$. In Fig. 6-4, each circle

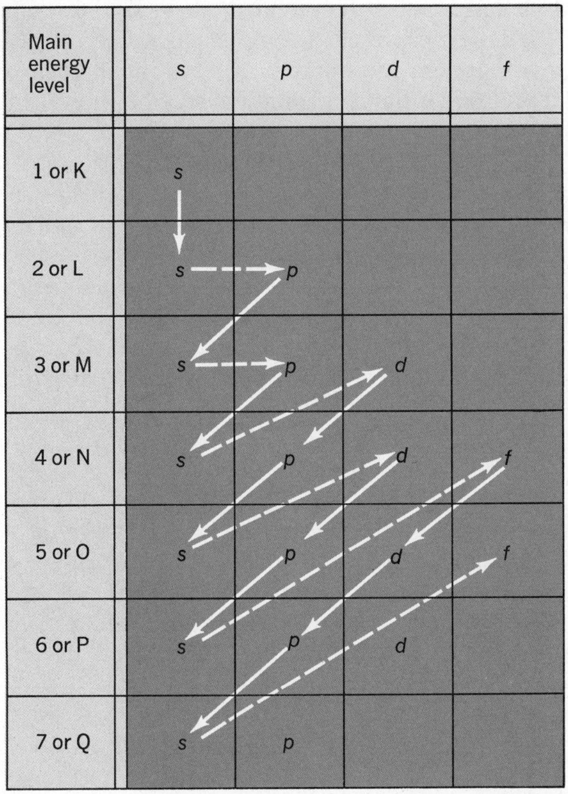

Main energy level	s	p	d	f
1 or K	s			
2 or L	s	p		
3 or M	s	p	d	
4 or N	s	p	d	f
5 or O	s	p	d	f
6 or P	s	p	d	
7 or Q	s	p		

Sequence: 1s 2s 2p 3s 3p 4s 3d 4p 5s 4d 5p 6s 4f 5d 6p 7s 5f

6-5 Memory aid for determining the ground state electronic configuration of a multi-electron atom.

represents an orbital capable of holding two electrons. Arrows pointing in opposite directions are used to denote electrons having opposite spin direction. A filled orbital is represented as ⇅.

There are several ways to represent the electron configuration of an atom. The starting point is to locate the element in the Periodic Table. The atomic number of the element represents the positive charge on the nucleus and the number of electrons outside the nucleus of a neutral atom of the element. Knowing the total number of electrons available, we can then distribute them by following the energy-level diagram or mnemonic diagram and applying the rules noted above. Consider nitrogen, atomic number 7, as an example, According to the atomic number, there are 7 electrons in a nitrogen atom. The first two electrons are located in the 1*s* orbital. The third and fourth electrons are located in the 2*s* orbital. The fifth, sixth, and seventh electrons occupy, respectively, the $2p_x$, $2p_y$, $2p_z$ orbitals. This distribution may be represented as $1s^2$, $2s^2$, $2p_x^1$, $2p_y^1$, $2p_z^1$ or as

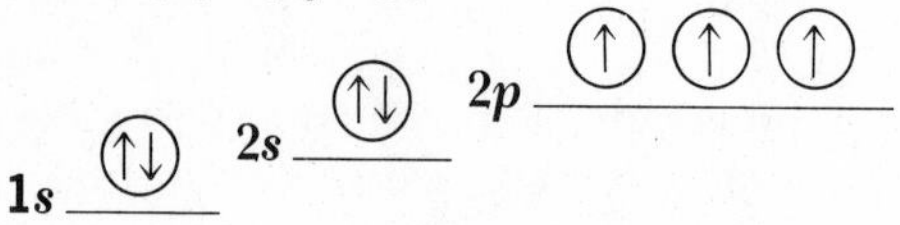

2p (↑) (↑) (↑)

2s (↑↓)

1s (↑↓)

where the letters *s* and *p* represent the orbital, the coefficients in front of the letters represent the principal energy levels and the superscripts represent the number of electrons in each orbital. If less detail is desired, the configuration may be shown as

$$1s^2 2s^2 2p^3$$

The imaginary process of building up the electron structures of the elements in the ground state is known as the *aufbau process.* Let us illustrate its application.

A hydrogen atom has atomic number 1. This indicates that this atom has 1 electron. The single electron occupies the lowest energy orbital (*s*) in the lowest energy level (n = 1). Accordingly, the electronic configuration of a hydrogen atom in the ground state is represented as

$$1s^1$$

Helium, atomic number 2, has 2 electrons. One of these electrons occupies the *s* orbital in the first level, as in the case of the hydrogen atom. The second electron also occupies the 1*s* orbital. Thus, the first two electrons form an electron pair and completely fill both the orbital and the first-energy level. These 2 electrons must have opposite spins since they occupy the same orbital. The configuration may be designated as

$$1s^2$$

Helium is chemically inert. This experimental fact suggests that the helium atom has a very stable structure. Consequently, we assume that 2 electrons complete the first energy level and represent a stable electronic configuration for the first energy level.

An atom of lithium, atomic number 3, contains 3 electrons. Since the first level is complete, the third electron is found in the lowest-energy orbital (*s*) of the second major energy level. The electronic configuration of lithium in the ground state is

$$1s^2 2s^1$$

Let us next write the electronic configuration of potassium, atomic number 19. Reference to Fig. 6-4 shows that for the ground-state atom, a 4*s* orbital is slightly lower in energy than the 3*d* orbital. Thus, the 19th electron is found in the 4*s* orbital, which is the lowest-energy orbital available. Starting with potassium, all of the atoms of the elements in Groups IA and IIA contain electrons in an outer *s* orbital before any electrons are found in the underlying *d* orbital. The configuration of a potassium atom is

$$1s^2 2s^2 2p^6 3s^2 3p^6 4s^1$$

The ten elements following calcium are known as the *transition elements*. The ground-state configurations of these atoms show that underlying *d* orbitals are occupied only after the *s* orbital in the next higher energy level is occupied. The orbital configurations of these elements are shown in Table 6-4. Note that chromium, atomic number 24, and copper, atomic number 29, are exceptions to the general trend. The exception consists of an unpairing of the 4*s* electrons, yielding a $3d^5$ or $3d^{10}$ configuration. The electronic configurations of these atoms are given below.

$$\text{Cr}\ 1s^2 2s^2 2p^6 3s^2 3p^6 3d^5 4s^1$$
$$\text{Cu}\ 1s^2 2s^2 2p^6 3s^2 3p^6 3d^{10} 4s^1$$

The detailed explanation for these irregularities is beyond the scope of this book. In general, it can be said that there is an unusual stability attached to a one-half or completely filled set of orbitals. A second transition series of ten elements begins after strontium, atomic number 38. In this and in the third series of transition elements, there are several exceptions to the usual pattern. In general, we shall not be concerned with the reasons for these irregularities. We shall assume that the arrangement of electrons implied from experiments is more stable than one which we would predict by following the rules.

To determine the ground-state electronic configuration of cadmium, atomic number 48, we may use the memory aid shown in Fig. 6-5. The 48 electrons may be distributed in the proper orbitals by following the arrows in the diagram. The two electrons of lowest energy are located in the 1*s* orbital. The two with the next higher energy are located in the 2*s* orbital. Continuing in this pattern, the fifth through tenth electrons occupy the $2p_x$, $2p_y$, and $2p_z$ orbitals. These are followed by two electrons in the 3*s* orbital, six in the 4*p* orbitals, two in the 4*s* orbital, ten in the 3*d* orbitals, six in the 4*p* orbitals, two in the 5*s* orbital, and ten in the 4*d* orbitals.

Using the energy-level diagram or the mnemonic, the configuration of a cadmium atom in the ground state is found to be

$$1s^22s^22p^63s^23p^64s^23d^{10}4p^65s^24d^{10}$$

Rather than show the exact sequence obtained by following the energy-level diagram, it is customary to show the ground-state configuration by grouping all of the orbitals in a given energy level together. Thus, the ground-state configuration of cadmium is usually written as

$$1s^22s^22p^63s^23p^63d^{10}4s^24p^64d^{10}5s^2$$

This order makes it easier to recognize the outer energy-level electrons.

The electrons in the highest or outermost energy level are the ones generally involved in chemical reactions and are, therefore, of special interest to us. They are sometimes referred to as ***valence electrons.*** The *number of valence electrons in an atom of any representative element (A Groups in the Periodic Table) is numerically equal to the Roman numeral at the top of the column.* For example, the atoms of the elements in Group VIIA all have 7 electrons in their outer energy level. With few exceptions, the atoms of the transition elements (B Group elements) have only 2 electrons in their outer energy level.

FOLLOW-UP PROBLEM

Use orbital notation to represent the groundstate electronic configuration of rubidium (at. no. 37), cesium (55), bromine (35), and iodine (53). What is the orbital configuration of the outer energy level in each of the atoms in Group IA of the Periodic Table? What is the outer energy-level configuration in each of the atoms in Group VIIA? Use the answers to the last two questions to suggest a reason for the similarities in the properties of the elements in a given group of the Periodic Table.

With the ability to represent properly the electronic configuration of atoms, we are now in a position to examine the relationship between the electronic structure of atoms and their location in the Periodic Table. We shall find that we can use the location of an element in the Periodic Table to help us write the orbital electronic configuration of its atoms.

PERIODICITY IN OUTER ENERGY-LEVEL ELECTRON CONFIGURATION

6-4 Electron Configuration of Outer Energy Levels Electric and spectrographic measurements reveal that in general the atoms of all elements in a given group of the Periodic Table have the same electron population and orbital configuration in their outer energy levels. The electron population and configuration of the atoms in three of the groups, as revealed by experimental measurements, are listed in Table 6-3. *Each group is composed of elements having similar properties, but each group has properties which differ from those of the other groups.*

The symbols of the noble gases in the configurations of Group IA elements are used to represent inner electron levels having noble-gas structures.

The data in Table 6-3 are summarized below.

1. The atoms of all Group-IA elements have one electron in the *s* orbital of the outer energy level.
2. The atoms of all Group-VIIA elements have two electrons in the *s* orbital and five electrons in the *p* orbitals of the outer level.
3. The atoms of all elements in Group O have completely occupied *s* and *p* orbitals in their outer levels. In general, these elements are characterized by chemical inactivity.
4. The atoms of all elements in Group IA have one electron more than the noble-gas atoms just preceding them in the Periodic Table. The elements in this group are all observed to be very reactive metals.
5. All the atoms of the elements in Group VIIA have one electron fewer than the noble-gas atoms which follow them in the Periodic Table. These elements are observed to be extremely reactive nonmetals.
6. The group number of the *A* families, known as the *representative elements,* is equal to the number of electrons in the outer energy level of the atoms in the family. For example, all atoms of elements in Group VIIA have seven electrons in their outer energy level.
7. The number of the period (row) in which an element is located is equal to the number of major energy levels occupied by electrons in the ground-state atoms of the elements.

TABLE 6-3
OUTER ENERGY-LEVEL ELECTRON CONFIGURATIONS

Group O Noble Gases
neon $2s^22p^6$
argon $3s^23p^6$
krypton $4s^24p^6$
xenon $5s^25p^6$
Group VIIA Halogens
fluorine $2s^22p^5$
chlorine $3s^23p^5$
bromine $4s^24p^5$
iodine $5s^25p^5$
Group IA Alkali Metals
sodium Ne $3s^1$
potassium Ar $4s^1$
rubidium Kr $5s^1$
cesium Xe $6s^1$

The observations above suggest these conclusions:

1. Similarities in chemical and physical properties are paralleled by and reflect similarities in electron configurations.
2. Completed *s* and *p* orbitals are associated with high stability.
3. Chemical behavior is related to electron configuration of the outer level.

Let us now see how the Periodic Table can be used as a guide to help us determine the electronic configuration of atoms.

6-5 Orbital Blocks of the Periodic Table A detailed examination of the experimentally determined outer energy-level electron configuration of the atoms in the Periodic Table reveals these facts:

1. The outermost (highest-energy) electrons in the atoms of Group IA and IIA elements are located in *s* orbitals.

2. The outermost (highest-energy) electrons in the atoms of Groups IIIA, IVA, VA, VIA, VIIA and O elements are located in *p* orbitals.

3. In general, the atoms of the elements in the B groups contain one or two electrons in the *s* orbital of the highest energy level but differ from one another in the number of electrons in an underlying *d* orbital. These elements are known as the transition elements. As you go from left to right across the fourth row (period 4) of the Periodic Table, the change in properties of the ***transition elements*** is not as drastic as in the elements of the A or representative groups. This is not totally unexpected since the main difference in electron configuration of the transition-element atoms is in the population of the underlying *d* orbital rather than an outer *s* or *p* orbital.

4. In general, the atoms of the elements in the separate section at the bottom of the Periodic Chart contain 2 electrons in the *s* orbital of the highest energy level but differ from one another in the number of electrons in an underlying *f* orbital that has a principal quantum number two less than that of the outer *s* orbital. These elements are known as the inner transition elements. As you go from left to right across these long rows there are marked similarities in the properties of the elements. This is to be expected since the main difference in the electron configuration of the atoms is in the electron population of an underlying *f* orbital.

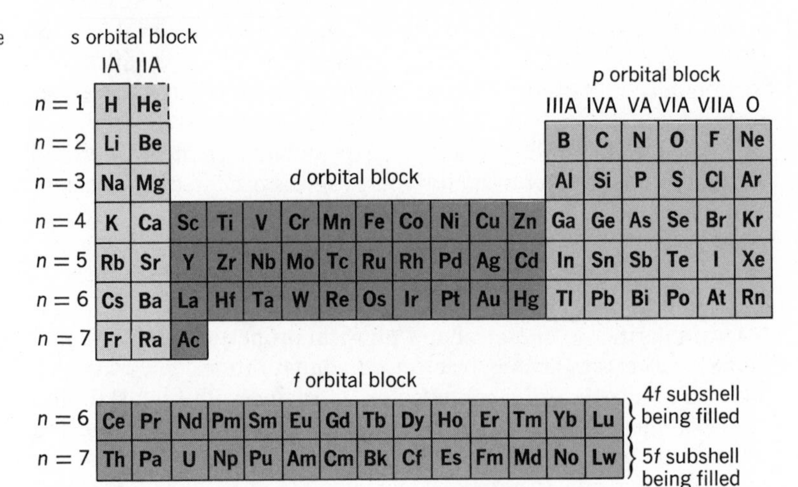

6-6 The *s*, *p*, *d*, and *f* orbital blocks of the Periodic Table.

The analysis above suggests that the long form of the Periodic Table may be dissected into four blocks: an *s*-orbital block, a *d*-orbital block, a *p*-orbital block, and an *f*-orbital block. These blocks are shown in Fig. 6-6. The *s* block is two elements wide, corresponding to the maximum allowable number of *s* electrons in a given energy level. The *p* block is six elements wide, corresponding to the six electrons which can occupy a set of *p* orbitals.

The *d* block is ten elements wide, corresponding to the ten electrons which can occupy a set of five *d* orbitals. The *f* block is fourteen elements wide, corresponding to the fourteen electrons which can occupy a set of seven *f* orbitals.

6-6 The Periodic Table as a Guide to the Electron Configuration of Atoms Examination of the electron configuration of the atoms in a given row of the Periodic Table reveals that each atom differs from its neighbors by one electron. Inspection of the orbital energy-level diagram shows that the orbitals containing the outer electrons have closely related energy values. The length of each row is determined by the relative energies of the orbitals. After each *p* orbital is fully occupied there is a relatively large energy gap. These gaps represent the difference in energy between two successive principal energy levels and are related to the great stability of the noble-gas atoms. Each noble-gas (except He) atom has completely occupied *p* orbitals and it is surrounded by a symmetrical charge cloud. The negligible electron affinities of these atoms suggest that there is no room for another electron. Thus, the next electron occupies an *s* orbital of a new higher energy level that is relatively distant from the completed *p* orbital. This corresponds to the beginning of a new row in the Periodic Table. The completion of a period in the table occurs when a *p* orbital of a noble-gas atom is completely filled.

Since a new period begins after each noble gas, the number of elements in a period is determined by difference in the electron populations of successive noble-gas atoms. For example, argon, atomic number 18, is the noble gas located at the end of Period 3. Krypton, atomic number 36, is another noble gas and it is found at the end of Period 4. The difference in electron population between argon and krypton is $36 - 18 = 18$. Therefore, Period 4 may be expected to contain 18 elements. It does. In Fig. 6-7, orbitals having closely-related energy values are grouped together. Examination of this figure reveals that there are only small energy differences between different orbitals in the same period. The relationship between the location of an atom in the Periodic Table and the electron configuration of its outer orbitals is shown in Fig. 6-8. Many of the relationships discussed above are depicted by this table.

Again, the imaginary "order of filling" concept serves a useful purpose because it helps rationalize the form of the Periodic Table and suggests a positive correlation between the electron configuration of an atom and its position in the table. This means we should be able to express in writing the electron configuration of an atom by noting its position in the Periodic Table. There are, of course, a number of deviations from the general pattern. This is to be expected since each atom is unique and has an electron configuration that is in no way controlled by its neighbors. With these limitations in mind we shall use our imaginary "order of filling" process to deduce the electron configuration of osmium, element

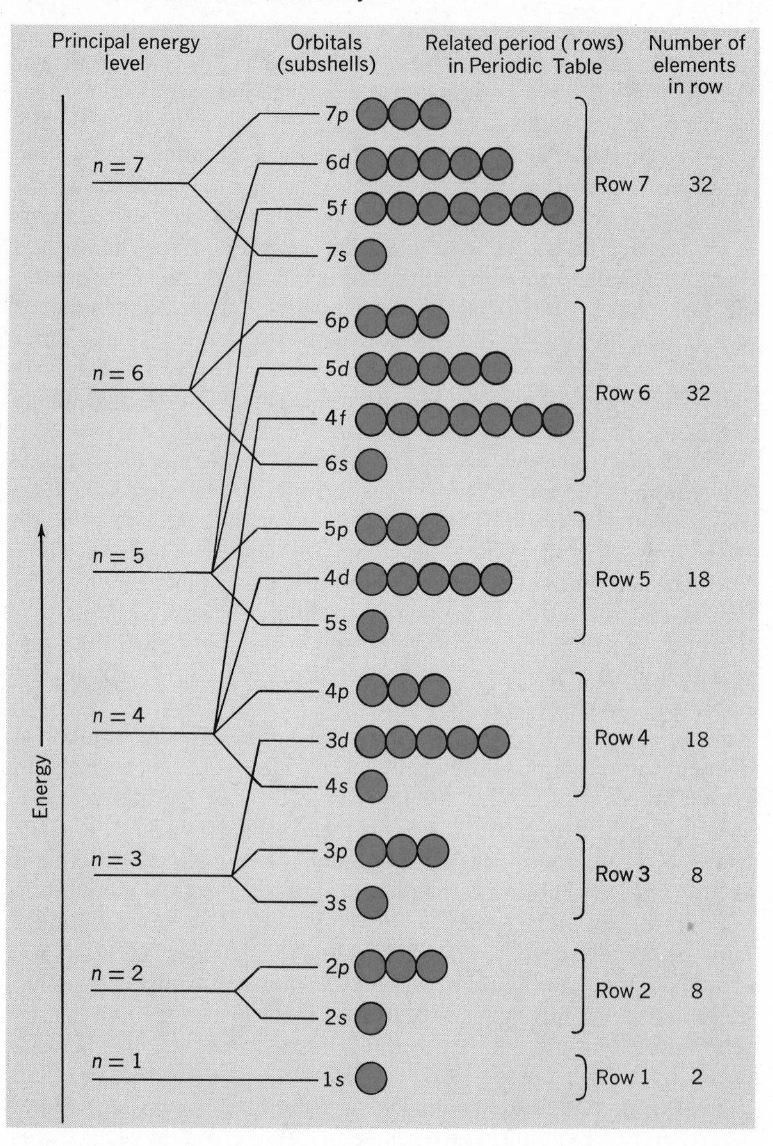

6-7 Energy-level diagram showing principal quantum levels, orbitals, related periods in the Periodic Table, and the number of elements in each row.

number 76, from the position of the atom in the Periodic Table.

Notice that osmium follows the lanthanide series of elements in which the $4f$ orbitals are completely filled. Since osmium is the sixth element in the $5d$ transition-element series, it is expected to have six electrons in the $5d$ orbitals. The $5d$ elements follow the $4f$ elements and the $6s$ elements in Period 6 of the table. Therefore, the three highest sublevels which contain electrons in an osmium atom are expected to have an electron configuration of $6s^2$, $4f^{14}$, $5d^6$. The remaining underlying orbitals are completely filled. Their configuration may be obtained by writing the completely filled orbitals of the atoms in the first five rows of the Periodic Table. These represent the Xe core and are $1s^2$, $2s^2$,$2p^6$, $3s^2$,$3p^6$,

TABLE 6-4
ELECTRONIC CONFIGURATION OF ELEMENT NUMBERS 1–36

Element	*Atomic Number*	*1s*	*2s*	*2p*	*3s*	*3p*	*4s*	*3d*	*4p*
H	1	↑	—	— — —	—	— — —	—		
He	2	↑↓	—	— — —	—	— — —	—		
Li	3	↑↓	↑	— — —	—	— — —	—		
Be	4	↑↓	↑↓	— — —	—	— — —	—		
B	5	↑↓	↑↓	↑ — —	—	— — —	—		
C	6	↑↓	↑↓	↑ ↑ —	—	— — —	—		
N	7	↑↓	↑↓	↑ ↑ ↑	—	— — —	—		
O	8	↑↓	↑↓	↑↓ ↑ ↑	—	— — —	—		
F	9	↑↓	↑↓	↑↓ ↑↓ ↑	—	— — —	—		
Ne	10	↑↓	↑↓	↑↓ ↑↓ ↑↓	—	— — —	—		
Na	11	↑↓	↑↓	↑↓ ↑↓ ↑↓	↑	— — —	—	— — — — —	
Mg	12	↑↓	↑↓	↑↓ ↑↓ ↑↓	↑↓	— — —	—	— — — — —	
Al	13	↑↓	↑↓	↑↓ ↑↓ ↑↓	↑↓	↑ — —	—	— — — — —	
Si	14	↑↓	↑↓	↑↓ ↑↓ ↑↓	↑↓	↑ ↑ —	—	— — — — —	
P	15	↑↓	↑↓	↑↓ ↑↓ ↑↓	↑↓	↑ ↑ ↑	—	— — — — —	
S	16	↑↓	↑↓	↑↓ ↑↓ ↑↓	↑↓	↑↓ ↑ ↑	—	— — — — —	
Cl	17	↑↓	↑↓	↑↓ ↑↓ ↑↓	↑↓	↑↓ ↑↓ ↑	—	— — — — —	
Ar	18	↑↓	↑↓	↑↓ ↑↓ ↑↓	↑↓	↑↓ ↑↓ ↑↓	—	— — — — —	
K	19	↑↓	↑↓	↑↓ ↑↓ ↑↓	↑↓	↑↓ ↑↓ ↑↓	↑	— — — — —	
Ca	20	↑↓	↑↓	↑↓ ↑↓ ↑↓	↑↓	↑↓ ↑↓ ↑↓	↑↓	— — — — —	
Sc	21	↑↓	↑↓	↑↓ ↑↓ ↑↓	↑↓	↑↓ ↑↓ ↑↓	↑↓	↑ — — — —	
Ti	22	↑↓	↑↓	↑↓ ↑↓ ↑↓	↑↓	↑↓ ↑↓ ↑↓	↑↓	↑ ↑ — — —	
V	23	↑↓	↑↓	↑↓ ↑↓ ↑↓	↑↓	↑↓ ↑↓ ↑↓	↑↓	↑ ↑ ↑ — —	
Cr	24	↑↓	↑↓	↑↓ ↑↓ ↑↓	↑↓	↑↓ ↑↓ ↑↓	↑	↑ ↑ ↑ ↑ ↑	
Mn	25	↑↓	↑↓	↑↓ ↑↓ ↑↓	↑↓	↑↓ ↑↓ ↑↓	↑↓	↑ ↑ ↑ ↑ ↑	
Fe	26	↑↓	↑↓	↑↓ ↑↓ ↑↓	↑↓	↑↓ ↑↓ ↑↓	↑↓	↑↓ ↑ ↑ ↑ ↑	
Co	27	↑↓	↑↓	↑↓ ↑↓ ↑↓	↑↓	↑↓ ↑↓ ↑↓	↑↓	↑↓ ↑↓ ↑ ↑ ↑	
Ni	28	↑↓	↑↓	↑↓ ↑↓ ↑↓	↑↓	↑↓ ↑↓ ↑↓	↑↓	↑↓ ↑↓ ↑↓ ↑ ↑	
Cu	29	↑↓	↑↓	↑↓ ↑↓ ↑↓	↑↓	↑↓ ↑↓ ↑↓	↑	↑↓ ↑↓ ↑↓ ↑↓ ↑↓	
Zn	30	↑↓	↑↓	↑↓ ↑↓ ↑↓	↑↓	↑↓ ↑↓ ↑↓	↑↓	↑↓ ↑↓ ↑↓ ↑↓ ↑↓	
Ga	31	↑↓	↑↓	↑↓ ↑↓ ↑↓	↑↓	↑↓ ↑↓ ↑↓	↑↓	↑↓ ↑↓ ↑↓ ↑↓ ↑↓	↑ — —
Ge	32	↑↓	↑↓	↑↓ ↑↓ ↑↓	↑↓	↑↓ ↑↓ ↑↓	↑↓	↑↓ ↑↓ ↑↓ ↑↓ ↑↓	↑ ↑ —
As	33	↑↓	↑↓	↑↓ ↑↓ ↑↓	↑↓	↑↓ ↑↓ ↑↓	↑↓	↑↓ ↑↓ ↑↓ ↑↓ ↑↓	↑ ↑ ↑
Se	34	↑↓	↑↓	↑↓ ↑↓ ↑↓	↑↓	↑↓ ↑↓ ↑↓	↑↓	↑↓ ↑↓ ↑↓ ↑↓ ↑↓	↑↓ ↑ ↑
Br	35	↑↓	↑↓	↑↓ ↑↓ ↑↓	↑↓	↑↓ ↑↓ ↑↓	↑↓	↑↓ ↑↓ ↑↓ ↑↓ ↑↓	↑↓ ↑↓ ↑
Kr	36	↑↓	↑↓	↑↓ ↑↓ ↑↓	↑↓	↑↓ ↑↓ ↑↓	↑↓	↑↓ ↑↓ ↑↓ ↑↓ ↑↓	↑↓ ↑↓ ↑↓

Periodic Table

		Representative elements		Transition elements — d					
	Outer orbital configuration	s^1	s^2	$d^1s^2f^x$	d^2s^2	(d^3s^2)*	(d^5s^1)*	d^5s^2	(d^6s^2)*
Period	Orbitals being filled	IA	IIA	IIIB	IVB	VB	VIB	VIIB	←
$n = 1$	1s	1.00797 **H** 1 (1)							
$n = 2$	2s2p	6.939 **Li** 3 (2, 1)	9.0122 **Be** 4 (2, 2)						
$n = 3$	3s3p	22.9898 **Na** 11 (2, 8, 1)	24.312 **Mg** 12 (2, 8, 2)						
$n = 4$	4s3d4p	39.102 **K** 19 (2, 8, 8, 1)	40.08 **Ca** 20 (2, 8, 8, 2)	44.956 **Sc** 21 (2, 8, 9, 2)	47.90 **Ti** 22 (2, 8, 10, 2)	50.942 **V** 23 (2, 8, 11, 2)	51.996 **Cr** 24 (2, 8, 13, 1)	54.9380 **Mn** 25 (2, 8, 13, 2)	55.847 **Fe** 26 (2, 8, 14, 2)
$n = 5$	5s4d5p	85.47 **Rb** 37 (2, 8, 18, 8, 1)	87.62 **Sr** 38 (2, 8, 18, 8, 2)	88.905 **Y** 39 (2, 8, 18, 9, 2)	91.22 **Zr** 40 (2, 8, 18, 10, 2)	92.906 **Nb** 41 (2, 8, 18, 12, 1)	95.94 **Mo** 42 (2, 8, 18, 13, 1)	[99*] **Tc** 43 (2, 8, 18, 13, 2)	101.07 **Ru** 44 (2, 8, 18, 15, 1)
$n = 6$	6s4f5d6p	132.905 **Cs** 55 (2, 8, 18, 18, 8, 1)	137.34 **Ba** 56 (2, 8, 18, 18, 8, 2)	Lanthanide Series 57–71	178.49 **Hf** 72 (2, 8, 18, 32, 10, 2)	180.948 **Ta** 73 (2, 8, 18, 32, 11, 2)	183.85 **W** 74 (2, 8, 18, 32, 12, 2)	186.2 **Re** 75 (2, 8, 18, 32, 13, 2)	190.2 **Os** 76 (2, 8, 18, 32, 14, 2)
$n = 7$	7s5f6d7p	[223] **Fr** 87 (2, 8, 18, 32, 18, 8, 1)	[226] **Ra** 88 (2, 8, 18, 32, 18, 8, 2)	Actinide Series 89–103					

Period	Orbitals being filled						
$n = 6$	4f	Lanthanide Series	138.91 **La** 57 (2, 8, 18, 18, 9, 2)	140.12 **Ce** 58 (2, 8, 18, 20, 8, 2)	140.907 **Pr** 59 (2, 8, 18, 21, 8, 2)	144.24 **Nd** 60 (2, 8, 18, 22, 8, 2)	[147*] **Pm** 61 (2, 8, 18, 23, 8, 2)
$n = 7$	5f	Actinide Series	[227] **Ac** 89 (2, 8, 18, 32, 18, 9, 2)	232.038 **Th** 90 (2, 8, 18, 32, 18, 10, 2)	[231] **Pa** 91 (2, 8, 18, 32, 20, 9, 2)	238.03 **U** 92 (2, 8, 18, 32, 21, 9, 2)	[237] **Np** 93 (2, 8, 18, 32, 23, 8, 2)

*Some variations in configuration of family members

6-8 The Periodic Table.

$4s^2,3d^{10},4p^6$, $5s^2,4d^{10},5p^6$. It is usually desirable to write the ground-state sequence so that the orbitals are grouped according to their principal quantum numbers. For osmium, the ground-state configuration could be written $1s^2$, $2s^2$, $2p^6$, $3s^2$, $3p^6$, $3d^{10}$, $4s^2$, $4p^6$, $4d^{10}$, $4f^{14}$, $5s^2$, $5p^6$, $5d^6$, $6s^2$. There are irregularities in the configurations of certain atoms which cannot be predicted by following the general pattern. These exceptions need not concern us at this point. For quick reference, the orbital configurations for the first 36 elements in the Periodic Table are given in Table 6-4.

The electron configurations of neutral gaseous atoms in the ground state serve a useful function in providing a basis for our

Of The Elements

Representative elements — Noble gas elements

(d^7s^2)*	(d^8s^2)*	$d^{10}s^1$	$d^{10}s^2$	s^2p^1	s^2p^2	s^2p^3	s^2p^4	s^2p^5	s^2p^6
VIII		IB	IIB	IIIA	IVA	VA	VIA	VIIA	0
									4.0026 **He** 2 (2)
				10.811 **B** 5 (2, 3)	12.01115 **C** 6 (2, 4)	14.0067 **N** 7 (2, 5)	15.9994 **O** 8 (2, 6)	18.9984 **F** 9 (2, 7)	20.183 **Ne** 10 (2, 8)
				26.9815 **Al** 13 (2, 8, 3)	28.086 **Si** 14 (2, 8, 4)	30.9738 **P** 15 (2, 8, 5)	32.064 **S** 16 (2, 8, 6)	35.453 **Cl** 17 (2, 8, 7)	39.948 **Ar** 18 (2, 8, 8)
58.9332 **Co** 27 (2, 8, 15, 2)	58.71 **Ni** 28 (2, 8, 16, 2)	63.54 **Cu** 29 (2, 8, 18, 1)	65.37 **Zn** 30 (2, 8, 18, 2)	69.72 **Ga** 31 (2, 8, 18, 3)	72.59 **Ge** 32 (2, 8, 18, 4)	74.9216 **As** 33 (2, 8, 18, 5)	78.96 **Se** 34 (2, 8, 18, 6)	79.909 **Br** 35 (2, 8, 18, 7)	83.80 **Kr** 36 (2, 8, 18, 8)
102.905 **Rh** 45 (2, 8, 18, 16, 1)	106.4 **Pd** 46 (2, 8, 18, 18, 0)	107.870 **Ag** 47 (2, 8, 18, 18, 1)	112.40 **Cd** 48 (2, 8, 18, 18, 2)	114.82 **In** 49 (2, 8, 18, 18, 3)	118.69 **Sn** 50 (2, 8, 18, 18, 4)	121.75 **Sb** 51 (2, 8, 18, 18, 5)	127.60 **Te** 52 (2, 8, 18, 18, 6)	126.9044 **I** 53 (2, 8, 18, 18, 7)	131.30 **Xe** 54 (2, 8, 18, 18, 8)
192.2 **Ir** 77 (2, 8, 18, 32, 15, 2)	195.09 **Pt** 78 (2, 8, 18, 32, 16, 2)	196.967 **Au** 79 (2, 8, 18, 32, 18, 1)	200.59 **Hg** 80 (2, 8, 18, 32, 18, 2)	204.37 **Tl** 81 (2, 8, 18, 32, 18, 3)	207.19 **Pb** 82 (2, 8, 18, 32, 18, 4)	208.980 **Bi** 83 (2, 8, 18, 32, 18, 5)	[210*] **Po** 84 (2, 8, 18, 32, 18, 6)	[210] **At** 85 (2, 8, 18, 32, 18, 7)	[222] **Rn** 86 (2, 8, 18, 32, 18, 8)

Inner transition elements — f orbital elements

150.35 **Sm** 62 (2, 8, 18, 24, 8, 2)	151.96 **Eu** 63 (2, 8, 18, 25, 8, 2)	157.25 **Gd** 64 (2, 8, 18, 25, 9, 2)	158.924 **Tb** 65 (2, 8, 18, 27, 8, 2)	162.50 **Dy** 66 (2, 8, 18, 28, 8, 2)	164.930 **Ho** 67 (2, 8, 18, 29, 8, 2)	167.26 **Er** 68 (2, 8, 18, 30, 8, 2)	168.934 **Tm** 69 (2, 8, 18, 31, 8, 2)	173.04 **Yb** 70 (2, 8, 18, 32, 8, 2)	174.97 **Lu** 71 (2, 8, 18, 32, 9, 2)
[242] **Pu** 94 (2, 8, 18, 32, 24, 8, 2)	[243] **Am** 95 (2, 8, 18, 32, 24, 9, 2)	[247] **Cm** 96 (2, 8, 18, 32, 25, 9, 2)	[249*] **Bk** 97 (2, 8, 18, 32, 27, 8, 2)	[251*] **Cf** 98 (2, 8, 18, 32, 28, 8, 2)	[254] **Es** 99 (2, 8, 18, 32, 29, 8, 2)	[253] **Fm** 100 (2, 8, 18, 32, 30, 8, 2)	[256] **Md** 101 (2, 8, 18, 32, 31, 8, 2)	[254] **No** 102 (2, 8, 18, 32, 32, 8, 2)	[257] **Lw** 103 (2, 8, 18, 32, 32, 9, 2)

"The brackets indicate that the mass number is that of the isotope with the longest half-life."

discussion in subsequent chapters of how atoms combine. Most of the situations we deal with involve atoms in the combined state. We have already noted that many solid compounds are composed of ions. Knowledge of the electron configuration of ions and the way they may be formed will help us to understand their properties, and the characteristics of the aggregates they form. For these reasons, let us see how the electron configuration of ions is related to the configuration of neutral gas atoms.

6-7 Formation of Positive Ions It is reasonable to assume that the formation of a positive gas ion is related to the ease with which an electron can be removed from a neutral gas atom. As stated

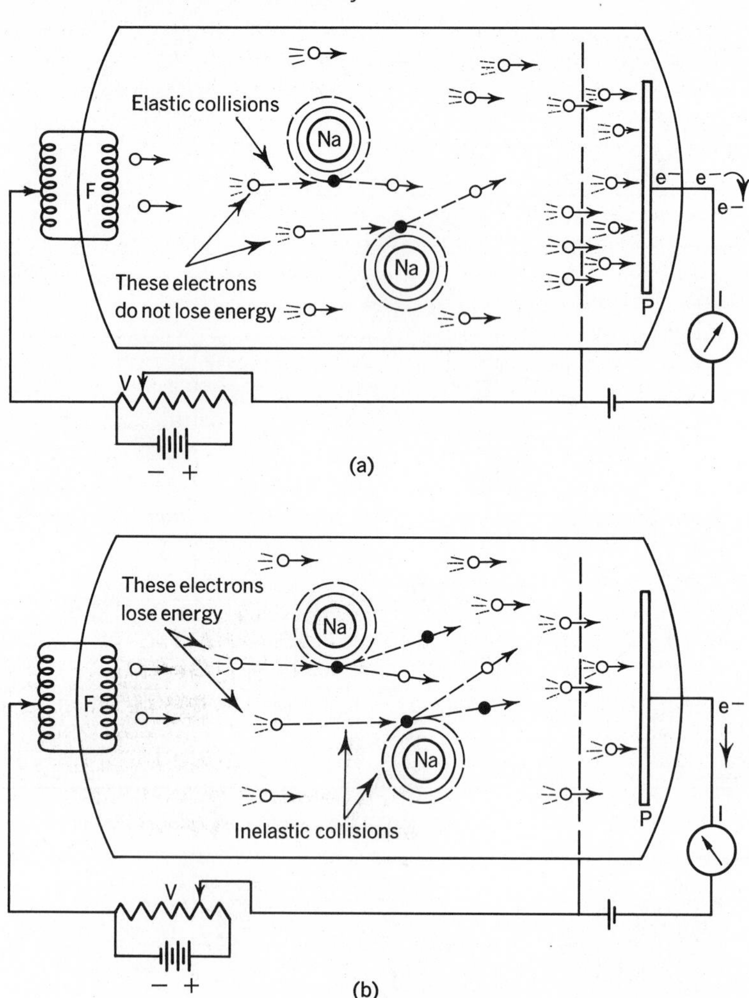

6-9 (a) As voltage (*V*) is increased, electrons from the filament (*F*) gain sufficient energy to reach the plate (*P*) and cause a rise in the plate current (*I*), but do not have enough energy to dislodge the outer electron of sodium atoms. No loss of energy is suffered in the elastic collisions with the sodium atoms. (b) When voltage reaches a critical value, bombarding electrons have enough energy to dislodge an electron from a sodium atom. Kinetic energy of bombarding electrons is distributed between the two electrons (the bombarding electron and the "target" electron). Neither has enough energy to reach the plate so the current drops. The ionization energy may be determined by analyzing a plot of voltage *vs* current.

earlier, the energy or work required to remove an electron from a gaseous atom is called the ***ionization energy.*** Ionization energies may be determined spectroscopically or in some cases, by means of electric measurements. The second method is easier to understand and is depicted schematically in Fig. 6-9. The experiment involves bombarding gaseous atoms with electrons in a cathode-ray-like tube. The kinetic energy of the bombarding electrons is related to and controlled by the applied voltage. Higher voltages increase the kinetic energies of the bombarding electrons. To determine the ionization energy of a gas atom, the voltage is increased until the kinetic energy of the bombarding electrons is equal to the energy needed to overcome the force of attraction between the nucleus and the easiest-to-remove electron of a gas atom. When this critical voltage is reached, positive gas ions are formed. The formation is signaled by a sudden change in current flow.

In theory, each atom has as many ionization energies as it has electrons. Experimental data show that the energy required to remove a second electron is always greater than that to remove the first. The removal of the first electron reduces the number of electrons and, consequently, the total electronic repulsion. This results in drawing the electron cloud closer to the nucleus (Fig. 6-10). There it is more compact and, because of the smaller radius, each electron in the cloud is subjected to a greater force of attraction by the nucleus. A greater force of attraction between an electron and its nucleus means that more energy is required to dislodge the electron, or to achieve ionization. Thus, the *second ionization energy* of a gaseous atom is always greater than the first. The first, second, and third ionization energies of elements 1 through 20 are listed in Table 6-5.

Analysis of the data in Table 6-5 leads to these observations.

1. In general, nonmetallic atoms have higher first-ionization energies than metallic atoms.

2. Certain gases such as helium (He), neon (Ne), and argon (Ar) have very high first-ionization energies. This indicates that the electrons are held very firmly. The compactness and symmetry of their electron clouds also prevent them from taking on electrons. In general, the gases helium, neon, argon, krypton, xenon, and

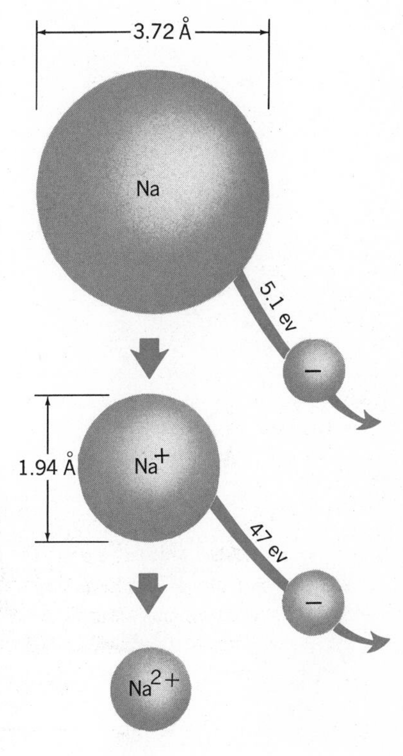

6-10 The removal of an outer electron from a sodium atom reduces the electron-to-proton ratio, the total electron repulsion, and the radius of the particle.

TABLE 6-5
IONIZATION ENERGY OF GASEOUS ATOMS

		Ionization Energies					
		I_1		I_2		I_3	
Atomic Number	Element	kcal/mole	ev	kcal/mole	ev	kcal/mole	ev
1	H	314	13.6				
2	He	567	24.6	1254	54.4		
3	Li	124	5.4	1744	75.5	2823	122
4	Be	215	9.3	420	18.2	3548	154
5	B	191	8.3	580	25.1	875	37.9
6	C	260	11.3	562	24.4	1104	47.9
7	N	335	14.5	683	29.6	1094	47.4
8	O	314	13.6	811	35.2	1267	54.9
9	F	402	17.4	807	35.0	1445	62.6
10	Ne	497	21.6	947	41.1	1500	64.0
11	Na	118	5.1	1091	47.3	1652	71.7
12	Mg	176	7.6	347	15.0	1848	80.1
13	Al	138	6.0	434	18.8	656	28.4
14	Si	188	8.1	377	16.3	772	33.4
15	P	254	11.0	453	19.6	696	30.2
16	S	239	10.4	540	23.4	807	35.0
17	Cl	300	13.0	549	23.8	920	39.9
18	Ar	363	15.8	637	27.6	943	40.9
19	K	99	4.3	734	31.8	1100	47.8
20	Ca	141	6.1	274	11.9	1181	51.3

radon are chemically unreactive. They are known as the noble gases. With the exception of helium, they all have two *s* electrons and six *p* electrons in the outermost energy level.

3. There is an unusually large gap between the energy required to remove the first electron and the second electron from lithium (Li), sodium (Na), and potassium (K) atoms. This suggests that it is relatively easy for these atoms to form ions having a 1+ charge. The low ionization energy is consistent with the fact that the single outer *s* electron in each of these atoms is located at a relatively long distance from the nucleus. It is also consistent with the relatively high chemical reactivity of these elements.

4. There is an unusually large gap between the energy required to remove the second electron and the third electron from beryllium (Be), magnesium (Mg), and calcium (Ca) atoms. This suggests that it is relatively easy for these atoms to form ions having a 2+ charge.

Let us examine the electronic configuration of a few of these simple positive ions.

6-8 Electron Configuration of Ions Examination of the electron configuration of many simple positive ions shows that they are *isoelectronic* (same number of electrons) with the atoms of the noble gases. For example, sodium *ions* (Na^+) and neon *atoms* both have 10 electrons and the same configuration. Let us examine the significance of this observation. Neon atoms have a ground-state configuration of $1s^2$, $2s^2$, $2p^6$. In neon atoms all available orbitals in the first and second energy levels are fully occupied. Note that the outer level has 8 electrons. This outer-level octet structure is characteristic of all noble-gas atoms except helium, which has only one principal energy level. Because noble gases are chemically unreactive and extremely stable, it is assumed that *there is a special stability associated with the outer-octet structure and the configuration of the noble-gas atoms.* Apparently the 8 electrons form a spherically symmetrical charge cloud about the nucleus.

Neutral sodium atoms have 11 electrons—one more than neon atoms. By losing one electron, sodium atoms achieve the electron configuration of neon and become stable sodium ions. Sodium ions have a nuclear charge of 11+ and only 10 electrons. This gives the ion an over-all charge of 1+. Accordingly, the electron configuration of Na^+ is the same as that of neon, $1s^2$, $2s^2$, $2p^6$.

It is evident from the relatively high ionization energies of fluorine, chlorine, oxygen, and sulfur that these and other nonmetals do not readily lose electrons and form positive ions. Rather, they are observed to form simple negative ions quite readily. In doing so, they achieve the electron configurations of the noble gases. Consider fluorine atoms which have one electron fewer than neon atoms. We would expect fluorine to gain one electron readily and thereby achieve the neon structure. The addition of one electron to a neutral fluorine atom produces a fluoride ion with a 1− charge. Thus fluoride ions (F^-), sodium ions (Na^+), and neon atoms (Ne) all have the same electron configurations.

This charge cloud is, in effect, impenetrable in that other atoms cannot "see" or "feel" the nucleus of a noble-gas atom. Hence, their stable structures are not disrupted by attracting the electrons of other atoms or by losing their own electrons. This suggests that other atoms might gain stability by acquiring the same electron configuration as a noble-gas atom.

These examples and other data indicate that the easiest-to-remove electrons are those with the highest principal quantum numbers. It is for this reason that we usually represent the ground-state configuration of an atom by grouping all orbitals in a given energy level together and arranging them in order of increasing principal quantum number.

It is apparent that positive ions have the same nuclear charge as the atoms from which they are formed but have fewer electrons. Let us write the electronic configuration of a titanium(II) ion (Ti^{2+}). The atomic number of titanium is 22. This means that the neutral gas atom has 22 electrons and a nuclear charge of 22+. Accordingly, the electron configuration of a titanium atom is

$$\mathbf{1s^2 2s^2 2p^6 3s^2 3p^6 3d^2 4s^2}$$

The 2+ charge on the ion indicates that the atom has lost two electrons. The two electrons which are lost come from the 4*s* orbital even though in our hypothetical process a *d* electron was the last to "enter" in the "building up" of the atom in the ground state. This behavior is explained in terms of the relative stabilities of electrons in the 4*s* and 3*d* orbitals. The relative stabilities are related to the energy changes associated with orbital overlap and increasing complexity of orbitals in the higher energy-levels. A detailed explanation is beyond the scope of this text. We may represent the configuration of Ti^{2+} as $1s^2 2s^2 2p^6 3s^2 3p^6 3d^2$, or

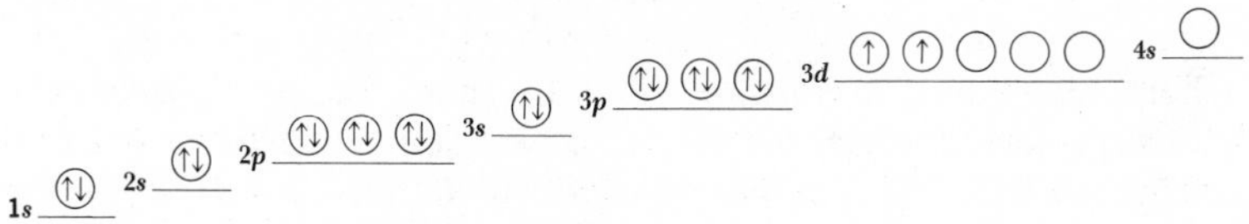

1*s* (↑↓) 2*s* (↑↓) 2*p* (↑↓)(↑↓)(↑↓) 3*s* (↑↓) 3*p* (↑↓)(↑↓)(↑↓) 3*d* (↑)(↑)( )( )( ) 4*s* ( )

FOLLOW-UP PROBLEM

Write the electron configuration for these species: (a) K^+, (b) Cl^-, (c) Ar, (d) Fe^{2+}, (e) Fe^{3+}, (f) Zn^{2+}.

You now have acquired sufficient background to go directly to Chapter 7 where we shall look for a relationship between the electronic configuration of atoms and the observed similarities in the properties of various groups of elements. Many of you however, may be curious as to how the modern quantum-mechanical theory originated and the nature of its theoretical basis. If you wish a greater insight into how contemporary scientists describe atomic and molecular systems, then you will want to read the remaining sections in this chapter.

THE EVOLUTION AND NATURE OF THE WAVE-MECHANICAL MODEL OF AN ATOM

6-9 "Matter Waves" In 1923, a young French scientist, Louis de Broglie, (1892–) supplied the creative spark which led to the development of this new model. De Broglie's imaginative proposal was that *matter, like light or any other type of radiant energy, has both particle and wave characteristics.* He reinforced this proposal by deriving an equation for the wavelength λ of a particle with mass m and velocity v. The equation was derived by equating the energy of a photon from Planck's equation $E = h\nu$, to the mass energy from Einstein's equation, $E = mc^2$. That is,

$$mc^2 = h\nu$$

Substituting c/λ for ν yields

$$mc^2 = \frac{hc}{\lambda}$$

Dividing both sides by mc^2 and multiplying both sides by λ yields

$$\lambda = \frac{h}{mc}$$

Substituting the velocity of a material particle v for the speed of light yields the de Broglie equation

$$\lambda = \frac{h}{mv}$$

where h is Planck's constant, m is the mass of the particle, and v is its velocity. Thus, de Broglie concluded that *every particle is associated with a wave, the frequency of the wave being dependent on the velocity of the particle.* At the time, de Broglie's concept of "matter waves" was looked upon with a great deal of skepticism by other scientists who had no evidence of the existence of such waves. However, even the skeptics were forced to accept the idea when two American scientists experimentally verified de Broglie's hypothesis.

In 1927, C. H. Davisson (1881–1958) and L. H. Germer (1896–) conducted experiments which involved the "shooting" of electrons at a sample of nickel. To their amazement, they obtained diffraction patterns similar to those obtained when X rays (a form of light) are diffracted by a crystal. Furthermore, the experimentally determined wavelength agreed with the value calculated by using the de Broglie equation. It has since been confirmed that beams of

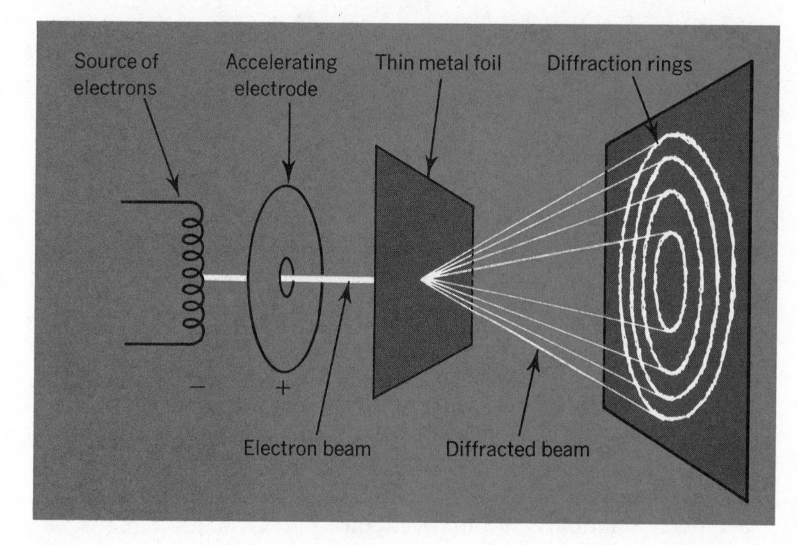

6-11 Schematic representation of the diffraction of electrons by a thin metal foil. The regular spacing of the atoms in a metallic crystal (foil) affects the beam of electrons in the same way that the slits in a diffraction grating or the pinhole in a piece of cardboard affect visible light (see suggested experiment on p. 170). The spacing of the atoms in a crystal is of the order of 1 Å. This approximates the wavelength of a 100-ev electron. The distance between the lines in an ordinary grating approximates the wavelength of visible light.

neutrons, hydrogen atoms, and other small particles exhibit wave characteristics. Wavelengths can be calculated theoretically for material objects ranging from electrons to elephants. The wavelengths associated with large objects, however, are so small that there is no physical method available for detecting or measuring them. Hence, they have no practical significance (Fig. 6-13).

Practical use of the wave character of electrons is shown in the operation of the *electron microscope.* This instrument, Fig. 6-14, uses a beam of electrons rather than light waves to produce the image of an object being viewed. The wavelength of "electron waves" is much smaller than that of visible light waves. The electron microscope, therefore, has a much greater resolving power than an ordinary microscope.

Although we can calculate a wavelength for "matter waves," such waves do not resemble electromagnetic waves, sound waves, water waves, or any other waves associated with our normal everyday experience. It was Max Born (1882–) who proposed the presently accepted interpretation of "matter waves." He suggested that these waves are waves of probability, indicating a region of space where we are most likely to find particles associated with these waves. We can consider electron waves as "pilot waves" which guide the motion of the particles (electrons) analogous to electromagnetic waves which guide the flow of "particles" (photons) of energy.

6-10 The Uncertainty Principle The inability to predict the exact location of atomic particles was rationalized by Werner Heisenberg (1901–). He proposed that the basis of the dilemma was the attempt to apply the rules and methods used in making observations on the macroscopic level to atomic phenomena. It is possible to carry out observations and measurements of objects common to everyday experience without disturbing the object being measured. For example, the exact path that a satellite follows or the spot that a rifle bullet hits can be precisely determined by

6-12 Electron interference pattern resulting from the diffraction of electrons by a crystal. This experimental observation provides evidence that beams of electrons behave as waves. (Bell Telephone Laboratories)

6-13 Even an elephant has a wavelength that can be calculated. In order for a 5000-lb elephant to have a measureable wavelength (0.01 Å), it would have to move at a velocity of about 10^{-23} in. per sec. At this rate, it would take 10^{15} years (million billion years) to go one inch.

applying classical mechanical laws. As long as the momentum (the product of mass and velocity) and the position of a satellite can be determined simultaneously, the path it will follow can be predicted. Since the position of the massive satellite is not changed by the impact of the light rays which strike it, the image we observe is the actual point from which the rays are reflected.

A different situation exists in the world of atomic particles. Investigating the behavior of an electron with photons of light is like investigating the working of a fine watch with a crowbar or pickax. For example, if we attempt to determine the position of an electron, the light reflected from a tiny electron has enough energy to alter the electron's path completely. Thus, *we cannot simultaneously measure the position and the momentum of an electron.* Since we cannot observe the true position of an electron, we are unable to describe its trajectory. Consequently, an exact orbit such as that described by the Bohr theory cannot be determined for electrons.

Werner Heisenberg summarized these ideas in his famous ***uncertainty principle,*** which states that *it is impossible to determine simultaneously the exact position and momentum of a single atomic particle.* The uncertainty principle may be expressed mathematically as

$$(\Delta p)(\Delta x) = h \qquad 6\text{-}1$$

where h equals Planck's constant. In this specialized instance, Δp represents the *uncertainty in momentum* and Δx *the uncertainty in position.* Equation 6-1 shows that the smaller the uncertainty in momentum or velocity ($\Delta p = m\,\Delta v$), the greater the uncertainty

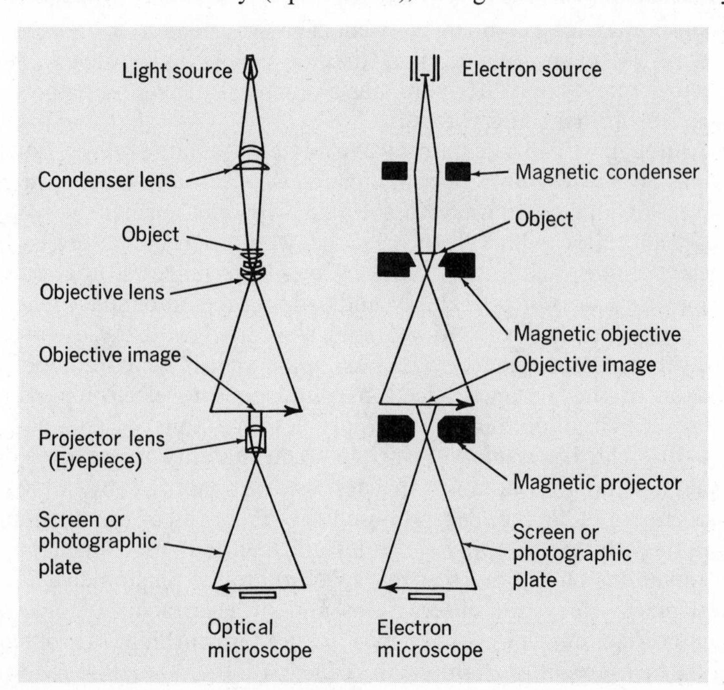

6-14 Schematic diagrams of the electron microscope and the optical (light) microscope, inverted to make comparison easier.

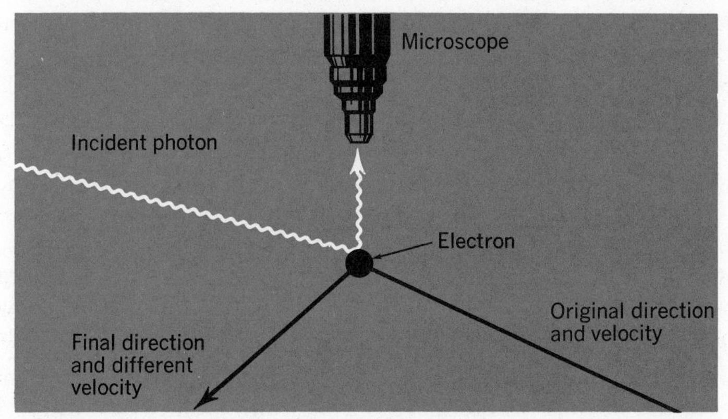

6-15 A photon of light used to observe and determine the position of an electron would change the momentum (velocity) and alter the path of an electron. Therefore, we cannot simultaneously determine the position and momentum of an electron.

in position. It can be shown that an uncertainty of as much as 1 percent in the velocity of an electron in a hydrogen atom yields an uncertainty of 3.7×10^{-6} cm in the position of an electron. This uncertainty is 700 times the actual radius of a hydrogen atom. More precise measurements of the velocity yield even greater uncertainty in the position of an electron. This example vividly illustrates that it is impossible to specify the location of an electron at any given time. Thus, Bohr's definite orbits have no real meaning and can never be experimentally demonstrated.

Since the value of h in Equation 6-1 is so small, we can ignore uncertainty on the macroscopic scale. It can be shown that an uncertainty of 10 mi/hr in the velocity of an ordinary automobile results in an uncertainty of only 1×10^{-35} cm in its position. This example indicates that the uncertainty principle does not apply to events on the macroscopic scale.

6-11 The Quantitative Basis for the Wave-Mechanical Model In 1926, Erwin Schroedinger (1887–1961), an Austrian scientist, furnished the quantitative basis for the new atomic model. Schroedinger, taking a cue from de Broglie, *visualized an atom as a positively charged nucleus surrounded by vibrating electron waves.* As a theoretical physicist, Schroedinger knew that a complete description of an atom required an equation for the waves associated with its electrons. The correct equation would have to yield solutions which would agree with experimental data. That is, it would have to predict energy states that agreed with those obtained spectroscopically. Schroedinger considered the pulsating electron waves surrounding the nucleus analogous to the periodic vibrations of water covering a uniformly flooded, spherical planet. Equations describing the wave motion on a hypothetical flooded planet had been worked out 100 years earlier by William Hamilton (1805–1865), an Irish mathematician, while he was working on the problem of predicting tides.

In 1926, Schroedinger adapted Hamilton's equations to an atomic system, introduced de Broglie's wavelength of an electron, and applied conditions corresponding to the "geometry" of an atom. The result was his famous *wave-mechanical equation, the solution of which described the behavior of a hydrogen atom in terms of its wave characteristics.* This equation, or a modification of it, provides the theoretical basis for the current wave-mechanical model of atomic and molecular structures. The wave equations describe an electron as a three-dimensional wave in the electric field of a positively charged nucleus. The wave equation is shown and briefly discussed in the sketch about Erwin Schroedinger in the margins of pp. 218 and 219.

The "geometry" of an atom or molecule refers to its dimensions and shape.

When the wave equation is solved for a hydrogen atom, the allowed energy values are in close agreement with the experimentally derived values. In addition, solutions to the equation provide data which enable scientists to identify regions of space (orbitals) about the nucleus where there is a high probability of locating an electron with a specific energy. Computers are used to solve the Schroedinger equation and to plot the probable locations of an electron as it moves about the nucleus.

The solution to the Schroedinger equation yields a quantity psi (ψ) whose square (ψ^2) can be used to determine the probability of finding an electron with a specific energy in a tiny unit volume of space at various distances from the nucleus. By determining the probability at various distances from the nucleus, it is possible to outline regions of space where there is, for example, 90 or 95 percent probability of finding a given electron.

The computer-produced plot is equivalent to plotting the results of a series of hypothetical experiments designed to determine the frequency of an electron's appearance at different points in the space surrounding the nucleus. If we plotted the results of many such experiments on a three-dimensional space graph, we would obtain an electron charge-cloud picture which would have the size, shape, and direction of the regions of space in which an electron could be found a given percentage of the time. These regions of space are called ***orbitals*** and are pictured as electron charge-clouds. The probability of finding an electron at various regions in the cloud is proportional to the density of the cloud in those regions.

Each orbital is associated with and defined by a set of integers called ***quantum numbers.*** These numbers are rather analogous to the three dimensions needed to describe anything occupying three-dimensional space. The three quantum numbers are derived from the mathematical solutions of the wave equations. These numbers are designated as n, l, and m_l. The quantum numbers can have only specific values and are interrelated. That is, the value of l is limited by the value of n and the value of m_l is limited by the value of l. The permitted values assigned to these numbers determine the specific (quantized) energies that a hydrogen atom can have. The quantum numbers are like an address for the electrons in an atom. To identify an electron in an atom completely, it is necessary to state the value of each of the quantum numbers. *No two electrons in a given atom can have the same set of quantum numbers.* This important principle, known as the *Pauli (Wolfgang*

ERWIN SCHROEDINGER
1887–1961

Schroedinger was an Austrian physicist who received his education at the University of Vienna and later served as professor of physics at Stuttgart, Breslau, Zurich, and the University of Berlin. He became a Fellow of Magdalene College, Oxford, and in 1940 went to Dublin, Ireland, as a professor of physics at the Institute of Advanced Studies. In 1933, Schroedinger was awarded the Nobel Prize for his contribution to the development of the wave-mechanical model of an atom.

Schroedinger expressed his ideas of the wave-particle duality of the electron proposed by de Broglie in the following statement: "The electron revolving in the atom is a disturbance proceeding along the electron orbit in the form of a wave." In 1926, Schroedinger developed the equation which bears his name. The equation is

$$\left(\frac{\partial^2\psi}{\partial x^2} + \frac{\partial^2\psi}{\partial y^2} + \frac{\partial^2\psi}{\partial z^2}\right) + \frac{8\pi^2 m}{h^2}(E - V)\psi = 0$$

where x, y, and z are the space coordinates of the electron with the nucleus at 0, 0, 0, m is the mass of the electron, E is the total energy of the electron-proton system, and V is the potential energy. The Greek letter ψ (psi) is called a "wave amplitude function" because in many solutions to the equation it exhibits a wave-like series of maxima and minima. It is analogous to the wave amplitude of an ordinary wave. The expression $\partial^2\psi/\partial x^2$ indicates the mathematical operation which must be carried out on ψ to solve the equation. (Continued in opposite margin.)

Pauli, 1900–1958) *exclusion principle,* underlies the electron arrangement about the nucleus.

The significant aspects of the quantum numbers are summarized below.

6-12 *n*, the Principal Quantum Number The *principal quantum number* is designated as *n*. This quantum number designates the principal energy level in which a given electron is located. The value of *n* determines to a large extent the energy of an electron and is related to the average distance of an electron from the nucleus. *n* is limited to any positive whole number value excluding zero. An electron with $n = 1$ is in the lowest energy-level of an atom. **The value of *n* also indicates the number of nodal surfaces associated with each orbital in a given energy level. Nodal surfaces** represent locations where the probability of finding the electron is zero.

Solutions of the Schroedinger equation yield expressions for the wave function, ψ. Each wave function is associated with a set of quantum numbers which can have only certain allowed values. The values determine the allowed energy states of the atom. Thus, each acceptable wave function, ψ, is related to a given energy state as well as to the space coordinates which describe the atomic system. Psi squared, ψ^2, is interpreted as being proportional to the probability of finding the electron in an element of unit volume about the nucleus. If we think of the electron in a specific energy state as being smeared out over a certain region of space about the nucelus, then ψ^2 may be interpreted as being proportional to the "electron density" of the electron charge cloud in a given volume element of the cloud.

A nodal surface for a three-dimensional vibrating object is analogous to a nodal point for a vibrating string. In Fig. 6-16(1) the string is fixed at both ends. Points *a* and *b* constitute the boundaries of the string system. When plucked, the string vibrates. The maximum displacement from the rest position at point *c* is called the amplitude of the vibration. There is no vibration at the ends. These points of zero displacement are called ***nodes.*** Only certain vibrations can occur. These are determined by boundaries of the system. A second mode of vibration is shown in Fig. 6-16(2). The vibration in this case has three nodes: one at each end and one in the center. Note that on either side of the central node the displacements of the curve are equal in magnitude but opposite in direction. The displacements are said to be *out of phase.* A change in phase always occurs at a nodal point. In an atom there

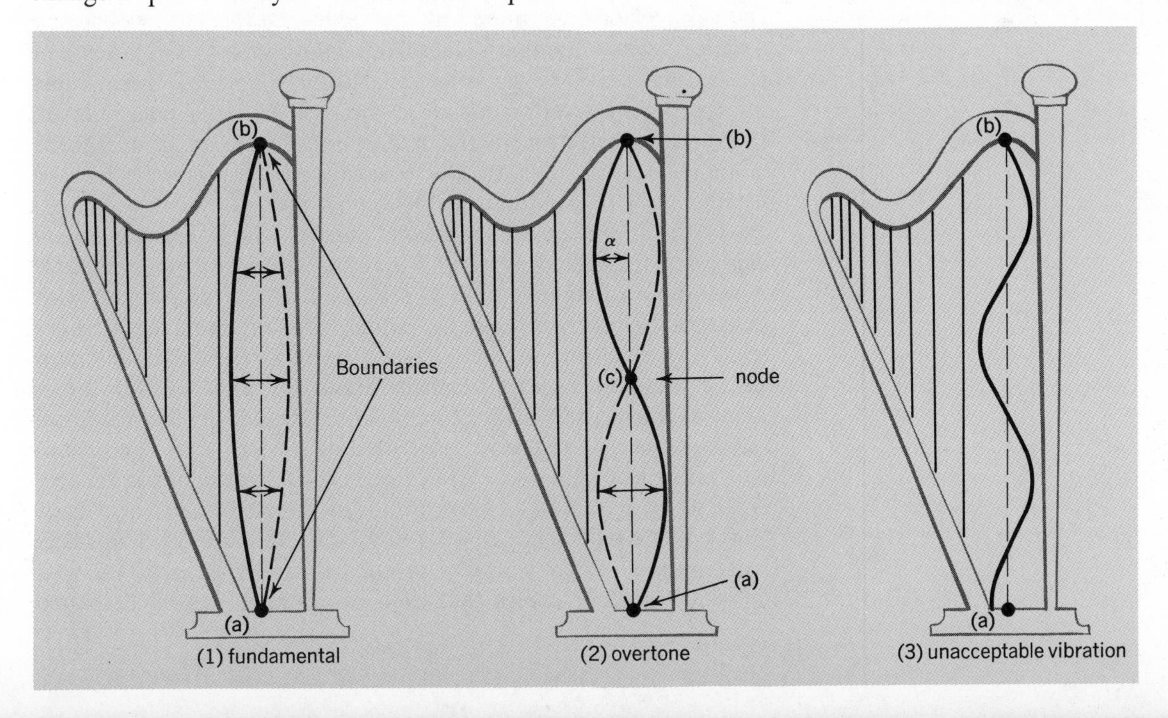

6-16 (1) and (2). Allowed vibrations of a string. Note in (2) that the string is displaced in opposite directions on either side of the node. The two displacements are said to be out of phase. Any vibrations that would require displacement of the string at the boundaries are not allowed. When the string in (1) is plucked, standing waves shown in (2) appear. That is, the waves in (2) appear to be stationary. The wave-mechanical model treats electron waves as standing waves of energy. (3) An unallowed vibration analogous to an unallowed set of quantum numbers for describing an electron.

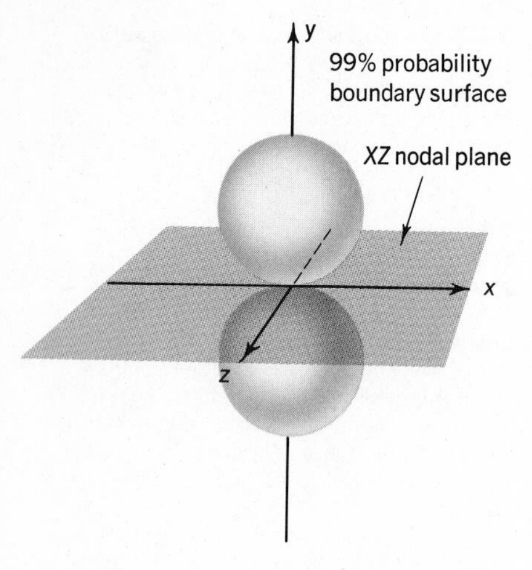

6-17 The p_y orbital showing the region of zero probability in the xz plane. The xz plane is called a nodal plane. What is the nodal plane for the p_z orbital?

is a nodal surface at infinity for all orbitals. Thus, in theory, all orbitals extend to infinity. We are concerned with the regions relatively close to the nucleus where there is a high probability of an electron being found. The most likely distance at which a given electron may be found is represented in terms of a probability graph to be discussed later. In addition to the number of nodal surfaces, n is equal to the number of different types of orbitals in the given energy level. For example, in the third energy level in which $n = 3$, there are three types of orbitals: *s*, *p*, and *d*.

6-13 *l*, the Angular Momentum Quantum Number Although the general energy position of an electron is represented by the principal quantum number, a portion of this energy can be associated with the orbital motion of the electron about the nucleus. This orbital motion is described by the ***angular-momentum quantum number*** which is designated as l. l is related to the angular momentum of the electron and the shape of the orbitals. The l number is also related to the position of the nodal surfaces. Interpretation of this quantity reveals that for *p* orbitals there is a nodal plane through the nucleus. *p* electrons have zero probability of being found in this plane. The value of l indicates the type of sublevel and orbital in which an electron is located. Sublevels and orbitals are designated as *s*, *p*, *d*, or *f*, depending on the value of l. The relationship is shown below.

Value of l	*Type of sublevel and orbital*
0	s
1	p
2	d
3	f

The value of l is determined by the value of n for that level. l may assume all whole number values from 0 through $n - 1$. Thus, when $n = 1$, l has only one value and that is 0. This means there is just one type of sublevel or orbital in the first or $n = 1$ energy level. Therefore, an electron in the first energy level can be identified as an *s* electron only. An electron with $n = 2$ might be located in either an *s* or *p* orbital in the second energy level.

6-14 *m*, the Magnetic Quantum Number ***The magnetic quantum number*** is designated as m_l. As a result of an electron's angular momentum, a magnetic field is produced which can interact with an external electric or magnetic field. Under the influence of these fields, the electrons in the orbitals or sublevels are oriented in certain preferred regions of space about the nucleus. The directional characteristics of these regions are related to the *x*, *y*, and *z* axes in space. These letters are used as subscripts on the orbital letters. For example, these are three *p* orbitals designated, respectively, as p_x, p_y, and p_z. The p_x orbital is directed along the *x* axis. The *yz* plane represents a nodal surface for this orbital. The directional nature of the *p* and *d* orbitals is shown in Fig. 6-2 and Fig. 6-3. Fig. 6-17 shows that each *p* orbital is dumbbell-shaped with two lobes separated by a *nodal plane*. It should be noted that an electron in a p orbital has an equal probability of being found in either lobe of the orbital. We shall find that the directional characteristics of the p and d orbitals play an important role in determining the shapes and properties of molecules. Fig. 6-1 shows that s orbitals are spherically symmetrical about the nucleus and are nondirectional. This means that the probability of finding an s electron at a given distance from the nucleus is identical in all directions from the nucleus. The area of high concentration of dots represents the region where an electron is most likely to be found.

The value of the magnetic quantum number is related to the *spatial orientations of the orbitals* which result when the atom is subjected to a magnetic field. The values allowed m_l depend on the value of l. Thus, m_l can assume all values from $+l$ through 0 to $-l$. For example, when $l = 0$, m_l can have only one value, 0. This means that there is one s orbital per energy level. When $l = 1$, m can have three values: $+1, 0, -1$. This means that there can be three p orbitals. It can be seen that for each value of l, there are $2l + 1$ different values of m_l. Thus, the number of each kind of orbital in a principal energy level is determined by the number of values that m_l has for a given value of l. It is left as an exercise for you to show that d orbitals can have five spatial orientations (5 orbitals per level) and that f orbitals can have seven spatial orientations (7 orbitals per level). The relationships among the three quantum numbers, the principal energy levels, and the orbitals are summarized in Table 6-6.

6-15 m_s, the Spin Quantum Number In order to explain finer details of atomic spectra, it was proposed that an electron has an angular momentum of rotation about its own axis. In other words,

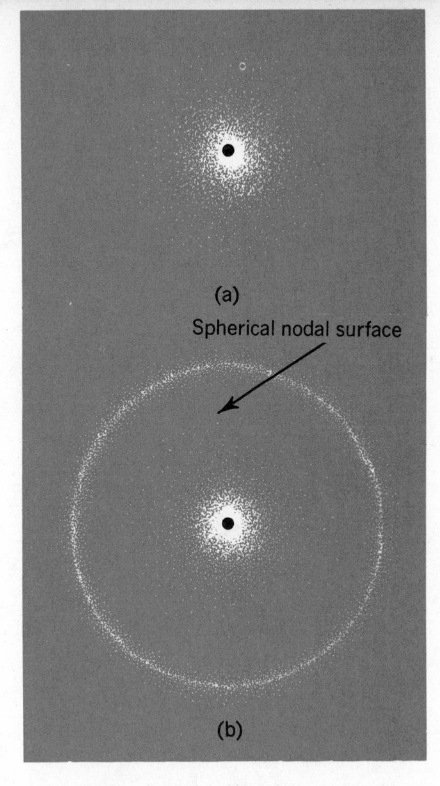

6-18 Diffuse charge cloud representation of orbital-electron charge distribution. This pattern represents the distribution of electron charge that would be obtained by observing the motion of an electron over a period of time. The stippling represents the probability density (ψ^2). The size of the cloud depends on the value of n. Thus the 2s cloud in (b) is larger than the ls cloud in (a). Note that the 2s orbital has a second nodal surface in addition to the one at infinity.

TABLE 6-6
QUANTUM-NUMBER AND ENERGY-LEVEL RELATIONSHIPS IN ATOMS

Principal Energy Level	Principal Quantum Number n	Angular-momentum Quantum Number l	Orbital or Sublevel Designation	Different Types of Orbitals per Principal Energy Level	Magnetic Quantum Number m_l	Number of Each Type of Orbital per Energy Level
1	1	0	s	1	0	1
2	2	0	s	2	0	1
		1	p		$+1, 0, -1$	3
3	3	0	s	3	0	1
		1	p		$+1, 0, -1$	3
		2	d		$+2, +1, 0, -1, -2$	5
4	4	0	s	4	0	1
		1	p		$+1, 0, -1$	3
		2	d		$+2, +1, 0, -1, -2$	5
		3	f		$+3, +2, +1, 0, -1, -2, -3$	7

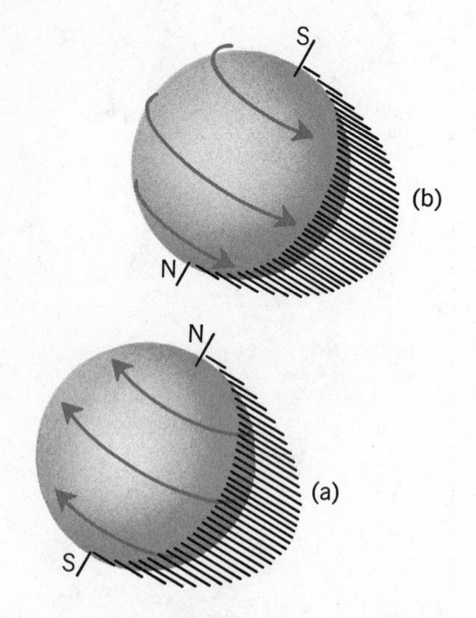

6-19 Electrons spinning about an axis produce a magnetic field which interacts with an externally applied field and the field produced by the electron's motion around the nucleus. The electrons in (a) and (b) have opposite spins and interact with the orbital magnetic field giving different energies. The differences can be determined spectroscopically.

it has a spin and behaves like a small magnet. It can spin in either of two directions and can have one of two possible values: $+\frac{1}{2}$ and $-\frac{1}{2}$. Thus, a fourth quantum number, called the ***spin quantum number, m_s***, distinguishes between the oppositely spinning electrons in an orbital (Fig. 6-19). See Table 6-7 below.

6-16 Pauli Exclusion Principle Since the values of the quantum numbers are restricted and interdependent, only certain combinations are possible. The Pauli exclusion principle states that no two electrons in an atom can have the same set of four quantum numbers. Thus, *no orbital can contain more than two electrons and the two electrons must have opposite spins.* Application of this principle determines and limits the maximum number of electrons that can occupy a given energy level. The following table illustrates the possible combinations of quantum numbers for electrons in a neon atom and is consistent with our finding that the first energy level can hold only two electrons and the second level only eight.

The values of the four quantum numbers in the first row of Table 6-7 constitute the "address" of the single electron in a hydrogen atom. The first- and second-row values are those which identify, respectively, the two electrons in a helium atom. The data in Table 6-7 show that *s* orbitals contain a maximum of two electrons and that *p* orbitals contain a maximum of six electrons. An extension of this table would show that the *d* orbitals in the third or higher energy level could contain a maximum of ten electrons and that *f* orbitals in the fourth or higher energy level could contain a maximum of 14 electrons. Application of the Pauli exclusion principle shows that the general formula for the maximum number of electrons that can be accommodated by a given energy level is $2n^2$, where n is the principal quantum number of the energy level. Thus, the fourth energy-level can hold a maximum of $2 \times 4^2 = 32$ electrons. See Table 6-7 below.

TABLE 6-7
ALLOWED QUANTUM NUMBERS OF ELECTRONS IN A NEON ATOM

Energy Level	n	l	Orbital	m_l	m_s	Possible Combinations of 4 Quantum Numbers (Maximum no. of electrons)
1	1	0	1s	0	$+\frac{1}{2}$	2
	1	0	1s	0	$-\frac{1}{2}$	
2	2	0	2s	0	$+\frac{1}{2}$	8
	2	0	2s	0	$-\frac{1}{2}$	
	2	1	2p	-1	$+\frac{1}{2}$	
	2	1	2p	-1	$-\frac{1}{2}$	
	2	1	2p	0	$+\frac{1}{2}$	
	2	1	2p	0	$-\frac{1}{2}$	
	2	1	2p	$+1$	$+\frac{1}{2}$	
	2	1	2p	$+1$	$-\frac{1}{2}$	

6-17 Energy Levels in Atoms Solutions to the Schroedinger equation yield quantum numbers which fix the energy states of a hydrogen atom. Support for the validity of the wave-mechanical model is found in the close agreement between the theoretically-calculated energy values for hydrogen and those determined experimentally from atomic spectra. The relative energy values for each state can be depicted by means of an energy-level diagram. The energy-level diagram for hydrogen is shown in Fig. 6-20. As may be seen from this figure all of the orbitals in any given principal quantum level represent the same energy: that is, they are *degenerate.*

Orbitals having the same energy are said to be degenerate **energy levels.**

The wave equation cannot be rigorously solved for atomic systems containing more than one electron. Therefore, approximation methods are used. Prediction of the behavior of the electrons in more complicated atoms and molecules is made by analogy to the solutions of the wave equation for a hydrogen atom. Since solutions have been verified experimentally, predictions made on this basis are valid. In multi-electron atoms, inter-electron repulsion is a factor so that the orbitals in a given energy level have different energy values. In these atoms, a 3*d* electron has more energy than a 3*p* electron which, in turn, has more energy than a 3*s* electron. The relative positions of the energy levels of a given atom depend on the nuclear charge and on the extent to which other orbitals are occupied. Therefore, energy-level diagrams of different atoms vary in terms of actual energies, but in a qualitative sense, they are similar and follow the general pattern shown in Fig. 6-4.

The electron configuration of atoms in the ground state may be written by using an energy-level diagram, Hund's rule, and the Pauli exclusion principle. The procedure was illustrated earlier in some detail.

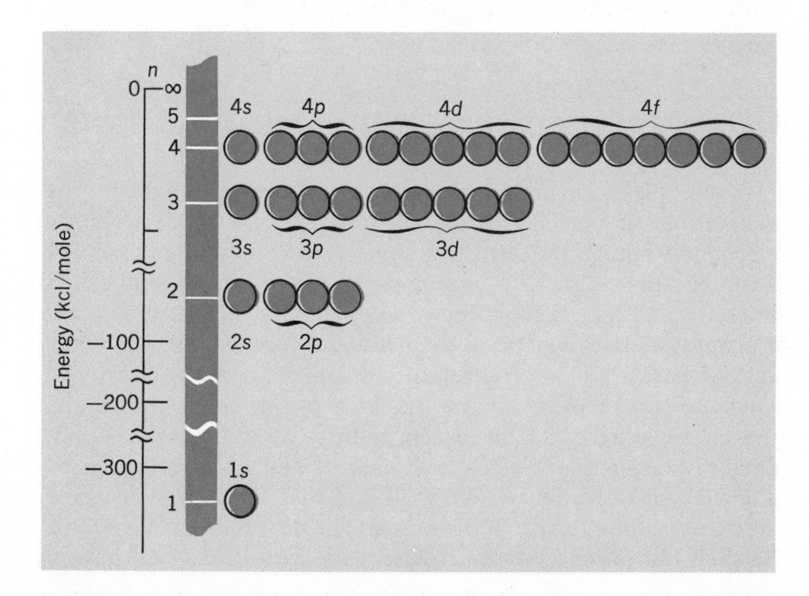

6-20 Energy-level diagram for hydrogen. The circles represent empty orbitals.

LOOKING AHEAD

The wave-mechanical model of an atom, evolved during the 1920's, has withstood the test of time and in the 1970's, is still the model that provides the tools and concepts whereby contemporary scientists describe atomic and molecular systems. From now on we shall draw on its resources to help us explain the observed properties of matter and behavior of the systems we investigate.

Up to this point, we have developed laws and principles which enable us to calculate masses and volumes of substances involved in chemical reactions. In addition, we have constructed models which help to explain the behavoir of gaseous molecules and which account for the attractive forces between atoms. In this chapter we discovered that similarities in the properties of elements are related to similarities in the outer energy-level electron configuration of their atoms. It was further observed that there is a correlation between the electron configuration of an atom and its location in the Periodic Table. References to the nature and specific properties of substances, however, have been rather incidental.

We are now ready to examine some specific properties of the elements and a few of their compounds. We shall use the Periodic Table as a guide in our search for trends in physical, chemical, and atomic properties. The knowledge we obtained through our study of electron configuration of atoms will help us to explain the observed trends. In the next chapter we shall also investigate in some detail atomic properties such as ionization energy and electron affinity, atomic and ionic radii, oxidation state, and electronegativity. An understanding of these concepts will provide background needed to explain properties of substances and will set the stage for the discussion of chemical bonding in Chapter 8 and properties of aggregates in Chapter 9.

Special Feature: Probability

Chance plays an important role in the events of nature. The makeup of our own minds and bodies is the result of a chance selection by nature of a few genes from an almost infinite possibility of combinations. Life-insurance companies consult actuarial charts which contain information about the life spans of a large number of people and use the data to predict your probable life span when you purchase a policy. No insurance company, however, can make a definite statement about the life span of any single individual. In a later chapter we shall discuss radioactive atoms which spontaneously disintegrate into a new kind of atom with an emission of some type of radiation. We shall find that for a large group of radioactive atoms, one-half disintegrate in a given interval of time. We shall also discover that a single radioactive atom will have a

50 percent chance of undergoing radioactive decay in a given time interval. This period of time is known as the half-life of that particular radioactive material.

Statements about the location of an electron in an atom, like those about the life-span of a person, or the disintegration of a radioactive atom, can be made only in terms of probability. Calculation of the probability of a given event occurring are based on the possible combinations which can lead to that event. For example, when a coin is flipped, there are only two ways that it can land: heads or tails. Therefore, there is a 1 to 1 chance (50 percent probability) of the coin coming up heads on a given toss. The odds offered by gamblers on craps, roulette, and other games of chance are based on the number of possible combinations that can lead to a given result.

Think of a third way a coin could land. What odds would you give that this event would occur?

One of the highlights of the year's sporting events is the World Series in baseball. This series affords some individuals the opportunity to capitalize on the odds associated with the possible outcome of the series. If we were to attempt to determine the probability of a given baseball team winning the World Series we would begin by analyzing all of the possible combinations of games that might be played. To win the Series one team must win 4 games. The Series could last just four games or it could end in game number 5, 6, or 7. If we assume that the players are equally skilled at baseball, we could then conclude that there are 70 possible outcomes of the Series. These are shown in Fig. 6-21. The 70 possible outcomes could be used as the basis for computing the probability of a given outcome. This, of course, does not imply that one can predict the outcome in a given instance. The many possible outcomes of the World Series seem rather complicated but in reality is very simple compared to the complexity in specifying the possible configurations of electrons within a molecule.

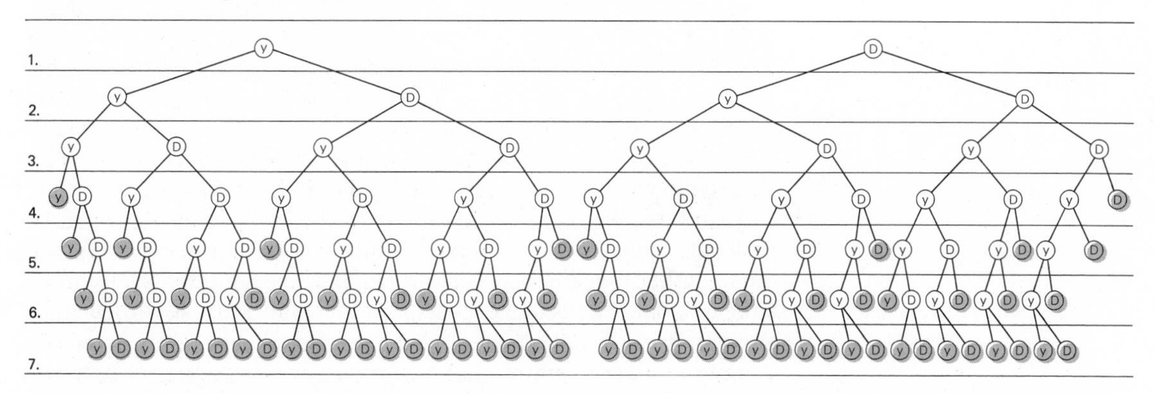

6-21 Possible outcomes of the Yankees and Dodgers in a World Series. The shaded circles show possible methods of terminating the series.

Consider H_2, the simplest neutral molecule. Suppose that we attempt to specify all of the possible positions that the two electrons may occupy in a hydrogen molecule. Let us assume that an electron may effectively fill a region within a hydrogen molecule which approximates a rectangular solid $1\,\text{Å} \times 1\,\text{Å} \times 2\,\text{Å}$ or two cubic

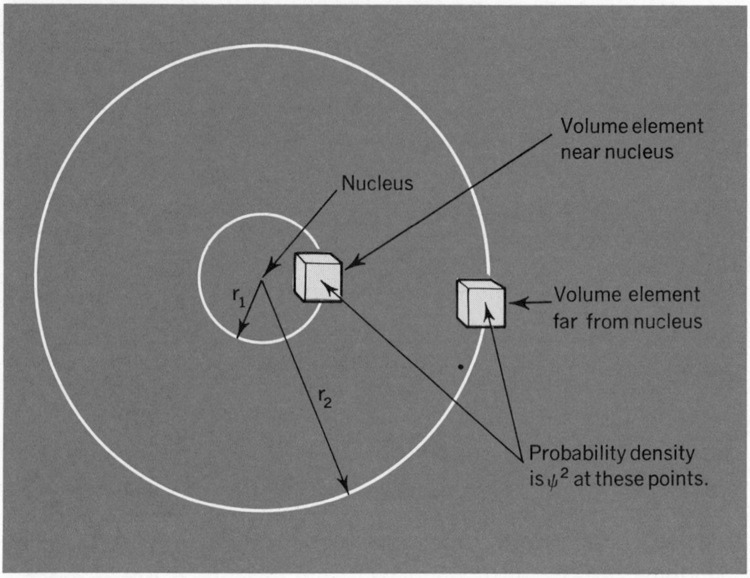

6-22 There is a greater probability of finding an electron in the cube (volume element) at r_1 than in the cube at r_2. The probability of an electron being in the cube is equal to the probability density at the point in the center of the cube multiplied by the volume of the cube (p. 228). If a H_2 molecule has a volume of 2 cubic Ångstroms, how many cubic "cells" each measuring 0.01 Å on a side are contained in the molecule?

Ångstroms. If we break this volume into volume-element cubes as shown in Fig. 6-22 where each cube is 0.01 Å on an edge, we find that each electron in the molecule could occupy any one of 2×10^6 (2 million) of these "cells" at any given instant. To specify completely the total possible combinations of possible positions for the two electrons, we must specify the position of one electron (2×10^6 possibilities) then at the same time specify the position of the second electron. There would be $(2 \times 10^6)(2 \times 10^6)$ or 4×10^{12} (4 trillion) possibilities. This is a very large number of possibilities. To appreciate how large this number is, let us imagine that we are to gather together enough encyclopedias to contain 4×10^{12} (4 trillion) words. Assume 10^3 (one thousand words on each page and 10^3 (one thousand) pages in each volume. We would need 4×10^6 or 4 million volumes. This great complexity represents the simplest of all molecules. Suppose we were to attempt to specify all of the possible combinations of position that the electrons could occupy in a benzene molecule. Benzene (C_6H_6) is also a relatively simple molecule. It has been calculated that there are 10^{300} different ways that the electrons could occupy the "cells" of a single benzene molecule. This is truly a large number. Again let us imagine we are to gather together enough encyclopedias to contain 10^{300} words. Take the largest distance ever seen by man with high-powered telescopes 2×10^9 light years) and make this the radius of a sphere. This sphere is the entire region of the known universe. If we were to pack this sphere solidly full of encyclopedias, it would contain only 10^{80} volumes with only 10^{86} words. You can see that to specify all of the possible positions of the electrons in a large molecule, would require a really large number, one that expands the imagination. When a scientist says **large** he is not fooling.

In this chapter we have referred to orbitals as regions of space about the nucleus in which there is a high probability of a given electron being found. Let us see how the probabilities are determined and represented. A simple analogy may help you understand the concept as applied to specifying the probable location of an electron.

Let us view an electron of a hydrogen atom as moving around the nucleus much the way a bee moves around its hive. Suppose we took a large number of photographs of the area including the bee and its hive. The bee would dart around the area collecting nectar from flowers but always returning to the hive which is a kind of focal point of its activity. Any given photograph could show the bee far from the hive but if a large number of photographs were considered, there is a great probability of finding the bee near the hive. It is important to remember, however, that probability tells us only where the bee is likely to be found in any given instant. It never tells us with certainty where the bee can be found. If we marked on a chart the locations of the flowers the bee visited, we would obtain a pattern similar to that shown in Fig. 6-23. This chart conveys probability information. It tells us that close to the hive the probability of finding the bee is high and far from the hive, it is low. The probability density (dots per cm^2 on a plane plot) is greatest near the hive.

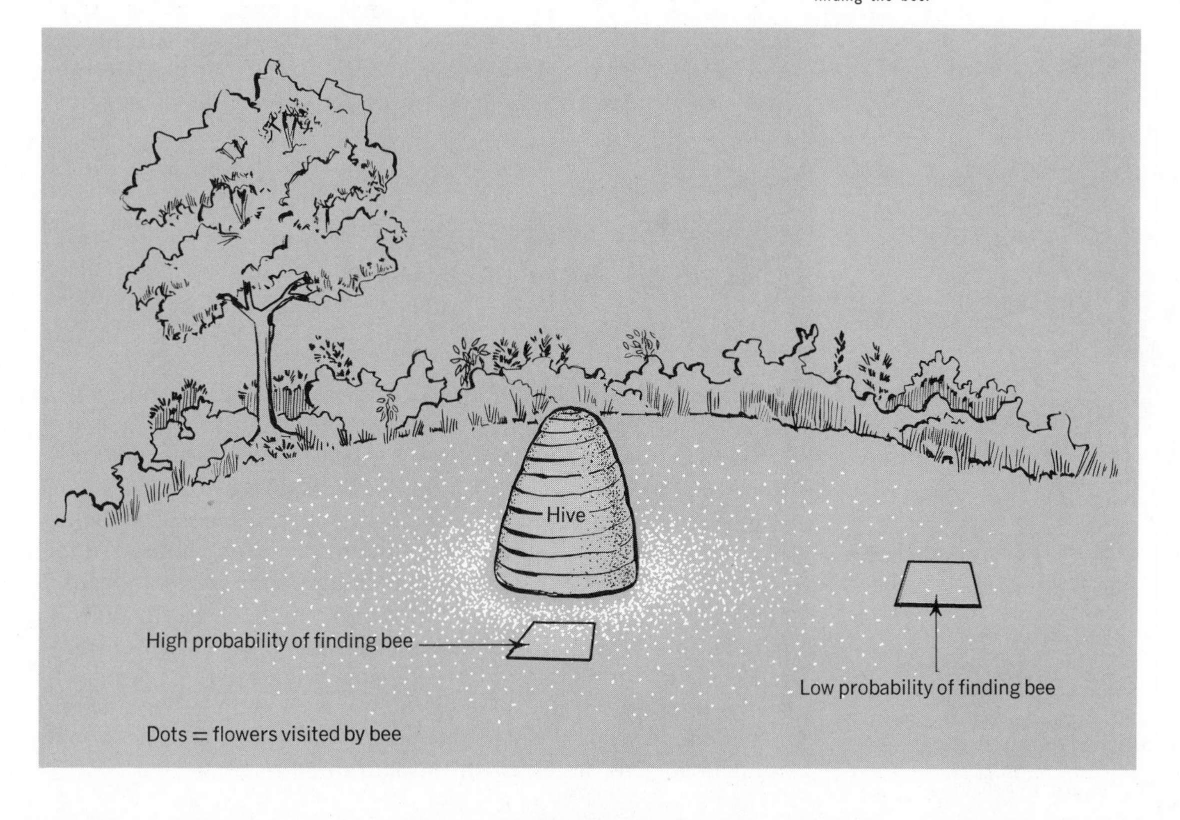

6-23 A bee visits more flowers per unit area near the hive than away from it. We can use the information shown on the diagram to outline an area within which there is a 95 percent probability of finding the bee.

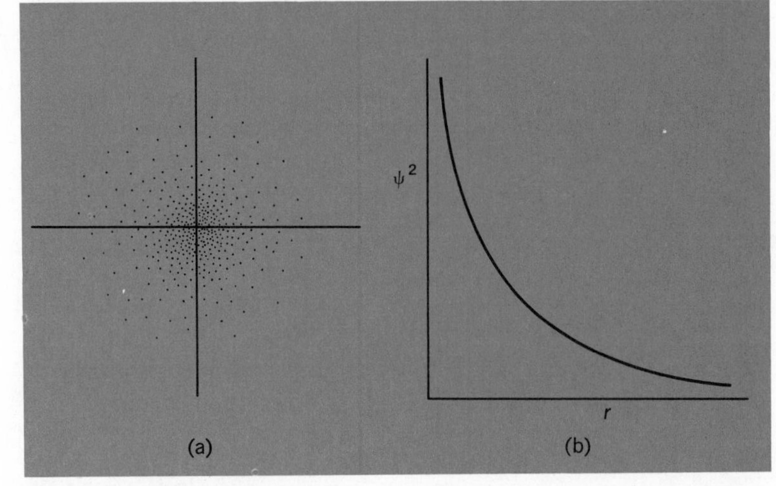

6-24 (a) Probability density ψ^2 represented by the density of the dots. (b) Probability density ψ^2 plotted against r for the ls orbital of a hydrogen atom. The probability curve never reaches zero. It extends throughout the universe. Approximately 99 percent of the probability for the ls electron of a hydrogen atom is contained within a spherical shell having a radius of approximately two Å.

Locating an electron around a nucleus of a hydrogen atom is somewhat like locating the bee around its hive. Obviously, we cannot photograph an electron as it moves about the nucleus. In order to plot the probable locations of an electron, we must rely on solutions to the equation which describes an electron's behavior. The equation was furnished by Erwin Schroedinger (p. 218). The solution to the equation yields a wave (probability) function, psi (ψ), whose value depends on the distance and direction of an electron from its nucleus. The probability function ψ is related to and can be used to calculate the probability of finding an electron in a tiny volume-element (ΔV) around a point. The value of psi squared (ψ^2) at a given point is interpreted as the *probability density* at that point. Probability density has the units of *per* cm^3 (cm^{-3}). The probability of an electron being in the tiny-volume element ΔV is represented by the equation

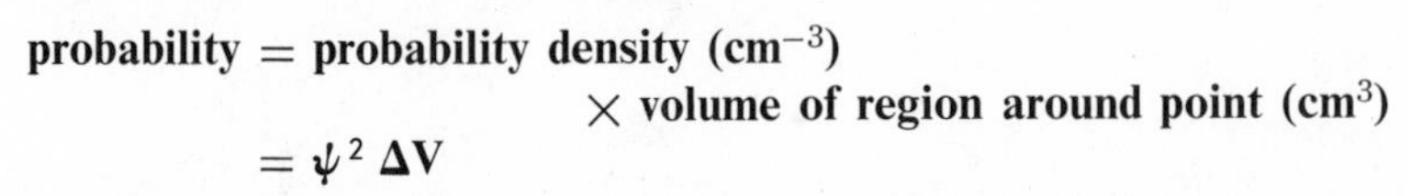

$$\textbf{probability} = \textbf{probability density } (\mathbf{cm^{-3}}) \times \textbf{volume of region around point } (\mathbf{cm^{3}}) = \boldsymbol{\psi}^2\,\boldsymbol{\Delta V}$$

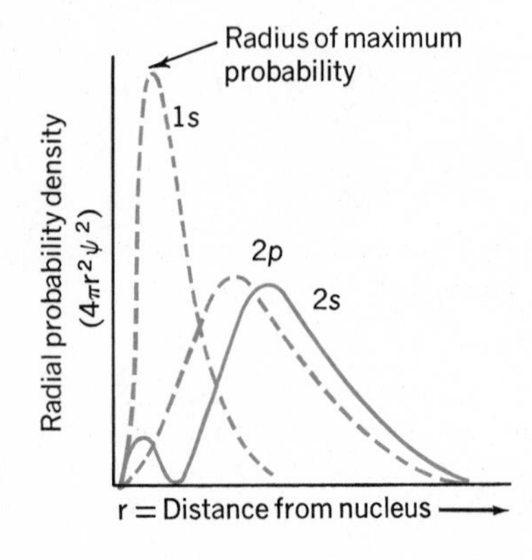

6-25 Radial probability density plotted vs radial distance from the nucleus. The shape of this curve reflects two factors: (1) the probability density ψ^2 at a point decreases as r increases, and (2) the number of points in the concentric spherical "electron shells" increases as the radius increases.

The probability distribution of an electron about the nucleus is obtained by calculating its probability of being found in many different regions. The charge cloud shown in Fig. 6-24 represents the probability distribution for the 1*s* electron in a hydrogen atom.

When we plot the probability density (ψ^2) against r for the 1*s* electron in a hydrogen atom, we obtain the graph shown in Fig. 6-24. Notice that ψ^2 reaches a maximum at the nucleus ($r = 0$) and decreases as we proceed away from the nucleus. To draw a parallel to the Bohr "shell" model of an atom, we can graphically represent the probability of an electron being found in infinitesimally thin concentric spherical shells surrounding the nucleus rather than in a small volume-element of the shell. The shells would be analogous to the layers of the skin of an onion.

These concentric shells have greater volumes as we proceed further out from the nucleus. You can see from Fig. 6-22, that there could be more cubes (volume elements) at r_2 than at r_1. Even though the probability of locating an electron in a single volume-element decreases as we proceed away from the nucleus, there is a radius where the probability of finding an electron is at a maximum simply because more volume-elements (cubes) exist at this radius than near the nucleus. For the 1*s* electron in a hydrogen atom this radius is 0.529 Å. This distance is known as the radius of maximum probability. The graphical data showing the radial probability density for the 1*s*, 2*p*, and 2*s* electrons are shown in Fig. 6-25. The radius of maximum probability for the 1*s* electron in hydrogen obtained from quantum-mechanical considerations is the same as the radius of the "orbit" calculated by Bohr. Whereas Bohr visualized an electron as being restricted to this distance from the nucleus, a quantum-mechanical electron might be found at any distance from the nucleus, as may be seen from the graphs in Figs. 6-24 and 6-25. In Fig. 6-25, it can be seen that the radius of maximum probability for a 2*p* electron is less than that for a 2*s* electron. However, the small hump in the 2*s* curve indicates that a 2*s* electron with no nodal surface at the nucleus penetrates closer to the nucleus than a 2*p* electron. Because of this, a 2*s* electron is more strongly attracted by the nucleus and is in a more stable, lower-energy state than a 2*p* electron. Thus, penetration effects along with the presence of nodal surfaces (locations of zero probability) help to explain the energy differences in the *s*, *p*, *d*, and *f* orbitals. For a given energy level, the orbital energies are $s < p < d < f$. Probability graphs convey valuable information which can be used to help explain observed characteristics of atoms and molecules.

6-26 Diagrams showing the analogy between the Bohr "electron shells" and the radius of maximum probability associated with the wave-mechanical model of an atom. Note that in (b) there is a spherical nodal surface between the radius of maximum probability for a ls electron and that for a 2s electron.

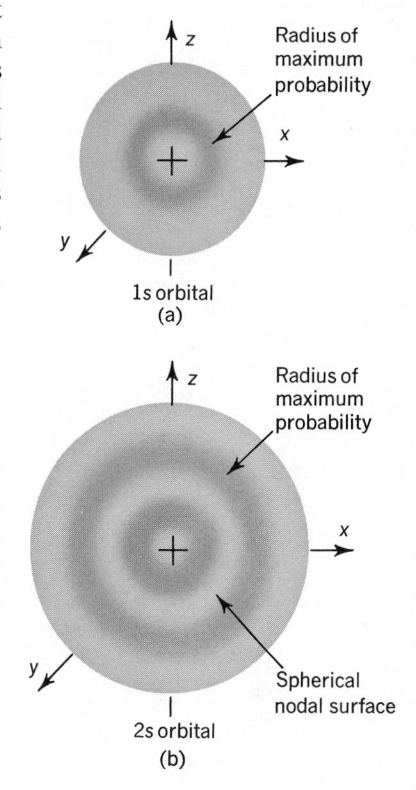

QUESTIONS

1. Calculate the maximum number of electrons with principal quantum number (a) 1, (b) 2, (c) 3, (d) 4.

2. How many electrons fill an orbital?

3. Describe the shapes of the s and *p* orbitals.

4. What is the maximum number of electrons that may be designated (a) 2*s*, (b) 2*p*, (c) 3*s*, (d) 3*d*, (e) 4*f*?

5. What is the physical significance in our model of the *x*, *y*, and *z* in the *p* orbital (p_x, p_y, p_z)?

6. Identify the elements whose neutral atoms have the electron configurations: (a) $1s^2 2s^2$, (b) $1s^2 2s^2 2p^3$, (c) $1s^2 2s^2 2p^6 3s^2$, (d) $1s^2 2s^2 2p^6 3s^2 3p^3$, (e) $1s^2$.

7. What kind of evidence was used by physicists and chemists to establish the relative energies of the energy levels?

8. What is the relationship between the principal quantum number of the last electron to "enter" an atom of an element and the period (row) in the Periodic Table in which we find the element?

9. Use the Periodic Table and write the electron configuration of (a) Sr, (b) the element with atomic number 52, (c) Ta, (d) Gd, (e) U.

10. Identify the group containing the element composed of atoms whose last electron (a) enters and fills an s orbital, (b) enters but does not fill an s orbital, (c) is

the first to enter a *p* sublevel, (d) is the next to the last in a given *p* sublevel, (e) enters and fills a given *p* sublevel, (f) is the first to enter a *d* sublevel, (g) half fills a *d* sublevel.

11. Explain why we find noble-gas properties upon completion of *p* orbitals rather than *s* orbitals.

12. Which of these represents the electron configuration of an excited atom: (a) $1s^2 2s^2 2p^6$, (b) $1s^2 2s^1 3s^1$, (c) $1s^2 2s^2 2p^2 3s^1$, (d) $1s^2 2s^1 4d^1$? Which is likely to give off a photon with the greatest energy on returning to the ground state?

13. Write the electron configuration for these substances, using the orbital notation described previously: (a) zinc atom, (b) vanadium atom, (c) chloride ion (Cl^-), (d) aluminum ion (Al^{3+}), (e) gallium (Ga), (f) bromine (Br).

14. (a) Write the electron configuration for argon. (b) Name two positive and two negative ions that have this configuration.

15. Chemists often consult tables of atomic radii in order to compare sizes of atoms. Why is it not strictly correct to say that a hydrogen atom has a radius of 0.53 Å?

16. (a) Is it possible to define the exact path of an electron in a *p* orbital? (b) Which atomic principle do we rely on to answer this question?

17. The Bohr theory allows only specific orbits, which means that an electron is at some exact distance from the nucleus. The Schroedinger model indicates that the distance of an electron from the nucleus can be defined only in terms of probability. Discuss the possible truth of the statement: An electron from an atom in the ink of this page may at this instant be on the moon.

18. What evidence is there to support the idea that matter has wave properties?

19. How does the wavelength of "matter waves" vary with (a) increasing mass, (b) decreasing velocity?

20. Show the permissible values of l and m for (a) $n = 1$, (b) $n = 2$, (c) $n = 3$.

21. Write a set of 4 permissible quantum numbers for the electron found in the outermost energy-level of a potassium atom (atomic number 19).

22. How many nodal surfaces are there for (a) a 2s orbital, (b) a $3p_x$ orbital?

23. What are the relative energies of the 3s, 3*p*, and 3*d* orbitals in a hydrogen atom? (b) Is the order the same for multi-electron atoms? Explain.

24. The radius of maximum probability for a 2s electron is greater than that for a 2*p* electron. Explain why a 2s electron is in a lower energy state.

7

Periodicity

Preview

In earlier chapters we noted that elements could be grouped in terms of similarities in their properties. In Chapter 6, we discovered that the common underlying factor responsible for the similarities in properties of the elements is the similarities in the electron configuration of the outer orbitals of their atoms.

We are now ready to investigate the periodicity of specific properties. This investigation will reveal the value of the Periodic Table as a predictive device which correlates a large number of apparently unrelated, isolated facts. An understanding of the use and limitations of the Periodic Table will greatly simplify your study of chemistry.

We shall first review the circumstances surrounding the origin of the Periodic Table. In this discussion, you will have a chance to assume the role of the man credited with the discovery of the Periodic Law when you are asked to predict the properties of three "unknown" elements. In the follow-up discussion you will learn how the Periodic Table can be used to help predict the formulas of compounds, the acid or basic nature of certain solutions, and the products of certain reactions.

In the second half of the chapter, we shall define a number of periodic atomic properties which are related to and can be explained in terms of the structure and electronic configuration of atoms. These properties are (1) ionization energy, (2) electron affinity, (3) atomic and ionic radii, (4) oxidation state, and (5) electronegativity.

You will be able to use these atomic properties throughout the rest of the text to help explain observed physical and chemical properties of substances.

At the end of this chapter you will find an extremely interesting and informative Special Feature on "The Origin of the Elements."

"Science is not everything. But science is very beautiful."

J. Robert Oppenheimer (1904–1967).

"I wished to establish some sort of system of simple bodies in which their distribution is not guided by chance, as might be thought instinctively, but by some sort of definite and exact principle."

—Dmitri Mendeleyev

PERIODICITY IN PHYSICAL AND CHEMICAL PROPERTIES

7-1 The Experimental Basis of the Periodic Table At first glance, the subject matter of chemistry may appear to be a confusing mass of unrelated data and facts. There are approximately

DMITRI MENDELEYEV
(1834–1907)

Dmitri Mendeleyev was born in Siberia, the youngest of seventeen children. His father was a high-school principal and his grandfather the publisher of the first newspaper in Siberia. A political exile in Siberia tutored Mendeleyev in science and sparked an interest in this field which lasted the remainder of his life. Mendeleyev graduated at the top of his class and received his doctor's degree in chemistry. He studied in Germany and France where he met and worked with Bunsen and other leading scientists of Europe. Mendeleyev returned to the University of St. Petersburg and became a professor of chemistry at the age of 32.

Mendeleyev's arrangement of the elements in the Periodic Table and his prediction of the existence and properties of three undiscovered elements is a striking example of creative imagination, courage, and confidence.

Within a very few years after Mendeleyev made his predictions, gallium, scandium, and germanium were discovered. The properties of these elements agreed so remarkably well with Mendeleyev's predictions that he quickly became the most famous chemist in the world.

From the time he first announced his periodic law until the present, elements have been discovered that have fitted into the periodic system of classification. The value of the Periodic Table as a device for prediction is illustrated by the following example.

During World War II, a new element, plutonium, not found in nature and which could be used as a source of energy for

one hundred elements and, in theory, millions of compounds each having different properties. The organization of these data and the formulation of broad generalizations and laws which brought order and unity to what were apparently thousands of unrelated, isolated facts, stand as milestones in the progress of the science of chemistry.

The evolution of the laws and generalizations related to the systematic organization of the elements began with a search for a pattern in the properties of the elements. Although a number of scientists, such as John Newlands (1838–1898) of England, and J. Lothar Meyer (1830–1895) of Germany, independently discovered regularities in the properties of the elements, Dmitri Mendeleyev of Russia is credited with the formulation of the original Periodic Law which led to the development of the Periodic Table.

Mendeleyev did not directly participate in the experimental isolation of any one element. He did spend many years reading, studying, and collecting data on each element from every available source. His dream was to bring order to the heterogeneous collection of data which had accumulated. In his search for some pattern, Mendeleyev wrote the name and properties of each element on a separate card and arranged the cards in seven groups, each group containing the cards of elements with somewhat similar properties. When he started with lithium (atomic mass 7) from the first group and followed with a card from each group in order of increasing atomic mass, a startling relationship was revealed. Apparently the properties of the elements repeated themselves at regular intervals. In other words, "the properties of the elements were periodic functions of their atomic masses." This was the original ***Periodic Law*** as presented by Mendeleyev. This discovery resulted in the preparation of a ***Periodic Table of the Elements*** in which the elements were arranged horizontally in order of increasing atomic masses. Elements having similar properties were grouped together in the same column.

The great Russian chemist's confidence in his discovery was demonstrated when he boldy predicted that three new elements would have to be discovered to fill the vacant spaces he had left in his Periodic Table. Using this law, Mendeleyev went so far as to predict the specific properties of the new elements. How well his predictions were fulfilled is shown in Table 7-1.

When the elements are arranged sequentially in order of increasing atomic masses, there are several inconsistencies in the locations of certain elements with respect to their properties. For example, the properties of potassium are rather similar to those of sodium. Yet, according to its atomic mass, potassium should be placed under neon, a noble gas. Iodine and tellurium are another pair of elements whose properties do not appear to be functions of their atomic masses. When the position of the elements in each of the pairs mentioned above is interchanged, the inconsistency disappears. The apparent irregularities in atomic mass sequence are explained by Moseley's discovery that "the properties of the elements are periodic functions of their *atomic numbers*." This means that the atomic number of an element is a more fundamental property than the atomic mass. Thus, *the elements in the modern version of the Periodic Table are arranged in order of their atomic numbers.*

TABLE 7-1

PREDICTED AND ACTUAL PROPERTIES OF GERMANIUM

Property	*Predicted* (X)	*Actual* (Ge)
Atomic mass	72	72.6
Density (g/ml)	5.5	5.35
Atomic volume	13	13.5
Color	dirty grey	greyish-white
Action on strong heating	yields XO_2	yields GeO_2
Effect of water	none	none
Effect of acids	slight	HCl has no effect
Effect of alkalis	slight	KOH has no effect
Properties of oxide	density 4.7 g/ml; more basic than SiO_2	density 4.07 g/ml; very slightly basic
Properties of chloride	XCl_4 boils below 100°C; density 1.9 g/ml at 0°C	$GeCl_4$ boils at 83°C; density 1.89 g/ml at 0°C

atomic bombs, was made in a chemical laboratory. In order to obtain it in the required degree of purity, it had to be separated from the other elements, by chemical means. Scientists had such confidence in the Periodic Law that they designed a process for the separation and purification of plutonium on the basis of its location in the Periodic Table. That is, from its location in the table and its relation to elements adjacent to it, they were able to predict its properties and the reactions it would undergo. This and many other examples illustrate the significance of Mendeleyev's discovery. It is surprising that he never received the Nobel Prize in chemistry. It was, instead, awarded to Henri Moissan of France who discovered fluorine. Most scientists of today would place Mendeleyev's discovery ahead of any work done by his contemporaries.

American scientists at Berkeley, California, named element number 101, mendelevium, in honor of Dmitri Mendeleyev, the founder of the periodic classification of the elements.

Even in a modern version of the Periodic Table there are difficulties. One problem is the location of hydrogen. In some tables hydrogen is placed in Group VIIA because in some ways it resembles the halogens. In some reactions, however, hydrogen behaves like the alkali metals. On this basis, it is sometimes placed in Group IA. In general, the element at the top of each *A* group in the table exhibits anomalous behavior with respect to that of the other elements in the group. For example, in some ways the behavior of lithium resembles that of magnesium more than it does that of sodium. These anomalies can be explained in terms of radii and other atomic properties and should not reduce our confidence in the validities of the Periodic Law and Table. The value of the Table as a classification device and as an instrument for predicting the chemical and physical properties of substances has been verified and effectively demonstrated in countless ways. It is, however, important to be aware of the limitations of any predictive device.

Perhaps it would be of interest to you to play the role of Mendeleyev and predict certain properties of elements X, Y, and Z located in the blank spaces of the Periodic Table illustrated in Fig. 7-1. Note that in this abbreviated table only the *A* groups (representative elements) are shown. The properties to be predicted are: atomic mass, density (g/ml), atomic radius (Å), melting point (°C), ionization energy (ev), phase (solid, liquid, gas), and formula of an oxide formed by the element. Your prediction of the properties of a given element should be largely based on the properties of its neighbors and the general trends observed as you go from top

IA	IIA	IIIA	IVA	VA	VIA	VIIA	O
3 5.39 1.52 Li(s) 0.53 Li_2O 6.9 180	4 9.32 1.11 Be(s) 1.85 BeO 9.0 1283	5 8.30 0.79 B(s) 2.34 B_2O_3 10.8 2300	6 11.3 0.77 C(s) 2.25 CO_2 12.0 3570	7 14.5 0.74 N(g) 1.25 X 10^{-3} N_2O_3 14.0 -210	8 13.6 0.73 O(g) 1.4 X 10^{-3} 16.0 -219	9 17.4 0.71 F(g) 1.8 X 10^{-3} 19.0 -218	10 21.6 Ne(g) 0.90 X 10^{-3} 20. -249
11 5.14 1.86 Na(s) 0.97 Na_2O 23.0 97.5	12 7.64 1.60 Mg(s) 1.74 MgO 24.3 650	13 5.98 1.43 Al(s) 2.70 Al_2O_3 27.0 660	14 8.15 1.17 Si(s) 2.33 SiO_2 28.1 1414	15 11.0 1.10 P(s) 1.8 P_2O_3 31.0 44	16 10.4 1.02 S(s) 2.07 SO_2 32.1 119	17 13.0 0.99 Cl(g) 3.2 X 10^{-3} Cl_2O 35.5 -101	18 15.8 Ar(g) 1.7 X 10^{-3} 39.9 -189
19 4.34 2.27 K(s) 0.86 K_2O 39.1 63.4	20 6.11 1.97 Ca(s) 1.55 CaO 40.1 850	31 6.00 1.22 Ga(s) 5.89 Ga_2O_3 69.7 30	Y	33 9.81 1.21 As(s) 5.72 As_2O_3 74.9 817	Z	35 11.8 1.14 Br(l) 3.1 Br_2O 79.9 -7.3	36 14.0 Kr(g) 3.7 X 10^{-3} 83.8 -157
X	38 5.69 2.15 Sr(s) 2.6 SrO 87.6 769	49 5.79 1.62 In(s) 7.3 In_2O_3 114.8 157	50 7.34 1.41 Sn(s) 7.3 SnO_2 118.7 232	51 8.64 1.41 Sb(s) 6.7 Sb_2O_3 121.8 631	52 9.01 1.3 Te(s) 6.2 TeO_2 127.6 450	53 10.5 1.33 I(s) 4.9 126.9 114	54 12.1 Xe(g) 5.9 X 10^{-3} 131.3 -112
55 3.89 2.65 Cs(s) 1.9 Cs_2O 132.9 28.7	56 5.21 2.17 Ba(s) 3.5 BaO 137.3 704	81 6.11 1.71 Tl(s) 11.9 Tl_2O_3 204.4 304	82 7.42 1.75 Pb(s) 11.4 PbO_2 207.2 328	83 7.29 1.55 Bi(s) 9.8 Bi_2O_3 209 271	84 8.43 1.4 Po(s) 9.2 210	85 --- 1.4 At(s) ---	86 10.8 Rn(g) 9.9 X 10^{-3} 222 -77

7-1 Use the data in this table and the concept of periodicity to predict the properties of elements *X*, *Y*, and *Z*.

Atomic number	Ionization energy (ev)	Atomic radius (Å)
	Symbol (Phase)	
Density (g/ml)		Formula of oxide
Atomic mass		Melting point (°C)

to bottom in a group of the Periodic Table. You can check your predictions by referring to the margin on p. 237. The code to the items in the table is given in the code square in the margin.

7-2 The Periodic Table as an Aid in Predicting Formulas and Names Most of you have probably already discovered that the Periodic Table is a valuable aid in predicting the formulas of compounds containing elements from different groups. Consider the oxides noted in Group IIIA (Fig. 7-1). Knowing the formula of one oxide containing a metal from this group enables us to predict a reasonable formula for an oxide containing any other metal in the same group. In general, we can represent the formula of such an oxide as M_2O_3 where *M* represents the symbol of any metal in the group.

We can extend this type of generalization to include other types of compounds. For example, knowing the formula and name of an acid containing chlorine, we could write a reasonable formula and name for an analogous acid containing bromine; specifically, HClO is the formula of hypochlorous acid. Therefore, we might predict that HBrO would be called hypobromous acid.

FOLLOW-UP PROBLEMS

1. The formula for a compound containing beryllium and chlorine is $BeCl_2$. (a) What is a formula for a compound containing strontium and chlorine? (b) What is a formula for a compound containing calcium and bromine? (c) The name of $BeCl_2$ is beryllium chloride. What is the name of the compounds in (a) and (b)? What is the general formula for a compound containing any metal (M) from Group IIA and any nonmetal (X) from Group IIA and any nonmetal (Z) from Group VIIA?

2. The formula for sulfuric acid is H_2SO_4. What would be the name and formula for an analogous acid formed by selenium (atomic number 34)?

7-3 Periodicity in the Acid or Basic Nature of Oxides The periodicity in the formulas of the oxides shown in Table 7-2 suggests that there may be a periodicity in their chemical behavior. Let us examine the reaction of various oxides with water and see whether such a chemical property is periodic in nature. When you test solutions prepared by dissolving oxides in water you will discover that some are basic and others are acid. In general, you will find that metallic oxides dissolve and form basic solutions by reaction with water, and that nonmetallic oxides dissolve and form acid solutions. Thus, oxides such as CaO and Li_2O which dissolve in water and yield basic solutions are called ***basic oxides*** or ***basic anhydrides.*** The products of the reactions are, respectively, calcium hydroxide [$Ca(OH)_2$] and lithium hydroxide (LiOH). In aqueous solution, these substances, known as strong bases, are completely dissociated. Hence, in ionic equations they are represented as ions. The net ionic equations for the reactions described above are

$$\mathbf{CaO(s) + HOH(l) \longrightarrow Ca^{2+}(aq) + 2OH^-(aq) \quad (basic)}$$
$$\mathbf{Li_2O(s) + HOH(l) \longrightarrow 2Li^+(aq) + 2OH^-(aq) \quad (basic)}$$

Chemical indicators such as litmus, phenolphthalein, or wide-range test papers can be used to determine the relative acidity of solutions.

Nonmetallic oxides such as CO_2 and SO_2 which dissolve in water and yield acid solutions are called ***acid oxides*** or ***acid anhydrides.*** The products of the reactions are carbonic acid (H_2CO_3) and sulfurous acid (H_2SO_3). The net ionic equations are

Strong acids and bases are completely dissociated into ions. A list of strong acids is provided on page 453.

$$\mathbf{CO_2(g) + 2H_2O(l) \rightleftharpoons H_3O^+(aq) + HCO_3^-(aq) \quad (acid)}$$
$$\mathbf{SO_2(g) + 2H_2O(l) \rightleftharpoons H_3O^+(aq) + HSO_3^-(aq) \quad (acid)}$$

In aqueous solution, weak acids and bases are largely molecular and furnish few ions. Carbonic and sulfurous acids are weak acids and are discussed at the bottom of page 453.

It should be noted that the formulas of the oxides in Table 7-2 are the simplest formulas. Molecular mass determinations and structural investigations show that the actual formulas of the phosphorous, arsenic, and antimony compounds are, P_4O_6, As_4O_6, and Sb_4O_6. A solution of P_4O_6 is called phosphorus acid (H_3PO_3). The equation for the reaction is

$$\mathbf{P_4O_6(s) + 10H_2O(l) \rightleftharpoons 4H_3O^+(aq) + 4H_2PO_3^-(aq) \quad (acid)}$$

FOLLOW-UP PROBLEM

(a) Write the name and formula of the acid formed by the reaction of P_4O_{10} with water. (b) Write a net ionic equation for the reaction.

Hydrochloric acid ($H_3O^+ + Cl^-$) and other strong acids are represented ions in ionic equations. When Cl^- and other unreactive "spectator" ions appear on both sides of an equation, they are deleted to yield a net ionic equation.

An oxide that does not dissolve in water can be classified according to its reaction with a known acid or base. For example, silicon dioxide (SiO_2) does not dissolve appreciably in water, but will react with a strong base. In this reaction it is behaving as an acid. Thus, it may be classified as an acid oxide.

Oxides which react with either acids or bases are called ***amphoteric oxides.*** For example, aluminum oxide does not dissolve appreciably in water. It does, however, dissolve in either hydrochloric acid or in sodium hydroxide.

$$Al_2O_3(s) + 6H_3O^+(aq) \longrightarrow 2Al^{3+}(aq) + 9H_2O(l)$$
$$Al_2O_3(s) + 2OH^-(aq) \longrightarrow 2AlO_2^-(aq) + H_2O(l)$$

In the first equation Al_2O_3 is behaving as a base and in the second equation as an acid. AlO_2^- is called the aluminate ion.

FOLLOW-UP PROBLEM

Zinc oxide (ZnO) is an amphoteric oxide. Write net ionic equations for the reaction of zinc oxide with (a) hydrochloric acid and (b) sodium hydroxide. The formula for a zincate ion is ZnO_2^{2-}.

Observations made on systems such as those described above are organized in Table 7-2. The periodicity in the acid or basic nature of the oxides is shown in Table 7-2. It can be seen that for the representative elements, *the acid nature of the oxides increases as you go from left to right across the Periodic Table and decreases as you go down a group of the table.* The oxides of the transition metals are not shown in the table above. A number of these elements form several oxides. When a given element forms more than one oxide, the acidity generally increases as the oxidation state of the metallic atom increases. In CrO, Cr_2O_3, and CrO_3, the oxidation states of chromium are, respectively, 2+, 3+, and 6+. Thus, CrO is basic, Cr_2O_3 is amphoteric, and CrO_3 is acid. In the next two chapters we shall attempt to relate the nature of the oxide to the nature of the bond between atoms and type of aggregate formed by the atoms.

The *oxidation state* of an atom in a molecule is the *apparent* charge carried by the atom. We may assume the oxidation state of oxygen in most oxides is 2−.

TABLE 7-2
PERIODICITY OF ACID AND BASIC OXIDES

Increasing acidity of oxides ↑							
	Li_2O basic	BeO amphoteric	B_2O_3 acid	CO_2 acid	N_2O_3 acid		F_2O acid
	Na_2O basic	MgO basic	Al_2O_3 amphoteric	SiO_2 acid	P_4O_6 acid	SO_3 acid	Cl_2O acid
	K_2O basic	CaO basic	Ga_2O_3 amphoteric	GeO_2 amphoteric	As_4O_6 amphoteric	SeO_3 acid	Br_2O acid
	Rb_2O basic	SrO basic	In_2O_3 basic	SnO_2 amphoteric	Sb_4O_6 amphoteric	TeO_3 acid	I_2O_5 acid
	Cs_2O basic	BaO basic	Tl_2O_3 basic	PbO_2 amphoteric	Bi_2O_3 basic		

Increasing basicity of oxides ←

FOLLOW-UP PROBLEM

Which would you predict to be the more acid oxide, Sb_4O_6 or Sb_2O_5?

7-4 Periodicity in Chemical Properties Our discussion of periodicity in formulas, and in the nature of oxides suggests that certain chemical properties are periodic in nature. Our observations of these systems lead us to suspect the elements in the same group might undergo similar chemical reactions. For example, knowing that magnesium reacts with hydrochloric acid yielding hydrogen gas and a solution of magnesium chloride leads us to predict that calcium would behave similarly. A laboratory experiment would confirm our prediction. The equation for the first reaction is

$$Mg(s) + 2H_3O^+(aq) \longrightarrow Mg^{2+}(aq) + H_2(g) + 2H_2O$$

Another reaction of interest to first-year chemistry students is that which occurs when metallic sodium is added to water. Most students are quite astonished to see a metal react rather violently with water. If your instructor demonstrates this reaction, you will see evidence that indicates the formation of hydrogen gas and a highly basic solution as products. The equation for the reaction is

$$2Na(s) + 2HOH(l) \longrightarrow 2Na^+(aq) + 2OH^-(aq) + H_2(g)$$

On the basis of this reaction you might predict that metallic potassium would react in the same manner. If this reaction is demonstrated, you will note that the hydrogen gas ignites instantly. This indicates that the reaction of potassium with water is more vigorous and exothermic than the one involving sodium.

37	4.18	2.48
	X(s)	
1.53		X_2O
85.5		38.8

32	7.88	1.22
	Y(s)	
5.32		YO_2
72.6		958

34	9.75	1.16
	Z(s)	
4.79		ZO_2
79.0		217

Properties of elements X, Y, and Z in Fig. 7-1.

7-2 Metallic sodium has a bright luster when freshly cut, but it soon corrodes in air. In order to exclude air and water, sodium and the other alkali metals are stored in the laboratory in bottles of kerosene or oil. (Chemistry: A First Course in Modern Chemistry, Garrett, Richardson, and Montague, Ginn and Company.)

FOLLOW-UP PROBLEMS

1. Write the net ionic equation for the reaction between metallic calcium and an aqueous solution of hydrochloric acid.

2. (a) Write an ionic equation for the reaction between potassium and water. (b) Is the resulting solution acid, basic, or neutral?

Observation of the last two demonstrations suggests that metallic potassium is more chemically reactive than metallic sodium. It is also apparent that, in general, metallic properties decrease in going from left to right across a row (period) and increase in going from top to bottom in a column (group) of the Periodic Table. Let us now seek an explanation for these and other observations in terms of atomic properties (properties of individual atoms). We may assume that atomic properties which can be used to explain periodicity of chemical and physical properties must also be periodic in nature. In turn, we would expect atomic properties to be related to the observed periodicity in the electronic configuration of atoms. We shall examine in some detail four of the atomic properties that are the most useful in explaining chemical and physical properties. These are

Electron affinity is a measure of the tendency of a gaseous atom to accept an additional electron.

1. Ionization energy and electron affinity,
2. Atomic and ionic radii
3. Oxidation state
4. Electronegativity

PERIODICITY IN ATOMIC PROPERTIES

7-3 The heat evolved when potassium metal reacts with water ignites the hydrogen gas produced by the reaction. (B. M. Shaub, Northampton, Mass.)

Perhaps the easiest way to detect regularities in the properties of atoms and/or elements is to plot each property against the atomic numbers of the atoms in the Periodic Table. Inspection of such a graph would enable us to identify properties that are periodic functions of the atomic number. A periodic property yields a cyclic graph; that is, the graph has maxima (highs) and minima (lows).

For each of the properties noted above, we shall prepare a Periodic Table showing the atomic number of each atom and the value of the atomic property. This will make it easy to detect trends in the properties. Using the data in the table, we shall plot the atomic property against the atomic number of the atom. The resulting graph will enable us to determine whether or not the property is a periodic function of the atomic number. We shall then seek an explanation for the observed periodicity and trends in terms of the structure and electron configuration of the atoms. Finally, we shall briefly relate the periodicity and trends in the atomic property to observed chemical and physical properties. Many occasions will arise when we shall make extensive use of these properties in explaining observations noted in subsequent chapters. Let us first consider ionization energy.

7-5 Ionization Energy The first ionization energy of an atom is the energy (work) required to remove the least tightly-held electron from an isolated atom in the vapor state, thus forming a positive ion. This endothermic process may be represented by a general equation as

$$\textbf{energy} + \textbf{X} \longrightarrow \textbf{X}^+ + \textbf{e}^-$$

As the attraction between an electron and the nucleus increases, the energy required to remove the electron, the ionization energy, also increases.

The first ionization energies of the elements are given in the Periodic Table shown in Fig. 7-5. The energy required to remove a second electron from the outer level of an atom (the second ionization energy) is always greater than that required to remove the first electron. The removal of the first (least tightly held) electron reduces the number of electrons and, consequently, the total electronic repulsion. This results in drawing the electron cloud closer to the nucleus. There it is more compact and, because of the smaller radius, is subjected to a greater force of attraction by the nucleus.

The data in Fig. 7-5 may be plotted as shown in Fig. 7-6. It can be seen that maxima (highs) in ionization energies occur at elements belonging to the noble-gas family and that minima occur

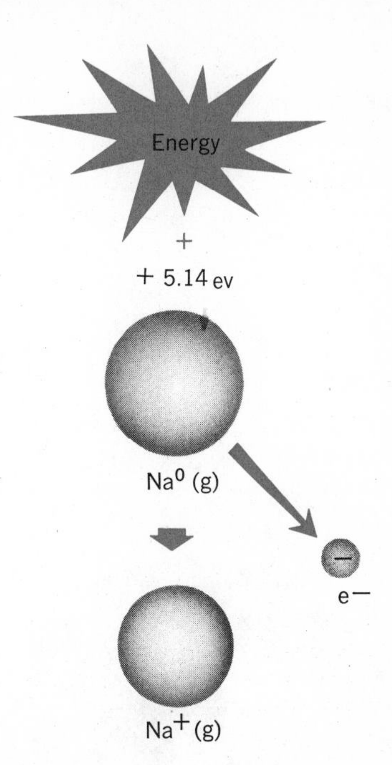

7-4 5.14 ev is required to remove one electron from a gaseous sodium atom. A sodium ion has a smaller radius than a sodium atom.

Periodic Table
First Ionization Energy (ev/atom)

	1A	2A	3B	4B	5B	6B	7B	8	8	8	1B	2B	3A	4A	5A	6A	7A	0
1	1 H 13.6																1 H 13.6	2 He 24.6
2	3 Li 5.39	4 Be 9.32											5 B 8.30	6 C 11.3	7 N 14.5	8 O 13.6	9 F 17.4	10 Ne 21.6
3	11 Na 5.14	12 Mg 7.64											13 Al 5.98	14 Si 8.15	15 P 11.0	16 S 10.4	17 Cl 13.0	18 Ar 15.8
4	19 K 4.34	20 Ca 6.11	21 Sc 6.6	22 Ti 6.8	23 V 6.7	24 Cr 6.8	25 Mn 7.4	26 Fe 7.9	27 Co 7.9	28 Ni 7.6	29 Cu 7.7	30 Zn 9.4	31 Ga 6.0	32 Ge 7.88	33 As 9.81	34 Se 9.75	35 Br 11.8	36 Kr 14.0
5	37 Rb 4.18	38 Sr 5.69	39 Y 6.5	40 Zr 7.0	41 Nb 6.8	42 Mo 7.1	43 Tc 7.3	44 Ru 7.4	45 Rh 7.4	46 Pd 8.3	47 Ag 7.6	48 Cd 9.0	49 In 5.79	50 Sn 7.34	51 Sb 8.64	52 Te 9.01	53 I 10.5	54 Xe 12.1
6	55 Cs 3.89	56 Ba 5.21	57 La 5.6	72 Hf 5.5	73 Ta 7.9	74 W 8.0	75 Re 7.9	76 Os 8.7	77 Ir 9.0	78 Pt 9.0	79 Au 9.2	80 Hg 10.4	81 Tl 6.11	82 Pb 7.42	83 Bi 7.29	84 Po 8.43	85 At 9.5	86 Rn 10.5
7	87 Fr	88 Ra 5.3	89 Ac 6.9															

58 Ce 5.6	59 Pr 5.7	60 Nd 5.5	61 Pm	62 Sm 5.7	63 Eu 5.8	64 Gd 6.1	65 Tb 6.0	66 Dy 6.8	67 Ho	68 Er 6.0	69 Tm 5.8	70 Yb 6.2	71 Lu 6.1
90 Th 7.0	91 Pa	92 U 6.1	93 Np	94 Pu 5.1	95 Am	96 Cm	97 Bk	98 Cf	99 Es	100 Fm	101 Md	102 No	103 Lw

7-5 Periodicity of ionization energies.

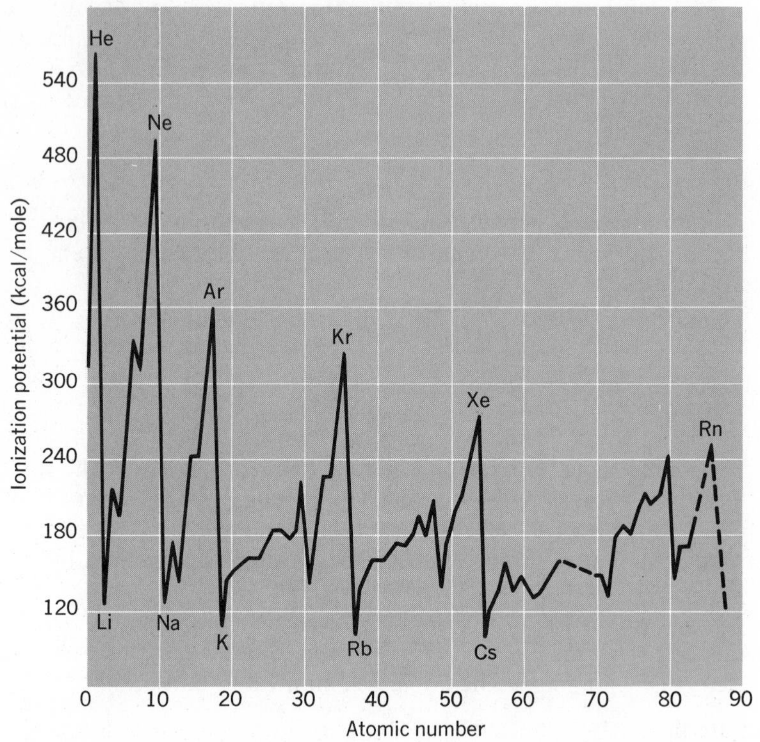

7-6 Graph showing the periodic nature of ionization energies.

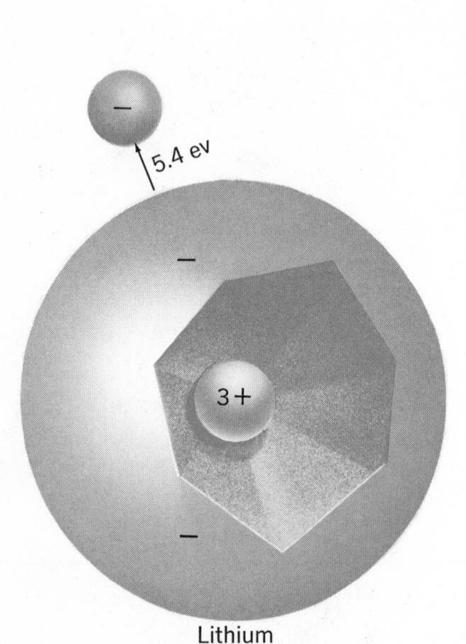

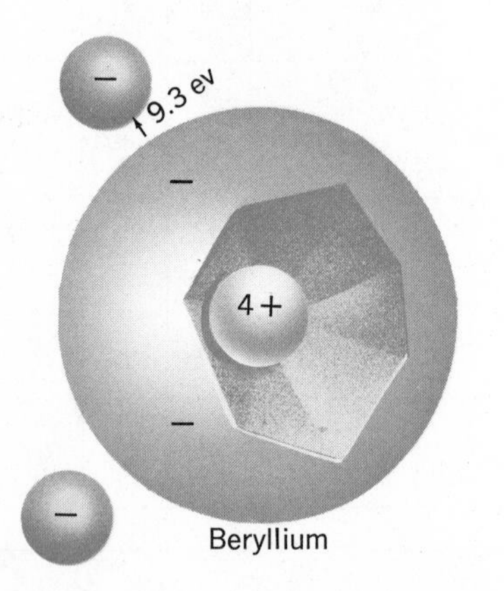

7-7 As the nuclear charge increases, the ionization energy of the atoms in the same period of the Periodic Table usually increases. Since the effective nuclear charge of beryllium is greater than that of lithium, beryllium's outer electron is held more firmly and more energy is required to remove it.

at elements belonging to the alkali metal family (Group IA elements). The cyclic nature of the graph indicates that *ionization energy is a periodic function of atomic number.*

In order to explain the trends and variations in ionization energy shown in Fig. 7-6, it is necessary to understand the factors which determine the quantity of energy required to remove an electron from an atom. Let us examine the four factors which interact and determine ionization energies.

1. *The magnitude of the positive charge on the nucleus; that is, the atomic number.* Other factors being constant, the greater the nuclear charge, the greater is the attraction for the outer-level electron. As the force of attraction increases, more energy is needed to remove the electron from the atom. Thus, ionization energy increases with increased nuclear charge provided that other factors which normally influence the ionization energy are not considered. The influence of nuclear charge on ionization energy is shown in Fig. 7-7.

2. *The radius of the atom.* The force of attraction between an electron and its nucleus is electrostatic. According to Coulomb's Law, the force varies inversely with the square of the distance between the charges.

$$\mathbf{F} \propto \frac{\mathbf{q_1 q_2}}{\mathbf{r^2}}$$

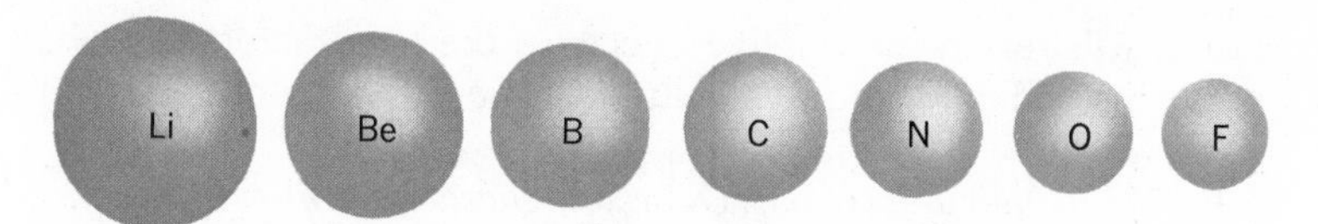

7-8 As the nuclear charge of the Period 2 elements increases, the added electrons go into the same principal energy level. The more effective nuclear charge causes the volume of the atoms to decrease.

This means that the force of attraction decreases as the radius increases. Hence, other factors being constant, the energy required to remove an electron decreases as the distance from the nucleus increases. In general, *the radii of atoms gradually decrease as we progress from left to right in a given row of the Periodic Table.* This is because the *atomic number (nuclear charge) increases.* The relative sizes of the atoms in Period 2 are illustrated in Fig. 7-8. As the nuclear charge increases, all complete inner-electron clouds are drawn closer to the nucleus. For example, the most probable distance of the 1*s* electrons from the nucleus of a krypton atom (36 protons) is less than that in a bromine atom (35 protons). The *radii of the atoms generally increase from top to bottom in a given column of the table.* As you go down a family, the number of energy levels increases. Each additional energy level of electrons screens the other electrons from the attractive force of the nucleus. Consequently, as the attractive force decreases, the outer electrons are not attracted as strongly. Therefore, they are not pulled in as closely as in the case of atoms having fewer energy levels of electrons. The atoms of the Group-IA alkali metals shown in Fig. 7-9 illustrate the relationship between the number of energy levels and the radii of atoms.

3. *Number of inner-level electrons underlying the outer energy level.* The inner-level electrons serve to shield the outer electron from the pull of the nuclear charge. This is known as the *screening effect.* Other factors being constant, as the number of inner energy levels increases, both the force of attraction and the ionization energy decrease. Thus, the first ionization energy of a cesium atom, shown in Fig. 7-10, is less than that of a sodium atom.

4. *The number of occupied orbitals in the outer levels.* The volume-probability graph in Fig. 7-11 shows that the radius of maximum probability for the 2*s* electron is, on the average, slightly farther from the nucleus than the 2*p* electron, but the small hump in the curve for the 2*s* electron close to the nucleus represents a greater electron density closer to the nucleus. Hence, the 2*s* electron is more penetrating than that for the 2*p* electron. This means it has a greater chance of being close to the nucleus and "feels" more of the nuclear charge than the 2*p* electron. In general, *s* electrons penetrate the inner-electron levels better than *p* electrons which, in turn, penetrate the inner-electron screen better than *d* electrons. This ability to penetrate plays an important role in determining atomic properties.

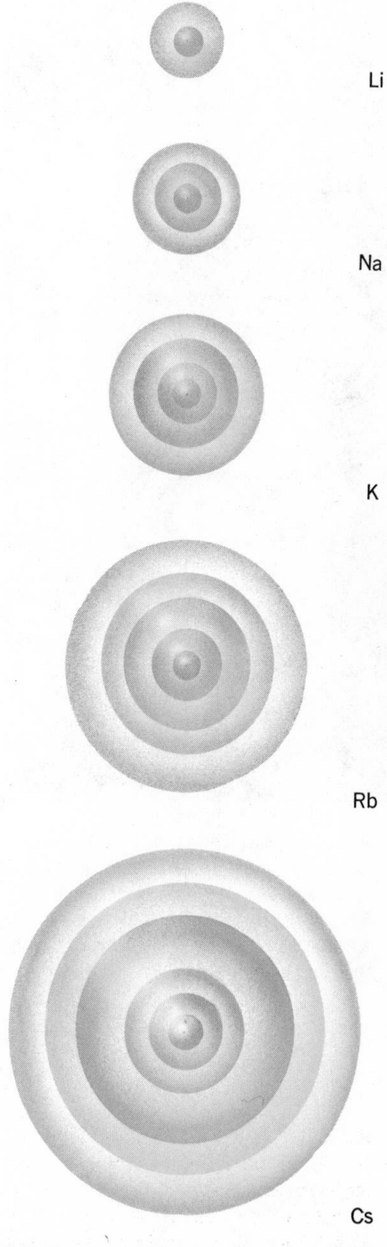

7-9 The atomic radii of the alkali metals. The radii increase as the number of energy levels increases.

Let us use the information given above to explain the trends shown by the graph in Fig. 7-6 and table in Fig. 7-5. This expla-

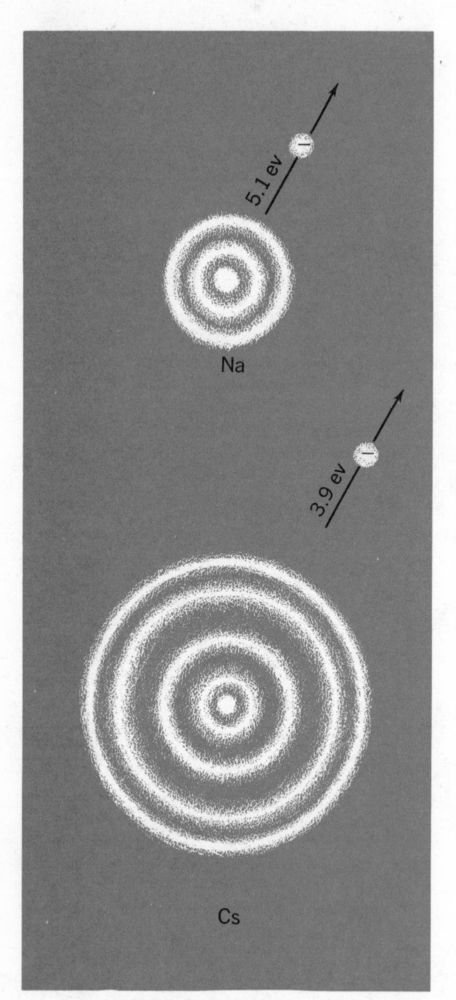

7-10 The outer electron in a cesium atom is shielded from the nuclear charge by five inner levels of electrons; the outer electron in a sodium atom by only two inner levels of electrons. Thus, a sodium atom has a more effective nuclear charge and a greater ionization energy.

nation will reveal the relative effects of the four factors discussed in the preceding section. The trends and explanations are summarized below.

1. *The noble gases, Group O, are all characterized by high ionization energies.* This indicates that the electrons are held very firmly and have very little tendency to give up electrons. The compactness and symmetry of their electron clouds prevent them from taking on electrons. This is in agreement with experimental observation of their chemical inertness.

2. *The alkali metals, Group IA, all have relatively low first-ionization energies,* separated by a relatively large energy gap from the second ionization energy (Table 7-3). This suggests the presence of one easy-to-remove electron and accounts for the 1+ charge of the ions formed by this group of elements. The low ionization energy is consistent with the fact that the single outer *s* electron is located at a relatively greater distance from the nucleus. In addition, the outer electron is well screened from the nuclear charge of electrons in inner energy levels.

3. *The alkaline-earth metals, Group IIA, have first and second ionization energies which are relatively low but still greater than those of the IA metals.* The first large energy gap occurs between the second and third ionization energies (Table 7-4). This indicates the presence of two relatively easy-to-remove electrons and is in agreement with the 2+ charge which characterizes the ions in this group of elements. These elements are observed to be *less reactive than the adjacent IA elements.* This is consistent with their higher ionization energies.

TABLE 7-3
FIRST AND SECOND IONIZATION ENERGIES FOR THE ELEMENTS OF GROUP IA

Element	Atomic Number	First Ionization Energy (kcal/mole)	(ev)	Second Ionization Energy (kcal/mole)	(ev)
Li	3	124	5.39	1744	75.7
Na	11	118	5.14	1091	47.4
K	19	99	4.34	734	31.9
Rb	37	96	4.18	634	27.6
Cs	55	90	3.89	579	25.1

TABLE 7-4
FIRST, SECOND, AND THIRD IONIZATION ENERGIES FOR THREE ELEMENTS IN GROUP IIA

Element	Atomic Number	First Ionization Energy (kcal/mole)	(ev)	Second Ionization Energy (kcal/mole)	(ev)	Third Ionization Energy (kcal/mole)	(ev)
Be	4	215	9.32	420	18.2	3548	154
Mg	12	176	7.64	347	15.1	1848	80.3
Ca	20	141	6.1	274	11.8	1181	50.9

The atoms in Group IIA have higher first ionization energies than their IA neighbors because there is a *more effective nuclear* charge and a smaller radius. The effective nuclear charge is greater because the presence of an electron in a given orbital does not provide an appreciable screening effect on another electron in that orbital. Therefore, the nuclear charge "felt" by the outer electrons of Group-IIA atoms is essentially larger by one unit (one more proton) than that "felt" by the outer electron in the IA atoms. Hence, the outermost electrons are held more firmly, and the ionization energy is greater than in the case of the IA atoms of the element preceding it in the table. When one of the outer *s* electrons is removed, the remaining electronic repulsion is reduced, and the remaining electron is drawn closer to the nucleus. Thus, more energy is required to remove the second electron than the first.

4. *In general, the ionization energy of the elements increases as we go across a given row of the table* (Table 7-5). The gradual increases are attributed to the increase in effective nuclear charge and a decrease in radius. That is, the positive charge in the nucleus increases, but in a given period the outer electrons are all in the same energy level and frequently in the same sublevel. Thus, the effective nuclear charge increases so that each additional electron experiences a greater attraction. Hence, it becomes progressively more difficult to remove each of these additional electrons. The few irregularities in the general trend are related to the extra stabilities of empty, half-filled, and filled sublevels. For example, aluminum has a lower first ionization energy than magnesium. The outer orbital configurations of the two atoms are

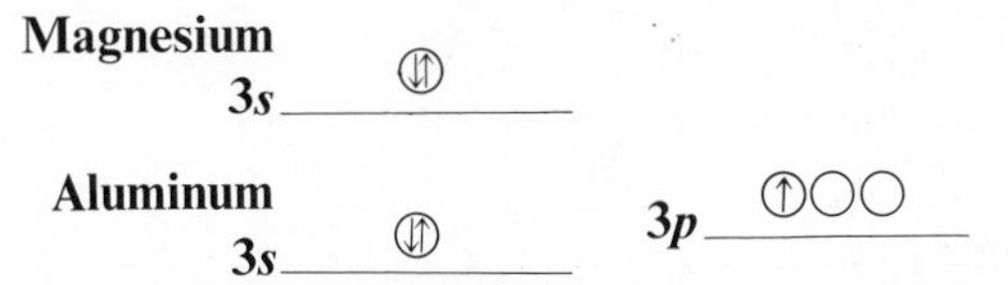

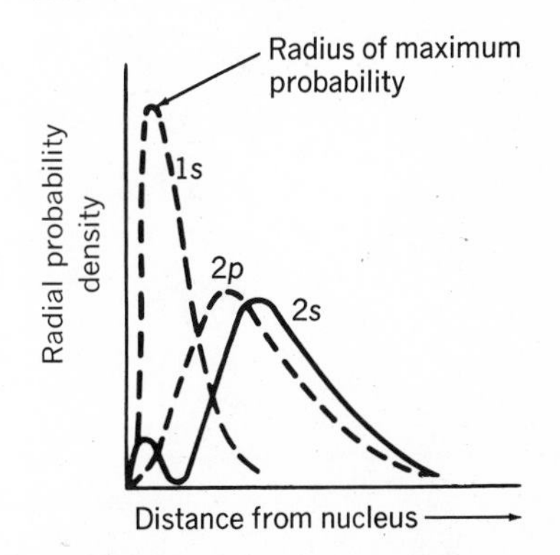

7-11 This graph shows how the probability of finding the electron in a small volume-element in the space around the nucleus varies with the distance of the small volume-element from the nucleus.

Remember that ionization energies are calculated for gaseous atoms.

7-12 The removal of the first electron from a calcium atom requires only 6.1 ev, resulting in the formation of a positive ion having a smaller radius than the original atom. The remaining 4s electron is more strongly attracted by the nucleus than it was originally. Thus, its removal requires an additional 11.8 ev. The third electron is still closer to the nucleus and constitutes part of the noble gas core. Since its removal requires 50.9 ev, we would not expect to find a stable Ca^{3+} ion. Note that after removal of each electron, the electron clouds representing the inner levels contract so they are closer to the nucleus. This contraction is not shown in the figure.

Ca (g) + 6.1 ev ⟶ Ca^{+} (g) + e^{-}

Ca^{+2} (g) + 50.9 ev ⟶ Ca^{+3} (g) + e^{-}

Ca^{+} (g) + 11.8 ev ⟶ Ca^{+2} (g) + e^{-}

TABLE 7-5
FIRST IONIZATION ENERGIES FOR THE ELEMENTS OF PERIOD 3

Element	Atomic Number	First Ionization Energy (kcal/mole)	First Ionization Energy (ev)
Na	11	118	5.14
Mg	12	176	7.64
Al	13	138	5.98
Si	14	188	8.15
P	15	254	11.0
S	16	239	10.4
Cl	17	300	13.0
Ar	18	363	15.8

It may be seen that the aluminum electron is an unpaired $3p$ electron whereas the magnesium electron must be removed from a filled $3s$ orbital. Also, the slight shielding of the $3p$ electron reduces the effective nuclear attraction in an aluminum atom. The net effect is a lower ionization energy for aluminum. This is also consistent with the evidence that the p sublevel is slightly higher in energy than is the s level for the same value of n.

The slight decrease in first ionization energy in going from nitrogen to oxygen may in part, be attributed to the extra stability of the half-filled p orbitals in the nitrogen atom.

Nitrogen **2s** ⇅ **2p** ↑ ↑ ↑

Intuitively, we would expect that the introduction of a second electron into one of the $2p$ orbitals results in an additional electrostatic repulsion which makes that electron easier to remove in spite of the increased nuclear charge.

5. *The ionization energy generally decreases as the atomic mass increases within a given main family of the representative elements* (Tables 7-3 and 7-4). Here, the effect of increased nuclear charge is less important than the large increase in atomic radius (Coulomb's Law) and the presence of additional inner layers of screening electrons. These latter factors serve to *reduce the effective nuclear charge.* Thus, the outer electron is held less firmly and a smaller quantity of energy is required to remove it.

Metallic and nonmetallic properties are discussed on p. 46.

6. Increases in ionization energy are paralleled by increases in nonmetallic characteristics. Consider the elements of Period 3. As you go from left to right across the table, the properties of the elements become progressively more nonmetallic and the ionization energies gradually increase. In Group VA, the element at the top, nitrogen, has nonmetallic properties and a relatively high first-ionization energy. As you go down the group, the elements become more metallic. Bismuth, at the bottom of the group, is definitely metallic with a low ionization energy compared to that of nitrogen.

7. *The ionization energies for the B families or the transition elements are larger than those for the corresponding elements in Group IA and IIA* which, like the transition elements, contain one or two electrons in an outer s orbital. This is reflected in the lower reactivities of the members of the B families. For example, potassium and copper are in the same period. Both have one $4s$ electron. Potassium, a Group IA element, has a lower ionization energy than copper, a Group IB element. As a consequence, potassium is more reactive than copper. The higher ionization energy of copper is a reflection of a more effective nuclear charge.

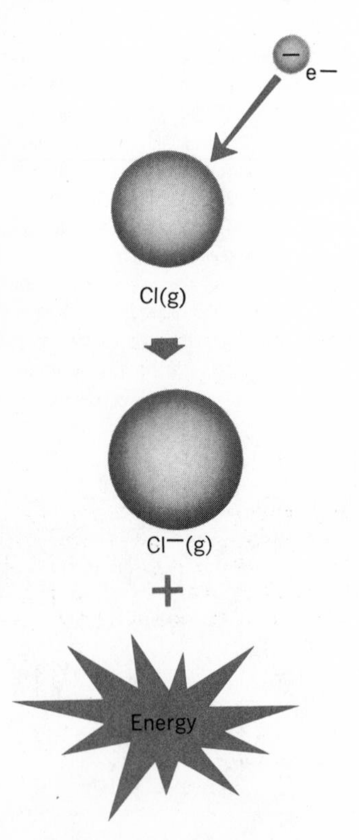

7-13 The potential energy of the atomic system decreases when a chlorine atom acquires an electron and forms a chloride ion.

7-6 Electron Affinity It is evident from the relatively high ionization energies of the halogens (Group VIIA) and other nonmetals that these elements do not readily lose electrons, thus forming positive ions. Rather, they are observed to form negative ions quite readily. In doing so, they achieve the electron configurations of the noble gases. The exothermic processes that the nonmetallic elements undergo may be represented as

$$\mathbf{X^\circ + e^- \longrightarrow X^- + energy}$$

The energy released when a neutral atom in the vapor state accepts an electron is known as the electron affinity. It is a measure of the tendency of an atom to accept another electron; that is, it determines how easily an atom forms a negative ion. Along with ionization energy, it is an important factor in determining chemical properties. Electron affinities are difficult to determine experimentally. Thus, values are available for only a relatively few elements.

From Table 7-6 it can be seen that Group VIIA elements have high electron affinities. This is to be expected because the addition of one electron to each of these atoms yields a stable noble-gas configuration. Inspection of the data in Table 7-6 shows that the electron affinity decreases in going from chlorine to bromine to iodine. *As the atomic number increases within a family, the atoms increase in size and the effective nuclear attraction decreases.* Thus, there is less tendency to attract an additional electron. The unexpected, low values for fluorine and other Period 2 elements may be related to the extremely small sizes of these atoms. In these cases, an additional electron produces a negative ion which has a high electron density; that is, a large number of electrons in a small volume. This would be a rather unstable condition and could be opposed by inter-electronic repulsion. Apparently, the repulsive forces are strong enough to result in an abnormally low electron affinity which does not follow the general trend.

The limited data in Table 7-7 show that, in general, electron affinity increases as you go from left to right across a period of the Periodic Table. Thus, carbon, compared with fluorine, has very little tendency to form negative ions.

TABLE 7-6
ELECTRON AFFINITIES FOR THE ELEMENTS OF GROUP VIIA

Element	*Atomic Number*	*Electron Affinity (kcal/mole)*	*(ev)*
F	9	80	3.5
Cl	17	87	3.8
Br	35	80	3.5
I	53	70	3.0

TABLE 7-7
ELECTRON AFFINITIES FOR THE ELEMENTS OF PERIOD II

Element	*Atomic Number*	*Electron Affinity (kcal/mole)*	*(ev)*
C	6	29	1.3
O	8	34	1.5
F	9	80	3.5

7-7 Atomic and Ionic Radii We have seen how atomic radius is a factor related to the magnitude of an atom's ionization energy. In addition, atomic radii are related to bond strength and play an important role in determining such physical and chemical properties as density, solubility, melting point, and acid strength. Since it is impossible to isolate an individual atom and to measure its properties, it is necessary to make measurements using large numbers of atoms. Thus, most atomic properties pertain to an atom as a part of a molecule or a crystal rather than to an atom itself. *Atomic* and *ionic radii* are examples of such properties. Since an electron may be found at almost any distance from the nucleus of an atom, the distance designated as the radius is arbitrary. Furthermore, the electron probability distribution will be affected by the presence of other atoms in a molecule or crystal. This means that the radius of an atom may vary depending on the condition under which the measurement is made. The best that we can do is to compare the relative sizes of atoms whose radii have been

Bond strength is described in terms of bonding energy. Bonding energy may be defined as the energy required to break all the bonds in a mole of gaseous molecules which are in their lowest energy states.

7-14 Radius *a* represents the covalent bond radius, one-half the distance between two nuclei within a molecule. Radius *b* represents the van der Waals radius, one-half the distance between nuclei of atoms in adjacent molecules.

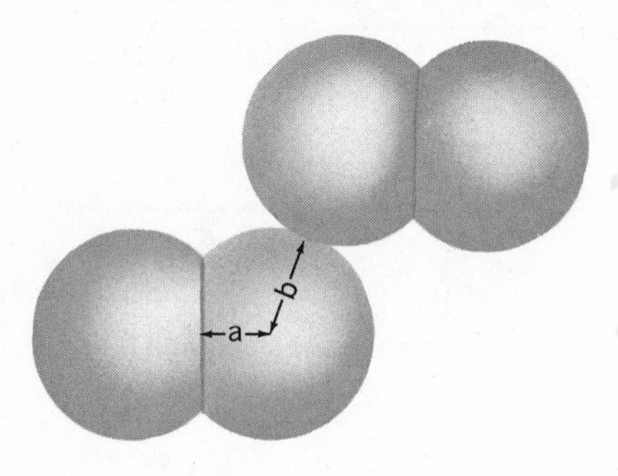

determined by similar methods. For example, the radius might be defined as one-half the distance between the nuclei of adjacent atoms bound together in a molecule. This value would differ from a radius defined as one-half the distance between the nuclei of adjacent atoms in two molecules, Fig. 7-14. If the type of radius is specified, valid comparisons various atomic properties can be made.

Interatomic distances can be determined experimentally by two general methods.

1. *Spectroscopic methods.* These methods involve the *absorption and emission of radiation.* Absorption of radiation results in a change in the energy state of an atom or molecule. By measuring the frequencies absorbed by molecules, scientists can calculate interatomic distances. That is, there are quantitative relationships between frequencies absorbed and distances between atoms in molecules. The frequencies of interest in the calculation of interatomic distances fall in the microwave (0.05 cm) region of the spectrum. Hence, this process is known as *microwave spectroscopy.*
2. Diffraction methods. These methods depend upon the diffraction of electron beams or X rays by atoms or ions in molecules and crystals.

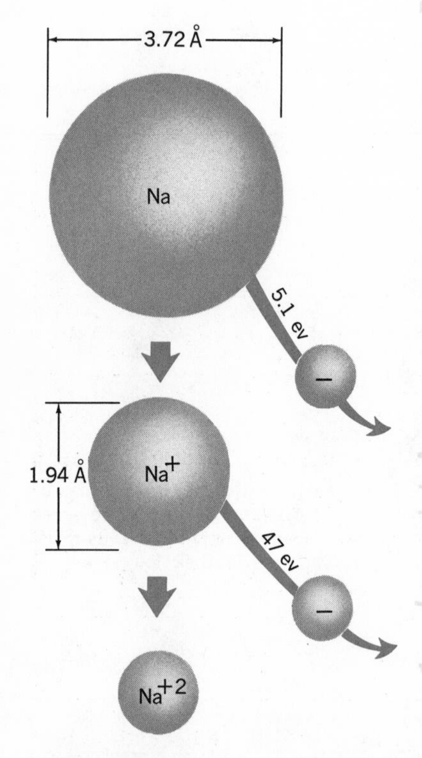

7-15 The removal of the outer electron from the sodium atom reduces the electron to proton ratio, the total electron repulsion, and the radius of the particle.

The periodic Table, Fig. 7-16, shows the covalent radii of atoms. These radii are plotted against atomic number in Fig. 7-17. The graph clearly shows the periodic character of atomic radii. It is of interest to note that J. Lothar Meyer, in the middle of the 19th century showed that atomic size (volume) is periodic in nature. Meyer determined atomic volumes by dividing the atomic mass of the elements by their density. He then plotted atomic volumes against atomic mass and obtained the graph shown in Fig. 7-18.

Reference to Fig. 7-18 reveals these trends:

1. In a given period of the Periodic Table, the atomic radius decreases as the atomic number increases. This is because the progressive addition of electrons to the same energy level is accompanied by an increase in the effective nuclear charge. The higher effective nuclear charge pulls the electrons in the outer energy level closer to the nucleus, resulting in a smaller atomic radius.
2. In general, the atoms in a given family increase in size as the atomic number increases. The number of major energy levels increases by one for each new period. The additional energy level more than compensates for the added nuclear charge, so the atomic radius increases.

Removing an electron from an atom reduces the interelectronic repulsion and causes the electron cloud to shrink. Adding an electron to an atom increases the interelectronic repulsion and causes the charge cloud to expand. This means that positive ions are always smaller than the atoms from which they are formed, and negative ions are always larger than the atoms from which they are formed. The data in Tables 7-8 and 7-9 confirm this observation.

It can be seen that the radius of a sodium ion (Na^+) is approximately one-half that of a sodium atom (Na^o). The data in Table 7-9 show that the radius of a fluoride ion (F^-) is approximately twice that of a fluorine atom (F^o). As would be expected, small positive ions have a strong attraction for electrons, and large negative ions have relatively weak attraction for electrons. The relative sizes of atoms and ions are pictorially represented in Fig. 7-19.

7-8 Oxidation States In an earlier section we noted that it is possible to predict a reasonable formula for a compound containing a given metal provided we know the formula of an analogous compound containing another metal from the same group in the Periodic Table. Because formulas are related to the combining powers of atoms we might suspect that the combining powers of atoms could also be predicted from their location in the Periodic Table. You may recall that the combining powers of atoms may be expressed in terms of the actual or apparent electric charges (oxidation states) they exhibit when combined with other atoms.

TABLE 7-8
ATOMIC AND IONIC RADII FOR METALLIC ELEMENTS

Element	*Atomic Radius, Å*	*Ionic Radius, Å*
GROUP IA		
Li	1.52	0.68
Na	1.86	0.95
K	2.27	1.33
Rb	2.48	1.48
GROUP IIA		
Be	1.11	0.31
Mg	1.60	0.65
Ca	1.97	0.99
Sr	2.15	1.13

7-16 Periodicity of atomic radii.

Periodic Table
Atomic Radii (Ångstroms)

	1A	2A	3B	4B	5B	6B	7B	8	8	8	1B	2B	3A	4A	5A	6A	7A	0
1	1 H 0.31																1 H 0.31	2 He
2	3 Li 1.520	4 Be 1.11											5 B 0.79	6 C 0.77	7 N 0.74	8 O 0.73	9 F 0.709	10 Ne
3	11 Na 1.858	12 Mg 1.598											13 Al 1.432	14 Si 1.17	15 P 1.10	16 S 1.02	17 Cl 0.99	18 Ar
4	19 K 2.272	20 Ca 1.974	21 Sc 1.606	22 Ti 1.448	23 V 1.311	24 Cr 1.249	25 Mn 1.29	26 Fe 1.26	27 Co 1.253	28 Ni 1.246	29 Cu 1.278	30 Zn 1.332	31 Ga 1.221	32 Ge 1.225	33 As 1.2	34 Se 1.16	35 Br 1.142	36 Kr
5	37 Rb 2.48	38 Sr 2.152	39 Y 1.776	40 Zr 1.590	41 Nb 1.429	42 Mo 1.362	43 Tc 1.325	44 Ru 1.325	45 Rh 1.345	46 Pd 1.376	47 Ag 1.444	48 Cd 1.490	49 In 1.62	50 Sn 1.41	51 Sb 1.4	52 Te 1.3	53 I 1.333	54 Xe
6	55 Cs 2.654	56 Ba 2.174	57 La 1.870	72 Hf 1.563	73 Ta 1.43	74 W 1.370	75 Re 1.370	76 Os 1.338	77 Ir 1.358	78 Pt 1.388	79 Au 1.442	80 Hg 1.502	81 Tl 1.71	82 Pb 1.750	83 Bi 1.548	84 Po 1.4	85 At 1.4	86 Rn
7	87 Fr	88 Ra	89 Ac 1.878															

58 Ce 1.825	59 Pr 1.820	60 Nd 1.814	61 Pm	62 Sm	63 Eu 1.994	64 Gd 1.786	65 Tb 1.762	66 Dy 1.752	67 Ho 1.743	68 Er 1.734	69 Tm 1.724	70 Yb 1.940	71 Lu 1.718
90 Th 1.798	91 Pa 1.606	92 U 1.38	93 Np 1.3	94 Pu 1.513	95 Am	96 Cm	97 Bk	98 Cf	99 Es	100 Fm	101 Md	102 No	103 Lw

TABLE 7-9
ATOMIC AND IONIC RADII FOR NONMETALLIC ELEMENTS
GROUP VIIA

Element	*Atomic Radius, Å*	*Ionic Radius, Å*
F	0.71	1.36
Cl	0.99	1.81
Br	1.14	1.95
I	1.33	2.16
GROUP VIA		
O	0.73	1.40
S	1.02	1.84
Se	1.16	1.98
Te	1.30	2.21

7-17 Graph showing the periodic nature of atomic radii.

In Chapter 6 and in this chapter we found that the charges exhibited by atoms are related to their electronic structures. Because electronic structures are periodic functions, oxidation states should also be periodic in nature. Let us now see how common oxidation states of an element can sometimes be predicted from its location in the Periodic Table.

It is necessary to recall that reactions between atoms generally involve electrons in their outer energy level(s). During a reaction chemical bonds are formed. We shall find that in the formation

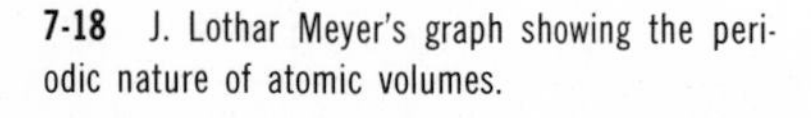

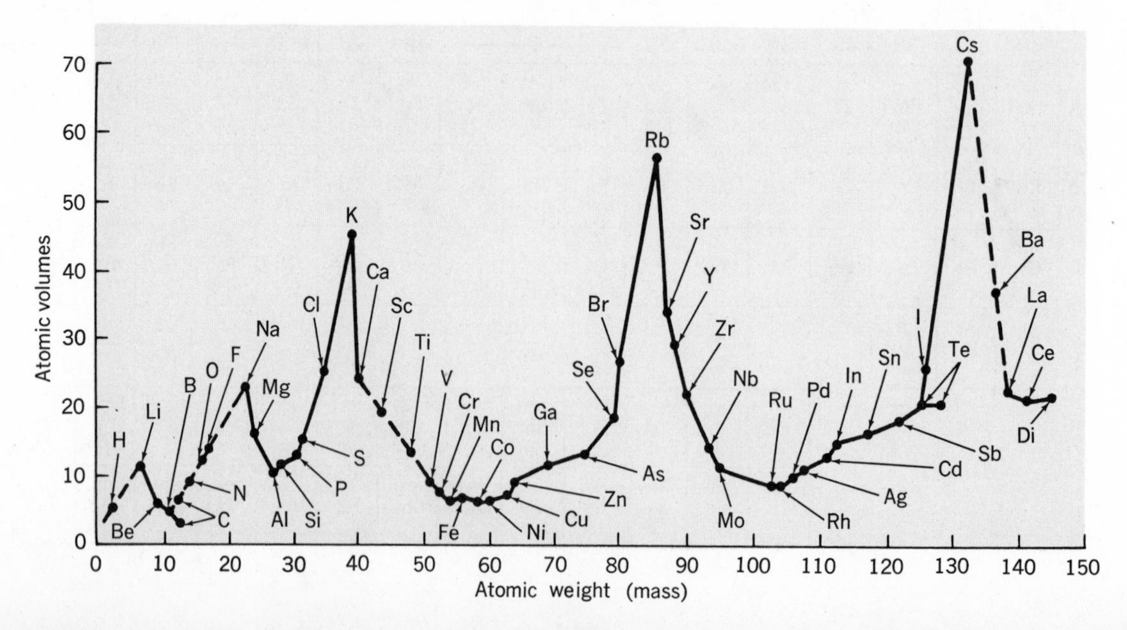

7-18 J. Lothar Meyer's graph showing the periodic nature of atomic volumes.

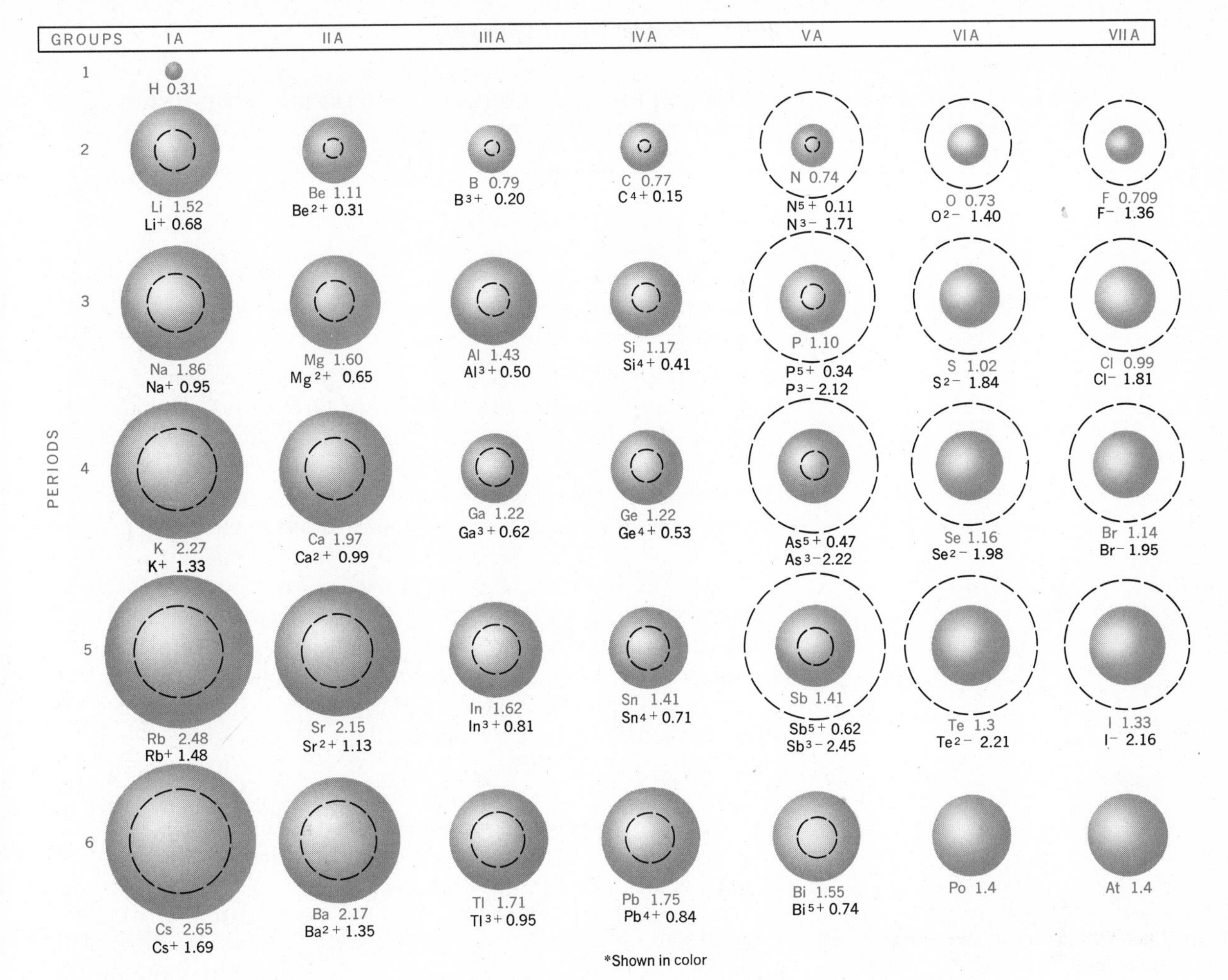

7-19 Pictorial representation of the relative atomic and ionic radii of the representative elements. The dashed lines represent the relative circumferences of the ions.

of bonds, electrons may be donated or accepted by an atom, depending on the relative ionization energies and electron affinities of the combining atoms. The combined atoms and ions of the representative elements often have the same electron configuration as the noble-gas atoms. To predict common oxidation states for atoms of the representative elements we must determine the number of electrons that, in theory, could be donated or accepted by the atom if it were to achieve a noble-gas configuration. It is found that, in general, electrons in an orbital tend to be donated in pairs. The common oxidation states of the representative elements listed in Table 7-10 were obtained by applying these rules. It should be noted that donation or acceptance of electrons by an atom does not necessarily imply the formation of ions. For example, in many of its compounds, chromium exhibits an oxidation state of 6+. In chemical reactions or in solutions, however, chromium is never found as a 6+ ion. The 6+ oxidation state suggests that chromium is donating 6 electrons to a bond holding a chromium atom to an atom having a greater attraction for electrons.

TABLE 7-10
COMMON OXIDATION STATES OF THE REPRESENTATIVE ELEMENTS

Group Number	*IA*	*IIA*	*IIIA*	*IVA*	*VA*	*VIA*	*VIIA*
Outer orbital configuration	s^1	s^2	s^2p^1	s^2p^2	s^2p^3	s^2p^4	s^2p^5
Common oxidation states	1+	2+	1+, 3+	2+, 4+	3+, 5+, 3−	2+, 4+, 6+ 2−	5+, 7+, 1−

As an example, let us illustrate the origin of the common oxidation states exhibited by the elements of Group VIA. Donation of one pair of p electrons gives rise to the 2+ state, donation of all four p electrons gives rise to the 4+ state, donation of all six outer electrons gives rise to the 6+ state, and the acquisition of two electrons thereby forming an outer octet results in the 2− oxidation state.

To help us represent and explain chemical bonds in the next chapter, we need a concise means of showing the distribution of the electrons involved in bond formation. One simple representation involves the use of electron-dot formulas. We shall introduce the concept at this point and then extend its application to chemical bonds in the next chapter.

In an electron-dot formula, dots or x's are used to represent the electrons in the outer energy level. These electrons are sometimes referred to as *bonding or valence electrons.* The nucleus and the inner level electrons are represented by the symbol of the element. Oxygen is element number 8 and is located in Period 2 of Group VIA. This information indicates that the oxygen atom has 8 electrons, 6 of which are in the outermost ($n = 2$) energy level. The outer level orbital configuration is $s^2p^2p^1p^1$ and the electron-dot formula is

The *oxidation state* of an atom in a molecule is the *apparent* charge carried by the atom. We may assume the oxidation state of oxygen in most oxides is 2−.

$$:\ddot{\mathrm{O}}\cdot$$

A pair of electrons in a given orbital is sometimes represented by a short line. Using this notation, the formula for the oxygen atom would be $|\overline{\mathrm{O}}\cdot$. The electron-dot formula for an oxide ion is

$$\left[:\ddot{\mathrm{O}}\,\overset{\times}{\times}\right]^{2-}$$

Electron-dot formulas are sometimes referred to as Lewis structures in honor of the American chemist, G. N. Lewis (1875–1946).

Note that the electron-dot formula of an ion is enclosed in brackets with a charge placed outside the bracket. Since all electrons are identical, the *dots* and x's are used only to identify their origin. In the case of an oxide ion, the x's represent electrons donated by hydrogen or some other atom.

FOLLOW-UP PROBLEM

Write an electron-dot formula for (a) a chlorine atom, (b) a chloride ion (Cl^-), (c) a magnesium atom (Mg), (d) a magnesium ion (Mg^{2+}).

In general, the transition elements (B families) have multiple oxidation states. This is because the *d* and adjacent *s* orbitals are very close in energy. The atoms of the Group IIB elements (Sc, Y, La) have only one common oxidation state, 3+. This corresponds to the donation of the single *d* and the pair of *s* electrons having very close energy values. For the rest of the transition elements, the lowest *common* oxidation state is usually 2+ and the highest is usually the same as the group number. In general, the group number indicates the highest positive oxidation state that the members of the family can theoretically attain. The table on p. 250 contains only the more common oxidation states of the elements. Less common states are known, however.

Sometimes it is instructive to relate the oxidation state of an atom in a given species to some property of a species. For example, we shall find in a later chapter that the relative strengths of some acids can be explained in terms of the oxidation state of their central atom. This means we must be able to calculate the oxidation state of a given atom from the formula of its compound. This can be done by assigning fixed oxidation states to all but the atom in question and conforming to the requirements listed below.

1. The sum of the positive and negative oxidation states in a neutral molecule is zero.
2. The algebraic sum of the positive and negative oxidation states of the atoms in a polyatomic ion is equal to the charge on the ion.

Oxidation states conventionally assigned to atoms combined in common chemical species are noted below:

1. Oxygen = 2− (except in peroxides where it is 1− or when combined with fluorine where it is 2+.)
2. Hydrogen = 1+ (except in metallic hydrides, where it is 1−).
3. Group IA elements = 1+
4. Group IIA elements = 2+.
5. Ions of the halogen atoms in binary ionic compounds = 1−.

Following these suggestions, we obtain 4+ as the oxidation state of carbon (C) in carbon dioxide (CO_2), 6+ as the oxidation state of sulfur (S) in sulfate ion (SO_4^{2-}), and 6+ as the oxidation state of chromium (Cr) in dichromate ion ($Cr_2O_7^{2-}$).

FOLLOW-UP PROBLEM

Calculate the oxidation state of the underlined element each of these species: (a) $\underline{S}O_2$, (b) $\underline{Mn}O_4^-$, (c) $\underline{Mn_2}O_7$, (d) $H\underline{Cl}O$, (e) $H\underline{Cl}O_2$, (f) $H\underline{Cl}O_3$, (g) $H\underline{Cl}O_4$, (h) $\underline{I}O_4^-$.

The use of oxidation states to predict and write formulas for chemical substances suggests that the chemical combination of atoms is the result of attractive forces between atoms. In the language of chemistry we say that *the atoms in an aggregate of atoms are held together by chemical bonds which originate from the attractive forces between atoms.*

In our study of chemical bonding we shall find that bonded atoms "share" outer-orbital electrons. The degree of sharing is determined by the relative attraction that two atoms have for electrons. Equal sharing occurs when bonding atoms are identical and have the same attraction for electrons. The bond between the atoms in this case is called a *nonpolar covalent bond.* When the difference in electron-attracting ability of two atoms is at a maximum, the electron sharing is at a minimum and the bond between them is called an *ionic bond.* In this case, the unequally shared electrons are so close to one atom that it acquires a negative charge while the other atom acquires a positive charge. The charged atoms are, of course, referred to as negative and positive ions. In between these extremes the covalent bonds have a partial ionic character depending on the difference in the electron-attracting ability of the atoms. For convenience, a bond with over 50 percent ionic character is referred to as ionic.

The characteristics of bonded species are related to the character of the bonds between the atoms. That is, the properties of covalently bonded aggregates differ from those of ionically bonded aggregates. It is apparent that it would be valuable to have a method for measuring the relative electron-attracting ability of bonded atoms, and hence, be able to identify the nature of the bond and the characteristics of the bonded species. Fortunately, there is an experimentally based periodic property that indicates the relative electron-attracting ability of bonded atoms. This property, known as ***electronegativity,*** will help us predict the nature of the bond between atoms when we investigate chemical bonding in the next chapter. The periodic nature of electronegativity suggests that the location of pairs of bonding atoms in the Periodic Table will help us predict the type of bond between them.

7-9 Electronegativity Electronegativity is a measure of the tendency of an atom in a molecule or aggregate of atoms to attract electrons. Atoms, such as fluorine, which have a *strong attraction for electrons* are said to be highly electronegative. Those atoms, such as cesium, which have little electron-attracting ability have low electronegativities.

A number of methods have been used to determine electronegativities of atoms, and a number of scales have been devised. One of the more commonly used is the scale developed by Linus Pauling (1901–), formerly a professor of chemistry at Caltech. Pauling's scale of electronegativities is based on energy changes accompanying chemical reactions. The electronegativity values obtained by Pauling are shown in the Periodic Table in Fig. 7-20. These values are plotted against atomic numbers in Fig. 7-21. The graph shows the periodic nature of this property.

Periodic Table
Electronegativities

	1A	2A	3B	4B	5B	6B	7B	8	8	8	1B	2B	3A	4A	5A	6A	7A	0
1	1 **H** 2.1																1 **H** 2.1	2 **He**
2	3 **Li** 1.0	4 **Be** 1.5											5 **B** 2.0	6 **C** 2.5	7 **N** 3.0	8 **O** 3.5	9 **F** 4.0	10 **Ne**
3	11 **Na** 0.9	12 **Mg** 1.2											13 **Al** 1.5	14 **Si** 1.8	15 **P** 2.1	16 **S** 2.5	17 **Cl** 3.0	18 **Ar**
4	19 **K** 0.8	20 **Ca** 1.0	21 **Sc** 1.3	22 **Ti** 1.5	23 **V** 1.6	24 **Cr** 1.6	25 **Mn** 1.5	26 **Fe** 1.8	27 **Co** 1.8	28 **Ni** 1.8	29 **Cu** 1.9	30 **Zn** 1.6	31 **Ga** 1.6	32 **Ge** 1.8	33 **As** 2.0	34 **Se** 2.4	35 **Br** 2.8	36 **Kr**
5	37 **Rb** 0.8	38 **Sr** 1.0	39 **Y** 1.2	40 **Zr** 1.4	41 **Nb** 1.6	42 **Mo** 1.8	43 **Tc** 1.9	44 **Ru** 2.2	45 **Rh** 2.2	46 **Pd** 2.2	47 **Ag** 1.9	48 **Cd** 1.7	49 **In** 1.7	50 **Sn** 1.8	51 **Sb** 1.9	52 **Te** 2.1	53 **I** 2.5	54 **Xe**
6	55 **Cs** 0.7	56 **Ba** 0.9	57 **La** 1.1	72 **Hf** 1.3	73 **Ta** 1.5	74 **W** 1.7	75 **Re** 1.9	76 **Os** 2.2	77 **Ir** 2.2	78 **Pt** 2.2	79 **Au** 2.4	80 **Hg** 1.9	81 **Tl** 1.8	82 **Pb** 1.8	83 **Bi** 1.9	84 **Po** 2.0	85 **At** 2.2	86 **Rn**
7	87 **Fr** 0.7	88 **Ra** 0.9	89 **Ac** 1.1															

58 **Ce** 1.1	59 **Pr** 1.1	60 **Nd** 1.1	61 **Pm** 1.1	62 **Sm** 1.1	63 **Eu** 1.1	64 **Gd** 1.1	65 **Tb** 1.2	66 **Dy** 1.2	67 **Ho** 1.2	68 **Er** 1.2	69 **Tm** 1.2	70 **Yb** 1.2	71 **Lu** 1.2
90 **Th** 1.3	91 **Pa** 1.5	92 **U** 1.7	93 **Np** 1.3	94 **Pu** 1.3	95 **Am** 1.3	96 **Cm** 1.3	97 **Bk** 1.3	98 **Cf** 1.3	99 **Es** 1.3	100 **Fm** 1.3	101 **Md** 1.3	102 **No** 1.3	103 **Lw** 1.3

7-20 Electronegativity is a periodic property.

The data also show that *electronegativity generally decreases as you go down a given family in the Periodic Table.* This observation may be explained largely by the increase in atomic radius as the atomic number increases within a family. Since the force of attraction between the nucleus of an atom and an electron from another atom varies inversely with the square of the distance between them, we would expect a large atom to have much less electron-attracting ability than a smaller one. Thus, a cesium atom would have less tendency to attract an electron of another atom than would a sodium atom.

It is the difference between the electronegativity of two atoms that can be used to identify the character of the bond. Large differences are associated with a high degree of ionic character. On Pauling's scale, a difference of 1.7 is often used to denote a 50 percent ionic character. It should be noted that the concept of electronegativity is somewhat controversial. Many chemists question the validity of the numerical values. Even the critics, however, admit they find the concept useful. The important thing is to be aware of the limitations and not expect 100 percent accuracy in the predictions you make using this concept. Remember that all predictive devices have shortcomings.

For quick reference, the general trends in the properties discussed in this chapter are summarized in Fig. 7-22.

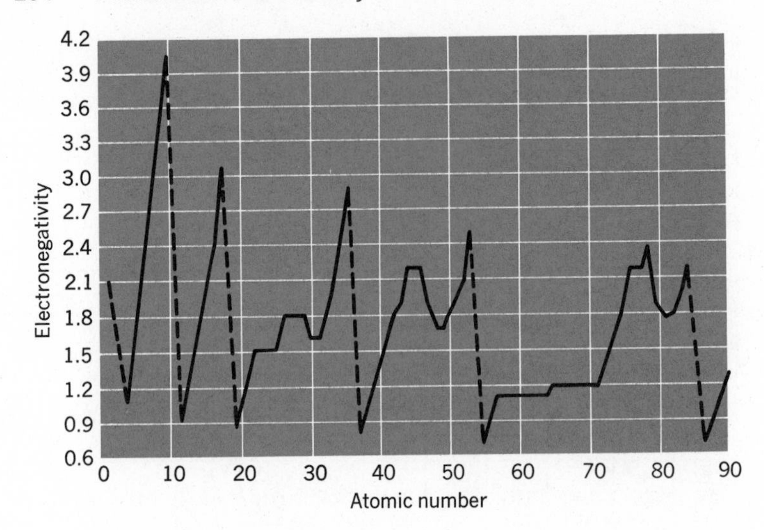

7-21 Graph showing the periodic nature of electro-negativity.

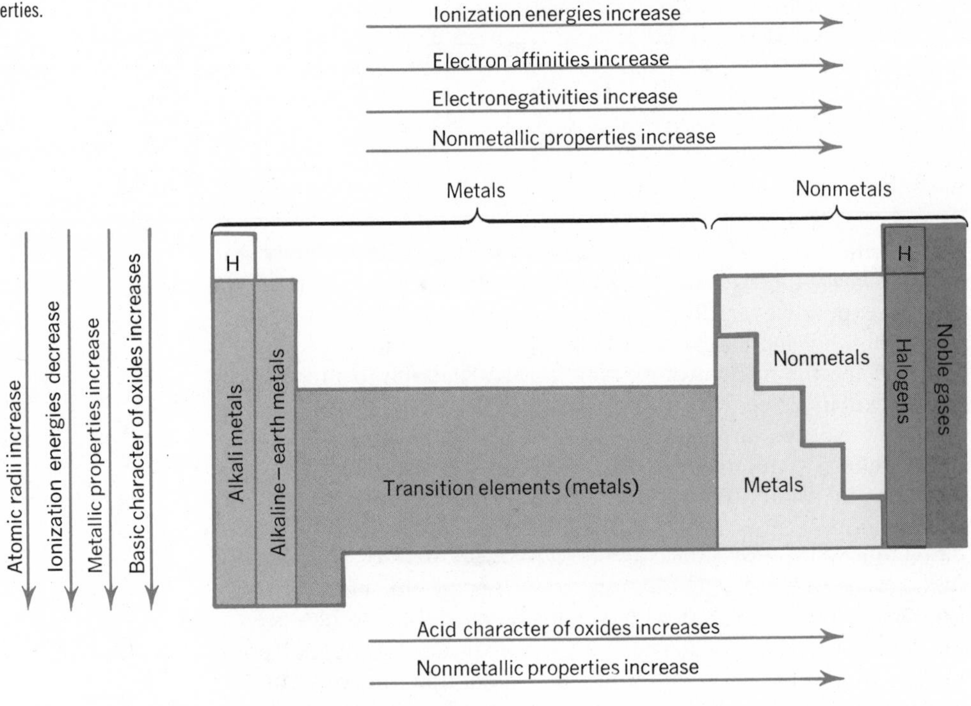

7-22 Composition of the Periodic Table and trends in periodic properties.

FOLLOW-UP PROBLEM

1. (a) Would you say the bond between potassium and chlorine would be largely ionic or largely covalent? (b) Would you predict the bond between sulfur and oxygen to be largely ionic or largely covalent? (c) Make a generalized statement about the nature of the bonds that exist between the atoms of Group I-A and Group VII-A elements. (d) Make a generalized statement about the nature of the bonds that exist between two nonmetallic atoms.

LOOKING AHEAD

With our ability to represent the electronic configuration of atoms, recognize the types and shapes of atomic orbitals, and use the concept of periodicity to generalize and make predictions, we are now ready to consider the formation and nature of chemical bonds.

One of our objectives is to discover and explain the reason that substances behave the way they do. Most of you, at some time in your lives, have been asked or have wondered about such basic questions as: Why are diamonds hard? Why is a candle soft, and why does it melt so easily? Why is honey sticky? Why can metals be drawn into wires and be used as electric conductors? Why doesn't oil or gasoline dissolve in water the way salt, sugar, and alcohol do?

The answers to these questions depend upon the nature of the substances involved. In turn, the nature of the substances depends upon the forces (bonds) which hold together the individual particles of which they are composed. To obtain the knowledge that will help us answer these and other basic questions, we shall focus our attention in the next chapter on chemical bonds. We shall use as background the orbital concept and our knowledge of the electronic configuration of atoms to help explain how the different types of bonds are formed, and to account for the shapes and characteristics of molecules. The concept of periodicity will enable us to extend and apply the principles of bond formation to a large number of substances.

Special Feature: The Origin of the Elements

An English physician, William Prout (1785–1850), was primarily interested in nutrition and digestion. He discovered that the stomach contained hydrochloric acid, a very corrosive acid capable of damaging some flesh and reacting with many metals. He was the first person to classify foods as fats, carbohydrates, and proteins. Prout, however, is best known for an idea that was totally outside his field of medicine. Just seven years after Dalton had published his atomic theory, Prout published an article suggesting that hydrogen seemed to be the common denominator of all the other elements. He pointed out that if hydrogen were assigned an atomic mass of 1, then elements could be considered as whole number multiples of hydrogen. All of the elements, he said, were simply conglomerates of hydrogen atoms. Further investigation of atomic masses led to the discovery that magnesium had an atomic mass of about 24.3 and chlorine about 35.5. Prout's idea did not seem to fit all cases and was, therefore, generally discarded.

For over 100 years, Prout's hypothesis was nearly dormant. In 1938, Hans Bethe (1906–) suggested that hydrogen in the sun reacted with carbon and produced helium with the carbon being regenerated. Thus, carbon was acting as a kind of catalyst and hydrogen was the fuel for the energy from the sun. Sir Ernest Rutherford of England had earlier demonstrated that it was possible to change (transmute) one element to another. Bethe's concept used Einstein's theory to explain the almost endless energy that is being generated by the sun. The four hydrogen atoms that ultimately become a helium atom in Bethe's process have a greater mass than the resulting helium atom. The difference in mass is equal to the energy produced. The energy equivalence of the mass may be calculated using Einstein's famous equation, $E = mc^2$. The almost perpetual emission of energy by the sun was a mystery that had plagued scientists for many generations. Bethe's theory not only explained this mystery but revived new interest in Prout's hypothesis. The vast amount of energy released by this process was first witnessed at close range by man in 1952 when the first experimental hydrogen bomb was exploded. This type of nuclear reaction is called a fusion reaction because it involves the "melting together" or fusing of lighter nuclei of atoms into heavier nuclei.

A star like our sun has a temperature of about $10^7\,°K$ in its core, and a density of about 100 g/ml. These conditions are sufficient to carry out the fusion process described by Bethe. By fusion, helium (element number 2) is formed from hydrogen (element number 1), and a constant flow of energy is released. If a star contains carbon, as does our sun, then the carbon acts as a catalyst for this reaction. If a star contains no carbon, the fusion reaction may proceed anyway and produce helium, but no element with an atomic number higher than that of helium. How then are the other elements produced? Spectroscopy (analysis of light) reveals that our sun contains not only carbon, but iron and other denser elements. How can we explain the existence of these elements in the sun?

Let us begin with an assumption, which most scientists accept, that the universe is 99 percent hydrogen and nearly 1 percent helium. Imagine a cold cloud of hydrogen gas in a galaxy being compressed by the starlight of surrounding stars into a rather dense cloud of swirling gas. This cloud contracts by its own gravitation force into a hot star in which hydrogen "burns" and forms helium. This is not a chemical, but rather, a nuclear "burning" process. This is a "main-sequence" star like our own sun but it explains only the formation of helium. The elements in our sun, like carbon and iron, must have been present as "contaminants" when the original cloud of hydrogen contracted and formed the star.

When about one-tenth of the hydrogen in a main-sequence star has been "burned," the star then has contracted so that the density of the core becomes about 1000 times as great as before. The temperature in the core increases to about $10^8\,°K$. The high temperature in the core expands the outer portion of this star, called the envelope, so that it is greately enlarged. The outer surface of this star is so expanded that the energy output per unit area of surface is less than for a main-sequence star and it appears red in color. This type of star is known as a red giant. The temperature and density in the core of a red giant are sufficiently great to cause the helium to fuse into larger aggregates. Neutrons produced in these reactions are absorbed by nuclei of other elements present. The absorption of a neutron by a nucleus increases its mass and may make it unstable. The unstable nucleus may emit a negative beta particle (high-speed electron) so the positive charge on the nucleus increases and a new element with a higher atomic number is formed. Thus, elements with the middle range of atomic numbers are produced. The nuclear processes in a red giant star produce such elements as carbon, nitrogen, iron, and these elements up to atomic number 82 and 83 which are lead and bismuth. How then are elements produced that have atomic numbers 84 and greater?

When the first experimental hydrogen bomb was exploded on the South Pacific Island of Bikini, in 1952, the uranium-238 that was a part of the triggering mechanism, absorbed a sufficient number of neutrons to form an element which was called californium. The particular isotope formed was californium-254. The half-life (time required for one-half of the sample to degenerate into other elements with lower atomic numbers) was found to be 55 days. Several other higher elements were also produced in this reaction. In nature, neither the main-sequence star or the red giant duplicate these conditions. The only time nature develops conditions sufficient for producing higher elements such as polonium, atomic number 84, and above, is in an exploding star called a supernova.

A supernova is a titanic explosion of a star. The radiation emitted from one of these explosions is enormous. In a few hours a supernova radiates as much light as our sun does in a million years. After a few dozen hours, the supernova fades and slowly diminishes in intensity. These blowups are as rare as they are spectacular. In our own galaxy, the Milky Way, there have been only three known supernovas in the past two thousand years.

In 1054, Chinese astronomers observed an extremely bright star in the sky. They kept an accurate log of its luminosity by comparing its brightness with other stars and with planets. Modern astronomers have analyzed their log and have concluded that this supernova diminished in intensity by one-half every 55 days. In 1572, the Danish astronomer, Tycho Brahe (1546–1601), observed a supernova which increased in intensity until it could be seen in the daytime. In 1604, Johann Kepler (1571–1630), also observed a supernova. In both of these cases the stars lost one-half of their luminosity each 55 days. Most astronomers agree that the Crab Nebula, which is 4×10^3 light years away, is the remains of the supernova observed by the Chinese astronomers in 1054. This

7-23 Modern astronomers believe that the Crab Nebula shown here is the result of the supernova observed by Chinese astronomers in 1054. This explosion is still taking place as gases are spreading out at the rate of 70 million miles per day. (Mount Wilson and Palomar Observatories)

explosion is still taking place and a gaseous shell is spreading out at the rate of 7×10^7 miles a day. The gases in the Crab Nebula are at a very high temperature and are emitting strong radio signals. Astronomers today are still intensely interested in this system. By turning their telescopes to other galaxies, astronomers can see supernovae each year. No opportunity, however, has ever existed to study one in our own galaxy with modern equipment. There is much yet to be learned about these colossal events.

The high temperatures and pressures in the core of a supernova are sufficient for producing the higher elements. The density of the gas in the interior of a supernova is calculated to be 1×10^8 g/ml. The temperature is 1×10^9 to 1×10^{10} degrees. There are high concentrations of neutrons so that nuclei can capture sufficient numbers of them and produce the higher elements such as uranium and californium. Many scientists believe that the Chinese astronomers who observed the supernova in 1054 were perhaps the real discoverers of californium which was not isolated and named until 1952. Is is possible, however, that the diminishing in luminosity of supernovae over a 55-day period and the half-life of californium being also 55 days is merely a coincidence. Many nuclear scientists and astronomers, however, feel that there may be a relationship.

The supernovae of past billions of years have filled the intergalactic spaces with all of the higher elements that they produced. The clouds of hydrogen that form new stars are therefore, "contaminated" with these materials. The Crab Nebula today is in the process of contaminating intergalactic space in this way. We feel that we can say that iron and other heavy elements in our own sun originated in this manner. By studying the concentrations of uranium and lead in deposits in the earth, scientists have concluded that our terrestrial uranium was formed in a great supernova about 6.6×10^9 years ago. The nuclear chemistry that the universe performs in stars and supernovae fuses hydrogen, the simplest form of matter, into all of the other more complicated elements. Modern scientific theory, therefore, tends to substantiate Prout's hypothesis that all elements are indeed conglomerates of hydrogen.

QUESTIONS

1. (a) Was the original Periodic Table developed on an experimental or theoretical basis? Explain. (b) How is the form of the Periodic Table explained on a theoretical basis?

2. In which section of the Periodic Table are the nonmetallic elements?

3. Find three places in the Periodic Table where Mendeleyev's generalization that the properties of the elements are a function of their atomic masses does not hold.

4. What is the family name of (a) Group IA, (b) Group IIA, (c) Group VIIA, (d) Group O, (e) the B-family?

5. Why do elements in the same group display rather similar chemical properties?

6. Calculations indicate that elements of numbers 110 and 114 could be quite stable. If scientists can make elements 110 and 114, what well-known elements would you expect them to resemble?

7. The formula for phospine is PH_3. Write the name and formula for an analogous compound containing arsenic.

8. The boiling point of CH_4 is $-161\,°C$ and the boiling point of GeH_4 is $-90\,°C$. Estimate the boiling point of SiH_4.

9. The boiling point of HI is $-35.5\,°C$ and that of HCl is $-85\,°C$. Estimate the boiling point of HBr.

10. The formula for calcium chromate is $CaCrO_4$. Use the Periodic Table to predict a formula for (a) barium molybdate and (b) strontium tungstate.

11. Write net ionic equations for (a) sulfur trioxide (SO_3) with water, (b) phosphorus(V) oxide (P_4O_{10}) with water (the products are $H_2PO_4^-$ and H_3O^+), (c) magnesium oxide with water, (d) strontium with hydrochloric acid.

12. (a) A water solution of HCl is called hydrochloric acid: what is the name of a water solution of HBr? (b) A solution of $HClO_3$ is called chloric acid; what is the name of a solution of $HBrO_3$?

13. Use the location of the elements in the Periodic Table to help write formulas for these compounds: (a) a chloride containing indium (number 49), (b) an oxide containing germanium (32), (c) a bromide containing barium (56); (d) an oxide containing osmium in its highest oxidation state.

14. What are the characteristics of an amphoteric oxide?

15. What are the characteristics of a periodic function?

16. Complete this sentence using the terms increase(s) or decrease(s). As you go from left to right across a row of the Periodic Table, the ionization energies ______, the metallic properties ______, the atomic radius ______, and the electronegativity ______.

17. Which is the most metallic element of Group VA? What is the basis of your choice?

18. Explain the general trend in ionization energies as you go down a given family in the Periodic Table.

19. Explain the general trend in atomic radii as you go across a row of the Periodic Table.

20. On the basis of ionic radii, predict in which of these ion pairs the force of attraction between ions is greater: Cs^+Cl^- or Na^+Cl^-?

21. (a) Would you predict the electron affinity of the noble-gas atoms to be high or low? Explain. (b) How are electron affinity and ionization energy related to the observed resistance of these atoms to enter into chemical combination with the atoms of other elements?

22. Why would you not expect sodium to exhibit a charge of 2+ in its compounds?

23. (a) Suggest a reason why, contrary to the trend, the first ionization energy for sulfur, element number 16, is less than that for phosphorus, element number 15. (b) Suggest a reason why, contrary to the trend, the first ionization energy for boron, element number 5, is less than that for beryllium, element number 4.

24. (a) The element X, combines with oxygen and forms an oxide, XO. The compound (XO) dissolves in water and forms a solution which changes the color of red litmus paper to blue. Is element X a metal or a nonmetal? (b) Is the compound (XO) an acid or basic anhydride? (c) Write a formula for a compound composed of elements X and chlorine.

25. Within the Group IA elements, rubidium has a lower first ionization energy than sodium. Explain.

26. Lithium has an atomic radius of 1.52 Å and an ionic radius of 0.65 Å; fluorine has an atomic radius of 0.71 Å and an ionic radius of 1.36 Å. Explain why Li decreases in size on becoming an ion while fluorine increases in size.

27. Which group of elements in the Periodic Table would you expect to have low first- and second-ionization energies with very high third-ionization energies?

28. Values for the first, second, and third ionization energies of an unknown element are found to be as follows: $E_1 = 9.32$ ev, $E_2 = 18.3$ ev, $E_3 = 153.9$ ev. (a) Identify this element as a metal or nonmetal. (b) In which group in the Periodic Table would you expect to find this element? (c) Would you expect the next ionization energy to be greater or smaller than 153.9 ev?

Ans. (a) Metal (Be), (b) Group II, (c) greater.

29. Discuss some of the problems scientists face when assigning atomic radii to atoms.

30. Using the table of ionization energies, explain why Na forms Na^+ ions while F forms F^- ions in ordinary chemical reactions.

31. Of the eight elements in Period III of the Periodic Table, which would you predict to (a) lose electrons and form ions, (b) gain electrons and form ions?

32. Explain the decrease in first ionization energy for elements in a group of the Periodic Table as the atomic number increases.

33. What is the relationship between first ionization energies and metallic and nonmetallic properties?

34. Use the Periodic Table to predict which atom in each of these pairs has the greater first-ionization energy: (a) Li, Ne, (b) K, Cs, (c) Na, Cl, (d) K, Cu.

35. The ionization energies of 5 elements (A, B, C, D, E) are listed below

ATOM	*Ionization energies (kcal/mole)* I_1	I_2	I_3
A	300	549	920
B	99	734	1100
C	118	1091	1652
D	176	347	1848
E	497	947	1500

Two of the elements are nonmetals and three are metals. One of the elements is a noble gas. Identify by letter: (a) the noble gas, (b) the reactive nonmetal, (c) the most reactive metal, (d) the atom(s) which exhibit(s) a common oxidation state of 2^+, (e) the atom(s) which exhibit(s) a common oxidation state of 1^+.

36. Use the Periodic Table to predict which species in each of these pairs has the smaller radius: (a) Na, Cl, (b) Na^+, Al^{3+}, (d) Na^+, Cs^+, (e) F^-, I^-. Justify each choice.

37. Use the Periodic Table (inside back cover) to predict which species in each of these pairs has the greatest electronegativity. Justify each choice. (a) K, Ca, (b) Li, N, (c) S, Se, (d) F, Cs, (e) K, Cu.

38. For each of these outer-shell electron configurations, give the probable oxidation state(s) of the element: (a) ns^1, (b) ns^2, np^5, (c) ns^2, (d) ns^2, np^6, (e) ns^2, np^4, (f) ns^2, np^2, (g) ns^2, np^3, (h) ns^2, np^1.

39. In Question 38, determine whether the atom is acting as a metal or a nonmetal.

40. What is the oxidation number of each of the underlined elements: (a) $\underline{P}_4$, $\underline{S}O_2$, $\underline{S}O_3$, (b) $\underline{N}_2O$, $\underline{N}O$, $\underline{N}_2O_3$, $\underline{N}_2O_4$, $\underline{N}_2O_5$, (c) $\underline{C}O$, $\underline{C}O_2$, (d) $H\underline{Br}$, $H\underline{Br}O$, $H\underline{Br}O_2$, $H\underline{Br}O_3$, (e) $H_2\underline{S}$, $H_2\underline{S}O_3$, $H_2\underline{S}O_4$?

41. Calculate the oxidation number of the underlined atom (a) $H_2\underline{F}_2$, (b) $H_2\underline{Te}$, (c) $\underline{Na}_2O_2$, (d) $\underline{N}H_3$, (e) $K\underline{Mn}O_4$, (f) $\underline{S}_8$, (g) $K_2\underline{Cr}_2O_7$, (h) $\underline{As}_4$, (i) $K_2H\underline{P}O_4$, (j) $\underline{C}_6H_{12}O_6$.

42. Separate these formulas into two groups (a) those representing compounds with predominantly ionic bonds, (b) those representing compounds with predominantly covalent bonds.

(a) Rb_2O	(f) SO_2	(k) XeF_4	(p) K_2S
(b) NO	(g) SiF_4	(l) IF_7	(q) ClF
(c) CO_2	(h) GeH_4	(m) CsF	(r) RbBr
(d) KI	(i) $CaCl_2$	(n) BeO	(s) P_2S_3
(e) KH	(j) IBr	(o) LiI	

8

Chemical Bonding

Preview

By this time you are, no doubt, aware of the countless chemical changes that occur every day in your environment. One of our goals is to interpret, explain, and predict chemical changes. To achieve this goal, it is necessary to gain an understanding of the principles that underlie chemical changes, and of the nature of the particles that make up chemical systems.

We now have acquired a background knowledge and an understanding of the structure and nature of atoms. Most chemical systems and changes involve molecules and polyatomic ions. For continued progress, therefore, it is necessary that we investigate the structure and nature of these species. You are already aware that molecules and polyatomic ions are aggregates of chemically-combined atoms. In this chapter, we shall discuss how atoms combine, the reason that they combine in certain ratios, and the characteristics of the species resulting from their combination. We can then use this knowledge in the following chapter to help explain the properties of systems involving *aggregates* of molecules and ions.

Using the language of chemistry, we say that the atoms which compose a molecule or polyatomic ion are held together by *chemical bonds*. Because chemical bonding is the theme of this chapter, we shall first define and describe the origin of "bonds." This leads us to mention briefly and show the interrelationship between two major types of bonds: *covalent* and *ionic*. For the rest of the chapter, we focus our attention on developing a covalent-bond model. This model will help us predict the shapes, electric nature, and other properties of covalent molecules.

In this chapter, we shall use the concept of periodicity to extend and apply the principles of bond formation to a large number of substances. In our development of bonding concepts we shall utilize ideas from several bonding theories. A bonding theory should account for polarity (electric nature) and shapes of molecules, as well as their stability, reactivity, and other properties. There is no single theory that accomplishes all of these objectives equally well. Each one has its own strengths.

We shall use the *Lewis scheme* (electron-dot formulas) to represent structures involving covalent bonds. This scheme will help us identify electron-pairs when we use the electrostatic repulsion concept to determine shapes of molecules and

"The fairest thing we can experience is the mysterious. It is the fundamental emotion which stands at the cradle of true art and true science. He who knows it not and can no longer wonder, no longer feel amazement, is as good as dead, a snuffed-out candle."

—Albert Einstein

ions. We shall use the *charge-cloud model* to emphasize the basic electrostatic nature of chemical bonding and to help predict shapes of molecules. The *valence-bond theory* of bonding supplies the concepts of *hybridization* and *resonance* which are useful for describing, explaining, and predicting geometry (shapes) and other characteristics of molecules.

THE ORIGIN AND NATURE OF CHEMICAL BONDS

8-1 Origin, Nature, and Classification of Chemical Bonds Structural investigation of most substances found in nature reveals that they are composed of an aggregate of atoms held together by electrostatic forces of attraction. For example, the oxygen (O_2) and nitrogen (N_2) gas in the air we breathe are composed of diatomic molecules. A great deal of energy is required to overcome the forces of attraction which bind together the atoms in these molecules. In nature, very few metals exist in the elemental state. The ones that do, such as gold and silver, are found in metallic crystals which are also aggregates of atoms held together by electrostatic forces of attraction. In nature, most metals are found in the form of their compounds. For example, the element sodium is found in a variety of naturally-occurring compounds, one of which is ordinary sodium chloride. Laboratory tests show that a large number of these compounds are crystalline ionic solids which consist of an aggregate of positive and negative ions held together by strong electrostatic forces of attraction.

The electrostatic forces of attraction which hold atoms, ions, and molecules together, are, in general, referred to as ***chemical bonds.*** In an earlier chapter, we learned that electrostatic forces of attraction are those that exist between oppositely charged bodies. Since the nuclei of atoms bear positive charges and the electrons carry negative charges, we can say that **all chemical bonds are formed as the result of the simultaneous attraction of two or more nuclei for electrons.**

electrostatic forces →

It is apparent from our examples and discussion that a bonded condition is usually a more energetically stable one than an unbonded one. That is, the energy required to overcome attractive forces (break bonds) between bonded particles raises the energy of the system. Thus, bonded atoms are in a lower energy, more stable state than unbonded ones. Therefore, it is reasonable to assume that when conditions are such that atoms can form bonds and thereby achieve structures that have lower energy, they do so.

To simplify and expedite our study of bonding, we shall classify chemical bonds in terms of the behavior and probable location of the bonding (outer energy level) electrons of the atoms involved. Such a classification will enable us to explain a large number of observations and predict the behavior of many substances in terms of a relatively few general principles. Clues to the type of bond that exists between the component particles of a substance are obtained by studying the properties of the substances and the structural characteristics of the particles.

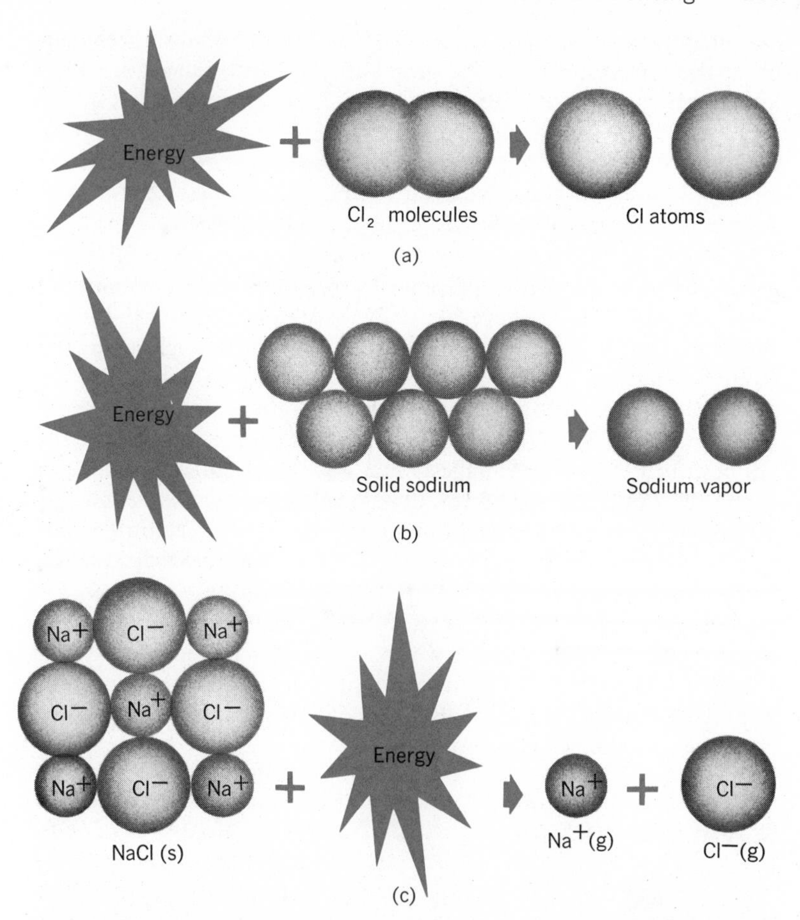

8-1 Energy is required to overcome the force of attraction (chemical bond) which binds atoms together in (a) molecules, (b) metallic crystals, and (c) ionic crystals. The separated atoms are, therefore, in a higher energy state than the bound atoms. In (a) the force of attraction is called a covalent bond, in (b) a metallic bond, and in (c) an ionic bond.

Examples:

(a) The dissociation of a chlorine molecule into chlorine atoms is an endothermic process which absorbs 2.34 ev of energy (58 kcal per mole of chlorine).

(b) The conversion of one mole of sodium atoms from the solid phase into the gaseous phase in which the atoms are separated requires 26 kcal of energy.

(c) Energy measuring 184 kcal is required to convert one mole of solid sodium chloride into individual gaseous sodium and chloride ions.

One of the most revealing properties of a substance is its electric conductivity in the solid and liquid phases. You discovered earlier that a large number of substances classified as ionic salts conduct an electric current in the liquid phase but not in the solid phase. This evidence coupled with data from X-ray diffraction studies of ionic crystals suggests that these compounds are composed of oppositely charged ions. The model which accounts for these observations is called an **ionic-bond model.** This model explains the formation of positive and negative ions in terms of electron transfer from an atom having a low electronegativity to one having a high electronegativity. The electrostatic force of attraction between the oppositely charged ions is called an **ionic bond.** We shall develop and investigate the nature and significance of the ionic bond model in the next chapter.

It can be shown experimentally that many pure substances such as sulfur (S_8), carbon tetrachloride (CCl_4), and water (H_2O) do not conduct an appreciable electric current in either the solid or the liquid phase. This observation suggests that neither the solid nor the liquid phase of these substances contains ions. Rather, it suggests that they are composed of neutral molecules. The physical and chemical properties of these and many other substances lead us to the conclusion that the atoms in these substances are held together by nonionic bonds. In 1916, an American chemist, Gilbert N. Lewis (1875–1946), suggested that atoms in molecules might be held together by an electrostatic force of attraction between atomic nuclei and *electron-pairs* shared by the bonding atoms. Lewis called an ***electron-pair bond*** between atoms a ***covalent bond.***

It is reasonable to assume that in molecules composed of different kinds of atoms with different electronegativities, the bonding electrons will not be shared equally between the atoms. In this situation, the bonding electrons are, on the average, closer to the atom having the greater attraction for electrons (higher electronegativity). Therefore the more electronegative atom acquires a *partial negative charge* and the less electronegative acquires a *partial positive charge.* This suggests that covalent bonds in molecules composed of different kinds of atoms have a ***partial ionic character*** and that there are no sharp boundary lines which separate a covalent bond from an ionic bond. That is, as the difference in electron-attracting ability (electronegativity) between atoms increases, the degree of sharing decreases, and the ionic character of the bond increases. Thus, we may consider the ionic bond as an extension of the covalent bond in which there is minimum sharing of bonding electrons. Because of this relationship, we shall first consider the covalent-bond model.

COVALENT BONDS

8-2 Energy Changes During Bond Formation Let us first consider the covalent bond formed by the interaction of two hydrogen atoms. The starting point is to examine the ground-state electron configuration of the atoms, the atomic properties related to the behavior of the electrons, and the forces which exist between the interacting atoms.

The ground state electronic configuration of a hydrogen atom is

$$1s^1$$

From this configuration it is seen that hydrogen atoms need only one electron to fill their 1*s* orbital and to attain the configuration of helium, the noble-gas atom which follows hydrogen in the Periodic Table.

Both hydrogen atoms have an electronegativity of 2.1, and an ionization energy of 13.6 ev. When the two atoms approach each other, the relative attraction of the two nuclei for their own (and for each other's) outer energy-level electrons (bonding electrons) determines the extent to which electrons will be transferred from one atom to the other. Since two hydrogen atoms approaching each other have the same attraction for electrons (same electronega-

TABLE 8-1
FORCES BETWEEN INTERACTING ATOMS

Attractive Forces Operate Between
1. The nucleus of an atom and its own electrons
2. The nucleus of one atom and the electron clouds of the other
Repulsive Forces Operate Between
1. The negative electron clouds of two different atoms
2. The positive nuclei of two different atoms

tivity), there is essentially no tendency for an electron to be transferred from one atom to the other.

As the atoms approach each other, attractive and repulsive forces operate to change the energy of the atomic system. These forces are listed in Table 8-1.

When the nuclei of the two hydrogen atoms are 0.74 Å apart, the atomic system has minimum potential energy. At this distance the atoms are close enough for their half-filled 1*s* orbitals to *overlap.* Thus, the volume represented by the 1*s* orbitals of the two atoms is now available to the bonding electrons. Since the region between the positive nuclei represents a region of low potential energy for the negative electrons, there is a high probability of finding the bonding electrons in the internuclear region. Thus, we can say that in general, an electron from each atom shares a region of space between the two nuclei. Sharing a pair of electrons results in a filled *s* orbital and a stable, helium-like electronic configuration for each atom. The covalent bond formed by the overlap of *s*-orbital electron clouds is sometimes referred to as a ***sigma (σ) bond.*** The electron-dot and line formula for the hydrogen molecule are shown.

H:H H—H

A pair of dots or a single line may be used to represent a covalent bond.

The graph in Fig. 8-5 shows how the potential energy varies as two hydrogen atoms approach each other. As a reference for comparision, the potential energy of the separate atoms is assumed to be zero. The graph shows that the potential energy of the combined atoms is lower than that of the separated atoms (the potential energy has a negative value). As the atoms approach each other, attractive forces operate and increase their velocity and kinetic energy. As the kinetic energy increases, the electric potential energy decreases. At the intermolecular distance represented by point B, the attractive force between the atoms is at a maximum, and the system has minimum potential energy. At internuclear distances less than that represented by point B, repulsive forces predominate, kinetic energy decreases, and potential energy increases. Finally, the atoms momentarily stop and then begin to move apart. Thus, the distance between the nuclei varies as the atoms in the molecule vibrate about the equilibrium bond length. In other words, the point on the curve that represents the internuclear distance moves back and forth about the equilibrium point.

Helium does not form He_2 molecules. The 1*s* orbitals of separate helium atoms already contain 2 electrons. When two helium atoms approach each other, repulsive forces exceed the attractive forces and the potential energy of the atomic system increases. An increase in potential energy is associated with a decrease in stability. Therefore, He_2 represents an unstable system. It is of interest to note that another more complicated bonding theory provides a method for predicting the relative stability of molecules.

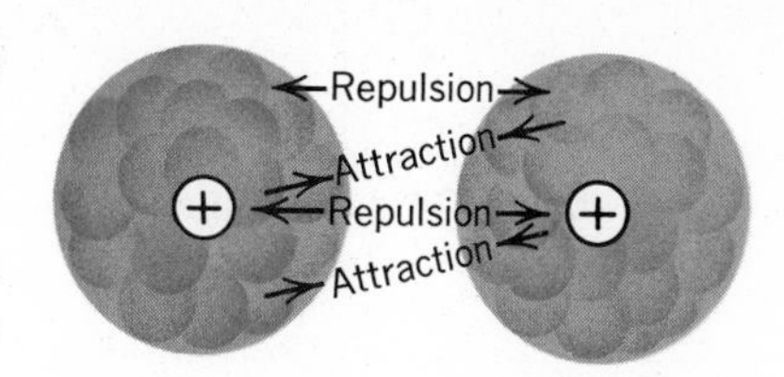

8-2 As two atoms approach each other, attractive and repulsive forces begin to operate. If the attractive forces predominate before the equilibrium bond distance is reached, as in the case of hydrogen, a bond is formed. If the repulsive forces predominate, as in the case of helium, no bond forms.

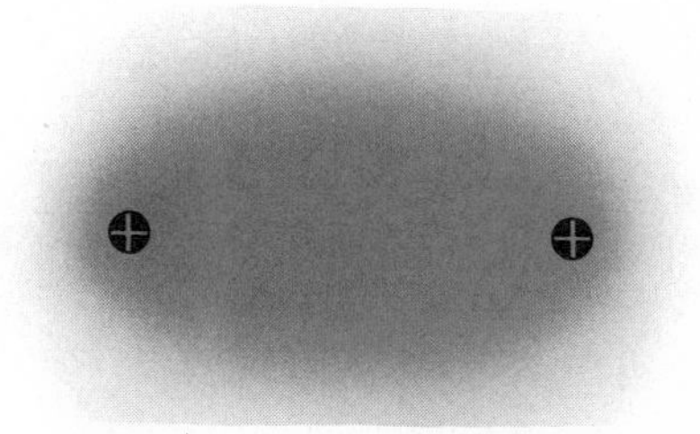

8-3 An electronic charge distribution representation of a hydrogen molecule. It can be seen that maximum electron density occurs in the internuclear region of the molecule.

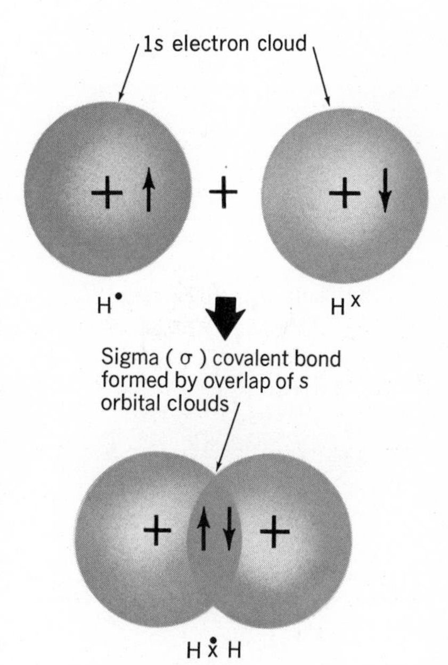

8-4 The covalent bond within an H_2 molecule is formed when the s orbitals of the two H atoms overlap and form a single electron cloud which is attracted to both nuclei. The arrows indicate that the paired electrons have opposite spins.

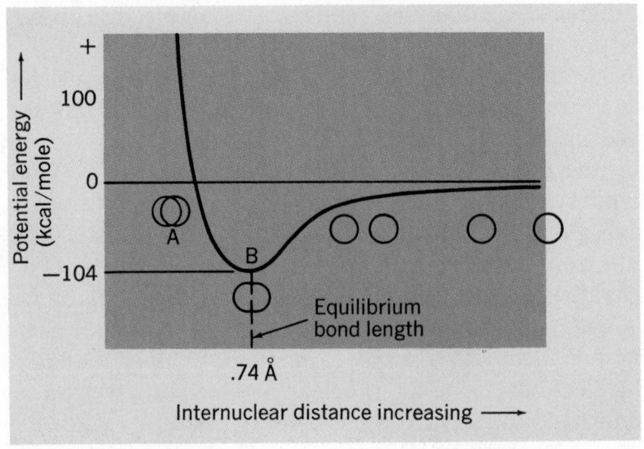

8-5 Graphical representation of potential energy changes which occur when two hydrogen atoms approach each other.

8-3 Nonpolar Bonds Diatomic molecules such as H_2, Cl_2, and N_2 are composed of identical atoms having the same attraction for electrons. Thus, we would expect the electronic charge distribution in these molecules to be symmetrical about the two nuclei. Experiments (Fig. 8-44) designed to measure the electric symmetry of molecules verify this expectation. Covalent bonds in which the bonding electrons are equally shared and symmetrically distributed are called ***nonpolar bonds.***

8-4 Polar Covalent Bonds: Partial Ionic Character Electric measurements made on diatomic molecules such as HF and HCl show that these molecules have an *electric disymmetry.* This suggests that the bonding electrons are unequally shared by the two atoms and unsymmetrically distributed in the molecule. Covalent bonds in which there is an unsymmetrical distribution of electrons are called ***polar bonds.*** Molecules in which there is an unsymmetrical distribution of electronic charge are called ***polar molecules.*** As a result of the unequal sharing, one atom acquires a partial positive charge symbolized by the Greek letter delta (δ^+), and the other acquires a partial negative charge (δ^-) equal in magnitude to the partial positive charge. Thus, diatomic polar molecules are neutral but have positive and negative regions called *electric poles.* This means that **all polar covalent bonds have a partial ionic character.** Let us examine hydrogen chloride (HCl) as an example of a diatomic polar molecule.

The electronegativities and ground state electronic configuration of hydrogen and chlorine atoms are given below.

Ground state configuration	Electronegativity
H: $1s^1$	2.1
Cl: $1s^22s^22p^63s^23p_x^23p_y^23p_z^1$	3.0

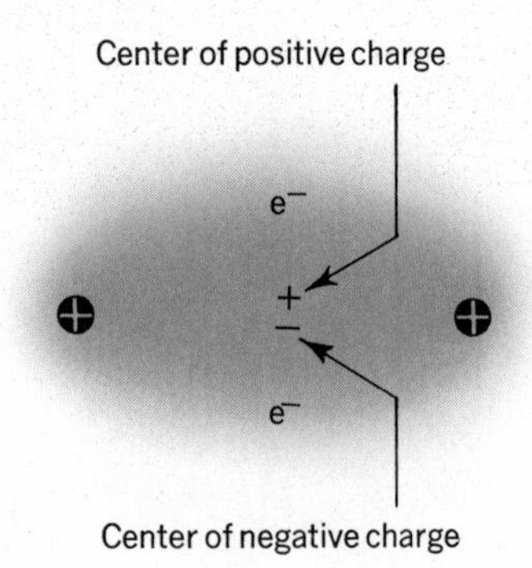

8-6 A nonpolar covalent bond results when two nuclei with the same number of protons simultaneously attract the electron cloud between them. In these molecules composed of identical atoms, the center of positive charge coincides with the center of negative charge.

The overlap of the half-filled $1s$ orbital of hydrogen and the half-filled $3p$ orbital of chlorine results in the formation of a covalent bond when the two atoms share the electron-pair in their internuclear region (Fig. 8-7). The covalent bond formed by the

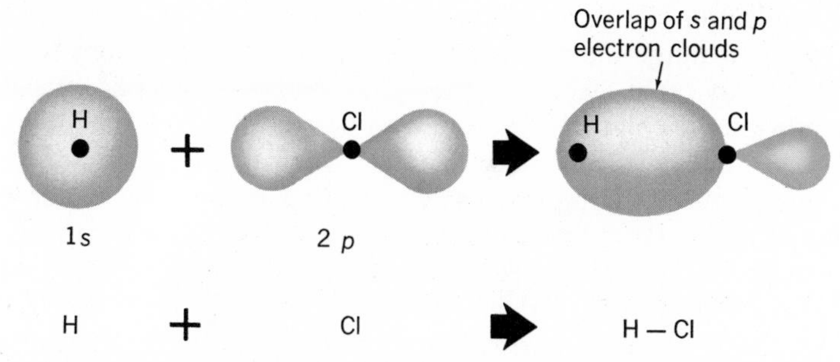

8-7 When a singly occupied s orbital from a hydrogen atom overlaps a singly occupied p orbital from a chlorine atom, a covalent bond is formed. Like s-s overlap, s-p overlap also forms a sigma (σ) covalent bond.

overlap of *s* and *p* electron clouds is also referred to as a *sigma bond.* Since chlorine has a much greater electronegativity than hydrogen, the electron charge density is greater near the chlorine part of the molecule. The result is that the chlorine end of the molecule acquires a partial negative charge (δ^-) while the hydrogen end acquires a partial positive charge (δ^+) (Fig. 8-8). Thus, hydrogen chloride (HCl) is a polar molecule. Electronegativity differences are sometimes used to denote the ***partial ionic character*** of a bond. On Pauling's scale, a difference of 1.7 is used to denote 50 percent ionic character. A larger difference suggests greater ionic character. Equations are available for calculating the percentage of ionic character in a bond. Calculations made for a few selected substances are tabulated in Table 8-2.

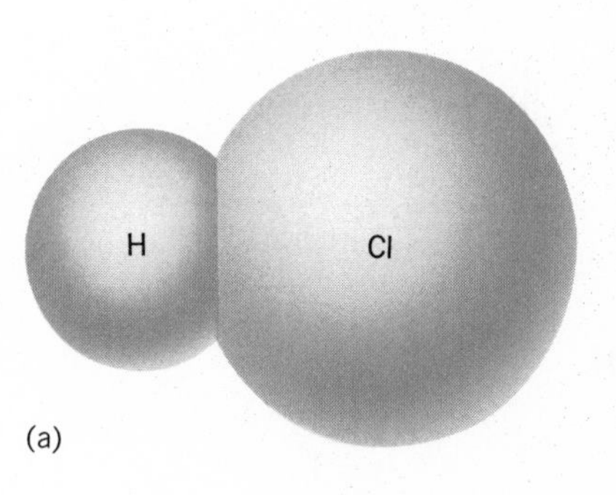

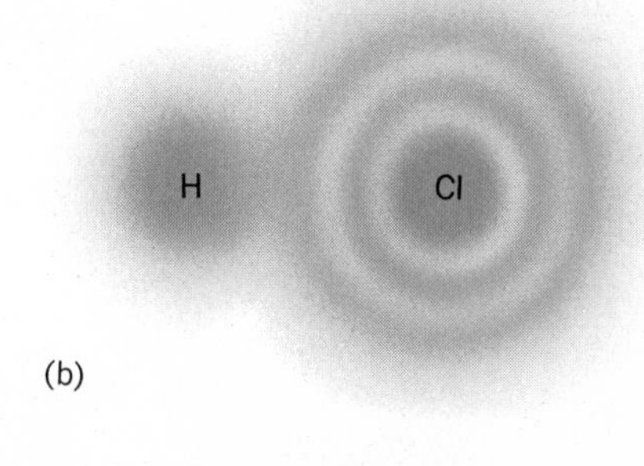

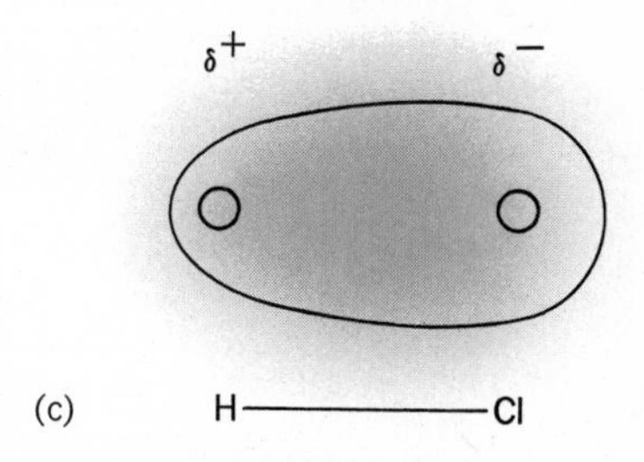

8-8 Three representations of an HCl molecule. (a) A space-filling model. (b) The greater electronegativity of the Cl atom is illustrated by the greater electron density around the Cl atom. (c) In polar molecules the center of positive charge does not coincide with the center of negative charge.

TABLE 8-2
PARTIAL IONIC CHARACTER OF SELECTED SUBSTANCES

Name	*Formula*	*Electro-negativity Difference*	*Partial Ionic Character*
Cesium fluoride	CsF	3.3	94%
Sodium chloride	NaCl	2.1	67%
Hydrogen chloride	HCl	0.9	19%
Hydrogen bromide	HBr	0.7	12%
Hydrogen iodide	HI	0.4	4%
Bromine chloride	BrCl	0.2	1%
Hydrogen	H_2	0.0	0%

ELECTRON-DOT (LEWIS) STRUCTURES

8-5 The Octet Rule and Structures It is observed that many stable molecules and polyatomic ions are composed of covalently bonded atoms having a noble-gas electron configuration. This configuration is sometimes referred to as an ***octet structure*** and the atoms that achieve this structure are said to follow the ***octet rule.*** The rule is frequently used when constructing electron-dot formulas (Lewis structures) to show the electron configuration and bonding of atoms in molecules. To construct an electron-dot formula for a molecule, we determine the total number of available outer electrons in the atoms composing the molecule and then, if possible, arrange them about the atoms so that each atom is surrounded by eight electrons. Shared electrons are common to two atoms and are counted as part of the outer-level population of each atom. The number of outer level (bonding or valence) electrons in the atoms of the representative elements is given by the group number in the Periodic Table. To construct the electron-dot formula for a chlorine molecule (Cl_2), we note that its atoms are in Group VIIA (contains seven valence electrons). It is apparent that each atom needs one electron to achieve an octet structure. By sharing a pair of electrons, each atom completes its outer octet. The electron-dot and line formulas for Cl_2 are given below.

$$:\ddot{\underset{\cdot\cdot}{Cl}}\overset{\times}{\underset{\times}{:}}\overset{\times\times}{\underset{\times\times}{Cl}}\overset{\times}{\underset{\times}{}} \qquad |\overline{\underline{Cl}}—\overline{\underline{Cl}}|$$

FOLLOW-UP PROBLEM

Write the electron-dot and line formulas for (a) F_2 and (b) OF_2.

8-6 Multiple Bonds In order to achieve the octet structure, it is necessary for some atoms to share two or even three pairs of electrons. Sharing two pairs of electrons produces a ***double bond;*** sharing three pairs of electrons produces a ***triple bond.*** The bonding in carbon dioxide (CO_2) illustrates the point. Experimental data to be discussed later suggest that the molecule is symmetrical and that two oxygen atoms are bonded to one carbon atom. From the location of the atoms in the Periodic Table, we can determine that the carbon atom (Group IVA) has four electrons available for bonding and each oxygen atom (Group VIA) has six electrons in its outer level. This makes a total of 16 electrons to be arranged so that, if possible, each atom has an outer octet. We may arrange the two oxygen atoms around the carbon atom as follows:

$$\overset{\times}{\underset{\times}{}}\underset{\times\times}{O}\overset{\times}{\underset{\times}{}} \qquad :C: \qquad \overset{\times}{\underset{\times}{}}\underset{\times\times}{O}\overset{\times}{\underset{\times}{}}$$

oxygen **carbon** **oxygen**

In order to obtain an octet structure, the carbon atom must share its 4 electrons with 4 from the two oxygen atoms. By obtaining a share in two of carbon's electrons each oxygen atom will complete its outer octet. The electron arrangement which involves two double bonds is represented by an electron-dot formula as

$$\overset{\times}{\underset{\times}{}}\underset{\times\times}{O}\overset{\times}{\underset{\times}{}}:C:\overset{\times}{\underset{\times}{}}\underset{\times\times}{O}\overset{\times}{\underset{\times}{}}$$

and by a line formula as

$$|\underline{O}=C=\underline{O}|$$

It is reasonable to assume that there is a greater electron density between the nuclei and hence, a greater attractive force between the nuclei and the shared electrons with four electrons than with two. Experimental data verify that greater energy is required to break double bonds than single bonds. Hence, *double bonds are stronger than single bonds.* Stronger bonds tend to pull atoms closer together; therefore, atoms joined by double bonds are closer together than those joined by single bonds. It should be noted that more than one arrangement of electrons that yields an octet structure can be drawn for CO_2. This phenomenon is discussed later in Section 8-8.

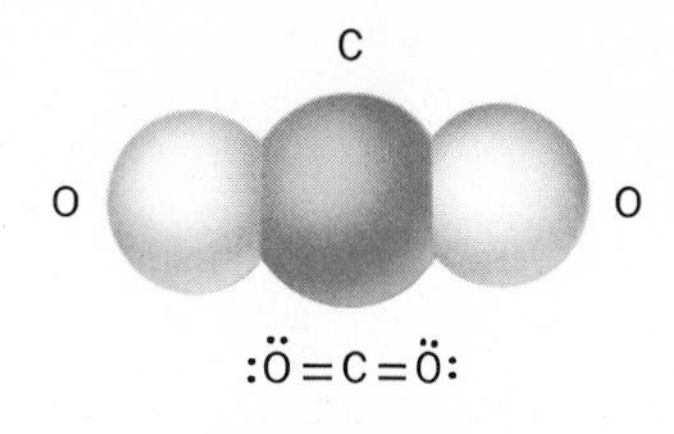

8-9 Space-filling model of carbon dioxide (CO_2).

FOLLOW-UP PROBLEM

The large amount of energy that is required to decompose a nitrogen molecule (N_2) into nitrogen atoms suggests the presence of a triple covalent bond. (a) Write the electron-dot and line formulas for N_2. (b) Are triple bonds stronger or weaker than double bonds? Explain. (c) Are triple bonds longer or shorter than double bonds? Explain.

The oxidation state of an atom in a simple binary molecule can sometimes be used with other information to help write the electron-dot formula for a molecule. Consider acetylene (C_2H_2). Experimental data show that the carbon atoms are bonded together and that one hydrogen atom is bonded to each carbon atom. In this molecule, the oxidation state of carbon is 1−. This suggests that each carbon atom shares one electron with a hydrogen atom. In common with other elements in Group IVA, each carbon atom has four valence electrons. This means that each carbon atom shares its other three electrons with the other carbon atom. In this way, each carbon atom achieves an outer octet structure and each hydrogen atom achieves a helium-like structure. The electron-dot and line formulas for acetylene are

Measurements show that C_2H_2, CH_4, C_2H_6, and C_2H_4 are nonpolar. You may assume that all valence electrons of carbon in these molecules are used in bond formation. Additional data related to these molecules are provided in the Table on p. 290.

$$H:C\vdots\vdots C:H \qquad H{-}C{\equiv}C{-}H$$

FOLLOW-UP PROBLEM

Write the electron-dot and line formulas for methane (CH_4), ethane (C_2H_6), and ethylene (C_2H_4).

8-7 Polyatomic Ions: Coordinate Bonds In an earlier chapter we noted that a large number of ionic substances contain polyatomic ions. Since there is no easy way to predict the formulas of these ions, we listed them and suggested that you memorize a number of them. We are now able to rationalize their formulas in terms of bonding concepts and electron-dot formulas. Let us first investigate the bonding in an ammonium ion (NH_4^+). We shall find that the bonding is characteristic of that in many polyatomic ions.

We can experimentally demonstrate the formation of the ammonium ion with a simple reaction involving ammonia gas (NH_3) and hydrogen chloride gas (HCl). When a bottle of hydrogen chloride is inverted over a bottle of ammonia and the glass cover plates are removed, a white smoke (powder) is instantly observed. The reaction is exothermic. When the white powder is analyzed it is found to be composed of two ions whose formulas are NH_4^+ and Cl^-. To explain the formation of ammonium ions, we start with the electron-dot formula of an ammonia molecule

$$\begin{matrix} & \mathrm{H} & \\ & \bullet\times & \\ \mathrm{H}\,\overset{\bullet}{\times} & \mathrm{N} & : \\ & \bullet\times & \\ & \mathrm{H} & \end{matrix}$$

It may be seen that NH_3 has an unshared pair of valence electrons which are available for bond formation. The attractive force operating between the nucleus of the hydrogen atom in a HCl molecule and the unshared electrons of a nitrogen atom in an NH_3 molecule lowers the energy and increases the stability of the system.

When the hydrogen atom in the HCl molecule breaks away from a chlorine atom, the hydrogen electron remains with the more electronegative chlorine. Thus, the hydrogen atom becomes a positively charged hydrogen ion, (H^+) and the chlorine, left with the extra electron, becomes a chloride ion, (Cl^-). An ammonium ion is formed when the *vacant s* orbital of the hydrogen ion overlaps the orbital of the nitrogen atom which contains the *unshared electron-pair.* A covalent bond in which one atom donates both bonding electrons is called a ***coordinate covalent bond*** (or simply a ***coordinate bond***). The formation of such a bond requires that one atom have an orbital containing an unshared pair of electrons and the other atom must have an empty orbital available for bonding. The atom furnishing the pair of electrons is called an ***electron-pair donor;*** the atom furnishing the empty orbital is called an ***electron-pair acceptor.*** The above reaction may be represented by the following electron-dot equation.

$$\begin{matrix} & \mathrm{H} & \\ & \bullet\times & \\ \mathrm{H}\overset{\times}{\bullet} & \mathrm{N} & : \\ & \times\bullet & \\ & \mathrm{H} & \end{matrix} + \mathrm{H}\overset{\bullet}{\times}\underset{\bullet\bullet}{\overset{\bullet\bullet}{\mathrm{Cl}}}: \longrightarrow \left[\begin{matrix} & \mathrm{H} & \\ & \bullet\times & \\ \mathrm{H}\times & \mathrm{N} & :\mathrm{H} \\ & \times\bullet & \\ & \mathrm{H} & \end{matrix}\right]^{+} + \overset{\bullet}{\times}\underset{\bullet\bullet}{\overset{\bullet\bullet}{\mathrm{Cl}}}:^{-}$$

The addition of a positively charged hydrogen ion to a neutral ammonia molecule yields a positively charged ammonium ion. In this reaction, NH_3 is an *electron-pair donor* and H^+ is an *electron-pair acceptor.* Since H^+ is a bare proton, we may also describe NH_3 as a ***proton acceptor*** and HCl as a ***proton donor.***

The coordinate covalent bond may be represented in a line formula by an arrow which points from the donor to the acceptor. The line formula for an ammonium ion is

$$\left[\begin{matrix} & \mathrm{H} & \\ & | & \\ \mathrm{H}- & \mathrm{N} & \rightarrow\mathrm{H} \\ & | & \\ & \mathrm{H} & \end{matrix}\right]^{+}$$

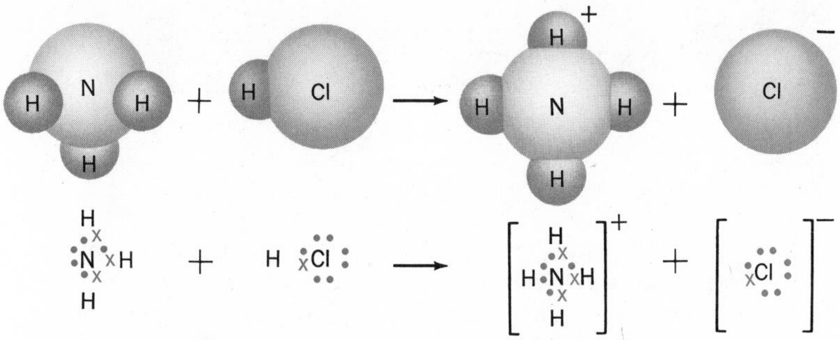

8-10 Pictorial representation of the reaction between gaseous HCl and gaseous NH_3, forming solid ammonium chloride (NH_4Cl).

The arrow does not imply that one of the covalent bonds in an ammonium ion is different from the others. Experimental data show that all four bonds are equivalent.

A general technique for writing the electron-dot or line formulas for many binary species is illustrated in the following examples. We shall assume that each atom except hydrogen must be surrounded by an octet of electrons.

Example 8-1
Write the electron-dot formula for a sulfite ion, SO_3^{2-}.

Solution
1. Count up the available valence electrons plus the electrons giving the ion its charge. Because both sulfur and oxygen are in Group VIA, there are $6 \times 4 = 24$ valence electrons plus two more for the 2− charge.
2. Arrange the 26 dots around the atoms so that, if possible, the octet rule is followed. Note that we shall later encounter many cases where this rule is not followed by the central atom. We shall find, however, that the rule can be invoked for all atoms in the second row of the Periodic Table.
3. The formula (SO_3^{2-}) implies that three oxygen atoms are bonded to one sulfur atom. Therefore, write the formula for a sulfur atom and surround it with the symbols for three oxygen atoms.

O O
S
O

4. Draw four lines (which represent eight electrons) from the central sulfur atom to the surrounding oxygen atoms. Recall that a single line represents two electrons common to two atoms. Since a sulfur atom must have eight electrons around it, we can show a single bond between the sulfur atom and each of two oxygen atoms, and a double bond between the sulfur and third oxygen atom.

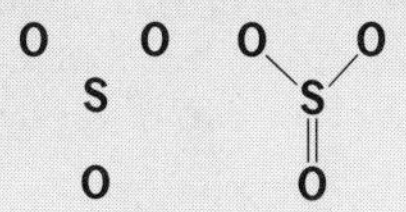

O O O O
S S
O O

5. Complete the outer octet of the oxygen atoms by using dots or lines

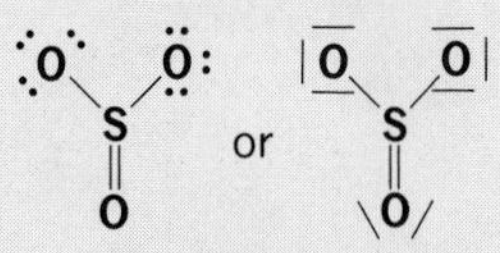

6. Count the total number of electrons in the tentative formula to see if this number agrees with the available electrons. Our electron-dot formula above shows a total of 24 electrons. The number available to be distributed in a sulfite ion, (SO_3^{2-}) is 26. To make room for the extra electrons we can shift the pair which represents one-half the double bond on to the oxygen atom. This converts the double bond to a single bond and leaves the sulfur atom with 6 electrons.

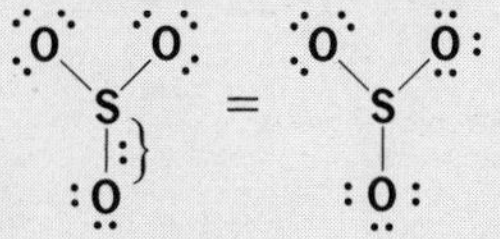

The other two electrons can then be placed as a lone pair on the sulfur atom

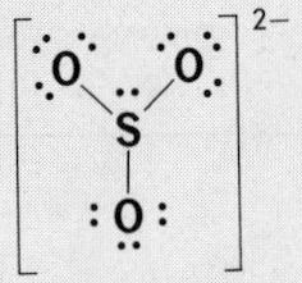

Note that the electron-dot formulas for ions should be bracketed and the charge shown on the outside of the bracket. The formula shows that three identical atoms are bonded to a central atom which still has an unbonded electron-pair.

Example 8-2
Write the electron-dot formula for sulfur trioxide (SO_3).

Solution
The steps are the same as in Example 8-1. In sulfur trioxide there are exactly 24 outer electrons to be distributed. This suggests that there is a double bond in the molecule and that its formula is

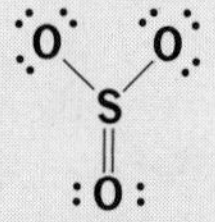

There are, however, two other ways to draw the electron-dot formula for SO_3. We could have placed the double bond between either of the other two oxygen atoms. The resulting formulas would be

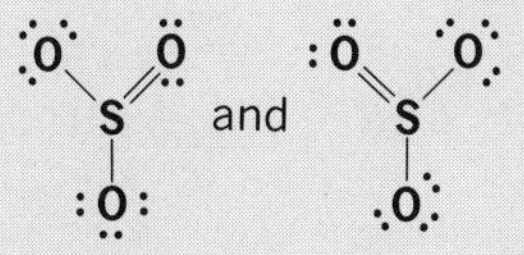

These three equivalent electron-dot formulas for SO_3 are known as **resonance structures.**

8-8 Resonance Structures Strangely enough, none of the three configurations shown above agree with experimental data which show that the three sulfur-oxygen bonds are of the same length and of the same strength. The above structures show a double bond between a sulfur atom and an oxygen atom. This would indicate that one oxygen atom is closer and more tightly bound to the sulfur atom than the other two singly-bonded oxygen atoms. The experimentally determined bond distances and strengths fall between those of single and double bonds. Therefore, the bonding must be intermediate between the electron-dot structures shown above. It is apparent that the SO_3 molecule cannot be accurately represented by a single electron-dot formula. The concept invoked to explain such a situation is known as ***resonance,*** and the three diagrams shown above are called ***resonance structures.*** These structures do not actually exist; they are imaginary. It is not implied that the SO_3 molecule resonates between these three structures. Sulfur trioxide is a real substance with a definite electronic structure. We may think of the real structure as a *hybrid* of the three imaginary electron-dot structures. It is conventional to show resonance structures joined by a double-headed arrow.

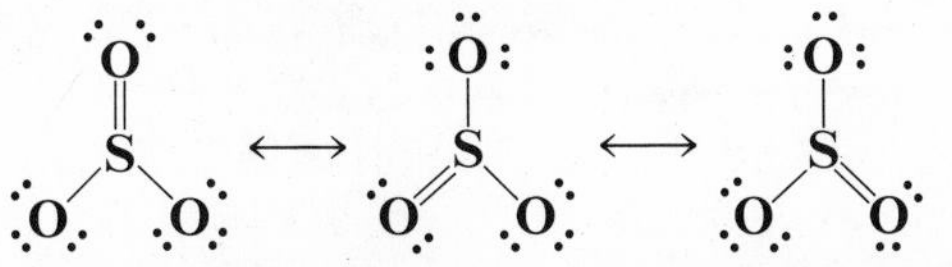

Sulfur trioxide is representative of a large number of substances for which no single, correct electron-dot formula can be written. These substances are classified as having structures which are ***resonance hybrids*** of the two or more structures which can be drawn. Substances for which resonance structures may be written are more energetically stable than would be predicted on the basis of any one of the individual structures. The increase in stability resulting from resonance is important in explaining bond strengths, heats of reaction, and the rates of chemical reactions.

You will no doubt encounter a number of stable species for which no satisfactory electron-dot structures, resonance or nonresonance, can be written. Conversely, you may be able to write Lewis structures for species that cannot exist. In either case, the problem can be traced to deficiencies in the Lewis scheme. Often application of other bonding theories will solve the problem.

FOLLOW-UP PROBLEM

Write electron-dot formulas for the species listed below. Assume that octet rule holds. Indicate which formulas are resonance structures. (a) carbon tetrachloride (CCl_4), (b) phosphine (PH_3), (c) hydroxide ion (OH^-), (d) phosphate ion (PO_4^{3-}), (e) nitrate ion (NO_3^-), (f) sulfur dioxide (SO_2), (g) carbonate ion (CO_3^{2-}), (h) ozone (O_3), (i) carbon monoxide (CO), (j) sulfate ion (SO_4^{2-})

8-9 Species Which Do Not Follow the Octet Rule Examination of the outer-electron configuration of the atoms in certain molecules show that not all atoms attain an octet structure. These include the species listed below.

1. Molecules in which more than four atoms are bonded to a central atom. An example is phosphorus pentachloride (PCl_5). In this molecule, a phosphorus atom is covalently bonded to five chlorine atoms. Since each bond involves two electrons, there are ten electrons in the outer energy level of the phosphorus atom. The Lewis structure is

```
        :Cl:
   :Cl   |   Cl:
       \ P /
       /   \
   :Cl       Cl:
```

2. Molecules containing an odd number of bonding electrons. An example of an *odd-electron* molecule is nitric oxide (NO). This molecule is a resonance hybrid whose resonance structures may be written as

$$:\ddot{N} = \ddot{O}\cdot \longleftrightarrow \cdot\ddot{N} = \ddot{O}:$$

3. Species that contain no multiple bonds (as shown by experiment) and whose central atom has fewer than four bonding electrons. An example is boron trifluoride (BF_3). Boron has three bonding electrons. Thus, in BF_3 there are only six electrons in the outer level of the boron atom as shown by its Lewis structure.

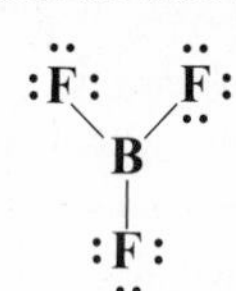

Application of the octet rule to BF_3 yields resonance structures involving a double bond. One structure is

```
:F:   F:
   \ //
    B
    |
   :F:
```

FOLLOW-UP PROBLEM

Write Lewis structures for the following species whose formulas do not follow the octet rule: (a) sulfur hexafluoride (SF_6), (b) nitrogen dioxide (NO_2), (c) beryllium fluoride (BeF_2).

A knowledge of the shapes and sizes of molecules and polyatomic ions will help us explain the observed properties of substances. The shapes and sizes of molecules are related to the geometric arrangement of their component atoms. Let us now see how our models

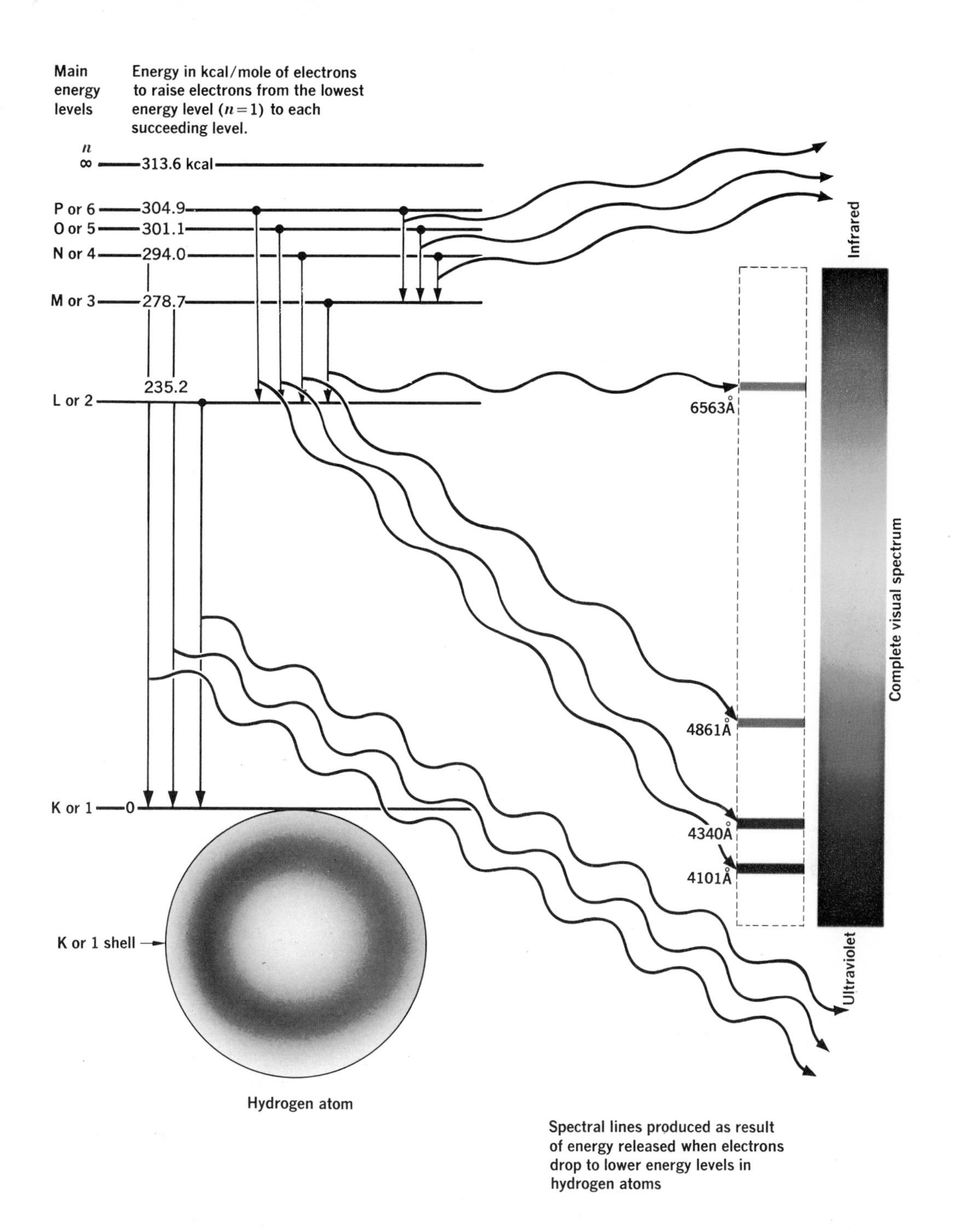

Spectral lines produced as result of energy released when electrons drop to lower energy levels in hydrogen atoms

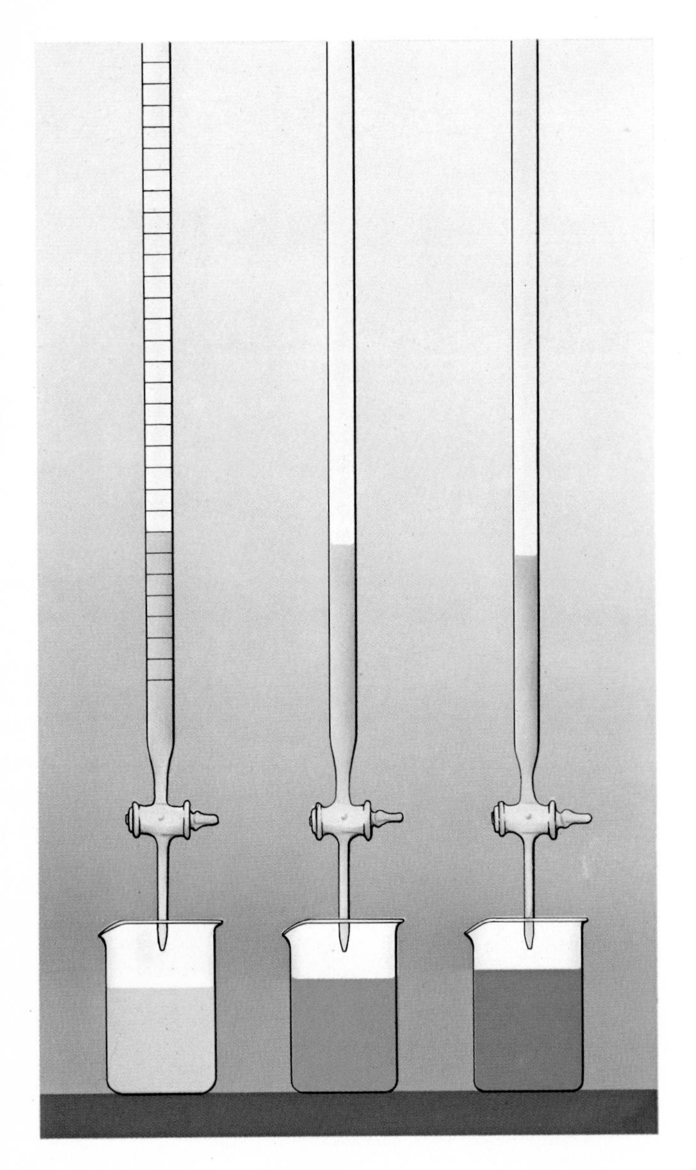

The titration of 50.00 ml of 0.100-*M* HCl with 0.100-*M* NaOH using bromthymol blue indicator. Prior to the endpoint, the indicator is yellow (left beaker). A sudden change in the color of the solution from yellow to green marks the endpoint of the titration (center beaker). One drop past the endpoint the color changes to blue (right beaker). The pH at various stages in the titration may be calculated or determined with a pH meter. The calculated titration data and titration curve for this reaction are shown to the right.

Changes in pH during the titration of 50.00 ml of 0.100-*M* HCl with 0.100-*M* NaOH

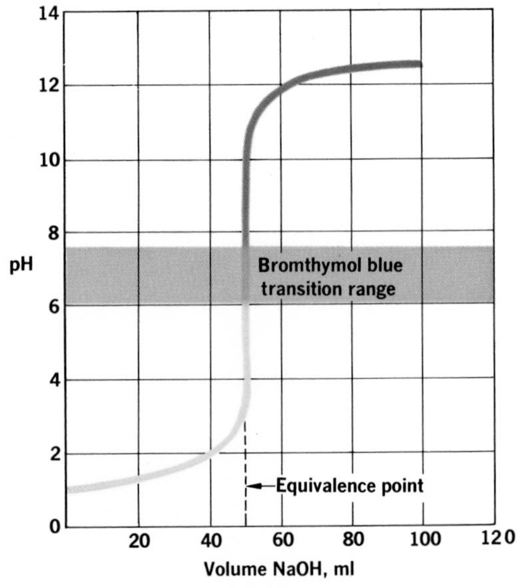

Volume NaOH added (ml)	pH
100.00	12.52
75.00	12.30
60.00	11.96
51.00	11.00
50.10	10.00
50.01	9.00
50.00	7.00
49.99	5.00
49.90	4.00
49.00	3.00
40.00	1.95
25.00	1.48
10.00	1.18
00.00	1.00

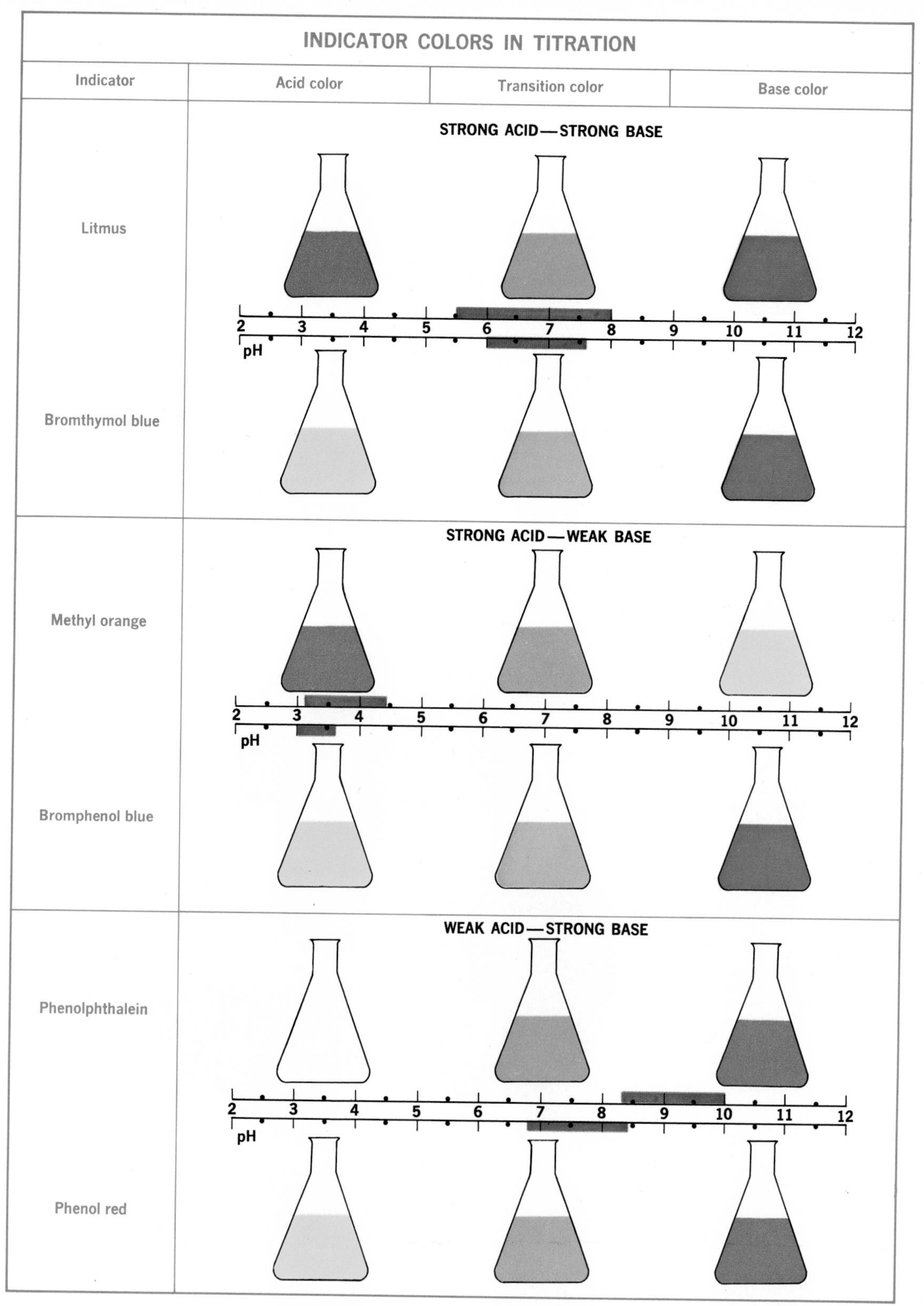

Color plate III

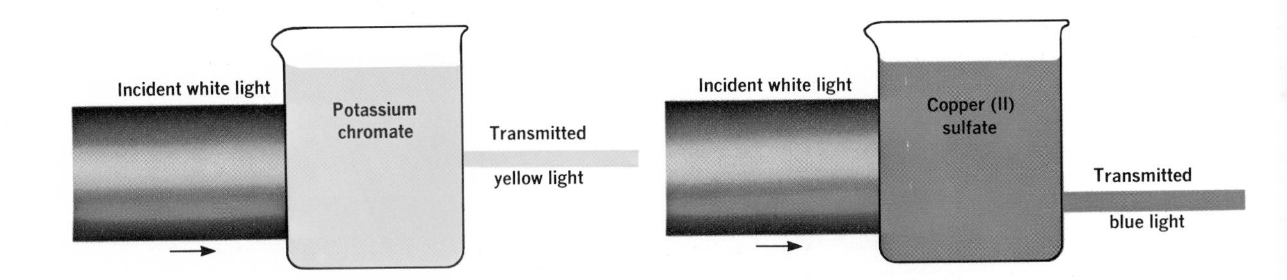

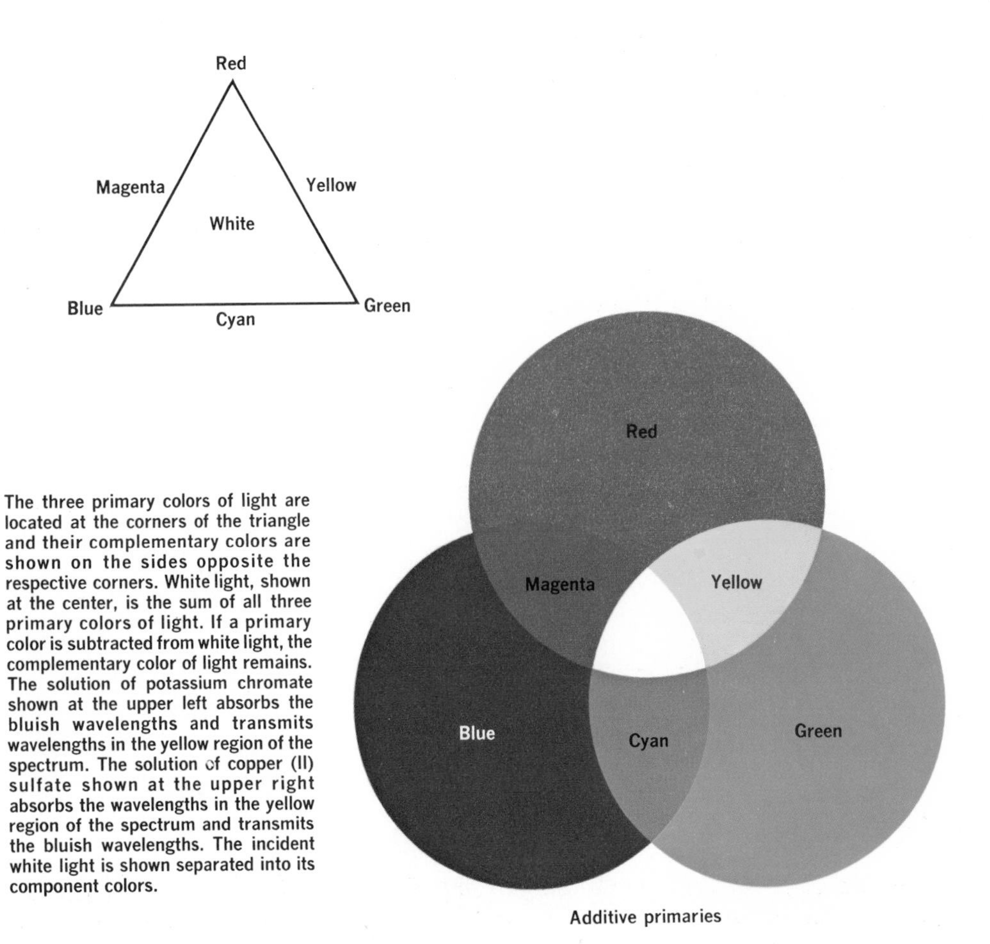

The three primary colors of light are located at the corners of the triangle and their complementary colors are shown on the sides opposite the respective corners. White light, shown at the center, is the sum of all three primary colors of light. If a primary color is subtracted from white light, the complementary color of light remains. The solution of potassium chromate shown at the upper left absorbs the bluish wavelengths and transmits wavelengths in the yellow region of the spectrum. The solution of copper (II) sulfate shown at the upper right absorbs the wavelengths in the yellow region of the spectrum and transmits the bluish wavelengths. The incident white light is shown separated into its component colors.

Additive primaries

Color plate IV

of electronic configuration and chemical bonding can be used to help predict the geometric arrangement of atoms in molecules and polyatomic ions.

SHAPES OF COVALENT MOLECULES AND POLYATOMIC IONS

8-10 Bond Angles and Bond Lengths The arrangement of atoms in a molecule or ion is described in terms of bond distances and bond angles. The average distance between the nuclei of two vibrating covalently bonded atoms defines the length of a covalent bond. These distances can be determined by the detailed analysis of molecular spectra. For example, spectroscopic data show that the average equilibrium distance (covalent bond length) between the nuclei in a hydrogen molecule (H_2) is 0.74 Å (Fig. 8-11). One-half of the internuclear distance represents the *covalent radius* of a hydrogen atom.

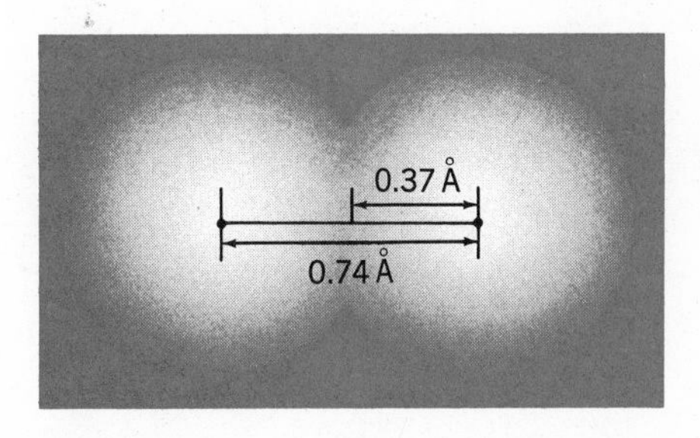

8-11 A cross-section of a hydrogen molecule (H_2). The covalent radius of a hydrogen atom within a hydrogen molecule is one-half the bond length (average distance between nuclei).

When a molecule is composed of just two atoms, it is not necessary to define a bond angle. When two or more atoms are bonded to a third atom, however, the bonds joining the two atoms to the third atom form an angle with each other which is called a bond angle. Bond angles are measured experimentally by X-ray diffraction or molecular spectroscopy. For example, spectroscopic data show that the vibrating hydrogen atoms in a water molecule form an average angle (H╲O╱H) of approximately 104.5° between each other (Fig. 8-12). Bond angles have been measured experimentally for many molecules. A few relevant ones are listed in Table 8-3.

TABLE 8-3
BOND ANGLES

Substance	*Formula*	*Angle Between Atoms Attached to Central Atom*
Water	H_2O	104.5°
Hydrogen sulfide	H_2S	92°
Ammonia	NH_3	107°
Methane	CH_4	109.5°
Boron trifluoride	BF_3	120°
Beryllium fluoride	BeF_2	180°

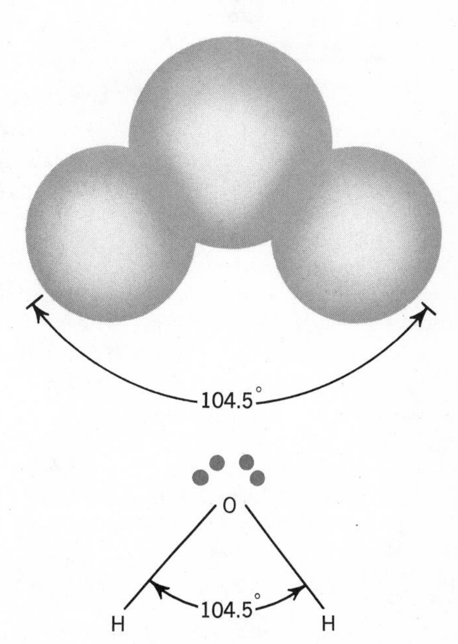

8-12 The angle formed by the imaginary lines joining the nuclei of the hydrogen atoms to the nucleus of the oxygen atom in a water molecule has been measured spectroscopically and found to be approximately 104.5°. Thus, H_2O is said to be an angular or bent molecule.

The concept of pure atomic orbitals (*s*, *p*, *d*, *f*) developed in Chapter 6 helped us to construct energy-level diagrams and furnished a basis for the concept of electron-pair bonds. Furthermore, it should help us to explain and predict the bond angles and shapes of covalent molecules. Consider a water molecule (H_2O) in which two hydrogen atoms are covalently bonded to an oxygen atom.

The shape of the molecule depends on the angle defined by imaginary lines joining between the nuclei of the two hydrogen atoms to the nucleus of the oxygen atom. We should be able to explain the experimentally observed 104.5° bond angle in terms of the directional characteristics of atomic orbitals. Let us first examine the electronic configuration of an oxygen atom in order to determine which orbitals are available for bond formation.

The ground state configuration for the central oxygen atom is

energy

2p (↑↓) (↑) (↑)

2s (↑↓)

1s (↑↓)

On the basis of this configuration we would predict that oxygen could form two covalent bonds by sharing two pairs of electrons with two hydrogen atoms. Each pair would consist of one *p* electron from oxygen and one *s* electron from hydrogen. This would complete the *s* orbitals of the hydrogen atoms and the *p* orbitals of the oxygen atom. The formation of the two covalent bonds requires that the *s* orbitals of each hydrogen atom overlap the *p* orbitals of the oxygen atom. Thus, the location of the hydrogen atoms depends upon the directional characteristics of the *p* orbitals of the oxygen atom. Since pure *p* atomic orbitals are at right angles to each other, we might predict the bond angle between the two hydrogen atoms to be 90° (Fig. 8-13).

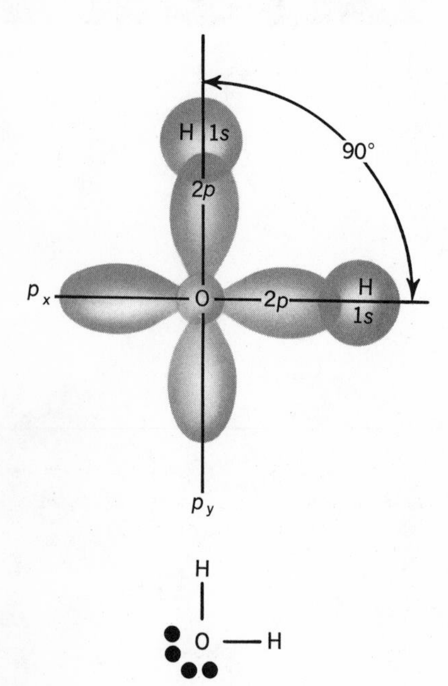

8-13 The hydrogen atoms in a water molecule would form a 90° angle if we assume that only pure *p* orbitals of oxygen are involved (p^2 bonding) and that there is no interaction between hydrogen atoms. The 1s, 2s, and $2p_z$ orbitals of oxygen are not shown.

Experiment shows the bond angle is 104.5°. This suggests that *the directional characteristics of orbitals in bonded atoms differ from those in isolated atoms.* That is, the orientation of the orbitals in a bonded atom is influenced by the presence of other atoms. We shall first explain observed bond angles in terms of the *electrostatic repulsion between valence electron charge clouds* and then later use the concept of *hybridization* to explain the same observations.

Keep in mind that the ideas expressed here are merely models to help us correlate experimental findings. The models are useful because they allow us to make predictions concerning molecular shapes which are reasonably correct in a large number of cases.

8-11 Electrostatic Repulsion of Electron-Pair Charge Clouds

Let us visualize either bonded or unbonded electron-pairs in the outer level of a bonded atom as negative charge clouds. Each negative charge cloud tends to repel all other charge clouds in the vicinity. To achieve a condition of minimum potential energy, it is necessary to locate the charge clouds so that they will be as far apart as possible. In this position the electrostatic repulsion between the clouds is reduced to a minimum. The position of the charge clouds on the central atom will determine the directional characteristics of the covalent bonds and shapes of the molecules which the atom forms.

The spatial orientation of the charge clouds depends on the number that are present and on their size. The number is equal to the total number of electron-pairs (bonded and unbonded) in the outer level of the central atom as indicated by the Lewis structure of the species. The relative size depends upon whether the electron-pair is a bonded or a lone (unbonded) pair. We would expect clouds associated with bonded electrons to be rather localized between nuclei and to take up less space than those associated with unbonded electrons.

Because this model applies best in the following systems, we shall consider at this point only those species in which a central atom is bonded, respectively, to two, three, four, or six identical atoms. Let us determine the nature of the geometric figure inscribed in an "atomic" sphere which joins, respectively, two, three, four, and six electron-pairs which are located as far apart as possible on the surface of an atomic sphere. We shall then use this information to describe the bond angles and shapes of some molecules and ions.

1. Mutual repulsion of two electron clouds force them to the opposite sides of the atomic sphere. Lines from the electron-pairs to the center of the sphere form a 180° angle with each other. These lines represent the directional characteristics of the covalent bonds. When atoms are bonded to these electron-pairs, the resulting molecule is said to be *linear* and the bond angle is 180°.

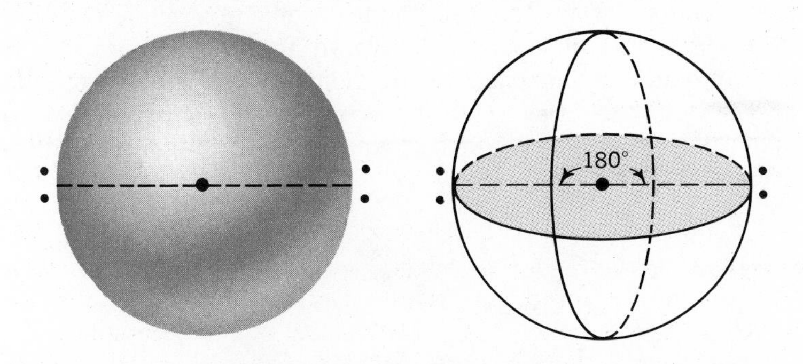

8-14 Minimum repulsion between two electron-pairs occurs when the pairs are located at the opposite ends of the diameter of a sphere.

2. Mutual repulsion of three identical electron clouds will direct them to the corners of an equilateral triangle where repulsive forces are at a minimum. Lines from the electron-pairs to the center of the sphere make an angle of 120° with each other. Atoms bonded to these electron-pairs lie in the same plane and make a bond angle of 120° with each other. Accordingly, the molecule is said to be *trigonal planar.*

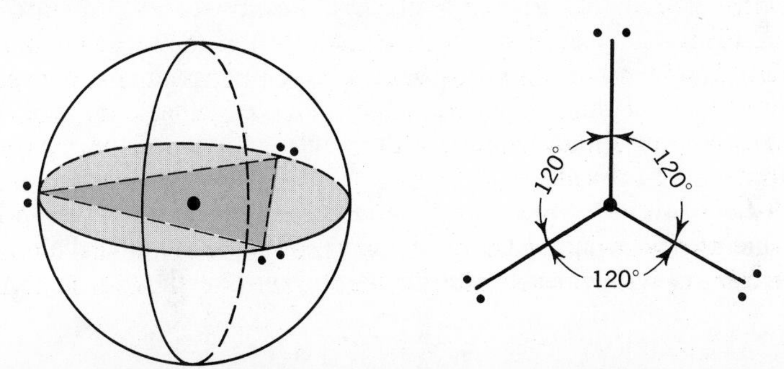

8-15 Minimum repulsion between three electron-pairs occurs when each of the pairs is located at the vertices of an equilateral triangle inscribed in a great circle of the sphere.

3. Mutual repulsion of four identical electron clouds directs them to the corners of an inscribed regular tetrahedron. A regular tetrahedron has four identical faces. Lines drawn from any two of the electron-pairs to the center of the sphere (and tetrahedron) make an angle of 109° 28′ with each other. This angle, usually rounded to 109°, is called the *tetrahedral angle.* When four identical groups are bonded to the four electron-pairs, the species is said to have a *tetrahedral shape.*

8-16 Four electron-pairs are the farthest apart on the surface of an atomic sphere at the vertices of an inscribed tetrahedron.

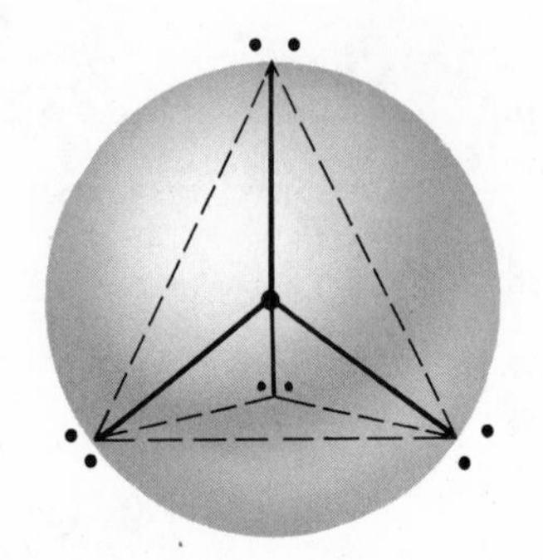

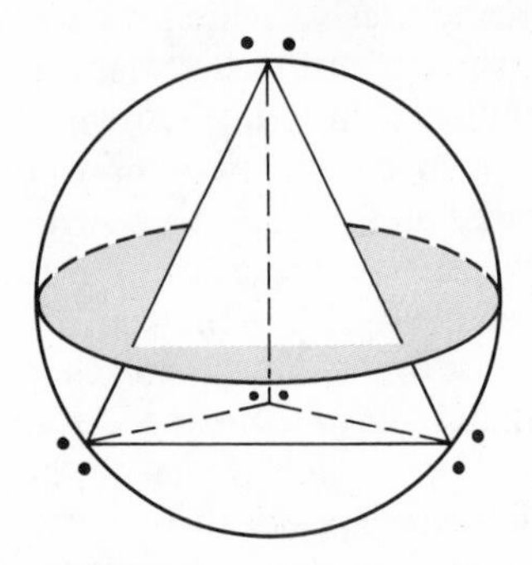

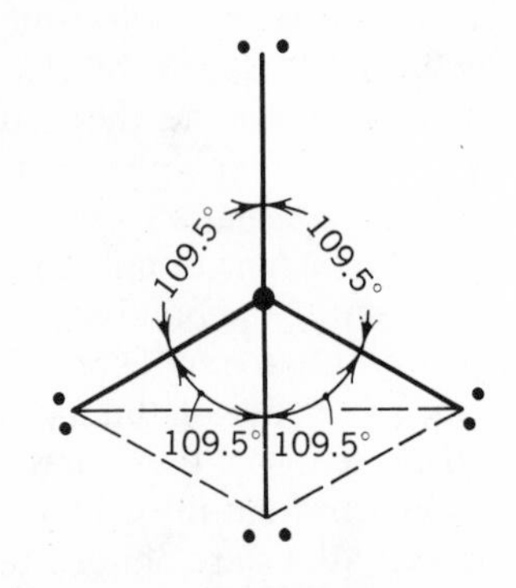

4. Mutual repulsion of six identical electron clouds directs them to the corners of an inscribed regular octahedron where they are as far apart as possible. Lines drawn from any two adjacent electron-pairs to the center of the atomic sphere make an angle of 90° with each other. Thus, molecules or ions composed of six identical groups bonded to a central atom are said to have an octahedral geometry.

8-17 Six electron-pairs are the farthest apart on surface of an atomic sphere at the vertices of an inscribed octahedron.

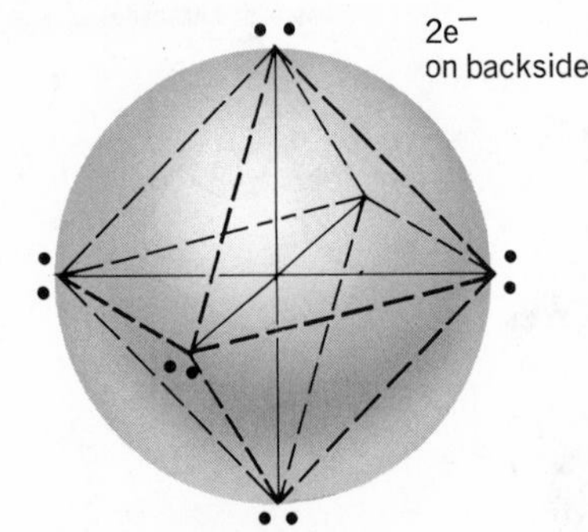

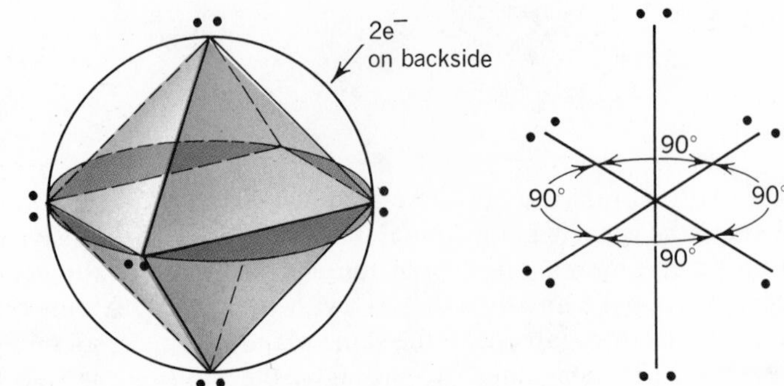

TABLE 8-4
ELECTRON-PAIR ARRANGEMENT

Pairs of Electrons on Surface of Sphere	*Geometric Shape of Inscribed Figure*
2	linear
3	trigonal planar
4	tetrahedral
6	octahedral

The electron-pair arrangements described above are summarized in Table 8-4.

Let us now use these generalities to help predict the bond angles and shapes of simple covalent molecules of binary compounds containing atoms of elements from each *A* group of the Periodic Table as the central atom.

8-12 Covalent Molecules Containing an Atom of a Group-IA Element The atoms in this group have only one bonding (valence) electron. A covalent molecule containing any of these atoms as a central atom would involve only one covalent bond. Therefore, no angle can be defined. Any diatomic molecule may be considered linear. In the next chapter we shall find that most compounds containing these elements are ionic rather than molecular.

8-13 Covalent Molecules Containing an Atom of a Group-IIA Element The atoms of these elements have only two bonding (valence) electrons and can, therefore, form a maximum of two regular covalent bonds. In beryllium fluoride (BeF_2) each of the two valence electrons in a beryllium atom is shared with an electron from a fluorine atom. Using the valence electron-pair repulsion theory applied to two electron pairs, we would predict that the angle between the fluorine atoms is 180° and that BeF_2 is a linear molecule. A *linear arrangement is characteristic of molecules in which the central atom uses all of its bonding electrons to bond to two identical atoms.* The symmetry of the BeF_2 molecule is shown in Fig. 8-18(a).

8-14 Covalent Molecules Containing an Atom of a Group-IIIA Element Boron trifluoride (BF_3) is a covalent molecule from Group IIIA of the Periodic Table. Elements in this group have three outer energy-level electrons and can form three regular covalent bonds. In a molecule such as BF_3, there are three electron-pairs in the valence level of the boron atom. Thus, all three fluorine atoms lie in the same plane at 120° from each other. The molecule is symmetrical (Fig. 8-18(b)) and said to be trigonal planar. *When all bonding electrons are used to bond three identical atoms to a fourth atom,* the molecule is trigonal planar and the bond angles are 120°.

8-15 Covalent Molecules Containing an Atom of a Group-IVA Element Methane (CH_4) is an example of a molecule in which four identical groups are bonded to a central atom. In methane, the outer level of the carbon atom contains four electron-pairs. This means the electrostatic repulsion will be at a minimum when the angle between the hydrogen atoms is 109.5°. Since the hydrogen atoms are located at the corners of a regular tetrahedron, the molecule is said to be tetrahedral. The symmetric methane molecule is shown in Fig. 8-18(c). In general, *when an atom of the representative elements uses all of its outer electrons to bond to four other atoms, the resulting molecule is tetrahedral.*

8-16 Covalent Molecules and Ions Containing an Atom of a Group-VA Element The atoms of these elements have five valence electrons and should be able to form five regular covalent bonds. The phosphorous atom in phosphorus pentachloride (PCl_5) is an example. Electrostatic repulsion is at a minimum when 5 electron pairs are located at the vertices of a trigonal bipyramid (Fig. 8-19). Because this type of molecule is relatively uncommon, we shall be concerned primarily with the very abundant and important molecules which contain nitrogen. We would not expect nitrogen to form five covalent bonds. The bonding level in nitrogen is the second quantum level which can accommodate a maximum of four

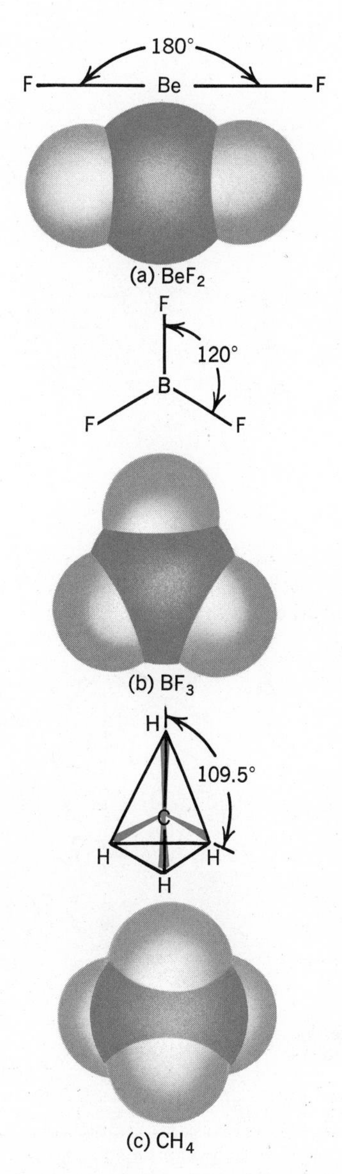

8-18 BeF_2, BF_3, and CH_4 are all symmetrical molecules. BeF_2 is a linear, BF_3 is a trigonal planar, and CH_4 is tetrahedral. All valence electrons of the central atoms are used in bond formation.

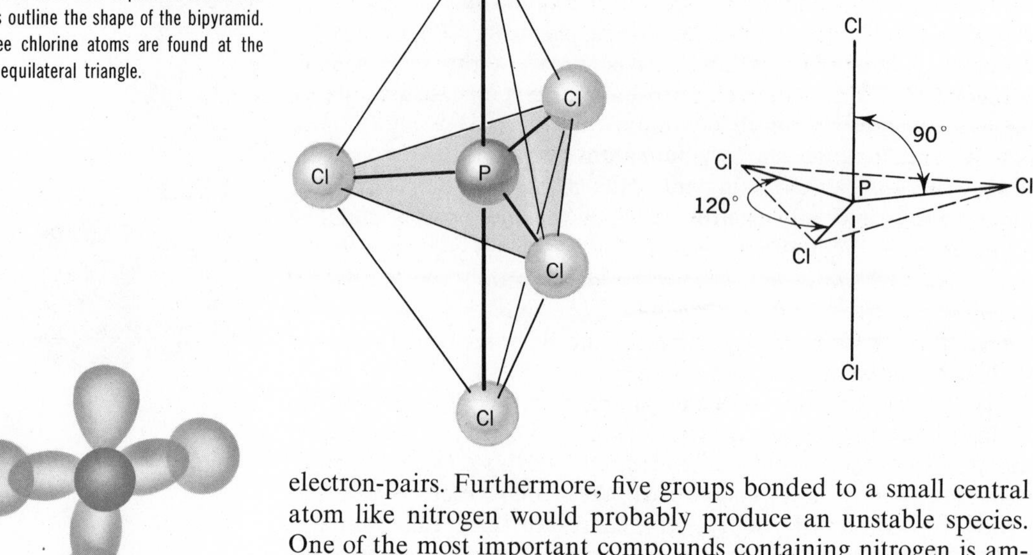

8-19 When the central atom uses all of its valence electrons to bond to five identical atoms, the resulting species has a trigonal bipyramidal shape. In this figure, the dark lines represent bonds and the white lines outline the shape of the bipyramid. Note that three chlorine atoms are found at the corners of an equilateral triangle.

electron-pairs. Furthermore, five groups bonded to a small central atom like nitrogen would probably produce an unstable species. One of the most important compounds containing nitrogen is ammonia (NH_3). Examination of the Lewis structure of NH_3 reveals that there are three pairs of shared electrons and one pair of unshared electrons (a lone pair) in the outer energy level of the nitrogen atom

$$\begin{array}{c} \mathrm{H} \\ \mathrm{:\ddot{N}:H} \\ \mathrm{\ddot{}} \\ \mathrm{H} \end{array}$$

If all pairs were equivalent, the four charge clouds would be identical. Mutual repulsion would force them apart until they made an angle of 109.5° with each other. The one large charge cloud associated with the lone pair, however, has a distorting effect on the tetrahedral angle and reduces it to about 107°. The result is a distorted tetrahedron with a lone pair of electrons at one apex.

Lines joining the three hydrogen atoms outline an equilateral triangle. The nitrogen atom is located above the center of the triangle. Lines radiating from the nitrogen atom to each hydrogen atom form an angle of about 107° with each other. An ammonia molecule shown in Fig. 8-20 is said to be pyramidal. In general, *when three atoms bond to a central atom, the molecule or ion will be pyramidal if the central atom contains one unbonded electron-pair.*

In an ammonium ion (NH_4^+), there are four identical groups with no unbonded electron pairs. Hence, we would predict that the ion is tetrahedral. Experimental data confirm this prediction.

(a)

N
H
H
H
(b)

(c)

8-20 (a) The structure of NH_3 is related to a tetrahedron. The unbounded pair of electrons exerts a greater repulsion on a bond pair than bond pairs do on each other. The result is a distorted tetrahedron. (b) Three hydrogen atoms form the base of a pyramid with a nitrogen atom at the apex. (c) Space-filling model for NH_3.

8-17 Covalent Molecules Containing an Atom of a Group-VIA Element According to our generality, atoms in this group should be able to form six covalent bonds. The exception is oxygen which, like nitrogen, is a small atom whose bonding level can contain only four electron-pairs. Let us apply our charge cloud concept to predict the shape of sulfur hexafluoride (SF_6), a molecule in which six fluorine atoms are bonded to one sulfur atom. The formula (SF_6) shows that six bonding positions are required. Since sulfur is in Period 3 of the Periodic Table, it has three energy levels which, in theory, can contain nine electron-pairs. In SF_6, six electron-pairs are used to form six covalent bonds. We have shown that, in this case, minimum repulsion occurs when the electron-pairs are located at the vertices of a regular octahedron (Fig. 8-22(a)). The bond angles between any adjacent fluorine atoms are 90°. The SF_6 molecule shown in Fig. 8-22(b) is octahedral and symmetrical. *Most species (molecules or ions) in which six atoms or ions are bonded to a central atom or ion are octahedral.*

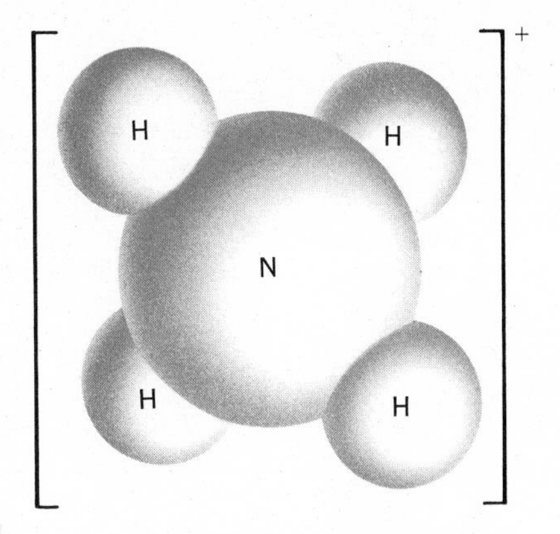

8-21 In NH_4^+, all bonding electrons are used to bond four identical atoms. An ammonium ion, therefore, is tetrahedral.

The most familiar and important molecule formed by an element of Group VIA is, of course, the water molecule. The Lewis formula shows that in a water molecule, the oxygen atom has two pairs of shared electrons and two pairs of unshared electrons in its outer energy level. If all pairs were equivalent, the four charge clouds would be identical, and the bond angle would be 109.5°. We would, however, predict that the larger clouds associated with the unbonded electron pairs would repel and force the bonding pairs closer together than 109.5°. This is confirmed by data which show that the bond angle in water is 104.5°. Note that the one lone pair in an ammonia molecule distorted the tetrahedral angle to a less extent than the two lone pairs in a water molecule. In general, when two atoms are bonded to a third atom, the resulting molecules are *angular (bent) if unbonded outer electrons are present in* the central atom. This shape is characteristic of triatomic molecules, containing as a central atom, the atoms of Group VIA.

8-18 Covalent Molecules Containing an Atom of Group-VIIA Element In theory, the larger atoms of this group can form seven covalent bonds. Molecules of the type IF_7 (Fig. 8-24) are known but are relatively unimportant in a beginning course. We shall concern ourselves mainly with those species in which two, three, or four atoms are bonded to a central atom. Examples involving these species have already been discussed.

The possible shapes of molecules and ions that are related to the possible arrangements of valence electron-pairs on the central atom are summarized in Tables 8-5 and 8-5A. In the Lewis structures, *A* represents the central atom and *B* represents the attached atoms. In the geometric figures the light colored lines outline three-dimensional geometric figures and unbroken dark lines represent bonds. Dashed lines are used to help give a three-dimensional effect. Note that the structures of all species in which there are four pairs of bonding electrons around the central atom are related to a tetrahedron. Similarly, the structures of all species in which there are five pairs of bonding electrons are shown experimentally as being related to the trigonal bipyramid. For example, chlorine

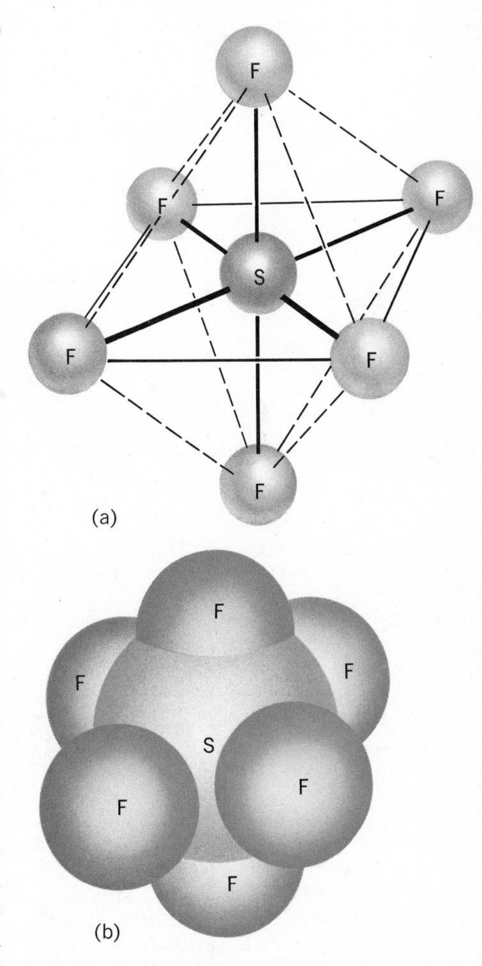

8-22 (a) In an SF_6 molecule, the fluorine atoms are located at the vertices of a regular octahedron. (b) Space-filling model of SF_6.

TABLE 8-5

POSSIBLE SHAPES OF SPECIES CONTAINING ATOMS OF REPRESENTATIVE ELEMENTS

No. of e^- pairs in valence level of bonded central atom	No. of atoms bonded to central atom	No. of unbonded electron pairs (lone pairs)	Lewis structure showing bonds and unbonded e^- pairs	Formula	Representation of shape	Name of shape	Example
2	2	0	B— A —B	AB_2	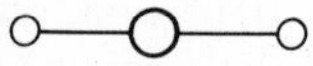	linear	BeF_2
3	3	0	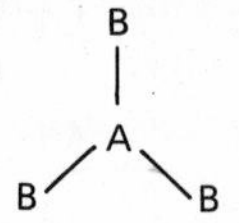	AB_3	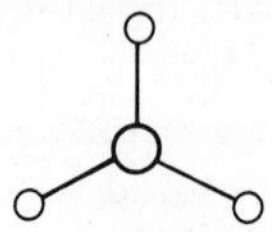	trigonal planar	BF_3
3	2	1		AB_2	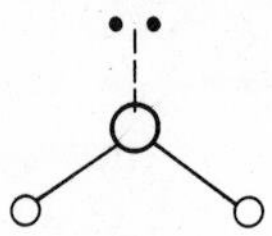	angular	$SnCl_2$
4	4	0	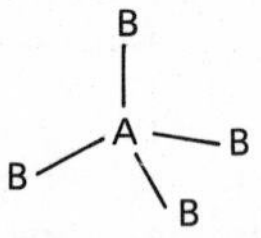	AB_4	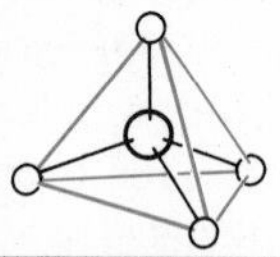	tetrahedral	CCl_4
4	3	1	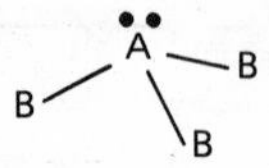	AB_3	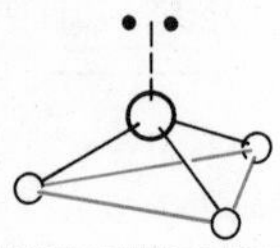	pyramidal	NH_3
4	3	0	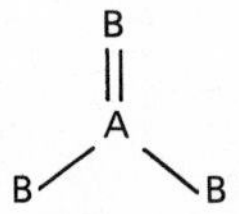	AB_3	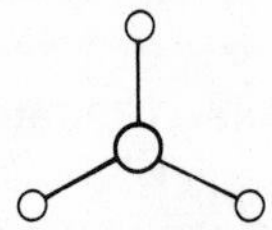	trigonal planar	SO_3
4	2	2	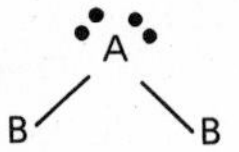	AB_2	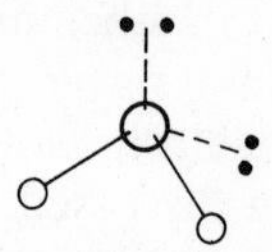	angular	H_2O
4	2	1	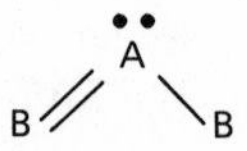	AB_2	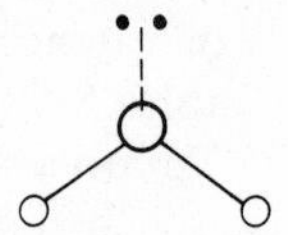	angular	SO_2
4	2	0	B═A═B	AB_2	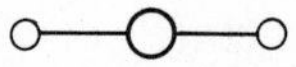	linear	CO_2

TABLE 8-5A

POSSIBLE SHAPES OF SPECIES CONTAINING ATOMS WITH 5, 6, or 7 ELECTRON-PAIRS

No. of e^- pairs in valence level of bonded central atom	No. of atoms bonded to central atom	No. of unbonded electron pairs (lone pairs)	Lewis structure showing bonds and unbonded e^- pairs	Formula	Representation of shape	Name of shape	Example
5	5	0	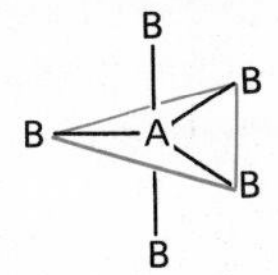	AB_5	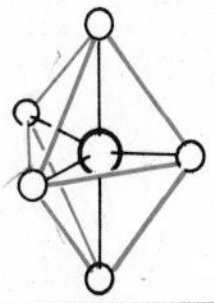	trigonal bipyramidal	PCl_5
5	4	1	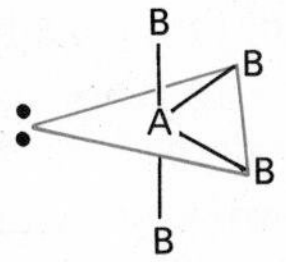	AB_4	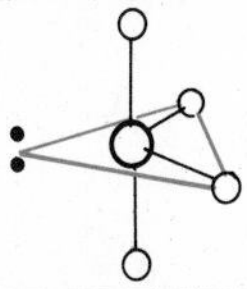	seesaw	SF_4
5	3	2	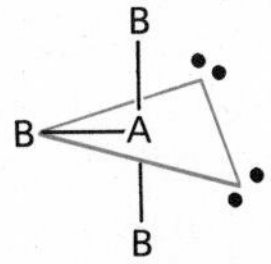	AB_3	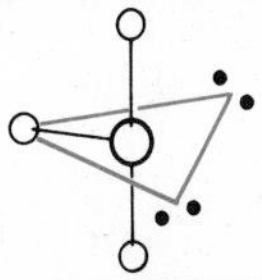	T-shaped	ClF_3
5	2	3	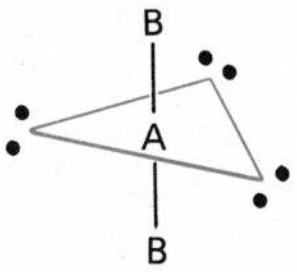	AB_2	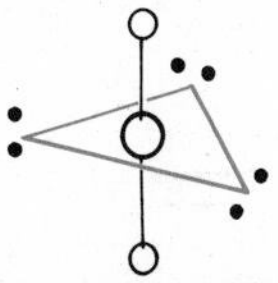	linear	XeF_2
6	6	0	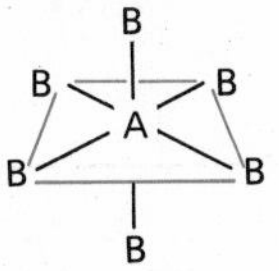	AB_6	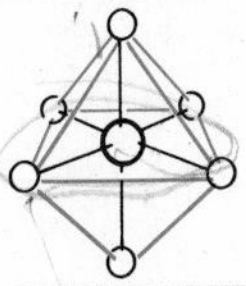	octahedral	SF_6
6	5	1	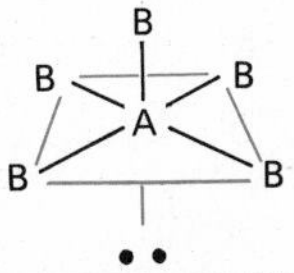	AB_5	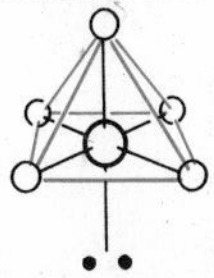	square-based pyramid	IF_5
6	4	2	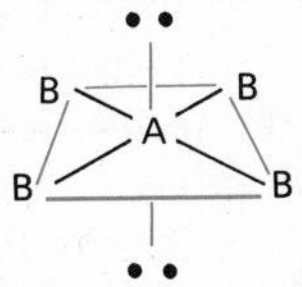	AB_4	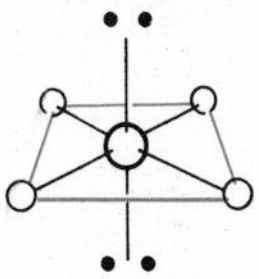	square planar	XeF_4
7	7	0	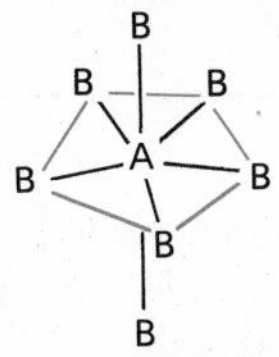	AB_7	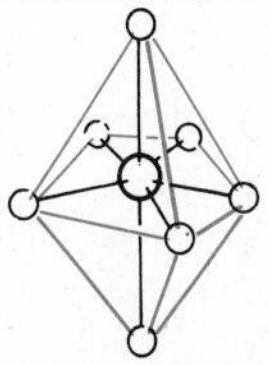	pentagonal bipyramid	IF_7

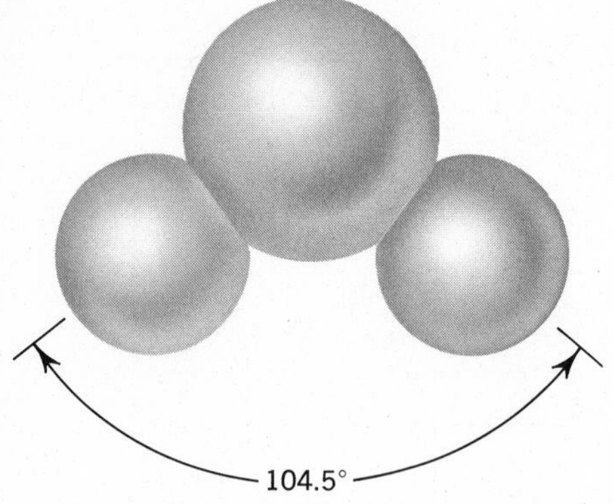

8-23 The bond angle between the hydrogen atoms and the oxygen atom in a water molecule is 104.5°. Explain why it is neither 90° nor 109.5°.

trifluoride (ClF_3) has five pairs of bonding electrons. Since only three positions on the bipyramid are occupied, the molecule is said to be T-shaped. Minimum repulsion is achieved when the lone pairs of electrons are in the central plane. See margin on p. 285.

When there are six pairs of bonding electrons the shape of the structure is related to the octahedron. When only five positions are occupied as in iodine pentafluoride (IF_5), the bonded atoms define a square-based pyramid. See margin on p. 285.

Note that in the pictorial representation of molecular shapes, multiple bonds are not shown. As far as shape is concerned, they have the same orientation as single bonds. You should also be aware that the shapes described in Tables 8-5 and 8-5A are idealized and represent general shapes. No bonding theory permits the accurate prediction of all bond angles because the angles are affected by the relative sizes of the atoms and their relative electronegativity. For example, in a T-shaped molecule it is unlikely that the top of the T will be precisely perpendicular to the stem.

The relationship between central atom (A) bonded atoms (B), and number of lone pairs (nonbonding electron-pairs) symbolized by E is summarized in the margin on p. 285.

FOLLOW-UP PROBLEM

Write electron-dot formulas for and predict the shapes of (a) BeH_2, (b) H_2S, (c) BCl_3, (d) H_3O^+, (e) SiH_4, (f) $Sb(OH)_6^-$, (g) ClO_4^-, (h) CO_3^{2-}, (i) $Al(H_2O)_6^{3+}$, (j) CO_2, (k) $SbCl_5$, (l) $TeCl_4$, (m) ClO_2^-, (n) BrF_3, (o) BrF_5, (p) NO_2^-, (q) N_2O.

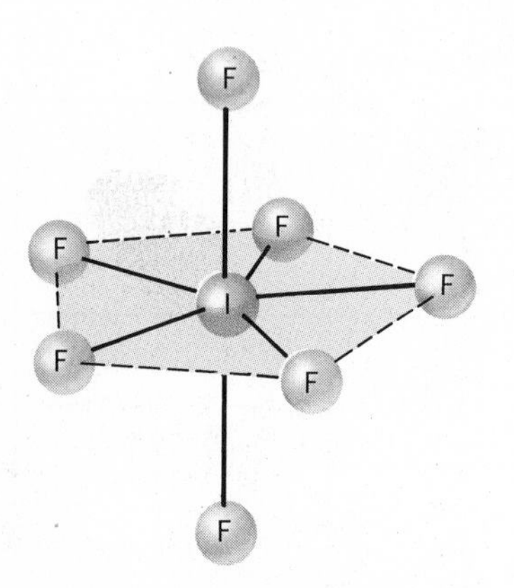

8-24 When the central atom uses all of its valence electrons to bond to seven identical atoms, the resulting species has a pentagonal bipyramidal shape. In Fig. I five fluorine atoms are found at the vertices of a regular pentagon. Lines drawn from the fluorine atoms at the poles to each of the atoms in the central plane outline the pentagonal bipyramid.

You may have noticed that we avoided identifying by name or letter any of the bonding orbitals used by the central atom. The reason is that the directional characteristics of pure atomic orbitals do not give rise to the observed bond angles. For example, we could not explain the 109.5° angles in CH_4 by assuming that one s and three p hydrogen-like orbitals in the carbon atom were used for bonding. The s orbital is nondirectional, and the three p orbitals form right angles with each other. Furthermore, the electron configuration of the atom in the ground state is not consistent with the observed bonding capacity of carbon in most of its compounds. The ground state configuration shows two unpaired electrons but in most of its compounds carbon exhibits a bonding capacity of four.

In order to explain experimentally observed bonding capacities and bond angles, and still retain the concept of pure atomic orbitals, scientists use a synthetic concept known as *hybridization.* Hybridization is one aspect of an important bonding theory known as the valence bond model. Let us briefly examine this helpful concept.

8-19 Hybridization Hybridization of pure atomic orbitals means that two or more different pure atomic orbitals can be mixed (hybridized) to yield two or more hybrid atomic orbitals which are identical. For example, mixing one *s* and one *p* orbital in an atom yields two *sp* hybrid orbitals. Whereas the individual *s* and *p* orbitals have different energies and directional characteristics, the two *sp* hybrids are identical in every respect. Mixing pure *s* and *p* atomic orbitals and yielding two *sp* hybrids is rather like mixing a gallon of pure red paint and a gallon of pure white paint to give two gallons of pastel pink paint. The analogy is illustrated in Fig. 8-25.

8-20 *sp* Hybrid Orbitals We shall illustrate application of the hybridization concept by showing how it can account for the experimentally observed bonding capacity of beryllium and the bond angles in BeF_2. The formula shows that a beryllium atom is bonded to two fluorine atoms and, therefore, has a bonding capacity of two. Spectroscopic measurements reveal that the bond angle between the fluorine atoms is 180° and that the bond lengths are equal. The ground state electronic configuration of beryllium is

$$1s^2 2s^2$$

It is apparent that this configuration is not consistent with the formation of two covalent bonds oriented at 180°. Our theory must account for these observations. In other words, there must be two bonds of equal strength directed at an angle of 180° with respect to each other.

To explain these observations, valence bond theory invokes the hybridization concept in which a specific number of pure atomic orbitals are hydridized (mixed) and yield an equal number of identical hybrids. If all valence electrons of the central atom are used in bond formation, we can assume that the angle between the hybrid orbitals is maximized. If not all valence electrons are used in bond formation, then the angle between hybrid orbitals may be slightly distorted.

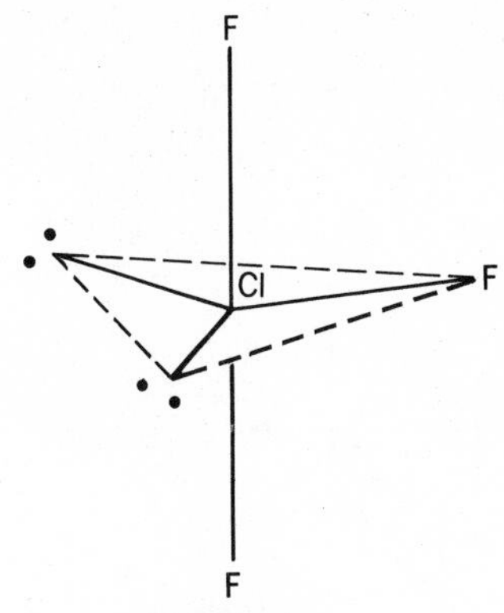

3 bond pairs
2 lone pairs

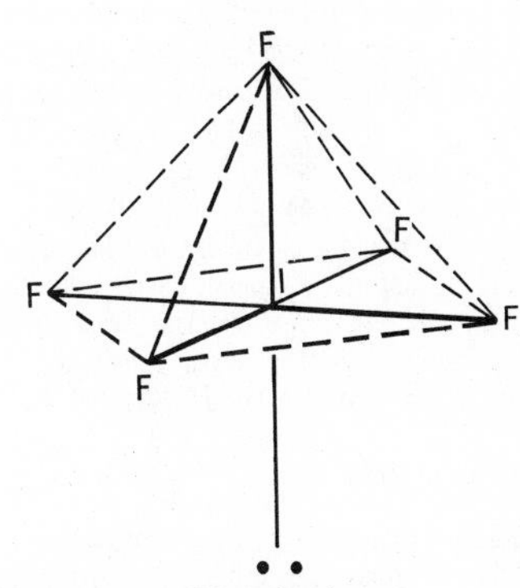

5 bond pairs
1 lone pair

Formula	*Shape*
AB_2	linear
AB_2E	angular
AB_2E_2	angular
AB_2E_3	linear
AB_3	trigonal planar
AB_3E	pyramidal
AB_3E_2	T-shaped
AB_4	tetrahedral
AB_4E	seesaw
AB_4E_2	square planar
AB_5	trigonal bipyramid
AB_5E	square-based pyramid
AB_6	octahedral
AB_7	pentagonal bipyramid

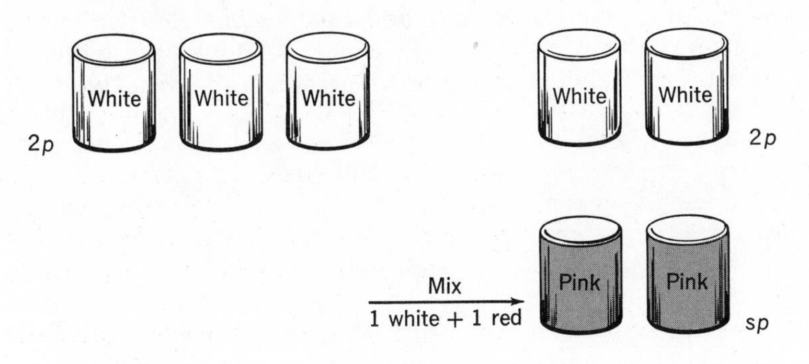

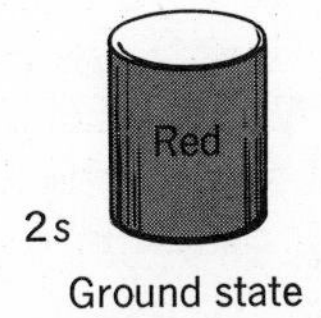

Ground state

Hybridized state

8-25 Hybridization analogy. Starting with one can of red paint and three cans of white paint one can of red paint is mixed with one can of white paint, two cans of pink paint are formed. This leaves no red paint and two cans of white paint. This is analogous to mixing one s and one *p* orbital to give two *sp* hybrid orbitals. This leaves two *p* and two *sp* orbitals available for bonding.

LINUS PAULING
1901–

Linus Pauling received his undergraduate education at Oregon State College and then went to the California Institute of Technology for his graduate studies. After receiving a Ph.D. in chemistry in 1925, he went to Europe for a year of post-doctoral study. His greatest contribution to chemistry was his explanation of the nature of the chemical bond. He was the first person to apply quantum mechanical principles to chemical reactions. From 1931 to 1933, Pauling published seven papers on "The Nature of the Chemical Bond." His book by the same title, first published in 1938, became a classic in its field.

He was the first to suggest a helical (spring-like) structure for protein molecules. Pauling first recognized the hydrogen bond as a unique type of bond different in nature and strength from the covalent bond and van der Waals forces.

Pauling was the first chemist to express the idea that chemical bonds could have a mixed character, including both a covalent and ionic nature. He developed an electronegativity scale and assigned electronegativity values to elements, which helped in predicting the degree of ionic or covalent character of their various compounds. He was the first to use the term *resonance* in chemical bonding.

For his work in chemistry, Pauling received the 1954 Nobel Prize in chemistry. Because of his concern about peace during the nuclear age and his efforts to inform the public of the potential danger of the military use of nucelar energy, he was awarded the 1962 Nobel Peace Prize. This made him the first man to receive two Nobel Prizes. He is now working as a research professor in the Center for the Study of Democratic Institutions, Santa Barbara, California.

A hypothetical sequence of steps can be used to illustrate the formation of hybrid orbitals in BeF_2. As would be expected, the approach of fluorine atoms causes a rearrangement of electrons and energy changes to take place in the beryllium atom. The ground state electron configuration of a beryllium atom is

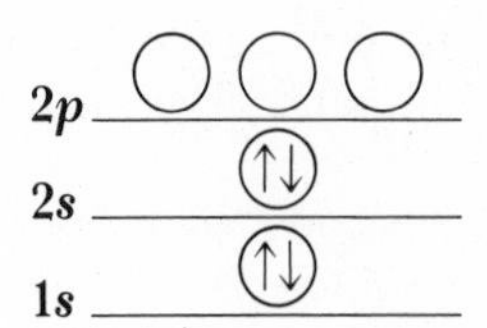

We may imagine that the energy released as fluorine atoms approach a beryllium atom promotes one of the *s* electrons into an empty *p* orbital. This provides two bonding orbitals, each with one electron. This intermediate configuration, which is strictly hypothetical, would be

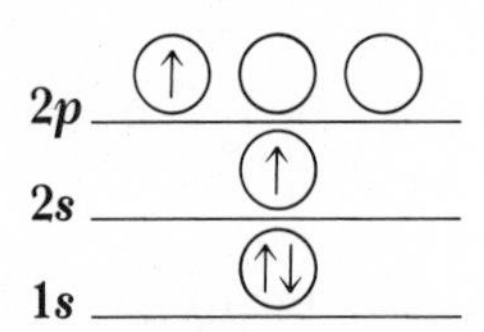

However, since electrons in *s* and *p* orbitals have different energies, bonds involving electrons in these pure atomic orbitals would not have the same strength. In order to make the bonding theory agree with experimental evidence, it is necessary to postulate that one *s* and one *p* orbital hybridize (mix) and yield two identical orbitals. We identify the hybrid by stating with superscripts the number of each pure orbital involved in the formation of the hybrid. Thus, the two hybrid orbitals formed by mixing one *s* and one *p* orbital are called *sp* hybrids. Each hybrid contains one electron and is, in nature, intermediate between a pure *s* and a pure *p* atomic orbital. The overall process would be represented schematically as

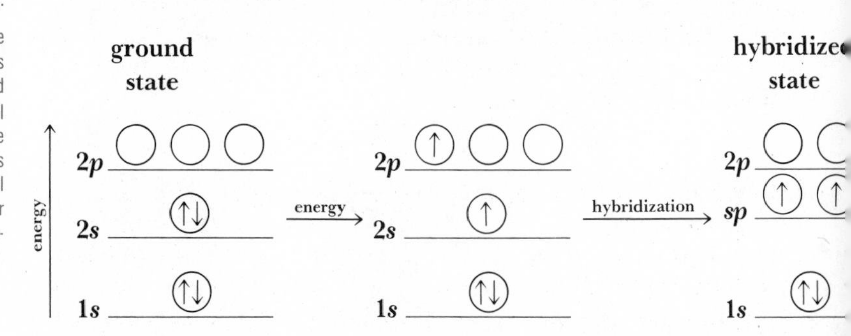

The formation of sp hybrid orbitals from pure orbitals is shown in Fig. 8-26.

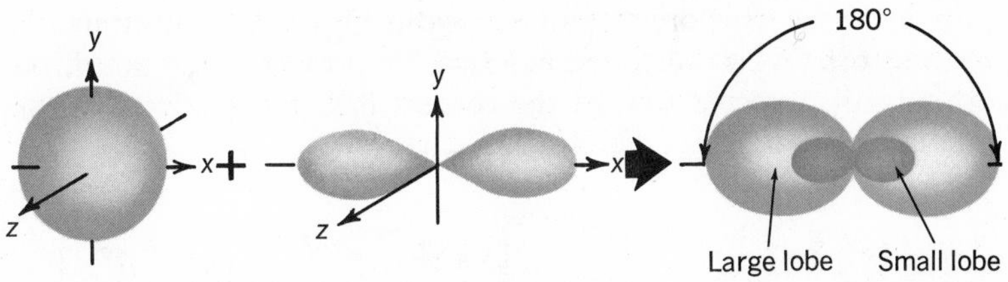

8-26 Two equivalent *sp* hybrid orbitals are formed by hybridizing one s and one *p* orbital. Each *sp* orbital consists of two lobes, one large and one small. The large lobe occupies a greater region of space than the pure *p* orbital. This permits a greater degree of overlapping with the orbital of another atom, resulting in the formation of a stronger bond.

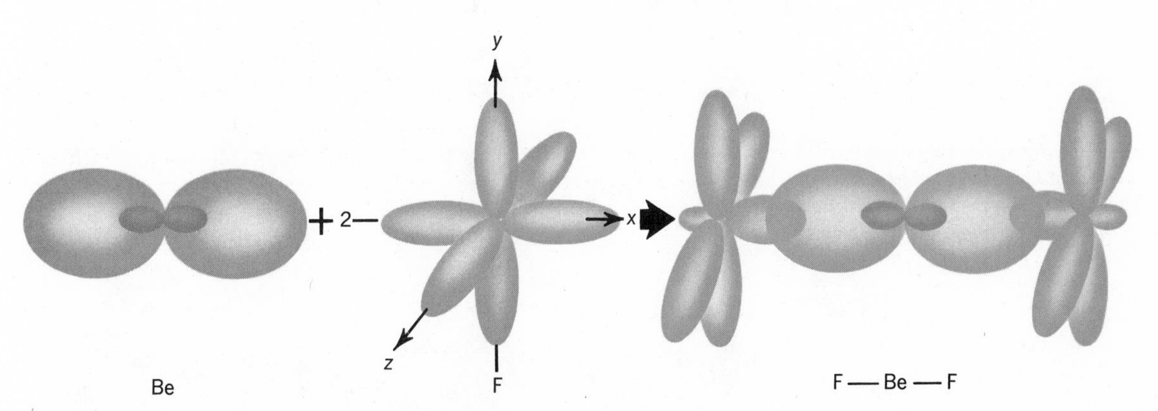

8-27 Representation of bond formation in BeF_2. The s orbitals of fluorine are not shown. The overlap of the p_x orbital of each fluorine atom with the *sp* orbitals of beryllium give rise to the *sigma*-type (σ) covalent bond.

Bonds are formed when a singly occupied *p* orbital of each fluorine atom overlaps a singly occupied *sp* orbital of a beryllium atom so that two electrons are common to both orbitals. The hybrid *sp* beryllium orbitals are directed at angles that keep them as far apart as possible (Fig. 8-26). Therefore, the bond angle in BeF_2 is 180°.

8-21 sp^2 Hybrid Orbitals Earlier we used the electrostatic repulsion model of chemical bonding to show that when *a central atom uses all its bonding electrons to bond three* other atoms, the molecule is planar and the bond angles are 120°. Let us apply the valence bond theory concept of hybridization to this situation. Consider a boron trifluoride molecule (BF_3). This molecule is nonpolar with three identical bonds and trigonal planar geometry. The ground state configuration of a boron atom is

$$1s^2 \quad 2s^2 \quad 2p^1$$

To exhibit a bonding capacity of 3 and form three identical bonds, we can imagine that a 2*s* electron is promoted to a 2*p* orbital whereupon hybridization occurs. The three resulting hybrid orbitals are identified as sp^2 orbitals. Their spatial orientation and the shape of a BF_3 molecule are shown respectively in Figs. 8-28 and 8-18.

8-22 sp^3 Hybrid Orbitals In methane (CH_4) one *s* and three *p* orbitals of the carbon atom may hybridize and give four identical sp^3 hybrid orbitals. The hypothetical steps involved in the forma-

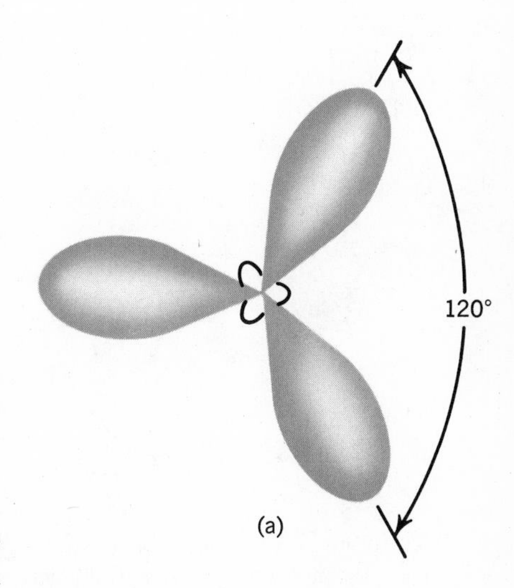

8-28 Spatial orientation of sp^2 orbitals.

tion of the four sp^3 orbitals are shown in Fig. 8-29. The formation of these orbitals is illustrated in Fig. 8-30. It can be seen that these orbitals are directed toward the corners of a regular tetrahedron and make an angle of 109.5° with each other (Fig. 8-30).

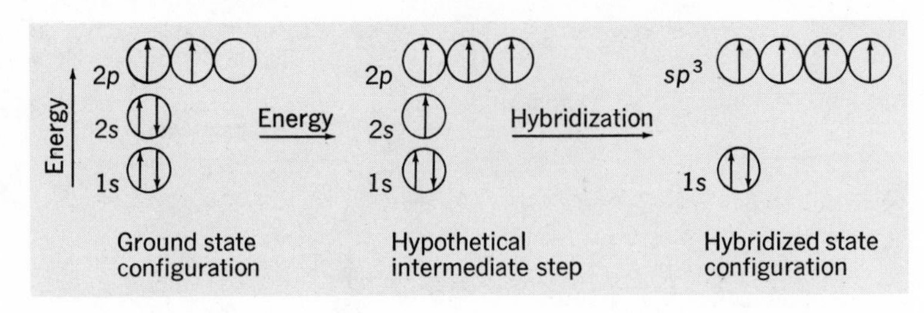

8-29 One pure s orbital and three pure p orbitals may hybridize and give four equivalent sp^3 orbitals with 75 percent p character and 25 percent s character.

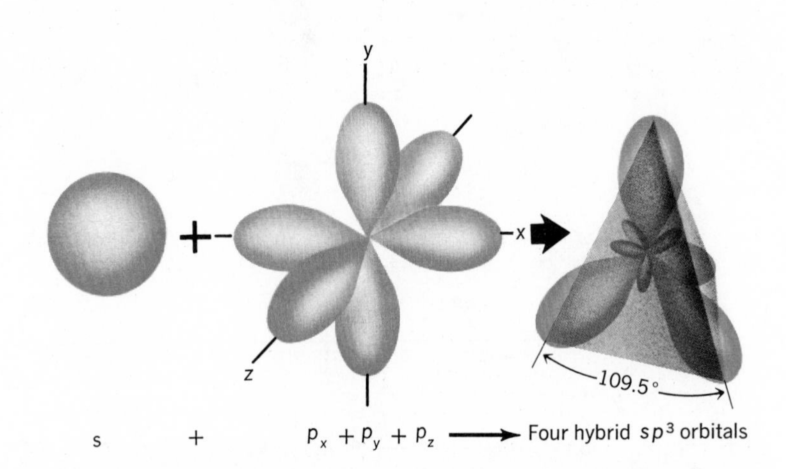

8-30 Hybridization of one s and three p orbitals yields four identical sp^3 orbitals, the larger lobes of which are directed toward the vertices of a regular tetrahedron.

8-23 Hybrid Orbitals in H_2O and NH_3 The sp^3 orbitals are the ones referred to in our discussion of the bonding in H_2O and NH_3. In H_2O and NH_3, however, the angles of the sp^3 orbitals are influenced and distorted by the presence of unbonded electron pairs. The ground state and hybridized electron configurations of an oxygen atom are

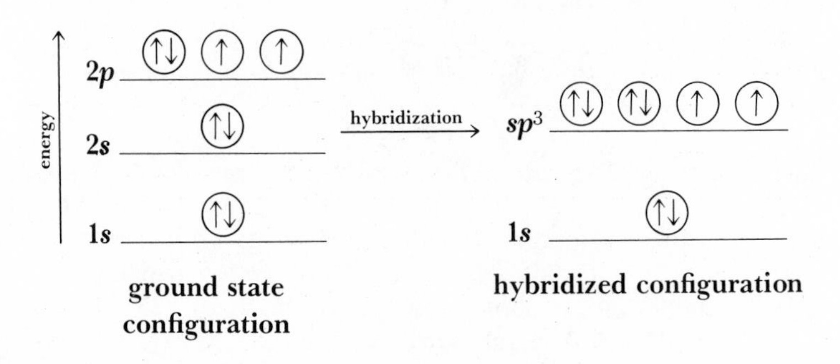

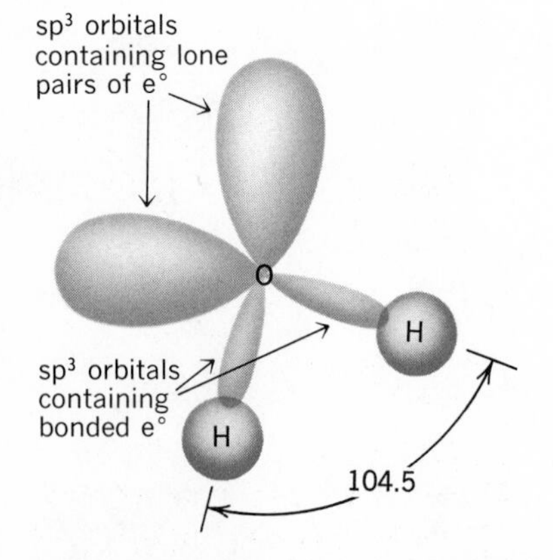

8-31 Orbital model for water. The larger electron clouds associated with lone pairs of electrons distort the tetrahedral angle from 109.5° to 104.5°.

The orbital model for water is shown in Fig. 8-31.

The ground state and hybridized electron configurations for a nitrogen atom are

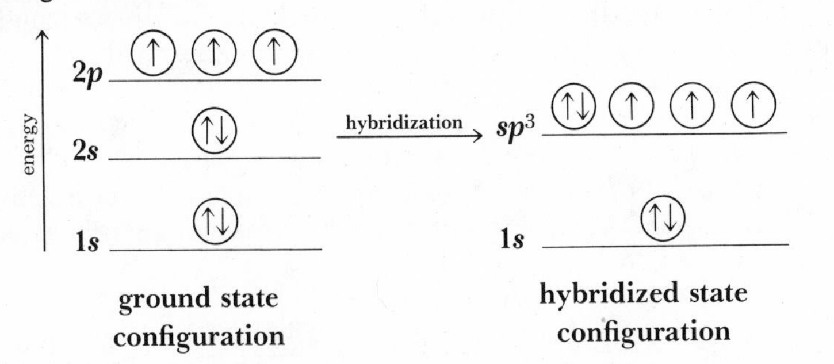

The orbital and space-filling models for NH_3 are shown respectively in Figs. 8-32 and 8-20.

8-24 sp^3d^2 Hybrid Orbitals When six identical bonds are formed, the central atom must use *d* orbitals. Consider a sulfur hexafluoride molecule (SF_6). The ground state configuration of a sulfur atom is

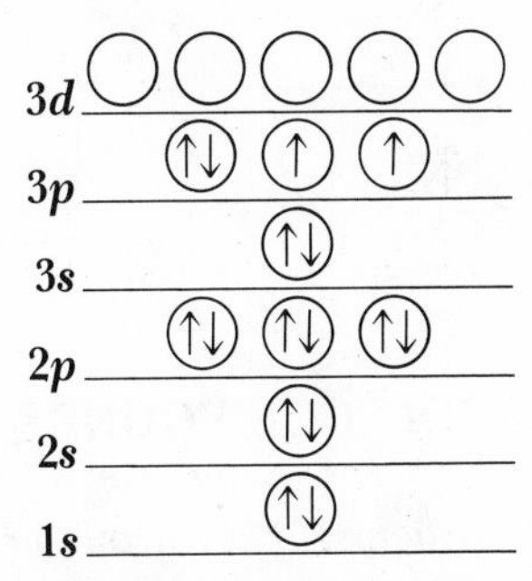

To provide the necessary bonding orbitals, an *s* and a *p* electron are promoted to *d* orbitals, whereupon sp^3d^2 hybridization occurs. The spatial orientation of the sp^3d^2 hybrid orbitals and the geometry of a SF_6 molecule are shown in Fig. 8-33.

8-25 Summary The relationships among the different types of hybrid orbitals and the characteristics of molecules are summarized in Table 8-6.

TABLE 8-6

RELATIONSHIPS AMONG TYPES OF HYBRID ORBITALS, FEATURES OF CENTRAL ATOM, AND CHARACTERISTICS OF RESULTING MOLECULE

No. of Bonds	No. of Unused e⁻ Pairs	Type of Hybrid Orbital	Angle Between Atoms Bonded to the Central Atom	Geometry of Molecule or Ion Formed	Example
2	0	sp	180°	linear	BeF_2
3	0	sp^2	120°	trigonal planar	BF_3
4	0	sp^3	109.5°	tetrahedral	CH_4
3	1	sp^3	90° to 109.5°	pyramidal	NH_3
2	2	sp^3	90° to 109.5°	angular	H_2O
6	0	d^2sp^3 or sp^3d^2	90°	octahedral	SF_6

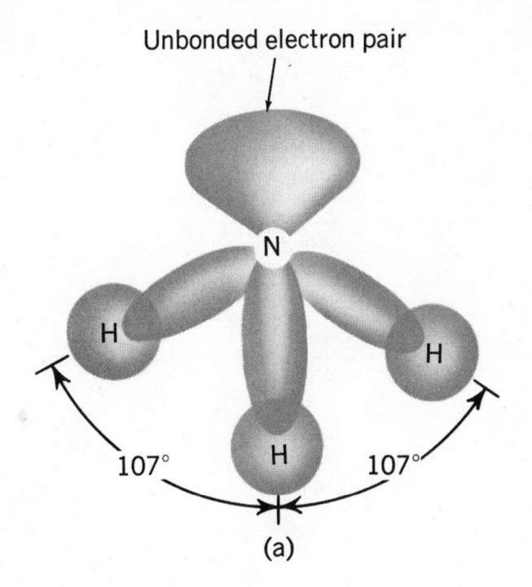

8-32 Orbital model for NH_3.

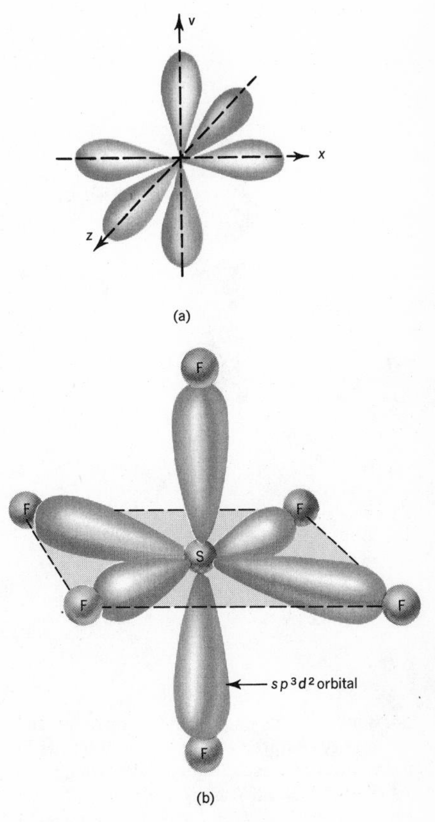

8-33 (a) Shape and spatial orientation of sp^3d^2 hybrid orbitals. In SF_6, four fluorine atoms are located at the corners of a square. The other two, located at the poles, make an angle of 90° with the central plane. See Fig. 8-22 for a space filling model.

To enhance your confidence in the concept of hybridization it is worth noting that it is based on quantum mechanical calculations. In other words, the hybrid orbital angles and relative strengths can be calculated. The calculated values agree closely with observed values. In general, hybrid orbitals permit a greater degree of overlap than the corresponding pure atomic orbitals. Hence, stronger bonds are formed by hybridization than with pure atomic orbitals. A summary of orbitals used for bond formation and their relative strengths based upon a value of 1.000 for a pure *s* orbital are given in Table 8-7.

TABLE 8-7
RELATIVE STRENGTHS OF BONDING ORBITALS

Orbitals Used	Number of Orbitals	Relative Strength
s	1	1.000
p	3	1.732
sp	2	1.932
sp^2	3	1.991
sp^3	4	2.000
d^2sp^3	6	2.923

FOLLOW-UP QUESTION

Identify the hybrid orbitals used by the central atom in these (a) BeH_2, (b) BCl_3, (c) H_3O^+, (d) SeH_4, (e) SeF_6, (f) NO_3^-, (g) SO_3^{2-}.

BONDING IN CARBON COMPOUNDS

When writing Lewis structures, we encountered two carbon compounds which involved multiple bonds. These compounds were ethylene (C_2H_4) and acetylene (C_2H_2), both of which are classified as organic compounds. Because of the importance of organic compounds, and to set the stage for the discussion of organic chemistry in a later chapter, we shall briefly examine the structural features of a few simple organic molecules. We have already considered methane (CH_4) commonly used as a fuel for heating and cooking. Three other carbon-hydrogen compounds closely related to methane are propane (C_3H_8), ethylene (C_2H_4), and acetylene (C_2H_2). Experimental data related to these molecules are in Table 8-8.

Bonding energy may be defined as the energy required to break all the bonds in a mole of gaseous molecules which are in their lowest energy states.

TABLE 8-8
CHARACTERISTICS OF C_3H_8, C_2H_4, and C_2H_2

	Formula	Carbon-carbon Bond Length (Ångstroms)	Carbon-carbon Bond Energy (kcal/mole)	C—H Bond Angles	Molecular Shape	Relative Reactivity
Propane	C_3H_8	1.54	88	109.3°		low
Ethylene	C_2H_4	1.34	167	120°	trigonal planar	moderately high
Acetylene	C_2H_2	1.20	230	180°	linear	high

Inspection of the data in Table 8-8 reveals that the carbon-carbon bond strength in acetylene is approximately one and one-half times that in ethylene and almost three times that in propane. This factor coupled with the corresponding changes in bond length suggests the presence of single bonds in propane, a double bond in ethylene, and a triple bond in acetylene. These structures provide an outer octet of electrons for each carbon atom and a helium configuration for each hydrogen atom. The line formulas of the molecules are listed below.

```
              H   H   H
              |   |   |
Propane    H—C—C—C—H
              |   |   |
              H   H   H

            H       H
             \     /
Ethylene      C=C
             /     \
            H       H

Acetylene  H—C≡C—H
```

8-26 Single Bonds: Propane The bond orientation (109.3°) in C_3H_8 shown in Fig. 8-34 may be explained by assuming that sp^3

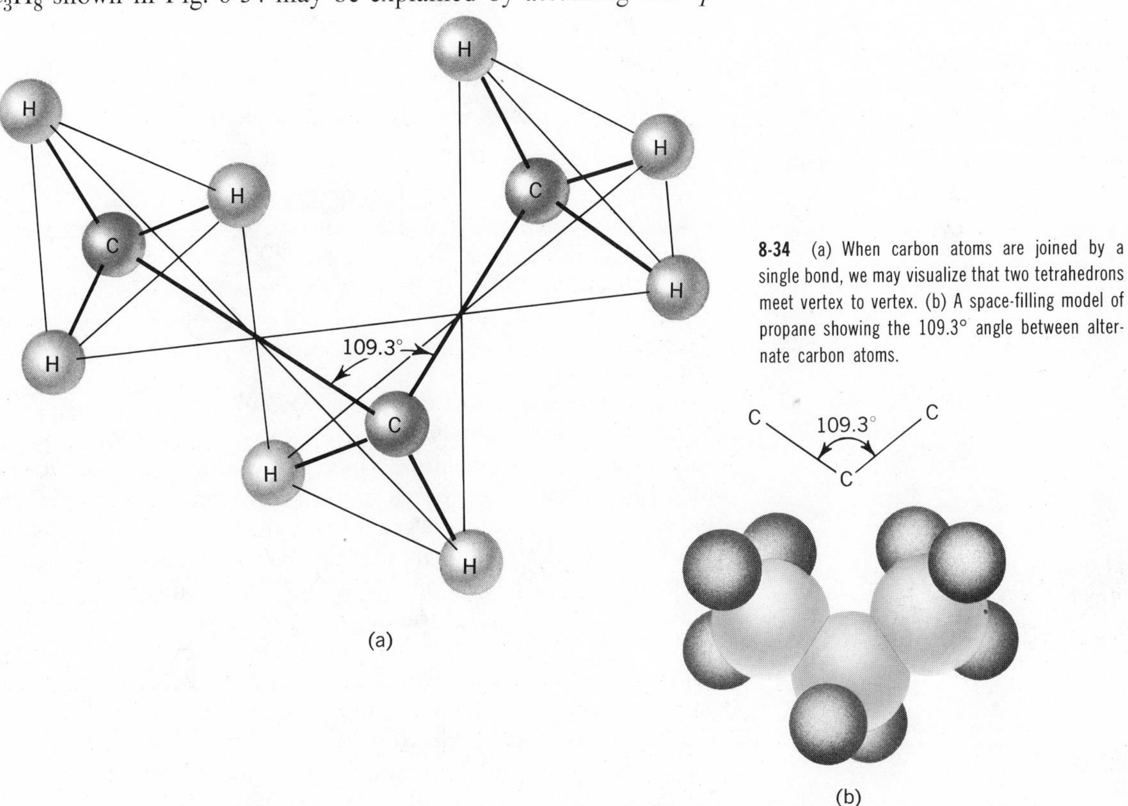

8-34 (a) When carbon atoms are joined by a single bond, we may visualize that two tetrahedrons meet vertex to vertex. (b) A space-filling model of propane showing the 109.3° angle between alternate carbon atoms.

hybrid orbitals are involved. One singly occupied sp^3 orbital from each adjacent atom overlaps endwise in the region between the carbon nuclei forming a strong sigma covalent bond. The other singly occupied sp^3 orbitals from each carbon atom overlap the singly occupied *s* orbitals of the hydrogen atoms and form additional sigma bonds.

8-27 Double Bonds: Ethylene The observed trigonal planar geometry of an ethylene molecule suggests the presence of sp^2 hybrid orbitals which form 120° angles with each other. The basic structure showing single bond orientation only of ethylene may be represented as

A double or triple bond involving, respectively, two or three electron pairs is counted as a single electronic charge cloud for purposes of determining spatial orientation.

H 120° H
C—C 120°
H 120° H

All of the atoms lie in the same plane and the angle between atoms is 120°. The bond between the carbon atoms is a strong sigma bond formed by the endwise or head-on overlap of two singly occupied sp^2 orbitals. The C—H bonds are also sigma bonds formed by the overlap of a carbon sp^2 orbital with a hydrogen *s* orbital. These bonds are represented in Fig. 8-35.

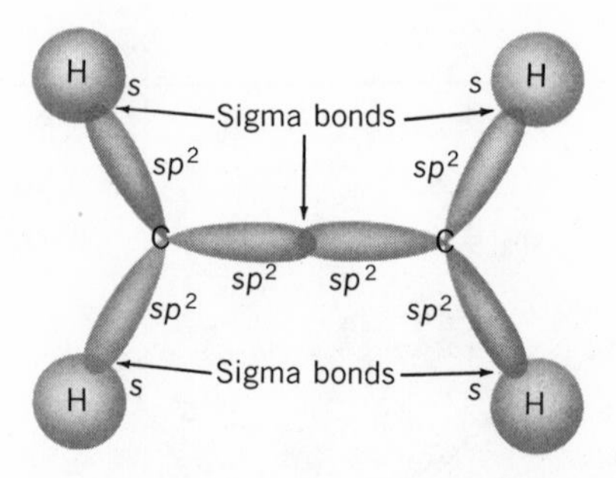

8-35 Five *sigma* bonds are formed in an ethylene molecule. The *sigma* bond between the two carbon atoms constitutes half of the double bond in ethylene. The p_y orbitals containing the fourth bonding electron of the carbon atom are not shown.

Since the formation of three singly occupied sp^2 hybrid orbital requires only two *p* orbitals and two electrons, each carbon atom still possesses an unused *p* electron. These electrons must account for the double bond nature of the carbon-carbon bond in ethylene. The moderately high reactivity (a general chemical property) suggests that some of the outer electrons are more available for reaction than those of ethane. These observations along with other data (spectroscopic) suggest that all of the electrons in the double bond are not localized in the region between the carbon nuclei where they would be rather confined and not readily available for reaction. It is the remaining *p* electrons which account for the greater reactivity of ethylene.

The fourth bonding electron of each carbon atom is located in a p_y orbital. A minimum energy condition occurs when the p_y orbitals on the two carbon atoms are parallel to each other.

or argon reported. From a table of ionization energies (Table 6-5) you can see that helium and argon have first ionization energies which exceed that of fluorine, while those of the other noble gases are lower. Many scientists predict that, even in the future, it is unlikely that stable compounds will ever be formed with helium or neon. The first ionization energies are so high that combination even with fluorine, the most reactive nonmetal, would seem to be improbable.

With a background knowledge which enables us to predict shapes of simple molecules, we can now investigate their polarity.

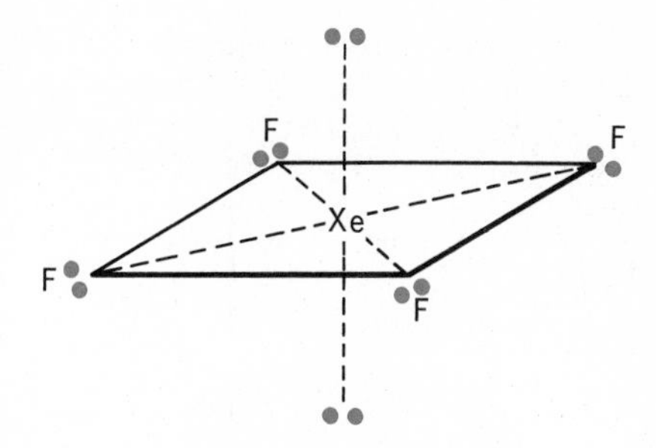

8-41 Electron-dot structure of XeF_4.

POLARITY OF BONDS AND MOLECULES

The melting points, boiling points, and other properties of pure substances are influenced by the shapes, sizes, and polarities of their component molecules. The shapes and polarities of molecules are important only in relation to other molecules. For example, the melting point of a solid such as sulfur is influenced by the shape of the S_8 molecules that compose it. There are, of course, wide variations in the properties of individual molecular substances. We would expect the properties of substances composed of polar covalent molecules to differ considerably from those composed of nonpolar covalent molecules. Properties such as melting points and boiling points depend on attractions between molecules. Thus, the properties of a macroscopic sample depend on the polarity of the individual molecules. Molecular polarity, in turn, depends on the shape of the molecule and the polarity of the bonds between the atoms which compose the molecule.

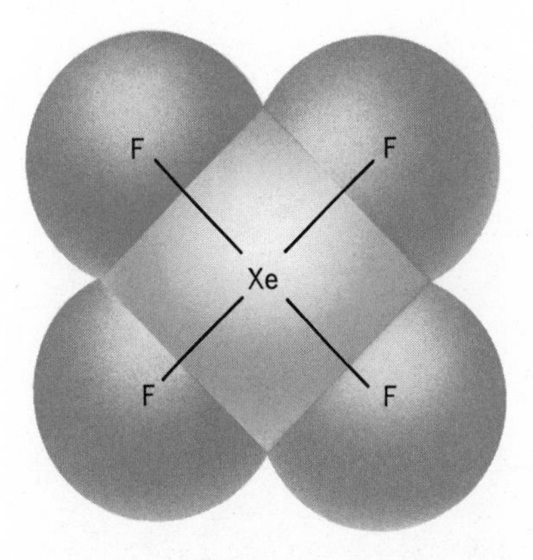

8-42 A space-filling model of the square planar XeF_4 molecule.

8-30 Bond Polarity The polarity of a bond is related to the difference in the electron affinities or electronegativities of the bonding atoms. The larger the difference, the greater is the polarity. When the difference is great enough, it is assumed that the "shared" electrons become the "property" of the more electronegative atom. In other words, an electron is transferred from one atom to another. This results in the formation of a positive and negative ion which carry charges equal to or multiples of the unit electronic charge. Since electron transfer in polar molecules is not complete, the atoms behave as though they carry a partial electronic charge. This partial charge, symbolized by the Greek letter delta (δ), represents the extent of electron transfer and may be used to calculate the percent ionic character in a bond. Let us now go from bond polarity to molecular polarity.

8-31 Molecular Polarity We have already learned that a *diatomic molecule is nonpolar if the atoms are identical and polar if the atoms are different* in electronegativity. When, however, two or more atoms are bonded to a third atom, the polarity of the molecule depends not only on bond distances but also on bond angles. Molecules in which the atoms attached to the central atom are

8-43 As the difference in electronegativity between two atoms increases the bond between them becomes more polar. When the atoms acquire a discrete charge, the bond is said to be ionic. The degree of distortion of an ion's electron cloud is related to the relative size and charge of the two ions.

identical and *symmetrically* arranged are nonpolar. This means that a molecule may be nonpolar even though all of the covalent bonds in the molecule are polar. For example, experiments designed to test the polarity of the molecules show that CH_4, BF_3, and BeF_2 are nonpolar. All the covalent bonds in each of these molecules hold together dissimilar atoms and are, therefore, polar.

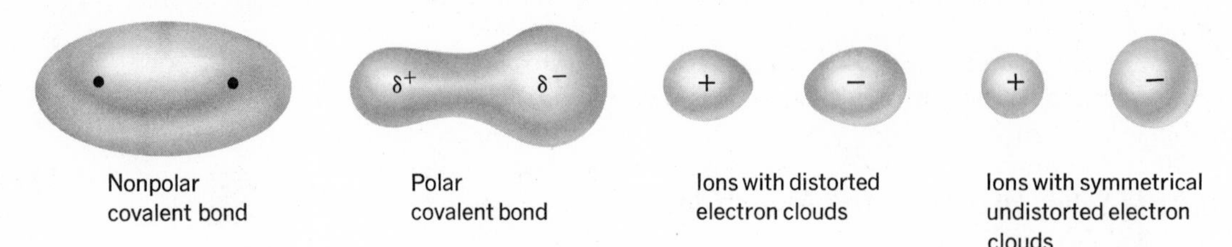

The polarity of a molecule is equal to the geometrical sum of the polarities of the individual bonds. Using the observed bonding angles (Table 8-3) and representing the relative polarity of each bond by the length of an arrow, it can be shown that the polar bonds in CH_4, BF_3, and BeF_2 cancel. The arrows in Fig. 8-45 show that the two bonds in BeF_2 are equal in polarity and opposite in direction. Thus, they cancel, and the molecule has a net polarity of zero. In other words, it is nonpolar.

On the other hand, the arrangement of atoms in water or in ammonia molecules is not symmetrical. As a result of the unsymmetrical distribution of electron charge, one end of these molecules is negative with respect to the other end. In other words, water and ammonia are dipoles (Fig. 8-46). The degree of molecular polarity has an important bearing on the behavior of substances. We shall see in Chapter 14 that the solubility of ionic and certain molecular compounds in water can be partially explained in terms of the high polarity of water molecules.

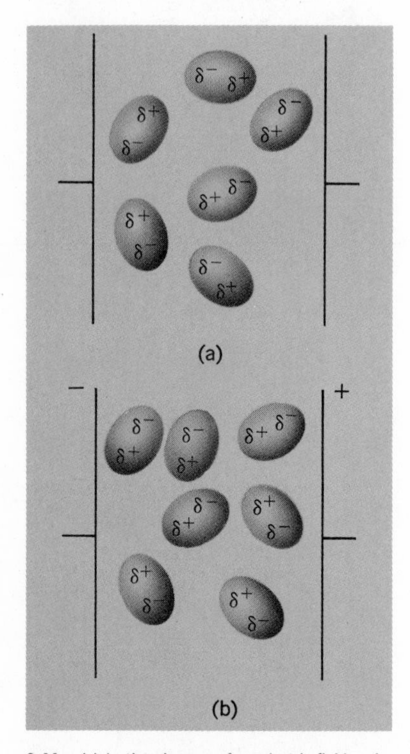

8-44 (a) In the absence of an electric field, polar molecules are randomly oriented. (b) In the presence of an electric field, the positive pole of the molecules is generally oriented toward the negative plate, and the negative pole of the molecules is generally oriented toward the positive plate. The more polar a molecule is, the greater is its effect on the electric field.

LOOKING AHEAD

In this chapter we have developed a covalent-bond model that enables us to predict with reasonable correctness the electric characteristics (polarity) and geometrical shapes of molecules. We concluded our discussion of covalent bonds by showing that all polar covalent bonds have a partial ionic character. We found that the degree of ionic character increases as the difference in the electronegativities of the bonding atoms increases.

In the next chapter we shall extend our discussion to include the bonds that exist between atoms having relatively low ionization energies (low electronegativities) and those having relatively high electron affinities (high electronegativities). These bonds are called ionic bonds and are associated with an aggregate of oppositely charged ions known as an ionic crystal. In our discussion we shall relate the observable properties of the aggregate to the nature of the ionic bond.

We shall then extend our discussion of aggregates to include liquids and solids composed of covalent molecules. This leads us to introduce and investigate the nature of intermolecular (between molecule) forces and bonds. Our understanding of the composition, shapes, and polarities of covalent molecules will help us explain intermolecular forces and the observable properties of molecular aggregates such as water and other familiar substances.

In addition to ionic and molecular aggregates, we shall briefly investigate the structure, bonding, and properties of atomic aggregates such as network solids and metallic crystals. This investigation will enable you to explain, among other things, why diamonds are hard and poor conductors of electricity while metals are ductile and good conductors of heat and electricity.

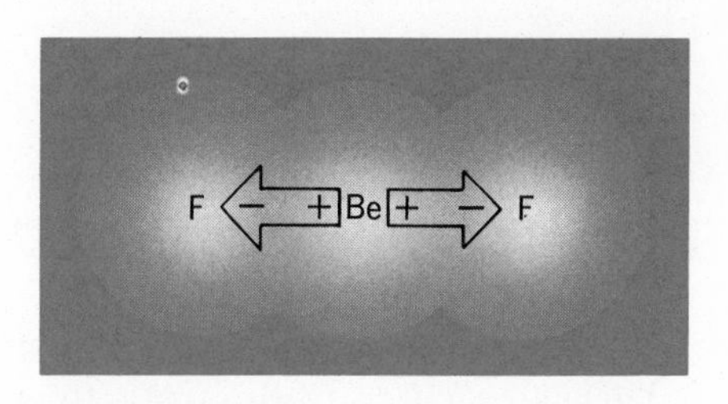

8-45 Nonpolar BeF_2 molecule. The polarities of the individual bonds are oriented at 180° from each other.

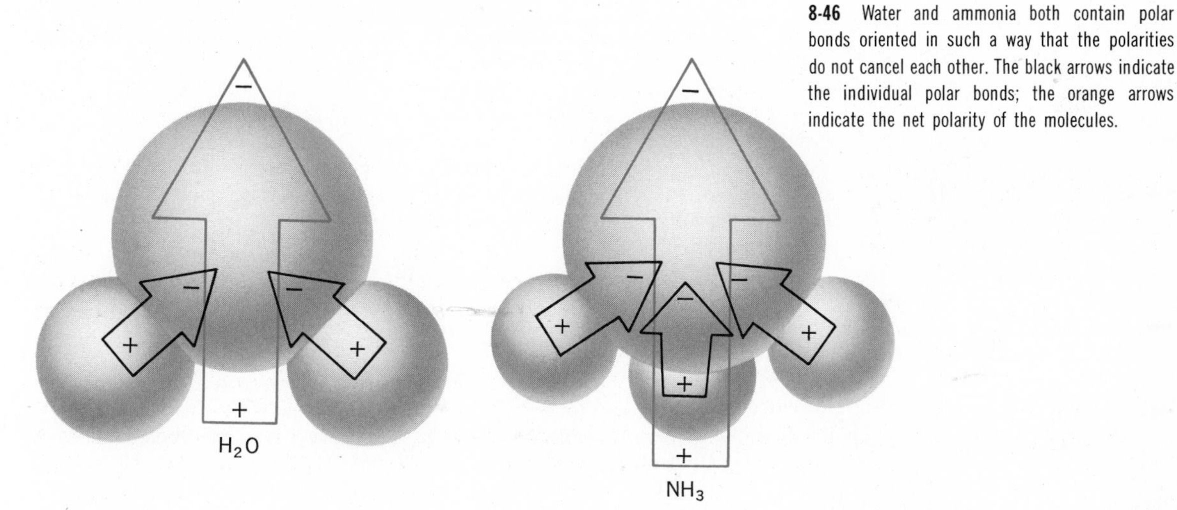

8-46 Water and ammonia both contain polar bonds oriented in such a way that the polarities do not cancel each other. The black arrows indicate the individual polar bonds; the orange arrows indicate the net polarity of the molecules.

QUESTIONS

1. Identify two repulsive forces and two attractive forces that influence the formation of chemical bonds between two atoms.

2. Energetically speaking, which is the more stable, hydrogen atoms or a hydrogen molecule? Explain.

3. Where are the bonding electrons most apt to be found in a diatomic covalent molecule such as H_2 or Cl_2?

4. Describe qualtitatively why two hydrogen atoms form an H_2 molecule while He remains monatomic.

5. Explain why it is not possible to describe the bonding of compounds in terms of purely ionic or purely covalent bonds.

6. Explain what is meant by a polar bond.

7. Which of these bonds is most polar and which is least: S—O, Cl—Cl, or Cl—O? Justify your arrangement.

8. Which has the greater degree of ionic character, a N—F bond or a Se—F bond? Explain.

9. Which bond is the more polar in each of these pairs: (a) H—O or S—O, (b) H—O or H—S, (c) B—O or C—O, (d) N—Cl or N—F, (e) Sn—Cl or P—Cl?

10. Which of the bonds in Problem 9 has the smallest partial ionic character?

11. (a) What is meant by an octet structure? (b) What evidence indicates that octet structures might be more stable than others?

12. Draw electron-dot structures of (a) SO_3^{2-}, (b) ClO_4^-, (c) ClO_3^-, (d) CCl_4, (e) N_2, (f) NCl_3, (g) BrCl, (h) BeO, (i) SeF_6, (j) OF_2, (k) HOCl, (l) HCN, (m) BrO_3^-, (n) SiO_2.

13. What reasons can you give for thinking that triple bonds are shorter and stronger than double or single bonds?

14. (a) In which is the bond distance the greater, C—N or C≡N? (b) In which is the bond energy (strength) the greater, C—N or C≡N?

15. Phosphorus forms PCl_3 and PCl_5. Nitrogen forms NCl_3 but not NCl_5. Explain why NCl_5 does not exist.

16. What is meant by a coordinate covalent bond?

17. (a) Draw a Lewis structure for an hydronium ion (H_3O^+) and identify the coordinate bond. (b) In this ion, how does the strength of the coordinate bond differ from the other bonds?

18. (a) What is meant by resonance? (b) In what situations does the concept apply?

19. (a) What experimental evidence indicates that the Lewis structure does not represent the structure of a nitrate ion accurately? (b) Draw two other Lewis structures for a nitrate ion. (c) What are the three structures called? (d) What is a nitrate ion called in valence-bond terminology? (e) Would a nitrate ion actually be more or less energetically stable than the structures represented in (a) and (b)?

20. Explain what is meant by a polar molecule.

21. Must a polar molecule contain a polar bond?

22. Is it possible for a nonpolar molecule to contain polar bonds? Explain using examples.

23. How can you explain the fact that HF is polar while H_2 and F_2 are not?

24. (a) Would you predict SiF_4 to be polar or nonpolar? Justify your answer. (b) Is the Si—F bond polar or nonpolar? Justify your answer. (c) Explain why your answers to (a) and (b) are consistent.

25. Since there is a difference in electronegativity between carbon and bromine, how can you explain the lack of polarity in the compound CBr_4?

26. What are the approximate bond angles in each: (a) NO_3^-, (b) SiF_4, (c) PH_3, (d) CO_3^{2-}, (e) SO_3^{2-}, (f) SO_3, (g) $Co(NH_3)_6^{2+}$?

27. Explain why you would expect a BeF_2 molecule to be linear but an SF_2 molecule to be angular.

28. Predict the shapes of (a) PH_3, (b) PH_4^+, (c) SeF_6, (d) SO_3^{2-}, (e) BrO_3^-, (f) CCl_4, (g) CaH_2, (h) H_2S, (i) $Fe(CN)_6^{3-}$, (j) PO_4^{3-}.

29. Which of the neutral molecules in Problem 28 are dipoles?

30. (a) How do you explain the existence of a Cl_2O molecule when both chlorine and oxygen have *common* negative oxidation states. (b) What is the shape of the molecule? (c) Is the molecule polar or nonpolar? (d) If polar, which end of the molecule has the partial negative charge? (e) What is the oxidation state of each atom in the molecule? (f) Draw the Lewis structure for the molecule.

31. (a) Do NH_3 and BF_3 have the same shape? Explain. (b) Can both NH_3 and BF_3 form coordinate covalent bonds? Explain any differences.

32. Aluminum(III) ions react with water molecules and form a species known as a complex ion. The formula of the ion is $Al(H_2O)_6^{3+}$. (a) How many water molecules are bonded to an Al^{3+} ion? (b) Draw a Lewis structure for the complex ion. (c) Which is the electron-pair acceptor, the Al^{3+} ion or the water molecules (H_2O)? (d) What is the shape of this ion?

33. What shapes are associated with these hybrid orbitals: (a) *sp*, (b) sp^2, (c) sp^3, (d) sp^3d^2?

34. (a) How would you designate the hybrid orbitals formed by "mixing" one *d*, one *s*, and two *p* orbitals? (b) How many of the hybrid orbitals would be formed by the mixing?

35. If nitrogen, carbon, and oxygen all form sp^3 orbitals, what reason can you advance for NH_3 and H_2O being polar while CH_4 is not?

36. The bond angles in NH_3, H_2O, and CH_4 are 107°, 104.5°, and 109.5°, respectively. How can these values be justified if sp^3 orbitals are involved in each case?

9

Aggregates: Bonding and Properties

Preview

"Imagination is more important than knowledge."

—Albert Einstein

Most of the pure substances which we use in our laboratory experiments are either solids or liquids at ordinary room conditions. The smallest measurable quantity of a solid or liquid contains countless atoms, molecules, or ions. Thus, the observed physical properties of a pure substance are not those of individual particles but, rather, are those of an aggregate of particles (atoms, molecules, or ions). The particles themselves are held together by attractive forces (chemical bonds). For example, we cannot say that a sodium atom has a melting point of 97.5°, that a sulfur atom is yellow and a carbon atom is black, or that a chlorine molecule has a density of 1.56 g/ml. It is true, however, that metallic crystals of sodium do melt at 97.5°C, that nonmetallic crystals of sulfur are yellow, and that liquid chlorine at −30°C has a density of 1.56 g/ml.

In this chapter, one of our objectives is to relate the physical properties of aggregates to the nature of the particles of which they are composed, and to the relative strength of the attractive forces which exist between the particles. We shall develop models of aggregates which will enable us to explain the wide variation in the physical properties of different pure substances, and qualitatively answer such questions as: What is the reason that the melting point of helium is less than −272°C and that of tungsten is greater than 3400°C? How does one explain that sodium (Na) conducts electricity in both the solid and liquid phases, but that sodium chloride (NaCl) conducts only in the liquid phase, and that chlorine (Cl_2) conducts in neither phase? Why does water (H_2O) have a boiling point of 100°C at 1 atm pressure but hydrogen sulfide (H_2S), a similar compound, a boiling point of −60.7°C? How does one explain the waxiness or softness of a candle, the brittleness of a salt crystal, and the ductility of a metal such as copper? What is the explanation of these properties?

To develop generalizations which we can apply to large numbers of substances, we shall classify substances in terms of the structural units which make up the aggregates and the type of bond or force that exists between the units. The classification scheme is summarized below.

1. *Ionic substances*. Examples are: NaCl, K_2SO_4, $Cu(NO_3)_2$, and BaO. The structural units which make up ionic solids are *oppositely charged ions*. The force of attraction between the ions is called an *ionic bond*.

JOHANNES D. VAN DER WAALS
1837–1923

Johannes van der Waals was a native of the Netherlands and attended the University of Leyden. He later taught at Deventer, The Hague, and finally became a professor of physics at the University of Amsterdam. Van der Waals was concerned about the deviations from Boyle's Law at higher pressures. His doctoral dissertation, "On the Continuity of the Gaseous and Liquid States," brought him immediate recognition. He introduced two factors in his *Equation of State* which accounted for the deviations in the behavior of gases. These factors were: the volume occupied by gaseous molecules themselves and the attractive forces between the molecules. The attractive forces between the molecules in a gas are called van der Waals forces. Since many chemical phenomena involving solids, liquids, and gases require the consideration of these intermolecular forces, the name of van der Waals appears frequently throughout every modern chemistry textbook.

Van der Waals' theoretical calculations and predictions stimulated James Dewar of England and others to attempt to achieve the liquefaction of the so-called "permanent gases" such as oxygen, nitrogen, and nitric oxide. As a result of their investigations, van der Waals made the interesting prediction that the pressure just inside a droplet of water is about 10,000 atmospheres. Van der Waals was awarded the 1910 Nobel Prize in physics for his efforts.

2. *Molecular substances*. The structural units of molecular solids are either *nonpolar* or *polar molecules*. Examples of the former are sulfur (S_8) and methane (CH_4). Examples of the latter are water (H_2O) and sugar ($C_{12}H_{22}O_{11}$). The attractive forces between the molecules themselves are called *van der Waals forces*.

3. *Network or macromolecular solids*. These substances may be visualized as a gigantic network of *atoms* held together by *covalent bonds*. Examples are diamonds (C) and quartz (SiO_2).

4. *Metallic substances*. Examples are sodium (Na), iron (Fe), and copper (Cu). The structural units which make up metallic solids are *atoms* (positive ions). The atoms are held together by *metallic bonds*.

Let us now examine the properties of ionic substances and then develop an ionic-bond model that will account for the properties observed.

IONIC SUBSTANCES AND IONIC BONDS

9-1 Properties of Ionic Substances The properties of sodium chloride and most other compounds known as salts cannot be explained in terms of molecules. The electrolysis of liquid sodium chloride described in Section 2-9 provides indirect evidence that the liquid is composed of charged particles. X-ray diffraction provides additional evidence that the basic structural units of solid crystals are oppositely charged ions. Laboratory experiments reveal that, in general, ionic substances are characterized by the properties listed below.

1. In the solid phase at room temperature they do not conduct an appreciable electric current.
2. In the liquid phase they are relatively good conductors of an electric current. The conductivity of ionic substances is much smaller than that of metallic substances.
3. They have relatively high melting and boiling points. There is a wide variation in the properties of different ionic compounds. For example, potassium iodide (KI) melts at 686°C and boils at 1330°C while magnesium oxide (MgO) melts at 2800°C and boils at 3600°C. Both KI and MgO are ionic compounds.
4. They have relatively low volatilities and low vapor pressures. In other words, they do not vaporize readily at room temperatures.
5. They are brittle and easily broken when a stress is exerted on them.
6. Those that are soluble in water form electrolytic solutions which are good conductors of electricity. There is, however, a wide range in the solubilities of ionic compounds. For example, at 25°C, 92 g of sodium nitrate ($NaNO_3$) dissolves in 100 g of water while only 0.0002 g of $BaSO_4$ dissolves in the same mass of water.

These observations suggest that most ionic crystals are energetically stable and that the ionic bond is a relatively strong chemical bond. Let us examine a typical ionic compound for clues that will help us identify which atoms in the Periodic Table are most likely to be found in ionic compounds and also help us to explain the relatively high stabilities of the compounds.

9-2 Ionic Bonds Consider sodium chloride (NaCl). When metallic sodium is heated in an atmosphere of chlorine gas, a white crystalline solid composed of positive sodium and negative chloride ions is formed. The existence of these ions in the crystal suggests that the net overall reaction involves the transfer of an electron from a neutral sodium atom to a neutral chlorine atom. We can use the concepts of ionization energy, electron affinity, and electronegativity to interpret this observation. We would expect atoms with high electron affinities or high electro-negativities to remove electrons from atoms with low ionization energies or low electronegativities. Thus, when chlorine atoms (electronegativity 3.0) collide with sodium atoms (electronegativity 0.9) there can be sufficient overlap of the electron clouds for the 3*s* electron of sodium to have a stronger attraction to the chlorine nucleus than to its own nucleus.

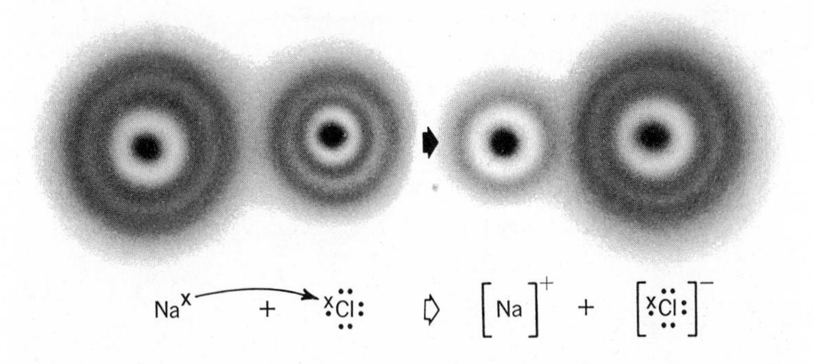

9-1 When sodium reacts with chlorine, ions are formed which have the same electron configuration as a noble gas. A Na^+ ion is smaller than a Na atom, and a Cl^- ion is larger than a Cl atom.

Thus, in the reaction (Fig. 9-1), the single electron in the 3*s* orbital of a sodium atom is transferred to the half-filled 3*p* orbital of a chlorine atom.

Each ion formed in the reaction has an octet structure and is an electrically charged particle surrounded by a nondirectional electric field which is uniformly distributed about the ion. In our model, *an ion may be visualized as a spherical particle whose charge is concentrated at the center of the sphere.* It should be noted that 118 kcal is absorbed when 1 mole of gaseous sodium atoms is converted to 1 mole of gaseous sodium ions whereas only 87 kcal is evolved when a mole of gaseous chlorine atoms is converted to gaseous chloride ions. This means that the gaseous ions have a greater potential energy (31 kcal/mole) and, thus, less stability than the gaseous atoms. This process would not tend to occur spontaneously unless there were a simultaneous change which lowered the potential energy of the system. It is the coalescing of positive and negative ions (Fig. 9-2) that leads to a lower energy, more stable aggregation called an *ionic crystal. The forces of attraction between the positive and negative ions in an ionic crystal constitute*

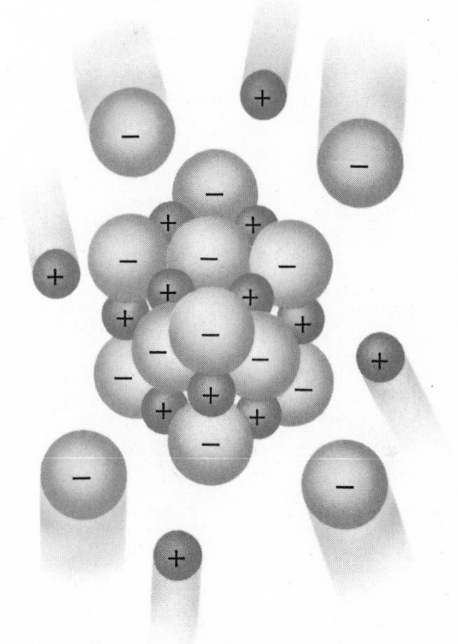

9-2 The potential energy of a system decreases when positive and negative ions coalesce and form stable aggregates called ionic crystals.

ionic bonds. This force of attraction between the ions accounts for the high stability of the crystal. This implies that *ionic bonds are formed in response to the attractive forces between ions rather than in response to any tendency to form an octet or noble-gas structure.*

Stability is a *relative* term which is related to a specific process. Because a relatively large amount of heat energy is required to disrupt their lattice structure, most ionic crystals are said to be stable with respect to their conversion into the liquid or vapor phases. It should be noted that certain ionic crystals containing polyatomic ions are *relatively* unstable with respect to their thermal decomposition into simpler pure substances.

The distance between the nuclei of adjacent ions in a crystal (***interionic distance***) is determined by the tendency of the outer electrons of the ions to repel one another. The positions of the ions may be determined by X-ray diffraction and are represented by lattice points in the crystal. The three-dimensional array of ***lattice points*** is known as the ***lattice structure*** of the crystal. The lattice structure of sodium chloride is shown in Fig. 9-3. The attractive forces between the particles restrict the particles to a vibratory motion at these points. The kinetic energies associated with the vibratory motion of the particles in a solid follow a characteristic distribution curve similar to that of gases.

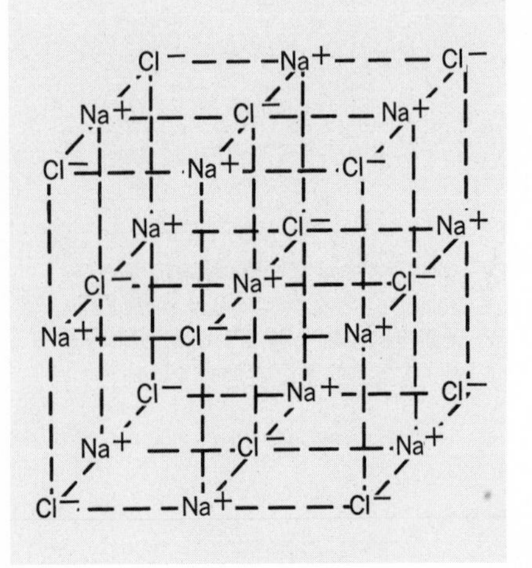

9-3 The lattice structure of the sodium chloride crystal. The symbols represent the location of the nuclei of the atoms in space.

The properties of ionic solids depend on the stability of the crystal. The stability of the crystal lattice is measured experimentally in terms of the ***crystal-lattice energy.*** This is the energy liberated when one mole of an ionic crystal is formed from gaseous ions. The crystal-lattice energy of NaCl is -182 kcal. The negative sign indicates that the process is exothermic and that the energy of the system decreases as the crystal forms. This relatively high value of the crystal energy of NaCl is the reason that NaCl is thermally stable and has a low vapor pressure.

9-4 Sodium chloride. (a) Ordinary table salt processed and purified by industry. (b) Rock salt crystals found in nature. (c) Laboratory grown cubic crystal of NaCl. (d) Model showing how Na^+ and Cl^- ions pack in a cubic crystal. (Photos for (a) and (b) by Donald Reddick).

Let us now examine the role played by ionic radius and charge in determining bond strength and crystal stability.

9-3 Factors Affecting Bond Strength and Crystal Stability We have indicated that ions tend to approach each other until the repulsion of their charge clouds causes the potential energy of the system to increase. In a crystal, the interionic distance is the distance between the nuclei of adjacent ions when the potential energy of the system is at a minimum (Fig. 9-5). This distance can be determined by X-ray measurements. By measuring the interionic distances in a number of different crystals formed by an element and considering various other factors, scientists have been able to determine what fraction of the internuclear distance is occupied by each ion. Thus, values for the radii of most ions have been calculated and tabulated.

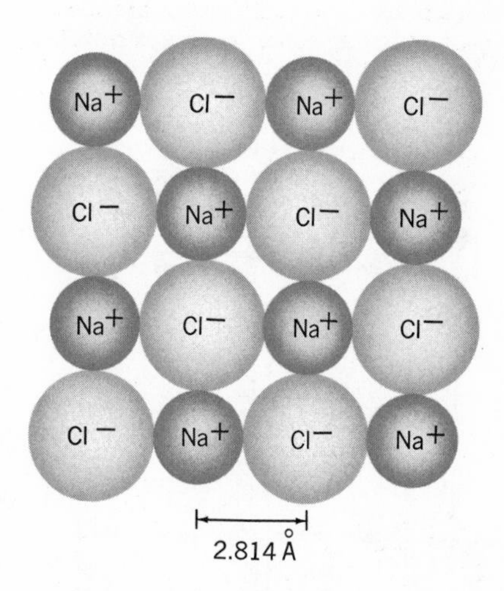

9-5 Sodium ions and chloride ions pack into a solid crystal in which each sodium ion is in contact with six chloride ions. The distance between the nucleus of any Na^+ ion and that of any adjacent Cl^- ion (interionic distance) is 2.814 Å.

The force of attraction between ions is determined largely by the radii of and the charges on the ions. Since the charge of an ion acts as though it were concentrated at the center of the ion, the force of attraction between ions may be expressed by the inverse square law

$$\mathbf{F = kq_1q_2/r^2}$$

where k is a constant of proportionality, q_1 and q_2 are the charges of the ions and r is the sum of the radii of the positive and negative ions ($r^+ + r^-$). One can see that the larger the ionic radius, the greater is the distance between the ions and the smaller is the force of attraction. The bond strength and the stability of the crystal are proportional to this force of attraction. *The greater the attraction, the stronger is the bond and the more stable is the crystal.* The role played by the ionic radius is illustrated by comparing sodium chloride with cesium chloride. A cesium ion is much larger than a sodium ion. The distance between the centers of the Cs^+ ions and Cl^- ions in CsCl is much greater than that between the Na^+ ions and Cl^- ions in NaCl (Fig. 9-6). Consequently the force of attraction between Cs^+ and Cl^- is much less than that between Na^+ and Cl^-. This suggests that the ionic bonds in CsCl are weaker than those in NaCl.

Weaker bonds correspond to smaller lattice energies and less stable structures. We may conclude correctly that CsCl is less stable than NaCl, and it is easier to separate a Cs^+ ion from Cl^- in a CsCl crystal than to separate Na^+ from Cl^- in NaCl. On this basis, we would predict that CsCl would have a lower melting point than NaCl.

TABLE 9-1
MELTING POINTS AND SOLUBILITIES OF TWO IONIC COMPOUNDS

	Melting point (°C)	Solubility (g/100 g H_2O, 0°C)
CsCl	646	161
NaCl	800	35.7

The dissolving of an ionic substance also involves the separation of ions. Other factors being equal, therefore, the solubility of CsCl should be greater than that of NaCl. The data in Table 9-1 show that this prediction is correct.

Experiments show that, in general, solutes composed of polar molecules or of ions are more soluble in polar solvents than in nonpolar solvents. Thus, sugar and salt dissolve in water but not appreciably in benzene. On the other hand, solutes composed of nonpolar molecules have a much greater tendency to dissolve in nonpolar solvents than in polar ones. These phenomena are discussed in Chapter 14.

The role played by the charge on the ions in determining bond strength and crystal stability may be shown by comparing magnesium oxide (MgO), and sodium chloride (NaCl). Since the force of attraction between charged particles is proportional to the magnitude of the charges, the attraction between doubly charged magnesium ions and doubly charged oxide ions is greater than that between singly charged sodium and chloride ions. The large lattice energy of MgO is a reflection of this large attractive force and is the main reason that MgO crystals are more stable than NaCl crystals. The melting points and solubilities of the two crystals are shown in Table 9-2.

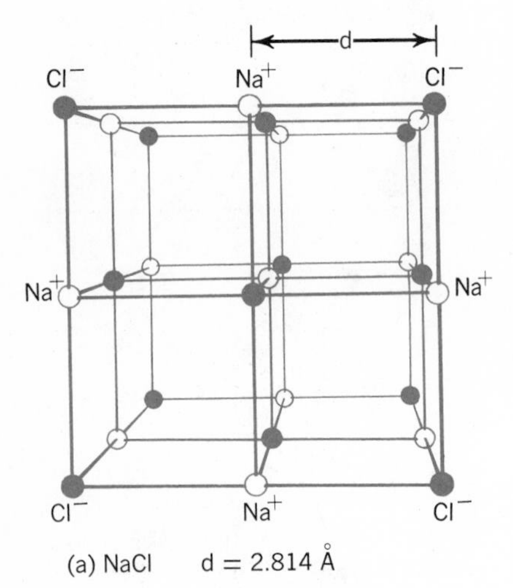

(a) NaCl d = 2.814 Å

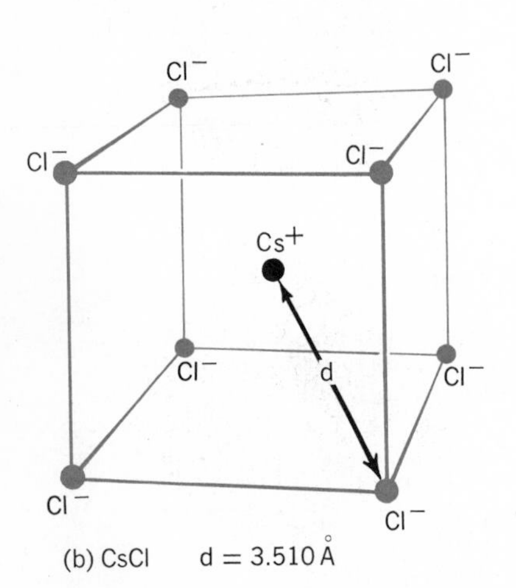

(b) CsCl d = 3.510 Å

9-6 Sodium chloride crystallizes in a face-centered structure in which each Na^+ ion is surrounded by six Cl^- ions. Cesium chloride crystallizes in a body-centered cubic structure in which each Cs^+ ion is surrounded by eight Cl^- ions as nearest neighbors.

FOLLOW-UP PROBLEM

Which of these substances has the highest melting point? Explain your choice. (a) BeO, (b) BaO, (c) KCl.

There is a wide range in the physical properties of ionic substances and these differences can sometimes be accounted for in terms of ionic radii and charge. In Chapter 14, we shall use these concepts to help us explain the differences in solubilities of certain ionic compounds. At this point, we are primarily interested in identifying ionic compounds and their general properties. We can use the Periodic Table to help us predict those elements which are most likely to be found in ionic compounds by identifying atoms that have low ionization energies or low electronegativities, and those that have high electron affinities or high electronegativities.

9-4 Use of the Periodic Table to Identify Ionic Substances In our study of periodic properties we found that all Group-IA elements have relatively low ionization energies and that all Group-VIIA elements have relatively high electron affinities. We would predict therefore that the elements from these two groups would tend to react and to form stable compounds in which the bonds have a high degree of ionic character. This is, to a smaller extent, also true for the elements in Groups IIA and VIA. In the binary ionic compounds containing elements from these groups, the ions have the characteristic octet structure of noble-gas atoms. It should be noted, however, that not all ions have this configuration. Notable exceptions are the ions formed by the atoms of the transition elements. These atoms may transfer their outer *s* electrons but because of the presence of underlying *d* orbital electrons, form ions having as many as eighteen electrons in the outer energy level.

9-5 Use of the Ionic-Bond Model to Interpret Behavior of Ionic Substances We can now explain some of the general properties of ionic crystals in terms of their composition and the strong electrostatic forces of attraction between positive and negative ions.

1. *At room temperature, solid ionic crystals do not conduct an appreciable electric current.* The electrons are firmly held by the

nuclei of the individual atoms, and the ions are bound together by strong electrostatic attractions. Even under the influence of an applied voltage, the ions are limited to vibratory motion and are unable to move about to any great extent; therefore, no current can flow.

2. Melted ionic substances are moderate conductors of electric current. When heated, the crystals reach a point where their ionic vibrations become so great that they break free from their fixed positions and begin to slide freely over each other, forming a liquid. The temperature at which this change occurs is the *melting point.* In the liquid phase, the mobile ions are free to move under the influence of an applied voltage and thus an electric current (transfer of charge) results. As noted earlier, ionic conduction differs in several ways from metallic conduction. In metals, the electric charge is transferred by electrons and because there is more resistance to the flow of ions than to the flow of electrons, metallic substances are much better electric conductors than ionic substances. Recall that a chemical change occurs when a direct current is passed through liquid sodium chloride. No such change is observed when a direct current is passed through a metallic conductor.

3. *Ionic crystals have relatively high melting and boiling points, low volatilities, and low vapor pressures.* Ions which are tightly bound by strong attractive forces require relatively large amounts of energy in order to escape from the crystal lattice. In general, we find that ionic crystals composed of small ions with high charges have the highest melting points and lowest vapor pressures. Thus, aluminum oxide (Al_2O_3) melts at 2045°C and has a vapor pressure of only one torr when heated to 2148°C.

4. *Ionic crystals are brittle and are easily broken when a stress is exerted on them.* Ionic crystals tend to be hard and brittle because of the nature of the packing. The oppositely charged ions surround each other in the crystal. A stress tending to distort the crystal brings similarly charged ions in contact with one another. The result is repulsion, instability, and the breakdown of the crystal. The ions of crystalline substances are arranged in a definite geometrical fashion characteristic of the particular substance.

These crystals have definite boundaries or planes which intersect at certain precise angles. Thus, when a crystalline solid is split, it fractures (cleaves) along one of these planes, leaving the faces of the crystal at an angle which is characteristic of the given crystal. As an example, this angle is 90° in a sodium-chloride crystal.

5. *Many ionic solids composed of polyatomic ions are thermally unstable.* It should be noted that ionic crystals containing a polyatomic ion sometimes decompose when heated. Many salts containing carbonate ions (CO_3^{2-}), bicarbonate ions (HCO_3^-), hydroxide ions (OH^-), and ammonium ions (NH_4^+) can be decomposed when heated to simpler compounds, one or more of which is a gas. General rules which might help you in predicting products of

TABLE 9-2
MELTING POINTS AND SOLUBILITIES OF TWO IONIC COMPOUNDS

	Melting point (°C)	Solubility ($g/100\ g\ H_2O$, 0°C)
MgO	2800	0.0006
NaCl	800	35.7

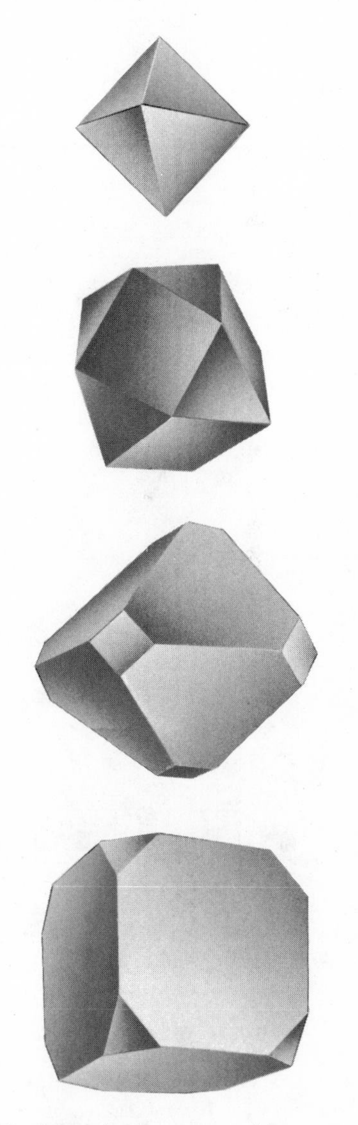

9-7 Shapes of some common crystals.

decomposition reactions are listed below.

1. Metallic carbonates and bicarbonates usually decompose and form carbon dioxide and a metallic oxide.
2. The decomposition of hydroxides yields oxide ions and water.
3. Ammonium compounds decompose and may produce ammonia gas (NH_3).
4. Nitrates often decompose and yield oxygen gas and an oxide of nitrogen.
5. Chlorates usually decompose and form a metallic chloride and oxygen gas.

The equations for the decomposition of several of these thermally unstable salts are given below.

$$CaCO_3(s) + \text{heat} \longrightarrow CaO(s) + CO_2(g)$$
$$2NaHCO_3(s) + \text{heat} \longrightarrow Na_2CO_3(s) + H_2O(g) + CO_2(g)$$
$$NH_4HCO_3(s) + \text{heat} \longrightarrow NH_3(g) + H_2O(g) + CO_2(g)$$
$$Ca(OH)_2(s) + \text{heat} \longrightarrow CaO(s) + H_2O(g)$$
$$2Cu(NO_3)_2(s) + \text{heat} \longrightarrow 2CuO(s) + 4NO_2(g) + O_2(g)$$
$$2KClO_3(s) + \text{heat} \longrightarrow 2KCl(s) + 3O_2(g)$$

FOLLOW-UP PROBLEM

Write equations for the thermal decomposition of (a) $BaCO_3$, (b) $Ca(HCO_3)_2$, (c) NH_4Cl, (d) $Mg(OH)_2$, (e) $Zn(NO_3)_2$.

In the last chapter we found that many common substances such as water (H_2O), carbon dioxide (CO_2), ammonia (NH_3), chlorine (Cl_2), and methane (CH_4) are composed of covalent molecules. In the solid phase, these substances are referred to as *molecular crystals*. The properties of these solids are related to the nature (polarity), shape, and arrangement of the basic structural units (molecules) which compose the crystal. Let us now look at some general characteristics of these solids and attempt to interpret them in terms of molecular characteristics and intermolecular forces.

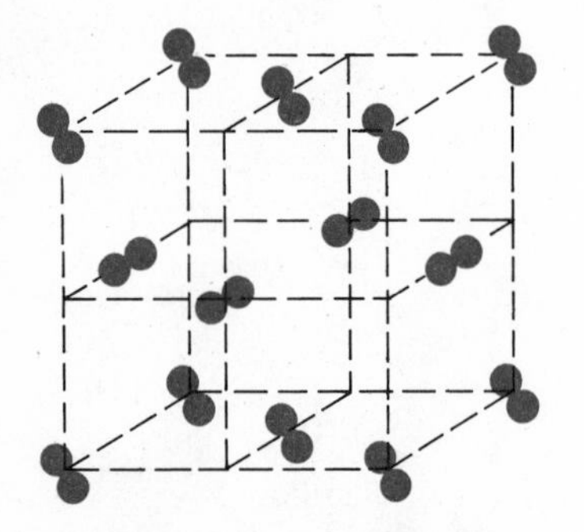

9-8 Orderly arrangement of I_2 molecules in a crystal of solid iodine. The above lattice structure shows the spatial arrangement of the molecules that occupy the lattice points but does not show accurately the size of the molecules.

MOLECULAR CRYSTALS

9-6 General Properties of Molecular Crystals and Liquids Experiments show that, in general, molecular liquids and crystals have these properties

1. Neither the liquids nor the solids conduct an electric current appreciably.
2. Many exist as gases at room temperature and atmospheric pressure and many solids and liquids are relatively volatile.

3. The melting points of solid crystals are relatively low.
4. The boiling points of the liquids are relatively low.
5. The solids are generally soft and have a waxy consistency.
6. A large amount of energy is often required to decompose the substance chemically into simpler substances.

The first observation suggests that *molecular crystals* are crystalline solids whose lattice points are occupied by neutral molecules. If the points are occupied by polar molecules, we may call the solid a ***polar molecular crystal.*** If the basic structural unit is a nonpolar molecule, then the crystal is called a ***nonpolar molecular crystal*** (Fig. 9-8 and Fig. 9-9).

The next four observations suggest that the intermolecular forces are relatively weak compared to the interionic forces in ionic substances (which are usually solids with relatively high melting and boiling points).

The last observation indicates that the *intra*molecular forces (covalent bonds) are much greater than the *inter*molecular forces. For example, at one atmosphere pressure, water undergoes a phase change from liquid to gas at 100°C (the boiling point), but liquid water does not decompose into H_2 and O_2 appreciably even at 2000°C.

As with ionic substances, there are also wide variations in the properties of individual molecular substances. For example, we would expect a nonpolar molecular solid such as methane (CH_4), Dry Ice (solid carbon dioxide, CO_2) or iodine (I_2) (Fig. 9-9) to be much more ***volatile*** (vaporize more readily) than a polar molecular solid such as ordinary ice (solid water). The attractive forces between nonpolar methane, iodine, or carbon dioxide molecules are much less than those between polar water molecules. It is apparent that properties such as melting points and boiling points depend largely on attractions between molecules and will, therefore, be influenced by the polarities of the individual molecules. As a second example of this difference, the melting point of solid NH_3, an aggregate composed of polar covalent molecules, is −77°C, while that of methane (CH_4), an aggregate composed of nonpolar molecules, is −182°C.

9-7 Effect of Molecular Shape on Properties In the case of *nonpolar* molecular crystals and liquids, the shapes of the molecules play an important role in determining the properties of the substance. To determine the effect of shape we can compare the properties of two different substances composed of nonpolar molecules that differ only in shape. Consider normal pentane and neopentane. These substances have identical molecular formulas (C_5H_{12}). They have, however, different properties and a different arrangement of atoms in their respective molecules, resulting in a difference in shape. Such compounds are known as ***isomers.*** The ability of carbon to bond with other carbon atoms, to hydrogen atoms, and also to atoms of many other elements in a wide variety of combinations

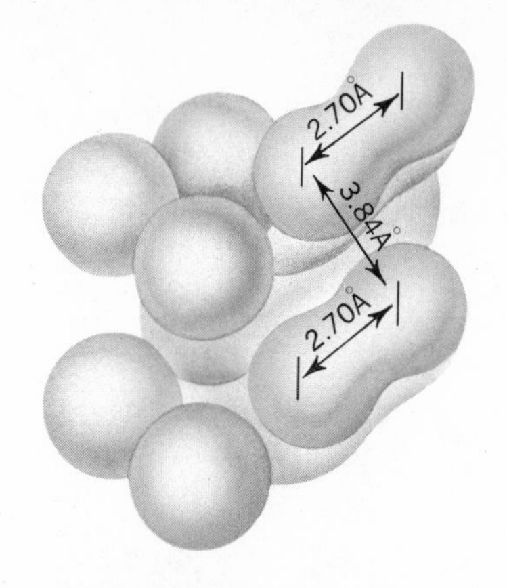

9-9 Section of crystal showing arrangement of I_2 molecules in iodine(s).

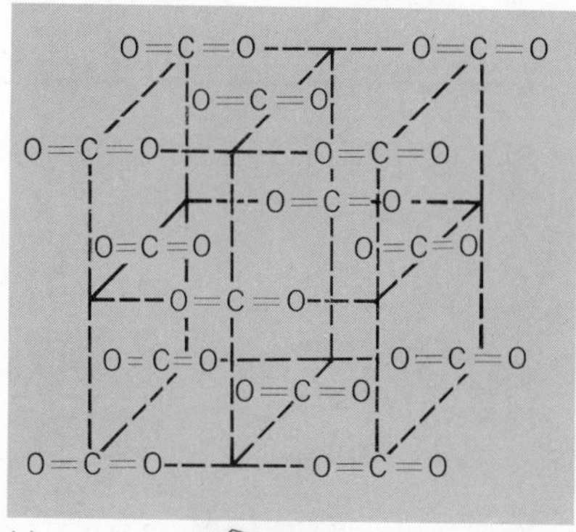

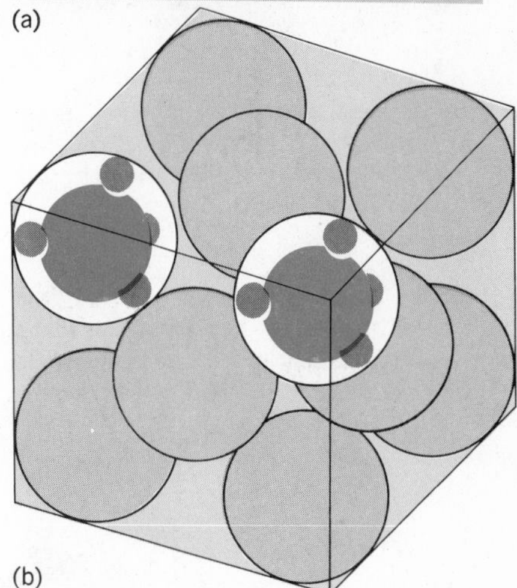

9-10 (a) Spatial arrangement of CO_2 molecules in solid CO_2 (Dry Ice). (b) Space-filling model of solid methane. Each section in the crystal lattice is occupied by a CH_4 molecule. For simplicity all except two CH_4 molecules are represented by spheres.

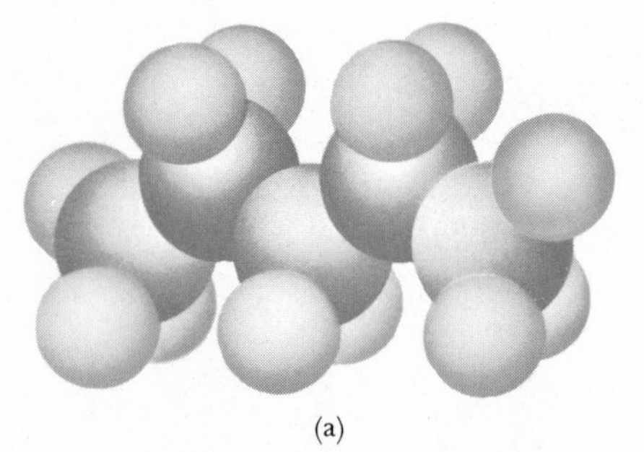
(a)

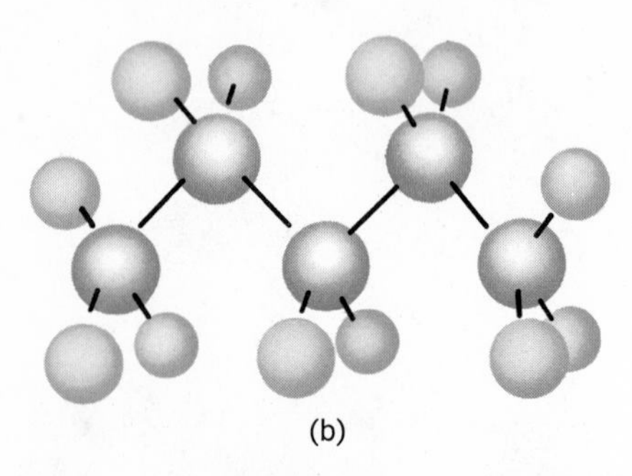
(b)

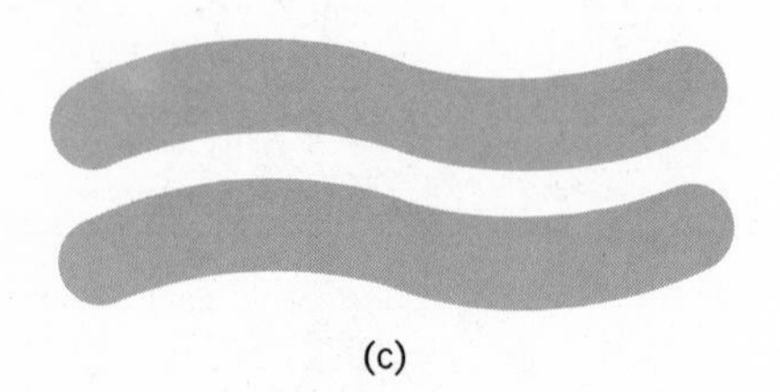
(c)

9-11 Three representations of normal pentane: (a) space-filling model, (b) ball and stick model, (c) silhouettes of two molecules showing large surface contact in liquid.

is largely responsible for the vast number of different compounds containing carbon (organic compounds). Because isomers are similar except for physical properties and shapes, it is reasonable to assume that the difference in properties is related to the difference in shape.

Normal pentane boils at 36°C and melts at −130°C, while neopentane boils at 9°C and melts at −20°C. Normal pentane is a linear, chainlike molecule with a zigzag shape, Fig. 9-11. Because there can be a relatively large surface contact between the molecules in the liquid, forces of attraction are large enough to give a relatively high boiling point. Because of its flexible, chainlike structure, however, it does not easily pack into a regular lattice, so the solid crystal has a relatively low melting point.

Neopentane is a compact, symmetrical, tetrahedral molecule, Fig. 9-12. Thus, it readily packs into a stable crystal lattice having a relatively high melting point. In the liquid state, the compact, ball-like molecules do not afford as much surface contact as the chainlike molecules of normal pentane. Hence, the forces of attraction between neopentane molecules are less than those between normal pentane molecules. Consequently the boiling point of neopentane is lower than that of normal pentane. The characteristics of the two compounds are compared in Table 9-3.

TABLE 9-3
CHARACTERISTICS AND PROPERTIES OF NEOPENTANE AND NORMAL PENTANE

Name	Molecular Shape	Surface Contact in Liquid	Crystal Packing in Solid	Melting Point (°C)	Boiling Point (°C)
Normal pentane C_5H_{12}	zigzag chainlike	large	noncompact, relatively unstable	−130	36
Neopentane C_5H_{12}	compact symmetrical, tetrahedral	small	compact, relatively stable	−20	9

9-12 Three representations of neopentane (tetramethyl methane): (a) space-filling model, (b) ball and stick model, (c) silhouettes of two molecules showing small surface contact in liquid.

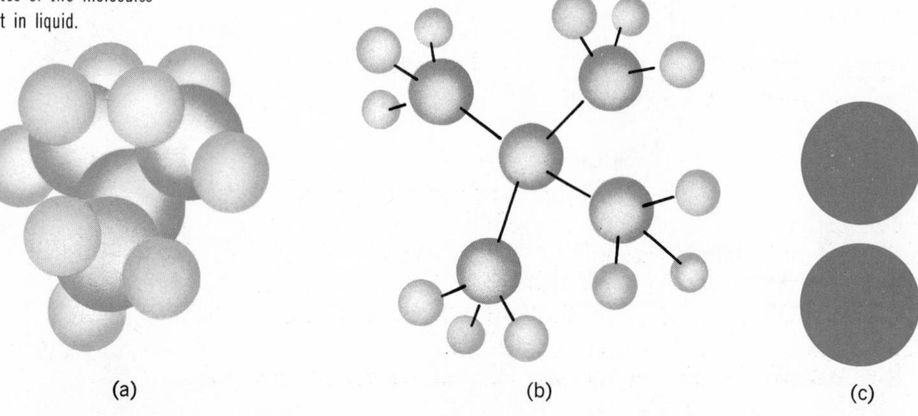
(a) (b) (c)

Another example of the importance of shape in determining behavior is found in sulfur. Ordinary sulfur, when heated, goes through a series of transformations that can be explained in terms of the puckered-ring structure of a sulfur molecule, whose molecular formula is S_8. This molecule is pictured in Fig. 9-13. At 120°C, sulfur melts, forming a clear, light yellow, free-flowing liquid (low viscosity). As the temperature rises from 120°C to 200°C, the viscosity increases to the consistency of molasses. The sulfur becomes dark and opaque at the same time. From 200°C to the boiling point of 440°C, the viscosity again decreases. This behavior is explained by assuming that just above the melting point, the S_8 rings gain enough kinetic energy to slip by and roll over one another easily. As the temperature continues to rise, the rings split into eight-atom chains which may then join and form extremely long-chain molecules of sulfur atoms (Fig. 9-14). The long chains easily become tangled and lose mobility. This represents the viscous stage. Above 200°C, the vibrational kinetic energy becomes sufficiently high so that the chains begin to break into smaller and smaller units and liquid sulfur becomes lower in viscosity as the temperature continues to rise.

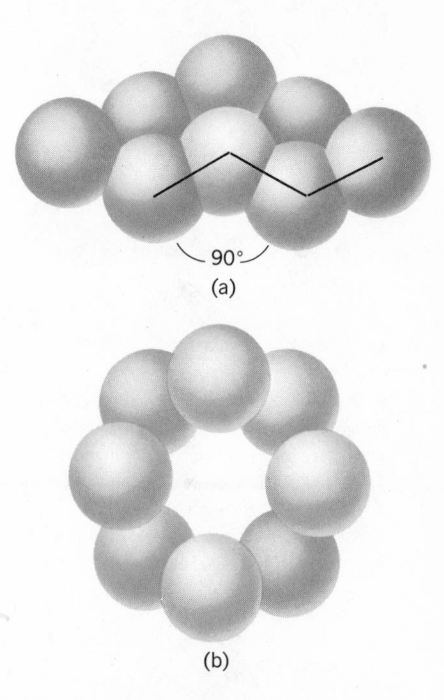

9-13 Two views of an S_8 molecule. Each lattice point in a sulfur crystal is occupied by an eight-membered ring of sulfur atoms. This puckered ring consists of atoms bonded to each other at 90° angles.

We have demonstrated that the melting and boiling points of molecular substances are influenced by the shape, size, and polarity of the molecules, and by the symmetry of the crystals. In general, we can say that the *melting and boiling points of molecular substances are much lower than those of ionic and metallic substances.* This implies that the forces of attraction between the molecules that occupy the lattice points of a molecular crystal are relatively weak. The weak forces between the molecules in molecular crystals and liquids are called ***van der Waals forces.*** Let us now examine the nature and origin of van der Waals forces.

INTERMOLECULAR FORCES OF ATTRACTION

Although it is proper to refer to all intermolecular forces as van der Waals forces, we shall classify intermolecular forces into the three categories.

1. Dipole forces of attraction
2. van der Waals forces
3. Hydrogen bonds

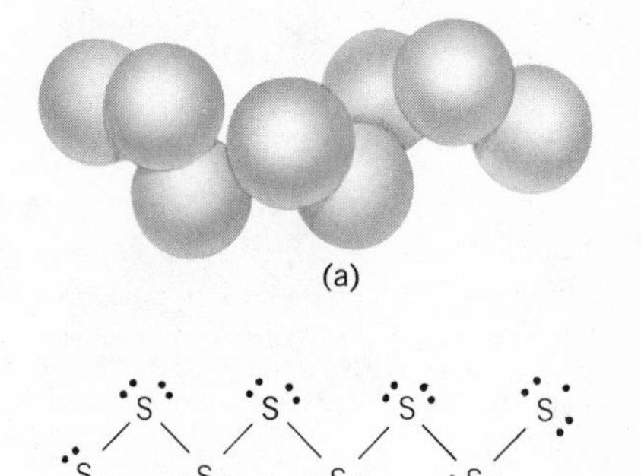

9-14 (a) When sulfur is heated above its melting point, the eight-membered puckered ring opens, forming straight chains of sulfur atoms which link together and form large molecules. (b) Electron-dot formula for S_8. Note that the half-filled orbitals at each end of the chain gives it the ability to bond covalently to the ends of other chains.

Hydrogen bonds and dipole forces exist between specific types of *polar molecules.* We shall use the name *van der Waals forces* to represent intermolecular forces between *nonpolar molecules.* Thus, van der Waals forces have the distinction of being the factor responsible for the existence of the liquid and solid phases of such substances as sulfur, iodine, and hydrogen.

9-8 Dipole Forces In the last chapter we found that polar

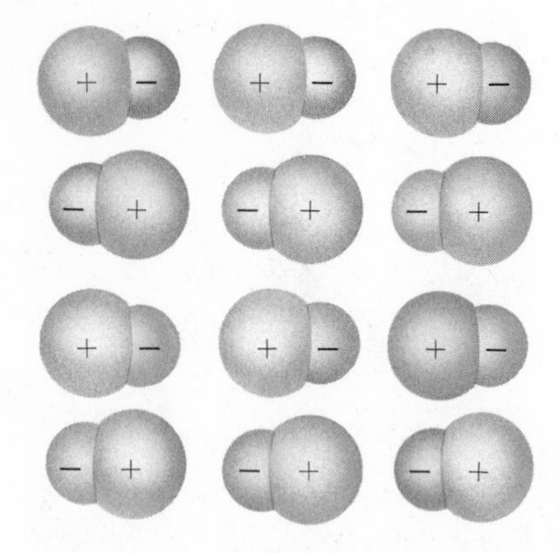

9-15 Dipole-dipole attraction. This type of intermolecular attraction occurs between molecules which have separate centers of positive and negative charge. That is, between polar molecules. In a crystal, the polar molecules are packed so that the positive pole of one molecule is close to the negative pole of an adjacent molecule.

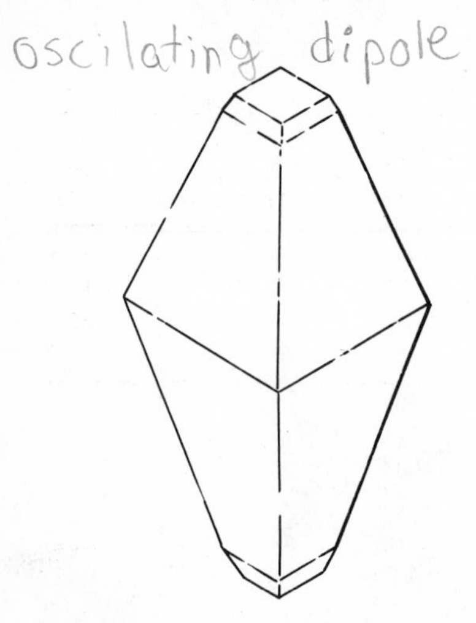

9-16 The most stable packing of the S_8 rings of sulfur at room temperature is in a rhombic crystal.

TABLE 9-4
BOILING POINTS OF RELATED MOLECULAR COMPOUNDS OF GROUP IVA

Formula	Number of Electrons	Boiling Point (°C)
CH_4	10	−161
SiH_4	18	−112
GeH_4	36	−90
SnH_4	54	−52

molecules are characterized by an unsymmetrical distribution of electronic charge that leads to partial positive and negative charges within the molecule. The molecules in which there is a separation of the centers of positive and negative charge are called ***dipoles.*** In polar molecular crystals, the dipoles line up so the positive pole of one molecule is next to the negative pole of another molecule. This is analogous to the behavior of small bar magnets. The force of attraction between polar molecules is called a ***dipole force.*** The forces of attraction between the partial positive and partial negative charges are relatively weak compared to that between the full charges carried by ions in ionic crystals. Thus, solid phosphorous trichloride (PCl_3), a polar molecular crystal, melts at −112°C and sodium chloride (NaCl), an ionic crystal, melts at 901°C.

9-9 van der Waals Forces The type of crystal formed by nonpolar molecular substances is largely a function of the shape of the molecules which make up the crystal. For example, sulfur molecules (S_8) tend to pack together so as to leave as little vacant space as possible. The most stable packing gives rise to *rhombic crystals* (Fig. 9-16). The *intra*molecular forces (covalent bonds) between the sulfur atoms are strong compared to the intermolecular forces (van der Waals forces) between molecules. Thus, sulfur crystals are brittle, readily crushed, and do not flow easily. There is no simple way to predict the structure of molecular crystals. One-half the distance between nuclei in adjacent molecules in a molecular crystal is known as the *van der Waals radius* of the atom. For comparison, both the van der Waals and covalent radii of chlorine are illustrated in Fig. 9-17. It is evident that the van der Waals radius is almost twice the covalent radius. This supports the observation that covalent bonds between atoms are hundreds of times stronger than van der Waals forces of attraction between molecules. Because atoms in molecules usually have no unpaired outer electrons, there can be no overlapping of orbitals and pairing of electrons thus forming strong covalent bonds. There are cases of odd-electron molecules such as NO_2, which contain an unpaired electron. Below a certain temperature two NO_2 molecules bond to each other by a covalent linkage, forming a *dimer,* (N_2O_4). These molecules in turn, if solidified, bond to one another by van der Waals forces.

Unlike dipole forces, van der Waals forces are rather difficult to visualize. They are caused, in part, by temporarily induced dipole effects caused by fluctuations in the density of the electron cloud surrounding an atom or molecule. These effects are illustrated in Fig. 9-18.

The regular rise in boiling points of closely related compounds such as CH_4, SiH_4, and GeH_4 (Table 9-4) suggests that the magnitude of the van der Waals force is largely determined by the number of electrons and the size and shape of the molecule. The greater the number of electrons and the larger the molecule, the stronger is the attractive force between the molecules. In large molecules, the electron cloud is farther from the nucleus and more easily distorted by an adjacent dipole than in smaller molecules.

It should be emphasized that the relative melting points of ionic and molecular crystals do not reflect the relative strengths of ionic and covalent bonds. Melting an ionic crystal partially overcomes the ionic forces which constitute an ionic bond. Melting a molecular crystal separates the molecules in the crystal but does not break the covalent bonds between the atoms in the molecules. The lower melting points of molecular crystals simply means that the force of attraction between molecules in a molecular crystal is less than that between ions in an ionic crystal.

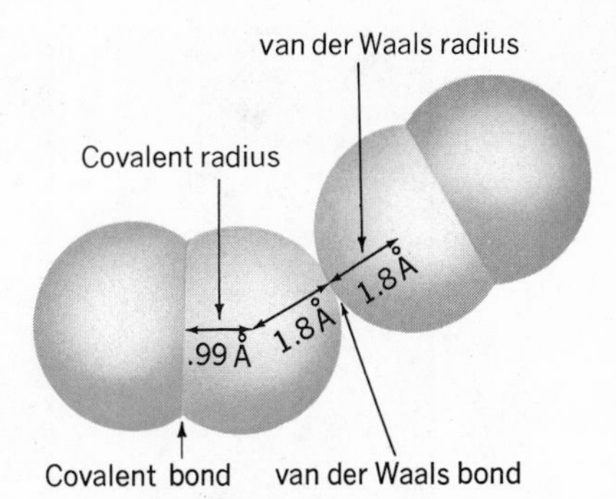

9-17 The van der Waals radius for chlorine is about twice the covalent radius. The shortest distance between the nuclei of two chlorine atoms in adjacent chlorine molecules is about 3.6 Å.

van der Waals forces or ordinary dipole forces are not strong enough to account for the properties of certain molecular crystals and liquids such as water (H_2O), ammonia (NH_3), hydrogen fluoride (HF), and other substances that contain hydrogen atoms bonded to these three highly electronegative atoms. In addition to ordinary van der Waals forces, molecules in these substances are bonded to one another by *hydrogen bonds*. We shall now investigate the origin and nature of these highly important intermolecular bonds.

TABLE 9-5
BOILING POINTS OF BINARY HYDROGEN COMPOUNDS

	Group VA			Group VIA			Group VIIA	
Formula	Number of Electrons	BP (°C)	Formula	Number of Electrons	BP (°C)	Formula	Number of Electrons	BP (°C)
SbH_3	54	−17	H_2Te	54	−1.8	HI	54	−35.5
AsH_3	36	−55	H_2Se	36	−42	HBr	36	−67
PH_3	18	−85	H_2S	18	−59.6	HCl	18	−85
NH_3	10	−33	H_2O	10	100	HF	10	19.4

9-10 Hydrogen Bonds Consider the boiling points of the hydrides of the elements in Groups VA, VIA, and VIIA which are listed in Table 9-5 and graphed with those of the hydrides of Group-IVA elements on p. 312.

Examination of the data in Table 9-5 reveals that the trend toward a lower boiling point with decreasing number of electrons suddenly reverses in the case of the smallest members of each group. This reversal is not observed in the analogous compounds of Group IVA listed in Table 9-4.

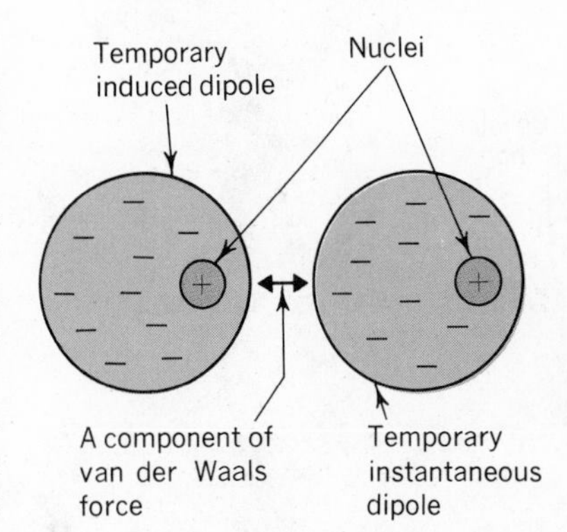

9-18 Fluctuation in the electron density of the molecule on the right causes a small instantaneous dipole. At that instant, an interaction with an adjacent molecule produces in it an induced dipole. At that instant there is a small attractive force operating between the two nonpolar molecules.

The unusually high boiling points of NH_3, H_2O, and HF lead us to believe that intermolecular forces in addition to van der Waals or ordinary dipole forces are operating between molecules in these substances. This additional force is explained by the attraction between the extra pair of electrons on the highly electronegative element of the molecule and the hydrogen atom of an adjacent molecule. A highly electronegative atom such as oxygen very strongly attracts the electrons bonding it to hydrogen atoms. In water, for example, the hydrogen atoms are almost like two exposed

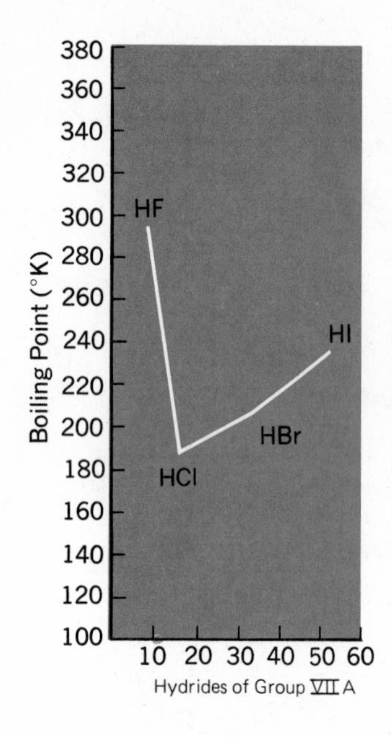

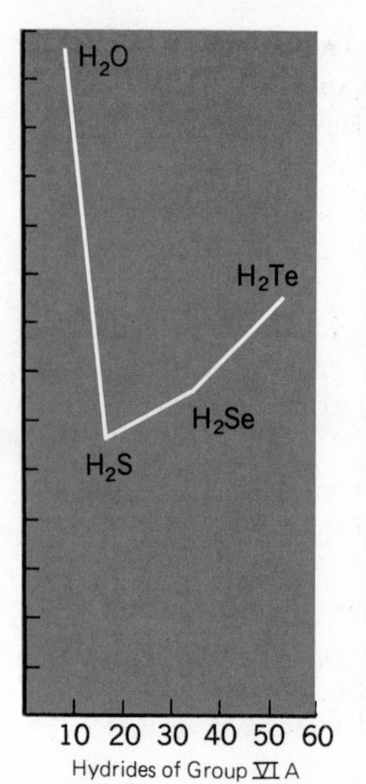

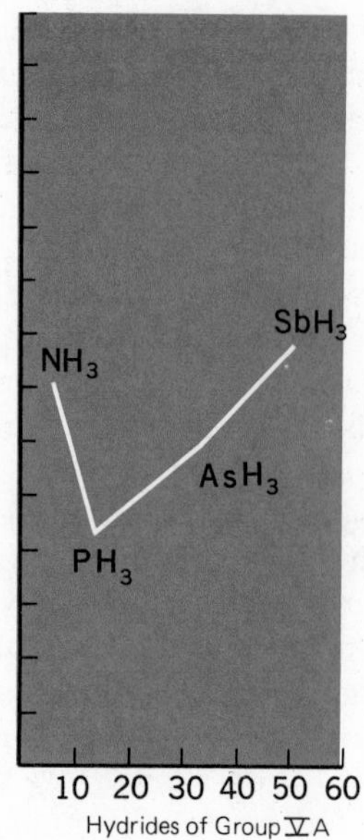

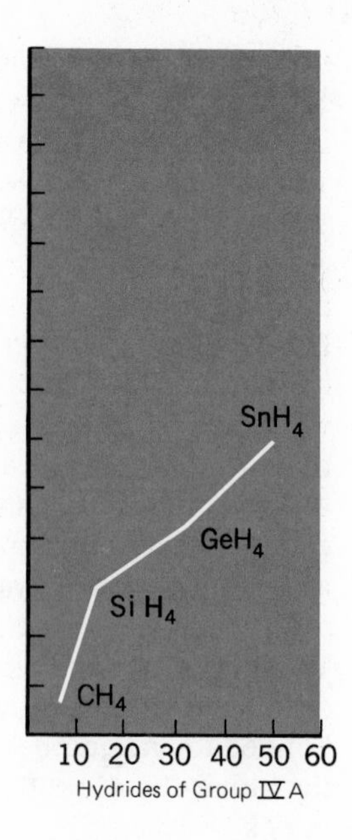

Atomic Number of Representative Element

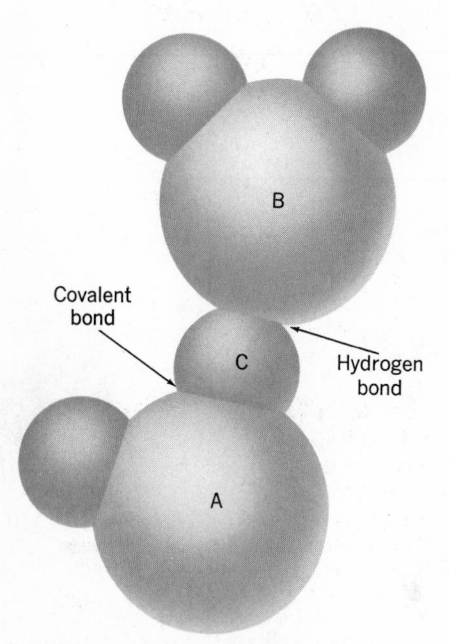

9-19 Hydrogen atom C, covalently bonded to oxygen atom A, is attracted to the unbonded electron-pair of the oxygen atom of water molecule B.

protons. Since the hydrogen atoms in these molecules are somewhat similar to positively charged particles, they exert an electrostatic force of attraction on electron pairs of other highly electronegative atoms. In water molecules, the hydrogen atoms are joined by strong covalent bonds to an oxygen atom in one molecule and are held by electrostatic attractions to an adjacent molecule (Fig. 9-19). In effect, hydrogen is acting as a bridge between two oxygen atoms. In fact, we may think of a ***hydrogen bond*** as an interaction between two molecules or two parts of a single molecule in which a proton is unequally shared by two atoms. We shall not be concerned at this point with intramolecular hydrogen bonds (formed between two parts of the same molecule). Hydrogen bonds do not involve the sharing of electrons, so they cannot be classified as covalent. They do not bind two ions together, so they are not ionic, and they are too electrostatic in nature to be van der Waals or ordinary dipole forces, hence, they deserve a category of their own.

The energies of hydrogen bonds are, in general, about ten times those of van der Waals forces and approximately one-tenth those of ionic or covalent bonds. The energy needed to break hydrogen bonds added to that required to overcome the van der Waals forces also present in water and related compounds, is reflected in the high boiling points of hydrogen-bonded compounds.

It is fortunate for life on earth that the boiling point of water does not follow the trend observed for the related compounds in Group VIA. If it did, no liquid water could exist at the temperatures found on this planet.

9-11 Properties of Water and Other Compounds Containing Hydrogen Bonds The presence of hydrogen bonds accounts for many of the unique properties of water. For example, *hydrogen bonds account for the density of ice being less than that of liquid water.* Because of the shape and the characteristics of water molecules, these bonds seem to form at preferred angles. In ice, and partially in liquid water, the water molecules are tetrahedrally oriented to each other (Fig. 9-20). The hydrogen bonds joining adjacent molecules are weaker and longer than regular covalent bonds within a water molecule, and give rise to the observed open, cage-like structure of ice. This open structure partially collapses when the ice melts so that the molecules in the liquid are closer in water than in ice. Hence the liquid has a greater density than the solid. Without hydrogen bonding, ice would sink to the bottom of our oceans and lakes, a process that would be disastrous to aquatic life and to the environment.

The density of most substances decreases when they are warmed. When water at 0°C is warmed, however, the density increases as

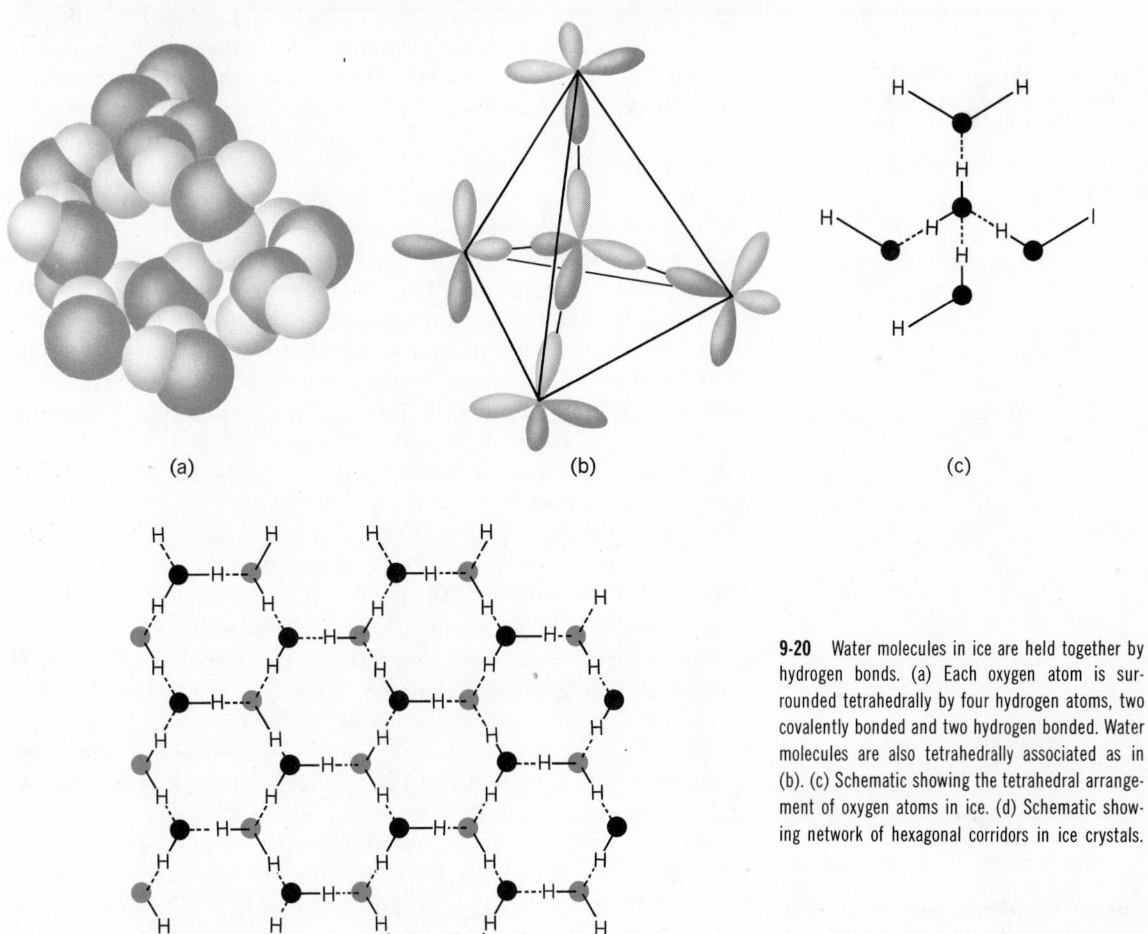

9-20 Water molecules in ice are held together by hydrogen bonds. (a) Each oxygen atom is surrounded tetrahedrally by four hydrogen atoms, two covalently bonded and two hydrogen bonded. Water molecules are also tetrahedrally associated as in (b). (c) Schematic showing the tetrahedral arrangement of oxygen atoms in ice. (d) Schematic showing network of hexagonal corridors in ice crystals.

more hydrogen bonds are broken because the molecules continue to come closer together. Above 4°C, the expansion caused by heating more than compensates for the contraction caused by the breaking of hydrogen bonds. As the temperature rises above 4°C, the density of water gradually decreases. This behavior is shown graphically in Fig. 9-21.

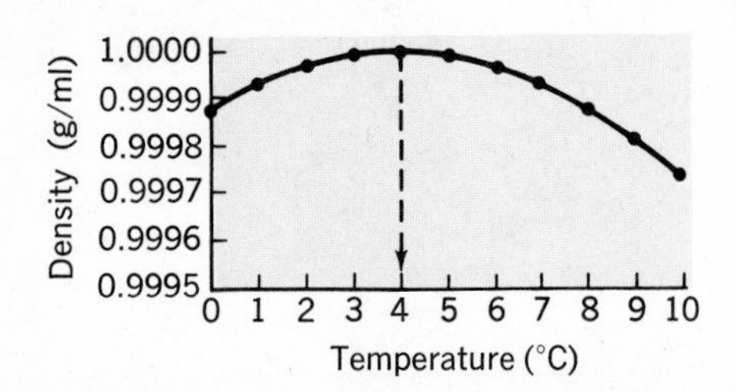

9-21 Density-temperature plot for water. Above 0°C, the open-cage structure of ice collapses; the density increases. Above 4°C, the increased molecular motion causes the molecules to move apart. Thus, the density tends to decrease.

The hydrogen-bond strength in ice is approximately 5 kcal/mole of water molecules. the heat required to melt ice at 0°C is only 1.4 kcal/mole. These figures show that less than 30 percent of the hydrogen bonds are broken when ice melts. This means that considerable hydrogen bonding remains in liquid water. It is hydrogen bonding that is responsible for the unusually high heat capacity of liquid water when compared to that of other molecular liquids. The high heat capacity means that a relatively large amount of heat must be added to raise the temperature of liquid water and conversely, a large amount of heat is released when it cools. Thus, water prevents temperature extremes in the surroundings and acts as a modifier of the planet's climate. The unusually high heat of vaporization of water (9.7 kcal/mole) is further evidence of the strength of hydrogen bonds and of the many hydrogen-bonded molecules which remain even at temperatures just below the boiling point of 100°C.

The molar heat of vaporization is the amount of heat required to vaporize a mole of liquid already at its boiling point.

There are many examples of the widespread occurrence and importance of hydrogen bonds. All living organisms contain hydrogen-bonded substances. Dr. Linus Pauling has shown how hydrogen bonding partially accounts for the structure of proteins, the building materials of animal tissue. The greater degree of hydrogen bonding in plants and wood makes their fibers more rigid than those of animal tissue. Hydrogen bonds in wood are oriented with the grain rather than across it. This means that a piece of wood is stronger if it is cut with the grain rather than against it. For example, a baseball bat is made by cutting with the grain of a tree, never across it.

Much of our clothing and food is composed of hydrogen-bonded materials. Hydrogen bonds are responsible for the stickiness of honey. Both the water and sugar molecules in honey are hydrogen-bonded substances. Sugar molecules have many hydroxyl (—OH) groups protruding at various angles. The oxygen in the —OH groups on one molecule is joined by hydrogen bonds to the hydrogen in neighboring molecules. Considerable energy is required to break a large number of such bonds. The stickiness of honey (viscosity or resistance to flow) is a reflection of the large number of hydrogen bonds which resist being broken.

Experimental data indicate that the solubility of a substance is enhanced when hydrogen bonding occurs between the solute and solvent particles. Thus, ammonium compounds tend to be quite soluble in water. This phenomenon is discussed in Chap. 14.

In view of the important materials whose properties are related to hydrogen bonding, it is apparent that this important intermolecular bond is in large measure responsible for the nature of life on earth.

In earlier chapters we learned that molecules were aggregates of a relatively few atoms bonded by covalent bonds. We called aggregates of molecules molecular crystals. When a large number (in the magnitude of one mole) of atoms are bonded together in a crystal by a network of covalent bonds, the solid is called a *network or macromolecular solid.* A number of very important familiar substances fall into this category. For this reason, we shall briefly examine the structure, bonding, and properties related to this type of aggregate.

NETWORK (MACROMOLECULAR) SOLIDS

9-12 Three-dimensional Network Solids Network crystals, or macromolecules, consist of a one-, two-, or three-dimensional network of atoms joined by single covalent bonds. These may be atoms of the same or different elements. Pure substances such as diamond, quartz, and silicon carbide (Carborundum®) are highly stable, extremely hard, poor electric conductors, and insoluble in most solvents. Furthermore, they have very high melting and boiling points. For example, carbon in the form of a diamond has the maximum rating on the Mohs (or any) scale of hardness, melting point greater than 3550°C, boiling point 4827°C, and density 3.5 g/cm^3. These properties suggest that the atoms in the crystal are joined by very strong bonds oriented in such a way as to yield a very rigid structure. Experimental data reveal that all C—C bond lengths in diamond equal 1.54 Å, all C—C—C bond angles are 109°28′, and all carbon-carbon bond strengths are equal. Using these data we can visualize a diamond crystal as a single macromolecule composed of a network of carbon atoms extending uniformly throughout the entire crystal. Each carbon atom must be at the center of a tetrahedron whose vertices are occupied by other carbon atoms. In this *three-dimensional network of atoms,* each carbon atom shares its four valence electrons with four other carbon atoms, thus filling all available sp^3 orbitals and providing a stable outer octet of electrons for each atom. The bonding electrons are tightly bound, and highly localized. Thus, diamond crystals are poor electric conductors and transparent. The network of atoms joined by strong, tetrahedrally oriented covalent bonds accounts for the extreme hardness of diamond crystals. The close packing of the small carbon atoms accounts for the high density compared to graphite. When the highly directional bonds in a diamond are distorted by stress, the crystal cleaves. This cleavage is related to the fact that new bonds cannot be formed easily with adjacent atoms which already have formed bonds in certain preferred directions.

Pure silicon has the same outer-level electronic configuration as carbon. Thus, both silicon and carbon form crystalline compounds with the rigid diamond structure. Silicon carbide (SiC), a common abrasive known as Carborundum®, and quartz, a form of silicon

The Mohs scale (left) is a numerical scale from 1 through 10, used in mineralogy to indicate relative hardness. The degrees of hardness are listed below in order. Knoop values (right), are obtained by pressing a tiny diamond wedge-like chisel into the sample with a known force and measuring the length of the impression.

Mohs	Knoop
1. talc ($3MgO \cdot 4SiO_2 \cdot H_2O$)	
2. gypsum ($CaSO_4 \cdot 2H_2O$)	32
3. calcite ($CaCO_3$)	135
4. fluorite (CaF_2)	163
5. apatite [$CaF_2 \cdot 3Ca_3(PO_4)_2$	430
6. feldspar ($K_2O \cdot Al_2O_3 \cdot 6SiO_2$)	560
7. quartz (SiO_2)	820
8. topaz (AlF_2SiO_4)	1340
9. sapphire (Al_2O_3)	1950
10. diamond (C)	7000

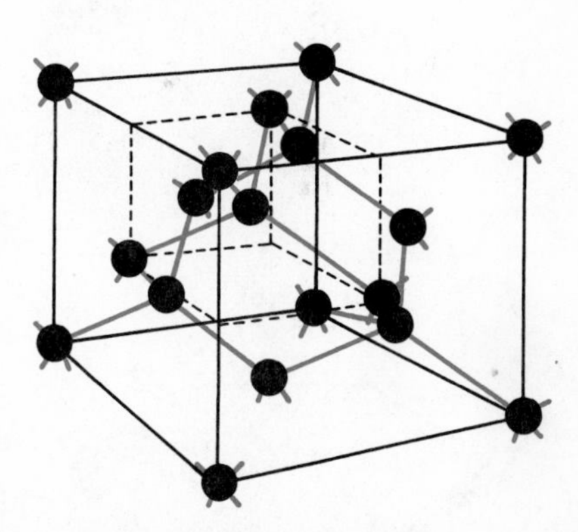

9-22 Carbon atoms in a diamond crystal. Each carbon atom is bonded covalently to four neighboring atoms at tetrahedral angles. The diamond structure is characterized by a three-dimensional network of atoms.

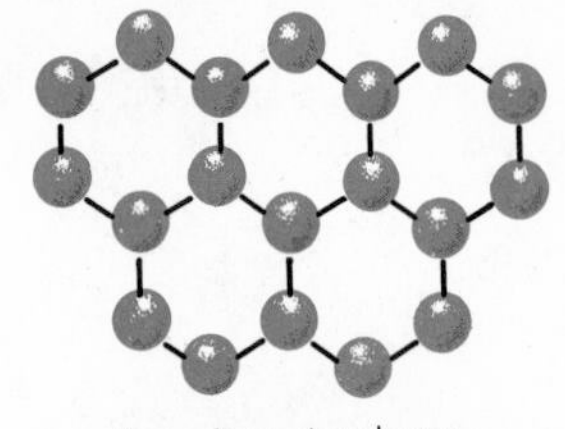

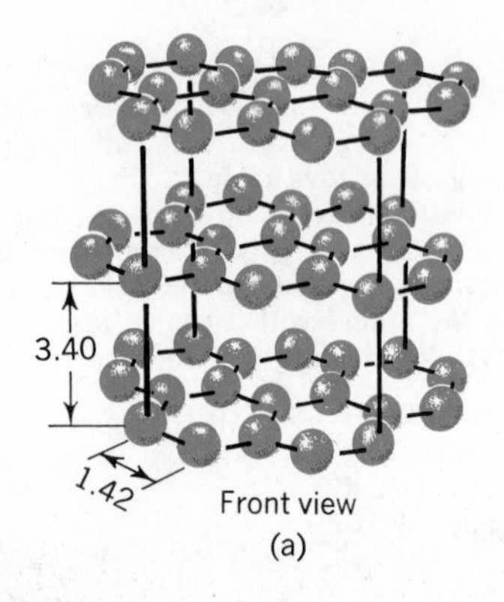

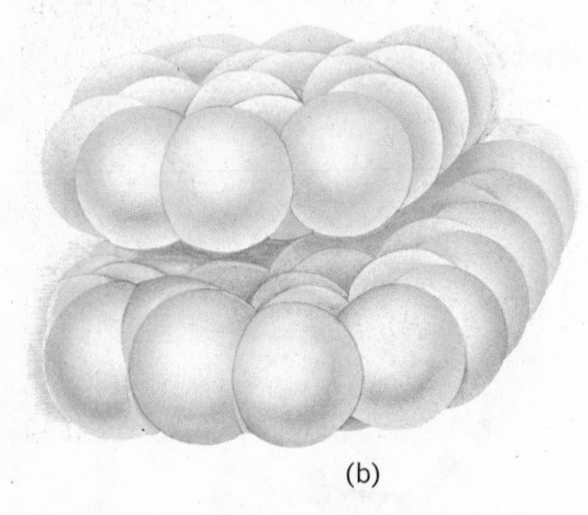

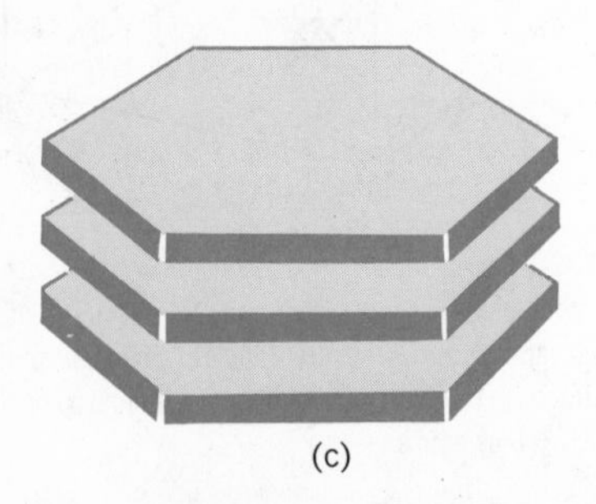

9-23 (a) Carbon atoms in the graphite crystal. Note that the distance between the atoms in adjacent layers is greater than that between carbon atoms in the same layer. (b) Space-filling model. (c) Layers of graphite which may be obtained by cleaving a graphite crystal with a razor blade.

dioxide $(SiO_2)_x$, are examples of three-dimensional covalent crystals. As a result of their structural similarity to diamond, they are characterized by extreme hardness and high melting points.

9-13 Two-dimensional Network Solids Graphite, another form of pure carbon, has properties entirely different from those of diamond. Graphite has a high melting and boiling point but is soft and a good conductor of electricity. These properties suggest that there is more than one kind of bonding in this crystal. The high melting point is a reflection of strong sigma covalent bonds and the electric conductivity is characteristic of delocalized mobile pi-electrons. The softness, however, implies that there are some very weak bonds in the crystal. Visual observation of graphite reveals that it has a layered structure. Apparently the layers are a *two-dimensional network of atoms.* The atoms in the individual layers are bonded to each other by strong covalent bonds. Only weak bonds or forces join adjacent layers so they move over each other rather easily. As a result of this softness and slipperiness, graphite is used as a lubricant. The softness of graphite is a reflection of this layered structure. When a film of gas such as oxygen, nitrogen, or air is adsorbed between the layers, the slipperiness of graphite increases.

Studies of graphite crystals reveal that the atoms in the individual layers form a hexagonal pattern which involves 120° bond angles. Furthermore, it is found that the distance between the layers is 3.40 Å while the distance between the atoms in a given layer is 1.42 Å. The average C—C single bond length is 1.54 Å and the average C=C double bond length is 1.34 Å. These data lead to the conclusion that there are rather weak bonds between the carbon atoms in adjacent layers. The bond distance between atoms within each layer is intermediate between a single and double bond length. This implies that the bonding within a layer involves resonance structures. The hexagon pattern of carbon atoms in the layers and the delocalized electrons can be explained in terms of sp^2 hybridization similar to that observed in ethylene. Within a layer the overlap of singly occupied sp^2 orbitals yields strong σ bonds between carbon atoms. Between adjacent layers, the unhybridized parallel p orbitals overlap laterally and form π bonds. The electrons forming the π bonds are delocalized and spread out over the hexagons where they are mobile and relatively free to move under the influence of electromagnetic radiation (light) or an externally applied electric voltage. In other words, the delocalized electrons account for the electric conductivity and the reflectivity (luster) of graphite. Alternatively, the weak forces of attraction between the layers are sometimes interpreted as *van der Waals attractions.*

9-14 One-dimensional Network Solids The analysis of asbestos-type minerals reveals that they are composed of silicon, oxygen, and certain metallic elements such as magnesium and calcium. The fibrous nature and other properties of these minerals lead us to conclude that they consist of a one-dimensional network of atoms. In these crystals the silicon and oxygen atoms form long chains with metal atoms attached to the silicon atoms. The atoms within the chain are held together by strong covalent bonds. The forces between the adjacent parallel chains are relatively weak, resulting in a substance with a thread-like structure.

We have now discussed ionic bonds, covalent bonds, and intermolecular bonds. As a result of our study we can make some generalizations.

1. Ionic bonds exist between atoms with low electronegativities and those of high electronegativities.
2. Polar covalent bonds exist between dissimilar atoms with different electronegativities.
3. Nonpolar covalent bonds exist between identical nonmetallic atoms.
4. Dipole forces exist between most polar molecules.
5. Hydrogen bonds exist between those polar molecules containing hydrogen covalently bonded to a highly electronegative atom such as fluorine, oxygen, or nitrogen.
6. van der Waals forces exist between all molecules but are the only forces that exist between nonpolar molecules.

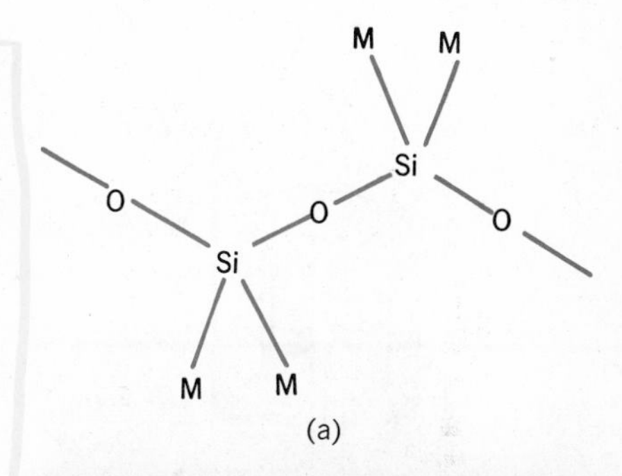

9-24 (a) Asbestos is a fiber-like mineral consisting of chains of alternating silicon and oxygen atoms covalently bonded. Metallic atoms such as calcium or magnesium are attached to the silicon atoms. (b) Asbestos crystals. (Johns-Manville)

Examination of the list above reveals that we have not accounted for the bonds that exist between metallic atoms. We shall now discuss the properties and bonding related to an aggregate of metallic atoms. Such an aggregate is called a *metallic crystal.*

METALLIC CRYSTALS AND METALLIC BONDS

Approximately 80 percent of all the elements are metals. The properties of individual metals vary to a great extent. Because of this they serve many useful purposes. Gold, silver, platinum, and iridium are used in valuable jewelry. Copper and aluminum are used in thousands of miles of wire which bring electricity to every part of the country, and iron is used as the structural backbone of our buildings and bridges. Although some of these metals have been known and used for thousands of years, it has been only in the last 100 years that scientists have developed theories which help explain many of their properties.

9-15 Properties of Metals Most metals are characterized by these properties:

1. *Luster or reflectivity.* Freshly cleaned metallic surfaces are good reflectors of light. Most metals reflect all frequencies of visible light and appear silvery white under white light.
2. *High electric conductivity.* Metals such as copper and aluminum are used in electric cables because of their outstanding ability to conduct an electric current. When a small difference in voltage

is maintained in a metal, a relatively large current flows through the metal. This flow of current represents the passage of electrons through the metal. The passage of current does not cause any change in the composition of the metal as it does when it passes through melted ionic substances or solutions.

3. *High heat conductivity.* Aluminum ware and copper plated pans have long been used by housewives as cooking utensils because of their outstanding ability to conduct heat. In contrast, most nonmetals (such as sulfur) and compounds (such as water) are poor conductors of heat and electricity. *The best electric conductors are the best heat conductors.* This implies that the two properties may be related to a common factor.

4. *Workability.* Most metals can be hammered into sheets (malleability), drawn into wires (ductility), or formed into various shapes without shattering. This is not true of most other pure substances.

5. *Electron emission caused by heat or light.* When metals are heated or when their surfaces are exposed to sufficiently short wavelengths of light, they emit electrons. Vacuum tubes and photoelectric tubes used in automatic door-openers and other electronic devices use this property of metals.

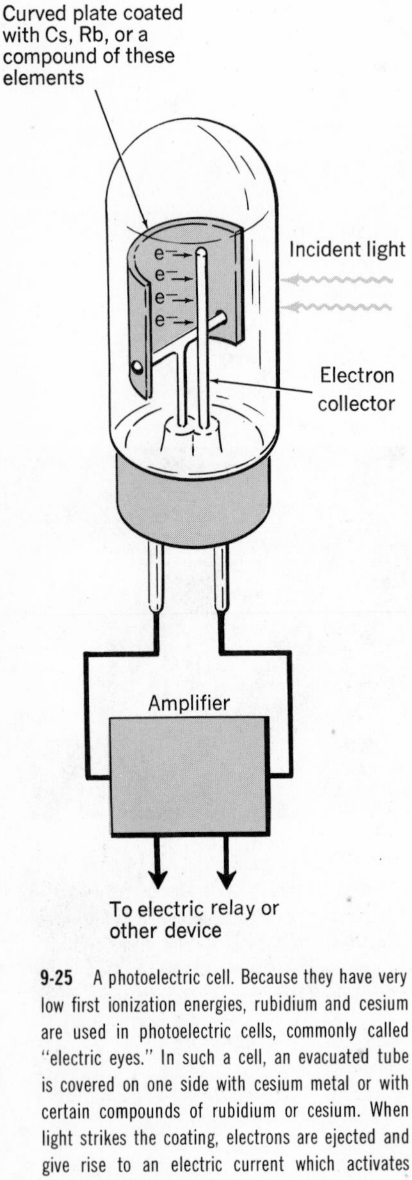

Curved plate coated with Cs, Rb, or a compound of these elements

Incident light

Electron collector

Amplifier

To electric relay or other device

9-25 A photoelectric cell. Because they have very low first ionization energies, rubidium and cesium are used in photoelectric cells, commonly called "electric eyes." In such a cell, an evacuated tube is covered on one side with cesium metal or with certain compounds of rubidium or cesium. When light strikes the coating, electrons are ejected and give rise to an electric current which activates mechanical devices.

These features cannot be accounted for in terms of pure covalent bonds, ionic bonds, or van der Waals forces. The existence of such bonds in metallic crystals can be discounted on the basis of these observations:

1. *Covalent bonds are directional with fixed lengths.* Such bonds would resist the deformation which takes place when a stress is applied during the working of metals. Also, the electrons in covalent bonds are tightly bound and highly localized. Consequently, they do not readily conduct electricity or heat or reflect light.

2. *Ionic bonds, although nondirectional, resist deformation while metallic bonds do not.* When stress is applied to an ionic substance, ions of like charge come into contact, causing a high-energy, unstable situation that causes the shattering of the crystal. Substances containing ionic bonds do not conduct an appreciable electric current nor do they reflect much light. The electrons are strongly attracted by the nuclei of the respective ions and are not free to move throughout the crystal under the influence of applied voltage or of light energy.

3. *van der Waals forces can be discounted as a significant factor because of the relatively weak nature of these forces.* The melting points of most metals are far too high to be explained in terms of the relatively weak van der Waals attractions.

9-16 Metallic Bonds Most observable properties of substances can be interpreted in terms of three atomic characteristics: outer-level electron configurations, ionization energies, and atomic radii.

In turn, these properties are related to the nature of the bond between atoms of the aggregate. A satisfactory model of metallic bonding must account for observed metallic properties. Let us first look for features common to all metallic atoms. Then we shall determine the manner in which these characteristics can be used to develop a model of metallic bonding which will explain the observed properties of metals. Examination of the electron configurations and the ionization energies of atoms of substance with metallic properties reveals two common features: (1) a number of *vacant orbitals* and (2) low *ionization energies.*

Consider the element sodium, whose atoms meet these requirements. A sodium atom has only one electron in its outer energy level and a low ionization energy, indicating that the 3*s* electron is rather loosely bound. X-ray diffraction shows that each sodium atom in the solid crystal is surrounded by eight other sodium atoms. Each has one electron and several vacant orbitals in the third energy level.

There are not enough valence electrons for each sodium atom to form a covalent bond with all of its neighbors. Each atom would need eight electrons to form covalent bonds with each of its eight neighbors. Because this is not the case, the single, loosely held electron from each atom remains delocalized in the region between the atoms. The space about each of the atoms is subject to the same positive nuclear charge. The space between the atoms, therefore, represents a region of uniformly low potential energy for the negative electrons. A given electron in the crystal can move easily through this region of low potential energy which extends throughout the crystal. In other words, the outer, loosely bound electrons of atoms in a metallic crystal are not localized but are free to move throughout the crystal.

In a metallic crystal, each atom contributes its outer electrons to the common pool of electrons which extends uniformly throughout the crystal. The contribution of an electron by each atom leaves a positive ion, or positive atomic kernel, occupying a lattice point of the crystal. *The electrostatic attraction between the mobile electrons and the atomic kernels gives rise to what is called the metallic bond.* The structure of a metallic crystal and the nature of the metallic bond is shown in Fig. 9-26 and is summarized below.

1. Metallic crystals consist of a three-dimensional, closely-packed latticework of atomic kernels surrounded by a sea of delocalized, mobile valence electrons.
2. The electrons can move throughout the crystal rather like gas molecules confined in a container. The electrons are held within the metal by the attraction of the positive atomic kernels.
3. The atomic kernels are held together by the electrostatic attraction of the electrons which move between them. It is this force of attraction which results in a metallic bond between the atoms of a metal.

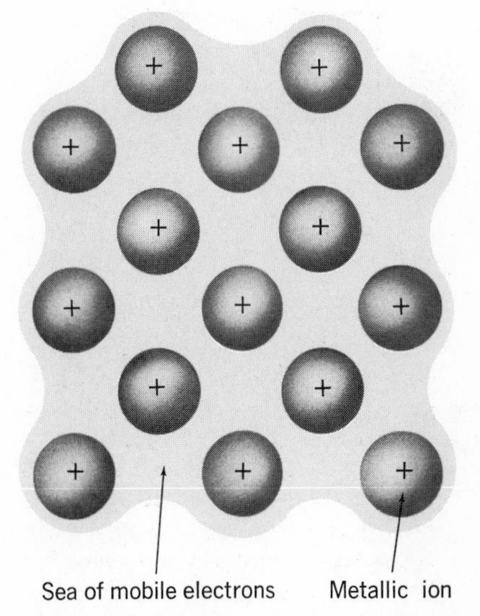

9-26 A metal consists of positive metallic ions arranged in a "sea" of highly mobile electrons. The metallic bond is the result of the attraction between the positively charged kernels and the surrounding electrons.

4. The strength of a metallic bond depends on the nuclear charge and number of electrons in the outer energy levels. As the number of outer electrons increases, the strength of the metallic bond increases. This trend would be expected, since a greater number of electrons would give each atomic kernel a higher charge and a greater share of bonding electrons. In general, *as the strength of the metallic bond increases, the melting and boiling points and the hardness of the metal increase.* Furthermore, the transition elements such as iron and nickel are harder and have higher melting points than sodium and other metals among the representative elements.

9-17 Use of Metallic-Bond Model to Interpret Properties of Metals Let us now use our simple model to explain observed properties of metals.

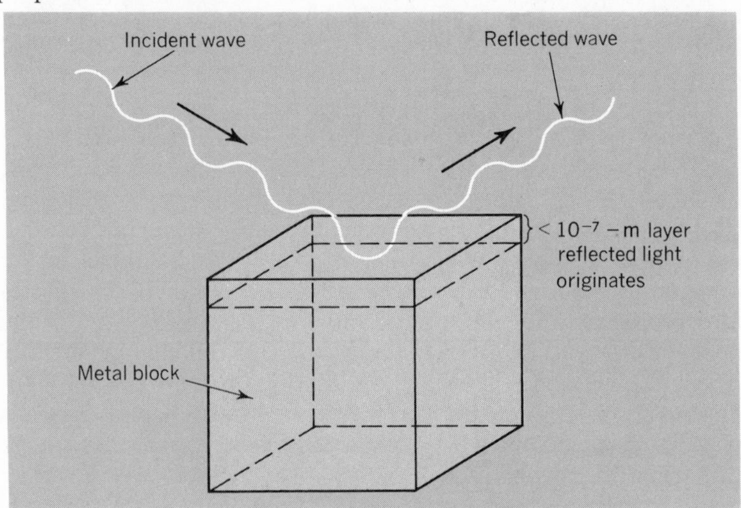

9-27 Reflection of light from a metal surface. Only the "free-moving" conduction electrons located in a small volume (thin layer) of the *surface* of a metal reradiate and reflect visible light that strikes the metal. Almost all the energy of the incident light goes into the reflected light. Thus, uncorroded, smooth surfaces of many metals are shiny and can be used as mirrors.

1. *Luster or reflectivity.* Nonlocalized, mobile electrons account for this and other properties. Light energy causes the outer electrons of metals to move back and forth with the same frequency as the light which strikes the metal. Oscillation of a charged particle involves acceleration, which results in radiation of electromagnetic energy. Since the radiated energy is the same frequency as the incident light, we see it as a reflection of the original light beam.

2. *Electric conductivity.* Mobile electrons account for the high electric conductivity of metals. The very tiny electrons can move very rapidly from one position to another under the influence of an applied voltage. At higher temperatures, the conductivity in general decreases because the increased vibration of the atomic kernels interferes with the motion of the electrons.

3. *Heat conductivity.* Metals conduct heat well, a fact that is also explained in terms of highly mobile electrons. These electrons may move rapidly from regions of high temperature, where they attain high kinetic energy, to cooler regions, where they transfer some of this kinetic energy to the crystal lattice. Heat is slowly trans-

ported throughout a crystal lattice by the vibrational motion of the atoms, molecules, or ions which occupy the lattice sites. Covalent and other crystals have tightly bound, localized bonding electrons. Heat conduction through such substances must depend solely upon the vibration of atoms or molecules about fixed positions in the crystal lattice colliding with adjacent atoms, with a resultant transfer of energy. They are therefore poor conductors.

4. *Workability.* Since metallic bonds are nondirectional, metal lattices can easily be deformed. Unlike ionic crystals, they do not shatter when a stress is applied. In metallic crystals, as one plane of atoms slides over another, there is no increase in electrostatic repulsion. The environment of each atom remains essentially unchanged, and the bonding electrons continue to exert uniform attractive force on the atomic kernels (Fig. 9-28).

5. *Electron emission caused by heat or light.* When enough heat energy is applied to a metal to overcome the attractive force between the atomic kernels and an outer electron, the electron is emitted from the metallic atom. When the frequency and, hence, the energy of the light that strikes the metal is great enough to overcome the attractive forces, the electron escapes from the metal with a resultant decrease in the energy of the incident photon. This is the photoelectric effect. This phenomenon is illustrated on p. 171.

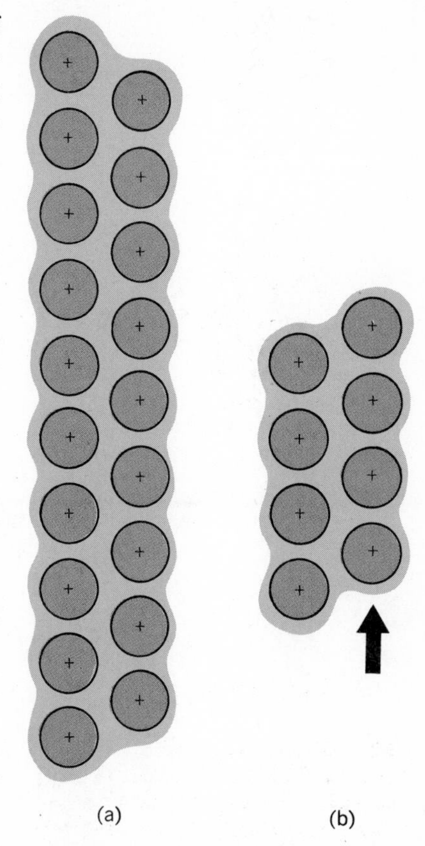

9-28 A metallic crystal, unlike an ionic crystal, can be "worked" into thin sheets or drawn into wires. This is because the positively charged atomic kernels can be moved within the "sea" of electrons without fundamentally altering their environment. Thus, crystal (a) may be converted into crystal (b) without cracking or shattering.

SUMMARY

In this chapter, we have related the different types of crystalline solids to the types of bonds or forces which bind together the particles in the crystal. We have also shown how the properties of these solids are related to bond type. Table 9-6 summarizes these relationships.

The changes in the properties of the elements from left to right in a period of the Periodic Table reflect the changes in types of chemical bonds. In viewing a row (period) of the Periodic Table, we see a range of dissimilar properties which is in sharp contrast to the rather similar properties encountered in a vertical group of elements. As we move from left to right in a row of elements, properties such as electric conductivity, melting points, boiling points, molar heats of fusion, and molar heats of vaporization undergo sharp increases and decreases. These abrupt changes in properties reflect the differences in the type of bonding that we find as the atomic number increases by one unit. Table 9-7 shows some of the properties of Period-3 elements.

Let us examine more closely the bonding in each of the elements in Table 9-7 and relate the type of bonding to some of the observed properties. Sodium, magnesium, and aluminum all form metallic crystals in which the atoms are held together by metallic bonds. We can attribute the respective increases in melting point, boiling point, heat of fusion, and heat of vaporization partially to the increase in strength of the metallic bond as we go from sodium to aluminum.

The sudden rise in the melting point of silicon is a reflection of a complete change in the type of bond found in silicon crystals. Evidence from a variety of sources indicates that silicon crystals are covalent crystals in which the silicon atoms are covalently bonded and arranged in a rigid three-dimensional network.

The sharp drop in the melting point, boiling point, heat of fusion, and heat of vaporization that occurs from silicon to phosphorus reflects another change in bonding. Phosphorus forms molecular crystals. The van der Waals attractions between the molecules that

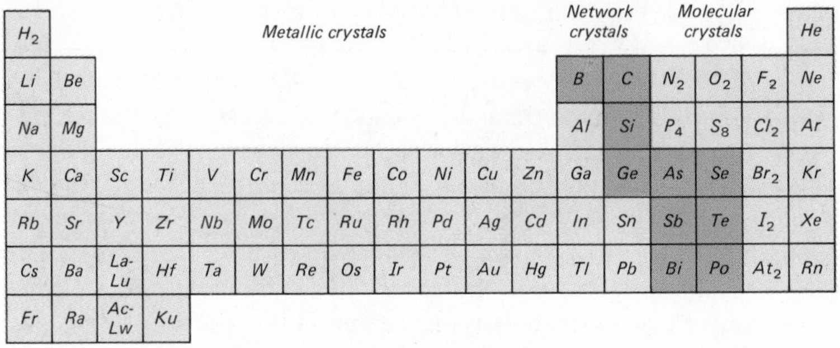

9-29 Types of crystals formed by elements in the Periodic Table. What type of forces bind together the atoms in elemental molecular crystals?

Metallic crystals · *Network crystals* · *Molecular crystals*

H_2																	He
Li	Be											B	C	N_2	O_2	F_2	Ne
Na	Mg											Al	Si	P_4	S_8	Cl_2	Ar
K	Ca	Sc	Ti	V	Cr	Mn	Fe	Co	Ni	Cu	Zn	Ga	Ge	As	Se	Br_2	Kr
Rb	Sr	Y	Zr	Nb	Mo	Tc	Ru	Rh	Pd	Ag	Cd	In	Sn	Sb	Te	I_2	Xe
Cs	Ba	La-Lu	Hf	Ta	W	Re	Os	Ir	Pt	Au	Hg	Tl	Pb	Bi	Po	At_2	Rn
Fr	Ra	Ac-Lw	Ku														

TABLE 9-6
PROPERTIES OF CRYSTALLINE SOLIDS

	Solid Ionic Crystals	*Solid Molecular Crystals*	*Solid Covalent Crystals*	*Solid Metallic Crystals*
Examples	NaCl CsF	Cl_2 H_2O CH_4	diamond (C) quartz (SiO_2) silicon carbide (SiC)	sodium iron copper
Particles occupying lattice points of crystal	positive ions or negative ions	molecules	atoms	positive ions or positive kernels of atoms
Type of bond or force between units occupying lattice points	ionic (electrostatic)	van der Waals, dipole, and sometimes hydrogen bonds	covalent bonds	metallic bonds
Relative melting point	high	low	very high	moderate to high
Electric conductivity in solid state	poor	poor	poor	good
Electric conductivity in liquid state	moderate	poor	poor	good
Heat conductivity	poor	poor	poor	good
Workability or hardness	hard and brittle	usually soft	hard	soft to hard, workable
Reflectivity or transparency	transparent	translucent transparent	transparent	opaque, good reflectors

TABLE 9-7
PROPERTIES OF ELEMENTS IN PERIOD 3 OF THE PERIODIC TABLE

	Na	Mg	Al	Si	P_4 (yellow)	S_8 (rhombic)	Cl_2	Ar
Melting Point (°C)	97.5	651	659.7	1420	44.1	112	−103	−189
Heat of Fusion (kcal/mole)	0.712	2.14	2.07	11.1	0.155	0.34	0.77	0.268
Boiling Point (°C)	889	1120	2357	2355	280	445	−34.1	−186
Heat of Vaporization (kcal/mole)	23.1	31.5	67.9	105	3	2.5	4.9	1.6
Metallic Properties	(diminishing) →							
Conductivity	+	+	+	*s*	−	−	−	−

(+ conducting, *s* semiconducting, − nonconducting)

occupy the lattice points of the crystal are small compared with the strong covalent bonds between the silicon atoms that occupy the lattice points of the covalent crystal. The covalent bonds between the phosphorus atoms in the P_4 molecules are not ruptured during melting. Therefore, individual phosphorus atoms are not present in the liquid state.

The molar heat of fusion is the amount of heat required to liquify one mole of solid already at its melting point.

The nature of the bonding in sulfur crystals is essentially the same as that in phosphorus crystals. Sulfur forms molecular crystals containing S_8 molecules. The S_8 molecules are much larger and contain many more electrons than P_4 molecules. Therefore we would expect the van der Waals forces to be greater and would predict that sulfur would have a higher melting point than phosphorus.

Both chlorine and argon molecules are small and symmetrical with filled outer orbitals. Thus, we predict that the melting points of these elements are low, a consequence of the small intermolecular forces of attraction between the molecules.

LOOKING AHEAD

In this chapter we have examined the general structure of aggregates, the nature of the bond that exists between the units which compose the aggregate, and the relationship between the type of bonding and the properties of the aggregate. Energy changes are involved when a solid aggregate melts or when a liquid aggregate solidifies or vaporizes.

In the next chapter, we shall examine the energy involved in phase changes and interpret our observations in terms of the structural characteristics of solids, liquids, and gases. This will enhance our understanding of the behavior of matter and provide background for our study in Chapter 11 of the energy associated with chemical reactions.

QUESTIONS

1. What factors determine which of two reacting atoms donates electrons and which accepts electrons?

2. List four characteristics of ionically bonded crystals.

3. Given these data:

	Solubility in Water (g/100 g H_2O, 25°C)	*Melting Point (°C)*
NaF	4.2	988
NaCl	35.7	801
NaBr	116	755
NaI	184	651

(a) Explain the decreasing melting point from NaF to NaI. (b) Explain the increasing solubility in the same order. (c) Cesium fluoride has a melting point of 682°C and a solubility of 367 g/100 g H_2O at 25°C. Explain the difference between the properties of CsF and NaF.

4. Lattice energy is related to melting point. What factors explain that sodium sulfide melts at 1180°C while magnesium sulfide melts at a temperature over 2000°C?

5. Arrange aggregates of these in order of decreasing % ionic character

LiI	BaO	$AlCl_3$	CsF	RbBr
K_2S	CaO	ClF	P_2S_3	F_2

6. What tests might you perform to determine whether a solid substance contains ionic bonds?

7. Explain the poor conductivity of heat and electricity exhibited by covalently bonded substances.

8. What is the reason for the generally low melting points of molecular solids?

9. (a) What are van der Waals forces, (b) what are the structural features of substances in which only van der Waals forces are of importance in their interparticle bonding?

10. Why do NH_3, H_2O, and HF have abnormally high boiling points when compared to their analogs PH_3, H_2S, and HCl?

11. The three compounds shown below are fairly similar. Thus we might expect them to have similar boiling points. Explain the increase in boiling points of the compounds that follow:

Formula	*Common Name*	*Chemical Name*	*BP(°C)*
C_3H_7OH	Rubbing alcohol	Propanol	82
$C_2H_4(OH)_2$	Antifreeze	Ethylene glycol	189
$C_3H_5(OH)_3$	Glycerin	Glycerol	290

12. (a) Which compound in Question 11 is the most viscous (has the least ability to flow)? (b) Which is the most fluid (least viscous)?

13. Explain why life might not be possible on the earth if water did not expand upon cooling from 4°C to 0°C and also upon freezing.

14. How do network crystals differ from molecular crystals (a) in structure, (b) in hardness and (c) in melting point?

15. What property of graphite is the result of the presence of delocalized electrons?

16. A metal is sometimes said to be composed of atoms in which there are more vacant orbitals than occupied orbitals in the outer energy level. Show how this holds for Li, Mg, and Al.

17. Show how the idea that a metal is composed of the kernels of atoms in a sea of mobile electrons helps explain a metal's (a) luster, (b) electric conductivity, (c) heat conductivity, (d) photoelectric effect, (e) thermoelectric effect, (f) malleability.

18. What is the reason for the generally low melting points of the Group-IA metals as compared with the high melting points of metals in the center of the transition series?

19. Classify these substances as ionic, solid molecular, solid covalent, or solid metallic crystals: (a) lattice composed of positive ions sharing electrons with neighboring positive ions, (b) lattice composed of atoms bonded covalently to neighboring atoms, (c) a solid only at extremely low temperatures, (d) a good conductor of heat and electricity, (e) extremely hard, yet very soluble in polar liquids, (f) extremely hard, not workable, and a poor conductor of heat and electricity, (g) a good electric conductor only when melted.

20. Classify these materials according to the type of bonds between particles, (a) paraffin, (b) wood, (c) sugar, (d) liquid nitrogen, (e) gasoline, (f) ammonia gas. Explain each answer.

21. (a) Explain how a nonpolar covalent molecule such as BeH_2 can be held together by electrostatic forces. (b) Explain how a steel cable can be strong even though it consists of neutral atoms adjacent to each other.

22. Explain why $C_{20}H_{40}$ is solid at 25°C while C_4H_8 is a gas at 25°C.

23. Germanium (Ge) is a solid whose atoms are all covalently bonded to each other, much like those of carbon atoms in a diamond crystal. Glycerol [$C_3H_5(OH)_3$] is an alcohol. Potassium chloride (KCl) is a white crystalline solid. Methane (CH_4) is a gas which can be liquefied only under high pressures and low temperatures. Rubidium metal is very malleable and is an excellent conductor of electricity. Which has (a) hydrogen bonding, (b) the greatest hardness in the solid phase, (c) Which has the highest melting point? (d) conductivity of electricity when melted, (e) the lowest molar heat of vaporization, (f) particles held together primarily by van der Waals forces, (g) nondirectional bonds (more than one answer)?

24. (a) Explain why pure carbon in the form of graphite is soft and has a high melting point. (b) Explain why pure carbon in the form of diamond is hard and has a high melting point. (c) Why is diamond a poor conductor of an electric current and graphite a good conductor?

25. Account for the sharp drop in melting point as you move in the Periodic Table from element 14, silicon, to element 15, phosphorus.

26. Explain why the electric conductivity changes as we move from left to right across a row of the Periodic Table.

27. Is the melting point of element 32, which is directly below silicon in the Periodic Table, low or high? Explain.

28. The boiling point of sulfur (S_8) is very high compared with that of chlorine (Cl_2). Explain this difference.

29. Aluminum has a heat of fusion of 2.07 kcal/mole, while the value for silicon is 11.1 kcal/mole. How are these values related to changes in bond types?

30. Elements A, B, C, and D have consecutive atomic numbers. Element D is a monatomic gas with low melting and boiling points. All efforts to form compounds of D in the laboratory have failed. (a) Which of the remaining elements, A, B, or C, has the strongest affinity for an additional electron? (b) A compound of an alkali metal M with element C has a formula of MC. Does this compound have ionic or covalent bonds? Predict other properties of MC such as melting point and solubility in water. (c) Write the formulas for the hydrides of elements A, B, and C. (d) Predict the shape of each hydride molecule. (e) Predict the conductivity of solid B.

31. Elements X, Y, and Z have consecutive atomic numbers. X is the only good conductor of electricity and heat, Y has an extremely high melting point, and Z has a low melting point and a molecular formula of Z_4. (a) What type of bonding exists between the atoms of element Y? (b) Which of the three elements is malleable? (c) Write formulas for the hydrides of these elements. (d) Which of these elements could be cut with a knife? (e) Which oxide of these elements dissolves in water and forms an acid?

32. Explain why metals are malleable while ionic substances are not.

33. Why is CO_2 a gas at 25°C and SiO_2 a high-melting-point crystalline substance?

34. (a) Which is the better conductor of a direct electric current, liquid KCl or metallic aluminum? (b) How is the electric charge transported in each case? (c) What effect does the passage of an electric current have on each?

35. Which compound in each of these has the higher boiling point? Explain your choice. (a) NH_3 or PH_3, (b) C_2H_6 or C_4H_{10}, (c) sulfur (S_8) or chlorine (Cl_2).

36. Which compound in each of these pairs has the higher melting point? (a) CaO or KI, (b) KCl or KI, (c) RbCl or ICl.

37. (a) Identify the inter- and intramolecular forces in methane (CH_4) by name and magnitude. (b) When a crystal of methane melts, which force is being overcome, the inter- or intramolecular?

10

The Behavior of Liquids and Energetics of Phase Changes

Preview

"Science is not technology, it is not gadgetry, it is not some mysterious cult, it is not a great mechanical monster. Science is an adventure of the human spirit. It is essentially an artistic enterprise, stimulated largely by curiosity, served largely by disciplined imagination, and based largely on faith in the reasonableness, order, and beauty of the universe of which man is a part."

Warren Weaver (1894–)
Ex-president of the Sloan Foundation
1960

In an earlier chapter, our study of gases provided us with a better understanding of the behavior of the molecules that make up a sample of gaseous matter. In the case of an ideal gas, it is possible to analyze the motion of a molecule and use well-established laws to develop the mathematical equations which are used to explain the behavior of a macroscopic sample of the gas. This accomplishment helped establish the validity and success of the Kinetic Molecular Theory.

We are now ready to extend our investigation to the more complex phases of matter: liquids and solids. In this chapter, our objectives are twofold. First we shall attempt to explain the characteristics of liquids and solids in terms of their general structures and the nature of their component particles. Unfortunately, the structure of liquids and solids is more difficult to study than that of gases. Unlike the motion of gaseous molecules, the motion of liquid molecules is difficult to analyze. This is because of the close packing and large number of collisions and interactions between molecules in liquids.

Secondly, we shall examine phase changes in matter for clues to specific factors which are responsible for the behavior of molecular systems. We shall find that the Kinetic Molecular Theory, developed during our study of gases, as well as our study of molecular characteristics and intermolecular forces in the last chapter, will help us understand the behavior of liquids and solids.

LIQUIDS

10-1 Effects of Intermolecular Distances and Attractions In Chapter 4 we developed the Kinetic Molecular Theory to help us describe the behavior of an ideal gas. According to this theory the molecules of an ideal gas occupy negligible volume and exert no attractive forces on each other. Real gases deviate from ideal behavior for two reasons. First, the molecules do have a finite volume. When a gas is compressed, the volume of the molecules in the gas becomes a significant fraction of the total volume. Second, gaseous molecules actually have a small attraction for one another. These attractive forces (van der Waals forces) are not very significant in most gases at high temperatures and low pressures. When, however, a gas is placed under great pressure and the temperature is lowered, these forces do become very significant. Under the proper conditions of temperature and pressure, all gases can be liquefied. High pressures tend to reduce the space between molecules and low temperatures reduce the kinetic energy of the molecules to a point where the attractive forces result in the liquefaction of a gas.

In a liquid, the volume of the molecules and the attractive forces between them are much more important than they are in a gas. In a gas, the molecules constitute far less than 1 percent of the total volume. When a gas is liquefied, the spaces between the molecules become diminished so that the molecules actually constitute approximately 70 percent of the total volume. Because the spaces between the molecules in a liquid are so greatly decreased, and because powerful electric forces as well as molecular motion exist within liquids, *a liquid expands and contracts only very slightly with a change in temperature.* For the same reason, liquids lack the compressibility typical of gases; the volume of a liquid can be reduced only very slightly, even under extreme pressure. Thus, we find that effects which are not evident in gases are of prime importance in liquids.

Because of the large attraction of the molecules of a liquid for one another, a given sample of liquid (at a given temperature) always occupies essentially a constant volume regardless of external pressure or size of the container. A 50-ml sample of mercury for example, occupies 50 ml of space whether it is in a 50-ml container or a 5-liter container. This is one of the differences which distinguishes a gas from a liquid. Gas molecules are far removed from each other and essentially independent, while molecules in the liquid phase are very much affected by the mutual attraction of their close neighbors. Thus, a liquid can be poured from one vessel to another as a body; however, the ease of transfer depends on the liquid's *viscosity* (internal resistance to flow). Liquids such as molasses have a high viscosity and are difficult to pour, especially at low temperatures where the attractive forces between the molecules become more important.

In 1827, the Scottish botanist, Robert Brown (1773–1858), noticed that small particles that were suspended in a liquid had a constant random motion. This motion is the result of collisions between the particles and the molecules of the liquid which are in continuous thermal motion. You can observe the effects of molecular motion in a liquid by placing a drop of ink in a beaker of cold water. The diffusion of the ink in the water indicates that the *molecules of a liquid are in motion*. The difference between this diffusion and that in gases is that the molecules of liquids cannot move very far before colliding with their neighboring molecules.

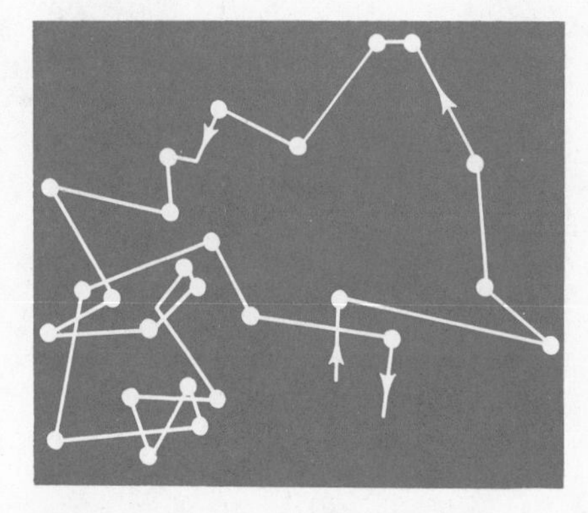

10-1 The zigzag motion of microscopic particles suspended in a liquid furnishes evidence of molecular motion in liquids and supports the Kinetic Molecular Theory.

Thus, liquids diffuse more slowly than do gases. Put an ink drop in a beaker of hot water and compare its rate of diffusion with that of one in cold water. This observation is consistent with our observations of gases, which indicate that the *molecules move more rapidly as the temperature increases.* See Fig. 10-2.

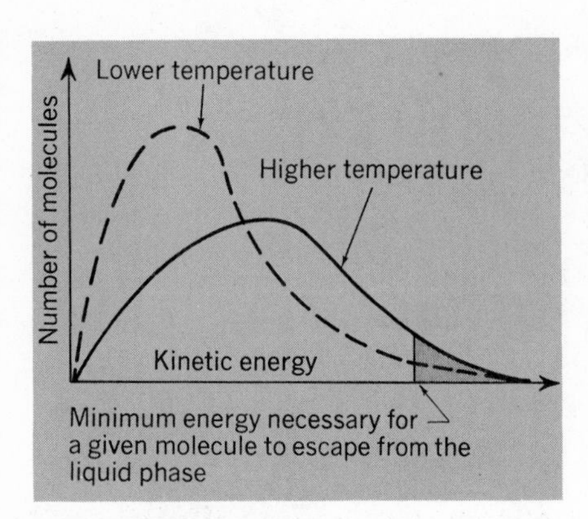

10-2 Energy distribution of molecules in a cold and in a warm liquid.

10-2 Evaporation When a liquid evaporates, the average kinetic energy of the molecules and the temperature of the liquid decrease. This implies that the strong attractive forces between the molecules of liquids may sometimes be overcome. Let us examine the process of evaporation in detail. The molecules in any one sample of liquid do not all possess the same kinetic energy. These molecules have an energy distribution, Fig. 10-2, as has a sample of gaseous molecules. The molecules in a sample of cold liquid have, on the average, less kinetic energy than those in a warmer sample. Assume that we could devise a method for momentarily obtaining a sample of water whose molecules have identical kinetic energy. They would soon be varied in energy as a result of the random collisions of the molecules. Molecules A and B might both collide with molecule C in such a way that molecule C would have its own kinetic energy plus a share of that from molecules A and B. Molecule C might then have enough energy to be included among those molecules at the extreme right of Fig. 10-2. Molecules A and B would be found toward the left in the same diagram. If molecule C happens to be near the surface of the liquid at the same moment that its kinetic energy reaches its highest point, it may be able to overcome attractive forces and escape into the gaseous phase. Part of its energy is used in overcoming attractive forces. If this single molecule does manage to escape, we say that it has undergone a *change of phase*. Since the faster-moving molecules are the ones which have enough energy to escape, the remaining molecules have a lower average kinetic energy. Thus, the temperature is lowered because of the removal of the higher-energy molecules.

We might compare this situation to another, more familiar one. If we removed the three heaviest players from a football team, the average weight of the team would be decreased even though none of the remaining players underwent any change. In the case of the molecular system, however, the Law of Conservation of Energy shows that the decrease in average kinetic energy of the system is balanced by the increase in potential energy of the molecules undergoing the change in phase.

By examining Fig. 10-2, we can easily see that a warmer sample of liquid would have a higher percentage of molecules in the high kinetic energy range and would, therefore, evaporate faster. We can observe this experimentally if we compare the evaporation rates of two samples of water at different temperatures.

The human body uses this mechanism for lowering its temperature. Evaporation of perspiration is one of the most effective methods of losing excess heat.

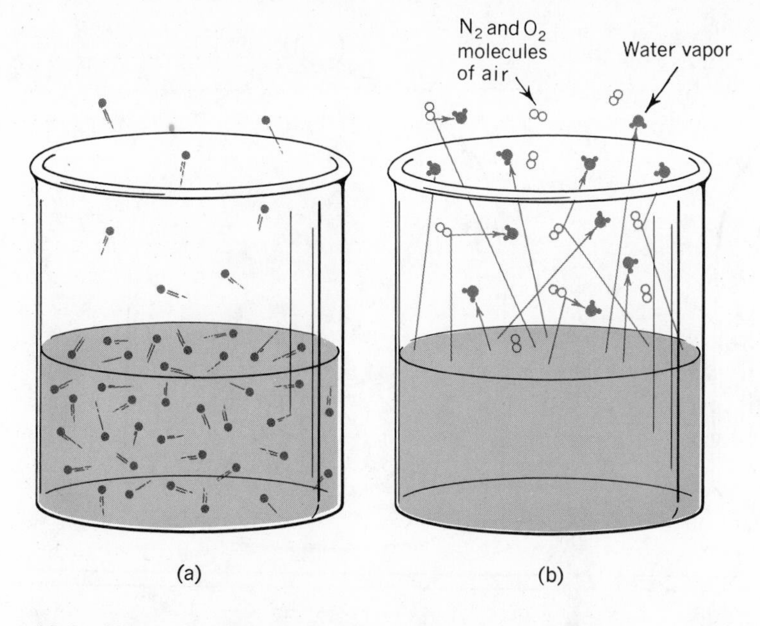

10-3 The fastest moving molecules near the surface of a liquid escape into the gaseous phase. The increase in potential energy experienced by those molecules undergoing a phase change is just balanced by the decrease in average kinetic energy of the molecules remaining in the liquid. Thus, the temperature of the remaining liquid is lowered. In (a) the molecules are evaporating into a vacuum. Thus, the rate of evaporation is greater than in (b) where they are evaporating into air. Explain. Would the presence of air currents in (b) affect the rate of evaporation? Explain.

10-3 Phase Equilibrium If we place some water in an open container and place a large bell jar over the container, careful measurement shows that the water level drops a small amount at first. This indicates that evaporation has begun. Continued observation of this system shows that evaporation appears to stop, and no further change is obvious. This experiment can be explained in terms of the Kinetic Molecular Theory. When the water is first placed in the container, some of the more energetic molecules near the surface escape into the gaseous phase. This process continues simultaneously with the reverse process, in which some of the molecules in the gaseous phase strike the surface of the liquid where strong forces of attraction can cause some of them to return to the liquid phase. After a short time, the rates of evaporation and condensation equalize. This condition is known as ***phase equilibrium.*** When, in a given closed (constant mass) system, opposing changes are taking place at equal rates, the system is said to have reached a state of ***dynamic equilibrium.*** Although the changes continue to take place, the quantities of materials and observable (macroscopic) properties remain constant.

If the bell jar were removed, the gaseous phase would partially escape. Then all of the water would eventually evaporate, and equilibrium conditions would never be reached. *The system must be closed before equilibrium can be established.* If we were to increase the temperature of the entire system, how would this affect the equilibrium? By examining Fig. 10-2, we can see that increasing the temperature shifts the average kinetic energy of the molecules to the right in both the liquid and gas phases. This elevation of temperature increases the average velocity of the molecules and decreases the opportunity for a gaseous molecule to be trapped

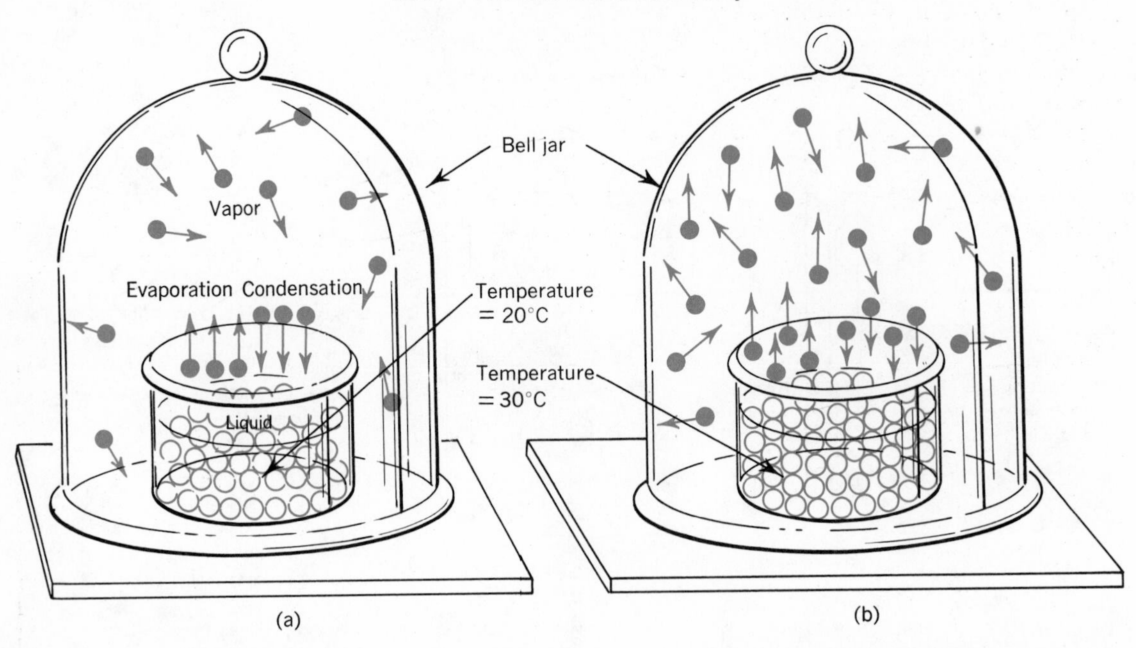

10-4 In a closed system, a dynamic equilibrium is established when the rate of evaporation is equal to the rate of condensation. When the temperature is increased as in (b), equilibrium is re-established but there are more molecules in the gaseous phase than in (a).

in the liquid phase, even with a substantial increase in impacts per unit area per unit of time. At the same time, it increases the chances that a molecule in the liquid phase will pass into the gaseous phase. The increase in temperature causes a reduction in the amount of liquid and an increase in the amount of gas. If we maintain the same system at a *higher temperature, a new equilibrium condition is established.* At this new equilibrium point, the rates of evaporation and condensation again become equal even though we now have more molecules in the gaseous phase than in the previous equilibrium.

10-4 Le Châtelier's Principle The behavior of the system described above illustrates what is known as Le Châtelier's Principle. The principle may be stated: *when a system at equilibrium is disturbed by application of a stress (change in temperature, pressure, concentration) it adjusts so as to minimize the stress and attain a new equilibrium position.* In the preceding example, adding energy to the system at equilibrium was a stress which disturbed the system. The stress was minimized when the system absorbed energy by increasing the amount of gas at the expense of decreasing the amount of liquid. In chemical language we say that the equilibrium system represented by the equation

$$\textbf{heat} + H_2O(l) \rightleftharpoons H_2O(g)$$

shifts to the right when the temperature is increased. The process represented above (to the right) is endothermic. It is worth noting that any system which is in a state of equilibrium shifts in the direction of the endothermic process when the temperature is increased.

10-5 Equilibrium Vapor Pressure The molecules in the vapor which are in equilibrium with a liquid at a given temperature exert a constant pressure. This pressure is known as the ***equilibrium vapor pressure*** of the liquid at that temperature. At constant temperature, the vapor pressure remains constant regardless of the size of the container as long as the liquid phase is present. Vapor pressures can be determined experimentally as shown in Figs. 10-5 and 10-6. The equilibrium vapor pressures for several substances are listed in Table 10-1.

TABLE 10-1
EQUILIBRIUM VAPOR PRESSURES AT 20°C

Substance	*Formula*	*Vapor Pressure (torr)*
Ethyl alcohol	CH_3CH_2OH	44
Benzene	C_6H_6	75
Carbon disulfide	CS_2	225
Diethyl ether	$CH_3CH_2OCH_2CH_3$	440
Mercury	Hg	0.00120
Water	H_2O	17.5

As shown earlier, the equilibrium vapor pressure increases with an increase in temperature. The vapor pressure of water at various temperatures is tabulated in Table 10-2.

The vapor pressure-temperature curve for water shown in Fig. 10-7 is obtained by plotting the data in Table 10-2. Any point on the line represents the pressure of water vapor that can exist in equilibrium with liquid water at the specified temperature in a closed system.

10-6 Boiling Point When a liquid is heated in an open container, the liquid and vapor are not in equilibrium. As the temperature is continuously increased, the vapor pressure also increases until it becomes equal to the pressure above the liquid. At this

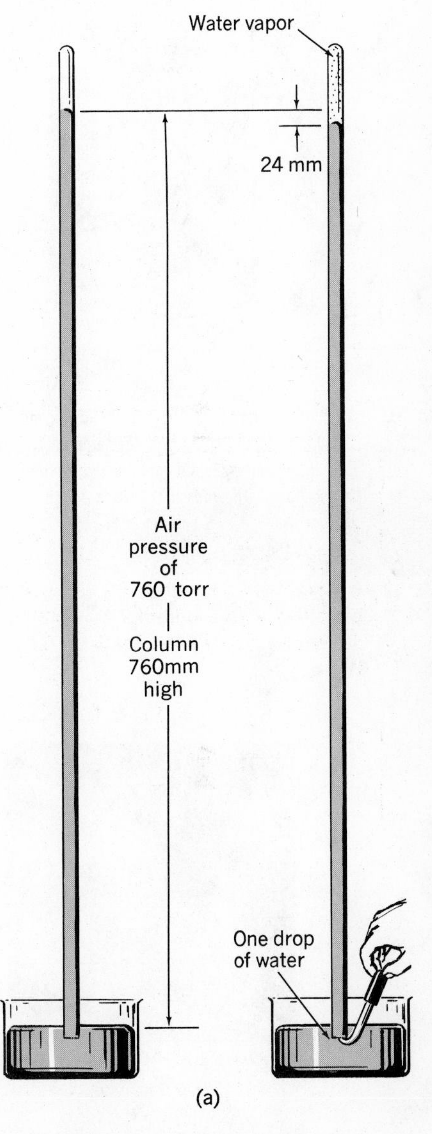

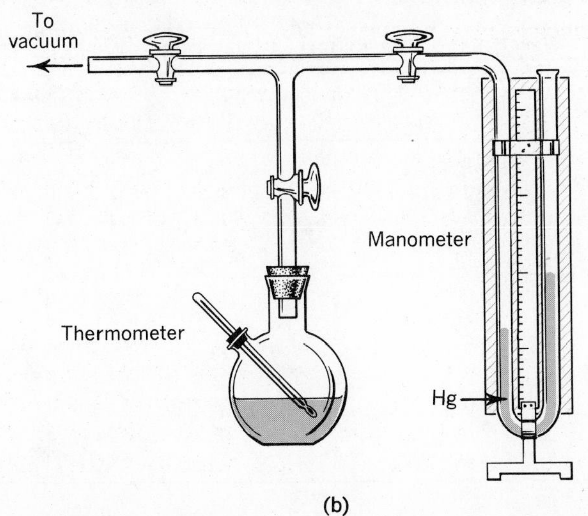

10-5 Apparatus for determining vapor pressure. (a) In the barometer, the water molecules escaping from a drop of liquid water exert enough pressure at 25°C to lower the mercury level by 24 mm. (b) A more sophisticated apparatus for determining vapor pressure so that the mercury in the manometer does not become contaminated with the liquid being tested.

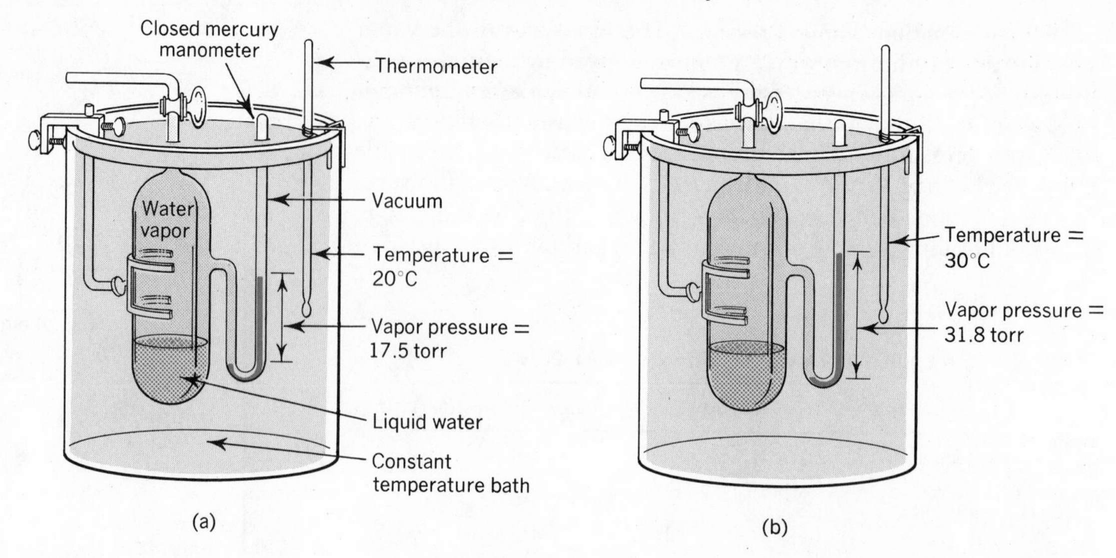

10-6 The vapor pressure of a liquid depends on the nature of the liquid and the temperature, not on the amount of liquid nor the size of the container. In (a) the pressure exerted by the vapor is just balanced by a column of mercury 17.5 mm high. At the higher temperature in (b), a column of mercury 31.8 mm high is required to balance the pressure exerted by the water vapor.

point, the average kinetic energy of the molecules is such that they are rapidly converted from the liquid to the vapor phase within the liquid as well as at the surface. The event is called ***boiling*** and the temperature at which the vapor pressure of a liquid is equal to the pressure above it is known as the ***boiling point*** of the liquid. Notice in Table 10-2 that the equilibrium vapor pressure of water at 100°C is 760 torr. We may then say that the boiling point of any liquid at 1 atm pressure is the temperature required for that liquid to exert a vapor pressure of 760 torr. If the pressure above a liquid is decreased, we expect it to boil at a lower temperature. This is verified by experimental evidence. The normal boiling points at 1 atm pressure for a few substances are listed in Table 10-3.

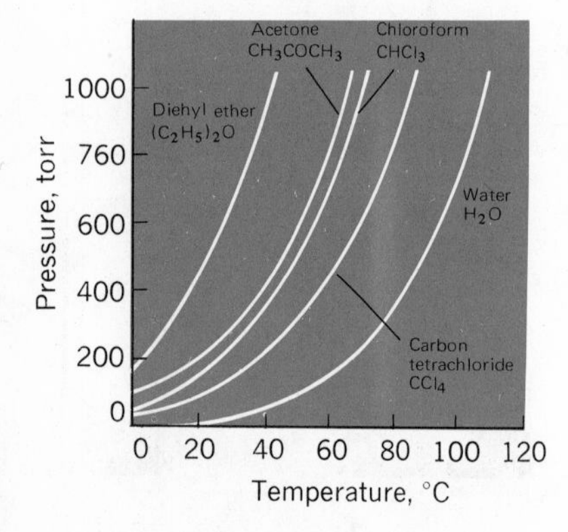

10-7 Equilibrium vapor pressures for several liquids. Use the graph to identify the normal boiling point of each of the liquids.

TABLE 10-2
WATER VAPOR PRESSURES AT VARIOUS TEMPERATURES

Temperature (°C)	*Pressure* (torr)	*Temperature* (°C)	*Pressure* (torr)	*Temperature* (°C)	*Pressure* (torr)
5	6.5	17	14.5	29	30.0
6	7.0	18	15.5	30	31.8
7	7.5	19	16.5	40	55.3
8	8.0	20	17.5	50	92.5
9	8.6	21	18.7	60	149.4
10	9.2	22	19.8	70	233.7
11	9.8	23	21.1	75	289.1
12	10.5	24	22.4	80	355.1
13	11.2	25	23.8	85	433.6
14	12.0	26	25.2	90	525.8
15	12.8	27	26.7	95	633.9
16	13.6	28	28.3	100	760.0

Comparison of the boiling points in Table 10-3, with the vapor pressures in Table 10-1, suggests that qualitatively, there is a relationship between the vapor pressure of a liquid at a given temperature and the boiling point of the liquid. That is, liquids with the highest vapor pressure at a given temperature have the lowest boiling points.

Vapor Pressure of Benzene

Temperature (°C)	Vapor Pressure (torr)
0	27
30	120
50	270
80	750
85	880
90	1020
100	1360

FOLLOW-UP PROBLEMS

1. The pressure above liquid water in a closed flask is 50 torr. At what temperature does the water boil? Refer to Table 10-2.

2. Water boils at 85°C at the summit of a mountain peak. What is the atmospheric pressure there?

3. (a) What is the approximate boiling point of benzene at 760 torr? The vapor pressure of benzene at various temperatures is listed in the margin. Plot vapor pressure on the ordinate and temperature on the abscissa. (b) What does the difference in boiling points between benzene and water suggest about the nature of their molecular structures and the nature of the bonds between the particles which compose the aggregates?

4. Does ethylene glycol have a higher or lower vapor pressure than water at 20°C? Answer in terms of differences in intermolecular bonding.

5. Explain in terms of intermolecular bonding, the difference in boiling points between glycerol [($C_3H_5(OH)_3$)] and carbon disulfide (CS_2).

10-7 Molar Heat of Vaporization Increasing the temperature of a liquid to the boiling point does not guarantee that all of the molecules will pass into the gaseous phase. We recall that a distribution of kinetic energy exists at each temperature. Even though we have a sample of liquid water at 100°C, we still have an energy distribution. If this sample of boiling water were permitted to stand at room temperature, some of the faster-moving molecules would escape. This would lower the average kinetic energy and the temperature. In time, the sample would cool to room temperature. In order to vaporize any given sample of water, we must continue

TABLE 10-3
BOILING POINTS OF SELECTED SUBSTANCES
(1 atm)

Substance	*Formula*	*Boiling Point (°C)*
Acetone	CH_3COCH_3	56.2
Alcohol (ethyl)	C_2H_5OH	78.5
Carbon tetrachloride	CCl_4	76.8
Carbon disulfide	CS_2	45.0
Chloroform	$CHCl_3$	61.2
Ethyl ether	$(C_2H_5)_2O$	34.6
Glycerol	$C_3H_5(OH)_3$	290 (decomposes)
Glycol (ethylene)	$C_2H_4(OH)_2$	198.0
Helium	He	−268.6
Mercury	Hg	356.6
Oxygen	O_2	−183.0
Water	H_2O	100.0

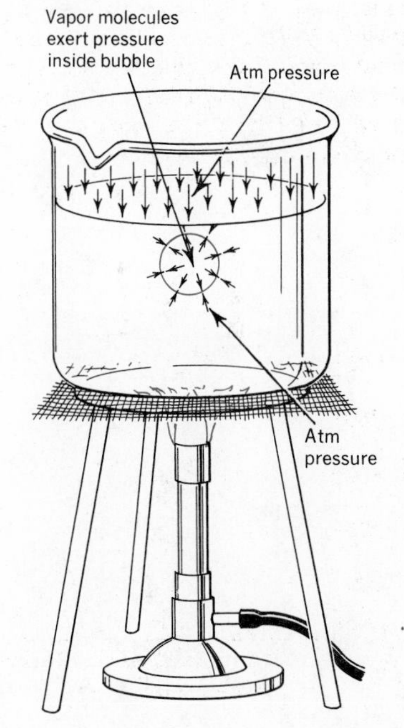

10-8 Boiling occurs when bubbles of vapor form in a liquid, rise to the surface, and escape. Boiling occurs when the vapor pressure of the liquid becomes equal to atmospheric pressure. If the atmospheric pressure exceeds the vapor pressure, the bubbles tend to collapse. If the vapor molecules inside the bubble exert a pressure greater than atmospheric, then the bubble rises to the surface.

to add a given amount of energy until all of the hydrogen bonds in the sample are broken. The added energy increases the potential energy of the water molecules being vaporized.

Let us examine quantitatively the amount of energy absorbed when a sample of water is heated. To determine the quantity of heat transferred during a change in a sample of matter, it is necessary to measure the temperature change (Δt) and the mass of the sample of matter undergoing the temperature change. In addition, it is necessary to know the ***specific heat capacity*** of the substance. *The specific heat capacity of a substance is the number of calories required to raise the temperature of one gram of that substance one Celsius degree.* The number of calories gained or lost by a given mass of substance may be expressed in formula form as

$$\textbf{cal lost or gained} = \textbf{g} \times \Delta \textbf{t} \times \textbf{specific heat capacity}$$

Assume that we have a one-mole sample of water (18 g) at 60°C and a heating coil capable of delivering 80 cal/min to our sample. In order to raise the temperature of the water to its boiling point (100°C at 1 atm), 720 cal is required. That is,

$$18 \text{ g } H_2O \times \frac{1.0 \text{ cal}}{\text{g } H_2O \text{ °C}} \times (100\text{°C} - 60\text{°C}) = 720 \text{ cal}$$

If we measure both the temperature and the elapsed time, we see that in 9 minutes the coil would provide sufficient heat to raise the temperature of the sample from 60°C to 100°C. At this point, we observe that the water begins to boil. Two additional hours would be required to vaporize this one-mole sample of water completely. This indicates that 9600 cal (80 cal/min × 120 min) is absorbed by the boiling water during the vaporization. Our thermometer, whose reading rose from 60°C to 100°C in a matter of 9 minutes, records 100°C during the entire two hours. Since the coil was delivering 80 cal/min to our sample during this period of time and no temperature change was noted, we can conclude that the energy is being absorbed by the molecules in order to bring about the change of phase. The absorbed energy is used in overcoming the attractive forces between the molecules. Since there is no change in average *kinetic energy* (temperature is constant), the energy being absorbed must be increasing the *potential energy* of the H_2O molecules being vaporized. In this experiment, we assume that all of the heat from the coil was transmitted to the water and that any heat loss to the environment during the heating period was negligible. If the water vapor were trapped and if the coil were to continue to supply 80 cal/min, the temperature would immediately rise above 100°C as soon as the liquid water was completely vaporized. A graph of the data from this experiment would appear as shown in Fig. 10-9. The plateau of the curve is characteristic of any pure substance (element or compound) which is being heated while undergoing a phase change. One might ask if there is actually any difference in the system at points A and B since

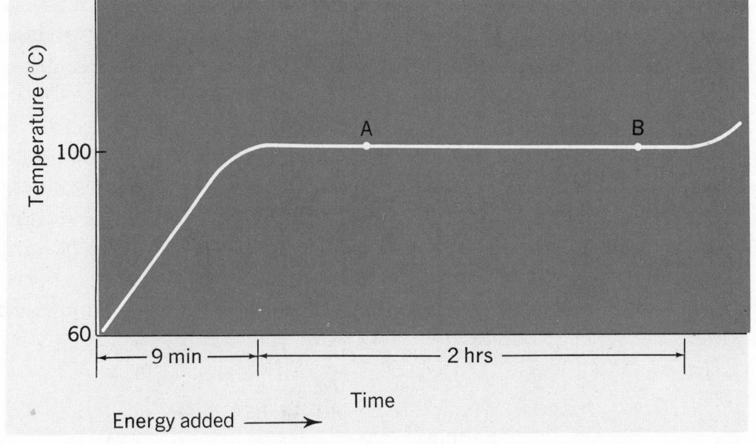

10-9 How does the system change from point A to point B?

the temperature is 100°C in both cases. There is only one difference: at point B, more of the liquid has been converted to the gaseous phase. Consequently, the entire system has more potential energy at point B.

The amount of heat required to vaporize a mole of liquid already at its boiling point is known as the ***molar heat of vaporization.*** This experiment indicates that the *molar heat of vaporization* for water is about 9,600 cal/mole. This means that 9.6 kcal is absorbed when a mole of liquid water is vaporized, and an equal amount of energy is liberated when the same amount of water vapor is condensed.

The internal energy stored in a substance is related to its ***heat content* or *enthalpy*** which is symbolized by H. Thus, the one mole of water vapor at 100°C has a greater heat content or enthalpy than one mole of liquid water at 100°C. The change in heat content or enthalpy that occurs when a system changes from one energy state to another is symbolized by ΔH and defined as

$$\Delta H = \begin{bmatrix}\textbf{Total enthalpy} \\ \textbf{of system in} \\ \textbf{final state}\end{bmatrix} - \begin{bmatrix}\textbf{Total enthalpy} \\ \textbf{of system in} \\ \textbf{initial state}\end{bmatrix} \qquad \text{10-1}$$

At a given temperature and pressure, one mole of substance has a characteristic enthalpy. It is, however, difficult to determine absolute values of enthalpies. For our purposes it is not necessary to know the absolute enthalpies. We are primarily interested in knowing the energy absorbed or liberated during a phase change or chemical reaction. We can measure the *energy changes* of a system. In the experiment described above, the enthalpy of one mole of water vapor was 9.6 kcal greater than the one mole of liquid water at 100°C. Therefore the change in enthalpy known as the molar heat of vaporization and symbolized by ΔH_v for the process

$$\textbf{1 mole } H_2O(l) = \textbf{1 mole } H_2O(g)$$

at 100°C is equal to 9.6 kcal. Note that ΔH is always positive for an endothermic process and negative for an exothermic process because it is defined as $H_{final} - H_{initial}$. In an exothermic process, the products in the final state have less enthalpy than the reactants in the initial state. Subtracting a number from a smaller one yields a negative difference.

Since a mole of any liquid substance contains 6.02×10^{23} molecules, the *molar heats of vaporization* for several liquids *provide us with a comparison of the relative attraction that the molecules of these liquids have for one another.* Table 10-4 shows the boiling points and molar heats of vaporization for several liquids. More precise experiments show that the molar heat of vaporization for water is 9.72 kcal/mole.

TABLE 10-4
BOILING POINTS AND MOLAR HEATS OF VAPORIZATION AT 1 ATM

Substance			
Name	*Formula*	*Boiling Point (°K)*	*Molar Heat of Vaporization (kcal/mole)*
Helium	He	4.22	0.020
Methane	CH_4	111.7	1.96
Chlorine	Cl_2	238.9	4.88
Diethyl ether	$(C_2H_5)_2O$	307.6	6.21
Water	H_2O	373	9.72
Mercury	Hg	630	14.2
Sodium	Na	1162	24.1
Sodium chloride	NaCl	1738	40.8

Example 10-1

A 2.00-kg sample of boiling water is heated by an electric hot plate which delivers 5.00 kcal/min. After 25.0 minutes, how many grams of $H_2O(l)$ remain? Assume that all energy was absorbed in the phase change.

Solution

1. Determine how many calories were delivered to the sample.

$$5.00 \text{ kcal/min} \times 25.0 \text{ min} = 125 \text{ kcal}$$

2. Find the number of moles of water vaporized.

$$\frac{1 \text{ mole}}{9.7 \text{ kcal}} \times 125 \text{ kcal} = 12.9 \text{ moles}$$

3. Find the number of grams of water vaporized.

$$18.0 \text{ g/mole} \times 12.9 \text{ moles} = 232 \text{ g}$$

4. Find the number of grams remaining.

$$2.00 \text{ kg} - 0.23 \text{ kg} = 1.77 \text{ kg} = 1770 \text{ g}$$

Alternate one-step solution

$$25.0 \text{ min} \times \frac{5.00 \text{ kcal}}{\text{min}} \times \frac{\text{mole } H_2O}{9.7 \text{ kcal}} \times \frac{18.0 \text{ g } H_2O}{\text{mole } H_2O}$$

$$= 232 \text{ g } H_2O \text{ vaporized}$$

FOLLOW-UP PROBLEMS

1. Find the number of kcal required to vaporize a 5.00-mole sample of water originally at 40.0°C.
Ans. 53.9 kcal.

2. When red-hot iron is dipped into a beaker of boiling water, 25 g of water is vaporized. How many kilocalories were given up by the iron? **Ans. 13 kcal.**

3. The molar heat of vaporization for ether (molecular mass 74 g) is 6.2 kcal/mole. How many grams of boiling ether can be vaporized by 0.50 kcal?
Ans. 6.0 g.

From Table 10-4, we see that the molar heat of vaporization of helium is very low compared to that of the other gases in this group. What is the fundamental reason for the low molar heat of vaporization of helium? Helium and the other noble gases consist of monatomic molecules with almost perfect symmetry and minimal surface contact. Because of these, the van der Waals forces between helium atoms are a minimum. The amount of energy required to pull apart one mole of these molecules in the liquid phase and to change them to the gaseous phase is very small.

Notice that the molar heat of vaporization of water is much higher than that of helium. Water molecules are nonsymmetrical particles which have considerable attraction for each other. These attractive forces are shown by the strong *surface tension* of water.

The phenomenon of surface tension is explained by the fact that molecules on the surface of a liquid are only partly surrounded by other molecules. At the surface of a sample of water, therefore, the hydrogen bonds act largely toward the interior. The molecules on the surface are consequently drawn inward (see Fig. 10-10). The surface tension of water enables us to float a needle on water and it is responsible for the almost spherical droplets of water which form when water is placed on a waxed surface. A sphere has the smallest surface area-to-volume ratio of any three-dimensional geometric figure. You have all seen spherical droplets of mercury roll around on a glass surface. This observation, supported by the fact that mercury has a high molar heat of vaporization, suggests that the intermolecular forces in liquid mercury are relatively strong and that mercury has a higher surface tension than water.

Diethyl ether ($CH_3CH_2{-}O{-}CH_2CH_3$) is only slightly polar and thus its dipole bonds are relatively weak. This accounts for the fact that this substance exhibits little surface tension and has a lower molar heat of vaporization than water. The strong odor emitted

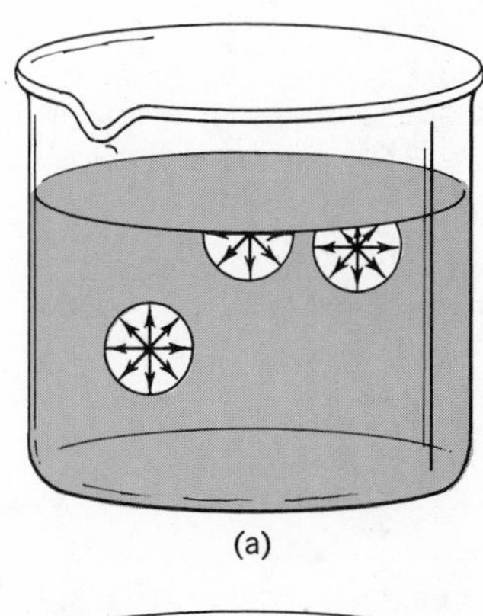

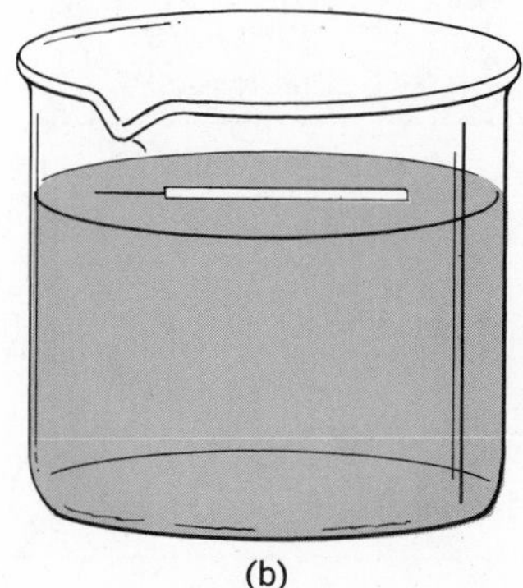

10-10 (a) The net forces operating on the surface molecules tend to be directed toward the interior of the liquid. They produce surface tension. (b) Because of surface tension, a steel needle can be floated on the surface of the water in a beaker.

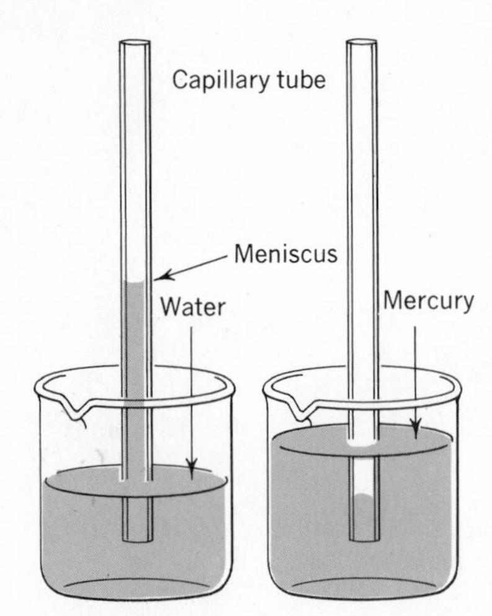

10-11 Capillary action. The relative forces of attraction between liquid molecules themselves as compared to those between liquid molecules and glass determine whether the liquid rises or falls in a capillary tube. The depressed surface of mercury in the tube is consistent with our observation that the attractive forces between mercury atoms are greater than those between water molecules. Mercury, unlike water, does not wet glass.

by ether is further evidence of its high vapor pressure and weak dipole bonds. We may make the generalization that *liquids and solids with strong odors are likely to have high vapor pressures and low molar heats of vaporization.*

A simple experiment compares the rate of vaporization of water with that of ether. Place a drop of each on the back of your hand. You will notice that the ether evaporates much more rapidly. The cooling sensation produced by the vaporization of the ether indicates that the strength of the attractive forces acting between ether molecules is less than that between water molecules.

10-8 Critical Temperature and Pressure We have noted that under proper conditions of temperature and pressure any gas can be liquefied. Perhaps you have wondered whether there are any conditions under which a gas cannot be liquefied or, put in another way, are there conditions under which it is impossible for a liquid to exist? We learned earlier that the average kinetic energy of a molecular system is directly proportional to the Kelvin temperature. It is reasonable, therefore, to assume that there is a temperature at which the kinetic energy of the molecules is so great that the attractive forces between molecules are too ineffective for the liquid phase to remain. The temperature above which the liquid phase of a substance cannot exist is called its *critical temperature.* Above its critical temperature, no gas can be liquefied regardless of the applied pressure. The critical temperature of helium is −267.9°C. Above this temperature, helium cannot be liquefied even if thousands of atmospheres of pressure are applied. The minimum pressure required to liquefy a gas at its critical temperature is called its *critical pressure.* Because there is no liquid phase above the critical temperature, the critical pressure is the maximum pressure that a vapor can exert when in equilibrium with the liquid phase. The critical pressure of helium is 2.26 atm. This means that a pressure of 2.26 atm or greater liquefies a sample of helium gas if the temperature of the gas is no higher than −267.9°C. A few gases are listed with their critical temperatures and critical pressures in Table 10-5.

TABLE 10-5
CRITICAL TEMPERATURES AND PRESSURES OF COMMON GASES

	Critical Temperature (°C)	(°K)	Critical Pressure (atm)
He	−267.9	5.1	2.26
H_2	−239.9	33.1	12.8
O_2	−118.8	154.2	49.7
Cl_2	144.0	417.0	76.1
C_4H_{10} (n-butane)	153.0	426.0	36.0
SO_2	157.2	430.2	77.7

From Table 10-5 it is evident that Cl_2, C_4H_{10}, and SO_2 could be stored as liquids at room temperature in strong steel tanks. The others could not be stored as liquids at room temperature at any pressure. Substances with low critical temperatures behave more "ideally" than those with higher critical temperatures. Thus, the behavior of helium approaches that of an ideal gas.

SOLIDS

10-9 General Characteristics of Solids In considering samples of gases, liquids, and solids, we understand that the *particles in gases have the highest degree of disorder.* Particles of solids are the most ordered, and particles of liquids are intermediate between them. The observation that solids are practically incompressible should not be surprising since the particles of a solid are packed more closely than particles in the liquid phase. Since the particles are fixed in a rather definite position, *solids maintain a definite shape.* This does not mean that their positions are absolutely fixed but rather that their motion is severely restricted. The degree of restriction varies greatly with different types of solids.

Particles in a solid may vibrate in position, and even diffusion may occur in some solids. If solid gold is clamped against solid lead for a long period of time, diffusion occurs at the junction of the two metals. No such diffusion is observed for such solids as table salt (NaCl) or diamond, a crystalline form of carbon. This suggests that as a class, metallic bonding is weaker than ionic bonding or the covalent bonds in network solids. In the two latter cases, we assume that particle *movement* is at a minimum.

When heated at certain pressures, some solids vaporize directly without passing through the liquid phase. This direct change from the solid to the gaseous phase is known as sublimation. Solid iodine, naphthalene, paradichlorobenzene (moth crystals), and solid carbon dioxide (Dry Ice) are examples of solids which sublime readily. All solids exhibit vapor pressures, but these values are usually extremely low when compared with those of liquids. As with liquids, the vapor pressure of a solid increases with increasing temperature. We would expect odorous solids, such as moth crystals, to have a higher vapor pressure than a solid which exhibits no odor; however, this is not necessarily the case. Dry Ice, for example, has a high vapor pressure and no odor. The sublimation process also explains the disappearance of snow on dry days at temperatures below the freezing point of water. Sublimation occurs when the vapor pressure of the solid exceeds the partial pressure of that substance in the atmosphere while at a temperature below its melting point (see discussion of triple point in Sect. 10-11).

When heated, most solids reach a point at which their atomic or molecular vibrations become so great that the particles break free from their fixed positions and begin to slide freely over each other, forming a liquid. The temperature at which the transition from the solid to liquid phase occurs is known as the ***melting point.*** Solids have a great range in melting points. Gallium metal melts at 29.8°C, and sodium chloride (table salt) melts at 801°C. This range in melting points indicates that the particles in some solids are bonded together much more strongly than others. A number of solids, such as glass, slowly soften upon being heated and never exhibit any definite melting point. These solids are called ***noncrystalline*** or ***amorphous*** solids. The atoms or molecules of amorphous solids are not arranged in the definite geometrical fashion characteristic of crystalline solids. Because amorphous materials are not pure substances, interparticle forces vary in type and strength.

10-12 If a sample of radioactive lead is clamped against a sample of nonradioactive lead, the radioactive atoms, after a period of time, can be detected in the nonradioactive sample.

10-10 Molar Heat of Fusion, ΔH_f Earlier, we saw that a definite amount of additional energy (the molar heat of vaporization)

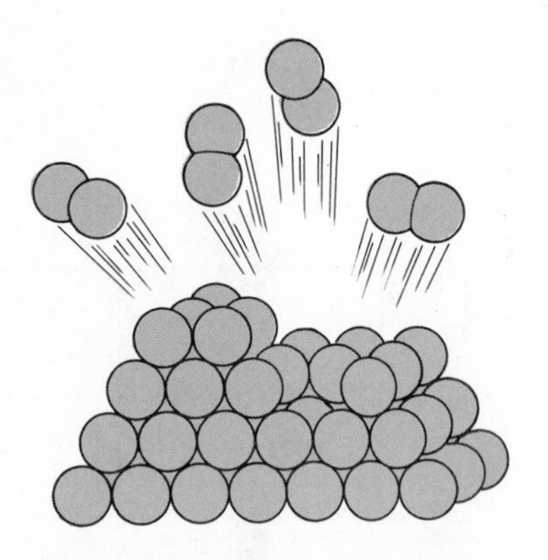

10-13 Sublimation of iodine. In this endothermic process, the iodine molecules leave the solid surface and enter directly the gaseous phase.

must be provided in order to vaporize a mole of liquid at its boiling point. The same general principle applies to the melting of solids. When a solid is warmed to its melting point, its atoms or molecules are actively vibrating in position. Additional energy must be added at the melting point before the solid can be liquefied. The amount of energy required for the liquefaction (fusion) of a mole of solid is called the ***molar heat of fusion.*** This amount of energy depends upon the nature of the solid and the types of bonds present. Solids such as acetamide or paradichlorobenzene which are easily melted, have lower molar heats of fusion than solids which are difficult to melt. An example of the latter is table salt which has a high molar heat of fusion. It is generally true that substances with low melting points also have low molar heats of fusions. Table 10-6 lists the molar heats of fusion for some solids.

You will notice in comparing Table 10-6 with Table 10-4 that the molar heats of fusion for these materials differ widely from their molar heats of vaporization. This can be partially explained by the observation that the melting of solids does not require that the atoms or molecules be pulled completely apart from one another, as in the case of vaporization, but only that they be freed from the crystalline structure so that they may slide over one another. Your knowledge of the nature of the chemical bonds in these materials should help you to understand the wide variety of melting points and molar heats of fusion noted above.

10-14 Some solids, such as gallium (above), melt at a low temperature, 29.8°C. Others, such as table salt (NaCl), melt at a relatively high temperature, 808°C. (Morton Salt Company)

TABLE 10-6
MOLAR HEATS OF FUSION

Substance			
Name	*Formula*	*Melting Point (°C)*	*Molar Heat of Fusion (H_f) (kcal/mole)*
Helium	He	−272.2	0.0050
Chlorine	Cl_2	−101	1.53
Water	H_2O	0	1.44
Sodium	Na	97.8	0.63
Sodium chloride	NaCl	801	6.8

The data in Tables 10-4 and 10-6 indicate that more energy is required to vaporize a mole of a certain liquid than is required to fuse (melt) a mole of the same material. In comparing the molar heat of fusion with the molar heat of vaporization, consider what happens as we melt a one-mole sample of ice. Assume that the temperature of the 18-g sample of ice is initially −20°C and that we have a heating coil which supplies exactly 40 cal/min to our sample. Ice has a specific heat of 0.5 cal/gC°. This means that 0.5 cal is required to warm 1 g of ice 1C°. Warming the 18-g sample from −20°C to 0°C requires 180 cal. At the rate of 40 cal/min, 180 cal would require 4.5 minutes. At this point, the temperature of the ice is 0°C. We would notice that for the next 36 minutes the thermometer remains in ice water at 0°C until all of the ice has melted. Then the temperature begins to rise above 0°C and the heating coil begins to warm the liquid water.

By plotting the time against the temperature in this experiment, we obtain the curve shown in Fig. 10-15. This curve is very similar to the one representing the vaporization of a mole of water. The characteristic plateau is present whenever thermal energy is added to a pure substance during a phase change. In comparing the melting of a mole of solid water to the vaporization of a mole of liquid water, notice that in the latter experiment the rate of heating was only 40 cal/min as compared to 80 cal/min in the vaporization experiment. Even at this reduced rate of heating, we see that 36 minutes is required for the fusion of a mole of solid ice, while two hours is required for the vaporization of the same amount of sample. From these experiments, we can conclude that much less energy is required for fusion than for vaporization. We can quantitatively calculate that at 40 cal/min for 36 minutes, 1.44 kcal was required the fusion of a mole of water, while almost 10 kcal was required for vaporization. We have discussed only the warming and vaporizing of water but, of course, these same quantities of energy must be given off if the process were reversed. We can summarize the energy changes by considering the conversion of a mole of water vapor at 100°C to a mole of ice at 0°C.

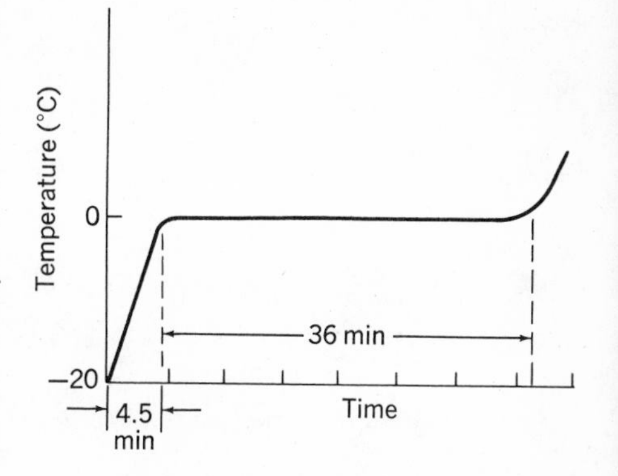

10-15 1440 calories is required to convert a mole of ice at 0°C into liquid water at 0°C.

Change	*Kilocalories Liberated*
18 g H_2O(g), 100°C ⟶ 18 g H_2O(l), 100°C	9.7
18 g H_2O)l), 100°C ⟶ 18 g H_2O(l), 0°C	1.8
18 g H_2O(l), 0°C ⟶ 18 g H_2O(s), 0°C	1.44

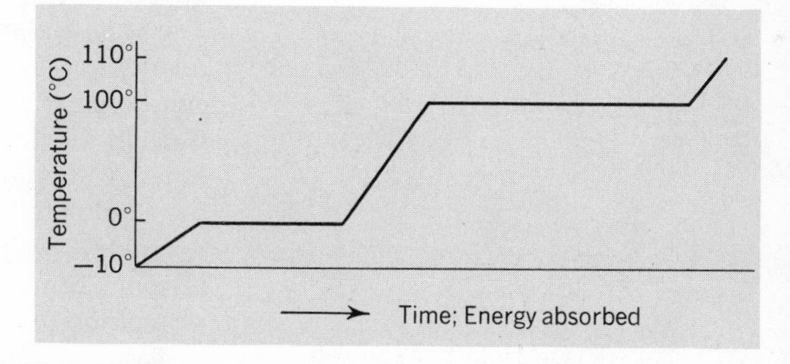

10-16 When water at −10°C is converted into steam at 110°C, two plateaus are noted, one for each phase change. The lengths of the plateaus are not drawn to scale.

Example 10-1

A 5.0-kg block of ice at −10.0°C is placed in a container of warm water. The entire block of ice is warmed to 0.0°C, and 4.0 kg of ice remains unmelted. At this point, how many calories were transferred from the warm water?

Solution

1. Determine the number of calories absorbed by the entire block of ice in warming from −10.0°C to 0.0°C. The specific heat capacity of ice is about 0.50 cal/g/C°

$$5.0 \text{ kg } H_2O(s) \times \frac{1000 \text{ g}}{1 \text{ kg}} \times 10.0C° \times \frac{0.50 \text{ cal}}{1 \text{ g } H_2O(s)/1C°}$$

$$= 25\,000 \text{ cal or } 25 \text{ kcal}$$

2. Find the total energy absorbed in the melting of 1.0 kg of ice. The heat of fusion of H_2O(s) = 1.44 kcal/mole.

$$1.0 \text{ kg} \times \frac{1000 \text{ g}}{1 \text{ kg}} \times \frac{1.44 \text{ kcal}}{1 \text{ mole } H_2O(s)} \times 1 \text{ mole}/18 \text{ g} = 80 \text{ kcal}$$

3. Total heat transferred

$$25 \text{ kcal} + 80 \text{ kcal} = 105 \text{ kcal}$$

FOLLOW-UP PROBLEMS

1. How many calories are required to melt 10 g of ice at 0°C? **Ans. 800 cal.**

2. A candle flame delivers 250 cal/min. If no loss, how long will it take this candle to vaporize 180 g of H_2O at 25°C? The molar heat of vaporization for water is 9.7 kcal/mole. **Ans. 442 min.**

3. A refrigeration unit is used to solidify 5.00 moles of $Cl_2(l)$ to $Cl_2(s)$ at its freezing point. How much heat must the refrigeration unit absorb from the chlorine? The molar heat of fusion for chlorine is 1.53 kcal/mole. **Ans. 7.65 kcal.**

PHASE DIAGRAMS

10-11 Phase Diagrams In our discussion of phase changes, we have thus far considered only the effect of temperature. Pressure also has an important influence on any system, especially one involving a gaseous substance. The effect of pressure on a system can be clearly illustrated by using carbon dioxide as an example. Carbon dioxide (Dry Ice) is an example of a material which sublimes readily at ordinary pressures. One might wonder if it would be possible to liquefy this compound under any conditions. Carbon dioxide (as well as iodine and moth crystals, which tend to sublime at room conditions) can be liquefied under the proper conditions of temperature and pressure. Phase diagrams can be constructed to show in which phase a substance exists at a given temperature and pressure. Let us interpret the partial phase diagrams for carbon dioxide shown in Fig. 10-17.

Line BD is essentially a vapor-pressure curve for the liquid phase (similar to Fig. 10-7). Any point on this line represents a temperature and pressure at which the liquid and vapor are in equilibrium.

Line AB is essentially a vapor-pressure curve for the solid phase. Any point on this line represents a temperature and pressure at which the solid and vapor are in equilibrium.

Line BC is a melting-point curve which shows how the melting point varies with the total pressure. Any point on this line represents a temperature and pressure at which the solid and liquid are in equilibrium.

At point B, where the liquid and solid vapor-pressure curves meet, the liquid and solid phases are in simultaneous equilibrium with the vapor. Therefore, the solid and liquid phases are also in equilibrium. Point B is known as the ***triple point.*** It is the only

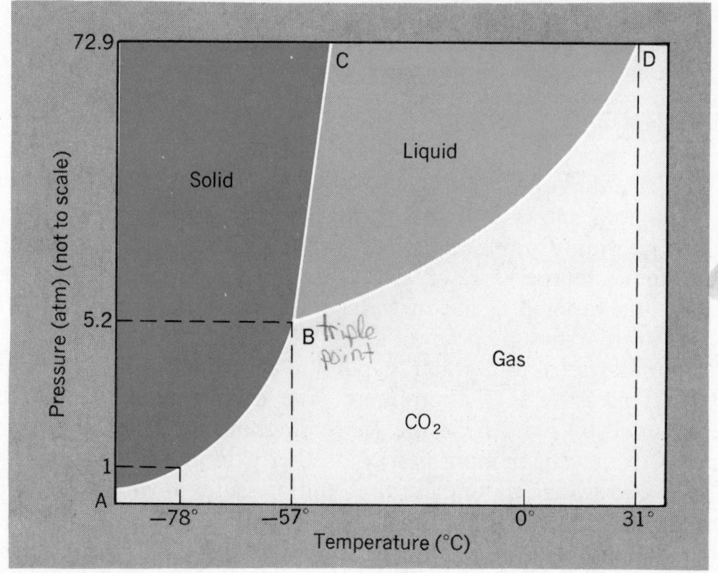

10-17 A partial phase diagram for CO_2.

temperature and pressure at which three phases of a pure substance can exist in equilibrium with one another in a system containing ***only*** the pure substance.

The definition of the normal *melting* or *freezing* point of a substance differs from that of the triple point in one way. At the normal melting point, the liquid and solid phases are in equilibrium with each other but *in the presence of air at one atmosphere of pressure.* According to this definition, the phase diagram for CO_2 indicates that CO_2 does not have a normal (at one atmosphere pressure) melting point. The vapor pressures of CO_2, I_2, and a few other solids equal one atmosphere at temperatures below their triple point. At one atmosphere pressure, therefore, they do not melt but pass directly from the solid to vapor phase. Fig. 10-17 indicates that in a closed system, solid carbon dioxide is in equilibrium with carbon dioxide vapor at $-78\,°C$. In an open system, solid carbon dioxide passes *directly* from the solid to vapor phase at a constant temperature of $-78\,°C$ until all of the solid phase has disappeared. It will, of course, sublime if the surrounding temperature is above $-78\,°C$. In order to liquefy carbon dioxide, the external pressure must be raised to a minimum of 5.2 atmospheres and the temperature must be higher than $-57\,°C$, and lower than $31\,°C$.

FOLLOW-UP PROBLEMS

1. Specify in each of these conditions which phases of carbon dioxide may exist in simultaneous equilibrium. If no equilibrium is possible, identify the existing phase. The critical temperature of CO_2 is $31\,°C$. (a) 1 atmosphere and $20\,°C$, (b) 1 atm and $-60\,°C$, (c) 1000 atm and $32\,°C$, (d) 5.2 atm and $-57\,°C$, (e) 1 atm and $-100\,°C$, (f) 60 atm and $0\,°C$.

2. (a) Which phase would increase if a system in equilibrium along line BC (Fig. 10-17) was subject to a slight increase in temperature? (b) Which phase would increase if a slight increase in pressure were applied to a system in equilibrium along line BC? (c) Which phase would increase if a system in equilibrium along line BD were subjected to a slight increase in temperature? (d) Which phase(s) would increase if the system at the triple point were subjected to a slight increase in pressure?

A phase diagram for water is shown in Fig. 10-18. Analysis of this diagram shows that in a closed system (and in the absence of air) ice, liquid water, and water vapor are in simultaneous equilibrium at a temperature of 0.01°C. At this temperature (the triple point), the vapor pressure of water is 4.5 torr (point B).

The normal melting point of ice is 0°C, the temperature at which ice, liquid water, and water vapor are in equilibrium when the external pressure is 1 atmosphere. The normal melting point is represented by point E on the phase diagram. All other points on line BC represent melting points at other external pressures. That is, they correspond to temperatures and pressures at which the solid and liquid are at equilibrium.

10-18 A partial phase diagram for water, distorted somewhat to distinguish the triple point from the freezing point.

It can be seen that the triple point for water is only slightly above its normal freezing point. It is interesting to note that in low temperature research (cryogenics), the temperature scale based on two fixed points (melting and boiling point of water) is not accurate enough because of the difficulty in obtaining an invariant value for the melting point. To resolve the problem an International Committee has defined a temperature scale based on the triple point of water which is assigned a value of 273.16°K.

We can use the phase diagram for water to determine the effect of increasing the external pressure on the melting point of ice. According to Fig. 10-18, when an external pressure is applied to the equilibrium system at the triple point, the vapor phase disappears. The remaining solid and liquid can exist in equilibrium only along line BC. Therefore, as the external pressure is increased, the

10-19 What would be the effect on these ice skaters if the freezing point of water, like that of most substances, increased with pressure?

temperature must be decreased if equilibrium is to be maintained. In other words, increasing the external pressure decreases the melting temperature. This behavior is uncommon among most familiar substances and is partially responsible for the movement of glaciers and the existence of the popular sport of ice skating.

The sublimation of snow (ice) referred to earlier does not represent an equilibrium system. The information conveyed by the phase diagram, however, indicates that in an open system, snow sublimes only when the partial pressure of water vapor in the air and the temperature are both below the triple-point values. In other words, sublimation occurs only on a very cold, dry day.

10-20 A valley glacier, Fer pèrcle in the Valais. Mt. Matterhorn is in the background. Glaciers move several inches to several feet per day. Can you explain how the movement of this glacier is related to the effect of pressure on the freezing point of water? (Photograph courtesy of Swiss Center).

Example 10-3
Explain what happens when a sample of water at -10°C is heated at a constant external pressure of 700 torr.

Solution
According to the phase diagram, water at -10°C and 700 torr is in the solid phase region. The effect of adding thermal energy to this system can be described by following a line parallel to line EF from left to right across the diagram. Between -10°C and line BC, the solid is being warmed. At the intersection with BC (slightly above 0°C) the solid melts at constant temperature. From this point on BC to the intersection of the 700-torr line with BD, the liquid is being warmed. At the intersection with BD (slightly below 100°C) the liquid vaporizes (boils). Beyond this point the water vapor is being heated.

FOLLOW-UP PROBLEM

Describe what happens when a sample of water at 380 torr is compressed at a constant temperature of 85°C.

FACTORS AFFECTING THE BEHAVIOR OF MOLECULAR SYSTEMS

10-12 Potential Energy of a Molecular System For conciseness and convenience, the phase changes discussed in this chapter are summarized in Table 10-7.

The energy changes associated with phase changes for water are represented schematically in Fig. 10-22. It is apparent that energy is an important factor that is related to the behavior of a molecular system. Energetically speaking, molecular systems behave in much the same way as mechanical systems. That is, *if the energy factor only is considered, then the system tends to reach a condition of minimum potential energy.* For example, water spontaneously runs

10-21 Water at the bottom of the falls has less potential energy than it does at the top. Thus it travels spontaneously "downhill" to a position of lower potential energy. Churchill Falls, Labrador (Annan Photo).

TABLE 10-7
SUMMARY OF PHASE CHANGES

Phase Change	*Name*	*Energy Change*
Solid to liquid	melting, fusion	endothermic
Liquid to solid	freezing, solidification	exothermic
Liquid to gas	vaporization, boiling	endothermic
Gas to liquid	condensation	exothermic
Solid to gas	sublimation	endothermic
Gas to solid	no special name	exothermic

downhill and rocks slide down the side of a mountain to the base and reach a condition of minimum potential energy. The energy used to lift the rocks back to a higher level (a nonspontaneous change) becomes potential energy stored in the rocks. Similarly, many substances such as melted sulfur spontaneously solidify at room temperature. This is consistent with our observation that the potential energy of the molecules in the solid phase is lower than that of the molecules in the liquid or gaseous phases. All substances, for example water, do not spontaneously form solids at all temperatures. Some factor other than the tendency to attain minimum potential energy must affect the behavior of molecular systems. To explore this factor, let us examine the behavior of a molecular system in which no appreciable energy change is involved.

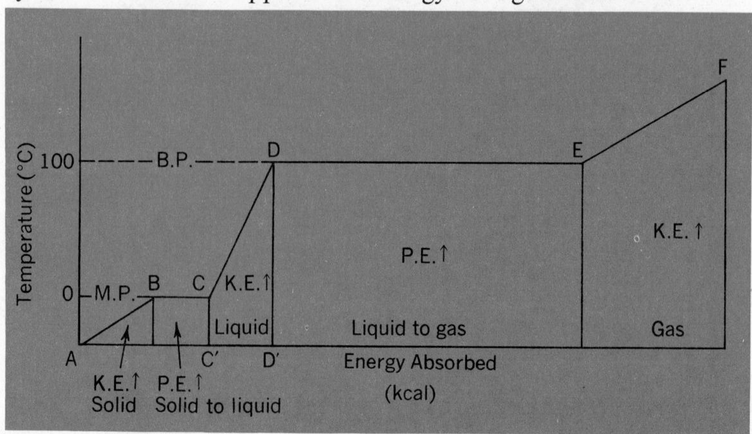

10-22 Temperature changes which occur when thermal energy (heat) is added at a constant rate to one mole of water. How many calories does the distance C′D′ represent? What should be the ratio in the lengths of lines DE/BC?

Range	*Types of Energy Change Involved*
AB	Average kinetic energy of molecules increasing rapidly, along with a small potential energy increase due to expansion of solid during heating.
BC	No change in average kinetic energy, but a considerable increase in potential energy due to the overcoming in part of the intermolecular forces of attraction.
CD	Large increase in average kinetic energy and slight increase in potential energy due to expansion of the liquid as it warms.
DE	No change in average kinetic energy, but a large change in potential energy due to the further overcoming of intermolecular forces *and* the large increase in average distance between adjacent molecules.
EF	Increase in average kinetic energy as the gas warms, but no increase in potential energy if the gas is confined at constant volume.

10-13 The Disorder of a Molecular System Consider two tanks connected by a valve. One tank contains helium gas and the other neon gas, both at the same temperature. Both gases are inert chemically. Now open the valve for a time. Then close it and examine the contents of the tanks. You find a mixture of both gases in each tank. The thermometer readings remain constant; thus there was no apparent energy transfer. The molecules, however, have spontaneously rearranged themselves in a more probable, disordered state than that which existed originally. There are many arrangements of particles which represent a disordered condition but only a few that correspond to an ordered one. Therefore, when changes occur, it is more probable for particles to achieve a disordered condition than an ordered one. A new deck of cards, arranged numerically and by suits, represents an ordered system. Once the cards are put into play, it is highly unlikely that they will spontaneously reach the original ordered arrangement.

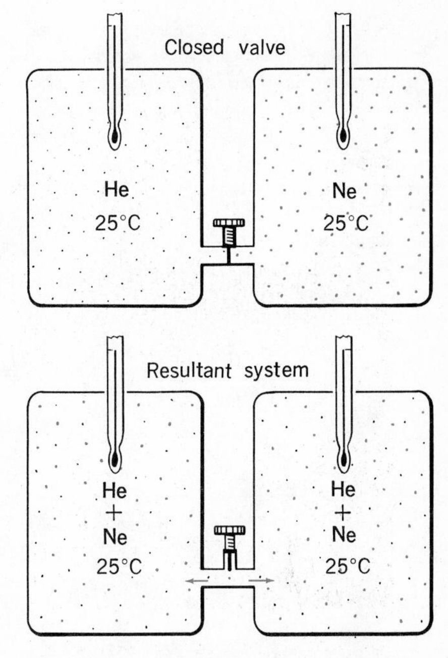

10-23 When no energy change is involved, molecular systems spontaneously seek a state of maximum disorder (entropy). Note that there is no measurable temperature change.

The randomness or disorder of a system is expressed in terms of a quantity called ***entropy,*** which is symbolized by S. A perfectly ordered system has zero entropy. The imaginary, ideal perfectly ordered system used as a standard for entropy measurement is a perfect crystal at absolute zero. At any temperature other than absolute zero, thermal energy gives rise to some disorder. The more disordered the system, the greater is its entropy. On the basis of the above example and many other experiments, it can be stated that *molecular systems favor the state of maximum entropy, provided that energy factors do not prohibit.* It is apparent that the tendency to achieve maximum entropy is not the only factor that affects the behavior of a molecular system. If it were, all substances would spontaneously change into the gaseous phase.

Fig. 10-24 shows the entropy changes that accompany the phase changes of a substance. At the melting point, the liquid phase of a substance has greater entropy than the solid phase. The gaseous phase has more entropy than the liquid phase at a given temperature. We shall find that processes involving an increase in entropy are favored over those involving a decrease. Thus, *spontaneous chemical changes are favored by a decrease in energy and an increase in entropy.* When the two factors oppose each other, the dominant one prevails. Since temperature is a measure of molecular motion, it is a partial basis in determining which of the two factors is favored. High temperatures indicate rapid molecular motion, and since rapid molecular motion results in greater disorder, *higher temperatures favor the entropy factor.* A more detailed account of this effect is given later. We can use the experimental observation that a liquid exerts a fixed vapor pressure at a given temperature to illustrate the compromise which is reached between the tendency of a system to reach a state of minimum energy and that to reach one of maximum entropy.

10-14 Equilibrium in Molecular Systems As stated above, in a closed system containing liquid water and water vapor at a con-

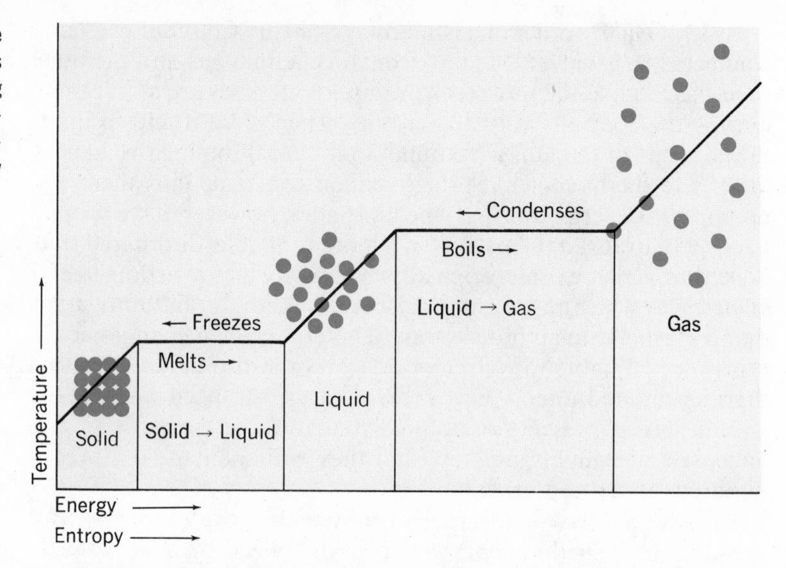

10-24 Entropy changes accompanying phase changes. The lower and upper horizontal lines represent, respectively, the melting and boiling points of the substance. At these constant temperatures, phase changes occur which are accompanied by an increase in both the potential energy and entropy of the system.

stant temperature, a dynamic equilibrium is established when the rate of the opposing processes, evaporation and condensation, become equal. When any system reaches an equilibrium state, the *observable macroscopic properties* (color, pressure, temperature, density) do not change. If the temperature is increased and maintained at a higher level, then a new equilibrium condition is established. At the new equilibrium, more molecules are in the gaseous phase than at the lower temperature. This is consistent with our observation that higher temperatures favor the more disordered state.

If the tendency to achieve a condition of maximum entropy were the only factor affecting the behavior of the system, the water would evaporate completely. If the tendency to achieve a condition of minimum energy were the only factor, then there would be no evaporation. Since both phases are present with no further observable changes occurring, a compromise between the two tendencies must have been reached. Thus, *in a system at equilibrium there is a compromise between the tendency to achieve a state of maximum entropy and a state of minimum energy. In other words, the behavior of a closed molecular system is determined by its tendency to achieve an equilibrium state.*

The preceding discussion suggests that processes and reactions which result in a decrease in the energy state combined with an increase in disorder (entropy) should be highly probable or spontaneous processes. For example, when alcohol is added to water, the entropy of the system after mixing is greater than before mixing. In addition, the only process involving energy is exothermic and the system achieves a lower energy state. With both the entropy and energy factors favoring the dissolving process, we would predict that the process would be highly favorable and that alcohol would be highly soluble in water. It is.

It also follows from the above discussion that processes which would result in an increase in energy and a decrease in entropy should be highly improbable and nonspontaneous. The following analogy illustrates the point. A glass bottle resting on a wall has much greater potential energy than on the ground below. It is unstable with respect to its potential energy. It could fall to the ground and break. In this way the glass system achieves a lower energy state and at the same time reaches a state of greater entropy (disorder). It is highly improbable that the broken glass will spontaneously reassemble itself as a bottle and rise to the top of the wall. This amusing situation which would simultaneously increase the energy and decrease the entropy of the system can be visualized by running a film of a spontaneous event backwards.

In a given chemical system, the energy and entropy may either increase or decrease. There are, however, several fundamental laws of nature that underlie the behavior of all processes.

1. *The Law of Conservation of Energy* which states that the *total amount of energy of the universe is constant.*
2. *The Second Law of Thermodynamics* which states that *the total amount of entropy in the universe is increasing.*

Although spontaneous changes in some chemical systems we investigate may involve a decrease in the entropy of the system, the second law indicates that the surroundings simultaneously must be increasing in entropy and always by a greater amount than the decrease in the system. There has been no observed exception to this fundamental law even though such an exception would still be in accord with the Law of Conservation of Energy.

The Law of Conservation of Energy permits the energy of a system to increase or to decrease but dictates that the accompany-

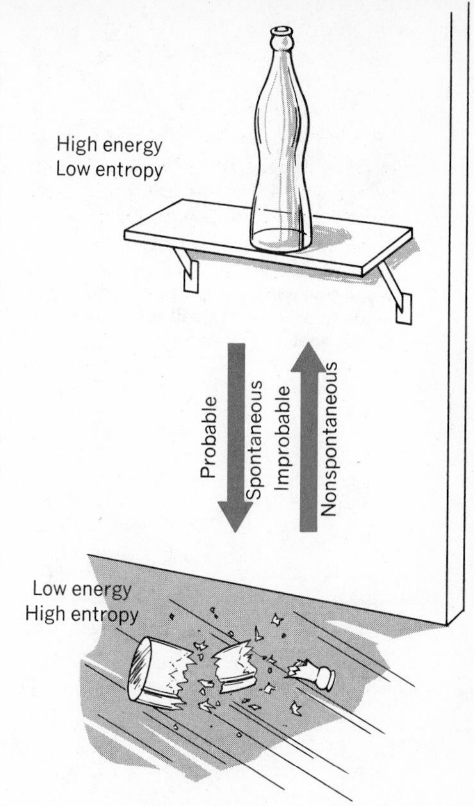

10-25 The bottle on the shelf has a higher potential energy than on the floor. When the bottle falls, the potential energy of the system decreases while the entropy increases. Thus the change is highly probable.

10-26 All systems undergoing a spontaneous change increase the disorder (entropy) of the universe.

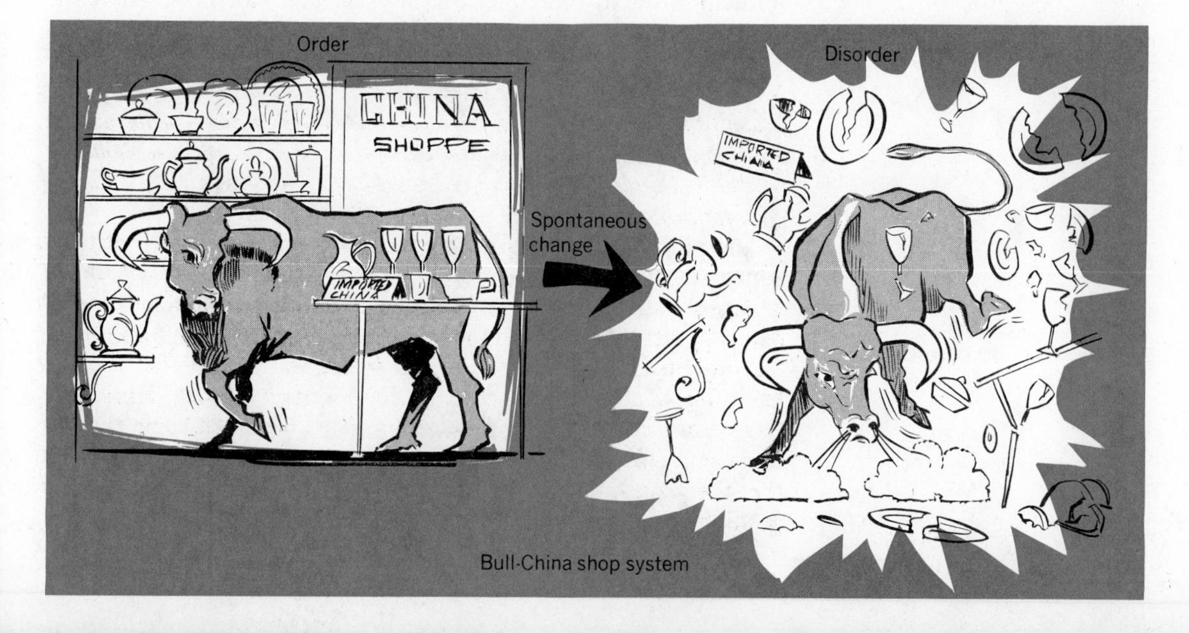

ing energy change in the surroundings must be such that the total energy of the universe is conserved. We shall find in Chapter 12 that the energy change during a reaction involves the interchange of kinetic and potential energy. We may use a mechanical system involving gravitational potential energy to represent this interchange.

A spontaneous change is a change which tends to proceed of itself in the direction of the equilibrium state.

In Fig. 10-27, Point C represents the lowest part of the arc swept out by the swing. At this point, the swing moves most rapidly. Thus, its kinetic energy is at a maximum while its potential energy is zero relative to the motion of the swing. At points A and E, the potential energy is at a maximum while the kinetic energy is equal to zero. As the swing moves downward from points A or E, it gains kinetic energy and loses potential energy. At points B and D, which are one-half the distance between the top and bottom of the arc, half of the maximum potential energy has been converted to kinetic energy. At these points, therefore, the potential and kinetic energy have the same value. In other words, at any point on the arc the sum of the kinetic and potential energies is constant. Since this is not an ideal system, frictional losses prevent the swing from returning to the same height on subsequent cycles. This means that some energy has been dissipated as heat evolved to the environment. The Law of Conservation of Energy requires that *the heat energy gained by the surroundings exactly equals the energy lost by the swing system.* We shall apply this principle when we study the energetics of chemical reactions in the next chapter.

LOOKING AHEAD

In this chapter, we found that a phase change involves an energy change. We were able to explain the observed energy changes in terms of attractive forces between molecules. Let us now turn our

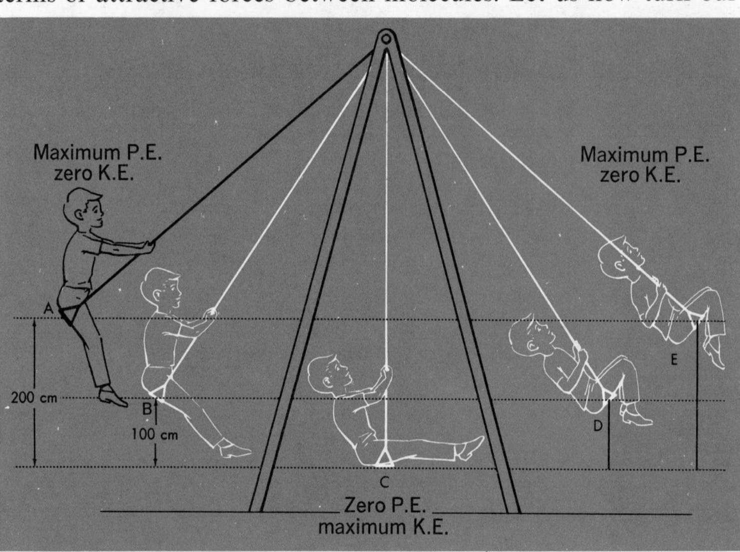

10-27 The total energy of this system is constant although it is continuously changing between the kinetic and potential forms.

attention to the energy aspect of chemical changes. The energy possessed by chemical substances and the energy changes associated with a chemical reaction are often just as important to the chemist as the mass of the chemicals and the products of the reaction. For example, in developing a rocket fuel, the amount of available energy is of primary interest. The energy evolved during a chemical reaction may be put to work lifting a rocket, spinning a turbine, or running an automobile.

Because of the importance of energy to scientists and because it will help us understand the behavior of material substances, we shall devote the next chapter to a discussion of the energy changes associated with chemical reactions. Our discussion will provide an opportunity for us to use and reinforce concepts discussed in earlier chapters such as: mole relationships in equations, structural and bonding concepts, and the nature and characteristics of aggregates.

QUESTIONS

1. Why is the structure of liquids more complex than that of gases?

2. Does a liquid with a high surface tension have a relatively low or high molar heat of vaporization? Explain.

3. In which of these closed containers is the vapor pressure the greatest? Explain your choice. (a) A one-liter container with five ml of water held at 20°C. (b) A two-liter container with 15 ml of water held at 30°C. (c) A three-liter container with 500 ml of water held at 25°C.

4. Two beakers, A and B, are filled with a mixture of ice cubes and water. An ice cube is added to beaker A, and a sample of warm water is added to beaker B. The temperatures of both are still 0°C. Explain.

5. When commercial gases such as acetylene and oxygen are purchased, pressures are carefully noted. Why are we not concerned with pressure when we purchase gasoline?

6. Is this statement true or false? If a sample of water is heated from 40°C to 80°C, all of the molecules have undergone an increase in velocity. Explain your answer.

7. Describe how to use a mercury barometer to measure the vapor pressure of a liquid.

8. Steam sterilizers (autoclaves) used in hospitals and laboratories attain temperatures of 250°F with boiling water. Explain how this is possible. Water ordinarily boils at 212°F.

9. At the top of a mountain, an egg is heated in boiling water for four minutes and is found to be uncooked. Explain.

10. Which would cause a more severe burn, 1 g of $H_2O(g)$ at 100°C or 1 g of $H_2O(l)$ at 100°C. Explain.

11. If water is poured into a flask and the air above it is removed with a vacuum pump, the water boils at room temperature. Explain.

12. When an aerosol bomb is used to spray an insecticide, the jet becomes cold. Explain.

13. (a) From where does the energy required to vaporize the refrigerant in a refrigerator come? (b) How does the gaseous refrigerant give up energy and become a liquid?

14. The noble gases such as helium and neon have low boiling points. Explain.

15. Which has a greater vapor pressure at 30°C: (a) water or ether, (b) perfume or mineral oil, (c) motor oil or gasoline, (d) mercury or table salt?

16. Sodium metal has a very high molar heat of vaporization and a very low molar heat of fusion. Which measurement better represents the attractive forces which operate between the atoms? Explain your answer.

17. According to Boyle's Law, an increase in the volume of a gas causes a decrease in gas pressure at constant temperature. Does doubling the volume of a vessel containing a liquid and its vapor at constant temperature cause a corresponding change in vapor pressure?

18. Container A is filled with $H_2O(s)$ at $-5°C$, container B with $H_2O(l)$ at 20°C, and container C with $H_2O(g)$ at 105°C. The pressure is one atm in all three cases. (a) Which system has the highest degree of entropy? (b) Which sample of molecules is the most compressible? (c) Which sample is the densest? (d) Which molecules have the greatest potential energy (e) Which molecules have the least entropy? (f) Which sample(s) has (have) a definite shape? (g) Which sample(s) has (have) a definite volume?

19. State for each of these situations (1) whether the system or surroundings gain energy and (2) whether the process is exothermic or endothermic. (a) An electric current converts water into H_2 (g) and O_2 (g). (b) Solid carbon dioxide is converted directly into the gaseous phase. (c) When ammonium chloride, a salt, is dissolved in water, the solution and beaker become cold. Consider the beaker as part of the surroundings. (d) A beaker containing water at 20°C is stirred until the temperature reaches 21°C. (e) A hydrogen molecule which is composed of two atoms held together by electric forces is split into two separate atoms. (f) Positively charged ions coalesce with negatively charged ions and form a solid ionic crystal. (g) Iodine vapor changes directly to a solid. (h) An electron is completely removed from a hydrogen atom. (i) Wood burning in a fireplace is converted into gases and leaves ashes. (j) A stick of dynamite explodes and changes into a number of gases.

20. What is the significance of the triple-point temperature?

21. How does the triple point differ from the normal melting point?

22. Refer to Fig. 10-18 and answer these questions. The critical temperature of water is 374°C. (a) What phases of water exist in equilibrium at point F? What is this temperature called? (b) What phase(s) of water exist(s) at 80°C and 380 torr? (c) What phases exist in equilibrium along line AB? (d) What phase(s) exist(s) at 10 atm and 500°C?

23. Identify the sign of ΔH for each of these phase changes: (1) (g) to (l), (2) (s) to (g), (3) (l) to (s).

24. Some people attempted to "economize" by wrapping the ice in the old "ice box" in several layers of newspaper. The ice didn't melt so fast. They bought ice less often. Comment on this "economy" measure.

25. Refer to Fig. 10-18 and explain what happens when: (a) a sample of water originally at 0°C is heated at a constant pressure of 380 torr. (b) A sample of water originally at a pressure of 4.5 torr is compressed at a constant temperature of 50°C.

26. Does a change in external pressure have a more pronounced effect on the melting point or on the boiling point of a substance?

27. Explain the conclusion that phase changes from ice to liquid water and from liquid water to water vapor depend more upon entropy than upon energy considerations.

28. (a) What changes in the entropy and energy of a system are associated with spontaneous processes? (b) What changes are associated with nonspontaneous processes?

29. In each of the reactions, state whether the entropy or the energy factor is the predominant force: (a) the evaporation of water, (b) the sublimation of snow, (c) the dissolving of ammonium chloride (the surroundings become cold), (d) the addition of water to builder's lime (the surroundings become hot).

30. Which are examples of dynamic equilibrium? (a) A beehive during the day. The number of bees in the hive and the number in the field are constant. (b) Liquid ether and its vapor in a closed vessel. The level of the liquid remains constant. (c) Water in a beaker and its vapor. The level of liquid is kept constant by adding water at the same rate at which it evaporates. (d) The players on the field and those on the bench in a football game. (e) The dissolved and undissolved solute in a saturated solution containing excess solid solute. (f) The liquid and vapor within a thermometer.

31. For each of these changes, state whether the entropy of the system increases or decreases. (a) Salt (NaCl) is dissolved in water and forms a solution. (b) Carbon dioxide (g) is dissolved in water and forms a solution. (c) Dry Ice is converted directly into carbon dioxide (g). (d) A stick of dynamite explodes and forms several gases. (e) Water is frozen into ice cubes in a refrigerator.

PROBLEM

1. The molar heat of fusion of water is 1.44 kcal at 0°C. Its molar heat of vaporization is 9.72 kcal at 100°C. Calculate the number of calories required to (a) melt 1.0 g of ice at 0°C, (b) completely vaporize 1.0 g of water at 100°C, (c) completely vaporize 18 g of ice starting at 0°C and ending with steam at 100°C.

Ans. (a) 80 cal/g, (b) 540 cal/g.

11

Energy Associated with Chemical Reactions

Preview

Of all the factors responsible for our present high economic standards, the use of energy is the most important. Nations that have developed the ability to control and convert available forms of energy into useful forms have the highest standard of living and wield the greatest influence in the world today.

"If, then, you ask me to put into one sentence the cause of that recent, rapid, and enormous change and the prognosis for the achievement of human liberty, I should reply, 'It is found in the discovery and utilization of the means by which heat energy can be made to do man's work for him.'"

Robert A. Millikan (1868–1953)

Only since 1800 has a large part of our work been performed by energy other than that supplied by the muscles of men and animals. In the 19th century, the steam engine was invented. This device converted heat energy into energy of motion. Thus, heat energy became a servant of man. Since then, scientists have been constantly seeking new energy sources and trying to develop practical ways to change available energy into forms which can be made to do useful work.

The three primary sources of available energy are solar energy from nuclear reactions taking place in the sun, nuclear energy released when the nuclei of atoms split in a fission reaction or merge in a fusion reaction, and chemical energy in matter which is recoverable through the rearrangement of atoms during a chemical reaction.

At present, steam power-plants represent the most important method of converting available energy into useful energy. In these plants, the internal energy of coal, oil, or natural gas is released as heat by burning. The internal energy or heat is converted into mechanical energy when the steam produced by heating water is used to turn a turbine. The turbine is coupled to a generator which converts mechanical energy to usable electric energy.

In this chapter, we shall be concerned with the energy absorbed or liberated in chemical reactions. Our objectives will be to describe its origin, its measurement, its calculation, and its theoretical and practical significance.

ORIGIN OF CHEMICAL ENERGY

Chemical energy in matter is one of the primary sources of energy available to man. There are countless examples of the conversion of chemical potential energy into heat or electric energy.

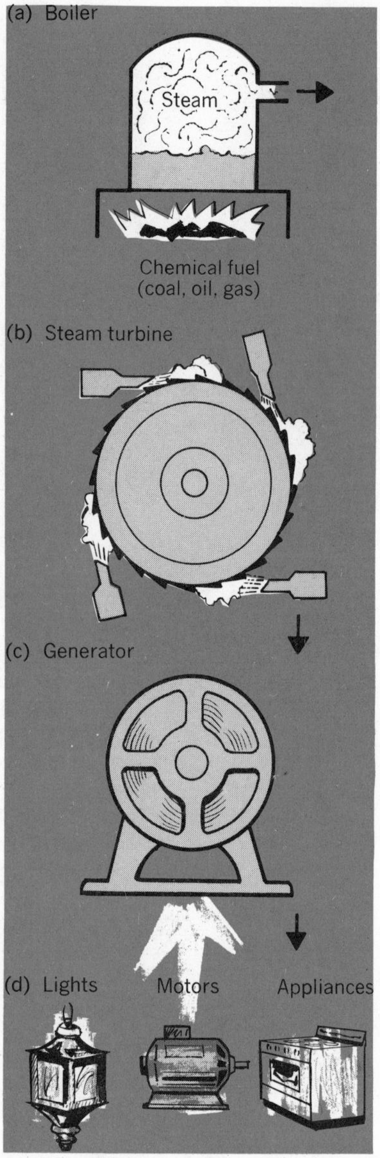

(a) Boiler

Steam

Chemical fuel (coal, oil, gas)

(b) Steam turbine

(c) Generator

(d) Lights Motors Appliances

11-1 Transformation of the chemical energy in a fuel into electric, heat, light, and mechanical energy. (a) The chemical energy in the fuel is evolved as heat energy during the combustion. The thermal energy added to water in the boiler converts it into steam. (b) The pressure of the hot steam on the blades of the rotor in the turbine causes it to turn. (c) In the generator or dynamo, the rotating shaft of the turbine causes a coil of wire to revolve within a large magnet, generating electric energy. (d) Electric energy is converted into light and heat energy in a light bulb, heat energy in a range, and mechanical energy in a motor.

In batteries, chemical energy of the reactants is converted into useful electric energy. Chemical energy stored in foods and released through biochemical processes maintains proper body temperature, promotes growth, and enables man to do work. In steam plants, chemical energy of fossil fuels is released as heat by burning.

The energy associated with a chemical reaction is often just as important to the chemist as the products. For example, in developing a rocket fuel, the amount of available energy is of primary interest. The energy evolved during a chemical reaction may be put to work lifting a rocket, spinning a turbine, or running an automobile.

In general, all chemical reactions either liberate or absorb heat. Let us first investigate the source of the energy liberated or absorbed during a chemical reaction.

The origin of chemical energy lies in the position and motion of atoms, molecules, and subatomic particles. The total energy possessed by a molecule is the sum of all the forms of potential and kinetic energy associated with it. Let us consider the kinetic and potential energy associated with a molecule of hydrogen chloride (HCl).

11-1 Electronic Energy, E_e Under the heading of electronic energy, we can group the kinetic energy of the electrons moving about the atomic nuclei and the potential energy resulting from the interaction between electrons and nuclei both within and between atoms. This interaction gives rise to what we have called chemical bonds. The energy associated with this interaction between positive and negative charges is called *bonding energy.*

The energy changes that occur during a chemical reaction are, to a large extent, the result of the potential energy changes that accompany the breaking of chemical bonds in the reactants and the formation of new bonds in the products. The energy absorbed or emitted in the ultraviolet and visible range of the electromagnetic spectrum corresponds to electron transitions between electronic energy levels. Thus, the wavelengths of the energy emitted or absorbed may be used to help identify the atoms which compose the molecule.

11-2 Rotational Energy, E_r A molecule may have rotational energy if it is rotating about an axis through its center of mass (Fig. 11-3). Both theory and experiment show that rotational energy, like electronic energy, is quantized. That is, there are only certain allowed rotational energy levels. Rotational quanta have a frequency which falls in the microwave region of the electromagnetic spectrum.

11-3 Vibrational Energy, E_v Another form of kinetic energy associated with an HCl molecule is the energy produced by vibration of the two atoms with respect to each other. The bond between the hydrogen atom and the chlorine atom is analogous to a spring. Attractive forces pull the atoms together until repulsive forces cause them to spring away from each other. This alternating attraction and repulsion results in a vibrational motion. The vibrational motion of an HCl molecule is depicted in Fig. 11-4. As the vibrational energy increases, the distance through which the atoms vibrate increases. The frequency of vibration does not change. When the energy reaches a certain value, the amplitude of vibration becomes great enough so that the attractive forces between the atoms are no longer effective, and the atoms fly apart. As with other types of energy, vibrational energy is quantized. Vibrational quanta have a frequency which falls in the infrared region of the electromagnetic spectrum.

11-4 Translational Energy, E_t This type of kinetic energy is associated with gas molecules as they move linearly from one point to another. The translational energy of a molecule is expressed as $KE = \frac{1}{2}mv^2$. At absolute zero, the translational motion of a molecule becomes zero, and thus the translational energy is also zero at this temperature. The translational motion of an HCl molecule is shown in Fig. 11-5.

11-5 Miscellaneous Energy, E_m This category includes several types of energy which may contribute to the total molecular energy, depending on the circumstances. Nuclear energy is related to the forces which hold the nucleus together and is important when considering nuclear reactions in Chapter 21. Furthermore, whenever externally applied electric or gravitational fields interact with a molecular system, additional energy effects arise and must be taken into account. In general, the molecular energies discussed above are determined spectroscopically. The actual methods and calculations involved are beyond the scope of this text.

The total internal energy of a molecule is the sum of all the forms of molecular energy discussed above. That is,

$$E = E_e + E_r + E_v + E_t + E_m$$

Vibrational, rotational, and translational motion all contribute to the kinetic energy of liquid and gaseous substances. In general, *the molecular kinetic energy of solids is associated with vibrational motion.* For example, the kinetic energy of water molecules in an ice crystal is the result of the vibrational motion about fixed points in the crystal as well as vibrations of the hydrogen and oxygen atoms within each molecule. Calculations show, however, that at ordinary temperatures, *the kinetic energy of molecules contributes little to the energy of a substance.* In elementary chemistry, we are mainly concerned with the electronic energy (E_e) involved in the making and breaking of chemical bonds.

It is apparent that it would be difficult to determine the total energy of a molecule or of a macroscopic system containing countless molecules. The energy which we measure experimentally is that associated with a system changing from some initial to some final state. In other words, we are interested in energy *changes* rather than in absolute total energy content. We are primarily concerned

11-2 Use and transformation of chemical energy. On December 21, 1968, Apollo 8 lifts off, carrying astronauts Frank Borman, Jim Lovell, and Bill Anders on man's historic first journey to the moon. Kerosene and liquid oxygen are shot into the combustion zone of this Saturn V rocket engine at a rate of 15 tons per second. The extremely fast reaction yields CO_2 and H_2O. The intense heat drives these exhaust gases downward and produces the tremendous thrust necessary to boost the massive rocket (massive as a Navy destroyer and as tall as a 36-story building) off the launching pad. Other fuels used in different stages of the rocket assembly and lunar modules are liquid hydrogen, hydrazine (N_2H_4), and dimethyl hydrazine ($N_2C_2H_8$). The relative mass, and the amount of energy released during the combustion are of prime importance in the selection of fuels for space flight.

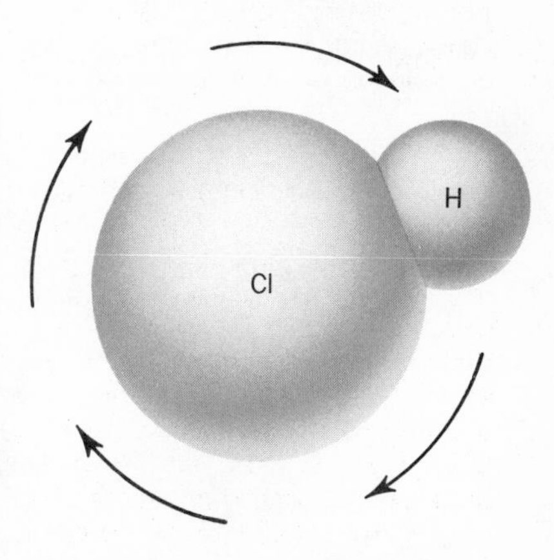

11-3 Rotational motion of an HCl molecule.

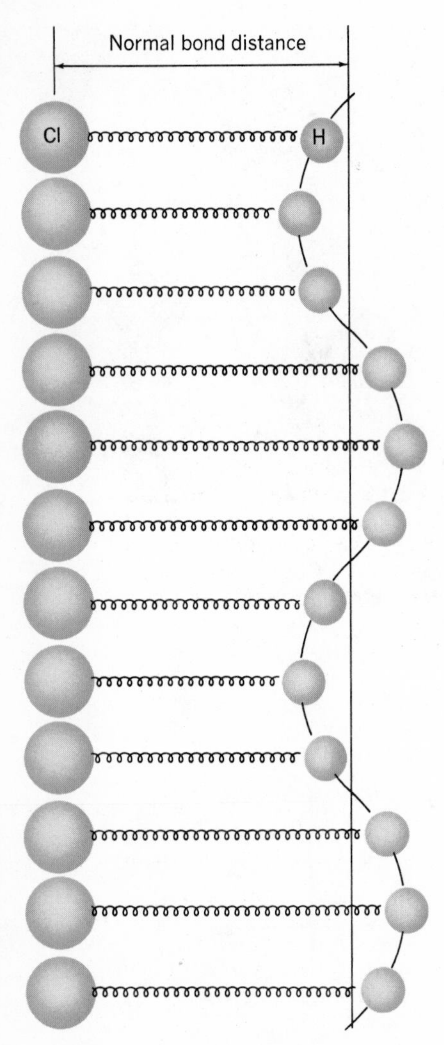

11-4 The vibrational motion of a HCl molecule involves alternately increasing and decreasing the bond distance between the hydrogen and chlorine nuclei.

with energy in the form of heat that is absorbed or liberated by a reaction that takes place in an open container at constant pressure. This quantity of energy is known as the *change in enthalpy* (heat content) of the chemical system and is symbolized by ΔH.

CHANGES IN ENTHALPY

11-6 Calorimetric Measurements In general, all reactions either liberate or absorb heat. The heat liberated or absorbed during a chemical reaction is called the ***enthalpy (heat) of reaction.*** The enthalpy of reaction is defined as

$$\Delta \mathbf{H} = \textbf{(enthalpy of products)} - \textbf{(enthalpy of reactants)}$$

The Greek letter Δ, pronounced "delta," is frequently used to denote a difference between the initial and final values of a variable in a system undergoing a change. The above definition of ΔH shows that if heat is liberated (flows out of the system), then the enthalpy (***heat content***) of the system decreases and ΔH has a negative value. In other words, ΔH is negative for any exothermic reaction. This is because the enthalpy (heat content) of the reactants is greater than the enthalpy (heat content) of the products. On the other hand, ΔH is positive for endothermic reactions because heat is absorbed so that the enthalpy of the products is greater than that of the reactants.

Experimentally, the sign of ΔH may be determined by noting the temperature change which the system and its immediate surroundings undergoes. If the temperature rises as in the case of combustion reactions, the system is losing some of its chemical potential energy by its conversion into kinetic energy of the system and surroundings. This means that ΔH has a negative value. Quantitatively, the process of determining heat changes associated with chemical reactions is known as ***calorimetery.*** Experimentally, it is possible to determine how much heat is liberated during a reaction by allowing the reaction to take place in a calorimeter (Fig. 11-6), a well-insulated vessel containing a liquid that absorbs the heat evolved by the reaction. The reaction is carried out in the inner reaction vessel. Heat evolved by the reaction is transferred to the weighed amount of water in the insulated outer vessel. The original and final temperatures of the water are read on the thermometer. Of course, any heat transferred to the calorimeter must be taken into account.

Example 11-1

A chemical reaction takes place in a calorimeter. The water surrounding the reaction vessel has a mass of 120 g. During the course of the reaction, the temperature changes from 25°C to 55°C. How many calories were absorbed by the water? The specific heat capacity of water is defined as 1.0 cal/g C°.

Solution
In order to raise the temperature of 1.0 g of water 1.0 C°, 1.0 cal is required. To raise the temperature of 120 g of water from 25°C to 55°C requires

$$120 \text{ g} \times \frac{1.0 \text{ cal}}{\text{g } H_2O \text{ C}°} \times (55°\text{C} - 25°\text{C}) = 3600 \text{ cal}$$

FOLLOW-UP PROBLEM

We wish to determine how much heat paraffin gives off on burning. We use a candle flame to heat some water in a calorimeter. These data are obtained.

Mass of water in calorimeter	350 g
Initial mass of candle	150 g
Final mass of candle	112 g
Initial temperature of water	15°C
Final temperature of water	23°C

Calculate (a) the temperature rise, (b) the calories absorbed by the water in the calorimeter, (c) the grams of paraffin burned, (d) the approximate value of the heat of combustion of paraffin in cal/g. Neglect the energy absorbed by calorimeter.

Ans. (a) 8.0 C°, (b) 2800 cal, (c) 38 g, (d) 74 cal/g.

11-7 State Properties and Changes in State Properties Enthalpy, *H*, is an example of a ***state property.*** This means that its value is determined by the state of the system. The *state of a system* is fixed when the temperature, pressure, number of moles, composition, and other properties of the system are specified. The energy of the system in Fig. 11-7b, because of its higher temperature, has a definite value which is greater than that of the system in Fig. 11-7a. The value of the energy of the system in the final state (b) does not depend on the method or path by which the energy was transferred. The difference in energy between the two states could be the result of adding heat to the system, doing mechanical work such as vigorously stirring the water, or using a combination of these or other methods (Fig. 11-8).

Changes in state properties, such as enthalpy, *are always independent of the path taken to change a system from some initial state to some final state* (Fig. 11-9). This means the net energy change must be the same regardless of the path. If it were not, it would be possible to put one amount of energy into a system to reach a higher energy state and then return to the initial state by a different path, obtaining more energy than was put into the system. This would mean that energy could be created, a phenomenon which has never been observed and one which contradicts the experience stated in the Law of Conservation of Energy.

The quantity of heat absorbed or liberated during a reaction varies with the temperature. Scientists have, therefore, adopted 25°C and 1 atmosphere pressure as ***standard state*** conditions for

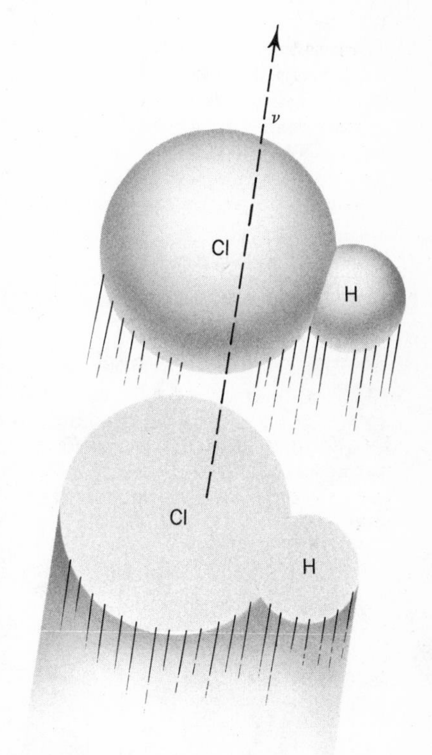

11-5 The translational energy of a HCl molecule is associated with the linear motion of the molecule through space. In this type of motion, the center of gravity of the molecule moves.

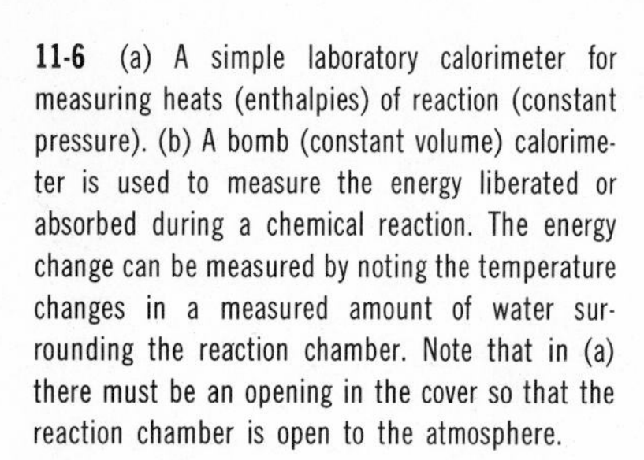

11-6 (a) A simple laboratory calorimeter for measuring heats (enthalpies) of reaction (constant pressure). (b) A bomb (constant volume) calorimeter is used to measure the energy liberated or absorbed during a chemical reaction. The energy change can be measured by noting the temperature changes in a measured amount of water surrounding the reaction chamber. Note that in (a) there must be an opening in the cover so that the reaction chamber is open to the atmosphere.

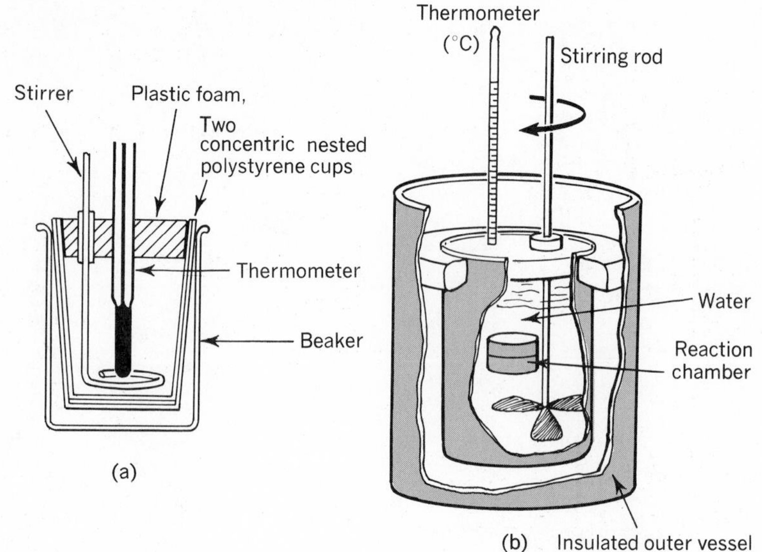

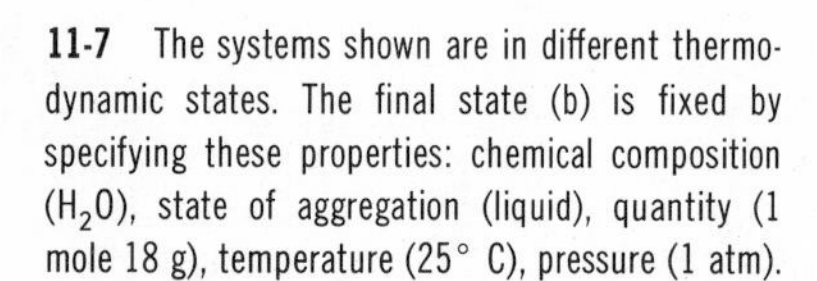

11-7 The systems shown are in different thermodynamic states. The final state (b) is fixed by specifying these properties: chemical composition (H_2O), state of aggregation (liquid), quantity (1 mole 18 g), temperature (25° C), pressure (1 atm).

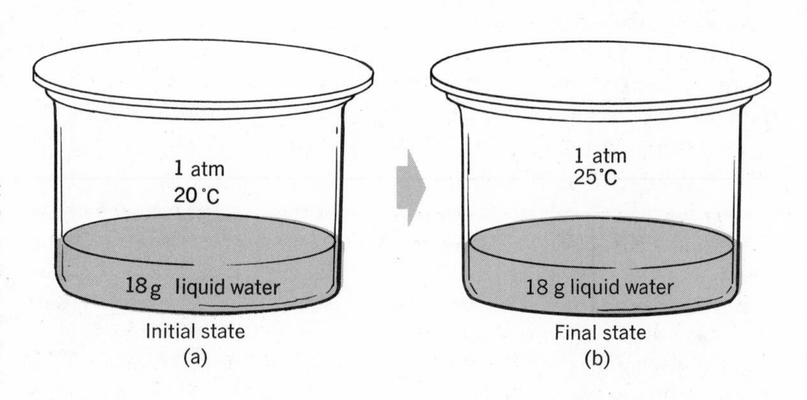

11-8 In the systems represented below, assume there is no heat transferred between the system and its surroundings. Processes carried out under these conditions are called *adiabatic* processes. The same change in state (energy change) is observed in each of the adiabatic systems shown. In (a) vigorous stirring increases the energy content of the water; in (b) electric work, represented by a current flowing through a resistance is used; in (c) energy is transferred by means of the work associated with the compression of a gas; in (d) the work done when two metal blocks are rubbed together is converted into energy; and in (e) energy is transferred from the hot metal block by heat "flow" from the region of high temperature (the block) to one of lower temperature (the water). The change in state is independent of the method used to bring it about.

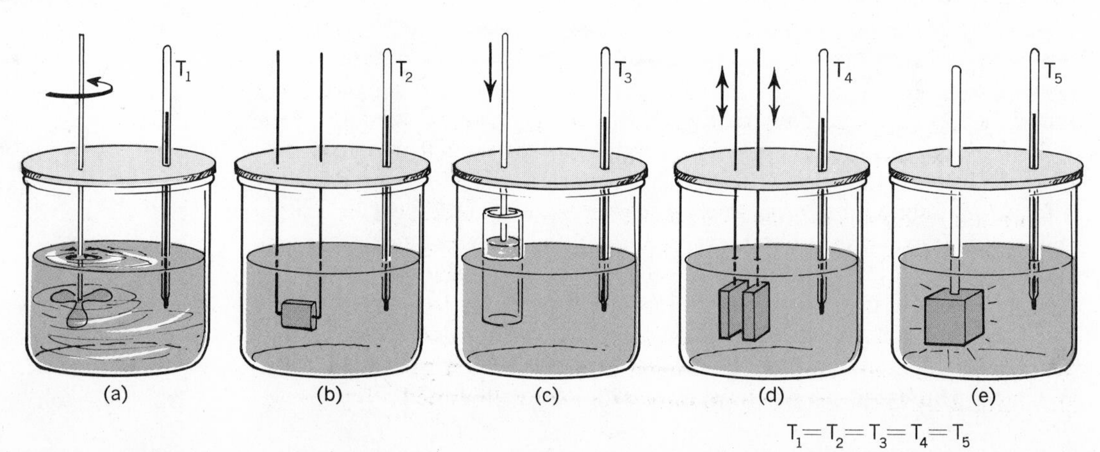

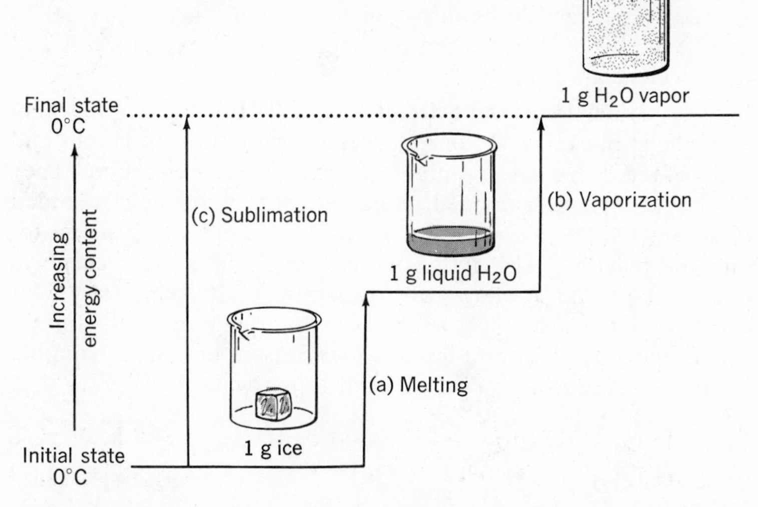

11-9 The conversion of ice at 0 °C into saturated water vapor at 0 °C may be accomplished by melting the ice and then vaporizing the water or by subliming the ice. The energy absorbed from the surroundings along path (c) is equal to the sum of the energies absorbed along paths a + b.

reporting heat data. To calculate the enthalpy of a reaction, it is necessary to write an equation for the reaction, similar to that used to calculate quantities of products formed. The standard enthalpy change, designated by $\Delta H°$, for a given reaction is usually expressed in kilocalories and depends on how the equation is written. For example, we may write the equation for the reaction of hydrogen with oxygen in two ways:

$$\mathbf{H_2(g) + \tfrac{1}{2}O_2(g) \longrightarrow H_2O(g) \qquad \Delta H° = -57.8\ kcal} \qquad 11\text{-}1$$

$$\mathbf{2H_2(g) + O_2(g) \longrightarrow 2H_2O(g) \qquad \Delta H° = -115.6\ kcal} \qquad 11\text{-}2$$

Experimentally, the change in enthalpy, $\Delta H°$, for the reaction is found to be -57.8 kcal per mole of $H_2O(g)$. The coefficients in Equation 11-2 are twice those in Equation 11-1 and represent the combustion of 2 moles of H_2. Therefore, the enthalpy change represented by Equation 11-2 is -115.6 kcal. Note that it is assumed that the initial and final states are measured at 25°C and 1.00 atm although the reaction occurs at a higher temperature.

FOLLOW-UP PROBLEM

How much heat is liberated when 40.0 g of $H_2(g)$ reacts with excess oxygen (g)? **Ans. 1156 kcal.**

Note that the physical state of each participant is indicated in Equations 11-1 and 11-2. *Because energy is involved in a phase change, the energy related to a given reaction depends on the physical state of the participating reagents.* For example, when hydrogen gas and oxygen gas react and form a mole of liquid water, 68.3 kcal is liberated. In contrast, when hydrogen gas and oxygen gas react and form a mole of water vapor, only 57.8 kcal is liberated.

The difference is the heat required to convert one mole of water from the liquid to the vapor phase. Because 10.5 kcal of energy is used to change the liquid water to a vapor, only (68.3 – 10.5) or 57.8 kcal is liberated when the hydrogen and oxygen react and form water vapor directly.

11-8 Principle of Additivity of Reaction Heats The preceding example implies that two or more separate chemical equations and heats of reaction can be manipulated like algebraic equations. They can be multiplied or divided by numerical factors and then added or subtracted to give a net equation and the heat of reaction for another reaction. The following example shows how chemical equations and $\Delta H°$ values can be manipulated algebraically. Our objective in this example is to combine the equations for the formation of gaseous water and liquid water so as to obtain an equation which represents the vaporization of liquid water.

$$H_2(g) + \tfrac{1}{2}O_2(g) \longrightarrow H_2O(g) \qquad \Delta H° = -57.8 \text{ kcal} \qquad 11\text{-}3$$
$$H_2(g) + \tfrac{1}{2}O_2(g) \longrightarrow H_2O(l) \qquad \Delta H° = -68.3 \text{ kcal} \qquad 11\text{-}4$$

Algebraically subtracting the second equation from the first yields

$$0 = H_2O(g) - H_2O(l) \qquad \Delta H° = 10.5 \text{ kcal}$$

Rearranging this equation so as to obtain a conventional expression for a reaction (no minus signs) we obtain

$$H_2O(l) \longrightarrow H_2O(g) \qquad \Delta H° = 10.5 \text{ kcal}$$

This equation shows that 10.5 kcal is required to convert one mole of liquid water at 25°C into one mole of water vapor. This value agrees with the experimentally determined molar heat of vaporization for water at this temperature.

Rather than subtracting one equation from another, it is more common practice to reverse one of the equations algebraically and then add them. This procedure prevents negative signs from appearing in the final equation. It must be remembered that when the equation for an exothermic reaction is reversed, the expression now represents an endothermic reaction. Therefore, the sign of $\Delta H°$ must be changed whenever the equation of a reaction is

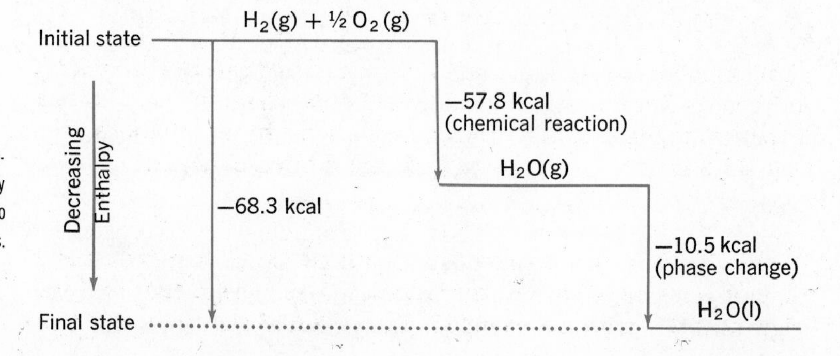

11-10 Energy diagram for the conversion of hydrogen gas and oxygen gas into liquid water by two different processes. The total heat evolved to the surroundings is the same for both pathways.

reversed. In the preceding example, reversing Equation 11-4 yields

$$H_2O(l) \longrightarrow H_2(g) + \tfrac{1}{2}O_2(g) \qquad \Delta H° = 68.3 \text{ kcal}$$

Adding

$$H_2(g) + \tfrac{1}{2}O_2(g) \longrightarrow H_2O(g) \qquad \Delta H° = -57.8 \text{ kcal}$$

yields

$$H_2O(l) + H_2(g) + \tfrac{1}{2}O_2(g) \longrightarrow H_2(g) + \tfrac{1}{2}O_2(g) + H_2O(g) \qquad \Delta H° = 10.5 \text{ kcal}$$

Simplification gives a net equation for the change

$$H_2O(l) \longrightarrow H_2O(g) \qquad \Delta H° = 10.5 \text{ kcal}$$

The principle underlying the preceding calculations is known as ***Hess' Law of Heat Summation*** or the ***Principle of Additivity of Reaction Heats.*** This principle states that *when a reaction can be expressed as the algebraic sum of two or more other reactions, the heat of the reaction is the algebraic sum of the heats of these reactions.* This principle is based upon the ***First Law of Thermodynamics,*** a fundamental law of nature which can be expressed in various ways. The simplest statement is that *the total energy of the universe is constant* and cannot be created or destroyed. Application of the law to the calculation of enthalpies of reactions means that the difference between the total enthalpy of specific reactants and specific products is constant regardless of the path by which the products are obtained from the reactants. There may be one or a dozen steps involved, but as long as the initial and final states of the reaction system are fixed, the enthalpy difference is constant.

Hess's Law enables us to determine heats of reaction that cannot be easily determined experimentally. This can be accomplished by combining experimentally determined heats of reaction for related reactions. For example, it is relatively simple to measure the heats of reaction when one mole of carbon reacts with oxygen and forms carbon dioxide and when one mole of carbon monoxide reacts with oxygen and forms carbon dioxide. The equations and heats of reaction are

$$C(s) + O_2(g) \longrightarrow CO_2(g) \qquad \Delta H_r° = -94.0 \text{ kcal} \qquad 11\text{-}5$$

$$CO(g) + \tfrac{1}{2}O_2(g) \longrightarrow CO_2(g) \qquad \Delta H_r° = -67.6 \text{ kcal} \qquad 11\text{-}6$$

Although the heat of reaction for the reaction between carbon and oxygen forming carbon monoxide cannot be accurately determined experimentally, it can be calculated by combining Equations 11-5 and 11-6 in such a way as to yield the desired equation

$$C(s) + \tfrac{1}{2}O_2(g) \longrightarrow CO(g)$$

Since CO(g) in the desired equation must be on the right-hand side, it is necessary to reverse Equation 11-6. This means that the sign of $\Delta H°$, must be changed. The negative sign indicates that the formation of CO_2 from CO is exothermic; therefore, the reverse reaction (decomposition of CO_2) must be endothermic and have a positive $\Delta H°$. The two equations which, when added, give the desired net reaction are

$$C(s) + O_2(g) \longrightarrow CO_2(g) \qquad \Delta H° = -94.0 \text{ kcal}$$
$$CO_2(g) \longrightarrow \tfrac{1}{2}O_2(g) + CO(g) \qquad \Delta H° = +67.6 \text{ kcal}$$

Addition yields

$$C(s) + \tfrac{1}{2}O_2(g) \longrightarrow CO(g) \qquad \Delta H° = -26.4 \text{ kcal}$$

The relationships are shown schematically in Fig. 11-11. Note again that it assumed that all reactants and products are changed to standard state for purposes of evaluating $\Delta H°$ values.

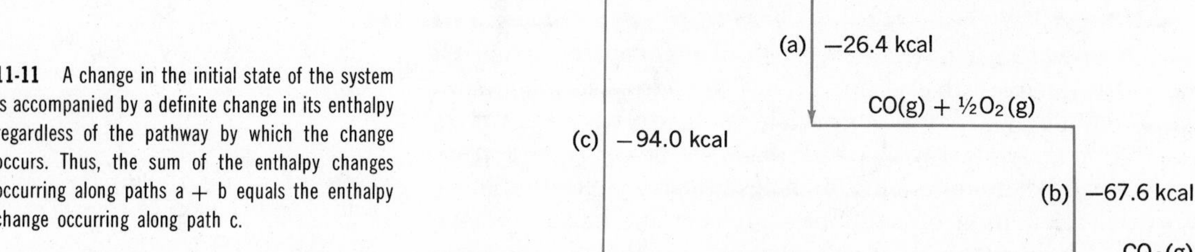

11-11 A change in the initial state of the system is accompanied by a definite change in its enthalpy regardless of the pathway by which the change occurs. Thus, the sum of the enthalpy changes occurring along paths a + b equals the enthalpy change occurring along path c.

FOLLOW-UP PROBLEM

The standard heat of combustion of liquid ethyl alcohol is −327 kcal/mole and that of acetic acid is −209 kcal/mole. The equations are

$$C_2H_5OH(l) + 3O_2(g) \longrightarrow 2CO_2(g) + 3H_2O(l) \qquad \Delta H° = -327 \text{ kcal}$$

$$HC_2H_3O_2(l) + 2O_2(g) \longrightarrow 2CO_2(g) + 2H_2O(l) \qquad \Delta H° = -209 \text{ kcal}$$

What is the heat of reaction for the oxidation of ethyl alcohol to acetic acid?

$$C_2H_5OH(l) + O_2(g) \longrightarrow HC_2H_3O_2(l) + H_2O(l)$$

Ans. $\Delta H° = -118$ **kcal.**

11-9 Standard Enthalpies (Heats) of Formation Although many enthalpies (heats) of reaction have been measured experimentally and tabulated in handbooks, there are numerous reactions whose enthalpies have not been determined or are not readily available. A simple method of calculating reaction enthalpies (heats) from readily available data would be desirable. We shall now develop the concept of enthalpy of formation of compounds and show how it can be used to calculate the enthalpy of a given reaction.

The molar enthalpy (heat) of formation for a compound is the enthalpy change which accompanies the formation of one mole of compound directly from its elements in their standard state. For example, the standard enthalpy (heat) of formation for calcium oxide, designated as ΔH_f°, is -151.9 kcal. This means that 151.9 kcal is liberated to the surroundings when one mole of calcium metal at standard state (25°C, 1 atm) reacts with one-half mole of oxygen gas at standard state and forms one mole of solid calcium oxide at standard state. The equation for the reaction is

$$\mathbf{Ca(s) + \tfrac{1}{2}O_2(g) \longrightarrow CaO(s) \qquad \Delta H_f^\circ = -151.9\ kcal}$$

Of course, during the reaction the product formed (CaO) is very hot and not at standard state. The 151.9 kcal represents the heat that must be transferred from the system when a mole of calcium oxide is formed and cooled to 25°C, the temperature of the original reactants. In general, the enthalpy of formation for a compound containing *n* elements may be represented

$$\Delta H_f^\circ = H^\circ_{\text{compound}} - \left[H^\circ_{\text{element 1}} + H^\circ_{\text{element 2}} + \cdots H^\circ_n\right] \qquad 11\text{-}7$$

At standard state every substance has a characteristic enthalpy (heat content) designated as $H°$. Since it would be very difficult to determine absolute values of enthalpies, scientists use *relative enthalpies.* Relative enthalpies are based upon the enthalpies of free elements which are *arbitrarily assigned an enthalpy of zero at 25°C and 1 atm pressure.* This means that the bracketed term in Equation 11-7 is assigned a value of zero and that the relative enthalpy $H^\circ_{\text{compound}}$ of a compound such as calcium oxide, is equal to its standard heat of formation, ΔH_f°. Standard enthalpies of formation for a number of compounds are listed in Table 11-1.

A positive enthalpy of formation indicates that the enthalpy of the compound is greater than the sum of the enthalpies of the elements from which it was formed. For example, when hydrogen gas and solid iodine react and form hydrogen iodide gas, 6.2 kcal of heat is absorbed from the surroundings. The thermochemical equation for the reaction is

$$\mathbf{6.2\ kcal + \tfrac{1}{2}H_2(g) + \tfrac{1}{2}I_2(s) \longrightarrow HI(g)}$$

Thus the standard enthalpy of formation of one mole of HI(g) is 6.2 kcal. This indicates that the enthalpy of one mole of HI(g) at 25°C and 1 atm is greater than the total enthalpy of one-half mole of $H_2(g)$ and one-half mole of I_2 also measured under these conditions. Thus energy must be added to H_2 and I_2 in order to form HI. ΔH_f° is positive because the chemical system gained energy in the formation of the compound. The positive value indicates a greater amount of energy is absorbed in breaking bonds than is liberated in the formation of bonds. The net result is an absorption of energy and an endothermic reaction.

A negative enthalpy of formation indicates that the enthalpy of the compound is less than the sum of the enthalpies of the elements

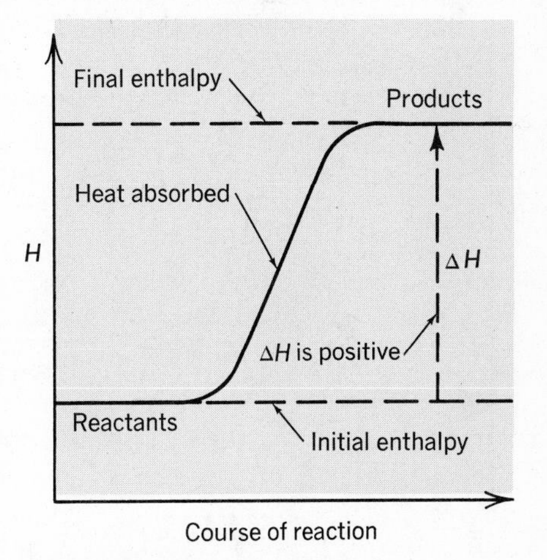

11-12 The enthalpy of a system increases when an endothermic reaction takes place. This means that the products have a greater enthalpy than the reactants.

TABLE 11-1
STANDARD ENTHALPIES OF FORMATION
(25°C, and 1.0 atm)

Compound	*Formation Reaction*	ΔH_f° (kcal/mole)
$Ag_2O(s)$	$2Ag(s) + \frac{1}{2}O_2(g) \longrightarrow Ag_2O(s)$	−7.3
$C_6H_6(g)$	$6C(s) + 3H_2(g) \longrightarrow C_6H_6(g)$	+19.82
$C_6H_6(l)$	$6C(s) + 3H_2(g) \longrightarrow C_6H_6(l)$	+11.63
$CaO(s)$	$Ca(s) + \frac{1}{2}O_2(g) \longrightarrow CaO(s)$	−151.9
$Ca(OH)_2(s)$	$Ca(s) + O_2(g) + H_2(g) \longrightarrow Ca(OH)_2(s)$	−235.8
$CH_3OH(g)$	$C(s) + 2H_2(g) + \frac{1}{2}O_2(g) \longrightarrow CH_3OH(g)$	−48.10
$CO(g)$	$C(s) + \frac{1}{2}O_2(g) \longrightarrow CO(g)$	−26.42
$CO_2(g)$	$C(s) + O_2(g) \longrightarrow CO_2(g)$	−94.05
$CuO(s)$	$Cu(s) + \frac{1}{2}O_2(g) \longrightarrow CuO(s)$	−37.1
$Cu_2O(s)$	$2Cu(s) + \frac{1}{2}O_2(g) \longrightarrow Cu_2O(s)$	−39.8
$Fe_2O_3(s)$	$2Fe(s) + \frac{3}{2}O_2(g) \longrightarrow Fe_2O_3(s)$	−196.5
$HBr(g)$	$\frac{1}{2}H_2(g) + \frac{1}{2}Br_2(l) \longrightarrow HBr(g)$	−8.66
$HCl(g)$	$\frac{1}{2}H_2(g) + \frac{1}{2}Cl_2(g) \longrightarrow HCl(g)$	−22.06
$HF(g)$	$\frac{1}{2}H_2(g) + \frac{1}{2}F(g) \longrightarrow HF(g)$	−129.0
$HI(g)$	$\frac{1}{2}H_2(g) + \frac{1}{2}I_2(s) \longrightarrow HI(g)$	+6.20
$H_2O(g)$	$H_2(g) + \frac{1}{2}O_2(g) \longrightarrow H_2O(g)$	−57.80
$H_2O(l)$	$H_2(g) + \frac{1}{2}O_2(g) \longrightarrow H_2O(l)$	−68.32
$H_2S(g)$	$H_2(g) + S(s) \longrightarrow H_2S(g)$	−4.82
$HgO(s)$	$Hg(l) + \frac{1}{2}O_2(g) \longrightarrow HgO(s)$	−21.7
$MgO(s)$	$Mg(s) + \frac{1}{2}O_2(g) \longrightarrow MgO(s)$	−143.8
$NaCl(s)$	$Na(s) + \frac{1}{2}Cl_2(g) \longrightarrow NaCl(s)$	−98.2
$NaClO_3(s)$	$Na(s) + \frac{1}{2}Cl_2(g) + \frac{3}{2}O_2(g) \longrightarrow NaClO_3(s)$	−85.7
$NH_3(g)$	$\frac{1}{2}N_2(g) + \frac{3}{2}H_2(g) \longrightarrow NH_3(g)$	−11.04
$NH_4Cl(s)$	$\frac{1}{2}N_2(g) + 2H_2(g) + \frac{1}{2}Cl_2(g) \longrightarrow NH_4Cl(s)$	−75.38
$NO(g)$	$\frac{1}{2}N_2(g) + \frac{1}{2}O_2(g) \longrightarrow NO(g)$	+21.60
$NO_2(g)$	$\frac{1}{2}N_2(g) + O_2(g) \longrightarrow NO_2(g)$	+8.09
$SO_2(g)$	$S(s) + O_2(g) \longrightarrow SO_2(g)$	−70.76
$SO_3(g)$	$S(s) + \frac{3}{2}O_2(g) \longrightarrow SO_3(g)$	−94.45

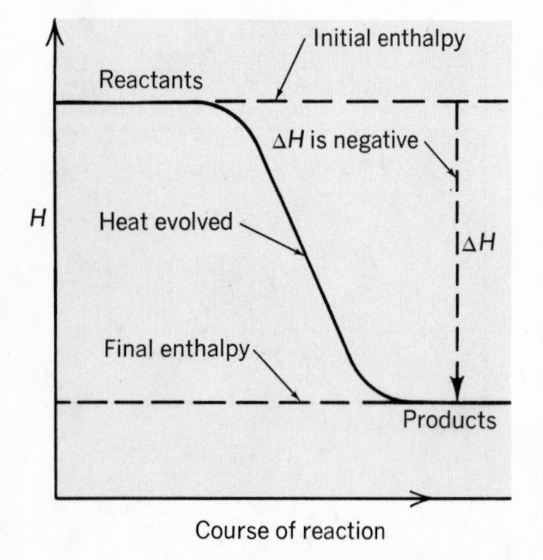

11-13 The enthalpy of a system decreases when an exothermic reaction takes place. This indicates that the reactants have a greater enthalpy than the products.

from which it was formed. For example, the standard enthalpy of formation for $H_2O(g)$ is −57.8 kcal/mole (see Table 11-1). This indicates that the enthalpy of a mole of $H_2O(g)$ is less than the total enthalpy of a mole of H_2 plus $\frac{1}{2}$ mole of O_2. It means that energy is liberated when H_2 and O_2 gas react and form $H_2O(g)$. ΔH_f° is negative because the chemical system loses energy in the formation of the compound. A negative value indicates that there are fewer and/or weaker bonds in the reactants than in the products. In other words, more energy is liberated when bonds are formed between H and O than is absorbed when bonds are broken within the H_2 and O_2 molecules. The net result is the liberation of energy and an exothermic reaction. The value of ΔH° depends on the relative numbers and strengths of the bonds in the reactant as compared with the molecules in the product. Now that we have related ΔH_f° qualitatively to bonding concepts, let us see how it can be used to calculate enthalpies of reactions.

11-10 Enthalpies (Heats) of Reactions in General We can use the First Law of Thermodynamics or Hess's Law with the heats of formation in Table 11-1 to calculate heats of reactions that are not found in tables. The technique is illustrated in Example 11-2.

Example 11-2

Calculate the standard state enthalpy change (ΔH_r°) when ammonia is oxidized.

$$NH_3(g) + \tfrac{7}{4}O_2(g) \longrightarrow NO_2(g) + \tfrac{3}{2}H_2O(g)$$

Solution

1. Refer to Table 11-1 and write the equation for the formation of each compound which appears in the above equation. Write the ΔH_f°(in general ΔH_r°) value after each equation. The three compounds of interest in this reaction are $NH_3(g)$, $NO_2(g)$, and $H_2O(g)$. The equations from Table 11-1 are:

$$\tfrac{1}{2}N_2(g) + \tfrac{3}{2}H_2(g) \longrightarrow NH_3(g) \qquad \Delta H_f^\circ = -11.0 \text{ kcal}$$
$$\tfrac{1}{2}N_2(g) + O_2(g) \longrightarrow NO_2(g) \qquad \Delta H_f^\circ = +8.1 \text{ kcal}$$
$$H_2(g) + \tfrac{1}{2}O_2(g) \longrightarrow H_2O(g) \qquad \Delta H_f^\circ = -57.8 \text{ kcal}$$

2. The molecules of reactants and products in the equations taken from the table must correspond to the molecules of reactants and products in the net equation for the desired reaction. If any do not, the equation and the sign of ΔH° should be reversed. The net equation shows $NH_3(g)$ as a reactant. Therefore, the equation for the formation of $NH_3(g)$ and the sign of ΔH° must be reversed. Rewriting the three equations gives

$$NH_3(g) \longrightarrow \tfrac{1}{2}N_2(g) + \tfrac{3}{2}H_2(g) \qquad \Delta H^\circ = +11.0 \text{ kcal}$$
$$\tfrac{1}{2}N_2(g) + O_2(g) \longrightarrow NO_2(g) \qquad \Delta H_f^\circ = +8.1 \text{ kcal}$$
$$H_2(g) + \tfrac{1}{2}O_2(g) \longrightarrow H_2O(g) \qquad \Delta H_f^\circ = -57.8 \text{ kcal}$$

3. Compare the coefficients of the compounds in the equations in Step 2 with those in the net equation. Multiply any equation and its ΔH° in Step 2 by a number that will make the coefficient of the compound equal to that in the net equation. Water has a coefficient of $\tfrac{3}{2}$ in the net equation. Therefore, multiply the third equation in Step 2 by a factor of $\tfrac{3}{2}$. Perform this multiplication and then add the three equations.

$$\tfrac{3}{2}H_2(g) + \tfrac{3}{4}O_2(g) \longrightarrow \tfrac{3}{2}H_2O(g)$$
$$\Delta H^\circ = \tfrac{3}{2}(-57.8) = -86.7 \text{ kcal}$$
$$NH_3(g) \longrightarrow \tfrac{1}{2}N_2(g) + \tfrac{3}{2}H_2(g) \qquad \Delta H^\circ = +11.0 \text{ kcal}$$

$$\tfrac{1}{2}N_2(g) + O_2(g) \longrightarrow NO_2(g) \qquad \Delta H^\circ = +8.1 \text{ kcal}$$

$$NH_3(g) + \tfrac{3}{2}H_2(g) + \tfrac{3}{4}O_2(g) + \tfrac{1}{2}N_2(g) + O_2(g) \longrightarrow$$
$$\tfrac{3}{2}H_2O(g) + \tfrac{1}{2}N_2(g) + \tfrac{3}{2}H_2(g) + NO_2(g) \qquad \Delta H_r^\circ = -67.6 \text{ kcal}$$

4. Simplifying gives

$$NH_3(g) + \tfrac{7}{4}O_2(g) \longrightarrow \tfrac{3}{2}H_2O(g) + NO_2(g)$$
$$\Delta H_r^\circ = -67.6 \text{ kcal/mole}$$

FOLLOW-UP PROBLEMS

1. Calculate ΔH° for the reaction for

$$CO_2(g) + H_2(g) \longrightarrow CO(g) + H_2O(g)$$

Ans. $\Delta H^\circ = 9.8$ kcal/mole.

2. Calculate the standard molar heat of combustion of ethyl alcohol.

$$C_2H_5OH(l) + 3O_2(g) \longrightarrow 2CO_2(g) + 3H_2O(l)$$

ΔH_f° for $C_2H_5OH = -66.36$ kcal

Ans. $\Delta H_r^\circ = -327$ kcal.

3. Calculate the standard molar heat of combustion of ethane $(C_2H_6)(g)$.

$$C_2H_6(g) + \tfrac{7}{2}O_2(g) \longrightarrow 2CO_2(g) + 3H_2O(l)$$

ΔH_f° for $C_2H_6 = -20.24$ kcal

Ans. $\Delta H_r^\circ = -373$ kcal.

An alternate and simpler method for calculating enthalpies of reaction is based on the definition of ΔH°: the enthalpy change is equal to the difference between the total enthalpies of the reactants and products. Because the enthalpies of formation are equal to the relative enthalpies of the compounds, we can say that at constant pressure the enthalpy (heat) of a reaction, ΔH_r°, is equal to the difference between the sum of the enthalpies of the formation of the products and that of the reactants. For a general reaction

$$\mathbf{aA + bB \longrightarrow cC + dD}$$

this relationship may be expressed as

Greek sigma (Σ) is read "the summation of."

$$\Delta H^\circ = \Sigma \Delta H_f^\circ \text{ products} - \Sigma \Delta H_f^\circ \text{ reactants}$$

or as

$$\Delta H^\circ = [c(\Delta H_f^\circ)C + d(\Delta H_f^\circ)D] - [a(\Delta H_f^\circ)A + b(\Delta H_f^\circ)B] \qquad 11\text{-}8$$

The use of this relationship is illustrated by examples.

Example 11-3
Calculate the enthalpy change, ΔH_r°, for the decomposition of sodium chlorate.

$$NaClO_3(s) \longrightarrow NaCl(s) + \tfrac{3}{2}O_2(g)$$

Solution
1. Obtain enthalpies of formation from Table 11-1

$$NaClO_3(s)\ \Delta H_f^\circ = -85.7 \text{ kcal/mole}$$
$$NaCl(s)\ \Delta H_f^\circ = -98.2 \text{ kcal/mole}$$
$$O_2(g)\ \Delta H_f^\circ = 0 \text{ kcal/mole}$$

2. Substitute enthalpies of formation in Equation 11-8

$$\begin{aligned}\Delta H_r^\circ &= [\Delta H_f^\circ(NaCl) + \Delta H_f^\circ(\tfrac{3}{2}O_2)] - [\Delta H_f^\circ(NaClO_3)]\\ &= -98.2 \text{ kcal} + 0 - [-85.7 \text{ kcal}]\\ &= -98.2 \text{ kcal} + 85.7 \text{ kcal}\\ &= -12.5 \text{ kcal}\end{aligned}$$

Example 11-4
Calculate the standard enthalpy change for this oxidation of ammonia.

$$4NH_3(g) + 5O_2(g) \longrightarrow 6H_2O(g) + 4NO(g)$$

Solution
1. Itemize the enthalpies of formation of all products and reactants, and multiply each value by the coefficient in the balanced equation.

$$\begin{aligned}4NH_3 &= 4(-11.0 \text{ kcal}) = -44.0 \text{ kcal}\\ 5O_2 &= 5(0) = 0\\ 6H_2O &= 6(-57.8 \text{ kcal}) = -346.8 \text{ kcal}\\ 4NO &= 4(21.6 \text{ kcal}) = 86.4 \text{ kcal}\end{aligned}$$

2. Substitute the values from Step 1 into Equation 11-8

$$\begin{aligned}\Delta H_r^\circ\\ &= [6\Delta H_f^\circ(H_2O) + 4\Delta H_f^\circ(NO)] - [4\Delta H_f^\circ(NH_3) + 5\Delta H_f^\circ(O_2)]\\ &= -346.8 \text{ kcal} + 86.4 \text{ kcal} - [-44.0 \text{ kcal}]\\ &= -216.4 \text{ kcal}\end{aligned}$$

FOLLOW-UP PROBLEMS

1. Use the standard enthalpies of formation in Table 11-1 and determine the change in enthalpy for each of these reactions.

(a) $2CO(g) + O_2(g) \longrightarrow 2CO_2(g)$ **Ans. −135 kcal.**
(b) $CH_4(g) + 2O_2(g) \longrightarrow CO_2(g) + 2H_2O(l)$
ΔH_f° for $CH_4 = -19.1$ kcal/mole **Ans. −212 kcal.**
(c) $2H_2S(g) + 3O_2(g) \longrightarrow 2H_2O(l) + 2SO_2(g)$
Ans. −269 kcal.

2. Calculate the enthalpy of formation for $C_2H_6(g)$ using information from Table 11-1 and the equation

$$2C_2H_6(g) + 7O_2(g) \longrightarrow 4CO_2(g) + 6H_2O(l);\quad \Delta H_r = -748 \text{ kcal}$$

Ans. $\Delta H_f^\circ = -19$ kcal/mole C_2H_6.

We noted earlier that energy changes which accompany a chemical reaction are the result of making and breaking chemical bonds. Furthermore, we implied that the magnitude and sign of ΔH° was related to the relative strength and number of bonds broken to the number and strength of bonds formed. This suggests that bond energies might be used to approximate enthalpy changes in chemical reactions. We shall now illustrate the validity of this statement.

BOND DISSOCIATION ENERGY

Because reactions involve the breaking and formation of chemical bonds, a knowledge of the strength of bonds is necessary if we are to estimate energies of reactions. The strength of the forces holding atoms together in a gaseous molecule is measured in terms of bond dissociation energy (more simply, *bond energy*).

11-11 Principle of Additivity of Bond Energies ***Bonding energy*** may be defined as the energy required to break all the bonds in a mole of gaseous molecules which are in their lowest energy states (usually room temperature). For example, 104 kcal is required to break the bonds in one mole of hydrogen molecules. Therefore, the H—H bond energy is said to be 104 kcal/mole. In other words, the bond energy per mole of H_2 is the enthalpy change for the reaction represented by the equation

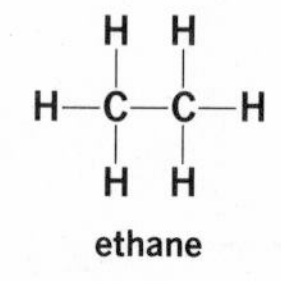

ethane

$$H_2(g) \longrightarrow H(g) + H(g) \qquad \Delta H° = 104 \text{ kcal/mole}$$

For substances whose molecules have more than one bond, the total bonding energy is the sum of the contributions from each bond in each of the molecules. For example, the bond energy for ethane molecules (see margin) is the sum of the energy of six C—H bonds and one C—C bond. This illustrates the Principle of Additivity of Bond Energies which states that the bond energy of a substance is the sum of all the bond energies of the bonds which make up the substance. Using the average bond energies given in Table 11-2, the bond energy, H_{be}, for ethane is calculated thus:

$$H_{be} = 6E_{C-H} + E_{C-C} = 6(99 \text{ kcal}) + 1(83 \text{ kcal}) = 677 \text{ kcal}$$

TABLE 11-2
AVERAGE BOND ENERGIES

Bond	Energy (kcal/mole)
H—H	104
C—H	99
C—C	83
C=C	146
C≡C	200
O—O	35
O=O	119
O—H	111
C—O	86
C=O	177
H—F	135
H—Cl	103
H—Br	87
H—I	71
N—N	39
N≡N	225
N—H	93
Cl—Cl	58
Br—Br	46
I—I	35

FOLLOW-UP PROBLEMS

1. Calculate the value of the H_{be} for

(a) methane H—C(H)(H)—H, (b) propane H—C(H)(H)—C(H)(H)—C(H)(H)—H.

11-12 Enthalpy Changes From Bond Energies The average values listed in Table 11-2 show relatively little variation from actual values and may be used to give good approximations when actual values are not available. For example, consider the reaction between hydrogen gas and chlorine gas which produces hydrogen chloride.

$$H_2(g) + Cl_2(g) \longrightarrow 2HCl(g)$$

This reaction, once initiated, is observed to be spontaneous and highly exothermic. The enthalpy change in the above reaction represents the difference between the total energy required to break the bonds in a mole of hydrogen molecules and a mole of chlorine molecules and the energy liberated when new bonds are formed in the two moles of hydrogen chloride molecules. The bond energies of chlorine molecules and hydrogen molecules are, respectively, 57.8 and 104.2 kcal/mole. The bond energy of HCl is 103.0 kcal. These data indicate that 162.0 kcal of energy should be absorbed in breaking bonds and 206 kcal should be liberated by bond formation. This information is summarized as follows:

Bonds Broken	*Energy Absorbed*	*Bonds Formed*	*Energy Evolved*
H—H	104.2 kcal		
Cl—Cl	57.8 kcal	2H—Cl	206.0 kcal
Totals	162.0 kcal		206.0 kcal

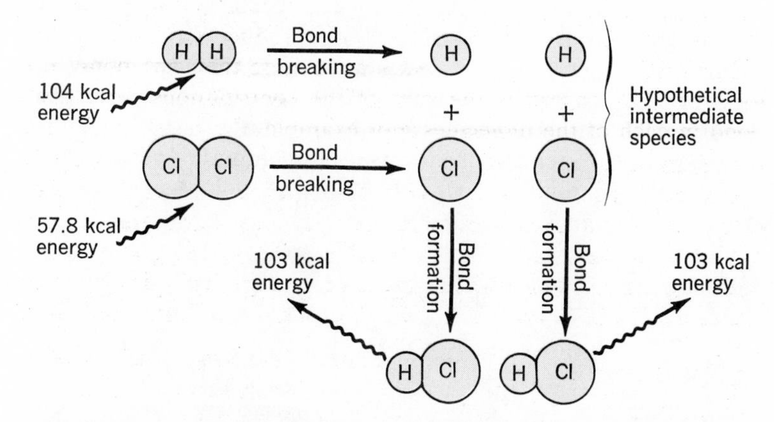

11-14 Diagram representing the net energy liberated when two moles of HCl are formed.

The difference between the energy evolved and that absorbed in the reaction is 206.0 − 162.0 = 44.0 kcal. This approximates the net energy evolved in the formation of two moles of HCl. Accordingly, the enthalpy of formation for HCl calculated from bond energies is −44.0 kcal/2 moles = −22.0 kcal/mole. This calculated value agrees very closely with the experimentally determined enthalpy of formation for HCl given in Table 11-1.

In general, single bonds between dissimilar atoms are stronger than those between similar ones. In our example it may be seen that the H—Cl bond is stronger than the average of H—H and Cl—Cl bonds. We could therefore predict qualitatively that the reaction would be exothermic and that $\Delta H°$ has a negative value.

It is also possible for an endothermic reaction to be spontaneous under a particular set of conditions. The industrial preparation of the fuel known as water gas is an example of such a reaction. *Water gas* is a gaseous mixture of carbon monoxide (CO) and hydrogen (H_2) produced when steam is passed over hot coke (carbon). The equation for the reaction is

$$H_2O(g) + C(s) \longrightarrow CO(g) + H_2(g)$$

The product of the reaction is a mixture of two combustible gases which may be burned as a fuel. The reaction takes place spontaneously at temperatures of around 600°C. Measurements show that 31.4 kcal is required per mole of carbon. This should agree with the value obtained by calculating H_r° from bond energies (Table 11-2). The thermochemical equation for the process is

$$\textbf{31.4 kcal} + H_2O(g) + C(s) \longrightarrow CO(g) + H_2(g) \qquad 11\text{-}9$$

Because the reaction is endothermic, the products have a greater enthalpy than the reactants. It is apparent that some factor other than the tendency to achieve a condition of minimum energy is responsible for the spontaneity of this reaction. This factor is, of course, the tendency of the system to achieve a condition of *maximum entropy.*

Examination of Equation 11-9 suggests that there is an increase in the entropy of the system during the reaction because one of the reactants is a solid whereas both products are gases. You learned in Chapter 10 that when a substance changes from the solid to the gaseous phase, there is an over-all increase in the entropy of the system at the temperature used. In the preparation of water gas, the increase in entropy is more important in determining spontaneity than the enthalpy factor. A follow-up discussion of entropy is provided in Chapter 13.

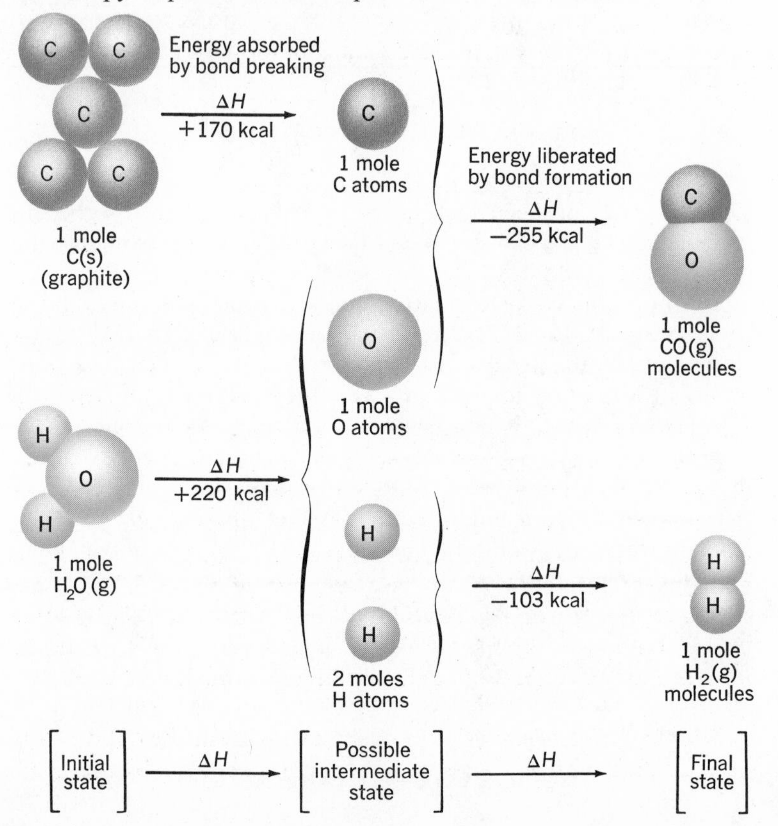

11-15 The bond-breaking and bond-forming energies for the reaction

$$C(s) + H_2O(g) \longrightarrow CO(g) + H_2(g)$$

are shown at the right. Use these values to calculate the theoretical enthalpy of reaction, ΔH_r.

FOLLOW-UP PROBLEM

Using the data in Table 11-2, calculate $\Delta H°$ for the reaction

$$N_2(g) + O_2(g) = 2NO(g)$$

The bond energy for N—O bond in NO is 150 kcal/mole.

Ans. $\Delta H° = 44$ **kcal.**

We have shown that bond energies can be used to approximate heats of reaction. They are, of course, related to the stabilities of substances. For example, one of the strongest covalent bonds is the triple bond between the nitrogen atoms in molecular nitrogen (225 kcal/mole). Thus, molecular nitrogen (N_2) is unusually stable and relatively unreactive. On the other hand, one of the weakest covalent bonds is that between fluorine atoms in molecular fluorine (F_2). This shows the reason that elemental fluorine is the most reactive nonmetallic element known.

11-13 Lattice Energy of an Ionic Crystal (Born-Haber Cycle) In the case of ionic crystals, "bond energy" is expressed in terms of crystal-lattice energy. The stability of the crystal lattice is measured in terms of the crystal-lattice energy which is defined *as the energy liberated when one mole of an ionic crystal is formed from gaseous ions.* This quantity has been measured experimentally for only a few substances. The value may be calculated from experimentally determined enthalpies of formation (ΔH_f°) plus other experimental data. We shall illustrate the calculation of crystal lattice energy for NaCl(s). The enthalpy of formation ΔH_f° for NaCl is -98 kcal/mole.

$$\mathbf{Na(s) + \tfrac{1}{2}Cl_2 \longrightarrow NaCl(s) + 98\ kcal} \qquad 11\text{-}10$$

When solid sodium and gaseous chlorine are mixed under proper conditions the reaction appears to take place in one simple step although the reaction is probably very complicated. Although the actual steps are not known, we can represent the formation of sodium chloride from its elements by a series of five hypothetical steps. All steps but one involve energy transfers which have been measured experimentally. To conform to the Law of Conservation of Energy, the five-step sequence must be energetically equal to the one-step process represented by Equation 11-10. This *hypothetical sequence* is known as the *Born-Haber cycle* and is represented by Fig. 11-16. By equating the enthalpy of formation to the sum of the enthalpy changes for each step in the cycle, we can calculate the crystal-lattice energy. Thus the analysis of the enthalpy of formation gives us a *measure of the stability of the crystal lattice.* It will also provide an insight into the driving force which accounts for the formation of ionic bonds. Let us examine in detail the energetics of each step in this Born-Haber cycle.

Step 1: The conversion of sodium atoms from the solid to the gaseous phase. This process, called sublimation, absorbs energy (is

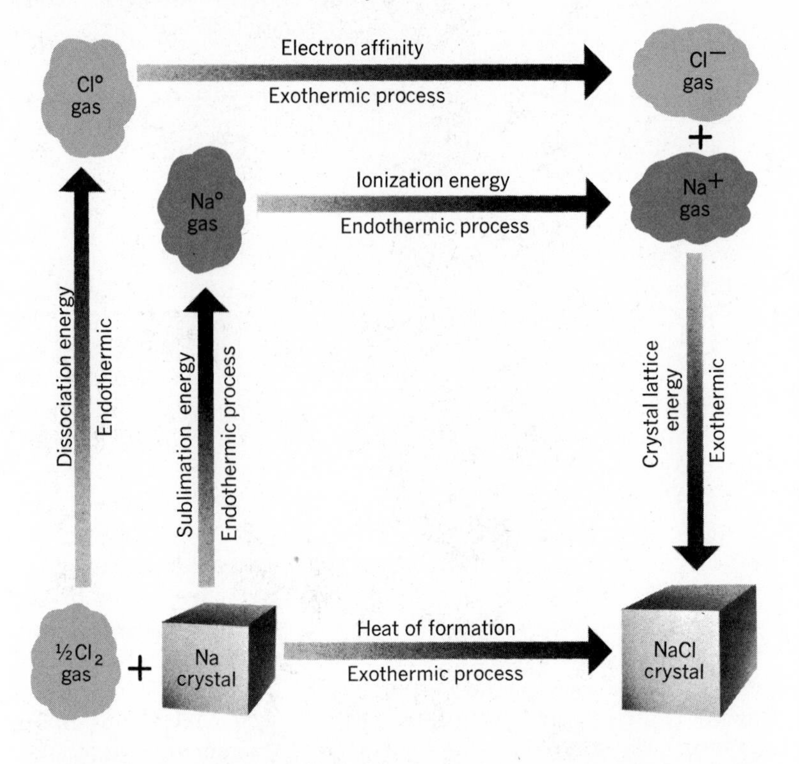

11-16 The Born-Haber cycle. When five of the six energy factors in the cycle are known, the sixth may be calculated. This is an application of the law of Conservation of Energy.

endothermic). The energy absorbed per mole, called the ***enthalpy of sublimation,*** ΔH_s°, has a value of 26 kcal/mole. The thermochemical equation representing the process is

$$\mathbf{26\ kcal + Na(s) \longrightarrow Na(g)}$$

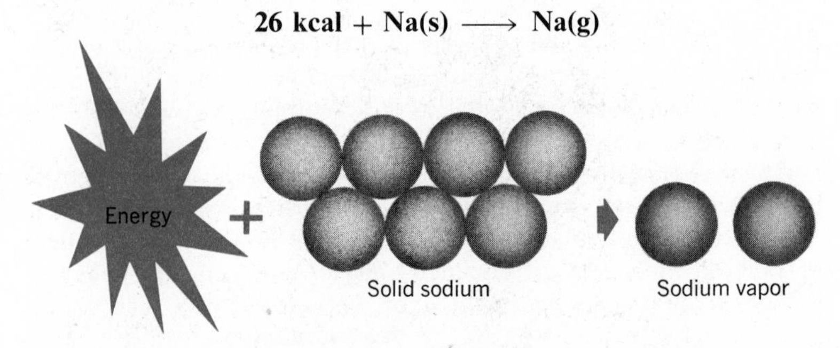

11-17 Sublimation of sodium metal. In order to convert one mole of metallic sodium crystals into sodium vapor, 26 kcal are required.

Step 2: The conversion of gaseous chlorine molecules into gaseous chlorine atoms. This process, called ***dissociation,*** absorbs energy. The energy required to break the bonds holding the chlorine atoms together in the molecule is called ***dissociation energy,*** ΔH_d°. ΔH_d° has a value of 58 kcal/mole of Cl_2. The following equation shows that only $\frac{1}{2}$ mole of chlorine is involved, so ΔH_d° for this reaction is 29 kcal per $\frac{1}{2}$ mole of Cl_2. The thermochemical equation for the reaction is

$$\mathbf{29\ kcal + \tfrac{1}{2}Cl_2(g) \longrightarrow Cl(g)}$$

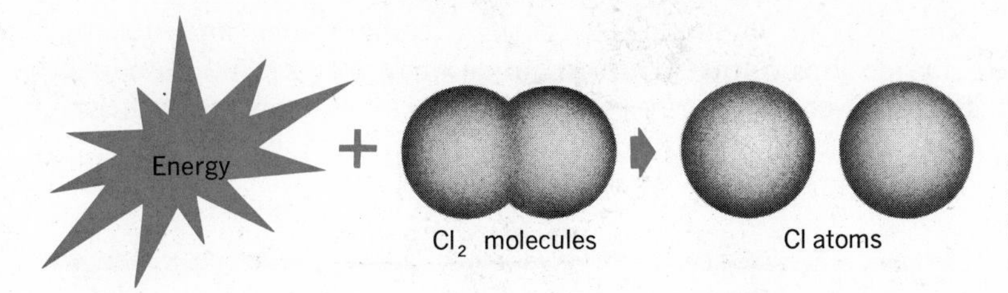

11-18 Energy is required to overcome the force of attraction (chemical bond) binding two chlorine atoms together in a molecule. The separate chlorine atoms are, therefore, in a higher energy state than the bound atoms.

Step 3: The conversion of sodium atoms in the gaseous phase into gaseous sodium ions. This process is called ***ionization.*** As you have already learned, this energy is known as the ***first ionization energy,*** ΔH_i°, and is the energy required to remove the most loosely bound electrons from one mole of sodium atoms. The value of ΔH_i° for the reaction represented below is 118 kcal/mole. The thermochemical equation is

$$\mathbf{118\ kcal + Na(g) \longrightarrow Na^+(g) + e^-}$$

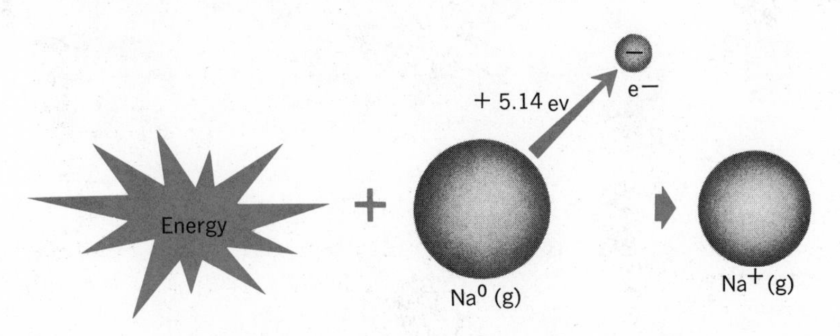

11-19 5.14 ev are required to remove one electron from a sodium atom. 5.14 ev per atom is equivalent to 118 kcal per mole of atoms. Can you determine the conversion factor?

Step 4: The conversion of gaseous chlorine atoms into gaseous chloride ions. This process is exothermic. The addition of an electron to a neutral atom involves attractive forces which result in a decrease in energy for this system. Therefore, energy is released to the surroundings. As discussed in Chapter 7, the energy released is known as the ***electron affinity,*** ΔH_e°. The electron affinity of an atom is the energy released when an electron is added to a gaseous atom in the ground state. The value of ΔH_e° for the equation below is -87 kcal/mole. The thermochemical equation is

$$\mathbf{Cl(g) + e^- \longrightarrow Cl^-(g) + 87\ kcal/mole}$$

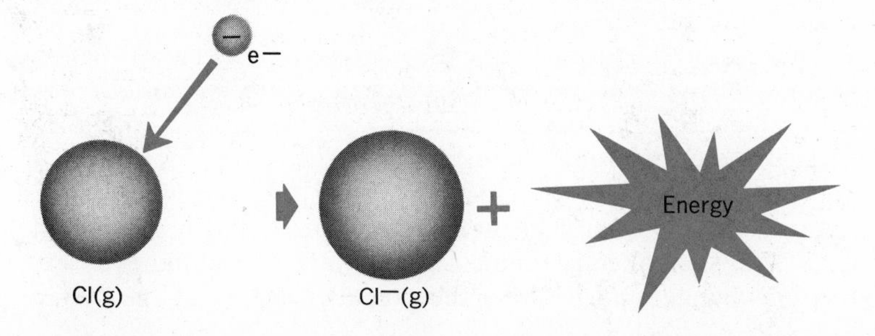

11-20 The potential energy of the system decreases when the chlorine atom acquires an electron to form the chloride ion.

Step 5: The combining of positive sodium ions and negative chloride ions forms solid sodium chloride as a crystal. Attractive forces lower the energy of the system. The reaction is exothermic and releases energy. The process is represented as

$$\mathbf{Na^+(g) + Cl^-(g) \longrightarrow Na^+Cl^-(s) + energy}$$

The energy released in this reaction is called the ***crystal-lattice energy,*** ΔH_c°. The value of ΔH_c° is a measure of the energy required to separate the ions in the crystal so that they may enter the gaseous phase. Thus, a substance with a large crystal energy is a solid with a relatively low vapor pressure.

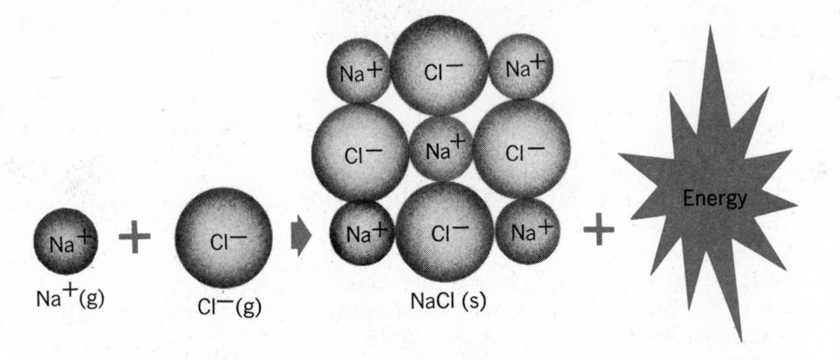

11-21 When oppositely charged ions in the gaseous phase condense to form a solid crystal, crystal lattice energy is released.

Applying Hess's Law to the Born-Haber cycle in Fig. 11-16, we can write

$$\Delta H_f^\circ = \Delta H_s^\circ + \Delta H_d^\circ + \Delta H_i^\circ + \Delta H_e^\circ + \Delta H_c^\circ$$

Solving this equation for ΔH_c°, we obtain

$$\Delta H_c^\circ = \Delta H_f^\circ - \Delta H_s^\circ - \Delta H_d^\circ - \Delta H_i^\circ - \Delta H_e^\circ$$

Substituting experimentally determined values for the terms on the right side of the preceding equation yields

$$\begin{aligned}\Delta H_c &= (-98\text{ kcal}) - (26\text{ kcal}) - (29\text{ kcal}) - (118\text{ kcal}) \\ &\qquad - (-87\text{ kcal}) \\ &= -271\text{ kcal} + 87\text{ kcal} = -184\text{ kcal}\end{aligned}$$

This relatively high value for the crystal lattice energy of NaCl is consistent with the fact that NaCl is thermally stable and has a low vapor pressure.

Analysis of the steps in the Born-Haber cycle shows that the gaseous ions with octet structures have a greater potential energy (31 kcal/mole) and, thus, less stability than the gaseous atoms. The 31 kcal/mole represents the difference between the ionization energy for sodium and the electron affinity for chlorine (118–87). Because there is no significant entropy increase this process would not occur spontaneously unless there were a simultaneous exothermic change which lowers the potential energy of the system.

It is the force of attraction between the ions in the crystal that supplies the required stability. This reasoning implies that the ionic bond is formed in response to the attractive forces between ions rather than in response to a tendency to form an octet structure.

Analysis of the individual steps of the cycle also suggests that an ionic substance such as Na^+Cl^- has a unique formula. The second ionization energy for sodium is over 1000 kcal/mole. The formation of Na^{2+} would require the absorption of this amount of energy and would produce an extremely high-energy, unstable system. The relatively small amount of crystal lattice energy which would be evolved when the ions condensed and formed a crystal of Na^+Cl^- would not be sufficient to reduce the energy enough to form a stable system.

FOLLOW-UP PROBLEM

1. (a) Using the data below, calculate the net energy liberated in the formation of one mole of solid potassium fluoride from its elements (the heat of formation). (b) What is the value of ΔH_f°? (c) Write an equation for the formation of one mole of solid KF, showing the heat term on the right-hand side of the equation. Heat of sublimation (ΔH_s°) of potassium = 21 kcal/mole. Ionization energy (ΔH_i°) of potassium = 99 kcal/mole. Heat of dissociation (ΔH_d°) of fluorine, F_2 = 38 kcal/mole. Electron affinity (ΔH_e°) of fluorine = −80 kcal/mole. Crystal-lattice energy (ΔH_c°) of KF = −193 kcal/mole.

Ans. (a) 134 kcal/mole is liberated, (b) $\Delta H_f^\circ = -134$ kcal/mole, (c) $K(s) + \frac{1}{2}F_2(g) \longrightarrow KF(s) + 134$ kcal.

LOOKING AHEAD

In this chapter we found that reactions such as

$$H_2(g) + \tfrac{1}{2}O_2(g) \longrightarrow H_2O(g) \text{ or } (l)$$

and

$$\tfrac{1}{2}N_2 + \tfrac{3}{2}H_2(g) \longrightarrow NH_3(g)$$

were both highly exothermic. They are also spontaneous. However, when hydrogen gas and oxygen gas, or hydrogen gas and nitrogen gas are mixed at room temperature and pressure, no measurable reaction occurs regardless of how long they are allowed to stand. This illustrates a very important point: in theory a reaction may be highly probable and spontaneous but its rate may be so slow that it cannot be detected thus making it impractical from an economic or utilitarian viewpoint. This suggests that the rate of a reaction has considerable importance.

In the latter part of this chapter we noted that reactions that appear to take place in a single step may actually involve a number of simpler steps. It turns out that the study of reaction rates known as *chemical kinetics* yields data which help scientists determine the mechanism (series of simple steps) of a reaction. Knowledge of reaction mechanisms helps scientists improve the efficiency of reactions.

A knowledge of reaction rate concepts will help us understand chemical reactions (changes). We shall, therefore, devote the next chapter to discussion of the simpler aspects of this important topic.

QUESTIONS

1. Discuss the idea that the standard of living in a society might be measured in units of horsepower/person or kcal/person available in that society.

2. (a) List five ways in which energy may be stored in a molecule. (b) Which of these are potential? (c) Which are kinetic?

3. Explain why the energy stored in chemical bonds is mostly potential rather than kinetic.

4. What is meant by a state property?

5. Show that, if one measures the pressure and temperature of a known mass of gas, the volume is also known.

6. Why is it difficult to measure the total energy of a chemical system?

7. Indicate how each of these affects the total energy, E, of a system. (a) Heat is transferred from the surroundings to the system. (b) Heat is transferred from the system to the surroundings. (c) Light energy is absorbed by the system. (d) Electric work is done by the system. (e) Mechanical work is done by the system. (f) The system contracts in response to external pressure. (g) Externally applied electric energy is used to decompose substances within the system. (h) The system expands against the atmosphere.
Ans. (a) increased, (d) decreased, (f) increased, (h) decreased.

8. For the reaction $N_2(g) + O_2(g) \longrightarrow 2NO(g)$, $\Delta H_r^\circ = 43.2$ kcal. (a) Which is greater, the total value of ΔH_f° for the reactants or ΔH_f° of the product? (b) Calculate ΔH_f° for the product. (c) How do the strengths of the bonds in a NO molecule compare with those in $N_2(g)$ and $O_2(g)$?

9. Which of these are exothermic reactions?

(a) $N_2 + 3H_2 \longrightarrow 2NH_3 \quad \Delta H_r^\circ = -22$ kcal

(b) $2HgO + 43.3\text{ kcal} \longrightarrow 2Hg + O_2$

(c) $Fe_2O_3 + 3CO \longrightarrow 2Fe + 3CO_2 + 6.6\text{ kcal}$

(d) $N_2 + O_2 \longrightarrow 2NO - 43.2\text{ kcal}$

10. Rewrite each of the equations above so that the coefficient in front of the first formula to the right of the arrow is 1. Show the enthalpy change associated with one mole of this substance.

11. Which substance in each of these pairs is the more thermally stable: (a) N_2 with $N_{be} = +225$ kcal or NCl_3 with $\Delta H_f^\circ = +111$ kcal, (b) F_2 with $H_{be} = +37$ kcal or HF with $\Delta H_f^\circ = -135$ kcal?

12. Table 11-1 gives the ΔH_f° of the hydrogen halides (HF, HCl, HBr, and HI). (a) Account for the change in ΔH_f° as the atomic number of the halogen increases. (b) Which of the halogens forms the strongest bond with hydrogen? (c) Which hydrogen halide most readily gives up hydrogen atoms in a chemical reaction?

13. Describe the changes in (a) enthalpy, ΔH°, and (b) entropy, ΔS, when ice melts spontaneously.

14. Is it possible for a reaction to be spontaneous (probable) and yet undergo no observable changes when the reactants are mixed? Give an example and explain your answer.

15. The equation representing the sublimation of I_2 is

$$I_2(s) \longrightarrow I_2(g) \qquad \Delta H_s^\circ = 14.9\text{ kcal}$$

(a) Is the sublimation or crystallization of I_2 favored by the tendency of a system to achieve a state of maximum entropy? (b) Is the sublimation or crystallization of I_2 favored by the tendency of a system to achieve a condition of minimum enthalpy? (c) At a given temperature, would solid iodine have a greater or smaller tendency than X_2 to vaporize if the ΔH_s° for X_2 is 18.7 kcal? Assume entropy changes are the same for both substances.

PROBLEMS

1. A calorimeter containing 1.00 liter of water at 23°C is warmed to 68°C when 5.00 g of butter is burned. Calculate (a) the total number of calories absorbed by the water, (b) the number of cal/g given off by the oxidation of butter fat, (c) the kcal/g.
Ans. (a) 4.5×10^4 cal, (b) 9000 cal/g, (c) 9.0 kcal/g.

2. A man is immersed in a tub containing 60.0 ℓ of water. The heat of his body raises the temperature of the water from 30.0°C to 31.5°C in one hour. (a) At what rate is the man giving off heat? (b) How many kcal would he give off in a day? (c) How many grams of fat ($\Delta H_{\text{combustion of fat}} = 9.5$ kcal/g) would he need per day to supply this much energy?
Ans. (a) 90 kcal/hr, (b) 2160 kcal/day, (c) 227 g.

3. Use these data to determine the heat of combustion of a candle made of paraffin:

Mass of water in calorimeter	340 g
Temperature of water at start	15.0°C
Temperature of water at end	31.2°C
Mass of candle at start	24.7 g
Mass of candle at end	23.6 g

4. (a) How many kcal of heat are evolved when 6.0 g of pure carbon is completely burned? (b) How many grams of carbon must be burned to furnish 9.40 kcal of heat? Use data in Table 11-1.

$$C(s) + O_2(g) \longrightarrow CO_2(g) \qquad \Delta H^\circ = -94.0 \text{ kcal}$$

5. How many grams of carbon would have to be burned in order to heat 1000 ℓ of water from 20.0°C to 100°C. Specific heat capacity of water is $\frac{1.00 \text{ cal}}{gH_2O°C}$. Assume the density of water is 1.0 g/ml. **Ans. 1×10^4g.**

6. When 1.37 g of Ba reacts with oxygen gas at 25°C and atmospheric pressure, there is a net release of 1.33 kcal after the BaO cools to a temperature of 25°C. What is ΔH_f° for BaO?

$$Ba(s) + \tfrac{1}{2}O_2(g) \longrightarrow BaO(s)$$

7. Given the equation

$$H_2(g) + \tfrac{1}{2}O_2(g) \longrightarrow H_2O(g) + 57.8 \text{ kcal}$$

(a) How much heat is evolved in the formation of 2.5 moles of H_2O? (b) How much heat is liberated when 1.75 moles of H_2 burn in excess oxygen? (c) Write a thermochemical equation corresponding to the reaction above for the combustion of 2 moles of H_2. (d) How much heat is absorbed when 4 moles of H_2O decompose in the reverse reaction? (e) How much heat is evolved for each gram of hydrogen burned? **Ans. (a) 145 kcal.**

8. Using Table 11-1, calculate ΔH_r° for these reactions. In each case, state whether the reaction is exo- or endothermic, rewrite the equation as a thermochemical equation to include the heat term, and indicate whether the products have a greater or smaller enthalpy than the reactants. ΔH_f° $NH_4Cl = -75.38$ kcal/mole.

(a) $SO_2(g) + \tfrac{1}{2}O_2(g) \longrightarrow SO_3(g)$
Ans. $\Delta H_r^\circ = -23.69$ kcal.

(b) $CaO(s) + H_2O(l) \longrightarrow Ca(OH)_2(s)$
Ans. $\Delta H_r^\circ = -15.6$ kcal.

(c) $N_2(g) + 3H_2(g) \longrightarrow 2NH_3(g)$
Ans. $\Delta H_r^\circ = -22.08$ kcal.

(d) $C_6H_6(l) + \tfrac{3}{2}O_2(g) \longrightarrow 6C(s) + 3H_2O(l)$
Ans. $\Delta H_r^\circ = -216.6$ kcal.

(e) $NH_3(g) + HCl(g) \longrightarrow NH_4Cl(s)$
Ans. $\Delta H_r^\circ = -42.3$ kcal.

9. Calculate the heat of reaction, ΔH_r°, for each of these reactions, using the data from Table 11-1 and this information

ΔH_f° (kcal/mole: $CH_4(g)$, −19.1; $H_2O(l)$, −68.3; $CO_2(g)$, −94.0;
$C_2H_2(g)$, +53.3; $C_2H_6(g)$, −20.2; $C_3H_8(g)$, −24.8.

(a) $CH_4(g) + 2O_2(g) \longrightarrow CO_2(g) + 2H_2O(l)$
(b) $C_2H_2(g) + 2H_2(g) \longrightarrow C_2H_6(g)$
(c) $C_3H_8(g) + 5O_2(g) \longrightarrow 3CO_2(g) + 4H_2O(l)$
(d) $H_2O(g) + C(s) \longrightarrow CO(g) + H_2(g)$
Ans. (b) $\Delta H_r = -73.5$ kcal, (d) $\Delta H_r = +31.38$ kcal.

10. Calculate ΔH_r° for

$$SO_2(g) + \tfrac{1}{2}O_2(g) = SO_3(g)$$

from these equations.

$S(s) + O_2(g) = SO_2(g) \qquad \Delta H_f^\circ = -70.96$ kcal
$S(s) + \tfrac{3}{2}O_2(g) = SO_3(g) \qquad \Delta H_f^\circ = -94.45$ kcal

11. ΔH_r° for the complete combustion of ethylene (C_2H_4) is −331.6 kcal

$$C_2H_4(g) + 3O_2(g) = 2CO_2(g) + 2H_2O(l) \qquad \Delta H = -331.6 \text{ kcal}$$

Use this information with the data in Table 11-1 to calculate the heat of formation, ΔH_f°, for ethylene.
Ans. $\Delta H_f^\circ = 7.0$ kcal/mole.

12. Calculate ΔH_r° for

$$C(s) + H_2O(g) = CO(g) + H_2(g)$$

using this information.

$$C(s) + O_2(g) = CO_2(g) \qquad \Delta H = -94.0 \text{ kcal}$$
$$CO(g) + \tfrac{1}{2}O_2(g) = CO_2(g) \qquad \Delta H = -67.6 \text{ kcal}$$
$$H_2(g) + \tfrac{1}{2}O_2(g) = H_2O(g) \qquad \Delta H = -57.8 \text{ kcal}$$

Ans. 31.4 kcal.

13. The heat of combustion, ΔH_c° for ethylene glycol [$(CH_2OH)_2$] is -281.9 kcal.

$$(CH_2OH)_2(l) + 2\tfrac{1}{2}O_2(g) = 2CO_2(g) + 3H_2O(l)$$

Use this information and the data in Table 11-1 to calculate ΔH_f° for $(CH_2OH)_2(l)$.

14. Calculate the heat of combustion for C_2H_6 from this information.

$$C_2H_4 + 3O_2 \longrightarrow 2CO_2 + 2H_2O \qquad \Delta H = -337.2 \text{ kcal}$$
$$C_2H_4 + H_2 \longrightarrow C_2H_6 \qquad \Delta H = -32.7 \text{ kcal}$$
$$H_2 + \tfrac{1}{2}O_2 \longrightarrow H_2O \qquad \Delta H = -68.3 \text{ kcal}$$

15. Calculate the heat of formation, ΔH_f°, for ethane (C_2H_6) from these data.

ΔH° of combustion for C_2H_6 = -372 kcal/mole
ΔH_f° for CO_2 = -94.0 kcal/mole
ΔH_f° for $H_2O(l)$ = -68.3 kcal/mole

16. Use bond energies listed in Table 11-2 to calculate ΔH° for the reaction

$$2H_2(g) + O_2(g) = 2H_2O(g)$$

Pg. 368

12

Rates of Chemical Reactions

Preview

". . . it is by logic that we prove, but by intuition that we discover."

—Jules Henri Poincaré (1854–1912)

Man has always sought methods for increasing or decreasing the rates of chemical processes. Sometimes we wish to increase the rate of a reaction so that the process is efficient, practical, and economical. This is usually the objective of the chemical industry which produces thousands of substances that enhance our standard of living and add to our well-being. In our daily activities, we often wish we could speed up and make more efficient many processes such as cooking various foods, cleaning dirty materials and surfaces, drying paints and finishes, or relieving pain. On the other hand, we sometimes wish to decrease the rate of undesirable processes such as spoilage of food, spontaneous ignition or explosion of fuels, and corrosion of metals.

Common sense and past experience can often be used to design methods for varying the rates of the reactions. Our ancestors learned long ago that many small twigs burned faster than a large log and that the higher temperature produced by the rapidly burning smaller pieces cooked their meat faster. Most of the early methods for increasing or decreasing rates of reactions were trial and error discoveries.

Although much progress has been made in the study of reaction rates, scientists do not always know exactly how or why a particular substance increases the rate of a certain reaction. They do not know the exact paths or sequence of steps that reacting molecules follow when they change into observed products. The study of reaction rates helps to determine the steps in a reaction path. The study of reaction rates, called *chemical kinetics,* is one of the frontiers of chemistry which provides an almost unlimited supply of problems and opportunities for those interested in chemical research.

Balanced equations for overall reactions do not tell the whole story. The key to the way in which reactions take place lies on the path between the initial reactants and the final product. This is the key that enables scientists to increase the efficiency of reactions.

The primary objective of a scientist who does basic research on reaction rates is to explain and describe macroscopic observations on rates of reactions at the microscopic level in terms of atoms, molecules, and ions. Since events on the microparticle level cannot be observed directly, the task of identifying the steps in a reaction is very difficult.

The identification and analysis of the intermediate steps (path) of a reaction are extremely complex and beyond the scope of this text. In this chapter we shall be primarily concerned with the

(a) *factors which affect the rate of a reaction,*
(b) *why some reactions take place rapidly and others slowly,*
(c) *how rates are measured and what is meant by a reaction mechanism,*
(d) *energy changes that occur during a reaction, and*
(e) *use of the molecular theory to explain reaction-rate phenomena.*

THE MEANING AND MEASUREMENT OF REACTION RATES

Methods of expressing rates of chemical reactions are analogous to expressing the rate at which an automobile travels. The average rate at which an automobile travels a certain distance may be expressed in miles per hour. Similarly, the rate of a chemical reaction may be expressed in the number of moles per second.

12-1 The Meaning of the Rate of a Reaction Let us use a few common reactions to illustrate the qualitative and quantitative meaning of reaction rate. For some reactions a qualitative evaluation of relative rates may be made by observing the *rate of appearance of a product or disappearance of a reactant.* For example, when a piece of magnesium ribbon is placed in a beaker of dilute hydrochloric acid, an extremely rapid evolution of hydrogen gas (a product) takes place as the magnesium (reactant) rapidly disappears. When a piece of iron is placed in the same acid, hydrogen gas is evolved slowly, and the iron disappears at a relatively slow rate. Evidently the rate of the first reaction is much faster than that of the second.

We could quantitatively express the rate of the above reaction in several ways. We might weigh the magnesium and record the number of seconds required for it to disappear completely. The average rate of the reaction could then be expressed as

$$\textbf{average rate} = \frac{\textbf{moles of magnesium consumed}}{\textbf{seconds required for reaction}}$$

For example, if 0.048 g (0.002 mole) of magnesium completely reacts in 20 seconds, then the average rate of reaction over the 20-second time interval is

$$\begin{aligned}\textbf{average rate} &= \textbf{0.002 mole/20 sec} = \textbf{0.0001 mole/sec}\\ &= \textbf{1.0} \times \textbf{10}^{-4}\ \textbf{mole/sec}\end{aligned}$$

The equation for the reaction is

$$\mathbf{Mg(s) + 2HCl(aq) \longrightarrow MgCl_2(aq) + H_2(g)}$$

This equation shows that one mole of magnesium yields one mole of hydrogen gas. Thus, the rate at which hydrogen appears is the same as the rate at which magnesium disappears.

Rates of reaction are usually expressed in terms of a change in concentration of one of the participants per unit time. Changes in concentration are symbolized by ΔC and changes in time by Δt; therefore, the rate of a reaction may be expressed as

$$\mathbf{r} = \Delta C/\Delta t \qquad 12\text{-}1$$

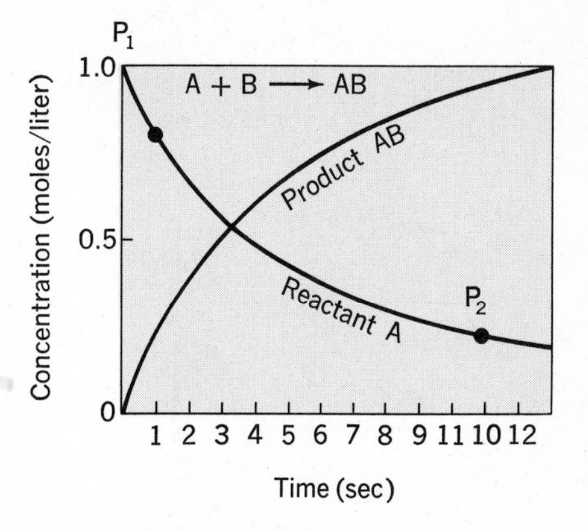

12-1 Graph showing how the concentrations of reactant A and product AB vary with time. The steepness of the curve (slope of tangent to the curve) at any point is a measure of the rate of the reaction at that point.

Experiments show that for most reactions, the concentrations of all participants change most rapidly at the beginning of the reaction. That is, the concentration of the products shows the greatest rate of increase, and the rate of decrease in the concentration of the reactants is highest at this point. This means that *the rate of a reaction changes with time.* Therefore, a rate must be identified with a specific time.

12-2 Methods of Measuring Reaction Rates The technique used to determine the difference in concentrations varies with the reaction and the available apparatus. For reactions in solution, small samples may be periodically withdrawn from the reaction mixture in order to determine the concentration by methods of quantitative analysis (titration). For reactions involving colored substances, *changes in the color* can be measured with special instruments and related to changes in concentration of the colored substances. A number of other properties such as *density and electric conductivity may vary with concentration.* These also may be measured and related to changes in concentration.

For many reactions involving gases in a closed system, changes in pressure may be *related to changes in the quantity of reactants or products.* The development of instruments which enable scientists to identify and measure the components of reaction rapidly, has facilitated research in the field of chemical kinetics. *Absorption spectrometers* and *gas chromatographs* are two instruments widely used to identify and analyze reaction mixtures. Spectrometers are discussed in Chapter 20, whereas, gas chromatography, a powerful analytical tool, was discussed in the Special Feature at the end of Chapter 2.

FACTORS AFFECTING REACTION RATES

There are several rather simple and interesting experiments that you may be able to perform or to see demonstrated that qualitatively illustrate the factors that affect the rate of a reaction. We shall briefly describe a few of these in case you don't have an opportunity to observe them directly. We shall first consider reactions in which the reactants and products are all in one phase. These are known as ***homogeneous reactions.*** Systems in which all participants are gases or in which all are dissolved in aqueous solution are *homogeneous*.

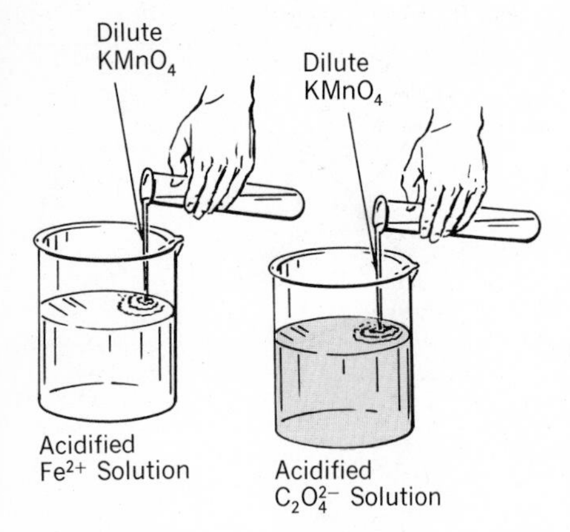

12-2 One drop of $KMnO_4$ colors an oxalate solution purple. A relatively large time interval (compared to the Fe^{2+} solution) is required for the reaction to take place and for the color to disappear.

12-3 The Nature of the Reactants Let us first add a solution of permanganate ions (MnO_4^-) dropwise to iron(II) ions (Fe^{2+}) in an acid solution, and then add permanganate ions dropwise to oxalate ions ($C_2O_4^{2-}$) in acid solution. We can qualitatively estimate the relative rates of reaction by noting the time required for the purple color of the permanganate solution to disappear. Observations reveal that in the first procedure, the permanganate solution is decolorized almost instantly. In the latter experiment, the color remains for a relatively long period of time. This means that permanganate ions reacted rapidly with iron(II) ions but slowly with oxalate ions. The only difference in the two reactions was the nature of one reagent. Iron(II) (Fe^{2+}) is a simple ion, whereas an oxalate ion ($C_2O_4^{2-}$) is polyatomic and contains covalent bonds which must be broken or weakened in the reaction process. In general, *reactions between simple ions* such as Ag^+ and Cl^- ions, which combine in a one-to-one mole ratio, are almost instantaneous. Experimental measurements show that most of these reactions occur in about one-millionth of a second. On the basis of this and many other similar experiments, we can conclude that the nature of the reactants affects the rate of a reaction, and that complicated species generally react more slowly than simple ions. You will find that there are several reasons why reactions between complicated molecules are usually slow compared with those between simple ions.

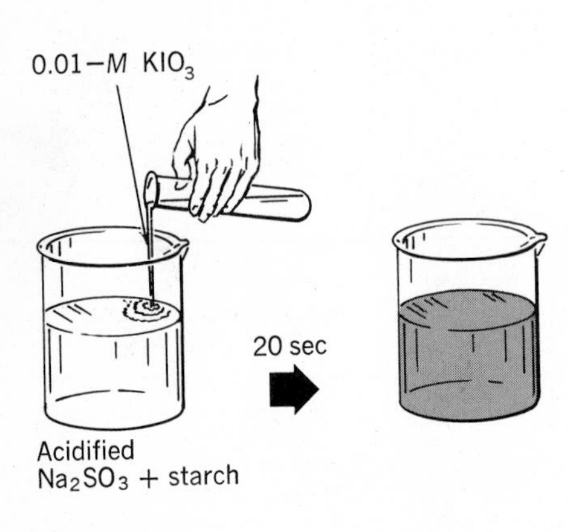

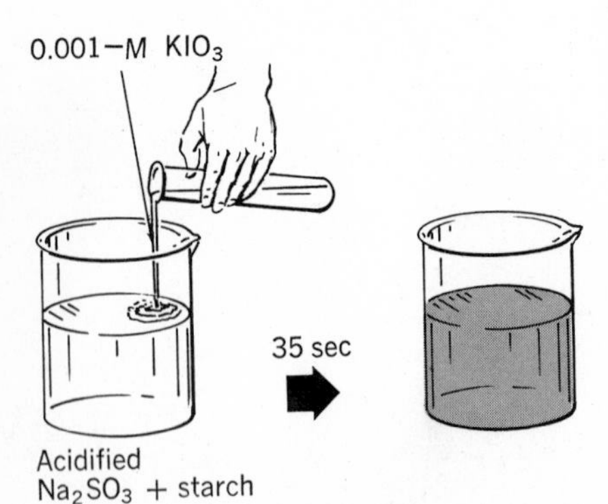

12-3 Reducing the concentration of a reactant reduces the rate of the reaction and increases the time required for its completion.

12-4 The Concentration of Reactants Let us now see how changing the concentration of a reactant affects the rate of a reaction. In this experiment we shall mix potassium iodate (KIO_3) solutions of varying concentration with an acidified Na_2SO_3 solution of fixed concentration and measure the time required for the added starch indicator to turn blue. We observe that diluting the KIO_3 solution increases the time required for the blue color to appear suddenly. In this case, decreasing the concentration decreases the rate of the reaction. Conversely, we find that increasing the concentration of KIO_3 increases the rate of the reaction. The effect of an increase or decrease in the concentration of oxygen on the rate of burning is familiar to most people. A piece of steel wool heated in air (20 percent oxygen by volume) burns slowly, but when heated in pure oxygen, it undergoes rapid combustion as evidenced by a dazzling shower of sparks.

12-5 Temperature The effect of changing the temperature on the rate of a reaction can be determined by warming the KIO_3 and Na_2SO_3 solutions and measuring the time required for the blue color to appear at different temperatures. Here, the experimental data show that the time required for the blue color to appear decreases as the temperature of the KIO_3 solution is increased. This is evidence that the *rate of the reaction increases as the temperature* *rises.* This knowledge is put to practical use at home when you store foods and milk in a refrigerator to slow down reactions which ordinarily result in spoilage and souring.

Pressure cookers are used to obtain higher temperatures in order that the reactions involved in cooking food will take place at a faster rate. In general, *a 10C° rise in temperature doubles or sometimes triples the rate of a reaction.* It should be noted that there are a few reactions whose rates decrease with increasing temperature. These exceptions can be explained, but the explanations need not concern us at this point.

12-6 Catalysts It is possible, in many reactions, to introduce a chemical substance that will change the rate of a reaction without its being consumed by the reaction. For example, if we add a small quantity of a manganese(II) compound to the oxalate solution described in Section 12-3, we find that the purple color of the permanganate solution disappears much more rapidly than without the manganese salt. Substances which affect reaction rates without being consumed by the overall reaction are called ***catalysts.*** Formulas of catalysts are not included in the equation for the overall reaction. Catalysts may be consumed during some intermediate step in a reaction and be regenerated in a subsequent step. The catalysts most familiar to you are the enzymes produced by living organisms which catalyze digestive and other biochemical processes in the body. For example, ptyalin in saliva increases the rate at which starch is converted to maltose. Without the catalyst, the conversion would take weeks and be of little biological value. We shall examine the behavior of some specific catalysts later.

12-7 Surface Area Assuming that many experiments similar to those described above show the same general effects, we can say that in homogeneous reactions there are at least four factors that affect the rate of a reaction. These factors are

1. Nature of the reactants (complexity of reacting species, nature and number of bonds).
2. The concentration of the reactants.
3. The temperature.
4. Catalysts.

To this list we can add a fifth factor that plays an important role in determining the rate of a reaction in which more than one phase is present in the system. Such a reaction is said to be ***heterogenous.*** Solid zinc reacting with hydrochloric acid, and oxygen reacting with iron, are two examples of heterogeneous reactions. The following equations for these reactions show the presence of more than one phase.

$$Zn(s) + 2HCl(aq) \longrightarrow ZnCl_2(aq) + H_2(g)$$
$$4Fe(s) + 3O_2(g) \longrightarrow 2Fe_2O_3(s)$$

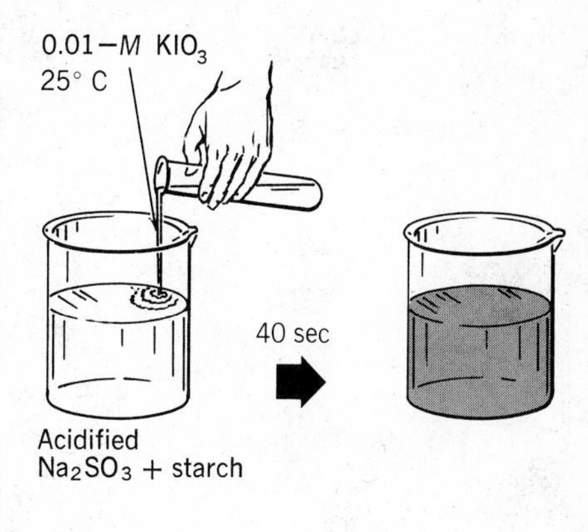

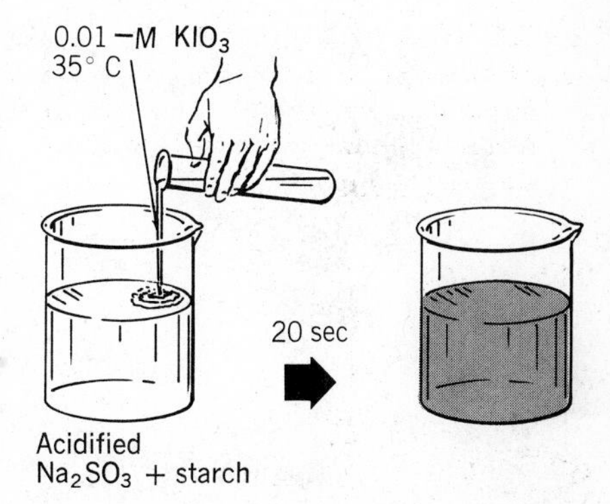

12-4 Warming the KIO_3 solution increases the rate of the reaction and decreases the time required for the appearance of the blue color.

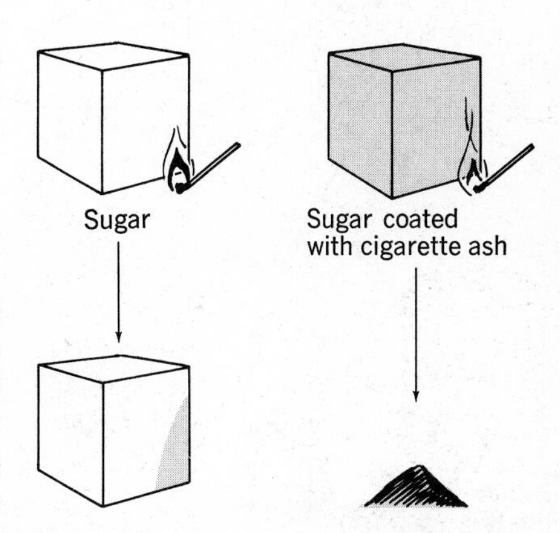

12-5 A sugar cube treated with cigarette ashes is readily ignited. The oxidation is catalyzed by the oxides in the ashes. An untreated sugar cube does not burn when placed in the flame of a match.

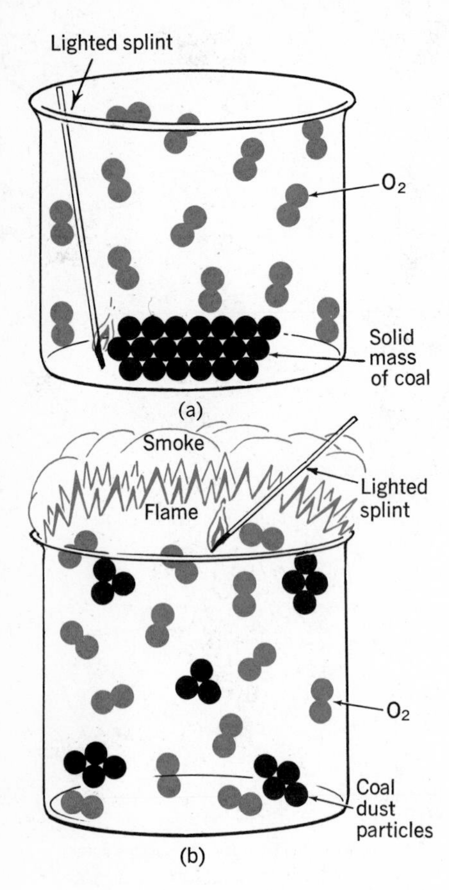

12-6 A heterogeneous reaction. The lump of coal represented in (a) has a relatively small surface area compared to the finely divided coal dust represented in (b). The reaction between carbon and oxygen in (a) takes place very slowly compared to (b).

These reactions take place at the surface which lies between the two phases of the system. *As the surface area increases, the rate of the reaction increases.* You have all observed the rapid burning of wood chips compared with the rather slow rate of burning of a log or the rapid solution of granulated sugar compared with the slow solution of lump sugar. The total surface area of the smaller particles is much greater than that of the larger ones. Since more surface area is available for reaction, the rate is greater.

We now need a model which will help us explain our macroscopic observations on reaction rates at the microscopic level of atoms, molecules, and ions. The model we shall use is known as the *Collision Theory of Reaction Rates.*

COLLISION THEORY OF REACTION RATES

12-8 Relationship Between Concentration of Reactants and Number of Molecular Collisions This theory makes the assumption that for a reaction to occur, there must be collisions between the reacting species. In the case of the decomposition of a single species, the theory assumes that the single species must have absorbed light or gained energy by contact with another microparticle. According to the collision theory, the rate of a reaction depends upon two factors: the *number of collisions* per unit time between the reacting particles and the *fraction of these collisions that are successful* or effective in producing new product molecules.

We shall first investigate the relationship between the concentration of the reactants and the number of collisions between them by considering a hypothetical reaction between A and B forming AB according to the equation

$$\mathbf{A + B \longrightarrow AB}$$

Let us put one molecule of A and one of B in a unit volume and assume that there is a chance for one collision to take place in unit time. If two molecules of A and one of B are placed in the container, then there will be two chances for the occurrence of a collision between A and B in the same time interval. When three molecules of A and one of B are placed in the container, each A has a chance to collide with the B molecule. Therefore, there are three possible (A-B) collisions per unit time. When two B molecules are placed in the container with three A molecules, there is a chance for six collisions in the same unit time. That is, each of the three A molecules has a chance of colliding with each of two B molecules. The possible collisions are illustrated in Fig. 12-7. It can be seen that the rate of collisions between A and B per unit time is proportional to the product of the number of molecules of A and the number of molecules of B. The number of molecules per unit volume may be expressed as a concentration term; thus, we may

[A] is read "the molar concentration of A."

say that the rate of collision (number per unit time) is proportional to the product of the concentrations of reactants A and B. That is,

$$\mathbf{r \propto [A][B]}$$

This agrees with our experimental observations that increasing the concentration of either reactant increases the rate of the reaction.

Earlier we noted that hydrogen gas (H_2) and oxygen gas (O_2) could be mixed and allowed to stand indefinitely without undergoing any apparent reaction regardless of their concentrations. As soon as a certain amount of energy is added to the system, however, the gases react violently and exothermically. Let us now explore the reason that so many reactions are slow and do not take place appreciably until external energy is supplied. We shall seek an answer to this problem on the molecular level in terms of the Collision Theory of Reaction Rates.

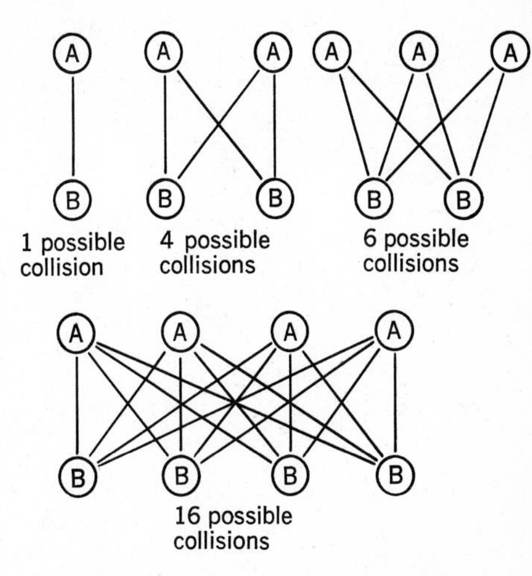

12-7 The number of possible collisions between reactant molecules A and B is proportional to the product of the number of molecules present.

12-9 Activation Energy In any sample of gas there are an enormous number of collisions. If each molecule in a mole (6×10^{23} molecules) of gas averages one billion (10^9) collisions per second, there would be 10^{32} collisions per second per mole of gas. If the rate of collisions per second were equal to the rate of reaction, then every reaction would be extremely rapid. Since many gaseous reactions are slow, it is apparent that only a fraction of the total collisions results in a reaction.

The tremendous number of collisions results in a wide range of velocities and kinetic energies. The distribution of kinetic energies is shown in Fig. 12-8. The area under the curve represents the total number of molecules. The activation energy, E_a, represents the minimum kinetic energy that must be possessed by a molecule in order for it to react. From the shape of the curve it can be seen that only a few molecules in a given sample have kinetic energies greater than E_a. The collision theory assumes that only collisions between molecules having a minimum kinetic energy of E_a are energetic enough to overcome repulsive forces between the electron clouds of the interacting molecules, and to weaken or break bonds, resulting in a reaction. Therefore, the main reason that H_2 and O_2 molecules react so slowly at room conditions is that relatively few of them have energies greater than E_a.

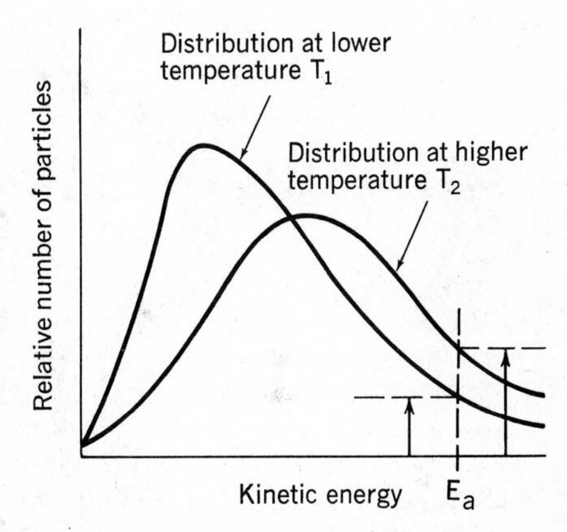

12-8 Distribution of molecular energies. As the temperature is increased, the number of molecules having energies greater than E_a is greatly increased.

Inspection of Fig. 12-8 reveals that at a higher temperature, T_2, there is a large increase in the number of molecules with energy

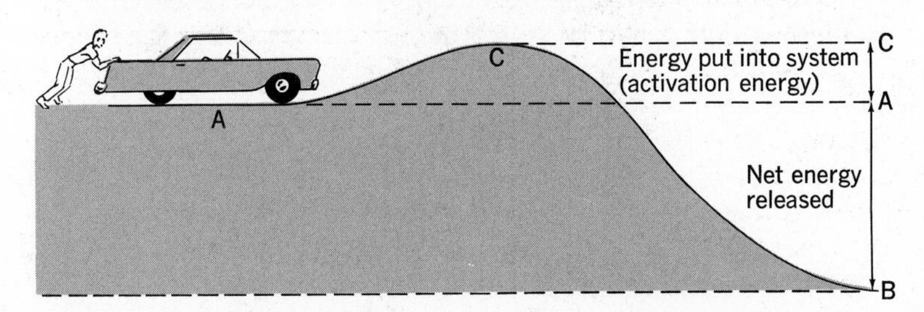

12-9 Activation energy analogy. When enough energy has been expended to reach crest C, the car rolls to the bottom of the hill B and releases energy. The energy put into the system (activation energy) is regained as the car rolls from level C to level A. The energy released as the car goes from level A to level B is analogous to the net energy evolved by an exothermic reaction.

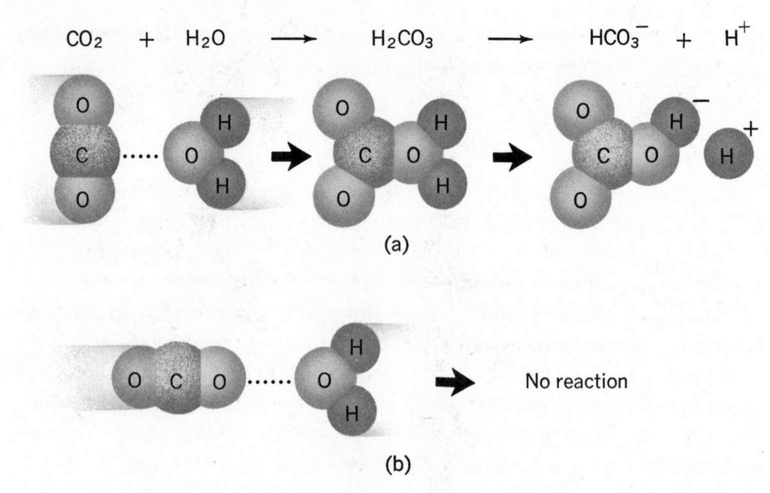

12-10 Orientation and reaction rate. In diagram (a), the reaction occurs because the molecules have the "correct" collision geometry and sufficient energy. In (b), there is no reaction because the molecules are not properly oriented.

E_a or greater. Thus, at a higher temperature, there is an increase in the number of effective collisions and a faster rate of reaction.

12-10 Collision Geometry A molecule possessing the activation energy associated with a given reaction does not necessarily react when it undergoes a collision. Even very high-speed collisions may not be effective if the colliding molecules are not oriented properly. For example, carbon dioxide molecules react with water molecules only when the two molecules come together in such a fashion that the carbon atom on a carbon dioxide molecule collides with the oxygen atom of a water molecule as shown in Fig. 12-10.

$$CO_2 + H_2O \longrightarrow H_2CO_3 \longrightarrow H^+ + HCO_3^-$$

When the oxygen atoms of CO_2 approach the oxygen atom of H_2O, the collision geometry is totally unfavorable. Hence no reaction

12-11 In (a) the player has not supplied enough energy. Hence, the ball falls short of the basket. This is analogous to the reactants in a reaction lacking enough activation energy to surmount the potential-energy barrier of the activated complex. In (b), the ball has enough energy to reach the height of the basket, but is not traveling in the correct direction to drop through the basket. This is analogous to the reactants lacking the proper collision geometry. In (c) the energy and orientation requirements are correct and the ball drops into the basket. This is analogous to a molecular collision which results in a reaction.

occurs. Thus, the geometrical shape and the collision geometry of reacting molecules affect reaction rates.

12-11 The Activated Complex When collision geometry is favorable and the colliding molecules have a kinetic energy at least equal to E_a, they interact and form a high-energy, unstable, transitory species known as an *activated complex*. An ***activated complex*** is a high-energy, unstable, short-lived configuration of reactant atoms. This unstable complex may decompose and form new, lower-energy, stabler products or revert to the reactant state.

The game of basketball furnishes an analogy for the path followed by reactant molecules as they undergo energy changes and form new substances. For this analogy, we shall let the ball represent a molecule and the basket represent the high-energy barrier between the reactant and product molecules. "Making a basket" corresponds to reactants forming products. In order for the ball to reach the basket, it must be given a certain minimum kinetic energy (analogous to activation energy). When enough energy is supplied for the ball to reach the level of the basket, it may teeter on the rim in an unstable position (analogous to the instability of the activated complex). It will have more potential energy in this position than it had in the hands of the player. The ball remains in this unstable position only momentarily and then either drops through the basket to a more stable position or falls back into the hands of the player, also a more stable position. No matter how much energy is supplied, the ball will not go through the basket unless it is properly oriented.

Molecules involved in a chemical reaction behave in much the same way. That is, they must possess a certain minimum kinetic energy which can be converted to the greater potential energy of the activated complex. They also must be properly oriented and able to enter into high-energy, unstable bonding configurations before forming stable products. In the next section we shall illustrate graphically the relationship of the activated complex to the reactant and product molecules.

ENERGY CHANGES DURING A REACTION

12-12 Graphical Representation of Energy Changes Let us now examine the potential-energy changes taking place when diatomic molecules A_2 react with molecules B_2 and form AB in a single-step reaction. The general equation representing this reaction is

$$A_2(g) + B_2(g) \longrightarrow 2AB(g) \qquad 12\text{-}2$$

These changes are represented graphically in Fig. 12-12. The total potential energy of the system is shown along the vertical axis. The horizontal axis is called the reaction coordinate. The points along this axis represent different stages during the progress of the reac-

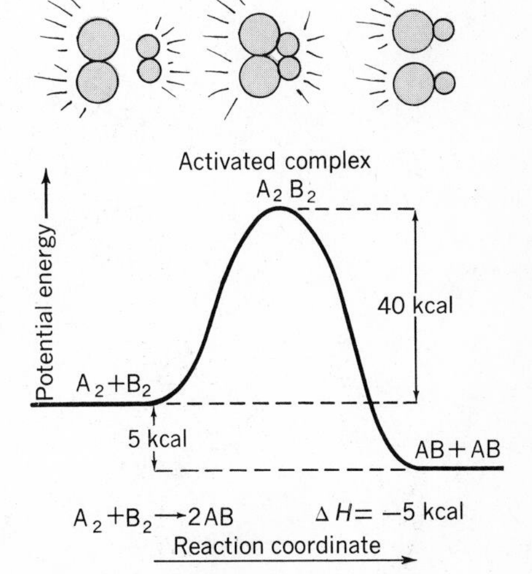

12-12 Potential energy changes occurring in a chemical system during an exothermic reaction.

tion. At each point, the interatomic distances of the interacting atoms change, and also the potential energy changes.

We shall start at the left side of the curve and follow the path of the reaction through the formation of the products. The height of the plateau at the left represents the total potential energy or enthalpy of the A_2 and B_2 molecules. Along this part of the path, the reactant molecules have kinetic energy but are so far apart that no potential energy changes occur.

When the A_2 and B_2 molecules are close enough for their electron clouds to exert a significant repelling effect, the molecules slow down, lose kinetic energy, and gain potential energy with an accompanying enthalpy increase. As kinetic energy is converted to potential energy, the curve shows a rise. From this point to the peak, the $A—A$ and $B—B$ bonds are lengthening and weakening as energy is absorbed. At the same time, new bonds are forming between A and B, a process which evolves energy. The continued rise of the curve shows that more energy is absorbed than liberated. At the peak where the potential energy is a maximum, the activated complex is formed. This unstable, short-lived complex is represented in Fig. 12-12 as A_2B_2.

The potential energy of the activated complex determines the activation energy for the reaction. It may be seen from the figure that the activation energy of 40 kcal/mole is the difference between the total potential energy of the separated reactant molecules and that of the activated complex. Activation energy always increases the potential energy of a reaction system.

Once formed, the activated complex either reverts to the original reactant molecules or follows the path to the right. From the peak down to the right plateau, $A—A$ bonds and $B—B$ bonds continue to lengthen, and the $A—B$ bonds become shorter and stronger. Bond formation now predominates so that more energy is liberated than absorbed. The potential energy of the system decreases as the potential energy of the activated complex is converted into the kinetic energy of the product molecules. When the plateau of the right side is reached, all the potential energy of the complex has been converted to the kinetic energy of the products. The height of the right plateau represents the total potential energy or enthalpy of the products.

The difference of 5 kcal/mole between the total potential energy of the products and that of the reactants is the enthalpy change or heat of reaction. Since the enthalpy of the products is less than that of the reactants, the reaction is exothermic. It should be noted that the *net energy* liberated is independent of the activation energy. Experimental measurements show that the activation energy is usually less than the bonding energy of the reacting molecules. This is attributed to the fact that new bonds are forming at the same time that old ones are weakening. This implies that it is not absolutely necessary for molecules to undergo violent collisions and break into individual atoms before a reaction can occur.

12-13 Relationship Between Activation Energies of Opposing Reactions It is observed that slow reactions generally have high activation energies, and fast ones have relatively low activation energies. An endothermic reaction always has a greater activation energy and a slower rate than the opposing exothermic reaction. Fig. 12-13 shows the path for the reaction between AB molecules in the formation of A_2 and B_2 molecules.

$$2AB(g) \longrightarrow A_2(g) + B_2(g) \qquad 12\text{-}3$$

This reaction is the reverse of that illustrated by the curve in Fig. 12-12. Note that the AB molecules must collide with a favorable orientation in order to form the same activated complex as A_2 and B_2 in the reverse reaction. The total potential energy of two AB molecules, however, is less than that of $A_2 + B_2$. The activation energy for Equation 12-3 is therefore greater than that for Equation 12-2.

Comparison of Fig. 12-12 with Fig. 12-13 shows that the AB molecules have a higher barrier to surmount than the A_2 and B_2 molecules. Thus, we would expect a smaller fraction of the AB molecules to get "over the top." In other words, there would be a smaller fraction of effective collisions between AB molecules than between A_2 and B_2 molecules per unit time.

Since the reactant AB molecules are at a lower potential energy level than the product molecules, Equation 12-3 represents an endothermic reaction. It is apparent that Equations 12-2 and 12-3 are opposing reactions. Since the endothermic reaction always has a higher activation energy, it is always slower at a given temperature than the opposing exothermic one. Increasing the temperature increases the rate of both endothermic and exothermic reactions. Experiments show, however, that an *increase in temperature affects the rate of the endothermic reaction more than that of the exothermic reaction.*

The fact that catalysts increase the rate of a reaction suggests that they may lower the activation energy of the system. Let us briefly examine the role of a catalyst in chemical reactions.

12-14 Effect of Catalysts on Activation Energy Increasing the temperature of a reaction system will usually increase the rate of the reaction, but it may also cause the decomposition of the reactants before they can react. In addition, increased temperature may result in the formation of unwanted (side) products. Fortunately, scientists have known for many years that certain catalysts are able to increase the rate of a reaction. For example, platinum gauze causes the reaction between hydrogen and oxygen gas to take place at room temperature. Without the platinum catalyst, there is no apparent reaction under these conditions. The discovery and use of catalysts have been significant factors in the growth of the petroleum and most other chemical industries. Research chemists are constantly working on the development of new catalytic agents.

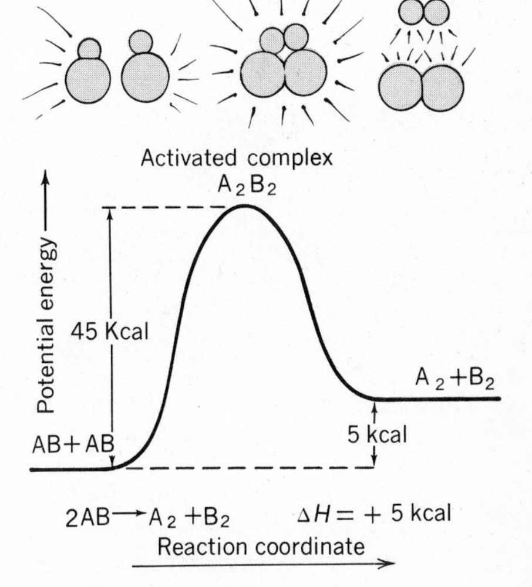

12-13 Potential energy changes occurring in a chemical system during an endothermic reaction.

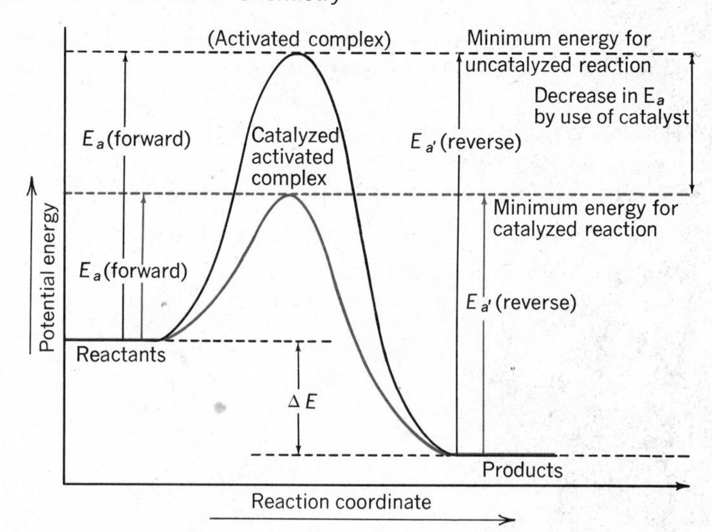

12-14 A catalyst lowers the energy barrier between the reactants and products. It is assumed that catalyst forms an intermediate activated complex with a lower activation energy than that formed without a catalyst.

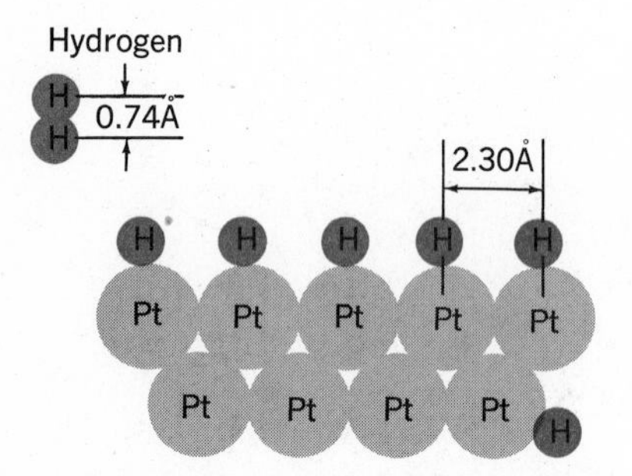

(a) Hydrogen adheres to surface. Bond between hydrogen atoms is greatly extended or broken.

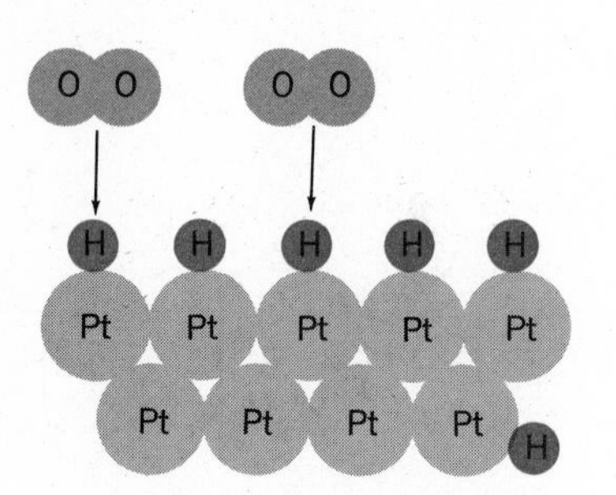

(b) Oxygen molecules approach surface containing adsorbed hydrogen atoms.

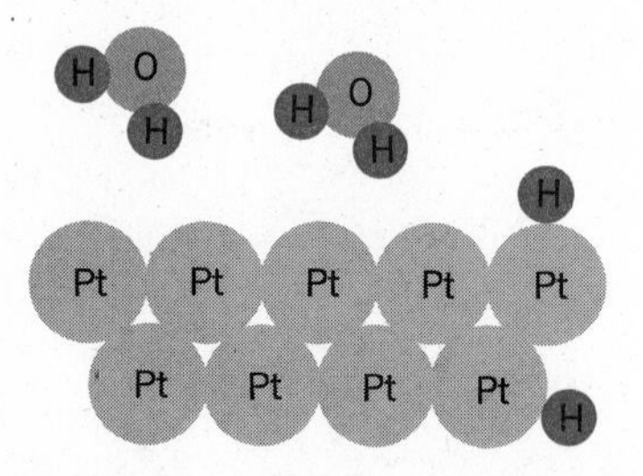

(c) As water molecules are formed the platinum surface is regenerated.

12-15 Platinum catalysis of the reaction between H_2 and O_2. Hydrogen is adsorbed on the surface of the platinum, and the covalent bond in the H_2 molecule is extended or broken. The reaction occurs more readily because the activation energy of a single H atom is less than that of H_2.

In general, catalysts provide a new reaction path in which a different, lower-energy, activated complex can form. They lower the activation energy and thereby increase the rate of the reaction. Fig. 12-14 shows the difference in activation energy between a catalyzed and noncatalyzed reaction. Addition of a catalyst is like lowering the basket in our basketball analogy. The drawing shows that both the forward and reverse reactions follow the same path. The activation energy is reduced to the same extent for both forward and reverse reactions.

12-15 Mechanisms of Catalytic Reactions The actual mechanisms of many catalytic reactions are not well understood. Solid catalysts have large surface areas and are capable of adsorbing the reactants on their surfaces. In some cases, one of the reactant molecules may readily react with the atoms of the catalyst and form an intermediate species which reacts readily with the second reagent and forms the desired product and regenerates the catalyst. When platinum is used to catalyze the reaction between H_2 and O_2, it is possible that hydrogen molecules are absorbed on the platinum. Oxygen reacts with the adsorbed hydrogen, forms water, and regenerates metallic platinum. Fig. 12-15 illustrates a reaction.

Many reactions are catalyzed by the presence of an acid. Here it is possible that hydrogen ions from the acid react with and modify the structure of one of the reactants so that it is more susceptible to reaction with the other reagent. This is believed to be the case in the acid-catalyzed decomposition of formic acid. The overall equation for the decomposition of formic acid is

$$\mathrm{HCOOH} \xrightleftharpoons{H^+} \mathrm{CO} + \mathrm{H_2O} \qquad 12\text{-}4$$

The steps involved in the formation of the intermediate and the regeneration of the H^+ ion catalyst are

$$\mathrm{HCOOH} + \mathrm{H^+} \rightleftharpoons \mathrm{HCOOH_2^+} \qquad 12\text{-}5$$

$$HCOOH_2^+ \rightleftharpoons HCO^+ + H_2O \qquad 12\text{-}6$$

$$HCO^+ \rightleftharpoons H^+ + CO \qquad 12\text{-}7$$

Reactions 12-5, 12-6, and 12-7 constitute the *mechanism* of Reaction 12-4. That is, they are believed to be the *elementary processes* involved in the overall reaction (12-4). An ***elementary process*** is a one-step process in which the product particles are in most cases the direct result of the collision of only two reactant particles. Three or more particle collisions are known but are rare. Elementary processes cannot be observed directly, but must be deduced from macroscopic observations made during studies of reaction rates.

REACTION MECHANISMS

As a result of rate studies, we can predict that gaseous reactions whose net (overall) equations show more than two or three reactant molecules do not represent elementary processes. This is logical since we would expect the simultaneous collision of more than two or three gaseous molecules to be a highly improbable event.

12-16 Rate-determining Step When a reaction is the result of a series of elementary processes, the *rate of the overall reaction is determined by the slowest reaction in the sequence.* The situation is rather analogous to an automobile assembly line. The rate at which completed automobiles roll off the assembly line depends on the rates at which several component parts are produced. If frames are produced at the rate of 100 per hour, bodies at 100 per hour, and engines at 95 per hour, the overall rate of production of completed automobiles is 95 per hour, the rate of the *slowest* step in the assembly line.

In general, overall reactions are rapid if the equations for them show that the simultaneous collision of two *simple* oppositely charged ions is all that is needed for a reaction to occur. Consider the following net ionic equation for the reaction between silver ions and chloride ions.

$$Ag^+(aq) + Cl^-(aq) \longrightarrow AgCl(s) \qquad 12\text{-}8$$

This equation shows that a two-particle collision between Ag^+ ions and the Cl^- ions could result in a reaction. Assuming that the reacting particles are uncombined ions, there are no strong bonds to be broken before the reaction can take place. We would predict, therefore, that the rate of reaction would be rapid. As you have no doubt observed, rapid rates are characteristic of precipitation reactions.

Rate studies can be used to identify and eliminate an incorrect mechanism, but cannot prove the correctness of one. For example, rate studies show that the reaction mechanism for the overall reaction

$$2NO(g) + 2H_2(g) \longrightarrow N_2(g) + 2H_2O(g) \qquad 12\text{-}9$$

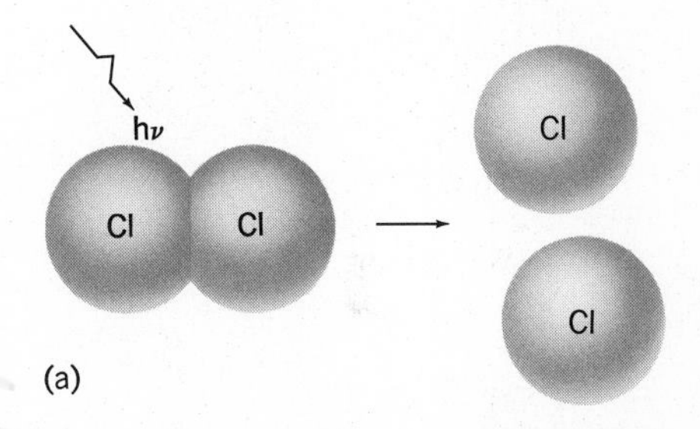

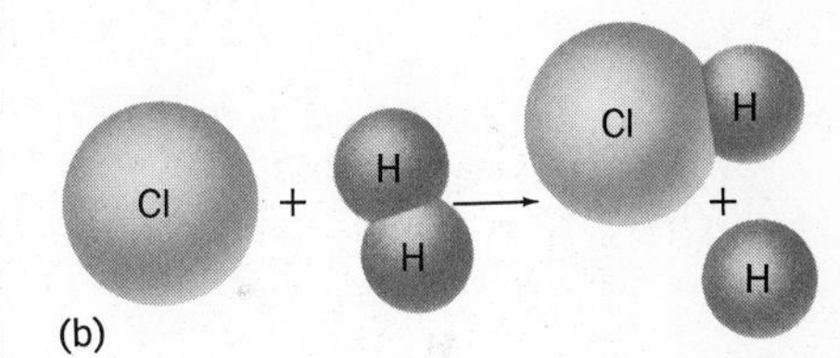

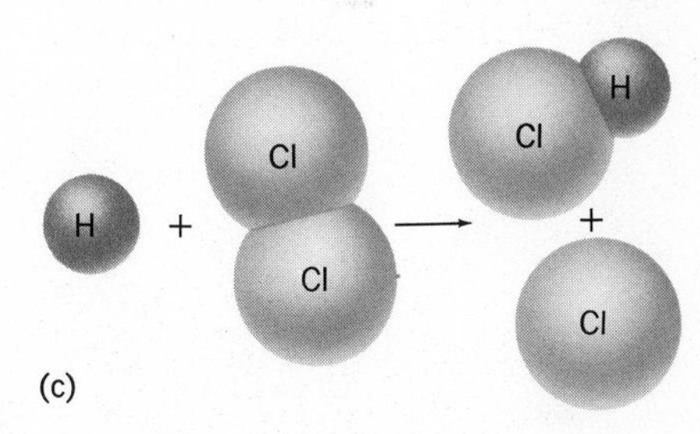

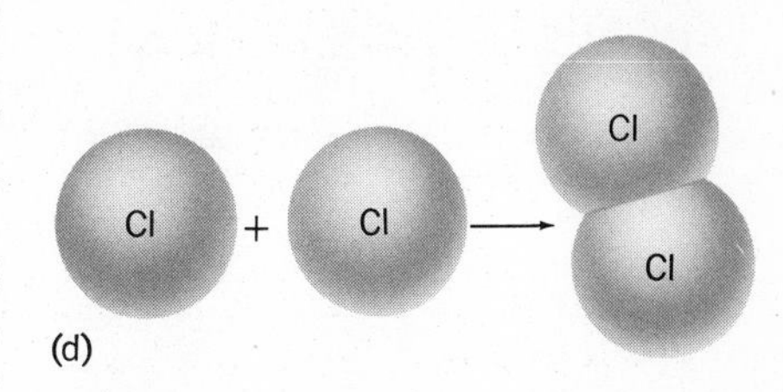

12-16 Reaction mechanism for the formation of HCl.

may involve a three-particle collision as an elementary process (***termolecular elementary process***) but not a four-particle collision. This indicates that Equation 12-9 cannot represent an elementary process. In other words, collisions between NO and H_2 molecules do not result directly in the formation of N_2 and H_2O molecules. We might guess that a termolecular elementary process such as

$$2NO + H_2 \longrightarrow N_2 + H_2O_2 \qquad 12\text{-}10$$

might be the rate-determining step in the reaction mechanism. A second faster step might be

$$H_2 + H_2O_2 \longrightarrow 2H_2O \qquad 12\text{-}11$$

The sum of the two elementary processes (12-10 and 12-11) gives the net equation

$$2NO + 2H_2 \longrightarrow N_2 + 2H_2O$$

Alternatively, we could have proposed a simpler, more probable sequence, consisting of a sequence of bimolecular steps.

FOLLOW-UP PROBLEM

Consider the reaction

$$4HBr(g) + O_2(g) \longrightarrow 2H_2O(g) + 2Br_2(g)$$

(a) Does the above equation represent an elementary process? Explain. (b) Experiments show that a change in the concentration of HBr has the same effect on the rate of the reaction as an identical change in the concentration of O_2. Propose a reaction mechanism for the overall reaction. Assume that HOOBr and HOBr are intermediates. (c) Identify the rate-determining step.

12-17 Mechanism for the Reaction of H_2 with Cl_2 An example of a relatively simple reaction mechanism that has been extensively studied is the formation of HCl from hydrogen gas and chlorine gas. This photochemical reaction is an example of a ***chain mechanism.*** When a mixture of these gases is heated or exposed to ultraviolet light, a violent reaction takes place. The reaction, however, does not occur directly between H_2 and Cl_2 molecules. It has been shown that the first step in the reaction mechanism occurs when Cl_2 molecules absorb a quantum of ultraviolet light.

$$Cl_2 + h\nu \longrightarrow 2Cl$$

This reaction is called a ***chain-initiating*** step. The chlorine atoms generated in this step are highly reactive species known as ***free radicals*** which react with H_2 molecules and form HCl molecules and H atoms.

$$:\ddot{\underset{..}{Cl}}\cdot + H_2 \longrightarrow HCl + H\cdot$$

The reactive H atoms are also free radicals which react with Cl_2 molecules and produce another molecule of HCl and a Cl atom.

$$\cdot H + Cl_2 \longrightarrow HCl + :\ddot{\underset{..}{Cl}}\cdot$$

The preceding free radical reactions are ***chain-propagating steps*** of the mechanism. The regenerated chlorine atom continues the process so that HCl is formed continuously until a *chain-breaking* reaction occurs. A chain-breaking reaction such as

$$:\ddot{\underset{\cdot\cdot}{Cl}}\cdot + :\ddot{\underset{\cdot\cdot}{Cl}}\cdot \longrightarrow Cl_2$$

or

$$H\cdot + \cdot H \longrightarrow H_2$$

is one which removes the reactive free radicals from the system. *The reaction rate is very slow in a dark room at room temperature.* This is because the bond energy of Cl_2 is relatively high. Considerable energy (heat or ultraviolet light), therefore, is required to rupture the bond and bring about the formation of the active species needed to initiate the reaction.

The clue to what occurs on the path between reactants and products, as represented by the equation for the overall reaction, is found in the so-called rate laws. These laws, which are determined experimentally, relate the rate of a reaction to the concentration of the participants. The laws can be used to help tell whether the equation for an overall reaction represents an elementary process and to help deduce a reaction mechanism.

REACTION RATE LAW

12-18 Law of Mass Action The relationship between the rate of a reaction and the masses (expressed as concentrations) of reacting substances was recognized in 1864 by two Norwegian chemists, Cato M. Guldberg and Peter Waage. The relationship is summarized in their famous ***Law of Mass Action*** which states that *the rate of a chemical reaction is proportional to the product of the concentrations of the reactants.* For a general reaction between A and B represented by

$$aA + bB \longrightarrow \ldots \qquad 12\text{-}12$$

the rate law expression is

$$r \propto [A]^m[B]^n$$

or

$$r = k[A]^m[B]^n \qquad 12\text{-}13$$

where [A] and [B] represent the molar concentrations of *A* and *B*, *m* and *n* are the powers to which the concentrations must be raised, and *k* is a constant of proportionality known as the *rate constant.* Data show that the rate constant is not affected by concentration changes but does vary with the temperature changes. The only valid way to obtain *m* and *n* is to use *experimental data.* They are derived by determining the effect of changing reactant concentration on the rate of the reaction. The exponents, *m* and *n*, may be zero, a fraction, or an integer. The sum of the exponents $m + n$, on the concentration factors in the rate law equation (12-13), is called the

total reaction order. For example, the experimentally derived rate law expression for the reaction

$$H_2(g) + I_2(g) \longrightarrow 2HI(g) \qquad 12\text{-}14$$

is

$$r = k[H_2][I_2]$$

Thus, the reaction is second order overall. That is, the sum of the exponents is $1 + 1 = 2$. This does not prove, however, that Equation 12-14 represents a *bimolecular elementary process* and that the collision of one H_2 molecule with one I_2 molecule produces HI molecules directly. Let us see how the exponents m and n can be determined experimentally.

12-19 Experimental Determination of Reaction Order Consider the reaction between nitric oxide (NO) and hydrogen (H_2) forming water vapor (H_2O), and nitrogen (N_2).

$$2NO(g) + 2H_2(g) \longrightarrow N_2(g) + 2H_2O(g) \qquad 12\text{-}15$$

As this reaction proceeds, four moles of reactant gas form three moles of product gas and the pressure drops. The rate of the reaction can be determined by following the pressure change.

For convenience and clarity, we are representing the rate in terms of moles/ℓ per unit time. To determine the relationship between the rate of the reaction and the concentration of NO, it is necessary to keep the concentration of H_2 and the temperature constant and to vary the concentration of NO. The initial rate of reaction may then be determined using various concentrations of NO. The concentration of NO and the temperature are then held constant, and the initial rates of reaction for various concentrations of H_2 are determined. The data in Table 12-1 show that increasing the concentration of either of the reactants increases the rate of the reaction.

TABLE 12-1
RATES OF REACTION BETWEEN NO AND H_2 AT 800°C

Experiment Number	NO (moles/ℓ)	H_2 (moles/ℓ)	Initial Rate of Reaction (moles/ℓ sec)
1	0.001	0.004	0.002
2	0.002	0.004	0.008
3	0.003	0.004	0.018
4	0.004	0.001	0.008
5	0.004	0.002	0.016
6	0.004	0.003	0.024

Examination of the data reveals that when the concentration of NO is held constant at 0.004 *M* and the concentration of hydrogen is doubled, the rate doubles from 8 to 16 units. When the concentration of hydrogen is tripled, the rate triples from 8 to 24 units. Mathematically, the rate of the reaction is directly proportional to the concentration of hydrogen. That is,

$$r \propto [H_2]$$

When the concentration of H_2 is held constant and that of NO is varied, the data show that the rate of the reaction is proportional to the square of the concentration of NO. Specifically, when the concentration of NO is doubled from 0.001 to 0.002 *M*, the rate is quadrupled from 2 to 8 rate units. When the concentration of NO is tripled from 0.001 *M* to 0.003 *M*, the rate increases by a factor of 9. Mathematically, this rate is proportional to the square of the concentration of NO,

$$r \propto [NO]^2$$

CATO MAXIMILIAN GULDBERG
1836–1902

PETER WAAGE
1833–1900

The accomplishments of Guldberg and Waage, as in the case of many scientists, required the ground work of other men who preceded them. Guldberg, a professor of applied mathematics at the University of Christiana (now Oslo) in Norway, and his brother-in-law, Peter Waage, had heard much about the work of Professor Berthelot of Paris. Berthelot had been doing some studies on the decomposition rates of esters (organic compounds) and had come to the conclusion that esters decomposed at a rate which depended upon the *amount* of remaining undecomposed ester.

Guldberg and Waage extended the work of Berthelot and interpreted the rate of a reaction in terms of *concentrations* rather than amounts. A summary of their work is expressed in the *Law of Mass Action* which states that the rate of a chemical reaction is proportional to the concentration of the reacting species. Their pamphlet, first published in Norwegian in 1863, was translated into French in 1867. They described their work by saying, "Investigations in this field are doubtless more difficult, more tedious, and less fruitful than those which now enjoy the attention of most chemists, namely the discovery of new compounds," but investigations such as these are necessary to bring chemistry "into the class with the truly exact sciences."

Since the rate is proportional to $[H_2]$ and $[NO]^2$, it is proportional to their product.

$$\mathbf{r} \propto [\mathbf{H_2}][\mathbf{NO}]^2$$

Mathematically, a proportionality can be changed to an equality by inserting a constant of proportionality.

$$\mathbf{r} = \mathbf{k}[\mathbf{H_2}][\mathbf{NO}]^2 \qquad 12\text{-}16$$

k is the ***specific rate constant*** for the above reaction at the temperature of the reaction. Equation 12-16 is the rate law expression for reaction 12-15. In Equation 12-16 the value of the exponents are $m = 1, n = 2$. Thus, the reaction represented by equation 12-15 is a *third order reaction.*

Equation 12-15 does not represent an elementary process. That is, the sum of the coefficients of the reactants (H_2 and NO) in Equation 12-15 is four while the sum of the exponents in Equation 12-16 is three. One of many possible mechanisms for this reaction was proposed on p. 392.

When the exponents, as determined experimentally, do not agree with the coefficients in the net equation, then the net equation must represent the sum of a series of elementary processes. This does not imply, however, that when they do agree the net equation represents an elementary process. Detailed studies of reaction rates at different temperatures often provide clues which help scientists to determine the nature of elementary processes.

In recent years, progress has been made in identifying many reaction mechanisms. The importance of kinetics cannot be overemphasized. It provides a description of what happens on the microparticle level during a reaction. Analyses and interpretation of the data help scientists find solutions to problems such as atmospheric pollution by smog, to understand reactions of biological systems, and to develop more efficient processes for producing desired products.

LOOKING AHEAD

We are now in the midst of our study of reaction principles. In Chapter 11 we considered the energy associated with reactions and in this chapter we investigated factors related to the rates of reactions. Because the equilibrium condition can be described on a molecular level in terms of the rates of opposing reactions, it is now logical to make a detailed study of molecular equilibrium. In this study we shall acquire the tools (principles and equations) that will enable us to predict the extent of a reaction. Most of the reactions we have studied in earlier chapters (precipitation reactions) are quantitative (essentially complete). In a large number of reactions, however, the reactants are not quantitatively converted into products. We shall focus our attention on these so-called "incomplete" reactions in the next chapter.

As you can imagine, the chemist is concerned with the efficiency of a reaction and with methods of controlling it so as to obtain a maximum yield of desired product. We shall devote the next chapter to developing equilibrium principles which will help us to determine the amount of product we can expect to obtain from a reaction, and to determine how changing various factors will affect the yield of product.

Special Feature: Atmospheric Pollution (Photochemical Smog)

Those who live in or have visited a number of large, highly industrialized population centers are familiar with the eye-smarting, cough-inducing, nauseating, suffocating, photochemical smog that plagues these areas many days each year. Rate studies have proved a valuable aid in deducing the mechanism of the reactions which give rise to the components of this environmental pollutant. It should be of interest as well as instructive to discover what scientists have learned about the components and nature of photochemical smog and atmospheric pollution.

and produce noxious products that irritate human membranes and destroy vegetation. Methods must be found to reduce hydrocarbon emission from automobiles and from other sources. This would help break the smog-producing chain of reactions. One approach involves developing catalysts which promote the decomposition of NO into N_2 and O_2 before the gas leaves the automobile.

Another atmospheric pollutant that causes some concern is the lead compounds produced during the combustion in automobile engines of gasoline containing tetraethyl lead [$(C_2H_5)_4Pb$], an antiknock compound. This pollutant is particularly hazardous since it is a cumulative poison. Approximately 40 percent of inhaled lead is absorbed and stored by the body, particularly in the aorta and in the bones. The absorption rate is much greater for inhaled lead than for that ingested with food. Some of the lead found in foods is believed to originate in the exhaust gases of automobiles. One solution to this problem is to find other antiknock substances that do not yield toxic or harmful products.

It is apparent that man is probably having a profound and unpredictable effect upon his own environment by polluting the atmosphere. Upsetting the balances established by nature over millions of years can lead to changes in climate and alteration of ecological balances of nature. The technology which has helped to pollute our environment and cause problems can and must find solutions to the problems. In some cases the technology to solve some pollution problems is already available. The implementation of these methods, however, requires time and money. As a citizen, your vote will help to establish regulatory laws as well as to appropriate funds which will help solve our pollution problems. It is important, therefore, that you have some knowledge and appreciation of the nature, causes, and possible solutions of environmental problems.

QUESTIONS

1. What is the meaning of *rate* as applied to a chemical reaction?

2. Write three simple rate expressions for the reaction

$$HgO(s) \longrightarrow Hg(\ell) + \tfrac{1}{2}O_2(g)$$

3. Explain why the rate of a simple chemical reaction such as

$$NO(g) + \tfrac{1}{2}O_2(g) \longrightarrow NO_2(g)$$

is likely to be most rapid at the beginning of the reaction.

4. (a) List four factors which affect the rate of a homogeneous reaction. (b) What additional factors must be considered for heterogeneous reactions?

5. Which of these reactions are likely to proceed rapidly once they have begun? Explain.
(a) $H_2(g) + Cl_2(g) \longrightarrow 2HCl(g)$
(b) $C_6H_{12}O_6(aq) + 6O_2(g) \longrightarrow 6CO_2(g) + 6H_2O(l)$
(c) $Cu^{2+}(aq) + S^{2-}(aq) \longrightarrow CuS(s)$
(d) A lump of iron in melted sulfur
(e) Powdered iron in melted sulfur

Ans. (a) Probable because reactants are gaseous, (d) not likely since reaction is limited to surface of iron.

6. What property might be measured in order to follow the rate of reaction for each of these reactions?
(a) $KI(aq) + \frac{1}{2}Cl_2(aq) \longrightarrow \frac{1}{2}I_2(s) + KCl(aq)$
(b) $PCl_5(g) \longrightarrow PCl_3(g) + Cl_2(g)$
(c) $2NO_2(g) \longrightarrow N_2O_4(g)$
(d) $Cu(s) + 2Ag^+(aq) \longrightarrow Cu^{2+}(aq) + 2Ag(s)$
(e) $Fe(s) + 2HCl(aq) \longrightarrow H_2(g) + FeCl_2(aq)$
Ans. (b) Pressure.

7. Use the Kinetic Molecular Theory to explain why reaction rates vary directly with temperature.

8. Explain why a mixture of hydrogen and chlorine gas is stable if kept in the dark, yet is likely to explode if exposed to sunlight.

9. If you wish to dissolve a lump of sugar in water, how might you increase the rate at which the sugar dissolves? Explain why each procedure you use is effective.

10. A mixture of natural gas and air does not react appreciably at room temperature. When a piece of platinum is inserted into the reaction vessel, the mixture explodes. Explain.

11. A fresh starch suspension shows no positive test for sugar. If boiled for a long time, a small quantity of sugar is detected. If boiled with acid, additional sugar is present. If saliva is added to the original mixture, even more sugar is formed in the same time. Explain.

12. What is meant by an elementary process?

13. What is a bimolecular process?

14. What is meant by the *mechanism of a reaction?*

15. (a) Describe the "reaction mechanism" in the "reaction" which gets the dinner dishes cleaned and into the cupboard. (b) Which is the rate-determining step? (c) Can the "reaction" be catalyzed? (d) How else can the "reaction" be speeded up?

16. Using collision theory, explain (a) the rapidity with which gas reactions usually occur, (b) the speed of ionic reactions, (c) the temperature dependence of the rate of chemical reactions.

17. Lavoisier was able to form the oxide of mercury by moderately heating liquid mercury in air. Upon heating the same oxide more strongly, he was able to decompose the oxide and re-form liquid mercury and oxygen. If the rate of reaction is increased by an increase in temperature, why wasn't the oxide of mercury formed more rapidly?

18. Distinguish between reaction rate and reaction-rate constant.

19. What is meant by activation energy?

20. Distinguish between activation energy and activated complex.

21. Compound A reacts with compound B and forms the products C and D. This reaction proceeds very slowly at first, then accelerates rapidly until virtually all A and B have been consumed. (a) Suggest a possible reason for this. (b) A highly unstable species ABCD is formed as an intermediate. What is ABCD called?
Ans. (a) One of the products C or D could be acting as a catalyst for this reaction. Thus, the more C and D produced, the more rapidly the reaction takes place. Such reactions are said to be autocatalytic. (b) An activated complex.

22. (a) Compare the kinetic energies of the reactants, activated complex, and products of a reaction when all are at the same temperature. (b) How do the potential energies compare?

23. Draw a representative reaction curve for an exothermic reaction. (a) Label the activation energy for the forward reaction, the activation energy for the reverse reaction, the enthalpy of the forward and reverse reactions. (b) Do the same for a representative endothermic reaction. (c) How do both curves change upon the addition of a catalyst?

24. On the reaction curve (Question 23) for an exothermic reaction, indicate the regions (a) where the bond-breaking energies predominate, (b) where the bond-forming energies predominate.

25. What is the reason the activation energy for a reaction is usually less than the bond energy of the reactants.

26. What would be the effect of an increase in temperature on (a) the rate constant of the forward reaction, (b) the rate constant of the reverse reaction?

27. How does a catalyst affect the enthalpy change of a reaction?

28. The mechanism of the reaction between nitrogen (II) oxide, (NO), and oxygen (O_2) is believed to take place in two steps

$$NO + O_2 \longrightarrow NO \cdot O_2 \text{(rapid)}$$
$$NO \cdot O_2 + NO \longrightarrow 2NO_2 \text{(slow)}$$

The exact nature of the bond between the NO and O_2 in $NO \cdot O_2$ is not understood. (a) Which of the steps above is the rate-determining step?

29. Suggest a mechanism for the decomposition of ozone (O_3) into oxygen (O_2). This reaction takes place in two steps.

Ans. $O_3 \longrightarrow O_2 + O$ **(rapid),** $O + O_3 \longrightarrow 2O_2$ **(slow).**

30. Propose a mechanism for each of these catalyzed reactions?

(a) $SO_2(g) + \frac{1}{2}O_2(g) \xrightarrow{\text{Pt black}} SO_3(g)$

(b) $H_2O_2(aq) \xrightarrow{\text{hemoglobin}} H_2O(l) + \frac{1}{2}O_2(g)$

31. Explain why an exothermic reaction once started is self-sustaining.

32. Explain why an endothermic reaction once started requires a continual supply of energy.

33. (a) What is meant by the order of a reaction? (b) Can the order be determined from the equation for the no overall reaction? (c) If the sum of the coefficients of the reactants in the equation equals the total order of a reaction, can it be assumed that the equation represents an elementary process? no

34. What is the order of these reactions with these rate-law expressions? (a) $r = k[A]^{1/2}[B]$, (b) $r = k[A][B]^2$.

a) 3/2 b) 3

13

Chemical Equilibrium

Preview

"The mind is but a barren soil—a soil which is soon exhausted, and will produce no crop, or only one, unless it be continually fertilized and enriched with foreign matter."

—Sir Joshua Reynolds (1723–1792)

Almost all physical and chemical changes which you encounter are reversible processes. Reversing a process converts some of the products back into reactants. Let us cite a few familiar examples. Water changes into ice when the temperature is lowered; increasing the temperature reverses the process. Chemicals which react and produce electricity in an automobile battery are consumed when the battery is furnishing current. When the battery is recharged, the process is reversed and the original chemicals are regenerated. Hemoglobin in blood cells combines with oxygen in the lungs and forms oxyhemoglobin. The oxyhemoglobin then releases oxygen for use in metabolic processes in different parts of the body.

In Chapter 10 we found that in a closed system at a given temperature, a phase equilibrium is established between liquid water and water vapor. Increasing the temperature increases the relative number of vapor molecules. Chemical changes (reactions) are rather similar to phase changes in this respect. Many of them reach an equilibrium state long before much product is formed. Fortunately, these reactions can be made to proceed in either direction by changing temperature, pressure, and other factors which affect the process. In other words, it is possible to control the direction and extent of reactions. Controlling reversible reactions is valuable to chemical manufacturers who produce drugs, plastics, or any one of thousands of products. Relevant factors are adjusted to achieve a maximum yield of desired products. In these processes, attempts are made to minimize the reverse reaction in which the product reverts to the reactants.

Application of the principles of *chemical equilibrium* allows scientists to control the direction and extent of a reaction. In this chapter, we shall first consider the qualitative aspects of equilibrium. This will help us to identify equilibrium systems. We shall then develop the *equilibrium law* which can be used to determine the extent of a reaction and to calculate the quantity of reactant and product which are present together at equilibrium.

Finally, we shall briefly examine the driving forces that are responsible for a reaction system's tendency to reach spontaneously the equilibrium state in which there is no net particle or energy flow.

CHARACTERISTICS OF THE EQUILIBRIUM STATE

13-1 Constancy of Observable Macroscopic Properties In our study of phase equilibria we found that an equilibrium can exist only in closed systems and that, when equilibrium is established, observable macroscopic properties of the system are constant. Let us see if there is a parallel in closed systems involving chemical reactions. We shall use the reaction which occurs between $H_2(g)$ and $I_2(g)$ at 448°C for our study. The equation for the reaction is

$$H_2(g) + I_2(g) \longrightarrow 2HI(g) \qquad 13\text{-}1$$

We can follow the progress of the reaction by noting the color of the system. All components of the system are colorless except violet iodine vapor (g). Thus, as the reaction proceeds, the intensity of the color diminishes because the concentration of I_2 decreases.

Let us place 1.00 mole each of H_2 and I_2 in a one-liter container and allow them to react at 448°C until no further changes in color or other observable properties are seen. How many moles of HI

13-1 (a) How do the number of moles of participants in the equilibrium system below compare with those in the nonequilibrium system? (b) Does this relationship between total moles of participants in the original and equilibrium state hold for all systems? (c) What is the percentage yield in the reaction depicted below?

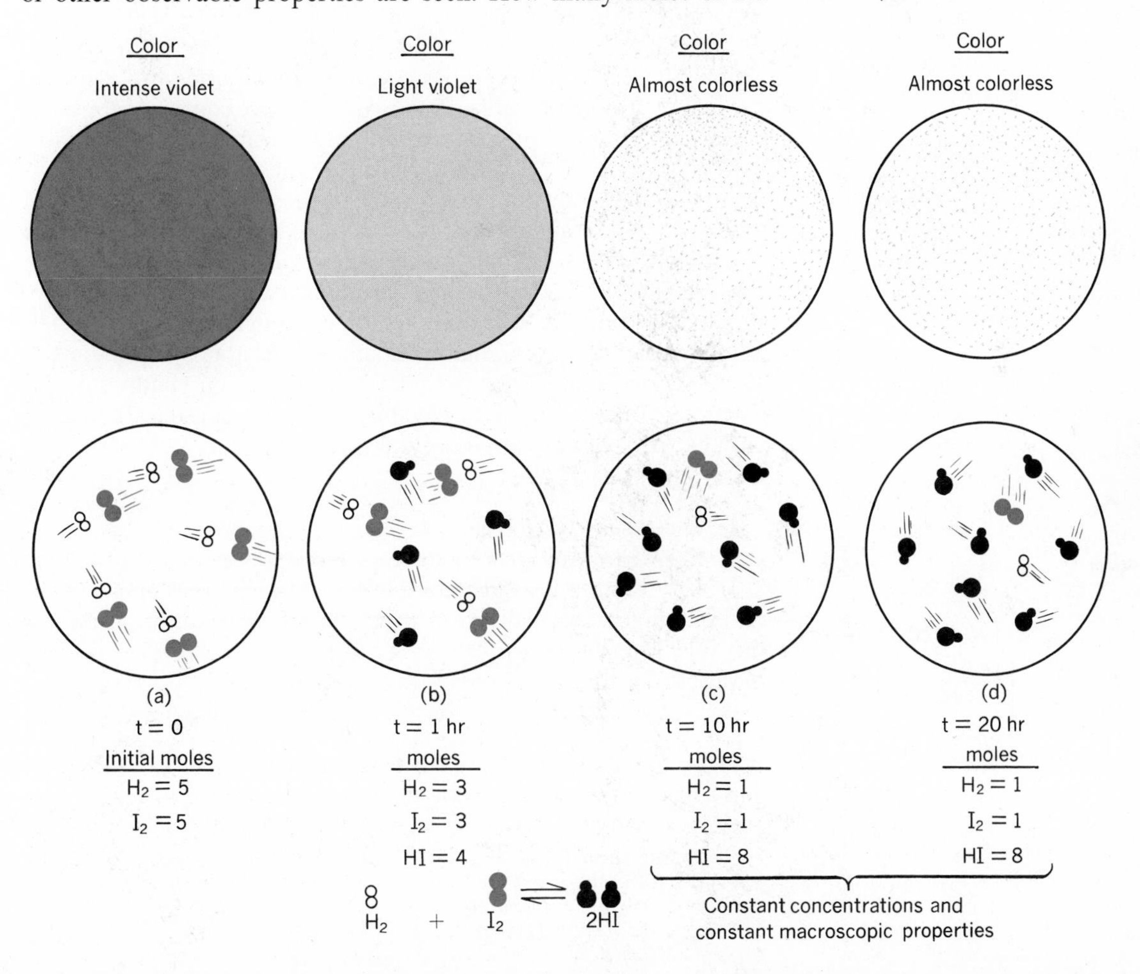

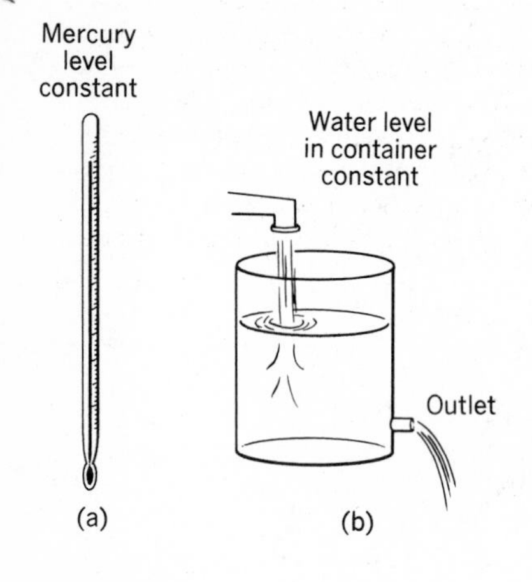

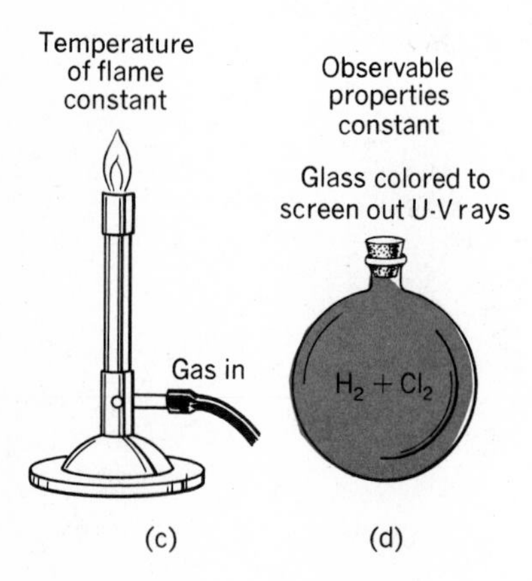

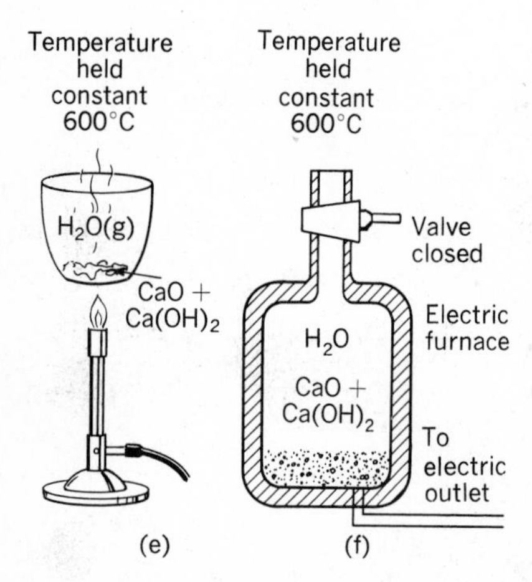

13-2 Which of the systems above are either at equilibrium or will reach an equilibrium state under the conditions noted? Refer to Section 12-9 for information related to (d).

would be present? Based on the mole relationship in the equation, you might assume that two moles of HI would be present. The answer, two moles, however, is incorrect because it does not take into account the fact that the system reaches a state of equilibrium in which the conversion of reactants to products is incomplete. When equilibrium is attained, analysis of the mixture shows that there are 0.220 moles each of unreacted H_2 and I_2 and 1.56 moles of HI present. Observation shows that this ratio of product to reactants remains constant at constant temperature regardless of how long the mixture is allowed to react. We may summarize this observation thus: *observable properties and concentrations of all participants become constant when a chemical system reaches a state of equilibrium.*

Just because the properties of a closed system appear to be constant does not necessarily mean that the system is at equilibrium. For example, a flask containing a mixture of H_2 and O_2 has constant properties but provides no evidence that it is at equilibrium. The violent explosion that occurs when the mixture is sparked certainly demonstrates that the system was not originally at equilibrium. It may be helpful, then, to distinguish between systems which have constant properties because two opposing processes are occurring at the same rate and those which have constant properties because the rate of any reaction among the components is too slow to be observable. The constancy-of-properties criterion for an equilibrium system can be applied only to those in which the rate at which possible changes occur is observable. For example, when the $H_2 - I_2 -$ HI equilibrium described previously is subjected to a slight increase in temperature, the intensity of the color changes slightly, but rapidly. When the system is brought back to the original temperature, the original intensity returns almost immediately. These observations indicate that, under the given conditions, the reaction is quite rapid and reversible. Thus, systems at equilibrium may be identified by applying these criteria

1. The system is closed.
2. The observable macroscopic properties are constant.
3. The reaction is sufficiently reversible so that observable properties change and then return to the original condition at an observable rate when a factor that affects the rate of the reaction is varied and then restored to its original value.

13-2 Spontaneous Approach to Equilibrium from Either Direction In contrast to the method used in Section 13-1, let us now place two moles of HI(g) in the same container, maintained at 448°C, and then analyze the mixture as before. The equation for the decomposition of HI(g) is

$$2HI(g) \longrightarrow H_2(g) + I_2(g) \qquad 13\text{-}2$$

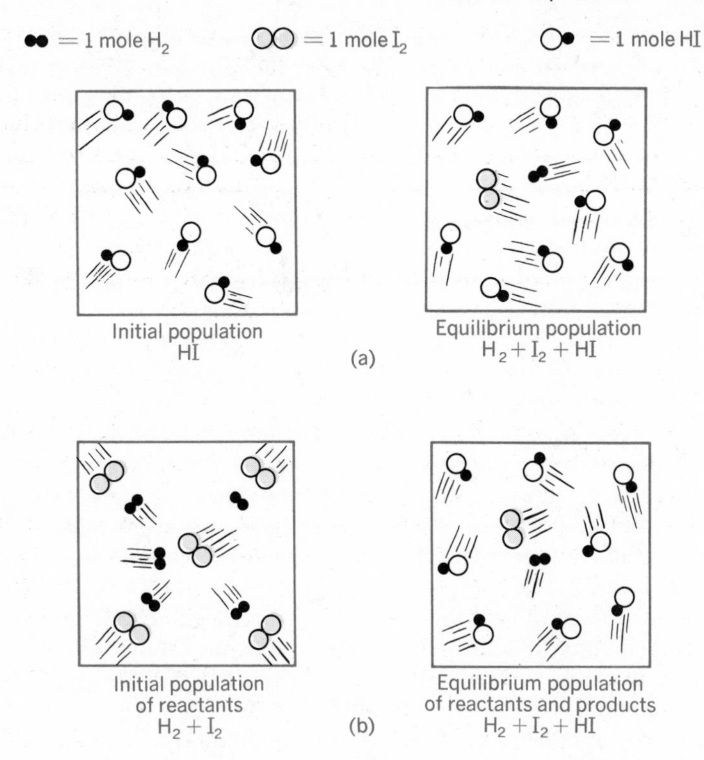

13-3 For a reversible reaction such as $H_2(g) + I_2(g) \rightleftharpoons 2HI(g)$ which takes place in a closed system at a given temperature, equilibrium can be approached from either direction. In (a) 10 moles of HI decompose and yield an equilibrium mixture containing 8 moles of HI, 1 mole of H_2, and 1 mole of I_2. In (b) 5 moles of H_2 and 5 moles of I_2 react and form the same equilibrium mixture.

As you may have guessed, the resulting concentrations at equilibrium are the same as those obtained when we started with one mole of $H_2(g)$ and with one mole of $I_2(g)$. The two reactions illustrate two characteristics of all reversible reactions in closed systems; they all *spontaneously tend to approach an equilibrium, and equilibrium can be approached from either direction.*

Equations 13-1 and 13-2 represent opposing reactions and can be represented by a single equation using double arrows to indicate the forward and reverse reactions. At equilibrium the reaction is reversible and is represented by

$$H_2(g) + I_2(g) \rightleftharpoons 2HI(g) \qquad 13\text{-}3$$

13-3 The Dynamic Nature of an Equilibrium System The constancy of observable properties in an equilibrium system does not imply that the reaction on the microscopic level has ceased. In our example, it can be shown that there is a continuous formation of HI after equilibrium has been established. The technique consists of substituting radioactive iodine, which can be traced, for nonradioactive iodine. Radioactive isotopes have the same chemical properties as their nonradioactive counterparts. Substitution can be accomplished without disturbing the equilibrium. Subsequent analysis reveals that HI in the equilibrium mixture contains radioactive iodine. In another equilibrium mixture, HI containing radio-

Addition of a *small* quantity of radioactive iodine to a system already at equilibrium does not affect significantly the equilibrium position.

active iodine can be substituted for some of the nonradioactive HI in the mixture. In this instance, some of the radioactive iodine from the HI becomes free I_2. This illustrates the dynamic nature of chemical equilibria *and leads us to conclude that the properties and concentration of an equilibrium system are constant because the rates of the forward and reverse reactions are equal* so that it *appears* that no reactions are occurring. Let us investigate the changes taking place as the reaction proceeds.

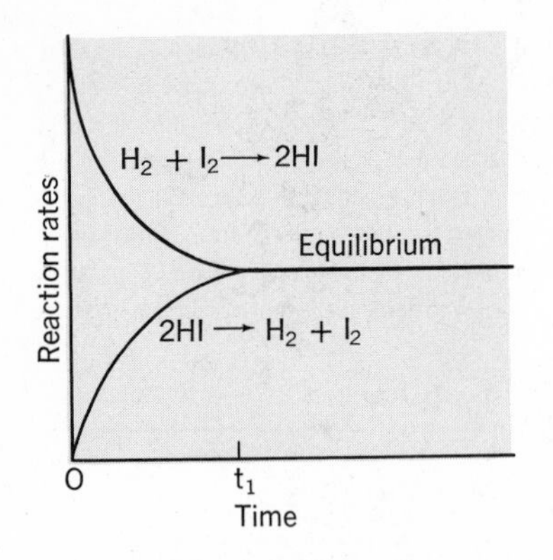

13-4 Reaction rates expressed as a function of time for the forward and reverse reactions occurring in the system $H_2 + I_2 \rightleftharpoons 2HI$. Equilibrium is established at time t_1.

As the reaction proceeds, the concentrations of all reactants change less rapidly than they did initially. (The decreasing concentration of H_2 and I_2, however, results in a decrease in the rate of the forward reaction, while the increasing concentration of HI results in an increase in the rate of the reverse reaction.) Finally, the two rates become equal and equilibrium is established. At this point the concentrations of all participants remain constant; that is, there is no net reaction. The reactants H_2 and I_2 are reacting and forming the product HI at exactly the same rate as HI is decomposing and forming H_2 and I_2. This is an example of ***dynamic equilibrium.*** The changing rates of the forward and reverse reactions are shown in graphical form in Fig. 13-4. It can be seen from the graph that neither rate becomes zero at equilibrium. They both reach the same value determined by the nature of the reactants and the conditions of the experiment.

13-4 Reversibility of Reactions at Equilibrium Once a system has reached equilibrium, any factor which causes a change in the rate of either the forward or the reverse reaction disturbs the equilibrium. The participants then no longer are in equilibrium with each other. *The system readjusts, however, so that the rates again*

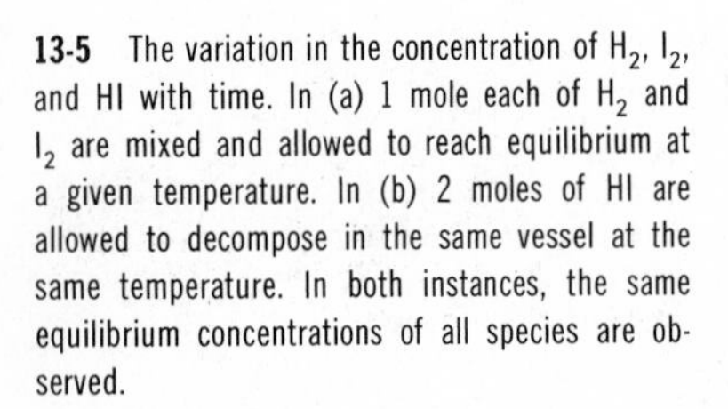

13-5 The variation in the concentration of H_2, I_2, and HI with time. In (a) 1 mole each of H_2 and I_2 are mixed and allowed to reach equilibrium at a given temperature. In (b) 2 moles of HI are allowed to decompose in the same vessel at the same temperature. In both instances, the same equilibrium concentrations of all species are observed.

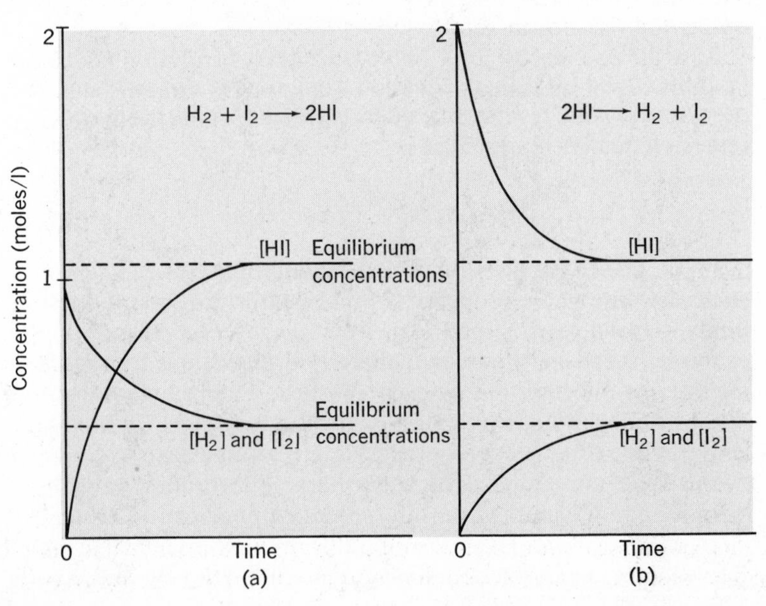

become equal, and a new equilibrium is established. For example, after an equilibrium has been established between H_2, I_2, and HI, let us increase the concentration of H_2. According to the collision theory, increasing the concentration of H_2 increases the forward rate, r_f of the reversible reaction

$$H_2(g) + I_2(g) \underset{r_r}{\overset{r_f}{\rightleftharpoons}} 2HI(g)$$

Thus, because the value of r_f temporarily exceeds r_r, the concentration of HI increases. As soon as the concentration of HI increases, r_r will also start to increase and continue to do so until $r_f = r_r$ and a new equilibrium is established. In this new equilibrium state the rates differ from their former values and the concentrations of HI and H_2 are greater than in the former state. However, because a small quantity of I_2 reacted with some of the added H_2, the concentration of I_2 is less in the new state than in the original. Conversely, the addition of HI to the equilibrium system reverses the reaction by increasing the reverse rate, r_r, and increasing the concentration of H_2 and I_2.

LE CHÂTELIER'S PRINCIPLE

13-5 Effect of Changing Concentration The principle underlying the behavior of the system described in the preceding paragraph was stated in 1888 by the French chemist, Henri Le Châtelier. Le Châtelier's principle states that *when a stress is applied to a system at equilibrium, the system readjusts so as to relieve or offset the stress.* The term *stress* applies to any imposed factor which upsets the balance in rates of the forward and reverse reactions. Such a factor must affect a *change in the concentration* of one or more of the participants involved in a chemical equilibrium. The three factors which may be varied in order to change the relative rates and disturb a gaseous system at equilibrium are changes in *concentration, temperature,* or *total pressure* brought about by a *volume change.*

In our example above, the stress was imposed by adding H_2 to the equilibrium mixture. The effect was offset to some extent when the system reacted and consumed some of the H_2 by forming more product. In chemical language we say that the equilibrium position is displaced or *shifted* toward the product side or to the right. The same shift toward the product side would occur if the concentration of a product were reduced. Increasing the concentration of a reactant or decreasing the concentration of a product causes r_f to exceed r_r temporarily. This has the immediate effect of increasing the concentration of the product molecules. The subsequent effect of this, of course, is to increase r_r because the formation of HI molecules results in more collisions between HI molecules. Additional

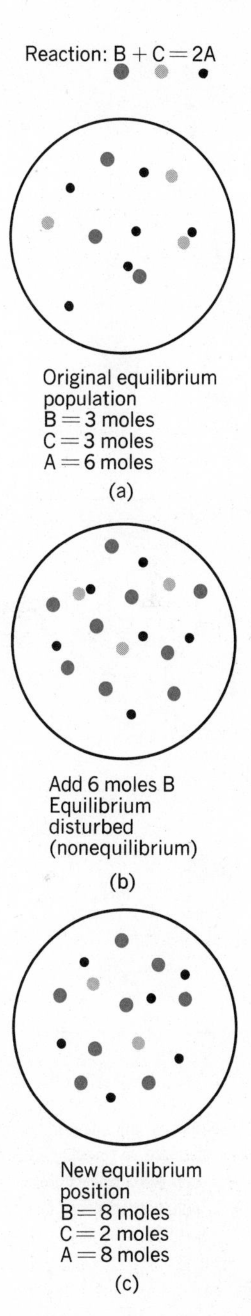

13-6 In order to obtain an additional 2 moles of product (A), it was necessary to add 6 moles of B to the original equilibrium mixture. Explain the relative number of moles of each participant in diagram (c). After you have read Section 13-15 you should be able to show that the mathematical relationship between the moles of participants in (c) is the same as that in (a).

collisions between HI molecules are thus reflected in an increase in the reverse rate, r_r. Ultimately, however, a new balance between r_f and r_r results.

13-6 Change in Total Volume (or Related Pressure) of an Equilibrium System A second stress that may be applied to an equilibrium system involving gases is a change in the volume of the system. To illustrate the effect of the change we shall use the Haber reaction for the formation of ammonia. The thermochemical equation for this reaction is

$$N_2(g) + 3H_2(g) \rightleftharpoons 2NH_3(g) + 22 \text{ kcal} \qquad 13\text{-}4$$

The stress imposed by the decrease in volume is actually a stress caused by the increase in concentration or in the total number of molecules per unit volume. The stress is relieved when the system reacts and reduces the number of molecules. The equation shows that a decrease in the total number of molecules occurs every time four molecules of reactant form two molecules of product. Thus, decreasing the volume results in an increased yield of NH_3 as the equilibrium shifts to the right.

Alternatively, we could interpret the shift caused by the volume decrease in terms of the pressure change that accompanies the decrease in volume. That is, the increase in pressure associated with a decrease in volume causes a shift that reduces the number of molecules and thus minimizes the effect of the change. Thus, we can say that a decrease in the total volume of a gaseous system *(or the accompanying increase in pressure) shifts an equilibrium in the direction of the fewer molecules as shown by the equation for the reaction.* In other words, to predict the effect of a volume or related pressure change, you count the number of product and reactant molecules (or moles) separately as shown by the equation for the reaction. If the number of product molecules (or moles) equals the number of reactant molecules (or moles) as in Equation 13-3, then a change in total volume or accompanying pressure does not affect the equilibrium.

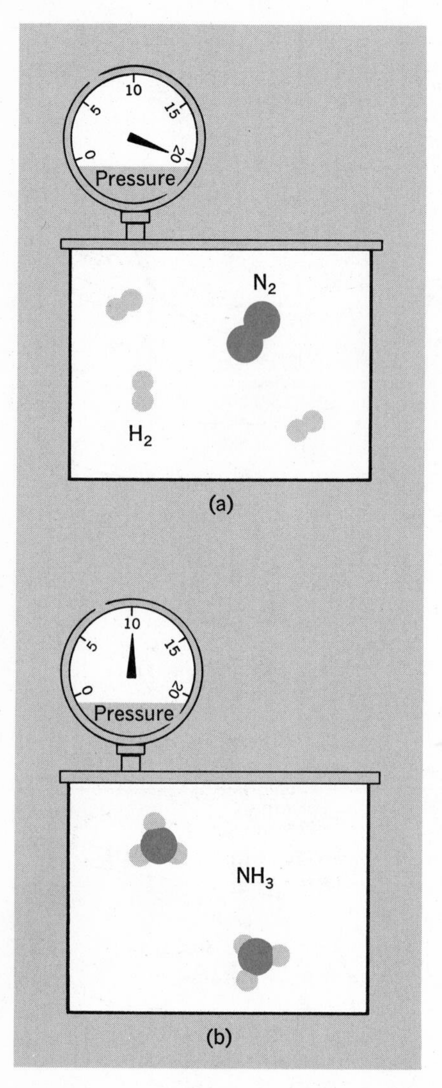

13-7 A decrease in the number of molecules results in a decrease in pressure. When three hydrogen molecules and a single nitrogen molecule combine and form two ammonia molecules, the pressure is reduced to one-half.

In terms of rates, increased pressure or decreased container volume means that molecules are closer together and collide more often. This results in an increase in both the forward and reverse rates of reactions. When the same number of molecules is shown on both sides of the balanced equation, as in Equation 13-3, r_f and r_r increase in the same ratio. Thus no net change in molecular population of the reactants or products occurs.

In the Haber process, when the volume is decreased (pressure increased), there is a greater increase in the number of collisions between reactant molecules per unit time than between product molecules.

In the Haber reaction, when the product molecules are fewer than the reactant molecules, the rate of the forward reaction is found to increase to a larger extent than that of the reverse reaction when the pressure increases. Then additional NH_3 molecules form and bring the reverse rate into balance with the forward rate.

It should be noted that when a noble gas or any gas which does not react with any of the participants is added to a gaseous system at equilibrium in a container of fixed volume, the total pressure increases. This, however, has no effect on the equilibrium because the concentrations or partial pressures of the reacting gases remain constant.

13-7 Change in the Temperature of a System at Equilibrium We can again use the Haber reaction to illustrate the effect of a temperature change on a system at equilibrium. Equation 13-4 indicates the reaction is exothermic (ΔH is negative). According to Le Châtelier's principle, a temperature increase should cause the system to react in a way that would absorb energy. The thermochemical equation reveals that the decomposition of ammonia absorbs heat. An increase in temperature, therefore, should cause an increase in the concentration of H_2 and N_2 and a decrease in the concentration of NH_3. In other words, the stress caused by increasing the temperature is relieved when heat is absorbed during the decomposition of NH_3. Increasing the temperature of this or any other system, which is exothermic in the forward direction, at equilibrium, therefore decreases the quantity of product. Do not say or visualize that heat is being added to the right side of the reaction. Remember, all participants in the system are present together at equilibrium and any change in the conditions affects them all. Some students like to think in terms of heat as a "product" in an exothermic reaction. Thus, an increase in heat would have the same effect as an increase in the "concentration" of a product.

In terms of rates, increasing the temperature increases both the forward and reverse rates. However, as mentioned earlier, because the reverse reaction is endothermic, increasing the temperature increases r_r to a greater extent than it does r_f. In restoring the balance in rates, the concentration of the reactants increases to the point where there are enough collisions for r_f to again equal r_r.

13-8 Effect of a Catalyst on an Equilibrium In Chapter 12 we illustrated that a catalyst generally increased the rate of the forward and reverse reactions equally by lowering the activation energy to the same extent for both reactions. Since the ratio of the forward to reverse rate remains constant, no net change occurs in the relative amounts of product and reactant present at equilibrium. Addition of a catalyst, therefore, does not affect the position of the equilibrium. *It causes the reaction system to reach equilibrium in a shorter period of time.*

The use of Le Châtelier's Principle as a predictive device to help us determine the effect of changes in reaction conditions on a system at equilibrium is illustrated in this example.

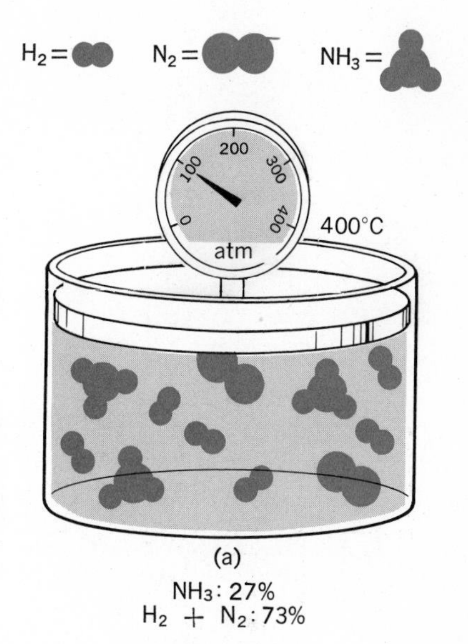

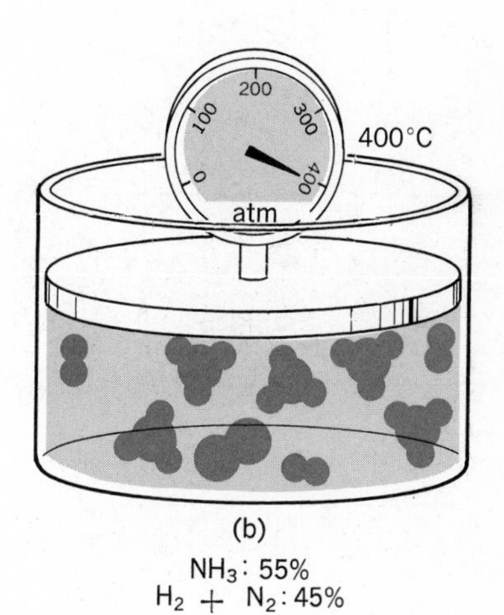

13-8 Effect on the equilibrium system $N_2 + 3H_2 \rightleftharpoons 2NH_3$ of decreasing the total volume. The increase in pressure associated with the decrease in volume causes a reduction in the total number of molecules but an increase in the number of ammonia molecules. By this action, the system minimizes the stress brought about by the increased pressure (decreased volume).

Example 13-1

This equation represents a gaseous system at equilibrium.

$$2SO_2(g) + O_2(g) \rightleftharpoons 2SO_3(g) + \text{heat}$$

Indicate the direction in which the equilibrium shifts when these changes are made. Briefly justify your answer. (a) The concentration of SO_2 is increased.

Ans. The equilibrium shifts to the right. The formation of more SO_3 uses up SO_2, and helps to relieve the stress imposed by the increased concentration of SO_2.

(b) The partial pressure of SO_3 is decreased.

Ans. The equilibrium shifts to the right. Decreasing the partial pressure is analogous to decreasing the concentration. The formation of additional SO_3 tends to relieve the stress by increasing its concentration (partial pressure).

(c) The temperature of the system is decreased.

Ans. The equilibrium shifts to the right and increases the concentration of SO_3 as it produces heat to compensate for the lowering of the temperature.

(d) The volume of the container is increased.

Ans. The equilibrium shifts to the left. Increasing the volume results in more space between the molecules. This stress is relieved by the formation of more molecules. For every two SO_3 molecules that decompose, three molecules of reactants form.

(e) Helium gas is added at constant volume so that the total pressure is increased.

Ans. There is no change. The volume of the container and the concentration of the reacting molecules are constant so that no stress is imposed, and no change occurs.

(f) Helium gas is added, but the total pressure is kept constant.

Ans. The equilibrium shifts to the left. If the total pressure is constant, then volume must increase and the reactant molecules are farther apart. Because this is the same stress situation as in (d), a shift that tends to increase the number of molecules takes place.

(g) A catalyst is added.

Ans. No shift occurs. Both the forward and reverse rates are increased by the same ratio. Equilibrium is established sooner, but the equilibrium concentrations remain the same.

13-9 Ammonia made by the Haber process is sometimes added directly to the soil as a source of fixed nitrogen. (W. R. Grace Chemical Co.)

FOLLOW-UP PROBLEM

This equation represents a gaseous system at equilibrium.

$$\text{heat} + 2H_2O(g) \rightleftharpoons 2H_2(g) + O_2(g)$$

Indicate in which direction the equilibrium shifts when these changes are made. Justify your answer in terms of Le Châtelier's principle. The answers are omitted. (a) The concentration of H_2 is increased. (b) The partial pressure (concentration) of H_2O is increased. (c) The concentration of O_2 is decreased. (d) The temperature is increased. (e) The volume of the container is decreased. (f) Helium gas is added at constant volume so that the total pressure is increased. (g) Helium gas is added, but the total pressure is kept constant. (h) A catalyst is added.

13-9 The Haber Process The Haber process for synthesizing ammonia from elemental hydrogen and nitrogen is one of the most important industrial processes in the world. It is the major source of "fixed" nitrogen in compounds which is an essential plant food applied in the form of chemical fertilizers. The importance of chemical fertilizers cannot be overemphasized when we view the expanding population and the problem of supplying adequate food. It is apparent that NH_3 will continue to play an important role in the production of the world's food supply.

On the other side of the coin, the development of this process in 1913 by the German chemist, Fritz Haber (1868–1934), was an important factor in World War I. The preparation of explosives used during this war required nitrogen compounds. Before the war, Germany imported natural nitrates from Chile. The blockade by the British early in the war cut off this supply. By converting the NH_3 produced by the Haber process to explosive nitrates, Germany was able to prolong the war. Let us examine the conditions under which this important chemical is produced.

In our discussion of Le Châtelier's Principle, we applied it separately to the temperature and pressure factors. The principle may indicate that, for some reactions, high pressure and low temperature are favorable equilibrium conditions for high yields of product. At low temperatures, reaction rates may be too slow to be practical. In these cases a compromise must be made between the temperature and pressure factors so that a good yield is obtained in a reasonable time. A number of important chemical manufacturing processes depend on such a compromise. One of them is the Haber process, Equation 13-4. Application of Le Châtelier's Principle shows that, in theory, the most favorable conditions for a high yield of NH_3 would be low temperature and high pressure. At low temperatures it takes so long for the system to reach equilibrium that the amount of NH_3 produced in a reasonable time is not economically practical. Even with the best catalyst known, the process is slow. It has been found that temperatures near 500°C must be used. Because increasing the temperature tends to decrease the yield of ammonia, the only solution to the problem is to run the process at a high pressure which gives a favorable yield in a reasonable period of time. The effect of different temperature and pressure combinations on the percent of ammonia at equilibrium is shown in Table 13-1. A schematic diagram of the process is shown in Fig. 13-10.

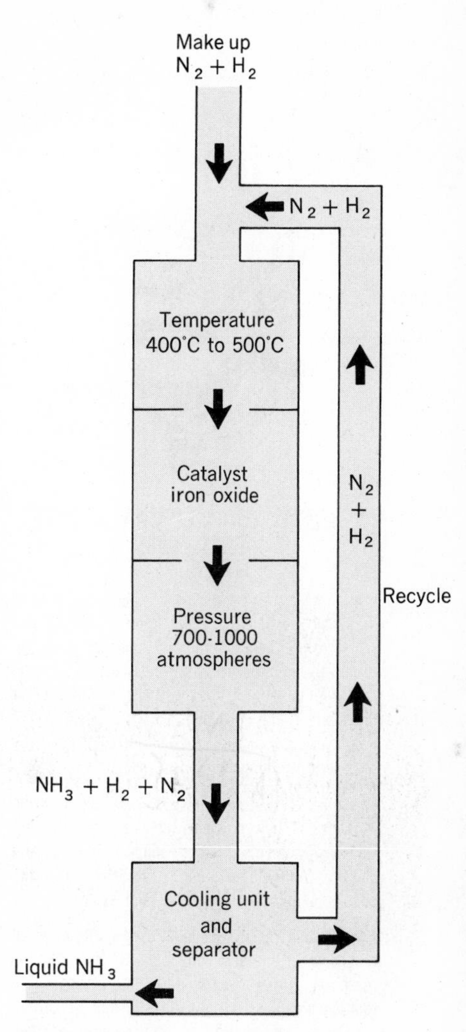

13-10 The Haber process. In this process, practical use is made of Le Châtelier's Principle. A satisfactory yield requires the use of high pressure, a relatively low temperature (500°C), a catalyst, and the continuous removal of a product. Unreacted hydrogen and nitrogen are recycled through the reaction chamber with any residual ammonia gas. Additional hydrogen and nitrogen are added as required. In the cooling unit only the $NH_3(g)$ condenses. This, of course, lowers the concentration of $NH_3(g)$ and shifts the equilibrium further to the product side as predicted by Le Châtelier's Principle.

TABLE 13-1
PERCENTAGE OF AMMONIA IN THE EQUILIBRIUM MIXTURE, HABER PROCESS

Temperature (°C)	*Pressure (atm)* 100	200	300	400	600	800	1000
200	82	90	92	96	97	98	99
300	52	64	71	76	82	88	92
350	38	51	59	65	73	80	88
400	25	39	48	55	65	73	80
500	11	20	27	32	43	51	58
600	5	8	14	16	24	28	32

HENRY LOUIS LE CHÂTELIER
1850–1936

Le Châtelier was trained as a mining engineer at the Ecole des Mines in Paris. After working two years in the field of mining engineering, he returned to his old school as Professor of Chemistry. In 1908, he accepted a position as Professor of Chemistry at the University of Paris, a position that was previously occupied by Henri Moissan, the discoverer of fluorine.

Le Châtelier was an energetic experimenter who worked on such problems as the calcining and setting of cements, the annealing of ceramics and glassware, the preparation of abrasives, and the development of fuels, glasses, and explosives.

Le Châtelier was an excellent teacher who stressed precision in measurement as well as scientific reasoning. He was particularly interested in the relationship of science to industry and in how to get the maximum yield from any chemical reaction. His research in this area led to his discovery of the principle for which he is best known. The "Le Châtelier Principle" deals with chemical equilibrium systems. It states that if a system at equilibrium is disturbed, changes will occur which tend to restore the original conditions of the equilibrium. Methods for controlling chemical reactions are based on this principle. The efficiency of many industrial processes has been greatly increased by the application of this principle.

It may be seen from the data in Table 13-1 that the conditions which would produce the optimum yield of NH_3 are 200°C and 1000 atm pressure. The rate of reaction at 200°C is so slow that the time required for the system to reach equilibrium is prohibitively long. Therefore, a temperature of 400 to 500°C and a catalyst are used to speed up the reaction so that equilibrium is established within a reasonable time interval. Although the increased rate of reaction is achieved at the expense of a reduced yield, the process is still economically feasible.

It can be seen from our examples that at a given temperature, different chemical systems reach an equilibrium state at varying degrees of "completeness." Let us now develop an expression which relates the concentrations of reaction participants after equilibrium is established. This expression, known as the ***Equilibrium Law*** expression, enables us to determine the extent of a reaction and to calculate the quantity of product present at equilibrium. We shall use rate concepts and microparticle behavior to derive a *quantitative equilibrium law.*

THE EQUILIBRIUM LAW

13-10 Derivation of the Equilibrium Law Expression, K_e Let us consider a reversible reaction involving two opposing elementary processes.

$$\mathrm{aA + bB \underset{r_r}{\overset{r_f}{\rightleftharpoons}} cC + dD} \tag{13-5}$$

The rate of the forward reaction is

$$r_f = k_f[\mathrm{A}]^{\mathrm{a}}[\mathrm{B}]^{\mathrm{b}} \tag{13-6}$$

and that of the reverse reaction is

$$r_r = k_r[\mathrm{C}]^{\mathrm{c}}[\mathrm{D}]^{\mathrm{d}} \tag{13-7}$$

The rate of the net reaction is

$$r_n = r_f - r_r$$

At equilibrium, there is no net reaction so

$$r_f - r_r = 0$$

and

$$r_f = r_r$$

Substituting the value of r_f and r_r from 13-6 and 13-7 into 13-8 we obtain

$$k_f[\mathrm{A}]^{\mathrm{a}}[\mathrm{B}]^{\mathrm{b}} = k_r[\mathrm{C}]^{\mathrm{c}}[\mathrm{D}]^{\mathrm{d}} \tag{13-8}$$

Combining the rate constants r_f and r_r yields another constant K.

$$\frac{k_f}{k_r} = \frac{[\mathrm{C}]^{\mathrm{c}}[\mathrm{D}]^{\mathrm{d}}}{[\mathrm{A}]^{\mathrm{a}}[\mathrm{B}]^{\mathrm{b}}} = \mathrm{K}_e \tag{13-9}$$

The quotient of the two rate constants is known as the ***equilibrium constant*** (K_e) for the reaction. It is agreed by convention that K_e is defined as k_f/k_r although it could just as well have been defined as k_r/k_f. Equation 13-9, known as the ***Equilibrium Law Expression,*** is a general expression of the ***Law of Mass Action.*** This law may be stated: in a system at equilibrium, at a fixed temperature, the product of the equilibrium concentration of the products divided by the concentrations of the reactants, each concentration being raised to a power equal to the coefficient of the substance in the equation, must be equal to a constant. Equation 13-9 is *strictly applicable only to systems of ideal gases and ideal solutions;* we shall assume that the expression applies to the systems we encounter in this text.

Equilibrium constants can be derived by applying rate concepts to equations for overall reactions. The net equation need not represent an elementary process. This means that the Equilibrium Law expression is independent of the actual mechanism. For example, the Equilibrium Law expression for

$$\mathbf{H_2(g) + I_2(g) = 2HI(g) + heat} \qquad 13\text{-}10$$

is

$$\mathbf{K}_e = \frac{\mathbf{[HI]^2}}{\mathbf{[H_2][I_2]}} \qquad 13\text{-}11$$

Experimental data show that at a given temperature the equilibrium concentration of the participants may vary but always in such a way as to satisfy Equation 13-11. That is, at constant temperature, K_e is constant and there are countless values for the concentrations of the participants that satisfy the equation. The concentrations in the equilibrium expression may be stated in moles per liter. In the case of gases, they may be given in terms of partial pressures which are proportional to the concentrations.

13-11 Effect of Temperature on K_e We can use the rate theory discussed in Chapter 12 to show that the equilibrium constant changes with temperature. Since

$$\mathbf{K}_e = \frac{\mathbf{k}_f}{\mathbf{k}_r}, \qquad 13\text{-}12$$

any factor which causes an unequal change in k_f and k_r causes a change in K_e. In Chapter 12 it was pointed out that increasing the temperature resulted in greater increases in the rate and the rate constant of an endothermic reaction than those for the opposing exothermic reaction. Because the forward reaction in Equation 13-10 is exothermic, increasing the temperature of the system causes a greater increase in k_r than in k_f (Equation 13-12); therefore, the ratio K_e decreases. Conversely, it can be shown that increasing the temperature causes K_e for an overall endothermic reaction to increase. Changes in K_e are accompanied by changes in equilibrium concentrations. Therefore, **changing the temperature of an equilibrium system changes both the concentration of the**

participants and the value of K_e. At constant temperature, changing the equilibrium concentrations does not affect K_e because rate constants are not affected by concentration changes. *When the concentration of one of the participants is changed, the concentration of the others varies in such a way as to maintain a constant value for K_e.*

By applying the principles of the Law of Mass Action to any equilibrium system, we can obtain an expression for the equilibrium constant K_e. For example, the equilibrium constant for the reaction

$$2SO_2(g) + O_2(g) \rightleftharpoons 2SO_3(g)$$

is

$$K_e = \frac{[SO_3]^2}{[SO_2]^2[O_2]}$$

FOLLOW-UP PROBLEM

Write the equilibrium law expression for these equilibrium reactions

(a) $3H_2(g) + N_2(g) \rightleftharpoons 2NH_3(g)$
(b) $2NO(g) + O_2(g) \rightleftharpoons 2NO_2(g)$

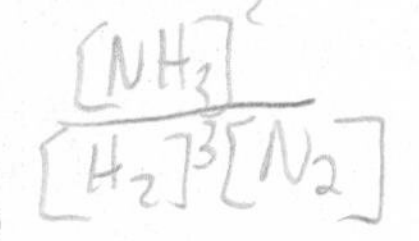

13-12 Relationship of K to Form of Equation for Reaction It may be seen from Equation 13-9 that the form and value of the equilibrium constant depends on the way the equation is written. For example, the $H_2 - I_2 - HI$ reaction system at 448°C might be described by any of these equations and constants.

$$H_2 + I_2 \rightleftharpoons 2HI; \quad K_1 = \frac{[HI]^2}{[H_2][I_2]} = 50.2 \qquad 13\text{-}13$$

$$\tfrac{1}{2}H_2 + \tfrac{1}{2}I_2 \rightleftharpoons HI; \quad K_2 = \frac{[HI]}{[H_2]^{1/2}[I_2]^{1/2}} = \sqrt{50.2}$$

$$2H_2 + 2I_2 \rightleftharpoons 4HI; \quad K_3 = \frac{[HI]^4}{[H_2]^2[I_2]^2} = (50.2)^2$$

$$2HI \rightleftharpoons H_2 + I_2; \ K_4 = \frac{[H_2][I_2]}{[HI]^2} = \frac{1}{50.2} \qquad 13\text{-}14$$

We can summarize the above relationships:

1. When the coefficients of a given reaction equation are multiplied by $\frac{1}{2}$, then the K for the new reaction equation is the square root of the original K.
2. When the coefficients of a given reaction equation are doubled, then the K for the new reaction equation is the square of the original K. In general, *when we multiply the terms in an equation by a certain value, we must change the K for the equation to a power equal to that value.*

3. *When a given reaction equation is reversed, then the K for the reverse reaction equation is the reciprocal of that for the original reaction equation.*

FOLLOW-UP PROBLEMS

1. Write the equilibrium law expression for these equilibrium reactions.

(a) $N_2(g) + 3H_2(g) \rightleftharpoons 2NH_3(g)$
(b) $2SO_2(g) + O_2(g) \rightleftharpoons 2SO_3(g)$
(c) $SO_2(g) + \frac{1}{2}O_2(g) \rightleftharpoons SO_3(g)$
(d) $SO_3(g) \rightleftharpoons SO_2(g) + \frac{1}{2}O_2(g)$
(e) $4HCl(g) + O_2(g) \rightleftharpoons 2H_2O(g) + 2Cl_2(g)$

2. The equilibrium constant for reaction 1(c) is 6.60. What is the equilibrium constant for reactions 1(b) and 1(d)? **Ans. 1(b) = 43.6, 1(d) = 0.152.**

13-13 Relationship of *K* to the Extent of a Reaction In general, the magnitude of *K* is a measure of the extent to which a given reaction has taken place before equilibrium was established. A large value for *K* indicates that, at equilibrium, there is a high concentration of products and a relatively low concentration of reactants. On the other hand, a small value for *K* indicates that, at equilibrium, the concentration of reactants is high compared with that of the products. A value for *K* equal to 1 for the reaction

$$A + B \rightleftharpoons C + D$$

in which the coefficients all have a value of 1, indicates that, at equilibrium, one half the mass of the reactants is converted to products. That is, the yield is 50 percent. *The larger the value of K is, the more complete the conversion of reactants to products. A large value of K for a forward reaction indicates a small K for a reverse reaction.* In general, a reaction with a *K* of about 1×10^{10} is "complete" enough to be considered quantitative. One with a *K* equal to 1×10^{-10} is relatively incomplete. That is, the equilibrium concentration of the products is extremely small and probably not measurable by ordinary means. Actually, the constants for reactions have a much greater range than this. For example, some reactions have a *K* as low as 1×10^{-50} which is about as close to zero as you can get. On the other hand, some reactions have a *K* greater than 1×10^{100}, a number so large it is difficult to comprehend and is essentially infinite. Using the $H_2 - I_2 - HI$ system, the forward reaction represented by Equation 13-13 has a *K* of 50.2 at 448°C. Starting with one mole each of H_2 and I_2 it can be shown that the equilibrium yield of HI is

$$\frac{\textbf{1.56 moles actual yield}}{\textbf{2.00 moles theoretical yield}}\textbf{(100)} = \textbf{78.0 percent.}$$

For the reverse reaction represented by Equation 13-14, *K* is $\frac{1}{50.2}$. Starting with two moles of HI, the equilibrium yield of H_2 and of I_2 is

$$\frac{\textbf{0.22 mole actual yield}}{\textbf{1.00 mole theoretical yield}}\textbf{(100)} = \textbf{22.0 percent.}$$

The yield for a reaction with a large value for K is greater than that for one with a small value of K.

13-14 Calculation of Equilibrium Constants and Concentrations We can now illustrate some quantitative aspects of equilibrium in homogeneous gaseous reactions.

1. Calculation of the equilibrium constant.
2. Calculation of the equilibrium concentration of a participant when K for the reaction and the concentration of the other participants are known.
3. Determination of the net direction of a reaction prior to establishing equilibrium.
4. Calculation of equilibrium concentrations when initial concentration and K for the reaction are known.
5. Calculation of the percentage dissociation and/or the percentage yield of a reaction.

Example 13-2

When 0.40 mole of PCl_5 is heated in a 10.0-ℓ container, an equilibrium is established in which 0.25 moles of Cl_2 is present.

$$PCl_5(g) \rightleftharpoons PCl_3(g) + Cl_2(g)$$

(a) What is the number of moles of PCl_5 and PCl_3 at equilibrium? (b) What are the equilibrium concentrations of all three components?

Solution

(a) The equation shows that for each mole of Cl_2 formed, 1 mole of PCl_3 is formed and 1 mole of PCl_5 is consumed. At equilibrium there are

$$0.25 \text{ moles } Cl_2 \times \frac{1 \text{ mole } PCl_3}{1 \text{ mole } Cl_2} = 0.25 \text{ moles } PCl_3$$

$$0.40 \text{ moles } PCl_5 - \left[0.25 \text{ moles } Cl_2 \times \frac{1 \text{ mole } PCl_5}{1 \text{ mole } Cl_2}\right] = 0.15 \text{ moles } PCl_5$$

(b) The equilibrium concentrations are equal to moles/liter

$$Cl_2 \quad \frac{0.25 \text{ moles}}{10.0\ \ell} = 0.025\ M$$

$$PCl_3 \quad \frac{0.25 \text{ moles}}{10.0\ \ell} = 0.025\ M$$

$$PCl_5 \quad \frac{0.15 \text{ moles}}{10.0\ \ell} = 0.015\ M$$

FOLLOW-UP PROBLEM

When 1.00 mole of NH_3 gas and 0.40 mole of N_2 gas are placed in a 5.0-liter container and allowed to reach equilibrium at a certain temperature, it is found that 0.78 mole of NH_3 is present. The reaction is

$$2NH_3(g) \rightleftharpoons 3H_2(g) + N_2(g)$$

(a) What is the number of moles of H_2 and N_2 at equilibrium? (b) What is the concentration in moles/ℓ of each component?

Ans. (a) H_2 = 0.33 moles, N_2 = 0.51 moles, (b) H_2 = 0.066*M*, N_2 = 0.10 M, NH_3 = 0.16 *M*.

Example 13-3

A mixture of H_2 and I_2 is allowed to react at 448°C. When equilibrium is established, the concentrations of the participants are found to be $[H_2] = 0.46$ mole/ℓ, $[I_2] = 0.39$ mole/ℓ, $[HI] = 3.0$ moles/ℓ. Calculate the value of K at 448°C from these data.

Solution

1. Write the equation and the equilibrium expression for the reaction

$$H_2(g) + I_2(g) \rightleftharpoons 2HI(g)$$

$$K = \frac{[HI]^2}{[H_2][I_2]}$$

2. Substitute the equilibrium concentrations in the equilibrium expression and solve for K.

$$K = \frac{[3.0]^2}{[0.46][0.39]} = 50$$

Example 13-4

Assume that in the analysis of another equilibrium mixture at 448°C, the equilibrium concentrations of I_2 and H_2 are both 0.50 mole/ℓ. What is the equilibrium concentration of HI?

Solution

Because at constant temperature the value of K does not change, the equilibrium concentration of HI must increase in order to compensate for the increased equilibrium concentrations of H_2 and I_2. The new equilibrium concentration for HI can be determined by substituting the known values for H_2 and I_2 into the equilibrium expression using the same value of K.

$$50 = \frac{[HI]^2}{[0.50][0.50]}$$

$$[HI]^2 = (50)(0.25) = 12.5$$

$$[HI] = 3.5 \text{ moles/ℓ}$$

Example 13-5

The equilibrium constant for the reaction represented by the equation is 50 at 448°C.

$$H_2(g) + I_2(g) \rightleftharpoons 2HI(g)$$

(a) How many moles of HI are present at equilibrium when 1.0 mole of H_2 is mixed with 1.0 mole of I_2 in a 0.50-ℓ container and allowed to react at 448°C? (b) How many moles of H_2 and I_2 are left unreacted? (c) If the conversion of H_2 and I_2 to HI were essentially complete, how many moles of HI would be present? (d) What is the percent yield of the equilibrium mixture?

Solution

(a) 1. Write the equation and equilibrium law expression for the reaction

$$K = \frac{[HI]^2}{[H_2][I_2]}$$

2. Express the initial concentrations of all participants in moles/ℓ.

$$H_2 = 1.0 \text{ mole}/0.50\ \ell = 2.0 \text{ moles}/\ell$$
$$I_2 = 1.0 \text{ mole}/0.50\ \ell = 2.0 \text{ moles}/\ell$$
$$HI = 0 \text{ moles}$$

3. Let x = the number of moles of H_2 per liter consumed in reaching equilibrium. The balanced equation shows that an equal number of moles per liter of I_2 are consumed and that the number of moles of HI formed is twice the number of moles of H_2 or I_2 consumed. Now write the equilibrium concentrations for each of the participants. The equilibrium concentrations of H_2 and I_2 are their initial concentrations minus the number of moles consumed. The equilibrium concentration of HI is the number of moles of HI formed ($2x$) per liter. The equilibrium concentrations are

$$[H_2] = 2.0 - x$$
$$[I_2] = 2.0 - x$$
$$[HI] = 2x$$

A better picture of the situation may be seen by writing the concentration of each species under its formula in the equation.

	$H_2(g)$	+	$I_2(g)$	$\rightleftharpoons$	$2HI(g)$
Initial conc.	2.0 moles/ℓ		2.0 moles/ℓ		0
Change in no. of moles during reaction	$-x$ moles		$-x$ moles		$2x$
Equilibrium conc.	$(2.0 - x)$ moles/ℓ		$(2.0 - x)$ moles/ℓ		$2x$ moles/ℓ

4. Substitute the equilibrium concentrations and the value of *K* in the equilibrium expression

$$K = \frac{[HI]^2}{[H_2][I_2]}$$

$$50 = \frac{(2x)^2}{(2.0 - x)(2.0 - x)} = \frac{(2x)^2}{(2.0 - x)^2}$$

This equation may be solved by taking the square root of both sides and solving for x.

$$\sqrt{50} = \sqrt{\frac{(2x)^2}{(2.0 - x)^2}}$$

$$7.1 = 2x/(2.0 - x)$$

$$14.2 - 7.1x = 2x$$

$$x = 14.2/9.1 = 1.6 \text{ moles}/\ell$$

5. Determine the equilibrium concentrations by substituting x in the expressions for equilibrium concentrations given in Step 3.

$$[H_2] = 2.0 - x = 2.0 - 1.6 = 0.40 \text{ mole}/\ell$$

$$[I_2] = 2.0 - x = 2 - 1.6 = 0.40 \text{ mole}/\ell$$

$$[HI] = 2x = 2(1.6) = 3.2 \text{ moles}/\ell$$

6. Determine the total number of moles of HI by multiplying the concentration of HI by the volume of the container.

$$3.2 \text{ moles}/\ell \times 0.50\ \ell = 1.6 \text{ moles HI}$$

(b) The number of moles of H_2 and I_2 left unreacted may be found by multiplying their equilibrium concentrations by the volume of the container.

$0.40 \text{ mole}/\ell \times 0.50\ \ell = 0.20$ mole of unreacted H_2 and I_2.

(c) The stoichiometric equation shows that 2 moles of HI form for each mole of H_2 and I_2 that react. Therefore, if the reaction were essentially complete, 1.0 mole of H_2 would react with 1.0 mole of I_2 and form 2.0 moles of HI.
(d) The percent yield of HI at equilibrium would be the actual equilibrium yield divided by the yield if the reaction were complete.

$$\% \text{ yield HI} = \frac{1.6 \text{ moles}}{2.0 \text{ moles}}(100) = 80\%$$

Example 13-6
How many moles of HI are present at equilibrium when 2.0 moles of H_2 is mixed with 1.0 mole of I_2 in a 0.50-ℓ container and allowed to react at 448°C? At this temperature *K* = 50.

Solution

1. Express the initial concentration of all participants in moles/ℓ.

$$[H_2] = 2.0 \text{ mole}/0.50\ \ell = 4.0 \text{ moles}/\ell$$
$$[I_2] = 1.0 \text{ mole}/0.50\ \ell = 2.0 \text{ moles}/\ell$$
$$[HI] = 0 \text{ moles}/\ell$$

2. Let x = the number of moles per liter of H_2 consumed in reaching equilibrium and express the equilibrium concentrations of all participants in terms of x.

$$H_2 = 4.0 - x$$
$$I_2 = 2.0 - x$$
$$HI = 2x$$

3. Substitute the equilibrium concentrations and the value of K in the equilibrium-law expression and solve for x.

$$50 = \frac{(2x)^2}{(4.0 - x)(2.0 - x)}$$

$$50 = \frac{4x^2}{(8.0 - 6.0x + x^2)}$$

$$400 - 300x + 50x^2 = 4x^2$$
$$46x^2 - 300x + 400 = 0$$

This equation is a quadratic having the form of

$ax^2 + bx + c = 0$ and it can be solved by

$$x = \frac{-b \pm \sqrt{b^2 - 4ac}}{2a}$$

In the example above,

$$a = 46$$
$$b = -300$$
$$c = 400$$

Substituting these values in the quadratic formula yields

$$x = \frac{+300 \pm \sqrt{(300)^2 - 4(46)(400)}}{92}$$

$$x = \frac{+300 \pm 127}{92}$$

$$x = 4.7 \text{ or } 1.9$$

The value of 4.7 for x has no meaning since it is larger than the original concentration of H_2. Using a value of 1.9 for x, the equilibrium concentration of HI is $2x$ or 3.8 moles/ℓ and the total number of moles equals

$$0.50\ell \times 3.8 \text{ moles}/\ell = 1.9 \text{ moles}$$

Example 13-7
(a) When 3.0 moles of HI, 2.0 moles of H_2, and 1.5 moles of I_2 are placed in a 1-ℓ container at 448°C, will a reaction occur? (b) If so, which reaction takes place?

Solution
Substitute the given concentrations in an expression similar to the equilibrium law expression and use the given experimental concentrations to determine the experimental concentration quotient Q. If the experimental quotient is exactly equal to *K*, the given concentrations are equilibrium concentrations, and no net reaction occurs. If the experimental quotient is greater than *K*, the numerator in the expression is larger than it would be at equilibrium. Therefore, the concentration of the product, HI, must decrease to reach the equilibrium concentration. On the other hand, if the experimental quotient is too small, the numerator is smaller than it would be at equilibrium. Therefore, the reactants, H_2 and I_2, would form additional product to increase the concentration of HI to an equilibrium value. Substituting given values yields

$$\text{experimental quotient } Q = \frac{(3.0)^2}{(2.0)(1.5)} = 9.0/3.0 = 3.0$$

Since the experimental quotient of 3.0 is less than the *K* value of 50, the reaction goes to the right and forms more HI and reduces the concentrations of H_2 and I_2.

FOLLOW-UP PROBLEMS

1. When 0.040 mole of PCl_5 is heated to 250°C in a 1.0-ℓ vessel, an equilibrium is established in which the concentration of Cl_2 is 0.025 mole/ℓ. Find the equilibrium constant, *K*, at 250°C for the reaction

$$PCl_5(g) \rightleftharpoons PCl_3(g) + Cl_2(g)$$

Note that the initial, not the equilibrium concentration, of PCl_5 is given. The amount of PCl_5 reacted is the same as the amount of Cl_2 formed. **Ans. 4.2×10^{-2}.**

2. Assume that the analysis of another equilibrium mixture of the system in Problem 1 shows that the equilibrium concentration of PCl_5 is 0.012 mole/ℓ and that of Cl_2 is 0.049 mole/ℓ. What is the equilibrium concentration of PCl_3 at 250°C? **Ans. 1.0×10^{-2} mole.**

3. How many moles of PCl_5 must be heated in a 1.0-ℓ flask at 250°C in order to produce enough chlorine to give an equilibrium concentration of 0.10 mole/ℓ? **Ans. 0.34 moles.**

4. Will there be a net reaction when 2.5 moles of PCl_5, 0.60 mole of Cl_2, and 0.60 mole of PCl_3 are placed in a 1-ℓ flask and heated to 250°C? If so, which reaction takes place? **Ans. Cl_2 and PCl_3 react and form more PCl_5.**

13-15 Quantitative Aspect of Le Châtelier's Principle (Concentration Changes) Let us now consider the effect of changing the concentration of a participant in a system already at equilibrium. Consider this system

$$SO_2(g) + NO_2(g) \rightleftharpoons SO_3(g) + NO(g) \qquad 13\text{-}16$$

whose K may be expressed as

$$\frac{[SO_3][NO]}{[SO_2][NO_2]} = K \qquad \text{13-17}$$

Suppose we add SO_2 to an equilibrium mixture of the above components. This addition produces nonequilibrium condition. Increasing the denominator in Equation 13-17 means that the experimental reaction quotient, Q, is less than K. Therefore, to re-establish equilibrium, the value of the products represented in the numerator must increase. In other words, the equilibrium position must shift to the right and increase the yield of products. The quantitative aspects of this change are illustrated in Example 13-8.

Example 13-8
At a given temperature, analysis of an equilibrium mixture represented by Equation 13-16, shows

$$[SO_2] = 4.0\ M,\ [NO_2] = 0.50\ M$$
$$[SO_3] = 3.0\ M,\ [NO] = 2.0\ M.$$

What is the new equilibrium concentration of NO when 1.5 moles of NO_2 is added to a liter of the first mixture?

Solution
1. Use the original equilibrium concentrations to calculate the equilibrium constant at the given temperature.

$$K = \frac{[SO_3][NO]}{[SO_2][NO_2]}$$
$$= \frac{(3.0)(2.0)}{(4.0)(0.50)} = 3.0$$

2. Let x = moles/ℓ of NO formed as a result of increasing the concentration of NO_2 from 0.50 M to 2.0 M. Express the second set of equilibrium concentrations in terms of x.

	$SO_2(g)$ +	$NO_2(g)$	$\longrightarrow$	$SO_3(g)$ +	$NO(g)$
First equil. conc.	4.0M	0.50M		3.0M	2.0M
Second equil. conc.	$(4.0 - x)$	$(0.50 + 1.5 - x)$		$(3.0 + x)$	$(2.0 + x)$

3. Substitute the new equilibrium concentrations in the expression for K and solve the resulting quadratic equation for x.

$$\frac{(3.0 + x)(2.0 + x)}{(4.0 - x)(2.0 - x)} = 3.0$$
$$2.0x^2 - 23x + 18 = 0$$
$$x = 0.75\ M$$

4. Calculate the new equilibrium concentration of NO.

$$2.0 + x = 2.0 + 0.75 = 2.8\ M$$

FOLLOW-UP PROBLEM

How many moles/ℓ of NO_2 would have to be added to the original equilibrium mixture in Example 13-8 to increase the equilibrium concentration of SO_3 from 3.0 to 4.0 moles at the same temperature?

Ans. 1.8 moles/ℓ.

EQUILIBRIA IN HETEROGENEOUS SYSTEMS INVOLVING GASES

13-16 Form of Equilibrium Constant for Systems Involving Solids In earlier chapters we encountered many reactions in which a solid was decomposed, yielding at least one gaseous product. As you know, systems in which more than one phase are present are called heterogeneous reaction systems. The form of the equilibrium constant for these reactions differs markedly from that for homogeneous reactions involving only gases. Consider the thermal decomposition of calcium carbonate in a closed container. The equation for the reaction is

$$CaCO_3(s) \rightleftharpoons CaO(s) + CO_2(g)$$

Experiments show that at a given temperature an equilibrium is established in which the concentration or pressure of CO_2 is constant. Furthermore, it is noted that the concentration of CO_2 is not affected by the quantity of the solid $CaCO_3$ or solid CaO in the container. As long as both solids are present, the concentration of CO_2 is constant at a given temperature. At 900°C, the concentration of $CO_2(g)$ is 1.1×10^{-2} moles/ℓ. Following our original

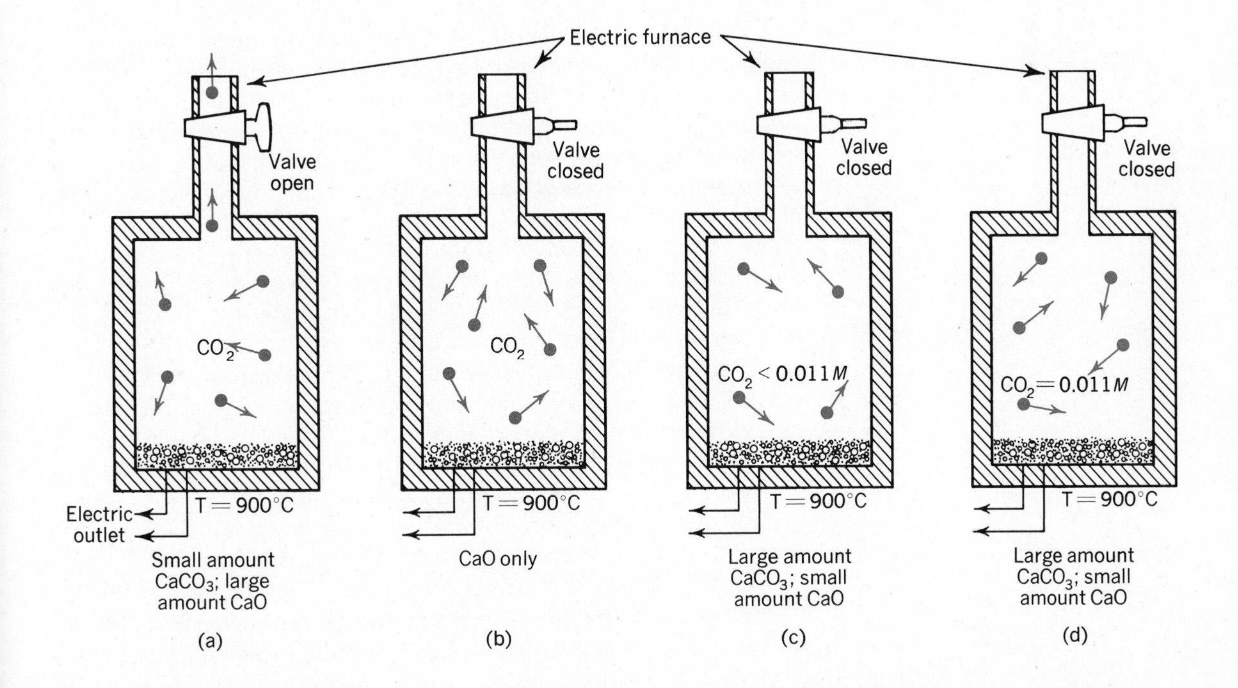

13-11 (a) Which of the systems shown below are in equilibrium? (b) What happens when the pressure (concentration) of CO_2 is increased in (d)? (c) Is it possible to bring the nonequilibrium systems shown below to an equilibrium state? In each case, indicate what can be done and what the effect on the quantities of CaO and $CaCO_3$ are.

Increasing the amount of solid present in a heterogeneous equilibrium system increases the total surface area at which both forward and reverse reactions are occurring. The total number of moles of solid reactant being converted into products per unit time increases. However, the total number of moles of product being converted back to solid reactant per unit time increases proportionally. This results in no net change in total amount of reactant or product. It is assumed that the change in total volume of solid is insignificant in relation to the container volume.

convention we would write the equilibrium constant expression for the reaction as

$$K_e = \frac{[CaO(s)][CO_2(g)]}{[CaCO_3(s)]}$$

Since K_e does not vary with the quantities of solid present, we can assume that they have a constant concentration. The concentration of a solid at a given temperature is fixed by its density (particles, grams or moles of solid per unit volume).

Since the concentrations of the solids are unchanging, they may be incorporated into the equilibrium constant, K_e, to give a new equilibrium constant, K.

$$[CO_2] = K_e \frac{[CaCO_3(s)]}{[CaO(s)]} = K$$

The value of K for this reaction at 900°C is, therefore, equal to the concentration of CO_2 only or 1.1×10^{-2}. We can summarize this discussion by saying that the equilibrium constant expression for a heterogeneous reaction involving gases does not include the concentrations of pure solids. For example, the K for the reaction

$$NH_4Cl(s) \rightleftharpoons NH_3(g) + HCl(g)$$

is

$$K = [NH_3][HCl]$$

FACTORS RELATED TO THE MAGNITUDE OF K_e

13-17 The Tendency or Drive Toward Minimum Energy (Enthalpy) and Maximum Disorder (Entropy) It is apparent that there is a wide variation in the degree of "completeness" of reactions. For some reactions, equilibrium is achieved when practically all of the reactants have been converted into products. These reactions are sometimes referred to as *complete.* For other reactions, the equilibrium state is achieved when only a negligible amount of the reactants has been converted into products. From a stoichiometric viewpoint, these substances do not react appreciably with one another. In between these extremes we find all degrees of "incompleteness" at equilibrium.

Let us now look for some fundamental reasons why certain reactions have larger equilibrium concentrations of products than others. What are the factors that are related to the magnitude of K_e and the completeness of a reaction?

Earlier we found that at a given temperature, the $H_2 - I_2 - HI$ system reached the same equilibrium state when we started with 2 moles of HI as when we started with one mole each of H_2 and I_2. We can express this observation by saying *in a closed system all reactions, unless prevented by exceedingly high activation energy barriers (slow rates), spontaneously approach an equilibrium state.*

It is apparent from our study of the rather inefficient spontaneous exothermic Haber process

$$N_2(g) + 3H_2(g) \rightleftharpoons 2NH_3(g) + \text{heat}$$

that equilibrium is established before the system reaches the lowest possible energy state represented by the complete conversion of N_2 and H_2 into NH_3. The "drive" toward the minimum energy is hampered by the opposing drive of the system to achieve a condition of maximum randomness (entropy). This drive favors the decomposition of NH_3 in which two molecules decompose and yield four molecules. Thus, we can say that the equilibrium state and the value of K represents a compromise between the drive of atoms and molecules in a reaction to achieve a state of minimum enthalpy and a state of maximum entropy.

We can use the thermal decomposition of $CaCO_3$ to illustrate the point further. The equation for the reaction is

$$CaCO_3(s) \rightleftharpoons CaO(s) + CO_2(g)$$

The equilibrium law expression for this equation is

$$K = [CO_2] \qquad 13\text{-}21$$

Equation 13-21 shows that the value of K is equal to the equilibrium concentration or partial pressure of CO_2. If the system were unhampered by a drive to achieve a minimum enthalpy condition, its drive to attain maximum entropy would result in the complete conversion of $CaCO_3(s)$ into $CaO(s)$ and gaseous CO_2. This is because the random distribution of gas molecules represents greater entropy than the orderly arrangement of ions in the crystal. The decomposition of $CaCO_3$ and the formation of CO_2, however, involves the absorption of energy and results in an increase in the enthalpy of the system. Thus, both the actual equilibrium concentration of CO_2 and the value of K represent a compromise between the drive of the system to remain in the lower enthalpy, more ordered, crystalline form and the drive to change into a higher enthalpy, less ordered, gaseous form. The relative importance of each drive depends on the temperature. At low temperatures the energy factor predominates. This means that the equilibrium favors the formation of the substances with the lowest enthalpy (heat content). At high temperatures, equilibrium favors the tendency toward maximum disorder (entropy).

This discussion suggests that the likelihood of a reaction occurring is related to the temperature effect and the relative enthalpy and entropy changes associated with the reaction. We would predict that reactions involving an *increase* in *entropy* and a *decrease* in *enthalpy* would be *highly probable* (*spontaneous*) while those involving a *decrease* in *entropy* and an *increase* in *enthalpy* would be *highly improbable* (*nonspontaneous*).

As we noted earlier, a decrease in the enthalpy of a system is associated with the tendency of the reacting particles to form strong bonds and with a negative value for ΔH. Qualitatively, it is often possible to predict the entropy change (sign of ΔS) by inspecting the equation for a reaction. Following is a list of conditions in which there is an increase in the entropy of a system.

1. When a gas is formed from a solid. An example is

$$CaCO_3(s) + heat \longrightarrow CaO(s) + CO_2(g)$$

2. When a gas is evolved from a solution. An example is

$$Zn(s) + 2H^+(aq) \longrightarrow H_2(g) + Zn^{2+}(aq)$$

3. When the number of moles of gaseous product exceeds the moles of gaseous reactant. An example is

$$2C_2H_6(g) + 7O_2(g) \longrightarrow 4CO_2(g) + 6H_2O(g)$$

4. When many crystals dissolve in water. An example is

$$NaCl(s) \longrightarrow Na^+(aq) + Cl^-(aq)$$

It should be noted that a decrease in entropy occurs when ions such as those shown in the above equation become hydrated. That is, there is a decrease in randomness of the water molecules as they cluster about the ions. The increase caused by the dissolution of the crystals usually predominates in the case of soluble salts.

FOLLOW-UP PROBLEM

For each of these processes, predict whether the entropy increases or decreases.

(a) $2H_2(g) + O_2(g) \longrightarrow 2H_2O(g)$
(b) $2SO_3(g) \longrightarrow 2SO_2(g) + O_2(g)$
(c) $MgCO_3(s) + 2H_3O^+ \longrightarrow Mg^{2+}(aq) + 3H_2O(l) + CO_2(g)$
(d) $Ag^+(aq) + Cl^-(aq) \longrightarrow AgCl(s)$
(e) $Cl_2(g) \longrightarrow 2Cl(g)$
(f) $NH_4NO_3(s) \longrightarrow NH_4^+(aq) + NO_3^-(aq)$
(g) $H_2O(l) \longrightarrow H_2O(g)$
(h) $Mg(s) + 2H_3O^+ \longrightarrow Mg^{2}(aq) + H_2(g) + 2H_2O$
(i) $2C_2H_2(g) + 5O_2(g) \longrightarrow 4CO_2(g) + 2H_2O(g)$
(j) $NH_3(g) + HCl(g) \longrightarrow NH_4Cl(s)$

Ans. (j) decrease.

13-18 The Second Law of Thermodynamics The drive of a system toward the equilibrium state does not necessarily imply that the energy (enthalpy) of the system is decreasing nor that the entropy of the system is increasing. We have already encountered examples of spontaneous endothermic reactions (enthalpy increases), and of spontaneous reactions in which the entropy of the system decreases. This further illustrates the competition between the drive toward maximum entropy and the drive toward minimum enthalpy of the same system. We need additional criteria for predicting the spontaneity of reactions in which the entropy and enthalpy factors oppose each other. These criteria must take into account temperature, entropy changes, enthalpy changes, and also relate to the ***Second Law of Thermodynamics*** which states that *the entropy of the universe increases for any spontaneous process.*

Applied to chemical systems, the second law indicates that the entropy of a system may increase or decrease but if it does decrease, then the entropy of the surroundings must increase to a greater extent so that the overall entropy change in the universe is positive. That is,

It can be shown that during the condensation of water vapor at a temperature below 100°C, the heat released by the process increases the entropy of the surroundings to a greater extent than the decrease in the entropy of the water molecules.

$$\Delta S_{universe} = \Delta S_{system} + \Delta S_{surroundings} > 0$$

where ΔS represents a change in entropy.

13-19 Change in Free Energy: Spontaneity of a Reaction The equation above indicates that we can use the total entropy change of the universe as a criterion to predict the spontaneity of a reaction. To do this, however, we have to calculate the entropy change in both the system undergoing the change and in the surroundings. A more specific criterion which we could apply directly to the chemical system without calculating the entropy change of the surroundings, would be of more use to us in predicting reaction spontaneity. This criterion, known as the change in the free energy of a system, is symbolized by ΔG and is related to the enthalpy change, entropy change, and temperature of a system. The relationship is given by the equation

$$\Delta G = \Delta H - T\,\Delta S \qquad 13\text{-}22$$

where T is the Kelvin temperature. Qualitatively, the sign of ΔG can be used to predict the spontaneity of a reaction at constant temperature and pressure. The criteria are listed below.

A spontaneous change is a change which tends to proceed of itself in the direction of the equilibrium state and one which increases the entropy of the universe.

1. If ΔG is negative, then the reaction is spontaneous (probable) as written. Thus a negative value for ΔG_{system} means that $\Delta S_{universe}$ must be positive.
2. If ΔG is positive, the reaction is improbable as written, but the reverse reaction is probable. Reversing the equation reverses the sign of ΔG.
3. If ΔG is zero, the system is at equilibrium and there is no net reaction. This means there is no net flow of particles or energy and no energy is available for work. When $\Delta G = 0$, then $\Delta H = T\,\Delta S$. Thus, at equilibrium, the entropy factor is balanced by the enthalpy factor.

Let us refer to Equation 13-22 and determine how different values of ΔH and ΔS affect the value of ΔG and the probability of spontaneous occurrence of a reaction.

Inspection of the equation shows that when ΔH is negative and ΔS is positive, ΔG is negative. Thus, **exothermic reactions which are accompanied by an increase in entropy of the system are probable.**

When ΔH is positive and ΔS is negative, ΔG is positive. This means that **endothermic reactions accompanied by a decrease in entropy are improbable.**

At very high temperatures, the sign and magnitude of ΔG and the spontaneity of a reaction are determined primarily by the change in entropy. When both ΔH and $T\,\Delta S$ are positive, ΔG may be either positive or negative. The reaction may or may not be spontaneous. The higher the temperature, the greater the chances are that the reaction will be spontaneous. A high value of T gives the $-T\,\Delta S$ term a large negative value which eventually overbalances a positive ΔH and gives ΔG a negative value.

At very low temperatures, the sign and magnitude of ΔG and the spontaneity of a reaction are determined primarily by the enthalpy change, ΔH. When T is very low, the $T\,\Delta S$ term is very small and has little influence on the value of ΔG. In other words, it will be overbalanced by a large negative ΔH. Large negative values of ΔH are associated with exothermic reactions. Therefore, when T is small, exothermic reactions are very probable and endothermic reactions with a positive ΔH are improbable.

The drive to achieve a state of minimum free energy may be interpreted as the driving force of a chemical reaction. In any reaction, there is a compromise between the tendency of a system to achieve a state of minimum enthalpy and that to achieve a state of maximum entropy. If the entropy change is small, then the primary driving force is to achieve minimum enthalpy (energetically stable products). If the reactants and products are energetically similar, then the primary driving force is to achieve greater entropy.

The effects described above are summarized below.

Thermal Effect	ΔH	ΔS	*Spontaneity of Reaction as Written*	*Comment*
Exothermic	−	+	probable	no exceptions
Exothermic	−	−	probable	at low temperatures
Endothermic	+	+	probable	at high temperatures
Endothermic	+	−	improbable	no exceptions

FOLLOW-UP PROBLEMS

Predict the probability of the following reactions by approximating the sign of ΔG. Classify each reaction as exothermic or endothermic.

(a) $H_2O(l) \longrightarrow H_2(g) + \frac{1}{2}O_2(g)$ $\quad \Delta H = +68$ kcal; $T\,\Delta S = +10$ kcal

(b) $C_6H_{14}(g) \longrightarrow 6C(s) + 7H_2(g)$ $\quad \Delta H = 39.95$ kcal; $T\,\Delta S = 40.05$ kcal

(c) $2Fe(s) + \frac{1}{2}N_2(g) \longrightarrow Fe_2N(s)$ $\quad \Delta H = -0.9$ kcal; $T\,\Delta S = -3.5$ kcal

(d) $HCl(g) + H_2O(l) \longrightarrow H_3O^+(aq) + Cl^-(aq)$ $\quad \Delta H = -18.0$ kcal; $T\,\Delta S = -9.4$ kcal

(e) $N_2(g) + 2O_2(g) \longrightarrow 2NO_2(g)$ $\quad \Delta H = +16.2$ kcal; $T\,\Delta S$: check equation

(f) $2C(s) + O_2(g) \longrightarrow 2CO(g)$ $\quad \Delta H = -26.4$ kcal; $T\,\Delta S$: check equation

Ans. (e) $\Delta S = -$, so $\Delta G = +$ and reaction is endothermic.

LOOKING AHEAD

From now on, most of the equilibrium systems we encounter involve solutions. This means we shall apply equilibrium principles to reactions and processes which occur in solution. To provide the proper background it is essential that we investigate the nature of solvents, the behavior of solutes, and the properties of solutions.

Most solutions of interest to us involve liquids as the solvent, and solids, liquids, or gases as the solute. The solvent is usually molecular in nature and the solute is either molecular or ionic. In the next chapter we shall examine the behavior of both types of solutes when they are added to water and form aqueous solutions. You will find that a number of familiar substances dissolve in or react with water and form equilibrium systems. We shall find that equilibrium principles can be used to determine the concentration of various species in these systems. In addition, we shall apply structural and bonding concepts to help us explain differences in the solubilities of substances.

Equilibrium principles serve as unifying threads for the next four chapters. In each chapter, the same principles are applied qualitatively and quantitatively to a different type of system. In the next chapter, the quantitative aspects of equilibrium between the ions and molecules in pure water are considered. Acid-base equilibria are then examined in some detail in Chapter 15. In Chapter 16 we shall focus our attention on the equilibria that exist in saturated solutions of slightly soluble salts. We shall conclude this series of equilibrium studies in Chapter 17 when we concentrate on equilibria in systems involving oxidation-reduction reactions.

It is suggested that you prepare yourself for the material to be covered in the following chapters by becoming as proficient as you can in working the problems involving equilibrium principles. It is recommended that, if time permits, you work all of the problems in this and succeeding chapters even if your instructors do not assign them. No musician or football player has ever been successful in his field without countless hours of practice. The same can be said for any successful chemist or chemistry student.

Special Feature: Energy and Entropy

There are few of us who, at one time or another, have not wished that we could predict future events. In science, the ability to predict the feasibility of processes and devices could have saved many wishful thinkers of the past countless hours of toil and wasted effort. The dreams of those who hoped to get something for nothing by constructing a perpetual motion machine were doomed to failure. Such a machine contradicts the experience of man with nature as expressed in fundamental laws or principles of nature. On the other

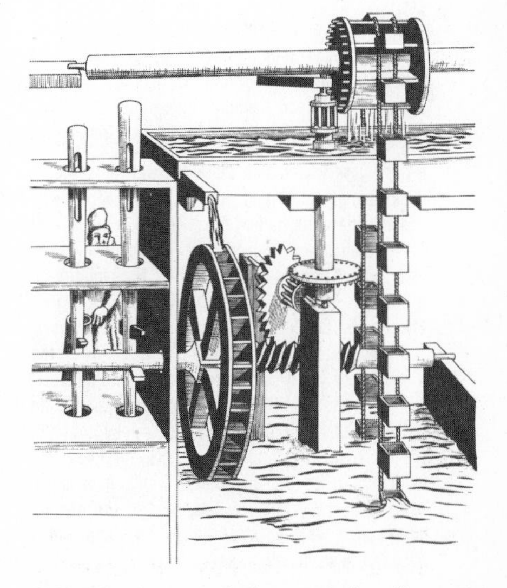

13-12 The inventor of the perpetual motion machine hoped to get something from nothing. Once set in motion, this machine was supposed to go on moving forever doing useful work, such as raising water or grinding corn, without the application of an external source of energy. What is wrong with it?

JOSIAH WILLARD GIBBS
1839–1903

Josiah Willard Gibbs is perhaps the most famous American scientist of the 19th century. He was a shy, unassuming professor of mathematical physics at Yale University from 1871 until his death in 1903. During this period he published many theoretical papers on thermodynamics, a branch of science concerned primarily with the energy transfer between a system and its surroundings. The laws of thermodynamics are applicable to all areas of science.

Gibbs prepared two papers covering about 400 pages which he published between 1873 and 1878 in the little known "Transactions of the Connecticut Academy of Sciences." For many years the work of this quiet and intensely religious Yale professor went unnoticed. Gibbs' paper was so thorough and the mathematical treatment so rigorous that few chemists who read it could understand it.

One of the first persons to recognize the value of Gibbs' work was James Clerk Maxwell of England. He explained Gibbs' concepts to van der Waals, who passed the information to a fellow professor at Amsterdam, Henrik Roozeboom. Roozeboom undertook many experiments to prove the validity of Gibbs' theories. It was not until the 1890's that Europe really recognized the value of Gibbs' work. Gibbs was described by Ostwald of Germany as the "founder of chemical energetics." Ostwald wrote: "The importance of the thermodynamic papers of Willard Gibbs can best be indicated by the fact that in them is contained, explicitly or implicitly, a large part of the discoveries which have since been made by various investigators in the domain of chemical and physical equilibrium and which have led to so notable a development in this field."

Once the value of his work was recognized, Gibbs received many honors and was paid tribute by scientists of all nations. He was, without doubt, one of the greatest theoretical scientists of all time. When visiting Yale University, Lord Kelvin, the British scientist, said: "By the year 2000 this university will be best known for having produced J. Willard Gibbs."

hand, the same principles assured modern scientists that their attempt to convert graphite into sparkling diamonds was theoretically possible and worth the effort. (See Fig. 3-9 p. 107.)

The key to predicting the feasibility of certain events and processes is energy. There are many facets of energy that enter the picture. Some of these are its conservation, its distribution (concentration in different parts of a system), its transfer, its transformation, and its conversion into work.

Energy is the prime mover of all material substances in the universe. It makes all work possible. In short, the *useful practical aspects of energy are realized by turning it into work.* An example is the conversion of energy from burning gasoline in the cylinders of an automobile into mechanical work represented by the motion of the pistons and of the automobile against frictional forces.

Work is a method of transferring energy from one part of a system to another part. This means that there is a "downhill" flow of energy from a region where there is a high concentration of energy (hot region) to one where there is a lower concentration (cold region). A cold object in contact with a hot one never becomes colder. In the automobile, the energy flows from the hot region (cylinders) to the cold water (radiator) region of the system.

It is impossible to convert all of the energy in a system into work. This would mean that all the energy would have to be concentrated in one part of the system—an unattainable event. The *entropy* change of a system during a spontaneous process may be thought of as a measure of the amount of energy that is unavailable for work.

The most fundamental law of nature, the Law of Conservation of Energy, dictates that no energy be lost by its conversion into work. Once it is used, however, it is distributed in such a way that it is less concentrated. There is a conservation of energy but less energy becomes *available for work.* In other words **the entropy (energy unavailable for work) of the universe increases whenever energy is transferred in a spontaneous process.** In terms of atoms and molecules we may think of the energy as being distributed among the quantized translational, rotational, and vibrational energy levels of molecules. We might refer to this as an *energy randomness.* When the faster-moving molecules of a system (representing a region of high energy concentration) collide with slow-moving ones, there is an energy transfer and redistribution so that, in general, the faster ones slow down and the slower ones speed up, thus decreasing the energy extremes and the amount of energy available for work.

Both *energy randomness* and *spatial* or *molecular randomness* (associated with increases in volume which provide more ways for molecules to be arranged) play a role in determining the spontaneity of a reaction. If, in accordance with the Second Law of Thermodynamics, the entropy of the universe is to increase, then **any spontaneous reaction in which the molecular randomness of the**

system decreases must be exothermic. The heat or thermal energy released may increase the temperature and energy randomness of the system and/or the energy randomness of the surroundings. The increase in energy randomness must be greater than the decrease in molecular disorder of the system.

The Second Law of Thermodynamics, which may be paraphrased: things in this world are becoming more mixed up; or, in the battle between order and disorder, disorder always wins, suggests that the universe is heading for an equilibrium state in which there is the most random distribution of energy and hence, maximum entropy. **As a system heads spontaneously for equilibrium, energy can be used to do work.** Once it reaches equilibrium, there is no net flow of energy or particles and no work can be done.

The criterion by which the spontaneity of a reaction may be judged was supplied in principle by Josiah Willard Gibbs, perhaps the most famous American scientist of the 19th century. He suggested that **any reaction which, in theory or practice, can do useful work at constant temperature and pressure is spontaneous.** In other words, any reaction which produces useful work is spontaneous and increases the entropy of the universe. The maximum useful electric work that can be obtained from a reaction is equal to a quantity called the change in free energy of the system (ΔG). When useful work is produced, the system loses an equivalent amount of energy so that there is a decrease in its free energy (G) and ΔG is negative. Thus, **a negative value for ΔG_{system} means that $\Delta S_{universe}$ must be positive.**

Useful work includes all work done by a system except the unavoidable work done by a system when it undergoes a volume change. When the volume of a system increases the system does mechanical work pushing back the atmosphere.

The tendency of a reacting system to achieve a state of minimum free energy (G) may be interpreted as the driving force of a chemical reaction.

Using the definition given above, a value for ΔG can be obtained by determining the maximum useful electric work obtainable from a reaction. This can be done by carrying out the reaction under the proper conditions in an electrochemical cell and measuring the voltage of the cell which in turn is related to ΔG and useful electric work. Electrochemical cells are discussed in Chapter 17.

QUESTIONS

1. Two colorless solutions are mixed in a stoppered flask. As the reaction proceeds, the resulting solution turns red, and a colorless gas is formed. After a few minutes, no more gas is evolved but the red color remains. What evidence is there that equilibrium has been established? pressure gage

2. An open flask contains a mixture of compound A and compound B. No observable reaction is taking place. May we conclude that this system has reached equilibrium? Explain.

3. Two gases are mixed and allowed to stand for a period of time. Continued observation reveals no change in the macroscopic properties of the system. This apparent constancy of properties might be caused by either the presence of an equilibrium condition or to a very slow reaction. How could you decide whether or not an equilibrium had been established?

4. Define equilibrium state in terms of (a) reaction rates, (b) change in observable properties, (c) energy and entropy states.

5. For each of these reactions, indicate the property which might be observed in order to determine when equilibrium has been reached.

(a) $PCl_5(g) \rightleftharpoons PCl_3(g) + Cl_2(g)$
(b) $CaCO_3(s) \rightleftharpoons CaO(s) + CO_2(g)$
(c) $H_2O(l) \rightleftharpoons H_2O(g)$
(d) $Cl_2(g) + 2HI(g) \rightleftharpoons I_2(s) + 2HCl(g)$
(e) $2HBr(g) \rightleftharpoons H_2(g) + Br_2(g)$

Ans. (c) pressure or temperature, (e) color.

6. What evidence is there to indicate that equilibrium is a dynamic state?

7. Explain why pressure can be considered as a concentration unit for gases.

8. Consider the reaction

$$N_2O_4(g) \rightleftharpoons 2NO_2(g)$$
$$\Delta H^\circ = 14 \text{ kcal}; K_e = 0.87 \text{ at } 55°C$$

What is the effect of each of these changes upon the concentration of N_2O_4 at equilibrium? (a) increasing the temperature, (b) increasing the volume, (c) adding more $NO_2(g)$ to the system without changing pressure or temperature, (d) adding He gas to the container, (e) adding a catalyst.

Ans. (b) Increased volume decreases the concentration of N_2O_4.

9. Answer questions (a, b, d, e) for $H_2O(g)$ in this reaction as you did for N_2O_4 in Problem 8.

$$H_2(g) + \tfrac{1}{2}O_2(g) \rightleftharpoons H_2O(g)$$
$$\Delta H^\circ = -57 \text{ kcal}; K_e = 1 \times 10^{40} \text{ at } 25°C$$

10. How can you increase the concentration of the product(s) in each of these reactions by varying the temperature and pressure (caused by volume change)?

(a) $4NH_3(g) + 5O_2(g) \rightleftharpoons 4NO(g) + 6H_2O(g)$ $\Delta H = -216$ kcal
(b) $Br_2(g) + Cl_2(g) \rightleftharpoons 2BrCl(g)$ $\Delta H = 3.5$ kcal
(c) $BaSO_4(s) \rightleftharpoons Ba^{2+}(aq) + SO_4^{2-}(aq)$ $\Delta H = 5800$ kcal

11. Write the equilibrium expressions (K) for each of these reactions.

(a) $H_2(g) + F_2(g) \rightleftharpoons 2HF(g)$
(b) $4NO(g) + 3O_2(g) \rightleftharpoons 2N_2O_5(g)$
(c) $BaCO_3(s) \rightleftharpoons BaO(s) + CO_2(g)$
(d) $Na_2CO_3 \cdot 10H_2O(s) \rightleftharpoons Na_2CO_3 \cdot H_2O(s) + 9H_2O(g)$

12. The equilibrium constants for three different reactions are (a) $K = 1.5 \times 10^{12}$, (b) $K = 0.15$, (c) $K = 4.3 \times 10^{-15}$. In which reaction is (a) the ratio of product to reactant large, (b) the ratio of product to reactant small?

13. Does the equilibrium constant for the reaction

$$Br_2(l) = Br_2(g)$$

increase or decrease as temperature increases? Explain.

14. Suggest four ways in which the equilibrium concentration of $NH_3(g)$ can be increased in a closed vessel if the only reaction is

$$N_2(g) + 3H_2(g) \rightleftharpoons 2NH_3(g)$$

15. The graph, Fig. 13-13, shows concentrations of all three species of the system

$$CO_{(g)} + Cl_{2(g)} \rightleftharpoons COCl_{2(g)}$$

plotted against time under a given set of conditions. (a) How much time was required for the system to reach equilibrium? (b) Approximate the value of K using the

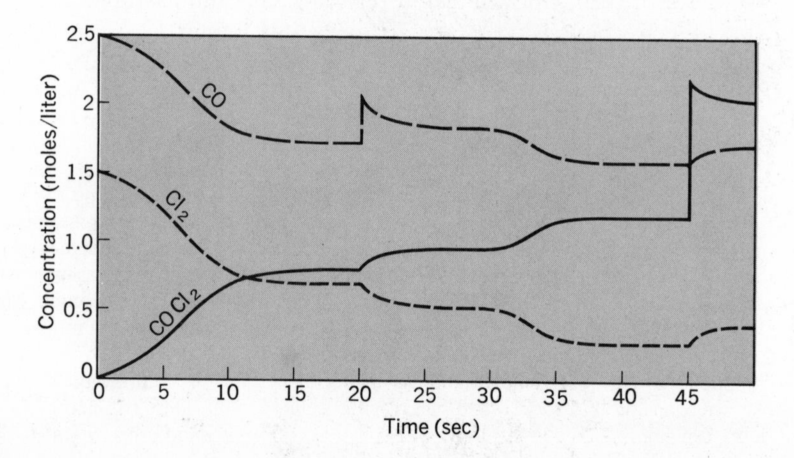

13-13 Graph of concentrations for CO, Cl_2, and $COCl_2$ for question 15.

concentrations at $t = 17$ sec. (c) Explain the changes 20 seconds after the initiation of the reaction. (d) What change in conditions might have been imposed on the system 30 seconds after the initiation of the reaction? (e) Are any events taking place between the interval of 15 sec and 20 sec? Explain. (f) What change may have taken place at $t = 45$ sec? (g) What differences would you have noted if a catalyst had been present during the entire course of this reaction? (h) List the changes you might impose on this system if you wanted to produce a maximum amount of $COCl_2$? (i) How could you account for the differences in the value of K at different points on the graph?

Ans. (a) 15 sec. (b) $K = 0.75$, (c) CO was introduced, (d) an increase in total pressure caused by a decrease in total volume, (e) in a state of dynamic equilibrium, microscopic changes are occurring, but there is no net change in amount of products or reactants, (f) a massive amount of $COCl_2$ was added to the system, (g) equilibrium would have become established more rapidly, (h) force more CO and Cl_2 into the reaction vessel, increase the pressure, remove $COCl_2$ as rapidly as it is formed, (i) changes in temperature.

PROBLEMS

1. A substance (CD) decomposes into C and D.

$$CD(g) \rightleftharpoons C(g) + D(g)$$

At the temperature of the experiment, 15.0 percent of CD is decomposed when equilibrium is established. (a) If the initial concentration of CD is 0.200 moles/ℓ, what are the equilibrium concentrations of CD, C, and D? (b) What is K for the reaction at this temperature?

2. A reaction may be represented by

$$A(g) + B(g) \rightleftharpoons AB(g)$$

At a given temperature 1.0 mole of A and 1.0 mole of B are placed in the 1.0-liter reaction vessel and allowed to reach equilibrium. Analysis revealed that the equilibrium concentration of AB was 0.40 molar. What percent of A had been converted to products?

3. Gas X_2 reacts with gas Y_2 according to the equation

$$X_2 + Y_2 \rightleftharpoons 2XY$$

0.50 mole each of X_2 and Y_2 are placed in a 1.0-liter vessel and allowed to reach equilibrium at a given temperature. The equilibrium concentration of XY is found to be 0.025 mole/ℓ. What is the equilibrium constant for this reaction?

4. Given the equilibrium reaction

$$A(g) \rightleftharpoons 2B(g) + C(g)$$

when 1.00 mole of A is placed in a 4.00-ℓ container at temperature T, the concentration of C at equilibrium is 0.050 mole/liter. What is the equilibrium constant for the reaction at temperature T?

5. The equilibrium constant for

$$2X(g) \rightleftharpoons Y(g) + Z(g)$$

is 3.0. How many moles of X are present at equilibrium when 1.00 mole each of Y and Z are placed in a 5.00-ℓ container? **Ans. 0.450 moles.**

6. Under a given set of conditions, an equilibrium mixture

$$SO_2(g) + NO_2(g) \rightleftharpoons SO_3(g) + NO(g)$$

in a 1.00-ℓ container was analyzed and found to contain 0.300 mole of SO_3, 0.200 mole of NO, 0.0500 mole of NO_2, and 0.400 mole of SO_2. Calculate the equilibrium constant for this reaction. **Ans. $K = 3.00$.**

7. At 55°C, the K for the reaction

$$2NO_2(g) \rightleftharpoons N_2O_4(g)$$

is 1.15. (a) Write the equilibrium expression. (b) Calculate the concentration of $N_2O_4(g)$ present in equilibrium with 0.50 mole/ℓ of NO_2. **Ans. 0.29 mole/ℓ.**

8. (a) Calculate the K for this reaction from the data $CO_2(g) + H_2(g) \rightleftharpoons CO(g) + H_2O(g)$ $[CO_2] = 1.17 \times 10^{-3}$ moles/ℓ, $[H_2] = 1.17 \times 10^{-3}$ moles/ℓ, $[CO] = 1.33 \times 10^{-3}$ moles/ℓ, $[H_2O] = 1.33 \times 10^{-3}$ moles/ℓ. **Ans. 1.29.**

9. One mole of NH_3 was injected into a 1-ℓ flask at a certain temperature. The equilibrium mixture

$$2NH_3 \rightleftharpoons N_2 + 3H_2$$

was then analyzed and found to contain 0.300 mole of H_2. (a) Calculate the concentration of N_2 at equilibrium. (b) Calculate the concentration of NH_3 at equilibrium. (c) Calculate the equilibrium constant for this system at this temperature and pressure. (d) Which way would the equilibrium be shifted if 0.600 mole of $H_2(g)$ were injected into the flask? (e) How would the injection of hydrogen into the flask affect the equilibrium constant? (f) How would the equilibrium constant be affected if the pressure of this system were suddenly increased?

Ans. (a) 0.100 mole of N_2/ℓ, (b) 0.800 mole of NH_3/ℓ, (c) 0.0042, (d) favoring the formation of NH_3, (e) no effect, (f) unchanged.

10. When 0.5 mole of CO_2 and 0.5 mole of H_2 were forced into a 1-liter reaction container, and equilibrium was established:

$$CO_2(g) + H_2(g) \rightleftharpoons H_2O(g) + CO(g)$$

Under the conditions of the experiment, $K = 2.00$. (a) Find the equilibrium concentration of each reactant and product. (b) How would the equilibrium concentrations differ if 0.50 mole of H_2O and 0.50 mole of CO had been introduced into the reaction vessel instead of the CO_2 and H_2?

11. An equilibrium mixture

$$H_2(g) + CO_2(g) \rightleftharpoons H_2O(g) + CO(g)$$

in a 10.00-ℓ container at a certain temperature was analyzed and found to contain $H_2(g) = 1.17$ moles, $CO_2(g) = 1.17$ moles, $H_2O(g) = 1.33$ moles, and $CO(g) = 1.33$ moles. (a) Calculate the equilibrium constant. (b) How would equilibrium quantity (moles) of H_2O be affected by an increase in the total volume of the system? (c) How would equilibrium concentration of water be affected by the increase in total volume? (d) How would the equilibrium constant be affected by an increase in total volume? (e) How many moles of water vapor would have to be injected into the original equilibrium mixture to increase the H_2 concentration to 0.150 moles/ℓ?

12. At 462°C, the reaction

$$(1)\ \text{heat} + 2NOCl(g) \rightleftharpoons 2NO(g) + Cl_2(g)$$

has an equilibrium constant, K of 8.0×10^{-2}.

(a) What is K at 462°C for the reaction

$$2NO(g) + Cl_2(g) = 2NOCl(g)$$

(b) What is K at 462°C for the reaction

$$NOCl(g) = NO(g) + \tfrac{1}{2}Cl_2(g)$$

Answer these true-or-false questions related to reaction (1) at 462°C. Do not write in this book.

(c) ______ The reaction is exothermic.

(d) ______ After equilibrium is established, increasing the concentration of NO causes an increase in the concentration of Cl_2.

(e) ______ After equilibrium is established, decreasing the volume of the container favors the formation of NOCl.

(f) ______ After equilibrium is established, increasing the temperature favors the formation of NO and Cl_2.

(g) ______ After equilibrium is established, decreasing the partial pressure of Cl_2 causes the equilibrium position to shift to the left.

(h) ______ Adding argon gas to the equilibrium system at constant total pressure will cause an increase in the yield of products.

(i) ______ Adding a catalyst decreases the time required for the reaction to reach equilibrium.

(j) ______ Adding more NOCl to the equilibrium system changes the value of K.

(l) ______ Increasing the temperature causes K to increase.

(m) ______ Adding a catalyst causes K to change.

(n) ______ The addition of a catalyst increases the yield of product.

(o) ______ At 462°C, two moles of NOCl react completely and form 3 moles of product (2 moles of NO and 1 mole of Cl_2).

(p) ______ Increasing the temperature increases the rate of the forward reaction.

(q) ______ Increasing the temperature increases the rate of the reverse reaction.

(r) ______ At a given temperature only one set of product and reactant concentrations satisfies K.

13. Consider this heterogeneous equilibrium

$$H_2(g) + S(s) \rightleftharpoons H_2S(g) + x\ \text{kcal}$$

At a given temperature, K for this reaction is 14.3. (a) What are the equilibrium concentrations of $H_2(g)$ and $H_2S(g)$ when 0.200 mole of each is allowed to react and reach equilibrium in a 2.00 ℓ container? (b) How is the magnitude of K affected by increasing the temperature of the system?

14

The Nature and Behavior of Solutes and the Properties of Solutions

Preview

We should so live and labor in our time that what came to us as seed may go to the next generation as blossom, and what came to us as blossom may go to them as fruit. That is what we mean by progress.

—Henry Ward Beecher (1813–1887)

Most reactions and processes encountered in industrial and school laboratories, as well as those associated with living organisms and the environment, occur in solutions. Reactions between substances involve collisions and interactions between molecules, atoms, or ions of the reacting chemicals.

In gaseous or liquid solutions, the solute particles are dispersed and mobile. Hence, there is adequate opportunity for collision and reaction. In the undissolved state there is minimum opportunity for contact between the particles of reactants. Thus, there is no apparent reaction taking place between the components of a dry Alka-Seltzer® tablet or of ordinary baking powder. A rapid reaction occurs, however, as soon as water is added to either of these dry mixtures. The addition of a solvent results in the dissolution of the crystals.

In general, aqueous solutions are prepared by adding either a molecular substance or an ionic crystal to water. In this chapter we shall investigate the behavior of various solutes when they are added to water and form aqueous solutions. We shall examine the nature of the solutions formed by the different kinds of solutes and use conductivity measurements to classify solutes as strong, weak, or nonelectrolytes.

The concept of solubility is of considerable interest to the practicing chemist as well as to the chemistry student. Both qualitative and quantitative analytical schemes which you may encounter are based on the solubilities of substances. Procedures for preparing and purifying substances are also developed, using solubility principles. Biological, as well as geological processes, involve the concept of solubility. We shall, therefore, qualitatively consider solubility principles and look for relationships between solubility and the nature of the solute and solvent. This discussion will provide an opportunity to apply again the concepts of enthalpy, entropy, and equilibrium.

An important section of this chapter involves the introduction of a rather generalized definition of acids and bases. In this section we shall apply equilibrium principles qualitatively to systems involving weak acids and bases, as well as to water. We shall find that the water equilibrium is an important factor in many reactions. The discussion of weak acid-base and water equilibria which also includes the concept of pH, will set the stage for a quantitative discussion of the same systems in Chapter 15.

DISSOLUTION OF IONIC CRYSTALS

14-1 Dissociation of Ionic Crystals Many of the reactions which we investigate involve solutions of dissolved salts. Thus, it is of considerable importance to understand the nature and behavior of salts when they are added to water. As you know, electric conductivity measurements can be used to reveal the nature of the dissolved particles. An apparatus for testing the conductivity of a solution is shown in Fig. 14-1. It consists of two platinum electrodes connected in series to an ammeter, an electric light bulb, and a source of electric energy. When the two electrodes are in

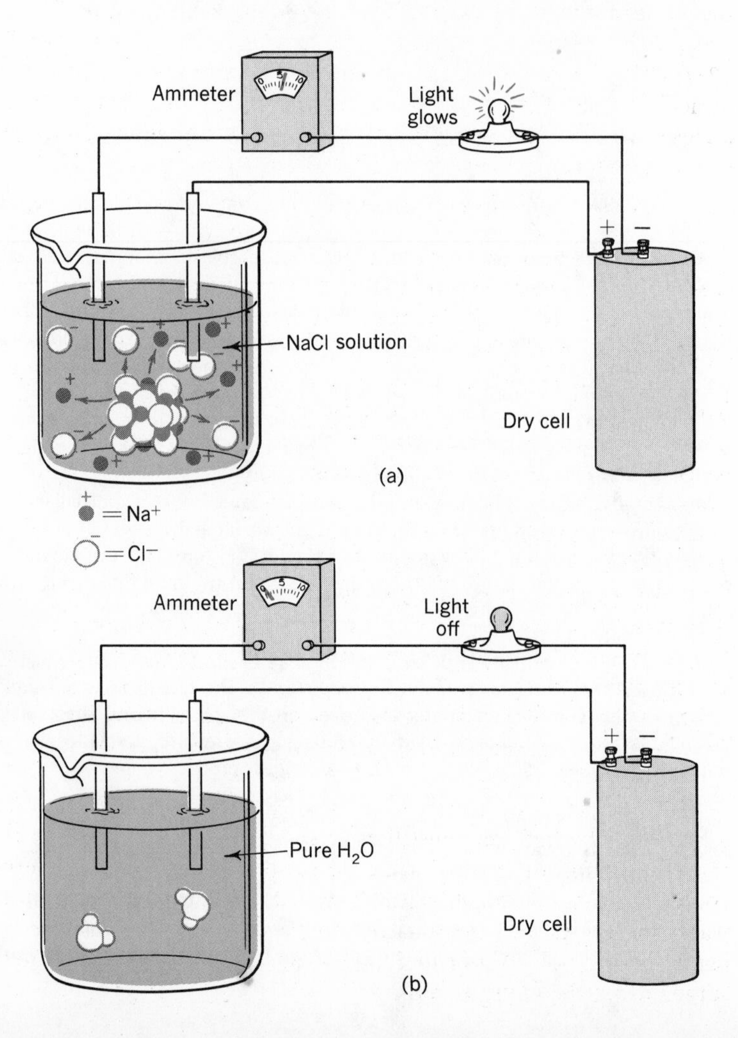

14-1 An apparatus such as this can be used to test the conductivity of solutions. Electrolytic solutions (a) contain ions which carry the charge between electrodes. Nonelectrolytes such as water (b) do not dissociate appreciably into ions and therefore conduct very little current. The ions in a solid ionic crystal are held together by strong ionic bonds and at room temperature do not possess enough kinetic energy to move freely under the influence of an applied voltage. Hence, when electrodes are placed on solid NaCl and other solid ionic crystals, no appreciable electric current is conducted.

contact, the light glows brightly and the ammeter shows a high reading as the electric charge flows through the completed circuit.

When the electrodes are placed in pure water, the light does not glow and no appreciable reading is observed on the meter. Apparently, water does not contain a sufficient number of electric charges to complete the circuit and permit the current to flow. Solid sodium chloride responds the same as water did.

Earlier (9-1) we found that solid sodium chloride does not conduct an appreciable electric current and that melted sodium chloride is a good conductor. These and other data (X-ray diffraction) suggest that NaCl crystals are an aggregate of ions. When the electrodes are immersed in an aqueous solution of an ionic crystal such as NaCl, the light glows brightly and a relatively high reading is observed on the ammeter. Because pure water does not conduct, we can assume that the current flow is caused by the presence of the solute. Since it has been shown that NaCl crystals are composed of ions, it is reasonable to assume that in a conducting solution, the electric charge is carried by ions. *Ionic crystals dissociate in water and yield hydrated ions.* The ions that are responsible for *electrolytic conduction* in a given salt solution may often be determined from the formula of the salt. Sodium chloride (NaCl) dissociates in water and yields $Na^+(aq)$ and $Cl^-(aq)$ ions. The *ionic dissociation* may be represented by

Dissociation is the antonym of *association.*

$$Na^+Cl^-(s) \longrightarrow Na^+(aq) + Cl^-(aq)$$

14-2 Classification of Solutes Solutes may be classified according to the extent to which they dissociate into ions in aqueous solution. Solutes which dissolve and form many ions are called ***strong electrolytes.*** When less than 50 percent of a dissolved solute exist as ions the solute is called a ***weak electrolyte.*** When less than 0.01 percent of a dissolved solute exists as ions, the solute is called a ***nonelectrolyte.*** Knowledge of whether a substance is a strong or weak electrolyte or a nonelectrolyte is valuable for predicting reactions and writing equations.

It can be demonstrated that *almost all salts are strong electrolytes.* This means that most dissolved salts exist largely in the form of ions. An ionic salt may not dissolve to an appreciable extent, but that quantity that does dissolve exists as hydrated ions in solution. For example, the solubility of silver chloride (AgCl) is approximately 0.0001 g/ℓ. Yet the small percentage that does dissolve exists in the form of ions. Let us now examine in some detail the dissolution of an ionic crystal and some of the factors related to its solubility.

14-3 Hydration Energy If an ionic crystal is to dissolve, the oppositely charged ions must be separated. This means that work must be done and energy must be supplied. Part of this energy comes from the interaction (attractive forces) between the polar water molecules and the ions of the solute. That is, energy is released as the polar water molecules interact with the electric field surrounding the charged ions.

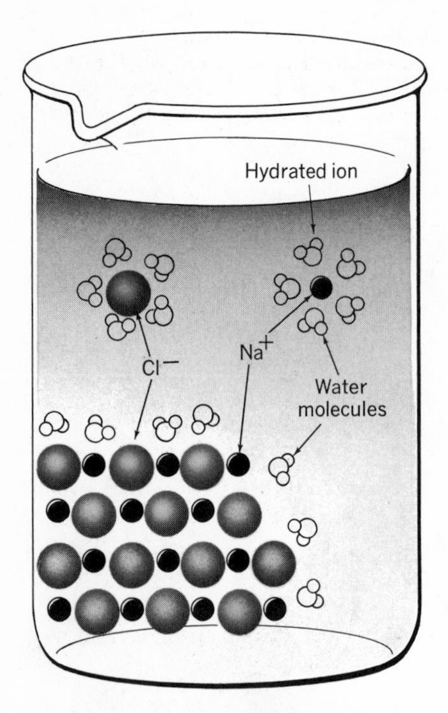

14-2 When an ionic crystal such as NaCl is added to water the Na^+ ions attract the negative ends of water molecules and the Cl^- ions attract the positive ends, resulting in dissociation.

TABLE 14-1
HYDRATION ENERGIES

Ion	*Radius Å*	*Hydration Energy (kcal/mole)*	*Ion*	*Radius Å*	*Hydration Energy (kcal/mole)*
Li^+	0.65	125	Na^+	0.96	100
Na^+	0.96	100	Mg^{2+}	0.65	464
Ca^{2+}	1.68	70	Al^{3+}	0.55	1122

Hydration energy decreases as radius increases.

Hydration energy increases with increasing charge and decreasing radius.

Water molecules are extremely polar. Hence, they attract and hold ions by relatively strong electrostatic forces. The positive pole (hydrogen end) of a water molecule attracts negative ions, and the negative pole (oxygen end) attracts positive ions. The *ion-dipole interaction* is called ***solvation*** or ***hydration.*** The energy released as a result of the process is known as the ***hydration energy.*** Hydration energies for the hydration of various ions are listed in Table 14-1. The hydration energy of a given ion depends on the strength of the electric field surrounding it. In turn, the magnitude of the field depends on the charge and radius of the ion. The ratio

ionic charge/ionic radius

is known as the ***charge density.*** The data in Table 14-1 can be used to show that charge density, field strength, and hydration energy *increase as the charge of an ion increases and as its radius decreases.* High hydration energies which represent a greater degree of interaction between the ions of a crystal and water molecules enhance the solubility of an ionic substance.

The tendency of an ionic salt to dissolve in water is related to the relative magnitudes of the hydration energy and the crystal lattice energy. Crystal lattice energy is the energy required to separate ions in a crystal into gaseous ions. A high crystal lattice energy reflects high crystal stability and indicates that much energy is required to separate the crystal into ions. Other factors being equal, *high crystal energies inhibit solubility, however, high hydration energies lower the energy of the system and enhance solubilities.*

TABLE 14-2
DIELECTRIC CONSTANTS

Solvent	*Formula*	*Constant*
Water	H_2O	80
Benzene	C_6H_6	2.3
Methanol	CH_3OH	33.6
Hexane	C_6H_{14}	1.9
Ammonia	NH_3	2.2
Ethylene glycol	$C_2H_4(OH)_2$	37
Glycerol	$C_3H_5(OH)_3$	42
Hydrogen fluoride (0°C)	HF	84
Hydrogen peroxide (0°C)	H_2O_2	84

14-4 Dielectric Effect As an ionic crystal dissociates, a number of water molecules surround each ion and form a protective or insulating mantle which partially neutralizes the charge on the ions and prevents them from recombining. The capsule of water molecules acts as an insulator and reduces the force of attraction between ions. The insulating effect is called the ***dielectric effect.*** It is described in terms of a *dielectric constant,* D, which is a measure of the ability of a substance to reduce the force of attraction between charged particles. The force of attraction is expressed by Coulomb's Law

$$F = k\frac{q_1q_2}{r^2}$$

where k is a constant of proportionality. The dielectric constant, D, can be considered a part of this constant of proportionality, k. It has been found that F varies *inversely* with D. We can therefore insert D in the above equation to show this relationship. The equation then becomes

$$F = k'\frac{q_1q_2}{Dr^2}$$

where k' is a new constant of proportionality whose value depends on the units used for the other factors in the equation.

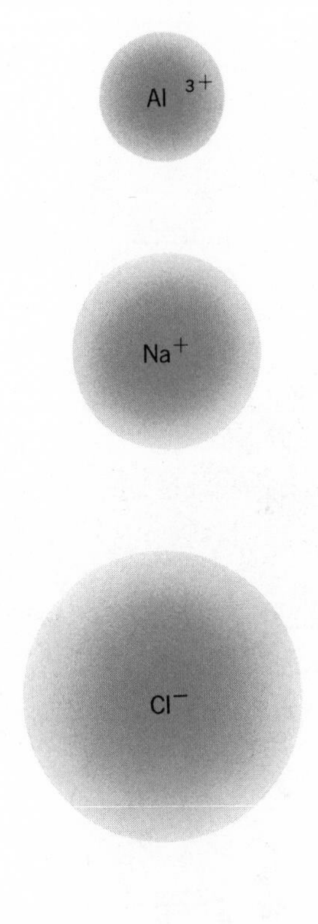

14-3 Of the three ions shown, it can be seen that the Al^{3+} ion has the greatest charge density and the Cl^- ion has the least charge density. Thus Al^{3+} ions are more strongly hydrated than the Na^+ or Cl^- ions.

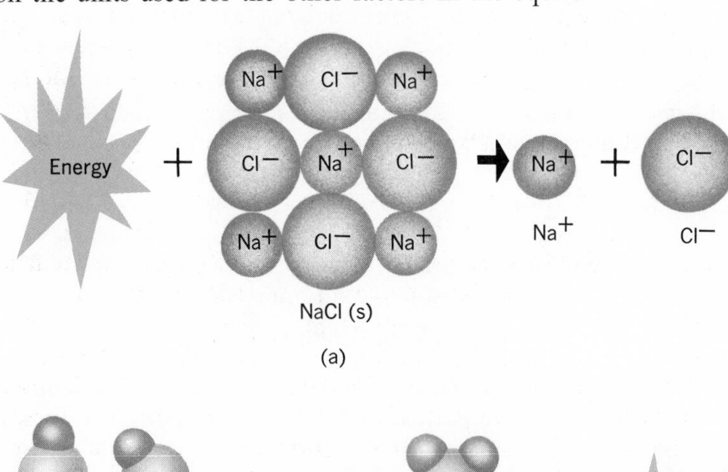

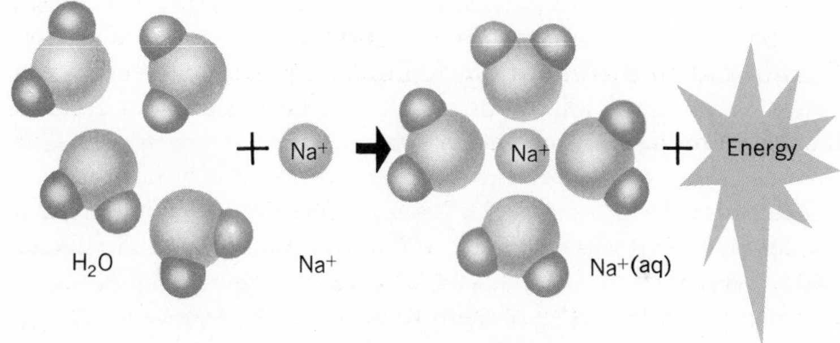

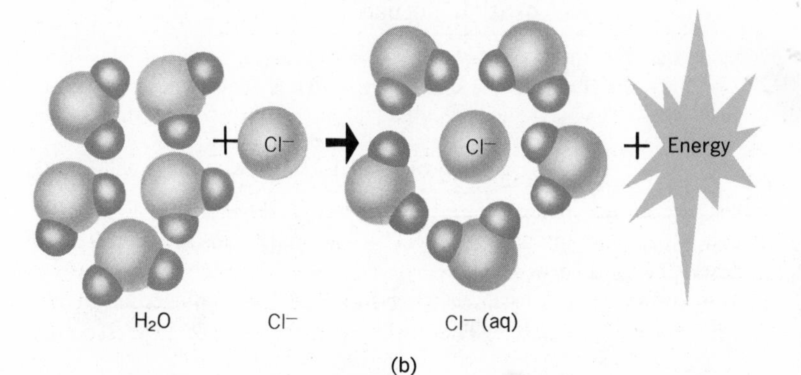

14-4 The dissociation of ions in an ionic crystal raises the energy of the system. This energy factor opposes dissolution of a crystal. (b) The hydration of ions liberates energy to the surroundings and lowers the energy of the system. The relative magnitude of the two energy factors determines the net heat change for the overall process.

The dielectric constant for water is approximately 80, which means that the force of attraction between charged particles in water is approximately $\frac{1}{80}$ of that in a vacuum or in air. In general, *ionic substances dissolve more rapidly in a solvent with a high dielectric constant than in one with a low constant.* The dielectric constants for various solvents are listed in Table 14-2, page 438.

The low dielectric constant of benzene is consistent with the fact that NaCl, an ionic solute, does not dissolve in benzene, a nonpolar solvent. Benzene molecules cannot effectively reduce the force of attraction between the ions in the crystal. Since there is little attraction (interaction) between the solute and solvent particles, dissolution does not occur.

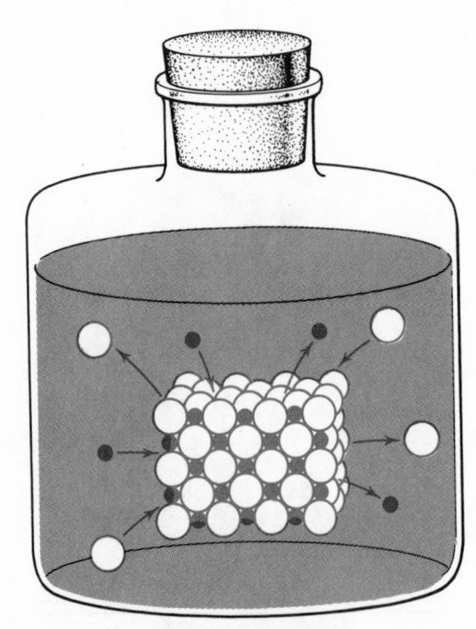

(a)

14-5 Solubility of Ionic Salts The actual solubility of a substance depends on the nature of the solute, the nature of the solvent, and other factors which are sometimes difficult to predict. The most reliable way to gain information regarding the solubilities of ionic compounds in water is to rely on experimental data from the chemical laboratory. Such experimental data provide a set of general solubility rules which are of value in predicting reactions and in writing equations. These rules are listed in Table 14-3.

14-6 Solubility Equilibrium The salts noted in Table 14-3 as slightly soluble do not dissolve to an appreciable extent in water. In general, they dissolve to an extent of less than 1 mg per liter of solution. The dissolving of any solute in any solvent provides another illustration of the *equilibrium concept.* When a solid is placed in water, solute particles immediately go into solution and may be dispersed throughout the solution. As the process continues, the concentration of solute in solution increases, and the rate at which solute particles return to the solid crystal (crystallize) increases. *Equilibrium is reached when the rate at which the solute particles are going into solution equals the rate at which they are returning to the solid state.* The process is shown in Fig. 14-5. When a solution reaches this state of dynamic equilibrium, it is said to be *saturated* with respect to the solute. At a given temperature, a saturated solution can hold no more solute. A solution which

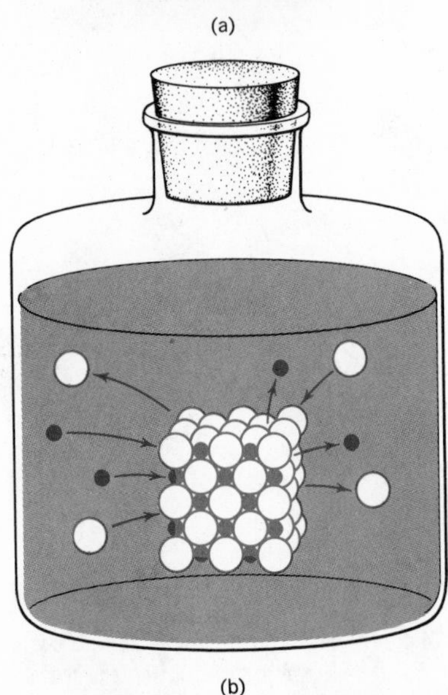

(b)

14-5 Solubility equilibrium. At a given temperature the ionic concentration is independent of the quantity of undissolved solid present. The ionic concentrations in (a) and (b) are the same. What second equilibrium exists in this container?

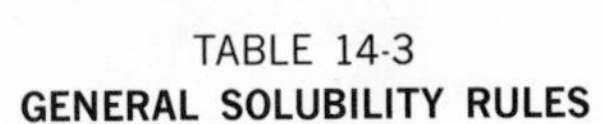

TABLE 14-3
GENERAL SOLUBILITY RULES

1. Most salts of Group I A cations and ammonium ions are soluble.
2. Most nitrates and acetates are soluble. $AgC_2H_3O_2$ is moderately soluble.
3. Chlorides, bromides, and iodides are soluble except the silver, mercury(1), and lead compounds ($PbCl_2$ is moderately soluble in hot water.)
4. Most sulfates are soluble except $BaSO_4$, $SrSO_4$, and $PbSO_4$ ($CaSO_4$ and Hg_2SO_4 are moderately soluble).
5. Most carbonates and phosphates are slightly soluble except those listed in Rule 1.
6. Most hydroxides are slightly soluble except those listed in Rule 1 and $Ba(OH)_2$. $Ca(OH)_2$ is moderately soluble.
7. Most sulfides are only slightly soluble except those listed in Rule 1. Magnesium, calcium, barium, and aluminum sulfides decompose in water.
8. Most silver salts are slightly soluble except $AgNO_3$ and $AgNO_2$. $AgClO_4$ and Ag_2SO_4 are moderately soluble.

is *unsaturated* at a given temperature is one which can dissolve more solute because an equilibrium does not exist.

When a solution holds more solute than would normally be dissolved in a saturated solution at a given temperature, it is said to be *supersaturated.* In supersaturated solutions, the solute particles are slow to achieve the geometric arrangement that is characteristic of their solid crystals. These solutions are unstable and usually return to the equilibrium state if sufficient time elapses. The solute can be made to crystallize more rapidly by *seeding* (adding a small crystal of solute) or by agitating the solution. Once it has begun, crystallization usually proceeds rapidly and equilibrium is attained in a short amount of time. The ***solubility*** of a substance is generally expressed in terms of the number of grams required to form a saturated solution with 100 g of water. Solubilities of specific substances may be found in a handbook of chemistry.

14-7 Energy and Entropy Changes Accompanying Dissolution of a Solid Solute Dissolution of a solid solute usually increases the entropy (disorder) of a system. In addition, energy *changes always accompany the dissolving of a solute.* It is not easy to predict whether the overall solution process will be endo- or exothermic. Energy is always required to separate particles that are attracted to each other, and such forces of attraction exist in any solid. If this were the only energy factor, then dissolution of a solid would always be endothermic. Attractive forces between solute and solvent, however, liberate energy so the net energy change during dissolution involves both an exothermic and endothermic factor.

The spontaneous solution of many solid solutes is an endothermic process and is an example of a spontaneous change which results in a condition of higher energy. This observation may seem contrary to the principle that spontaneous changes take place in the direction of decreased energy content. The observation is understandable when we recall that *spontaneous changes are favored by an increase in entropy.* Certainly, the system is likely to be more disordered when the solute is dispersed throughout the solution than when it is packed in the ordered arrangement of a solid crystal even though the water molecules which hydrate each ion are in a more ordered state. It should be noted, however, that the dissolution of a gaseous solute generally decreases both the entropy and energy of the system. That is, the disorder of a gas is generally greatly decreased by its dissolution. In general, spontaneous dissolution processes must *be interpreted in terms of both energy and entropy concepts.* At a given temperature, the equilibrium concentrations of the ions represent a compromise between the tendency of the system to achieve a condition of maximum entropy and minimum energy. The equilibrium that exists in a saturated solution of silver chloride (AgCl) may be represented as

$$\mathbf{AgCl(s) \rightleftharpoons Ag^{+}(aq) + Cl^{-}(aq)}$$

At 25°C the concentration of silver ions in a saturated solution of silver chloride is approximately 1.4×10^{-3} g per liter of water.

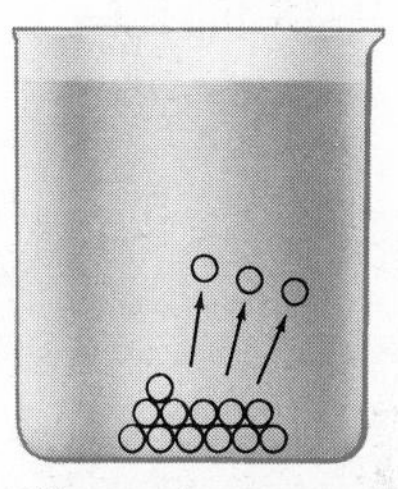

The tendency toward maximum entropy favors dissolution (dissolving)

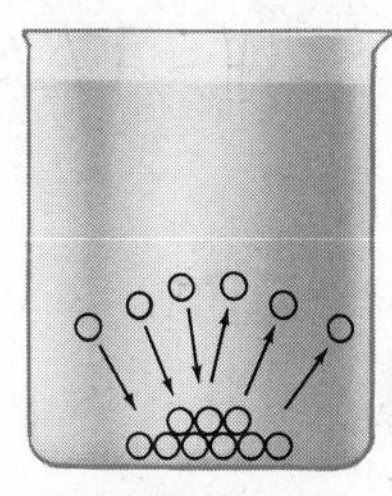

At equilibrium the solubility is fixed when the tendency toward minimum energy is balanced by the tendency toward maximum entropy

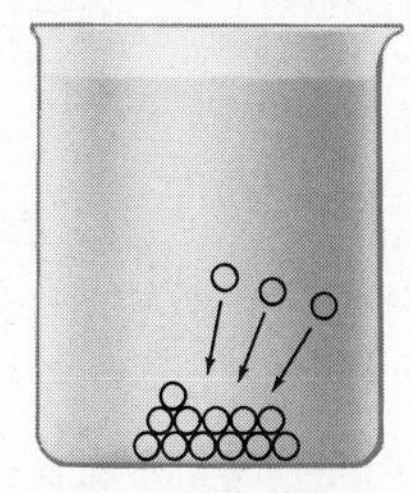

The tendency toward minimum energy favors crystallization

14-6 At a given temperature, the solubility of a solid is fixed by energy and entropy factors.

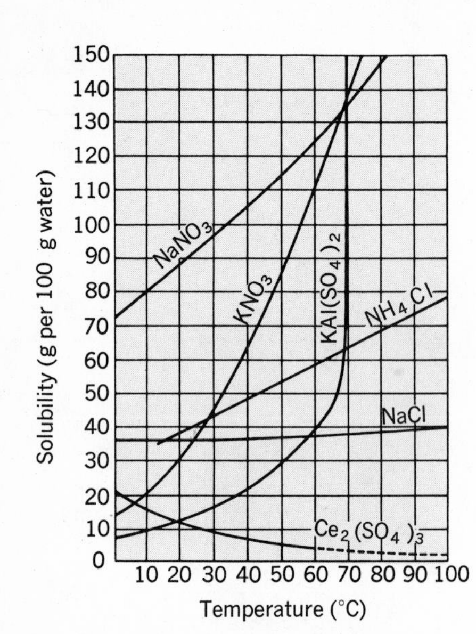

14-7 Solubility curves. Every point on each curve represents the number of grams of solute that dissolves in 100 g of water at a given temperature.

recall ΔG = ΔH − TΔS The more −ve ΔG the greater the tendency for spontaneity ie more −ve ΔH, more +ΔS. As T increases |−TΔS| increases.

Ionic salts having low solubilities are of considerable interest to us in chemical analysis as well as in natural processes. One such salt is calcium carbonate, the building material of stalagmites and stalactites. Laboratory observations indicate that $CaCO_3(s)$ does not dissolve to an appreciable extent in water. The dissolution of $CaCO_3$ is represented by the equation

$$\mathbf{CaCO_3(s) \rightleftharpoons Ca^{2+}(aq) + CO_3^{2-}(aq)}$$

We can use tabulated enthalpy and entropy data to gain insight and interpret our observations. These data reveal that the dissolution process is exothermic ($\Delta H < 0$) and there is a rather large decrease in entropy ($\Delta S < 0$). The negative value for ΔH indicates that the hydration energy of the ions must be greater than the crystal lattice energy. Because the enthalpy term tends to lower the energy of the system and make the process spontaneous, the entropy factor must be responsible for the low solubility. Qualitatively, without the help of the data, we might have used the equation to predict that the entropy would increase as the solid dissolves. The data, however, contradict this; so we must conclude that the overall decrease in entropy is connected to the reduction in the disorder of the water molecules when they are bound to the ions.

14-8 Effect of Temperature Change on Solubility The solubility of a salt may be varied by changing any of the factors which disturb the equilibrium. For example, changing the temperature is a stress that disturbs a system at equilibrium. Consider silver nitrate, a soluble salt, which at room temperature, dissolves to the extent of about 1200 g per liter of water. Experiment shows that the dissolution process is endothermic. Increasing the temperature of a saturated solution of silver nitrate originally at room temperature represents a stress which is relieved when more silver nitrate dissolves. Thus at a higher temperature, a new equilibrium is established in which there is a higher concentration of hydrated silver ions. Since the energy of the system increases, the high solubility of $AgNO_3$ must be attributed to the large increase in the entropy of the system. The variation of solubility with temperature is often represented in graphical form. Solubility graphs, similar to that represented in Fig. 14-7, enable us to determine quickly the effect of a temperature change on the solubility of a compound. They also readily allow us to compare the solubilities of different compounds at various temperatures.

FOLLOW-UP PROBLEMS

1. Write equations that represent the dissolution of these soluble salts (a) KBr, (b) $BaCl_2$, (c) $Ca(NO_3)_2$, (d) $(NH_4)_2Cr_2O_7$. (Place the formula for the solid solute on the left and those for the hydrated ions on the right side of the equation.)

2. Which compound in each of these pairs is more soluble? (a) Na_2CO_3 and $PbCO_3$, (b) $ZnSO_4$ and $BaSO_4$, (c) NH_4Cl and AgCl, (d) $AlPO_4$ and $AlCl_3$.

3. Refer to Fig. 14-7 and answer these questions. (a) How many grams of NaCl can be dissolved in 100 g of water at 0°C? (b) How many grams of KNO_3 can be dissolved in 1 liter of water at 45°C (assume density of water is 1 g/ml). (c) What mass of water at 50°C is needed to dissolve 90 g of $KAl(SO_4)_2$?

14-9 Bond Formation and Solubility When a number of ionic salts, such as copper sulfate, are crystallized from a water solution by the evaporation of solvent, a definite number of water molecules become incorporated as part of the ionic crystal. Such a crystal is called a *hydrate.* For example, when copper (II) sulfate, a white solid is dissolved in water, a blue solution is obtained. The hydrated copper (II) ions are responsible for the blue color. When the solid is crystallized from this solution, blue crystals of hydrated copper sulfate are formed. Analysis of this compound reveals that there are five water molecules associated with each copper sulfate ion-pair, and that four of the water molecules are bound to the Cu(II) ion and one is bound to the sulfate ion. One bonding model depicts the water molecules forming coordinate covalent bonds with the Cu^{2+} ions. The fifth water molecule is probably bonded by interaction between the positive end of the water dipole and the negatively charged sulfate ion.

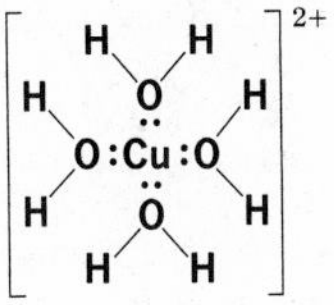

The electron-dot formula of a hydrated copper ion is shown in the margin. The four pairs of bonding electrons are furnished by the oxygen atoms in the water molecules. Thus a water molecule is an electron-pair donor and a Cu(II) ion is an electron-pair acceptor. The formula of the hydrate is written $CuSO_4 \cdot 5H_2O$. The hydrate is called copper(II) sulfate pentahydrate. The color of many compounds formed by the transition metals is caused by the interaction of the electrons of the ion with the electrostatic force fields of the species which coordinate with the ion. A more detailed discussion of this phenomenon is found in Chapter 19. The formulas and names of some common hydrates are listed in Table 14-4. Adding water to anhydrous copper(II) sulfate releases a vast amount of heat. The heat of hydration of Cu(II) is over 500 kcal/mole. It is the stabilizing effect of this large hydration energy that accounts for the relatively high solubility of copper sulfate. Because of the fact that a number of ions of the transition elements form stable coordinate bonds with water, they are often more soluble than corresponding compounds formed by Group I-A and II-A elements. Thus, $CuSO_4$ is much more soluble than K_2SO_4. As a result of these observations, we can conclude that the forma-

```
[    H   H        ] 2+
[     \ /         ]
[ H    O     H    ]
[  \   ..   /     ]
[   O : Cu : O    ]
[  /   ..   \     ]
[ H    O     H    ]
[     / \         ]
[    H   H        ]
```

TABLE 14-4
COMMON HYDRATES

Formula	*Name*	*Color*
$CuSO_4 \cdot 5H_2O$	copper(II) sulfate pentahydrate	blue
$CoCl_2 \cdot 6H_2O$	cobalt(II) chloride hexahydrate	red
$NiCl_2 \cdot 6H_2O$	nickel(II) chloride hexahydrate	green
$BaCl_2 \cdot 2H_2O$	barium chloride dihydrate	colorless
$MgSO_4 \cdot 7H_2O$	magnesium sulfate heptahydrate	colorless
$Na_2CO_3 \cdot 10H_2O$	sodium carbonate decahydrate	colorless
$Ba(OH)_2 \cdot 8H_2O$	barium hydroxide octahydrate	colorless
$AlCl_3 \cdot 6H_2O$	aluminum chloride hexahydrate	colorless
$CaCl_2 \cdot 2H_2O$	calcium chloride dihydrate	colorless

SVANTE AUGUST ARRHENIUS
1859–1927

When Svante Arrhenius presented his doctoral dissertation at the University of Upsala, Sweden, he was afraid it would not be accepted because he knew that it contained some unorthodox ideas. In his dissertation, Arrhenius pointed out that neither pure water nor dry salt would conduct an electric current but that salt dissolved in water produced a solution that would conduct electricity. He explained this observation by suggesting that "molecules" of salt broke up into charged particles or "ions" which were able to transport an electric charge.

To the chemists of Upsala this was a radical and unlikely idea. They grudgingly granted him a doctor of philosophy degree, although they indicated that they could not accept the idea of ions. They argued that sodium chloride dissolved in water gives no visible evidence of the greenish chlorine gas that should accompany the decomposition of sodium chloride. Arrhenius replied that the ion of chlorine had different properties from the chlorine atom. Faraday had previously used the term ion, but he had imagined that ions were produced by electric current. Arrhenius claimed that ions were already present in a salt solution prior to the passage of current through the solution and that these ions were responsible for conducting the current. He even ventured to say that chemical reactions involving these salts were merely combinations of these ions. He also defined acids and bases in terms of his ions.

Since Arrhenius could not get support for his theories in Sweden, he sent his dissertation to several German chemists, most of whom rejected the idea of ions. However, one influential German chemist, Wilhelm Ostwald, read the dissertation and immediately recognized the value of Arrhenius's idea. Ostwald and Arrhenius championed the cause of ions for several years against the strongest opposition such as that of Mendeleyev, who also opposed the idea of ions. In 1903, Arrhenius was awarded the Nobel Prize in chemistry for his work. He was later appointed Director of the newly founded Nobel Institute for Physical Research at Stockholm.

tion of stable coordinate bonds between the ions of a salt and water molecules produces hydrates and enhances solubility.

On the other hand, the relatively large degree of covalent character between metal and nonmetal atoms in certain binary salts can help to explain the very low solubilities of a number of compounds formed by transition elements. Solid compounds formed by elements which differ only slightly (less than 1.7) in electronegativity frequently have a low solubility in water. For example, the difference in electronegativity between mercury and sulfur is less than 0.7. This indicates that the bonds in solid HgS have a large degree of covalent character. Thus there is a greater electronic charge density between the nuclei of the atoms than in ionic compounds whose bonds have little covalent character. The attraction between the nuclei of the atoms and the electron cloud between them adds to any residual ionic attraction. Thus, the atoms are tightly bonded and the substance is slightly soluble. This is consistent with our observation that the solubility of HgS is approximately 1×10^{-25} moles/ℓ.

The unusually high solubilities of ammonium compounds can be partially explained in terms of hydrogen bonds formed between the hydrogen atoms of ammonium ions and water molecules. On the other hand, the low solubilities of solid metallic hydroxides such as $Fe(OH)_3$ and $Be(OH)_2$ can be partially explained in terms of hydrogen bonding within the solid crystal. In these solids, the positive ions are small enough to permit the formation of hydrogen bonds between the hydrogen atom of one hydroxide ion and the oxygen atom of its neighbor. The additional bonding increases the stability of the crystal and decreases the solubility.

We have explored the dissolution of ionic crystals in as much detail as is practical. Let us now turn our attention to the dissolution of molecular substances—first nonelectrolytes.

DISSOLUTION OF NONELECTROLYTES

14-10 Dissolution of Polar Covalent Molecular Species A large number of molecular solids and liquids such as sugar and alcohol are composed of polar covalent molecules. These are soluble to varying degrees in polar covalent solvents such as water. Solution of these substances is generally not the result of a chemical reaction but of *dipole-dipole* interaction between solute and solvent. For example, there are attractive forces between the opposite poles of ordinary (ethyl) alcohol molecules and water molecules. The interaction is great enough for the two substances to be completely *miscible* (soluble in all proportions).

Energetically speaking, the degree to which a molecular substance composed of covalent molecules dissolves in a molecular solvent depends on the attraction between the solute and solvent molecules relative to that between the solvent molecules them-

$$\Delta S = S_{final} - S_{initial} > 0$$

Acetone + Chloroform → Acetone-chloroform solution + Energy

$$\Delta H = H_{final} - H_{initial} < 0$$

14-8 When acetone and chloroform are mixed, the enthalpy of the system decreases and the entropy increases. As a result the two liquids dissolve in each other in all proportions. Note the molecular geometry and slight polarity of each molecule.

ΔH is −ve or only slightly +ve solution will take place because ΔS is almost always +ve in solution processes
ΔG = ΔH TΔS
−ve −TΔS > +ΔH
∴ ΔG is −ve

selves. If the two attractions are of approximately the same magnitude, solution takes place. For example, when liquid chloroform is mixed with liquid acetone, a relatively small quantity of heat is liberated (less than 0.5 kcal/mole). This indicates that the energy of the solution is slightly less than that of the two pure substances, and suggests that the solute-solvent attractive forces are slightly greater than the solute-solute forces. That is, more energy is liberated through solute-solvent interaction than is absorbed in separating solute molecules or solvent molecules. Although the energy effect is relatively minor, it favors the mutual solubility of acetone and chloroform. The entropy effect therefore, is largely responsible for the observed high solubility. The dispersion of chloroform molecules throughout the acetone increases the entropy of the system. The negative value for ΔH° and the positive value for ΔS° suggests a spontaneous process and a high solubility. This is consistent with the observation that acetone and chloroform are completely miscible. Large negative values for $\Delta H^\circ_{solution}$ and large positive values for $\Delta S^\circ_{solution}$ are associated with high solubilities.

The formation of intermolecular hydrogen bonds between the solute and solvent molecules enhances the solubility of a solute. This explains why some molecules with a low polarity dissolve in water while ones with higher polarities do not. For example, nitrobenzene ($C_6H_5NO_2$) is more polar than phenol (C_6H_5OH) but phenol is more soluble in water than nitrobenzene. The greater solubility of phenol is caused by the formation of hydrogen bonds between the oxygen atom of water and the hydrogen atom of the *hydroxyl group* in a phenol molecule (Fig. 14-9). Unlike an hydroxide ion, an hydroxyl group does not carry a charge and is not a free species in solution. A hydroxyl group has one fewer electron than an hydroxide ion. Thus, an hydroxyl group shares an electron and forms a covalent bond with the carbon atom in carbon compounds. In general, organic (carbon) compounds, such as phenol, sugar, alcohol, and glycerol which contain one or more hydroxyl groups (—OH) are quite soluble in water.

decrease the P.E. of the system

Phenol — Hydrogen bond — Water

14-9 The solubility of phenol in water may, in large part, be attributed to the hydrogen bonds formed between the oxygen atom of water molecules and the hydrogen atoms of the hydroxyl groups (OH) from phenol molecules.

hydroxide $:\ddot{O}:H^-$

hydroxyl $\cdot\ddot{O}:H$

14-11 Dissolution of Nonpolar Covalent Molecular Species Nonpolar covalent substances such as methane (CH_4) and benzene (C_6H_6) are composed of nonpolar covalent molecules. Their structures do not permit appreciable hydrogen bonding; thus, such compounds do not readily dissolve in polar solvents such as water. The energy required to rupture the hydrogen bonds in water is large and would not be balanced by energy released through formation of hydrogen bonds between solute and solvent and by the entropy increase of the system. $\Delta H°$, therefore, would have a large positive value and the process, energetically speaking, would be nonspontaneous (improbable).

Many molecular crystals composed of nonpolar covalent molecules, which do not dissolve appreciably in water, dissolve in nonpolar solvents. For example, naphthalene (moth balls, $C_{10}H_8$) a solid composed of nonpolar molecules, dissolves in nonpolar solvents such as benzene (C_6H_6). The van der Waals forces between the naphthalene molecules and benzene molecules are about the same as those between benzene molecules. Hence, there are no strong linkages to be disrupted in either the solvent or the solute, and no appreciable energy effects have to be considered. Because of the fact that the dissolution of the naphthalene increases the entropy of the system, the drive toward maximum entropy predominates, and the crystal dissolves. On the basis of observations such as these we can conclude that nonpolar solutes generally dissolve in nonpolar solvents.

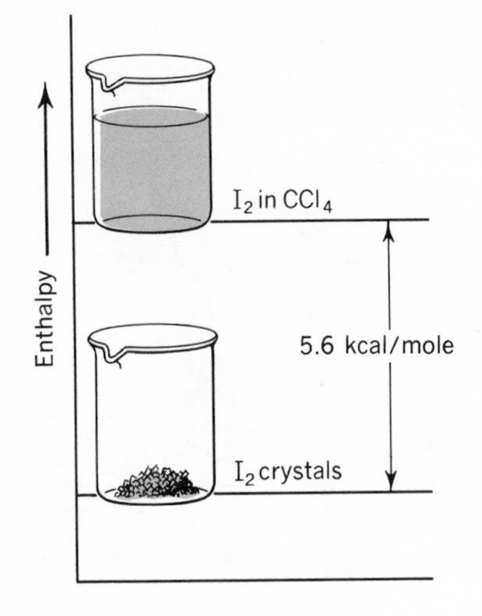

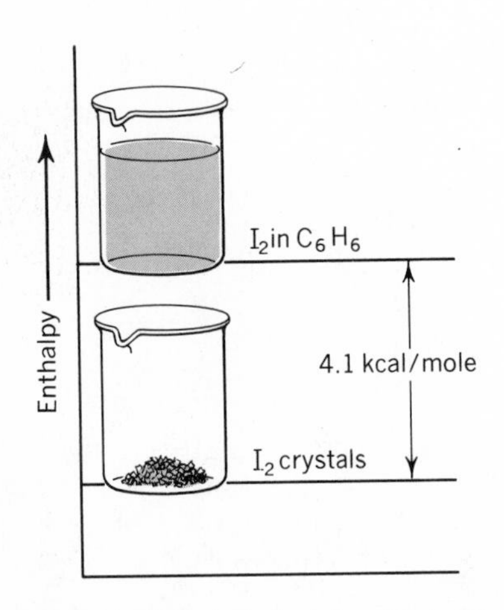

14-10 The higher heat of solution of I_2 in CCl_4 raises the potential energy of the system to a value higher than that of the I_2-benzene system. This energy effect opposes dissolution so that I_2 is more soluble in benzene than in carbon tetrachloride.

If we assume that the increase in entropy (randomness) is the same for the dissolution of a given nonpolar solute in two different solvents under the same conditions, then the relative solubilities in the two liquids can be interpreted in terms of the energy factor as represented by the enthalpy (heat) of solution ($\Delta H°_{solution}$). The solubility is greater in the system which absorbs the smaller amount of energy. For example, the enthalpy of solution of I_2 in carbon tetrachloride is 5.6 kcal/mole while in benzene the heat of solution is 4.1 kcal/mole. This suggests that there is more interaction between iodine and benzene than between iodine and carbon tetrachloride, and that iodine would therefore be more soluble in benzene than in CCl_4. See Fig. 14-10.

In general, nonpolar and polar molecular liquids are **immiscible.** That is, when mixed they do not dissolve in one another and produce a one-phase system. When immiscible liquids are mixed and allowed to separate, an equilibrium is established in which both phases are saturated solutions with respect to the other component. These phenomena are illustrated in Fig. 14-11.

14-12 Solubility of Gases in Liquids There are many examples of gases dissolved in liquids. Carbonated beverages and the numerous products packaged in spray cans contain gases dissolved in liquids. The small solubility of air in water is great enough to sustain marine life, and even the small solubility of air in human blood is enough to present a hazard to deep-sea divers.

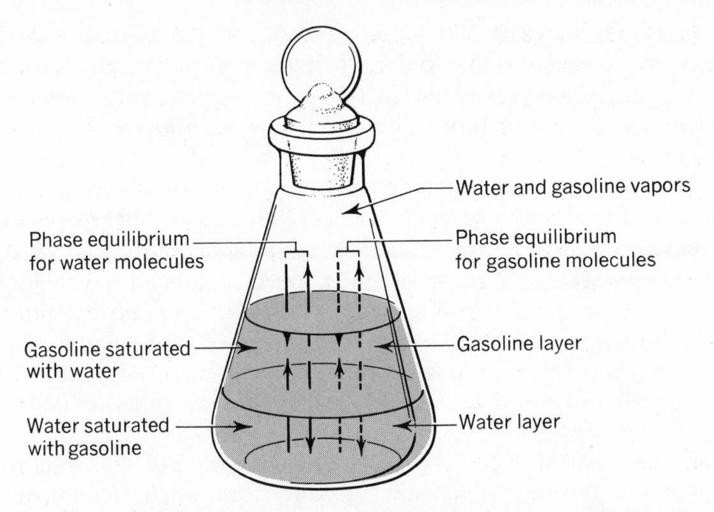

14-11 When water, a polar liquid, and gasoline, a mixture of nonpolar liquids are mixed, a two-phase (two-layered) system results. Each liquid layer is saturated with the other liquid. The double arrows represent the equilibria (movement of the molecules across the boundaries) that exist between the molecules in the different phases.

14-12 For most gases, both the energy and entropy are lower in solution than in the gas phase. Higher temperatures, therefore, favor the gas phase. This means that most gases become less soluble as the temperature rises.

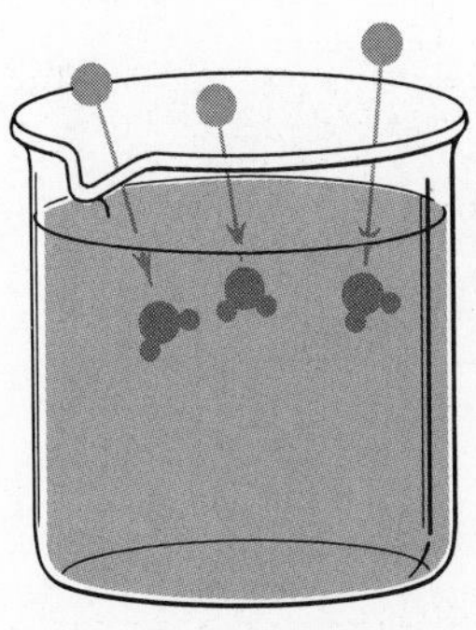

The solubility of gases, like those of solids and liquids, is related to enthalpy and entropy changes that occur during the solution process. Molecules are more disordered in the gas phase than in a solution. This suggests that unlike solids and liquids, *the drive to achieve greater disorder inhibits the dissolution of a gas in a liquid.*

In general, there is little attraction between molecules in the gas phase. When gas molecules are brought into contact with a liquid, they usually interact with solvent molecules and lower the energy of the system. *In most cases,* therefore, *the drive to reduce the energy of a system favors dissolution* of a gas in a liquid. In other words, the dissolution of a gas in a liquid is usually exothermic. If we assume that the entropy of solution for similar gases is about the same, then the gas with the largest negative $\Delta H°$ solution is the most soluble.

Application of Le Châtelier's Principle to the solubility equilibrium represented by

$$Cl_2(g) \rightleftharpoons Cl_2(aq) + 6 \text{ kcal}$$

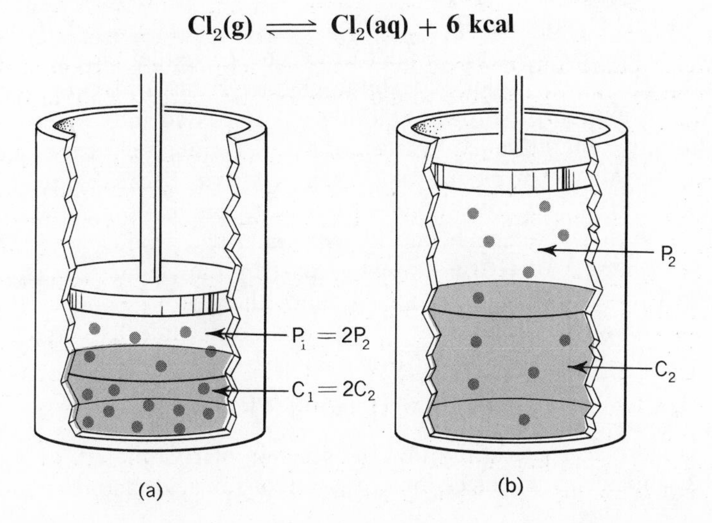

14-13 The partial pressure of O_2 (P) in (a) is twice that in (b). Therefore the concentration (C) of dissolved oxygen in (a) is twice that in (b).

reveals that increasing the temperature favors the more disordered state and decreases the solubility of chlorine in water. The fine bubbles of dissolved air which rise to the surface when you heat a pan of water also reflect the decreasing solubility of air in hot water.

The effect of changing the pressure on a liquid containing a dissolved gas can be summarized in ***Henry's Law:*** *the solubility of a gas in a liquid is proportional to the partial pressure of the gas on the surface of the liquid.* Application of this law indicates that doubling the number of molecules of a given gas above a liquid, doubles the number of gas molecules dissolved in the liquid. Since gas pressure is proportional to the number of molecules, doubling the number of specific gas molecules in a closed container doubles its partial pressure.

If you test the electric conductivity of a variety of substances you find that certain molecular substances composed of polar covalent molecules are, like ionic salts, electrolytes. Let us examine the behavior of this important type of compound when it is added to water.

14-14 In going from *a* to *b*, does the entropy of the system increase or decrease?

14-15 When hydrogen chloride dissolves in water, protons are transferred from the HCl molecules to the water molecules.

DISSOCIATION OF POLAR COVALENT MOLECULES

14-13 Hydronium Ions Consider hydrogen chloride (HCl), a gas composed of polar covalent molecules. Neither this gas nor liquid hydrogen chloride exhibit characteristic acid properties. When, however, hydrogen chloride gas is passed into water, the resulting solution, called hydrochloric acid, conducts electricity well and has all the properties of an acid solution. Because liquid or gaseous HCl does not contain ions, it is reasonable to assume that the molecules interact chemically with water and produce ions. The high solubility of HCl is caused in part by the chemical reaction which produces soluble products. Since there is no evidence that liquid or solid HCl consists of ions, it is not technically correct to show the dissociation of HCl in water as

$$\mathbf{HCl \longrightarrow H^+(aq) + Cl^-(aq)}$$

It is more accurate to say that the reaction of HCl molecules with H_2O molecules produces ions responsible for acid characteristics.

A hydrogen ion is a bare proton and has a diameter about $\frac{1}{100\,000}$ as large as other small positive ions. Because of its unit charge and small radius, this ion is surrounded by a strong electric field. Therefore, it is highly reactive and is strongly attracted to a highly polar molecule such as water. In other words, free hydrogen ions cannot exist in water and cannot be responsible for acid properties.

The reaction of HCl with water may more accurately be represented as

$$H_2O + HCl \longrightarrow H_3O^+(aq) + Cl^-(aq)$$

The reaction consists of a breaking away of a proton (H^+) from a chlorine atom. A stable coordinate bond is formed when a proton (hydrogen ion) shares a pair of electrons furnished by an oxygen atom of the highly polar water molecule. A hydrated proton, called an hydronium ion, is usually represented as H_3O^+ although it is also probably present to some extent as $H_5O_2^+$, $H_7O_3^+$, and $H_9O_4^+$. Spectroscopic evidence confirms its existence. We can conclude on the basis of experimental evidence that hydronium ions impart acid characteristics to aqueous solutions. *Most solutions formed by the reaction of polar molecular compounds with water are observed to have either acid or basic properties.*

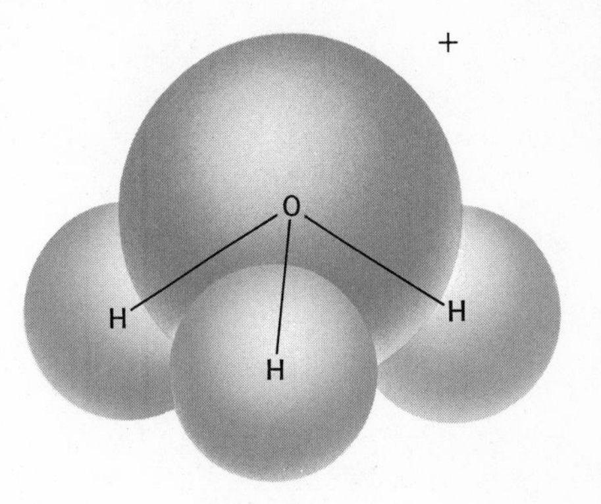

H_3O^+ ion

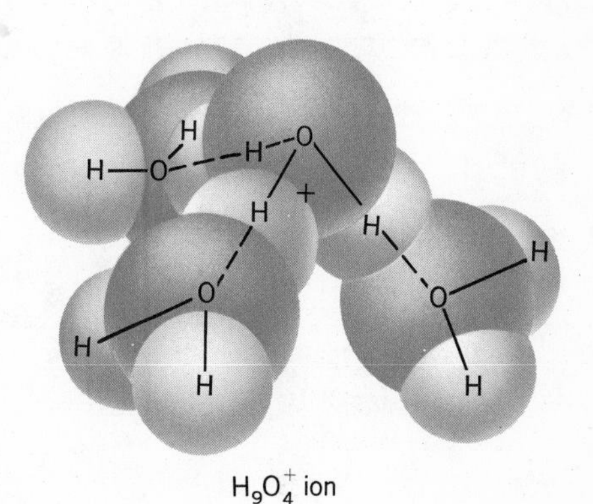

$H_9O_4^+$ ion

14-14 The Brønsted-Lowry Definition of Acids and Bases A number of nitrogen compounds react with water and yield basic solutions. For example, ammonia (NH_3) a polar covalent molecule, reacts with water and forms ammonium and hydroxide ions. The reaction of NH_3 in H_2O is represented as

$$NH_3 + H_2O \rightleftharpoons NH_4^+(aq) + OH^-(aq)$$

In 1923, J. N. Brønsted (1879–1947) of Denmark and T. M. Lowry of England independently recognized that the loss and gain of hydrogen ions (protons) might furnish a basis for classifying substances as acids or bases. A neutral hydrogen atom (H) has a nucleus with one proton and a single electron moving about the nucleus. When covalently bonded with highly electronegative atoms, a hydrogen atom behaves as if it has a partial positive charge because the bonding electrons are drawn toward the more electronegative atom. This often permits another polar molecule or negative ion to remove the hydrogen nucleus (proton). In this context, therefore, we refer to a hydrogen ion as a proton. Accord-

14-16 The conductance of an ammonia-water solution is caused by the presence of NH_4^+ and OH^- ions. Conductance measurements show that NH_3 dissociates to a much smaller extent than HCl.

ingly, Bronsted and Lowry defined *a base as a proton acceptor* and *an acid as a proton donor.* This conceptual definition permits us to classify NH_3 as a base and water as an acid. In the reaction above, the formation of NH_4^+ and OH^- ions can be explained by assuming that a hydrogen ion (a proton) is transferred from a water molecule to an ammonia molecule. The gain of a positive hydrogen ion by a neutral ammonia molecule converts it into a positive ammonium ion. The simultaneous loss of a proton by a neutral water molecule converts it into a negative hydroxide ion.

An aqueous solution of ammonia is sometimes referred to as ammonium hydroxide (NH_4OH). This usage is a matter of convenience since such a compound has never been isolated. A better notation for this mixture is $NH_3(aq)$.

weak acid does not give away protons readily

14-15 Strong and Weak Bases An aqueous solution of ammonia is a poor conductor of an electric current compared to a solution of sodium hydroxide. This means that in the ammonia solution there is a relatively small ratio of ions to molecules. In other words, the reaction (degree of dissociation) of NH_3 in water is small compared to that of NaOH. A base which is only slightly dissociated in aqueous solution is called a ***weak base;*** one which is highly dissociated is called a ***strong base.*** Aqueous ammonia ($NH_3(aq)$ is classed as a weak base. The common strong inorganic bases are the hydroxides of the Group I-A metals plus calcium hydroxide [$Ca(OH)_2$] and barium hydroxide [$Ba(OH)_2$]. Other hydroxides of Group II A are too slightly soluble to provide a significant concentration of OH^- ions in water. The conductivity and other properties of solutions of strong bases indicate that the solutes are completely dissociated into ions.

The double arrow used in the equations showing the reaction or dissociation of a weak base (Fig. 14-16) indicates that there is a dynamic equilibrium between the species on the product and reactant sides of the equation. The single arrows are used in equations showing the dissociation of a strong base, indicating that for practical purposes, the dissociation is complete. Thus it should be emphasized that a *water solution of a strong base does not contain molecules of the base.* Rather, it consists of positive metallic ions and negative hydroxide ions dispersed throughout the solution. On the other hand, a water solution of a weak base contains more neutral molecules of the base than product ions.

14-16 Strong and Weak Acids The relatively high conductivity and other properties of hydrochloric acid solution indicate that HCl reacts (dissociates) completely and forms a solution of ions. It is, therefore, classified as a ***strong acid.*** Weak acids, like weak bases, are equilibrium mixtures consisting of a higher concentration of reactant molecules than product ions. The implication is that atoms in the molecules of weak acids or bases are more firmly bonded to each other than in those which dissociate extensively in water. We have already noted that many organic polar molecular substances containing hydrogen and/or hydroxyl groups are nonelectrolytes and do not form ions in aqueous solution. Examples are sugar ($C_{12}H_{22}O_{11}$), ethyl alcohol (C_2H_5OH), and glycerol [$C_3H_5(OH)_3$]. For convenience, the formulas of most polar molecular substances that form acid solutions often begin with the element hydrogen. The hydrogen atoms represented first in the formula are referred to as *acid hydrogens.*

A common weak acid is acetic acid ($HC_2H_3O_2$), an acid found in vinegar. Its structural formula is $CH_3\overset{\overset{\displaystyle O}{\|}}{C}—OH$. This formula is commonly abbreviated as HOAc. The dissociation of acetic acid in aqueous solution may be represented as

$$\textbf{1. } CH_3\overset{\overset{\displaystyle O}{\|}}{C}—OH + H_2O \rightleftharpoons H_3O^+(aq) + CH_3\overset{\overset{\displaystyle O}{\|}}{C}—O^-(aq)$$

$$\textbf{2. } HC_2H_3O_2 + H_2O \rightleftharpoons H_3O^+(aq) + C_2H_3O_2^-(aq)$$

$$\textbf{3. } HOAc + H_2O \rightleftharpoons H_3O^+(aq) + OAc^-(aq)$$

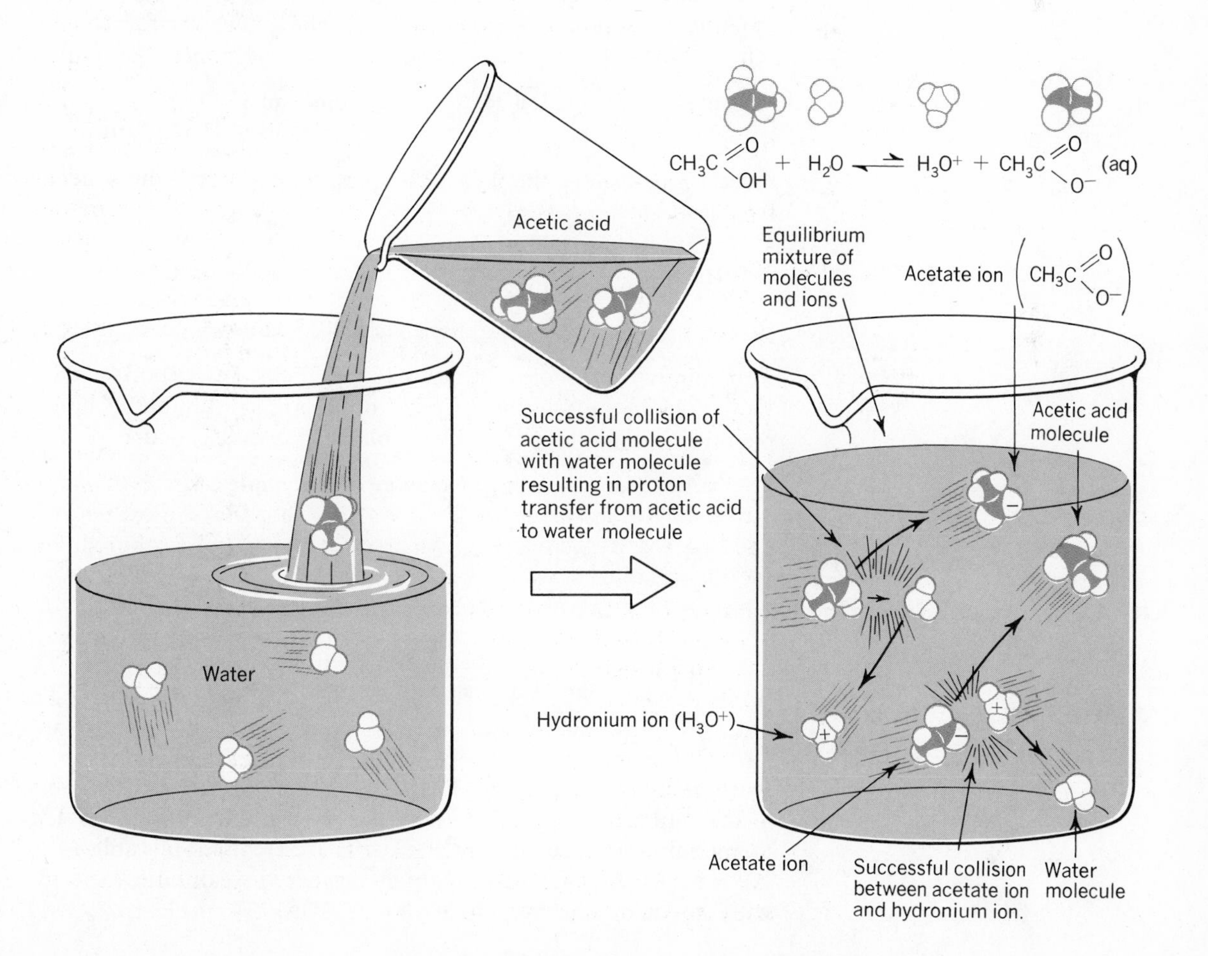

14-17 When acetic acid is added to water, collisions between $HC_2H_3O_2$ and water molecules occur. When the molecules are oriented properly and have enough kinetic energy, a proton transfer occurs resulting in the formation of acetate ions ($C_2H_3O_2^-$) and hydronium ions (H_3O^+). In a weak acid, the acid proton is rather tightly bound so few collisions are energetic enough to result in proton transfer. Collisions between acetate ions and hydronium ions can also result in a proton transfer. This results in the formation of acetic acid molecules and water molecules (lower right). An equilibrium is established when the rates of the two proton-transfer reactions are equal. When equilibrium is reached at a fixed temperature, there is a constant number of all species ($HC_2H_3O_2$, $C_2H_3O_2^-$, H_3O^+, H_2O).

Molarity (M) is discussed in Sections 3-5 and 3-6.

In this dissociation, the acid hydrogen is the one bonded to the very electronegative oxygen atom. The bonds in the methyl (CH_3^-) group are not polar enough for those hydrogens to be acid. Electric measurements show that the degree of dissociation of acetic acid is low. In dilute solutions, it is about one percent. This means that only one molecule in every 100 molecules is separated by the water into ions. As would be predicted by Le Châtelier's Principle, dilution with water increases the degree of dissociation of weak electrolytes in water. A 0.10-*M* solution of hydrochloric acid is much more acid than a 0.10-*M* solution of acetic acid. That is, the hydrochloric acid contains a higher percent of dissociated HCl molecules.

In order to calculate the ionic and molecular concentration in a solution of a weak electrolyte such as acetic acid, we must know the original concentration of the solute and its percentage dissociation. The percentage dissociation can be determined by conductance or freezing-point measurements (Sec. 14-29, page 467). These measurements show that a 0.100-*M* solution of acetic acid is 1.3 percent dissociated at room temperature. Equations 1, 2, and 3 p. 451 show that for each mole of acetic acid which dissociates, one mole of H_3O^+ ions and one mole of $C_2H_3O_2^-$ ions are formed. In other words,

$$\textbf{moles of dissociated } HC_2H_3O_2 = \textbf{moles of } H_3O^+$$
$$= \textbf{moles of } C_2H_3O_2^- \textbf{ formed}$$

Since 1.3 percent of the 0.10 moles per liter dissociate, the concentration of each ion is

$$0.013 \times \frac{0.10 \text{ moles } HC_2H_3O_2}{\text{liter}} = 0.0013 \text{ moles } H_3O^+/\text{liter}$$
$$= 0.0013 \text{ moles } C_2H_3O_2^-/\text{liter}$$

The number of moles of acetic acid molecules (undissociated $HC_2H_3O_2$) in the solution is equal to the original number of moles dissolved minus the number of moles which dissociated.

$$0.100 \text{ moles} - 0.0013 \text{ moles} = 0.0987 \text{ moles, } HC_2H_3O_2$$

FOLLOW-UP PROBLEM

The equation for the dissociation of formic acid (H—C(=O)—OH, $HCHO_2$) is

$$HC(=O)\text{—}OH + H_2O \rightleftharpoons H_3O^+ + HC(=O)\text{—}O^-$$

A 0.60-*M* solution of the acid is 1.8 percent dissociated. Calculate the molar concentration of each species in the equilibrium system.

Ans. H_3O^+ **and** $CHO_2^- = 0.011$ ***M***, $(HCHO_2) = 0.59$ ***M***.

It is often convenient to know the names and formulas of the commonly encountered strong acids. These are listed in Table 14-5. All acids other than these may be considered weak unless specifically noted.

TABLE 14-5
STRONG ACIDS

Formula	Name	Ions Formed in Aqueous Solution	
$HClO_4$	perchloric acid	H_3O^+ (or H^+)	ClO_4^-
HCl	hydrochloric acid	H_3O^+	Cl^-
HBr	hydrobromic acid	H_3O^+	Br^-
HI	hydriodic acid	H_3O^+	I^-
HNO_3	nitric acid	H_3O^+	NO_3^-
H_2SO_4	sulfuric acid	H_3O^+	HSO_4^- (high concentration) SO_4^{2-} (low concentration)

14-17 Polyprotic Acids Acids such as H_2SO_4 which have more than one proton (hydrogen ion) per molecule that can be transferred to solvent molecules are called *polyprotic* acids. The protons which can be donated are usually shown first in the formula. For example, acetic acid ($HC_2H_3O_2$) is monoprotic and phthalic acid ($H_2C_8H_4O_4$) is diprotic or polyprotic. The hydrogen atoms in acetate ions ($C_2H_3O_2^-$) and phthalate ions ($C_8H_4O_4^{2-}$) are not transferable and, therefore, are nonacid hydrogens.

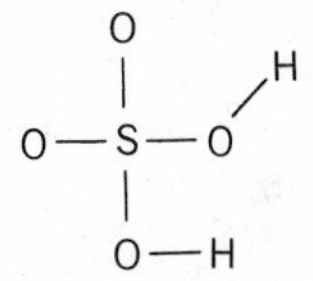

When the solution of a weak, polyprotic acid such as phosphoric acid (H_3PO_4) is analyzed, it is found to be an equilibrium mixture of H_2O, H_3PO_4, $H_2PO_4^-$, HPO_4^{2-}, PO_4^{3-}, and $H_3O^+(H^+)$. The observed concentrations of each species can be explained by assuming that H_3PO_4 dissociates in a series of steps and that there is an equilibrium between the species involved in each step. The steps are

$$H_3PO_4 + H_2O \rightleftharpoons H_3O^+ + H_2PO_4^-$$
$$H_2PO_4^- + H_2O \rightleftharpoons H_3O^+ + HPO_4^{2-}$$
$$HPO_4^{2-} + H_2O \rightleftharpoons H_3O^+ + PO_4^{3-}$$

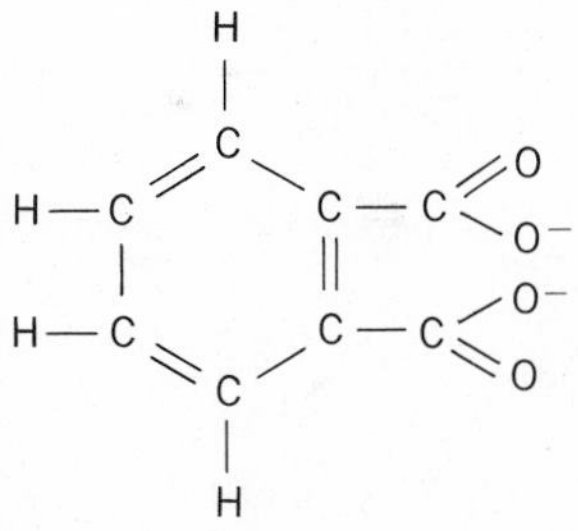

The first step of dissociation always proceeds to a greater extent than subsequent steps. This means that in phosphoric acid, most of the hydrogen ions in the solution are derived from the primary dissociation of H_3PO_4. The concentration of PO_4^{3-} ions is observed to be extremely low.

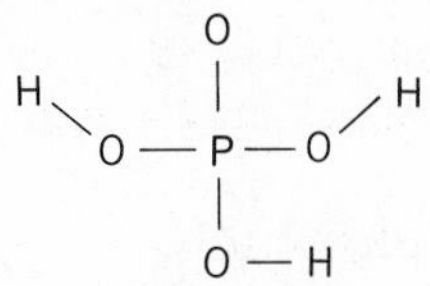

For two commonly encountered weak acids, the molecular species have never been isolated as pure compounds. These are carbonic acid (H_2CO_3) and sulfurous acid (H_2SO_3). The carbonic acid system may be represented as

$$\underbrace{CO_2 + 2H_2O}_{\text{“}H_2CO_3\text{”}} \rightleftharpoons H_3O^+ + HCO_3^-$$

and the sulfurous acid system as

$$\underbrace{SO_2 + 2H_2O}_{\text{“}H_2SO_3\text{”}} \rightleftharpoons H_3O^+ + HSO_3^-$$

Hydrogen sulfate (sulfuric acid) molecule.

Phthalate ion.

Hydrogen phosphate (phosphoric acid) molecule.

FOLLOW-UP PROBLEMS

Write equations which represent the dissociation of each of these acids or bases in aqueous solution. Use a single arrow in the case of a strong acid or base, and a double arrow to represent the equilibrium condition that exists in the solution of a weak acid or base. Show each step of dissociation for polyprotic acids. (a) KOH, (b) H_3AsO_4, (c) $HClO_4$, (d) HCN, (e) $C_6H_5NH_2$ (a weak base).

SUMMARY OF TYPES OF SOLUTES AND SOLUBILITIES OF SOLUTES

14-18 General Rules It is difficult to establish rules based on theory or principles that enable us to predict the relative solubilities of specific compounds. The solubility of a substance depends on many factors which are themselves difficult to predict.

In general, we can say that solutes are most apt to dissolve in solvents with similar characteristics; that is, ionic and polar solutes dissolve in polar solvents and nonpolar solutes dissolve in nonpolar solvents. For convenience our observations related to the solubility of substances are summarized below.

14-19 Types of Solutes and Their Behavior in Aqueous Solution

1. Polar molecular solutes such as sugar, alcohol, and glycerol dissolve but do not dissociate appreciably in aqueous solution. They form strong hydrogen bonds with water but are nonelectrolytes. The solute particles are molecules.

2. Polar molecular solutes such as hydrogen chloride (HCl) and hydrogen iodide (HI) dissociate completely in aqueous solution. The solute particles of these strong acids are ions.

3. Polar molecular solutes such as ammonia (NH_3) and acetic acid ($HC_2H_3O_2$) are partially dissociated in aqueous solution. The solute particles of these weak acids and bases are a mixture of molecules and ions at equilibrium with each other. The molecular species predominates.

4. Ionic crystals such as sodium chloride (NaCl), potassium bromide (KBr), and copper(II) sulfate ($CuSO_4$) are completely dissociated in aqueous solution. The solute particles of these salts are ions, usually hydrated. The solubilities of these salts vary over a wide range. In a saturated solution of a salt containing excess undissolved solute an equilibrium exists between the ions in solution and the solid.

In the preceding sections, we became acquainted with substances which react with water and yield acid or basic solutions. Let us now develop a method for expressing the relative acidity of a solution and for calculating the $[H_3O^+]$ or $[OH^-]$ concentration in a solution.

TABLE 14-6
APPROXIMATE pH OF SOME COMMON LIQUIDS

Liquid	*pH*
1.0 *M* HCl	0.1
0.1 *M* HCl	1.0
Gastric juice	2.0
Lemon juice	2.3
Vinegar	2.8
0.1 *M* $H \cdot C_2H_3O_2$	2.9
Soft drinks	3.0
Apple juice	3.1
Grapefruit juice	3.1
Orange juice	3.5
Tomato juice	4.2
Banana (fluid)	4.6
Rainwater	6.2
Milk	6.5
Pure water	7.0
Egg	7.8
0.1 *M* $NaHCO_3$ (baking soda)	8.4
Sea water	8.5
Milk of magnesia [$Mg(OH)_2$]	10.5
0.1 *M* NH_3(aq)	11.1
0.1 *M* NaOH	13.0
1.0 *M* NaOH	14.0

RELATIVE ACIDITY AND BASICITY OF SOLUTIONS

14-20 The pH Scale Every aqueous solution is either acid, basic, or neutral. The qualitative relationship between the concentrations of hydrogen (hydronium) ion and hydroxide ion in each of these solutions is given below.

neutral solution	$[H_3O^+] = [OH^-]$
acid solution	$[H_3O^+] > [OH^-]$
basic solution	$[H_3O^+] < [OH^-]$

The brackets denote molar concentration.

The acidity of a solution is a vital factor which determines the outcome of many reactions. The success or failure of many analytical and synthetic processes depends on the hydrogen-ion concentration. For example, you would not expect to remove carbonate ions from a solution by trying to precipitate it out in an acid medium. People who have swimming pools carefully control the acidity so as to avoid undesirable reactions with the plumbing and plaster and to expedite reactions with other chemicals. The food, drug, petroleum, and most chemical industries continually monitor the acidity of the solutions being used in the preparation of their products.

A large number of reactions involve extremely small concentrations of hydrogen ions. For example, most swimming pool owners try to maintain a hydronium-ion concentration of approximately 0.000 000 04 M or (4×10^{-8} M). As you can see, this is an awkward number. In 1909, Sören P. L. Sörenson (1868–1939), a Danish chemist, proposed a more concise method for expressing the acidity of a solution. Sörensen's proposal was that the acidity of a solution be expressed in terms of a quantity known as pH (potency of hydrogen).

The pH scale is a numerical scale which, for most applications, extends from 0 through 14. The numbers on the scale represent the relative acidity of solutions and can be converted to actual hydronium-ion concentrations. The midpoint of the scale is taken as 7.0. At 25°C a solution with a pH of 7.0 contains equal concentrations of H_3O^+ and OH^- ions and is neutral. A solution with a pH less than 7.0 has a greater H_3O^+ than OH^- ion concentration and is acid. Solutions with pH values above 7.0 have greater OH^- than H_3O^+ ion concentrations and are basic. The pH scale is represented in Fig. 14-18. The pH of a number of common liquids is given in Table 14-6.

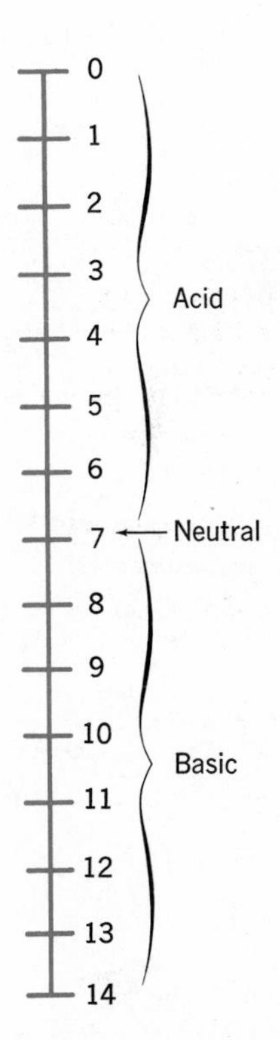

14-18 The pH scale. Most solutions with which we are concerned have pH values that fall between 0 and 14. Some very acid solutions, however, have negative pH values and some very basic solutions have pH values greater than 14.

The actual relation between the pH and the molarity of the H_3O^+ ion is given by

$$[H_3O^+] = 10^{-pH}\ M$$

Thus, the H_3O^+ ion concentration of a solution with a pH of 3.0 is $10^{-3.0}$ *M* or 0.001 *M*. The relation between pH and the molar concentration of the H_3O^+ ion expressed as a power of ten or a decimal fraction is shown in Table 14-7.

TABLE 14-7
$[H_3O^+]$ vs pH

Molarity of H_3O^+ (decimal fraction)	*Molarity of H_3O^+ (power of 10)*	*pH*	*Condition of Solution*
1 *M*	10^0	0	acid
0.1 *M*	10^{-1}	1	acid
0.000 000 1 *M*	10^{-7}	7	neutral
0.000 000 001 *M*	10^{-9}	9	basic

FOLLOW-UP PROBLEMS

1. What is the pH of 0.010-*M* hydrochloric acid?

2. What is the $[H_3O^+]$ concentration of acetic acid whose pH is 4.0?

The pH of a solution may be determined by the use of an electronic instrument known as a pH meter or by the use of chemical indicators. Acid-base ***indicators*** are dyes which undergo slight changes in molecular structure and color when the pH value of a solution changes. Specific colors correspond to certain pH values. Litmus and phenolphthalein are two chemical indicators. Litmus is red in a solution whose pH is less than 5 and blue in one whose pH is greater than 7. Between 5 and 7 it maintains an intermediate color (reddish purple or bluish purple). Above pH 8.2, phenolphthalein is red; below 8.2 colorless.

The color changes and pH range for a number of common indicators are shown on color plate III between pp. 274 and 275.

14-21 The pOH Scale In working with basic solutions, it is sometimes useful to refer to the OH^--ion concentration. For this reason a scale similar in structure to the pH scale is often used to express relative basicity. This is known as the ***pOH scale.*** The relationship between the pOH and molarity of OH^- ions is given by

$$[OH^-] = 10^{-pOH}\,M$$

A solution in which the $[OH^-] = 1 \times 10^{-4}\,M$ has a pOH of 4. If the solution has a pOH of 5, then the $[OH^-] = 1 \times 10^{-5}\,M$.

A useful relationship between pH and pOH in any aqueous solution at 25°C is

$$pH + pOH = 14 \qquad 14\text{-}1$$

Thus a solution with a pH of 7 has a pOH of 7. These statements suggest that in any aqueous solution there is a relationship between the molar concentration of the H^+ (or H_3O^+ ions) and that of the OH^- ions.

FOLLOW-UP PROBLEM

1. (a) What is the pOH of a 0.010-*M* NaOH solution? (b) What is the pH of the solution?

14-22 The Dissociation Constant of Water Very sensitive measurements reveal that pure water dissociates slightly into ions. Apparently, collisions between water molecules may result in a proton transfer from one molecule to another. The equation for the self-dissociation of water may be written

$$\mathbf{HOH + HOH \rightleftharpoons H_3O^+ + OH^-}$$

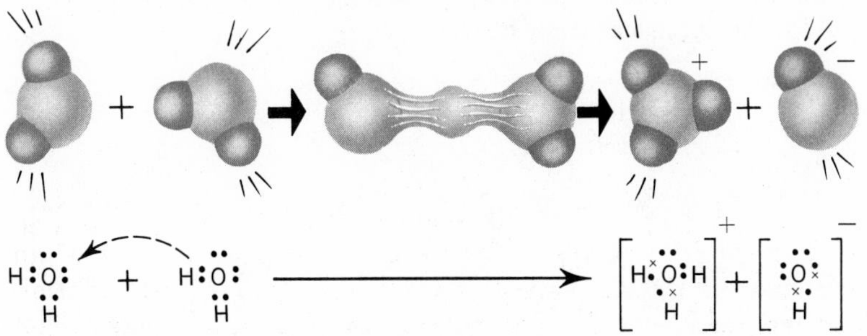

14-19 Proton transfer between water molecules.

This equation indicates that in pure water there is an equilibrium between the ions and molecules. As in all equilibrium systems, therefore, the concentration of all species is constant at a given temperature. The equilibrium law applied to the equilibrium between liquid water molecules and the ions yields a constant known as the *dissociation constant of water,* K_w.

$$\mathbf{K_w = [H_3O^+][OH^-]} \qquad 14\text{-}2$$

Because the concentration of water is essentially constant, H_2O is not included in the expression for K_w. Measurements show that in pure water at 25°C, the concentration of $[H_3O^+]$ and OH^- ions in pure water is 1×10^{-7} *M*. Substituting these values in Equation 14-2 shows that $K_w = 1 \times 10^{-14}$. In any aqueous solution at 25°C, the product of the $[H_3O^+]$ and the $[OH^-]$ must equal 1×10^{-14}. Equation 14-2 can be used to calculate the H_3O^+ ion or the OH^- ion concentration of a solution if one of the quantities is known. For example, the OH^--ion concentration of a solution having a H_3O^+ of 1×10^{-3} *M* is calculated thus:

$$\mathbf{[1 \times 10^{-3}][OH^-] = 1 \times 10^{-14}}$$

$$\mathbf{[OH^-] = \frac{1 \times 10^{-14}}{10^{-3}} = 1 \times 10^{-11}\ \mathit{M}}$$

TABLE 14-8
RELATIONSHIP BETWEEN $[H_3O^+]$, $[OH^-]$, pH, and pOH

$[H_3O^+]$	pH	pOH	$[OH^-]$	Condition of Solution
10^{-3} *M*	3	11	10^{-11} *M*	acid
10^{-7} *M*	7	7	10^{-7} *M*	neutral
10^{-9} *M*	9	5	10^{-5} *M*	basic

This relationship between pH, pOH, H_3O^+, and OH^- is shown in Table 14-8.

Example 14-1
What are the H_3O^+-ion and OH^--ion concentrations of a solution prepared by adding 1 ml of 1-*M* HCl to 9 ml of water? Assume that volumes are additive and that the 1-*M* HCl dissociates completely.

Solution
1. Calculate the molarity of the solution resulting from the addition of HCl to water.

$$\text{mmoles of solute} = MV = \frac{1 \text{ mmole}}{\text{ml}} \times 1 \text{ ml} = 1 \text{ mmole}$$

$$\begin{aligned} \text{ml of solution} &= 1 \text{ ml} + 9 \text{ ml} = 10 \text{ ml} \\ M &= \text{millimoles solute/ml solution} \\ &= 1 \text{ millimole/10 ml} = 0.1\ M \\ &= 1 \times 10^{-1}\ M \\ [HCl] &= [H_3O^+] \\ [H_3O^+] &= 1 \times 10^{-1}\ M \end{aligned}$$

2. Substitute $[H_3O^+]$ in Equation 14-2 and solve for $[OH^-]$.

$$1.0 \times 10^{-14} = [1 \times 10^{-1}][OH^-]$$

$$[OH^-] = \frac{1 \times 10^{-14}}{1 \times 10^{-1}} = 1 \times 10^{-13}\ M$$

FOLLOW-UP PROBLEM

What are the H_3O^+ ion and OH^- ion concentrations of a solution made by adding 1 ml of 0.1-*M* NaOH to 9 ml of water?

Ans. $[OH^-] = 1 \times 10^{-2}$ *M*, $[H_3O^+] = 1 \times 10^{-12}$ *M*.

14-23 The Definition of pH and pOH As long as concentrations can be converted to whole-number powers of ten, it is relatively simple to convert hydronium ion concentrations to pH and vice versa. In practice, solutions are not so accommodating, and actual concentrations usually yield pH values that are not whole numbers. In order to work with fractional pH values, we must use the general definition of pH given by Sörenson. That is

$$\mathbf{pH = -\log [H_3O^+]} \qquad 14\text{-}3$$

Similarly, pOH is defined as

$$\mathbf{pOH = -\log [OH^-]} \qquad 14\text{-}4$$

The use of Equations 14-3 and 14-4 is illustrated in this example.

Example 14-2
Find the pH, pOH, and $[OH^-]$ of a 0.000 10-*M* HCl solution. HCl is 100 percent dissociated.

Solution
1. Express the concentration of HCl in exponential form.

$$0.000\,10\ M = 1.0 \times 10^{-4}\ M = 10^{-4}$$

Since HCl dissociates to give one mole of H_3O^+ ion for each mole of HCl, the concentration of H_3O^+ ion equals the original concentration of HCl.

$$[H_3O^+] = 1.0 \times 10^{-4}\ M$$

2. Substitute the $[H_3O^+]$ in Equation 14-3 and solve for pH.

$$pH = -\log [1.0 \times 10^{-4}] = -[0 - 4] = 4.0$$

3. Substitute the pH obtained in Step 2 in Equation 14-1 and solve for pOH.

$$pOH = 14.0 - 4.0 = 10.0$$

4. The $[OH^-]$ may be obtained by substituting the H_3O^+ in Equation 14-2

$$[OH^-] = \frac{1 \times 10^{-14}}{1 \times 10^{-4}} = 1 \times 10^{-10}\ M$$

Example 14-3
Find the pH of a 0.00325-*M* NaOH solution. NaOH is 100 percent dissociated.

Solution
1. The $[OH^-]$ is equal to the [NaOH]. Express the $[OH^-]$ in exponential form.

$$[OH^-] = 0.00325\ M = 3.25 \times 10^{-3}\ M$$

2. Calculate the pOH, using Equation 14-4.

$$pOH = -\log (3.25 \times 10^{-3}) = -(\log 3.25 + \log 10^{-3})$$

Use a log table or slide rule to find the log of 3.25. The log is 0.51.

$$pOH = -(0.51 - 3.00) = -(-2.49) = 2.49$$

A short-cut calculation may be made thus: if

$$[OH^-] = 3.25 \times 10^{-3}\ M$$

then $pOH = 3 - \log 3.25 = 3 - (0.51) = 2.49$.

3. Calculate the pH by using Equation 14-1

$$pH = 14.00 - 2.49 = 11.51$$

Example 14-4
What is the hydrogen (hydronium)-ion concentration of a solution that has a pH of 2.6?

Solution
1. Substitute the pH value in Equation 14-3 and solve for $[H_3O^+]$

$$[H_3O^+] = 10^{-2.6}$$

2. The expression $10^{-2.6}$ can be written as $10^{-2} \times 10^{-0.6}$. Logarithm tables may be used to find the value of 10 raised to a fractional power. The fractional power, however, must be positive. Therefore, it is necessary to convert $10^{-2} \times 10^{-0.6}$ to its equivalent, which is

$$10^{-3} \times 10^{+0.4}$$

In other words, $-3 + (+0.4)$ is the same as $-2 + (-0.6)$. The value of $10^{+0.4}$ may be obtained by locating 0.4 in a logarithm table or on a slide rule and reading the number (antilog) which corresponds to it. Since 10^1 is 10, then $10^{+0.4}$ must be a number between 1 and 10. Reference to a table of logarithms shows that $10^{+0.4}$ equals 2.5. Therefore, the $[H_3O^+]$ of a solution with a pH of 2.6 is $2.5 \times 10^{-3}\ M$.

FOLLOW-UP PROBLEMS

1. Determine the $[OH^-]$, $[H_3O^+]$, pOH, and pH of a 0.001-*M* KOH solution. KOH is 100 percent dissociated.
Ans. $[OH^-] = 1 \times 10^{-3}\ M$, $[H_3O^+] = 1 \times 10^{-11}\ M$, **pOH = 3, pH = 11.**

2. What is the $[H_3O^+]$, $[OH^-]$, pH, and the pOH of a 0.045 0-*M* HCl solution?
Ans. $[H_3O^+] = 4.5 \times 10^{-2}\ M$, $[OH^-] = 2.2 \times 10^{-13}\ M$, **pOH = 12.65, pH = 1.35.**

3. What is the $[H_3O^+]$ of a solution having a pH of 3.4?
Ans. $4.0 \times 10^{-4}\ M$.

COLLIGATIVE PROPERTIES OF SOLUTIONS

In Chapter 10 we found that pure substances in the liquid phase have fixed boiling and freezing points at standard pressure. Let us now investigate how the addition of a solute to a pure liquid solvent changes the properties of the liquid. The *vapor pressure, boiling point,* and *freezing point of a solution are colligative properties and depend upon the total number of solute particles rather than upon their nature.* The solute may consist of ions, molecules, or a mixture of both. All have essentially the same effect as far as colligative properties are concerned. At this point we shall confine ourselves to nonvolatile solutes which do not dissociate. ***Nonvolatile solutes*** do not vaporize readily and do not contribute appreciably to the vapor pressure of the solution. Since the boiling point and freezing point of a solvent depend upon the vapor pressure, we shall first consider how the addition of a solute causes a variation in the vapor pressure of pure water.

14-24 Vapor Pressure of Solutions The vapor pressure of a liquid is related to the tendency of the molecules to escape from a solution. The relative escaping tendencies of the molecules can be demonstrated qualitatively by placing two beakers containing equal volumes of pure water and of sugar solution under a bell jar and observing the change in liquid levels. After a period of time, the volume of the pure solvent is observed to decrease while that of the solution increases. Apparently, solvent molecules are being transferred through the vapor phase from the pure solvent to the solution. In other words, the escaping tendency (vapor pressure) of water molecules of the pure water is greater than that of the solution. Ultimately, all of the pure solvent will be transferred to the solution. Qualitatively, we can describe this observation by assuming that, upon dissolving, the polar sugar molecules interact with the polar water molecules and disrupt the linkages between them. Sugar molecules replace some of the water molecules in the surface layer and, thus, reduce the number of water molecules in contact with the water vapor above the solution. The proportion and escaping tendency of the water molecules is reduced, and the *vapor pressure of the solution is,* therefore, *less than that of pure water.*

At a given temperature in a closed system, the vapor pressure of the solvent in the solution reaches a constant value. At this temperature, the escaping tendency of the solvent molecules in the solution is balanced by the tendency of the solvent molecules in the vapor to return to the solution. In other words, an equilibrium exists between the solvent in the solution and in the vapor state.

We can see that the transfer of solvent to solution under the bell jar in Fig. 14-20 is consistent with Le Châtelier's Principle because in this system, addition of the solute to one container reduces the escaping tendency of the solvent. This is a stress which can be

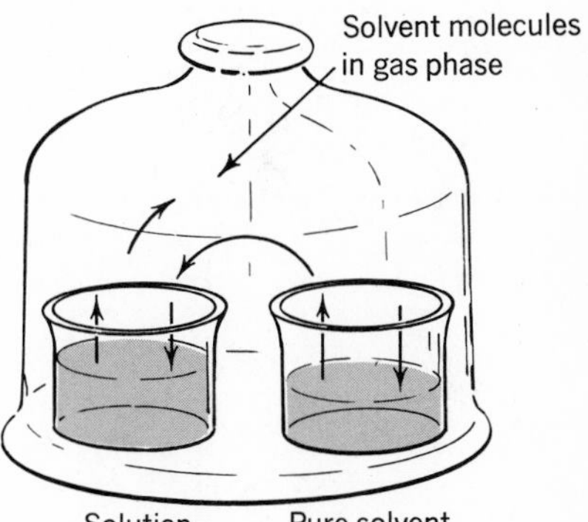

14-20 When equal volumes of pure solvent and solution are placed in separate beakers under a bell jar, the volume of the solvent decreases while that of the solution increases. In the initial and intermediate states, the system is not in equilibrium because the vapor pressure of the solvent is greater over the pure solvent than over the solution.

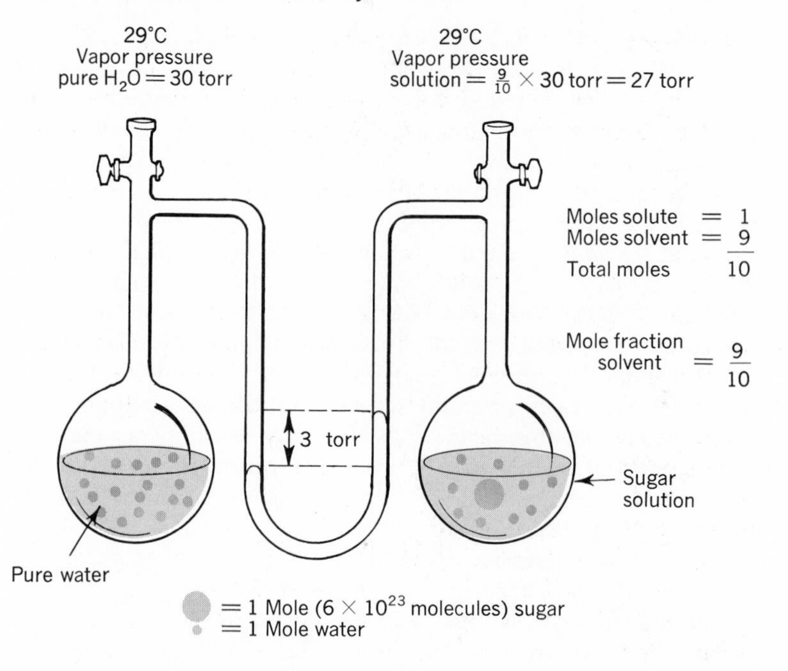

14-21 After air is evacuated from the system, the difference in the mercury levels is equal to the difference in the vapor pressures of the solvent and solution. The vapor pressure of a solution of a nonvolatile nonelectrolyte may be calculated by using the expression

$$\text{vapor pressure of solution} = \text{mole fraction of solvent} \times \text{vapor pressure of pure solvent}$$

relieved by increasing the amount of solvent in the solution. Additional solvent molecules are provided by condensation of the vapor. This lowers the vapor pressure in the system and disturbs the equilibrium between the pure solvent in the second container and its vapor. The liquid in the second vessel will then be evaporating faster than its vapor is condensing.

It can be shown that the vapor pressure of a solution depends on the molecular (mole) fraction of the solvent molecules in the solution. As the concentration of the nonvolatile solute increases, the vapor pressure of the solution decreases.

14-25 Boiling Points of Solutions We have defined the boiling point as *the temperature at which the vapor pressure of the liquid is equal to the atmospheric or external pressure. Decreasing the vapor pressure of a solvent by the addition of a nonvolatile solute causes an increase in the boiling point.* The percentage of water molecules in the surface layer is reduced by the addition of the solute. The average kinetic energy of the water molecules must be increased if a higher percentage of them are to escape from the surface into the vapor phase. In order to increase the average kinetic energy, it is necessary to raise the temperature of the solution. The relationship between the vapor pressure and the temperature of a pure solvent and a solution is seen by inspecting the graph in Fig. 14-22. The heavy line in the graph is a plot showing how the vapor pressure of pure water changes with temperature. The dashed line is a plot showing the relationship between the vapor pressure of a water solution of a nonvolatile solute and the temperature.

From the graph it can be seen that the boiling point of pure water, BP_w corresponds to a vapor pressure of 760 torr. Addition of a solute produces a solution and reduces the vapor pressure to point *C*. Because the vapor pressure is now below 760 torr, the solution does not boil. In order to raise the vapor pressure of the solution to 760 torr where boiling does occur, it is necessary to increase the temperature from 100°C to BP_s, the boiling point of the solution. The change in boiling point is usually designated as ΔBP or ΔT_b.

14-26 Freezing Points of Solutions The decrease in the vapor pressure of a solvent resulting from the addition of a nonvolatile solute causes a corresponding *decrease* in its freezing point. If a liquid is to change into a crystalline solid at the freezing point, the molecules of the liquid must lose energy and attain a more ordered crystalline pattern. Solute molecules do not fit into the same crystal pattern as solvent molecules. Consequently, they interfere with the process of crystallization. That is, they lower the rate of crystallization. Solute molecules do not, however, affect the rate of melting. The addition of a solute to an ice-liquid water system at 0°C disturbs the equilibrium and causes the ice to melt. Because the solution has greater disorder (entropy) than the pure solvent, the average kinetic energy of the solvent molecules must be reduced below that observed at the freezing point of the solvent. That is, to restore order to a greatly disordered system requires a greater reduction of energy than to restore it to a less disordered system. A lower average kinetic energy is reflected in a lower temperature. At the lower temperature, water molecules are moving more slowly and, therefore, will have a higher probability of re-

PETER J. W. DEBYE
1884–1966

Peter Debye was one of many European scientists who fled to America prior to World War II. He was a native of Holland and received his diploma in engineering at Aachen in 1905, his Ph.D. at Munich in 1908. He held professorships at Zurich, Utrecht, Göttingen, and Leipzig. He has published over 50 papers in Dutch, German, and English on such subjects as X-ray scattering, electric movements of molecules, and magnetic cooling. He is best remembered for his work on the theory of electrolytes. Debye believed that ions of salt when dissolved in water arrange themselves in solution in much the same manner as they are arranged in the solid state, although farther apart. He theorized, "If the solvent already contained ions of a foreign electrolyte, an ionic cloud of opposite charge can form around every ion as it leaves the solid phase and passes into solution." He explained the function of the polar water molecules in the process of dissolving an ionic solid. His 1929 book entitled *Polar Molecules* was published in English. Debye was awarded the 1936 Nobel Prize in chemistry for his research on ionization.

When Debye came to America in 1940, he accepted a position as professor of chemistry at Cornell University, where he remained as professor, then department chairman, and, finally, professor emeritus until his death in 1966.

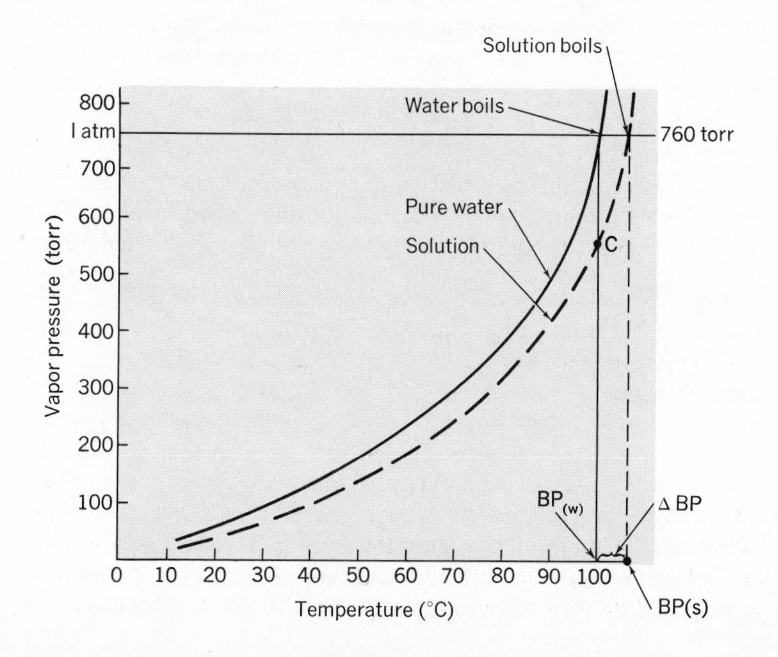

14-22 Vapor pressure-temperature graph illustrating the elevation of the boiling point of a solvent by the addition of a solute.

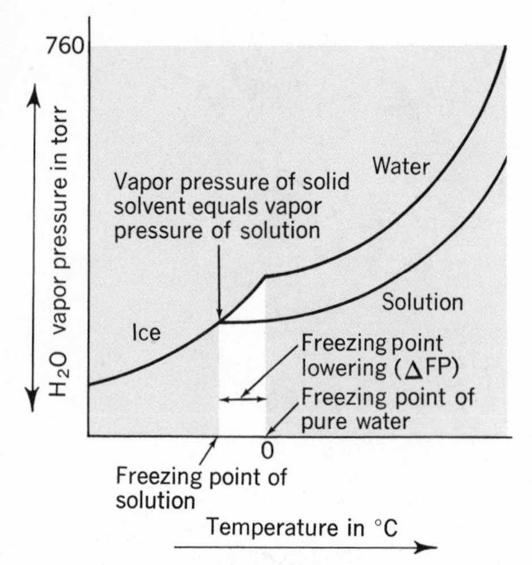

14-23 Graph showing freezing point depression caused by lowering the pressure of the solvent vapor by adding a solute.

turning to the solid phase and re-establishing an equilibrium condition in which the rate of melting equals the rate of crystallization. Thus the *freezing point of the solution is less than that of the pure solvent.* The difference between the freezing point of the solution and that of the pure solvent is designated as ΔFP or ΔT_f.

The depression of the freezing point that occurs when a nonvolatile solute is added to water is illustrated in Fig. 14-23. The freezing point of pure water is 0°C. This is the temperature at which the solid and liquid phases are in equilibrium at 1 atmosphere pressure. The freezing point of the solution is the temperature at which the vapor pressure of the solvent in the solution equals the vapor pressure of the solid solvent (ice). The vapor pressure of the solvent in the solution at a given temperature is lower than that of pure ice at 0°C. This means that ice does not form in a solution until the temperature of the system has been lowered below 0°C. In Fig. 14-23, the freezing point of the solution is the temperature at which the vapor pressure curve of the solid solvent intersects that of the solution curve. As the solution freezes, the solvent molecules are removed as a solid, the concentration of the solution increases, and the freezing point of the solution continues to decrease. It should be noted that freezing points vary slightly with changes in external pressure. At this point, however, we shall not be concerned with this phenomenon.

14-27 Molal Concentration It can be shown mathematically that the changes in boiling and freezing points that occur when a solution is formed by the addition of a solute to a solvent are proportional to the number of moles of solute particles per kilogram (1000 g) of solvent. The ratio of the moles of solute to a kilogram of solvent is called ***molality*** and symbolized by *m*. Thus molality is a concentration unit defined as *moles of solute per kilogram of solvent.* It may be expressed in formula form as

$$m = \frac{\text{moles of solute}}{\text{kilograms of solvent}} \qquad 14\text{-}5$$

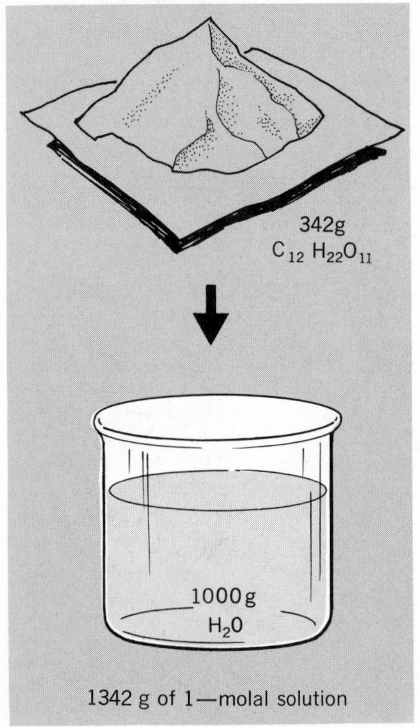

14-24 A 1-*m* solution of sugar is prepared by adding 342 g of sugar to 1 kg of water. The mass of the solution is 1342 g but there is no way to predict its exact volume. Molality is a temperature-independent method of expressing the concentration of a solution. To compare with molarity, see Fig. 3-3, p. 95.

Note that molality is based on the *mass of solvent* used and does not vary with temperature. The relationship between the molality and the change in boiling and freezing points may be represented as

$$\Delta T_b \propto m \quad \text{and} \quad \Delta T_f \propto m$$

or

$$\Delta T_b = k_b\, m \qquad 14\text{-}6$$
$$\Delta T_f = k_f\, m \qquad 14\text{-}7$$

where ΔT_b is the boiling-point elevation, ΔT_f is the freezing point depression, m is the molality or molal concentration of the solution, and k_b and k_f are experimentally determined constants known, respectively, as the ***molal boiling-point elevation constant*** and ***molal freezing point depression constant*** of the solvent. The derivation of Equations 14-6 and 14-7 involves several simplifying assumptions. The equations are, therefore, approximations which hold for dilute solutions only.

Using solutions of known concentrations, it is found that k_f for water is $\frac{(1.86°C)(\text{kg water})}{(\text{mole solute})}$. That is, when one mole of a nonelectrolyte such as sugar is dissolved in 1000 g (1 kg) of water, the freezing point of the solution is $-1.86°C$. The boiling point of the same solution is found to be 100.51°C. The value of k_b for water is, therefore $\frac{(0.51°C)(\text{kg water})}{(\text{mole solute})}$. The boiling points, freezing points, and constants for several common solvents are listed in Table 14-9.

TABLE 14-9
MOLAL BOILING-POINT AND FREEZING-POINT CONSTANTS

Solvent	*Boiling Point (°C)*	k_b *(°C)(kg solvent) / (mole solute)*	*Freezing Point 0°C*	k_f *(°C)(kg solvent) / (mole solute)*
Water	100.0	0.51	0	1.86
Benzene	80.2	2.53	5.48	5.12
Chloroform	60.2	3.63	−63.4	4.7
Carbon tetrachloride	76.5	5.03	−27.5	
Naphthalene	210.8		80.2	6.8
Camphor	208	5.95	180.0	40.0

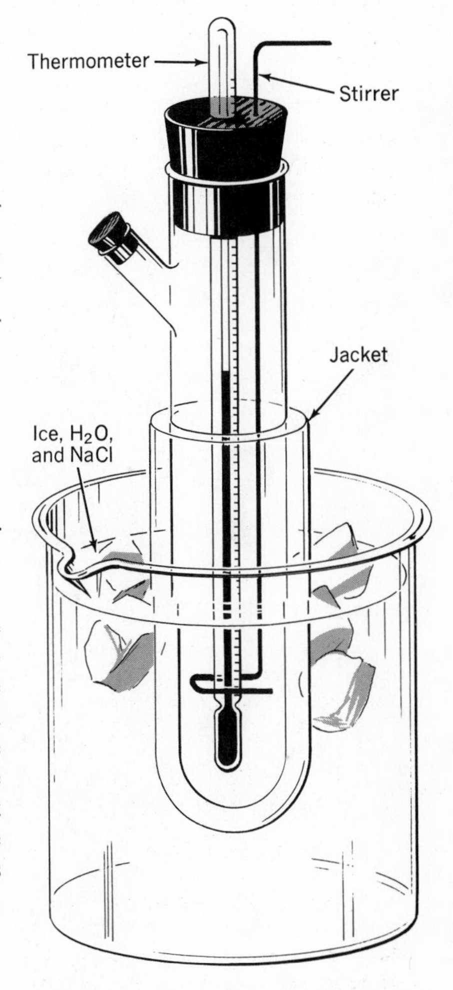

14-25 Determination of the approximate molecular masses of certain solids. A solution containing a measured mass of solute is placed in the central tube. The assembly is immersed in the cooling bath, and the freezing point is noted.

14-28 Molecular Masses of Solutes Boiling-point elevation and freezing-point depression can be used to determine the molecular masses of soluble, nonvolatile nonelectrolytes. The procedure consists of dissolving a weighed quantity of solute in a weighed quantity of solvent whose k_b or k_f is known. The change in the freezing or boiling point of the solution is then determined. The change in freezing or boiling point and the appropriate constant may be substituted into Equations 14-6 or 14-7 to obtain the molality of the solution. The molality, mass of solute and mass of solvent are related to the molecular mass of the solute as follows:

$$\textbf{molality} = \frac{\textbf{g of solute}}{\textbf{(molecular mass of solute)(kg of solvent)}}$$

or

$$\textbf{molecular mass of solute} = \frac{\textbf{g of solute}}{\textbf{(molality)(kg of solvent)}}$$

Equations 14-6 and 14-7 can be expanded to include the factor of molecular mass.

$$\Delta t_b = k_b m$$

$$m = \frac{\Delta t_b}{k_b}$$

$$\frac{\text{g solute}}{\left(\begin{array}{c}\text{molecular mass}\\ \text{of solute}\end{array}\right)(\text{kg of solvent})} = \frac{\Delta t_b}{k_b}$$

$$\begin{array}{c}\text{molecular mass}\\ \text{of solute}\end{array} = \frac{(\text{g solute})(k_b)}{(\Delta t_b)(\text{kg of solvent})} \qquad 14\text{-}8$$

$$\begin{array}{c}\text{molecular mass}\\ \text{of solute}\end{array} = \frac{(\text{g solute})(k_f)}{(\Delta t_f)(\text{kg of solvent})} \qquad 14\text{-}9$$

The several simplifying assumptions that are made in deriving Equations 14-6 and 14-7 also apply to Equations 14-8 and 14-9. Therefore their usage yields approximate molecular masses. Strictly speaking, these equations hold only for *ideal solutions.* Ideal solutions are those which are formed with no absorption or liberation of heat. The use of the equations and concepts discussed in this section is illustrated by these examples.

Example 14-5
Calculate the molality of a solution made by dissolving 18 g of glucose (m.m.180) in 200 g of water.

Solution
1. Determine the number of moles of glucose in 18 g.

$$18 \times 1 \text{ mole}/180 \text{ g} = 0.10 \text{ mole}$$

2. Calculate the kilograms of solvent in 200 g of water.

$$200 \text{ g} \times 1 \text{ kg}/1000 \text{ g} = 0.20 \text{ kg}$$

3. Substitute the moles of solute from Step 1 and the kilograms of solvent from Step 2 in Equation 14-5 and solve for m.

$$m = 0.100 \text{ mole}/0.200 \text{ kg} = 0.500 \text{ mole/kg}$$

Alternate solution, using conversion factors:

$$\frac{18.0 \text{ g glucose}}{200 \text{ g water}} \times \frac{1000 \text{ g water}}{1 \text{ kg water}} \times \frac{1 \text{ mole glucose}}{180 \text{ g glucose}} = \frac{0.500 \text{ mole glucose}}{\text{kg water}}$$

FOLLOW-UP PROBLEM

Calculate the molality of a solution made by dissolving 68.4 g of sugar (m.m. 342) in 800 g of water.

Ans. 0.250 *m*.

Example 14-6
When 0.50 g of an organic compound was dissolved in 100.0 g of benzene, the boiling point of the solution was found to be 80.40°C. What is the molecular mass of the compound? The boiling point of pure benzene is 80.20°C and $k_b = 2.53$.

Solution
1. Determine the kilograms of solvent and the boiling-point elevation.

$$\text{kg solvent} = 100 \text{ g} \times 1 \text{ kg}/1000 \text{ g} = 0.10 \text{ kg}$$
$$\Delta T_b = 80.40°\text{C} - 80.20°\text{C} = 0.20°\text{C}$$

2. Substitute values in Equation 14-8 and solve for molecular mass.

$$\text{m.m.} = \frac{0.50 \text{ g} \times 2.53°\text{C/mole/kg}}{0.20°\text{C} \times 0.10 \text{ kg}} = 63 \text{ g/mole}$$

FOLLOW-UP PROBLEM

A 1.50 g sample of urea dissolved in 105.0 g of water produces a solution which boils at 100.120°C. What is the molecular mass of urea from these data?

Ans. 60.8 g/mole.

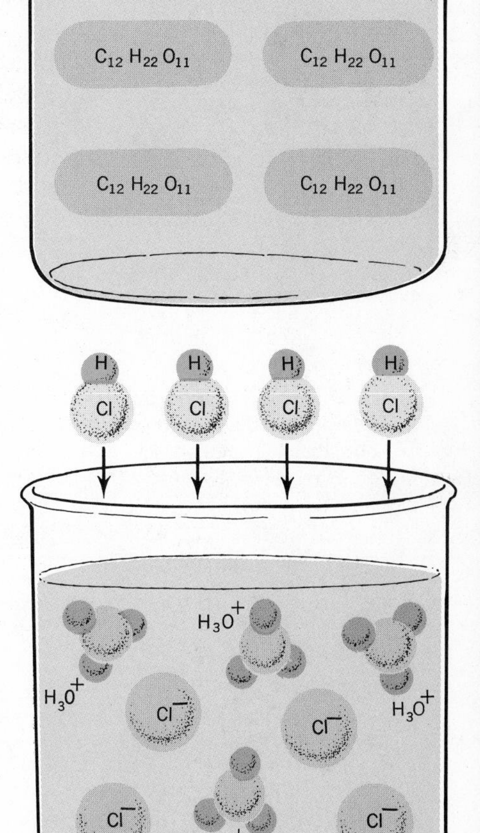

14-26 Four sugar molecules yield four particles in solution. Four HCl molecules yield eight particles: four hydronium ions and four chloride ions. Thus the colligative properties of a specific sugar solution differ from those of an acid solution, both with the same concentration properties.

DEGREE OF DISSOCIATION OF ELECTROLYTES

14-29 Percentage Dissociation from Freezing-Point Depression

Freezing point depression and boiling-point elevation measurements can be used to determine the degree of dissociation of electrolytes. When the freezing points of solutions of electrolytes such as HCl or NaCl are measured, it is found that one molecular (formula) mass of substance per kilogram of water causes an "abnormal" decrease in the freezing point; that is, a freezing point lower than −1.86°C. It is apparent that one formula mass of an electrolyte furnishes more than one mole of dissolved particles. Thus, one formula mass of NaCl in a kilogram of water should furnish a total of two moles of particles; that is, one mole of Na^+ ion and one mole of Cl^-, and yield a solution with a freezing point of

$$\frac{\textbf{2 moles}}{\textbf{kg water}} \times \frac{\textbf{(−1.86°C)(kg water)}}{\textbf{(mole of solute)}} = -3.72°\text{C}$$

The actual freezing point of −3.37°C indicates that NaCl behaves as though it were 81 percent dissociated, although it is completely dissolved and X-ray studies show that the crystal is 100 percent ionic.

The apparent degree of dissociation for a given solution is represented by the Greek letter alpha, α, and may be expressed as

$$\alpha = \frac{\textbf{moles of compound which dissociate}}{\textbf{moles of compound used}} \qquad 14\text{-}10$$

The percentage dissociation is equal to $\alpha \times 100$.

Example 14-7
A solution is prepared by adding 0.500 mole of KBr to 1000 g (1000 ml) of water. Experimental measurements reveal that the solution behaves as though 0.395 mole of KBr had dissociated. (a) What is the apparent fraction or degree of dissociation? (b) What is the apparent percentage of dissociation?

Solution
1. Substitute the data in Equation 14-10 and solve for α.

$$\alpha = 0.395/0.500 = 0.790$$

2. A fraction or degree may be converted to a percentage by multiplying by 100. Thus the apparent percentage of dissociation is $0.790 \times 100 = 79.0$ percent.

FOLLOW-UP PROBLEM

A solution is prepared by dissolving 0.200 mole of $Mg(NO_3)_2$ in 1000 g (1000 ml) of water. Measurements show that the solution acts as though 0.157 mole of $Mg(NO_3)_2$ had dissociated. Calculate (a) the degree of dissociation (α), (b) the percentage of dissociation.
Ans. (a) 0.785, (b) 78.5 percent.

Electric measurements as well as freezing point data reveal that the apparent degree of dissociation increases with dilution. The effect of diluting a KCl solution is illustrated by the data in Table 14-10.

The trend of the data in Table 14-10 suggests that in very dilute solutions, the solute approaches 100 percent dissociation. That is, it appears to be completely dissociated.

TABLE 14-10
APPARENT PERCENTAGE DISSOCIATION OF KCl SOLUTIONS

Concentration (molarity)	*Freezing Point °C*	*Apparent Percentage Dissociation*
0.50	−1.67	79.6
0.05	−0.175	88.2
0.005	−0.0182	95.6

14-30 Debye-Hückel Theory of Interionic Attraction The apparent lack of complete dissociation was explained in 1923 by Peter Debye and E. Hückel in their ***Theory of Interionic Attraction.*** They assumed that ionic solids completely dissociated into ions and that apparent deviations were caused by interactions between oppositely charged ions. Calculations using equations associated with the theory justify their assumptions. The main points of their theory are:

1. *Ionic crystals dissociate in water and yield solutions of ions.*
2. *Strong interionic attractions between ions in solution result in*

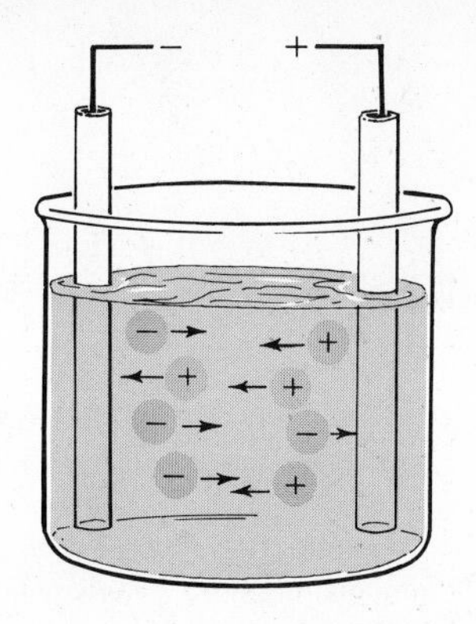

14-27 The cluster of oppositely charged ions around a given ion reduces its mobility (produces a drag effect), which decreases the conductance of the solution to a value below that which would be observed if the ion were unhampered.

a clustering action which decreases the effective concentration of the individual ions. That is, because of electrostatic attractions, ions in solution tend to be surrounded by a cluster of ions having an opposite charge. Consider a 0.1-*M* NaCl solution. If all the ions resulting from the complete dissociation of NaCl in this solution acted as individuals, then the effective concentration of both the Na^+ and Cl^- ions would be 0.1-*M*. In other words, each ion pair in a crystal of dissociated NaCl should form one Na^+ ion and one Cl^- ion. Therefore, 0.1 mole of NaCl would form 0.1 mole of Na^+ ion and 0.1 mole of Cl^- ion. However, because of residual attractions between ions in solution, some of them cluster together and act as a group rather than as individuals. Hence, the effective concentration of each ion is less than 0.1-*M*.

3. *The interionic attractive forces decrease as the solution is diluted because the ions are farther apart.* This means that the apparent degree of dissociation always increases with dilution. That is, dilution tends to break up the clusters so that more ions can act as individuals.

4. *The magnitude of the interionic attraction and the degree of clustering actions are related to the charge on the ions.* For example, there is greater interionic attraction and less apparent dissociation in a solution of $MgSO_4$ than in an NaCl solution of equal concentration. The double charge on Mg^{2+} ions and SO_4^{2-} ions results in greater attraction than the single charges on the Na^+ and Cl^- ions.

It should be noted that in precise determinations of percentage dissociation, interionic attractions must be accounted for.

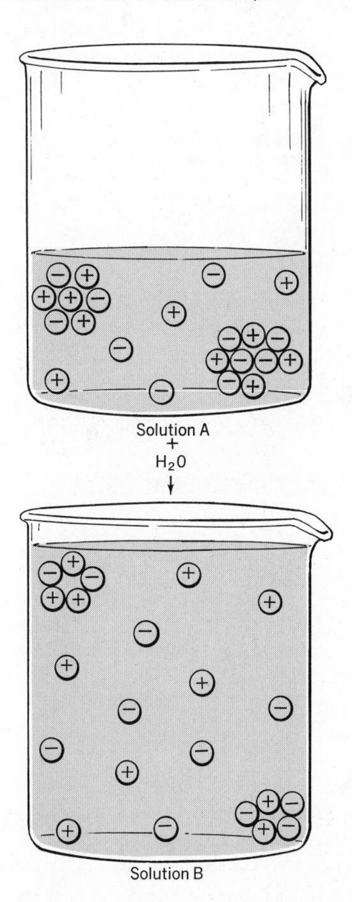

14-28 In concentrated solution A, oppositely charged ions tend to cluster in groups. Dilution (B) breaks up the clusters. The properties of solution A indicate the presence of four ion pairs. Solution B contains the same number of ions but behaves as though there were seven ion pairs.

LOOKING AHEAD

In this chapter we laid the foundations for an in-depth study of acid-base systems. The Brønsted-Lowry theory of acids and bases, the concept of K_w and pH, and the qualitative aspects of equilibria in systems of weak acids were introduced to set the stage for the next chapter.

In the next chapter we shall extend the Brønsted-Lowry concept to include ions as well as molecules. We shall also investigate the quantitative aspects of acid-base equilibria. Also, we shall apply the equilibrium law to systems involving weak acids and bases to obtain a dissociation constant that we can use to calculate H_3O^+ concentrations and, hence, the pH of the solution, percentage dissociation of the solute, and the extent of acid-base reactions. Application of these concepts and calculations will be made in our discussion of volumetric analysis.

QUESTIONS

1. Distinguish between electrolytic and nonelectrolytic solutions?

2. What are the characteristics of water molecules which make them such a good solvent for ionic substances?

3. What properties of ammonia molecules indicate that liquid ammonia might be a fairly good solvent for ionic substances?

4. What property of water is responsible for preventing the oppositely charged ions in a solution from recombining?

5. How can you distinguish between solutions of polar molecular substances in water which form ions and those which do not?

6. These reactions might be considered as hypothetical steps for the dissolution of NaCl(s) in water.

(1) 183 kcal + NaCl(s) $\longrightarrow$ Na^+(g) + Cl^-(g)
(2) Na^+(g) + H_2O(l) $\longrightarrow$ Na^+(aq) + 95 kcal
(3) Cl^-(g) + H_2O(l) $\longrightarrow$ Cl^-(aq) + 87 kcal

(a) Which reaction represents the crystal lattice energy of sodium chloride? (b) Which reaction(s) represent(s) the hydration energy? (c) Which reactions are exothermic? (d) Which reactions are endothermic? (e) Does a large or small temperature change occur when NaCl(s) is dissolved in water? Explain your answer.

7. Which of these ions has the greatest hydration energy?

	Ionic Charge	Ionic Radius (Å)
(a)	A^+	0.95
(b)	B^+	1.35
(c)	C^{2+}	0.65
(d)	D^{3+}	0.55
(e)	E^-	1.81

8. The dielectric constants for some substances are water, 80; liquid ammonia, 18; ethyl alcohol, 25; carbon tetrachloride, 2; and ether, 4. Which of these substances are good solvents for (a) ionic solutes, (b) nonpolar solutes?

9. Name and identify each of these salts as soluble or slightly soluble (insoluble) in water. Write equations which show the equilibrium that exists in saturated solutions of slightly soluble salts. (a) NaBr, (b) $MgCO_3$, (c) NH_4Cl, (d) AgI, (e) K_2CrO_4, (f) KOH, (g) $Fe(OH)_3$, (h) CuS, (i) $Cd(NO_3)_2$, (j) $Ba_3(PO_4)_2$, (k) $CoCl_2$, (l) $NiSO_4$, (m) NH_4NO_3.

10. Which compound in each of these pairs is the more soluble? (a) K_2CO_3 and $MnCO_3$, (b) $CuSO_4$ and $SrSO_4$, (c) NH_4I and AgI, (d) $FePO_4$ and $FeCl_3$, (e) $Cu(OH)_2$ and KOH (f) CdS and Na_2S.

11. Refer to the solubility graph, Fig. 14-7, and answer these questions: (a) How many grams of sodium chloride can dissolve in 0.500 kg of water at 0°C? (b) How many grams of KNO_3 can dissolve in a liter of water at 25°C? (c) Which of the substances shown in the graph is the most soluble at 50°C? (d) Which is least soluble at 50°C? (e) Which shows the least change in solubility between 0° and 100°C? (f) What mass of water at 30°C is needed to dissolve 50 g of NaCl? (g) What is the molarity of a saturated solution of KNO_3 at 20°C?

12. Show by writing equations the dissociation of each of these ionic substances in water. (a) KCl, (b) $Ca(C_2H_3O_2)_2$, (c) NH_4Br, (d) $NaNO_3$, (e) Li_2CO_3, (f) $MgSO_4$.

13. When solid anhydrous sodium sulfate is placed in water, heat is evolved. (a) Does the system have higher or lower enthalpy than in its original state? Explain. (b) Is it possible to predict accurately the direction of the entropy change? Explain.

14. Cesium iodide and lithium fluoride are both ionic compounds with general formulas M^+X^-. On the basis of the different charge densities of the ions, what are some of the differences in properties of these two compounds?

15. Fluoride ions and iodide ions both have a charge of −1. Which ions have the greater charge density? Explain.

16. (a) Explain why a compound like glycerol $C_3H_5(OH)_3$ is miscible with water. (b) Give several reasons why sugar is relatively soluble in water.

17. (a) What two factors help explain the solution of polar solutes in polar solvents? (b) Give an example of each factor.

18. Draw electron-dot structures for the species in these equations

(a) $HCl + H_2O \longrightarrow H_3O^+ + Cl^-$
(b) $H_2O + H_2O \longrightarrow H_3O^+ + OH^-$
(c) $NH_4^+ + H_2O \longrightarrow NH_3 + H_3O^+$

19. These substances are all water soluble in varying degrees. Which ones dissolve and form ions? Which dissolve in molecular form? (a) CsI, (b) CH_3OH, (c) KNO_3, (d) HCl, (e) $C_3H_5(OH)_3$, (f) NH_4NO_3, (g) $C_6H_{12}O_6$(glucose).

20. Which aqueous solutions in Question 19 would be classed as electrolytes?

21. Ethanol or ethyl alcohol (CH_3CH_2OH) is soluble in both water and carbon tetrachloride although water and carbon tetrachloride are immiscible. (a) What features of ethanol molecules are responsible for their miscibility in both CCl_4 and H_2O? (b) Which exhibits the greater entropy, a mixture of 5 ml of CCl_4 and 5 ml of H_2O, or a mixture of 5 ml of C_2H_5OH and 5 ml of H_2O?

22. When chloroform ($CHCl_3$) is dissolved in acetone ($CH_3\overset{O}{\overset{\|}{C}}CH_3$), energy is evolved. As more acetone is added, more energy is released. This energy is sometimes referred to as enthalpy of dilution, (a) Considering the bonding and molecular geometry of these species, why should there be any energy change during this process? (b) Explain, in terms of energy, why the miscibility of these liquids is favored.

23. When iodine is dissolved in carbon tetrachloride (CCl_4), 5.6 kcal/mole is absorbed. When iodine is dissolved in alcohol, 1.5 kcal/mole is absorbed. In which of the two solvents is iodine more soluble? Explain your choice. Assume that dissolution in both solvents results in relatively the same increase in entropy.

Ans. I_2 is more soluble in alcohol.

24. (a) Does the dissolution of a gas in a liquid increase or decrease the entropy of the system? Explain. (b) Does the tendency of a system to achieve maximum entropy favor or oppose the dissolution of a gas? Explain. (c) When most gases are dissolved in a liquid, energy is liberated. Give two reasons why the solubility of most gases in a liquid decreases when the temperature is increased.

25. Explain why butane ($CH_3CH_2CH_2CH_3$) is not water soluble while butanol ($CH_3CH_2CH_2CH_2OH$) is water soluble.

26. Name and identify each of these as strong or weak acid or base in aqueous solution. (a) HCl, (b) NH_4OH, (c) $H_2C_2O_4$, (d) KOH, (e) NaOH, (f) H_2S, (g) HBr, (h) HCN, (i) $Ca(OH)_2$, (j) H_3BO_3, (k) H_2CO_3, (l) H_2F_2, (m) $HClO_4$, (n) $Ba(OH)_2$, (o) H_2SO_3, (p) $HC_7H_6O_2$, (q) CsOH.

27. Write equations which represent the dissociation of each of these acids or bases in aqueous solution. Use a single arrow to show the dissociation of strong acids and bases, and a double arrow to represent the equilibrium condition that exists in the solution of a weak acid or base. Show each step of dissociation for polyprotic acids. (a) $Ba(OH)_2$, (b) HNO_3, (c) $HC_2H_3O_2$, (d) H_2CO_3, (e) $H_2C_2O_4$.

28. (a) Compare the relative strength of the bond holding the acid hydrogen in a weak acid to that holding it in a strong acid. (b) Explain how your answer is consistent with the observed dissociation of the two kinds of acid.

29. Which of these are soluble in water and which are soluble in CCl_4? (a) $CHCl_3$(l), (b) H_2S(g), (c) ICl(s), (d) HBr(g), (e) SiO_2(s), (f) CH_3OH(l).

30. (a) Explain what may happen when a diver who has been working under water at a great depth ascends too rapidly to the surface. (b) What measures are taken to help prevent the "bends?"

31. The spontaneous dissolution of sodium chloride is represented by the equation

$$NaCl(s) = Na^+(aq) + Cl^-(aq)$$

The heat of solution is

$$\Delta H = +0.93 \text{ kcal/mole}$$

(a) Is the decrease in disorder of the water molecules when they interact with the ions more or less than the increase in disorder resulting from the dissociation of the crystal? Explain. (b) Is more or less heat absorbed in separating ions than is liberated in hydrating them? Explain. (c) Would increasing the temperature increase or decrease the solubility of NaCl? Explain. (d) Which factor is responsible for the spontaneous dissolution of NaCl, the change in enthalpy or the change in entropy?

32. Why does the presence of nonvolatile solute particles lower the vapor pressure of a liquid?

33. Why is the boiling point of a liquid elevated when a solid is dissolved in it?

34. Why is there a discrepancy between the apparent degree of dissociation and the theoretical concept that ionic compounds are completely dissociated in solution?

35. Explain why it is necessary to add salt to an icy slush in order to freeze a custard into ice cream in an old-fashioned ice-cream freezer.

36. Which would you expect to exhibit the greater degree of dissociation, $CuSO_4$ or NaCl? Explain in terms of Debye-Hückel theory.

PROBLEMS

1. A solution is prepared by adding 18.0 g glucose ($C_6H_{12}O_6$) to 32.0 g of water. What is the molality of the solution?

2. An aqueous solution contains 50.0 percent of glycerol ($C_3H_8O_3$) by mass and has a density of 1.13 g/ml. What is the molarity of the solution?

3. Twenty-five g of a nondissociated compound whose molecular mass is 100 is dissolved in 500 g of water. Calculate (a) the boiling point of the solution, (b) the freezing point.

4. Calculate the freezing point and the boiling point of a water solution which contains 90 g of glucose ($C_6H_{12}O_6$) in 250 g of water.
Ans. FP = −3.72°C, BP = 101.02°C

5. A compound contains 40 percent carbon, 6.67 percent hydrogen, and 53.33 percent oxygen by mass. Forty-five grams of this compound, dissolved in 1000 g of water, boils at 100.26°C. Calculate (a) the simplest formula, (b) the molecular mass, (c) the molecular formula.
Ans. (a) $(CH_2O)_x$.

6. The molal-freezing point constant for naphthalene is 6.80. When 2.56 g of powdered sulfur is dissolved in 100 g of naphthalene, the freezing point of the mixture is lower than the freezing point of naphthalene by 0.680°C. Calculate the molecular formula for sulfur.
Ans. S_8.

7. A nonelectrolyte has a simplest formula CH_2O. A solution made by dissolving 4.0 g of the compound in 100 g of water freezes at −0.61°C. What is the formula of the compound?

8. A solution made by dissolving 0.823 g of phosphorus in 30.0 ml of carbon disulfide (CS_2) boils at 46.71°C. The density of CS_2 is 1.263 g/ml and the boiling point of pure CS_2 is 46.20°C. Use these data and determine the molecular formula of phosphorus. k_b = 2.91°C kg solvent/mole solute
Ans. P_4.

9. When 850 mg of compound **x** was dissolved in 50.0 g of H_2O, the freezing point of the solution was −0.186°C. (a) What is the molality of the solution? (b) If the simplest formula of the compound is C_2H_2O, what is the true formula?

10. Pure benzene freezes at 5.50°C. Calculate the apparent molecular mass of a compound if a solution of 10.0 g of the compound in 100.0 g of benzene freezes at 5.00°C. The molal freezing-point constant for benzene is 5.10 C°.

11. What is the apparent degree of dissociation of a solution prepared by dissolving 0.0500 mole of $Ca(NO_3)_2$ in 500 g of water if the solution behaves as if only 0.0785 mole of solute had dissociated per kg of water?
Ans. 0.785 percent.

12. A 0.200-*M* solution of KOH behaves as if the solute were 76.0 percent dissociated. (a) calculate the number of moles per liter which are *apparently* undissociated. (b) What is the apparent concentration of K^+ ions?
Ans. (a) 0.048 mole/ℓ.

13. A solution of NaCl conducts electricity as though 0.340 mole had dissociated per ℓ. α for NaCl = 0.680. What is the molarity of the solution?

14. (a) If magnesium chloride were completely dissociated when dissolved in water, what would be the boiling point of a 1.0-*M* solution? (b) Is the actual value greater or less than the value predicted in (a)? Explain.

15. If a 0.5-*M* solution of BaI_2 is totally dissociated into Ba^{2+} ions and I^- ions, what is the molar concentration of each ion? Review Sections 3-5 and 3-6.

16. One hundred ml of 0.40-*M* HNO_3 is mixed with 400 ml of 0.50-*M* $Ca(NO_3)_2$. Assume both compounds are 100 percent dissociated. Find the concentration of these ions in solution (a) $H^+(H_3O^+)$, (b) Ca^{2+}, (c) NO_3^-.

17. What is the molar concentration of these solutions? (a) A silver nitrate solution containing 1.0 mg Ag^+ per ml of solution. (b) A cadmium chloride solution containing 10 mg Cd^{2+} per ml of solution. (c) A magnesium sulfate solution containing 100 mg of $MgSO_4$ per ml of solution.
Ans. (b) 0.089 *M*

18. How many mg of metallic ions in 1.0 ml of these solutions? (a) 0.10-*M* $LiNO_3$, (b) 0.20-*M* $CuSO_4 \cdot 5H_2O$, (c) 0.50-*M* $Al(NO_3)_3$.

19. Calculate the pH of a solution in which the $[H_3O^+]$ equals (a) 1.0 *M*, (b) 0.1 *M*, (c) 0.01 *M*, (d) 1.0×10^{-7} *M*, (e) 3.0×10^{-7} *M*, (f) 4.0×10^{-10} *M*.

20. What is the pOH of each of the solutions in Question 19?
Ans. (b) 13, (d) 7, (f) 4.6.

21. Calculate the $[H_3O^+]$ and the pH of HCl solutions that are (a) 1.0 *M*, (b) 6.0 *M*, (c) 12 *M*. Assume 100 percent dissociation.

22. What is the pOH, pH, and $[H_3O^+]$ of a 0.01-*M* KOH solution?

23. What is the pH of solutions having $[OH^-]$ equal to (a) 0.1 *M*, (b) 0.001 *M*, (c) 0.002 *M*, (d) 1×10^{-5} *M*, (e) 7.0×10^{-1} *M*, (f) 3×10^{-12} *M*?

Ans. (b) 11, (d) 9, (f) 2.48.

24. What is the pH and pOH of a solution made by adding 400 ml of distilled water to 100 ml of 0.010-*M* HNO_3? Assume volumes are additive.

25. What is the H_3O^+ ion concentration solutions whose pH and pOH values are (a) pH = 5.0, (b) pOH = 3.0, (c) pH = 3.7, (d) pOH = 9.5. **Ans. (d) 3.2×10^{-5} M.**

26. What is the OH^- ion concentration and pH of these solutions? (a) 5.6 mg KOH dissolved in 100 ml of solution? (b) 74 mg $Ca(OH)_2$ dissolved in 2.0 liters of solution. Assume 100 percent dissociation.

Ans. (a) $[OH^-] = 1 \times 10^{-3}$ M, pH = 11.

27. The ΔH for the dissociation of water has a positive value. At 25°C, the K_w for water is 1×10^{-14}. (a) Is water dissociated to a greater or less degree at 100°C than at 25°C? (b) Is K_w higher or lower at 100°C? (c) Is the pH of boiling water greater or smaller than 7? (d) Is boiling water acid, basic, or neutral?

smaller

28. A 0.10-*M* solution of acetic acid ($HC_2H_3O_2$) is 1.3 percent dissociated. What is the pH of this solution?

$CH_3COOH_{(aq)} \rightleftharpoons CH_3COO^-_{(aq)} + H^+_{(aq)}$

.1 m

.1 − x   x   x

$\frac{1.3}{100} \times .1 \text{ mole} = x$

$x = 1.3 \times 10^{-3}$

$pH = -\log[H^+]$

$= -\log 1.3 \times 10^{-3}$

$= -\log_{10} 10^{.114} \times 10^{-3}$

$= -\log_{10} 10^{-2.886}$

$= -(-2.886)$

$= +2.89$

15

Acid-Base Systems and Ionic Equilibria

Preview

"It is not an overstatement to say that civilization depends for its survival on the joint wisdom of scientists and humanists."

—Harold F. Walton, Professor of Chemistry, University of Colorado

One type of reaction that has long occupied the attention of the chemist is the acid-base reaction. As a result of studying these systems, the concept of acids and bases has evolved from a very limited definition to a highly generalized one which includes a tremendous number of substances.

In an earlier chapter, we defined acids operationally as substances whose water solutions *conduct an electric current, turn blue litmus red, react with carbonates and form carbon dioxide gas, and neutralize bases*. Later these ideas were extended and acids were defined as *proton donors*.

Bases were operationally defined earlier as substances whose water solutions *turn red litmus blue, conduct an electric current, feel slippery, and neutralize acids*. Then later, bases were defined as *proton acceptors*.

In addition to the definitions, we found that acids and bases are classified as either *weak* or *strong*, depending respectively, on whether they were weakly or highly dissociated. Recall that solutions of weak acids and bases are identified as equilibrium systems.

The concepts of K_w, pH, and pOH were introduced to help us designate the relative acidity of a solution and relate it to the hydrogen (hydronium) or hydroxide-ion concentration.

After developing equilibrium concepts in the two previous chapters, we are now equipped to calculate the concentrations of various species in solutions of weak acids and bases, and estimate the relative "completeness" and suitability of a reaction as a basis for an analytical process. We shall first investigate the Brønsted-Lowry concept of acids and bases in more detail and finally extend it to include ions as well as molecules. After we consider the quantitative aspects of acid-base equilibria, we shall introduce the requirements, techniques, and calculations of volumetric analysis. In this section you will have a chance to review and extend stoichiometric principles and calculations introduced earlier.

We shall then plot curves which enable us to follow changes that occur in the acidity of a solution during an acid-base analysis. Calculating points needed to plot such a curve provides an opportunity for us to apply both stoichiometric principles and equilibrium calculations.

After briefly discussing the theory of chemical indicators, we shall attempt to show a relationship between the molecular structures and relative strengths of acids. The chapter ends with a brief discussion of a highly generalized concept known as the Lewis Theory of Acids and Bases.

BRØNSTED-LOWRY THEORY OF ACIDS AND BASES

15-1 Definition, Classification, and Characteristics of Brønsted-Lowry Acids and Bases The Brønsted-Lowry definition of an acid as a proton donor and a base as a proton acceptor implies that, in general, we should not classify a given compound as an acid or base unless we specify the reaction in which it is a participant. Consider the role of water in its reactions with HCl gas, and with NH_3 gas. The equations for these reactions, depicted in terms of space-filling models and with electron-dot formulas in Fig. 14-15 and 14-16, are

$$H_2O(l) + HCl(g) \rightleftharpoons H_3O^+ + Cl^-(aq) \qquad 15\text{-}1$$

$$H_2O(l) + NH_3(g) \rightleftharpoons NH_4^+ + OH^-(aq) \qquad 15\text{-}2$$

In equation 15-1, H_2O accepts a proton from HCl and behaves as a Brønsted base. In the reaction with NH_3, H_2O donates a proton and behaves as a Brønsted acid. It is apparent that the *role played by water depends upon the reaction in which it is involved.*

Classifying a chemical species as a Brønsted acid or base is like classifying the behavior of a person. As long as the person is isolated and inactive, he defies classification. He may be classified as a criminal or philanthropist only when he interacts with other people. Like a person's actions, the behavior of a chemical species varies with the environment. Thus, in one equilibrium system, water gives up protons and behaves as an acid, while in another system it takes on protons and acts as a base. Substances which are capable of acting as an acid in one situation and a base in another are said to be ***amphiprotic.*** Ammonia (NH_3) is another species that is amphiprotic. In Equation 15-2, NH_3 accepts a proton and behaves as a base. In some reactions, NH_3 acts as a Brønsted acid when it donates a proton and is converted into an amide ion (NH_2^-).

A little reflection will reveal that the formulas of *all Brønsted acids contain the element hydrogen, and that the structures of all Brønsted bases include an available, unshared electron-pair.* The electron-dot formula for methane (CH_4) shown in the margin, reveals that this substance could act as a Brønsted acid but not as a Brønsted base.

H
H:C:H
H

Using the Brønsted concept, **ions, as well as molecules, may behave as acids or bases.** For example, when an ionic salt such as Na_2CO_3 is added to water, the solution is found to contain small concentrations of HCO_3^- ions and OH^- ions. The presence of these ions can only be explained by assuming that CO_3^{2-} ions accept protons from water according to the equation

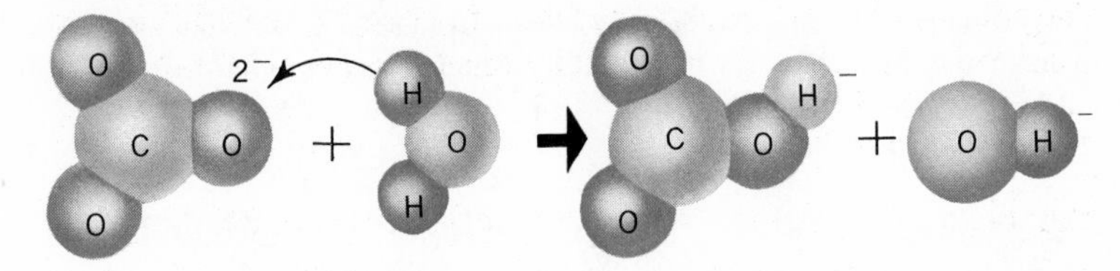

15-1 Reaction between water and carbonate ions.

$$CO_3^{2-} + H_2O(l) \rightleftharpoons HCO_3^-(aq) + OH^-(aq) \quad 15\text{-}3$$

Na^+ ions which are a part of the system but do not participate in the reaction ("spectators") are not shown in net ionic equations.

In this reaction CO_3^{2-} ions are acting as a Brønsted base. This classification of CO_3^{2-} ions is justified experimentally by the ability of an ion to "neutralize" acids. A solution of Na_2CO_3 is quite basic as evidenced by its ability to turn red litmus blue; thus, it must contain a significant concentration of OH^- ions.

15-2 Conjugate Acid-Base Pairs Examination of Equations 15-1, 15-2, and 15-3 reveals that in a reaction in which molecules or ions act as a Brønsted base, there must be some molecules or ions that act as an acid by losing protons. Furthermore, it can be seen that the base which accepts protons becomes a Brønsted acid on the product side of the equation. The species acting as a Brønsted acid on the reactant side loses protons and becomes a Brønsted base on the product side. Consider the reaction between cyanide ions (CN^-) and water, which is represented by the equation

$$\underset{\text{Brønsted base}}{CN^-} + \underset{\text{Brønsted acid}}{HOH} \rightleftharpoons \underset{\text{Brønsted acid}}{HCN} + \underset{\text{Brønsted base}}{OH^-}$$

The cyanide ions (CN^-) are obtained from a soluble ionic salt such as NaCN or KCN.

In this reaction, the CN^- ions, acting as a Brønsted base, accept one proton each and become HCN, a Brønsted acid on the product side of the equation. Water, acting as a Brønsted acid, donates one proton per molecule and becomes OH^- ions, a Brønsted base on the product side of the equation.

In general, *Brønsted acid-base reactions are equilibrium systems involving two acid-base pairs.* The equations below show that the members of an acid-base pair can be formed from each other by the transfer of a proton. The pairs are known as a ***conjugate acid-base pair.*** Proton transfer reactions are sometimes called ***protolysis*** or ***protolytic reactions.*** The proton transfer between a single conjugate acid-base pair and the general relationship between the two conjugate acid-base pairs involved in a Brønsted acid-base reaction is shown by the next equation. Addition of the two pairs gives the overall equation

$$\textbf{Pair 1} \quad \text{acid}_1 \rightleftharpoons H^+ + \text{base}_1$$

$$\textbf{Pair 2} \quad \text{base}_2 + H^+ \rightleftharpoons \text{acid}_2$$

$$\textbf{Overall reaction} \quad \text{acid}_1 + \underbrace{\text{base}_2 \rightleftharpoons \text{acid}_2}_{\text{conjugate pair 2}} + \text{base}_1$$

(conjugate pair 2: base_2 and acid_2; conjugate pair 1: acid_1 and base_1)

The species participating in the following Brønsted acid-base reactions are labeled to show the relationship between the conjugate acid-base pairs.

Ammonium ions (NH_4^+) are obtained from a soluble ionic salt such as NH_4NO_3. The nitrate ions (NO_3^-) which are part of the system but do not participate in the reaction are not shown in an ionic equation.

$$NH_4^+(aq) + HOH(l) \rightleftharpoons H_3O^+(aq) + NH_3(aq) \qquad 15\text{-}4$$

conjugate acid of the base NH_3	conjugate base of the acid H_3O^+	conjugate acid of the base H_2O	conjugate base of the acid NH_4^+

$$CH_3COO^-(aq) + HOH(l) \rightleftharpoons CH_3COOH(l) + OH^-(aq) \qquad 15\text{-}5$$

conjugate base of the acid CH_3COOH	conjugate acid of the base OH^-	conjugate acid of the base CH_3COO^-	conjugate base of the acid H_2O

Suggest a source of acetate ions (CH_3COO^-) for a reaction represented by Equation 15-5.

The conjugate acid-base concept indicates that all four species are present at equilibrium (the original acid and base, and their respective conjugates). The relative concentration of each in the system depends on the nature of the reacting species which we shall discuss in Sect. 15-3.

15-3 Equilibria in Brønsted Acid-Base Systems; Relative Strength of Acids and Bases In acid-base reactions, equilibrium favors the production of the weaker acid and base. **The stronger the reacting acid and base, the more complete the reaction.** In terms of the Brønsted concept, a ***strong base*** is one which readily accepts protons, and a ***strong acid*** is one which readily releases protons. For example, a perchloric acid solution is an excellent conductor of an electric current. This indicates that it is highly dissociated and readily gives up protons to water molecules. The equilibrium may be represented as

$$\underset{\text{acid}}{HClO_4(l)} + \underset{\text{base}}{H_2O(l)} \rightleftharpoons \underset{\text{acid}}{H_3O^+(aq)} + \underset{\text{base}}{ClO_4^-(aq)} \qquad 15\text{-}6$$

However, since dissociation is essentially complete, the equilibrium is displaced almost completely to the right. That is, at equilibrium, the ratio of products to reactants is very large. In general, in aqueous solutions, strong acids are not assigned equilibrium constants. Because the reaction shown in Equation 15-6 is essentially complete, it is usually written with a single arrow.

$$HClO_4 + H_2O \longrightarrow H_3O^+ + ClO_4^-$$

We may think of a Brønsted acid-base reaction as a competition between the two bases in the system for protons. The stronger base "wins" the competition and forces the equilibrium in the direction of the weaker acid and base. In Equation 15-6, two bases, H_2O and ClO_4^- ions, are competing for protons. H_2O, the stronger base, "wins" and the equilibrium is shifted to the right in the direction of the weaker base, ClO_4^- ions. $HClO_4$, the stronger acid, has a greater tendency to give its protons to H_2O than H_3O^+ ions have to give their protons to ClO_4^- ions. The equilibrium is displaced

TABLE 15-1
RELATIVE STRENGTHS OF ACIDS AND BASES

Acid	*Formula*	*Conjugate Base*	*Formula*
Perchloric	$HClO_4$	perchlorate ions	ClO_4^-
Hydriodic	HI	iodide ions	I^-
Hydrochloric	HCl	chloride ions	Cl^-
Nitric	HNO_3	nitrate ions	NO_3^-
Sulfuric	H_2SO_4	hydrogen sulfate ions	HSO_4^-
Hydronium ions	H_3O^+	water	H_2O
Sulfurous	$H_2SO_3(H_2O + SO_2(g))$	hydrogen sulfite ions	HSO_3^-
Hydrogen sulfate ions	HSO_4^-	sulfate ions	SO_4^{2-}
Phosphoric	H_3PO_4	dihydrogen phosphate ions	$H_2PO_4^-$
Hydrofluoric	HF	fluoride ions	F^-
Nitrous	HNO_2	nitrite ions	NO_2^-
Acetic	$HC_2H_3O_2$	acetate ions	$C_2H_3O_2^-$
Carbonic	$H_2CO_3(H_2O + CO_2(g))$	hydrogen carbonate ions	HCO_3^-
Hydrogen sulfide	H_2S	hydrogen sulfide ions	HS^-
Hydrogen sulfite ions	HSO_3^-	sulfite ions	SO_3^{2-}
Ammonium ions	NH_4^+	ammonia	NH_3
Hydrogen carbonate ions	HCO_3^-	carbonate ions	CO_3^{2-}
Hydrogen sulfide ions	HS^-	sulfide ions	S^{2-}
Water	H_2O	hydroxide ions	OH^-
Hydroxide ions	OH^-	oxide ions	O^{2-}
Ammonia	NH_3	amide ions	NH_2^-
Hydrogen	H_2	hydride ions	H^-

Decreasing acid strength (downward) — Decreasing base strength (upward)

so far to the right that the reaction represented by Equation 15-6 is essentially complete.

In contrast to $HClO_4$, HCN is very weakly dissociated.

$$\mathbf{HCN(g) + H_2O(l) \rightleftharpoons H_3O^+(aq) + CN^-(aq)} \qquad 15\text{-}7$$

This indicates that the equilibrium represented by Equation 15-7 is reached with the majority of the HCN molecules unreacted. The H_3O^+ ions are a much stronger acid than HCN, and CN^- ions are a much stronger base than H_2O. This means that the CN^- ions "win" the competition for protons and force the equilibrium to the left in the direction of the weaker acid and base. Thus, the H_3O^+ ion concentration of a 0.1-*M* HCN solution is much less than that of a 0.1-*M* $HClO_4$ solution. The displacement of the equilibrium position toward the reactant side is suggested by the relative length of the two arrows.

Inspection of Equations 15-6 and 15-7 shows that in a Brønsted acid-base reaction, the *stronger acid* has the weaker conjugate base. In Equation 15-6, ClO_4^- ions are the conjugate base of $HClO_4$. Since the base, H_2O, "won" the competition for the protons, it is stronger than ClO_4^- ions. Thus, H_3O^+, the conjugate acid of H_2O, is weaker than $HClO_4$, the conjugate acid of ClO_4^- ions. In Equation 15-7, the CN^- ions "won" the competition, and are, therefore, a stronger base than water. Accordingly, HCN, the conjugate acid of CN^- ions, is weaker than H_3O^+ ions. On the basis of these observations, we may conclude that **the stronger a Brønsted acid, the weaker its conjugate base.**

A list of acids in order of decreasing acid strength is given in Table 15-1. The conjugate bases of each acid are also shown.

Note that $HClO_4$, a very strong acid, is the conjugate acid of ClO_4^- ions, a very weak base. The perchlorate ions, ClO_4^-, are such a weak base that any base below it in the table can take the H^+ ions away from the ClO_4^- ions. Thus, water (H_2O) completely removes protons from $HClO_4$ molecules and forms hydronium ions (H_3O^+). On the other hand, *the conjugate base of a very weak acid is always a strong base.* Water is a very weak acid; hence OH^- ions, the conjugate base of H_2O, is very strong and removes protons from any acid stronger than water. You can use Table 15-1 to determine qualitatively the extent to which a given acid and base react. In general, the strong acids in the upper part of the left-hand column have the greatest tendency to react with the strong bases in the lower part of the right-hand column. The products are a weaker acid and base. For example, H_3O^+ ions have a relatively great tendency to react with NH_3 and form NH_4^+ and H_2O but a still greater tendency to react with OH^- ions and form H_2O.

ION CONCENTRATION AND pH

15-4 Solutions of Strong Acids and Strong Bases In the last chapter we identified six strong acids, $HClO_4$, HI, HBr, HCl, HNO_3, and H_2SO_4. If we use a general symbol, X, to represent the anion of these acids, then the general equation for their dissociation may be written

$$\mathbf{HX + H_2O \longrightarrow H_3O^+(aq) + X^-(aq)}$$

In this strong-acid system we assume that the equilibrium is completely displaced to the right; that is, the dissociation is complete. In other words, all strong acids are assumed to be completely converted to H_3O^+ ions. This means that H_3O^+ ions represent the strongest acid that can exist in significant concentration in aqueous solution.

The strongest base found in water solution is OH^- ions. The oxide ions (O^{2-}), amide ions (NH_2^-), and other bases stronger than OH^- ions react completely with water and produce OH^- ions.

$$\mathbf{O^{2-} + H_2O \longrightarrow 2OH^-}$$

and

$$\mathbf{NH_2^- + H_2O \longrightarrow OH^- + NH_3(g)}$$

In general, it is a relatively easy task to calculate the hydrogen ion concentration and pH of a solution of a strong acid or base. Since a strong, monoprotic acid is completely dissociated in dilute solution, the H_3O^+ ion concentration essentially equals the original concentration of the solute. For example, in a $1.0 \times 10^{-3}\ M$ (0.0010-*M*) HCl solution, the different species are

$$\mathbf{[H_3O^+] = 1.0 \times 10^{-3}\ \mathit{M},\quad [OH^-] = 1.0 \times 10^{-11}\ \mathit{M},\ [Cl^-] = 1.0 \times 10^{-3}\ \mathit{M}}$$

The pH of the solution is 3.0 and the pOH is 11.0. In this solution, the H_3O^+ ions which result from the dissociation of water did not significantly affect the pH of the solution. For this reason, we did not consider the dissociation of water in our calculations. Actually, the H_3O^+ from the HCl shifts the water equilibrium

$$\mathbf{HOH + HOH \rightleftharpoons H_3O^+ + OH^-}$$

to the left and decreases the dissociation of water. This means the $[H_3O^+]$ from the water is considerably less than 10^{-7} *M* and insignificant when compared to 10^{-3} *M* from the HCl.

15-5 Weak Monoprotic Acids Application of the Law of Chemical Equilibrium to the equilibrium system which exists in the solution of a weak acid yields a constant which expresses quantitatively the relative strength of the weak acid. A general equation for dissociation of a weak monoprotic acid is

$$\mathbf{HX + H_2O \rightleftharpoons H_3O^+ + X^-} \qquad 15\text{-}8$$

where X^- represents the anion of an acid. The equilibrium constant expression for Equation 15-8 is

$$K = \frac{[H_3O^+][X^-]}{[HX][H_2O]} \qquad 15\text{-}9$$

In dilute solutions of acids and bases which we encounter, the concentration of water is essentially constant (approximately 55.4 *M*) and may be combined with *K* and give what is called the dissociation constant, K_d. In the case of acids, the constant may be identified as K_a. Thus,

$$K_a = \frac{[H_3O^+][X^-]}{[HX]}$$

TABLE 15-2

DISSOCIATION CONSTANTS FOR SELECTED WEAK ACIDS (25°C)

Acid	Reaction	Conjugate Base Ions	K_a
Perchloric acid	$HClO_4 + H_2O \longrightarrow H_3O^+ + ClO_4^-$	perchlorate	
Hydriodic acid	$HI + H_2O \longrightarrow H_3O^+ + I^-$	iodide	
Hydrobromic acid	$HBr + H_2O \longrightarrow H_3O^+ + Br^-$	bromide	
Hydrochloric acid	$HCl + H_2O \longrightarrow H_3O^+ + Cl^-$	chloride	
Nitric acid	$HNO_3 + H_2O \longrightarrow H_3O^+ + NO_3^-$	nitrate	
Sulfuric acid	$H_2SO_4 + H_2O \longrightarrow H_3O^+ + HSO_4^-$	hydrogen sulfate	
Iodic acid	$HIO_3 + H_2O \longrightarrow H_3O^+ + IO_3^-$	iodate	1.6×10^{-1}
Benzoic acid	$HC_6H_5CO_2 + H_2O \longrightarrow H_3O^+ + C_6H_5CO_2^-$	benzoate	6.6×10^{-3}
Hydrofluoric acid	$HF + H_2O \longrightarrow H_3O^+ + F^-$	fluoride	6.8×10^{-4}
Nitrous acid	$HNO_2 + H_2O \longrightarrow H_3O^+ + NO_2^-$	nitrite	5.1×10^{-4}
Formic acid	$HCHO_2 + H_2O \longrightarrow H_3O^+ + CHO_2^-$	formate	2.0×10^{-4}
Acetic acid	$HC_2H_3O_2 + H_2O \longrightarrow H_3O^+ + C_2H_3O_2^-$	acetate	1.8×10^{-5}
Hypochlorous acid	$HOCl + H_2O \longrightarrow H_3O^+ + OCl^-$	hypochlorite	3.0×10^{-8}
Boric acid	$H_3BO_3 + H_2O \longrightarrow H_3O^+ + H_2BO_3^-$	dihydrogen borate	5.8×10^{-10}
Ammonium ion	$NH_4^+ + H_2O \longrightarrow H_3O^+ + NH_3$	ammonia (molecules)	5.6×10^{-10}
Hydrocyanic acid	$HCN + H_2O \longrightarrow H_3O^+ + CN^-$	cyanide	4.0×10^{-10}
Water	$H_2O + H_2O \longrightarrow H_3O^+ + OH^-$	hydroxide	1.0×10^{-14}

The dissociation constants for a number of acids are listed in Table 15-2. Note that there are no K_a values listed for the acids stronger than H_3O^+. Because of their very high degree of dissociation, the denominator in Equation 15-9 would approach zero and result in infinitely large K_a. The magnitude of the constants reflects the relative tendencies of the species on the left to give up protons to water.

We can use K_a to help determine the equilibrium concentrations of the species present in the solution of a weak acid. Knowledge of the H_3O^+ ion concentration enables us to calculate the pH and pOH of the solution. These calculations are illustrated.

Example 15-1

Calculate (a) the $[H_3O^+]$, (b) the pH, and (c) the percentage dissociation for 0.100-*M* acetic acid at 25°C. K_a for CH_3COOH is 1.8×10^{-5}.

Solution

(a) Write the equation and the equilibrium law expression for the reaction

$$H_2O(l) + CH_3COOH(l) \rightleftharpoons H_3O^+(aq) + CH_3COO^-(aq)$$

$$K_a = \frac{[H_3O^+][CH_3COO^-]}{[CH_3COOH]}$$

Let x = the number of moles/ℓ of CH_3COOH that dissociate and reach equilibrium, and express the equilibrium concentration of each species in terms of x. The balanced equation shows that every mole of CH_3COOH which dissociates yields 1 mole of H_3O^+ and 1 mole of CH_3COO^- ions. If we assume that the H_3O^+ from the dissociation of H_2O is negligible relative to that from the acid, then the concentration of these H_3O^+ ions and CH_3COO^- ions is equal to the moles/ℓ of CH_3OOH which dissociate.

$$[H_3O^+] = x \qquad [CH_3COO^-] = x$$

We can check the validity of this assumption by comparing the calculated H_3O^+ ion concentration with 1×10^{-7}, the maximum concentration of H_3O^+ ion in pure water. The equilibrium concentration of molecular or undissociated CH_3COOH is actually the original molar concentration minus the moles per liter dissociated.

$$[CH_3COOH] = 0.10 - x$$

Now substitute the equilibrium concentrations in the dissociation-constant expression

$$1.8 \times 10^{-5} = \frac{[x][x]}{[0.10 - x]}$$

Solving this equation yields a quadratic expression. Fortunately, there are valid simplifications we can make that save time. In solutions of fairly concentrated (0.01 *M* or greater) weak acids, there is very little dissociation so that the equilibrium concentration of the molecular species is very nearly equal to the original concentration (C_o). That is,

$$[0.10 - x] \cong [0.10]$$

In general, we can omit x when subtracting it from or adding it to a value which is large compared to itself. Again, we can make the simplifying assumption, solve the problem, and then compare the calculated value of x with the original concentration. If x does not exceed five percent of the number, then our assumption is valid. If x is greater than five percent, then we must solve either by a quadratic equation or by other methods. With the simplifying assumptions, the equilibrium law expression becomes

$$1.8 \times 10^{-5} = \frac{(x)(x)}{0.10}$$

Solving for x we obtain

$$x^2 = 1.8 \times 10^{-6}$$
$$x = 1.3 \times 10^{-3} = [H_3O^+]$$

This value is much greater than 1×10^{-7}. Therefore, it was valid to assume that the $[H_3O^+]$ from the dissociation of water is negligible. To check the validity of the assumption that $C_o - x$ $[C_o - x] \cong [C_o]$ we must find whether or not

$$\frac{x}{C}(100) < 5 \text{ percent}$$

$$\frac{x}{C_o}(100) = \frac{1.3 \times 10^{-3}}{1.0 \times 10^{-1}}(100) = 1.3 \text{ percent}$$

Since $1.3 < 5$, our assumption was valid. For all problems in this text you may assume that $[C_o - x] = [C_o]$. Occasionally this assumption may not be valid but it still saves time.

(b) Substitute the value of $[H_3O^+]$ in the equation which defines pH

$$\begin{aligned} \text{pH} &= -\log 1.3 \times 10^{-3} \\ &= 3 - \log 1.3 \\ &= 2.9 \end{aligned}$$

(c) The percentage of the original acid which dissociates may be expressed as

$$\% \text{ dissociation} = \frac{\text{moles}/\ell \text{ which dissociate}}{\text{original concentration}}(100)$$

Substitute x, the moles/ℓ which dissociate, and the original concentration (C_o) in the expression given above.

$$\text{percentage dissociation} = \frac{1.3 \times 10^{-3}}{1.0 \times 10^{-1}}(100) = 1.3 \text{ percent}$$

FOLLOW-UP PROBLEMS

1. Calculate (a) the $[H_3O^+]$, (b) the pH, and (c) the percentage dissociation for 0.50-*M* HCN at 25°C. See Table 15-2 for value of K_a. **Ans. (a) 1.4×10^{-5}, (b) 4.9, (c) 2.8×10^{-3} percent.**

2. Calculate (a) the $[H_3O^+]$ and (b) the pH of a 0.10-*M* solution of NH_4^+ (derived from a salt such as NH_4NO_3). **Ans. (a) 7.5×10^{-6} *M*.**

15-6 Weak Bases Calculations involving the dissociation constant for weak bases are parallel to those for weak acids. The constant for a weak base may be determined experimentally or be derived by combining the water constant, K_w, with the dissociation constant of the conjugate acid. Consider a solution made by adding ammonia (NH_3) to water. The reaction may be represented by

$$NH_3(g) + H_2O(l) \rightleftharpoons NH_4^+(aq) + OH^-(aq) \qquad 15\text{-}10$$

The equilibrium law expression for this reaction is

$$K_b = \frac{[NH_4^+][OH^-]}{[NH_3]}$$

At 25°C, the value of K_b is found to be 1.8×10^{-5}. Let us see how we could calculate the value of this constant, K_b, by combining the water constant, K_w, with K_a for NH_4^+ ions, the conjugate acid of NH_3. The fact that we can obtain K_b for Equation 15-10 by combining K_W and $K_{NH_4^+}$ implies that we can obtain Equation 15-10 by combining equations from Table 15-2.

$$NH_4^+ + H_2O \rightleftharpoons H_3O^+ + NH_3 \qquad K_{NH_4^+} = K_a = 5.6 \times 10^{-10}$$
$$2HOH \rightleftharpoons H_3O^+ + OH^- \qquad K_w = 1.0 \times 10^{-14}$$

We must first reverse the top equation to get NH_3 on the left-hand side so that when we add the equations, it appears as a reactant. The constant for the reverse reaction is the reciprocal of that for the forward reaction.

Adding

$$NH_3 + H_3O^+ \rightleftharpoons NH_4^+ + H_2O \qquad K = \frac{1}{K_a} = \frac{1}{5.6 \times 10^{-10}}$$

to

$$2H_2O \rightleftharpoons H_3O^+ + OH^- \qquad K_w = 1.0 \times 10^{-14}$$

yields

$$NH_3 + H_2O \rightleftharpoons NH_4^+ + OH^-$$

The net equation is the same as Equation 15-10. It can be shown that the constant for the net equation is the product of the constants for the equations which were added. Thus the dissociation constant for any of the conjugate bases listed in Table 15-2 may be obtained by applying the expression

$$K_b = \frac{1}{K_a} \times K_w = \frac{K_w}{K_a}$$

In the case of NH_3, the value of the constant is

$$K_b = \frac{1.0 \times 10^{-14}}{5.6 \times 10^{-10}} = \frac{10 \times 10^{-15}}{5.6 \times 10^{-10}} = 1.8 \times 10^{-5}$$

The magnitude of this constant shows that NH_3 is a *relatively* weak base. That is, it does not have a great tendency to accept protons from water. In solution it exists largely in *molecular form.* This is consistent with our observation that NH_3 solutions are relatively poor conductors of an electric current. The constant may be used to determine the concentration of the OH^- and other species in an NH_3 solution of given concentration.

A class of organic compound known as *amines* contains nitrogen and behaves much like NH_3 in aqueous solution. These compounds are often referred to as ***organic bases.*** One of the simplest is methyl amine (CH_3NH_2). When added to water the reaction is

$$CH_3NH_2(g) + HOH(l) \rightleftharpoons CH_3NH_3^+(aq) + OH^-(aq)$$

Application of the equilibrium law yields

$$K_b = \frac{[CH_3NH_3^+][OH^-]}{[CH_3NH_2]}$$

It may be seen that the equations and form of the constants for the organic bases are parallel to those of the inorganic base, NH_3. The constants for a number of weak bases are listed in Table 15-3.

TABLE 15-3
DISSOCIATION CONSTANTS FOR SELECTED WEAK BASES

Name	Formula	Reaction	Dissociation Constant, K_b, (25 °C)
Trimethylamine	$(CH_3)_3N$	$(CH_3)_3N + H_2O \rightleftharpoons (CH_3)_3NH^+ + OH^-$	6.5×10^{-5}
Ethanolamine	$HOC_2H_4NH_2$	$HOC_2H_4NH_2 + H_2O \rightleftharpoons HOC_2H_4NH_3^+ + OH^-$	3.2×10^{-5}
Ammonia	NH_3	$NH_3 + H_2O \rightleftharpoons NH_4^+ + OH^-$	1.8×10^{-5}
Hydrazine	N_2H_4	$N_2H_4 + H_2O \rightleftharpoons N_2H_5^+ + OH^-$	1.7×10^{-6}(20 °C)
Hydroxylamine	$HONH_2$	$HONH_2 + H_2O \rightleftharpoons HONH_3^+ + OH^-$	1.1×10^{-8}(20 °C)
Pyridine	C_5H_5N	$C_5H_5N + H_2O \rightleftharpoons C_5H_5NH^+ + OH^-$	1.8×10^{-9}
Aniline	$C_6H_5NH_2$	$C_6H_5NH_2 + H_2O \rightleftharpoons C_6H_5NH_3^+ + OH^-$	4.3×10^{-10}

15-7 Solutions Containing Anions Which Behave as Brønsted Bases According to the Brønsted concept, ions as well as molecules, may act as acids or bases. When dealing strictly with aqueous solutions, the strength of an ion as an acid or base depends on its ability, respectively to release a proton to water or accept one from water. This ability is reflected in the magnitude of the constant for its reaction with water. For example, cyanide ions (CN^-) from an ionic salt such as NaCN act as a Brønsted base when they accept protons from water and form a basic solution.

$$CN^-(aq) + H_2O(l) \rightleftharpoons HCN(aq) + OH^-(aq) \qquad 15\text{-}11$$

The constant for this reaction is

$$K_{CN^-} = K_b = \frac{[HCN][OH^-]}{[CN^-]}$$

As in the case for NH_3, this K is obtained by combining K_a for HCN and K_w.

$$K_b = \frac{K_w}{K_a} = \frac{10 \times 10^{-15}}{4.0 \times 10^{-10}} = 2.5 \times 10^{-5}$$

The reaction of ions with water is sometimes referred to as ***hydrolysis.*** This means the ions disturb the water equilibrium by reacting with either hydronium ions or hydroxide ions which are in equilibrium with water molecules. Decreasing the concentration of either the hydronium or hydroxide ions results in a solution which will be either basic or acid. The constants for these reactions are often called *hydrolysis constants* and are symbolized as K_h rather than K_b. We shall, however, use K_b to be consistent with the discussion above.

Base constants are used in the same way as acid constants to determine the equilibrium concentrations of the species in the system.

Example 15-2

What is the $[OH^-]$ of a 0.10-*M* solution of NaCN? See Eq. 15-11.

Solution

Let $x = [OH^-] = [HCN]$

then $(0.10 - x) = [CN^-]$

Substitute these equilibrium concentrations in the equilibrium law expression and solve for x.

$$2.5 \times 10^{-5} = \frac{(x)(x)}{(0.10 - x)}$$

Deleting x from the denominator, we obtain

$$2.5 \times 10^{-5} = \frac{x^2}{0.10}$$

$$x = [OH^-] = 1.6 \times 10^{-3}\ M$$

$$pOH = 3 - \log 1.6 = 2.8;\ pH = 11.2$$

FOLLOW-UP PROBLEM

Calculate the pH of a 0.10-*M* $NaC_2H_3O_2$ solution. K_a for $HC_2H_3O_2 = 1.8 \times 10^{-5}$.

Ans. 8.9.

The answers to the preceding example and follow-up problem indicate that a 0.10-*M* NaCN solution is more basic than a 0.10-*M* $NaC_2H_3O_2$ solution. This is consistent with the general principle that *the stronger a Brønsted acid, the weaker is its conjugate base.* Specifically, $HC_2H_3O_2$ with a K_a of 1.8×10^{-5} is a stronger acid than HCN which has a K_a of 4.0×10^{-10}. Therefore, $C_2H_3O_2^-$ ions, ($K_b = 5.6 \times 10^{-10}$), are a weaker base than CN^- ions ($K_b = 2.5 \times 10^{-5}$). In terms of hydrolysis, we can say *an anion related to a weaker acid hydrolyzes to a greater extent (forms a more basic solution) than one related to a stronger acid.*

The constants for a number of ions which behave as Brønsted bases are listed in Table 15-4. The magnitude of the constant reflects the relative tendency of the species on the left to accept a proton from water.

15-8 Solutions Containing Ions Which Do Not Hydrolyze Appreciably Anions related to very strong acids (those with an infinitely large K) have little tendency to accept protons and do not hydrolyze to any extent; thus, they do not disturb the water equilibrium. We may, therefore, classify Cl^-, NO_3^-, Br^-, I^-, ClO_4^-, and SO_4^{2-} ions as extremely weak bases ($K_b = 0$) and assume that they do not appreciably affect the acidity of an aqueous solution.

Most weakly hydrated metallic cations of the singly charged representative elements such as Na^+, K^+, and Rb^+ do not react appreciably with water and, therefore, do not affect the acidity of a solution. The same is essentially true of the larger, doubly-charged ions of Group IIA. The ions of the elements in these groups

TABLE 15-4
SELECTED ANIONS WHICH BEHAVE AS BRØNSTED BASES

Name of Ion	Formula	Reactions: Base		Reactions: Conjugate Acid	$K_b = \frac{K_w}{K_a\text{(conjugate acid)}}$
Oxide	O^{2-}	$O^{2-} + H_2O$	$\rightleftharpoons$	$OH^- + OH^-$	very large
Amide	NH_2^-	$NH_2^- + H_2O$	$\rightleftharpoons$	$NH_3 + OH^-$	very large
Sulfide	S^{2-}	$S^{2-} + H_2O$	$\rightleftharpoons$	$HS^- + OH^-$	7.7×10^{-2}
Phosphate	PO_4^{3-}	$PO_4^{3-} + H_2O$	$\rightleftharpoons$	$HPO_4^{2-} + OH^-$	2.1×10^{-2}
Arsenate	AsO_4^{3-}	$AsO_4^{3-} + H_2O$	$\rightleftharpoons$	$HAsO_4^{2-} + OH^-$	3.3×10^{-3}
Hypoiodite	IO^-	$IO^- + H_2O$	$\rightleftharpoons$	$HIO + OH^-$	4.3×10^{-4}
Carbonate	CO_3^{2-}	$CO_3^{2-} + H_2O$	$\rightleftharpoons$	$HCO_3^- + OH^-$	1.8×10^{-4}
Cyanide	CN^-	$CN^- + H_2O$	$\rightleftharpoons$	$HCN + OH^-$	2.5×10^{-5}
Hypobromite	BrO^-	$BrO^- + H_2O$	$\rightleftharpoons$	$HBrO + OH^-$	4.8×10^{-6}
Hypochlorite	ClO^-	$ClO^- + H_2O$	$\rightleftharpoons$	$HClO + OH^-$	3.3×10^{-7}
Sulfite	SO_3^{2-}	$SO_3^{2-} + H_2O$	$\rightleftharpoons$	$HSO_3^- + OH^-$	1.6×10^{-7}
Acetate	$C_2H_3O_2^-$	$C_2H_3O_2^- + H_2O$	$\rightleftharpoons$	$HC_2H_3O_2 + OH^-$	5.6×10^{-10}
Benzoate	$C_7H_5O_2^-$	$C_7H_5O_2^- + H_2O$	$\rightleftharpoons$	$HC_7H_5O_2 + OH^-$	1.5×10^{-10}
Formate	CHO_2^-	$CHO_2^- + H_2O$	$\rightleftharpoons$	$HCHO_2 + OH^-$	5.6×10^{-11}
Cyanate	CNO^-	$CNO^- + H_2O$	$\rightleftharpoons$	$HCNO + OH^-$	5.0×10^{-11}
Fluoride	F^-	$F^- + H_2O$	$\rightleftharpoons$	$HF + OH^-$	2.9×10^{-11}
Nitrite	NO_2^-	$NO_2^- + H_2O$	$\rightleftharpoons$	$HNO_2 + OH^-$	2.2×10^{-11}

have relatively low charge densities; hence, they are not greatly affected by the electric field of water. Ion-dipole interaction does, however, result in a loosely held mantle of water molecules about these ions. Because the number of water molecules in the cluster is not accurately known, the formulas of these ions are usually written as $M^+(aq)$. Using these generalizations, we would correctly predict that aqueous salt solutions such as NaCl, KI, $BaCl_2$, and $RbNO_3$ are essentially neutral.

FOLLOW-UP PROBLEM

1. Without making calculations, predict the relative acidity (pH =, >, or <7) of these aqueous solutions: (a) 0.1-*M* KCl, (b) 1-*M* K_2CO_3, (c) 0.1-*M* NH_4NO_3, (d) 0.1-*M* NaF, (e) 0.1 *M* Na_2SO_3.

15-9 Solutions Containing Hydrated Metallic Ions Which Behave as Brønsted Acids It is experimentally observed that *metallic salts such as $Fe(NO_3)_3$, $Al(NO_3)_3$, and $SnCl_4$,* when added to water produce *acid solutions.* The metallic ions in these salts have a small radius and a large positive charge which results in a *high charge density.* These ions are readily hydrated in water. For example, Al^{3+} bonds to six water molecules and forms hexaaquoaluminum (III) ions $[Al(H_2O)_6^{3+}]$. Naming and bonding of this type of ion, known as a *complex ion,* is discussed in Chapter 19. This, and similar hydrated metallic ions, act as Brønsted acids and give up protons to water to a significant extent. The equation is

$$\mathbf{Al(H_2O)_6^{3+} + H_2O \rightleftharpoons H_3O^+ + [Al(H_2O)_5OH^{2+}]} \quad 15\text{-}12$$

Apparently one of the protons with a partial positive charge in the attached water molecules is repelled by the high charge on the metallic ion and is thus more easily transferred to a molecule of the solvent. The K_a for Equation 15-12 is approximately 1×10^{-5}. This indicates that $Al(H_2O)_6^{3+}$ has an acid strength approximately equal to that of acetic acid. In general, the metallic ions of the transition elements have relatively high charge densities. Soluble salts composed of these ions plus an anion which does not hydrolyze, yield acid solutions. A list of cations which behave as Brønsted acids is provided in Table 15-5.

TABLE 15-5
SELECTED CATIONS WHICH BEHAVE AS BRØNSTED ACIDS

Acid		*Conjugate Base*	K_a
$Bi(H_2O)_6^{3+} + H_2O$	$\rightleftharpoons$	$H_3O^+ + Bi(H_2O)_5(OH)^{2+}$	1×10^{-2}
$Fe(H_2O)_6^{3+} + H_2O$	$\rightleftharpoons$	$H_3O^+ + Fe(H_2O)_5(OH)^{2+}$	6.0×10^{-3}
$Hg(H_2O)_4^{2+} + H_2O$	$\rightleftharpoons$	$H_3O^+ + Hg(H_2O)_3(OH)^+$	2×10^{-3}
$Cr(H_2O)_6^{3+} + H_2O$	$\rightleftharpoons$	$H_3O^+ + Cr(H_2O)_5(OH)^{2+}$	1×10^{-4}
$Al(H_2O)_6^{3+} + H_2O$	$\rightleftharpoons$	$H_3O^+ + Al(H_2O)_5(OH)^{2+}$	1.4×10^{-5}
$Cu(H_2O)_4^{2+} + H_2O$	$\rightleftharpoons$	$H_3O^+ + Cu(H_2O)_3(OH)^+$	1×10^{-8}
$NH_4^+ + H_2O$	$\rightleftharpoons$	$H_3O^+ + NH_3$	5.6×10^{-10}
$Zn(H_2O)_4^{2+} + H_2O$	$\rightleftharpoons$	$H_3O^+ + Zn(H_2O)_3(OH)^+$	2.5×10^{-10}

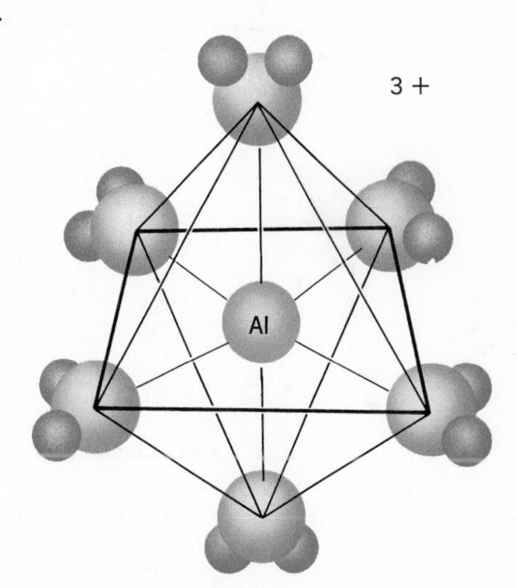

15-2 An $Al(H_2O)_6^{3+}$ ion. The six water molecules (ligands) are located at the apices of an inscribed octahedron.

15-10 Solutions Containing Both an Anion and a Cation Which Hydrolyze When the cation and anion of a salt both hydrolyze, the solution may be neutral, acid, or basic. In the case of a salt such as $NH_4C_2H_3O_2$, which contains a cation and anion that both hydrolyze, the condition of the solution depends on the extent to which each ion reacts.

$$\underset{\text{(acid)}}{NH_4^+} + HOH \rightleftharpoons H_3O^+ + NH_3 \quad K_a = 5.6 \times 10^{-10}$$

$$\underset{\text{(base)}}{CH_3COO^-} + HOH \rightleftharpoons OH^- + CH_3COOH \quad K_b = 5.6 \times 10^{-10}$$

The constant reveals that by coincidence the cation and anion hydrolyze to the same extent. Therefore, the H_3O^+ ions produced by the first reaction are completely neutralized by the OH^- ions from the second reaction. The net result is a neutral solution. If the strengths of the acid and base were different, the solution would not be neutral.

So far we have discussed aqueous solutions of four types of salts:

1. *Those in which there is neither cation nor anion hydrolysis* (no proton transfer). Solutions of these salts are neutral.
2. *Those in which the anion accepts a proton from water.* Solutions of these salts are basic. See Table 15-4.
3. *Those in which there is cation hydrolysis.* Solutions of these salts are acid. This is characteristic of many salts containing NH_4^+ ion or metallic ions having a high charge density. See Table 15-5.
4. *Those in which there is both cation and anion hydrolysis.* Solutions of these salts may be neutral, acid, or basic, depending on the relative extent of anion and cation hydrolysis.

15-11 Solutions Containing an Amphiprotic Anion Certain anions associated with polyprotic acids contain acid hydrogen and may behave as either a Brønsted acid or a Brønsted base in aqueous solution. Consider sodium bicarbonate ($NaHCO_3$), a so-called acid salt containing HCO_3^- ions. These ions may react with water in two ways.

$$\text{As a base } HCO_3^- + HOH \rightleftharpoons H_2CO_3 + OH^-$$

$$K_b = \frac{[H_2CO_3][OH^-]}{[HCO_3^-]}$$

$$\text{As an acid } HCO_3^- + HOH \rightleftharpoons H_3O^+ + CO_3^{2-}$$

$$K_a = \frac{[H_3O^+][CO_3^{2-}]}{[HCO_3^-]}$$

The acidity of a solution of $NaHCO_3$ depends on which of the two reactions occurs to the greater extent. If $K_b > K_a$, the solution is basic; if $K_a > K_b$, the solution is acid. The K_a for HCO_3^- ions is equal to the second dissociation constant (K_2) in the stepwise dissociation of H_2CO_3.

$$H_2CO_3 + H_2O \rightleftharpoons H_3O^+ + HCO_3^- \qquad K_1 = 4.3 \times 10^{-7}$$
$$HCO_3^- + H_2O \rightleftharpoons H_3O^+ + CO_3^{2-} \qquad K_2 = 5.6 \times 10^{-11}$$

The K_b for HCO_3^- is equal to

$$\frac{K_w}{K_a \text{ (for conjugate acid)}}$$

Since the conjugate acid of HCO_3^- ion is H_2CO_3, the K_b for HCO_3^- is $K_w/K_1(H_2CO_3)$. The values are

$$K_b = \frac{K_w}{K_1 \text{ for } H_2CO_3} = \frac{10 \times 10^{-15}}{4.3 \times 10^{-7}} = 2.3 \times 10^{-8}$$
$$K_a = K_2 \text{ for } H_2CO_3 = 5.6 \times 10^{-11}$$

Since $K_b > K_a$, we would predict that an aqueous solution of $NaHCO_3$ is basic. This prediction is confirmed by tests with indicators.

The hydronium-ion concentration of a solution of an amphiprotic anion may be calculated by simultaneously solving a number of equations which describe the equilibrium system. These equations and the solution to the problem may be found in almost any quantitative-analysis textbook. Here we are interested only in the qualitative aspects of the reactions and the nature of the solution formed when certain commonly encountered salts are added to water. The acid or basic characteristics of aqueous solutions of sodium salts containing an amphiprotic anion are noted in Table 15-6. Constants for the stepwise dissociation of polyprotic acids are listed in Table 15-7. As a follow-up problem you may wish to write the equations which show the acid or basic nature of the solutions listed in Table 15-6. As a further exercise you may wish to use the data in Table 15-7 to express the K_a and K_b values for the ions listed in Table 15-6. Comparison of the two values will help explain the general nature of the solutions noted in the table.

TABLE 15-6
AMPHIPROTIC ANIONS AND CHARACTERISTICS OF THEIR AQUEOUS SOLUTIONS

Name of Ion	*Formula*	*Nature of Solution*
Bisulfate	HSO_4^-	acid
Bicarbonate	HCO_3^-	basic
Dihydrogen phosphate	$H_2PO_4^-$	acid
Monohydrogen phosphate	HPO_4^{2-}	basic
Bioxalate	$HC_2O_4^-$	acid
Biphthalate	$HC_8H_4O_4^-$	acid
Bisulfide	HS^-	basic
Bisulfite	HSO_3^-	acid
Monohydrogen arsenate	$HAsO_4^{2-}$	basic
Dihydrogen arsenate	$H_2AsO_4^-$	acid

TABLE 15-7
DISSOCIATION CONSTANTS FOR SELECTED POLYPROTIC ACIDS (25°C)

Name of Acid	*Reaction for K_1*	K_1	K_2	K_3
Sulfuric acid	$H_2SO_4 + H_2O \rightleftharpoons H_3O^+ + HSO_4^-$	very large	1.3×10^{-2}	
Oxalic acid	$H_2C_2O_4 + H_2O \rightleftharpoons H_3O^+ + HC_2O_4^-$	5.4×10	5.4×10^{-5}	
Sulfurous acid	$H_2SO_3 + H_2O \rightleftharpoons H_3O^+ + HSO_3^-$	1.7×10^{-2}	6.2×10^{-8}	
Phosphoric acid	$H_3PO_4 + H_2O \rightleftharpoons H_3O^+ + H_2PO_4^-$	7.1×10^{-3}	6.3×10^{-8}	4.8×10^{-13}
Carbonic acid	$H_2CO_3 + H_2O \rightleftharpoons H_3O^+ + HCO_3^-$	4.3×10^{-7}	5.6×10^{-11}	
Hydrosulfuric acid	$H_2S + H_2O \rightleftharpoons H_3O^+ + HS^-$	1.0×10^{-7}	1.3×10^{-13}	
Arsenic acid	$H_3AsO_4 + H_2O \rightleftharpoons H_3O^+ + H_2AsO_4^-$	6.0×10^{-3}	1.0×10^{-7}	3.0×10^{-12}
Malonic acid	$H_2C_3H_2O_4 + H_2O \rightleftharpoons H_3O^+ + HC_3H_2O_4^-$	1.5×10^{-3}	3.0×10^{-6}	
Phthalic acid	$H_2C_8H_4O_4 + H_2O \rightleftharpoons H_3O^+ + HC_8H_4O_4^-$	1.3×10^{-3}	3.9×10^{-6}	
Maleic acid	$H_2C_4H_2O_4 + H_2O \rightleftharpoons H_3O^+ + HC_4H_2O_4^-$	1.4×10^{-2}	8.6×10^{-7}	

Up to this point we have discussed principles, techniques, and equations, the use of which enable us to calculate the pH and equilibrium concentration of the species present in solutions of (1) strong acids, (2) strong bases, (3) many weak monoprotic acids, (4) many weak Brønsted bases.

We are now ready to consider solutions containing more than one solute.

15-12 Solutions Containing a Weak Acid and Its Salt (Conjugate Base) or Weak Base and Its Salt (Conjugate Acid) When we add sodium acetate, a salt and conjugate base of acetic acid, to a solution of acetic acid, a weak acid, tests by indicators reveal that the pH of the solution increases. Let us use equilibrium principles to explain this observation. The principal equilibrium is

$$\mathbf{HC_2H_3O_2(l) + H_2O(l) \rightleftharpoons H_3O^+(aq) + C_2H_3O_2^-(aq)}$$

The formula for the acetate ion is sometimes abbreviated as OAc^- and the formula acetic acid is sometimes abbreviated as HOAc.

According to Le Châtelier's Principle, increasing the concentration of $C_2H_3O_2^-$, a product species, shifts the equilibrium to the left, causing an increase in the $HC_2H_3O_2$ concentration and a corresponding decrease in the H_3O^+ ion concentration. This means the solution becomes more basic and has a higher pH. In general, *the addition of a salt of a weak acid decreases the* $[H_3O^+]$ *and increases the pH of the solution.*

When NH_4Cl, a salt and conjugate acid of weak base (NH_3), is added to a solution of NH_3, the solution is observed to become more acid. The addition of NH_4^+ causes the equilibrium

$$\mathbf{NH_3(g) + H_2O(l) \rightleftharpoons NH_4^+(aq) + OH^-(aq)}$$

to shift to the left. This reduces the OH^- concentration so that the pH of the solution decreases. In general, *the addition of a salt of a weak base to a solution of the weak base decreases the* $[OH^-]$ *and decreases the pH of the solution.*

In the next example we shall demonstrate quantitatively the change in pH that occurs when a salt of a weak acid is added to a solution of the weak acid.

Example 15-3

Calculate the change in pH which occurs when 1.0 mole of sodium acetate is added to a liter of 1.0-*M* acetic acid. Assume no change in volume. $K_a = 1.8 \times 10^{-5}$

Solution

The pH of a 1.0-*M* HOAc solution is 2.4. The equation for the principal equilibrium is

$$HOAc(l) + H_2O(l) \rightleftharpoons H_3O^+(aq) + OAc^-(aq)$$

$$K_a = \frac{[H_3O^+][OAc^-]}{[HOAc]} = 1.8 \times 10^{-5}$$

1. Let $x = [H_3O^+]$

2. Express the equilibrium concentration of each of the other species in terms of x. One mole of OAc^- ion is added in the form of a completely dissociated salt, and x moles of OAc^- are derived from the dissociation of HOAc. Thus,

$$[OAc^-] = (x + 1.0)M$$

The equilibrium concentration of HOAc is the original molar concentration minus the number of moles dissociated per liter.

$$[HOAc] = 1.0 - x$$

Because HOAc is only slightly dissociated, x is relatively small, and $1.0 - x$ is essentially equal to 1.0. On the other hand, NaOAc is highly dissociated so that we may consider all the OAc^- ions to be derived from the salt. Thus, the OAc^- ion concentration of $(1.0 + x)M$ is essentially equal to 1.0 *M*. Therefore, at equilibrium, the concentrations are assumed to be

$$[OAc^-] \cong [HOAc] \cong 1.0\ M$$

3. Substitute the equilibrium concentrations in the equilibrium law expression and solve for x.

$$1.8 \times 10^{-5} = \frac{[x](1.0)}{(1.0)}$$

$$x = [H_3O^+] = 1.8 \times 10^{-5}$$

Because x is small compared to 1.0 *M*, the assumptions that $(1.0 + x) = 1.0$, and $(1.0 - x) = 1.0$ are valid.

4. Substitute x in the expression which defines pH.

$$\text{pH} = -\log 1.8 \times 10^{-5} = 5 - \log 1.8$$
$$= 4.8$$

5. The change in pH is ΔpH

$$\Delta\text{pH} = 4.8 - 2.4 = 2.4$$

Thus the $[H_3O^+]$ changes by a factor of approximately 250 (antilog of 2.4).

FOLLOW-UP PROBLEM

Calculate the pH of a solution containing 5.7 moles of sodium acetate (NaOAc) per liter of 0.10-*M* acetic acid solution. Assume no volume change. **Ans. 6.5.**

If we make the assumption that x, the $[H_3O^+]$, is small in comparison to the molar concentration of the acid and salt, as in the case of Example 15-3, then we can express the $[H_3O^+]$ of a solution composed of a weak acid and its salt (conjugate base) as

$$[H_3O^+] = K_a \frac{[C_{\text{acid}}]}{[C_{\text{salt (conjugate base)}}]} \qquad 15\text{-}13$$

A parallel formula for solutions composed of a weak base and its salt is

$$[OH^-] = K_b \frac{[C_{base}]}{[C_{salt\ (conjugate\ acid)}]} \qquad 15\text{-}14$$

Interpretation of Equations 15-13 and 15-14 reveals that the H_3O^+ of a solution containing a weak acid or base plus their respective salt depends upon the ratio of acid molarity to salt molarity. Furthermore, *dilution of such a solution does not change the pH as it does in the case of a solution of one component.*

FOLLOW-UP PROBLEMS

1. Calculate the pH of a solution 1.0 *M* in NH_3 and 1.0 *M* in NH_4Cl. $K_{NH_3} = 1.8 \times 10^{-5}$. **Ans. 9.2.**

2. How many grams of NaOAc must be added to 0.500 ℓ of 0.50-*M* HOAc to yield a solution with a pH of 5.0? $K_a = 1.8 \times 10^{-5}$. **Ans. 37 g.**

15-13 Buffer Solutions We have shown that the acidity (pH) of a solution containing a salt and its conjugate weak acid is not affected by dilution. Furthermore, the acidity (pH) of such a solution is, within limits, not greatly changed by the addition of strong acids or strong bases. Solutions with these characteristics are called ***buffer solutions.*** They play an important role in chemical processes where it is essential that a fairly constant pH be maintained.

In many industrial and physiological processes specific reactions occur at some optimum pH value. When the pH varies to any extent from this value, undesirable reactions and effects may result. For example, the pH of our blood is approximately 7.3 or 7.4. If this drops below 7.0, the results may be fatal. Fortunately, our blood contains a buffering system which maintains the acidity at the proper level. If it were not for the protection of the buffering system, we could not eat and absorb many of the acid juices and foods in our diet. Also, certain diseases such as diabetes would be much more serious.

Buffer solutions are equilibrium systems which, within limits, resist changes in acidity and maintain constant pH when acids or bases are added to them. A typical laboratory buffer may be prepared by mixing equal molar quantities of a weak acid such as $HC_2H_3O_2$ and its salt, $NaC_2H_3O_2$. When a strong base such as NaOH is added to the buffer, the acetic acid reacts with (consumes) most of the excess OH^- ion. The OH^- reacts with the H_3O^+ ion in the following system.

$$\mathbf{H_2O + HC_2H_3O_2 \rightleftharpoons H_3O^+ + C_2H_3O_2^-}$$

Reducing the H_3O^+ ion concentration causes a shift to the right, forming additional $C_2H_3O_2^-$ ion and H_3O^+ ion. For practical purposes, each mole of OH^- added consumes a mole of $HC_2H_3O_2$ and produces a mole of $C_2H_3O_2^-$. In other words, K for the reaction

$$OH^- + HC_2H_3O_2 \rightleftharpoons C_2H_3O_2^- + HOH$$

is very large. It is equal to

$$K = \frac{K_a}{K_w} = \frac{1.8 \times 10^{-5}}{10^{-14}} = 1.8 \times 10^9$$

When a strong acid such as HCl is added to the buffer, the hydrogen (hydronium) ions react with the $C_2H_3O_2^-$ ions of the salt and form more undissociated $HC_2H_3O_2$.

Sodium and chloride ions do not participate in Bronsted acid-base reactions and, therefore, are not shown in net ionic equations for the reactions.

$$H_3O^+ + C_2H_3O_2^- \rightleftharpoons HC_2H_3O_2 + H_2O;$$

$$K = \frac{1}{K_a} = \frac{1}{1.8 \times 10^{-5}} = 5.6 \times 10^4$$

The above equations show that for practical purposes, each mole of H^+ ion added consumes a mole of $C_2H_3O_2^-$ and produces a mole of $HC_2H_3O_2$. Proportional increases and decreases in the concentrations of $C_2H_3O_2^-$ and $HC_2H_3O_2$ do not significantly affect the acidity of the solution.

It is left as a follow-up problem for you to verify these statements quantitatively.

1. The addition of 1 millimole of HCl to 10 ml of pure water changes the pH from 7.0 to 1.0, or 6 pH units.
2. The addition of 1 millimole of HCl to 10 ml of a buffer 1.0-*M* in $HC_2H_3O_2$ and 1.0-*M* in $NaC_2H_3O_2$ changes the pH from 4.8 to 4.7, or 0.1 pH unit.

As you would expect, there is a limit to the quantity of H^+ or OH^- that a buffer can absorb without undergoing a significant change in pH. For example, addition of 1 mole of HCl to a liter of buffer solution containing 0.5 mole of NaOAc completely consumes the buffer and results in a drastic change in pH.

VOLUMETRIC ANALYSIS

15-14 Requirements and Objectives In many acid-base reactions, the equilibrium is displaced almost completely toward the product side. That is, the K for the reaction is exceedingly large. These reactions may be considered quantitative and can be used as the basis for the analysis of the amount of acid or base in a given sample. The process is called ***volumetric analysis.*** Any reaction considered for volumetric analysis must meet certain requirements. These requirements are

1. Only a single, specific reaction must take place between the unknown substance and the known substance used for the analysis.

2. The unknown substance must react completely and rapidly with the added standard reagent. In other words, the reaction must be quantitative.

3. An indicator or method must be available to signal when all of the unknown substance has reacted with the added standard reagent.

In a volumetric analysis, the usual objective is to determine the mass or percentage of a qualitatively identified component (the desired substance) in a sample whose quantitative composition is unknown. If the sample is a solution, the objective may be to determine its molar concentration.

15-15 Acid-Base Titration For example, the concentration of an acid solution may be determined by measuring the volume of a base of known concentration needed to react completely with a specific volume of the solution.

The application of molarity to reactions in solutions is discussed in Section 3-8, p. 100.

The solution of known concentration is known as a ***standard solution.*** The process of adding the standard solution from a graduated tube in controlled amounts is called ***titration.*** The graduated tube is known as a ***buret.*** In theory, the standard base is added until the amount of base is chemically equivalent to the amount of acid in the unknown sample. In the case of a strong acid-strong base titration, this is the point at which the number of hydroxide ions from the added base solution equals the number of hydronium ions furnished by the acid solution. In practice, the point at which chemically equivalent quantities of reactants are present, the ***stoichiometric point,*** may be estimated by the use of ***chemical indicators*** which change color at or very near the pH of the stoichiometric point. The color change of an indicator takes place at the ***endpoint*** of the titration. The indicator endpoint should, in theory, coincide with the stoichiometric point of the reaction. In a given acid-base titration, the stoichiometric point occurs at a specific pH, therefore, an indicator is chosen whose color change occurs as close as possible to this pH.

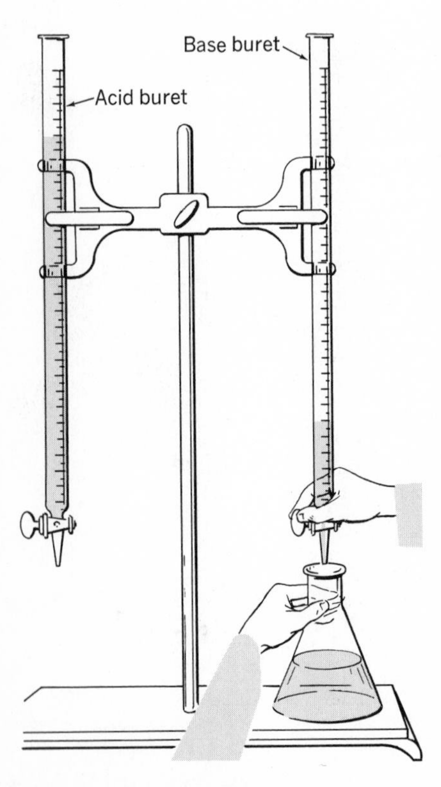

15-3 Burets used for titration experiments.

At the endpoint of an acid-base titration, the moles of standard base required to react with the desired substance may be calculated from the molarity and volume of standard solution used. The mole relationship between the desired substance (acid) and standard reagent (base) as shown by the equation for the reaction, may be used to calculate the quantity (moles) of acid in the sample. This value, divided by the volume of acid obtained from the buret readings gives the molarity of the acid.

Example 15-4

A 0.660-*M* NaOH solution is used to determine the molar concentration of a H_2SO_4 solution. What is the molarity of the acid, 20.0 ml of which is just neutralized by 36.0 ml of the standard base? The equation for the reaction is

$$2NaOH(aq) + H_2SO_4(aq) \longrightarrow Na_2SO_4(aq) + 2H_2O(l)$$

Solution

1. Calculate the millimoles (or moles) of NaOH used.

$$MV = \left(0.660 \frac{\text{mmole}}{\text{ml}}\right)(36.0 \text{ ml}) = 23.8 \text{ mmole NaOH}$$

2. Use the coefficients in the balanced equation to derive a conversion factor that relates mmole acid/mmole base. Use this factor to determine the mmole acid in the sample.

$$\frac{1 \text{ mmole acid}}{2 \text{ mmole base}} \times 23.8 \text{ mmole base} = 11.9 \text{ mmole acid}$$

3. Use the definition, $M = \frac{\text{mmole}}{\text{ml}}$, to determine the molarity of the acid.

$$M_{\text{acid}} = \frac{11.9 \text{ mmole}}{20.0 \text{ ml}} = 0.595\ M$$

The net ionic equation for the reaction between aqueous sodium hydroxide and aqueous sulfuric acid or between any strong acid and strong base is

$$H_3O^+ + OH^- \longrightarrow 2H_2O$$

Because sulfuric acid is a diprotic acid, it furnishes two moles of H_3O^+ ion per mole of acid. Thus, two moles of sodium hydroxide are required to react with one mole of sulfuric acid (H_2SO_4).

In net ionic equations for acid-base reactions, the formulas of weak acids and bases are written in molecular form and those of strong acids and bases in dissociated (ionic) form. Unchanged ions such as Na^+, K^+ and NO_3^- are not shown. A summary of rules for writing net ionic equations is given in Sect. 16-6.

FOLLOW-UP PROBLEMS

1. What is the molarity of a hydrochloric acid solution, 30.0 ml of which is just neutralized by 48.0 ml of 0.100-*M* NaOH? **Ans. 0.160 *M*.**

2. How many milliliters of 0.100-*M* HCl are required to neutralize 25.0 ml of 0.100-*M* $Ba(OH)_2$? **Ans. 50.0 ml.**

15-16 Standard Solutions Standard solutions play an important role in volumetric analysis. There are two common methods for preparing a standard solution.

1. Direct preparation by dissolving an accurately weighed pure dry substance in pure water and diluting to a known volume in a volumetric flask.
2. Reaction of an unstandardized solution with an accurately weighed quantity of a primary standard. ***Primary standards*** are stable, nonhygroscopic (nonwater absorbing) substances of known composition and high purity that react rapidly and quantitatively with the solution to be standardized. The number of moles in a weighed quantity of a solid primary standard is given by

$$\textbf{No. of moles} = \frac{\textbf{grams of solid}}{\textbf{gram molecular mass of solid}} \qquad 15\text{-}15$$

The primary standard is dissolved in a convenient amount of water which need not be measured precisely. An indicator is added, and the solution of the primary standard is titrated with the solution to be standardized. The coefficients in the balanced equation for the reaction are used to determine the moles of reagent in the solution being standardized.

The reaction in Example 15-5 involves H_3O^+ ions, not HCl molecules. This information is conveyed by the net ionic equation for the reaction.

$$Na_2CO_3(s) + 2H_3O^+ \longrightarrow 2Na^+(aq) + 3H_2O + CO_2(g)$$

Example 15-5

A hydrochloric-acid solution is standardized using pure Na_2CO_3 (m.m. = 106.0) as a primary standard. What is the molarity of the acid if 30.00 ml of the acid is required to react completely with a 0.500-g sample of Na_2CO_3?

$$Na_2CO_3(s) + 2HCl(aq) \longrightarrow 2NaCl(aq) + H_2O(l) + CO_2(g)$$

Solution

1. Calculate the moles of Na_2CO_3 in 0.500 g.

$$0.500 \text{ g } Na_2CO_3 \times \frac{1 \text{ mole } Na_2CO_3}{106.0 \text{ g } Na_2CO_3} = 0.004\,72 \text{ moles } Na_2CO_3$$

2. Calculate moles HCl needed to react with 0.004 72 moles Na_2CO_3.

$$0.004\,72 \text{ moles } Na_2CO_3 \times \frac{2 \text{ moles HCl}}{1 \text{ mole } Na_2CO_3} = 0.009\,44 \text{ moles HCl}$$

3. Calculate the molarity of the HCl.

$$M = \frac{\text{moles HCl}}{\text{l solution}} = \frac{0.009\,44 \text{ moles}}{0.030\,0 \text{ liters}} = 0.315\ M$$

FOLLOW-UP PROBLEM

A sodium hydroxide solution is standardized by reaction with benzoic acid, $HC_7H_5O_2$ (m.m. = 122.0). A 2.00-g sample of benzoic acid requires 35.00 ml of NaOH to reach the endpoint. What is the molarity of the base? Benzoic acid is a monoprotic acid. **Ans. 0.468 *M*.**

15-17 Percentage Purity of a Sample A common objective of a volumetric analysis is to determine the percentage purity of a substance. The procedure is essentially the same as in the standardization process. For example, we can use the HCl solution standardized in Example 15-5 to analyze an impure sample of Na_2CO_3. In the calculation for this type of analysis, we convert the moles of unknown into mass units (grams or milligrams) and then apply the general formula

$$\textbf{percentage of component} = \frac{\textbf{grams of component}}{\textbf{grams of sample}}\textbf{(100)} \qquad 15\text{-}16$$

Example 15-6

What is the percentage of Na_2CO_3 in an impure sample if 25.00 ml of 0.315-*M* HCl is required to react completely with a 0.600-g sample of the impure salt? The impurities do not react with HCl.

$$2HCl(aq) + Na_2CO_3(s) \longrightarrow 2NaCl(aq) + CO_2(g) + H_2O(l)$$

Solution

1. Calculate the millimoles (mmoles) of HCl used.

$$\text{mmoles} = MV = (0.315 \text{ mmoles/ml})(25.00 \text{ ml}) = 7.88 \text{ mmoles}$$

2. Use the coefficients in the equation to calculate the millimoles of Na_2CO_3 which react with 7.88 mmoles HCl.

$$7.88 \text{ mmoles HCl} \times \frac{1 \text{ mmoles } Na_2CO_3}{2 \text{ mmoles HCl}} = 3.94 \text{ mmoles } Na_2CO_3$$

3. Calculate the mg of Na_2CO_3 in the impure sample.

$$3.94 \text{ mm } Na_2CO_3 \times \frac{106 \text{ mg } Na_2CO_3}{1 \text{ mm } Na_2CO_3} = 418 \text{ mg } Na_2CO_3$$

4. Find the percentage of Na_2CO_3 in the impure sample.

$$\text{Percent } Na_2CO_3 = \frac{418 \text{ mg } Na_2CO_3}{600 \text{ mg sample}}(100) = 69.7\%$$

FOLLOW-UP PROBLEM

What is the percentage of acetic acid ($HC_2H_3O_2$) in a sample of vinegar if 35.00 ml of 0.468-*M* NaOH solution is required to neutralize a 25.00-ml sample of vinegar which has a density of 1.06 g/ml? Acetic acid is monoprotic and has a mm 60.0. **Ans. 3.70 percent.**

Chemical indicators play an important role in volumetric analysis. The choice of an indicator which yields a recognizable color change at the proper pH is essential if significant endpoint errors are to be avoided. Let us briefly consider the behavior and characteristics of acid-base indicators.

INDICATORS

15-18 Behavior and Choice of Indicators Acid-base indicators are weak acids or bases which establish an equilibrium between their molecular and ionic forms. The molecular form has a color different from that of the ionic form. Changes in acidity cause a shift in the equilibrium which favors one species over another. For example, phenolphthalein is a weak acid whose formula and equilibrium equation may be represented as

$$\underset{\text{colorless}}{\text{HPh}} + \text{B}^- \rightleftharpoons \text{HB} + \underset{\text{magenta}}{\text{Ph}^-}$$

The molecular form (HPh) is colorless, and the ionic form (Ph^-) is magenta. If an acid is added, it donates protons to the ionic form of the indicator and shifts the equilibrium to the left. Phenol-

Test your understanding of indicator behavior: The acid color of a certain indicator, HIn, is yellow and the basic color is blue. When this indicator is added to a certain acid, HB, the solution turns blue. The equation for the reaction may be written

$$HB + In^- \rightleftharpoons HIn + B^-$$

which is the stronger acid, HB or HIn?

phthalein is colorless in acid solution. When a base is added, it reacts with the HPh and produces colored Ph^- ions. As a result, in basic solution phenolphthalein is magenta or reddish. Note that the acid must be stronger than HPh to shift the equilibrium to the left and the base must be stronger than Ph^- to shift it to the right.

When a few drops of phenolphthalein indicator are added to an unknown acid sample (which is stronger than HPh), and OH^- ions are then added, the solution remains colorless until the added OH^- ions have reacted with essentially all of the stronger acid. Then the weaker acid HPh donates protons to OH^-.

Phenolphthalein has a K value equal to 1×10^{-9}. This small value indicates that HPh is a very weak acid; therefore, the indicator does not react appreciably with the OH^- ions being added until the H_3O^+ ions from the unknown acid have been changed to a very small concentration comparable to that of the $[H_3O^+]$ ions in a solution of HPh in water by itself. Phenolphthalein is a satisfactory indicator for the titration of a weak acid such as HOAc with OH^- because HOAc is much stronger than the acid, HPh. For a given titration, an indicator should be picked whose K is equal to $[H_3O^+]$ at the stoichiometric point of the titration, because at this pH the indicator is at its color transition point such that $[HIn] = [In^-]$. Thus,

$$[H_3O^+] = K\frac{[HIn]}{[In^-]} = K \text{ in this case.}$$

A number of common indicators with their color changes and pH ranges are listed in Table 15-8. Note that we use some indicators (thymol blue) which undergo two color transitions because they are diprotic (two acid hydrogens). Thus, as is the case of any

From Table 15-8 it can be seen that methyl orange is red in a solution having a pH of 3.2 or less and yellow in a solution having a pH of 4.4 or greater. The transition of intermediate color of orange is observed when the pH falls between these extremes. At an intermediate pH, $[HIn] = [In^-]$. That is, the concentration of the red-colored species is equal to that of the yellow-colored species so the solution appears orange. Thus methyl orange can be used for a titration reaction in which the pH of the solution at the stoichiometric point is between 3.2 and 4.4. The K_a for methyl orange at 25°C is 4×10^{-4}. Thus, ideally the solution at the stoichiometric point should have a $[H_3O^+] = 4 \times 10^{-4}$ M and a pH of 3.4.

TABLE 15-8
ACID-BASE INDICATORS

	Transition Range, pH	*Color Change*	
		Acid	*Base*
Methyl violet	0.5–1.6	yellow	blue
Thymol blue	1.2–2.8	red	yellow
Thymol blue	8.0–9.6	yellow	blue
Methyl orange	3.2–4.4	red	yellow
Bromcresol green	3.8–5.4	yellow	blue
Methyl red	4.8–6.0	red	yellow
Chlorophenol red	5.2–6.8	yellow	red
Bromthymol blue	6.0–7.6	yellow	blue
Phenol red	6.6–8.0	yellow	red
Neutral red	6.8–8.0	red	yellow-orange
Phenolphthalein	8.2–10.0	colorless	magenta
Thymolphthalein	9.4–10.6	colorless	blue
Alizarin yellow	10.1–12.0	yellow	red

diprotic acid, they give up their first and second proton at different points on the pH scale because H_2In is a stronger Brønsted acid than HIn^-.

TITRATION CURVES

15-19 Titration of Strong Acid with Strong Base We have now acquired the principles and tools needed to determine (a) the completeness of a reaction, (b) the composition of a solution at various stages during a titration, and (c) the pH of a solution at various stages during a titration. We can apply these principles by constructing a titration curve in which the pH of the solution during the titration is plotted against the volume of standard solution added. The resulting curve shows how rapidly the pH changes with the volume of standard solution and provides a clue to the practicability or feasibility of the titration. To define such a curve we need to calculate the pH of a solution at 10 to 12 different stages during the titration. In general, there are 4 significant stages which require 4 types of calculations. These are

1. The pH of the solution before the addition of the standard solution.
2. The pH at the stoichiometric point.
3. The pH of the solution between the start and stoichiometric point.
4. The pH of the solution after the stoichiometric point.

The type of calculation depends on whether a strong or weak acid is being titrated with a strong base. In the case of a strong acid, no equilibrium constants are involved. Stoichiometric principles can be used to calculate the $[H_3O^+]$ at any point during the titration by dividing the moles of remaining unreacted H_3O^+ ions by the total volume of solution at that time. In the titration of solutions of HCl with solutions of NaOH, the reaction is complete and at the stoichiometric point the solution contains only water and NaCl. Since neither Na^+ nor Cl^- ions hydrolyze, the only equilibrium reaction is the dissociation of water so the pH of the solution is 7. This and additional points are tabulated and plotted on color plate II opposite p. 274. As a follow-up problem you may wish to verify the pH values shown in this table.

It can be seen from the data and the graph that the addition of 0.02 ml of base (one small drop) near the stoichiometric point causes a pH change of 4 units. This means the $[H_3O^+]$ changes by a factor of 10^4. A rapid change in pH with a small change in volume of standard solution is desirable if the indicator is to give a sharp color change at the endpoint. The graph reveals that any indicator which has a detectable color change occurring in the range pH = 3.0 to pH = 10.0 would be suitable for the titration of a strong acid with a strong base. Three indicators that have commonly been used for this type of titration are

1. Bromthymol blue: range = 6.0–7.6
2. Phenolphthalein: range = 8.0–9.6
3. Methyl orange: range = 3.1–4.4

15-20 Titration of Weak Acid with Strong Base In the case of the titration of a weak acid with a strong base, we must use the equilibrium constant for the dissociation of the weak acid. Consider the titration of 50.0 ml of 0.100-*M* HOAc with 0.100-*M* NaOH. The equation for the reaction is

$$\mathbf{HOAc(l) + OH^-(aq) = H_2O(l) + OAc^-(aq)} \qquad 15\text{-}17$$

Stoichiometric calculations show that 50.00 ml of 0.100-*M* NaOH is needed to reach the stoichiometric point. The $[H_3O^+]$ at four different stages during the titration are calculated as described below.

1. Before the addition of any base. Use the method described on p. 481. For the titration of 0.100-*M* HOAc, the expression is

$$[H_3O^+] = \sqrt{C_oK_a}$$
$$= \sqrt{0.100 \times 1.8 \times 10^{-5}}$$

C_0 represents the original concentration of the acid (HOAc). See p. 482.

2. Between the start of the titration and the stoichiometric point, the titration solution is a buffer system containing a weak acid (HOAc) and its salt (conjugate base), OAc^-. The simplified equation in this region is

$$[H_3O^+] = K_a \frac{[C_o]}{[C_{salt\ (conjugate\ base)}]}$$

$$H_3O^+ = 1.8 \times 10^{-5} \frac{[HOAc]}{[OAc]}$$

3. At the stoichiometric point, the solution contains sodium and acetate ions. We may consider the pH to be the same as if we had prepared an aqueous NaOAc solution having the same concentration as the endpoint solution.

$$\text{mmole } OAc^- = \text{mmole original HOAc}$$
$$= mV = (0.100\ M)(50.0\ \text{ml}) = 5.00\ \text{mmole}$$
$$\text{ml solution} = 50.0\ \text{ml} + 50.0\ \text{ml} = 100.0\ \text{ml}$$
$$[OAc^-] = \frac{5.00\ \text{mmole}}{100.0\ \text{ml}} = 0.0500\ M$$

Since $[OAc^-]$ ions are a Brønsted base, and the principal equilibrium is

$$OAc^-(aq) + HOH(l) \rightleftharpoons HOAc(l) + OH^-(aq),$$

the hydroxide ion concentration is

$$[OH^-] = \sqrt{C_bK_b}$$

$$\text{where } K_b = \frac{K_w}{K^a} = \frac{10^{-14}}{1.8 \times 10^{-5}} = 5.6 \times 10^{-10}, \text{ and } C_b = 0.0500\ M$$

As a follow-up problem you may wish to show that the pH at this point is 8.7

4. After the stoichiometric point. In this region, the acidity of the solution is largely determined by the excess OH^- from the sodium hydroxide. The amount of OH^- furnished by the hydrolysis of the OAc^- ions is insignificant. The normally small degree of hydrolysis is repressed even further by the excess OH^-. Perhaps you can show that the pH after the addition of 75.0 ml of 0.100-*M* OH^- is 12.3.

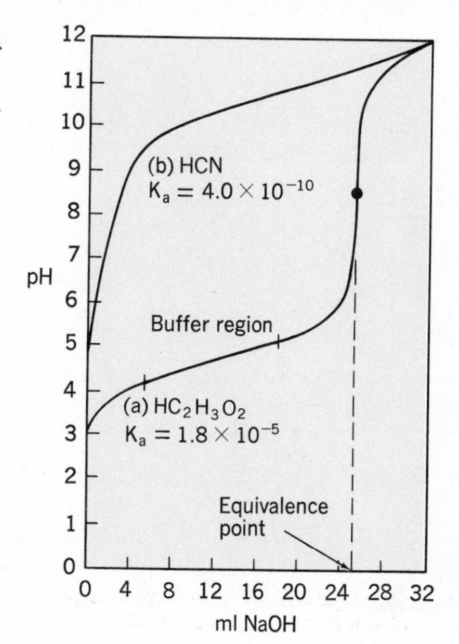

15-4 (a) Calculated curve for the titration of 0.100-*M* $HC_2H_3O_2$ (a weak acid) with 0.100-*M* NaOH (a strong base). (b) Calculated curve for the titration of 0.100-*M* HCN with 0.100-*M* NaOH.

The curve for this titration is shown in Fig. 15-4(a). Note that the vertical section of the curve is not nearly as long as that of the strong acid-strong base curve shown on color plate II.

This means that near the stoichiometric point the change in pH for a given amount of added base is less for strong base-weak acid titration than for a strong acid-strong base titration. Thus, indicator endpoints are not as sharp for the former titration as for the latter. The reaction between OH^- and HOAc produces OAc^-, which in the presence of HOAc constitutes a buffer solution. The gradual rather than sudden increase in pH may be attributed to the action of the buffer solution. This drastically reduces the choice of indicator. The stoichiometric point pH of 8.4 suggests that phenolphthalein could be used for this titration.

The choice of an indicator which yields a recognizable color change at the proper pH is essential if the color change is to represent the correct stoichiometric point.

FOLLOW-UP PROBLEM

1. (a) Calculate the pH at various stages during the titration of 50.00 ml of 0.100 *M* HCN with 0.100-*M* NaOH. $K_a = 4.0 \times 10^{-10}$. This curve is shown in Fig. 15-4. (b) Does there appear to be a relationship between the K_a of an acid and the slope of the titration curve near the stoichiometric point? Compare with the curve in Fig. 15-4a and on color plate II. (c) Is it feasible to use a chemical indicator to titrate HCN with OH^- ions? Explain.

STRUCTURAL INTERPRETATION OF RELATIVE ACID STRENGTHS

15-21 Relation of Relative Bond Strengths to Relative Acid Strengths One of our objectives in this text is to explain and predict the behavior of substances in terms of the particles which compose them. We have used experimentally determined dissociation constants to represent the relative strengths of acids. Let us now attempt to relate the relative strengths of acids to structural and bonding concepts. We shall first consider the oxyacids formed by the halogens. The oxyacids and oxyanions of chlorine are listed in Table 15-9. The structural formulas are shown in Fig. 15-5.

The structural formulas for the oxyacids shown in Fig. 15-5 reveal that all of the acids listed in the table may be written as hydroxy compounds. It may be seen that the hydrogen atom in each of these acids is covalently bonded to an oxygen atom which, in turn, is covalently bonded to a halogen atom. A simple oxyacid (hydroxy acid) may be represented by the general formula M—O—H, where M represents a nonmetallic atom other than oxygen or hydrogen.

TABLE 15-9
OXYACIDS AND OXYANIONS FORMED BY CHLORINE

Oxidation State of Central Atom	*Oxyacid*	*Oxyanion*	*Sodium salt*
+1	HClO hypochlorous	ClO^- hypochlorite	NaClO hypochlorite
+3	$HClO_2$ chlorous	ClO_2^- chlorite	$NaClO_2$ chlorite
+5	$HClO_3$ chloric	ClO_3^- chlorate	$NaClO_3$ chlorate
+7	$HClO_4$ perchloric	ClO_4^- perchlorate	$NaClO_4$ perchlorate

The strength of an oxyacid depends on the relative strengths of the M—O and the O—H bonds. The more closely the bonding electron pairs are drawn toward the oxygen, the weaker the O—H bond will be, and the greater will be the tendency for the acid molecule to transfer a proton to a solvent molecule. That is, relatively strong oxyacids are characterized by relatively weak O—H bonds. The relative strengths of the oxyacids formed by the halogens are indicated by the magnitude of the dissociation constants listed in Table 15-10.

Examination of the data in Table 15-10 reveals that *the strength of the chlorine oxyacids increases as the number of unprotonated oxygen atoms increases.* The additional oxygen atoms have a high electron affinity. This means that the electrons bonding the hydro-

TABLE 15-10
DISSOCIATION CONSTANTS FOR HALOGEN OXYACIDS

Chlorine Oxyacids	*K*	*Bromine Oxyacids*	*K*	*Iodine Oxyacids*	*K*
HClO	5.6×10^{-8}	HBrO	2×10^{-9}	HIO	1×10^{-11}
$HClO_2$	1×10^{-2}				
$HClO_3$	large	$HBrO_3$	large	HIO_3	large
$HClO_4$	very large				

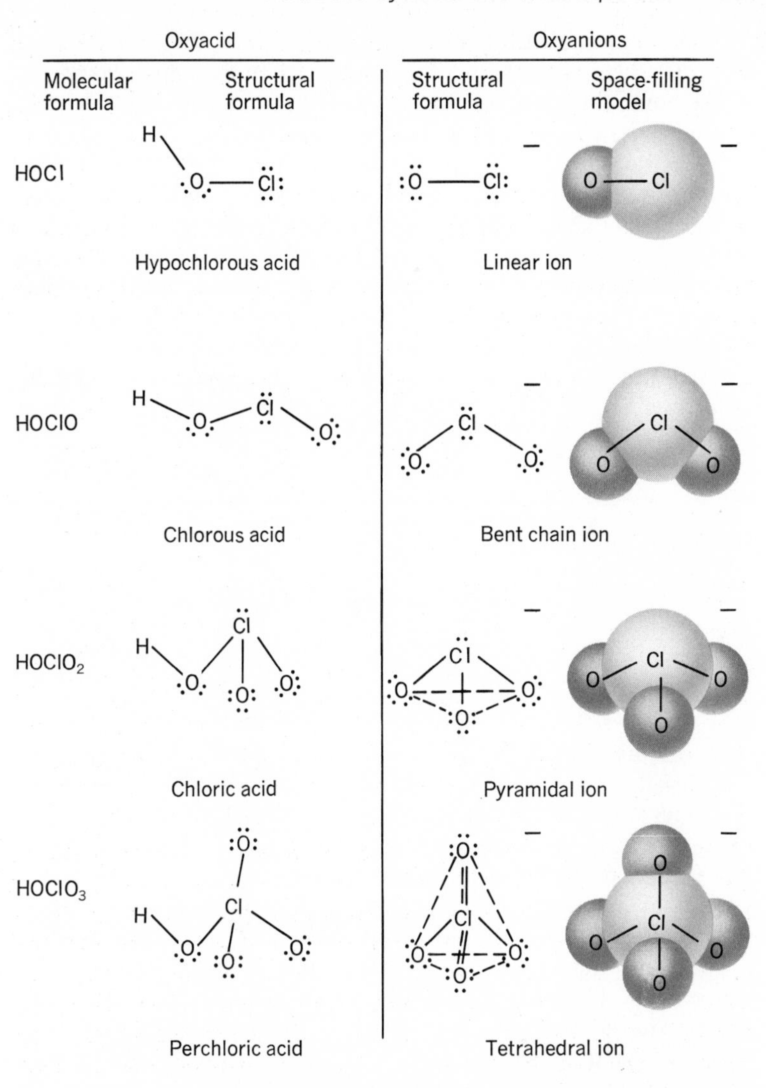

15-5 Formulas and structures of the oxyacids and oxyanions of chlorine.

gen to the oxygen atom in the O—H group tend to be pulled away from the hydrogen, thereby weakening the O—H bond. The same phenomenon may be interpreted in terms of the oxidation number of the central atom. The increase in number of oxygen atoms is paralleled by an increase in the oxidation number of the central atom. Examination of the electron-dot structure of hypochlorous acid (HClO or ClOH) in Fig. 15-5 reveals that only one outer electron of the chlorine atom is used to form the covalent bond with oxygen. Because oxygen is more electronegative than chlorine, the oxidation number of chlorine is +1. Now look at the line formula for perchloric acid [$ClO_3(OH)$]. It can be seen that the chlorine atom shares a pair of its own electrons with each of three additional oxygen atoms. This means that each chlorine atom is sharing a total of 7 electrons with the 4 oxygen atoms. Hence, in $ClO_3(OH)$, the chlorine atom has an oxidation number of +7.

Chlorine atoms with the higher positive oxidation number tend to attract the electrons away from oxygen atoms and weaken the O—H bond. This means that the O—H bond in perchloric acid is weaker than that in hypochlorous acid.

When the central atoms in two oxyacids have the same oxidation number, the relative strengths of the two acids may be predicted on the basis of the relative electronegativities of the two central atoms. The atom with the greater electronegativity will have the greater attraction for the electrons of the oxygen atom. On this basis, we would predict that HClO should be a stronger acid than HBrO. The data in Table 15-10 confirm our prediction.

The trend in the relative strengths of the oxyacids formed by the elements of Group VIA parallels the trend of those of Group VIIA.

FOLLOW-UP PROBLEM

Use oxidation numbers and/or electronegativity to predict the relative strengths of these groups of acids: (a) H_2SO_4 and H_2SO_3, (b) H_2TeO_3, H_2SO_3, H_2SeO_3.

According to the above principles, we would predict that HNO_3 would be the strongest oxyacid formed by the elements of Group VA. This may be explained by the fact that nitrogen has the highest electronegativity of any element in the group. In addition, HNO_3 is the only acid of the group in which 2 oxygen atoms are bonded to the nonmetallic atom but not to hydrogen.

O=N(–O:)–O–H

nitric acid molecule

For a given polyprotic acid, the strength of its anions as Brønsted acids decreases as the charge on the anion increases. Thus, $H_2PO_4^-$ is a stronger acid than HPO_4^{2-} because it is easier to remove a proton from a singly-charged anion than from a doubly-charged one.

The dissociation constants for the binary acids of Group VIA are

$$H_2O + H_2O \rightleftharpoons H_3O^+ + OH^- \qquad K = 1.0 \times 10^{-14}$$
$$H_2O + H_2S \rightleftharpoons H_3O^+ + HS^- \qquad K = 1.1 \times 10^{-7}$$
$$H_2O + H_2Se \rightleftharpoons H_3O^+ + HSe^- \qquad K = 2.0 \times 10^{-4}$$
$$H_2O + H_2Te \rightleftharpoons H_3O^+ + HTe^- \qquad K = 2.3 \times 10^{-3}$$

The relative strength of the binary acids of Group VIA elements increases as the atomic number increases. This reflects a decrease in the strength of the bond between hydrogen and the nonmetallic atom.

All of the binary acids formed by the elements of Group VIIA are completely dissociated in aqueous solution except hydrogen fluoride. As with the Group-VIA elements, we can explain this observation in terms of decreasing H—X bond strength as the atomic number of X increases. We can also use the thermodynamic concepts of enthalpy and entropy to help us explain relative strength of acids. These are discussed briefly in Section 18-9.

LEWIS THEORY OF ACIDS AND BASES

15-22 Definition and Characteristics of Lewis Acids and Bases **Lewis acids are electron-pair acceptors and Lewis bases are electron-pair donors.** Both the Arrhenius and Brønsted theories limit the classification as acids to substances which contain hydrogen. This restriction does not allow a number of substances which experimentally behave similarly to be designated as acids. In 1923, Gilbert N. Lewis, an American chemist, proposed a more general theory of acids and bases. The Lewis definitions are: an acid is an electron-pair acceptor; a base is an electron-pair donor; an acid-base reaction involves the formation of a coordinate covalent bond between the electron-pair donor (base) and the electron-pair acceptor (acid). These definitions indicate that any substance with an *unshared pair of electrons* in the outer energy level can act as a *Lewis base* and any substance with an *available empty orbital* can act as a Lewis acid.

Lewis acids need not contain hydrogen. A little reflection will reveal that a substance that qualifies as a Brønsted base will also be a Lewis base. That is, in order to accept a proton, a substance must donate a pair of electrons. Thus, NH_3 is acting as both a Brønsted and a Lewis base when it reacts with water and forms NH_4^+.

Note that when we apply the Lewis theory to a Brønsted acid-base reaction the *proton itself* is the Lewis acid, rather than the species which donates it.

15-23 Species Which Behave as Lewis Acids The most familiar Lewis acids may be grouped into the categories listed below.

1. *Molecules having a central atom with an incomplete octet or containing multiple bonds (except between carbon atoms).* Examples of Lewis acid-base reactions are:

(a) $BF_3 + NH_3 \longrightarrow H_3NBF_3$

Lewis acid — **Lewis base** — **coordination compound containing coordinate bond**

(b) $SO_3 + \left[:\ddot{\underset{\cdot\cdot}{O}}:\right]^{2-} \longrightarrow \left[\begin{array}{c} O \\ \cdot\cdot \\ O:S:O \\ \cdot\cdot \\ O \end{array}\right]^{2-}$

Lewis acid — **Lewis base** — **ion containing coordinate bond**

2. *Molecules with a central atom capable of expanding its outer octet by using empty d orbitals.* An example is

GILBERT NEWTON LEWIS
1875–1946

G. N. Lewis received a Ph.D. in chemistry from Harvard University in 1899, studied in Leipzig and Göttingen in Germany, and in 1905 joined the Research Laboratory of Physical Chemistry at the Massachusetts Institute of Technology. He remained there until 1912, at which time he accepted a position as professor of chemistry at the University of California. During his 30-year career at this university, he helped build one of the finest chemistry departments to be found in any university. More Nobel Prize laureates now work at the University of California than at any other single university in the world. Although Lewis never received a Nobel Prize himself, he trained a number of students who did receive this high honor. This is a reflection of the inspirational teaching and leadership abilities of this outstanding scientist.

Lewis received much recognition and many honors for his work on the electronic theory of the covalent bond and the electronic theory of acids and bases. Both are powerful tools which enable the chemist to correlate and explain many observations and phenomena associated with chemical reactions. For many years, chemists had been able to identify acids and bases by their specific properties. Until 1923, however, there were no theories that adequately explained the acid and basic behavior of many substances. This was, in part, caused by the belief that one particular element or group of elements was responsible for the observed properties. Lewis recognized the limitations of this belief and stated, "To restrict the group of acids to those substances which contain hydrogen interferes as seriously with the systematic understanding of chemistry as would the restriction of the term 'oxidizing agent' to substances containing oxygen."

As a result of his search for a more fundamental property common to all acids or bases, Lewis developed his generalized electronic theory. This theory permits the inclusion of a vast number of reactions within the scope of acid-base phenomena and has, therefore, helped chemists to systematize the study of chemical reactions.

$$\underset{\text{Lewis acid}}{SnCl_4} + \underset{\text{Lewis base}}{2Cl^-} \rightleftharpoons \underset{\text{complex ion containing coordinate bond}}{SnCl_6^{2-}}$$

Because in $SnCl_4$ molecules a tin atom has a complete outer octet, it must use empty *d* orbitals in its fifth energy level to accept electron-pairs from chloride ions.

3. *Simple Cations.* All simple cations have an empty orbital in their outer energy level and can, in theory, act as Lewis acids. For example, Zn^{2+} ions have 10 electrons in the 3*d* energy level but none in the 4th level. The empty orbitals in the 4th level are available for coordinate bond formation. In aqueous solution, Zn^{2+} ions react with four water molecules and form $Zn(H_2O)_4^{2+}$ ions. The four water molecules each share a pair of their electrons with a Zn^{2+} ion. Thus, the water molecules act as Lewis bases and the Zn^{2+} ions act as Lewis acids. The hydrated ion $Zn(H_2O)_4^{2+}$ has a tetrahedral configuration (sp^3 hybrid orbitals). The reaction may be represented by the equation

$$Zn^{2+}(aq) + 4H_2O(l) \rightleftharpoons Zn(H_2O)_4^{2+}(aq)$$

In the next chapter, we shall find that the reaction of cations as Lewis acids helps us to explain a number of experimental observations related to the solubilities of precipitates. For this reason, let us examine their nature and behavior in a little more detail.

The *strength of cations as acids increases with increasing charge density.* Thus, zinc ions with a large charge and a small radius have a high charge density and are a much stronger Lewis acid than K^+ ions. In Chapter 14 you learned that charge density depends on the charge and radius of an ion. The charge density of an ion with a given radius increases as the charge increases. The charge density also increases as the radius of an ion with a given charge *decreases. The ionic radius decreases as you go from left to right and increases from top to bottom in the Periodic Table.* This means that the *acid strength of cations* having the same charge generally *increases* as you go from *left* to *right* and *decreases* as you go from *top* to *bottom* in the Periodic Table.

15-24 Comparison of Lewis Theory with Brønsted-Lowry Theory and Arrhenius Theory Covalent bonds can be classified according to the mechanism of formation as either simple covalent or coordinate covalent, depending on whether each atom contributed an electron to the pair or one atom contributed both electrons. In a given molecule, the bonds are identical.

When a covalent bond is broken in a chemical reaction, the shared pair of electrons must either stay with one of the atoms or be divided between the two atoms. Likewise, when a covalent bond is formed in a reaction, each atom must either contribute one electron to the shared pair, or one atom must contribute both. Thus all reactions which involve the breaking or forming of covalent bonds may be grouped in one of two categories: those in which the *bonding electrons remain intact* and those in which the *bonding electrons are divided between two atoms.* The breadth and unifying characteristic of the Lewis Theory enable us to classify all *reactions in which the bonding pair of electrons remains intact* as *acid-base reactions.* Reactions in which the electron pair is divided so that temporarily each atom has an unpaired electron are known as *free-radical reactions.* An example of this type of reaction was found in Chapter 12 when we discussed the reaction mechanism of

$$H_2(g) + Cl_2(g) \longrightarrow 2HCl(g)$$

Many oxidations occurring in the gas phase are of this type.

In summary, in going from the Arrhenius to the Brønsted and finally to the Lewis theory, we have gone from a very restrictive, limited definition to a very general theory. The Arrhenius theory has only one kind of acid (H^+) and one kind of base (OH^-) and was restricted to aqueous solutions. The Brønsted theory has only one kind of acid (must contain a proton) but many bases (any proton acceptor) and can be applied to aqueous or nonaqueous systems. The Lewis theory has many acids (any electron-pair acceptor) and many bases (any electron-pair donor). This information is summarized in Table 15-11. It should be emphasized that each of the theories has certain advantages and useful applications. They all help us to systematize the study of chemical reactions.

TABLE 15-11
SUMMARY OF ACID-BASE THEORIES

Theory	*Acid*	*Base*
Arrhenius (in water only)	Specific: substances which furnish hydrogen ions (protons) in water solution	Specific: substances which furnish hydroxide ions in water solution
Brønsted (in any solvent)	Specific: substances which donate protons (H^+)	General: substances capable of accepting protons
Lewis (in any solvent)	General: electron-pair acceptors (may or may not contain hydrogen)	General: electron-pair donors

LOOKING AHEAD

We have now considered the qualitative and quantitative aspects of molecular equilibria and acid-base equilibria. With this background, we are now equipped to study the solubility equilibria that exist in saturated solutions of slightly soluble substances.

In the next chapter we shall use equilibrium principles to help us predict whether a precipitate forms when we mix ionic solutions, and whether a precipitate dissolves when we add specific acids, bases, and other reagents to the solution. Application of the principles will be illustrated when we discuss qualitative and quantitative analysis.

QUESTIONS

1. Write formulas for the conjugate bases for each of these acids (a) HCl, (b) CH_4, (c) HSO_3^-, (d) H_2SO_4, (e) NH_3, (f) $HClO_4$.

2. Show how each of these acids react with water and forms a conjugate acid-base pair (a) HCl, (b) HNO_3, (c) H_2SO_4, (d) $HClO_4$, (e) H_2S, (f) H_3PO_4.

3. Draw the electron-dot structures for these species and show that each has an unshared pair of electrons. (a) NH_3, (b) CH_3^-, (c) H_2O, (d) CH_3OH, (e) Cl^-, (f) SO_4^{2-}, (g) S^{2-}, (h) NH_2^-, (i) HSO_3^-.

4. (a) Write the reaction of water with each of the species listed in Question 3. (b) Label the acid-base pairs formed.

5. Use Table 15-1 to predict whether a reaction between these pairs occurs to any appreciable extent. Identify the reacting Brønsted acids and bases. (a) $HCl + H_2O$, (b) $H_2O + H_2SO_4$, (c) $HSO_4^- + H_3O^+$, (d) $HS^- + H_3O^+$, (e) $CH_3COOH + H_2SO_4$, (f) $HClO_4 + OH^-$, (g) $HCO_3^- + OH^-$, (h) $NH_3 + HSO_3^-$.

6. The formation of products is strongly favored in this acid-base system:

$$HX + B^- \rightleftharpoons HB + X^-$$

(a) Identify the bases competing for protons. (b) Which base is stronger? (c) Which is the weaker acid, HX or HB? (d) Does the K for this system have a large or small value? (e) How is the equilibrium affected by the addition of the soluble salt NaB?

7. Write the equation for the reaction of each of these ions with water. Experiments show that (b), (c), and (d) form acid solutions. (a) HCO_3^-, (b) $H_2PO_4^-$, (c) HSO_4^-, (d) NH_4^+, (e) HS^-, (f) HPO_4^-, (g) S^{2-} (h) CO_3^{2-}.

8. Which of these 1.0-*M* solutions is (a) basic or (b) acid? Explain. (a) Na_2CO_3, (b) Na_2S, (c) $FeCl_3$, (d) $(NH_4)_2SO_4$, (e) $Al_2(SO_4)_3$, (f) $MgSO_4$, (g) $KHCO_3$, (h) $AgNO_3$, (i) NH_4I, (j) $NaHSO_4$.

9. (a) What are buffer solutions? (b) Why are they useful?

10. Write the equilibrium expression for the indicator HIn. When $[In^-]$ and $[HIn]$ are equal, how do the values for K_a and $[H^+]$ compare?

11. Separate solutions of two acids HX and HY have the same pH. Does this mean that the molar concentration of the acids is identical? Explain.

12. Boric acid is a much weaker acid than acetic acid. Is a solution of sodium borate more or less basic than a solution of sodium acetate? Explain.

13. A solution of sodium cyanide has a much higher pH than a solution of sodium fluoride. On the basis of this observation, is the value of K for HCN larger or smaller than the K for HF? Explain.

14. Suggest a reason why the molarity of a NaOH solution changes when it is exposed to air for an extended period of time.

15. Does the neutralization of an acid with a standard base always provide a measure of the hydronium-ion concentration of the original acid solution? Explain.

16. Use Le Châtelier's Principle to explain the change in pH observed when NH_4Cl is added to an NH_3 solution.

17. What are the general requirements for a volumetric analysis?

18. What is the difference between the terms "stoichiometric point" and "endpoint?"

19. Use the concept of charge density to explain that HI is a stronger acid than HF.

20. Explain the increasing acid strength of the hydrogen halides as one goes from fluoride to iodide.

21. Explain the differences in acid strength of the oxyacids containing chlorine.

22. On the basis of the different oxidation states of the central atom, which of the following oxyacids is the stronger, (a) HXO_3 or (b) HXO_4? Explain.

23. Which is the stronger acid, an aqueous solution of H_2Se or one of H_2S? Explain.

24. Account for H_2SO_4 being a stronger acid than H_2TeO_4.

25. Which is the stronger acid, H_3AsO_4 or H_3SbO_4? Explain.

26. (a) With the aid of electron-dot structures, show how phosphine (PH_3) and ammonia both act as Brønsted bases when added to water. (b) Name the ions formed.

27. Draw the Lewis structures for each of these compounds (a) BF_3, (b) NH_3, (c) NCl_3.

28. Which of these could act as Lewis acids but not as Bronsted acids? (a) HCl, (b) H_2SO_4, (c) SO_3, (d) HSO_3^-, (e) BF_3, (f) CH_3COOH, (g) $SnCl_4$, (h) SeF_4.

29. In these reactions, identify the Lewis acid and the Lewis base.

(a) $Fe^{3+} + 6H_2O \rightleftharpoons Fe(H_2O)_6^{3+}$

(b) $BF_3 + NH_3 \rightleftharpoons F_3BNH_3$

(c) $BF_3 + F^- \rightleftharpoons BF_4^-$

(d) $H^+ + Cl^- \rightleftharpoons HCl$

30. KOH, H_3AsO_4, and $HClO_4$ all contain one or more OH groups. Explain in terms of bonding principles, why KOH is a base whereas H_3AsO_4 and $HClO_4$ are acids in water.

PROBLEMS

1. Calculate the $[H_3O^+]$, $[OH^-]$, pH, and pOH of these solutions (a) 1.0-*M* HCl, (b) 0.50-*M* HNO_3, (c) 0.0020-*M* $HClO_4$, (d) 1.5×10^{-4}-*M* KOH, (e) a solution prepared by dissolving 0.040 g NaOH in 2.0 ℓ of solution, (f) a solution prepared by diluting 1.0 ml of 0.20-*M* HCl to a total volume of 5.0 liters, (g) a solution made by dissolving 0.10 mole Na_2O in 1.0 ℓ of solution.
Ans. (d) $[H_3O^+] = 6.7 \times 10^{-11}$ *M*, $[OH^-] = 1.5 \times 10^{-4}$ *M*, pH = 10.2, pOH = 3.8, (g) $[H_3O^+] = 5.0 \times 10^{-14}$, $[OH^-] = 0.20$ *M*, pOH = 0.70, pH = 13.30.

2. Calculate the $[H_3O^+]$, pH, and percentage dissociation of these solutions.(a) 1.0-*M* HCN, (b) 0.001-*M* HCN, (c) 1.0-*M* HF, (d) 0.50-*M* HNO_2, (e) 0.5-*M* $HCHO_2$, (g) 0.50-*M* H_3BO_3.
Ans. (d) $[H_3O^+] = 1.6 \times 10^{-2}$ *M*, pH = 1.8, 3.2%.

3. A solution of hydrofluoric acid contains 2.0 g of HF per liter and has a pH of 2.2. What is the dissociation constant for HF?

4. A weak acid, HX, is a weak monoprotic acid. A 0.100-*M* solution is 6.0 percent dissociated. What is the dissociation constant for the acid?

5. A 1.0×10^{-3}-*M* solution of a weak acid, HX, is 20.0 percent dissociated. (a) What is the pH of the solution? (b) What is the concentration of X^-? (c) What is the dissociation constant for the acid?

6. Hypobromous acid (HBrO) has a dissociation constant of 2.0×10^{-9}. A solution of HBrO has a pH of 4.8. What is the molarity of the solution?

7. Calculate the $[OH^-]$, pOH, and pH of these solutions. (a) 1.0-*M* NH_3, (b) 0.10-*M* aniline ($C_6H_5NH_2$), (c) 5.0×10^{-2}-*M* hydrazine (N_2H_4), (d) 0.20-*M* hydroxylamine (NH_2OH), (e) 1.5-*M* trimethylamine $[(CH_3)_3NH_2]$.
Ans. (b) $[OH^-] = 6.6 \times 10^{-6}$ *M*, pOH = 5.18, pH = 8.82, (d) $[OH^-] = 2.3 \times 10^{-5}$ *M*, pOH = 4.64 pH = 9.36.

8. Calculate the $[OH^-]$, pOH and pH of these solutions. (a) 0.10-*M* Na_2SO_3, (b) 0.50-*M* KCN, (c) 1.0-*M*, Na_2CO_3, (d) 0.05-*M* $NaC_7H_5O_2$, (e) 0.2-*M* NaClO.

9. (a) What is the pH of a solution made by combining 0.60 moles of acetic acid with 0.40 moles of sodium acetate in enough water to make one liter of solution? (b) What is the pH of this solution if four additional liters of water is added to it?

10. What concentration of sodium acetate is required to prepare a solution in which the pH is 5.0 and the acetic acid is 0.10-*M*? **Ans. 0.18 *M*.**

11. How many grams of NH_4Cl must be added to 0.500 ℓ of 1.0-*M* NH_3 solution to yield a solution with a pH of 9.0? Assume no change in volume occurs.

12. A buffer solution is prepared by adding 1.0 mole of NH_4Cl to 1 liter of a solution containing 1.0 mole NH_3. (a) What is the pH of the solution? (b) What is the pH of the solution resulting from the addition of 1.0 millimole of HCl to 10.0 ml of the buffer? Assume no volume change occurs. (c) What is the pH of the solution resulting from the addition of 1.0 mmole NaOH to 10.0 ml of the buffer? Assume no volume change. (d) How many ml of 6-*M* HCl would be required to change the pH of one liter of buffer by 1 pH unit?
Ans. (b) 9.17, (d) 140 ml.

13. How many milliliters of 0.200-*M* NaOH are required to neutralize 50.0 ml of 0.100-*M* HCl?

14. What is the molarity of an H_2SO_4 solution, 25.0 ml of which is completely neutralized by 45.0 ml of 0.100-*M* NaOH solution?

15. A solution is prepared by dissolving 0.0370 g of $Ba(OH)_2$ and 0.855 g-of KOH in 50.0 ml of solution. How many ml of 0.500-*M* HCl are required to react with a 25.0-ml sample of this solution?

16. What is the molarity of a NaOH solution if 32.20 ml is needed to titrate a 1.10-g sample of potassium biphthalate ($KHC_8H_4O_4$)?

17. What is the percentage of $KHC_8H_4O_4$ in an impure sample, 1.00 g of which requires 25.0 ml of 0.100-*M* NaOH for neutralization?

18. How many ml of 0.100-*M* NaOH are required to react completely with 0.400 g of oxalic acid dihydrate ($H_2C_2O_4 \cdot 2H_2O$)?

19. How many ml of 1.50-*M* H_2SO_4 are required to neutralize a solution containing 32.0 g of NaOH?

20. What volume of 0.020-*M* H_2SO_4 is required to react completely with 20.0 ml of 0.0400-*M* NaOH?

21. A 15.20-ml sample of vinegar has a specific gravity of 1.060 and requires 42.40 ml of 0.3460-*M* NaOH for titration to the endpoint. What is the percentage by mass of acetic acid ($HC_2H_3O_2$) in the vinegar?

22. An impure sample of $Ba(OH)_2$ weighing 0.500 g was added to 50.00 ml of 0.100-*M* HCl. The excess HCl was then titrated with 7.50 ml of 0.200-*M* NaOH. What was the percentage of $Ba(OH)_2$ in the sample?

Ans. 59.8%.

23. Calculate the value of *K* for these reactions. Use the value of *K* to predict whether the reaction is quantitative and suitable as a basis for a quantitative analysis. Only one proton is removed from H_3BO_3. (a) $HF + OH^-$, (b) $H_3BO_3 + OH^-$.

24. Fifty ml of 0.200-*M* NH_3 solution is titrated with 0.200-*M* HCl solution. *K* for NH_3 is 1.8×10^{-5}. (a) Calculate the pH value of the solution at the start of the titration. (b) Calculate the pH of the solution when the base is one-half neutralized (one-half way to the equivalence point). (c) Calculate the pH at the equivalence point. (d) Calculate the pH after 75.0 ml of the acid has been added.

25. Calculate the pH of the solution resulting from these reactions (a) 30.0 ml of 0.200-*M* NaOH and 30.00 ml of 0.200-*M* HCl. (b) 30.0 ml of 0.200-*M* NaOH and 40.0 ml of 0.200-*M* HCl. (c) 50.0 ml of 0.10-*M* HCl and 50 ml of 0.10-*M* $NaC_2H_3O_2$. (d) 25.0 ml of 0.10-*M* HCl and 50.0 ml of 0.10-*M* $NaC_2H_3O_2$. (e) 50.0 ml of 0.10-*M* NH_4Cl with 50.0 ml of 0.10-*M* NaOH.

26. 30.0 ml of 0.150-*M* HCOOH is titrated with 0.150-*M* NaOH. (a) Calculate the pH of the solution after 15.0 ml of base has been added. (b) Calculate the pH of the solution after 30.0 ml of base has been added. (c) Calculate the pH after 40.0 ml of base has been added.

27. A 20.0 ml sample of a weak acid, HX, requires 50.0 ml of 0.050-*M* NaOH to reach the endpoint. After the addition of 30.0 ml of the base, the pH of the solution was 5.00. What is the dissociation constant for HX?

Ans. $K = 1.5 \times 10^{-5}$.

28. Refer to Table 15-8 and choose a suitable indicator for each of these titrations. (a) H_2SO_4 and NaOH, (b) $Ca(OH)_2$ and CH_3COOH, (c) HCl and $NH_3(aq)$.

29. For each of these titrations, specify the indicator(s) from Column II that should be used. State your reason, but do not make any quantitative calculations.

I *Titration*	*II* *Indicator pH range*
(a) $NaOH + KHC_8H_4O_4$	(a) neutral red (6.8–8.0)
(b) $NaOH + HCl$	(b) bromcresol green (3.8–5.4)
(c) $NaHCO_3 + HCl$	(c) o-cresolphthalein (8.2–9.8)

16

Equilibria in Saturated Solutions of Slightly Soluble Substances

Preview

Up to this point, we have applied the equilibrium law to gaseous and acid-base systems. In this chapter we shall extend the quantitative aspects of equilibrium principles to systems involving slightly soluble substances. There are several questions related to precipitates (slightly soluble substances) that are of interest to us. Some of these are

1. How can the approximate solubility of a precipitate be calculated?
2. How can we predict whether or not a precipitate forms when two solutions are mixed?
3. Which reagents or methods can be used to dissolve specific precipitates?

In this chapter we shall introduce the concept of the complex ion and use it with Lé Châtelier's Principle to help us predict the effect that specific reagents have on solubility of various precipitates. You will have an opportunity to apply the principles and concepts developed in this chapter if you perform a qualitative analysis as part of your laboratory program. We shall also have an opportunity to review stoichiometric calculations when we use precipitation reactions as a basis for a quantitative volumetric analysis.

I have steadily endeavoured to keep my mind free so as to give up any hypothesis, however much beloved (and I cannot resist forming one on every subject) as soon as facts are shown to be opposed to it . . . I cannot remember a single first-formed hypothesis which had not after a time have to be given up or be greatly modified.

—Charles R. Darwin (1809–1882)

THE SOLUBILITY PRODUCT AND PRECIPITATE FORMATION

16-1 Ionic Concentration and Approximate Solubility of Slightly Soluble Salts Application of the equilibrium law to the equilibrium that exists between a slightly soluble solid and its ions in a saturated solution yields an equilibrium expression which can be put to practical use by the chemist. This expression can be used for calculating ionic concentrations and approximate solubilities, for determining whether precipitates form when two solutions are mixed, and as an aid in devising methods for separating mixtures of ions in solutions.

Consider a saturated solution of silver chloride. The equilibrium can be represented as

$$\mathbf{AgCl(s) \rightleftharpoons Ag^+(aq) + Cl^-(aq)} \qquad 16\text{-}1$$

The equilibrium law expression for this reaction is

$$\mathbf{K_{sp} = [Ag^+][Cl^-]} \qquad 16\text{-}2$$

Notice that the K_{sp} does not contain a term representing the "concentration" of solid AgCl. K_{sp} is known as the *solubility-product constant.* The K_{sp} values for many slightly soluble substances have been determined and tabulated in handbooks. A brief list is given in Table 16-1.

TABLE 16-1
SOLUBILITY-PRODUCT CONSTANTS
(K_{sp} at 25°C)

Salt	K_{sp}	*Salt*	K_{sp}
$AgC_2H_3O_2$	2.5×10^{-3}	FeS	3.7×10^{-19}
AgBr	7.7×10^{-13}	$Fe(OH)_3$	6.0×10^{-38}
Ag_2CO_3	8.2×10^{-12}	HgS	3×10^{-53}
AgCl	1.8×10^{-10}	$MgCO_3$	2.5×10^{-5}
Ag_2CrO_4	1.1×10^{-12}	MgC_2O_4	8.6×10^{-5}
AgCN	1.6×10^{-14}	$Mg(OH)_2$	6×10^{-12}
AgI	8.3×10^{-17}	MnS	1.4×10^{-15}
Ag_2S	1.6×10^{-49}	NiS	1.8×10^{-21}
$Al(OH)_3$	3×10^{-33}	$PbCl_2$	1.6×10^{-5}
$BaCO_3$	4.9×10^{-9}	$PbCrO_4$	1.8×10^{-14}
$BaCrO_4$	1.2×10^{-10}	$Pb(IO_3)_2$	2.6×10^{-13}
$BaSO_4$	1.1×10^{-10}	PbI_2	7.1×10^{-9}
$CaCO_3$	4.8×10^{-9}	$PbSO_4$	1.6×10^{-8}
CaC_2O_4	2.3×10^{-9}	PbS	8.4×10^{-28}
CaF_2	4.9×10^{-11}	$SrCO_3$	7×10^{-10}
$CaSO_4$	2.6×10^{-5}	$SrCrO_4$	3.6×10^{-5}
CdS	1.0×10^{-28}	$SrSO_4$	7.6×10^{-7}
CoS	1.0×10^{-21}	TlBr	3.6×10^{-6}
CuCl	3.2×10^{-7}	TlCl	1.9×10^{-4}
$Cu(OH)_2$	1.6×10^{-19}	TlI	8.9×10^{-8}
CuS	8.5×10^{-45}	$Zn(OH)_2$	2×10^{-14}
Cu_2S	1.6×10^{-48}	ZnS	4.5×10^{-24}
$Fe(OH)_3$	6.0×10^{-38}		

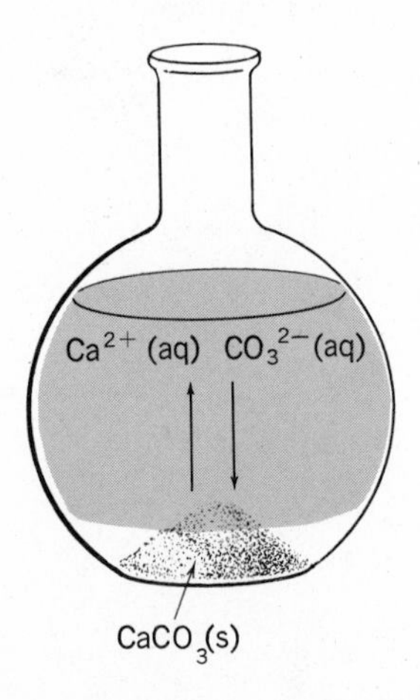

16-1 In a saturated calcium carbonate solution, solid $CaCO_3$ is at equilibrium with the $Ca^{2+}(aq)$ and the $CO_3^{2-}(aq)$ ions. Addition of solid $CaCO_3$ does not change the concentration of the ions.

The K_{sp} for AgCl is given in Table 16-1 as 1.8×10^{-10}. This small constant shows that the concentrations of the Ag^+ ions and the Cl^- ions are very small in a saturated solution of AgCl. In other words, the equilibrium represented by Equation 16-1 is displaced to the left. This means that equilibrium is established before a significant amount of AgCl dissolves. Another way of expressing this would be to say that Ag^+ ions and Cl^- ions can exist together at equilibrium in the same solution only when their concentrations are low enough so that the product of their concentrations does not exceed the K_{sp} value. When the original concentrations exceed the equilibrium value indicated by the K_{sp}, they are then reduced by the crystallization of solid AgCl. The concentration of the Ag^+ ions and the Cl^- ions in a saturated solution of AgCl can be calculated easily, using the solubility-product constant.

In solving problems using solubility-product constants and other equilibrium constants, we shall assume that the solutions involved resemble *ideal solutions.* That is, we shall not take into account the deviations from ideality caused by interionic attractions caused by the presence of other ions in the solution. The attraction between ions in a solution hinders their movement so that they are less "active." This effect causes their *effective concentration* (called their *activity*) to be lower than the measured or calculated molar concentrations might indicate. In precise measurements, activities rather than molar concentrations are used. The activity of a species is equal to the product of its concentration and an experimentally determined correction factor known as an activity coefficient. For example, the activity of Ag^+ ions in a saturated solution of AgCl is

$$\mathbf{a}_{Ag^+} = \mathbf{f}_{Ag^+}[\mathbf{Ag}^+]$$

where a_{Ag^+} *is the activity of* Ag^+ *ions, and* f_{Ag^+} *is the activity coefficient* whose value can be determined indirectly by experimental methods. In our calculations involving slightly soluble salts, we shall use molarities and, unless specifically noted, shall also assume that the dissolved species do not undergo reactions with the solvent. It is important to be aware of the assumptions and the limitations of our calculations, but it is not important at this point to apply the above refinements. Example 16-1 illustrates how the K_{sp} can be used to calculate ion concentrations and approximate solubilities.

Example 16-1

Calculate the concentration of Ag^+ ions and Cl^- ions in a saturated solution of the salt at 25°C. What is the approximate solubility of AgCl in moles/ℓ at this temperature?

1. Write the equation and the K_{sp} expression for the reaction

$$AgCl(s) \rightleftharpoons Ag^+ + Cl^-$$
$$K_{sp} = [Ag^+][Cl^-]$$

2. Let x equal the number of moles of AgCl that must dissolve in a liter of solution to attain equilibrium (this represents the solubility of AgCl in moles/ℓ). The equation shows that for each mole of silver chloride that dissolves, one mole of Ag^+ ions and one mole of Cl^- ions go into

solution. Therefore, the equilibrium concentrations of Ag^+ ions and Cl^- ions is also equal x.

3. Substitute the equilibrium concentrations of the ions and the given value of the K_{sp} in the equilibrium-law expression and solve for x.

$$1.8 \times 10^{-10} = x \times x$$
$$x^2 = 1.8 \times 10^{-10}$$
$$x = 1.3 \times 10^{-5} \text{ moles}/\ell$$

Therefore 1.3×10^{-5} moles of AgCl dissolves in one liter of solution, and the equilibrium concentrations of the Ag^+ ions and the Cl^- ions are each 1.3×10^{-5} moles/ℓ.

FOLLOW-UP PROBLEM

Determine the solubility of AgI in (a) moles/ℓ and (b) g/ℓ at 25°C

Ans. (a) 9.1×10^{-9} moles/ℓ, (b) 2.1×10^{-6} g/ℓ.

16-2 The Common Ion Effect Le Châtelier's Principle indicates that it should be possible to vary the solubility of a solid by varying the concentration of the ions at equilibrium with the solid. For example, to decrease the concentration of Ag^+ ions in a saturated solution of AgBr, we can add excess Br^- ions in the form of a soluble salt such as NaBr. This excess produces a stress which the equilibrium system

$$\textbf{AgBr(s)} \rightleftharpoons \textbf{Ag}^+\textbf{(aq)} + \textbf{Br}^-\textbf{(aq)}$$

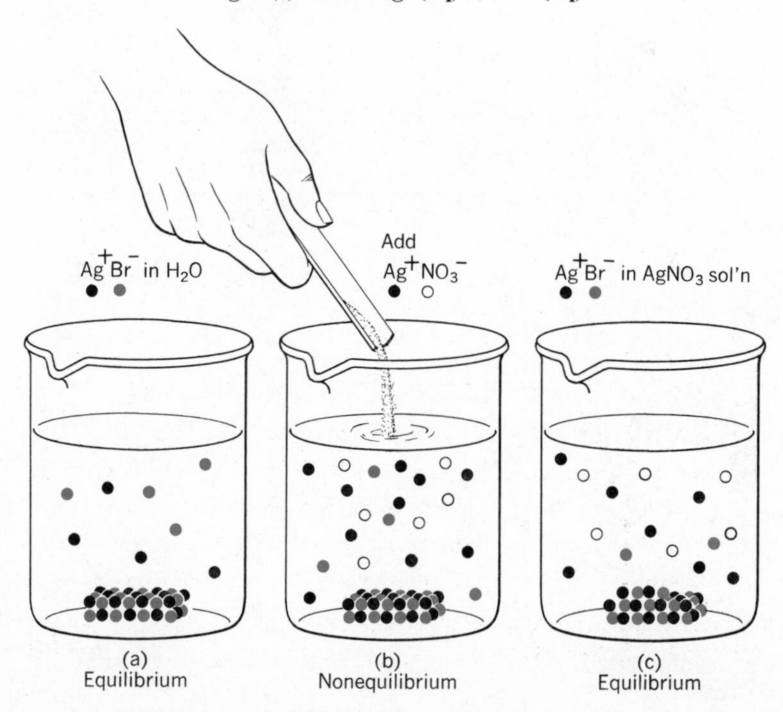

16-2 The common ion effect. Addition of a soluble silver salt to a saturated solution of AgBr increases the Ag^+ ion concentration to a value which exceeds the equilibrium value. To restore equilibrium, additional AgBr precipitates. At the new equilibrium position, the concentration of Ag^+ is greater, and the Br^- concentration is less than in the original saturated solution. However, the product of the $[Ag^+]$ and $[Br^-]$ in solution is constant.

adjusts by shifting to the left as it consumes Br^- ions and forms additional solid AgBr. Each Br^- ion that reacts ties up a Ag^+ ion, thus reducing the concentration of Ag^+ until the product of the ion concentrations satisfies the relationship

$$K_{sp} = [Ag^+][Br^-] = 4.8 \times 10^{-13}$$

The action decreases the solubility of AgBr. In other words, AgBr is less soluble in NaBr than in pure water. The same results could be accomplished by adding excess Ag^+ in the form of soluble $AgNO_3$. The phenomenon is known as the ***common-ion effect.*** Both NaBr and $AgNO_3$ have an ion in common with AgBr; hence, the name common-ion effect.

Example 16-2

Compare the molar solubility of AgBr in pure water and in 0.10-*M* NaBr solution. K_{sp} of AgBr at 20°C is 4.8×10^{-13}

Solution

1. Let $x = [Ag^+]$ and $[Br^-]$ in a saturated aqueous solution of AgBr.

$$\begin{array}{ccc} AgBr(s) \rightleftharpoons & Ag^+(aq) & + \; Br^-(aq) \\ & x & x \end{array}$$

2. Substitute and solve for x in the K_{sp} expression.

$$4.8 \times 10^{-13} = x \times x$$
$$x = \sqrt{48 \times 10^{-14}}$$
$$x = [Ag^+] = [Br^-] = \text{molar solubility in water}$$
$$= 6.9 \times 10^{-7}\ M$$

3. In 0.10-*M* NaBr, the $[Br^-] = 0.10\ M$. The small quantity of Br^- furnished by the dissolution of AgBr is insignificant. In pure water it was only 6.9×10^{-7}. Since dissolution is repressed by the presence of the common ion, it will be even less than this value. Therefore, it may be neglected.

Let $x = [Ag^+]$ = molar solubility of AgBr,

then $$[Br^-] = (x + 0.10) \cong 0.10$$

$$[Ag^+][Br^-] = K_{sp}$$
$$(x)(0.10) = 4.8 \times 10^{-13}$$
$$x = \frac{4.8 \times 10^{-13}}{0.10} = 4.8 \times 10^{-12}\ M$$

Notice that $0.10 + 4.8 \times 10^{-12} \cong 0.10$ so our assumption is valid. The answer indicates that AgBr is about 10^5 times more soluble in pure water than in 0.10-*M* NaBr.

FOLLOW-UP PROBLEM

Compare the molar solubility of PbI_2 in pure water and in 0.10-*M* NaI. See Table 16-1 for K_{sp}.

Ans. Solubility in water = 1.2×10^{-3} *M*. Solubility in 0.10-*M* NaI = 7.1×10^{-7} *M*. Solubility in water = 1.7×10^3 times as great.

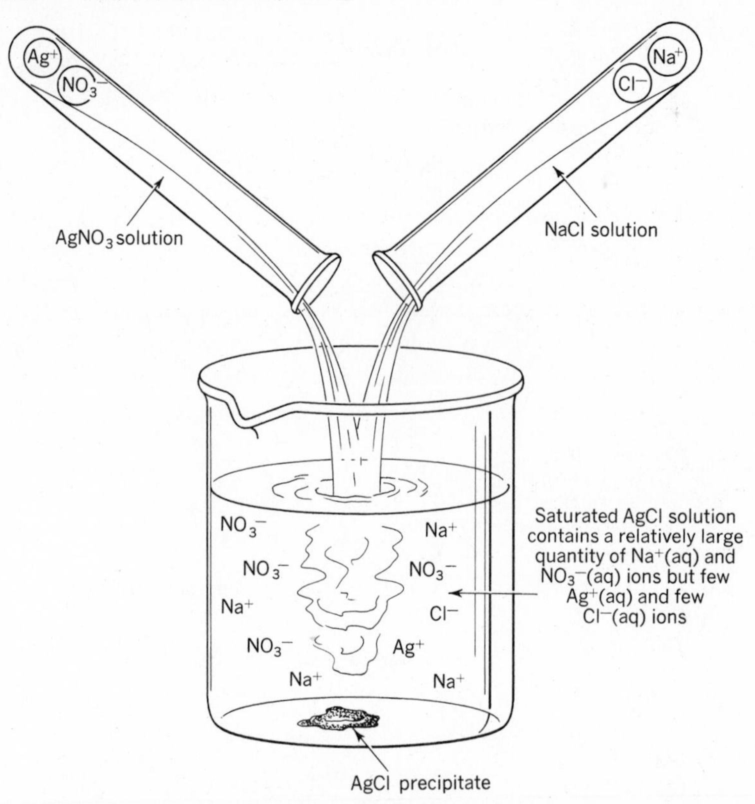

16-3 When $AgNO_3$ and NaCl solutions are mixed, a precipitate of AgCl forms until the Ag^+ and Cl^- ion concentrations are reduced to the equilibrium values. That is, until their product does not exceed the K_{sp} value at the given temperature.

16-3 Predicting Precipitate Formation Let us next see how the K_{sp} expression can be used to determine whether or not a precipitate forms when two solutions are mixed. The K_{sp} puts an upper limit on the product of the concentration of the two ions. *When the product of the concentration of two ions exceeds the value of K_{sp}, they do not exist in equilibrium. They form a precipitate and reduce their concentrations to an equilibrium value.*

To determine whether the K_{sp} is exceeded, we can substitute the ion concentrations in an expression similar to the K_{sp} expression and determine an experimental ion product. If the experimental ion product exceeds the given K_{sp}, then a precipitate forms. If the experimental ion product is less than the K_{sp}, no precipitate forms.

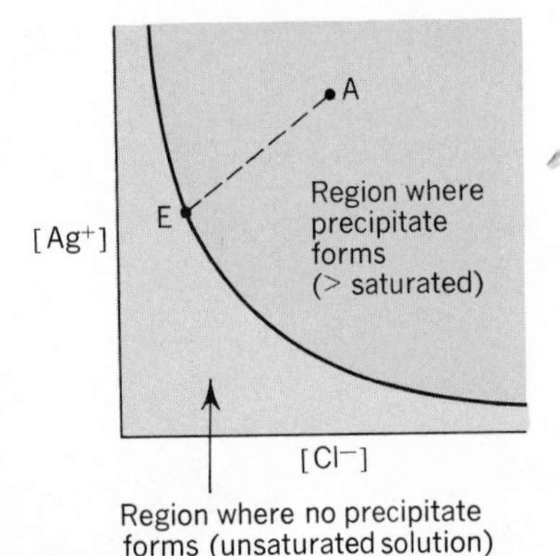

16-4 Graphic representation showing inverse relationship between $[Ag^+]$ and $[Cl^-]$. Any combination of concentrations, other than those on the curve are nonequilibrium. If we assume original concentrations at point A, then precipitation occurs following the dashed line AE until equilibrium point E is attained.

Example 16-3

Does a precipitate of AgCl form when 1 ml of 0.1-*M* $AgNO_3$ is added to a beaker containing 1 liter of tap water with a Cl^- ion concentration of 1×10^{-5} *M*?

Solution

1. Determine the concentration of Ag^+ ions and Cl^- ions.

$$[Cl^-] = 1 \times 10^{-5}\ M$$

$$[Ag^+] = \frac{\text{millimole } Ag^+}{\text{ml solution}} = \frac{1 \text{ ml } (0.1 \text{ millimole/ml})}{1000 \text{ ml} + 1 \text{ ml}}$$

$$= \text{approx. } 0.0001\ M = 1 \times 10^{-4}\ M$$

2. Substitute the ion concentration in an expression similar in form to the K_{sp} expression and solve for an experimental ion product.

$$\text{Experimental ion product} = [Ag^+][Cl^-] = (1 \times 10^{-4})(1 \times 10^{-5}) = 1 \times 10^{-9}$$

From Table 16-1, the K_{sp} for AgCl is 1.8×10^{-10}. Therefore, the experimental ion product exceeds the K_{sp}. That is, $1 \times 10^{-9} > 1.8 \times 10^{-10}$. This indicates that a precipitate of AgCl forms.

FOLLOW-UP PROBLEM

Does a precipitate of AgI form when 10 ml of 0.1-*M* $AgNO_3$ is added to 90 ml of a solution containing 1×10^{-10}-*M* KI?

Ans. Yes.

Example 16-4

(a) What is the highest concentration of I^- ions that can exist in equilibrium with 0.010-*M* Ag^+ ions? The K_{sp} for $AgI = 8.3 \times 10^{-17}$. (b) How many grams of KI are present in one liter of a solution with the I^- ion concentration found in (a)?

Solution

(a) 1. Write the equation and the equilibrium law expression for the reaction

$$AgI(s) \rightleftharpoons Ag^+(aq) + I^-(aq)$$
$$K_{sp} = [Ag^+][I^-]$$

2. Substitute the given values for the K_{sp} and the Ag^+ ion concentration in the K_{sp} expression and solve for $[I^-]$.

$$8.3 \times 10^{-17} = (1.0 \times 10^{-2})[I^-]$$
$$[I^-] = \frac{8.3 \times 10^{-17}}{1.0 \times 10^{-2}} = 8.3 \times 10^{-15}\ M$$

If $[I^-]$ exceeds 8.3×10^{-15} *M*, a precipitate of AgI forms.
(b) KI is a soluble salt. The formula for KI shows that one mole of KI furnishes one mole of I^- ions. Therefore, it takes 8.3×10^{-15} moles of KI to furnish 8.3×10^{-15} moles of I^- ions.

Multiply the number of moles of KI needed by the formula mass of KI.

$$\frac{8.3 \times 10^{-15}\ \text{moles}}{\ell} \times \frac{166\ \text{g}}{\text{mole}} = 1.4 \times 10^{-12}\ \text{g}/\ell$$

FOLLOW-UP PROBLEMS

1. What is the highest concentration of I^- ions that can exist in equilibrium with 0.001-*M* Tl^+ ions at 25°C?

2. Calculate the $[Cl^-]$ required to begin precipitation of these metal ions from solutions containing one g of metal ions per liter of solution (a) Ag^+, (b) Cu^+, (c) Pb^{2+}.

Ans. (a) 1.9×10^{-8} *M*, (b) 2.0×10^{-5} *M*, (c) 5.8×10^{-2} *M*

When a precipitating agent is slowly added to a solution that contains more than one ion that precipitates, then the compound whose solubility product is exceeded first precipitates first. For example, if I^--ion solution is slowly added to a solution containing 0.01-*M* Ag^+ and 0.01-*M* Pb^{2+}, AgI precipitates as soon as the $[I^-]$ exceeds

$$\frac{K_{sp}}{[Ag^+]} = \frac{8.3 \times 10^{-17}}{1 \times 10^{-2}} = 8.3 \times 10^{-15}\ M$$

The lead ions do not precipitate until $[I^-]$ exceeds

$$\sqrt{\frac{K_{sp}}{Pb^{2+}}} = \sqrt{\frac{7.1 \times 10^{-9}}{10^{-2}}} = 8.4 \times 10^{-4}\ M$$

At this point, when PbI_2 starts to precipitate the amount of Ag^+ left in solution is very small. When $[I^-]$ reaches 8.4×10^{-4} *M* which is needed to precipitate Pb^{2+}, then

$$[Ag^+] = \frac{K_{sp}}{[I^-]} = \frac{8.3 \times 10^{-17}}{8.4 \times 10^{-4}} = 9.9 \times 10^{-14}\ M$$

FOLLOW-UP PROBLEM

A solution contains 0.010-*M* Ag^+ ions and 0.010-*M* Sr^{2+} ions. (a) Which ion precipitates first when dilute K_2CrO_4 is slowly added to the mixture? (b) What percentage of the ion that is precipitated first remains unprecipitated when the second ion begins to precipitate?

Ans. (a) Ag^+, (b) 0.17 percent.

In the last chapter we found that the $[OH^-]$ of a weak base could be regulated to a certain degree by the addition of a salt of a weak base. The next example illustrates how the $[OH^-]$ of an ammonia solution can be adjusted to a predetermined value by the addition of the proper quantity of NH_4Cl.

Example 16-5

How many grams of NH_4Cl must be added to 250 ml of a solution which is 0.50-*M* NH_3 to prevent precipitation of $Mg(OH)_2$ when the base is added to 0.020-*M* $Mg(NO_3)_2$?

$$K_{sp} = 6.0 \times 10^{-12},\ K_b = 1.8 \times 10^{-5}$$

Solution

1. Use the K_{sp} expression to calculate the $[OH^-]$ which can just exist in equilibrium with 0.020-*M* Mg^{2+}.

$$Mg(OH)_2(s) \rightleftharpoons Mg^{2+} + 2OH^-$$

$$K^{sp} = [Mg^{2+}][OH^-]^2$$

$$[OH^-] = \sqrt{\frac{6.0 \times 10^{-12}}{2.0 \times 10^{-2}}}$$

$$= 1.7 \times 10^{-5}\ M$$

The $[OH^-]$ from the dissociation of aqueous NH_3 must be limited to 1.7×10^{-5} *M*; otherwise the K_{sp} of $Mg(OH)_2$ becomes exceeded and a precipitate of $Mg(OH)_2$ forms. The dissociation of aqueous NH_3 can be repressed by the addition of NH_4^+.

2. Use the K_b expression to calculate the NH_4^+ which can exist in equilibrium with 1.7×10^{-5}-*M* OH^- ions and 0.50-*M* NH_3 solution

$$NH_3 + H_2O \rightleftharpoons NH_4^+ + OH^-$$

$$K_b = \frac{[NH_4^+][OH^-]}{[NH_3]}$$

$$[NH_4^+] = \frac{(1.8 \times 10^{-5})(5.0 \times 10^{-1})}{1.7 \times 10^{-5}} = 5.3 \times 10^{-1}\ M$$

3. Convert moles/liter of NH_4^+ into grams per 250 ml of NH_4Cl

$$\frac{5.3 \times 10^{-1}\ \text{moles}}{\ell} \times 0.250\ \ell \times \frac{53.5\ \text{g}}{\text{mole}} = 7.1\ \text{g}$$

FOLLOW-UP PROBLEM

What is the maximum concentration of Mg^{2+} ions that can exist in equilibrium in a solution that is 0.050-*M* in NH_3 and 0.050 *M* in NH_4^+? $K_{NH_3} = 1.8 \times 10^{-5}$, K_{sp} $Mg(OH)_2 = 6 \times 10^{-12}$?
Ans. 1.9×10^{-2} *M*.

DISSOLVING PRECIPITATES

16-4 General Method Application of the Le Châtelier Principle to a solubility equilibrium indicates that a slightly soluble solid can be dissolved by decreasing the concentration of one or more of the ions in equilibrium with the solid. Consider a saturated $CaCO_3$ solution.

$$\mathbf{CaCO_3(s) \rightleftharpoons Ca^{2+}(aq) + CO_3^{2-}(aq)} \qquad 16\text{-}3$$

The dissolution of $CaCO_3$ can be enhanced by adding any reagent that ties up and decreases the concentration of either Ca^{2+} ions or CO_3^{2-} ions below the equilibrium value in the saturated solution. That is, a decrease in the concentration of either ion is accompanied by dissolution of the solid and an increase in the concentration of the other ion. Dilution with water decreases the concentration of both ions but has little effect on the dissolution of compounds with very small K_{sp} values.

There are five ways (in addition to dilution) by which the concentration of an ion in equilibrium with a slightly soluble solid may be reduced. These are: *formation of a weakly dissociated species, formation of a volatile substance, formation of a precipitate which is less soluble than the original solid, formation of a complex ion, and oxidation or reduction by a redox agent.* There is some overlapping in these methods. We shall discuss the first four.

16-5 Formation of a Weakly Dissociated Species and /or a Gas

Experimentation reveals that $CaCO_3$ dissolves in hydrochloric acid, forming a gas and soluble products. Specific tests show that the gas is carbon dioxide (CO_2) and that the solution contains calcium (Ca^{2+}) and chloride ions (Cl^-).

In order to determine the reason that hydrochloric acid (HCl(aq)) dissolves $CaCO_3$, we must decide whether the H_3O^+ ions or Cl^- ions of the reagent are able to tie up either the Ca^{2+} or CO_3^{2-}. By now you should be able to predict that H_3O^+ from a strong acid reacts with CO_3^{2-} ions and forms weakly dissociated H_2CO_3 which decomposes and forms CO_2 gas and H_2O. To summarize, we can say that H_3O^+ ions *reduce the concentration of the* CO_3^{2-} below its equilibrium value by forming CO_2 (a gas) and H_2O (a weakly dissociated species). This shifts the equilibrium represented by Equation 16-3 to the right, resulting in the dissolution of additional $CaCO_3$ and an increase in the Ca^{2+} ion concentration. Many slightly soluble salts of weak acids may be dissolved by strong acids. To write a net ionic equation for the reaction of HCl with $CaCO_3$, follow the rules listed here.

16-6 Writing Net Ionic Equations

1. Write the formulas of any strong acid or strong base in dissociated (ionic) form. Check lists of acids and bases if necessary.
2. Write the formulas of weak acids, weak bases, water, gases, and other weakly dissociated species in combined (molecular) form.
3. Write the formulas of soluble ionic salts in dissociated (ionic) form. Refer to the general solubility rules (p. 92) to help identify soluble and slightly soluble substances.
4. Write the formulas of slightly soluble ionic salts in undissociated form and identify them as slightly soluble by writing (s) after the formula.
5. Identify and write the formulas for the products.
6. Delete from the equation any ions that appear on both sides.
7. Use the inspection method to balance atoms and ionic charges. The atoms on the reactant side must be equal in number and type to those on the product side. Similarly, the net ionic charge on the reactant side must equal that on the product side.

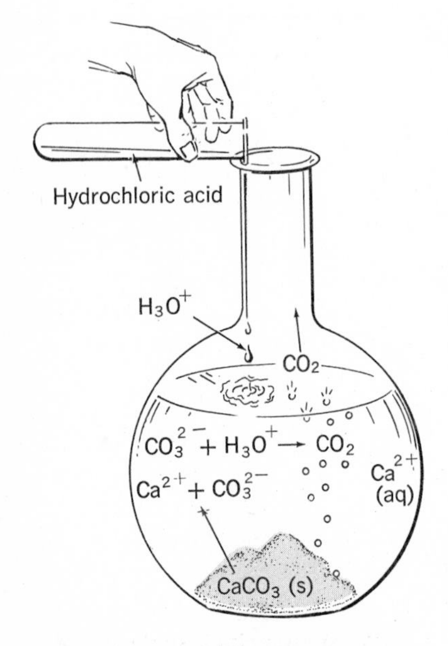

16-5 When hydrochloric acid is added to a saturated $CaCO_3$ solution, the H_3O^+ ions react with the CO_3^{2-} ions and form CO_2 gas which escapes from the solution. Additional solid $CaCO_3$ then dissolves and reestablishes equilibrium conditions.

In the reaction of hydrochloric acid with $CaCO_3$, the participants and their characteristics are listed below.

Reactants Hydrochloric acid, a strong acid which, in aqueous solution, exists in ion form ($H_3O^+ + Cl^-$). Calcium carbonate, a slightly soluble salt in equilibrium with its ions.

Products Calcium chloride, a soluble ionic salt which, in aqueous solution, exists in ionic form ($Ca^{2+} + 2Cl^-$). Carbonic acid (H_2CO_3), a weak, unstable acid which decomposes into water (H_2O) and a gas (CO_2).

Application of these rules leads to the ionic equation

$$CaCO_3(s) + 2H_3O^+(aq) + 2Cl^-(aq) \longrightarrow Ca^{2+}(aq) + CO_2(g) + 3H_2O(l) + 2Cl^-(aq)$$

The net ionic equation is

$$CaCO_3(s) + 2H_3O^+ \longrightarrow Ca^{2+}(aq) + CO_2(g) + 3H_2O(l)$$

16-7 Formation of Weak Acids In the reaction of calcium carbonate with hydrochloric acid, the ionic equation revealed that the hydronium ions from a strong acid reacted with carbonate ions from an ionic salt and produced carbonic acid. In general, *we can say that the reaction of a strong acid with an ionic salt containing the negative ion of a very weak acid produces the weak acid and goes essentially to completion.* The weaker the acid formed, the more nearly complete is the reaction. In other words, the attractive force binding the hydrogen atom in the weak acid is stronger than in a strong acid; hence, the weak acid has less tendency to dissociate. Some of the negative ions which, for practical purposes, react completely with a strong acid and form weak acids, are listed in Table 16-2.

The reaction of ammonium ions from a salt such as NH_4Cl with a strong base such as NaOH, produces the weak base, NH_3. The equation is

$$NH_4^+(aq) + OH^-(aq) \longrightarrow NH_3(g) + H_2O$$

TABLE 16-2
NEGATIVE IONS WHICH FORM WEAK ACIDS

Anion		*Weak acid formed*	
Formula	*Name*	*Formula*	*Name*
CN^-	cyanide	HCN	hydrocyanic
S^{2-}	sulfide	H_2S	hydrosulfuric
SO_3^{2-}	sulfite	H_2SO_3	sulfurous
CO_3^{2-}	carbonate	H_2CO_3	carbonic
NO_2^-	nitrite	HNO_2	nitrous
BO_3^{3-}	borate	H_3BO_3	boric
ClO^-	hypochlorite	HClO	hypochlorous
$C_2H_3O_2^-$	acetate	$HC_2H_3O_2$	acetic
$C_7H_5O_2^-$	benzoate	$HC_7H_5O_2$	benzoic
F^-	fluoride	HF	hydrofluoric

FOLLOW-UP PROBLEMS

1. Write the ionic equations for these reactions. (a) Sodium cyanide with hydrochloric acid. (b) Potassium carbonate with hydrobromic acid. (c) Sulfuric acid with ammonium borate (complete reaction). (d) Hydrochloric acid with sodium phosphate, forming NaH_2PO_4. (e) Ammonium nitrate with sodium hydroxide. (f) Iron (II) sulfide with hydrochloric acid. (g) Silver nitrate ($AgNO_3$) with hydrosulfuric acid (H_2S).

2. Which of these substances dissolve(s) to an appreciable extent in 1-*M* nitric acid (HNO_3)? (a) $PbCO_3$, (b) $SrSO_4$, (c) $Mg(OH)_2$, (d) AgCl, (e) $CuCrO_4$, (f) SrF_2.

16-8 Dissolution of Precipitates by Complex Ion Formation A number of slightly soluble salts do not dissolve readily in acids. Unlike the slightly soluble carbonates and hydroxides, the silver halides (AgCl, AgBr, AgI) cannot be dissolved to an appreciable extent by the addition of a strong acid (high H_3O^+ concentration). For example, AgCl does not dissolve in nitric acid (HNO_3). In order to dissolve AgCl, the equilibrium

$$\mathbf{AgCl(s) \rightleftharpoons Ag^+(aq) + Cl^-(aq)}$$

must be shifted to the right. This can be accomplished by adding a reagent that reacts with and reduces the concentration of either Ag^+ or Cl^- below their equilibrium values.

In Chapter 15 we noted that Ag^+ ions and other positively charged ions with small radii behave as Lewis acids (electron-pair acceptors). This suggests that the addition of a Lewis base (electron-pair donor) such as NH_3 or CN^- ions might form a coordinate bond with silver ions and remove them from solution. Thus, when excess aqueous NH_3 or CN^- (in the form of NaCN) is added to a saturated solution of AgCl, the solid dissolves. Analysis of the ammoniacal solution identifies a species having the formula

There is no simple way to determine the formula of a complex ion. You must either memorize the formulas of a few common ones, or refer to a table such as 16-4.

$$\mathbf{Ag(NH_3)_2^+.}$$

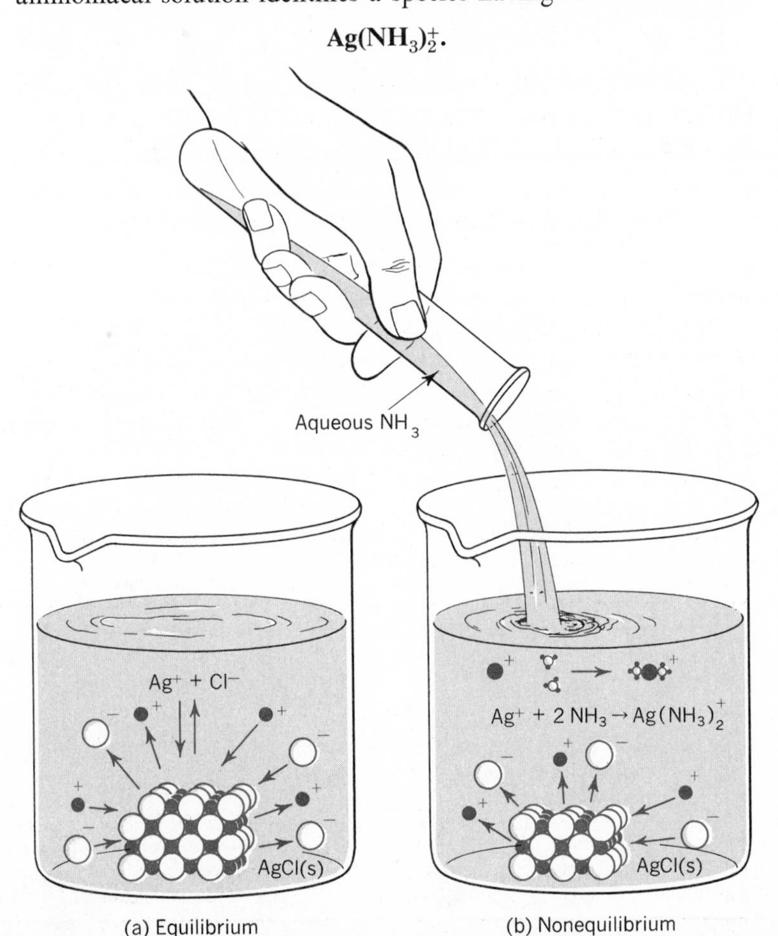

16-6 When ammonia is added to a saturated AgCl solution, the NH_3 reacts with the Ag^+ and forms diamminesilver(I) ions. Reducing the number of Ag^+ ions makes a nonequilibrium condition (b). Additional solid AgCl dissolves to reestablish equilibrium conditions.

Analysis of the cyanide solution shows the presence of a species having the formula

$$\mathbf{Ag(CN)_2^-}$$

Experiments reveal that in aqueous solution these two polyatomic ions are weakly dissociated. In this respect they resemble weak acids and weak bases. These polyatomic ions which dissociate into their component species are called ***complex ions.*** The electron-donor groups (Lewis bases) attached by coordinate bonds to the central positively charged ion are often called ***ligands.*** Thus, silver chloride dissolves in NH_3 or CN^- solutions because these two Lewis bases are able to reduce the $[Ag^+]$ below the equilibrium concentration in a saturated AgCl solution. This represents a stress which is offset when the equilibrium

$$\mathbf{AgCl(s) \rightleftharpoons Ag^+(aq) + Cl^-(aq)}$$

shifts to the right, thereby increasing the $[Cl^-]$ and increasing the solubility of AgCl. The overall equations representing the dissolution of AgCl in excess NH_3 and excess CN^- are, respectively,

$$\mathbf{AgCl(s) + 2NH_3(aq) \rightleftharpoons Ag(NH_3)_2^+(aq) + Cl^-(aq)} \quad 16\text{-}6$$

$$\mathbf{AgCl(s) + 2CN^-(aq) \rightleftharpoons Ag(CN)_2^-(aq) + Cl^-(aq)} \quad 16\text{-}7$$

A special nomenclature has been developed for complex ions. The names of the two complex ions formed in the preceding equations are

$\mathbf{Ag(NH_3)_2^+}$	$\mathbf{Ag(CN)_2^-}$
diamminesilver(I) ion	**dicyanoargentate(I) ion**

The naming of complex ions is discussed in Section 19-10. In the meantime, a list of Lewis bases (ligands) which react with small positive ions and form complex ions is provided in Table 16-3. A list of complex ions commonly encountered in laboratory work is given in Table 16-4. You may assume that these weakly dissociated species are formed when excess Lewis base is added to a solution containing the given positive ion.

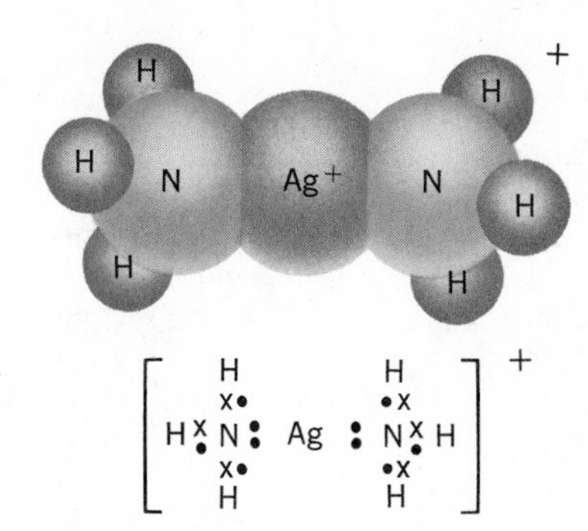

16-7 The space-filling model shows the linear shape of the diamminesilver(I) ion. The bonds are coordinate covalent bonds in which ammonia molecules are electron-pair donors.

TABLE 16-3
COMMON LEWIS BASES (LIGANDS)

Ligand	*Name of Ligand*
H_2O	aquo
NH_3	ammine
CN^-	cyano
Cl^-	chloro
OH^-	hydroxo
SO_4^{2-}	sulfato
NO_3^-	nitrato
$C_2O_4^{2-}$	oxalato
NO_2^-	nitro
$S_2O_3^{2-}$	thiosulfato

The names of all negative complex ions end with the suffix -ate. A Roman numeral in parentheses is used to indicate the oxidation state of the central positive ion.

TABLE 16-4
COMPLEX IONS

Positive Ion	*Lewis Base (Ligand)*	*Formula of Complex Ion*	*Name of Complex Ion*	*Equilibrium*
Ag^+	NH_3	$Ag(NH_3)_2^+$	diamminesilver(I)	$Ag(NH_3)_2^+ \rightleftharpoons Ag^+ + 2NH_3$
Ag^+	CN^-	$Ag(CN)_2^-$	dicyanoargentate(I)	$Ag(CN)_2^- \rightleftharpoons Ag^+ + 2CN^-$
Ag^+	Cl^-	$AgCl_2^-$	dichloroargentate(I)	$AgCl_2^- \rightleftharpoons Ag^+ + 2Cl^-$
Cu^{2+}	NH_3	$Cu(NH_3)_4^{2+}$	tetramminecopper(II)	$Cu(NH_3)_4^{2+} \rightleftharpoons Cu^{2+} + 4NH_3$
Cu^{2+}	CN^-	$Cu(CN)_4^{2-}$	tetracyanocuprate(II)	$Cu(CN)_4^{2-} \rightleftharpoons Cu^{2+} + 4CN^-$
Cd^{2+}	NH_3	$Cd(NH_3)_4^{2+}$	tetramminecadmium(II)	$Cd(NH_3)_4^{2+} \rightleftharpoons Cd^{2+} + 4NH_3$
Cd^{2+}	CN^-	$Cd(CN)_4^{2-}$	tetracyanocadmate(II)	$Cd(CN)_4^{2-} \rightleftharpoons Cd^{2+} + 4CN^-$
Co^{3+}	NH_3	$Co(NH_3)_6^{3+}$	hexamminecobalt(III)	$Co(NH_3)_6^{3+} \rightleftharpoons Co^{3+} + 6NH_3$
Al^{3+}	OH^-	$Al(OH)_4^-$	tetrahydroxoaluminate(III)	$Al(OH)_4^- \rightleftharpoons Al(OH)_3(s) + OH^-$
Zn^{2+}	NH_3	$Zn(NH_3)_4^{2+}$	tetramminezinc(II)	$Zn(NH_3)_4^{2+} \rightleftharpoons Zn^{2+} + 4NH_3$
Zn^{2+}	OH^-	$Zn(OH)_4^{2-}$	tetrahydroxozincate(II)	$Zn(OH)_4^{2-} \rightleftharpoons Zn(OH)_2(s) + 2OH^-$

FOLLOW-UP PROBLEM

Write net ionic equations for these reactions which involve the formation of a complex ion. In each case, name the complex ion, using the pattern given in Table 16-4. (a) copper(II) hydroxide with excess aqueous ammonia. (b) Nickel(II) hydroxide with excess sodium cyanide. The complex product is $Ni(CN)_4^{2-}$. (c) Silver bromide with excess sodium thiosulfate ($Na_2S_2O_3$). The complex product is $Ag(S_2O_3)_2^{3-}$. (d) Zinc hydroxide with excess sodium hydroxide.

16-9 Amphoteric Hydroxides Reference to the general solubility rules or table of solubility products reveals that a large number of metallic hydroxides are only slightly soluble. In general, we would expect metallic hydroxides to dissolve in acids. Adding H_3O^+ ion to the equilibrium system

$$M(OH)_n(s) \rightleftharpoons M^{n+} + nOH^- \qquad 16\text{-}8$$

would reduce the OH^- concentration and shift the equilibrium to the right.

A special type of metallic hydroxides dissolves in either an acid or a base. These are known as ***amphoteric hydroxides.*** Aluminum hydroxide [$Al(OH)_3$] represents this type of compound. When an aluminum salt such as $Al(NO_3)_3$ is added to water, it reacts and forms hexaaquoaluminum(III) ions [$Al(H_2O)_6^{3+}$]. Addition of three moles of OH^- ions to one mole of the hexahydrated aluminum ions causes a precipitate of slightly soluble aluminum hydroxide to form. The reaction is represented as

$$Al(H_2O)_6^{3+}(aq) + 3OH^-(aq) \longrightarrow Al(H_2O)_3(OH)_3(s) + 3H_2O(l) \qquad 16\text{-}9$$

Equation 16-9 is sometimes written without showing the hydrated species as

$$Al^{3+}(aq) + 3OH^-(aq) \longrightarrow Al(OH)_3(s)$$

When excess sodium hydroxide is added to solid $Al(OH)_3$, it dissolves and forms soluble tetrahydroxoaluminate(III) ions [$Al(OH)_4^-$]. The reaction may be represented as

$$Al(OH)_3(s) + OH^-(aq) \longrightarrow Al(OH)_4^-(aq)$$

When hydrochloric acid is added to solid $Al(OH)_3$, it dissolves and forms the soluble dihydroxoaluminum(III) ion [$Al(OH)_2^+$] or $Al(OH)^{2+}$ ions, depending on the concentration of the acid.

$$Al(OH)_3(s) + H_3O^+(aq) \longrightarrow Al(OH)_2^+(aq) + 2H_2O(l)$$

Zinc, chromium, and tin are three other common metallic ions that form amphoteric hydroxides. The amphoteric behavior of certain metallic hydroxides can be partially explained in terms of the relative strength of the bond between the metal atom and the oxygen atom as compared with that between the oxygen atom and the hydrogen atom in a metallic hydroxide having the general formula M—O—H. In most *amphoteric hydroxides, the M—O and O—H bonds are of approximately equal strengths.* When a strong base is added to an amphoteric hydroxide, the hydroxide behaves as an acid; that is, the O—H bond of the amphoteric hydroxide is broken and the hydrogen from M—O—H combines with the hydroxide of the strong base. On the other hand, the addition of a strong acid to an amphoteric hydroxide results in the cleavage of the M—O bond. In this case, M—O—H behaves as a base. Frequently, *it is possible to relate the acid, basic, or amphoteric nature of M—O—H to the electronegativity and charge density of the metal atom M.* The relationship was discussed in Sect. 15-21.

The formation and subsequent dissolution of $Al(OH)_3$ observed when a strong base is added to a solution of an aluminum salt illustrates the importance of controlling the concentration of the precipitating agent (ion). Dissolution of the $Al(OH)_3$ precipitate may be minimized by using a weak base such as aqueous NH_3 as the precipitating agent. In excess NH_3, $Al(OH)_3$ precipitates but does not redissolve and form $Al(OH)_4^-$. Since $NH_3(aq)$ is a rather weak base, it does not yield a large enough OH^- ion concentration to form the complex ion.

FOLLOW-UP PROBLEM

Write net ionic equations which show (a) the reaction of aqueous ammonia with a chromium(III) nitrate solution forming the slightly soluble hydroxide, (b) the subsequent reaction of the slightly soluble hydroxide with excess sodium hydroxide forming a soluble complex ion, and (c) the reaction of the slightly soluble hydroxide with hydrochloric acid.

ENVIRONMENTAL APPLICATIONS OF LE CHÂTELIER'S PRINCIPLE

16-10 Limestone Caverns Although calcium carbonate is quite insoluble, $K_{sp} = 5 \times 10^{-9}$, it does dissolve to a significant extent in the presence of CO_2. The formation of beautiful limestone caverns is a result of the dissolving and recrystallizing of $CaCO_3$. The equilibrium involved in the formation of these caverns is an interesting example of Le Châtelier's Principle. As water with dissolved CO_2 seeps over $CaCO_3$, these reactions take place.

$$CO_2 + H_2O \rightleftharpoons HCO_3^- + H^+ \qquad 16\text{-}10$$

$$CO_3^{2-} + H_2O \rightleftharpoons HCO_3^- + OH^- \qquad 16\text{-}11$$

$$CaCO_3(s) \rightleftharpoons Ca^{2+}(aq) + CO_3^{2-}(aq) \qquad 16\text{-}12$$

Equation 16-10 shows that as the CO_2 concentration is increased, the H^+ ion concentration is also increased. The H^+ ions react with the OH^- in Equation 16-11 and shift the equilibrium to the right. The resulting decrease in CO_3^{2-}-ion concentration forces the equilibrium for Equation 16-12 to the right and causes dissolution of the $CaCO_3$. The overall equation which shows the conversion of insoluble $CaCO_3$ into *soluble* $Ca(HCO_3)_2$ is

$$CaCO_3(s) + CO_2(aq) + H_2O \rightleftharpoons Ca^{2+}(aq) + 2HCO_3^-(aq)$$

16-8 Stalactites and stalagmites form in Carlsbad Caverns in New Mexico. (Sante Fe Railway)

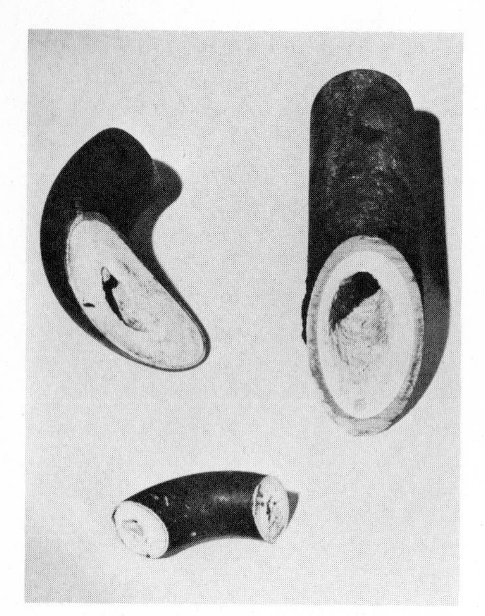

16-9 Hard-water scale consists of insoluble compounds of calcium, magnesium, and iron. These deposits clog pipes and decrease water flow. (The Permutit Company)

The end result is a solution of $Ca(HCO_3)_2$ which trickles down from the ceiling onto the floors of the caverns. When evaporation takes place, stalagmites of solid $CaCO_3$ build up from the floor and stalactites form from the ceiling.

$$Ca(HCO_3)_2(aq) \rightleftharpoons CaCO_3(s) + H_2O(g) + CO_2(g)$$

16-11 Water Softening The presence of Ca^{2+}, Mg^{2+}, or Fe^{2+} ions in water gives water an undesirable characteristic called *hardness.* The degree of hardness is usually expressed in terms of parts per million (ppm) of $CaCO_3$ regardless of what other minerals are present. Water for domestic use exceeding 200 ppm of mineral is usually considered hard water. The presence of these calcium and magnesium compounds in hard water causes pipes to become coated inside with insoluble $CaCO_3$ (Fig. 16-9). The familiar bathtub ring is formed when soap is used in hard water. This scum is actually a precipitate of calcium stearate.

$$\underset{\text{soap}}{2Na(stearate)(aq)} + Ca^{2+}(aq) \longrightarrow \underset{\text{scum}}{Ca(stearate)_2(s)} + 2Na^+(aq)$$

Naturally occurring minerals, called *zeolites,* have long been used in water softening. The zeolites are minerals containing Na, Si, Al, and O atoms bonded in such a way that the Na^+ ions can be readily exchanged for the Ca^{2+} ions in the sample of hard water.

$$2Na(zeolite) + Ca^{2+}(aq) \rightleftharpoons Ca(zeolite)_2 + 2Na^+(aq)$$

Parts per million (ppm) is a unit used to express the concentration of very dilute solutions.

$$\text{ppm} = \frac{\text{mgs of impurity}}{\text{kilograms of solution}}$$

Zeolites are called *ion exchangers* because the essential process involves the exchange of calcium ions for sodium ions. By adding brine (concentrated NaCl solution) to the $Ca(zeolite)_2$, the above reaction can be reversed, thus regenerating the initial Na(zeolite), which can be reused in a water-softening tank. This reversal is an application of Le Châtelier's principle.

Synthetic ion exchange resins which have been developed by chemists are much superior to the natural zeolites. These resins consist of *giant hydrocarbon polymers with anionic and cationic exchange groups.* Cationic exchange groups contain covalently bonded groups such as SO_3^-, which are capable of holding a loosely bonded H^+ ion. If the concentration of the Ca^{2+} ion is increased, the SO_3^- group releases hydrogen ions $(H_3O)^+$ and the Ca^{2+} ions attach themselves to the resin. Anion exchange groups consist of covalently bonded groups like $N(CH_3)_3^+$, which are capable of holding loosely bonded OH^- ions. If negative ions like Cl^- ions come into contact with the resin, the OH^- ions are released and the Cl^- ions attach themselves to the resin. Resins such as this can absorb ions and release only H_3O^+ and OH^- ions, which combine and form water. The following equation shows the removal of $CaCl_2(aq)$ from solution. *R* represents the resin.

$$H_2R(OH)_2 + Ca^{2+} + 2Cl^- \longrightarrow CaRCl_2 + 2H_2O$$

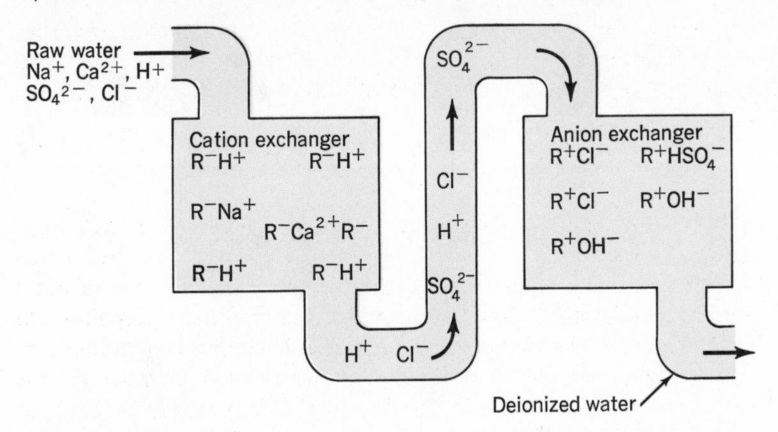

16-10 An ion-exchange system. In the first tank, all positive ions in the water are exchanged for hydrogen ions. In the second tank, all negative ions are exchanged for hydroxide ions. The final deionized water contains no ions except those from the slight dissociation of water itself.

Water purified with these resins is often as pure as carefully distilled water.

VOLUMETRIC ANALYSIS (PRECIPITATION METHODS)

A number of important volumetric analyses are based on ionic reactions that yield precipitates. These are known as volumetric precipitation methods. These methods are widely used in the analysis of substances containing silver or anions such as chloride, bromide, iodide, thiocyanate, and others. For the large majority of precipitation reactions we encounter, however, there are no satisfactory endpoint indicators. Let us examine one of these classical volumetric precipitation methods.

16-12 Volhard Method One method of endpoint detection involves the formation of a colored complex ion, $Fe(SCN)^{2+}$. In this method, known as the ***Volhard method,*** a standard solution or potassium ammonium thiocyanate is used to titrate solutions containing silver ions. A soluble compound containing Fe(III) ions is used as an indicator. The titration reaction is

$$\mathbf{Ag^+(aq) + SCN^-(aq) \longrightarrow AgSCN(s)}$$

At the stoichiometric point, the silver ions have been quantitatively removed from solution in the form of the slightly soluble white precipitate, AgSCN. The SCN^- ions in the next drop of standard solution react with the Fe(III) ions from the indicator and form the red complex ion, $Fe(SCN)^{2+}$, which imparts a reddish tinge to the solution.

The Volhard method is also used to determine the percentage of I^-, Br^-, Cl^-, AsO_4^{3-}, CNO^- and other anions in an unknown. In these analyses, a measured excess of standard silver nitrate is added to the unknown solution. The silver ions form a precipitate with the anion and the excess of silver ions is determined by back titration with a standard thiocyanate solution.

Example 16-6
Silver-copper alloys can be analyzed for the percentage of silver by dissolving the alloy in nitric acid, obtaining silver and copper ions. Standard ammonium thiocyanate added to the solution reacts only with silver ions according to the equation

$$Ag^+(aq) + SCN^-(aq) \longrightarrow AgSCN(s).$$

Iron(III) ions added as an indicator change color as soon as all of the Ag^+ ions have reacted. Assume that a sample of silver-copper alloy weighing 0.360 g dissolved in nitric acid requires exactly 30.00 ml of 0.100-*M* NH_4SCN solution to react completely with the Ag^+ ions. What is the percentage of silver in the alloy?

Solution
1. Calculate the number of moles of SCN^- ions used in the analysis.

$$\text{moles} = MV = 0.100 \text{ mole}/\ell \times 0.0300\ \ell = 0.00300 \text{ mole}$$

2. Obtain the mole ratio of Ag^+ ions to SCN^- ions from the equation.

$$Ag^+/SCN^- = 1:1$$

3. Multiply the mole ratio from Step 2 by the moles of SCN^- ions from Step 1.

$$\frac{1 \text{ mole } Ag^+}{1 \text{ mole } SCN^-} \times 0.00300 \text{ mole } SCN^- = 0.00300 \text{ mole } Ag^+$$

4. Multiply the number of moles of Ag^+ ions from Step 3 by the mass of one mole of Ag to obtain the grams of Ag in the sample.

$$0.00300 \text{ mole Ag} \times \frac{108 \text{ g Ag}}{1 \text{ mole } Ag^+} = 0.324 \text{ g Ag}$$

5. Divide the grams of Ag in the sample by the mass of the sample to obtain the percentage of Ag in the sample.

$$\text{Percentage Ag} = \frac{0.324 \text{ g Ag}}{0.360 \text{ g sample}} \times 100 = 90.0\% \text{ Ag}$$

FOLLOW-UP PROBLEM

1. A certain dental alloy containing silver is dissolved in nitric acid and titrated with 0.120-*M* NH_4SCN in the presence of Fe(III) indicator. If 35.4 ml of the standard SCN^- is required to titrate the silver from a 0.800-g sample of alloy, what is the percentage of silver in the alloy? **Ans. 57.3 percent Ag.**

2. A sample containing Br^- ions weighs 2.500 g and is analyzed by the Volhard method. Exactly 50.00 ml of 0.2400-*M* $AgNO_3$ is added to the sample. After completely precipitating the Br^- ions as AgBr, the excess $AgNO_3$ is titrated with 5.00 ml of 0.1200-*M* KSCN. What is the percentage of Br^- in the sample? **Ans. 36.4 percent Br^-.**

QUALITATIVE ANALYSIS

16-13 General Principles The separation and identification of ions in a solution provide an interesting illustration and application of acid-base, solubility, and complex-ion equilibria. It also provides us with an opportunity to learn a great deal about the descriptive chemistry of the elements. There are a number of rather extensive schemes designed for laboratory work in qualitative analysis. All of them involve the systems and principles we have studied up to this point.

One method of separating and identifying ions in solution is based on the differences in the solubilities of their compounds. The wide range of K_{sp} values shown in Table 16-1 is evidence of wide differences in the solubilities of salts.

By controlling the concentrations of H_3O^+ ions and other reagents, it is possible to add a reagent which precipitates specific ions and leaves others in solution. For example, the data in Table 16-1 show that $BaSO_4$ is only slightly soluble. Since there are no K_{sp} values listed for aluminum, magnesium, and zinc sulfate, it is implied that these compounds are relatively soluble. Thus, it should be possible to separate Ba^{2+} ions from a solution containing Al^{3+} ions, Zn^{2+} ions, and Mg^{2+} ions by adding a solution of some soluble sulfate such as Na_2SO_4. A small amount of SO_4^{2-} ions from this solution causes the K_{sp} for $BaSO_4$ to be exceeded. The $BaSO_4$ precipitates so that it can be filtered from the solution. The Al^{3+}, Zn^{2+}, and Mg^{2+} ions remain in solution and can be separated by adding other reagents.

LOOKING AHEAD

Up to this point we have been primarily concerned with reactions which do not involve the transfer of electrons from one reactant to another. We may classify most reactions we have studied into these categories

1. Formation of a precipitate (precipitation reactions). These are reactions between ions that occur in response to the strong electrostatic forces of attraction between oppositely charged ions and generally result in the formation of slightly soluble ionic solid. An example is

$$Ag^+(aq) + I^-(aq) \rightleftharpoons AgI(s)$$

2. Formation of a weakly dissociated species or gas. These reactions occur when strong covalent bonds are formed between reactant ions. Examples are

$$S^{2-}(aq) + 2H_3O^+(aq) \rightleftharpoons H_2S(g) + 2H_2O(l)$$
$$NH_4^+(aq) + OH^-(aq) \rightleftharpoons NH_3(g) + H_2O(l)$$

3. Formation of a complex ion. These reactions occur when stable coordinate bonds are formed between positive ions and electron-donor species called ligands. Examples are

$$Cu^{2+}(aq) + 4NH_3(aq) \rightleftharpoons Cu(NH_3)_4^{2+}(aq)$$
$$Cu(OH)_2 + 3CN^-(aq) \rightleftharpoons Cu(CN)_3^-(aq) + 2OH^-(aq)$$

There is one more highly important type of chemical reaction that we must consider before we conclude our study of reactions and reaction principles. This is the oxidation-reduction reaction. Oxidation-reduction reactions take place because one reactant has a greater attraction for electrons than another. This suggests that oxidation-reduction reactions involve an electron transfer between reacting species. Thus, in a way, oxidation-reduction reactions involving electron transfer are analogous to acid-base reactions involving proton transfer.

We shall devote the next chapter to a detailed investigation of oxidation-reduction reactions. You will find that many familiar phenomena can be explained in terms of oxidation-reduction principles. All combustion processes, battery reactions, electroplating operations, and corrosion processes involve oxidation-reduction reactions.

In this chapter, we indicated that oxidation reactions could be used to dissolve certain precipitates and implied that oxidation-reduction (redox) reactions play an important role in qualitative analysis schemes. In addition, many quantitative analysis procedures are based on redox reactions. An understanding of oxidation-reduction reactions and equilibria is essential to a well-rounded background of chemical knowledge. Hopefully, we shall supply that background in the next chapter.

Special Feature: Chemistry in Crime Detection

The problem of scientific crime detection involves almost all phases of science. The modern crime laboratory employs the skills of medical doctors, biologists, chemists, and physicists. Specialists in these areas who work specifically in solving problems which arise in connection with the administration of justice are called ***forensic scientists.*** The chemists can rightfully claim to be a central member of the team of forensic scientists because a majority of cases requiring scientific investigation involve some chemical testing. The forensic chemist must not only obtain evidence that is satisfying to himself, but he must collect and prepare evidence that is convincing enough to be used in a court of law. The types of crimes committed are so varied and the list of types of evidence needed is so exhaustive, that the forensic chemist must be versatile. He must use all types of modern instruments as well as classical methods of detecting the presence of various chemicals in any case being investigated.

It has only been during the last 50 years that instruments have been developed to help the chemist in his work. Prior to this time chemists were limited to traditional methods and traditional equipment very much like the methods and equipment employed by first-year chemistry students in their laboratory investigation of chemical principles. Classical methods of analysis involve the use of this type of equipment.

A famous classical method known as the Marsh test was developed in 1836 for the detection of arsenic in suspected poison cases. Arsenic compounds were commonly used by poisoners in the 18th and 19th centuries. Because many arsenic compounds are white and relatively tasteless, they could easily be slipped into a victim's food or beverage. Perhaps for this reason, arsenic compounds were sometimes used as a poison by women. In 1873 forensic chemists used the Marsh test to obtain evidence to convict Mary Ann Cotton of England of poisoning 24 people with arsenic. Because arsenic remains in the victim's body almost indefinitely, it is possible to exhume bodies and perform this test on the contents of the stomach or liver. The apparatus used in this test is shown in Fig. 16-11.

Poisons such as arsenic and lead compounds, when administered to a victim in small quantities over a long period of time are

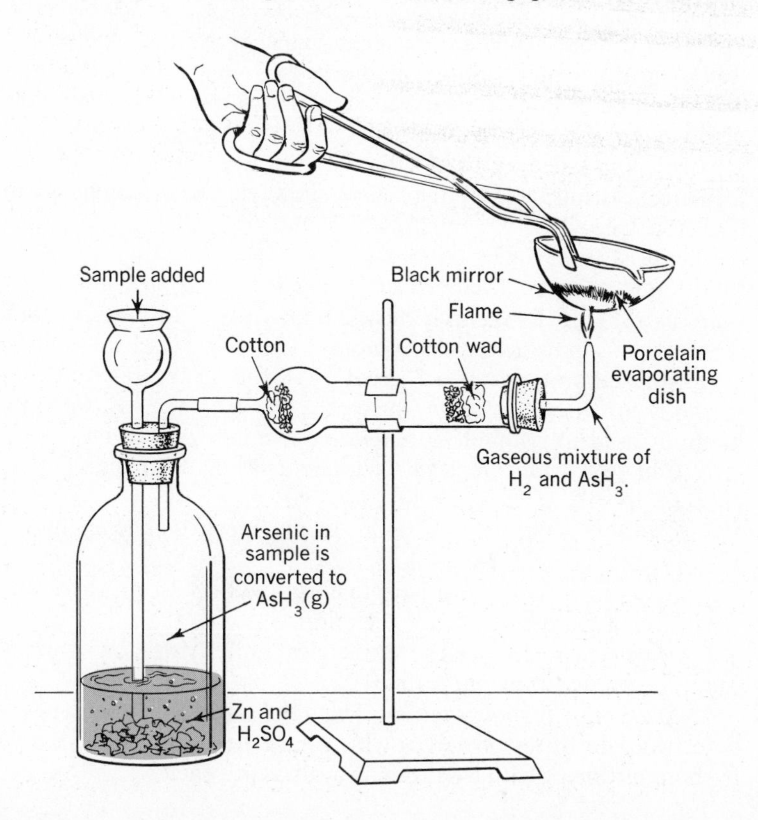

16-11 The classical test for arsenic known as the Marsh test was developed in 1836. The formation of a black mirror on the evaporating dish confirms the presence of arsenic in the sample.

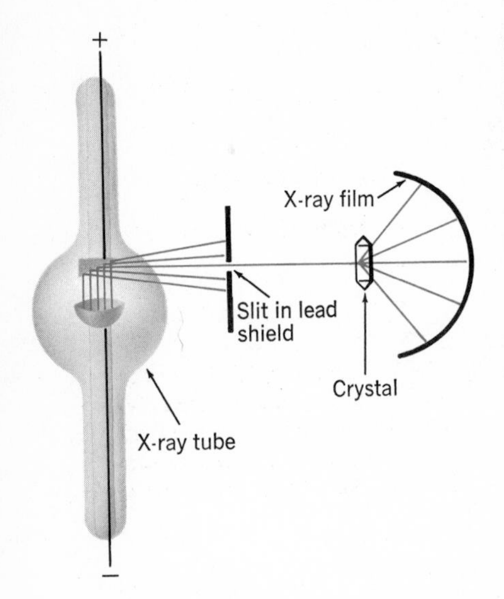

16-12 X-ray diffraction use to determine crystal structure, furnishes evidence used in the detection of fake jewelry. The beam of x rays is scattered (diffracted) by the atoms in the crystal. The diffraction pattern recorded on x-ray film gives important clues to the arrangement of the atoms in the crystal. See margin note below.

absorbed in the body in different ways. Arsenic is deposited in the hair and nails, while lead is deposited in the bones. It is interesting to note that modern chemists have analyzed a sample of hair which belonged to Napoleon I and found arsenic present. This evidence tends to support the claim he made before his death, that he was being poisoned. The analysis by German chemists of a bone from the mummified body of Pope Clement II, who died in 1047, indicated that his body contained the same amount of lead as present-day victims known to have been fatally poisoned by lead. Analysis of human bones recovered from ancient Rome reveals a very high lead content. Literature of that era indicates that a common practice existed of drinking wine which had been heated in lead pots. Some historians have suggested that "lead poisoning" could have been a contributing reason for the fall of the Roman Empire. Most cases of lead poisoning that have been reported recently involve small children gnawing furniture containing lead paints.

Organic poisons which contain no metallic residue are much more difficult to detect than inorganic poisons such as lead or arsenic. Poisons, like strychnine (derived from the dried seeds of a tree found in southeast Asia), cocaine, nicotine, and morphine, are examples of the organic poisons which are difficult to detect. These poisons can be detected, however, by doing a paper or gas chromatographic analysis on a blood, urine, or tissue sample. Chromatography was discussed in a Special Feature in Chapter 2. Probably the most difficult poison to detect is insulin. This poison leaves no detectable residue in the body but causes death by a condition known as hypoglycemia which is an excessively low blood-sugar level. An analysis of the blood can give a clue that a person was killed by insulin but cannot give absolutely positive evidence. "Insulin murderers" have, however, been convicted in both the United States and England, but not solely on the basis of chemical tests.

Atoms have diameters of approximately 10^{-8} cm and are much too small to appreciably disturb visible light waves that have wavelengths of the order of 10^{-5} cm. This means that the waves of visible light can pass through rows of atoms without being diffracted (disturbed). This is analogous to the passing of ocean waves through a picket fence. The direction and motion of the wave is not affected by the pickets in the fence. However, X rays have wavelengths of approximately 10^{-8} cm, which is about the diameter of an atom. Hence, *rows of atoms cause changes in the wave patterns of the X rays* that strike them. That is, X rays are diffracted by rows of atoms (Fig. 16-13). Diffraction occurs at certain angles, depending on the arrangement of the atoms in the crystal. The beam of diffracted X rays strikes an X-ray film, producing spots on the film (Fig. 16-14).

Although classical methods of chemical analysis are still sometimes used by the forensic chemist, the majority of his work is done with rather sophisticated instruments. The work load in forensic chemistry laboratories is so great that the faster instrumental methods are used almost exclusively. The primary instrumental methods used in criminal investigation are itemized below.

1. *Gas, paper, and thin-layer chromatography.* See Special Feature, Chapter 2.

2. *Ultraviolet and infrared spectroscopy.* Spectroscopic analysis is discussed in Sect. 20-35.

3. *X-ray diffraction.* See p. margin notes and Figs. 16-12, 16-13, and 16-14. for a discussion of this technique.

4. *Mass spectroscopy.* The principles underlying this type of analysis are discussed in Sect. 20-37.

5. *Neutron activation analysis.* This method enables scientists to detect minute traces of metals which have been made radioactive by bombardment with neutrons in a nuclear reactor.

6. *Electron probe analysis.* In this type of analysis a sample is bombarded by a finely focused beam of electrons. The bombarding electrons dislodge electrons in the inner energy levels of the atoms in the sample. When outer electrons drop down to fill the vacancy, an X-ray photon of characteristic frequency is emitted. Thus, the atoms (elements) in a sample of paint or explosive residue can be identified from the characteristics of the emitted X ray.

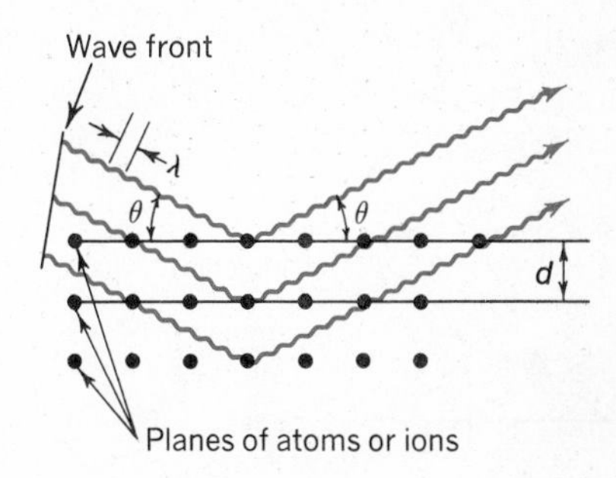

16-13 Schematic diagram of the diffraction of x rays by atoms in three successive planes. When the wavelength (λ) of the x rays is known, and the angle of diffraction (θ) is precisely measured, the distance (d) between the planes of atoms can be calculated.

The methods shown in the list above are by no means the only methods used by the forensic chemist because the nature of various crimes is such that specialties in virtually all fields of science may be needed in his investigation. One must keep in mind that evidence obtained by laboratory analysis is at best only supportive to a case under investigation and rarely is used by itself as the sole evidence in a case. Let us examine a few types of problems faced by the forensic chemist and the methods used to help solve the problem.

One of the most valuable and effective instruments available to the forensic chemist is the gas chromatograph. Gas chromatography has been used effectively in criminal cases to separate and identify the components of a mixture. In one case, a murder victim was strangled and burned inside an automobile. A remnant of his clothing was analyzed by gas chromatography which indicated that his clothing had been soaked with a flammable solvent. Identification of the solvent was the factor that solved the case. In another case, a hit-and-run driver killed a young person who had been riding a motorcycle. The analysis of a speck of paint on the bumper of a suspect's car by gas chromatography revealed that the speck came from the motorcycle in question. In another recent case, a number of people in Tijuana, Mexico died mysteriously from some poison that could not be readily identified. Gas chromatography revealed that the poison was Parathion, an organo-phosphorous ester, which was used as an insecticide. The poison had contaminated some flour in a warehouse and was transmitted to bakery goods. Over 20 people were believed to have been killed before the source of the poison was located.

Neutron activation analysis is one of the newest but most sensitive methods of analysis. Analysis of the residue on the back of a man's hand can tell not only if he has fired a revolver, but how many times and the kind of ammunition used. This method was employed in the analysis of the hair of Napoleon I. A few hairs clutched in the hand of a murdered Canadian woman were examined by neutron activation analysis. The result of this analysis conclusively identified the hair as that of one of the suspects.

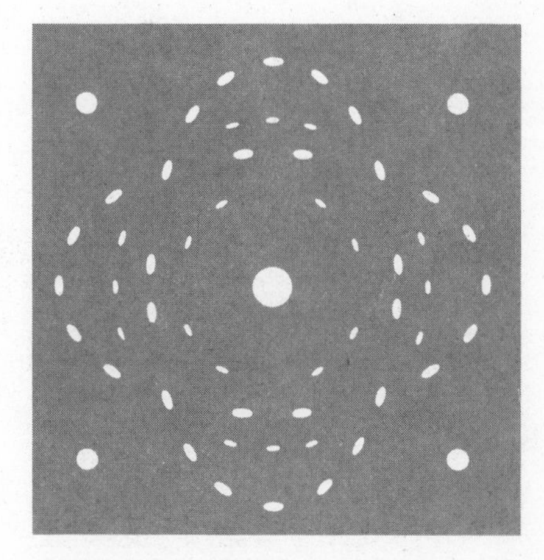

16-14 Diffraction pattern for sodium chloride. Each spot on the film represents the diffraction from a set of parallel planes of atoms of ions. A mathematical analysis of the spot pattern enables scientists to determine the arrangement of planes and distances between them. From the arrangement and intersections of the planes, the positions of the atoms or ions in the crystal can be deduced.

The identification of drugs is another area of great concern to the forensic chemist. The proliferation of illegal drugs is so great that the chemist must keep abreast with new methods of detection and identification. Both spectrophotometry and chromatography are important in drug identification. Hallucinogenic drugs like

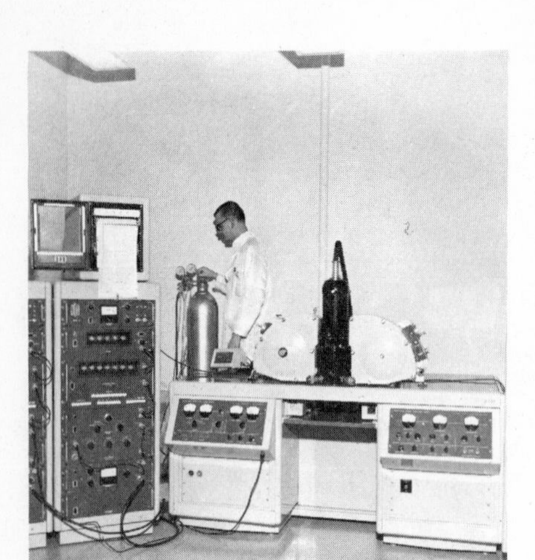

16-15 A modern crime laboratory. The instrument shown is an electron-probe analyzer. (Illinois Bureau of Identification)

lysergic acid diethylamide (LSD), as well as barbiturates and amphetamines, can be detected by ultraviolet or infrared spectrophotometry. These drugs tend to fluoresce which is really only a clue as to their chemical composition. Further tests with either classical analysis or instrumentation confirm the exact identity of the drug.

Ultraviolet and infrared light have been used to solve many cases involving documents and forgeries. Examination of documents with ultraviolet or infrared light often reveal attempts at alterations. Burned or charred documents which are otherwise not at all legible can be made legible using infrared photography. Fig. 16-16 shows a suicide note that was made legible by infrared photography.

Microscopic examination of evidence has long been used in criminal investigation. The "scratch marks" in a bullet reveal the weapon from which it was fired. These marks, from a bullet that is known to have been fired from a weapon in question, can be viewed in a comparison microscope. In one murder case, the victim was found in a barnyard. He had been shot but the bullet had passed through his body. Investigators searched the area and found a spent bullet that came from a suspect's rifle. The suspect claimed that he had used the rifle to shoot rats in the area. Microscopic examination revealed that one of the "rats" must have been wearing a sweater identical to that of the murder victim as the bullet had carried with it a small fiber which matched the sweater of the murdered man.

The X ray was adapted to police investigation soon after it was invented by Roentgen. In 1908 a jewelry theft in Berlin was solved when the expensive furniture of a couple was X rayed and found to contain a cache of stolen jewels. X rays, infrared and ultraviolet analysis are used to identify authentic "old master" paintings. Some clever painting frauds are done on genuinely old but previously unused canvases, and framed in 300-year old wood frames. Other fraudulent paintings are done over genuinely old but worthless paintings. Even these cleverly executed fakes can be exposed by combining the use of modern scientific instruments with a knowledge of art.

16-16 Infrared photography was used to make this suicide note legible. (Courtesy of FBI)

X-ray diffraction analysis is used in the detection of fake jewelry. Synthetic jewels have been manufactured to simulate diamonds and other gems. Some of these imitation gems are so nearly like the natural gem that even experts are sometimes baffled. X-ray diffraction analysis gives positive evidence regarding the composition of the "gems" in question.

Population increase, coupled with increased urbanization, may lead to increases of crime. The forensic scientists, including the forensic chemists, will have to continue to develop and use new discovery techniques and instruments in their constant struggle to cope with criminal ingenuity.

QUESTIONS

1. (a) Write net ionic equations for any of these reactions which result in the formation of a precipitate. If no reaction occurs, write N. R. Assume equal volumes of 0.1-*M* solutions are mixed. (b) Write the K_{sp} expressions for each precipitate. (1) silver nitrate and sodium cyanide (2) potassium chloride and ammonium sulfate (3) mercury(I) nitrate and potassium bromide (4) iron (III) chloride and aqueous ammonia (5) calcium nitrate and sodium oxalate (6) bismuth nitrate and aqueous hydrogen sulfide (7) lead nitrate and ammonium iodide.

2. Write net ionic equations for these acid-base reactions which result in the formation of a weakly dissociated species.
(a) $NaOH(aq) + HNO_3(aq)$
(b) $NaCN(s) + H_2SO_4(l)$
(c) $K_3BO_3(s) + HClO_4(aq)$
(d) $CaSO_3(s) + HCl(aq)$
(e) $(NH_4)_2CO_3(s) + HBr(aq)$
(f) $FeS(s) + HI(aq)$
(g) $NH_4Cl(s) + KOH(s)$
(h) $CuO(s) + HCl(aq)$
(i) $Cl_2O_7(l) + H_2O(l)$
(j) $MgCO_3(s) + HI(aq)$
(k) $Mg_3N_2(s) + H_2O(l) \longrightarrow Mg(OH)_2(s) + NH_3(g)$
(l) $NH_4OH(aq) + H_2SO_4(l)$
(m) $Cd(OH)_2(s) + HNO_3(aq)$
(n) $Ca(HCO_3)_2(aq) + HCl$
(o) $KOH(aq) + H_3PO_4(aq)$ (assume two acid hydrogens react)

3. Which of these substances dissolve to a significant extent in dilute (1-*M*) HNO_3? Justify your answers. (a) $BaCO_3$, (b) $BaSO_4$, (c) AgOH, (d) AgI, (e) $PbCrO_4$, (f) BaF_2.

4. Is $CaCO_3$ more or less soluble in a 6-*M* Na_2CO_3 solution than in pure water? Explain in terms of Le Châtelier's Principle.

5. Explain in terms of Le Châtelier's Principle why $AgC_2H_3O_2$ ($K_{sp} = 4 \times 10^{-3}$) can be dissolved in an excess of each of these reagents (a) H_2O, (b) $HNO_3(aq)$, (c) $NH_3(aq)$.

6. (a) Is AgCN more or less soluble in a 6.0-*M* NaCN solution than in pure water? Explain. (b) Is $Mg(OH)_2$ more or less soluble in excess NH_4Cl solution than in pure water? Explain.

7. Why are all slightly soluble carbonates appreciably soluble in dilute strong acids?

8. AgCl is appreciably soluble in 6-*M* aqueous NH_3, but AgI is not. Explain.

9. When 1-*M* Na_2CO_3 is added to a solution of $Fe(NO_3)_3$, a precipitate of $Fe(OH)_3$ is observed Explain.

10. What is an amphoteric oxide or hydroxide?

11. Could an amphoteric hydroxide be readily soluble in NaOH and yet no more soluble in aqueous NH_3 than in H_2O? Explain.

12. Write net ionic equations for these reactions which involve the formation of a complex ion. (a) $Zn(OH)_2(s)$ + excess NH_3, (b) $Cu(OH)_2(s)$ + excess NaCN, (c) $ZnO(s)$ + excess KOH, (d) $AgCl(s)$ + excess HCl, (e) $AlCl_3$ added to water.

13. (a) Write net ionic equations for those of these reactions which result in the formation of a complex ion. Refer to Table 16-4.

(1) silver chloride(s) + excess NaCN(aq)
(2) iron chloride(aq) + excess $NH_3(aq)$
(3) copper(II) hydroxide(s) + excess $NH_3(aq)$
(4) chromium(III) hydroxide(s) + excess NaOH(aq)
(5) silver nitrate + excess KI(aq)

14. Copy and complete the table by writing the name and formula of the substance formed. If no reaction occurs, write N. R. *Do not write in this book.*

Ion	*Excess NaOH*	*Excess $NH_3(aq)$*
Ni^{2+}		
Cu^{2+}		
Zn^{2+}		
Al^{3+}		
Ca^{2+}		

15. Show by writing six chemical equations, the amphoteric character of (a) $Al(OH)_3$, (b) ZnO, (c) $Sn(OH)_4$. In each case, indicate whether the amphoteric substance is acting as an acid or as a base.

16. A solution may contain Ba^{2+} and/or Al^{3+}. Explain how you would prove the presence or absence of these ions.

17. An unknown white crystalline substance dissolves in water and yields a solution that turns blue litmus red. When $AgNO_3(aq)$ is added to a portion of the sample, a white precipitate is formed which does not dissolve in nitric acid. Addition of barium nitrate(aq) to a separate portion yields a white precipitate. State which of these ions are present, absent, or inconclusive Na^+, NO_3^-, Cl^-, CO_3^{2-}, S^{2-}, SO_4^{2-}, NH_4^+.

PROBLEMS

1. Calculate the K_{sp} for each of the salts whose solubility is listed below.
(a) $CaSO_4 = 5.0 \times 10^{-3}$ moles/ℓ
(b) $MgF_2 = 2.7 \times 10^{-3}$ moles/ℓ
(c) $AgC_2H_3O_2 = 1.02$ g/100 ml
(d) $SrF_2 = 12.2$ mg/100 ml

2. Calculate (a) the solubility in moles/ℓ of each of these salts and (b) the concentration of the cations in mg/ml in each of the saturated solutions
(a) AgCN, $K_{sp} = 2 \times 10^{-12}$
(b) $BaSO_4$, $K_{sp} = 1.5 \times 10^{-9}$
(c) FeS, $K_{sp} = 3.7 \times 10^{-19}$
(d) $Mg(OH)_2$, $K_{sp} = 9 \times 10^{-12}$
(e) Ag_2S, $K_{sp} = 1.6 \times 10^{-49}$
(f) CaF_2, $K_{sp} = 4.9 \times 10^{-11}$.

3. Consider these slightly soluble salts:
(1) PbS $K_{sp} = 8.4 \times 10^{-28}$
(2) $PbSO_4$ $K_{sp} = 1.8 \times 10^{-8}$
(3) $Pb(IO_3)_2$ $K_{sp} = 2.6 \times 10^{-13}$.
(a) Which is the most soluble? (b) Calculate the solubility in moles/ℓ of $PbSO_4$. (c) How many grams of $PbSO_4$ dissolve in one liter of solution? (d) How can you decrease the concentration of Pb^{2+}(aq) in a saturated $PbSO_4$ solution? (e) What is the concentration in moles/ℓ of PbS in a saturated solution of the salt?
Ans. (c) 0.039 g/ℓ, (e) 2.9 × 10^{-14} moles/ℓ.

4. For each of these substances, calculate the milligrams of metallic ion that can remain at equilibrium in a solution having a $[OH^-] = 1.0 \times 10^{-4}$ *M*.
(a) $Cu(OH)_2$, $K_{sp} = 1.6 \times 10^{-9}$
(b) $Fe(OH)_3$, $K_{sp} = 6.0 \times 10^{-38}$
(c) $Mg(OH)_2$, $K_{sp} = 6.0 \times 10^{-12}$
Ans. (b) 3.4 × 10^{-24} mg/ml.

5. Calculate the $[Ag^+]$ needed to begin precipitation of each of these anions from solutions containing one mg of anion per ml of solution. (a) Br^-, (b) S^{2-}, (c) BrO_3^-, (d) CrO_4^{2-}, (e) IO_3^-. K_{sp} $AgBrO_3$ 6×10^{-5}, $AgIO_3$ 3.1×10^{-8}.
Ans. (a) 6.2 × 10^{-11} *M*.

6. How many mg of TlI can dissolve in 500 ml of (a) water, (b) 0.1-*M* $TlNO_3$, (c) 0.02-*M* KI?

7. What is the solubility in moles per liter of AgBr in a solution resulting from the addition of 50.0 ml of 0.01-*M* $CaBr_2$ to 50.0 ml of 0.008-*M* $AgNO_3$?

8. Fifty ml of 0.10-*M* $AgNO_3$ is added to 150 ml of 0.10-*M* $CaCl_2$. What is the concentration of each ion in the resulting solution?

9. In which of these reactions does a precipitate form?
(a) 10.0 ml of 0.01-*M* $AgNO_3$ + 10.0 ml of 0.10-*M* Na_2SO_4. K_{sp} for $Ag_2SO_4 = 1.2 \times 10^{-5}$
(b) 1 mg of $MgCl_2$ + 1 liter of 0.01-*M* $Na_2C_2O_4$
(c) 1 ml of 0.1-*M* $Ca(NO_3)_2$ + 1 liter of 0.01-*M* HF
(d) 1 ml of 0.1-*M* $Ca(NO_3)_2$ + 1 liter of 0.01-*M* NaF
(e) 5 ml of 0.004-*M* $AgNO_3$ + 15 ml of a solution containing 1.5 mg Br^- ions.
Ans. c, d, e.

10. How many mg of Pb^{2+} must be present in 10.0 ml of a 0.135-*M* NaCl solution for $PbCl_2$ to precipitate?

11. A solution contains 0.01-*M* $TlNO_3$ and 0.01-*M* $AgNO_3$. (a) Which compound precipitates first when NaI is slowly added to 100 ml of this solution? (b) How many mg of this ion remain unprecipitated when the second compound begins to precipitate?
Ans. (b) 1.0 × 10^{-7} mg.

12. A liter of solution contains 100 mg of Ba^{2+} and 10.0 g of Sr^{2+}. Within what range must the concentration of CrO_4^{2-} be to precipitate Ba^{2+} without precipitating any Sr^{2+}?

13. Does a precipitate of $Mg(OH)_2$ form when 10.0 ml of a 0.10-*M* NH_3 solution containing 3.0 g of NH_4Cl is mixed with 10.0 ml of 0.10-*M* $MgCl_2$ solution?

14. Does a precipitate of $Mg(OH)_2$ form when 10.0 ml of 0.050-*M* aqueous NH_3 is mixed with 10.0 ml of 0.15-*M* $MgCl_2$?
Ans. yes.

15. Barium nitrate reacts with potassium sulfate solution and forms insoluble $BaSO_4$. How many milliliters of 0.40-*M* $Ba(NO_3)_2$ solution are required to precipitate effectively the sulfate ions in 25 ml of 0.80-*M* K_2SO_4 solution?
Ans. 50 ml.

16. What volume of 0.25-*M* KCl solution is required to precipitate effectively the silver ions from 160.0-ml sample of 0.60-*M* $AgNO_3$ solution?

17. What mass of silver chloride can be precipitated from a silver nitrate solution by 200 ml of a solution of 0.50-*M* $CaCl_2$?
Ans. 28.6g

18. A sample of an unknown chloride weighing 0.210 g is treated with 30.00 ml of 0.100-*M* $AgNO_3$. The precipitate is filtered and the excess silver in the filtrate is titrated with 5.65 ml of 0.0900-*M* KSCN. What is the percentage of Cl^- in the sample?

19. A solution of KI contains 0.200 g KI in 50.00 ml of solution. If 25.00 ml of 0.100-*M* $AgNO_3$ is added to the KI solution, how many ml of 0.110-*M* KSCN are needed for back titration?

17

Oxidation-Reduction Reactions and Electric Energy Associated with Chemical Reactions

Preview

"The world little knows how many of the thoughts and theories which have passed through the mind of a scientific investigator have been crushed in silence and secrecy by his own severe criticism and adverse examinations; that in the most successful instances not a tenth of the suggestions, the hopes, the wishes, the preliminary conclusions have been realized."

—Michael Faraday (1791–1867)

Every time you turn on a flashlight or press the starter of your automobile you are using electric energy liberated by the chemical reactions taking place in a battery. Spontaneous reactions which liberate electric energy and nonspontaneous reactions which consume electric energy are all part of a large and important class of reactions known as *oxidation-reduction reactions*. These are frequently referred to as *redox reactions*.

A number of parallels may be drawn between the acid-base reactions examined in Chapter 15 and the oxidation-reduction reactions presented in this chapter. An acid "neutralizes" a base in an acid-base reaction, and an oxidizing agent "neutralizes" a reducing agent in a redox reaction. A large number of redox reactions involve an electron transfer in the same general way that a large number of acid-base reactions involve a proton transfer. Both acid-base and oxidation-reduction reactions are equilibrium reactions. The relative strength of an acid is indicated by the magnitude of its dissociation constant. We shall find that the relative strength of an oxidizing or reducing agent is indicated by the magnitude of the voltage associated with a reaction. Furthermore, you will find that voltages associated with redox reactions are numerically related to the equilibrium constants for redox reactions. An important feature of cell voltages is that they can be used to predict whether a given redox reaction will be spontaneous or nonspontaneous.

In this chapter, we shall first examine and learn to balance equations for general spontaneous oxidation-reduction reactions in which the reactants are in the same vessel. In many of these exothermic reactions, electrons are apparently transferred directly from one chemical species to another. The decrease in potential energy (enthalpy) of these redox systems appears as heat given off to the surroundings.

In other systems, the reactants are in separate compartments linked together by an electric conductor. The decrease in potential energy of these systems (called *voltaic, galvanic,* or electrochemical cells) appears in large part as usable electric energy. You will learn how to calculate the voltage for different electrochemical cells and how to relate this voltage to the spontaneity of a reaction and to the equilibrium constant for a reaction.

After examining a number of applications of electrochemical cells, we shall consider redox reactions which take place in electrolytic cells. You will find that these reactions are the reverse of those taking place in electrochemical cells. That is, the reactions are nonspontaneous, so electric energy must be externally supplied in order that the reaction can take place. The electrolysis of water and the charging of an automobile battery are examples of reactions taking place in an electrolytic cell. Unlike the potential energy of the chemical system in a voltaic cell, that of the reaction system in electrolytic cells increases. This means that the products of electrolytic reactions have higher enthalpy than the reactants.

After discussing applications of electrolytic cells, we shall use the laws proposed by the great English chemist, Michael Faraday, to calculate the quantity of product yielded by a given current flowing through an electrolytic cell.

The study of redox reactions taking place in electrochemical or electrolytic cells is known as *electrochemistry*. You will find that the study of electrochemistry provides an opportunity to use and relate a large number of major chemical concepts which we have already encountered.

OXIDATION-REDUCTION REACTIONS AND EQUATIONS

17-1 Oxidation Numbers Like acid-base concepts, oxidation has evolved from a limited concept to a general one. Originally, this term was applied only to reactions in which a metal or a nonmetal combined with oxygen. It soon became apparent, however, that there was a factor in the reactions of metals with oxygen that was common to many other reactions. This factor is the loss of electrons by metallic atoms, forming metallic ions. To illustrate this observation, we use the concept of oxidation number developed earlier in this book. For example, in the oxidation reaction

$$2Mg(s) + O_2(g) \longrightarrow 2MgO(s)$$

each magnesium atom loses two electrons and becomes a magnesium ion (Mg^{2+}) while each oxygen atom gains two electrons and becomes an oxide ion (O^{2-}). Experiments reveal that when magnesium reacts with any acid or any nonmetal, its atoms always lose two electrons per atom and become Mg^{2+} ions. These observations show that reactions involving oxygen constitute only a small segment of the total number of reactions in which a substance loses electrons.

To help systematize the study of chemistry, chemists classify all reactions in which a substance loses electrons as ***oxidation reactions.*** Electrons are not transferred to the surroundings in ordinary chemical reactions. This means that the *electrons lost by one species must be gained by another*. The process in which an atom or ion gains an electron is known as ***reduction.*** The loss of electrons by one species and the simultaneous gain by another constitutes an ***oxidation-reduction or redox reaction.*** In Equation 17-1, the two electrons lost by a magnesium atom are gained by an oxygen atom. Thus, magnesium is oxidized and oxygen is reduced. These changes may be represented by the equations which deal with each element separately.

$$\mathbf{Mg^0 - 2e^- \longrightarrow Mg^{2+}} \qquad 17\text{-}1$$
$$\mathbf{\tfrac{1}{2}O_2^0 + 2e^- \longrightarrow O^{2-}}$$

In these equations the oxidation numbers of the atoms are written as superscripts.

Since oxygen is responsible for the oxidation of magnesium, it is called the *oxidizing agent.* Magnesium brings about the reduction of oxygen and is called the *reducing agent.*

$$\mathbf{Mg + \tfrac{1}{2}O_2 \longrightarrow MgO}$$

(Mg → MgO: $-2e^-$; $\tfrac{1}{2}O_2$ → MgO: $+2e^-$)

reducing agent (Mg) **oxidizing agent** ($\tfrac{1}{2}O_2$)

Although we can account for the formation of ions by assuming electron transfer, it should be noted that the electrons in magnesium oxide are more tightly packed around the nucleus than they are in magnesium metal. In magnesium crystals the distance between nuclei is 3.2 Å; in magnesium oxide, the distance between magnesium nuclei is 2.9 Å.

Notice that **oxidizing agents** take electrons and, consequently, become reduced, while **reducing agents give up electrons** and become oxidized. As a result of the reaction, the oxidation number of an oxidizing agent decreases while that of a reducing agent increases.

In terms of oxidation numbers, ***oxidation*** may be defined as an *increase* in oxidation number and ***reduction*** as a *decrease* in oxidation number. The reactions show that in Equation 17-1, the oxidation number of Mg increases from 0 to 2+, while the oxidation number of each oxygen atom in O_2 decreases from 0 to 2−. This means that if we are to recognize oxidation-reduction equations, we must be able to determine the oxidation numbers of atoms.

For example, we identify the reaction

$$\mathbf{2H_2 + O_2 \longrightarrow 2H_2O}$$

as an example of oxidation-reduction because the oxidation numbers change, not because there is a complete transfer of electrons from hydrogen to oxygen. It is true that the oxygen atoms in H_2O have a greater share in the bonding electrons than hydrogen, but this does not mean that the hydrogen in a water molecule has a 1+ charge. Rather, it has an arbitrarily assigned oxidation number of 1+. Thus, the concept of oxidation number enables us to identify readily the *apparent electron transfer* that occurs in oxidation-reduction (redox) reactions.

Oxidation numbers for atoms in covalent molecules cannot be determined experimentally. They are based on arbitrarily chosen standards. In covalent bonds the electrons are generally arbitrarily assigned to the more electronegative atom. The rules for assigning and determining oxidation numbers are summarized below. You will need these rules to help you balance oxidation-reduction equations by the method described in Section 17-2.

SUMMARY OF RULES FOR DETERMINING OXIDATION NUMBERS

1. The oxidation number of an atom in a molecule of an element is zero. Thus, the oxidation number of oxygen in O_2 is zero.

2. The oxidation number of a monatomic ion is the charge on the ion. The oxidation number of a calcium ion is 2+.

3. Oxidation numbers conventionally assigned to atoms combined in common chemical compounds are these:

(a) Oxygen = 2− (except in peroxides, where it is 1−). For example, the oxidation state of oxygen in SO_2, $KClO_3$, and $KMnO_4$ is 2−. In Na_2O_2 and H_2O_2, the single oxidation number of oxygen is 1−.

(b) Hydrogen = 1+ (except in metallic hydrides, where it is 1−). For example, the single oxidation number of hydrogen in H_2O, H_2O_2, NH_3, and $HC_2H_3O_2$ is 1+. In LiH, it is 1−.

(c) Group IA elements = 1+.

(d) Group IIA elements = 2+.

(e) Ions of the halogen atoms in binary ionic compounds (halides) = 1−. The halogen ions in NaF, KBr, and CsI all have an oxidation number of 1−.

4. When necessary, use charges on the polyatomic ions given in Tables 2-8 and 2-9. Some of the more common polyatomic ions and their charges are SO_4^{2-}, OH^{1-}, NO_3^{1-}, CO_3^{2-}, $C_2H_3O_2^{1-}$.

5. The sum of the positive and negative oxidation numbers in a compound is zero. The single oxidation number of a specified atom in a compound is determined thus:

(a) Assign a common oxidation number (rules 1, 2, 3) to all but the specified atom. To illustrate, let us determine the oxidation number of chromium (Cr) in potassium dichromate ($K_2Cr_2O_7$). The first step is to assign oxygen an oxidation number of 2− and potassium an oxidation number of 1+.

(b) Algebraically add the *total* oxidation numbers of all but the unknown atom. The total oxidation number of an atom in a formula is the assigned single oxidation number multiplied by the subscript of the atom in the formula. For example, the total oxidation number of hydrogen in H_2O is 2(H) = 2(1+). In the case of $K_2Cr_2O_7$, the total oxidation number of oxygen is 7(2−) = 14− and that of potassium is 2(1+) = 2+. The algebraic sum is (2+) + (14−) = 12−.

(c) The total oxidation number of the atom in question is the electric charge which must be assigned to the specified atom in the formula to give the formula an overall zero charge. If the designated atom has a subscript in the formula, the total oxidation number must be divided by the subscript to yield the single oxidation number. In $K_2Cr_2O_7$ the total charge contributed by 2 chromium atoms must be 12+ in order to balance the 12− charge contributed by the other two atoms. Therefore, the oxidation number of chromium in $K_2Cr_2O_7$ is 12+/2, or 6+.

6. The algebraic sum of the positive and negative oxidation numbers of the atoms in a polyatomic ion is equal to the charge on the ion. For example, in an hydroxide ion (OH^-) the oxidation number of oxygen is 2−, that of hydrogen is 1+. These add and give an overall charge of 1−. You can also use this principle to determine the single oxidation number of an atom in a polyatomic ion such as a dichromate ion ($Cr_2O_7^{2-}$). You should obtain a value of 6+ for the single oxidation number of Cr in $Cr_2O_7^{2-}$.

Since the assignment of oxidation numbers is somewhat arbitrary, a certain amount of ambiguity is inherent in the rules given above. That is, it is possible for two people to obtain two different values for the oxidation number of a given atom in a given compound. This, however, is a minor flaw and relatively unimportant. The use of this set of rules is illustrated in Example 17-1.

Example 17-1
What is the single oxidation number of manganese (Mn) in (a) manganese dioxide (MnO_2), (b) manganese heptoxide (Mn_2O_7), (c) manganate ions (MnO_4^{2-}), (d) potassium permanganate ($KMnO_4$)?

Solution
(a) Assign an oxidation number of 2− to oxygen. Multiply the single oxidation number of oxygen by 2 to obtain the total oxidation number (total charge of oxygen), 2(2−) = 4−. Since MnO_2 is a neutral substance, determine what positive charge is required to exactly balance the 4− charge contributed by oxygen. The answer is 4+. That is, 4− from oxygen added to 4+ from manganese gives 0 overall charge for MnO_2, (4−) + (4+) = 0. (b) Assign a 2− oxidation number to oxygen. Multiply by 7 to obtain the total oxidation number of the oxygen, 7(2−) = 14−. Mn_2O_7 has no overall charge. Manganese must therefore contribute a total charge of 14+ to the formula in order to achieve an overall 0 charge. Since there are two manganese atoms per formula, divide 14+ by 2 to obtain the single oxidation number of the atom, 14+/2 = 7+. (c) Assign a 2− oxidation number to oxygen. Multiply by 4 to obtain the total oxidation number of oxygen, 4(2−) = 8−. Since a manganate ion bears an overall charge of 2−, determine what electric charge must be added to 8− to yield a 2− overall charge on the ion (8−) + x = 2−; x = 6+. Therefore, manganese in MnO_4^{2-} has an oxidation number of 6+. (d) Assign an oxidation number of 2− to oxygen and 1+ to potassium. Determine the total charge on oxygen and potassium, oxygen = 4(2−) = 8−, potassium = 1+. Determine the total charge contributed by oxygen and potassium, (8−) + 1 = 7−. Since the formula has an overall charge of 0, manganese must contribute a 7+ charge in order to balance exactly the 7− charge. Hence, its oxidation number in $KMnO_4$ is 7+.

FOLLOW-UP PROBLEM

Determine the single oxidation number of the underlined atom in each of these species (a) $\underline{C}rO_4^{2-}$, (b) $\underline{N}H_2^-$, (c) $\underline{C}_2H_2$, (d) $\underline{P}_4O_{10}$, (e) $\underline{Fe}_3O_4$, (f) $\underline{S}_2O_3^{2-}$, (g) $\underline{S}O_4^{2-}$, (h) $\underline{N}_2O_3$.

By examining the formulas of the substances in the equation for a chemical reaction, we can calculate the oxidation number of specified atoms or ions. Thus, we can determine whether the oxidation number of a species has increased or decreased during the course of a reaction. We shall first consider redox reactions in which the reactants are both added to the same reaction vessel. In these reactions, the electrons are transferred directly from one species to another, and the decrease in potential energy of the system appears as heat energy given off to the surroundings. Our primary objective will be to supply the coefficients which balance the equations for the reactions. For balancing these equations, we shall use a method based on *changes in oxidation number.*

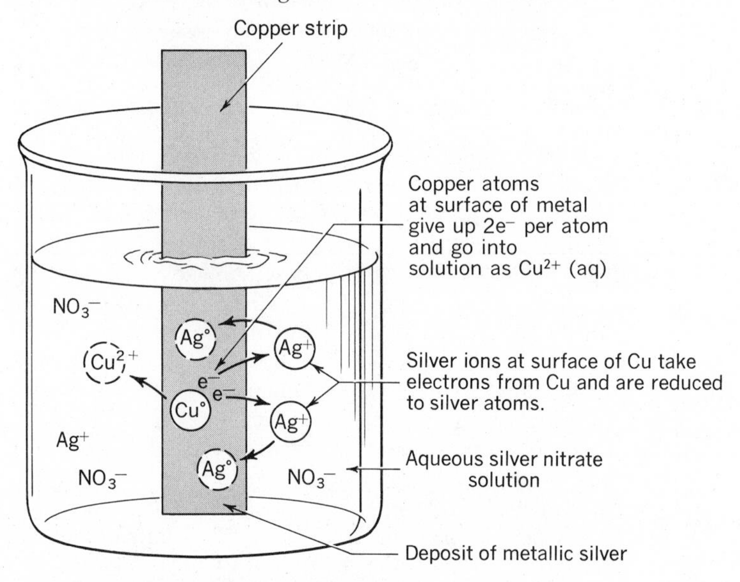

17-1 A simple redox reaction. In this reaction, Ag^+ ions oxidize Cu atoms to Cu^{2+} ions. Thus, silver ions are a stronger oxidizing agent than copper ions. All ions in water solution are hydrated (aq).

17-2 The Change in Oxidation Number Method of Balancing Redox Equations One of the simpler types of redox reactions involves the reactions of neutral atoms or molecules with ions. For example, when a piece of copper (Cu) is placed in a colorless solution of silver nitrate (a soluble ionic solid), metallic silver (Ag) forms on the surface of the copper and the solution acquires a light blue color. Analysis reveals that the blue color is that of hydrated copper(II) ions. On the basis of these observations we can write

$$\mathbf{Cu(s) + Ag^+(aq) \longrightarrow Cu^{2+}(aq) + Ag(s)} \qquad 17\text{-}2$$

The nitrate ions are not involved. To balance an oxidation-reduction equation, we apply the following criteria.

1. The total atoms on the left side of the equation must equal the total atoms on the right. In other words, there must be a mass or material balance.
2. The net ionic charge on the left must equal the net ionic charge on the right. This means there must be an electric balance.
3. The total increase in oxidation number experienced by the reducing agent must equal the total decrease in oxidation number experienced by the oxidizing agent. In other words, the electrons lost or apparently lost by the reducing agent must equal those gained by the oxidizing agent. The four coefficients needed to balance Equation 17-2 can be obtained by equalizing the changes in oxidation numbers (Criterion 3). To accomplish this, we first determine the oxidation number, and change in oxidation number for each agent.

increase of 2, oxidation: Cu 0 → 2+

$$Cu(s) + Ag^+(aq) \longrightarrow Cu^{2+}(aq) + Ag(s)$$

decrease of 1: Ag 1+ → 0

We then multiply the two changes and the species at the end of each line by whatever factors are needed to make the increase equal the decrease. In this case the factors are 1 for the copper species and 2 for the silver species.

$$\underset{\text{reducing agent}}{Cu(s)} + \underset{\text{oxidizing agent}}{2Ag^+(aq)} \longrightarrow Cu^{2+}(aq) + 2Ag(s)$$

FOLLOW-UP PROBLEM

Balance these equations.

(a) $Al(s) + Ag^+(aq) \longrightarrow Al^{3+}(aq) + Ag(s)$

(b) $Al(s) + Cu^{2+}(aq) \longrightarrow Al^{3+}(aq) + Cu(s)$

(c) $Cl_2(g) + Br^-(aq) \longrightarrow Br_2(l) + Cl^-(aq)$

Our observations, used to write Equation 17-2 and our analysis of the reaction in terms of oxidation numbers, indicate that copper metal (Cu) is oxidized and silver ions (Ag^+) are reduced. In this reaction Ag^+ is the oxidizing agent and Cu(s) is the reducing agent. If we now place a strip of silver metal in a solution of copper(II) nitrate, no reaction as observed to occur.

$$Ag(s) + Cu^{2+}(aq) \longrightarrow \text{no reaction} \qquad 17\text{-}3$$

We can interpret our observations of the two experiments represented by Equations 17-2 and 17-3 thus

1. Silver ions can oxidize (remove electrons) from copper atoms (Equation 17-2).
2. Copper(II) ions do not oxidize silver atoms (Equation 17-3).
3. Silver ions are a better oxidizing agent than copper ions.
4. Copper atoms can reduce silver ions (Equation 17-2).
5. Silver atoms do not reduce copper(II) ions (Equation 17-3).
6. Copper is a better reducing agent than silver.

We can rank silver ions and copper ions in order of their oxidizing ability. Similarly, we can rank copper atoms and silver atoms in order of their reducing ability. The oxidizing ability of Ag^+ is related to its tendency to gain electrons and become Ag°. Therefore, we can represent (rank) the relative oxidizing ability of the two ions in terms of equations

$$\textbf{1. } Ag^+ + e^- \rightleftharpoons Ag(s)$$
$$\textbf{2. } Cu^{2+} + 2e^- \rightleftharpoons Cu(s)$$

The relative order of these equations shows that Ag^+ have a greater tendency to be reduced (gain electrons) than Cu^{2+}. This means that Ag^+ are a stronger oxidizing agent than Cu^{2+}. We can use this listing to predict that Ag^+ can oxidize Cu(s) which is below Ag(s) on the right side of the equation, but that Cu^{2+} cannot oxidize Ag(s). In general, a given oxidizing agent can spontaneously oxidize the reduced form of another oxidizing agent weaker than itself.

In a similar manner, we can represent the relative reducing ability of the two metals (atoms) in terms of the equations

$$\textbf{1. } Cu(s) \rightleftharpoons Cu^{2+} + 2e^-$$
$$\textbf{2. } Ag(s) \rightleftharpoons Ag^+ + e^-$$

These equations show that Cu(s) has a greater tendency to be oxidized (lose electrons) than Ag(s). Cu(s), therefore, is a stronger reducing agent than Ag(s). Notice that the last two equations may be obtained by reversing the first two. The first two are written as reduction processes; the second two as oxidation processes.

Note again that *reduction processes,* represented by the first two equations, always result in a *decrease* in the oxidation number of a species. Thus, silver is reduced from a 1+ to a zero oxidation number. *Oxidation processes,* represented by the second set of equations, always result in an *increase* in the oxidation number of a species. The equation reveals that copper is oxidized from a zero to 2+. Metals do not exhibit negative oxidation numbers; that is, neutral metallic atoms do not gain electrons. Therefore, metals do not act as oxidizing agents.

As a result of many experiments similar to those described above, tables of oxidizing and reducing agents listed in order of their relative strengths have been compiled. A list of agents is given in Table 17-1. In this table, chlorine gas is the strongest oxidizing agent. This means that Cl_2 can oxidize or take electrons from any species below it on the right side of the equal signs. Note that the reactions of elements in this table are all written as reduction processes. In our study of acids and bases, it was useful to express $H^+(aq)$ as H_3O^+; in redox systems, this notation is of little value. For this reason, H_3O^+ will be generally noted as H^+ or $H^+(aq)$ in this chapter. Although other ions are also hydrated, we shall, for simplicity, omit the (aq) notation.

The name single replacement was derived from the form of an equation which suggests that one element replaces an ion of a second element from the compound containing the second element. Thus, zinc is said to replace hydrogen ions from hydrochloric acid,

$$Zn(s) + 2HCl(aq) \longrightarrow H_2(g) + ZnCl_2(aq).$$

The information in Table 17-1 may be used to write what are sometimes referred to as *single replacement reactions.* A rule of thumb often used by students to predict and write equations for these reactions may be stated. *Any species on the left side of the arrows reacts spontaneously with any species on the right side which is below it in the series.* For example, H^+ ions react with metallic zinc and form H_2 and Zn^{2+}.

$$\mathbf{2H^+(aq) + Zn(s) \longrightarrow H_2(g) + Zn^{2+}}$$

TABLE 17-1
IONS, ATOMS, AND MOLECULES WHICH BEHAVE AS OXIDIZING AND/OR REDUCING AGENTS

	Oxidized Form		Reduced Form	
Strongest oxidizing agent $\longrightarrow$	$Cl_2 + 2e^-$	$\longrightarrow$	$2Cl^-$	
	$Br_2(l) + 2e^-$	$\longrightarrow$	$2Br^-$	
	$Hg^{2+} + 2e^-$	$\longrightarrow$	$Hg(l)$	
	$Ag^+ + e^-$	$\longrightarrow$	$Ag(s)$	
Oxidizing strength increases ↑	$Fe^{3+} + e^-$	$\longrightarrow$	Fe^{2+}	Reducing strength increases ↓
	$I_2(s) + 2e^-$	$\longrightarrow$	$2I^-$	
	$Cu^{2+} + 2e^-$	$\longrightarrow$	$Cu(s)$	
	$Sn^{4+} + 2e^-$	$\longrightarrow$	Sn^{2+}	
	$2H^+ + 2e^-$	$\longrightarrow$	$H_2(g)$	
	$Pb^{2+} + 2e^-$	$\longrightarrow$	$Pb(s)$	
	$Sn^{2+} + 2e^-$	$\longrightarrow$	$Sn(s)$	
	$Ni^{2+} + 2e^-$	$\longrightarrow$	$Ni(s)$	
	$Cd^{2+} + 2e^-$	$\longrightarrow$	$Cd(s)$	
	$Fe^{2+} + 2e^-$	$\longrightarrow$	$Fe(s)$	
	$Zn^{2+} + 2e^-$	$\longrightarrow$	$Zn(s)$	
	$Al^{3+} + 3e^-$	$\longrightarrow$	$Al(s)$	
	$Mg^{2+} + 2e^-$	$\longrightarrow$	$Mg(s)$	$\longleftarrow$ Strongest reducing agent

FOLLOW-UP PROBLEMS

Use the techniques described above and the equations in Table 17-1 and write net ionic equations for these reactions (all aqueous solution except (g)). If no reaction occurs, write *no reaction.*

(a) Chlorine + tin(II) nitrate $\longrightarrow$
(b) Iron(II) nitrate + potassium iodide $\longrightarrow$
(c) Silver and hydrochloric acid $\longrightarrow$
(d) Cadmium + sulfuric acid $\longrightarrow$
(e) Zinc + nickel nitrate $\longrightarrow$
(f) Iron(II) nitrate + aluminum nitrate $\longrightarrow$
(g) Bromine(l) + magnesium $\longrightarrow$

Many polyatomic ions are also common oxidizing agents. Equations involving these agents are more complicated than those involving only simple ions and atoms.

Consider the reaction of sodium sulfite (Na_2SO_3) with potassium permanganate ($KMnO_4$) in an acid solution. When purple $KMnO_4$ solution is added slowly to colorless Na_2SO_3 solution, the color disappears until stoichiometric quantities are present. Ion tests reveal that sulfate ions (SO_4^{2-}) and manganese(II) ions are formed. Since both Na_2SO_3 and $KMnO_4$ are soluble ionic compounds, we may write the skeleton equation as

$$MnO_4^- + SO_3^{2-} \longrightarrow Mn^{2+} + SO_4^{2-}$$

To balance this equation, we apply these rules:

1. Identify the elements which change oxidation numbers. Connect the oxidized and reduced forms of each by lines and write the oxidation numbers above both symbols. The information in Tables 17-2 and 17-3 will help you identify the elements which commonly undergo a change in oxidation number.

$$\overset{7+}{Mn}O_4^- + \underset{4+}{S}O_3^{2-} \longrightarrow \overset{2+}{Mn}{}^{2+} + \underset{6+}{S}O_4^{2-}$$

2. Write the increase in oxidation number experienced by the reducing agent and the decrease in oxidation number experienced by the oxidizing agent.

decrease of 5

$$\overset{7+}{Mn}O_4^- + \underset{4+}{S}O_3^{2-} \longrightarrow \overset{2+}{Mn}{}^{2+} + \underset{6+}{S}O_4^{2-}$$

increase of 2

3. Equalize the decrease in oxidation number with the increase. In this case, the net change in each must be 10. Multiply the species

decrease of 10

$$2MnO_4^- + 5SO_3^{2-} \longrightarrow 2Mn^{2+} + 5SO_4^{2-}$$

increase of 10

joined by the lines by the necessary factors, two for the manganese and five for the sulfur.

TABLE 17-2
COMMON OXIDIZING AGENTS

Oxidizing Agent	*Formula (underlined element is reduced)*	*Oxidation State of Element Which is Reduced*
Permanganate ions	$\underline{Mn}O_4^-$	7+
Nitrate ions	$\underline{N}O_3^-$	5+
Dichromate ions	$\underline{Cr}_2O_7^{2-}$	6+
Chlorate ions	$\underline{Cl}O_3^-$	5+
Cerium(IV) ions	Ce^{4+}	4+
Iron(III) ions	Fe^{3+}	3+
Chlorine	Cl_2	0
Bromine	Br_2	0
Metallic ions in higher oxidation state	metallic ions X^{n+}	higher state n+

TABLE 17-3
COMMON REDUCING AGENTS

Reducing Agent	*Formula (underlined element is oxidized)*	*Oxidation State of Element Which is Oxidized*
Metallic atoms	Zn, Na	0
Nonmetallic atoms	C, S, P	0
Metallic ions in lower oxidation states	metallic ions (X^{n+})	lower state n+
Nonmetallic ions	I^-, Br^-, Cl^-	1−
Hydrogen sulfide	$H_2\underline{S}$	2−
Oxalic acid	$H_2\underline{C}_2O_4$	3+
Molecules or polyatomic ions containing atoms in lower oxidation states	$\underline{C}O$, $\underline{S}O_3^{2-}$, $\underline{N}O$	lower oxidation state

4. If the solution is noted as acid then add H^+ to balance the ionic charge. If solution is basic, add OH^- to balance the charge. The net charge on the left side of the above equation is 12−; that on the right side is 6−. Because the solution is acid, we must add $6H^+$ to the left side.

$$\mathbf{2MnO_4^- + 5SO_3^{2-} + 6H^+ \longrightarrow 2Mn^{2+} + 5SO_4^{2-}}$$

5. Add H_2O to balance hydrogen atoms. In the above equation, addition of $3H_2O$ to the right side will balance the 6 hydrogen ions on the left.

$$\mathbf{2MnO_4^- + 5SO_3^{2-} + 6H^+ = 2Mn^{2+} + 5SO_4^{2-} + 3H_2O}$$

6. Check by counting oxygen atoms on each side. Note that it is often advantageous to convert a formula equation into an ionic one before balancing it. This can be accomplished by removing inactive ions such as Na^+ and K^+. If it is desired to convert a balanced ionic equation into an equation showing neutral formulas, simply add the same number and kind of those ions to each side of the equation.

FOLLOW-UP PROBLEMS

Balance these equations. For each reaction, indicate (a) the oxidizing agent, (b) the reducing agent, (c) the element being oxidized, (d) the element being reduced. Write all equations as net ionic equations.

(a) $BrO_3^- + I^- + H^+ \longrightarrow Br^- + I_2 + H_2O$

(b) $SeO_4^{2-} + Cl^- + H^+ \longrightarrow SeO_3^{2-} + Cl_2 + H_2O$

(c) $Al(s) + NO_3^- + OH^- + H_2O \longrightarrow NH_3 + AlO_2^-$

(d) $Zn(s) + NO_3^- + H^+ \longrightarrow Zn^{2+} + NH_4^+ + H_2O$

(e) $KMnO_4 + KNO_2 + H_2SO_4 \longrightarrow MnSO_4 + H_2O + KNO_3 + K_2SO_4$

(f) $ClO_2(g) + SbO_2^- \longrightarrow ClO_2^- + Sb(OH)_6^-$ (basic)
(g) $Cr_2O_7^{2-} + I^- \longrightarrow Cr^{3+} + I_2$ (acid)
(h) $CN^- + CrO_4^{2-} \longrightarrow CNO^- + Cr(OH)_3(s)$ (basic)
(i) $I_2 + ClO_3^- \longrightarrow IO_3^- + Cl^-$ (acid)
(j) $MnO_4^- + NH_3 \longrightarrow MnO_2(s) + NO_3^-$ (basic)
(k) $Mg(s) + ReO_4^- \longrightarrow Mg^{2+} + Re^-$ (acid)
(l) $Ag(s) + NO_3^- \longrightarrow Ag^+ + NO$ (acid)
(m) $Fe(CN)_6^{3-} + Cr_2O_3 \longrightarrow Fe(CN)_6^{4-} + CrO_4^{2-}$ (basic)
(n) $MnO_4^- + C_2O_4^{2-} \longrightarrow CO_2 + Mn^{2+}$ (acid)

17-3 The Half-cell Method of Balancing Redox Equations It is possible to balance oxidation-reduction equations without using the concept of oxidation numbers. The method, known as the ***half-cell method,*** involves writing two half-cell reactions and then adding them to obtain a net equation for the overall reaction. One of the half-cell reactions is an oxidation; the other is a reduction.

In order to construct half-cell reactions, it is necessary to know both the oxidized and reduced forms of both agents. The oxidized and reduced forms of a number of common agents are listed in Table 17-4. Note that the oxidizing strength of a species and the product formed often depend on whether the solution is acid or basic. In Table 17-4, the strongest oxidizing agents are located at the upper left and the strongest reducing agents are at the lower right.

TABLE 17-4
OXIDIZED AND REDUCED FORMS OF SOME COMMON REDOX AGENTS IN AQUEOUS SOLUTION

Oxidized Form	Reduced Form
Co^{3+}	Co^{2+}
H_2O_2 (acid)	H_2O
MnO_4^- (acid)	Mn^{2+}
BrO_3^- (acid)	Br^-
ClO_4^- (acid)	Cl^-
Cl_2	Cl^-
$Cr_2O_7^{2-}$ (acid)	Cr^{3+}
O_2 (acid)	H_2O
$Br_2(l)$	Br^-
NO_3^- (acid)	$NO(g)$
ClO^- (basic)	Cl^-
Hg^{2+}	$Hg(l)$
Ag^+	$Ag(s)$
Fe^{3+}	Fe^{2+}
MnO_4^- (basic)	$MnO_2(s)$
$I_2(s)$ (acid)	I^-
Cu^+	$Cu(s)$
$O_2(g)$	OH^-
Cu^{2+}	$Cu(s)$
Cu^{2+}	Cu^+
$S(s)$ (acid)	$H_2S(g)$
H^+ (acid)	H_2
Pb^{2+}	$Pb(s)$
Ni^{2+}	$Ni(s)$
Co^{2+}	$Co(s)$
Cd^{2+}	$Cd(s)$
Fe^{2+}	$Fe(s)$
$S(s)$ (basic)	S^{2-}
Cr^{3+}	$Cr(s)$
Zn^{2+}	$Zn(s)$
H_2O (basic)	$H_2(g)$
Mn^{2+}	$Mn(s)$
Al^{3+}	$Al(s)$
Mg^{2+}	$Mg(s)$
Na^+	$Na(s)$
Ca^{2+}	$Ca(s)$
Cs^+	$Cs(s)$
K^+	$K(s)$

(Left arrow: Decreasing strength as oxidizing agent; right arrow: Increasing strength as reducing agent)

The data in Table 17-4 can, like that in Table 17-1, be used to predict whether a redox reaction is apt to occur between two species. It also enables us to predict products of the reaction. In general, a species on the left reacts spontaneously with a species below it on the right. The further apart the two species are, the more tendency there is for the reaction to occur. For example, Cu^{2+} has more tendency to react with Mg(s) than with Pb(s). Cu^{2+} does not react spontaneously with Ag(s). In addition to its use for predicting the feasibility of reactions, the data in Table 17-4 enable us to construct half-cell equations. Then combine half-cell equations and obtain an equation for any desired redox reaction. By following a proper sequence of steps, we can write an equation for almost any redox reaction.

We shall illustrate the sequence of steps by using the half-cell method to balance an equation for the reaction between permanganate ions (MnO_4^-) and iron(II) in an *acid solution.* In an acid solution, MnO_4^- acts as an oxidizing agent and is reduced to Mn^{2+} ions by iron(II), the reducing agent. Iron(II) is oxidized by MnO_4^- ions to Fe(III). The skeleton equation for this reaction is

$$MnO_4^- + Fe^{2+} \longrightarrow Mn^{2+} + Fe^{3+} \qquad 17\text{-}4$$

Step 1 Divide the skeleton equation into two half-cell equations. Each half-cell equation must show either an oxidizing or a reducing agent on the left side and the product(s) associated with its reduction or oxidation on the right side. The unbalanced half-cell expressions for Equation 17-4 are

$$\textbf{Reduction} \quad MnO_4^- \longrightarrow Mn^{2+} \qquad 17\text{-}5$$

$$\textbf{Oxidation} \quad Fe^{2+} \longrightarrow Fe^{3+} \qquad 17\text{-}6$$

We shall first balance equation 17-5 and then follow the same steps to balance Equation 17-6.

Step 2 Balance all atoms other than hydrogen and oxygen. In Equation 17-5 the manganese atoms are already in balance.

Step 3 Count the number of oxygen atoms and balance them by adding the proper number of H_2O molecules to whichever side is deficient in oxygen atoms. In Equation 17-5 there are 4 oxygen atoms on the left side. Four H_2O molecules must be added to the right side of the equation to supply the needed oxygen atoms.

$$MnO_4^- \longrightarrow Mn^{2+} + 4H_2O \qquad 17\text{-}7$$

Step 4 Count the number of hydrogen atoms. Balance them by adding the proper number of H^+ ions to whichever side is deficient in hydrogen. Since there are 8 hydrogen atoms on the right side, add 8 H^+ ions to the left side of Equation 17-7.

$$8H^+ + MnO_4^- \longrightarrow Mn^{2+} + 4H_2O \qquad 17\text{-}8$$

Step 5 Bring the two sides of Equation 17-8 into electric balance. This may be done by counting the total ionic charge on each side and adding electrons to whichever side needs additional negative charge to bring it into balance with the opposite side. In Equation 17-8 the ionic charge on the right side is $+2$; on the left, it is $+7$. Adding $5e^-$ to the left side reduces the charge to $+2$.

$$5e^- + 8H^+ + MnO_4^- \longrightarrow Mn^{2+} + 4H_2O \qquad 17\text{-}9$$

Step 6 Repeat Steps 2 through 4 for Equation 17-6. This yields

$$Fe^{2+} \longrightarrow Fe^{3+} + e^- \qquad 17\text{-}10$$

Step 7 Balance the number of electrons in Equations 17-9 and 17-10. The number of electrons must be the lowest common multiple of the numbers shown in the two equations. The lowest common multiple of e^- and $5e^-$ is 5. Multiply each factor in each half-cell reaction by the number that makes the number of electrons in the two half-cell reactions equal. Thus, multiply Equation 17-9 by 1 and Equation 17-10 by 5.

$$5e^- + 8H^+ + MnO_4^- \longrightarrow Mn^{2+} + 4H_2O \qquad 17\text{-}11$$

$$5Fe^{2+} \longrightarrow 5Fe^{3+} + 5e^- \qquad 17\text{-}12$$

Step 8 Algebraically add the two half-cell reactions, right and left sides separately. The net equation is

$$8H^+ + MnO_4^- + 5Fe^{2+} \longrightarrow Mn^{2+} + 5Fe^{3+} + 4H_2O$$

In order to obtain an equation for a redox reaction in *basic solution,* it is necessary to add an extra step. The purpose of the extra step is to eliminate the H^+ ions, since equations for reactions in basic solution cannot show the presence of excess H^+ ions. In order to illustrate the additional step, we shall write an equation for the reaction between permanganate ions and cyanide ions

SIR HUMPHREY DAVY
1778–1829

Humphrey Davy attended school until he was 16 years old. However, he was the kind of person who never stopped experimenting and studying so, in effect, his education continued throughout his entire life. When his father died, he was apprenticed to a surgeon-apothecary during which time he had an opportunity to do some experimentation. At this time, he became interested in the writings of Casper Neumann, an apothecary and chemistry professor in Berlin. When Davy was released from his indenture as an apprentice apothecary, he became superintendent of the Medical Pneumatic Institution of Bristol. This organization was devoted to the study of the medical value of various gases. His first discovery was the effect of nitrous oxide, which became known as laughing gas. He "breathed 16 quarts of the gas in 7 minutes" and became "completely intoxicated" with it.

In 1801, the Royal Institution in London engaged him as a lecturer. Davy, while on a lecture tour, met Faraday, who came to hear his lecture. Faraday took copious notes, made exact drawings of Davy's apparatus, and sent these to Davy as part of his application for a job as assistant. Davy hired Faraday at once. Later, when asked what his greatest discovery was, he replied, "Michael Faraday." Faraday accompanied Davy to France on a lecture tour and worked with him closely for many years. However, when Faraday became a member of the Royal Society, it was against the will of Davy, who had become jealous of his successful assistant.

Davy is remembered for his discovery of sodium and potassium. He obtained these metals by electrolyzing their fused salts. He had previously tried electrolyzing their water solutions which yielded only hydrogen gas. His persistence was rewarded when he was able to separate the globules of pure metal. He described the potassium as particles which, when thrown into water, "skimmed about excitedly with a hissing sound, and soon burned with a lovely lavender light." Dr. John Davy, Humphrey's brother, said that Humphrey "danced around and was delirious with joy" at his discovery.

(CN^-) in basic solution. In this reaction, MnO_4^- is reduced to MnO_2 and CN^- ions are oxidized to cyanate ions (CNO^-). The skeleton equation is

$$MnO_4^- + CN^- \longrightarrow MnO_2(s) + CNO^-$$

Step 1 Same as acid solution.

$$MnO_4^- \longrightarrow MnO_2(s) \qquad \text{17-13}$$
$$CN^- \longrightarrow CNO^-$$

Step 2 Same as acid solution.

$$MnO_4^- \longrightarrow MnO_2(s) + 2H_2O$$

Step 3 Same as acid solution.

$$4H^+ + MnO_4^- \longrightarrow MnO_2(s) + 2H_2O \qquad \text{17-14}$$

Step 3a (reactions in basic solution) Remove the H^+ ions from the half-cell reaction. This can be done without changing the present balance by adding the same number of OH^- ions to both sides of the equation. In Equation 17-14, 4 OH^- ions are required to neutralize 4 H^+ ions. To maintain balance, add 4 OH^- ions to both sides.

$$\begin{array}{l} 4H^+ + MnO_4^- \longrightarrow MnO_2(s) + 2H_2O \\ 4OH^- \qquad\qquad\qquad\qquad\qquad\qquad + 4OH^- \\ \hline 4HOH + MnO_4^- \longrightarrow MnO_2(s) + 2H_2O + 4OH^- \end{array}$$

This operation results in the appearance of water molecules on both sides of the equation. Simplification yields

$$2H_2O + MnO_4^- \longrightarrow MnO_2 + 4OH^-$$

Step 4 Same as acid solution.

$$3e^- + 2H_2O + MnO_4^- \longrightarrow MnO_2(s) + 4OH^- \qquad \text{17-15}$$

Step 5 Repeating Steps 2 through 4 for Equation 17-13 yields

$$2OH^- + CN^- \longrightarrow CNO^- + H_2O + 2e^- \qquad \text{17-16}$$

Step 6 Balance the electrons by multiplying Equation 17-15 by 2 and Equation 17-16 by 3.

$$6e^- + 4H_2O + 2MnO_4^- \longrightarrow 2MnO_2(s) + 8OH^- \qquad \text{17-17}$$
$$6OH^- + 3CN^- \longrightarrow 3CNO^- + 3H_2O + 6e^- \qquad \text{17-18}$$

Step 7 Adding Equations 17-17 and 17-18 and collecting like terms gives

$$H_2O + 2MnO_4^- + 3CN^- \longrightarrow 2MnO_2(s) + 2OH^- + 3CNO^-$$

If valid half-cell equations are not required, then the removal of H^+ ions from the equations may be postponed until the last step.

FOLLOW-UP PROBLEMS

1. Use the half-cell reaction method to write equations for these reactions.

	Solution
(a) $NO_3^- + Bi \longrightarrow Bi^{3+} + NO_2$	(acid)
(b) $Cr_2O_7^{2-} + I^- \longrightarrow Cr^{3+} + I_2$	(acid)
(c) $Cr^{3+} + ClO_3^- \longrightarrow ClO_2 + Cr_2O_7^{2-}$	(acid)
(d) $CHCl_3 + MnO_4^- \longrightarrow Cl_2 + CO_2 + Mn^{2+}$	(acid)
(e) $MnO_4^- + NO_2^- \longrightarrow MnO_2(s) + NO_3^-$	(basic)
(f) $ClO^- \longrightarrow Cl^- + ClO_3^-$	(basic)
(g) $Al + NO_3^- \longrightarrow Al(OH)_4^- + NH_3$	(basic)
(h) $O_2 + H_2O + I^- \longrightarrow I_2 + OH^-$	(basic)

2. Use the data in Table 17-4 to write net ionic equations for these reactions in aqueous solutions (except e).
(a) Hydrogen peroxide (H_2O_2) with manganese(II) sulfate ($MnSO_4$) (acid).
(b) Potassium dichromate with $H_2S(g)$ (acid).
(c) Sodium perchlorate with iron(II) nitrate (acid).
(d) Metallic sodium plus water (basic).
(e) Oxygen (O_2) + hydrogen sulfide (H_2S) (acid).

ELECTROCHEMICAL CELLS

17-4 Meaning of Terms about Electricity Perhaps you have wondered what is meant by the term *half-cell reaction.* The term is related to the fact that oxidation-reduction reactions can be used as a source of electric energy when they occur in an electrochemical cell such as an ordinary flashlight cell. An experimental electrochemical cell is composed of two half-cells, each half-cell involving a different reaction. The process (oxidation or reduction) occurring in each half-cell is known as a half-cell reaction. When a flashlight cell is operating, an oxidation half-cell reaction is occurring at the surface of one electrode (the zinc container), while a reduction process is occurring at the other electrode (central rod). The sum of the equations for the two half-cell reactions constitutes the equation for the overall cell (oxidation-reduction) reaction.

Most of you know that an ordinary flashlight cell has a voltage of about 1.5 *volts* and produces a current which may be measured in *amperes.* Chances are, however, that you would have difficulty explaining what is meant by the terms *volt* and *ampere.* Because these and other terms associated with electric phenomena are used to describe cell characteristics, we shall attach a meaning to them in order to provide background for later discussions.

When a flashlight is turned on, energy is being transferred. The source of this energy is the chemical reaction taking place in the cell. That is, chemical energy is being transformed into electric energy which is being transferred and is doing electric work when it is used to operate any electric device. The work is done by the cell as it forces electrons through the wires and other parts of the *circuit.* The ***voltage*** of the cell is a measure of its ability to do electric work. The current is measured in terms of the number of electrons which flow through a circuit per second. An ***ampere*** may be defined as a flow of 6.25×10^{18} electrons per second. The electric charge carried by this quantity of electrons is called a ***coulomb.*** An ***ampere*** is also defined in practical units as *one coulomb per second.* The

A circuit is the external path that electrons follow from one electrode of a cell through a conductor plus any devices operated by the cell back to the other electrode of the cell.

Important relationships between energy units are listed below.
1 calorie = 4.18×10^7 ergs
1 joule = 10^7 ergs
1 calorie = 4.18 joules
1 electron volt (ev) is the amount of energy a single electron absorbs when it passes through a potential difference of one volt.
1 ev = 1.6×10^{-12} ergs
1 ev = 3.8×10^{-20} cal

energy transferred or work done by 6.25×10^{18} electrons (1 coulomb) is called one *joule.* The *volt* may be defined as the energy transferred or work done by the cell when it drives one electron through the *circuit.* In other words,

$$1 \text{ volt} = \frac{1 \text{ joule}}{\text{coulomb}(6.25 \times 10^{18} \text{ electrons})}$$

$$= \frac{1.6 \times 10^{-19} \text{ joules}}{\text{electron(elementary charge)}}$$

Thus the electric energy supplied is equal to the voltage times the charge. This means the energy transferred when 1 volt drives one electron equals 1.6×10^{-19} joules, or 1.6×10^{-19} joules $\times$ $\frac{10^7 \text{ erg}}{\text{joule}} = 1.6 \times 10^{-12}$ ergs. We shall now examine the principles underlying the operation of an electrochemical cell and attempt to show how it can be used to determine the spontaneity of a reaction, the equilibrium constant, and other factors associated with reactions.

17-5 Components and Operation of an Electrochemical Cell Electron transfer between an oxidizing agent and a reducing agent occurs when the reaction is carried out in electrochemical cells. These cells are devices in which oxidation-reduction reactions take place indirectly. That is, the oxidation and reduction reactions each take place at separate surfaces called ***electrodes.*** In this way, the chemical energy of the participants in a reaction may be converted to and evolved largely as electric energy.

For example, zinc metal reacts with aqueous silver nitrate. The equation for the reaction is

$$Zn(s) + 2Ag^+ \rightleftharpoons Zn^{2+} + 2Ag(s)$$

When the reaction takes place in a single beaker, the two reactants are in direct contact with each other so that the *decrease in potential energy of the system appears as heat given off to the surroundings.* When, however, the reaction is carried out in an *electrochemical cell,* the reactants are *physically separated. Electron transfer takes place through a wire which connects the electrodes.* The decrease in potential energy of the system appears, in part, as electric energy, which can be used to light a small bulb or produce a reading on an ammeter (an instrument used for measuring an electric current). Let us examine the construction and operation of the electrochemical cell shown in Fig. 17-2.

The overall electrochemical cell is composed of ***two half-cells.*** These consist of two beakers, one of which contains a 1-M zinc-ion solution and a strip of zinc metal; the other a 1-M silver-ion solution and a strip of silver. The metal electrodes are connected

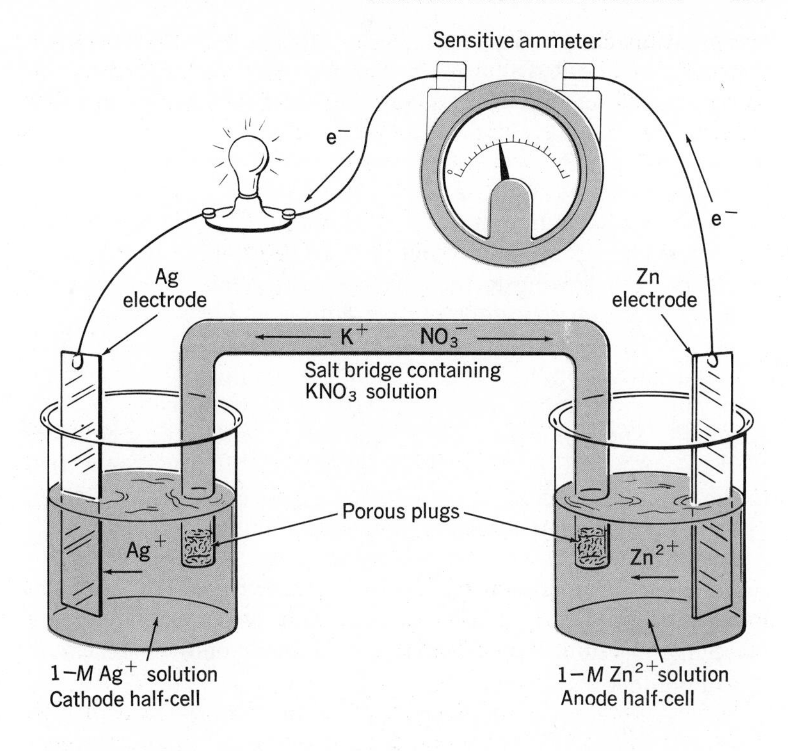

17-2 A simple electrochemical cell.

through the meter by a *metallic conductor* (wire). The two solutions are then connected by a ***salt bridge*** containing an *electrolytic solution,* such as KNO_3.

When the electrodes are connected through an ammeter, there is ample evidence that a reaction is taking place. The deflection of the ammeter needle indicates that electric current is passing through the meter. The zinc strip starts to disintegrate, and the concentration of the Zn^{2+} ions in solution increases. The silver strip increases in mass as silver metal deposits on its surface. Simultaneously, the concentration of Ag^+ ions in solution decreases. As the reaction proceeds, the meter deflection falls off, showing a decrease in the flow of current. The deposition of metallic silver on the silver strip resulting from the reduction of Ag^+ ions in solution constitutes evidence that electrons are flowing from the zinc strip through the wire to the silver strip.

The chemical reactions taking place at the surface of each electrode may be represented by equations. The *oxidation reaction* taking place at the *zinc electrode* is

$$\mathbf{Zn(s) \longrightarrow Zn^{2+} + 2e^-} \qquad 17\text{-}19$$

The *reduction reaction* taking place at the *silver electrode* is

$$\mathbf{Ag^+ + e^- \longrightarrow Ag(s)} \qquad 17\text{-}20$$

By definition, the electrode at which *reduction* occurs is the ***cathode.***

MICHAEL FARADAY
1791–1867

Michael Faraday, the son of a London blacksmith, received only a very elementary education and never received any formal training in science. He was apprenticed to a bookbinder in 1804, and although he did not enjoy this work, he became an expert in a very short time. During his apprenticeship, he read every book he could. His contact with scientific publications stimulated his interest in science.

Many of the terms associated with electrochemistry can be traced to Faraday. He used the term *electrode* to replace the word *pole.* He coined the words *electrolyte, electrolysis, anode, cathode, ion, anion,* and *cation.* He mistakenly believed that ions were created at the electrode during electrolysis. His contributions in the field of physics probably exceeded those in the field of chemistry. He believed that since electricity could produce a magnetic field, a moving magnet must be capable of producing an electric current. His experiments laid the groundwork for transformers and electric motors. Faraday did not believe, as Newton did, that a force could act through a distance but rather that a "field" existed. He expressed his idea of a field when he wrote, "In this view of the magnet, the medium or space around it is as essential as the magnet itself, being a part of the true and complete magnetic system." Many scientists feel that his concept of the "field theory" was his single most important contribution.

Like Einstein, Faraday believed that there must be some fundamental relationship between gravitational and electromagnetic forces. He performed many experiments to discover the nature of this relationship. He wrote: "Here end my trials for the present. The results are negative. They do not shake my strong feeling of the existence of a relation between gravity and electricity, though they give no proof that such a relation exists."

Oxidation occurs at the a***node.*** Thus, the zinc strip is the anode and the silver strip is the cathode.

The migration of ions through the salt bridge completes the electric circuit and preserves the electric neutrality of the solutions. The salt bridge is necessary to prevent the solutions from becoming electrically charged. If the salt bridge were missing, zinc ions resulting from the dissolution of the zinc strip would accumulate around the electrode, give the solution a positive charge, and prevent electrons from flowing to the silver compartment. In the cathode compartment, the deposition of silver ions would leave the solution negatively charged with an excess of negative ions. The salt bridge allows negative ions (anions) to move toward the anode compartment and positive ions (cations) to move toward the cathode compartment. The diffusion of ions maintains the electric neutrality of the solutions. Electrons are then free to flow from the anode to the cathode in the external circuit.

The overall oxidation-reduction reaction taking place in an electrochemical cell is the sum of two half-cell reactions. The reactions represented by Equations 17-19 and 17-20 are called ***half-cell reactions.*** In order to obtain the overall equation for the reaction taking place in the cell illustrated in Fig. 17-3 these half-cell reactions must be added, as shown below.

Anode reaction (oxidation)

$$\mathbf{Zn(s) \longrightarrow Zn^{2+} + 2e^-}$$

Cathode reaction (reduction)

$$\mathbf{2Ag^+ + 2e^- \longrightarrow 2Ag(s)}$$

Total cell reaction

$$\mathbf{Zn(s) + 2Ag^+ \longrightarrow 2Ag(s) + Zn^{2+}} \qquad 17\text{-}21$$

Note that Equation 17-20 was multiplied by a factor of two before being added to Equation 17-19. This was done to *equalize the number of electrons gained by the silver and the number lost by the zinc.* Equation 17-21 represents the overall or net reaction.

In theory, any oxidation-reduction reaction can be separated into half-cell reactions. The half-cell reactions can be balanced and then combined, yielding an equation for the overall oxidation-reduction reaction. The half-cell reaction method of balancing redox equations was illustrated in Sect. 17-3.

17-6 The Meaning and Measurement of Electric Potential and Potential Difference In Fig. 17-2, observation that electrons flow through the meter from the zinc to the silver electrode indicates that the silver metal-silver ion system in the one beaker has more attraction for electrons than the zinc metal-zinc ion combination in the other beaker. The electron-attracting ability of a half-cell is called its ***electric potential, E.*** The difference in electron-attracting ability between two half-cells is a measure of the *tendency for the overall cell reaction to take place.* The difference between the electric potentials of two half-cells is called the *potential difference* and is usually expressed as a *voltage* (see definition on p. 555).

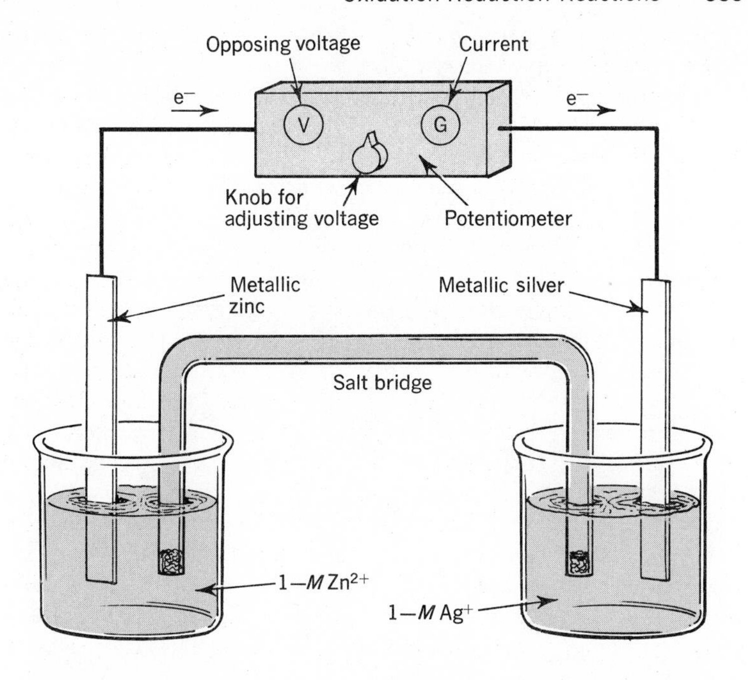

17-3 The potentiometer contains a variable voltage source which opposes the spontaneous flow of electrons as shown in the cell above. The voltage-adjust knob on the potentiometer is turned until the current meter (galvanometer, G) reads zero. At this point, there is no current flowing in the external circuit and the voltage read on the meter is equal numerically to the cell potential.

Precise differences in potential between two electrodes are measured with a *potentiometer.* This instrument is a source of opposing voltage adjusted so that no current flows through the cell whose voltage is being measured. Any current drawn from the cell during measurement results in a slight voltage drop and, therefore, a low reading of the cell potential. A current through the cell also changes the composition of the solution. For convenience in laboratory experiments, high resistance *voltmeters* which draw an insignificant current are used to measure cell potentials. A voltmeter connected to the electrodes of the zinc-silver electrochemical cell shown in Fig. 17-4 reads approximately 1.56 v. This means that the Ag-Ag^+ half-cell has 1.56 v more electron-attracting ability than the Zn-Zn^{2+} half-cell. If the potential of the zinc half-cell were known, then the potential of the silver half-cell could be determined by adding 1.56 v to the potential of the zinc half-cell. The absolute potential of an isolated half-cell, however, like the total energy content of a chemical species, cannot be measured experimentally. *Only the difference in potential between two half-cells can be measured.*

A volt is defined as the potential difference necessary to cause one ampere of current to flow through a conductor which has a resistance of one ohm.

17-7 The Standard Hydrogen Half-cell If the potential of a standard half-cell is assumed to be zero, then the experimentally measured difference in potential between the standard and a second half-cell is the relative potential of the second half-cell. A useful table of relative half-cell potentials has been developed by determining the voltages of hundreds of half-cells coupled to a *standard reference half-cell with an arbitrarily assigned potential of zero volts.*

The cell would operate at other concentrations. A 1-*M* concentration of dissolved species is part of the arbitrary choice of *standard state conditions* for cells. The other factors are 1.00 atm pressure for gases and 25°C.

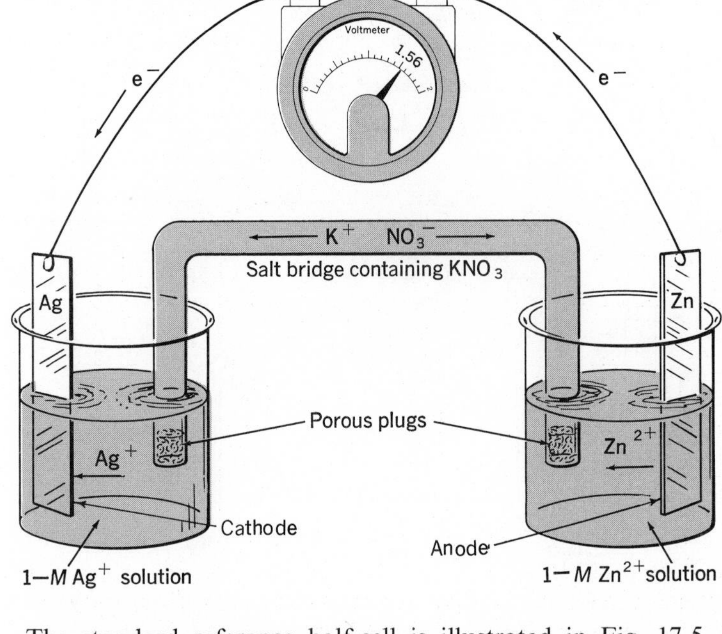

17-4 The voltage (1.56 v) of this cell is a measure of the tendency for the reaction

$$Zn(s) + 2Ag^+(aq) \longrightarrow Zn^{2+}(aq) + 2Ag(s)$$

to take place when the ion concentrations are 1-*M* and the temperature is 25 °C.

The standard reference half-cell is illustrated in Fig. 17-5. It consists of a platinum electrode in a solution in which the $[H_3O^+]$ is 1 *M* with hydrogen gas at a pressure of 1 atm bubbled over the electrode. Constant temperature is maintained at 25 °C.

17-8 The Silver-Hydrogen Cell Fig. 17-6 shows a standard hydrogen half-cell coupled by a circuit containing a voltmeter to a standard silver half-cell in which the $[Ag^+]$ is 1 *M*. The voltmeter reading shows that there is a potential difference of 0.80 v between the hydrogen half-cell and that of the silver-silver ion half-cell. The potential of the standard hydrogen half-cell is arbitrarily taken as zero. Therefore, the potential of the silver half-cell is 0.80 v relative to that of the standard.

The experiments reveal that electrons flow from the hydrogen half-cell to the silver half-cell and that the acidity of the solution in the hydrogen half-cell increases. In other words, the silver half-cell has an electron-attracting ability which is 0.80 v greater than that of the hydrogen standard half-cell. The reduction of silver ions at the silver electrode implies that the *electron-attracting ability of a half-cell is related to the reduction process.* Therefore, the voltage, 0.80 v, is called the ***reduction potential.*** The tendency for silver ions to be reduced to silver atoms in the half-cell

$$\mathbf{Ag^+ + e^- \rightleftharpoons Ag(s)} \qquad 17\text{-}22$$

is greater by 0.80 v than the tendency for H^+ ions to be reduced to H_2 in the half-cell

$$\mathbf{2H^+ + 2e^- \rightleftharpoons H_2(g)}$$

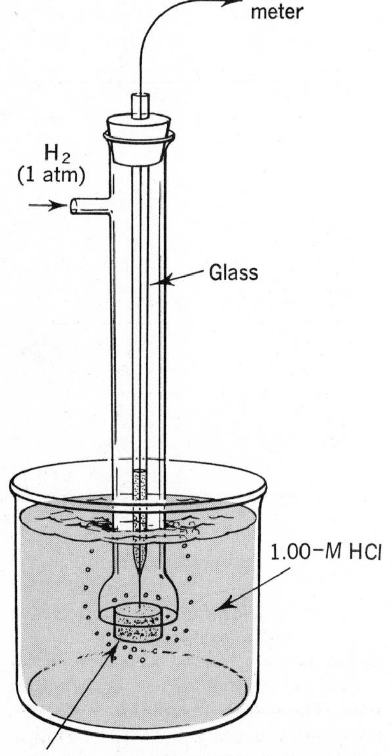

17-5 A standard hydrogen electrode.

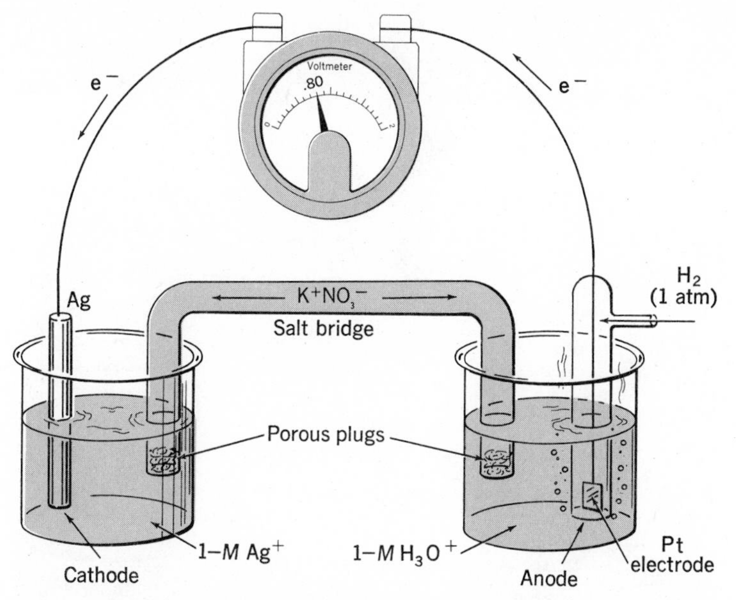

17-6 A silver-hydrogen cell.

17-9 The Zinc-Hydrogen Cell Fig. 17-7 shows that, when a standard hydrogen and a standard zinc half-cell are coupled, the voltmeter reads 0.76 v. Experiments show that the acidity of the solution in the hydrogen half-cell decreases and that the electrons are flowing from zinc through the external circuit to the hydrogen electrode. There they reduce H^+ ions to H_2 gas. This means that the tendency for zinc ions to be reduced to zinc atoms in the half-cell

$$\mathbf{Zn^{2+} + 2e^- \rightleftharpoons Zn(s)}$$

is 0.76 v less than the tendency for H^+ ions to be reduced to H_2 in the half-cell.

$$\mathbf{2H^+ + 2e^- \rightleftharpoons H_2(g)}$$

Since the electron-attracting ability of a hydrogen half-cell is represented by a potential of zero volts, that of the zinc half-cell would be −0.76 v.

The electron-attracting ability of a silver half-cell is 0.80 v more than that of a hydrogen half-cell, and the electron-attracting ability of a zinc half-cell is 0.76 v less than that of a hydrogen half-cell. Therefore, the electron attracting-ability of a silver half-cell must be 1.56 v more than that of a zinc half-cell. This relationship is represented in Table 17-5. The cell is illustrated in Fig. 17-4.

TABLE 17-5
HALF-CELL POTENTIALS

Half-cell	*Relative Electron Attracting Ability*
Silver	0.80 v
Hydrogen	0.00 v
Zinc	−0.76 v

17-10 The Zinc-Silver Cell When a standard silver and a standard zinc half-cell are coupled, the voltmeter reads 1.56 v (Fig. 17-4). This is the potential difference between the two half-cells in their standard states and is called the *standard cell potential*, E°_{cell}. The voltage represents the tendency for the cell reaction

$$\mathbf{2Ag^+ + Zn(s) \rightleftharpoons 2Ag(s) + Zn^{2+}} \qquad \text{17-23}$$

TABLE 17-5A
CELL POTENTIALS

Cell	*Anode*	*Cathode*	*Potential*
Hydrogen-silver	H_2	Ag	0.80 v
Zinc-hydrogen	Zn	H_2	0.76 v
Zinc-silver	Zn	Ag	1.56 v

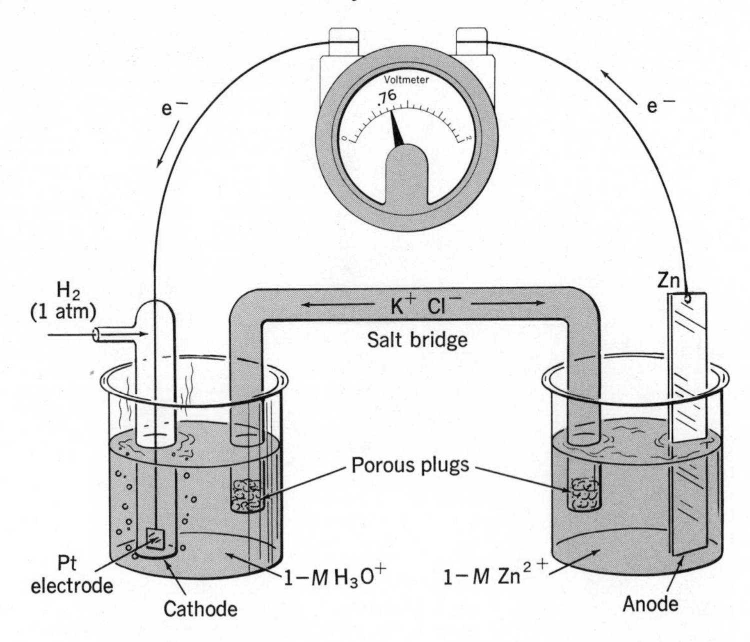

17-7 A zinc-hydrogen cell.

to take place when all participants are at standard state (25°C, 1 atm, 1-molar concentrations). The value of the ***standard potential*** can be thought of as a *measure of the driving force of the reaction.* The *positive sign of the cell voltage indicates that the reaction with which it is associated tends at standard-state conditions to proceed spontaneously as written.* A negative voltage indicates that the reverse reaction tends to proceed spontaneously at standard state. This does not mean that when a reaction has a negative voltage, no reaction occurs. It simply means that the products are not formed in their standard-state concentrations. We shall see later that the magnitude and sign of the voltage is related to the magnitude of the equilibrium constant. A negative voltage is associated with a small equilibrium constant.

17-11 Reduction Half-cell Reactions and Potentials We may use standard half-cell reactions and voltages to write equations for redox reactions and to calculate voltages associated with each reaction. Standard reduction potentials for a large number of half-cell reactions are given in Table 17-6.

Let us now examine, interpret, and apply the data given in Table 17-6. All cell-reactions in this table are written so that the *forward reaction (to the right) is a reduction.* Strong oxidizing agents are associated with reduction half-cell reactions having high reduction potentials. Thus, the strongest oxidizing agent, F_2, is at the top left. The poorest is the lithium ion, Li^+, at the bottom left. *The oxidizing agents are listed in decreasing order of oxidizing strength.*

The tendency for the reduction half-cell reaction to proceed to the right is indicated by the magnitude of the half-cell potential.

TABLE 17-6
STANDARD REDUCTION POTENTIALS IN VOLTS AT 25°C

Oxidizing Agents			*Reducing Agents*	E^0
$F_2(g)$	$+ 2e^-$	$\rightleftharpoons$	$2F^-$	+2.85
Co^{3+}	$+ e^-$	$\rightleftharpoons$	Co^{2+}	+1.82
Ce^{4+}	$+ e^-$	$\rightleftharpoons$	Ce^{3+}	+1.61
$BrO_3^- + 6H^+$	$+ 5e^-$	$\rightleftharpoons$	$\frac{1}{2}Br_2 + 3H_2O$	+1.52
$MnO_4^- + 8H^+$	$+ 5e^-$	$\rightleftharpoons$	$4H_2O + Mn^{2+}$	+1.52
Au^{3+}	$+ 3e^-$	$\rightleftharpoons$	Au	+1.50
$ClO_4^- + 8H^+$	$+ 8e^-$	$\rightleftharpoons$	$4H_2O + Cl^-$	+1.39
$Cl_2(g)$	$+ 2e^-$	$\rightleftharpoons$	$2Cl^-$	+1.36
$Cr_2O_7^{2-} + 14H^+$	$+ 6e^-$	$\rightleftharpoons$	$7H_2O + 2Cr^{3+}$	+1.33
$MnO_2(s) + 4H^+$	$+ 2e^-$	$\rightleftharpoons$	$2H_2O + Mn^{2+}$	+1.23
$O_2(g) + 4H^+$	$+ 4e^-$	$\rightleftharpoons$	$2H_2O(l)$	+1.23
$IO_3^- + 6H^+$	$+ 5e^-$	$\rightleftharpoons$	$\frac{1}{2}I_2 + 3H_2O$	+1.20
$SeO_4^{2-} + 4H^+$	$+ 2e^-$	$\rightleftharpoons$	$H_2SeO_3 + H_2O$	+1.15
$Br_2(l)$	$+ 2e^-$	$\rightleftharpoons$	$2Br^-$	+1.06
ICl_2^-	$+ e^-$	$\rightleftharpoons$	$\frac{1}{2}I_2 + 2Cl^-$	+1.06
$NO_3^- + 4H^+$	$+ 3e^-$	$\rightleftharpoons$	$2H_2O + NO(g)$	+0.96
$HO_2^- + H_2O(l)$	$+ 2e^-$	$\rightleftharpoons$	$3OH^-$	+0.88
$AuCl_4^-$	$+ 3e^-$	$\rightleftharpoons$	$Au + 4Cl^-$	+0.86
$Cu^{2+} + I^-$	$+ e^-$	$\rightleftharpoons$	CuI	+0.86
Hg^{2+}	$+ 2e^-$	$\rightleftharpoons$	$Hg(l)$	+0.85
$\frac{1}{2}O_2(g) + 2H^+(10^{-7}M)$	$+ 2e^-$	$\rightleftharpoons$	$2H_2O$	+0.82
Ag^+	$+ e^-$	$\rightleftharpoons$	$Ag(s)$	+0.80
Fe^{3+}	$+ e^-$	$\rightleftharpoons$	Fe^{2+}	+0.77
$O_2 + 2H_3O^+$	$+ 2e^-$	$\rightleftharpoons$	$H_2O_2 + 2H_2O$	+0.68
$MnO_4^- + 2H_2O$	$+ 3e^-$	$\rightleftharpoons$	$4OH^- + MnO_2(s)$	+0.59
$H_3AsO_4 + 2H^+$	$+ 2e^-$	$\rightleftharpoons$	$H_3AsO_3 + H_2O$	+0.56
$I_2(s)$	$+ 2e^-$	$\rightleftharpoons$	$2I^-$	+0.54
I_3^-	$+ 2e^-$	$\rightleftharpoons$	$3I^-$	+0.54
Cu^+	$+ e^-$	$\rightleftharpoons$	$Cu(s)$	+0.52
$Ag(NH_3)_2^+$	$+ e^-$	$\rightleftharpoons$	$2NH_3 + Ag$	+0.37
Cu^{2+}	$+ 2e^-$	$\rightleftharpoons$	$Cu(s)$	+0.34
$Hg_2Cl_2(s)$	$+ 2e^-$	$\rightleftharpoons$	$2Hg(l) + 2Cl^-$	+0.28
$IO_3^- + 3H_2O$	$+ 6e^-$	$\rightleftharpoons$	$I^- + 6OH^-$	+0.26
$AgCl(s)$	$+ e^-$	$\rightleftharpoons$	$Ag(s) + Cl^-$	+0.22
$SO_4^{2-} + 4H^+$	$+ 2e^-$	$\rightleftharpoons$	$H_2SO_3 + H_2O$	+0.17
Cu^{2+}	$+ e^-$	$\rightleftharpoons$	Cu^+	+0.15
Sn^{4+}	$+ 2e^-$	$\rightleftharpoons$	Sn^{2+}	+0.15
$S + 2H^+$	$+ 2e^-$	$\rightleftharpoons$	H_2S	+0.14
$S_4O_6^{2-}$	$+ 2e^-$	$\rightleftharpoons$	$2S_2O_3^{2-}$	+0.08
$AgBr(s)$	$+ e^-$	$\rightleftharpoons$	$Ag(s) + Br^-$	+0.071
$Ag(S_2O_3)_2^{3-}$	$+ e^-$	$\rightleftharpoons$	$Ag + 2S_2O_3^{2-}$	+0.01
$2H^+$	$+ 2e^-$	$\rightleftharpoons$	$H_2(g)$	0.00
Hg_2I_2	$+ 2e^-$	$\rightleftharpoons$	$2Hg + 2I^-$	−0.04
HgI_4^{2-}	$+ 2e^-$	$\rightleftharpoons$	$Hg + 4I^-$	−0.04
Pb^{2+}	$+ 2e^-$	$\rightleftharpoons$	$Pb(s)$	−0.13
Sn^{2+}	$+ 2e^-$	$\rightleftharpoons$	$Sn(s)$	−0.14
$AgI(s)$	$+ e^-$	$\rightleftharpoons$	$Ag(s) + I^-$	−0.15
Ni^{2+}	$+ 2e^-$	$\rightleftharpoons$	$Ni(s)$	−0.24
$H_3PO_4 + 2H^+$	$+ 2e^-$	$\rightleftharpoons$	$H_3PO_3 + H_2O$	−0.27
$PbCl_2$	$+ 2e^-$	$\rightleftharpoons$	$Pb + 2Cl^-$	−0.27
Co^{2+}	$+ 2e^-$	$\rightleftharpoons$	$Co(s)$	−0.28

Increasing oxidizing ability ↑ (from $AgBr(s)$ up to $IO_3^- + 6H^+$)

Increasing reducing ability ↓ (from $3H_2O$ down to Br^-)

TABLE 17-6 CONTINUED

Oxidizing Agents			*Reducing Agents*	E^0
$PbBr_2(s)$	$+ 2e^- \rightleftharpoons$	$Pb +$	$2Br^-$	−0.28
$Ag(CN)_2^-$	$+ 2e^- \rightleftharpoons$	$Ag +$	$2CN^-$	−0.31
Tl^+	$+ e^- \rightleftharpoons$		Tl	−0.34
$Hg(CN)_4^{2-}$	$+ 2e^- \rightleftharpoons$	$Hg +$	$4CN^-$	−0.37
$PbI_2(s)$	$+ 2e^- \rightleftharpoons$	$Pb +$	$2I^-$	−0.37
Cd^{2+}	$+ 2e^- \rightleftharpoons$		$Cd(s)$	−0.40
$2H_2O$	$+ 2e^- \rightleftharpoons$		$H_2(g) + 2OH^-(10^{-7}M)$	−0.41
$PbSO_4(s)$	$+ 2e^- \rightleftharpoons$	$Pb(s) +$	SO_4^{2-}	−0.41
Fe^{2+}	$+ 2e^- \rightleftharpoons$		$Fe(s)$	−0.44
$Cd(NH_3)_4^{2+}$	$+ 2e^- \rightleftharpoons$	$Cd +$	$4NH_3$	−0.60
$SO_4^{2-} + 4H_2O$	$+ 8e^- \rightleftharpoons$	$8OH^- +$	S^{2-}	−0.68
$Ag_2S(s)$	$+ 2e^- \rightleftharpoons$	$2Ag +$	S^{2-}	−0.69
$HgS(s)$	$+ 2e^- \rightleftharpoons$	$Hg +$	S^{2-}	−0.72
Cr^{3+}	$+ 3e^- \rightleftharpoons$		Cr	−0.74
Zn^{2+}	$+ 2e^- \rightleftharpoons$		$Zn(s)$	−0.76
$Fe(OH)_2(s)$	$+ 2e^- \rightleftharpoons$	$Fe(s) +$	$2OH^-$	−0.88
$SO_4^{2-} + H_2O$	$+ 2e^- \rightleftharpoons$	$SO_3^{2-} +$	$2OH^-$	−0.93
$PbS(s)$	$+ 2e^- \rightleftharpoons$	$Pb +$	S^{2-}	−0.95
$CNO^- + H_2O$	$+ 2e^- \rightleftharpoons$	$CN^- +$	$2OH^-$	−0.97
$Zn(NH_3)_4^{2+}$	$+ 2e^- \rightleftharpoons$	$Zn +$	$4NH_3$	−1.01
Mn^{2+}	$+ 2e^- \rightleftharpoons$		Mn	−1.18
$CdS(s)$	$+ 2e^- \rightleftharpoons$	$Cd +$	S^{2-}	−1.21
$Zn(CN)_4^{2-}$	$+ 2e^- \rightleftharpoons$	$Zn +$	$4CN^-$	−1.26
ZnS	$+ 2e^- \rightleftharpoons$	$Zn +$	S^{2-}	−1.44
Al^{3+}	$+ 3e^- \rightleftharpoons$		$Al(s)$	−1.66
Mg^{2+}	$+ 2e^- \rightleftharpoons$		$Mg(s)$	−2.34
$Mg(OH)_2(s)$	$+ 2e^- \rightleftharpoons$	$Mg +$	$2OH^-$	−2.69
Na^+	$+ e^- \rightleftharpoons$		$Na(s)$	−2.71
K^+	$+ e^- \rightleftharpoons$		$K(s)$	−2.92
Li^+	$+ e^- \rightleftharpoons$		$Li(s)$	−3.04

↑ Increasing oxidizing ability

Increasing reducing ability ↓

Most of the E^0 values in this table are taken from *The Oxidation States of the Elements and Their Potentials in Aqueous Solutions,* 2 ed, Wendell M. Latimer, Prentice Hall, Inc., 1952.

A positive potential for a reduction half-cell reaction means that the oxidizing agent is a stronger oxidizing agent than hydrogen ions, $H^+(aq)$. A negative potential means that the oxidizing agent is a weaker oxidizing agent than hydrogen ions. Thus, Ag^+ ions are a stronger oxidizing agent than H^+ ions, but Zn^{2+} ions are a weaker oxidizing agent than H^+ ions.

It should be noted that there are various ways to list the reduction half-cell reactions and potentials. In the *Handbook of Chemistry and Physics,* they are listed alphabetically. In some texts the reactions with the largest negative reduction potential are at the top of the list. Regardless of the order, *the oxidized species of the half-cell reaction having the highest positive reduction potential is the best oxidizing agent.* It can, therefore, oxidize the reduced species of any half-cell reaction having a lower positive reduction potential. Conversely, *the reduced species of the half-cell reaction having the most negative reduction potential is the best reducing agent.* Consider this alphabetical arrangement of half-cell reduction reactions.

$$\mathbf{Al^{3+} + 3e^- \longrightarrow Al(s) \quad -1.66\ v}$$
$$\mathbf{Cl_2(g) + 2e^- \longrightarrow 2Cl^- \quad +1.36\ v}$$

Applying the rule noted above, we can say that in the standard state chlorine (Cl_2) oxidizes Al(s) and forms Al^{3+}. The chlorine at the same time is reduced to chloride ions (Cl^-). Of the four species above, Al(s) is the best reducing agent and $Cl_2(g)$ is the best oxidizing agent.

17-12 Oxidation Half-cell Reactions and Potentials Note that the reduction half-cell reactions in Table 17-6 may be converted to oxidation half-cell reactions by reversing the species on the reactant and product side. Reversing

$$\mathbf{F_2(g) + 2e^- \longrightarrow 2F^- \qquad E° = 2.85\ v} \qquad 17\text{-}24$$

yields

$$\mathbf{2F^- \longrightarrow F_2(g) + 2e^- \qquad E° = -2.85\ v} \qquad 17\text{-}25$$

Reversing Equation 17-24 requires that the sign of the standard potential be changed. The negative value of E°, the standard oxidation potential for Equation 17-25, indicates there is relatively little tendency for F^- ions to give up electrons and become elemental fluorine. In other words, fluoride ions (F^-) are a poor reducing agent. They are the poorest reducing agent in Table 17-6. The strongest reducing agent in Table 17-6 is metallic lithium Li(s). The magnitude of the voltage for the reaction

$$\mathbf{Li(s) \rightleftharpoons Li^+ + e^- \qquad E° = 3.04\ v} \qquad 17\text{-}26$$

indicates a strong tendency for Li(s) to be oxidized. Hence, lithium is a powerful reducing agent. It can be seen that *the strength of the reducing agents* in Table 17-6 increases as you go down.

Equations 17-25 and 17-26 are written as oxidations. The table in some texts lists half-cell reactions as oxidations, so that the strongest reducing agents are at the upper left of the table. In these tables, the relative strengths of the reducing agents are indicated by ***oxidation potentials*** which are numerically equal to the reduction potentials in Table 17-6 but have opposite signs. A few oxidation half-cell reactions and their oxidation potentials are listed in Table 17-7. In Table 17-6, a fluorine half-cell is at the top with a high positive potential while in Table 17-7 it is at the bottom with a large negative potential. The important point is that *a positive potential for an oxidation half-cell reaction means the reducing agent is a stronger reducing agent than* $H_2(g)$.

STANDARD-STATE CELL POTENTIALS

17-13 The Calculation and Significance of Standard-state Cell Potentials A net equation for an oxidation-reduction reaction can be obtained by combining any two half-cell reactions in Table 17-6

TABLE 17-7
STANDARD OXIDATION POTENTIALS

Reducing Agents		Oxidizing Agents		E^0
Li(s)	$\rightleftharpoons$	Li^+	$+ e^-$	3.04
Zn(s)	$\rightleftharpoons$	Zn^{2+}	$+ 2e^-$	0.76
Fe(s)	$\rightleftharpoons$	Fe^{2+}	$+ 2e^-$	0.44
$H_2(g)$	$\rightleftharpoons$	$2H^+$	$+ 2e^-$	0.00
Cu(s)	$\rightleftharpoons$	Cu^{2+}	$+ 2e^-$	−0.34
Ag(s)	$\rightleftharpoons$	Ag^+	$+ e^-$	−0.80
$2Cl^-$	$\rightleftharpoons$	$Cl_2(g)$	$+ 2e^-$	−1.36
$2H_2O(l)$	$\rightleftharpoons$	$O_2(g) + 4H^+$	$+ 4e^-$	−1.23
$Mn^{2+} + 4H_2O$	$\rightleftharpoons$	$MnO_4^- + 8H^+$	$+ 5e^-$	−1.52
$2F^-$	$\rightleftharpoons$	$F_2(g)$	$+ 2e^-$	−2,85

Increasing reducing ability ↑ (upward); Increasing oxidizing ability ↓ (downward)

or Table 17-7 so that the electrons algebraically cancel. One method is to write *one half-cell reaction as a reduction and the other as an oxidation, and then add the two.* Because all reactions in Table 17-6 are written as reductions, one of them must be reversed before adding. Generally, the half-cell reaction with the lower reduction potential is reversed and then added to the one with the higher reduction potential. This is because the one with the higher reduction potential has the greater tendency to take place and because addition then yields a spontaneous reaction (one with a positive voltage).

Consider these half-cell reactions taken from Table 17-6.

$$\mathbf{Ag^+ + e^- \rightleftharpoons Ag(s) \qquad E^\circ = 0.80\ v} \qquad 17\text{-}27$$

$$\mathbf{Cu^{2+} + 2e^- \rightleftharpoons Cu(s) \qquad E^\circ = 0.34\ v} \qquad 17\text{-}28$$

Before adding, Equation 17-27 must be multiplied by 2 in order to balance the electrons, and Equation 17-28 must be reversed to represent an oxidation. The two new equations become

$$\text{Cathode,} \quad \mathbf{2Ag^+ + 2e^- \rightleftharpoons 2Ag(s) \qquad E^\circ = 0.80\ v} \qquad 17\text{-}29$$

$$\text{Anode,} \quad \mathbf{Cu(s) \rightleftharpoons Cu^{2+} + 2e^- \qquad E^\circ = -0.34\ v} \qquad 17\text{-}30$$

Note that we did not multiply the potential of the silver half-cell reaction by two. This is not necessary since the potential is not a function of the number of electrons. Adding Equations 17-29 and 17-30 yields

Cell reaction,

$$\mathbf{2Ag^+ + Cu(s) \rightleftharpoons Cu^{2+} + 2Ag^\circ(s) \qquad E^\circ_{cell} = 0.46\ v} \qquad 17\text{-}31$$

The *positive sign of the cell voltage indicates that the reaction tends to proceed spontaneously at standard state conditions.* When standard state half-cell reactions are taken from a table of reduction potentials, the cathode reaction at standard state will always be the one having the higher reduction potential.

Note that the same reaction and cell voltage would have been obtained using the oxidation half-cell reactions and potentials in Table 17-7 where the copper and silver half-cell reactions are

$$\mathbf{Cu(s) \rightleftharpoons Cu^{2+} + 2e^- \qquad E° = -0.34\ v}$$

$$\mathbf{Ag(s) \rightleftharpoons Ag^+ + e^- \qquad E° = -0.80\ v}$$

Before adding, we must reverse the lower equation and multiply it by 2. The two equations then become

Anode, $\mathbf{Cu(s) \rightleftharpoons Cu^{2+} + 2e^- \qquad E° = -0.34\ v}$

Cathode, $\mathbf{2Ag^+ + 2e^- \rightleftharpoons 2Ag(s) \qquad E° = +0.80\ v}$

Addition yields

Cell reaction

$$\mathbf{2Ag^+ + Cu(s) \rightleftharpoons Cu^{2+} + 2Ag(s) \qquad E°_{cell} = 0.46\ v}$$

When standard state half-cell reactions are taken from a table of oxidation potentials, then the anode reaction at standard state is the half-cell reaction having the higher oxidation potential. Regardless of which table is used, the anode and cathode can be identified by applying the rule that *oxidation occurs at the anode and reduction at the cathode.* Once the electrodes have been identified, the direction of electron flow and ion migration can be determined.

An electrochemical cell is often represented schematically by using single lines to separate an electrode from a solution and a double line to represent the porous partition or salt bridge which separates two solutions. Using this notation, the copper-silver cell discussed above would be represented as

$$\mathbf{Cu(s)\,|\,Cu^{2+}(1\mathit{M})\,||\,Ag^+(1\mathit{M})\,|\,Ag(s)}$$

By convention, the two couples are usually shown with anode on the left and cathode on the right.

The species listed in Table 17-6 are arranged so that *any oxidizing agent (on the left side) in the table tends to react spontaneously at standard state with any reducing agent (on the right side) which is below it in the table.* Thus, Br_2 oxidizes Cu, I^-, and Zn but not Cl^-. In this Table, the farther apart the oxidizing and reducing agents are, the greater will be the tendency for the reaction to take place. In general, the greater the distance between the agents, the greater is the difference in their electric potentials and the greater is the driving force of the reaction.

$E°$ values can be used to predict whether a reaction is probable but not whether it will actually take place at an appreciable rate. Thus, a cell voltage, $E°_{cell}$, can be used to predict whether or not a certain reaction is probable or spontaneous. It does not indicate, however, whether a reaction rate is fast enough for the reaction to be detected nor does it allow us always to predict which of several possible products are actually obtained.

Many oxidizing agents can be reduced in steps, yielding products having various oxidation states. For example, HNO_3 is a powerful oxidizing agent which can be reduced to NO_2, N_2O_3, NO, N_2O, N_2, or NH_4^+ ions. Reduction potentials for each of the above reductions are available in handbooks. The reduction of HNO_3 to N_2 has the highest potential difference while the reduction to NO_2 has the lowest value. Thus, it would seem that HNO_3 would have the greatest tendency to be reduced to N_2. In other words, N_2 is the most stable of the reduction products, and its formation from HNO_3 would represent the greatest decrease in energy. Actually, at room temperature, NO_2 is usually observed. This is because the reduction reaction which produces NO_2 has a lower activation energy and proceeds at a faster rate than the conversion to N_2. In general, the reaction with the fastest rate is the one that is observed. At higher temperatures, the yield of N_2 increases. Thus, temperature plays an important role in determining which product is formed.

FOLLOW-UP PROBLEMS

1. Which of these equations represent reactions that tend (a) to proceed spontaneously to the right (standard state conditions); (b) tend to proceed spontaneously to the left (standard state); (c) do not occur?

(a) $2Ag^+ + Cu(s) \rightleftharpoons Cu^{2+} + 2Ag(s)$
(b) $Zn(s) + Cu^{2+} \rightleftharpoons Cu(s) + Zn^{2+}$
(c) $I_2(s) + 2Cl^- \rightleftharpoons Cl_2(g) + 2I^-$
(d) $Fe(s) + Cl_2(g) \rightleftharpoons 2Cl^-(g) + Fe^{3+}$
(e) $MnO_4^- + 8H^+ + Cu^{2+} = Mn^{2+} + 4H_2O + Cu(s)$

Ans. (a), (b), and (d) to right, (c) to left, (e) no reaction.

2. What are the E°_{cell} values for these oxidation-reduction reactions? Which, if any, proceed spontaneously to the left?

(a) $5Fe^{2+} + MnO_4^- + 8H^+ \rightleftharpoons 5Fe^{3+} + Mn^{2+} + 4H_2O$
(b) $Sn^{2+} + Cl_2(g) \rightleftharpoons Sn^{4+} + 2Cl^-$
(c) $3Br_2(l) + 2Cr^{3+} + 7H_2O = Cr_2O_7^{2-} + 14H^+ + 6Br^-$

Ans. (a) 0.75 v, (b) 1.21 v, (c) −0.27 v (to left).

NONSTANDARD-STATE CELL POTENTIALS

17-14 Effect of Changing Concentration on Cell Potential You are probably aware that the voltage of a battery does not remain constant. After continued use, the voltage gradually decreases. Let us examine the changes that take place in the zinc-silver cell as it discharges. We can use Le Châtelier's principle to predict how these changes affect the voltage of the cell.

The equilibrium for the cell reaction

$$2Ag^+ + Zn(s) \rightleftharpoons 2Ag(s) + Zn^{2+}$$

shows that as the reaction proceeds, the concentration of Ag^+ ions decreases and the concentration of Zn^{2+} ions increases. According to Le Châtelier's principle, decreasing the concentration of the Ag^+ ions or increasing the concentration of the Zn^{2+} ions causes the equilibrium to shift to the left. Since a shift to the left opposes the cell reaction, the voltage decreases.

Originally, the electron attracting ability of the Ag-Ag^+ half-cell was much greater than that of the Zn-Zn^{2+} half-cell, and the electrons traveled from the zinc electrode to the silver electrode. As the concentration of the Ag^+ ions in the cathode compartment decreases, the electron-attracting ability of the cathode half-cell decreases. At the same time, the increasing concentration of the Zn^{2+} ions in the anode compartment causes an increase in the electron-attracting ability of the anode half-cell. When the electron-attracting abilities (potentials) of the two half-cells equalize, there is no further net reaction. The system is at equilibrium. Thus, *at equilibrium, the cell voltage is zero, and the ion concentrations are equilibrium concentrations.* A voltage of zero indicates that the cell is "dead."

Application of Le Châtelier's Principle indicates that increasing the concentration of the reactant in the cathode solution increases the cell voltage. This implies that changing concentrations may cause a change in the spontaneity of a reaction. That is, if concentration changes cause E°_{cell} to change from a positive to a negative value, then the reverse reaction becomes spontaneous.

17-15 The Nernst Equation In 1889, Walter Nernst (1864–1941), a German scientist, developed a mathematical relationship which enables us to calculate cell voltages and the direction of a spontaneous reaction at other than standard-state concentrations. For a general oxidation-reduction reaction

$$aA + bB \rightleftharpoons cC + dD$$

the ***Nernst equation,*** which has been experimentally verified, has the form

$$E_{cell} = E^\circ_{cell} - \frac{0.059}{n} \log \frac{[C]^c\,[D]^d}{[A]^a\,[B]^b} \qquad 17\text{-}32$$

where E°_{cell} is the standard-state cell voltage, n is the number of electrons exchanged in the equation for the reaction and 0.059 is a combination of constants at 298°K. The concentrations of ions in the logarithm term are expressed in terms of *molarity*. Concentrations of solids and liquid solvents are considered constant and are not included in the expression.

The value 0.059 is obtained by substituting numerical values in the expression $\frac{2.3\,RT}{nF}$, where *R* is the universal gas constant expressed in joules/mole K°, *T* is the Kelvin temperature (298°K), *F* is the charge on a mole of electrons (96 500 coulombs), *n* is the number of moles of electrons transferred, and 2.3 is the factor used to convert natural logs to base-10 logs. You can verify the value of 0.059 by performing the indicated operations.

Example 17-2

Calculate the cell voltage of

$$Zn(s)|Zn^{2+}(0.001\,M)||Ag^+(0.1\,M)|Ag(s)$$

Solution

1. Identify the cathode and anode half-cells and write the half-cell reactions. Then add the anode and cathode reactions to obtain the cell reaction and the standard-state cell potential.

Cathode	$2Ag^+ + 2e^- \longrightarrow 2Ag(s)$	$E^\circ = +0.80v$
Anode	$Zn(s) \longrightarrow Zn^{2+} + 2e^-$	$E^\circ = +0.76v$
Cell	$Zn(s) + 2Ag^+ \rightleftharpoons Zn^{2+} + 2Ag$	$E^\circ = 1.56v$

2. Substitute the known values in the Nernst equation and solve for E_{cell}.

$$E_{cell} = E^\circ_{cell} - \frac{0.059}{n} \log \frac{[Zn^{2+}]}{[Ag^+]^2}$$

$$E_{cell} = 1.56 - \frac{0.059}{2} \log \frac{[10^{-3}]}{[10^{-1}]^2}$$

$$= 1.56 - 0.03 \log 10^{-1}$$
$$= 1.56 + 0.03$$
$$= 1.59 \text{ v}$$

FOLLOW-UP PROBLEMS

1. Calculate the voltage for this cell.

$Zn(s)|Zn^{2+}(0.10\ M)||Cu^{2+}(0.010\ M)|Cu(s)$

Ans. 1.07 v.

2. (a) What is the cell voltage for this cell?

$Co(s)|Co^{2+}(0.1\ M)||Ni^{2+}(0.001\ M)|Ni(s)$

(b) In which direction does the reaction

$$Co(s) + Ni^{2+} \rightleftharpoons Co^{2+} + Ni(s)$$

with the above concentrations proceed spontaneously?
(c) In which direction does the reaction proceed spontaneously at standard-state concentrations?

Ans. (a) −0.02 v, (b) to left, (c) to right.

EQUILIBRIUM CONSTANTS FOR REDOX REACTIONS

17-16 Determination of K_{eq} from E°_{cell} Values We have suggested that both the standard-state voltage and the equilibrium constant for a reaction are measures of the tendency for a reaction to take place. It seems reasonable that the two factors should be related. To develop this relationship, recall that when the cell voltage is zero, the cell reaction is at equilibrium. Thus, the existing concentrations are equilibrium concentrations, and the logarithm term in Equation 17-32 becomes *log* K_e. At equilibrium, when the cell voltage is zero, Equation 17-32 can be written

$$0 = E^\circ_{cell} - \frac{0.059}{n} \log K_e \qquad 17\text{-}33$$

$$\log K_e = \frac{nE^\circ_{cell}}{0.059} \qquad 17\text{-}34$$

$$K = 10^{\frac{nE^\circ_{cell}}{0.059}} \qquad 17\text{-}35$$

Equation 17-35 reveals that the equilibrium constant for any oxidation-reduction reaction can be calculated using standard-state half-cell reduction potentials. The same principles apply to redox reactions which involve a direct transfer of electrons as to those which occur in electrochemical cells. Equation 17-35, shows that a reaction with a positive $E°$ has a $K_e > 1$ while a reaction with a negative $E°$ has a $K_e < 1$. These examples illustrate the calculation and use of equilibrium constants and cell potentials.

Example 17-3

Calculate the equilibrium constant for the reaction between silver nitrate and metallic zinc. Is the reaction essentially complete or incomplete?

Solution

1. Write the equation for the reaction

$$2Ag^+(aq) + Zn(s) \rightleftharpoons Zn^{2+}(aq) + 2Ag(s) \qquad E^\circ_{cell} = 1.56 \text{ v}$$

2. Substitute values in Equations 17-33, 17-34 or 17-35

$$0 = 1.56 - 0.03 \log K_e$$

$$-1.56 = -0.03 \log K_e$$

$$\log K_e = \frac{-1.56}{-0.03} = 52$$

$$K_e = 1 \times 10^{52}$$

$$10^{52} = \frac{[Zn^{2+}]}{[Ag^+]^2}$$

The large value of K_e indicates that the reaction is essentially complete.

FOLLOW-UP PROBLEM

Calculate the equilibrium constant for the reaction between metallic zinc and copper(II) sulfate. Is the reaction complete?

Ans. 5×10^{36}, reaction is complete.

Many equilibrium mixtures contain extremely large and extremely small concentrations of ions. Chemical analysis of such a mixture would present a formidable problem. Thus, measurement of cell voltages provides a relatively simple way to measure ion concentration and equilibrium constants. The electrometric (potentiometric) method is widely applied in determining solubility-product and equilibrium constants in which the equilibrium concentrations of the ions are very small. Consider the solubility-product constant for AgCl. In a saturated solution of AgCl the ion concentrations are so small that accurate measurement by ordinary chemical analysis is difficult. However, by measuring the voltage of an appropriate cell, we can calculate the K_{sp} for AgCl. One standard half-cell of such a cell consists of a metal electrode in a saturated solution of a slightly soluble salt of the metal. The other half-cell could be a standard-state half-cell of the same metal-metallic ion couple, or any other reference half-cell whose accurate E° value is known. See Fig. 17-8 on the next page.

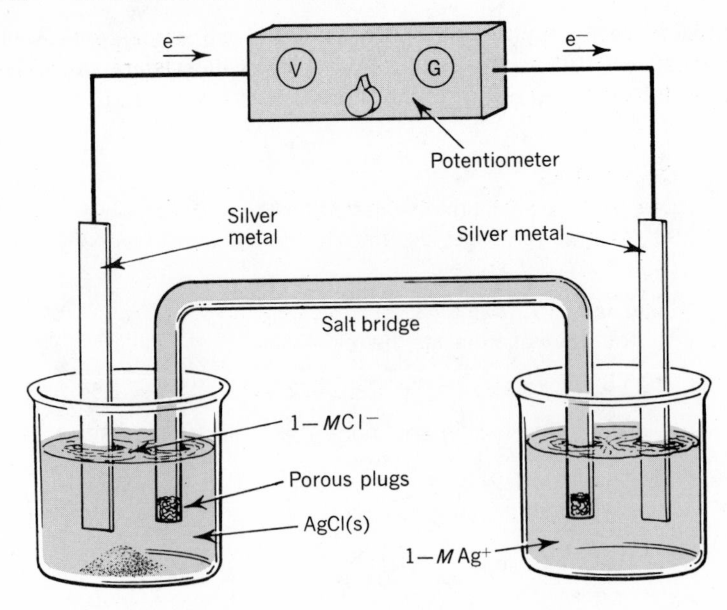

17-8 Electrochemical cell for determining solubility product (K_{sp}) for AgCl. In this cell, the anode consists of a piece of metallic silver immersed in a 1-*M* Cl^- ion solution containing excess solid AgCl. The cathode is a standard Ag(s) — Ag^+ half-cell. The schematic for the cell is

$$Ag(s) | AgCl(s) | Cl^-(1M) || Ag^+(1M) | Ag(s)$$

The anode half-cell potential at standard state is −0.22 v for the reaction.

$$Ag(s) + Cl^- \rightleftharpoons AgCl(s) + e^-$$

The standard state cathode potential is 0.80v for the reaction,

$Ag^+ + e^- \rightleftharpoons Ag(s)$. Thus the overall cell potential is 0.58 v for the reaction,

$$Ag^+ + Cl^- \rightleftharpoons AgCl(s).$$

At equilibrium, $E_{cell} = 0$ and

$$0 = 0.58\text{ v} - \frac{0.059}{1}\log\frac{1}{[Ag^+][Cl^-]}$$

$$\begin{aligned} -0.58 &= 0.059 \log [Ag^+][Cl^-] \\ -0.58 &= 0.059 \log K_{sp} \\ \log K_{sp} &= -9.8 \\ K_{sp} &= 10^{-9.8} \\ &= 10^{0.2} \times 10^{-10} \\ &= 1.6 \times 10^{-10} \end{aligned}$$

STANDARD STATE FREE ENERGY CHANGE

17-17 Calculation of Standard State Free Energy Change From Standard State Cell Potentials. In Chapter 13 we found that the change in free energy (ΔG) associated with a reaction could, like the standard-state potential for a reaction, be used as a criterion of reaction spontaneity. That is, a spontaneous reaction has a positive E_{cell} value and is associated with a decrease in free energy (ΔG is negative). These relationships suggest that cell potentials (E_{cell}) must be related to changes in the free energy (ΔG) of a reacting system. The relationship between standard state free energy change ($\Delta G°$) and standard cell potential ($E°_{cell}$) is given by

$$\Delta G° = -nFE°_{cell} \qquad 17\text{-}36$$

when n is the number of electrons transferred in the equation for the cell reaction, $E°$ is the standard-state cell potential expressed in volts, and F is the charge carried by a mole of electrons or 96 500 coulombs.

We may interpret Equation 17-36 by saying the ΔG is *numerically equal to the useful electric work which a chemical system can do on its surroundings.* Electric work is done when a charge is moved through a voltage difference. In the operation of an electrochemical cell, the voltage or potential difference between two half-cells drives the electrons through the circuit, where they perform electric work.

Equation 17-36 enables scientists to determine values for changes in standard-state free energy by measuring standard-cell potentials. In Chapter 13, we introduced an equation which related the change in free energy to enthalpy and entropy changes in a system. Since

E°_{cell} and ΔG° are related, we may assume that enthalpy and entropy changes also affect cell voltages. These effects are discussed in Sect. 18-11.

Example 17-4
Calculate ΔG° for the zinc-silver cell at standard-state conditions.

1. Obtain the required half-cell reduction potentials from Table 17-6 and determine E°_{cell} by using the relation

$$E^\circ_{cell} = E^\circ_{cathode} + E^\circ_{anode}$$
$$E^\circ = 0.80 + 0.76 = 1.56 \text{ v} = 1.56 \text{ joules/coul}$$

2. Substitute all known values in Equation 17-36 and solve for ΔG°.

$$\begin{aligned}\Delta G^\circ &= -nFE^\circ_{cell}\\ &= -2(96\,500 \text{ coul})(1.56 \text{ joules/coul})\\ &= -3.02 \times 10^5 \text{ joules}\\ &= 3.02 \times 10^5 \text{ joules} \times \frac{1 \text{ kcal}}{4180 \text{ joules}}\\ &= -72 \text{ kcal}\end{aligned}$$

FOLLOW-UP PROBLEM

Calculate ΔG° for the reaction between zinc and copper(II) using the data in Table 17-6. **Ans. −50.8 kcal.**

APPLICATIONS OF PRINCIPLES UNDERLYING THE OPERATION OF ELECTROCHEMICAL CELLS

There are many applications of electrochemical principles and voltaic cells. The research chemist may be primarily interested in using them to determine the spontaneity of reactions, the values for equilibrium constants, or the concentration of ions. Manufacturing chemists apply electrochemical principles when they use pH meters to monitor the acidity of solutions. The engineer is concerned with electrochemical principles and processes when he devises methods to prevent corrosion. The same principles are being used by scientists who are engaged in developing fuel cells that will more efficiently convert chemical energy into electric energy. All of you depend on electrochemical cells every time you drive your car or press the button on a flashlight. Let us briefly consider a few of these applications.

17-18 Storage Batteries One of the most familiar uses of electrochemical cells is in automobile storage batteries. The 12-v bat-

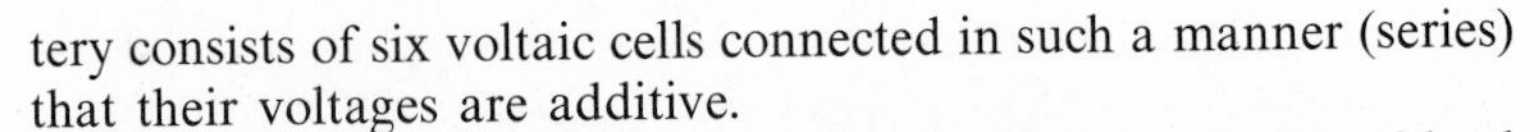

tery consists of six voltaic cells connected in such a manner (series) that their voltages are additive.

The anodes are made of spongy lead and the cathodes of lead dioxide (PbO_2). The electrodes are in a solution of sulfuric acid. When the cells are producing electricity, the reactions are

Anode,

$$\mathbf{Pb(s) + HSO_4^- \rightleftharpoons PbSO_4(s) + H^+ + 2e^-} \quad 17\text{-}37$$

Cathode,

$$\mathbf{PbO_2(s) + HSO_4^- + 3H^+ + 2e^- \rightleftharpoons PbSO_4(s) + 2H_2O} \quad 17\text{-}38$$

Addition of Equations 17-37 and 17-38 gives the overall cell reaction.

$$\mathbf{Pb(s) + PbO_2(s) + 2H^+ + 2HSO_4^- \rightleftharpoons 2PbSO_4(s) + 2H_2O} \quad 17\text{-}39$$

Equation 17-39 shows that HSO_4^- ions from the electrolyte react with the Pb^{2+} ions formed by the oxidation of the Pb anode and the reduction of the PbO_2 cathode. The lead sulfate deposits on both electrodes.

During the discharge of the cell, the concentration of H_2SO_4 decreases as sulfate ions precipitate out and as the H^+ ions form water. As the concentration of the sulfuric acid decreases, the density of the solution decreases. Thus, the condition of the battery can be easily checked by measuring the density of the sulfuric acid solution with a hydrometer. A low density (specific gravity) indicates a partially discharged cell. The sulfuric acid solution in a fully charged battery has a density of 1.250 g/ml to 1.300 g/ml. The acid solution in a discharged battery has a density of 1.100 g/ml to 1.150 g/ml.

The battery can be recharged by connecting it to an external source of direct current and reversing the flow of electrons. This reverses the reactions, forming lead on the anode and lead dioxide on the cathode, and increases the concentration of the sulfuric acid. The charging reaction is

$$\mathbf{2PbSO_4(s) + 2H_2O \rightleftharpoons Pb(s) + PbO_2(s) + 2H^+ + 2HSO_4^-}$$

17-19 Dry Cells and Mercury Cells Another type of cell used frequently is the ordinary dry cell used to supply electric energy for flashlights and portable radios.

The outside terminal is attached to a zinc container which acts as the anode. The center terminal is connected to a rod composed of carbon and manganese dioxide which acts as the cathode. The zinc container is filled with a moist paste composed of zinc chloride, ammonium chloride, water and an inert filler such as asbestos. A porous liner prevents the zinc from coming into direct contact with the paste and serves as a salt bridge. A number of reactions may occur when the cell is operating. The operating voltage for a dry cell is usually between 0.9 and 1.4 volts. When only a small current

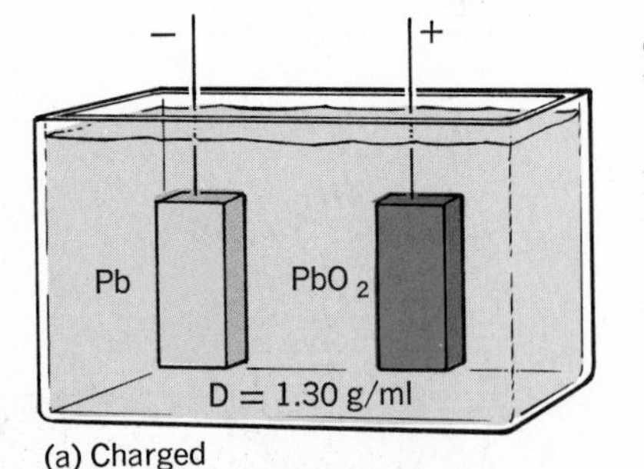

(a) Charged

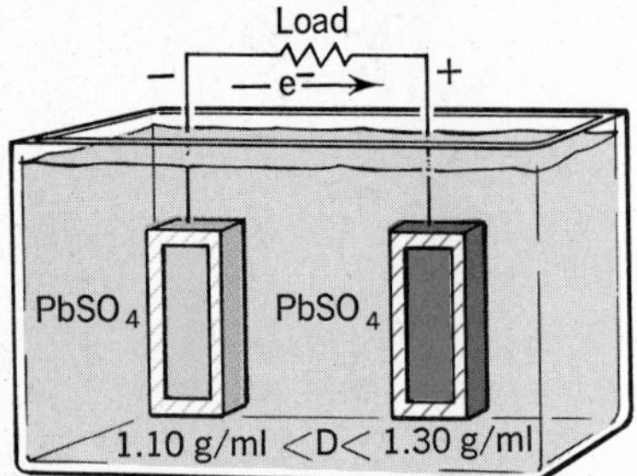
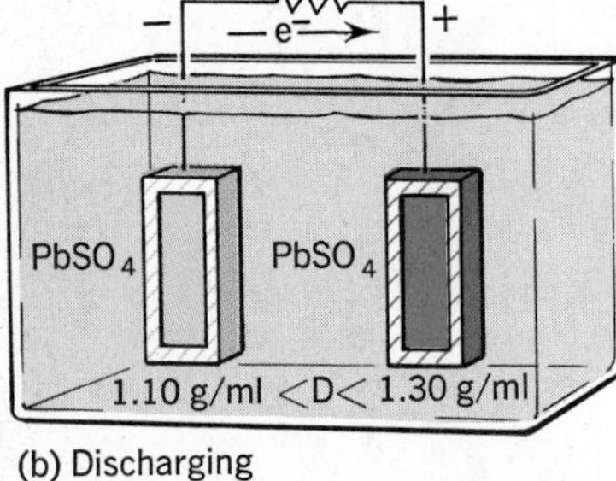

(b) Discharging

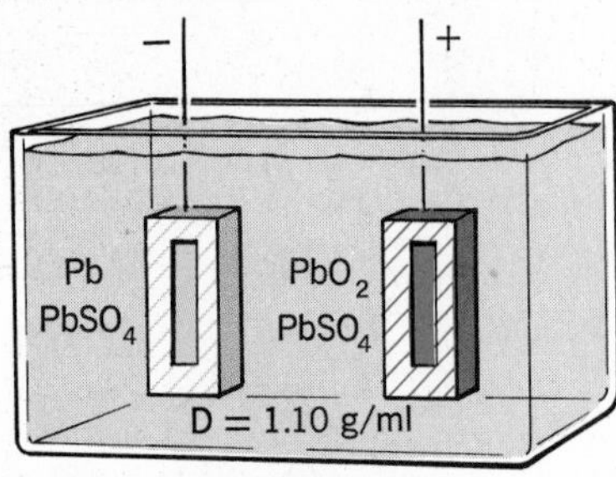

(c) Discharged

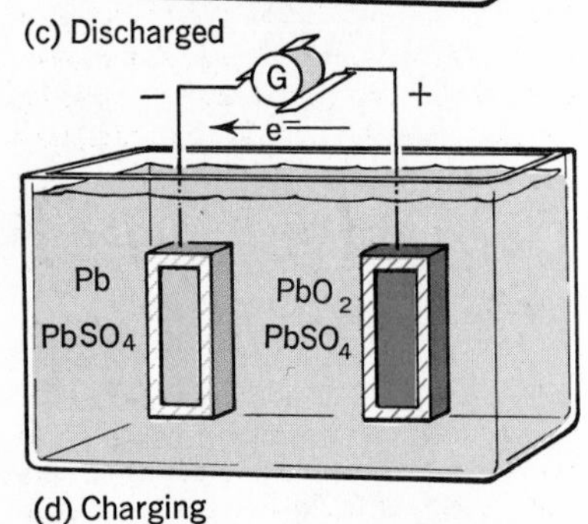

(d) Charging

17-9 Changes occurring during the operation of a storage cell. (a) A fully charged cell, (b) a discharging cell, (c) a fully discharged cell, (d) a storage cell being charged by the use of an external source of direct current.

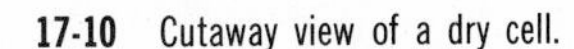

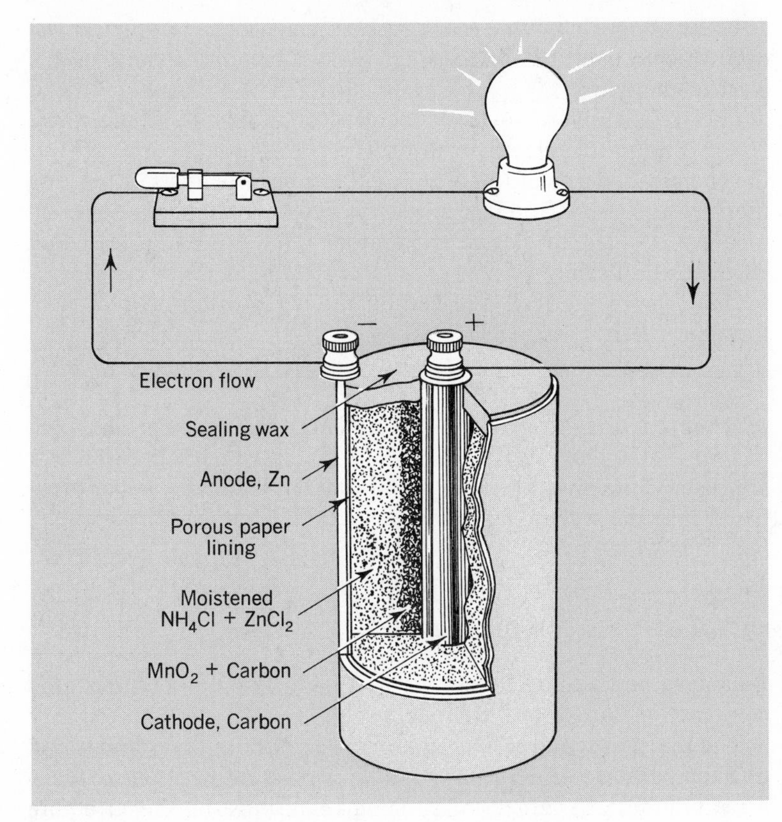

17-10 Cutaway view of a dry cell.

is flowing, the reactions are believed to be these:

Cathode,

$$2MnO_2 + 2NH_4^+ + 2e^- \longrightarrow Mn_2O_3 + 2NH_3 + H_2O$$

Anode,

$$Zn + 4NH_3 \longrightarrow Zn(NH_3)_4^{2+} + 2e^-$$

Another cell frequently used in miniature batteries to supply energy for transistor radios, movie cameras, hearing aids, and compact devices is the mercury cell. This cell is extremely important to the thousands of people who depend on electronic devices called *pacemakers* to keep their hearts beating with the proper rhythm. Most of these pacemakers, which are about the size of a cigarette package and usually implanted in the chest or abdominal wall, are powered by a set of mercury batteries. In a mercury cell, the anode is amalgamated zinc, Zn(Hg), the cathode is mercury(II) oxide and the electrolyte is KOH – ZnO. The operating voltage for a mercury cell is 1.30 volts and is derived from the reactions given below.

Anode,

$$Zn + 2OH^- \longrightarrow ZnO + H_2O + 2e^-$$

Cathode,

$$HgO + H_2O + 2e^- \longrightarrow Hg + 2OH^-$$

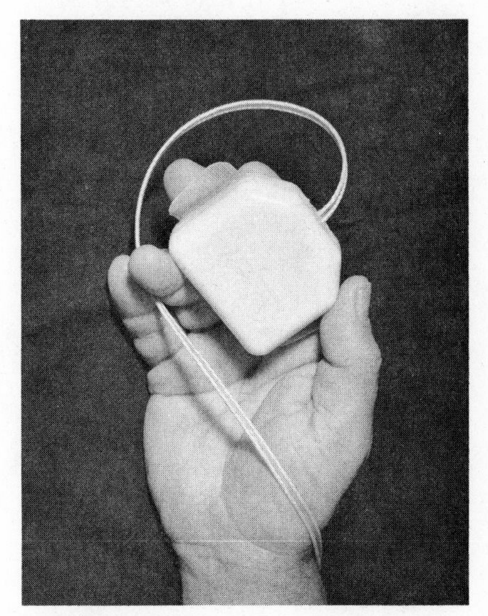

17-11 The heart has a natural pacemaker which generates electric impulses at an average rate of about 70 per minute. When injury, disease, or birth defects impair the operation of the natural pacemaker, an artificial pacemaker such as that shown above may be implanted in a patient's body. The mercury batteries in a pacemaker must be replaced about every two years. Research scientists are currently working on atomic-powered pacemakers that could work effectively for 20 years. (Huberland, N.Y.)

An amalgam is a mercury alloy. As a relevant sidelight, it is interesting to note that silver and gold are easily amalgamated, and once alloyed with mercury, cannot readily be restored to their original condition. This suggests that jewelry should be removed when working with mercury.

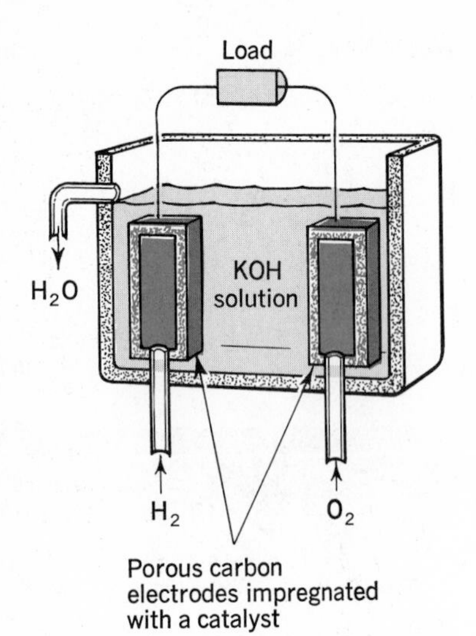

17-12 Schematic representation of a hydrogen-oxygen fuel cell.

17-20 Fuel Cells In recent years, scientists have designed *fuel cells* which convert the chemical energy of a fuel directly into usable electric energy.

Fuel cells convert about 75 percent of the available chemical energy into usable electric energy. The conventional conversion of the chemical energy of fuels into electric energy consists of burning the fuel and using the heat energy to produce steam for spinning turbines coupled to electric generators. This process is less than 40 percent efficient.

One type of fuel cell is based on the combustion of hydrogen, forming water.

$$2H_2(g) + O_2(g) \rightleftharpoons 2H_2O(l) \qquad 17\text{-}40$$

A schematic diagram of the experimental arrangement is shown in Fig. 17-12. The porous chambers through which the gases flow are placed in a solution of KOH or NaOH. Catalysts are impregnated into the walls of the porous carbon chamber. The half-cell reactions which occur at the electrodes are

Anode $$H_2(g) + 2OH^- \rightleftharpoons 2H_2O + 2e^- \qquad 17\text{-}41$$

Cathode $$O_2(g) + 2H_2O + 4e^- \rightleftharpoons 4OH^- \qquad 17\text{-}42$$

Balancing and adding Equations 17-41 and 17-42 gives the overall cell reaction shown by Equation 17-40.

The reaction requires a continuous flow of both gases and is run at a temperature of about 250°C and a pressure of 50 atm. There are a number of engineering problems that must be solved before fuel cells become practical sources of electric energy. Many of these problems will no doubt be overcome. Fuel cells have already played an important role in some of the manned space flights.

17-21 Corrosion The presence of impurities, oxygen, and moisture is responsible for the corrosion of certain metals. Corrosion of metals such as iron is a problem which confronts us all. Any iron object, from a garden hoe to the largest bridge, is subject to corrosion.

Experimental evidence indicates that the corrosion of iron is an electrochemical phenomenon. Very pure samples of iron seem to resist corrosion. In contrast, when a piece of iron containing specks of impurities such as copper is exposed to moist air, the iron becomes pitted by rust spots. Rust is a reddish-brown hydrated compound of varying composition having the formula $Fe(OH)_3 \cdot xH_2O$. Therefore, rust spots indicate the location at which pure iron has been oxidized. *These rust spots represent the anode half-cells of electrochemical cells.*

Such impurities as copper serve as the cathode half-cells and are located in unaffected areas adjacent to the rust spots. The electrons travel through the iron from the anode to the cathode. A film of moisture serves as the medium through which ions travel to complete the circuit.

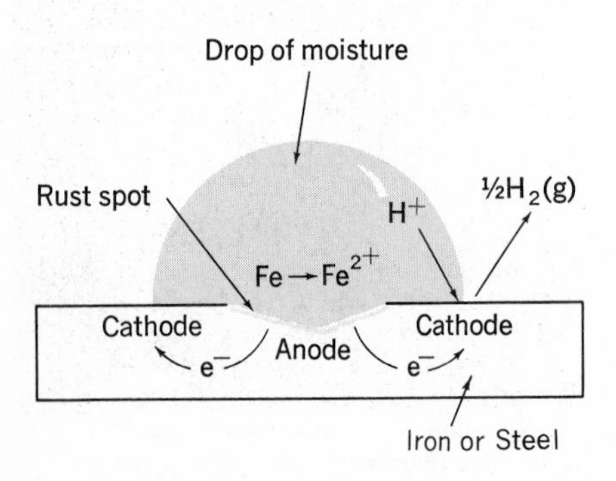

17-13 Corrosion of impure iron in contact with moisture.

The electrochemical theory of corrosion is supported by the observation that iron exposed to perfectly dry air does not corrode. Three half-cell reactions which would represent possible reactions in this complex mechanism are

Anode $\quad Fe(s) \rightleftharpoons Fe^{2+} + 2e^-$ $\quad$ 17-43

Cathode
(neutral solution) $\frac{1}{2}O_2(g) + H_2O(l) + 2e^- \rightleftharpoons 2OH^-$ $\quad$ 17-44

Cathode
(acid solution) $\quad 2H^+ + 2e^- \rightleftharpoons H_2(g)$ $\quad$ 17-45

The overall cell reaction can be obtained by adding Equations 17-43 and 17-44 or 17-43 and 17-45. The former addition gives

$$Fe(s) + \tfrac{1}{2}O_2(g) + H_2O(l) \rightleftharpoons Fe(OH)_2(s)$$

Further oxidation of $Fe(OH)_2$ by oxygen yields $Fe(OH)_3$. Examination of Equations 17-43, 17-44, and 17-45 shows that some of the factors involved in corrosion are the *concentration of oxygen, the acidity, and the presence of moisture.*

To minimize corrosion, protective coatings are applied to prevent the direct contact of moisture and oxygen with the metal. Electrochemical principles can also be applied to inhibit corrosion. This is known as *cathodic protection.* In cathodic protection, iron is made the cathode half-cell so that it will not lose electrons and dissolve. This can be accomplished by attaching a more active metal such as magnesium to the iron or by connecting the iron to the cathode of an external power source. These devices are frequently used to protect underground pipes and tanks and even the hulls of ships. Steel hulls of ships are especially vulnerable to corrosion because the bronze propeller acts as a cathode, salt water is a salt bridge, and the hull as an anode. This problem has been solved by attaching large pieces of magnesium metal to the hull. The magnesium metal, being more easily oxidized than iron, acts as an anode, making the steel hull the cathode. Thus, the magnesium metal is corroded and dissolved, but the steel hull remains unaffected. Underground pipes given cathodic protection remain unaffected by corrosion for many years.

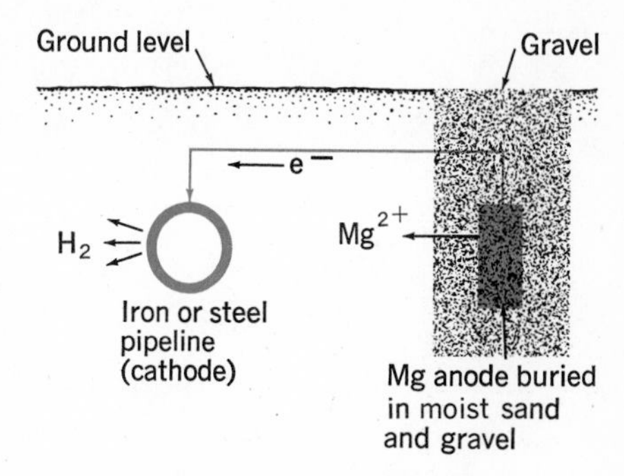

17-14 Underground iron pipes can be cathodically protected from corrosion by connecting the pipe to a more easily oxidized metal such as zinc or magnesium.

17-22 The pH Meter The development of reference electrodes which maintain a constant half-cell potential independent of solution concentration, and indicator electrodes which respond to changes in the concentration of specific ions, make it possible to use cell potentials to determine the concentration of a specific ion in a solution. The potential difference between the indicator and reference electrode as read on a meter can be related in a simple fashion to the concentration of the specific ion. One of the most familiar applications of this relationship is found in the use of pH meters to determine the $[H^+]$ of a solution. The potential of a cell changes when the concentration of a reaction participant is varied. When H^+ ions are a reactant, both the electrode voltage and the

17-15 Chunks of magnesium anode are connected to the hulls of steel ships to protect the hull cathodically from corrosion.

pH change when the H^+-ion concentration is varied. This implies that *the pH of a solution is related to the voltage of a cell in which H^+ ions are a reactant.*

The potential of the standard hydrogen electrode varies when $[H^+]$ is changed. When this half-cell is coupled to a reference half-cell that maintains a constant potential, the variation in the overall cell potential is a measure of the variation in the H^+ ion concentration of the solution. With the use of the Nernst equation, it can be shown that a change of one pH unit (10-fold $[H^+]$ change) causes a change in the overall cell potential of 0.059 v.

A pH meter is an electronic instrument that measures the potential difference between a reference half-cell and a half-cell whose potential changes when $[H^+]$ varies. The dial of the meter is often graduated in both pH and millivolt units. The hydrogen electrode is not a convenient half-cell to use for most laboratory pH measurements. Instead, pH meters are generally equipped with a standard reference electrode and a rugged, portable "glass" electrode whose half-cell potential is sensitive to changes in $[H^+]$. An experimental arrangement for determining the pH of a solution is shown in Fig. 17-16.

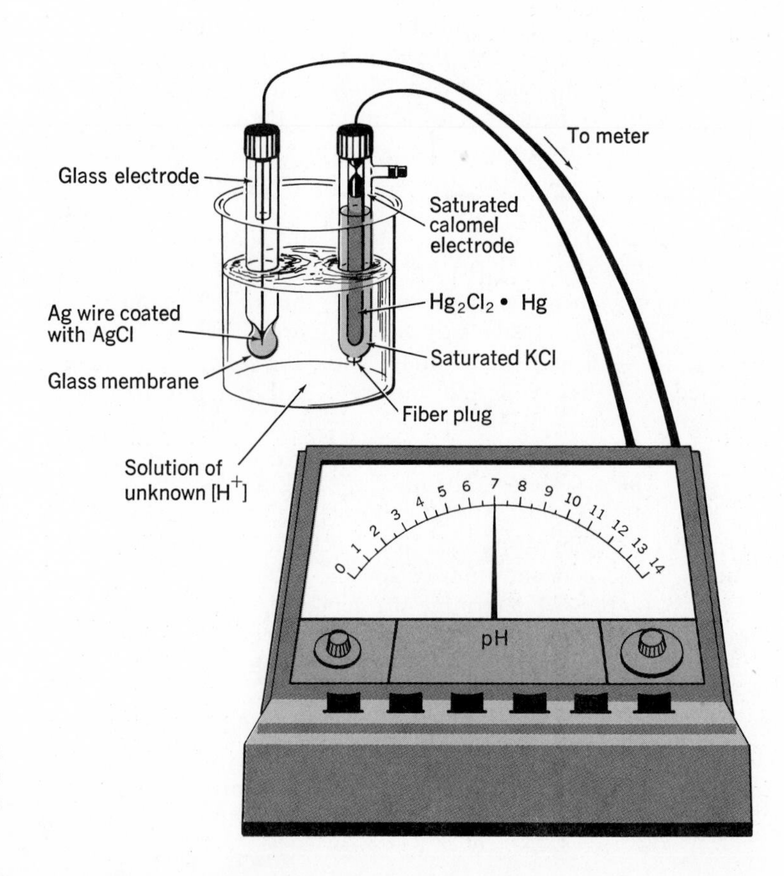

17-16 A pH meter. The calomel electrode is a reference electrode which maintains a constant potential. The potential across the glass electrode varies with the $[H^+]$ of the solution.

The saturated calomel (Hg_2Cl_2) electrode shown in Fig. 17-16 is commonly used as a reference electrode. The half-cell reaction for this electrode is

$$Hg_2Cl_2 + 2e^- \rightleftharpoons 2Hg + 2Cl^- \qquad E = 0.242 \text{ v}$$

In the so-called normal calomel electrode, the KCl is 1.0 *M* and the half-cell potential is 0.280 v.

ELECTROLYTIC CELLS

In electrochemical cells, spontaneous chemical reactions take place, liberate electric energy, and result in a decrease in the chemical energy of the system. Cells in which nonspontaneous reactions take place by the addition of electric energy to the system undergo an increase in potential energy. They are known as electrolytic cells. The decomposition of water by electrolysis is an example of an electrolytic cell reaction. The reaction

$$2H_2O(l) \rightleftharpoons 2H_2(g) + O_2(g)$$

is a *nonspontaneous reaction.* Electric energy must be added to the system to bring about the decomposition. The products, H_2 and O_2, possess more potential energy than the H_2O. In theory, the electric energy put into the system can be recovered by allowing the H_2 and O_2 to react as in a fuel cell.

17-23 Electrolysis of Melted Binary Salts In the electrolysis of melted salts, cations are reduced at the cathode, and anions are oxidized at the anode. The products of such an electrolysis are generally *metals at the cathode* and *nonmetals at the anode.* Let us examine the operation of an electrolytic cell containing melted sodium chloride, schematically illustrated in Fig. 17-18 on p. 576.

The external source of electric energy may be a dry cell or any source of direct current. The terminals of the battery are connected to inert electrodes A and B by copper wire. The electrodes dip into melted sodium chloride.

As long as the temperature remains above 800°C, the melting point of sodium chloride, the sodium and chloride ions are able to move freely and transport an electric charge through the liquid. When the circuit is completed, the electric field produced by the battery forces electrons through the wires in the direction shown by the arrows. That is, electrons are forced toward electrode A and pulled away from B. Positive sodium ions in the liquid are attracted to electrode A, where they pick up one electron per ion and are reduced to neutral sodium atoms. Chloride ions migrate toward the positive electrode, where there is a deficiency of electrons. There they give up one electron per ion to the electrode and become neutral chlorine atoms. Two chlorine atoms combine and form diatomic molecules of Cl_2 gas. The electrode reactions are

Cathode $\quad Na^+(l) + e^- \rightleftharpoons Na(l)$

Anode $\quad Cl^-(l) \rightleftharpoons \frac{1}{2}Cl_2(g) + e^-$

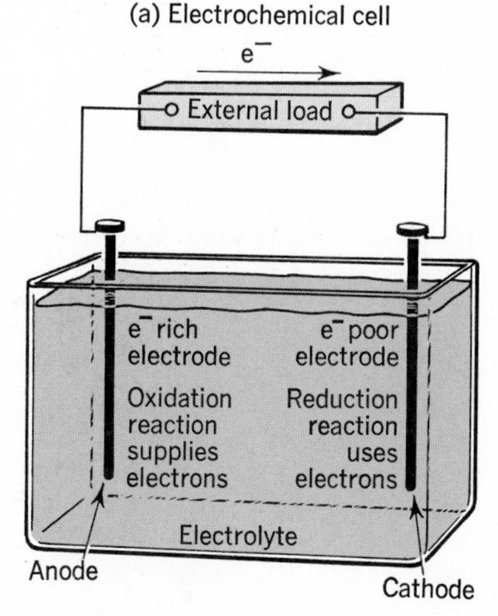

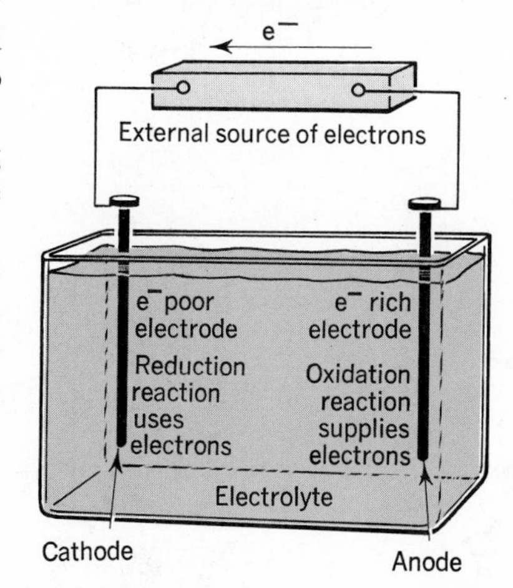

17-17 In electrochemical cells, chemical energy is converted into electric energy. In electrolytic cells, electric energy is converted into chemical energy.

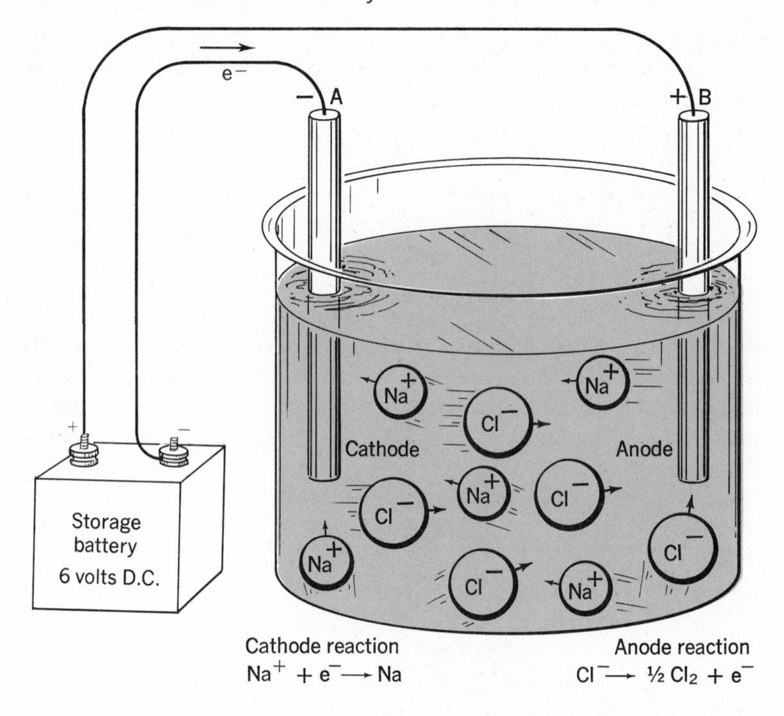

17-18 The electrolysis of melted sodium chloride.

The overall reaction can be obtained by adding the anode and cathode reactions. The result is

$$\textbf{electric energy} + Na^+(l) + Cl^-(l) \rightleftharpoons Na(l) + \tfrac{1}{2}Cl_2(g)$$

17-24 Electrolysis of Aqueous Solutions When inert electrodes are used in the electrolysis of water solutions, two possible reactions at each electrode must be considered. At the anode, these are the *oxidation of water*

$$H_2O(l) \rightleftharpoons \tfrac{1}{2}O_2(g) + 2H^+ + 2e^-$$

and the oxidation of a nonmetallic ion (X^-)

$$2X^- \rightleftharpoons X_2 + 2e^-$$

Experimentally, it is found that for fairly concentrated aqueous solutions of Cl^-, Br^-, or I^- ions, the free halogen is discharged at the anode rather than oxygen. Note that oxidation potentials indicate that it should be easier to oxidize H_2O ($[H^+] = 1 \times 10^{-7}\ M$) than to oxidize Cl^- and Br^- ions. That is, the *oxidation potential* for the water half-cell reaction is higher than that for the Cl^- or Br^- ion half-cell. This observation suggests that other factors must be considered in this instance. Since a discussion of these factors is beyond the scope of this text, we shall use experimental observations as a basis for our statement. When the salt solution contains sulfate ions as the anion, water is oxidized to oxygen. This is because the oxidation of SO_4^{2-} ions requires a much greater potential (driving force) than the oxidation of water.

At the cathode, the two possible reactions are the *reduction of a metallic ion* (M^{2+})

$$M^{2+} + 2e^- \rightleftharpoons M$$

and the *reduction of water*

$$2H_2O + 2e^- \rightleftharpoons H_2(g) + 2OH^-(10^{-7}\,M) \qquad E = -0.41\text{ v}$$

The species with the highest possible reduction potential should be the easiest to reduce. In general, therefore, metallic ions such as Cu^{2+} and Ag^+ which have reduction potentials greater than that for water should be reduced to the metal at the cathode. On the other hand, water is reduced to hydrogen gas at the cathode in solutions containing Na^+, K^+, Li^+, and other ions which have large negative reduction potentials. When an aqueous solution of Na_2SO_4 is electrolyzed using inert electrodes, hydrogen gas is observed at the cathode and oxygen gas at the anode. These observations may be interpreted in terms of the relevant half-cell reactions given below.

$$2H_2O + 2e^- \rightleftharpoons H_2(g) + 2OH^-(10^{-7}\,M) \quad E = -0.41\text{ v} \qquad 17\text{-}46$$

Use the Nernst equation to verify that E = −0.41 v.

$$Na^+(aq) + e^- \rightleftharpoons Na(s) \quad E^\circ = -2.71\text{ v} \qquad 17\text{-}47$$

$$2H_2O \rightleftharpoons O_2(g) + 4H^+(10^{-7}\,M) + 4e^- \quad E = -0.815\text{ v} \qquad 17\text{-}48$$

$$2SO_4^{2-} = S_2O_8^{2-} + 2e^- \quad E^\circ = -2.01\text{ v} \qquad 17\text{-}49$$

Equations 17-46 and 17-47 represent the possible cathode reactions. Equation 17-46 which represents the reduction of water has a higher reduction potential than Equation 17-47. Therefore, water is easier to reduce than $Na^+(aq)$. On this basis, we may assume that Equation 17-46 represents the cathode reaction.

Equations 17-48 and 17-49 represent possible anode reactions. Equation 17-48 which represents the oxidation of water, has a higher oxidation potential than Equation 17-49. Therefore, water is easier to oxidize than SO_4^{2-} ions. On this basis, we may assume that Equation 17-48 represents the anode reaction. The overall reaction may be obtained by multiplying Equation 17-46 by 2 and adding it to Equation 17-48. The result is

$$6H_2O \rightleftharpoons 2H_2(g) + O_2(g) + 4OH^- + 4H^+$$

The Na^+ ions and SO_4^{2-} ions maintain electric neutrality around the electrode as the water is electrolyzed.

The problem of predicting products is complicated even further by using electrodes that may react. Consider the electrolysis of a copper(II) sulfate solution using copper metal as the electrodes. This electrolytic cell is illustrated in Fig. 17-19 on the next page. At the cathode, the possible reactions are

$$Cu^{2+} + 2e^- \rightleftharpoons Cu(s) \quad E^\circ = 0.34\text{ v} \qquad 17\text{-}50$$

$$2H_2O + 2e^- \rightleftharpoons H_2(g) + 2OH^-(10^{-7}\,M) \quad E = -0.41\text{ v}$$

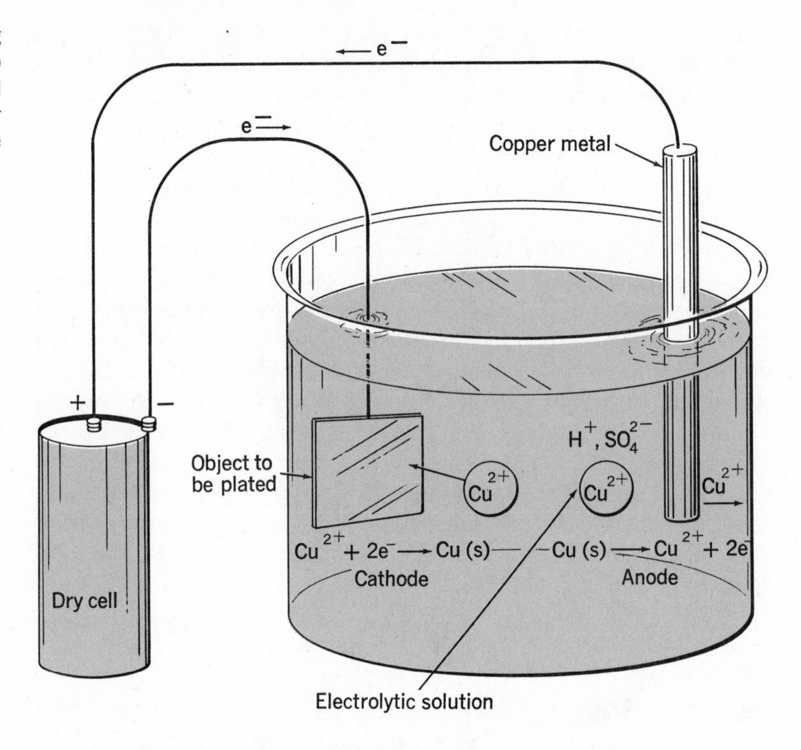

17-19 An electrolytic cell used for electroplating copper. The object to be plated is connected to the negative terminal of the electrochemical cell which serves as the source of electrons. The copper anode of the electrolytic cell is connected to the positive terminal of the electrochemical cell.

The half-cell potentials indicate that Equation 17-50 is the cathode reaction. At the anode there are now three possibilities: the oxidation of H_2O, the oxidation of SO_4^{2-}, and the oxidation of copper metal. The half-cell potentials of

$$\mathbf{Cu(s) = Cu^{2+} + 2e^-} \qquad \mathbf{E° = -0.34\ v} \qquad 17\text{-}51$$
$$\mathbf{2H_2O = O_2(g) + 4H^+ + 4e^-} \qquad \mathbf{E° = -0.815\ v}$$
$$\mathbf{2SO_4^{2-} = S_2O_8^{2-} + 2e^-} \qquad \mathbf{E° = -2.01\ v}$$

indicate that copper metal is the easiest of the three species to oxidize. Therefore, Equation 17-51 represents the anode reaction. The cell described above is characteristic of many electroplating cells. In these cells, the object to be plated is made the cathode, a bar of the plating metal is made the anode, and the solution contains a soluble salt of the plating metal. In a copper-plating cell, Cu^{2+} ions from the solution gain electrons and are deposited on the cathode as copper atoms. As fast as Cu^{2+} ions are removed from solution, copper atoms from the anode bar lose electrons and go into solution as Cu^{2+} ions. In this way, the copper(II) concentration of the solution remains constant at an optimum value. At this point, we should note that it is not always possible to predict exactly what products will be discharged in an electrolytic cell by referring to reduction or oxidation potentials. Such factors as rates, current levels, concentration, polarization, and overvoltage must also be considered.

17-25 Industrial Applications of Electrolysis Electrolytic processes are widely used to prepare metals in a highly pure state. The production of pure copper and pure magnesium are familiar examples. Approximately two million tons of 99.95 percent pure copper are electrolytically refined each year. In this process, impure copper slabs known as *blister copper* act as anodes in an electrolytic cell. The cathodes are thin sheets of pure copper and the electrolyte is a copper(II) sulfate solution containing sulfuric acid. During electrolysis the copper in the impure anode goes into solution as Cu^{2+} ions. The Cu^{2+} ions migrate to the cathode and plate out as pure copper. Impurities such as silver and gold which have more negative oxidation potentials than copper are not oxidized but fall to the bottom of the tank as the anode disintegrates. In this way valuable metals are salvaged in the "anode mud." Impurities which have higher oxidation potentials than copper are also oxidized and dissolved as ions in the solution. These ions have lower reduction potential than Cu^{2+} and, therefore, do not plate out unless most of the copper(II) ions have been removed from the solution. This, of course, is avoided by the use of impure copper anodes.

One important industrial process for preparing metallic sodium involves the electrolysis of melted NaCl. The reaction is carried out in a ***Downs cell*** similar to that illustrated in Fig. 17-21. In this cell, the electrodes are separated by an iron screen which permits

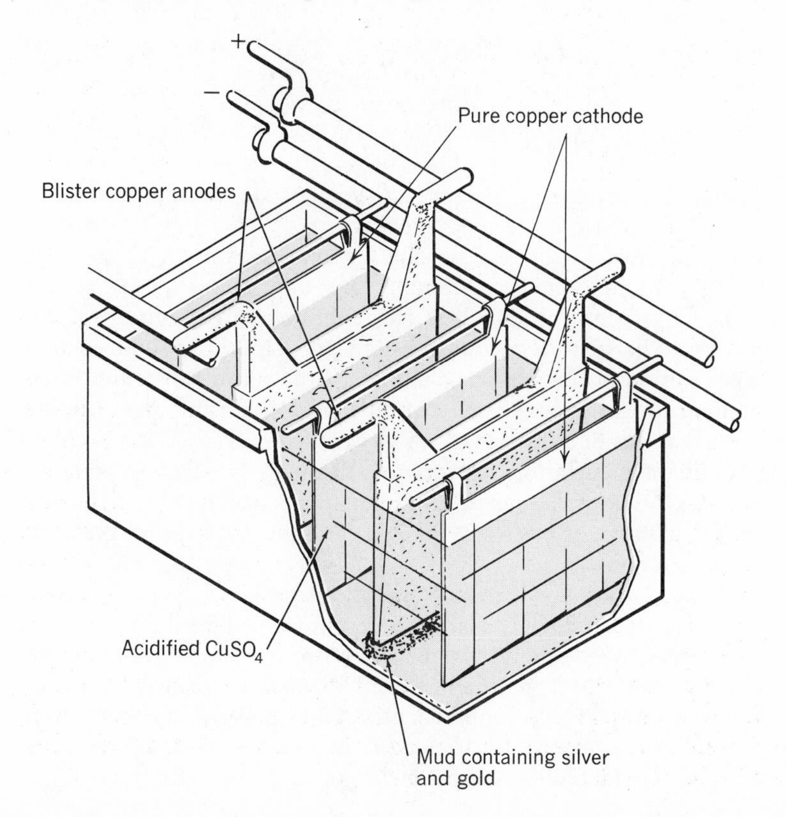

17-20 The final step in the refining of copper is an electrolytic process. The reduction potential of the copper half-cell is less than that of water. Hence, the Cu^{2+} ions are reduced rather than the water.

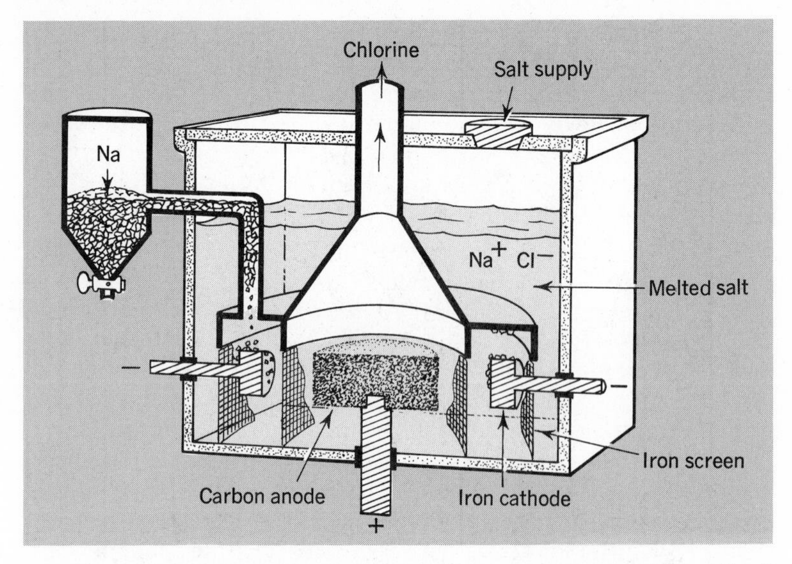

17-21 Melted sodium chloride is electrolyzed in a Downs cell. Chlorine gas is formed over the central positive carbon electrode, and melted sodium metal is produced at the negative iron electrode.

passage of ions but prevents the products from interacting. All of the alkali metals may be prepared by the electrolytic decomposition of their melted salts. This was the method used over 150 years ago by Sir Humphrey Davy, the discoverer of the elements sodium, potassium, and lithium.

Magnesium is primarily used as a structural material. Magnesium-aluminum alloys are utilized extensively in the aircraft and space industries because of their strength and light weight. Although magnesium is a strong reducing agent, it resists corrosion by forming a layer of basic carbonate, $MgCO_3 \cdot Mg(OH)_2$, on the surface which excludes further corrosion. Large quantities of magnesium are obtained from the sea. The last step in the production of magnesium from the sea is the electrolysis, in airtight electrolytic cells of melted $MgCl_2$. At the cathode, magnesium ions (Mg^{2+}) are reduced to metallic magnesium. The chlorine produced at the anode is used to produce the hydrochloric acid required for the process. The quantity of magnesium and chlorine produced in a given period of time depends on the current flow. In the industrial preparation of magnesium, currents of 50 000 amperes are frequently used. See flow chart on p. 611.

17-26 Quantitative Aspects of Electrolysis The masses of products liberated at the electrodes of an electrolytic cell are related to the quantity of electricity passed through the cell and the electrode reactions.

In the electrolysis of melted NaCl, one mole of electrons (called one *Faraday* of electric charge) is required to reduce one mole of Na^+ ions at the cathode. At the same time, one mole of electrons must be removed from one mole of Cl^- ions at the anode, forming one-half mole of chlorine molecules. Thus, *one Faraday discharges one mole of metallic sodium and one-half mole of chlorine gas.* The mass of a metal discharged by one Faraday depends on the number

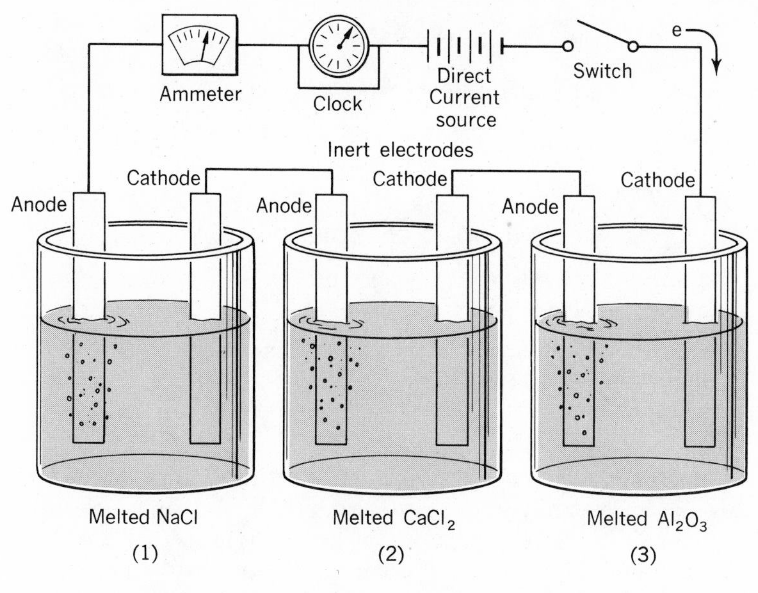

17-22 These three electrolytic cells are connected in series to a source of direct current (battery). It can be seen from the data in the Table that the amounts of different elements discharged at the cathode by the same quantity of electric charge form whole-number mole ratios. Determine these ratios from the data.

Cell	Metal discharged at cathode	Current (amp)	Time (sec)	Coulombs	Mass of metal discharged (g)	Molecular mass of metal
(1)	Na	2.0	48 250	96 500	23	23
(2)	Ca	2.0	48 250	96 500	20	40
(3)	Al	2.0	48 250	96 500	9	27

The mass (grams) of a chemical substance discharged at an electrode by one Faraday of electricity is called a gram-equivalent mass.

of electrons required to reduce each ion. The following cathode reactions show that one Faraday reduces one mole of sodium ions, one-half mole of calcium ions, and one-third mole of aluminum ions.

$$\mathbf{Na^+(l) + e^- \rightleftharpoons Na(l)}$$
$$\mathbf{Ca^{2+}(l) + 2e^- \rightleftharpoons Ca(s)}$$
$$\mathbf{Al^{3+}(l) + 3e^- \rightleftharpoons Al(l)}$$

The number of electrons associated with the discharge of any species at the electrode of an electrolytic cell may be determined from the electrode reaction.

Example 17-5

How many Faradays are required to reduce 2.93 g of nickel ions from melted $NiCl_2$?

Solution

1. Write the equation for the reduction of Ni^{2+}.

$$Ni^{2+}(l) + 2e^- \rightleftharpoons Ni(s)$$

2. Determine the number of moles of nickel in 2.93 g.

$$2.93\text{ g} \times 1\text{ mole}/58.7\text{ g} = 0.0500\text{ mole}$$

3. Multiply the number of moles of nickel by the number of Faradays per mole as shown by the number of electrons in the reduction equation.

$$0.0500\text{ mole} \times 2\text{ F/mole} = 0.100\text{ F}$$

FOLLOW-UP PROBLEM

How many Faradays are required to reduce 10 g of calcium from melted calcium chloride?

Ans. 0.500 F.

Experimentally, the quantity of electric charge transferred through an electrolytic cell is determined by measuring the time and magnitude of the current flow. The time is measured in seconds with a stopwatch or timer. The current(I) is measured in amperes with an ammeter. The product of the current and time is equal to the quantity of electric charge expressed in terms of coulombs, Q. That is,

$$\mathbf{Q = It} \qquad 17\text{-}52$$

Equation 17-52 indicates that one coulomb is the quantity of electric charge transferred when a current of one ampere flows for one second. One coulomb is the charge carried by 6.28×10^{18} electrons. It can be shown that one Faraday is equal to approximately 96 500 coulombs. The relationship between coulombs and Faradays is

$$\mathbf{F = coulombs \times \frac{1\ Faraday}{96\,500\ coul}} \qquad 17\text{-}53$$

Example 17-6
How much time is required to reduce 20.0 g of silver from a solution of silver nitrate, using 15.0 amp of current?

Solution
1. Determine the number of Faradays required to reduce 1 mole of silver ions by writing the equation for the reduction of silver ions.

$$Ag^+(aq) + e^- \longrightarrow Ag(s)$$

2. Determine the number of moles of silver in 20.0 g.

$$20.0\text{ g} \times \text{mole}/108\text{ g} = 0.185\text{ mole}$$

3. Determine the number of Faradays required to reduce 0.185 mole of silver. Because one silver ion requires one electron, one mole of silver ions requires one Faraday of electric charge, and 0.185 mole requires 0.185 Faraday.

4. Determine the number of coulombs in 0.185 Faraday. Equation 17-53 shows that Faradays can be converted to coulombs by multiplying by 96 500.

$$0.185 \text{ F} \times 96\,500 \text{ coul/F} = 17\,900 \text{ coul}$$

5. Solve Equation 17-51 for *t* and substitute the known values.

$$t = \frac{17\,900 \text{ coul}}{15.0 \text{ coul/sec}} = 1190 \text{ sec}$$

NB

Alternate Solution

$$\frac{20.0 \text{ g Ag}}{15.0 \text{ amp}} \times \frac{\text{mole Ag}}{108 \text{ g Ag}} \times \frac{\text{F}}{\text{mole Ag}} \times \frac{96\,500 \text{ coul}}{\text{F}} \times \frac{\text{ampsec}}{\text{coul}} = 1190 \text{ sec}$$

FOLLOW-UP PROBLEM

A current of 5.0 amp is passed through melted KCl for 1000 sec. How many grams of potassium are produced?

Ans. 2.03 g K.

Example 17-7

(a) How many grams of copper are deposited from a plating bath by a current of 3.0 amperes flowing for a period of 2.0 hours? (b) How many liters of oxygen gas will be evolved at the anode if the temperature is 27°C and the pressure is 740 torr? (c) If the cell originally contained 1.0 liter of 0.20-*M* $CuSO_4$ and a platinum (inert) anode, what is the concentration of $CuSO_4$ after the electrolysis assuming no volume change? (d) If the anode is a bar of copper, what is the concentration of $CuSO_4$ after the electrolysis?

Solution

(a) The electrode reaction is

$$Cu^{2+}(aq) + 2e^- \rightleftharpoons Cu(s)$$

Calculate the coulombs passed through the cell.

$$Q = It = 3.0 \text{ amps} \times 2.0 \text{ hrs} \times \frac{3600 \text{ sec}}{\text{hr}} = 1.8 \times 10^4 \text{ coul}$$

Calculate the Faradays of electricity.

$$F = 1.8 \times 10^4 \text{ coul} \times \frac{\text{F}}{9.65 \times 10^4 \text{ coul}} = 0.186 \text{ F}$$

Calculate the grams of Cu deposited. The electrode reaction shows there are 2 F per mole (63.6 g) of copper.

$$0.186 \text{ F} \times \frac{63.6 \text{ g Cu}}{2\text{F}} = 0.58 \text{ g Cu}$$

(b) The electrode reaction is

$$2H_2O = O_2(g) + 4H^+ + 4e^-$$

Calculate the mass of O_2 evolved by one F. From the electrode reaction 4F (4 moles of e^-) are associated with the discharge of one mole of O_2.

$$\frac{32 \text{ g } O_2}{\text{mole } O_2} \times \frac{\text{mole } O_2}{4\text{F}} = \frac{8.0 \text{ g } O_2}{\text{F}}$$

Calculate the mass of O_2 evolved by passage of 0.186 F (from Part a).

$$0.186 \text{ F} \times \frac{8.0 \text{ g } O_2}{\text{F}} = 1.5 \text{ g } O_2$$

Convert 1.5 g O_2 into liters at 27°C and 740 torr.

$$1.5 \text{ g } O_2 \times \frac{\text{mole } O_2}{32 \text{ g } O_2} \times \frac{22.4 \ \ell \ O_2}{\text{mole } O_2} \times \frac{300°\text{K}}{273°\text{K}} \times \frac{760 \text{ torr}}{740 \text{ torr}} = 1.2 \ \ell \ O_2$$

(c) Calculate the original moles of Cu^{2+}.

$$\text{moles} = MV = (0.20 \ M)(1.0) = 0.20 \text{ moles}$$

Calculate the moles of Cu deposited.

$$0.186 \text{ F} \times \frac{\text{mole Cu}}{2\text{F}} = 0.093 \text{ moles Cu}$$

Calculate the moles of Cu^{2+} remaining.

$$0.20 \text{ moles} - 0.09 \text{ moles} = 0.11 \text{ moles}$$

Calculate the molarity of the solution.

$$M = \frac{\text{moles } Cu^{2+}}{\ell \text{ sol'n}} = \frac{0.11 \text{ moles}}{1.0 \ \ell} = 0.11 \ M$$

(d) The concentration will be the same as the original concentration. The anode furnishes Cu^{2+} to the solution as fast as the Cu^{2+} is removed.

FOLLOW-UP PROBLEM

(a) How many liters of H_2 are evolved at 20°C and 770 torr when 2.0 amps is passed for 3.0 hrs through a cell containing sulfuric acid?

Ans. 2.7 ℓ.

LOOKING AHEAD

We have now concluded our study of reaction principles. We have developed and used energy concepts, rate phenomena, and equilibrium principles to help explain observations made on acid-base systems, systems involving slightly soluble substances, and redox systems.

We are now ready to study the properties and behavior of specific elements and groups of elements. This phase of our study is called descriptive chemistry. You will find that the concepts developed in the "principle" chapters will help us explain and understand the descriptive chemistry of the elements.

In the following chapter, we shall consider the nonmetals of Group VIIA in the Periodic Table. Metallic elements are discussed in Chapter 19.

QUESTIONS AND PROBLEMS

1. Relate these concepts to one another. (a) oxidizing agent, (b) oxidized substance, (c) reducing agent, (d) reduced substance.

2. For each state whether the change is an oxidation or a reduction. (a) MnO_4^- becomes MnO_4^{2-}, (b) K becomes K^+, (c) N_2 becomes NH_3, (d) NH_3 becomes N_2O, (e) P_4O_{10} becomes P_4O_6, (f) SO_4^{2-} becomes SO_3^{2-}, (g) $HClO_4$ becomes HCl and H_2O, (h) O_2 becomes O^{2-}, (i) $Cr_2O_7^{2-}$ becomes Cr^{3+} and H_2O.

Ans. (b) Oxidation, (d) oxidation.

3. Find the oxidation numbers of the underlined elements in these formulas. (a) H$\underline{Cl}$O, (b) K$\underline{Cl}O_3$, (c) $\underline{Mn}O_2$, (d) $\underline{Pb}O_2$, (e) $\underline{Pb}SO_4$, (f) $K_2\underline{S}O_4$, (g) $\underline{N}H_4^+$, (h) $Na_2\underline{O}_2$, (i) $\underline{Fe}O$, (j) $\underline{Fe}_2O_3$, (k) Na$\underline{I}O_4$, (l) $\underline{Fe}_3O_4$, (m) $\underline{Cr}_2O_7^{2-}$, (n) $\underline{Mn}O_4^{2-}$, (o) $\underline{N}O_3^-$, (p) $\underline{Cl}O_3^-$.

Ans. (c) 4+, (f) 6+, (l) $\frac{8}{3}$+, (p) 5+.

4. Balance these skeleton equations (ions are hydrated). Identify the oxidizing and reducing agent in each. Change to net ionic form before balancing.

(a) $I_2 + H_2S \rightleftharpoons H^+ + I^- + S$
(b) $Cu + HNO_3 \rightleftharpoons Cu(NO_3)_2 + NO_2 + H_2O$
(c) $H_2SO_4 + Zn \rightleftharpoons S + H_2O + ZnSO_4$
(d) $CuO + NH_3 \rightleftharpoons Cu + H_2O + N_2$
(e) $H_2O + ClO_3^- + SO_2 \rightleftharpoons SO_4^{2-} + H^+ + Cl^-$
(f) $K_2Cr_2O_7 + HCl \rightleftharpoons KCl + CrCl_3 + H_2O + Cl_2$
(g) $MnO_4^- + CH_3OH + H^+ \rightleftharpoons MnO_2 + CH_2O + H_2O$
(h) $P + HNO_3 + H_2O \rightleftharpoons H_3PO_4 + NO$
(i) $HIO_3 + H_2SO_3 \rightleftharpoons I_2 + H_2SO_4 + H_2O$
(j) $Zn + HNO_3 \rightleftharpoons Zn(NO_3)_2 + NH_4NO_3 + H_2O$

5. Balance these equations by the half-cell method. Show both half-cell reactions and identify them as oxidation or reduction.

(a) $SO_3^{2-} + MnO_4^- + H^+ \rightleftharpoons Mn^{2+} + SO_4^{2-} + H_2O(l)$
(b) $Cu(s) + NO_3^- + H^+ \rightleftharpoons Cu^{2+} + NO(g) + H_2O(l)$
(c) $Cl_2(g) + OH^- \rightleftharpoons Cl^- + ClO_3^- + H_2O(l)$
(d) $Cl_2(g) + H_2O \rightleftharpoons Cl^- + ClO^- + H^+$
(e) $SO_4^{2-} + I^- + H^+ \rightleftharpoons S^{2-} + I_2(s) + H_2O(l)$
(f) $NH_4NO_2(s) \rightleftharpoons N_2(g) + H_2O(g)$
(g) $Zn(s) + SO_4^{2-} + H^+ \rightleftharpoons S(s) + Zn^{2+} + H_2O(l)$
(h) $MnO_4^- + C_2O_4^{2-} + H^+ \rightleftharpoons MnO_2(s) + CO_2(g) + H_2O(l)$
(i) $SO_3^{2-} + IO_3^- + H^+ \rightleftharpoons I_2(s) + SO_4^{2-} + H_2O(l)$
(j) $ClO_3^- \rightleftharpoons Cl^- + ClO_4^- + O_2(g)$

6. Show that each of these reactions is a redox reaction by identifying the oxidation and the reduction.

(a) $2Ba(s) + O_2(g) \rightleftharpoons 2BaO(s)$
(b) $2NH_3(g) + 3\frac{1}{2}O_2(g) \rightleftharpoons 2NO_2(g) + 3H_2O(g)$
(c) $C_2H_6(g) + 3\frac{1}{2}O_2(g) \rightleftharpoons 2CO_2(g) + 3H_2O(g)$
(d) $ClO_3^-(s) + 5Cl^- + 6H^+ \rightleftharpoons Cl_2(g) + 3H_2O$
(e) $NH_4NO_2(s) \rightleftharpoons N_2(g) + 2H_2O$

7. Balance these equations. All ions are in solution.

(a) $ClO_3^- + Cr^{3+} \longrightarrow ClO_2 + Cr_2O_7^{2-}$ (acid)
(b) $H_2O_2 + Co(OH)_2 \longrightarrow Co_2O_3 + H_2O$ (basic)
(c) $Fe_3O_4 + H_2O_2 \longrightarrow Fe^{3+} + H_2O$ (acid)
(d) $W + NO_3^- \longrightarrow WO_3 + NO_2$ (acid)
(e) $Cu(NH_3)_4^{2+} + CN^- \longrightarrow CNO^- + Cu(CN)_3^{2-} + NH_3$ (basic)
(f) $Cr_2O_7^{2-} + C_2H_4O \longrightarrow HC_2H_3O_2 + Cr^{3+}$ (acid)
(g) $H_2O_2 + SCN^- \longrightarrow HCO_3^- + HSO_4^- + NH_4^+ + H_2$ (acid)
(h) $ClO^- + Mn(OH)_2 \longrightarrow MnO_2 + Cl^-$ (basic)
(i) $As_2S_3 + ClO_3^- \longrightarrow Cl^- + H_3AsO_4 + S$ (acid)
(j) $Cu(NH_3)_4^{2+} + S_2O_4^{2-} \longrightarrow SO_3^{2-} + Cu + NH_3$ (basic)
(k) $CrI_3 + KOH + Cl_2 \longrightarrow K_2CrO_4 + KIO_4 + KCl + H_2O$ (basic)
(l) $K_4Fe(CN)_6 + KMnO_4 + H_2SO_4 \longrightarrow Fe_2(SO_4)_3 + K_2SO_4 + CO_2 + HNO_3 + MnSO_4 + H_2O$ (acid)

Read 551-561

8. For each of these combinations of reactants, write an equation for a likely reaction. Assume 1 *M* concentration of ions, 1.00 atm pressure and 25°C for gases.

(a) $Mg(s) + Al^{3+}$
(b) $Hg^{2+} + Ag(s)$
(c) $Cl^- + I_2(s)$
(d) $Fe(s) + Sn^{2+}$
(e) $Al(s) + Zn^{2+}$
(f) $Zn(s) + Br_2(l)$
(g) $Sn(s) + H^+$
(h) $Cu(s) + I^-$
(i) $H_2(g) + Cl_2(g)$
(j) $H_2(g) + Ag^+$

9. Would it be practical to store a 0.5-*M* $Fe_2(SO_4)_3$ solution in an aluminum container?

10. Could you safely store 1-*M* $AlCl_3$ solution in an iron container?

11. Explain why a salt bridge is necessary in an electrochemical cell.

12. Define (a) anode, (b) cathode.

13. If instead of the hydrogen electrode being used as the standard for E° values, the silver-silver ion electrode were used, what would be the standard reduction potentials, E°, for these half-cell reactions?

(a) $F_2(g) + 2e^- \rightleftharpoons 2F^-$
(b) $O_2(g) + 4H^+ + 4e^- \rightleftharpoons 2H_2O$
(c) $Fe^{3+} + e^- \rightleftharpoons Fe^{2+}$
(d) $2H^+ + 2e^- \rightleftharpoons H_2(g)$
(e) $Mg^{2+} + 2e^- \rightleftharpoons Mg(s)$

14. An iron-nickel electrochemical cell uses a salt bridge to join a half-cell containing a strip of iron in a 1.0-*M* solution of Fe^{2+} to a half-cell which contains a strip of nickel in 1.0-*M* Ni^{2+} solution. A voltmeter connects the two metal strips. (a) in which cell does reduction occur? (b) Write the half-cell reactions involved. (c) Which metal is the anode? (d) In which direction are electrons passing through the voltmeter? (e) What is the expected voltmeter reading? (f) What would be the effect on the voltmeter reading if the Fe^{2+} concentration were increased? (g) What would be the effect on the reading if the $[Ni^{2+}]$ were increased; decreased? (h) What is the voltmeter reading when the cell reaches equilibrium? (i) What is the value of ΔG when the cell reaches equilibrium? (j) Calculate K_e for the reaction from E° values.
Ans. (d) From the iron to the nickel, (g) voltage of cell increases.

15. Observations (1) *A* reacts spontaneously with 1-*M* BNO_3, 1-*M* $D(NO_3)_2$, and dilute sulfuric acid. *A* does not react with 1-*M* $C(NO_3)_2$. (2) *B* does not react spontaneously with any of the 1-*M* solutions above nor with dilute sulfuric acid. (3) *C* reacts spontaneously with dilute sulfuric acid and with 1-*M* solutions of all the other metallic salts. (4) *D* reacts spontaneously with 1-*M* BNO_3. It does not react with dilute sulfuric acid.

(a) Use the observations and arrange the five reduction half-cell reactions in order with the one having the largest positive reduction potential listed first.

$A^{2+} + 2e^- = A(s)$
$B^+ + e^- = B(s)$
$C^{2+} + 2e^- = C(s)$
$D^{2+} + 2e^- = D(s)$
$2H^+ + 2e^- = H_2(g)$

(b) Which metal is the best reducing agent? (c) Which ion is the best oxidizing agent?

16. Write an equilibrium expression for the reaction

$$Zn(s) + 2Ag^+ \longrightarrow 2Ag(s) + Zn^{2+}$$

Ans. $K_e = [Zn^{2+}]/[Ag^+]^2$.

17. Assuming that K_e problem 16 has a high value. Is the Ag^+ ion concentration high or low in a "dead" cell containing an excess of zinc metal?

18. (a) Which of these at standard state is the best oxidizing agent? I_2, I^-, Au, Au^{3+}, Mg, Mg^{2+}? (b) which is the best reducing agent?

19. Calculate the voltage of a standard cell with these half-cell reactions

$$Ni(s) \longrightarrow Ni^{2+} + 2e^-$$
$$2e^- + Cl_2(g) \longrightarrow 2Cl^-$$

20. In a copper-zinc cell these half-cell reactions occur.

$$Cu^{2+} + 2e^- \longrightarrow Cu(s)$$
$$Zn(s) \longrightarrow Zn^{2+} + 2e^-$$

What would be the effect upon the voltage if (a) sulfide ions (S^{2-}) were added to the Cu^{2+}-ion compartment, (b) $CuSO_4$ were added to the Cu^{2+}-ion compartment, (c) sulfide ions were added to the Zn^{2+}-ion compartment, (d) the size of the zinc electrode were doubled, (e) water were added to both compartments?

21. In the equation

$$\tfrac{1}{2}Cl_2(g) + Br^- \longrightarrow \tfrac{1}{2}Br_2(l) + Cl^- \qquad E° = 0.30\ v$$

(a) Calculate the equilibrium constant for the cell reaction at equilibrium. (b) What is the voltage of the cell if the concentration of Br^- is increased tenfold?

Ans. $K_e = 1 \times 10^5$.

22. Calculate the standard-state cell potentials ($E°_{cell}$) for these reactions

(a) $Cl_2 + Ni(s) = Ni^{2+} + Cl^-$
(b) $2Ce^{4+} + 2I^- = 2Ce^{3+} + 5I_2$
(c) $Sn^{4+} + Cd(s) = Sn^{2+} + Cd^{2+}$
(d) $Br_2 + 2Fe^{2+} = 2Br^- + 2Fe^{3+}$

23. Calculate the equilibrium constant for each of the reactions in Problem 22.

24. Calculate the potential for each of these cells. Identify the anode in each case.
(a) $Fe(s)|Fe^{2+}(0.10\ M)||Cd^{2+}(0.0010\ M)|Cd(s)$
(b) $Mg(s)|Mg^{2+}(10^{-3}\ M)||Ag^{+}(2.0\ M)|Ag(s)$
(c) $Pt(s),\ H_2(1\ atm)|H^{+}(1.0 \times 10^{-4}\ M)||$
$Cl^{-}(1.0\ M),\ Hg_2Cl_2(sat)|Hg$
(d) $Pt(s),\ H_2(1\ atm)|H^{+}(0.10\ M)||$
$H^{+}(1.0 \times 10^{-5}\ M)|H_2(1\ atm),\ Pt(s)$

25. Name two substances which oxidize Ag to Ag^+.

26. Can iodine oxidize (a) Fe^{2+} to Fe^{3+}, (b) Sn to Sn^{2+}? Use data in Table 17-6 to help explain your answers.

27. Use the data in Table 17-6 to help explain in general terms the reason that hydrogen can be prepared in the laboratory by reacting zinc metal with hydrochloric acid, but not by copper with hydrochloric acid.

28. Use the data in Table 17-6 to help explain in general terms the reason that hydrogen is produced when zinc reacts with hydrochloric acid, but NO, NO_2, or NH_4^+ when zinc reacts with nitric acid.

29. Explain why silver dissolves in 1-*M* nitric acid but not in hydrochloric acid.

30. Concentrated HCl (aq) can be used to clean oxidized copper (copper containing a coat of copper(II) oxide) without attacking the metallic copper. Explain.

31. Which of the half-cell reactions listed below oxidize $H_2(g)$ to $2H^+$? (b) Which oxidize I^- but not Cl^-? (c) Which is the strongest reducing agent?
(1) $MnO_2(s) + 4H^+ + 3e^- \longrightarrow 2H_2O + Mn^{2+}$ 1.23 volts
(2) $Ag^+ + e^- \longrightarrow Ag(s)$ 0.80 volts
(3) $Cu^{2+} + 2e^- \longrightarrow Cu(s)$ 0.34 volts
(4) $Pb^{2+} + 2e^- \longrightarrow Pb(s)$ −0.13 volts
(5) $Mg^{2+} + 2e^- \longrightarrow Mg(s)$ −2.34 volts

32. Which of the reactions below (a) goes spontaneously at standard state to the right as written, (b) goes spontaneously to the left (reverse reaction is spontaneous), (c) is an impossible reaction (does not take place in either direction)? Assume standard-state conditions and concentrations.
(1) $BrO_3^- + H_3O^+ + I^- \longrightarrow I_2 + Br^- + H_2O$
(2) $Cu^{2+} + Fe^{2+} \longrightarrow Fe^{3+} + Cu(s)$
(3) $MnO_4^- + H_3O^+ + Co^{2+} \longrightarrow Co(s) + H_2O + Mn^{2+}$
(4) $Cr_2O_7^{2-} + H_3O^+ + Fe^{2+} \longrightarrow 2Cr^{3+} + H_2O + Fe^{3+}$
(5) $Fe^{2+} + I_2 \longrightarrow Fe^{3+} + I^-$
(6) $Cr_2O_7^{2-} + 14H^+ + 6Ce^{3+} \longrightarrow 2Cr^{3+} + 7H_2O + 6Ce^{4+}$
The half-cell potential for $BrO_3^- + 6H_3O^+ + 6e^- \longrightarrow 9H_2O + Br^-$ is 1.44 v.

33. Write equations for reactions that occur when equal volumes of these 2-*M* solutions are mixed. If no reaction occurs, write N.R. (1) Sulfurous acid and potassium permanganate; (2) Iron(III) nitrate, potassium bromide, and potassium iodide; (3) Sodium chloride and aqueous bromine; (4) Potassium dichromate, hydrogen peroxide, and sulfuric acid.

34. In the equations

$$Au^{3+} + 3e^- \longrightarrow Au(s) \qquad E° = 1.50\ v$$
$$AuCl_4^- + 3e^- \longrightarrow Au(s) + 4Cl^- \qquad E° = 1.00\ v$$

(a) Which of these species is the best oxidizing agent? (b) Explain why an $AuCl_4^-$ half-cell has a lower reduction potential than an Au^{3+} half-cell? (c) Could Au be plated better from an aqueous solution of $AuCl_4^-$ or from one of Au^{3+}? (d) At which electrode would the gold be deposited? **Ans. (a) Au^{3+}, (d) cathode.**

35. How does decreasing the pH (increasing the acidity) of a nitrate solution affect the reduction potential and the oxidizing strength of the nitrate system? (b) Write the formulas of two other oxidizing agents whose strength is affected by $[H^+]$. (c) Write the formulas of two oxidizing agents whose strength is not affected by the $[H^+]$.

36. Why are the alkali metals stored under kerosene?

37. Why is it not possible to prepare an alkali metal by electrolyzing an aqueous solution of ions of that metal?

38. Explain why magnesium metal does not corrode as rapidly as a less active metal such as iron.

39. Comment on the statement: Knowing the composition of a solution and the electrode potentials, it is possible to predict the products of electrolysis.

40. (a) Explain why the silver and gold found in blister (impure) copper do not go into solution when the anode disintegrates during the electrolytic refining of copper. (b) If Ag^+ ions were present in the electrolyte, would they interfere with the production of pure copper cathodes? Explain. (c) If Al^{3+} ions were present, would they interfere with the production of pure copper cathodes? Explain.

41. What advantage do fuel cells have over conventional methods of generating electric energy?

42. What is meant by cathodic protection against corrosion?

43. Why should aluminum nails be electrically insulated when used to hold sheet iron in place?

44. (a) What metal might you fasten to the hull of an aluminum boat to give it cathodic protection in sea water? (b) Name two metals you would not use.

45. (a) Why is it difficult or practically impossible to determine accurate values for solubility products by direct chemical analysis of equilibrium solutions? (b) Explain why measurement of cell potentials is an ideal method to measure these constants.

46. Predict the principal product discharged at each electrode during the electrolysis of these 1-*M* solutions. Assume platinum electrodes are used. (1) KI, (2) H_2SO_4, (3) HCl.

47. (a) Draw a diagram for an experimental setup that demonstrates how a copper-plated spoon could be plated with silver. (b) Identify the cathode and anode. (c) How much silver will be plated out of solution by a current of 1.0 amp flowing for 27 hours?

Ans. (c) 108 g.

48. How many coulombs are required to electroplate (a) 7.00 g of lithium, (b) 20.0 g of calcium, (c) 17.3 g of chromium from a solution of Cr^{3+}, (d) 27.0 g of Al?

Ans. (c) 96 500 coul.

49. By the electrolysis of water, 11.2 liters of oxygen at STP was prepared. (a) How many coulombs were required? (b) If a current of 0.5 amp was used, how long did it take?

50. (a) If there are 96 500 coul in a Faraday and 1 Faraday consists of 1 mole of electrons, calculate the charge on 1 electron (in coulombs). (b) How does this value compare with the value given in the Appendix?

51. If a constant current passing through a copper sulfate solution plated 2.00 g of copper in 90.0 minutes, what was the current in amperes?

52. How many kg of magnesium can be produced per hour by a plant which uses a current of 50 000 amperes to electrolyze melted $Mgcl_2$?

53. Electrolysis of 1.500 liter of a 0.500-*M* copper(II) sulfate solution is accomplished by passing 5 amp through the cell for 4 hr. Inert electrodes are used. Assume no change in the volume of the solution during electrolysis. (a) During electrolysis, does the potential energy of the chemical system increase or decrease? Explain. (b) Write an equation for the half-cell reaction occurring at the anode. (c) As electrolysis proceeds, does the solution become more acid, less acid, or maintain a constant pH value? (d) How many electrons pass through the system during the electrolysis? (e) How many grams of copper are deposited? (f) How many liters of gas are collected by displacement of water at 26°C and 750 torr? Water vapor pressure at 26°C is 25 torr. (g) What is the molar concentration of Cu^{2+} at completion of the electrolysis?

Ans. (g) 0.25 M.

18

The Halogens

Preview

"Rather than love, than money, than fame, give me truth."
—Henry David Thoreau (1817–1862)

A large part of the first 17 chapters in this text has been devoted to the development of fundamental concepts and principles of chemistry. The development and illustration of these principles have required the introduction of much descriptive chemistry. Frequently, we used experimental data as a basis for developing a principle or theory. We then used the theory to explain the observed properties of substances. Thus, descriptive and theoretical chemistry are closely integrated.

Most of the products used by all of us in our daily activities are in some way related to the chemical industry. It is safe to say that almost nothing we use comes directly from nature without first being processed by some branch of the chemical industry. Thus, it is the application of chemical principles by the chemical and other manufacturing industries to the processing of raw materials that is indirectly responsible for many of the materials which help us maintain our high standard of living.

In this chapter, we shall concentrate on the chemistry of the *halogens*, a family of five nonmetals with closely related properties. We shall interpret the properties of these elements in terms of periodic atomic properties such as electronegativity, ionization energy, and atomic radius. The concepts and techniques used to describe the chemistry of the halogens can be extended and used to predict the behavior of the elements found in the other groups of the Periodic Table. Much of the descriptive chemistry of oxygen, sulfur, and other nonmetals has already been covered in earlier chapters.

In the Special Feature at the end of this chapter we shall examine several industrial processes which yield materials essential to the welfare and economy of the nation.

THE ELEMENTS OF GROUP VIIA

18-1 Properties and Uses of the Halogens Fluorine, chlorine, bromine, iodine, and astatine, known collectively as the halogens, are reactive nonmetals. As a result of their reactivity, they are found in nature only in the form of compounds. Compounds containing these elements have a great variety of uses in our modern society. Freon®, a volatile, inert liquid, widely used as a refrigerant, and Teflon®, a heat- and corrosion-resistant plastic used for bearings, pipes, and many kinds of containers, are fluorine compounds. Ordinary salt, insecticides (DDT), bleaches, and many acids are chlorine compounds. Photographic film, anti-knock gasolines, and a number of drugs contain bromine compounds. Thyroxin, an iodine compound found in the human thyroid gland, plays a vital role in body metabolism. Although the chemistry of the halogens is extensive and rather involved, much of it can be explained in terms of atomic properties, bonding characteristics, and energy concepts discussed in the preceding chapters.

Data on the chemistry of the halogens are summarized in Table 18-1. Let us examine some of the properties and reactions of the halogens and use the summarized data to help us interpret our observations.

The outer level electronic configuration of the halogen atoms has one fewer electron than their adjacent noble-gas neighbors. In their reactions, the halogen atoms have a *strong tendency to achieve the noble-gas configuration by acquiring one more electron.* This may be accomplished by the formation of a negative ion (*called a halide*) during a reaction with a highly electropositive metal such as sodium or potassium or by the formation of a single covalent bond during a reaction with a less electropositive atom such as hydrogen.

High electronegativities, high electron affinities, high ionization energies, and large standard reduction potentials all reflect the relative ease with which halogen atoms (X) form halide ions (X^-).

TABLE 18-1
STRUCTURE AND PROPERTIES OF THE HALOGENS

	Fluorine	*Chlorine*	*Bromine*	*Iodine*
Color in gaseous phase	pale green-yellow	yellow-green	reddish brown	violet
Standard phase	gas	gas	liquid	solid
Atomic number	9	17	35	53
Electron configuration	$2s^22p^5$	$3s^23p^5$	$4s^24p^5$	$5s^25p^5$
Ionization energy (kcal) ΔH°	402	300	273	241
Electron affinity (kcal) ΔH°	−80	−85	−80	−73
Atomic radius (Å)	0.72	0.99	1.14	1.33
Ionic radius (Å)	1.33	1.81	1.96	2.20
Melting point (°C)	−223	−102	−7.3	113
Boiling point (°C)	−188	−34.6	58	183
Dissociation or bond energy $X_2 \longrightarrow 2X$ (kcal/mole)	38	57.6	45.4	35.5
$\Delta H_{reduction}$ (kcal/mole)	−366	−286	−270	−244
Reduction potential $X_2 + 2e^- \longrightarrow 2X^-$ (volts)	2.87	1.36	1.06	0.54
Electronegativity	4.0	3.0	2.8	2.5
$\Delta H_{hyd.}$ of X^- (kcal)	−122	−87	−81	−72
ΔH_f of HX (kcal)	−64	−22	−8.66	+6.2

The data in Table 18-1 show that the ability of a halogen atom to form a halide ion decreases as the radius of the atom increases. Iodine atoms have a radius of 1.33 Å and the $I_2 + 2e^- \longrightarrow 2I^-$ half-cell has a reduction potential of 0.54 v. Fluorine atoms have a radius of 0.72 Å and the $F_2 + 2e^- \longrightarrow 2F^-$ half-cell has a reduction potential of 2.87 v. The smaller fluorine atoms have more tendency to become halide ions than do the larger iodine atoms.

The *halogens* are *relatively strong oxidizing agents.* The data in Table 18-1 show that *within the family, the oxidizing strength of the halogens decreases as the atomic number increases.* Thus, fluorine, with the highest reduction potential, is the strongest oxidizing agent while iodine, with the lowest reduction potential, is the weakest.

The halogen atoms may form single covalent bonds and achieve noble-gas electron configurations by sharing electrons with either nonmetallic atoms, or with metallic atoms having intermediate electronegativity values. Metallic atoms such as beryllium and aluminum, which have electronegativity values of 1.5, combine with chlorine, bromine, and iodine, forming chlorides, bromides, and iodides which are largely covalent in nature. *All of the halogens* in the free elemental state *form diatomic molecules.* Thus, both atoms in the diatomic molecule achieve the electron configuration of a noble gas. The dissociation energies listed in Table 18-1 show that the halogen molecules are thermally stable. It may be seen, however, that the thermal stability of I_2 is considerably less than that of Cl_2. The trends in dissociation energy and other properties may be interpreted in terms of electronegativity values, atomic radii, and other atomic properties discussed in Chapter 7.

18-2 The Physical Phases of the Halogens The melting and boiling points listed in Table 18-1 show that, at room temperature, *fluorine* and *chlorine* are *gases, bromine* is a *liquid,* and *iodine* is a *solid.* Using our knowledge of periodicity, we would predict that astatine would be a darkly colored solid, more dense and less soluble in water than iodine. We would also predict that it would have a smaller ionization energy and higher melting point than iodine. The *physical phases of the halogens reflect the van der Waals forces of attraction* between their molecules. These forces, discussed in Section 9-9, tend to increase with atomic number. Iodine molecules have 106 electrons, while chlorine molecules have only 34 electrons. Thus, the van der Waals forces of attraction are greater between iodine molecules than between chlorine molecules. The higher melting and boiling points of iodine are a reflection of the stronger van der Waals forces.

PREPARATION OF THE HALOGENS

18-3 Chlorine and Bromine The halogens may be prepared by oxidizing the halide ions either electrolytically or, with the exception of fluoride ions, by means of oxidizing agents. Chlorine was first prepared by the Swedish chemist, K. W. Scheele (1741–1786), by the reaction of MnO_2 with concentrated hydrochloric acid (12*M*). The equation for the reaction, which is still commonly used for the *laboratory preparation of chlorine is*

$$\mathbf{MnO_2(s) + 4HCl(aq) \xrightarrow{heat} Cl_2(g) + MnCl_2(aq) + 2H_2O(l)}$$

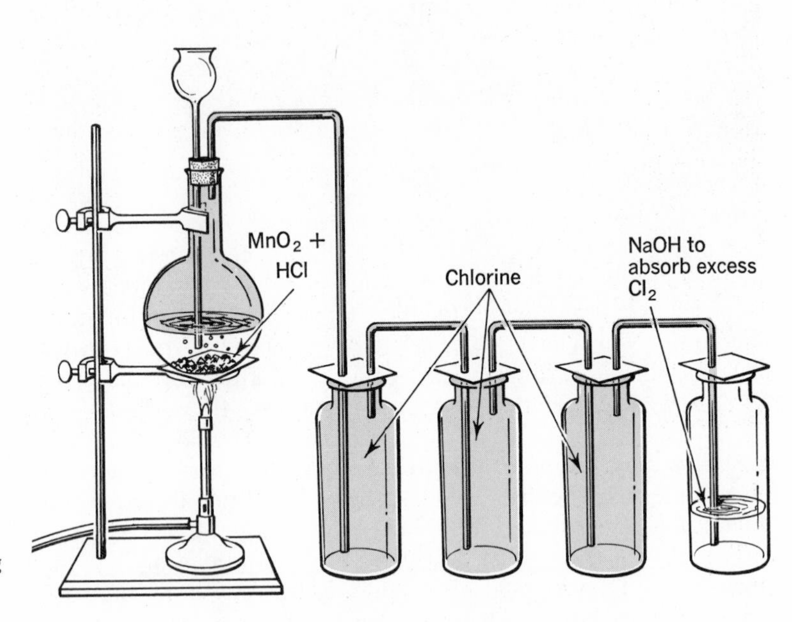

18-1 Apparatus for generating and collecting chlorine.

In this reaction, *MnO_2* serves as the *oxidizing agent.* The reaction may be carried out in an apparatus similar to that illustrated in Fig. 18-1. *Chlorine,* which is more dense than air and moderately soluble in water, is *collected by the upward displacement of air* in the collection bottles. In the laboratory preparation of chlorine, NaCl or other alkali chlorides may be used as a source of Cl^- ions. In these reactions, sulfuric acid is generally used to provide an acid solution. Any oxidizing agent in Table 17-6 which has a reduction potential greater than that of chlorine may be used in these reactions. For example, $KMnO_4$ or $K_2Cr_2O_7$ are strong enough to oxidize chloride ions to elemental chlorine. The ionic equation for the reaction between $K_2Cr_2O_7$ and NaCl is

$$\mathbf{Cr_2O_7^{2-}(aq) + 6Cl^-(aq) + 14H^+(aq) \longrightarrow 3Cl_2(g) + 2Cr^{3+}(aq) + 7H_2O(l)}$$

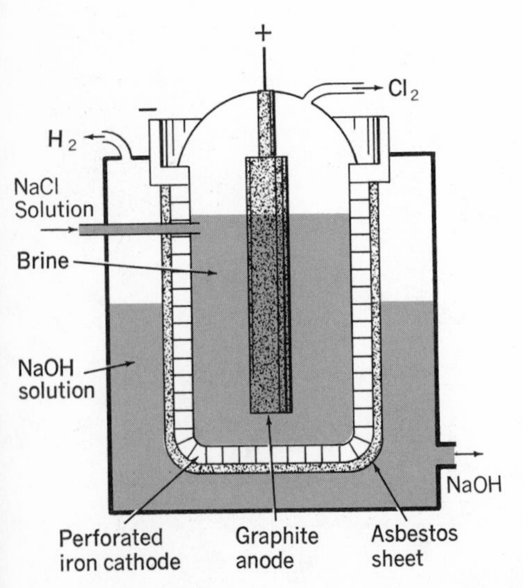

18-2 A cell for producing chlorine.

Chlorine is prepared industrially by the electrolysis of melted NaCl in a *Downs cell* (Fig. 17-21) or by the electrolysis of a sodium chloride solution (brine) in cells of various designs (Fig. 18-2). These reactions were described and illustrated in Section 17-23. The chlorine from the commercial electrolytic cells is liquefied and stored or transported in steel tanks. Large quantities of chlorine

18-3 Rows of Hooker cells. Electrolysis of concentrated NaCl solution produces a continuous flow of Cl_2, H_2, and NaOH from these cells. (Rotkin, Photography for Industry)

are used by the chemical industry to produce plastics, insecticides, bleaches, solvents such as carbon tetrachloride (CCl_4), and many other chlorine compounds.

The laboratory preparation of bromine involves the same general principles as the preparation of chlorine. Less powerful oxidizing agents may be used. That is, any oxidizing agent that lies above the $Br_2 + 2e^- \longrightarrow 2Br^-$ half-cell in Table 17-6 oxidizes Br^- ions. For example, Cl_2 may be used to oxidize Br^- ions. The reaction is

$$\mathbf{Cl_2(g) + 2Br^-(aq) \longrightarrow Br_2(l) + 2Cl^-(aq)}$$

18-4 Iodine Iodine may be prepared in the laboratory by gently warming a mixture of KI, MnO_2, and H_2SO_4 in an apparatus similar to that shown in Fig. 18-4. During the heating, purple vapors of iodine rise from the reaction mixture and condense as a solid on the cold surface of the porcelain dish without passing through the liquid phase. The equation for the reaction is

$$\mathbf{2KI(s) + MnO_2(s) + 3H_2SO_4(l) \longrightarrow I_2(s) + 2KHSO_4(aq) + MnSO_4(aq) + 2H_2O(l)}$$

Close examination of the solid residue on the bottom of the dish shows that iodine is a shiny, dark, flaky solid with an almost metallic appearance. As a solid, *iodine has a relatively high vapor pressure.* When the dark crystals of solid iodine are warmed, a cloud of violet iodine vapor is produced. At ordinary pressures, iodine *sublimes;* that is, it passes directly from the solid to the vapor phase. X-ray diffraction studies indicate that solid iodine consists of a regular arrangement of I_2 molecules (Fig. 9-8).

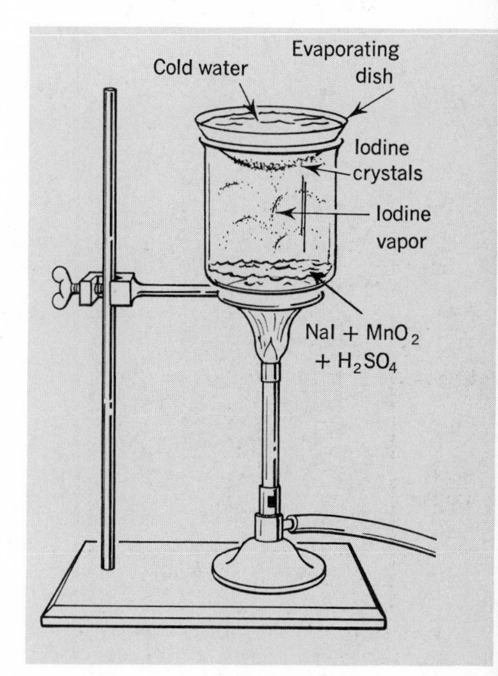

18-4 Apparatus for preparing iodine crystals.

It is estimated that iodine constitutes 3×10^{-5} percent by mass of the earth's crust (including the oceans). Iodine occurs principally as sodium iodate in $NaNO_3$ deposits found in Chile, S.A. Iodine in $NaIO_3$ is in the 5+ oxidation state; thus, in order to obtain the free element, a reducing agent must be used. In the industrial preparation of I_2 from $NaIO_3$, sodium hydrogen sulfite ($NaHSO_3$) is used as the reducing agent. The equation for the reaction is

$$\mathbf{2NaIO_3(aq) + 5NaHSO_3(aq) \longrightarrow I_2(s) + 3NaHSO_4(aq) + 2Na_2SO_4(aq) + H_2O(l)}$$

The net ionic equation is

$$\mathbf{2IO_3^-(aq) + 5HSO_3^-(aq) \longrightarrow I_2(s) + 3HSO_4^-(aq) + 2SO_4^{2-}(aq) + H_2O(l)}$$

Iodine and iodine compounds find some use as antiseptics. In addition, iodine in the form of iodides is an important component of the human diet. The thyroid gland produces an iodine compound called *thyroxin,* which has a regulatory function in the oxidation process that occurs in the cells. The absence of iodine in our diet causes a condition called a *goiter.* This condition is manifested as an enlargement of the thyroid gland resulting from its apparent attempt to produce more thyroxin. The addition of 0.02 percent NaI to table salt (iodized salt) is enough to prevent this condition. Radioactive iodine added in small amounts to the bloodstream of animals or humans is soon found to be concentrated in the thyroid gland. This method is sometimes used in diagnosing disorders of the thyroid gland because the radioactive iodine can easily be detected by using an electronic counting device.

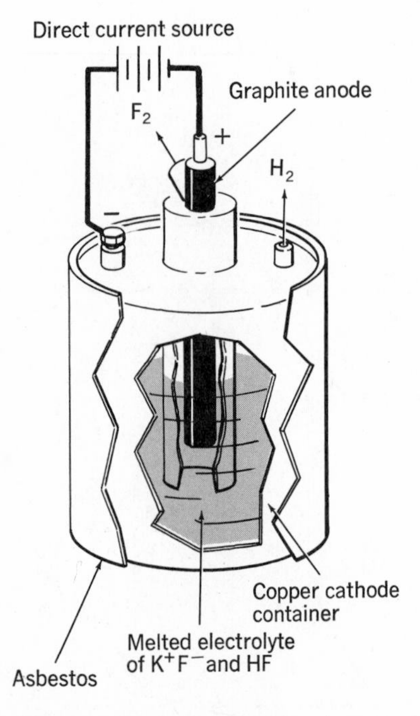

18-5 Electrolytic cell for preparing elementary fluorine. The reactions which take place in this cell are Cathode $2H^+ + 2e^- \longrightarrow H_2$ Anode $2F^- \longrightarrow F_2 + 2e^-$

18-5 Fluorine *Fluorine is the most reactive of all nonmetals.* It is so reactive that it ignites most metals and even ordinary rock (quartz) at room temperature. The position of the fluorine half-cell at the top of Table 17-6 indicates that it is the *strongest common oxidizing agent known.* This means that fluoride ions (F^-) cannot be oxidized by other chemical oxidizing agents. The stability of the ionic fluorides is attributed to the small dissociation energy of F_2 molecules and the small radius of F^- ions. In general, the bond energies of fluorine compounds are greater than those of any of the corresponding compounds formed by the other halogens. Thus, both metallic and nonmetallic fluorides are energetically more stable than any of the other corresponding halides.

Because of its great oxidizing strength, *elemental fluorine* can be prepared only electrolytically. In a practical sense, it was first produced in 1886 by Henri Moissan (1852–1907) of France. He electrolyzed a melted mixture of potassium fluoride and hydrogen fluoride in a crucible made of a platinum-iridium alloy, one of the few materials which could withstand the powerful oxidizing action of the fluorine produced by the process. Fluorine is now commercially produced in large quantities in copper furnaces by essentially the same process. It is transported in steel-alloy tanks. Steel in the presence of fluorine becomes coated with a thin layer of iron(III) fluoride which resists further attack by the elemental fluorine.

Before World War II, fluorine was scarcely more than an academic curiosity. The demand for fluorine in the processing of uranium for the atomic bomb stimulated fluorine research which ultimately resulted in the large-scale production of the element and the development of a number of useful fluorine products. The Freons, such as CCl_2F_2 and $CHClF_2$, are widely used as refrigerants. These compounds are inert and volatile and have largely replaced the more reactive, toxic refrigerants such as ammonia, sulfur dioxide, and methyl chloride. In addition, the Freons are used as propellants in many types of aerosol spray cans for spraying paint, insecticides, hair dressings, and shaving cream. *Teflon,* a slippery-surfaced plastic, which is resistant to heat and the action of many corrosive chemical agents, is also a fluorocarbon. Because of its properties, this versatile plastic finds applications that range from mechanical bearings and bushings to coatings for greaseless frying pans. It is widely used as a structural material in industry wherever resistance to heat and chemical corrosion is required. Teflon is a *polymer* consisting of long chains of carbon atoms covalently bonded to fluorine atoms. *Polymers* are giant molecules whose long chains are composed of recurring units called *monomers.* A monomer unit in Teflon may be represented as shown in the margin. *n* represents the number of identical units that compose the polymer.

18-6 The reaction of fluorine gas with water is so violent that its solubility cannot be determined. The flame furnishes visual evidence of the large amount of energy evolved. (Pennsalt Chemicals Corporation)

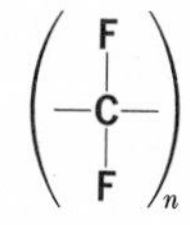

Thus far we have examined and explained some of the properties, methods of preparation, and uses of the halogens. Let us next investigate and explain the properties and behavior of some of the halogen acids.

HYDROGEN HALIDES AND HALOGEN OXYACIDS

18-6 Hydrogen Fluoride In the laboratory, the *hydrogen halides* (HF, HCl, HBr, and HI) can be prepared by adding a concentrated nonvolatile acid such as H_2SO_4 or H_3PO_4 to one of their salts. Thus, HF may be prepared in the laboratory by adding concentrated sulfuric acid to a fluoride salt such as CaF_2.

$$\mathbf{CaF_2(s) + H_2SO_4(l) \longrightarrow 2HF(g) + CaSO_4(s)}$$

A water solution of HF is known as hydrofluoric acid. A unique property of hydrofluoric acid is its ability to *etch glass.* The markings on the graduated glass apparatus in your laboratory locker were probably etched by using hydrofluoric acid. The etching is the result of the reaction between HF and the silicon dioxide (SiO_2) in the glass.

$$\mathbf{SiO_2(s) + 4HF(aq) \longrightarrow SiF_4(g) + 2H_2O(l)}$$

Hydrofluoric acid cannot be stored in glass containers. This acid was originally stored in wax bottles but is now packaged in Teflon or other corrosion-resistant plastic containers. Solutions of HF must be handled with care because they produce sores on the skin which heal very slowly.

Commercially, except HF, the hydrogen halides are produced to some extent by direct combination of the halogen with H_2.

$$\tfrac{1}{2}H_2(g) + \tfrac{1}{2}F_2(g) \rightleftharpoons HF(g) + 64 \text{ kcal}$$

This reaction is highly exothermic. It occurs with explosive violence, even at low temperatures. The formation of the energetically stable product, HF, results in the equilibrium being displaced far to the right. H_2 and Cl_2 do not react at room temperature unless activated by ultraviolet radiation. Once initiated, this reaction also proceeds with explosive violence. The data in Table 18-1 show that ΔH_f° for this reaction is -22 kcal/mole of HCl. Also ΔH_f° for HBr is -8.66 kcal/mole and ΔH_f° for HI is 6.2 kcal/mole. There is much less tendency for HBr and HI to be formed from their elements at standard state than there is for HF and HCl to be formed. Hydrogen gas reacts appreciably with iodine gas only in the presence of a catalyst and at an elevated temperature. Under these conditions the ΔH_r for the reaction $\tfrac{1}{2}H_2(g) + \tfrac{1}{2}I_2(g)$ is -5 kcal/mole. *The progressive decrease in the energy evolved as we go from HF to HI reflects the decrease in the tendency of the halogens to react with H_2 and the decrease in the strength of the bond between the hydrogen and halogen atoms.*

Hydrogen fluoride gas contains components which have a molecular weight of about 120, suggesting that its molecular formula is H_6F_6. Chemists believe that it exists in a six-membered ring of fluorine atoms with hydrogen atoms bridged between each pair of fluorine atoms (Fig. 18-7). This structure is caused by the *hydrogen bonding* that exists between the H atom of one HF molecule and the fluorine atom of the neighboring HF molecule. The species, H_2F_2, H_3F_3, H_4F_4, and H_5F_5 have also been identified in hydrogen fluoride gas.

18-7 Hydrogen Chloride Hydrogen chloride may be prepared at room temperature by the reaction of NaCl with concentrated H_2SO_4.

$$NaCl(s) + H_2SO_4(l) \rightleftharpoons HCl(g) + NaHSO_4(s)$$

The reaction mixture may be heated in order to evaporate the HCl and force the equilibrium to the right. Concentrated H_2SO_4 is diprotic (dibasic) and only slightly dissociated. The dissociation expression is

$$2H_2SO_4 \rightleftharpoons H_3SO_4^+ + HSO_4^-$$

Thus, at moderate temperatures, only a single proton is furnished in the form of $H_3SO_4^+$ by a H_2SO_4 molecule. The HSO_4^- ions combine with Na^+ ions and form $NaHSO_4$. Removal of a second proton

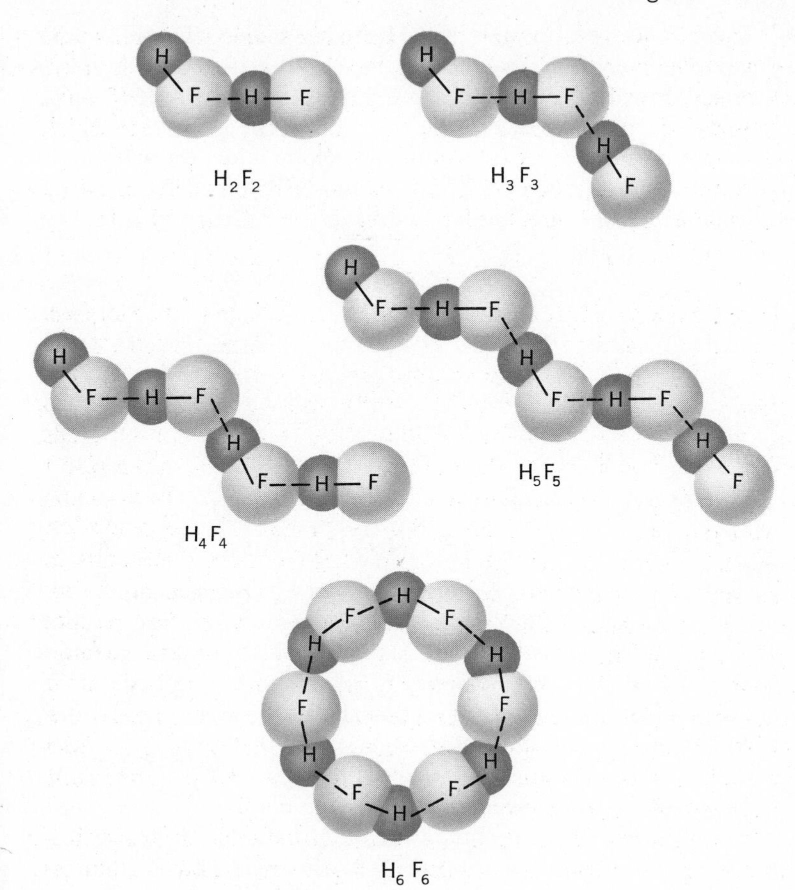

18-7 Hydrogen fluoride tends to form polymers with formulas of H_2F_2, H_3F_3, H_4F_4, H_5F_5, and H_6F_6. This clustering is caused by hydrogen bonding between neighboring HF molecules.

from HSO_4^- ions requires a higher temperature. At higher temperatures, the second reaction occurs.

$$\mathbf{NaCl(s) + NaHSO_4(s) \longrightarrow HCl(g) + Na_2SO_4(s)}$$

Even though HCl is a stronger Brønsted acid than HSO_4^-, the volatility of HCl prevents the reverse reaction from taking place to any extent.

Hydrogen chloride gas is colorless, has a penetrating pungent odor, and is extremely irritating to mucous membranes. It is very soluble in water. *A solution of hydrogen chloride in water is called hydrochloric acid.* Its nontechnical name is muriatic acid. Some of this acid is produced commercially by burning hydrogen gas in chlorine and then absorbing the hydrogen chloride in water. About one-half of the production of HCl is a by-product of the chlorination of hydrocarbons.

$$\mathbf{H_2(g) + Cl_2(g) \longrightarrow 2HCl(g)}$$
$$\mathbf{HCl(g) + H_2O(l) \longrightarrow H_3O^+(aq) + Cl^-(aq)}$$

18-8 Hydrogen Bromide and Hydrogen Iodide The reduction potential for concentrated H_2SO_4 is such that this acid is not a strong enough oxidizing agent to oxidize HF to F_2 or HCl to Cl_2. It is, however, powerful enough to oxidize HBr to Br_2 and HI to I_2. Therefore, if concentrated sulfuric acid is added to a bromide or iodide, the resulting hydrogen halide (HBr or HI) is contaminated with the halogen and the sulfur dioxide (SO_2) formed when the H_2SO_4 is reduced.

$$H_2SO_4(l) + 2HBr(aq) \longrightarrow Br_2(g) + SO_2(aq) + 2H_2O(l)$$
$$H_2SO_4(l) + 2HI(aq) \longrightarrow I_2(s) + SO_2(g) + 2H_2O(l)$$

To overcome this difficulty, a nonvolatile acid such as phosphoric acid (H_3PO_4), which is a weaker oxidizing agent than concentrated H_2SO_4 may be used.

$$H_3PO_4 + NaBr(s) \longrightarrow H_2PO_4^- + HBr(g) + Na^+$$
$$H_3PO_4 + NaI(s) \longrightarrow H_2PO_4^- + HI(g) + Na^+$$

18-9 Strength of Acids Formed by the Hydrogen Halides All of the hydrogen halides except hydrogen fluoride, form strong acids. Hydrofluoric acid is a weak acid with a dissociation constant (K_a) of 6.8×10^{-4}. In the series of hydrogen halides, the H—F bond is much stronger than that in the other molecules. This is caused primarily by the small radius of fluorine atoms. Energetically, the dissociation of HF is favored less than that of other acids.

The small F^- ions have a greater charge density and are, therefore, more highly hydrated than the other halide ions. This represents a greater degree of order for the solvent and means that the hydration effect decreases the entropy of the HF system more than that of the others. Hence, the entropy factor also operates to make HF a weaker acid than the others.

18-10 The Oxidizing Ability of the Halogen Oxyacids and Oxyanions The relatively large reduction potentials shown in Tables 18-2A and 18-2B indicate that all of the *oxyanions* and *oxyacids* formed by chlorine are, in theory, rather powerful oxidizing agents in either acid or basic solutions. In addition, all species except perchlorate ions (ClO_4^-) can act as reducing agents. For example, as a reducing agent ClO_3^- is oxidized to ClO_4^- and as an oxidizing agent ClO_3^- can be reduced to ClO_2, $HClO_2$, HClO, Cl_2 or Cl^-. The extent to which an oxyanion or oxyacid is reduced depends upon the strength and concentration of the reducing agent, the pH of the solution, the temperature, and other factors.

In certain reactions, a single species may become oxidized and reduced. These are called *self-oxidation-reduction* (*disproportionation*) *reactions*. Thus in acid solution, chlorate ions may undergo simultaneous oxidation and reduction. The equation for this disproportionation is

$$7ClO_3^- + 2H_3O^+ \longrightarrow Cl_2 + 5ClO_4^- + 3H_2O$$

In ClO_3^- ions, chlorine has an oxidation number of 5+; in ClO_4^- ions, its oxidation number is 7+; in Cl_2, the oxidation number is zero. Thus, in the reaction above, chlorine is oxidized from a 5+ to a 7+ state, and reduced from a 5+ to a 0 oxidation state. The overall equation and the standard potential for the reaction may be obtained by combining the two half-cell reactions which may be obtained from Table 18-2A or by constructing and combining your own half-cell reactions. The half-cell reactions are

$$\mathbf{2ClO_3^- + 12H_3O^+ + 10e^- \rightleftharpoons Cl_2(g) + 18H_2O \quad E° = 1.46\ v}$$
$$\mathbf{5ClO_3^- + 15H_2O \rightleftharpoons 5ClO_4^- + 10H_3O^+ + 10e^- \quad E° = -1.19\ v}$$
$$\mathbf{7ClO_3^- + 2H_3O^+ \rightleftharpoons Cl_2(g) + 5ClO_4^- + 3H_2O \quad E° = 0.27\ v}$$

Disproportionation may, in theory, occur when two half-cell reactions containing the same species can be combined and yield a positive cell voltage. Whether the reaction will actually occur depends on the rate and mechanism of the reaction. The data in Tables 18-2A and 18-2B show that the pH of a solution greatly affects the reduction potentials and, therefore, the tendency for a given species to undergo disproportionation. It can be seen from the reduction potentials in Tables 18-2A and 18-2B that the oxyacids and oxyanions of the halogens are not as powerful oxidizing agents in basic solutions as they are in acid solutions.

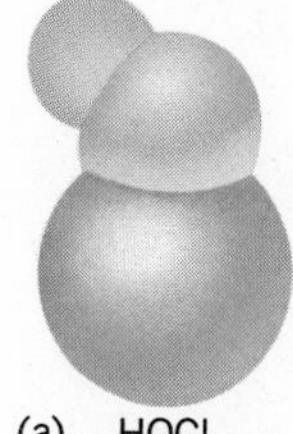

(a) HOCl (HClO)

(b) (HOClO) ($HClO_2$)

(c) ($HOClO_2$) ($HClO_3$)

(d) ($HOClO_3$) ($HClO_4$)

18-8 Space-filling models of the four oxyacids of chlorine. (a) hypochlorous acid, (b) chlorous acid, (c) chloric acid, (d) perchloric acid.

TABLE 18-2A

REDUCTION POTENTIALS FOR HALOGEN OXYANIONS AND OXYACIDS IN AQUEOUS ACID SOLUTION

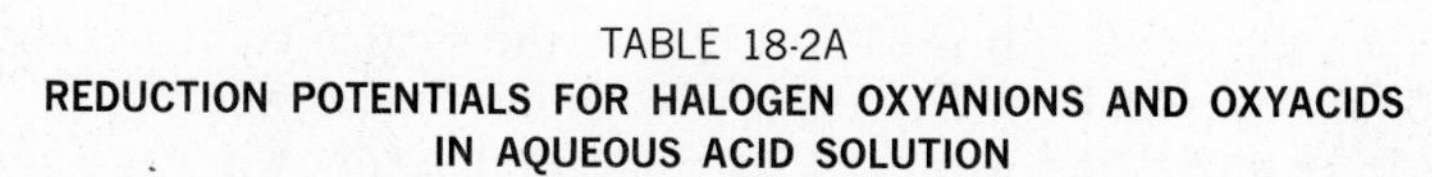

	Reducing Agents					
Oxidizing agents	*$Cl^-(aq)$*	*$Cl_2(g)$*	*$HClO(aq)$*	*$HClO_2(aq)$*	*$ClO_2(g)$*	*$ClO_3^-(aq)$*
$ClO_4^-(aq)$	+1.39	+1.39	+1.35	+1.20	+1.17	+1.19
$ClO_3^-(aq)$	+1.44	+1.46	+1.42	+1.21	+1.14	
$ClO_2(g)$	+1.51	+1.55	+1.52	+1.28		
$HClO_2(aq)$	+1.54	+1.65	+1.64			
$HClO(aq)$	+1.50	+1.63				
$Cl_2(g)$	+1.36					

TABLE 18-2B

REDUCTION POTENTIALS FOR HALOGEN OXYANIONS AND OXYACIDS IN AQUEOUS BASIC SOLUTION

	Reducing Agents					
Oxidizing agents	*$Cl^-(aq)$*	*$Cl_2(g)$*	*$ClO^-(aq)$*	*$ClO_2^-(aq)$*	*$ClO_2(g)$*	*$ClO_3^-(aq)$*
$ClO_4^-(aq)$	+0.56	+0.45	+0.45	+0.34	+0.07	+0.36
$ClO_3^-(aq)$	+0.63	+0.48	+0.50	+0.33	−0.50	
$ClO_2(g)$	+0.86	+0.73	+0.83	+1.16		
$ClO_2^-(aq)$	+0.78	+0.58	+0.66			
$ClO^-(aq)$	+0.89	+0.42				
$Cl_2(g)$	+1.36					

Two widely used salts of the chlorine oxyacids are prepared by disproportionation reactions. Sodium hypochlorite (NaOCl), a constituent of a variety of products sold as bleaching agents, is prepared by bubbling chlorine gas into a solution of NaOH. The equation for the disproportionation of Cl_2 in the basic solution is

$$Cl_2(g) + 2OH^-(aq) \rightleftharpoons Cl^-(aq) + ClO^-(aq) + H_2O(l)$$

Sodium chlorate ($NaClO_3$) may then be produced by heating the hypochlorite solution. This action results in disproportionation of ClO^- ions.

$$3ClO^-(aq) \rightleftharpoons ClO_3^-(aq) + 2Cl^-(aq)$$

When hypochlorites are acidified, they are converted to hypochlorous acid (HClO).

$$H_3O^+(aq) + ClO^-(aq) \rightleftharpoons HClO(aq) + H_2O(l)$$

HClO may also be produced by bubbling chlorine gas into water

$$Cl_2(g) + 2H_2O(l) \rightleftharpoons H_3O^+(aq) + Cl^-(aq) + HClO(aq)$$

HClO is a powerful oxidizing agent (E° = 1.63 v) which decomposes into HCl and O_2

$$2HClO(l) \longrightarrow 2HCl(aq) + O_2(g)$$

When the solution is used as a bleach, the oxygen combines with certain dyes and oxidizes them to colorless compounds.

It is interesting to note that some brands of swimming-pool chlorine are really sodium hypochlorite (NaClO) in a form more concentrated than in Purex®. Also, pool acid is concentrated hydrochloric acid.

18-11 Factors Affecting the Magnitude of Standard Reduction Potentials for the Halogen-Halide Half-cells In Chapter 11 we noted and demonstrated that the net enthalpy change for a process could be interpreted in terms of a number of intermediate processes each involving an energy change (Born-Haber cycle). In terms of enthalpy changes, the reduction potential for the conversion of a nonmetallic element to a hydrated ion represented by the general equation

$$X_2(g) + 2e^- \rightleftharpoons 2X^-(aq) \qquad 18\text{-}1$$

may be interpreted in terms of these factors

1. $\Delta H_{sublimation}$ or $\Delta H_{vaporization}$, depending on whether the element is a solid or liquid. In the case of a gas this term is not relevant.
2. $\Delta H_{dissociation}$
3. $\Delta H_{electron\ affinity}$
4. $\Delta H_{hydration}$
5. $\Delta H_{H_2 \rightarrow 2H^+(aq) + 2e^-}$

The overall enthalpy change, $\Delta H_{reduction}$, associated with the reduction potential (E°) for Equation 18-1 is

$$\Delta H_{red} = (\Delta H_s \text{ or } H_v) + \Delta H_D + 2\Delta H_E + 2H_H + \Delta H_{H_2 \rightarrow 2H^+(aq)+2e^-}$$

As ΔH_{red} becomes more negative, the tendency for the reaction to occur becomes greater and E° increases.

In the sequence above, ΔH_E and ΔH_H represent exothermic processes (ΔH is negative). Therefore, increases in hydration energies and electron affinities (ΔH more negative) are paralleled by a decrease in $\Delta H°_{red}$ and an increase in reduction potential, E°. On the other hand, increases in dissociation, sublimation or vaporization energies tend to increase $\Delta H°_{red}$ and decrease reduction potentials. The last term, representing the energy to convert $H_2 \longrightarrow 2H^+(aq)$ and to furnish electrons for the reduction process, refers to the reference electrode (standard hydrogen half-cell) and is constant for any half-cell we consider.

To the enthalpy considerations we must add the effect of entropy changes on standard reduction potentials. The change from a gaseous phase to an aqueous phase results in a decrease in entropy. In addition, the attraction between the ions and the water molecules results in a more ordered condition of the solvent. These unfavorable entropy changes oppose the favorable enthalpy changes so that, in general, the reduction potential is not as large as ΔH_{red} would indicate.

The enthalpy changes related to the overall reduction potential can be expressed in terms of a cyclic process consisting of several energy steps analogous to those found in the Born-Haber cycle illustrated in Fig. 11-16. A sketch of the energy cycle which is related to the reduction potential for the

$$2e^- + Cl_2(g) \longrightarrow 2Cl^-(aq)$$

half-cell is shown in Fig. 18-9.

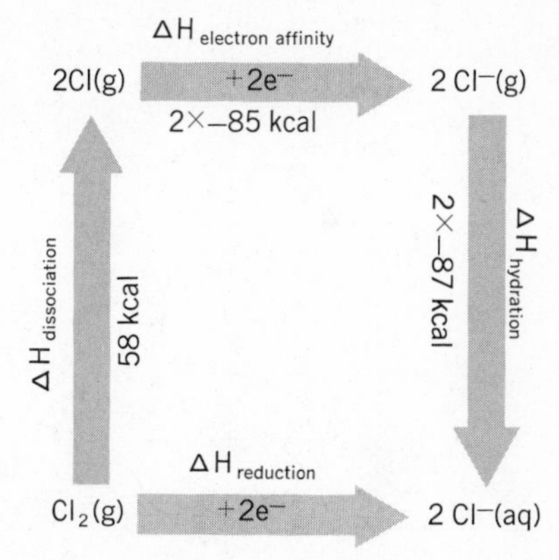

18-9 Born-Haber cycle showing enthalpy changes related to the reduction potential of the chlorine-chloride half-cell.

The conversion of chlorine atoms to chloride ions and the hydration of chloride ions are exothermic processes which tend to decrease ΔH_{red} and to increase the reduction potential. The dissociation of Cl_2 molecules is endothermic and tends to increase ΔH_{red}. The overall enthalpy change, ΔH_{red}, is equal to the sum of steps a, b, and c. Thus,

$$\Delta H_{red} = 58 \text{ kcal} + (-174 \text{ kcal}) + (-170 \text{ kcal}) = -286 \text{ kcal}$$

The data in Table 18-1 show that a fluorine half-cell has a very high reduction potential. The energy cycle (Fig. 18-10) shows that the high reduction potential of the fluorine-fluoride half-cell is associated with a large negative value of ΔH_{red} and is, in large part, caused by the small dissociation energy and the large hydration energy. The large hydration energy is, in turn, related to the

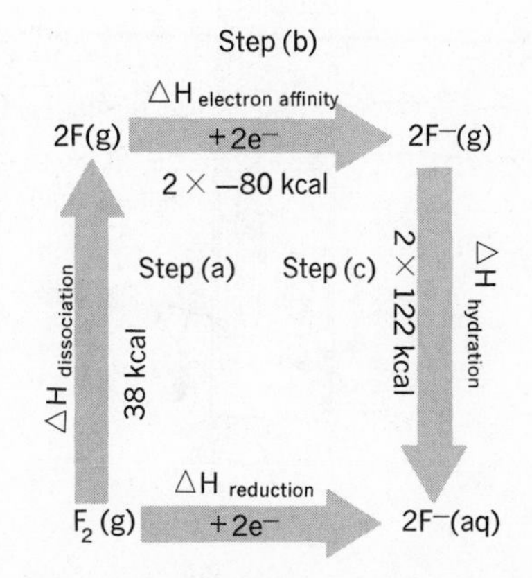

18-10 Born-Haber cycle showing enthalpy changes related to the reduction potential of the fluorine-fluoride half-cell.

small radius of fluoride ions. The data in Fig. 18-10 show that the overall enthalpy change (ΔH_{red}) for the half-cell reaction

$$2e^- + F_2(g) \longrightarrow 2F^-(aq)$$

is 38 kcal + (−244 kcal) + (−160 kcal) = −366 kcal. The increase in ΔH_{red} from the fluorine half-cell to the iodine half-cell is paralleled by a decrease in the reduction potentials. Unfavorable entropy changes oppose the favorable enthalpy changes so that the reduction potential is not as large as ΔH_{red} would indicate.

FOLLOW-UP PROBLEM

(a) Draw an energy cycle and calculate ΔH_{red} for

$$2e^- + Br_2(l) \longrightarrow 2Br^-(aq)$$

and

$$2e^- + I_2(s) \longrightarrow 2I^-(aq)$$

using these data.

Bromine $\Delta H_{ea} = -80$ kcal; $\Delta H_{hyd} = -81$ kcal;
$\Delta H_{diss} = 45.4$ kcal/mole;
$\Delta H_{vap} = 7$ kcal/mole

Iodine $\Delta H_{ea} = -73$ kcal; $\Delta H_{hyd} = -72$ kcal;
$\Delta H_{diss} = 35.5$ kcal/mole,
$\Delta H_{sub} = 10$ kcal/mole

(b) Is the trend in ΔH_{red} consistent with the trend in reduction potential? Explain

Ans. (a) ΔH_{red} for Br_2 half-cell = −270 kcal/mole, ΔH_{red} for I_2 half-cell = −244 kcal/mole, (b) Yes, as ΔH_{red} increases, E° decreases.

LOOKING AHEAD

In this and earlier chapters we have covered some of the chemistry of the important nonmetallic elements as well as that of the metals in the groups of representative elements (A groups). We have used the *light metals* of Groups IA and IIA to demonstrate periodic properties on a number of occasions. This is because there are more similarities between the members of the Group IA and Group IIA metals than between the members of any nonmetallic family.

We have had little opportunity and made little effort to discuss the chemistry of the *heavy metals* of the transition series in the Periodic Table. We shall find in the next chapter that the properties of the transition metals differ considerably from those of the representative metals. The explanation of many of these properties provides an opportunity to use and extend structure and bonding concepts discussed in earlier chapters. You will find the study of these important elements an unique experience.

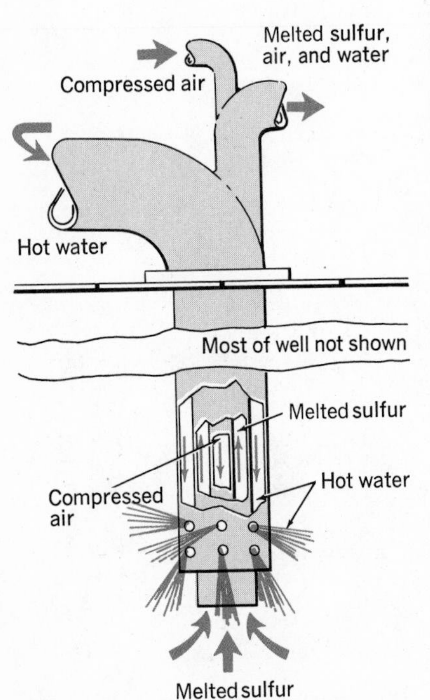

18-11 The Frasch process is a means of melting sulfur and bringing it to the surface. Why isn't this method used to mine other mineral deposits?

Special Feature: Industrial Chemistry

THE FRASCH PROCESS

The importance of sulfur to the chemical industry and the economy of the nation cannot be overemphasized. It is required for the synthesis of compounds used in the production of materials vital to our economy and way of life. Sulfur is the raw material

18-12 Sulfur brought to the surface by the Frasch process is about 99.9% pure. It is stored in huge bins such as that shown. It is then blasted and loaded into railroad cars or barges for shipment. (Freeport Sulphur Co.)

required for the production of sulfuric acid, the so-called "king of the chemicals." Over 20 million tons of sulfuric acid is produced each year in the United States. The fertilizer, steel, and petroleum industries consume enormous quantities of this acid. Large quantities of sulfur are also used to synthesize chemicals used in the production of synthetic fabrics, rubber, fungicides, drugs, and other useful products. We shall first examine one method for obtaining pure sulfur and then investigate the preparation and properties of its most common compound.

Elemental sulfur is often found in huge underground deposits several hundred feet below the earth's surface. The main deposits in the United States are found in Texas and Louisiana. This sulfur is recovered as a 99.9-percent pure element by the *Frasch process.* The underground sulfur layer is reached by using a drilling rig similar to those used for drilling oil wells. *Three concentric pipes* are lowered into the sulfur deposits. The arrangement is illustrated in Fig. 18-11. Water, superheated under pressure to a temperature above the melting point of sulfur, is pumped down the outermost pipe and melts the sulfur. Compressed air is forced down the small central pipe and causes the melted sulfur to form a low-density froth which is easily forced to the earth's surface through the intermediate pipe. On the surface, the sulfur is allowed to solidify in huge storage bins where it can be blasted and loaded into railroad cars and ships for marketing.

Sulfur, like oxygen, exists in several allotropic forms. Whereas the allotropes of oxygen are different molecular forms of the same element, the allotropes of sulfur are *different crystalline forms* of the same element. The two common crystalline allotropic forms of sulfur are *rhombic* and *monoclinic* sulfur. When you examine sulfur in the laboratory, you find that crystals of rhombic sulfur are diamond-shaped and that monoclinic crystals are long and needle-like (Fig. 18-13).

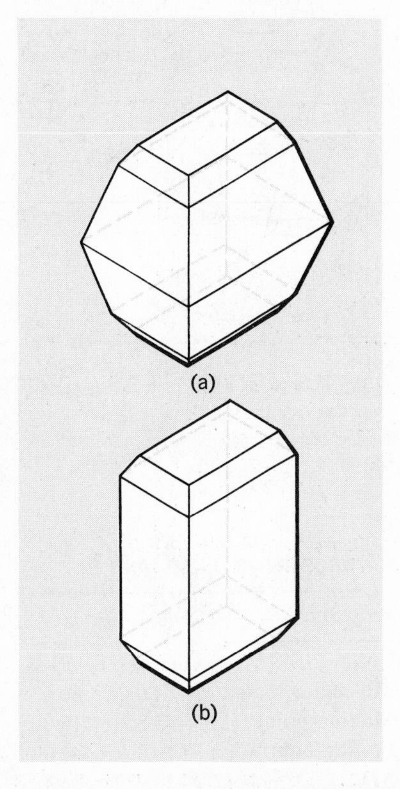

18-13 Crystalline allotropes of sulfur (a) rhombic, (b) monoclinic or prismatic.

18-14 Flow diagram for contact process.

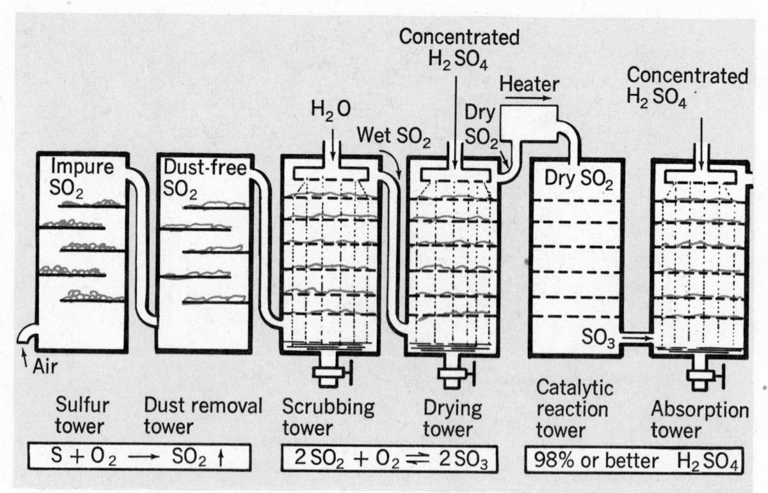

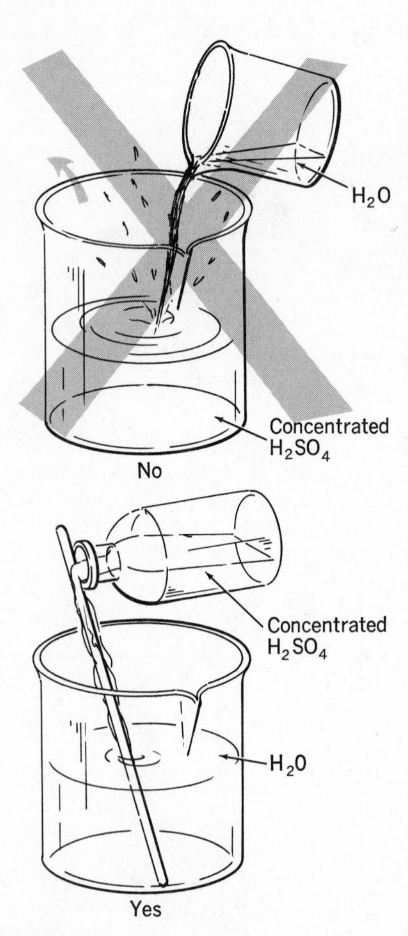

18-15 In order to dilute concentrated sulfuric acid safely, add the acid slowly to the water with constant stirring. If water is added to the acid, the heat evolved may cause the acid to spatter.

TABLE 18-3
PROPERTIES OF SO_2 AND SO_3

Property	SO_2	SO_3
Color	colorless	colorless
Molecular mass	64.0	80.0
Melting point (°C)	−75.5	16.8
Boiling point (°C)	−10.0	44.6
ΔH_f° (kcal/mole)	−70.7	−94.4
ΔG_f° (kcal/mole)	−72.0	−88.5

Under ordinary room conditions, *rhombic sulfur is the stable crystalline form.* X-ray analysis and electron diffraction measurements reveal that a crystal cell contains 16 S_8 molecules. These molecules consist of eight sulfur atoms covalently bonded to each other in a puckered-ring structure (Fig. 9-13). Since each sulfur atom shares a pair of electrons with each of two other sulfur atoms, it completes its outer octet and achieves the noble-gas electronic configuration. The properties of sulfur were interpreted in terms of this structure in Sect. 9-9.

The thermodynamic properties, ΔH_f° and ΔG_f°, indicate that SO_2 is very stable with respect to its decomposition into S and O_2. Very little decomposition takes place up to 2000°C. Sulfur trioxide (SO_3) involved in the contact process discussed below, is formed by the catalytic exidation of SO_2. In the presence of catalysts such as vanadium pentoxide or finely divided platinum, SO_2 combines with oxygen and yields sulfur trioxide (SO_3). The equation is

$$SO_2(g) + \tfrac{1}{2}O_2(g) \longrightarrow SO_3(g) \qquad \Delta H_f^\circ = -94.4 \text{ kcal}$$

The sign of ΔH_f° shows that the reaction is exothermic. The resonance structures of SO_3 are shown in the margin. It can be seen from the electron-dot formulas that all six of the bonding electrons in sulfur atoms are shared with the more electronegative oxygen atoms. Thus, in SO_3, sulfur has an oxidation number of 6+. Some of the properties of SO_2 and SO_3 are compared in Table 18-3.

Sulfur trioxide is a very reactive substance. When added to water the reaction is violent and exothermic. The equation for it is

$$SO_3(l) + H_2O(l) \longrightarrow H_2SO_4(l) \qquad \Delta H = -21.3 \text{ kcal}$$

THE CONTACT PROCESS

Approximately 20 million tons of H_2SO_4 is produced each year in the United States by the contact process. In this process, melted sulfur is burned in a furnace, forming SO_2. The SO_2 gas from the combustion chamber is contaminated with dust and other impurities which may interfere with the catalytic action in the reaction chamber and, thus, lower the efficiency of the process. To purify the gas, it is passed through a series of towers. After the dust particles are removed in the first tower, the gas is passed through the scrubber and drying towers where the SO_2 is first washed with water, then with concentrated sulfuric acid to dry it. The SO_2 then passes into a reaction chamber. There, in the presence of the catalyst mentioned above, it reacts efficiently and rapidly with oxygen and produces SO_3. Reaction occurs at the surface of the catalyst when the two gases make contact with the catalyst; hence, the names *contact catalyst* and *contact process.* The catalytic action is possibly caused by adsorption of the gases on the surface of the catalyst, thus providing more favorable collision geometry, and thus a lower activation energy for the reaction.

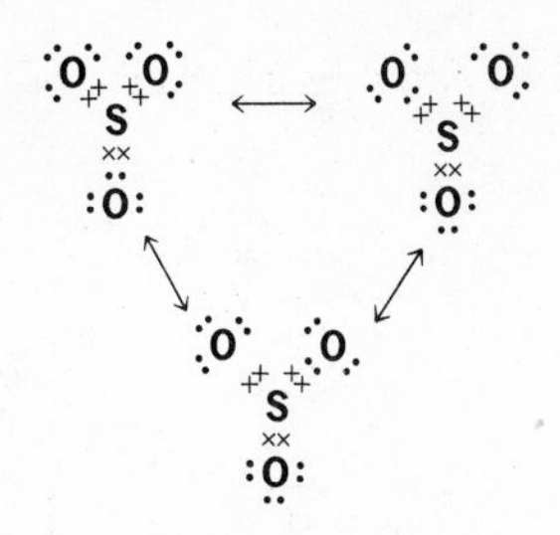

A high yield of over 90 percent is achieved when the temperature is controlled at 450°C. To maintain this temperature, it is necessary to make provision for removing some of the heat resulting from the highly exothermic reaction. The SO_3 gas is cooled and passed into absorption towers where it is absorbed by concentrated H_2SO_4. This reaction forms pyrosulfuric acid ($H_2S_2O_7$), sometimes called *oleum* or *fuming sulfuric acid.* The pyrosulfuric acid is then added to somewhat dilute sulfuric acid yielding sulfuric acid of the desired concentration. The equations for the principal reactions which occur during the contact process are

$$S(l) + O_2(g) \longrightarrow SO_2(g)$$
$$2SO_2(g) + O_2(g) \longrightarrow 2SO_3(l)$$
$$SO_3(l) + H_2SO_4(l) \longrightarrow H_2S_2O_7$$
$$H_2S_2O_7(l) + H_2O(l) \longrightarrow 2H_2SO_4(l)$$

Actually, SO_3 is not very soluble in water. This is the reason that it is necessary to pass the SO_3 into concentrated H_2SO_4. For each mole of H_2SO_4 used in the contact process, two moles are produced.

Pure sulfuric acid is a colorless, viscous liquid that freezes at 10.37°C and begins to boil at 290°C. It has a *strong attraction for water* and forms a number of stable hydrates with the evolution of a considerable amount of energy. It is often used in chemical laboratories as a *dehydrating agent.* For example, moist gases may be dried by bubbling them through concentrated sulfuric acid. It also removes hydrogen and oxygen from many organic compounds in the same ratio as they appear in water. For example, addition of concentrated sulfuric acid to sugar produces a black mass of porous carbon.

$$C_{12}H_{22}O_{11} \xrightarrow{H_2SO_4} 12C + 11H_2O$$

Hot concentrated sulfuric acid is a strong oxidizing agent; dilute sulfuric acid is a poor oxidizing agent. When hot, concentrated

18-16 The reaction between concentrated sulfuric acid and sugar is extremely exothermic. Explain the appearance of the large bulk of porous carbon.

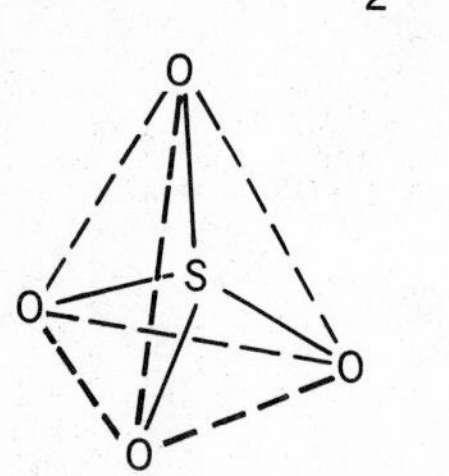

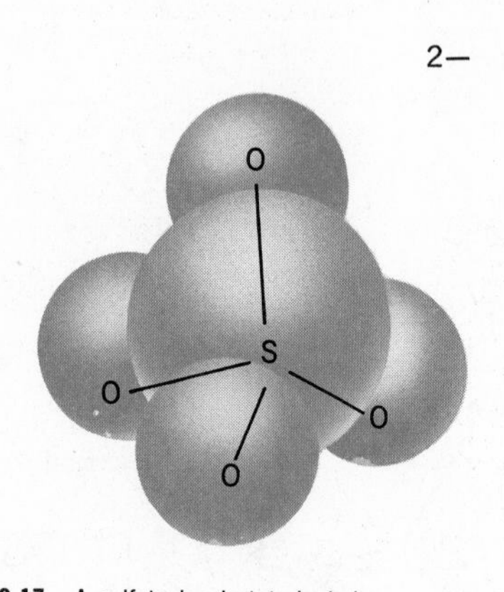

18-17 A sulfate ion is tetrahedral.

H_2SO_4 acts as a powerful oxidizing agent and dissolves metals such as copper.

$$\mathrm{Cu + 5H_2SO_4 \longrightarrow Cu^{2+} + SO_2 + 2H_3O^+ + 4HSO_4^-}$$

In dilute solution, the first stage of dissociation

$$\mathrm{H_2O^+ + H_2SO_4 \longrightarrow H_3O^+ + HSO_4^-}$$

is *essentially complete.* The dissociation constant for the second stage

$$\mathrm{H_2O^+ + HSO_4^- \rightleftharpoons H_3O^+ + SO_4^{2-}}$$

is 1.3×10^{-2}. Thus, H_2SO_4 is a strong acid with respect to the first step of its dissociation. In dilute solution, H_2SO_4 does not oxidize copper. The reduction potential for

$$\mathrm{HSO_4^- + 3H^+ + 2e^- \longrightarrow SO_2(g) + 2H_2O}$$

is only 0.18 v. Therefore, hydrogen sulfate ions (HSO_4^-) are not a powerful enough oxidizing agent to oxidize copper, which has a reduction potential of 0.34 v for the half-cell reaction

$$\mathrm{Cu^{2+} + 2e^- \longrightarrow Cu}$$

Dilute H_2SO_4, however, does oxidize metals such as zinc and magnesium, which have much lower reduction potentials. In these reactions, it is the

$$\mathrm{2H^+ + 2e^- \longrightarrow H_2}$$

half-cell reaction that is involved so that H_2 gas is usually observed as the product. As the concentration of the acid is increased, both zinc and magnesium react with H_2SO_4 and produce SO_2 and H_2S: This is another example of the role that concentration and rate play in determining the product of a reaction.

THE HABER AND OSTWALD PROCESSES

One of the most important nitrogen compounds produced by the chemical industry is ammonia. The economic importance of NH_3 is illustrated by the fact that over 8 million tons of it is produced each year in the United States. Most of this is used as fertilizer or in the production of synthetic fertilizers. The synthesis of ammonia by the Haber process was discussed and illustrated in Sect. 13-9. We found that the reaction

$$\mathrm{3H_2(g) + N_2(g) \rightleftharpoons 2NH_3(g)} \qquad \Delta H = -22 \text{ kcal/mole}$$

could be made economically feasible by using a catalyst and adjusting the temperature and pressure of the system so as to achieve a maximum yield. Some of the properties of ammonia are listed in Table 18-4.

TABLE 18-4
PROPERTIES OF AMMONIA

Color	colorless
Odor	pungent
Solubility in water (20°C, 760 torr)	690 ml/ml H_2O
Melting point (°C)	−77.7
Boiling point (°C)	−33.4

Large quantities of ammonia are used to produce nitric acid (HNO_3) by the ***Ostwald process.*** The first step in this process is the catalytic oxidation of ammonia, forming nitric oxide (NO).

$$4NH_3(g) + 5O_2(g) \xrightleftharpoons[Pt]{1000°C} 4NO(g) + 6H_2O(g)$$

After the nitric oxide is cooled, it is passed into an oxidation chamber where it is mixed with oxygen and spontaneously reacts, forming nitrogen dioxide (NO_2).

$$2NO(g) + O_2(g) \rightleftharpoons 2NO_2(g)$$

Because the reaction is exothermic, high temperatures shift the equilibrium to the left and reduce the yield of NO_2. For this reason, the NO is cooled before entering the reaction chamber.

In the final step, NO_2 gas is introduced into an absorption tower where it reacts with water and forms nitric acid.

$$3NO_2(g) + 3H_2O(l) \rightleftharpoons 2H_3O^+(aq) + 2NO_3^-(aq) + NO(g)$$

It can be seen that in the above reaction NO_2 simultaneously undergoes oxidation to NO_3^- ions and reduction to NO gas. The equilibrium is shifted to the right by maintaining a high concentration of oxygen which reacts with the NO and forms additional NO_2.

Nitrate ions in acid solution (nitric acid) are relatively strong oxidizing agents. They can be reduced to compounds containing nitrogen in any of its many oxidation states by varying the reducing agent and the concentration of the acid. Some relevant half-cell reactions and their reduction potentials are

$$2NO_3^- + 12H^+ + 10e^- \longrightarrow N_2(g) + 6H_2O \qquad E° = 1.25 \text{ v}$$
$$2NO_3^- + 10H^+ + 8e^- \longrightarrow N_2O(g) + 5H_2O \qquad E° = 1.12 \text{ v}$$
$$NO_3^- + 4H^+ + 3e^- \longrightarrow NO(g) + 2H_2O \qquad E° = 0.96 \text{ v}$$
$$NO_3^- + 3H^+ + 2e^- \longrightarrow HNO_2 + H_2O \qquad E° = 0.94 \text{ v}$$
$$NO_3^- + 10H^+ + 8e^- \longrightarrow NH_4^+ + 3H_2O \qquad E° = 0.88 \text{ v}$$
$$NO_3^- + 2H^+ + e^- \longrightarrow NO_2(g) + H_2O \qquad E° = 0.79 \text{ v}$$

It can be seen that all of the reactions listed above contain H^+ ions. Since H^+ ions appear on the left side of each equation, an increase in the acidity drives the reaction to the right and increases the reduction potential (and spontaneity). Reducing the concentration H^+ ions from 1-*M* to 1×10^{-7} *M* (neutral solution) causes the reduction potential of the reaction

$$NO_3^-(aq) + 2H^+(aq) + e^- \longrightarrow NO_2(g) + H_2O(l)$$

to decrease from 0.79 v to 0.01 v

The fact that all of the reduction potentials above are positive indicates that NO_3^- ions are a stronger oxidizing agent than H^+ ions. This helps explain why a metal like copper dissolves in nitric

Wilhelm Ostwald
1853–1932

Wilhelm Ostwald had a remarkable capability for recognizing the talents and ideas of other people. He received his education at the University of Dorpat in Estonia and eventually became professor of chemistry at the University of Leipsig. While at Leipsig he founded the first physical chemistry laboratory which attracted students from all over the world. The American students who studied under Ostwald were particularly fond of him because he was the only scientist in Europe who, for many years, recognized the value of contributions made by the American scientist, Willard Gibbs of Yale.

When Van't Hoff of Holland published his theories on stereochemistry, which became known as "space chemistry" because of its emphasis on the three-dimensional arrangement of atoms in space, he was severely ridiculed by the majority of the scientific community. It was Ostwald who immediately recognized the value of the idea and came to Van't Hoff's defense. Van't Hoff's studies on solutions and their colligative properties indicated that various salts must "dissociate" in solution and that each "molecule" of salt must act as though it were composed of separate particles. Ostwald was able to correlate the ideas of Van't Hoff and Arrhenius and find agreement in the experimental data obtained by both men. Ostwald recognized that the better a solution conducts an electric current, the more abnormally great are its colligative properties.

When Arrhenius was being attacked as a "foolish school boy" for his theory of ions, it was Ostwald who said, "Let us attack them," by publishing data on ions in the "Zeitschrift," a technical publication. Thus, Ostwald and his friends, Arrhenius and Van't Hoff, carried on a struggle to get the scientific community to accept the concept of the ion.

Although Ostwald did his most important work in the field of electrochemistry and catalysis, he is best known for the process he developed for converting ammonia into nitric acid. He passed ammonia and air over a platinum catalyst and produced oxides of nitrogen which react with water and produce nitric acid.

18-18 An important source of NaCl is the vast underground salt deposits in Louisiana. The mining operations are similar to those used to recover coal. Most of the 20 million tons of NaCl obtained from underground deposits or from salt lakes each year in the United States is used to manufacture other sodium compounds. The sodium chloride is usually converted into NaOH by the electrolysis of an aqueous sodium chloride solution or into Na_2CO_3 by the Solvay process.

acid but not in hydrochloric acid. The hydrogen ions from hydrochloric acid are not a strong enough oxidizing agent alone. In nitric acid, nitrate ions in the presence of hydrogen ions act as a strong oxidizing agent.

$$Cu(s) + 2NO_3^-(aq) + 4H^+(aq) \longrightarrow Cu^{2+}(aq) + 2NO_2(g) + H_2O(l)$$

From the magnitude of the E° values for the nitrate half-cell reactions, you might think that the reaction between a strong reducing agent, such as zinc, and nitric acid, would produce N_2 as a product. It is true that reduction to nitrogen is, in theory, the most probable reaction; that is, it has the largest reduction potential. When zinc reacts with concentrated HNO_3, however, NO_2 is formed. Here again, the rate of reaction is the critical factor. The reaction that proceeds at the fastest rate occurs. The rate is determined by the temperature, the concentration of the reactants, and the nature (strength) of the reducing agent.

THE SOLVAY PROCESS

Sodium carbonate, frequently called *sal soda* or *washing soda*, is used extensively in the manufacture of glass, soap, paper, textiles, rubber, and other products which require a soluble base. A relatively small percentage of the six million tons of Na_2CO_3 used annually in the United States is extracted from natural salt lakes. About 90 percent is synthesized by the Solvay process.

The ***Solvay process*** is an excellent example of the efficient use of raw material in chemical manufacturing. The only raw materials consumed in the overall process are NaCl, $CaCO_3$ (limestone), and fuel. In this process, CO_2 obtained by the thermal decomposition of $CaCO_3$, and ammonia are dissolved in cold, concentrated NaCl solution (brine). Sodium bicarbonate is not very soluble in cold brine and precipitates out of solution. It is removed by filtration and converted to Na_2CO_3 by heating.

$$CaCO_3(s) \xrightarrow{\text{heat}} CaO(s) + CO_2(g) \qquad 18\text{-}2$$

$$NH_3(g) + CO_2(g) + Na^+(aq) + Cl^-(aq) + H_2O(l) \longrightarrow NH_4^+(aq) + Cl^-(aq) + NaHCO_3(s) \qquad 18\text{-}3$$

$$2NaHCO_3(s) \xrightarrow{\text{heat}} Na_2CO_3(s) + CO_2(g) + H_2O(l) \qquad 18\text{-}4$$

Some of the $NaHCO_3$ produced in Equation 18-3 is sold after refining, as baking soda or used as an ingredient in baking powder.

The most expensive reactant in the above sequence is NH_3. However, only a few pounds of NH_3 is required for each ton of $NaHCO_3$ produced. Most of the NH_3 is recovered by treating the

NH_4Cl, a by-product from Equation 18-3, with $Ca(OH)_2$, a strong base. The net ionic equation is

$$NH_4^+(aq) + OH^-(aq) \xrightarrow{\text{heat}} NH_3(g) + H_2O(l)$$

The $Ca(OH)_2$ is obtained by slaking (adding water to) CaO, a product of Equation 18-2.

$$CaO(s) + H_2O(l) \longrightarrow Ca(OH)_2(s)$$

The CO_2 from Equation 18-4 is recycled and used as a reactant in Equation 18-3. Thus, the only by-product which is not recycled into the process is $CaCl_2$. A flow diagram for the process is illustrated in Fig. 18-19.

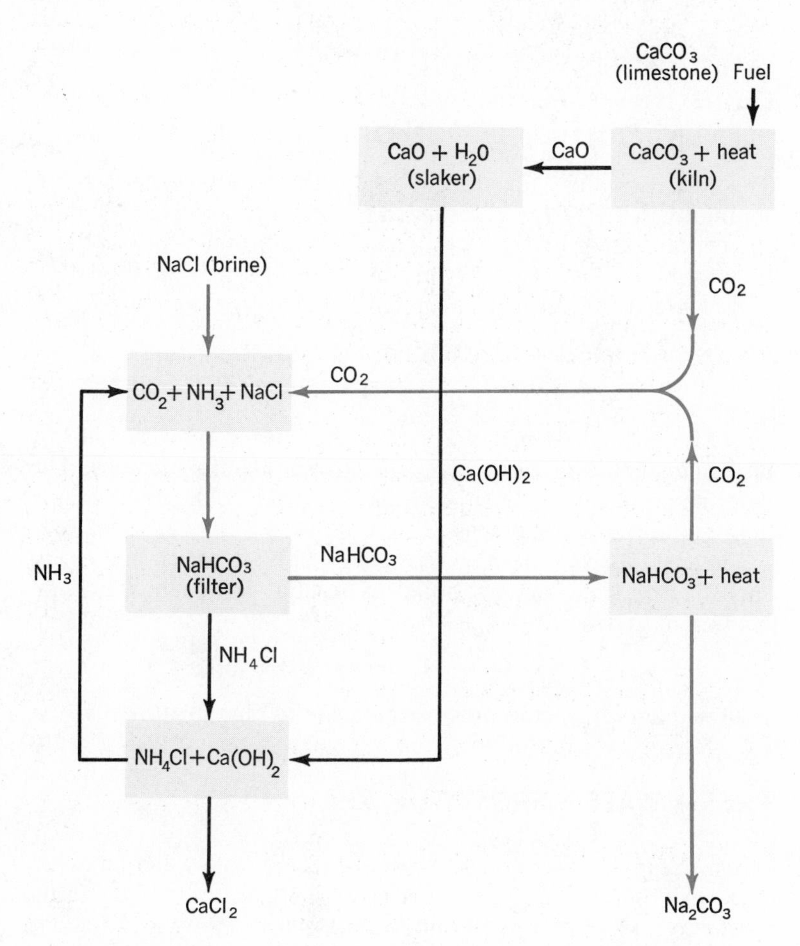

18-19 Flow chart for the Solvay process.

TABLE 18-5
MINERALS IN SEA WATER

Element	Tons per Cubic Mile
Chlorine	89 500 000
Sodium	49 500 000
Magnesium	6 400 000
Sulfur	4 200 000
Calcium	1 900 000
Potassium	1 800 000
Bromine	306 000
Carbon	132 000
Strontium	38 000
Boron	23 000
Silicon	14 000
Fluorine	6 100
Argon	2 800
Nitrogen	2 400
Lithium	800
Rubidium	570
Phosphorus	330
Iodine	280
Barium	140
Indium	94
Zinc	47
Iron	47
Aluminum	47
Molybdenum	47
Selenium	19
Tin	14
Copper	14
Arsenic	14
Uranium	14
Nickel	9
Vanadium	9
Manganese	9
Titanium	5
Antimony	2
Cobalt	2
Cesium	2
Cerium	2
Yttrium	1
Silver	1
Lanthanum	1
Krypton	1
Neon	0.5
Cadmium	0.5
Tungsten	0.5
Xenon	0.5
Germanium	0.3
Chromium	0.2
Thorium	0.2
Scandium	0.2
Lead	0.1
Mercury	0.1
Gallium	0.1
Bismuth	0.1
Niobium	0.05
Thallium	0.05
Helium	0.03
Gold	0.02

(From *Scientific American*, September, 1969)

MAGNESIUM FROM THE SEA

The oceans contain an almost inexhaustible supply of various minerals and raw materials but it is often difficult or uneconomic to extract them. Each cubic mile of sea water contains about 165 million tons of dissolved minerals. Table 18-5 shows the number of tons of dissolved mineral in a cubic mile of sea water.

Magnesium is the third most abundant element present in sea water. It is a very light structural metal with thousands of uses. Alloys of magnesium and aluminum have been used for small hand tools and even for structural parts of space vehicles.

In the industrial process for obtaining magnesium from sea water, lime (CaO) added to the sea water produces OH^- ions which react with Mg^{2+} ions and form insoluble $Mg(OH)_2$. Relevant reactions are

$$CaO(s) + H_2O(l) \rightleftharpoons Ca(OH)_2(s)$$
$$Ca(OH)_2(s) \rightleftharpoons Ca^{2+}(aq) + 2OH^-(aq) \qquad K_{sp} = 1.3 \times 10^{-6}$$
$$Mg^{2+}(aq) + 2OH^-(aq) \rightleftharpoons Mg(OH)_2(s) \qquad K_{sp} = 9 \times 10^{-12}$$

The relative values for the K_{sp} show that $Mg(OH)_2$ is far less soluble than $Ca(OH)_2$, and that Mg^{2+} ions react with and reduce the concentration of the OH^- ions in equilibrium with $Ca(OH)_2$, thus forming $Mg(OH)_2$. The precipitate of $Mg(OH)_2$ is filtered off and changed to $MgCl_2$ by treatment with hydrochloric acid.

$$2H_3O^+ + 2Cl^-(aq) + Mg(OH)_2(s) \longrightarrow 4H_2O(l) + Mg^{2+}(aq) + 2Cl^-(aq)$$

The $MgCl_2$ is dried in large vats, using energy from the sun. It is then melted and electrolyzed in airtight electrolytic cells. The chlorine produced at the anode is used to produce the hydrochloric acid required for the process.

The CaO used in the process is obtained by decomposing oyster shells. These shells are composed of calcium carbonate which, on heating, is converted to CaO.

$$CaCO_3(s) \longrightarrow CaO(s) + CO_2(g)$$

A flow diagram for the process is shown in Fig. 18-20.

FRESH WATER FROM THE SEA

The oceans are the ultimate source of all the fresh water on earth. The growing need for fresh water is forcing man to build desalting plants to take fresh water from the oceans. In many arid regions of the earth where no other fresh water is available, desalting plants have been successfully operated. One of the most successful types of desalination systems is the multistage flash distillation unit. The sea water is introduced into the unit where the pressure has been lowered so that the water boils instantly (flashes) into steam. This

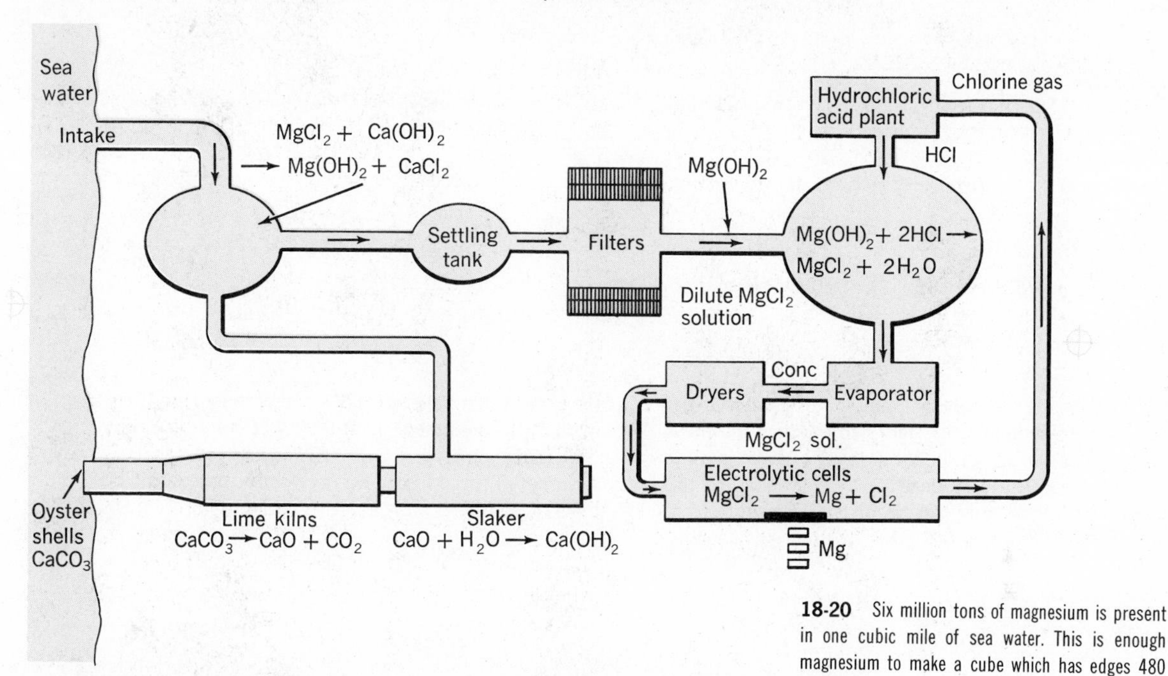

18-20 Six million tons of magnesium is present in one cubic mile of sea water. This is enough magnesium to make a cube which has edges 480 ft long.

evaporation causes the brine to be cooled and it is passed into a second chamber with even lower pressure. More water is evaporated from the brine at this stage, and the process is again repeated. A schematic representation of the entire process is shown in Fig. 18-21.

The multistage flash process and other desalination methods all require large amounts of energy. For this reason, combination nuclear power and desalting plants have been designed which use ocean water as a cooling agent, and which produce both electric energy and fresh water.

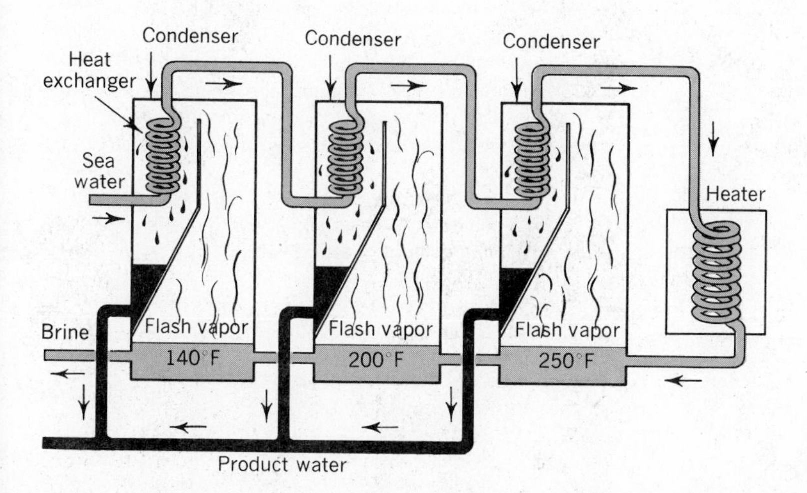

18-21 Multistage flash distillation process. Steam is condensed into fresh water within pipes in the heat exchanger. The pipes are cooled outside by the incoming flow of salt water. The heat evolved by the condensation process is absorbed by the incoming sea water which is thereby raised to its boiling point. (U.S. Dept. of the Interior: Office of Saline Water)

FOLLOW-UP PROBLEMS

1. The Ostwald process involves the conversion of ammonia to nitrates through these steps

(1) $4NH_3(g) + 5O_2(g) \xrightleftharpoons{\text{catalyst}} 4NO(g) + 6H_2O(g)$ $\quad \Delta H = -216.2$ kcal

(2) $NO(g) + \frac{1}{2}O_2(g) \rightleftharpoons NO_2(g)$ $\quad \Delta H = 13.5$ kcal

(3) $3NO_2(g) + H_2O(l) \rightleftharpoons 2HNO_3(l) + NO(g)$ $\quad \Delta H = -16.2$ kcal

(a) Describe the conditions of temperature and pressure you would recommend in Step 1 to produce the maximum amount of NO(g). (b) What mass of ammonia gas would be required to produce 1.00 ton of HNO_3 by this process? (c) How many grams of NH_3 are required to prepare 600 ml of 0.400-*M* HNO_3 by the Ostwald process?

2. These equations summarize the major reactions taking place in the contact process

$$S_8 + 8O_2 \longrightarrow 8SO_2$$
$$SO_2 + \tfrac{1}{2}O_2 \xrightarrow{\text{catalyst}} SO_3$$
$$H_2SO_4 + SO_3 \longrightarrow H_2S_2O_7$$
$$H_2S_2O_7 + H_2O \longrightarrow 2H_2SO_4$$

(a) How many moles of H_2SO_4 are prepared from each mole of S_8 molecules? (b) If 32.0 g of sulfur is burned, how many liters of SO_2 at STP would be produced? Assume SO_2 is an ideal gas. (c) What volume would one mole of SO_2 gas occupy at 800°C and 760 torr? (d) How many grams of pyrosulfuric acid ($H_2S_2O_7$) could be produced from 16.0 g of sulfur?

QUESTIONS AND PROBLEMS

1. Explain why the halogens are not found free in nature.

2. Use the orbital model of the atom to explain the diatomic structure of halogen molecules.

3. Draw electron-dot structures to illustrate the possible oxidation states of chlorine, bromine, or iodine.

4. What is the reason that fluorine exists in only the 1− and 0 oxidation states?

5. Explain the increasing acid strength of the hydrogen halides as one goes down the group from fluorine to iodine.

6. Which noble gas is iso-electronic with each halide ion?

7. Why are the ionic radii of the halide ions greater than the atomic radii of the atoms from which they are formed?

8. What are Freons?

9. Write an equation which represents the action of hydrofluoric acid on glass.

10. Explain the trend of decreasing oxidizing strength with increasing atomic number in the Group-VII elements.

11. Describe the laboratory setup you might use to prepare and collect chlorine gas from sodium chloride.

12. Complete and balance equations for the reactions which take place between

(a) $Cl^-(aq) + MnO_2(s) + H^+ \longrightarrow$

(b) $Br^-(aq) + MnO_4^-(aq) + H^+ \longrightarrow$

(c) $I^-(aq) + Cr_2O_7^=(aq) + H^+ \longrightarrow$

13. Given three colorless solutions, containing $Cl^-(aq)$, $Br^-(aq)$, or $I^-(aq)$, and a solution of chlorine gas in water. How would you identify each of the ions in solution?

14. Explain the differences in acid strength of the oxyacids containing chlorine (Sect. 18–10).

15. Can iodine oxidize (a) Fe^{2+} to Fe^{3+}, (b) Sn to Sn^{2+}, (c) Cr to Cr^{3+}? Explain.

16. Write the equations which illustrate the refining of magnesium from sea water.

19

Transition Elements and Coordination Compounds

Preview

> "Men love to wonder, and that is the seed of science."
>
> —Ralph Waldo Emerson (1803–1882)

The metals of Group IA and IIA are, with one exception, nonstructural metals. The *alkali* and *alkaline-earth metals* themselves have very few familiar applications. The compounds they form, however, are among the most common substances used by all of us. Ordinary salt (NaCl), chalk and marble ($CaCO_3$), gypsum ($CaSO_4 \cdot 2H_2O$), washing soda (Na_2CO_3), baking soda ($NaHCO_3$), and soap are a few of the familiar useful compounds of these elements.

In contrast, most of the compounds of the transition elements are not familiar household products, but the transition metals themselves are used to construct bridges and buildings or to fabricate appliances, utensils, or other common products.

The gold, silver, platinum, and iridium in valuable jewelry, the copper in the thousands of miles of wire bringing electricity to every part of the country, and the iron used as the structural backbone of our buildings and bridges are a few of the metals which appear in the Periodic Table in a series known as the transition elements. These are the elements located between Groups IIA and IIIA in the long form of the Periodic Table. Although a number of these metals and their compounds have been known and used for thousands of years, it has been only in the last 100 years that scientists have developed theories which explain many of their properties and reactions.

As we have implied, the properties of the transition metals differ considerably from those of the metals in Groups IA and IIA. We shall find that most of these differences can be attributed to the presence and behavior of *d*-orbital electrons. We shall, therefore, first investigate and review the electron configuration of these elements and then extend our model of chemical bonding in order to interpret their properties.

With this background, we shall take a brief look at the chemistry of complex ions and coordination compounds. This very important and rapidly expanding area of chemical knowledge has many significant applications in our contemporary world.

TABLE 19-1
PROPERTIES OF THE ROW 4 TRANSITION ELEMENTS AND CALCIUM

Element	*Ca*	*Sc*	*Ti*	*V*	*Cr*	*Mn*
Atomic number	20	21	22	23	24	25
Electron configuration, M^0	(Ar) $4s^2$	(Ar) $3d^1 4s^2$	(Ar) $3d^2 4s^2$	(Ar) $3d^3 4s^2$	(Ar) $3d^5 4s^1$	(Ar) $3d^5 4s^2$
Electron configuration, M^{2+}	(Ar) $4s^0$	(Ar) $3d^1$	(Ar) $3d^2$	(Ar) $3d^3$	(Ar) $3d^4$	(Ar) $3d^5$
Electron configuration, M^{3+}	—	(Ar) $3d^0$	(Ar) $3d^1$	(Ar) $3d^2$	(Ar) $3d^3$	(Ar) $3d^4$
Common oxidation numbers	2+	3+	2+, 3+ 4+	2+, 3+ 4+, 5+	2+, 3+ 6+	2+, 3+ 4+, 5+ 6+, 7+
1st ionization energy (kcal)	140	151	158	155	156	171
Atomic radius of M^0 (Å)	1.74	1.44	1.32	1.22	1.17	1.17
Ionic radius of M^{2+} (Å)	0.94	—	0.90	0.88	0.84	0.80
Ionic radius of M^{3+} (Å)	—	0.81	0.76	0.74	0.69	0.66
Melting point (°C)	850	1522	1725	1900	1875	1247
Packing in metallic crystal	fcc	fcc	hcp	bcc	bcc	bcc
Coordination number in crystal	12	12	12	8	8	8
Density (g/ml)	1.55	3	4.49	5.98	6.9	7.4
Reduction potential (volts) $M^{2+} + 2e^- \rightarrow M(s)$	−2.87	—	−1.6	−1.2	−0.91	−1.18

Finally, to enhance your background of descriptive and technological chemistry, we shall consider the metallurgical process by which iron ore is converted into the metal. Space will not permit us to discuss the metallurgy of other metals; however, we shall cite special properties and uses of several of them.

It should be noted that, at present, there are several models used by chemists who attempt to explain and corrleate the properties of the transition metals. They include the *valence-bond* theory, the *crystal-field* theory, and an extension of the latter, the *ligand-field* theory. In this text we shall use the valence-bond theory as originally postulated by Dr. Linus Pauling in his well-known treatise, the *Nature of the Chemical Bond,* (Freeman Co., San Francisco, California). No attempt is being made here to claim that one theory is superior to another. Again, this illustrates clearly the spirit of science in which a "no-holds-barred" approach is used to grapple with the unsolved problems of science.

THE TRANSITION ELEMENTS

19-1 General Characteristics In contrast to the representative metals of Groups IA and IIA, most of the transition metals are characterized by these properties.

TABLE 19-1 (CONT.)

Element	Fe	Co	Ni	Cu	Zn
Atomic number	26	27	28	29	30
Electron configuration, M^0	(Ar) $3d^64s^2$	(Ar) $3d^74s^2$	(Ar) $3d^84s^2$	(Ar) $3d^{10}4s^1$	(Ar) $3d^{10}4s^2$
Electron configuration, M^{2+}	(Ar) $3d^6$	(Ar) $3d^7$	(Ar) $3d^8$	(Ar) $3d^9$	(Ar) $3d^{10}$
Electron configuration, M^{3+}	(Ar) $3d^5$	(Ar) $3d^6$	(Ar) $3d^7$	—	—
Common oxidation numbers	2+, 3+	2+, 3+	2+, 3+	1+, 2+	2+
1st ionization energy (kcal)	182	181	176	178	217
Atomic radius of M^0 (Å)	1.16	1.16	1.15	1.17	1.25
Ionic radius of M^{2+} (Å)	0.76	0.74	0.72	0.72	0.72
Ionic radius of M^{3+} (Å)	0.64	0.63	—	—	—
Melting point (°C)	1528	1490	1452	1083	419.5
Packing in metallic crystal	bcc fcc	fcc hcp	fcc	fcc	hcp
Coordination number in crystal	12	12	12	12	12
Density (g/ml)	7.9	8.8	8.90	8.94	7.13
Reduction potential (volts) $M^{2+} + 2e^- \rightarrow M(s)$	−0.44	−0.28	−0.25	+0.34	−0.76

1. Hard and relatively dense
2. Relatively high melting and boiling points
3. Several oxidation states
4. Paramagnetic compounds.
5. Colored compounds
6. Weaker reducing agents than the metals of Group IA and Group IIA.

Let us seek an explanation for these properties which distinguish the transition metals from the representative metals. We shall use the data in Table 19-1 as a guide. In this table we have listed a few important properties of elements 20 to 30. Calcium, a non-transition element, is included for comparison with the transition elements.

19-2 Electronic Configuration of Atoms, and Bonding in Crystals It can be seen that the outermost energy levels of all atoms in the *first transition series have either one or two electrons.* Their first ionization energies indicate that the *outer electron* of each atom is *loosely bound* compared with that of a nonmetallic atom such as chlorine, whose first ionization energy is 300 kcal/mole. The outer *s* electrons are *delocalized* and able to spread out over many atoms throughout the metal. This delocalization gives rise to the characteristic metallic properties of electric conductivity and light reflectivity displayed by these elements.

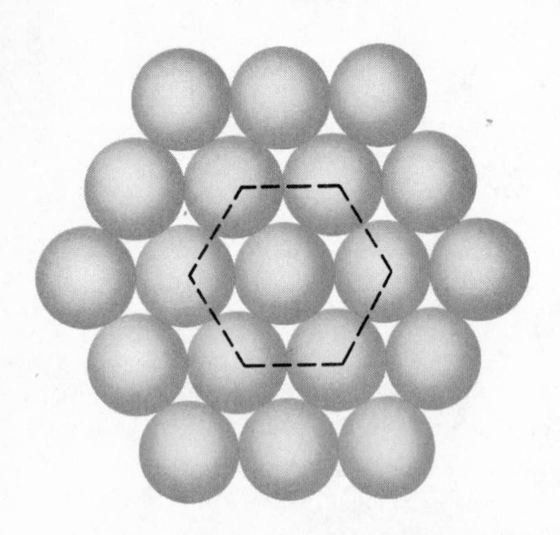

19-1 Each sphere in a closest-packed structure is surrounded and touched by 12 other spheres. A sphere in any given layer touches six in this layer (see figure), three in the layer above (not shown here), and three in the layer below (not shown). Closest-packed structures with a coordination number of 12 are the most efficient way that identical-sized spheres can be packed.

With the exception of copper and zinc, all of the atoms have unfilled *d* orbitals. In a metal, the partially filled *d* orbitals of adjacent atoms may overlap. The subsequent pairing of electrons results in covalent bond formation. Additional bonds cause an increase in the overall bond strength. The hardness of these metals is partially a reflection of the covalent bonding resulting from the sharing of *d* electrons. The increase in the melting points of the elements up through vanadium and chromium may, in general, be attributed to additional covalent bond formation resulting from the larger number of unpaired *d* electrons.

19-3 Density and Hardness of Crystals Atoms of most of the transition metals have *small radii* and pack in *closest-packed structures* with a coordination number of 12. They are, therefore, relatively dense compared with the metals of Group IA and Group IIA whose larger atoms pack in body-centered structures with a coordination number of 8. Thus, we may conclude that *partial covalent bonding* involving *d*-orbitals and closest-packed crystal structures are responsible for the hardness and relatively high melting points of most transition metals.

Coordination number is the number of the nearest-neighbor atoms which surround a given atom

19-4 Multiple Oxidation States Displayed by Atoms It can be seen in Table 19-1 that many of the transition elements exhibit several oxidation states. There is a gradual increase in the maximum oxidation state from Sc^{3+} on the left to Mn^{7+} in the center and then a gradual decrease to Zn^{2+} on the right.

This trend parallels the number of electrons available for bond formation. Energy-level diagrams show that there is only a small difference between the energies of electrons in an outermost *s* orbital and a *d* orbital in the next lower level. This means that both *s* and *d* electrons are available to participate in chemical reactions.

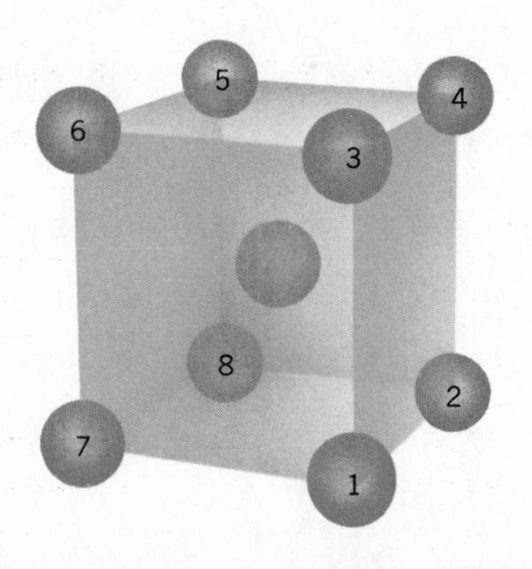

19-2 In a body-centered cubic crystal, the center atom in a unit cell is surrounded by 8 atoms all at the same distance. Accordingly, the coordination number is 8. Less efficient packing of atoms in this structure gives rise to softer metals with lower densities. A unit cell is the smallest representative building block of a crystal lattice structure.

In a reaction, scandium, with two 4*s* and one 3*d* electron fairly close in energy, loses all three and achieves a stable noble-gas core of electrons and a 3+ oxidation state. Manganese, with two 4*s* and five unpaired 3*d* electrons, readily loses the 4*s* electrons and forms Mn^{2+}. By sharing the *s* and all or some of its *d* electrons, manganese can exhibit a maximum oxidation state of 7+ or almost any positive state between 2+ and 7+.

Zinc, on the extreme right, has completely filled *d* orbitals. Its single oxidation state of 2+ corresponds to the loss of the outer two *s* electrons. Thus, we can conclude that the multiple *oxidation states of transition elements can be attributed to the availability of d electrons for bond formation.*

19-5 Magnetic Properties Atoms, ions, and molecules with unpaired electrons are attracted to a magnet. This property is known as ***paramagnetism.*** The greater the number of unpaired electrons, the greater will be the paramagnetism. Substances in which all electrons are paired are repelled by a magnet. These substances are said to be ***diamagnetic.*** By determining the magnetic properties of a substance, scientists can determine the number of unpaired electrons and use this information to help determine molecular structures and the electronic structures of certain complex ions. It is largely the presence of unpaired *d* electrons that gives rise to the large number of paramagnetic transition elements and compounds.

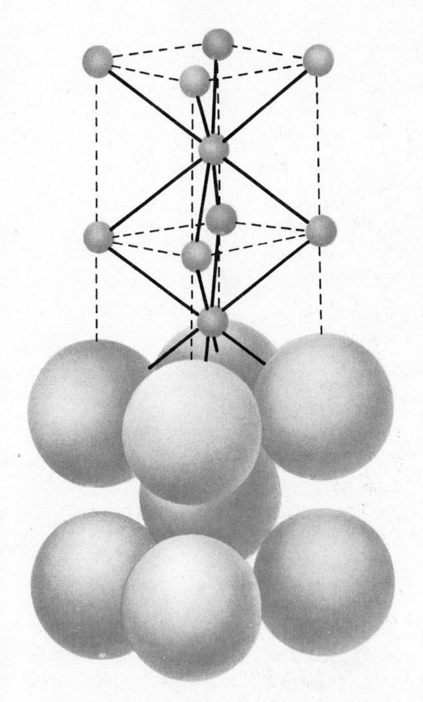

19-3 Space-filling model of a section of a body-centered crystal (bcc). In the overall crystal, every atom occupies the center of some unit cell. In theory, packing thousands of identical cells together generates an observable perfect crystal.

19-6 Formation of Colored Compounds The large number of colored compounds formed by the transition elements can be attributed to the presence of unpaired *d* electrons. Experimental evidence indicates that in the presence of certain Lewis bases (ligands) there may be slight energy differences among the five *d* orbitals in a given shell (principal energy level) of a transition metal ion. In many cases, visible light, which has a relatively long wavelength, is energetic enough to promote an electron from a lower level *d* orbital to a higher level *d* orbital in the same shell. In other words, certain wavelengths of light are absorbed. The transmitted light that characterizes the substance is the complementary color of that absorbed by the atom. A color triangle showing complementary colors is illustrated in Color Plate IV, between p. 274 and p. 275. Using the triangle, we can predict that a solution which absorbs light in the yellow portion of the spectrum appears blue to the eye. In other words, white light from which the yellow wavelengths have been removed appears blue to the eye.

Aqueous solutions of vanadium provide a good illustration of how unpaired *d* electrons give rise to colored substances. The formulas, colors, and structures of four vanadium ions are shown in Table 19-2. Solutions of $[V(OH)_4]^+$ are practically colorless. In this ion, vanadium is in the 5+ oxidation state. Since vanadium atoms have two *s* electrons and three *d* electrons, a 5+ state means that there are *no unpaired electrons.* Therefore, the substance *does not appreciably absorb light in the visible range,* and appears essentially colorless.

Solutions of $[V(OH)_2]^{2+}$ are blue. In this ion, vanadium is in the 4^+-oxidation state in which there is one unpaired *d* electron. This electron can be excited to a higher *d* level by low-energy photons which have frequencies in the yellow region of the spectrum. Thus, the light transmitted by the solution appears blue.

Solutions containing $[V(H_2O)_6]^{3+}$ appear green. Vanadium is in the 3+ state and has two unpaired *d* electrons. These two can be excited to higher energy *d* orbitals and can absorb light energy of frequencies that result in the transmission of green light. When more than one *d* electron can be excited, there will be interactions

TABLE 19-2
CHARACTERISTICS OF VANADIUM IONS

Formula	*Outer Level Electron Configuration*	*Oxidation State*
$[V(OH)_4]^+$	$3d^0 4s^0$	5+
$[V(OH)_2]^{2+}$	$3s^1 4s^0$	4+
$[V(H_2O)_6]^{3+}$	$3d^1 d^1 4s^0$	3+
$[V(H_2O)_6]^{2+}$	$3d^1 d^1 d^1 4s^0$	2+

Formula	*Color*	*Reduction Potential $E°$ (volts)*
$[V(OH)_4]^+$	light yellow	$+1.0$ to $V(OH_2)^{2+}$
$[V(OH)_2]^{2+}$	blue	$+0.36$ to $V(H_2O)_6^{3+}$
$[V(H_2O)_6]^{3+}$	green	-0.25 to $V(H_2O)_6^{2+}$
$[V(H_2O)_6]^{2+}$	violet	-1.2

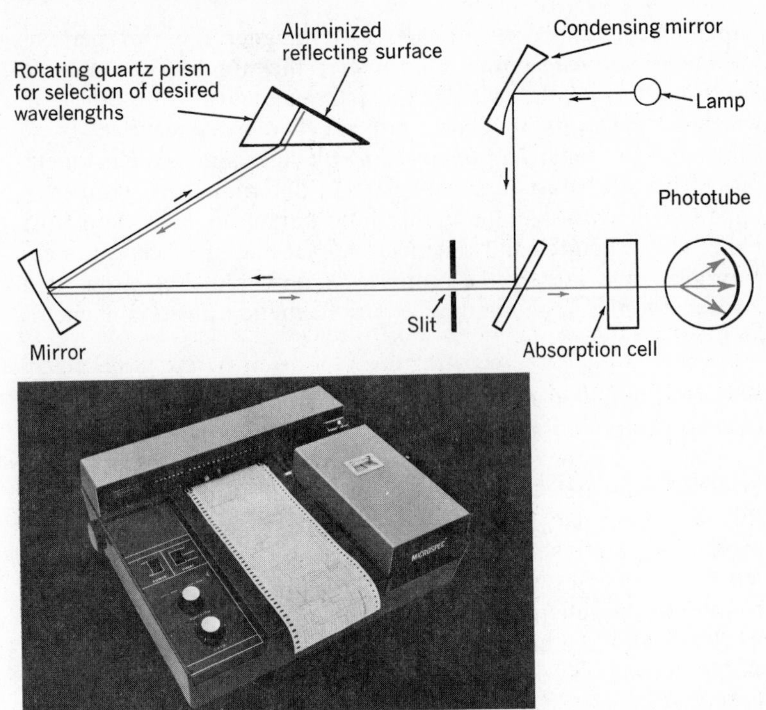

19-4 An absorption spectrophotometer is shown in the photo. The diagram is a simplified schematic of a spectrophotometer. (Beckman Instruments, Inc.)

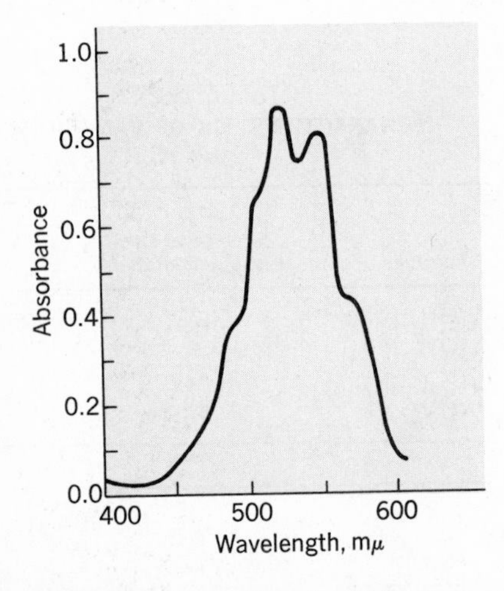

19-5 Absorption spectrum for $KMnO_4$. No two substances have the same absorption spectrum. Therefore, absorption spectra can be used to help identify a substance. The amount of radiation absorbed by a solution is related to its concentration. This relationship serves as a basis for quantitative analysis by absorption spectrometry.

between electrons so that absorption is more difficult to interpret. The frequencies of radiation absorbed by different solutions can be determined by using an *absorption spectrometer*. These instruments enable students to determine the absorbance of a solution for different wavelengths of radiation. The absorbance is usually plotted against wavelength. The wavelength at which maximum absorbance occurs can then be read from the graph. A typical absorption spectrum is shown in Fig. 19-5.

Finally, solutions containing $[V(H_2O)_6]^{2+}$ appear violet. Vanadium, in this ion, has an oxidation state of 3+ and has three unpaired *d* electrons. In this case the excitation of three electrons to higher-energy *d* orbitals results in absorption of light of still different frequencies. The overall absorbed frequencies are such that violet frequencies are transmitted by the solution.

The reduction potentials listed in Table 19-2 show that Zn ($E° = -0.76$ v) should be able to reduce each of the colored ions except $[V(H_2O)_6]^{2+}$ to a lower oxidation state. Experimentally, when a solution of $[V(OH)_4]^+$ containing metallic zinc is agitated, the color of the solution is observed to change from yellow to blue to green to violet.

19-7 Reducing Ability of Elements In theory, most of the transition metals are good reducing agents. The reduction potentials listed in Table 19-1 indicate that, with the exception of copper, the transition metals should spontaneously dissolve in acid (1-*M* H^+) solution. In practice, however, a number of them appear to be relatively unreactive. In some cases this is because the rate of the reaction is so slow that it cannot be detected while in other cases a metal such as chromium forms a thin layer of an oxide (Cr_2O_3) which protects it from the action of an oxidizing agent. For this reason, chromium is widely used as an ornamental protective coating for other metals.

The transition metals are not as strong reducing agents as calcium or the other Group-IA and Group-IIA metals, thus their reduction potentials are not as negative. The relatively large hydration energies of the small transition metal ions are opposed by the large heats (enthalpies) of vaporization of the stable crystals. To a large extent, this accounts for the more positive reduction potentials and lower reducing strengths of the transition metals.

COMPLEX IONS AND COORDINATION COMPOUNDS

19-8 Formation, Composition, and Application Another distinguishing characteristic of most transition elements is their ability to form complex ions and coordination compounds. In Section 16-8 we noted that most complex ions are electrically charged chemical species composed of a central atom or ion covalently bonded to a number of other atoms, ions, or molecules known as ligands. An example of complex ion formation is

$$\underset{\text{central ion}}{Cu^{2+}} + \underset{\text{ligand}}{4NH_3} \longrightarrow \underset{\text{complex ion}}{[Cu(NH_3)_4]^{2+}}$$

Because of its charge, a complex ion can combine with oppositely charged ions and form ionic crystals. For example, $Cu(H_2O)_4^{2+}$ is a complex ion which crystallizes from an aqueous sulfate solution as an ionic coordination compound, $CuSO_4 \cdot 5H_2O$.

Coordination compounds are prevalent both in nature and in chemical laboratories. Dyes containing coordination compounds were used thousands of years ago. The red color of blood is caused by the presence of hemoglobin, a coordination compound containing Fe(II). Chlorophyll, which is found in plants, is a coordination compound similar in structure to hemoglobin but containing Mg(II) instead of Fe(II). A number of substances incorporated into fertilizers and foods are coordination compounds.

The process of complex-ion formation is widely used to tie up and control the concentration of specific metallic ions which may cause harmful or undesirable effects in foods, beverages, drugs, cosmetics, and numerous industrial preparations. The process also has many important applications in analytical chemistry.

The *unfilled* d *orbitals,* the *relatively high positive charges,* and the *small radii* of the ions of the transition elements lead to the formation of a large variety of complex ions. Ions with a large positive charge density (intense electric fields) have a strong tendency to interact with ligands such as H_2O, NH_3, and CN^- all of

ALFRED WERNER
1866–1919

Werner was born in Alsace, between Germany and France, and studied with Berthelot in Paris prior to accepting a position as "extraordinary professor of Chemistry" at the University of Zurich. Werner was a very demanding teacher who required from his students the same endurance and determination that he himself possessed. He had a fantastic memory and was able to remember the detailed properties of thousands of compounds that he worked with.

Werner's great contribution to chemistry was the coordination theory of valency. Prior to 1893, the prevalent idea was that atoms had a fixed and invariable valency (combining power) and that atoms bonded into straight chains. Werner's great idea that a central "privileged" atom could coordinate a group of atoms or molecules about it came to him "like a flash" at two o'clock in the morning. He awoke to write down his theory and by five o'clock in the afternoon "the essential features of the coordination theory had been worked out."

Werner recognized that atoms could have multiple valencies (combining capacities) and that the existence of isomers of inorganic compounds was possible. He devised a method of classifying complex inorganic compounds and explained how atoms other than carbon were responsible for optical isomerism. In his honor, such ions as $[Co(NH_3)_4Cl_2]^+$, $[Fe(NH_3)_5Cl]^+$, and others have been called Werner complexes.

Werner stands with Kekule and Van't Hoff as one of the three great architects of the structure theory. Werner was awarded the 1913 Nobel Prize "in recognition of his works on linking up the atoms within the molecule, whereby new light has been opened up, especially within the realm of inorganic chemistry."

which have highly electronegative atoms and unshared electron pairs. In general, the smaller the positive ion and the larger the charge, the greater will be the tendency to form stable complexes. Thus, the relatively large ions of the Group-IA and -IIA elements do not have as great a tendency to form stable complex ions as do the smaller, more highly charged ions of the transition elements. The general relationship between the charge/radius ratio of an ion and its tendency to form complex ions is illustrated in Table 19-3.

The formation of one common type of complex ion can be demonstrated by adding colorless, anhydrous copper(II) sulfate to water. The resulting blue color of the solution is caused by the complex ion formed between water and copper(II) ions.

$$\underset{\text{white}}{\mathbf{Cu^{2+}}} + \mathbf{4H_2O} \longrightarrow \underset{\text{blue}}{\mathbf{[Cu(H_2O)_4]^{2+}}} \qquad 19\text{-}1$$

TABLE 19-3
RELATIONSHIP BETWEEN CHARGE TO RADIUS RATIO OF IONS AND THEIR COMPLEXING TENDENCY

Ion	*Radius (Å)*	*Charge/Radius Ratio*	*Tendency to Form Complexes and Behavior in Aqueous Solution*
K^+	1.33	0.71	no complexes; weak electrostatic forces between the ion and mantle of water molecules surrounding it
Hg^{2+}	1.10	1.8	few weakly bonded complexes
Cd^{2+}	0.97	2.0	several complexes, ions weakly hydrated
Co^{2+}	0.72	2.8	generally complexed, ions hydrated
Fe^{3+}	0.64	4.7	always complexed, ions hydrated
Pt^{4+}	0.65	6.2	always complexed, ions hydrated

The tetraaquocopper(II) ions absorb light frequencies in the yellow region of the spectrum, so the solution appears blue. *Hydration reactions* similar to that represented by Equation 19-1 are typical of the ions of the transition elements. The green color of aqueous solutions of Ni(II) ions, the blue of Cr(III), and the red of Co(II) ions result from the presence of *aquo complexes* of these ions.

19-9 Coordination Number and Coordination Sphere The number of ligands attached directly to the central species is known as the *coordination number of the complex*. Thus, the coordination number for $[Cu(H_2O)_4]^{2+}$ is 4. Most of the transition element complexes have coordination numbers of 2, 3, or 6, with 6 being the most common.

The central ion and the ligands coordinated about it are always enclosed in square brackets and constitute what is called the *inner coordination sphere*. Outside this sphere other less strongly held groups may be attached. Conductivity measurements, freezing-point measurements, chemical tests, and other instrumental measurements can be used to determine which groups are within and which are outside with the inner coordination sphere.

For example, all of the chloride in $CoCl_3 \cdot 6NH_3$ can be precipitated as AgCl by addition of $AgNO_3$. In addition, conductance measurements indicate the presence of four moles of ions per mole of compound. Other experiments show that one mole of this compound furnishes three moles of Cl^- ions and one mole of $[Co(NH_3)_6]^{3+}$ ions. This suggests that the chlorine is in the form of readily available ions which are not a part of the inner coordination sphere. Therefore the correct formula is $[Co(NH_3)_6]Cl_3$.

When $CoCl_3 \cdot 5NH_3$ is subjected to the same tests, only two thirds of the chlorine is precipitated by $AgNO_3$. Furthermore, conductance tests indicate that three moles of ions are formed per mole of compound. On this basis the correct formula is $[Co(NH_3)_5Cl]Cl_2$.

19-10 Nomenclature Because of the large number of complicated coordination compounds, it has been necessary to develop a systematic method for naming them. Complex species may be a *cation* such as $[Cu(H_2O)_4]^{2+}$, an anion such as $[Fe(CN)_6]^{4-}$, or a neutral molecule such as $[Cr(NH_3)_3Cl_3]$. Many different electronically satisfied entities may act as electron-pair donor ligands. The names of some common ligands are given in the margin. It may be seen that the names of negatively-charged ligands end in *o*. The name of the molecule is generally used for neutral ligands. Water and ammonia are the two important exceptions. The rules listed below enable you to name a large number of common complex substances.

COMMON LEWIS BASES (LIGANDS)

Ligand	*Name of Ligand*
H_2O	aquo
NH_3	ammine
CN^-	cyano
Cl^-	chloro
OH^-	hydroxo
SO_4^{2-}	sulfato
NO_3^-	nitrato
$C_2O_4^{2-}$	oxalato
NO_2^-	nitro
$S_2O_3^{2-}$	thiosulfato

1. Species making up the complex which are enclosed in the brackets are named in this order: negatively-charged ligands, neutral ligands, central ion. The number of each ligand present is indicated by the Greek prefixes *di-* for two, *tri-* for three, *tetra-* for four, *penta-* for five, and *hexa-* for six.
2. The oxidation state of the central ion is written as a Roman numeral and enclosed in parentheses following the name.
3. If the complex species is negative, the name of the central atom ends with the suffix *-ate*.

Example 19-1
Name these substances.

(a) $[Co(NH_3)_6]^{3+}$ **Ans. hexamminecobalt(III) ion.**

(b) $[Fe(CN)_6]^{4-}$ **Ans. hexacyanoferrate(II) ion.**

(c) $[PtCl_2(NH_3)_2]$ **Ans. dichlorodiammineplatinum(II).**

(d) $[CrCl_2(NH_3)_4]Cl$ **Ans. dichlorotetramminechromium(III) chloride.**

FOLLOW-UP PROBLEMS

1. Name these substances.
(a) $[Cu(CN)_3]^{2-}$, (b) $[HgCl_4]^{2-}$ (c) $[Co(NO_2)_6]^{3-}$, (d) $[PtCl_2(H_2O)_2]$, (e) $[CoCl_2(NH_3)_4]_2SO_4$, (f) $[Ni(H_2O)_6]^{2+}$

19-11 Relationship Between Coordination Number and Shape In earlier chapters we discussed the formation and chemistry of a number of complex ions. We can explain the observed geometry of these ions by using *Pauling's valance bond theory* and the concept of *hybridization.* The valence bond theory assumes that as the ligands (electron-pair donors) approach the central ion, hybridization of the atomic orbitals of the ion takes place. This results in an equal number of identical hybrid orbitals. A coordinate covalent bond is formed when the bonding orbital of the ligand, which must contain two electrons, overlaps the empty hybrid orbital of the central ion. The number of ligands attached to the central ion (the coordination number) is related to the geometry of the complex species. We shall consider only the complexes which have coordination numbers of 2, 4, and 6.

It has been experimentally determined that whenever six ligands are joined to a central ion, the complex is always octahedral. Complexes with a coordination number of 2 are linear. Two possible geometrical arrangements exist for complexes having coordination number four. These are (1) a tetrahedron in which the central ion is at the center of the tetrahedron and the four ligands are at the apices, and (2) a square in which the central ion is in the center and a ligand is at each of the four vertices. The hybrid orbitals and the shapes of the ions associated with each coordination number are shown in Table 19-4.

TABLE 19-4
SHAPES OF COMPLEX IONS

Coordination Number	Hybrid Orbitals Involved	Shape of Resulting Complex
2	*sp*	linear
4	sp^3	tetrahedral
4	dsp^2	square planar
6	d^2sp^3 or sp^3d^2	octahedral

19-12 Linear Complexes Let us consider examples of each type of complex. Experiments show that diamminesilver(I) ions, $[Ag(NH_3)_2]^+$, have a coordination number of 2 and are linear. The electron configurations of the silver atom, silver ion, and the silver ion in the complex are shown here. The 4*s*, 4*p*, and 4*d* orbitals in the silver ion are filled, but there are no electrons in the 5*s* or 5*p* orbitals. When the ligands approach, hybridization of the 5*s*

Ag	5*s* ①	5*p* ○○○
Ag^+	5*s* ○	5*p* ○○○
$[Ag(NH_3)_2]^+$	5*sp* ①①	5*p* ○○

H N Ag+ N H H H H H +

[H:N:Ag:N:H with H above and below each N] +

19-6 $Ag(NH_3)_2^+$ is a linear ion.

Zn	4s (↑↓)	4p ○○○
Zn^{2+}	4s ○	4p ○○○
$[Zn(NH_3)_4]^{2+}$		sp^3 (↑↓)(↑↓)(↑↓)(↑↓)

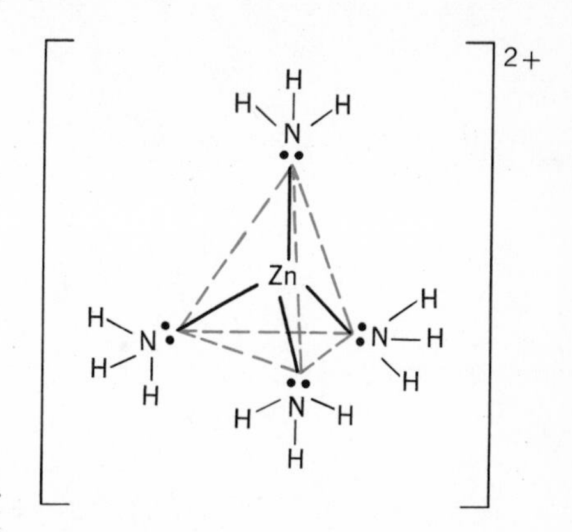

19-7 $Zn(NH_3)_4^{2+}$ is a tetrahedral ion. A zinc ion is located at the center of the tetrahedron with ammonia molecules at the vertices.

and one 5*p* orbital yields two equivalent *sp* orbitals available for bond formation. The four bonding electrons in the complex are those furnished by two NH_3 molecules. The two covalent bonds are formed when each *sp* hybrid of the silver overlaps the single available sp^3 (or *p*) orbital of NH_3. A space-filling model is shown in Fig. 19-6.

19-13 Tetrahedral Complexes A tetramminezinc(II) ion, $[Zn(NH_3)_4]^{2+}$, has a coordination number of 4 and is tetrahedral. All *3s*, *3p*, and *3d* orbitals of Zn^{2+} are filled. The electron configuration of a zinc ion shown above reveals that the empty *4s* and *4p* orbitals are used to bond four NH_3 molecules. Since experiments show that all four bonds are identical, hybridization is invoked so that four sp^3 orbitals are made available for bonding.

The eight bonding electrons in the complex ion are furnished by four ammonia molecules.

19-14 Square Planar Complexes Experiments show that a tetramminecopper(II) ion has a coordination number of 4 and is *square planar*. Furthermore, measurement of magnetic properties reveals the presence of one unpaired electron. The electron configuration of the atom, ion, and copper(II) ion in the complex are shown here. To account for dsp^2 square-planar geometry and a magnetic moment equivalent to one unpaired electron, it is necessary that one 3*d* electron be promoted to a 4*p* level. This makes

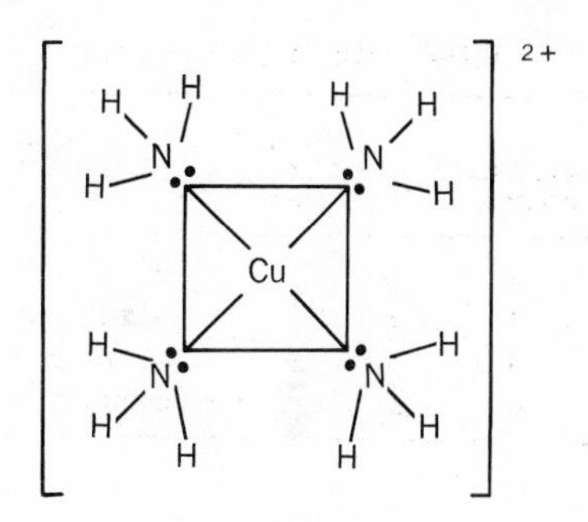

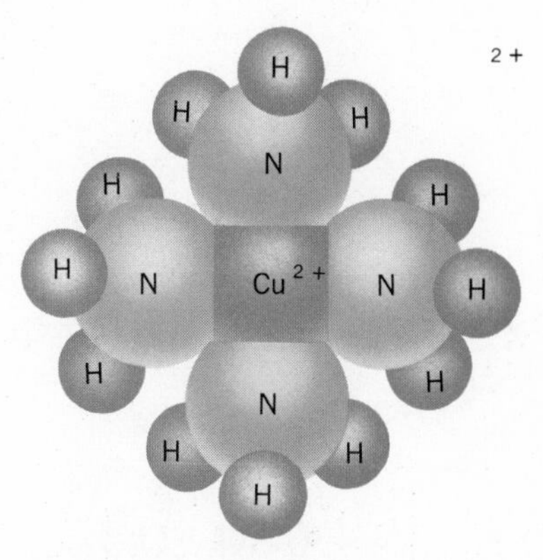

19-8 The square planar $Cu(NH_3)_4^{2+}$ ion. The lines directed to the corners of the square represent dsp^2 orbitals.

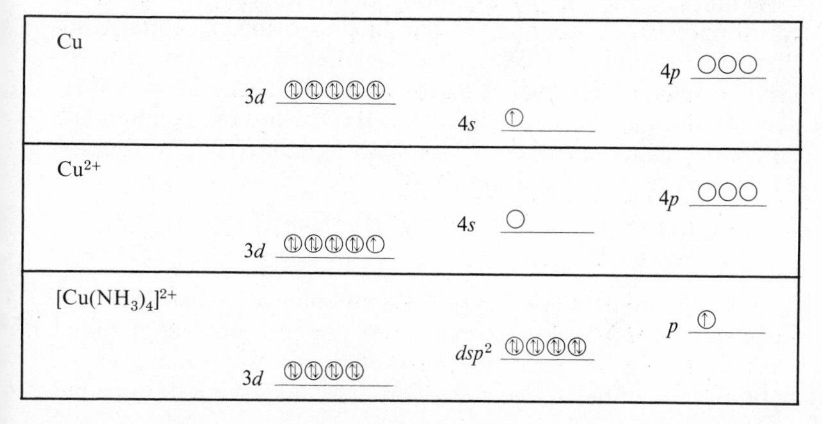

Cu	3d (↑↓)(↑↓)(↑↓)(↑↓)(↑↓)	4s (↑)	4p ○○○
Cu^{2+}	3d (↑↓)(↑↓)(↑↓)(↑↓)(↑)	4s ○	4p ○○○
$[Cu(NH_3)_4]^{2+}$	3d (↑↓)(↑↓)(↑↓)(↑↓)	dsp^2 (↑↓)(↑↓)(↑↓)(↑↓)	p (↑)

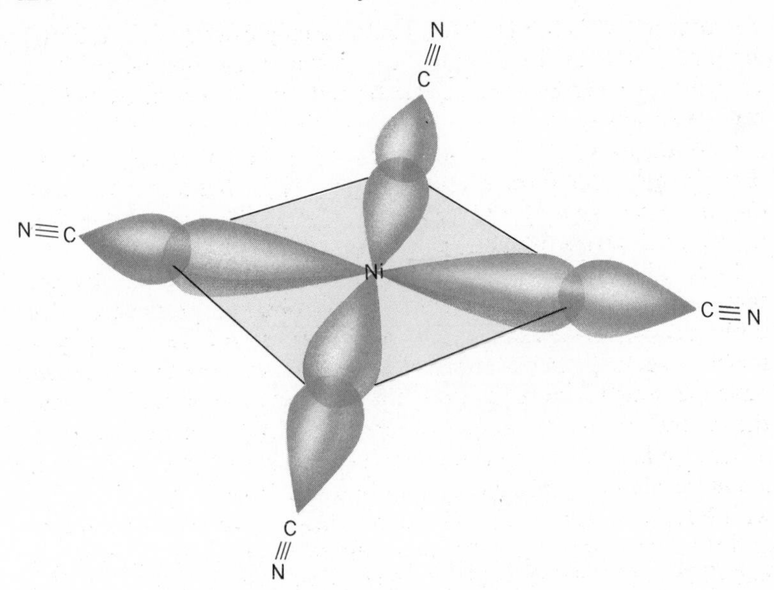

19-9 In a square-planar $Ni(CN)_4^{2-}$ ion, four equivalent dsp^2 hybrid orbitals of nickel are used. Show the configuration of the outer orbital electrons in this diamagnetic species.

the lowest energy and strongest orbitals available for bonding. The eight bonding electrons shown in the dsp^2 orbitals are furnished by four NH_3 molecules.

In contrast to a tetramminecopper(II) ion, the square planar complexes of nickel are not paramagnetic. This means that the electron configuration of the complex has no unpaired electrons. Sketch the configuration of a tetramminenickel(II) ion $[Ni(NH_3)_4]^{2+}$ to test your understanding of the procedure given above.

19-15 Octahedral Complexes The ligands in complexes having a coordination number of 6 are arranged octahedrally about the central atom. The hybrid orbitals involved in bonding are either d^2sp^3 or sp^3d^2. The former involve inner d orbitals and are called ***inner orbital complexes.*** The latter involve outer d orbitals and are known as ***outer orbital complexes.***

In general, weak Lewis bases such as fluoride ions and water form outer orbital complexes. Strong Lewis bases such as CN^- ions form inner orbital complexes. The ligands of outer orbital complexes are readily replaced by stronger Lewis bases. You can observe a ligand exchange reaction by adding concentrated NH_3 solution to an aqueous solution of Cu(II). The instant the ammonia enters the solution, the deep blue color of $[Cu(NH_3)_4]^{2+}$ appears. The ligand exchange reaction is

$$\underset{\text{light blue}}{[Cu(H_2O)_4]^{2+}} + 4NH_3 \longrightarrow \underset{\text{deep blue}}{[Cu(NH_3)_4]^{2+}} + 4H_2O$$

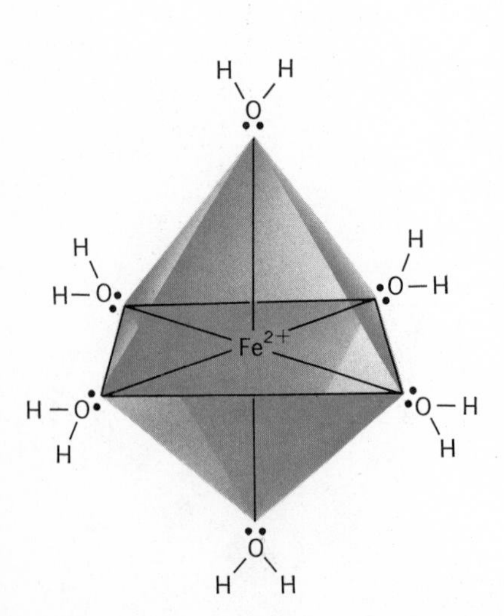

19-10 The octahedral structure of a $Fe(H_2O)_6^{2+}$ ion. The lines radiating from the Fe^{2+} ion represent sp^3d^2 hybrid orbitals.

Hexaaquoiron(II), $[Fe(H_2O)_6]^{2+}$, is an outer orbital complex. Its magnetic properties correspond to the presence of four unpaired electrons. To account for the unpaired electrons we may assume hybridization of the $4s$, $4p$, and $4d$ orbitals and leave four unpaired electrons in the *3d* orbital. The electron configuration of Fe, a Fe(II) ion, and Fe(II) in the complex are shown in the box below. The 12 bonding electrons in the sp^3d^2 orbitals are furnished by the six water molecules.

Cyanide ions (CN^-), a strong base, readily replace water in $[Fe(H_2O)_6]^{2+}$ and form $[Fe(CN)_6]^{4-}$ which is diamagnetic. Thus, *the magnetic property of a complex is a clue as to whether the substance is an inner or outer complex.*

19-16 Geometric Isomers Some complexes have exactly the same molecular or ionic formula but different properties. The differences in properties can be attributed to differences in geometric structure. These substances, known as ***geometric isomers,*** have the same formula but different arrangements of ligands about the central ion. For example, when $[Pt(NH_3)_4]^{2+}$ reacts with HCl, a light yellow compound, having the formula $[Pt(NH_3)_2Cl_2]$ and a square planar configuration, is obtained. When $[PtCl_4]^{2-}$ reacts with NH_3, a darker yellow compound with square planar geometry and the same formula is obtained. The latter compound, however, has different chemical and physical properties than the first. By studying the reactions of the two compounds with different chemical agents and measuring their polarities (dipole moments), scientists have been able to show that the ligands are arranged differently about the central atom in the two compounds.

In the case of the light yellow compound, the two ammonia molecules are opposite each other. The ammonia molecules in the dark yellow compound are found to be adjacent to each other. The prefix *cis* is used to designate the geometric isomer in which *similar ligands are next to each other.* The prefix *trans* is used when the *similar ligands are not adjacent to each other.* Thus, the light yellow complex is called *trans*-dichlorodiammineplatinum(II), and the dark yellow isomer is called *cis*-dichlorodiammineplatinum(II). See Figs. 19-11 and 19-12.

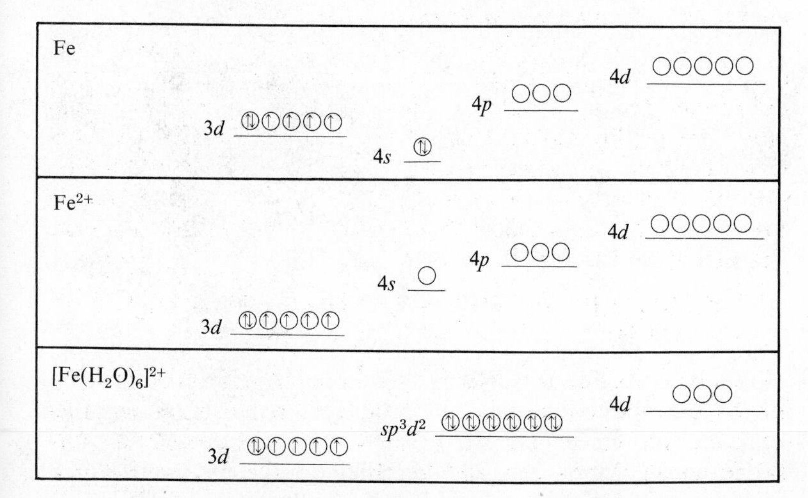

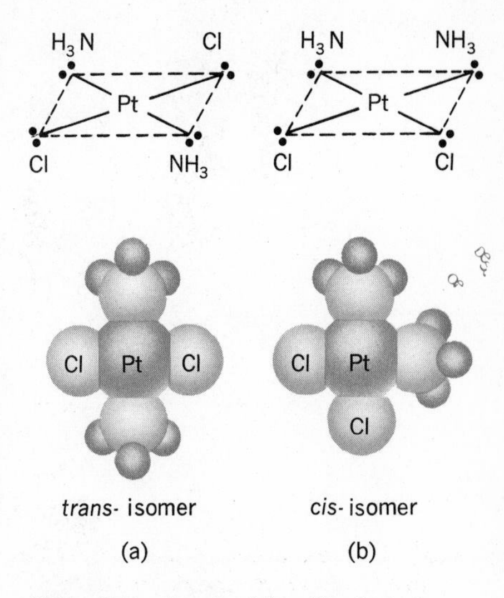

19-11 Dichlorodiammineplatinum(II) exists in the two geometric forms shown. Complex ions or molecules which contain two different kinds of ligands and which have square planar or octahedral geometry may form *cis*- or *trans*-isomers.

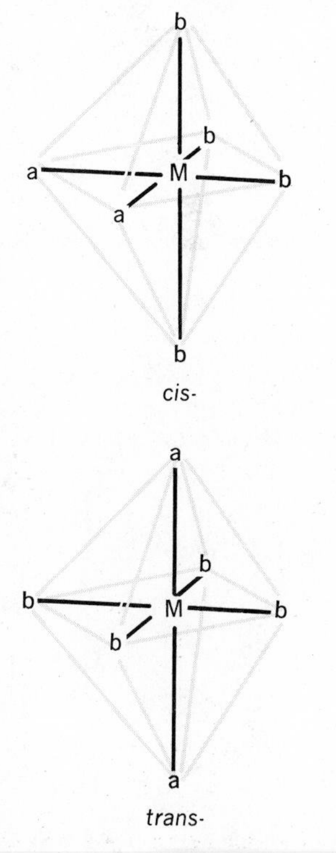

19-12 General structural formulas for the *cis*-*trans* isomers of octahedral complexes.

19-13 The heme molecule (left, above) is a chelate with an atom of iron at its center. This molecule, present in the red blood cells, carries oxygen to the cells of the body.

19-14 The chlorophyll-A molecule (right, above) has a structure somewhat similar to that of heme, except that the central atom is magnesium.

19-15 The color blocks show the six bonding sites on an EDTA ion. Each of these bonding sites has an electron pair which can be donated to a metallic ion.

19-17 Chelates The molecules and ions of some complexing agents are able to form two or more bonds with a central metallic ion. These substances are known as ***chelating agents.*** They "grasp" a metallic ion so that it is enclosed in a ring-like structure. They act rather like the claws of a crab clamping onto a food morsel. By this action, chelating agents are able to deactivate and control metallic ions. Traces of metallic ions in foods and other products often accelerate undesirable oxidation reactions which impair the color, flavor, clarity, and stability of the product. Chelating agents are often added in small quantities to deactivate the metallic ions.

A chelating agent that has two complexing groups is called a ***bidentate;*** ethylenediammine, $H_2NCH_2CH_2NH_2$, is an example. Each of the two nitrogen atoms has an unused pair of electrons and is capable of bonding to the same metallic ion. The reaction between ethylenediammine and tetramminecopper(II) may be represented as

$$[Cu(NH_3)_4]^{2+} + 2H_2N{-}CH_2CH_2{-}NH_2 \longrightarrow [Cu(H_2N{-}CH_2CH_2{-}NH_2)_2]^{2+} + 4NH_3$$

It can be seen that a Cu^{2+} ion is incorporated into the structure of two bidentate molecules. The product is a chelate and is called ethylenediamminecopper(II).

The molecules or ions of other chelating agents may form as many as six or eight bonds with a single positive ion. Heme, the active participant in hemoglobin, and chlorophyll (Figs. 19-13 and 19-14) are metal derivatives of a quadridentate ligand having four nitrogen atoms joined in a ring. *Ethylenediamminetetraacetic acid (EDTA)* is a widely used *hexadentate.* One ion of EDTA is able to deactivate all the bonding sites of metallic ions that form octahedral complexes. The formula of EDTA (Fig. 19-16) shows that it has four oxygen and two nitrogen atoms which can bond to metallic ions. When EDTA chelates Fe^{3+}, the oxygen and nitrogen atoms are located at the corners of an octahedron with an Fe^{3+} ion at the center.

Standard EDTA solutions are widely used in volumetric analysis to determine the *hardness* of water. In this analysis, the EDTA complexes Ca^{2+} and Mg^{2+}, ions which are responsible for the hardness of water. When all of the metallic ions have been chelated, the next drop of EDTA causes an indicator to change color.

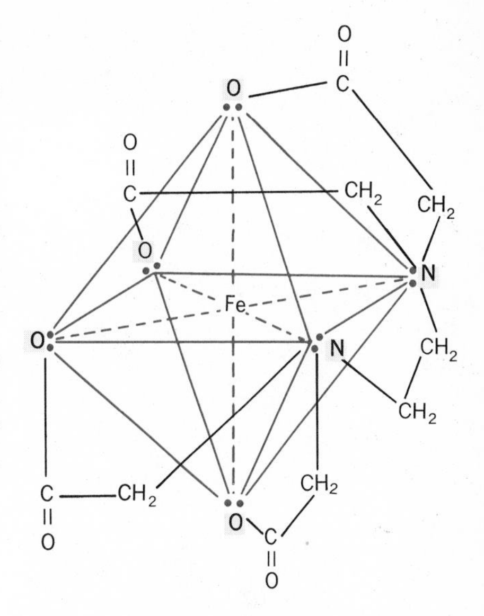

19-16 The EDTA ion surrounds an Fe^{3+} ion like an octopus. Each of the six bonding sites of an EDTA ion furnishes an electron pair at each vertex of the octahedral structure of an EDTA ion.

METALLURGY OF TRANSITION ELEMENTS

Only a few of the transition elements such as silver, gold, and platinum appear as free elements in nature. Iron, copper, zinc, tin, and other important structural metals are obtained by the chemical processing of ores.

Most ores contain the transition elements in the form of *oxides, sulfides,* or *carbonates.* Metals in these compounds are in their oxidized states. Therefore, after processing and purifying the ore, the metals are obtained by reducing it to the elemental state.

19-18 Flotation Cells The first step in processing an ore is to remove undesirable impurities called *gangue.* Gangue generally consists of sand, clay, and other worthless silicate materials. There are a number of methods for removing gangue and concentrating the ore. The most common method is known as *flotation.*

In this process, pulverized ore, frothing agents, and wetting agents known as collectors are added to a flotation tank filled with water. The ore is kept suspended by air bubbles forced through the mixture. This action is similar to that of a whirlpool bath. The collector molecule has a polar and a nonpolar end. The polar end attaches itself to the ore and the nonpolar end keeps the water from wetting the ore. The bubbles of air attached to the ore rise to the surface as a froth and are brushed off with a paddle wheel into a tank. The froth containing the sulfide, carbonate, or oxide of the metal is then dried and converted to a form which can be easily reduced.

19-19 Reduction to Metallic State Concentrated ore is reduced to the metallic state by chemical or electrolytic reduction. Some sulfides such as HgS and Cu_2S can be converted directly to the metal by controlled heating in air.

$$\mathbf{HgS + O_2 \longrightarrow Hg + SO_2}$$
$$\mathbf{Cu_2S + O_2 \longrightarrow 2Cu + SO_2}$$

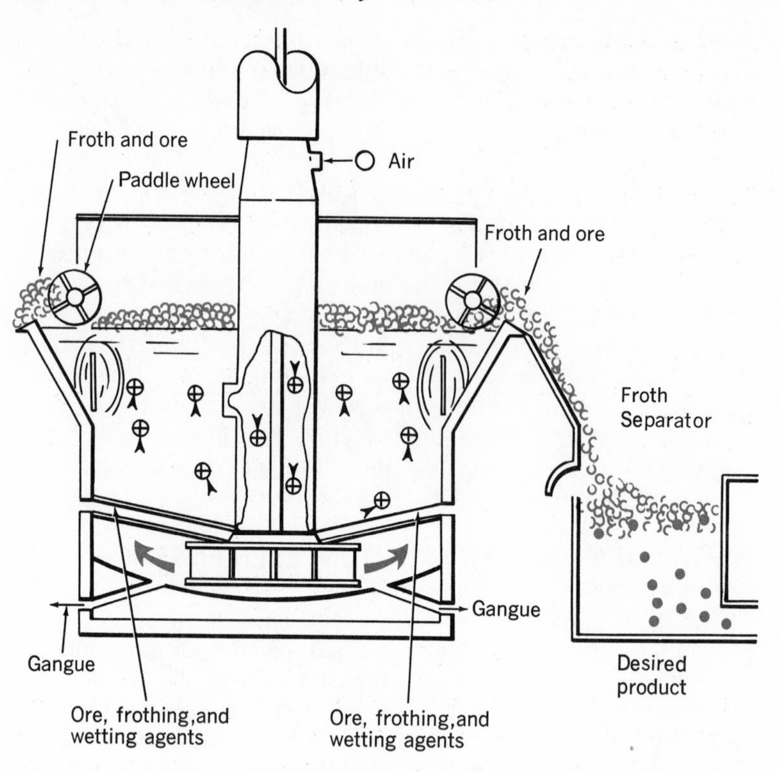

19-17 An ore-flotation cell. Special wetting agents are used for each ore, depending upon which minerals and impurities are present in the ore sample.

In general, it can be shown, using energy concepts, that oxides are more easily reduced than sulfides or carbonates. Therefore, the latter two ores are generally roasted to convert them into oxides. The equation for the conversion of ZnS into ZnO is

$$2ZnS + 3O_2 \longrightarrow 2ZnO + 2SO_2$$

The heat evolved per 8.00 g of oxygen used in forming an oxide is an indication of the thermal stability of the oxide with respect to its decomposition into its elements. Thus, the oxides which evolve the least amount of heat per 8.00 g of oxygen used in their formation are the easiest to decompose by the application of heat. A number of common metallic oxides are listed in Table 19-5 in decreasing order of thermal stability.

In general, an oxide which has a positive ΔH_f°, such as Au_2O_3, spontaneously decomposes at room temperature. The method of reducing an oxide to the metal depends on the stability of the compound. The data in the table show that only HgO and Ag_2O can be appreciably reduced at moderate temperatures (up to 500°C). The oxides of iron, tin, zinc, and lead can be reduced by carbon or carbon monoxide. In most cases, carbon is converted to CO by heating in a stream of oxygen.

$$2C + O_2 \xrightarrow{\text{heat}} 2CO$$

The carbon monoxide then acts as a reducing agent and converts the metallic oxide to the free metal. The reduction of iron oxide may be represented as

$$Fe_2O_3 + 3CO \rightleftharpoons 2Fe + 3CO_2$$

The equilibrium is displaced to the right by maintaining an excess of CO. The CO_2 formed in the reaction can be reduced to CO by passing it over the hot carbon.

$$CO_2 + C \xrightarrow{\text{heat}} 2CO$$

Some substances require a stronger reducing agent than carbon or CO. In these cases, aluminum, magnesium, or calcium may be used. For example, chromium(III) oxide is reduced by heating with aluminum.

$$Cr_2O_3 + 2Al \xrightarrow{\text{heat}} 2Cr + Al_2O_3$$

Titanium, which in recent years has become one of the most important structural metals, is prepared by reducing the tetrachloride with magnesium.

$$TiCl_4 + 2Mg \longrightarrow 2MgCl_2 + Ti$$

When heated to 1000°C in a vacuum, the $MgCl_2$ and any excess magnesium are vaporized. The titanium is then melted and removed. This metal, because of its resistance to corrosion at extremely high temperature, is used in jet and rocket engines. It is also used in structures which require high tensile strength.

TABLE 19-5

THERMAL STABILITIES OF SOME COMMON OXIDES

Oxides Reduced to Metal by	Oxide	ΔH_f° per Mole of Oxide	Kcal Evolved per 8.00 g of Oxygen
electrolysis of melted salt	Na_2O	− 99.5	49.7
	CaO	−157.5	78.8
	MgO	−146.1	73.05
	Al_2O_3	−380.0	63.3
aluminum	TiO_2	−218	54.5
carbon	Cr_2O_3	−270	45.0
	ZnO	− 83.2	41.6
hydrogen	SnO	− 68.4	34.2
	Fe_2O_3	−198.5	33.1
	CdO	− 60.9	30.5
	NiO	− 58.4	29.2
	CoO	− 57.2	28.6
	PbO	− 52.4	26.2
	CuO	− 38.5	19.3
heat	HgO	− 21.6	10.8
	Ag_2O	− 7.3	3.6
unstable at room temperature	Au_2O_3	11.0	− 1.8

(Arrow at left, pointing upward: Decreasing tendency to be reduced)

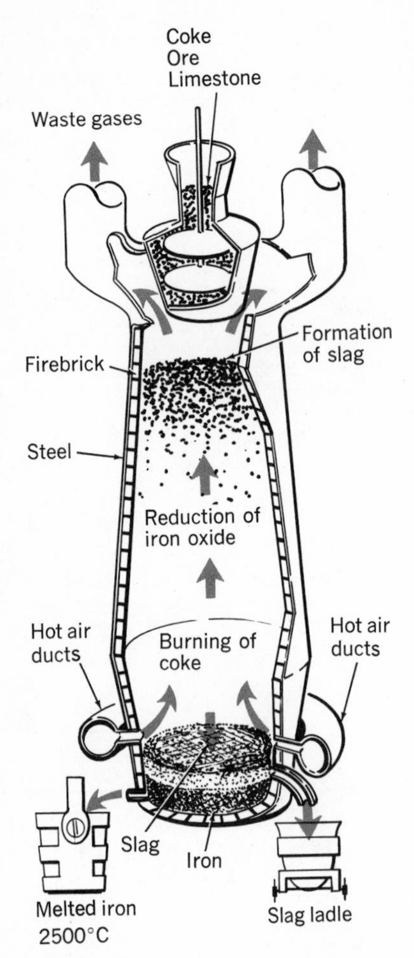

19-18 A blast furnace. These furnaces are kept in continuous operation as the raw materials are constantly fed in and the products are continuously drawn off.

Electrolytic methods are used for reducing a few compounds of the transition elements to the metallic state. The reduction of aluminum ore and other active-metal compounds is electrolytic.

19-20 Steelmaking One of the most important milestones in man's progress was his discovery of a method for obtaining metallic iron from its ore. There is evidence that metallic iron was produced over 4000 years ago; however, it was not until the 16th century that it came into widespread use. In 1855, Sir Henry Bessemer's discovery of a practical method for making steel ushered in the industrial revolution and changed our country from an agrarian-based to an industrial-based society.

The production of steel involves two major steps: the conversion of the *raw iron ore into an impure pig iron* containing 5 to 6 percent impurities and the conversion of *pig iron into steel* by removing impurities and adding necessary alloying substances.

The first step takes place in a *blast furnace.* The second occurs in a *basic oxygen furnace,* an *open-hearth furnace,* or an *electric furnace.*

Reactions occurring in a blast furnace convert iron ore into impure pig iron. A blast furnace is a gigantic chimney-like structure nearly 100 ft high and 20 to 25 ft in diameter. It is made of steel and lined with firebrick.

Iron ore, limestone, and coke are dumped in at the top of the stack and hot air is blown up through the mixture from the bottom. At the bottom of the furnace, where the temperature is approximately 1400°C, coke burns and forms CO_2 which is immediately reduced by the hot carbon to carbon monoxide. The carbon monoxide rises and cools to about 500°C as it nears the top of the furnace. Here hematite (Fe_2O_3) is reduced to Fe_3O_4.

$$3Fe_2O_3 + CO \rightleftharpoons 2Fe_3O_4 + CO_2$$

When the Fe_3O_4 reaches a lower level and a temperature of about 650°C, it is reduced by CO to FeO. About half way down the furnace at 800°C, the FeO is reduced to Fe. When the limestone reaches the middle of the furnace, it forms CaO and CO_2.

$$CaCO_3 \rightleftharpoons CaO + CO_2$$

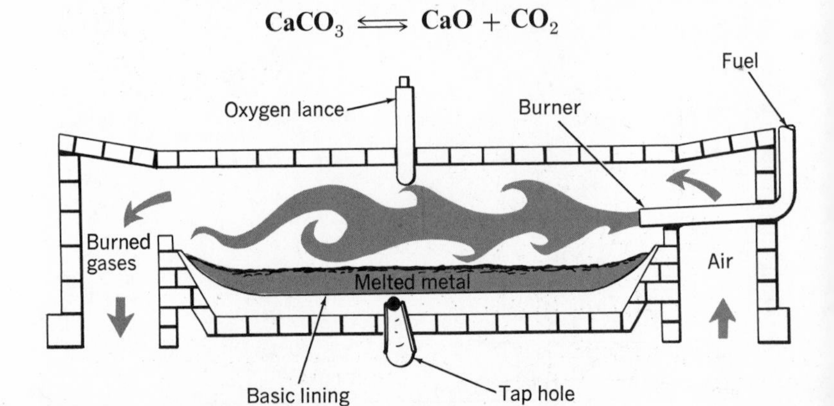

19-19 The open-hearth furnace.

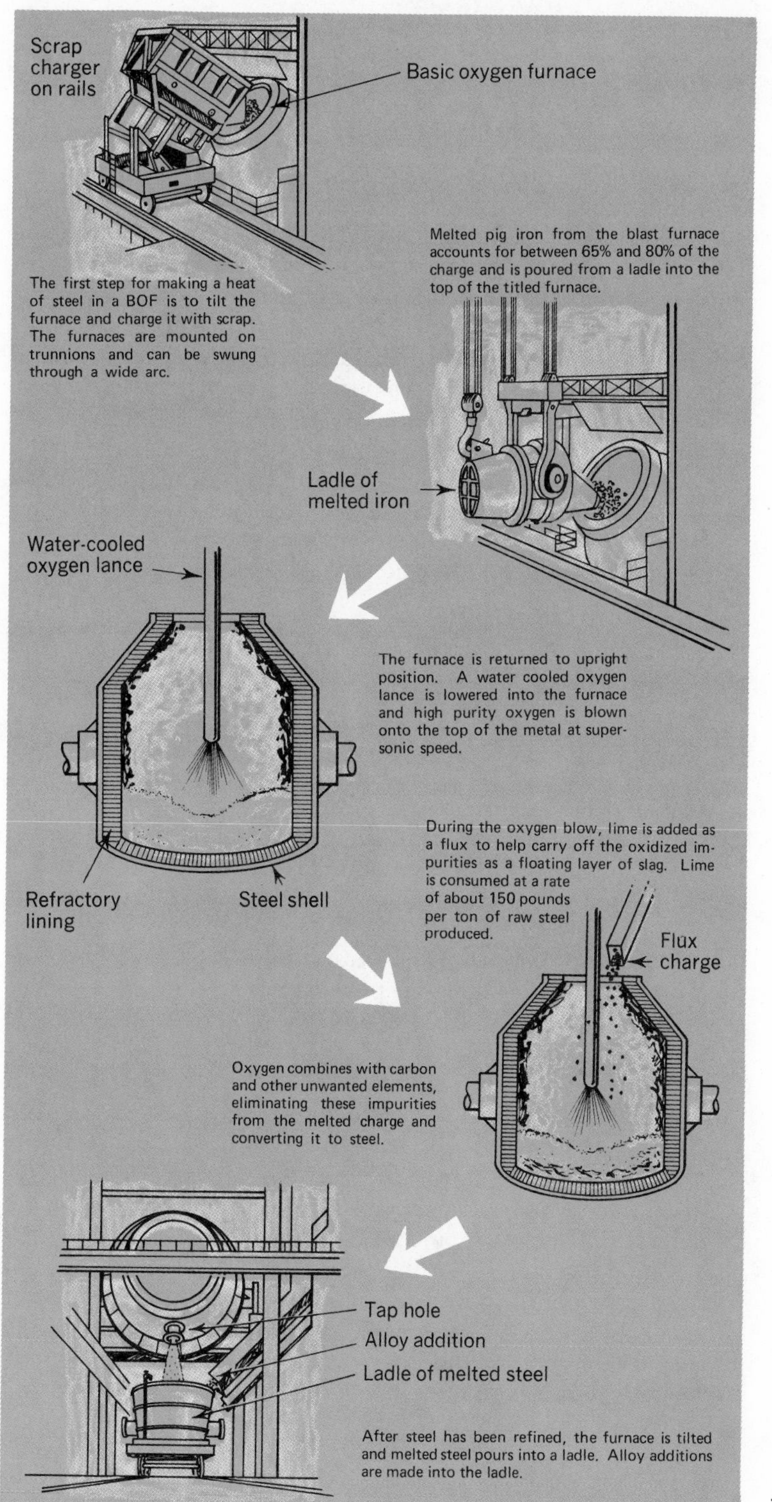

19-20 The basic oxygen furnace.

At about three-fourths of the way down, the temperature is roughly 1000°C and hot enough for the CaO to combine with the silica impurities.

$$CaO + SiO_2 \longrightarrow CaSiO_3$$

Impure calcium silicate is known as *slag*. Both the slag and pig iron settle as melted, immiscible liquids at the bottom of the furnace. The denser liquid iron forms the bottom layer. The iron and slag are drawn off at regular intervals. Blast furnaces in the United States yield approximately 200,000 tons per day.

The iron from a blast furnace contains carbon, phosphorus, sulfur, and silicon impurities. It is brittle and has poor structural characteristics. Most of it is converted to steel in basic oxygen furnaces.

Crude iron is converted to steel in basic oxygen, open-hearth, or electric furnaces by removing impurities and adding alloying metals. In an open-hearth furnace, pig iron from the blast furnace and rusty scrap iron are melted. The impurities of carbon, sulfur, phosphorus, and silicon are oxidized.

$$2Fe_2O_3 + 3C \longrightarrow 4Fe + 3CO_2$$
$$2Fe_2O_3 + 3S \longrightarrow 4Fe + 3SO_2$$
$$10Fe_2O_3 + 12P \longrightarrow 20Fe + 3P_4O_{10}$$
$$2Fe_2O_3 + 3Si \longrightarrow 3SiO_2 + 4Fe$$

The acid oxides of sulfur and phosphorus react with the basic lining (CaO) of the furnace and form a slag which floats on the melted iron. Controlled amounts of carbon and of alloying metals such as manganese, nickel, vanadium, chromium, molybdenum, and tungsten are added, depending on the type and characteristics of the steel desired.

With the help of modern analytical instruments, computers, and improved techniques employing pure oxygen, steelmaking has been streamlined into a relatively fast and efficient process necessary to meet the demands of modern technology.

19-21 Over 100 000 tons of steel have been used in Apollo ground-support applications. Super strong steels are used to construct crawler-transporters for moving space vehicles to the launch pad. Special exotic steels, unknown ten years ago, are used in constructing fuel storage systems, service gantries, umbilical towers, and other structures associated with the space program. (Courtesy United States Steel)

Looking Ahead

Other than hydrogen, there is one nonmetallic element found in more compounds than any other element. This element is carbon. The implication is that the chemistry of carbon and its compounds is broader in scope and complexity than that of any other element. Therefore, it is reasonable to devote a full chapter to the discussion of carbon and its compounds.

The study of most carbon compounds is called organic chemistry. A detailed study of organic chemistry would require years of concentrated effort and involve many volumes the size of this text. In our limited space we can introduce only a few concepts related to this important area of chemistry. In the next chapter, we shall examine the most important classes of organic compounds and types of organic reactions. Hopefully, this will provide insight into the nature, scope, applications, importance, and potential of this field of chemistry which is closely related to the chemistry of living organisms and responsible for a vast number of synthetic products essential to the well-being and comfort of mankind.

QUESTIONS AND PROBLEMS

1. Use atomic numbers to identify three different series of transition elements in the Periodic Table.

2. What characteristic properties distinguish the transition elements from the Group-IA and -IIA metals?

3. How does the atomic theory account for the high density of the metals in the center of the transition series?

4. Using the orbital model of an atom, explain (a) transition metals are generally hard, (b) transition metals have high melting points, (c) paramagnetic properties increase in the middle of the transition series, (d) many transition metals form colored compounds, (e) transition metals may have many oxidation states.

5. Explain why a water solution of copper sulfate is used to filter out heat waves (red end of the spectrum).

6. What color would you expect a solution to appear if the solute absorbed light in the blue area of the spectrum?

7. What is the meaning of each of these. (a) coordination number, (b) ligand, (c) coordination compound, (d) structural isomerism, (e) chelate, (f) bidentate ligand?

8. Name these substances. (a) $Cu(Cl_4)^{2-}$, (b) $Ag(NH_3)_2^+$, (c) $Fe(CN)_6^{4-}$, (d) $Cu(NH_3)_4SO_4$, (e) $Al(H_2O)_6Cl_3$, (f) $Fe_4[Fe(CN)_6]_3$, (g) $Pt(C_2O_4)_4^{4-}$, (h) $Ag(Cl)_2^-$, (i) $NaAg(Cl)_2$, (j) $[Cr(Cl)_2(H_2O)_4]_3[Ni(CN)_6]$.

9. Write formulas for (a) hexaaquochromium(III) chloride, (b) chloro-pentaaquochromium(III) chloride, (c) sodium tetra-chlorodiaquochromate(III), (d) tetramminecobalt(II) hexacyanoferrate(II), (e) diamminesilver(I) hexacyanomanganate(II).

Ans. (c) $Na[CrCl_4(H_2O)_2]$, (e) $[Ag(NH_3)_2]_4Mn(CN)_6$.

10. What is the shape of a complex ion in which the number of ligands attached to the central ion is (a) six, (b) four, (c) two?

Ans. (b) Tetrahedral or square planar.

11. Name the hybrid orbitals involved with each of these shapes of a complex (a) linear, (b) tetrahedral, (c) square planar, (d) octahedral.

12. Chromic ions (Cr^{3+}) form a complex ion by reacting with ammonia

$$Cr^{3+} + 6NH_3 \rightleftharpoons [Cr(NH_3)_6]^{3+}$$

(a) Use diagrams to show the shape of this ion, (b) name the compound, $Cr(NH_3)_6Cl_3$.

13. Use tetramminecopper(II) as an example to show that a copper ion is a Lewis acid and an ammonia molecule a Lewis base.

14. (a) Draw the electron configuration of the neutral cobalt atom. (b) Draw the configuration of the 3*d* electrons in a Co^{3+} ion. (c) How does the configuration in (b) change if the Co^{3+} complex is not paramagnetic? (d) If the complex formed by the Co^{3+} ion is paramagnetic, which orbitals are used for bonding? **Ans. (d) sp^3d^2.**

15. Using your answers to 14, explain why $Co(NH_3)_6^{3+}$ is not paramagnetic and $Co(F_6)^{3-}$ is.

16. (a) What are geometric isomers? (b) Draw the structural formulas of cis- and trans- dichlorodiamminecobalt(II).

17. Two distinct compounds with different melting points have the formula $[Cr(H_2O)_4Cl_2]Cl$. Draw possible structural formulas for these two compounds.

18. Consider the structural isomers $[Cr(H_2O)_6]Cl_3$, $[CrCl(H_2O)_5]Cl_2 \cdot H_2O$, $[CrCl_2(H_2O)_4]Cl \cdot 2H_2O$ (a) Which

of these has the lowest freezing point? Assume one mole of compound is dissolved in one kg of water. (b) Which of the compounds has the lowest conductance (the smallest number of ions)? (c) Which yields the greatest mass of AgCl when excess $AgNO_3$ is added to the solution?

19. (a) Show how ethylenediammine can complex with a tetrammineplatinum(II) ion which has a square planar structure. (b) Show how ethylenediammine can complex with cis-diamminedichloroplatinum(II) but not with the transisomer.

20. (a) Name two chelating agents. (b) How can they be used in the laboratory?

21. Which groups of metals in the Periodic Table are (a) reduced only by electrolysis, (b) reduced by carbon as well as by electrolysis, (c) found free in nature?

22. Write equations for these metallurgical reactions (a) the reduction of iron ore by carbon monoxide, (b) the reduction of chromium(III) oxide by aluminum, (c) the roasting of zinc carbonate, (d) heating a mixture of mercury(II) sulfide and mercury(II) oxide, (e) the reduction of titanium tetrachloride by magnesium.

23. How are limestone impurities in iron ore removed during the reduction process?

24. Distinguish between pig iron and steel. What advantage does steel have over pig iron in making (a) hammers, (b) griddles?

20

Organic Chemistry

Preview

Until 1828, chemists believed that certain materials such as sugar, silk, vinegar, and oil could be produced only by living organisms. Such materials were called *organic*. Chemists knew that organic substances were easily decomposed by heat into inorganic substances, but they had not been able to convert inorganic into organic substances.

Synthetic organic chemistry was born in 1828 when Friedrich Wöhler (1800–1882), a German chemist, converted an inorganic chemical called ammonium cyanate (NH_4CNO) into urea [$(NH_2)_2CO$], a substance with the same composition isolated from urine and recognized as organic by all chemists of that era. Since that time, approximately one million organic compounds have been synthesized. This is many times more than all the inorganic substances known at present. Modern drugs, synthetic fabrics, building materials, dyes, insecticides, and many other materials are products of synthetic organic chemistry. In 1965, Dr. Robert Woodward (1917–), a Harvard professor, was awarded the Nobel Prize in chemistry for outstanding achievement in the field of synthetic organic chemistry. During the last quarter century, Dr. Woodward has synthesized such complicated substances as quinine and chlorophyll.

Organic substances have one feature in common. They all contain carbon atoms. Thus, organic chemistry is the chemistry of carbon compounds. The fact that carbon forms the backbone of more compounds than all of the other elements put together suggests that carbon has some very unusual, almost unique, characteristics.

The most significant feature of carbon atoms is their ability to form covalent bonds with other carbon atoms as readily as they do with other kinds of atoms. This ability leads to the formation of chain-like molecules which may contain even thousands of carbon atoms. When the ends of a chain are joined, cyclic or ring structures are obtained. In theory, there is no limit to the number of carbon compounds which can be formed.

Many organic compounds have the same formula but entirely different properties and structures. An example is C_2H_6O, the formula for both ethyl alcohol

"Organic chemistry just now is enough to drive one mad. To me it appears like a primeval tropical forest full of the most remarkable things, a dreadful endless jungle into which one does not dare enter, for there seems no way out."

—Friedrich Wöhler

and dimethyl ether. Such sets of compounds are known as *isomers*. In theory, thousands of isomers are possible for a substance with a relatively simple molecular formula. To distinguish one isomer from another, we must use structural formulas. To simplify the study of organic compounds, we group them into relatively few categories in terms of their structures and compositions and use systematic methods for naming them.

We can barely scratch the surface of organic chemistry in a single chapter. The best we can do is to give an insight into the nature and scope of the subject.

We shall first examine the structure and bonding characteristics of a carbon atom. To systematize our study of the vast number of carbon compounds, we shall group them into four broad categories on the basis of their skeletal carbon structures. The four categories are *aliphatic, alicyclic, aromatic,* and *heterocyclic* compounds. We shall further classify the members in each category in terms of their *reactive or functional groups*. We will begin by discussing the nomenclature (naming) of the *saturated hydrocarbons*. The members of this series contain no reactive groups but the names of the members of this series serve as a basis for naming the members of other series of organic compounds.

After discussing the nomenclature of the saturated hydrocarbons, we shall consider the preparation, reactions, and uses of some major types of organic compounds. These types are *alkenes* and *alkanes, alcohols* and *ethers, aldehydes* and *ketones, carboxylic acids* and *esters, amines* and *amides*. You will find that most organic reactions such as oxidation, reduction, alkylation, acylation, and others can be classified as one of three reaction types: *addition, substitution,* and *elimination*.

The chapter concludes with brief discussions of biochemistry, synthetic polymers, and modern analytical techniques. You will find that acid-base concepts, bonding concepts, and reaction principles are common threads which link organic to inorganic chemistry.

GENERAL CHARACTERISTICS AND CLASSIFICATION OF ORGANIC COMPOUNDS

20-1 Bonding in Organic Compounds and General Nature of Organic Reactions A carbon atom in its ground state has an electron configuration of $1s^2$, $2s^2$, $2p^2$. By applying the concept of orbital hybridization, we can explain the ability of carbon atoms to form four covalent bonds. When undergoing reaction, the ground-state electron configuration can be altered by the promotion of a 2s electron to the empty 2p energy state and the formation of four equivalent sp^3 energy levels or orbitals. The four sp^3 orbitals are the result of the mixing or hybridizing of the s orbital and three p orbitals. The large number of organic compounds can be attributed to the rather unique ability of carbon atoms to form as many

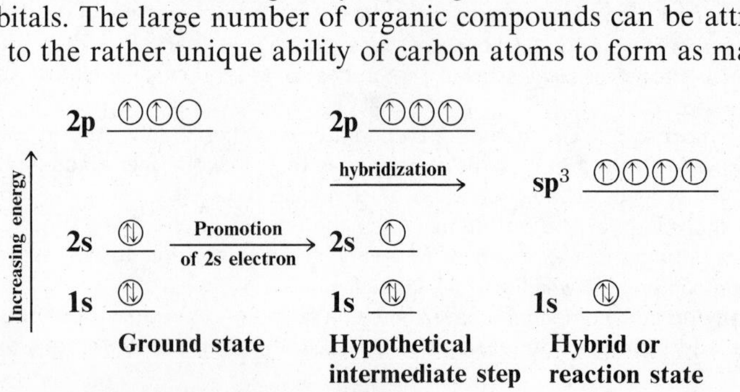

Increasing energy

2p ⓘⓘ○ 2p ⓘⓘⓘ

hybridization → sp^3 ⓘⓘⓘⓘ

2s ⇅ Promotion of 2s electron → 2s ⓘ

1s ⇅ 1s ⇅ 1s ⇅

Ground state | Hypothetical intermediate step | Hybrid or reaction state

as four strong covalent bonds with other carbon atoms and to form long chains of C—C bonds. Because the electronegativity of carbon (2.5) is near that of hydrogen, oxygen, nitrogen, and sulfur, it also forms covalent bonds with these elements. Most organic compounds are covalent in nature rather than ionic. This means that the properties of most organic substances can be interpreted in terms of van der Waals forces, dipole interactions, and hydrogen bonding. The melting points of a series of similar organic compounds, listed in Table 20-2 support this statement.

Large, bulky molecules composed of many covalently bonded atoms drastically affect the rate of most organic reactions. Because of the nature of the intramolecular bonding and the complexity of organic molecules, *organic reactions are usually slower* than inorganic ones. The reason for this is the fact that the barriers to making and breaking covalent bonds are greater than those associated with ionic bonds. In addition, organic reactions may involve a number of side reactions which yield a number of products. This gives the chemist the additional task of separating and purifying the various products which may be present in the final complex reaction mixture. Both intermolecular forces and reaction rates are affected by the shapes of molecules involved. Because sp^3 hybridization is assumed to be involved in the formation of a large number of organic molecules, the shapes of many organic molecules are related to the tetrahedral bond angle of 109.50°.

The simplest organic compound is methane (CH_4). If we could examine this molecule, we would find that it has the shape of a tetrahedron with a carbon atom in the center and a hydrogen atom at each apex. It is convenient to represent the structure of methane as shown in the margin. These formulas are misleading in that they imply that all five atoms are in the same plane. Three-dimensional models and diagrams, such as those shown in Fig. 20-1, suggest the correct spatial orientation of the atoms and the shapes of the molecules.

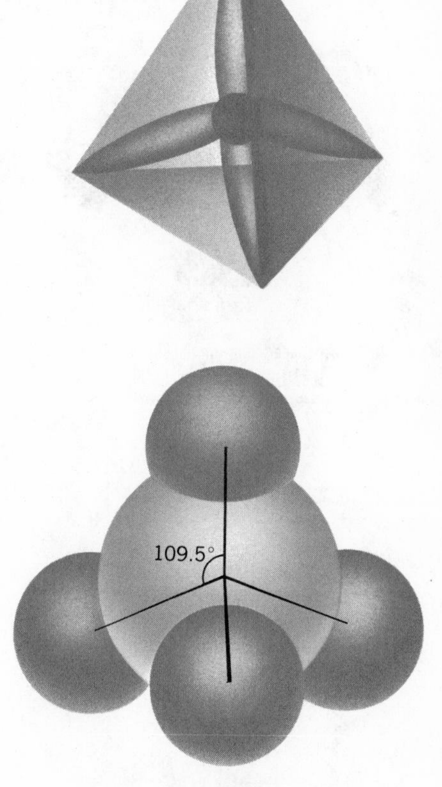

20-1 In methane, the sp^3 bonding orbitals of a carbon atom are directed toward the corners of a tetrahedron. In the space-filling model, all of the H—C—H bond angles are equal to 109.5°.

20-2 General Classification of Organic Compounds The strength of the C—C bond is so great that, during many reactions, the carbon skeleton of the molecule remains intact even though other groups (—Cl, —OH, —NO_2) attached to the molecule may be changed. This characteristic makes it convenient to classify organic compounds in terms of their *skeletal carbon structure* and the *reactive groups* (*functional groups*) located within the molecule.

The four general skeletal structures associated with organic compounds are called *aliphatic, alicyclic, aromatic,* and *heterocyclic.* ***Aliphatic*** compounds are composed of carbon atoms linked in an open chain structure. The chain may have branches but does not form any closed loops. For example, propane (C_3H_8) and isooctane (C_8H_{18}) are aliphatic compounds. Their two-dimensional structures are shown in the margin. The three-dimensional space-filling model of propane shown in Fig. 20-2 enables you to see that the bond angle between alternate carbon atoms is about *109.5°*.

```
  H          H
H:C:H     H—C—H
  H          H
     methane

   H  H  H
H—C—C—C—H
   H  H  H
    propane

   H  H  H  H  H  H  H
H—C—C—C—C—C—C—C—H
   H  |  H  H  H  H  H
   H—C—H
      H
     isooctane
```

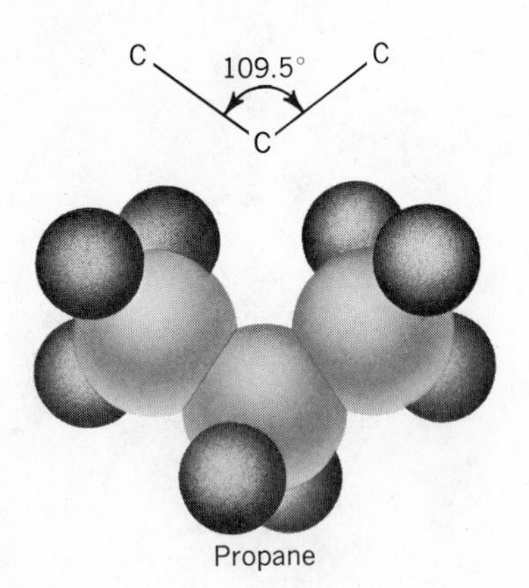

20-2 A space-filling model of propane (C_3H_8) showing the 109.5° bond angle between alternate carbon atoms.

Alicyclic compounds are compounds in which the carbon atoms are arranged in a ring structure. The rings may contain three, four, five, or a larger number of atoms. Cyclopropane (C_3H_6), sometimes used as an anesthetic, is an example of an alicyclic compound. Its structure is shown in the margin.

cyclopropane

In general, ***aromatic compounds*** have structures which are related to that of benzene (C_6H_6), the simplest aromatic compound. These compounds are characterized by molecules which contain a ring of six carbon atoms. The ring structure is usually represented by showing alternate single and double bonds, although it has been demonstrated that it is a delocalized electron system with six identical C—C bonds. The structural formula for benzene is shown in the margin.

benzene

TABLE 20-1

TYPES OF ORGANIC COMPOUNDS AND THEIR FUNCTIONAL GROUPS

Class	*Functional Group*	*Example*	*Name*
Alkanes	—C—C— (with single bonds)	$H_3C—CH_3$	ethane
Alkenes	>C=C<	$H_2C=CH_2$	ethene
Alkynes	—C≡C—	HC≡CH	ethyne (acetylene)
Alcohols	—OH	$CH_3C(H)(H)—OH$	ethanol
Ethers	—C—O—C—	$CH_3C(H)(H)—O—C(H)(H)CH_3$	ethyl ether
Aldehydes	—C(=O)—H	$CH_3C(=O)—H$	ethanal
Ketones	—C—C(=O)—C—	$CH_3C(=O)—CH_3$	propanone (acetone)
Carboxylic acids	—C(=O)—OH	$CH_3C(=O)—OH$	ethanoic acid (acetic acid)
Esters	—C(=O)—O—	$CH_3C(=O)—O—CH_3$	methyl ethanoate (methyl acetate)
Amines	—N— (with bond)	$CH_3C(H)(H)—N(H)—H$	ethyl amine
Amides	—C(=O)—NH_2	$CH_3C(=O)—NH_2$	acetamide

Heterocyclic compounds are composed of molecules which incorporate atoms such as oxygen, nitrogen, sulfur, phosphorus, and others as part of the ring system. *Nucleic acids,* important components of biological cells, contain heterocyclic ring systems. Ethylene oxide (C_2H_4O), a substance used to fumigate foodstuffs, is an example of a heterocyclic compound. Its structural formula is shown in the margin.

H H
H—C—C—H
O
ethylene oxide

20-3 Functional Groups Organic compounds can be classified on the basis of their functional groups. It has been found that organic compounds containing the same types of functional (reactive) groups behave in much the same manner. That is, the size and shape of the carbon-atom skeleton does not affect the chemical behavior of the substance to any great extent. This observation allows us to systematize the study of organic chemistry by classifying substances in terms of their functional groups. But, no classification system is perfect. There are many variations in composition and structure so that it is often difficult to classify a substance rigidly as belonging to a particular group. We shall first itemize the important classes of organic compounds and their functional groups in Table 20-1 and then discuss each class individually.

TYPES, PREPARATION, REACTIONS, NOMENCLATURE, AND USES OF ORGANIC COMPOUNDS

20-4 Saturated Hydrocarbons and Alkyl Groups The simplest organic, aliphatic compounds which have no reactive groups are the *saturated* hydrocarbons. These substances are known as alkanes or paraffin hydrocarbons. The molecules of these hydrocarbons contain only hydrogen and carbon atoms bonded by *single covalent bonds*. This means that the four bonding positions of each carbon atom in the molecule are occupied by either a hydrogen atom or another carbon atom. The members of this group have the general formula C_nH_{2n+2}. Methane (CH_4) is the first member of this series.

Closely related to each of the saturated hydrocarbon molecules are the corresponding *alkyl groups* or radicals. Alkyl radicals are simply the neutral species formed when one hydrogen atom is removed from the parent hydrocarbon. For example, the removal of a hydrogen atom from methane produces the methyl group ($—CH_3$). The methyl and other carbon radicals are *neutral species*. That is, they have the same number of electrons as protons. The dash (—) in this case simply indicates the presence of an unpaired electron and of a single bonding position. The methyl group could also be represented as shown in the margin. The names of many organic compounds are based on the names of the first ten members of the paraffin series and their corresponding alkyl groups. See Table 20-2 on the next page.

H
H—C—
H
methyl group

FOLLOW-UP PROBLEM

1. Using Tables 20-1 and 20-2, name these organic compounds (a) CH_3Br, (b) C_2H_5Cl, (c) CH_3OH.

2. Write the molecular formula for (a) methyl iodide, (b) propyl bromide, (c) butyl alcohol.

TABLE 20-2

η-ALKANES AND ALKYL RADICALS

Alkane	*Formula*	*Melting Point (°C)*	*Phase at Room Temperature*	*Alkyl Group*	*Formula*
Methane	CH_4	−183	gas	methyl	CH_3^-
Ethane	C_2H_6	−172	gas	ethyl	$C_2H_5^-$
Propane	C_3H_8	−187	gas	propyl	$C_3H_7^-$
Butane	C_4H_{10}	−135	gas	butyl	$C_4H_9^-$
Pentane	C_5H_{12}	−130	liquid	pentyl(amyl)	$C_5H_{11}^-$
Hexane	C_6H_{14}	−94	liquid	hexyl	$C_6H_{13}^-$
Heptane	C_7H_{16}	−91	liquid	heptyl	$C_7H_{15}^-$
Octane	C_8H_{18}	−57	liquid	octyl	$C_8H_{17}^-$
Nonane	C_9H_{20}	−54	liquid	nonyl	$C_9H_{19}^-$
Decane	$C_{10}H_{22}$	−30	liquid	decyl	$C_{10}H_{21}^-$

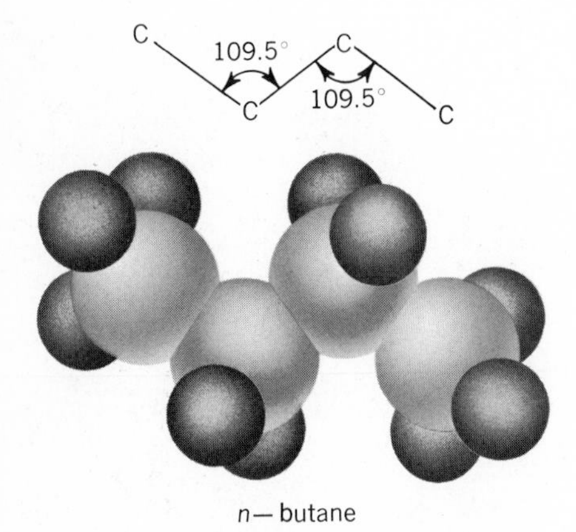

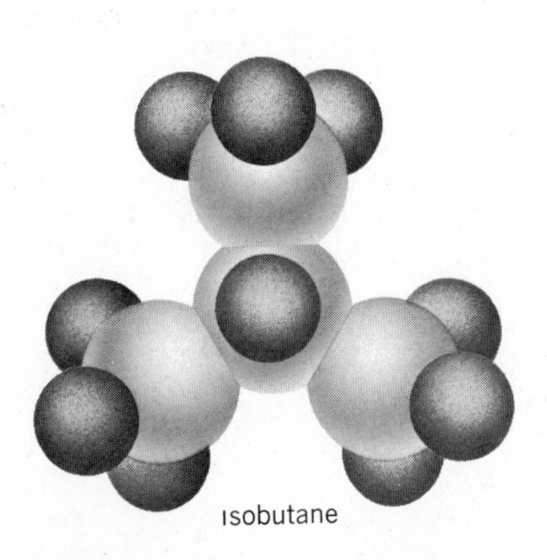

20-3 Butane and isobutane are structural isomers. Their formulas are the same, but the spatial arrangement of their atoms is different. In isobutane, the central carbon atom is bonded to three other carbon atoms.

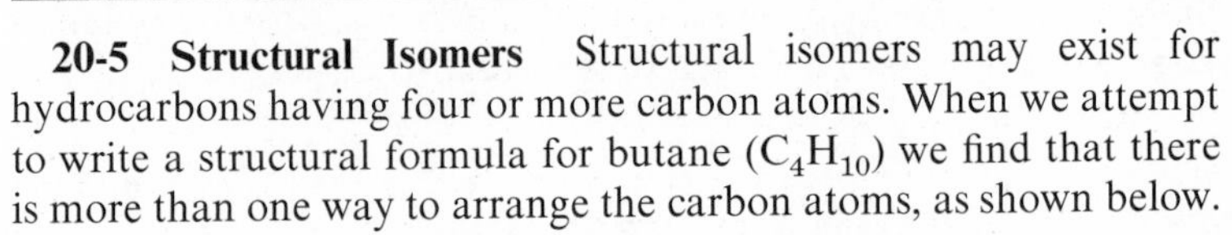

20-5 Structural Isomers Structural isomers may exist for hydrocarbons having four or more carbon atoms. When we attempt to write a structural formula for butane (C_4H_{10}) we find that there is more than one way to arrange the carbon atoms, as shown below.

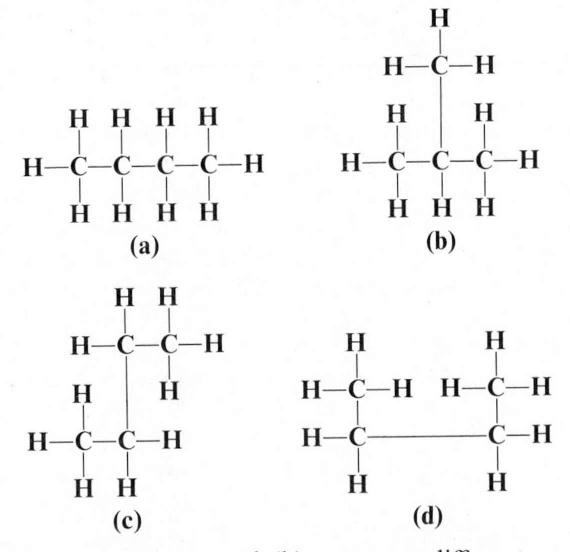

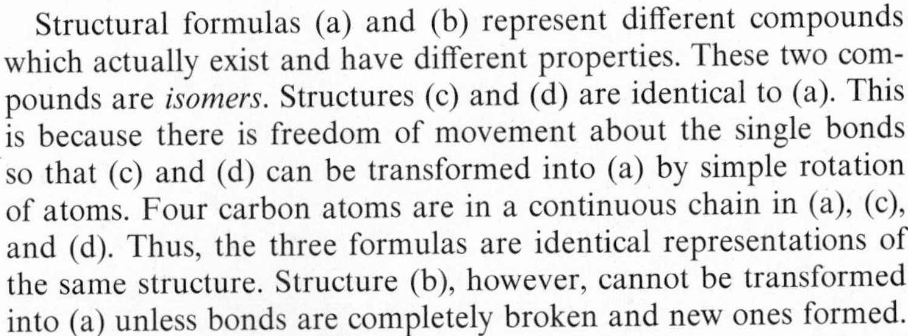

Structural formulas (a) and (b) represent different compounds which actually exist and have different properties. These two compounds are *isomers*. Structures (c) and (d) are identical to (a). This is because there is freedom of movement about the single bonds so that (c) and (d) can be transformed into (a) by simple rotation of atoms. Four carbon atoms are in a continuous chain in (a), (c), and (d). Thus, the three formulas are identical representations of the same structure. Structure (b), however, cannot be transformed into (a) unless bonds are completely broken and new ones formed.

Structure (a) is called *normal butane* or *n-butane.* The compound represented by structure (b) is called isobutane. The prefix iso- is used to represent the presence in the molecule of the alkyl group shown in the margin. This alkyl group is sometimes written as $(CH_3)_2CH—$. It is characterized by the substitution of a methyl group for one of the hydrogen atoms attached to the second carbon atom.

```
      H
      |
H3C—C—
      |
     CH3
```
alkyl group

FOLLOW-UP PROBLEMS

1. Write a structural formula for isopropyl bromide.

2. Name the substance whose formula is

```
H3C  H
   \ |
    C—OH
   /
H3C
```

20-6 Systematic Names of Organic Compounds Systematic names for organic compounds can be derived from structural formulas and vice versa. A systematic method for naming organic compounds has been developed by the International Union of Pure and Applied Chemistry (IUPAC). A number of organic compounds are still called by their common or "trivial" names. We shall emphasize the IUPAC system, but we shall also use common names whenever they are still in widespread use.

The basic part of the *systematic name* of an organic compound is the name of the longest carbon chain in the molecule. The chain names are those of the alkanes listed in Table 20-2. The carbon atoms of the longest chain are then numbered, beginning at the end closer to any side branches of the main chain. The side branches are named as alkyl groups and their positions identified by the number of the carbon atom in the chain to which they are attached. The systematic name of isobutane (structure shown in margin) is 2-methylpropane. The formulas and systematic names of the three isomers of pentane (C_5H_{12}) are shown in the margin.

We can also write the structure of a molecule from its name. The name 2,3,3-trimethylpentane gives us the information: the longest chain has five carbon atoms (pentane), there are three methyl side groups (trimethyl), and one attached to the second carbon atom and two attached to the third carbon atom in the five-carbon chain. The formula for this compound is shown in the margin.

```
      CH3
      |
CH3—CH—CH3
 1    2   3
```
isobutane

$CH_3CH_2CH_2CH_2CH_3$
pentane

```
 CH3
 |
CH3CHCH2CH3
```
2-methylbutane

```
    CH3
    |
CH3CCH3
    |
    CH3
```
2,2-dimethylpropane

```
  H3C CH3
   |   |
CH3CHCCH2CH3
       |
      CH3
```
2,3,3-trimethylpentane

FOLLOW-UP PROBLEMS

1. Name compounds (a) and (b) shown below

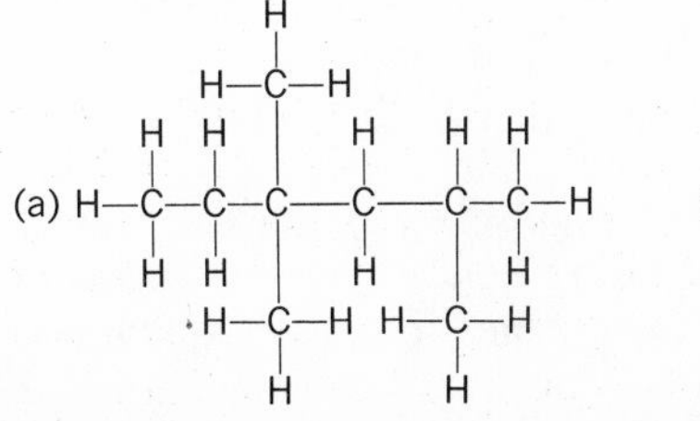

```
          H
          |
        H—C—H
     H    |    H  H
     |    |    |  |
(b) H—C———C———C——C—H
     |    |    |  |
     H    |    H  H
        H—C—H
          |
          H
```

2. Write structural formulas for (a) 2-iodo-3,3-dimethyl pentane, (b) 2-methyl-3-chloro butane.

20-7 Sources, Reactions, and Uses of Saturated Hydrocarbons Relatively speaking, the alkanes are chemically unreactive. Most alkanes are stable with respect to reaction with other substances. This is because their molecules do not contain any functional (reactive) groups. All carbon atoms in an alkane are linked to hydrogen or to other carbon atoms by relatively strong bonds. Most of the alkanes can be made to react with oxygen at high temperatures. These reactions are very exothermic, so saturated hydrocarbons are used as fuels. One of the most common fuels is methane in natural gas. *The complete combustion of any hydrocarbon yields carbon dioxide and water.*

$$CH_4(g) + 2O_2(g) \longrightarrow CO_2(g) + 2H_2O(g)$$

Incomplete combustion yields carbon monoxide and water, or even carbon and water.

$$CH_4(g) + \tfrac{3}{2}O_2(g) \longrightarrow CO(g) + 2H_2O(g)$$
$$CH_4(g) + O_2(g) \longrightarrow C(s) + 2H_2O(g)$$

In addition to their use as fuels, the alkanes find some application as solvents and cleaning agents.

Natural gas, petroleum, and coal furnish basic raw materials for synthetic organic chemistry. Alkanes as well as most other hydrocarbons, are derived from natural sources. They are found in natural gas, petroleum, coal, and plants. Petroleum is a mixture containing hundreds of different compounds. A large number of these are saturated aliphatic hydrocarbons. Gasoline, kerosene, lubricating oils, and greases are separated from petroleum by *fractional distillation* (p. 37). The gasoline fraction is greatly increased by the cracking process, which is carried out in catalytic cracking towers. In this process, catalysts, high pressures, and elevated temperatures are used to crack long-chain hydrocarbon molecules into smaller fragments which can then be used as components of gasoline.

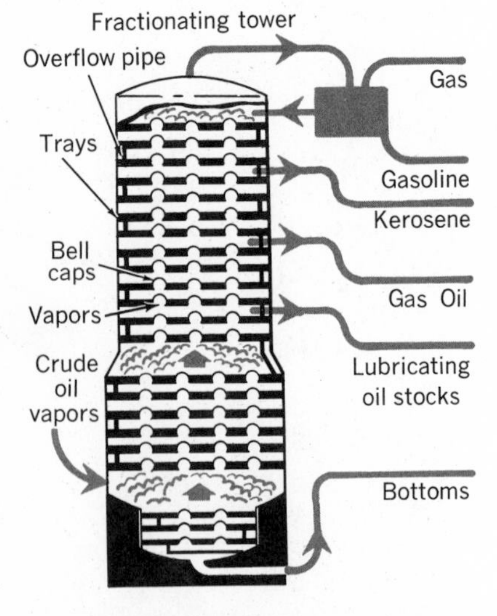

20-4 A fractionating tower in which the components of crude oil are separated on the basis of their different boiling points. The lower-boiling components condense near the top of the tower; the higher boiling near the bottom.

20-8 Alkenes The alkenes, sometimes called olefins, are unsaturated hydrocarbon molecules containing double bonds. The general formula for members of this series is C_nH_{2n}. The simplest member of the series is C_2H_4, commonly called ethylene. The second member, C_3H_6, is commonly called propylene. The structural formulas for these two compounds are shown below.

```
  H H               H
  | |               |
H—C=C—H       H—C=C—C—H
                | | |
                H H H
ethylene        propylene
```

The systematic base name for the alkenes is derived in the same way as it is for the alkanes. The ending *-ane* which characterizes the alkanes is changed to *-ene* for the alkenes. Thus, the systematic name for C_2H_4 is *ethene* and that for C_3H_6 is *propene.* For larger members of the series, the position of the double bond must be noted. The longest chain of carbon atoms is numbered from the end that is closer to the double bond. The position and names of any alkyl groups attached to the main chain are placed before the double bond number. Thus, the compound whose structure is shown in the margin is named 3-ethyl-l-hexene. Aliphatic hydrocarbons which contain two double bonds are called alkadienes. One of the most common alkadienes, used in preparing synthetic rubber, is 1, 3-butadiene, whose formula is shown in the margin. A commonly used *alkenyl group* or radical is the vinyl group, whose formula is also given in the margin.

$$\begin{array}{c} \quad\quad\quad\quad\quad\quad CH_2CH_3 \\ \quad\quad\quad\quad\quad\quad | \\ CH_3CH_2CH_2CHCH{=}CH_2 \end{array}$$

3-ethyl-1-hexene

$$CH_2{=}CH{-}CH{=}CH_2$$

1,3-butadiene

$$CH_2{=}CH{-}$$

vinyl group

FOLLOW-UP PROBLEM

Write structural formulas for (a) 2-chloro-3-methyl-2-pentene, (b) 3,3-dimethyl-1-pentene.

20-9 Addition Reactions Because they have a *double bond, alkenes are more reactive than alkanes.* One of the main types of organic reactions characteristic of alkenes, as well as of other types of compounds, is the addition reaction. In *addition reactions* two or more atoms are added to an unsaturated molecule. An *unsaturated hydrocarbon* is one which does not contain the maximum possible number of hydrogen atoms. The simplest addition reaction is the catalytic addition of hydrogen to the carbon-carbon double bond. This is one method for preparing an alkane from an alkene.

$$\begin{array}{c} \text{H}\ \ \text{H} \\ |\ \ \ | \\ \text{H}{-}\text{C}{=}\text{C}{-}\text{H} \end{array} + \text{H}_2 \longrightarrow \begin{array}{c} \text{H}\ \ \text{H} \\ |\ \ \ | \\ \text{H}{-}\text{C}{-}\text{C}{-}\text{H} \\ |\ \ \ | \\ \text{H}\ \ \text{H} \end{array}$$

With the proper catalyst, a number of unsaturated compounds can be hydrogenated at room temperature and at hydrogen pressures of 1 to 5 atm.

Chlorine and bromine also add to the site of the double bond. For example, bromine reacts with ethene (ethylene) and forms 1, 2-dibromoethane.

$$CH_2{=}CH_2 + Br_2 \longrightarrow \begin{array}{c} CH_2CH_2 \\ |\quad\ | \\ Br\ \ Br \end{array}$$

During the reaction described above, the red color of the bromine disappears. Because alkanes do not readily react with bromine, this halogen can be used to distinguish between an alkane and an alkene. With alkanes, the red color of the bromine persists when it is mixed with the hydrocarbon. The greater reactivity of ethene and the characteristics of the addition compounds can be explained in terms of the nature of the double bond (Sect. 8-27).

20-5 Catalytic cracking units. These units produce short-chain hydrocarbons, and aromatic hydrocarbons. (Humble Oil and Refining Co.)

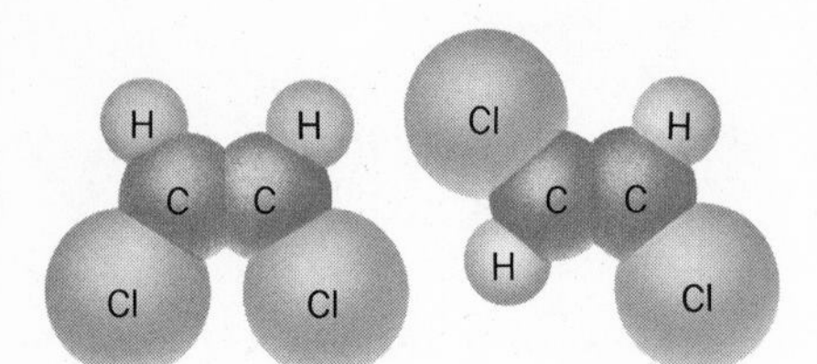

(a)

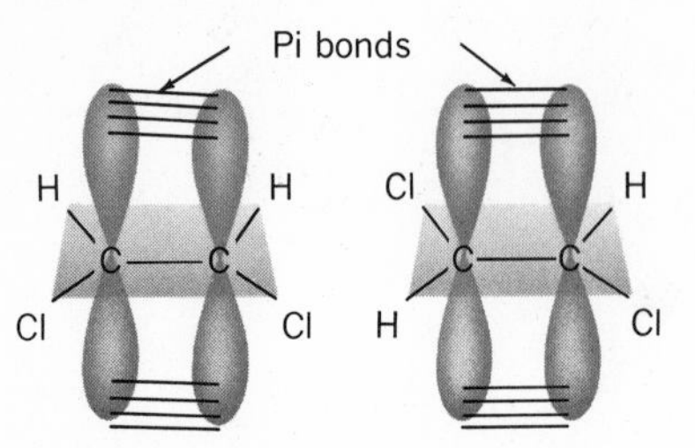

cis-dichloroethene trans-dichloroethene

(b)

20-6 From the schematic diagram (b) of dichloroethene, it can be seen that rotation of the two carbon atoms in opposite directions would extend and destroy the pi bonds. The space-filling model in (a) shows the geometric relationship of the atoms in the isomers.

Because of the double bond, alkenes have a much more rigid structure than alkanes. For example, the two CH_3—groups in ethane can be rotated in opposite directions. However, the CH_2—groups in ethene cannot be rotated in opposite directions. This rotation would destroy the π bond and would require a significant input of energy.

The rigid structure associated with double bonds gives rise to *cis* and *trans* isomerism (Fig. 20-6). For example, 1,2-dichloroethene can exist as *cis* and *trans geometric* isomers whose structural formulas are shown in the margin. The *cis* isomer is polar, but the *trans* isomer is nonpolar. The symmetrical arrangement of the chlorine atoms in the *trans* isomer results in the molecule having an electric symmetry and, hence, a nonpolar nature.

Cl Cl
C=C
H H
***cis*-dichloroethene**
(a)

H Cl
C=C
Cl H
***trans*-dichloroethene**
(b)

FOLLOW-UP PROBLEM

Describe a test that distinguishes butene from butane.

Now let us consider 1,2-dichloroethane, whose possible structural formulas are shown in the margin. Structures (a), (b), and (c) are different ways of representing the same compound. These are equivalent forms because there is *freedom of rotation about the single bond.* Such rotation converts any one of the forms into any of the others. Thus, 1,2-dichloroethane does not exhibit *cis-trans* isomerism. No geometric isomers exist for 1,2-dichloroethane.

Cl H
H—C—C—H
H Cl
(a)

H H
H—C—C—H
Cl Cl
(b)

H Cl
H—C—C—H
Cl H
(c)

Alkenes are often used as intermediates in the production of other chemicals. Many of the alcohols and chlorinated hydrocarbons can be made from the appropriate alkene. The addition of Cl_2 to the double bond in ethene yields 1,2-dichloroethane.

$$H_2C{=}CH_2 + Cl_2 \longrightarrow H_2\overset{Cl}{\overset{|}{C}}{-}\overset{Cl}{\overset{|}{C}}H_2$$

The addition of water in the presence of H_2SO_4 and a catalyst yields ethyl alcohol (C_2H_5OH).

$$H_2C{=}CH_2 + H_2O \xrightarrow[\text{catalyst}]{H_2SO_4} H_2\overset{H}{\overset{|}{C}}{-}\overset{OH}{\overset{|}{C}}H_2$$

Alkenes can also be an intermediate in the manufacture of polymers. *Polymers* are large molecules made by a repetitive connection of many small, molecular units called *monomers.* By connecting thousands of ethene molecules ($CH_2{=}CH_2$), we obtain the polymer, polyethylene $(—CH_2—CH_2—CH_2—CH_2—CH_2—)_n$, where *n* indicates the number of monomer units present. By polymerizing various derivatives of ethene, many different kinds of polymers can be produced. Tetrafluorethene, $CF_2{=}CF_2$, can be polymerized, forming Teflon $(—CF_2—CF_2—)_n$, a plastic with widespread usage.

20-10 Alkynes Alkynes are unsaturated hydrocarbons containing a triple bond. Acetylene (C_2H_2) is the simplest member of this series. The structural formula for acetylene is H—C≡C—H. The general formula for the members of this series is C_nH_{2n-2}. To name the alkynes systematically, the *-ane* of the corresponding alkane is replaced with *-yne*. Thus, the systematic name for acetylene is ethyne. Most of the simple members of this series are named as derivatives of acetylene. For example,

$$CH_3CH_2C{\equiv}CCH_2CH_3$$

is called diethylacetylene. Its systematic name is 3-hexyne.

Ethyne (acetylene) is manufactured industrially from CaO and C (coke) by the process:

$$CaO + 3C \xrightarrow{heat} CaC_2 \text{ (calcium carbide)} + CO$$

$$CaC_2 + 2H_2O \longrightarrow Ca(OH)_2 + C_2H_2$$

Most other alkynes can be produced as derivatives of acetylene.

FOLLOW-UP PROBLEM

Write two names for $CH_3C{\equiv}CCH_2CH_3$.

The carbon-carbon triple bond, composed of a sigma and two pi bonds, is shorter and stronger than the carbon-carbon double or single bond. The chemistry of acetylene and other members of this series is a reflection of the characteristics of the carbon-carbon triple bond (Sect. 8-28).

Acetylene is very reactive chemically. The pi electrons of the triple bond are readily attacked by reagents such as the halogens and hydrogen halides. Addition reactions involving these reagents result in the destruction of weak pi bonds and the formation of strong sigma bonds (σ bonds). For example, in the bromination of acetylene represented by the equation

$$2Br_2 + H{-}C{\equiv}C{-}H \longrightarrow H{-}\underset{\substack{|\\Br}}{\overset{\substack{Br\\|}}{C}}{-}\underset{\substack{|\\Br}}{\overset{\substack{Br\\|}}{C}}{-}H$$

four strong σ C—Br bonds are formed, while two weak π bonds in the triple bond and the two Br—Br bonds are destroyed. The product represents a more stable system than the reactants. It should be noted, however, that the combination of two π bonds and one σ bond forms a strong triple bond in acetylene which may require considerable activation energy to break.

Acetylene is widely used as a fuel in welding torches and as a raw material for the synthesis of a variety of organic chemicals.

FRIEDRICH WÖHLER
1800–1882

Wöhler studied medicine at the University of Heidelberg in Germany and received his medical degree in 1823, but he liked chemistry better than medicine. He wrote to Berzelius in Sweden, asking for permission to study with him. Berzelius welcomed him, but when Wöhler took his first chemical preparation to show Berzelius, the old master retorted, "Doctor, that was fast but bad." From Berzelius, Wöhler learned the importance of careful analysis, a technique which was to serve him for the rest of his life. The year that Wöhler spent in Stockholm with Berzelius was a most valuable experience and started a friendship that lasted for many years. After returning to Germany, Wöhler taught at a technical school in Berlin. He was then called to the University of Göttingen as professor of chemistry. He stayed there for the rest of his life.

A prevalent idea at this time was that naturally occurring compounds from plants or animals were produced by some mysterious force which was called a *vital force*. It was supposed that man would never be able to synthesize these compounds artificially. Wöhler's old teacher Berzelius strongly believed this, as did most chemists from Germany, France, and England. While in medical school, Wöhler had experimented with the waste products in urine and had often seen the long, thin crystals of urea. When, upon heating ammonium cyanate, he produced these same crystals, he was astounded. After carefully repeating the experiment and analyzing the crystals, he wrote to his old friend Berzelius, "I must tell you that I can prepare urea without requiring a kidney of an animal, either man or dog." With the old vital-force theory exploded, numerous other chemists began to synthesize organic compounds with great success.

The properties of acetylene explain its use as a fuel. The heat (enthalpy) of combustion of acetylene is 300 kcal/mole.

$$C_2H_2 + \tfrac{5}{2}O_2 \longrightarrow 2CO_2 + H_2O + 300 \text{ kcal}$$

This highly exothermic reaction is partially responsible for the very high temperature of the oxyacetylene flame (2700°C) used in welding.

The enthalpy of formation (ΔH) for acetylene is 54 kcal/mole.

$$2C(s) + H_2(g) \longrightarrow C_2H_2 - 54 \text{ kcal}$$

This suggests that acetylene is *thermally unstable.* Experimentally, this is verified by the observation that at pressures greater than 2 atm, it may explode violently. It is extremely soluble in acetone. Acetylene is therefore safely stored at a pressure of 15 atm in steel cylinders filled with a porous material, such as asbestos, which has been soaked with acetone.

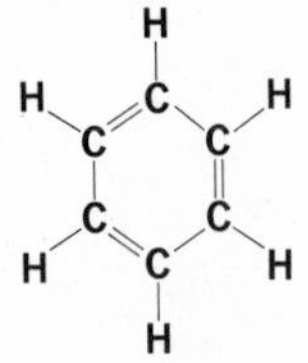

benzene

"Atoms were gamboling before my eyes, twisting and twining in snakelike motion. Then one of the snakes seized its own tail."

Description by August Kekulé (1824–1896) of a dream that led to his hypothesis of the ring structure of benzene.

20-11 Benzene and Aromatic Hydrocarbons Most aromatic compounds contain one or more benzene rings. Each ring contains six carbon atoms and several double bonds, as shown in the structural formulas in the margin. Although benzene and other aromatic compounds are unsaturated, *they do not undergo the addition reactions typical of the aliphatic unsaturated compounds and are not as reactive.* To help us explain these differences we shall first examine the structure of a benzene molecule in more detail.

A benzene molecule can be considered to be a resonance hybrid. X-ray diffraction and infrared spectroscopy studies reveal that the six carbon atoms in benzene form a *regular hexagon.* The molecule is *planar,* all C—C bonds are identical, and all bond angles are 120° angles. No single electron-dot formula can be drawn which is consistent with these data. Therefore, we must consider benzene to be a resonance hybrid. The two most important resonance structures are shown below. We shall represent a benzene ring by means of the simplified structural formulas shown below.

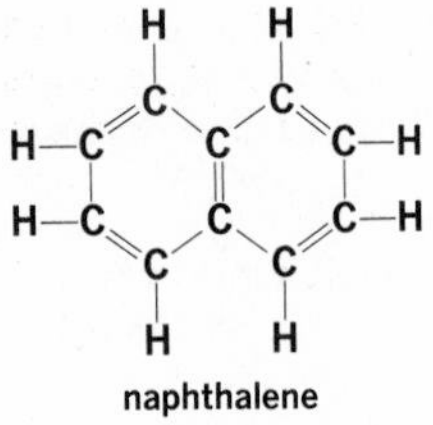

naphthalene

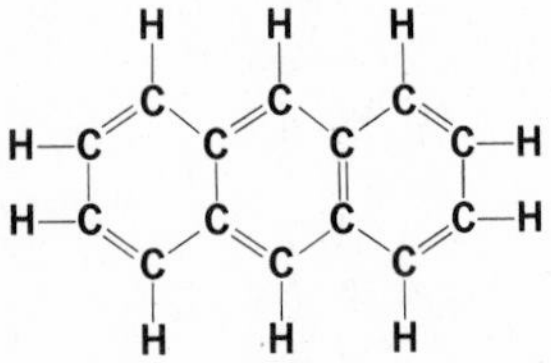

anthracene

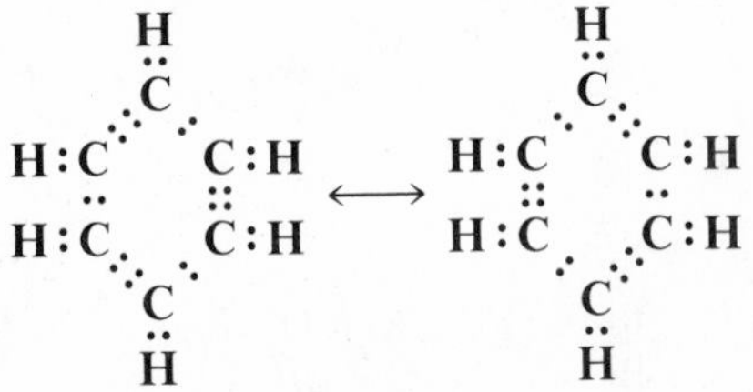

Resonance structures

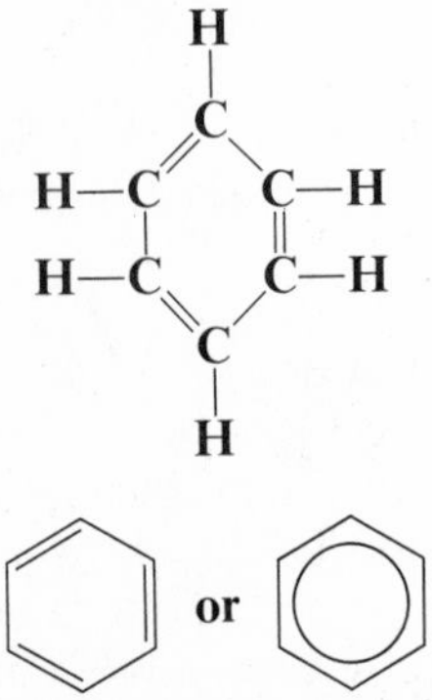

or

Simplified structural formulas

Because all four bonding electrons of each carbon atom are involved in either C—C or C—H bonds, we can assume that sp^2 hybrid orbitals, each containing an electron, are involved. Two of the sp^2 orbitals of each carbon atom overlap (endways) an sp^2 orbital from an adjacent carbon atom. This results in the formation of a strong single covalent σ bond between both carbon atoms. The unhybridized p_y orbitals of each carbon atom which contain a single electron are perpendicular to the plane of the sp^2 hybrid orbitals and parallel to each other (Fig. 20-7). The lateral overlap of p_y orbitals produces π bonds. The electrons involved in π-bonds are delocalized and spread out over the entire molecule and form the doughnut-shaped lobes illustrated in Fig. 20-8. The resonance hybrid model for benzene is partially explained in terms of the six delocalized electrons and justified on the basis of experimentally derived bond energies.

Unsaturated aromatic hydrocarbons are less reactive than unsaturated aliphatic hydrocarbons. We can use the delocalization of the π electrons and the concept of resonance to account for the extra stability of a benzene molecule. That is, the experimentally derived bond strength in benzene is actually much greater than that calculated by assuming the presence of three double bonds. The difference between actual and theoretically calculated bond energies is called *resonance energy*. It is the *stabilizing effect of this resonance energy* that causes unsaturated aromatic rings to be more stable than unsaturated aliphatic molecules that have true double bonds. *The six equivalent C—C bonds in benzene are intermediate in nature between a single and a double C—C bond.*

The simplest aromatic compounds are named either by their common names or systematically as derivatives of benzene. The structural and molecular formulas and the common and systematic names for several simple aromatic hydrocarbons are given below.

In the systematic scheme, the relative positions of the groups attached to the benzene ring are indicated by numbering the carbon atoms beginning with the carbon atom next to the original group and going around the ring so as to give the lowest numbers

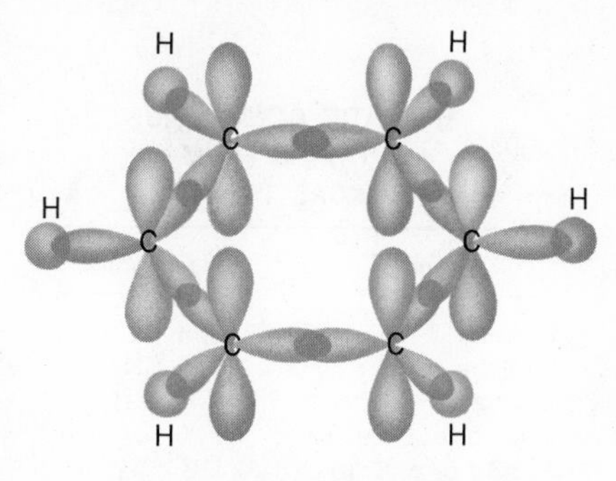

20-7 Schematic diagram for benzene showing sigma bonds between adjacent carbon atoms and between each carbon and two hydrogen atoms.

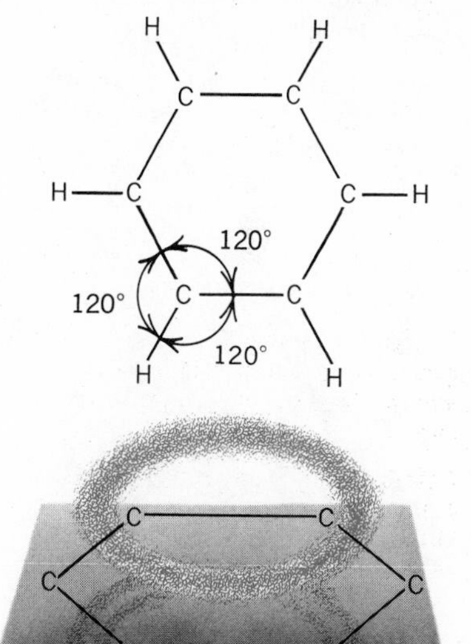

20-8 Bonding in benzene shows the charge cloud formed by the electrons of the laterally overlapping p_y orbitals consists of doughnut-shaped lobes above and below the plane of the carbon ring.

Molecular formula	C_7H_8	C_8H_{10}	C_8H_{10}	C_8H_{10}
Structural formula	CH_3	CH_3, CH_3	CH_3, CH_3	CH_3, CH_3
Common name	**toluene**	**ortho-xylene**	**meta-xylene**	**para-xylene**
Systematic name	**methyl-benzene**	**1,2-dimethyl-benzene**	**1,3-dimethyl-benzene**	**1,4-dimethyl-benzene**

(for example, 1,2- rather than 1,6-). In the conventional system, the relative positions of two groups are indicated by using the terms *ortho* (*o*-), *meta* (*m*-), *and para* (*p*-) to indicate, respectively, adjacent (1,2-), alternate (1,3-), and opposite (1,4-) orientations of two attached carbon atoms or other groups.

An important source of benzene and related compounds is coal tar, a black gummy material obtained when soft coal is heated in the absence of air. Distillation and chemical processing of coal tar yields a large number of aromatic compounds which serve as raw materials in the synthesis of many industrial and commercially important products. The formulas and names of certain aromatic compounds obtained from coal tar are given in the margin.

Two of the most commonly encountered aromatic hydrocarbon radicals (aryl groups) are the phenyl and the benzyl groups whose formulas are also shown in the margin.

AROMATIC COMPOUNDS OBTAINED FROM COAL TAR

Name	Formula
Benzene	C_6H_6
Toluene	$C_6H_5CH_3$
Xylene (*o*- or 1,2-; *m*- or 1,3-; *p*- or 1,4-)	$C_6H_4(CH_3)_2$
Naphthalene	$C_{10}H_8$
Phenol	C_6H_5OH
Cresol (*o*- or 1,2-; *m*- or 1,3-; *p*- or 1,4-)	$C_6H_5 \cdot CH_3 \cdot OH$
Naphthol (α- or β-)	$C_{10}H_7OH$

FOLLOW-UP PROBLEMS

1. Give two names for each. (a) [benzene ring with –Cl], (b) [benzene ring with –Cl, –Cl on adjacent positions], (c) [benzene ring with –Cl, –Cl in meta positions], (d) [benzene ring with Cl, Cl in para positions], (e) [benzene ring with –CH_3, –Cl on adjacent positions].

2. Write structural formulas for these compounds. (a) 2,4,6-trichlorotoluene, (b) meta-dibromobenzene, (c) diphenyl methane, (d) biphenyl, (e) benzyl chloride.

20-12 Substitution Reactions Substitution reactions are a major type of organic reaction. When a reagent such as Br_2 reacts with aromatic hydrocarbons, it does not attack the double bond as it does in the alkenes or alkynes. Rather, it displaces one of the hydrogen atoms from the benzene ring. This fundamental class of organic reaction is known as a substitution reaction. The net equation for $FeBr_3$-catalyzed reaction between benzene and bromine is

$$C_6H_6 \xrightarrow[(FeBr_3)]{Br_2} C_6H_5Br + HBr$$

bromobenzene

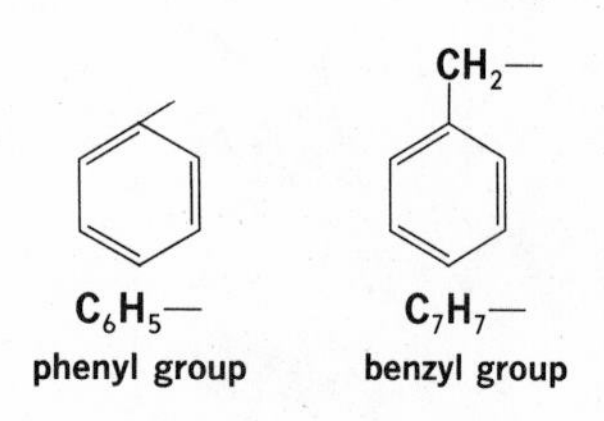

C_6H_5— phenyl group C_7H_7— benzyl group

Another rather important type of substitution reaction is that of nitration. Benzene can be nitrated by reacting it with a mixture of concentrated nitric acid and concentrated sulfuric acid. The equation for the net reaction is

$$C_6H_6 \xrightarrow[(H_2SO_4)]{HNO_3} C_6H_5NO_2 + H_2O$$

nitrobenzene

The reaction mechanism involves *electrophilic* (*electron-loving*) *nitronium* ions (NO_2^+), attacking the delocalized π-electrons of the benzene ring and the subsequent loss of an H^+ ion to an HSO_4^- ion. Thus H_2SO_4 is regenerated and not consumed by the reaction. The equation for the formation of NO_2^+ ions is

$$H_2SO_4 + HNO_3 \longrightarrow H_2NO_3^+ + HSO_4^- \longrightarrow H_2O + NO_2^+ + HSO_4^-$$

When HNO_3 accepts a proton from H_2SO_4, it forms an intermediate, unstable product, $H_2NO_3^+$, which decomposes and gives nitronium ions and water. The nitronium ions can then react with an organic compound such as benzene and form nitrobenzene. In the nitration process, a proton from benzene is transferred back to HSO_4^- ions, regenerating H_2SO_4. Thus one function of the sulfuric acid is that of a *catalyst*. It is also a powerful *dehydrating agent* and removes water from the system. In the reaction with concentrated sulfuric acid, nitric acid is a proton acceptor and thus acts as a Brønsted base. In most of its reactions, however, HNO_3 is a strong acid.

When nitrobenzene is further nitrated, a mixture of products is obtained. The principal product is *meta*-dinitrobenzene, whose structural formula is shown in the margin. These observations help to illustrate two important points: groups attached to the benzene ring have a "directing effect," and organic reactions often involve side reactions which may significantly reduce the yield of the desired product.

NO_2 NO_2
m-dinitrobenzene

The high proportion of *meta*-dinitrobenzene resulting from the two successive nitration reactions indicates that the nitro group ($—NO_2$) is a *meta-directing group*. It is beyond our scope to discuss the complexities of the substituent-directing abilities of different groups attached to the benzene ring. In general, groups such as nitro ($—NO_2$), carboxyl (—COOH), and aldehyde (—CHO) which tend to *withdraw electrons* from the benzene ring are meta-directing groups. Groups such as hydroxyl (—OH), amino ($—NH_2$), alkoxy ($—OCH_3$), and the halogens except fluorine act as a *source of electrons* and are *ortho- or para-directing groups.*

We have illustrated both addition and substitution reactions involving the halogens. The product in each instance was a halogen derivative of a hydrocarbon. These halogen derivatives are not generally found in nature, but many are synthesized. They are useful as solvents and have many other technological and practical applications. Examples are chloroform ($CHCl_3$), an anesthetic; a mixture of chlorinated alkanes, a dry-cleaning solvent; Freon (CCl_2F_2), a refrigerant; paradichlorobenzene ($C_6H_4Cl_2$), a moth repellent; and dichlorodiphenyl-trichloroethane (DDT), an insecticide.

20-13 Ethers Alcohols, ethers, aldehydes, ketones, acids, and esters can all be considered as oxygen derivatives of the hydrocarbons. We shall first consider alcohols and ethers, organic molecules in which all bonds to the oxygen atoms are single bonds. The relationship can be seen by examining the structural formulas for dimethyl ether and ethyl alcohol, shown in the margin. These

```
    H     H
    |     |
H—C—O—C—H
    |     |
    H     H
```
dimethyl ether

```
   H H
   | |
H—C—C—OH
   | |
   H H
```
ethyl alcohol
ethanol

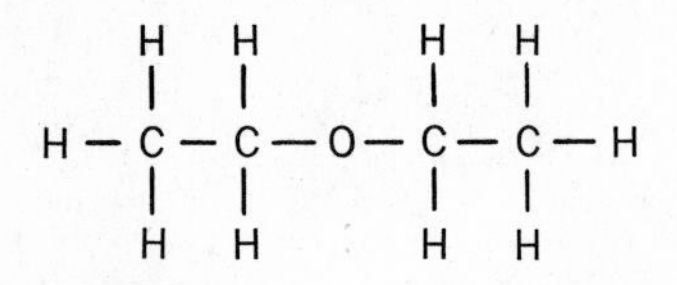

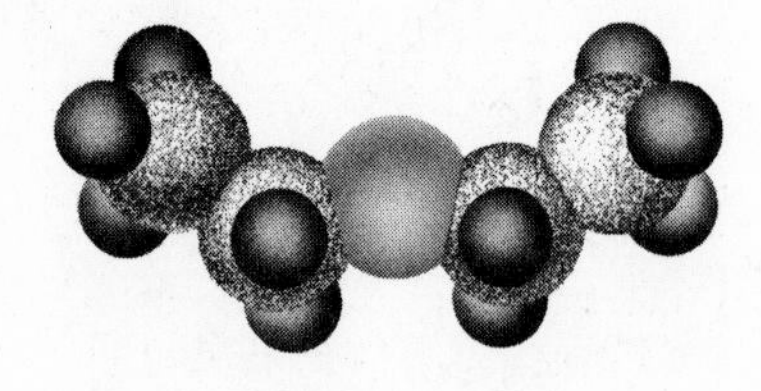

20-9 Diethyl ether.

substances are *functional isomers.* That is, there is the same number and kind of atoms in each molecule but they are arranged so as to give different functional groups.

In ethers, an oxygen atom is singly bonded to two carbon atoms. Thus, an ether can be viewed as a derivative of an alcohol in which an alkyl group has replaced a hydrogen of the hydroxyl group. Also, we can view an ether as being a derivative of water. Substitution of two alkyl groups for two hydrogen atoms in water yields an ether. The use here of the term *derivative* does not imply that ether is experimentally produced from water as a raw material.

In an ether, there is an oxide linkage (—O—) which is analogous to the methylene linkage (—CH_2—) in a saturated hydrocarbon. Replacement of a methylene group in a saturated hydrocarbon by an oxide linkage yields an ether structure. We might therefore expect the properties of ethers to resemble somewhat those of the corresponding hydrocarbons isoelectronic to them (same number of electrons). An examination of their properties confirms our expectations. Ethers are observed to be relatively unreactive and to have boiling points close to those of related hydrocarbons. Ethers are mainly used as solvents for organic reactions.

20-14 Classification of Alcohols Alcohols are characterized by the hydroxyl group (—OH). In alcohols, the oxygen atom is singly bonded to a hydrogen atom on one "side" and a carbon atom on the other "side." A *hydroxyl group is a neutral species.* Unlike hydroxide ions, it does not exist as a stable unit in aqueous solution, nor does it exist as a negatively charged unit in the liquid or solid phase of a pure substance.

Alcohols can be considered as derivatives of water. Substitution of an alkyl group for a single hydrogen atom in water produces an alcohol. Alternatively, an alcohol can be viewed as related to hydrocarbons, by the replacement of hydrogen by a hydroxyl (—OH) group. The systematic name is obtained by replacing the final *-e* of the alkane with *-ol.* The common name is obtained by adding the word *alcohol* to the name of the alkyl group attached to the hydroxyl group.

Formula	CH_3OH	$CH_3CH_2CH_2OH$	$CH_3CH(OH)CH_3$
Systematic name	**methanol**	**1-propanol**	**2-propanol or isopropanol**
Common name	**methyl or wood alcohol**	**normal propyl alcohol**	**isopropyl alcohol**

In general, the boiling points of the aliphatic alcohols are higher than those of the corresponding alkanes. This is caused by the *hydrogen bonding* between the hydrogen atom of one hydroxyl group and an oxygen atom on the hydroxyl group of a neighboring alcohol molecule.

Alcohols can be classified as primary, secondary, or tertiary. The classification is based on the location of the hydroxyl group. If the hydroxyl group is attached to a primary alkyl group, then the alcohol is a *primary alcohol.* A *primary alkyl group* is a carbon atom attached to only one other carbon atom. Ethanol and 1-propanol are primary alcohols. When the hydroxyl group is attached to a carbon atom which is bonded to two other carbon atoms, then the alcohol is called a *secondary alcohol.* Isopropanol, whose formula is shown in the margin, is a secondary alcohol. If the carbon atom bonded to the hydroxyl group is also bonded to three other carbon atoms, then the alcohol is a *tertiary alcohol.* An example of a tertiary alcohol is $(CH_3)_3COH$, whose structural formula is shown in the margin. This alcohol is commonly called *tert*-butyl alcohol. Its systematic name is 2-methyl-2-propanol. As we shall observe, the structural differences between primary, secondary, and tertiary alcohols cause differences in chemical behavior.

Alcohol molecules may contain one, two, three, or many hydroxyl groups. Alcohols having two hydroxyl groups attached to the carbon skeleton are called *glycols.*

Ethylene glycol (1,2-ethanediol, CH_2OHCH_2OH) is a dihydroxy alcohol commonly used as an antifreeze. Its structure is shown in the margin.

The simplest trihydroxy alcohol $[C_3H_5(OH)_3]$ is called glycerol (glycerin). Its structure is shown in the margin.

$$\begin{array}{ccccccc} & & H & & OH & & H \\ & & | & & | & & | \\ H & - & C & - & C & - & C & - & H \\ & & | & & | & & | \\ & & H & & H & & H \end{array}$$

isopropanol

$$\begin{array}{ccc} & & CH_3 & & \\ & & | & & \\ CH_3 & - & C & - & OH \\ & & | & & \\ & & CH_3 & & \end{array}$$

tert-butyl alcohol

$$\begin{array}{ccccc} & & OH & & OH \\ & & | & & | \\ H & - & C & - & C & - & H \\ & & | & & | \\ & & H & & H \end{array}$$

ethylene glycol

$$\begin{array}{ccccccc} & & OH & & OH & & OH \\ & & | & & | & & | \\ H & - & C & - & C & - & C & - & H \\ & & | & & | & & | \\ & & H & & H & & H \end{array}$$

glycerol

FOLLOW-UP PROBLEM

Consider ethanol, ethylene glycol, and glycerol. (a) Which has the lowest boiling point? (b) Which has the greatest viscosity? (c) Which ones are miscible with water? Explain your answers.

20-15 Preparation and Reactions of Alcohols Large quantities of methanol and ethanol are synthesized for use as starting materials in organic syntheses. Over a billion pounds of methanol are synthesized each year. Most of this is produced by the catalytic hydrogenation of carbon monoxide.

$$2H_2 + CO \xrightarrow{(cat.)} CH_3OH$$

The reaction takes place efficiently at 400°C and 200 atm. Methanol is extremely toxic. It is sometimes called wood alcohol because it can be obtained as a by-product in the destructive distillation of wood.

Ethanol, ordinary grain alcohol found in intoxicating beverages, can be prepared by fermenting sugar or starch solutions in the presence of yeast. This and certain other alcohols can also be produced by the hydration of alkenes in the presence of an acid.

$$\begin{array}{ccccccc} H - & C = C & - H + H_2O & \xrightarrow{H_2SO_4} & H - & \overset{OH}{\underset{H}{C}} - \overset{H}{\underset{H}{C}} & - H \\ & | \quad | & & & & & \\ & H \quad H & & & & & \end{array}$$

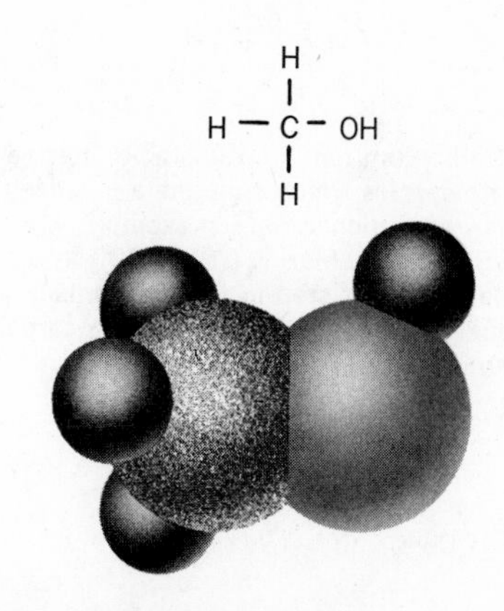

20-10 Methyl alcohol (methanol).

The reactivity of alcohols and the reaction product are largely determined by the location of the hydroxyl group. The most common reactions which alcohols undergo are *substitution, elimination,* and *oxidation*. The hydroxyl group in primary, secondary, and tertiary alcohols can be replaced by various reagents. For example alkyl halides are prepared by substituting a halogen atom for the hydroxyl group. To prepare ethyl bromide, react HBr or PBr_3 with ethyl alcohol.

$$CH_3CH_2OH + HBr \longrightarrow CH_3CH_2Br + H_2O$$
$$3CH_3CH_2OH + PBr_3 \longrightarrow 3CH_3CH_2Br + H_3PO_3$$

As with most organic reactions, side reactions occur. Reagents are chosen so as to minimize side reactions and to allow easy separation of products. Although this method can be used for primary, secondary, or tertiary alcohols, the mechanisms and rates may differ considerably.

FOLLOW-UP PROBLEM

A student preparing butyl bromide (C_4H_9Br) reacts 75.0 g of potassium bromide with 0.500 mole of butyl alcohol (C_4H_9OH) in the presence of excess H_2SO_4. (a) What is the theoretical yield of this reaction? (b) What is the percentage yield if he isolates 55.0 g of pure C_4H_9Br? The equation for the reaction is

$$C_4H_9OH + KBr + H_2SO_4 \longrightarrow C_4H_9Br + KHSO_4 + H_2O$$

Ans. (a) 68.5 g, (b) 80.2%.

Another major type of organic reaction is the *elimination* reaction. When alcohols undergo elimination reactions, the elements of water are removed (dehydration). This reaction is one method used to prepare alkenes. Ethene can be prepared from ethanol by heating the alcohol in the presence of an acid catalyst. Temperature control is an important factor in this reaction

$$CH_3CH_2OH \xrightarrow{H_2SO_4} CH_2{=}CH_2 + H_2O$$

Secondary and tertiary alcohols are more easily dehydrated than primary alcohols. These differences in rates are explained in terms of the ease with which certain intermediates (carbonium ions) are formed in the reaction mechanism.

A carbonium ion is an unstable intermediate species which contains a positively charged carbon atom. For example, when tertiary butyl chloride [$(CH_3)_3CCl$] hydrolyzes, the first step in the mechanism is believed to be the formation of a carbonium ion.

$$H_3C{-}\overset{\displaystyle CH_3}{\underset{\displaystyle CH_3}{\overset{|}{\underset{|}{C}}}}{-}Cl \longrightarrow H_3C{-}\overset{\displaystyle CH_3}{\underset{\displaystyle CH_3}{\overset{|}{\underset{|}{C^+}}}} + Cl^-$$

carbonium ion

When the conditions of temperature and concentration are properly adjusted (lower H_2SO_4 concentration, higher alcohol concentration), ethyl ether rather than ethene is the chief product. Again, the results can be explained in terms of the reaction mechanism. The overall net equation is

$$\begin{matrix} CH_3CH_2OH \\ + \\ CH_3CH_2OH \end{matrix} \xrightarrow[130^\circ C]{H_2SO_4} \begin{matrix} CH_3CH_2 \\ \quad\diagdown \\ \quad O + H_2O \\ \quad\diagup \\ CH_3CH_2 \end{matrix}$$

In this reaction, the net result is the elimination of one molecule of water from two molecules of alcohol.

Primary and secondary alcohols can be oxidized by a number of oxidizing agents. One of the most common is $K_2Cr_2O_7$. The oxidation of a primary alcohol proceeds in steps. The alcohol is first converted into the corresponding aldehyde. Unless the aldehyde is immediately removed from the reaction mixture, it is further oxidized to the corresponding carboxylic acid. For example, the one-step oxidation of ethanol yields ethanal (acetaldehyde).

$$\begin{array}{rcl} 3(CH_3CH_2OH & \longrightarrow & CH_3CHO + 2H^+ + 2e^-) \\ 6e^- + 14H^+ + Cr_2O_7^{2-} & \longrightarrow & 2Cr^{3+} + 7H_2O \\ \hline 3CH_3CH_2OH + 8H^+ + Cr_2O_7^{2-} & \longrightarrow & 3CH_3CHO + 2Cr^{3+} + 7H_2O \end{array}$$

The equation shows that one-third mole of dichromate ions is required per mole of alcohol.

The equation for the oxidation of ethanol to ethanoic acid (acetic acid) is

$$3CH_3CH_2OH + 16H^+ + 2Cr_2O_7^{2-} \longrightarrow 3CH_3COOH + 4Cr^{3+} + 11H_2O$$

In this reaction, two-thirds of a mole of dichromate ions is required to oxidize one mole of ethanol. Sometimes an oxidation such as that above is represented in abbreviated form as

$$CH_3CH_2OH \xrightarrow{(O)} CH_3CHO$$

The oxidation of a secondary alcohol produces a ketone. For example, the oxidation of isopropanol by $K_2Cr_2O_7$ produces propanone (acetone).

$$CH_3-\underset{\displaystyle H}{\overset{\displaystyle CH_3}{\underset{|}{\overset{|}{C}}}}-OH \xrightarrow[H_2SO_4]{K_2Cr_2O_7} (CH_3)_2C{=}O$$

Tertiary alcohols have no hydrogen atom on the hydroxylated carbon and *cannot be oxidized without destroying (degrading) the carbon skeleton.*

Note that compounds containing a hydroxyl group attached to a benzene (aromatic) ring are weakly acid and do not display properties characteristic of alcohols. As a class, they are called *phenols.* The simplest member of the series is phenol (C_6H_5OH) whose common name is carbolic acid. Phenol, a very poisonous compound, is used in the preparation of disinfectants and antiseptics. Its use by the surgeon, Joseph Lister, in 1867 ushered in the era of antiseptic medicine and was a milestone in the history of surgery. Phenol is also widely used as a raw material in the synthesis of drugs and dyes. Aspirin, one of the most widely used pain and fever suppressors, is prepared from phenol. Phenol is also used in the manufacture of Bakelite resins.

FOLLOW-UP PROBLEM

(a) Write an equation for the oxidation of normal propanol to propionaldehyde, using dichromate ions in acid solution as the oxidizing agent. (b) Write an equation for the oxidation of normal propanol to propionic acid. (c) Draw structural formulas for the organic products in each of the reactions.

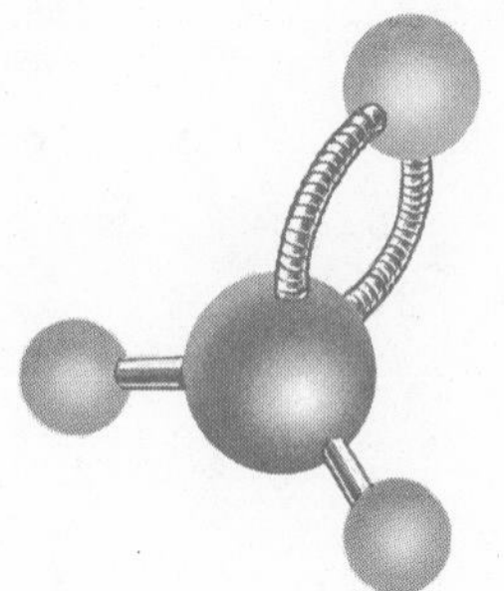

20-11 Formaldehyde.

20-16 Aldehydes and Ketones Aldehyde and ketone molecules are characterized by the presence of a carbonyl group. A *carbonyl group* consists of an oxygen atom doubly bonded to a carbon atom. The general formulas for aldehydes and ketones are shown in the margin. *R* represents an alkyl or aryl group. Aldehydes are systematically named by substituting the suffix *-al* for the final *-e* in the name of the corresponding saturated hydrocarbon. The location of side groups is indicated by numbers which begin with 1 at the carbonyl group. Examples are

Molecular formula	HCHO	CH_3CHO	$CH_3CH_2CH(CH_3)CHO$
Structural formula	H—C(=O)—H	H—C(H)(H)—C(=O)—H	H—C(H)(H)—C(H)(H)—C(H)(CH$_3$)—C(=O)—H
Common name	formaldehyde	acetaldehyde	
Systematic name	methanal	ethanal	2-methyl butanal

R—C(H)=O
aldehyde

R—C(R)=O
ketone

H—C(H)(H)—C(=O)—C(H)(H)—H
acetone

H—C(H)(H)—C(H)(H)—C(H)(CH$_3$)—C(=O)—C(H)(H)—H
3-methyl-2-pentanone

Simple ketones can be named by stating the names of the two alkyl groups attached to the carbonyl group and adding the word *ketone.* Acetone is the common name for dimethyl ketone, (CH_3COCH_3), whose structural formula is shown in the margin. The IUPAC system is used for more complicated ketones. In this method, the suffix *-one* is substituted for the final *-e* in the name of the corresponding hydrocarbon. The location of any side branches is indicated by numbers which begin at the end nearer the carbonyl group. For example, the structural formula and name of $CH_3CH_2CH(CH_3)COCH_3$ can be expressed as shown in the margin.

The carbon-oxygen double bond is more stable with respect to chemical reactivity than the alkene double bond. Data show that the carbonyl carbon and the two attached atoms are in the same plane and that the bond angle between the attached atoms is approximately 120°C. This suggests sp^2 hybridization on both the carbon and oxygen atoms and a double bond composed of a σ and a π bond.

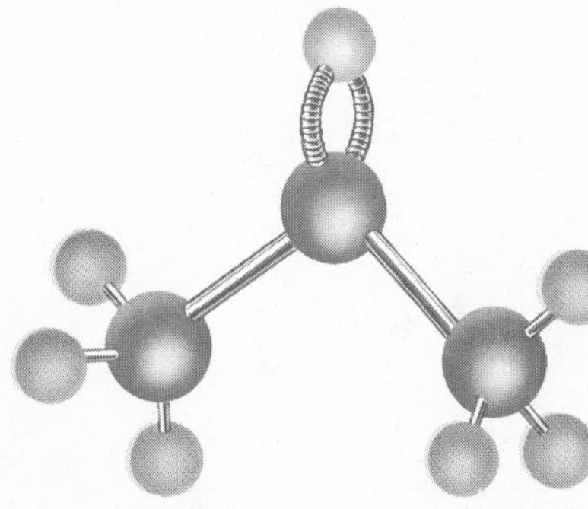

20-12 Acetone.

The double bond in the carbonyl group is analogous to that in the alkenes. Thus we might expect aldehydes and ketones to un-

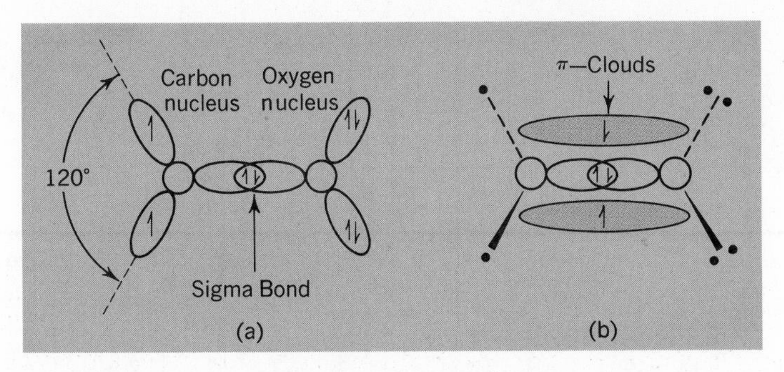

20-13 (a) "Top view" of sp^2 orbitals used by carbon and oxygen atoms in the carbonyl group. The overlap of the orbitals along the internuclear axis gives rise to a strong sigma bond. (b) "Side view" showing pi clouds formed by lateral overlap of parallel p orbitals on carbon and oxygen atoms. The delocalized electron system in the carbonyl group increases its stability.

dergo addition reactions similar to those of the alkenes. It turns out that there are more differences than similarities in the reactions of the two groups. Hydrogen can be added to both the olefinic double bond and the carbonyl double bond. The carbonyl group, however, does not undergo addition reactions with the halogens or halogen halides. This is caused, in large part, by the *greater stability of the carbon-oxygen double bond.* The greater stability can be attributed to the *polarity of the carbonyl group* and its *resonance hybrid characteristic.*

A carbonyl group in either ketones or aldehydes can be converted to a hydroxy group by catalytic hydrogenation or other reduction processes. For example, isopropyl alcohol can be prepared by the catalytic hydrogenation of acetone.

$$H_3C-\overset{\overset{O}{\|}}{C}-CH_3 + H_2 \xrightleftharpoons{\text{catalyst}} CH_3-\underset{\underset{H}{|}}{\overset{\overset{OH}{|}}{C}}CH_3$$

The process is exothermic and accompanied by a decrease in volume of the gases. Thus equilibrium principles suggest that the reaction would be more efficient at low temperatures and high pressures.

Aldehydes and ketones can be prepared by the oxidation of primary and secondary alcohols. This reaction was briefly discussed in the preceding section. Methanal (formaldehyde), the most common aldehyde, is used extensively as a preservative for biological specimens and as a starting material in the manufacture of Bakelite plastics. Formaldehyde is prepared commercially by the oxidation of methanol vapor with air in the presence of a suitable catalyst. The equation for the reaction is

$$CH_3OH(g) + \tfrac{1}{2}O_2(g) \xrightarrow[250\text{-}400^\circ C]{\text{catalyst}} CH_2O(g) + H_2O(g)$$

$$H-\underset{\underset{H}{|}}{\overset{\overset{H}{|}}{C}}-OH + \tfrac{1}{2}O_2 \longrightarrow H-\underset{\underset{H}{|}}{C}{=}O + H_2O$$

Aldehydes are easily oxidized to acids. Air oxidizes benzaldehyde to benzoic acid. Even a mild oxidizing agent such as a silver salt in an alkaline solution oxidizes an aldehyde. A reaction used as a test for the presence of aldehydes, called Tollen's test, uses a silver salt dissolved in excess aqueous ammonia. When this reagent and an aldehyde react in a clean glass vessel, metallic silver forms a mirror on the walls of the vessel. The equation is

$$\mathrm{H{-}\overset{\overset{\displaystyle O}{\|}}{C}{-}H + 2AgOH \longrightarrow H{-}\overset{\overset{\displaystyle O}{\|}}{C}{-}OH + 2Ag(s) + H_2O}$$

Copper II salts in alkaline solution form brick-red Cu_2O when heated with aldehydes. This is the basis for Benedict's and Fehling's tests.

Ketones cannot be oxidized without breaking C—C bonds and destroying the carbon skeleton. Both ketones and aldehydes are widely used as solvents and as starting materials in syntheses.

$$\mathrm{{-}\overset{\overset{\displaystyle O}{\|}}{C}{-}OH}$$

carboxyl group

20-17 Organic Acids Organic acids contain the carboxyl group. A *carboxyl group* consists of a carbonyl group with a hydroxyl group attached to the carbon atom. Carboxylic acids are represented by the general formula, R—COOH. The structural formula for the carboxyl group is shown in the margin. In water solution, the carboxyl group dissociates so that *such acids behave as weak acids.* Their dissociation can be represented as

$$\mathrm{H_2O + R{-}C{\overset{\displaystyle O}{\underset{\displaystyle OH}{\lessgtr}}} \rightleftharpoons \left[R{-}C{\overset{\displaystyle O}{\underset{\displaystyle O^-}{\lessgtr}}}\right] + H_3O^+}$$

This equilibrium is displaced to the left.

Many of these acids are still known by common names which were derived from the sources from which they were first obtained. For example, formic acid (HCOOH) was first obtained by the distillation of ants (L. *formica,* ant). It is found in the venom of ants and other insects. Acetic acid (CH_3COOH) is found in vinegar (L. *acetum,* vinegar). The odor of rancid butter is caused by the presence of butyric acid ($CH_3CH_2CH_2COOH$).

The systematic names for these acids are obtained by replacing the final *-e* of the hydrocarbon name with the suffix *-oic* and adding the word acid. The systematic names for formic, acetic, and butyric acid are, respectively, methanoic, ethanoic, and butanoic acid. Acids which contain two carboxylic groups are known as *dicarboxylic acids.* Their systematic names use the suffix *-dioic.* Thus, the systematic name for oxalic acid (HOOCCOOH) is ethanedioic acid. Aromatic acids such as benzoic acid (C_6H_5COOH) are named systematically by adding *carboxylic acid* to the name of the ring to which the acid groups are attached. Thus, benzoic acid is named benzene carboxylic acid. See the structural formula in the margin.

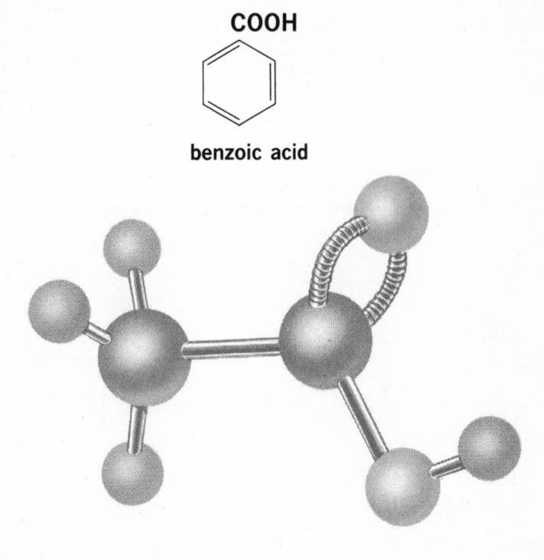

20-14 Acetic acid.

Carboxylic acids display the same Brønsted acid characteristics as inorganic acids but as a class they are very weak. Like inorganic acids, organic acids *affect indicators, react with active metals and carbonates,* and *are neutralized by bases.* The equation for the reaction of acetic acid with NaOH is

$$\mathrm{Na^+ + OH^- + CH_3C{\overset{\displaystyle O}{\diagup\!\!\diagup}}\!\!\underset{\displaystyle OH}{\diagdown}} \longrightarrow \mathrm{CH_3C{\overset{\displaystyle O}{\diagup\!\!\diagup}}\!\!\underset{\displaystyle O^-}{\diagdown} + Na^+ + H_2O}$$

Evaporation of the solution yields the salt, sodium acetate, whose structural formula is shown in the margin.

$$\mathrm{CH_3\overset{\displaystyle O}{\overset{\|}{C}}{-}ONa}$$

sodium acetate

A number of aliphatic acids which can be obtained from vegetable oils and certain animal fats are known as *fatty acids.* Examples of saturated and unsaturated fatty acids are

NAME	STEARIC ACID	LINOLEIC ACID
Formula	$CH_3(CH_2)_{16}COOH$	$CH_3(CH_2)_4CH{=}CHCH_2CH{=}CH(CH_2)_7COOH$
Source	beef tallow	cottonseed oil

Stearic acid with no double bonds, melts at approximately 70°C, and is a solid fatty acid. Linoleic acid with two double bonds melts at −9.5°C, and is a liquid fatty acid. These examples illustrate the point that the melting point of a fatty acid is related to the amount of unsaturation.

20-18 Esters Organic acids react with alcohols and form esters and water. The overall net reaction involves the elimination of the elements of water from the reactant molecules. The reaction, which is often catalyzed by an inorganic acid, can be represented by the general equation

$$\mathrm{RC{\overset{\displaystyle O}{\diagup\!\!\diagup}}\!\!\underset{\displaystyle OH}{\diagdown} + R'OH \xrightleftharpoons{H_2SO_4} RC{\overset{\displaystyle O}{\diagup\!\!\diagup}}\!\!\underset{\displaystyle OR'}{\diagdown} + H_2O} \qquad \text{20-1}$$

where R represents one alkyl or aryl group and R′ represents, a second alkyl or aryl group. R and R′ may represent the same or different groups. It has been proved experimentally that in many such reactions involving primary alcohols the *oxygen atom is removed from the acid rather than the alcohol molecule.*

$$\mathrm{R\overset{\displaystyle O}{\overset{\diagup\!\!\diagup}{C}}{-}\boxed{OH + H}OR' \longrightarrow R\overset{\displaystyle O}{\overset{\diagup\!\!\diagup}{C}}{-}O{-}R' + H_2O}$$

$$\textbf{acid} + \textbf{alcohol} \longrightarrow \textbf{ester} + \textbf{water}$$

Esterification reactions are reversible and reach an equilibrium long before all of the reactants are consumed. Essentially complete conversion of one reactant can be achieved by using a large excess of the other reagent and providing for the removal of water. These actions shift the equilibrium to the right and increase the yield of the ester.

Esters are named as derivatives of an alcohol and an acid. The name of an ester is derived by first naming the alkyl or aryl group from the alcohol (R′). The second part of the name is obtained by replacing the *-ic* of the acid name by the suffix *-ate.* For example, *ethyl* alcohol reacts with *acetic* acid (ethanoic acid) and forms ethyl acetate (ethyl ethanoate) which is a common ingredient in nail-polish removers.

$$\underset{\text{(ethanoic acid)}}{\underset{\text{acetic acid}}{CH_3\overset{\overset{\displaystyle O}{\|}}{C}-\boxed{OH}}} + \underset{\text{alcohol}}{\underset{\text{ethyl}}{C_2H_5O\boxed{H}}} \rightleftharpoons \underset{\text{(ethyl ethanoate)}}{\underset{\text{ethyl acetate}}{CH_3\overset{\overset{\displaystyle O}{\|}}{C}-OC_2H_5}} + H_2O$$

Esters can be prepared by the reaction of an alcohol with either a carboxylic acid or an acyl halide. The reaction with a carboxylic acid was shown above to reach an equilibrium state. The reaction with an acyl halide goes rapidly to completion because gaseous HCl is one of the products and can be easily removed from the system. One of the most widely used acyl halides is acetyl chloride (CH_3COCl). Acyl halides can be prepared by the reaction of a carboxylic acid with PCl_3 or PBr_3.

$$3CH_3\overset{\overset{\displaystyle O}{\|}}{C}-OH + PCl_3 \longrightarrow \underset{\text{chloride}}{\underset{\text{acetyl}}{3CH_3\overset{\overset{\displaystyle O}{\|}}{C}-Cl}} + H_3PO_3$$

In this reaction, the *hydroxyl from the carboxyl group is replaced by a halogen. The name of the acyl group is derived from the name of the parent acid.* The formulas of three acyl groups and their related acids are listed below.

ACID	ACYL GROUP	FORMULA
Formic	formyl	HCO—
Acetic	acetyl	CH_3CO—
Propionic	propionyl	CH_3CH_2CO—

Acyl halides are extremely reactive and can be used for introducing the acyl group, $R-\overset{\overset{\displaystyle O}{\|}}{C}-$, into substances having a replaceable hydrogen. Thus, methyl acetate can be prepared by reacting methyl alcohol with acetyl chloride.

$$CH_3\overset{\displaystyle O}{\overset{\|}{C}}-\boxed{Cl}(l) + CH_3O\boxed{H}(l) \longrightarrow CH_3\overset{\displaystyle O}{\overset{\|}{C}}-OCH_3 + HCl(g)$$

OH O C—OH

salicylic acid

Esters are often volatile liquids with pleasant odors. Specific esters and mixtures of esters produce the scents of flowers, fruits, perfumes, and flavorings. Ethyl butyrate has a pineapple odor, amyl acetate a banana odor, and methyl salicylate is wintergreen oil.

FOLLOW-UP PROBLEMS

1. Write equations and name the organic product for these reactions (a) methanol and formic acid, (b) methanol with salicylic acid whose formula (see margin) is C_6H_4—OH—COOH. One product of this reaction is oil of wintergreen.

2. Aspirin, known chemically as acetyl salicylic acid, may be prepared by the reaction of acetyl chloride with salicyclic acid (see problem 1). Write an equation for this reaction and show the structural formula for aspirin.

20-19 Fats and Fatty Acids Fats and oils (liquid fats) are usually esters of long-chain fatty acids and glycerol, a trihydroxy alcohol. The esters of glycerol can be named as esters, or they can be named as derivatives of the fatty acid by replacing the suffix *-ic* of the acid name with the suffix *-in.* For example, glyceryl tristearate (tristearin) is an ester formed by the reaction of stearic acid and glycerol.

$$\begin{array}{lcl} C_{17}H_{35}-\overset{\displaystyle O}{\overset{\|}{C}}-OH + HO-CH_2 & & C_{17}H_{35}-\overset{\displaystyle O}{\overset{\|}{C}}-O-CH_2 \\ C_{17}H_{35}-\overset{\displaystyle O}{\overset{\|}{C}}-OH + HO-CH & \longrightarrow & C_{17}H_{35}-\overset{\displaystyle O}{\overset{\|}{C}}-O-CH + 3H_2O \\ C_{17}H_{35}-\overset{\displaystyle O}{\overset{\|}{C}}-OH + HO-CH_2 & & C_{17}H_{35}-\overset{\displaystyle O}{\overset{\|}{C}}-O-CH_2 \end{array}$$

Solid fats are generally derivatives of saturated fatty acids. Some important *saturated fatty acids* are stearic acid [$CH_3(CH_2)_{16}COOH$] whose glyceryl ester is present in butter and lard, and palmitic acid [$CH_3(CH_2)_{14}COOH$] whose ester is present in large amounts in palm oil, butter, and lard. Two important *unsaturated fatty acids* are oleic acid [$CH_3(CH_2)_7CH{=}CH(CH_2)_7COOH$] whose glyceryl ester is the main constituent of olive oil, and linoleic acid [$CH_3(CH_2)_4CH{=}CHCH_2CH(CH_2)_7COOH$] whose ester is present in large amounts in sunflower-seed and soybean oils. Thus, margarine, advertised to be "low in saturated fats" would contain large amounts of the glyceryl esters of oleic, linoleic, and other unsaturated fatty acids. Waxes are often esters of long-chain alcohols and fatty acids. Both the alcohol and acid components have more than 20 carbon atoms.

Catalytic hydrogenation of liquid vegetable oils produces solid fats. Vegetable oils are generally mixtures of the esters of unsaturated fatty acids and glycerol. The fatty acids have 16 to 18 carbon atoms. These can be converted to solid fats by *catalytic hydrogenation.* For example, triolein, an unsaturated liquid fat (mp 16°C), can be converted by catalytic hydrogenation into tristearin (mp 70°C), a solid fat. The reaction involves the addition of hydrogen to the double bond. Cottonseed and other oils are commercially hydrogenated on a large scale, producing solid cooking fats.

20-20 Soaps and Synthetic Detergents Esters react with OH^- ions in a reaction called saponification or hydrolysis. These reactions are essentially the *reverse* of the esterification reaction given by Equation 20-1. During saponification, the ester is converted to an alcohol and the salt of fatty acid.

$$R\overset{\displaystyle O}{\overset{\|}{C}}{-}OR' + Na^+OH^- \rightleftharpoons R\overset{\displaystyle O}{\overset{\|}{C}}{-}O^-Na^+ + R'OH$$

$$C_3H_5(O{-}\overset{\displaystyle O}{\overset{\|}{C}}{-}C_{17}H_{35})_3 + 3NaOH \longrightarrow C_3H_5(OH)_3 + 3C_{17}H_{35}COONa$$

The alkaline hydrolysis of a fat produces a salt of a fatty acid known as a soap. For example, the reaction between sodium hydroxide and stearin yields sodium stearate, a useful soap.

$$\begin{array}{l} H{-}\overset{\displaystyle H}{\overset{|}{C}}{-}O{-}\overset{\displaystyle O}{\overset{\|}{C}}{-}C_{17}H_{35} \\ \quad\ \ | \\ H{-}C{-}O{-}\overset{\displaystyle O}{\overset{\|}{C}}{-}C_{17}H_{35} \\ \quad\ \ | \\ H{-}\underset{\displaystyle H}{\underset{|}{C}}{-}O{-}\overset{\displaystyle O}{\overset{\|}{C}}{-}C_{17}H_{35} \end{array} + 3NaOH \longrightarrow \begin{array}{c} CH_2OH \\ | \\ CHOH \\ | \\ CH_2OH \end{array} + 3CH_3(CH_2)_{16}\overset{\displaystyle O}{\overset{\|}{C}}{-}ONa$$

In this process, the fat and alkali are mixed and heated. Addition of Na^+ ions from saturated brine (common ion effect) causes the soap to float to the top, where it is drawn off.

$$\mathrm{RC(=O)\!-\!ONa \rightleftharpoons RC(=O)\!-\!O^- + Na^+}$$

In another method, fats are hydrolyzed by "live" steam, producing glycerol and fatty acids. The two products are then separated, and the fatty acids are reacted with NaOH, forming soap.

$$\mathrm{C_3H_5(OC(=O)\!-\!C_{17}H_{35})_3 + 3H_2O \longrightarrow C_3H_5(OH)_3 + 3C_{17}H_{35}\!-\!COOH}$$

$$\mathrm{C_{17}H_{35}COOH + NaOH \longrightarrow C_{17}H_{35}COONa + H_2O}$$

Ordinary soaps have the disadvantage of producing an insoluble scum when used in "hard" water. Formation of the scum involves a reaction between the carboxylate anions of the soap and Ca^{2+} ions or Mg^{2+} ions in the hard water. The cleansing ability of the soap is destroyed by this reaction. This disadvantage is overcome by the synthetic detergents.

Synthetic detergents are sulfonic esters of long-chain alcohols. For example, the 12-carbon alcohol, commonly called lauryl alcohol (dodecanol), reacts with concentrated sulfuric acid and produces lauryl hydrogen sulfate.

$$\underset{\text{lauryl alcohol (dodecanol)}}{\mathrm{C_{12}H_{25}OH}} + \underset{\text{sulfuric acid}}{\mathrm{HO\!-\!S(OH)(=O)\!=\!O}} \longrightarrow \underset{\text{lauryl hydrogen sulfate}}{\mathrm{C_{12}H_{25}O\!-\!S(OH)(=O)\!=\!O}} + \mathrm{H_2O}$$

Lauryl hydrogen sulfate is then neutralized with NaOH, forming the sodium salt, sodium lauryl sulfate.

$$\mathrm{C_{12}H_{25}O\!-\!S(OH)(=O)\!=\!O + NaOH \longrightarrow C_{12}H_{25}OS(O^-Na^+)(=O)\!=\!O + H_2O}$$

Calcium and magnesium salts of the detergent anions are soluble and, therefore, do not precipitate when the detergent is used in hard water.

The detergent or cleansing actions of soaps and other detergents is related to the structure of their component particles. The anions of both substances consist of a polar group attached to a long, nonpolar hydrocarbon chain. The polar group, because of H-bonding with the water, is hydrophilic (water-loving), and the nonpolar chain is hydrophobic (water-fearing), because it cannot

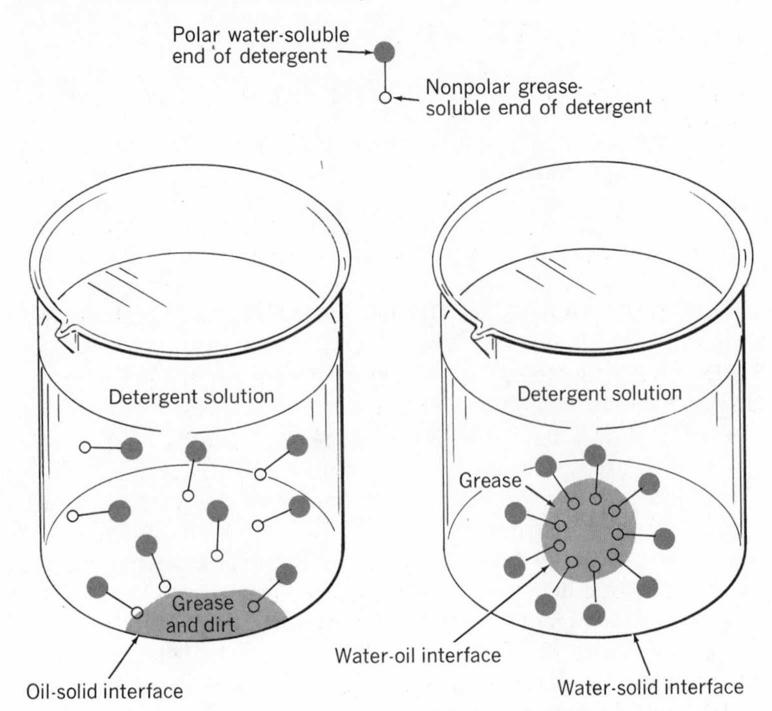

20-15 The removal of oil from a solid surface by a detergent solution is favored because the energy state of the system is lowered. In the presence of a detergent, the surface energy between oil and a solid surface is greater than that between the oil and water phase plus that of the water and solid phase. Once the grease globules have been removed from the solid surface, the polar "heads" of the detergent molecules repel so the grease droplets cannot coalesce. Once solubilized, the grease may be washed away.

overcome the intermolecular H-bonding of H_2O. Detergent action takes place at the interface or surface between oil droplets and water. The detergent anions orient themselves at this interface with their nonpolar ends clustered about and within the dirt-containing oil phase (weak, intermolecular der Waals forces), while their polar ends are attached to groups of water molecules. Thus, the covalent nonpolar oil and grease become dispersed in the water and can subsequently be washed away.

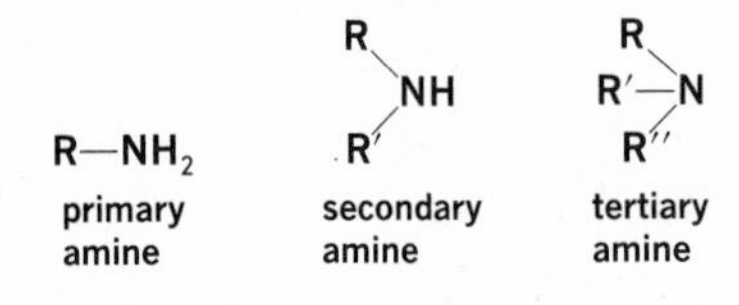

20-21 Amines Amines are nitrogen-containing organic derivatives of ammonia (NH_3). They are classified as primary, secondary, or tertiary depending on the *number of alkyl or aryl groups attached to the nitrogen atom.* The general formulas are shown in the margin.

There are several ways of naming amines. The common names of the simple amines are derived by adding the suffix *-amine* to the name(s) of the attached alkyl groups. For example, $CH_3CH_2NH_2$ is named ethylamine and $CH_3CH_2NHCH_2CH_3$ is diethylamine. Primary amines can be prepared by the alkylation of ammonia. This reaction involves the substitution of an alkyl or substituted alkyl group for one hydrogen atom in ammonia. The most common alkylating agents are the alkyl halides. The reaction of ammonia with an alkyl halide yields an ammonium salt which can be converted to an amine by reaction with OH^- ions. The synthesis of ethylamine beginning with ethyl chloride is represented at the top of the next page.

$$NH_3 + C_2H_5Cl \longrightarrow C_2H_5NH_3^+ + Cl^-$$

$$\underset{\text{ethylammonium chloride}}{C_2H_5NH_3^+Cl^-} + Na^+ + OH^- \longrightarrow \underset{\text{ethylamine}}{C_2H_5NH_2} + Na^+ + Cl^- + H_2O$$

This reaction produces a mixture of primary, secondary, and tertiary amines; thus it is not practical unless methods are available for separating the products.

Amines can form stable salts with strong acids. Amines have an *unshared pair of electrons* and, therefore, *act as Lewis or Bronsted* bases. This is illustrated by the reaction of methylamine with hydronium ions from hydrochloric acid.

$$\underset{\text{methyl amine}}{H_3C{-}\ddot{N}(H){-}H} + H^{+}{:}\ddot{Cl}{:}^- \longrightarrow \underset{\text{methylammonium chloride (methylamine hydrochloride)}}{H_3C{-}NH_2{-}H^+Cl^-}$$

This reaction is analogous to that which occurs between ammonia and hydrochloric acid. The salt is named as either a substituted ammonium salt or as the acid addition product of the amine.

aniline

The widely used trivial name for the simplest aromatic amine is aniline ($C_6H_5NH_2$). Aniline and other aromatic amines are used as raw materials in the synthesis of many dyes and drugs.

Aniline may be prepared by reducing nitrobenzene with zinc or iron and a trace of acid.

$$2Fe + \underset{\text{nitrobenzene}}{C_6H_5{-}NO_2} + 4H_2O \xrightarrow{H^+} \underset{\text{aniline}}{C_6H_5{-}NH_2} + 2Fe(OH)_3$$

20-22 Amino Acids Amino acids contain both the amine and carboxyl group and have the general formula

$$R{-}\underset{H}{\overset{NH_2}{C}}{-}C{\overset{O}{\underset{OH}{}}}$$

Because the amine group has basic characteristics and the carboxyl group has acid characteristics, amino acids have the characteristics of both and are thus *amphoteric* substances. That is, they react with either an acid or a base.

Name	*Structure*
Glycine (aminoacetic acid)	$H-C(NH_2)(H)-COOH$
Alanine	$H_3C-C(NH_2)(H)-COOH$
Valine	$CH_3C(H)(CH_3)-C(NH_2)(H)-COOH$
Phenylalanine	$C_6H_5-CH_2C(NH_2)(H)-COOH$
Cysteine	$HS-CH_2C(NH_2)(H)-COOH$

Twenty-one amino acids have been found to be the building blocks of natural proteins but not all of them are found in any single protein. The structure and names of five common amino acids are given in the margin.

Amino acids are solids which are reasonably soluble in water and have relatively high melting points. These properties suggest that amino acids have a salt-like structure. Infrared spectra and other experimental data reveal that crystals of amino acids are composed of *dipolar ions.* Thus, glycine can be formulated as an "inner salt," and represented as $H_3\overset{+}{N}CH_2COO^-$. Ions which carry both a positive and negative charge are sometimes called *zwitterions.* In aqueous solution this ion may behave as an acid and donate a proton to water.

$$H_3\overset{+}{N}CH_2COO^- + H_2O \longrightarrow H_3O^+ + H_2NCH_2COO^-$$

or it may behave as a base and accept a proton from water

$$H_3\overset{+}{N}CH_2COO^- + H_2O \longrightarrow H_3\overset{+}{N}CH_2COOH + OH^-$$

The species which predominates in solution is determined by the pH of the solution. Application of equilibrium principles reveals that in strongly acid solutions, the cationic form predominates ($H_3\overset{+}{N}CH_2COOH$), while in strongly basic solutions the anionic form predominates ($H_2NCH_2COO^-$). At some intermediate pH, most of the amino acid is in the dipolar form ($H_3\overset{+}{N}CH_2COO^-$). The dipolar nature of the amino acids accounts for many of their properties, such as their relatively high melting and boiling points.

20-16 Dr. S. E. Dixon and students of Ontario Agricultural College, Canada, work on an experiment with insect hormones. (Hughes)

FOLLOW-UP PROBLEM

A substance containing an amino acid may be analyzed by reactions with HNO_2. In this reaction the nitrogen in the amino acid is converted into $N_2(g)$. The equation for the reaction of glycine with HNO_2 is

$$CH_2(NH_2)CO_2H + HNO_2 \longrightarrow CH_2(OH)CO_2H + H_2O + N_2(g)$$

If a 0.600-g sample of a material containing glycine evolves 40.0 ml of N_2 at STP, what is the percentage of glycine in the material? **Ans. 22.3%.**

20-23 Amides Amides can be considered as either derivatives of ammonia and amines or of carboxylic acids. The general formula for a primary amide is shown in the margin. Their names are derived from the name of the carboxylic acid by changing the *-oic* or *-ic* of the acid name to the suffix *-amide.* For example, CH_3CONH_2 can be called acetamide or ethanamide.

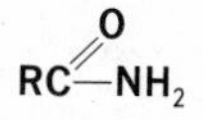

$$RC(=O)-NH_2$$

Amides can be prepared by the acylation of ammonia. Thus, acetamide can be prepared by reaction of ammonia with acetyl chloride.

$$CH_3\overset{O}{\overset{\|}{C}}-Cl + 2NH_3 \longrightarrow CH_3\overset{O}{\overset{\|}{C}}-NH_2 + NH_4^+ + Cl^-$$

Two moles of NH_3 are required for the reaction. One mole is required to produce a mole of amide, and one mole of NH_3 is required to neutralize the HCl formed during the reaction. Although amides appear to be related to ammonia and to amines, they neither react with acids and form salts nor do they display other basic properties. This can be attributed to the *electron-withdrawing tendency of the acyl group* ($R-\overset{O}{\overset{\|}{C}}-$ from an acid) and the *resonance hybrid characteristic of the amide.* Two resonance structures are shown in the margin.

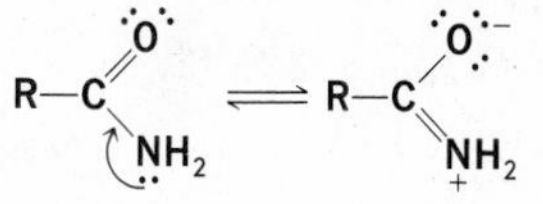

$$R-C(=\ddot{O}:)-\ddot{N}H_2 \rightleftharpoons R-C(-\ddot{O}:^-)=\overset{+}{N}H_2$$

FOLLOW-UP PROBLEM

Acetanilide, used in fever-suppressing drugs, is prepared by reacting aniline with acetyl chloride. Write an equation for this reaction. The formula of the salt formed as a by-product is $C_6H_5-NH_3^+Cl^-$

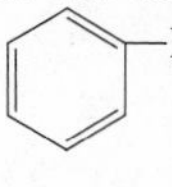

A special type of amide known as a peptide is formed during the bonding of two amino acids. During this process a water molecule is eliminated between the amino group of one amino acid and the carboxyl group of another amino-acid molecule. The simplest dipeptide is glycylglycine ($NH_2CH_2CONHCH_2COOH$) whose structural formula is shown in the margin. The amino acids are said to be joined by *peptide linkages.* The linking together of many amino acids produces a *polypeptide.* Proteins, discussed in the next section, are examples of complex polypeptides.

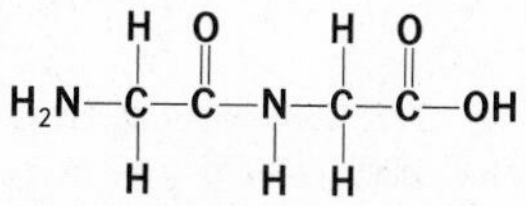

$$H_2N-\underset{H}{\overset{H}{C}}-\overset{O}{\overset{\|}{C}}-\underset{H}{N}-\underset{H}{\overset{H}{C}}-\overset{O}{\overset{\|}{C}}-OH$$

20-24 Structure and Function of Proteins Proteins are complex polymers of amino acids present in all forms of living material. Muscle, cartilage, hair, and nails, as well as enzymes, are all composed of protein. Proteins are not only complex molecules but are also very large molecules. Insulin, one of the smallest, has a molecular mass of about 6000. The influenza virus, one of the largest, has a molecular mass of about 400 000 000.

As stated above, *amino acids are linked together by peptide bonds, forming proteins.* A peptide bond occurs between the carboxyl group (—COOH) of one amino acid and the amino group (—NH_2) of a second amino acid. The carboxyl group of the second amino acid can then bond to the amino group of a third amino acid. This process can continue until extremely large molecules of protein are formed. The formation of a peptide bond between two glycine molecules is illustrated below.

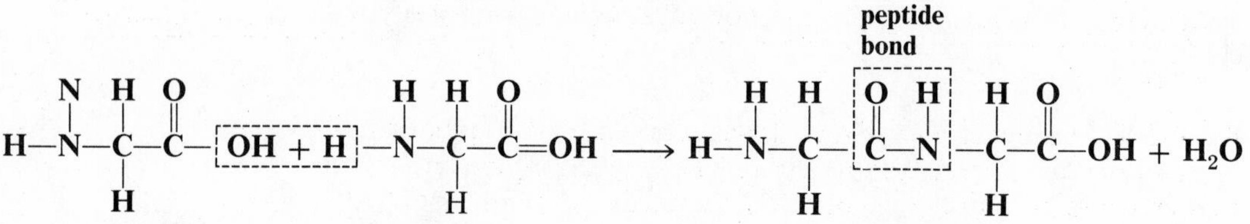

```
                                                              peptide
                                                               bond
                                                             ┌────────┐
   N  H  O                  H  H  O                   H  H   │ O  H   │ H  O
   |  |  ||                 |  |  ||                  |  |   │ ||  |  │ |  ||
H—N—C—C—┆OH + H┆—N—C—C=OH  ⟶  H—N—C—│C—N│—C—C—OH + H2O
      |    └──────┘           |                          |   └────────┘ |
      H                       H                          H             H
   glycine                 glycine                        glycylglycine
```

If this dipeptide, glycylglycine, reacts with the amino group of a third molecule of glycine, a *tripeptide* results.

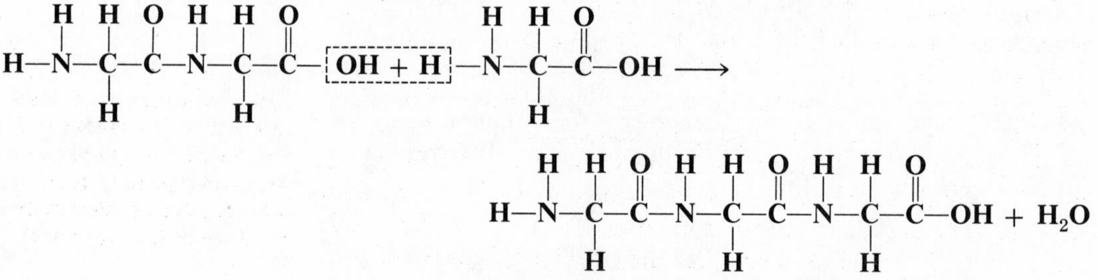

```
   H  H  O  H  H  O                  H  H  O
   |  |  ||  |  |  ||                |  |  ||
H—N—C—C—N—C—C—┆OH + H┆—N—C—C—OH  ⟶
      |        |   └──────┘           |
      H        H                      H

                    H  H  O  H  H  O  H  H  O
                    |  |  ||  |  |  ||  |  |  ||
                 H—N—C—C—N—C—C—N—C—C—OH + H2O
                       |        |        |
                       H        H        H
```

This tripeptide is called glycylglycylglycine. The formation of peptide bonds can continue until a molecule containing hundreds or thousands of amino acids is formed. Such a molecule is called a *polypeptide* or *protein* and can be represented in general by

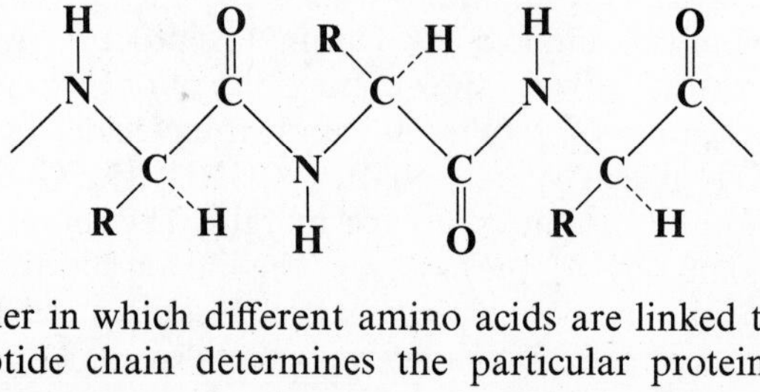

```
     H          O       R   H        H          O
     |          ||       \ ⁄         |          ||
     N          C         C          N          C
   ⁄   \      ⁄   \     ⁄   \      ⁄   \      ⁄   \
         C          N          C          C
       ⁄  `       |          ||       ⁄  `
      R    H      H          O       R    H
```

The order in which different amino acids are linked together in a polypeptide chain determines the particular protein which is

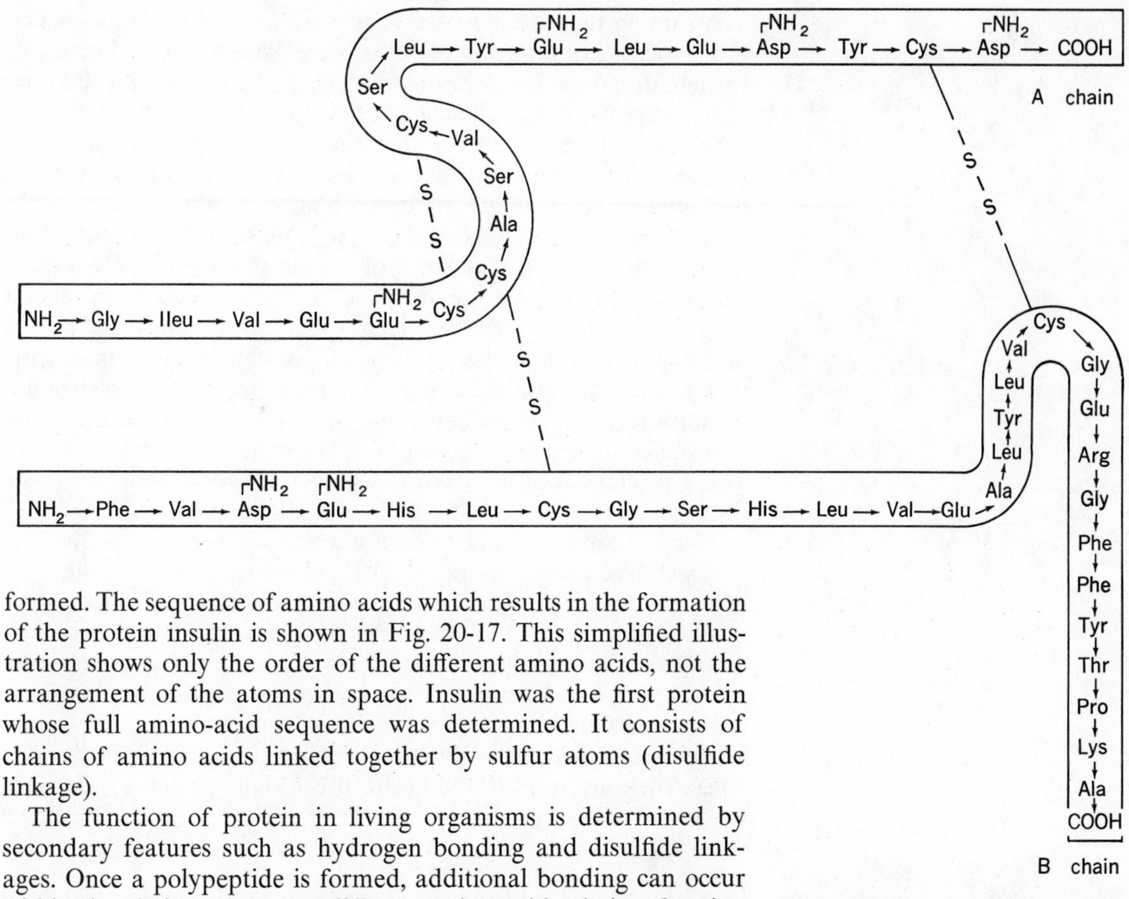

┌NH₂ ┌NH₂ ┌NH₂

Leu → Tyr → Glu → Leu → Glu → Asp → Tyr → Cys → Asp → COOH

Ser

Cys ← Val

Ser

Ala

Cys

┌NH₂

NH_2 → Gly → Ileu → Val → Glu → Glu → Cys

A chain

S—S S—S S—S

Cys → Gly → Glu → Arg → Gly → Phe → Phe → Tyr → Thr → Pro → Lys → Ala → COOH

Val ← Leu ← Tyr ← Leu ← Ala

┌NH₂ ┌NH₂

NH_2 → Phe → Val → Asp → Glu → His → Leu → Cys → Gly → Ser → His → Leu → Val → Glu

B chain

20-17 The structure of beef insulin. The various components of this protein, shown as Val, Gly, Cys, Phen, Ala, and so on, represent amino acids which are bonded together by peptide linkages. This protein consists of separate chains bonded together by sulfur atoms (disulfide linkages).

formed. The sequence of amino acids which results in the formation of the protein insulin is shown in Fig. 20-17. This simplified illustration shows only the order of the different amino acids, not the arrangement of the atoms in space. Insulin was the first protein whose full amino-acid sequence was determined. It consists of chains of amino acids linked together by sulfur atoms (disulfide linkage).

The function of protein in living organisms is determined by secondary features such as hydrogen bonding and disulfide linkages. Once a polypeptide is formed, additional bonding can occur within the chain or between different polypeptide chains, forming the final protein structure. Hydrogen bonding also can occur between H of the —NH group and the O of the C—O group. A disulfide bond, S—S, can be formed when 2 cysteine molecules approach one another. Electrostatic attractions and van der Waals forces, in addition, tend to stabilize the structural features of protein.

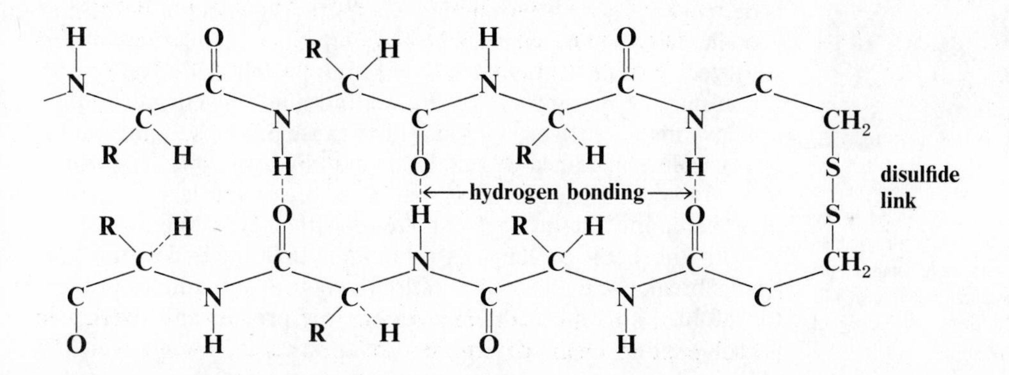

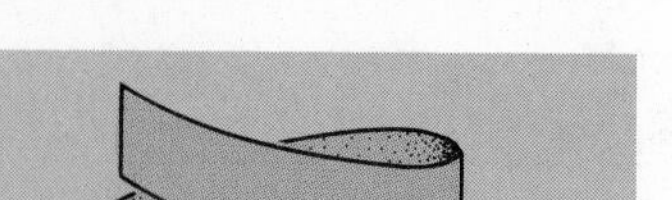
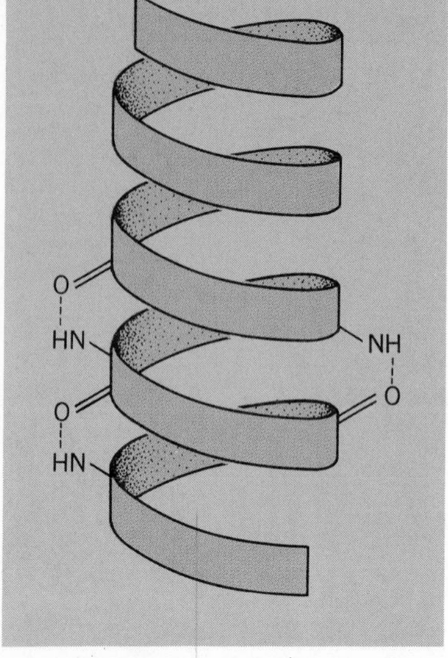

20-18 Schematic diagram showing the α-helical structure of protein. The helix is held in place by hydrogen bonds between the hydrogen of one amino acid and an oxygen atom of another acid further along the chain.

Assuming that the most stable structure would be one in which there was maximum hydrogen or secondary bonding, Dr. Linus Pauling proposed that protein molecules have a spiral staircase type of structure which he called an α-helix (Fig. 20-18). Several of these α-helix structures group together form a bundle of α-helices, resulting in a quarternary structure (Fig. 20-19). This type of aggregation adds to the stability of the protein. The protein hemoglobin consists of four closely packed polypeptide chains.

There are two general types of proteins: *fibrous* and *globular.* Those proteins which are involved in the structural aspects of animals are fibrous proteins which consist of many parallel chains of proteins forming a fiber. This type of protein is the main constituent of hair, nails, muscle, and other structural material. Thus the flexible nature of animal muscle fibers and membranes can be interpreted in terms of the useful properties of fibrous protein. Those proteins such as enzymes which are involved with metabolism are more spherical and are called globular proteins.

20-25 Hydrolysis and Denaturation of Proteins Proteins can be hydrolyzed forming amino acids. Digestion reverses the synthetic process by which the protein is formed. Hydrochloric acid in the gastric juices aids the *hydrolysis* process which breaks proteins down into smaller peptide chains or amino acids. The molecules which are formed by these reactions are small enough to diffuse through the intestinal wall and enter the blood stream. These amino acids or short peptide chains are carried to the cells where they are used to synthesize the specific proteins which are needed in the body. The equation for a typical hydrolysis reaction is shown below.

$$\mathrm{H{-}\overset{H}{\underset{}{N}}{-}\overset{H}{\underset{H}{C}}{-}\overset{O}{\overset{\|}{C}}{-}\overset{H}{N}{-}\overset{H}{\underset{H}{C}}{-}\overset{O}{\overset{\|}{C}}{-}\overset{H}{N}{-}\overset{H}{\underset{H}{C}}{-}\overset{O}{\overset{\|}{C}}{-}OH \xrightarrow[(HCl)]{2H_2O} 3HN{-}\overset{H}{\underset{H}{C}}{-}\overset{O}{\overset{\|}{C}}{-}OH}$$

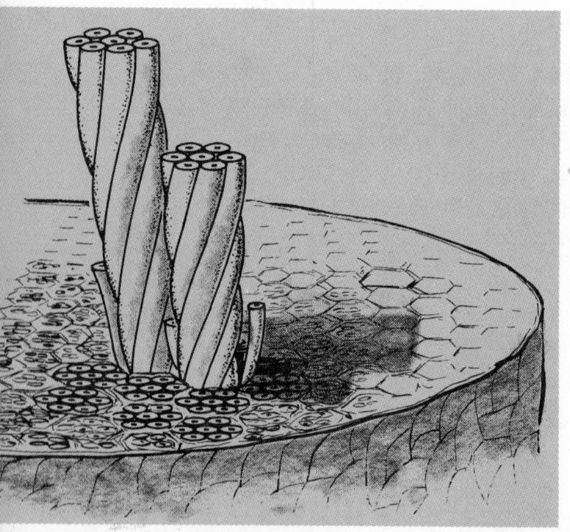

20-19 The molecular structure of fibrous protein such as that found in hair. Each strand in the seven-strand cables is a polypeptide chain having an α-helical structure. Explain in terms of the helical structure and hydrogen bonds why it is possible to stretch hair to almost twice its normal length.

When proteins are treated with *heat* or certain substances such as *alcohol* or *heavy metals,* they lose their specific properties. This is known as *denaturation.* When a protein is denatured, its function is destroyed. The frying of an egg illustrates the effect of heat upon protein. *The protein coagulates irreversibly.* Surgical instruments are sterilized at high temperature so that the proteins of any bacteria which may be present are denatured and made inactive. The ions of heavy metals such as Hg^{2+} and Pb^{2+} cause proteins to precipitate out of solution. If heavy metals are ingested, the antidote which is recommended is egg white or milk. Both contain large amounts of protein, and the heavy metals react with these proteins instead of with the body proteins. It is thought that these denaturation agents break the H-bonds and cause the helical structure of protein to unfold. This destroys the activity of the protein and renders it useless.

20-26 Carbohydrates Carbohydrates are important biochemical foodstuffs which are produced by plants as primary products of the process of photosynthesis. Carbohydrates provide much of the structural basis of plants and animals and include such organic compounds as *sugars, starch,* and *cellulose.* The breakdown of carbohydrates to simpler components is a major source of energy for living organisms. Carbohydrates are typified by the general formula $C_n(H_2O)_n$. We shall first investigate the process of photosynthesis, by which carbohydrates are formed, and the function of carbohydrates as a fuel source. The three general types of carbohydrates which we shall discuss are *monosaccharides, disaccharides,* and *polysaccharides.*

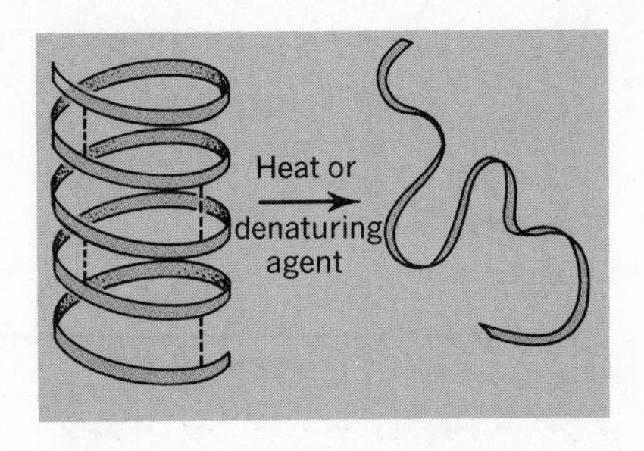

20-20 Denaturation of a protein. Heat and denaturing agents disrupt hydrogen bonds and other linkages and allow the polypeptide chains to uncoil. The long, uncoiled chains may become tangled and coagulate, forming an insoluble mass.

Carbohydrates are synthesized by green plants from carbon dioxide and water in the presence of chlorophyll and sunlight. The products of this synthesis are *simple sugars* which have the general formula $C_6H_{12}O_6$. The overall equation for the reaction, which does not include the intermediate products, is represented as

$$\underset{\text{obtained from air and soil}}{6CO_2 + 6H_2O} \xrightarrow[\text{chlorophyll}]{\text{energy from sunlight}} \underset{\text{simple sugar in plant tissue}}{C_6H_{12}O_6} + \underset{\text{released to air}}{6O_2}$$

Through photosynthesis, energy from the sun is stored in chemicals on earth. Thus, light energy is converted into chemical energy. Life as we know it on earth depends on this process. *Photosynthesis is a spontaneous, endothermic process which produces a more ordered system.* That is, the simple sugar, glucose, is more ordered than the reactants, carbon dioxide and water. On the surface, this seems to violate the general principle that reactions involving an increase in energy and a decrease in entropy are not probable.

The answer is that the sun must be considered as part of the system. The decrease in order in the sun caused by the emission of light is greater than the increase in order in the plant. Therefore, there is an overall net increase in entropy of the universe so the reaction continues spontaneously in the forward direction.

Simple sugars called monosaccharides are the building blocks of carbohydrates. The simplest carbohydrates generally contain chains of 3 to 8 carbon atoms. *These simple sugars contain several hydroxyl groups.* Those sugars which also contain an *aldehyde group* are referred to as polyhydroxy-aldehydes or *aldoses.* Those sugars containing a ketone group are called polyhydroxy-ketones or *ketoses.* The formulas and structures of some of the simple carbohydrates are shown at the top of the next page.

The compounds galactose, glucose, and fructose are examples of sugars which contain the same number of C, H, and O atoms. They differ in the spatial orientation of their atoms and are, therefore, *isomers.*

$$
\begin{array}{c}
\mathrm{H} \\
| \\
\mathrm{C{=}O} \\
| \\
\mathrm{H{-}C{-}OH} \\
| \\
\mathrm{HO{-}C{-}H} \\
| \\
\mathrm{HO{-}C{-}H} \\
| \\
\mathrm{H{-}C{-}OH} \\
| \\
\mathrm{CH_2OH}
\end{array}
\qquad
\begin{array}{c}
\mathrm{H} \\
| \\
\mathrm{C{=}O} \\
| \\
\mathrm{H{-}C{-}OH} \\
| \\
\mathrm{H{-}C{-}OH} \\
| \\
\mathrm{H{-}C{-}OH} \\
| \\
\mathrm{CH_2OH}
\end{array}
\qquad
\begin{array}{c}
\mathrm{H} \\
| \\
\mathrm{C{=}O} \\
| \\
\mathrm{HC{-}OH} \\
| \\
\mathrm{HO{-}C{-}H} \\
| \\
\mathrm{HC{-}OH} \\
| \\
\mathrm{HC{-}OH} \\
| \\
\mathrm{CH_2OH}
\end{array}
\qquad
\begin{array}{c}
\mathrm{CH_2OH} \\
| \\
\mathrm{C{=}O} \\
| \\
\mathrm{HO{-}C{-}H} \\
| \\
\mathrm{H{-}C{-}OH} \\
| \\
\mathrm{H{-}C{-}OH} \\
| \\
\mathrm{CH_2OH}
\end{array}
$$

D-galactose **D-ribose (aldose)** **D-glucose (aldose)** **D-fructose (ketose)**

The structural formulas show these sugar molecules with a straight-chain configuration. This simple structure enables us to identify galactose, ribose, and glucose as polyhydroxyl-aldehydes and fructose as a polyhydroxy-ketone. Galactose, glucose, and fructose are *hexoses* and ribose is a *pentose.* Actually, studies show that these substances have *heterocyclic structures.* The cyclic structure for glucose is shown in the margin on p. 671.

Glucose is an important monosaccharide found in the juices of fruits, in honey, and in blood. It is found in relatively large amounts in all plants and is probably the most abundant single, organic compound. Galactose does not exist as a separate species in living organisms but is a component of other complex carbohydrates such as lactose (milk sugar). Fructose, the sweetest of all sugars, is found in some fruits as well as in blood.

Five-carbon sugars, ribose, and deoxyribose, are components of *nucleic acids,* the "supervisors" of all biochemical activity, and of *adenosine triphosphate* (ATP), the compound which supplies the actual energy needed for the body's endothermic reactions.

Disaccharides are formed when two molecules of monosaccharide combine and lose a molecule of water. The most common example is the formation of sucrose from a molecule of glucose and one of fructose combined in the reaction

$$\underset{\text{glucose}}{C_6H_{12}O_6} + \underset{\text{fructose}}{C_6H_{12}O_6} \longrightarrow \underset{\text{sucrose}}{C_{12}H_{22}O_{11}} + H_2O \qquad \text{20-2}$$

Sucrose is ordinary table sugar, produced commercially by processing sugar cane or sugar beets.

Hydrolysis of a disaccharide is the reverse of Equation 20-2. In the laboratory, sucrose can be hydrolyzed by heating it in an acid solution. The same result is accomplished in the body in the absence of high temperature but with the aid of enzymes.

$$\underset{\text{sucrose}}{C_{12}H_{22}O_{11}} + H_2O \xrightarrow[\text{or enzyme}]{H^+} \underset{\text{glucose}}{C_6H_{12}O_6} + \underset{\text{fructose}}{C_6H_{12}O_6}$$

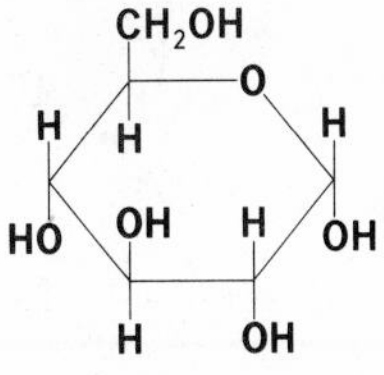

Disaccharide molecules such as sucrose, lactose, and maltose are too large to diffuse through cell membranes and be absorbed. They are first hydrolyzed to monosaccharides in the small intestine by the action of three separate enzymes, *sucrase, lactase,* and *maltase.* The smaller molecules diffuse into intestinal cells and, after a series of reactions, diffuse into the blood stream. As the blood circulates through the liver, some monosaccharides are synthesized into glycogen, a complex carbohydrate which is stored in the liver. Other carbohydrates are carried by the blood to other tissues or oxidized in various cells to carbon dioxide and water yielding immediate energy. Efficient oxidation of glucose in the cells requires the presence of a hormone called insulin. When there is a deficiency of insulin, the level of blood sugar rises. This happens in a condition known as diabetes.

glucose

Polysaccharides are high molecular mass polymers composed of many monosaccharide units. Polysaccharides are formed by a continuation of the process by which the disaccharides are formed. The conversion of glucose into polysaccharides proceeds in the body by a series of steps which are controlled by enzymes. In the body, carbohydrates are stored in the form of glycogen, a polysaccharide containing over 5000 glucose units. The general structural formula for a polysaccharide is

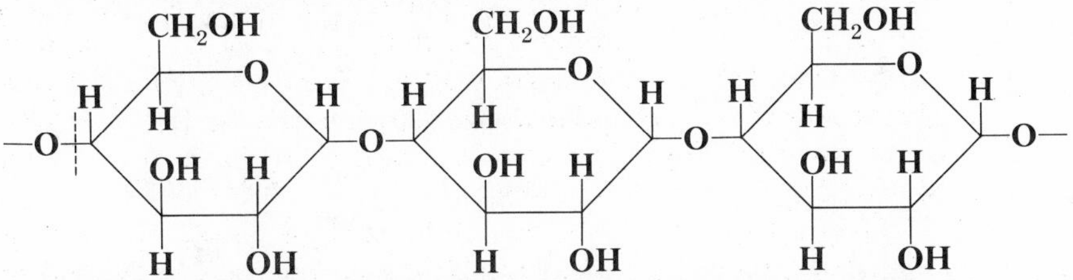

Cellulose, the fibrous framework of plant structure, is another polysaccharide containing thousands of glucose units. The ring structure keeps the unit rigid. Additional units added to opposite ends of this structure make a rigid, rodlike molecule. Large numbers of these molecules cluster together and form cellulose fibers. The large extent of hydrogen bonding between adjacent hydroxyl groups of the close-packed chains is responsible for the rigidity of cellulose (wood) fibers. The glucose in cellulose is not available for human nutrition because there are no enzymes in the human digestive system that can degrade the cellulose into glucose units. All polysaccharides, however, can be hydrolyzed into simple sugars by heating in an acid solution.

Starch is a common polysaccharide stored in plants such as rice, potatoes, and wheat. Starch is an important food source for men and animals, but it must be hydrolyzed to monosaccharides before it can be absorbed. The hydrolysis of starch begins in the mouth, where the digestive enzyme, ptyalin, catalyzes the hydrolysis of starch forming maltose, a dissacharide. Hydrolysis continues in the stomach and small intestine, as in the case of all carbohydrates.

20-27 Carbohydrate Metabolism Energy is liberated when carbohydrates are oxidized in plants and animals. These reactions proceed through a series of steps and require the presence of enzymes. The overall equation for the *catabolism* (oxidiation) of glucose is

$$C_6H_{12}O_6 + 6O_2 \longrightarrow 6CO_2 + 6H_2O + \text{energy}$$

The energy evolved is used to synthesize more complex molecules such as proteins, to maintain body temperature, and to support the activities of the organism.

The large amount of energy produced by the oxidation of simple carbohydrates may be stored in compounds which have very high-energy phosphate bonds. The most important of these is ATP (adenosine triphosphate) which has two *very high-energy* phosphate bonds. When energy is required for an endothermic process, it is supplied by the hydrolysis of ATP to adenosine diphosphate (ADP). The reaction is

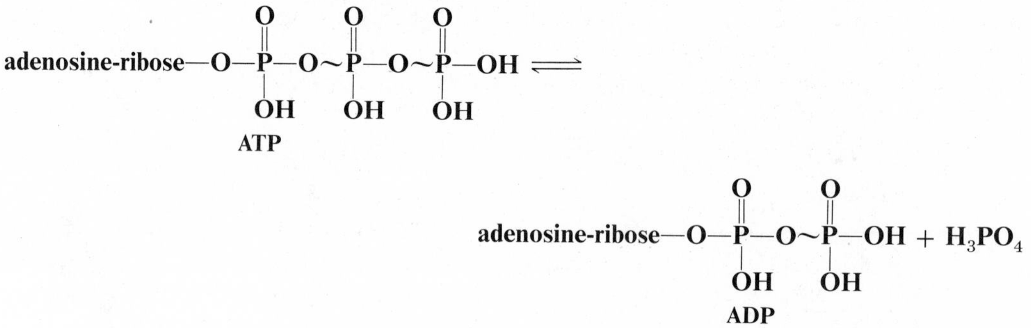

$$\underset{\text{ATP}}{\text{adenosine-ribose}{-}O{-}\overset{\overset{O}{\|}}{\underset{\underset{OH}{|}}{P}}{-}O{\sim}\overset{\overset{O}{\|}}{\underset{\underset{OH}{|}}{P}}{-}O{\sim}\overset{\overset{O}{\|}}{\underset{\underset{OH}{|}}{P}}{-}OH} \rightleftharpoons$$

$$\underset{\text{ADP}}{\text{adenosine-ribose}{-}O{-}\overset{\overset{O}{\|}}{\underset{\underset{OH}{|}}{P}}{-}O{\sim}\overset{\overset{O}{\|}}{\underset{\underset{OH}{|}}{P}}{-}OH} + H_3PO_4$$

where the *wriggle symbol* (~) designates a high-energy phosphate bond. This reaction is reversible so that ADP may be converted to ATP when sufficient energy is liberated by an exothermic reaction to reverse the process. ATP is just one of many energy-rich compounds. The energy which it releases can be used to "drive" biochemical reactions which require energy.

SYNTHETIC POLYMERS

20-28 Classification of Polymers Knowledge of the general nature of natural polymers such as proteins and carbohydrates led chemists to seek methods for synthesizing polymers which could produce fibers and other products that were similar to those found in nature. The wide variety of synthetic fabrics and plastics available to all of us is a result of that search. Some familiar polymers are synthetic and natural rubber, nylon, Dacron, Orlon, Teflon, polyethylene, and numerous other plastics.

Polymers which contain only one kind of fundamental repeating unit are called *homopolymers;* those containing different kinds but similar repeating units are called *copolymers.* Polymers can be further classified as either *addition* or *condensation* polymers, depending on their mode of formation. *Condensation* polymers are derived as the result of a reaction in which some small molecule such as H_2O or NH_3 is eliminated from the reactant monomer molecules when they condense and form the polymer. The formation of a protein is an example.

20-29 Addition Polymers Addition polymers are formed when monomer units add to each other. No by-products are formed in addition polymerization. One of the simplest addition homopolymers is polyethylene. This is the plastic used to make "squeeze bottles" and plastic pails. Polyethylene is formed by the addition of thousands of ethylene molecules (C_2H_4).

$$nCH_2{=}CH_2 \longrightarrow -(-CH_2CH_2-)_n-$$

In general, a reagent known as an *initiator* is required if addition polymerization of ethylene and other alkenes is to occur. Peroxides are often used as initiators. The products formed by the decomposition of peroxides readily add to the C=C bond, forming a neutral, short-lived, reactive species known as a *free radical.* The reactive free radical adds to a new C=C bond. This process continues and forms a giant molecule. Teflon, polystyrene, Orlon, and Plexiglas (Lucite) are other examples of commonly used addition polymers.

20-30 Condensation Polymers Nylon, one of the first synthetic fibers produced, is a polyamide and a condensation copolymer. It is produced by the reaction between adipic acid and hexamethylenediamine. The reaction can be represented as

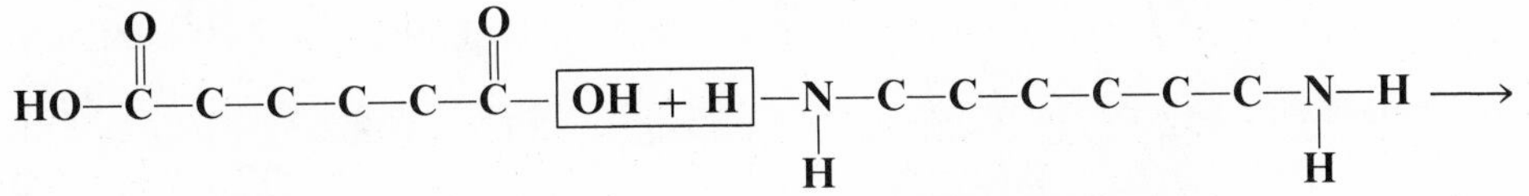

$$HO-\overset{\overset{O}{\|}}{C}-C-C-C-C-\overset{\overset{O}{\|}}{C}-\boxed{OH + H}-\underset{\underset{H}{|}}{N}-C-C-C-C-C-C-\underset{\underset{H}{|}}{N}-H \longrightarrow$$

adipic acid **1,6-diaminehexane (hexamethylenediamine)**

$$\left[-\underset{\underset{O}{\|}}{C}-C-C-C-C-\overset{\overset{O}{\|}}{C}-\underset{\underset{H}{|}}{N}-C-C-C-C-C-C-\overset{\overset{H}{|}}{N}-\right]_n + H_2O$$

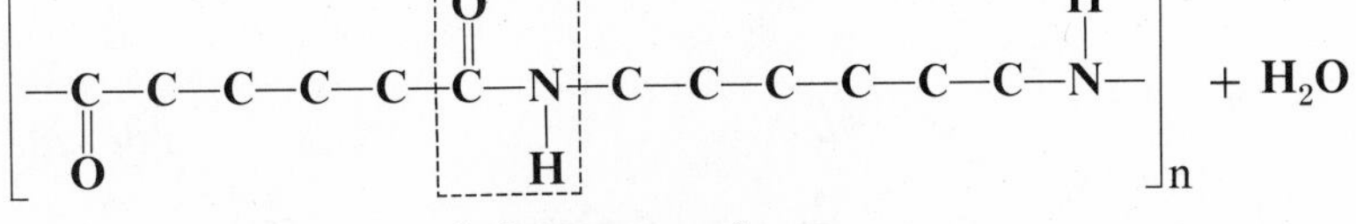

(nylon monomer)
hexamethylenediamineadipic acid monomer

The molecular mass of nylon may be as great as 25 000. The elimination of a water molecule enables the chain of the organic acid to link up covalently with that of the organic base (amine). Continuation of the condensation reaction is possible because both the acid and base molecules have reactive groups at both ends of the chain.

The characteristics of nylon and other plastics reflect the architecture of the individual molecules and the aggregates of molecules

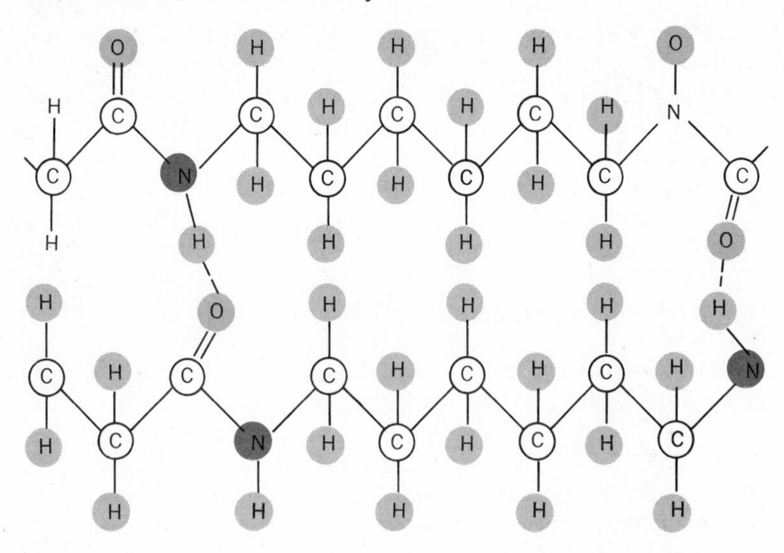

20-21 In fabrics, adjacent nylon molecules are aligned and linked together by hydrogen bonds.

which compose the plastic. One possible arrangement of nylon molecules is shown in Fig. 20-21. The diagram implies that in nylon fabrics and cords the molecules are highly ordered. There is a relatively strong attraction between the hydrogen in an NH group in one chain and oxygen in the adjacent chain. This close packing of chains and attractive force produces a plastic that, when stretched, has crystalline characteristics, considerable strength, and resistance to dissolution and chemical attack. Both nylon and polymers of the polyethylene type, however, will soften and melt at high temperatures.

20-31 Thermosetting Plastics Crosslinking the adjacent molecular chains in a plastic produces hard, thermosetting substances. Copolymers formed by the condensation of phenol or urea with formaldehyde produce *thermosetting plastics,* which are rigid and do not melt at high temperatures. The rigidness and resistance to heat are caused by the three-dimensional, highly cross-linked network architecture of the plastic. In turn, this structure is a reflection of the composition of the polymer molecules. A probable cross-linked structure for Bakelite, a hard thermosetting plastic made from phenol (C_6H_5OH) and formaldehyde (HCHO) and is

OH OH OH

$-CH_2-$ $-CH_2-$ $-CH_2OH$

CH_2 CH_2

$-CH_2-$ $-CH_2-$ $-CH_2OH$

OH OH OH

20-32 Elastomers Elastomers are rubber-like plastics. Isobutene (C_4H_8), an unsaturated hydrocarbon, under certain conditions polymerizes yielding butyl rubber, a synthetic material having physical properties similar to those of natural rubber.

$$\underset{\text{isobutylene}}{n\,CH_2{=}\overset{\displaystyle CH_3}{\underset{\displaystyle CH_3}{\overset{|}{\underset{|}{C}}}}} \longrightarrow -\left(CH_2\overset{\displaystyle CH_3}{\underset{\displaystyle CH_3}{\overset{|}{\underset{|}{C}}}}-\right)_n-$$

Energetic effects are such that the long-chain isobutene polymer molecules favor a randomly coiled arrangement. The intermingling of many randomly coiled molecules is like that of cooked spaghetti. The side branches prevent the close packing which is possible with unbranched chains. In addition, the lack of cross-linking between chains and lack of highly polar groups on the chain result in relatively weak van der Waals forces of attraction between the chains. The aggregate of the weakly cross-lined molecules can be deformed by a stress.

For example, when a piece of rubber is stretched, the molecules uncoil and elongate in the direction of stretch. In the stretched condition, the molecular chains assume a more ordered, "crystalline-like" arrangement. When the stress is removed, the strands tend to return to the more probable, randomly coiled arrangement. The elasticity depends on temperature. At low temperatures the kinetic energy of the molecules is reduced so that attractive forces between chains become more effective. This causes the rubber to assume a more rigid structure.

The rigidity and hardness of natural rubber can be varied by adding different quantities of sulfur to hot rubber. The process, accidentally discovered by Charles Goodyear (1800–1860) in 1838, is called *vulcanization.* In this reaction, the sulfur atoms form cross links between adjacent chains and thus, strengthen the rubber. When vulcanized rubber is stretched and then released, it returns to its original shape. The sulfur cross links keep the long-chain molecules from separating and tend to "pull" them back into their original positions. The hardness of the rubber increases as the sulfur content increases. Hard rubber such as that used to manufacture battery cases contains up to 35 percent sulfur.

From these examples, it can be seen that the observable properties of a plastic or synthetic fiber depend on a number of factors such as the *composition of the monomer molecule chains,* and the *orientation (alignment) of the chains in the product.* By varying these factors, an organic chemist skilled in synthetic polymers can tailor-make a plastic with almost any required characteristics. In some cases, he can improve on natural polymers. For example, synthetic butyl rubber molecules have few reactive sites and afford small opportunity for cross-linking. Hence, this synthetic rubber is stronger and more resistant to chemical attack than natural rubber.

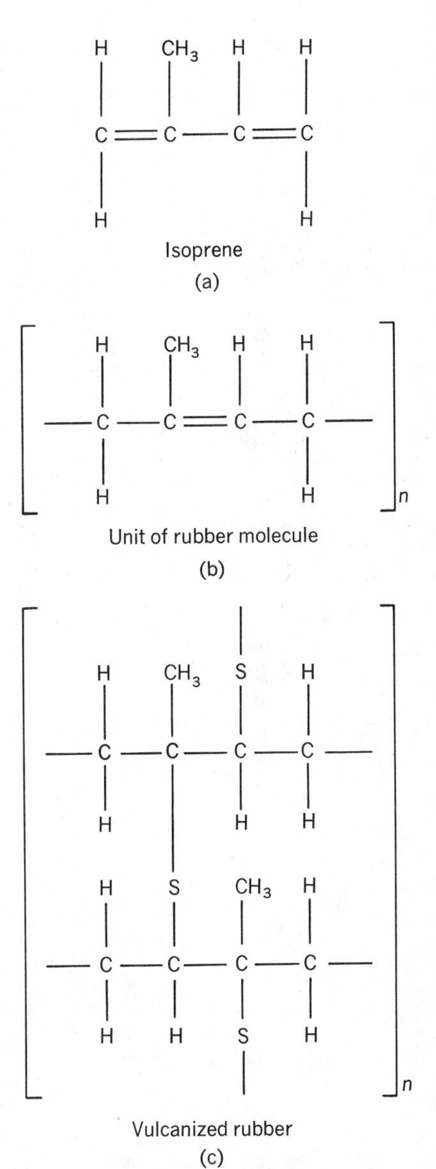

20-22 The structural unit of natural rubber is isoprene (a). When thousands of these units are linked together, they form natural rubber (b). Vulcanization results in a crosslinked structure (c).

MODERN ANALYTICAL TECHNIQUES IN ORGANIC CHEMISTRY

20-33 A Perspective Knowledge of the chemical composition and molecular structure of organic compounds is essential for synthesizing plastics, drugs, and other useful products. In order to reproduce natural products such as quinine and vitamin B-12, the chemist must know the chemical composition and the molecular structure of the substance. Prior to 1920, the determination of the molecular composition and architecture of organic substances was achieved largely through conventional chemical analysis. It is far more difficult to determine the molecular architecture than the chemical composition.

In general, the determination of structural characteristics was accomplished by interpreting the reactions of a substance and chemically identifying the fragments of an organic molecule which remained after a destructive reaction. This type of analysis not only requires a relatively large amount of sample, but also results in the destruction of the sample.

Since 1930 much progress has been made in developing instruments which enable scientists to determine *composition, molecular masses,* and the *architecture of organic molecules.* Most instrumental analyses require only extremely small samples (0.1 g or less) and are nondestructive. This means that the same small sample can be analyzed several times, using different instruments. Each instrumental analysis reveals different characteristics. The combined data obtained from the different sources give a composite picture of the molecule.

A presentation of the principles which underlie most instrumental analyses is beyond the scope of this text. We shall, however, identify and briefly discuss a few of the most important instrumental techniques used in organic analysis. One of the first modern analytical tools used by the organic chemist is the *infrared spectrometer.* Let us briefly examine some of the principles underlying analysis by *infrared spectroscopy.*

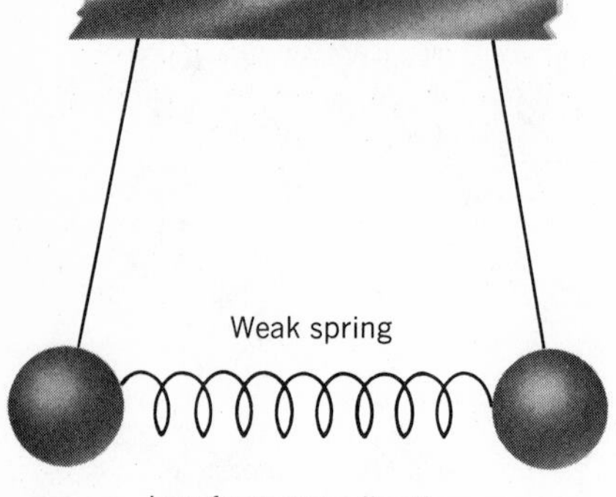

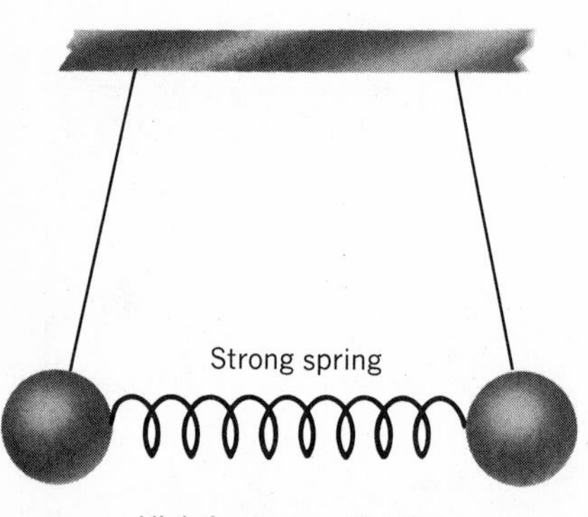

22-23 Spheres with identical masses connected by a strong spring vibrate with a greater frequency than when they are connected by a weak spring. The strength of the spring is analogous to the strength of the covalent bond between the atoms.

20-34 Infrared Spectroscopy Samples of matter absorb different wavelengths of light to different degrees. The petals of a red flower absorb green and blue wavelengths and reflect red wavelengths. Green leaves absorb red wavelengths and we see them as green. The molecules produced by nature thus appear to us as a myriad of colors. The wavelengths of light absorbed and reflected by these materials give important clues to the composition and structure of the molecules comprising these materials. The *differential absorption of various wavelengths of infrared energy* give chemists an insight into the structure of organic molecules.

You should recall that infrared "light" falls in the portion of the electromagnetic spectrum just above (longer wavelengths) that of the visible light. This region of the spectrum includes radiant energy with wavelengths between 2.5 and 25 microns. A micron is

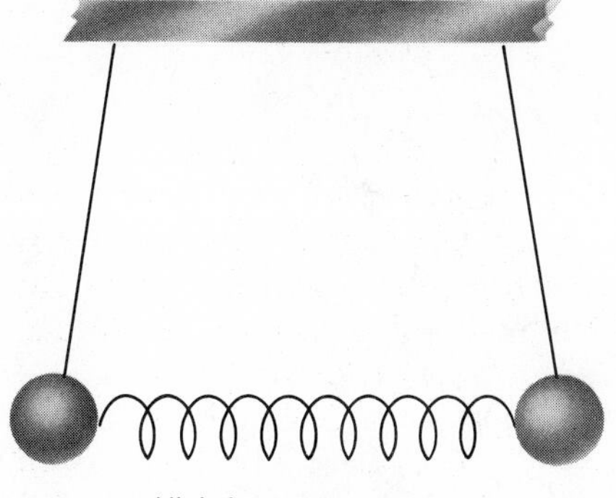

20-24 Two small spheres (atoms) held together by a spring (bond) of given strength vibrate more rapidly than two large spheres held together by the same spring (same size and strength).

one millionth of a meter or 1×10^{-4} cm. Most organic and many inorganic molecules absorb energies in this region. Let us examine the nature of this absorption.

The natural frequencies of energy waves which molecules absorb depends upon the masses of the atoms, the molecular shape, and the strengths of the chemical bonds. An organic molecule can be considered a three-dimensional array of nuclei held together by covalent bonds. In 1914, Niels Bjerrum (1879–1958), a Danish chemist, showed that the absorption of infrared energy by molecules could be correctly explained by viewing the nuclei as masses and the bonds as springs holding the masses together.

In a purely mechanical model, the stronger the spring, the higher the frequency of vibration. The greater the masses on the ends of the spring, the slower the vibration (Fig. 20-24). This behavior of a mechanical system is analogous to the behavior of atoms in a molecule. The various types (modes) of vibration or organic molecules can be described as stretching, bending, wagging, twisting, and rocking motions. We shall consider only the most common vibrational modes.

The *stretching* motion is an alternate shortening and lengthening of the bond (Fig. 20-24). This mode requires only a diatomic molecule. The *bending* mode which is present in water molecules requires at least three atoms. In this mode, the two hydrogen atoms engage in a flapping motion much as a bird flaps its wings (Fig. 20-25). In this case, the bond angle is alternately increased and decreased.

Organic functional groups exhibit specific infrared absorption frequencies. The C—H bond always absorbs infrared energy in the region of 3 microns. The C═O double bond absorbs in the region of 5.5 to 6 microns. The OH group has a stretch absorption at about 3.3 microns. An organic compound containing a triple bond absorbs energies of higher frequencies (greater energy per photon) than one containing a double bond. The double bond, in turn, absorbs higher frequencies than does the single bond. This is consistent with our spring analogy. The *triple bond is the strongest of*

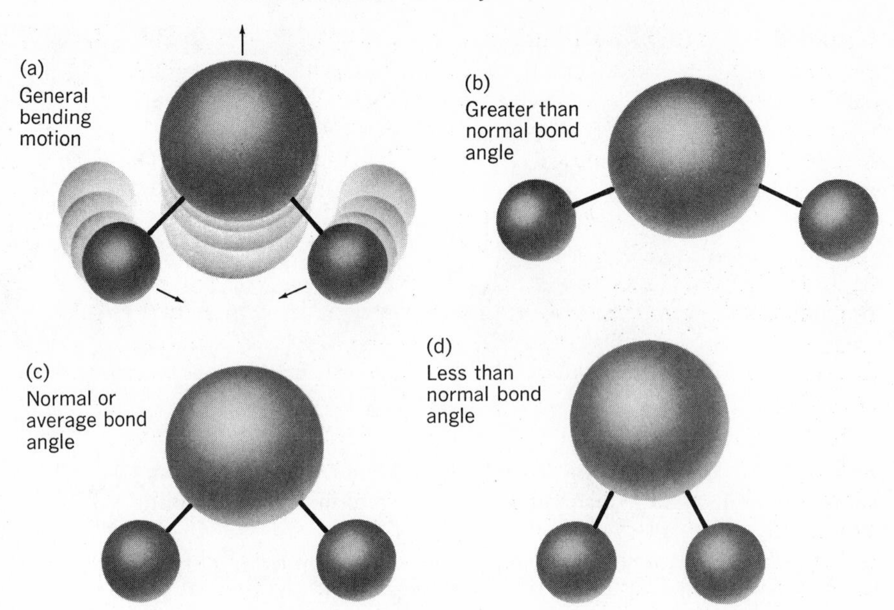

20-25 The flapping (bending) motion shown above involves the alternate increasing and decreasing of the bond angle.

the three bonds and is analogous to the strongest spring. HCl absorbs higher frequency radiation than DCl, (D, deuterium). The difference in absorption is fundamentally caused by the differences in the masses of hydrogen and deuterium.

The infrared spectrometer employs an infrared source, usually an electrically heated ceramic rod of silicon carbide. The infrared energy from the glowing rod is collected by mirrors and passed through the sample and reference cells (Fig. 20-26). The two beams,

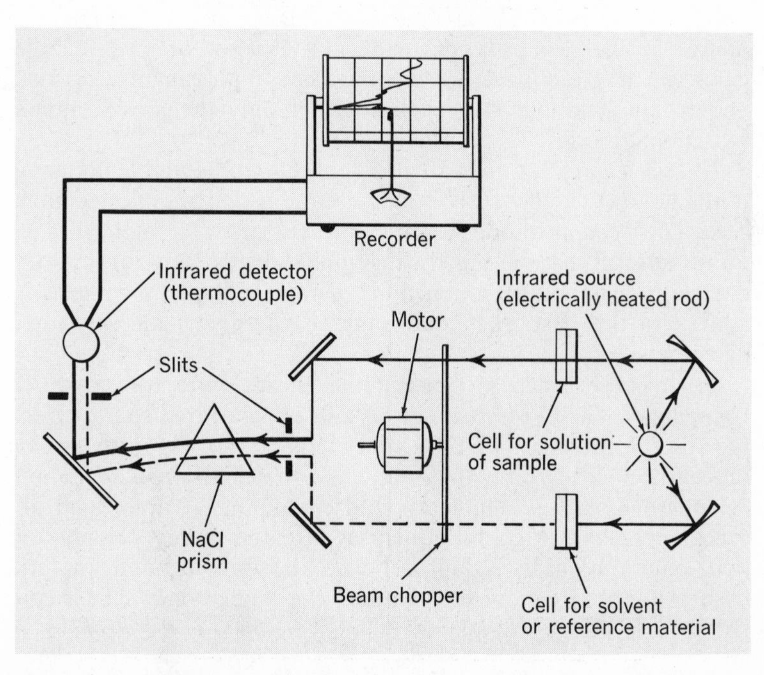

20-26 Schematic diagram for a double beam infrared spectrometer. This type of instrument allows for the difference in absorption of the solute and solvent to be measured. A diffraction grating is sometimes used in place of the prism.

separated in time by the beam chopper, are passed through the *monochromator* (a prism of pure NaCl) and impinge alternately on the detector. Differences in absorption between the sample and reference material give rise to a signal from the detector which is recorded as the transmittance of the sample. As the monochromator selects various wavelengths, a continuous automatic recorder indicates the percent transmission through the sample. Infrared spectrograms such as those shown in Fig. 20-28 are the final record from an infrared spectrometer. The regions showing 100 percent transmission indicate that no infrared energy is absorbed for those frequencies. In general, *infrared spectrograms allow chemists to determine the presence of many functional groups but often do not clearly indicate their relative positions in the molecule.*

20-35 Nuclear Magnetic Resonance Nuclear magnetic resonance (nmr) spectrometers yield data which enable scientists to determine the relative number of hydrogen atoms in a molecule as well as the environment of each hydrogen atom. Different functional groups characteristically influence the absorption behavior of the adjacent hydrogen atoms and, thus, can be identified by the influence which they exert on a hydrogen atom. By analyzing a nmr spectrogram (Fig. 20-31), the chemist can tell whether a specific hydrogen atom is on an alkyl group, an aromatic ring, or next to some electronically influential group such as a double bond or a carbonyl group. The absorption peaks corresponding to the hydrogen atoms next to each of these groups appear in a different region of the spectrogram. The theory of nuclear magnetic resonance is extremely complex. It is based on the ability of nuclei

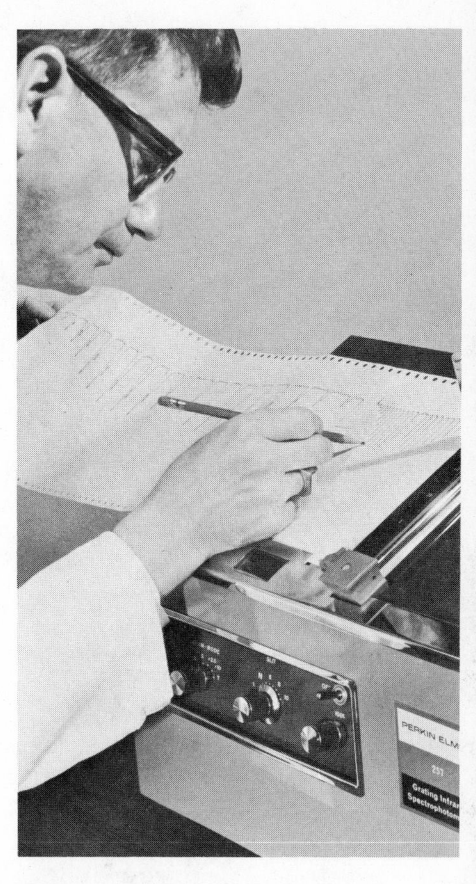

20-27 An infrared spectrophotometer. This important tool of the analytical chemist furnishes data (infrared spectrogram) which enable him to determine the arrangement of atoms in molecules. (Perkin-Elmer Corp.)

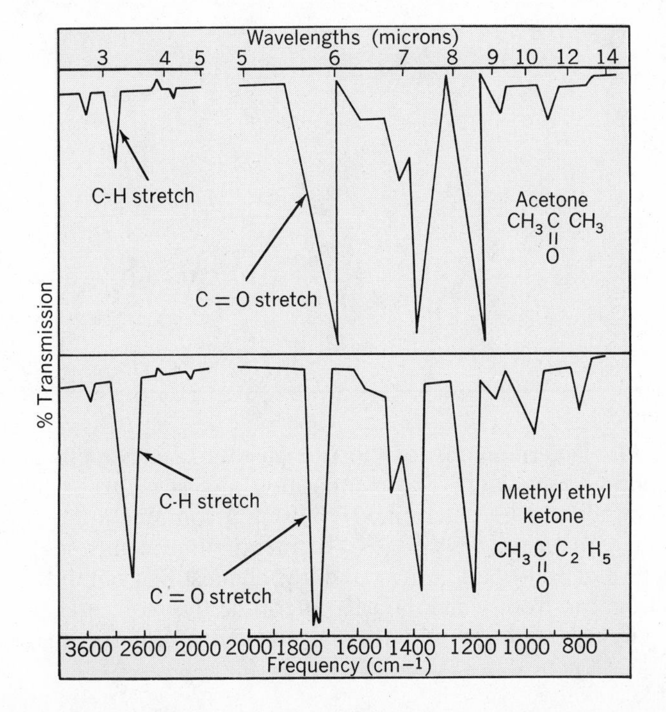

20-28 The infrared spectrograms for acetone and methyl ethyl ketone. Note that both compounds are ketones and both show absorption peaks for the C=O group. (D. P. Stevenson, Shell Development Co.)

R. BRUCE MERRIFIELD
(1921–)

Dr. R. Bruce Merrifield, professor of biochemistry at Rockefeller University, has been described as one of the greatest chemical innovators of our time. His innovation, the automatic protein synthesizer, has made it possible for chemists to synthesize enzymes, hormones, and other molecules of biological interest in the space of a few weeks or months. Prior to the development of his method, it was not uncommon for a biochemist to spend several years on the synthesis of just one of these substances.

Dr. Merrifield was born in Ft. Worth, Texas but spent most of his early life in California. He began his study of chemistry at Pasadena Junior College, transferred to UCLA and earned his B.S. While there he had a part-time job at Amino Acid Manufactures, under the direction of Prof. Max S. Dunn of UCLA. In so doing, Dr. Merrifield developed his interest in the chemistry of amino acids and proteins.

At UCLA he was awarded his Ph.D. in 1949, after having done much research on the techniques of microbiological assay of purines and pyramidines. The day after his graduation he was married and, on the following day, he and his wife, set out for New York and his new job at what was then known as the Rockefeller Institute for Medical Research.

At Rockefeller, Dr. Merrifield began working under the inspiration of the late Dr. D. W. Wolley—a great man in his own right, directing several research teams while being totally blind. Dr. Merrifield was involved in the separation and the determination of the sequence of protein growth factors, and he also synthesized several complex polypeptides containing as many as forty amino acids. It was his dissatisfaction with the very tedious methods of synthesis then available that caused him to search for a better one.

to absorb energy in the radio-frequency range while under the influence of a powerful magnetic field. The radio frequency depends on the strength of the magnetic field. The nmr spectrometer currently used employs a radio frequency of 60 megacycles (megahertz).

20-36 Mass Spectroscopy Exact formulas of molecules and fragments of molecules can often be determined from a mass spectrogram. In the late 1950's, mass spectroscopy matured as a valuable aid to structure determination of complex organic compounds. In the mass spectrometer, a volatile compound is bombarded by an electron beam. The resulting positively charged ion fragments are accelerated and passed through a magnetic field. This ion beam impinges upon a collector which is linked electronically to a recorder which produces the spectrogram. Analysis of the spectrogram (Fig. 20-30) reveals the exact mass of the particles responsible for each peak and therefore, the exact number and kinds of atoms composing the particles. Data such as that listed in Table 20-3 help the analyst identify the fragments.

TABLE 20-3
COMMON FRAGMENTS AND THEIR MOLECULAR MASSES

m.m.	*Fragment*	*m.m.*	*Fragment*
26	$C{\equiv}N$	44	$CH_2{-}\underset{\mid \atop H}{C}{=}O + H$, CH_3CHNH_2, CO_2
27	C_2H_3	45	$\underset{\mid \atop CH_3}{C}HOH$, CH_2CH_2OH
28	C_2H_4, CO, N_2	46	NO_2
29	C_2H_5, CHO	47	CH_2SH, CH_3S
30	CH_2NH_2, NO	48	$CH_3S + H$
31	CH_2OH, OCH_3	|	
|		54	$CH_2CH_2C{\equiv}N$
33	SH	55	C_4H_7
34	H_2S	56	C_4H_8
35	Cl	57	C_4H_9, $C_2H_5C{=}O$
36	HCl	58	$CH_3{-}\overset{O \atop \parallel}{C}{-}CH_2 + H$, $C_2H_5CHNH_2$, $(CH_3)_2NCH_2$, $C_2H_5NHCH_2$
|			
39	C_3H_3		
40	$CH_2C{\equiv}N$		
41	C_3H_5, $CH_2C{\equiv}N + H$		
42	C_3H_6		
43	C_3H_7, $CH_3C{=}O$		

20-37 Determination of the Composition and Structure of an Unknown Compound The composition and structure of organic compounds can be determined by using a combination of instrumental and chemical analysis. To identify the composition and structure of an organic compound, the chemist first obtains a mass spectrogram, from which he can determine the molecular mass of the parent compound and the masses of different fragments obtained by electron bombardment. From the molecular mass of the parent compound, he can obtain the probable *simplest formula* from a table relating masses to formulas. Thousands of fragments such as methyl (CH_3—), methylene (CH_2—), and phenyl (C_6H_5—) groups can be identified by the same general method.

An infrared spectrogram of the compound then reveals the presence of hydroxyl, carbonyl, and other functional groups. The proton distribution revealed by a nmr spectrogram then aids in locating the position of the functional groups and alkyl radicals. Finally, an ultraviolet spectrogram helps identify more complex ring structures. *Identification of a compound can often be confirmed by comparing the spectrogram of an unknown with that of a known sample.*

Let us imagine that the upper spectrograms shown in Fig. 20-28, 20-30, and 20-31, are obtained as data during the analysis of an unknown compound. In order to determine the composition and structure of our unknown compound, we begin with a *carbon-hydrogen analysis,* using an apparatus similar to that shown in Fig. 20-29. In this type of analysis, all of the carbon and hydrogen of the unknown are converted, respectively, to carbon dioxide and water. These gases are absorbed when they pass through tubes containing specific absorbents. The masses of carbon and hydrogen in the original sample may be determined from the mass of carbon dioxide and water absorbed. The mass of oxygen in the sample is equal to the difference between the mass of the original sample and the combined masses of the carbon and hydrogen. This chemical analysis enables us to determine the percentages of carbon, hydrogen, and oxygen in the compound. Suppose the analysis shows that the composition is

carbon	62.0% by mass
hydrogen	10.4% by mass
oxygen	27.6% by mass

From these data we can determine that the simplest formula is C_3H_6O. The true formula is $(C_3H_6O)_n$ and the molecular mass is $n \times 58$.

In 1959 he began work on the automatic synthesis of proteins, a method in which an amino acid is anchored chemically onto beads of polystyrene. Then other amino acids can be added, one by one, in automatically controlled steps to build the desired peptide linkages. His first success was a modest one, that of the production of a dipeptide consisting of alanine and leucine. Following this, he prepared a tetrapeptide, and then bradykinin, a peptide containing nine amino acids. His automatic synthesizer is now being produced commercially and has been developed to a point where one can add six amino acids in a period of 24 hours, the addition of each requiring ninety separate steps—all preprogrammed on a revolving drum which looks like something out of an old player piano.

Two substances synthesized by his method include the hormone insulin and the enzyme ribonuclease. Although more than five thousand separate operations were required to assemble the 51 amino acids of insulin, most of them were performed automatically under the control of the drum programmer, so that it was possible for one man to carry out the synthesis in only a few days! The synthesis of ribonuclease required the chemical linking of 124 amino acids, and for this feat Dr. Merrifield in 1969 was given the Albert Lasker Award for Basic Medical Research.

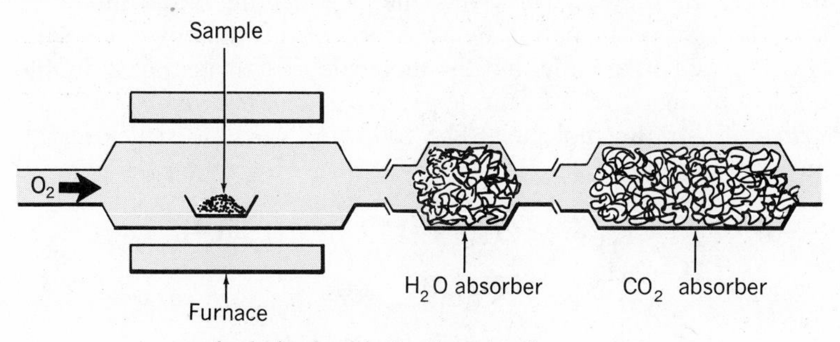

20-29 Apparatus for a combustion analysis.

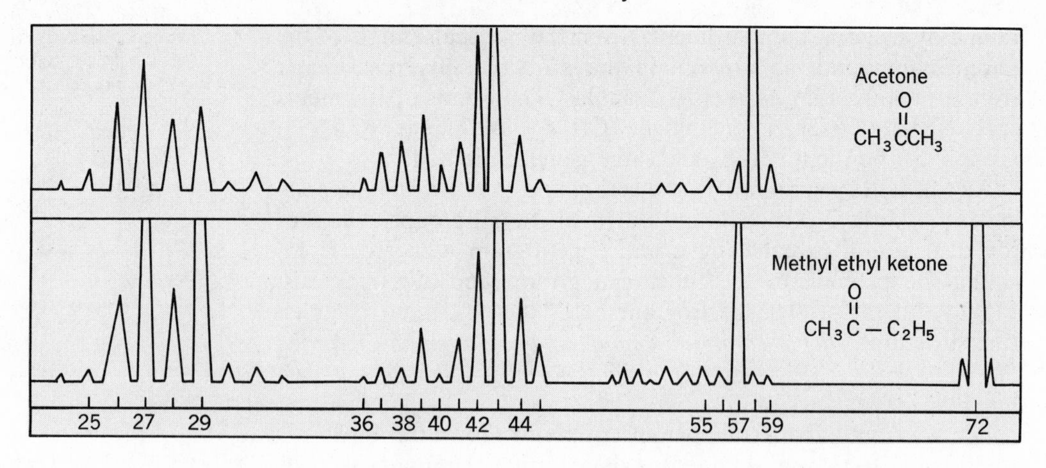

20-30 Mass spectra of acetone and methyl ethyl ketone. The peaks represent ion fragments, which can be identified by comparing the mass numbers on the spectrogram with those listed in data such as those in Table 20-3 (D. P. Stevenson, Shell Development Co.)

The next step is to obtain a *mass spectrogram* for the substance. The mass spectrogram for our compound, shown in Fig. 20-30, contains an intense peak at a mass number of 58. The peak with the highest mass number corresponds to the mass of the parent molecule less one electron. Thus, the true formula of our unknown is C_3H_6O. Since the parent hydrocarbon of this unknown molecule is C_3H_8, we can assume that the molecule contains either a double bond or a ring structure. Our next move is to *postulate some possible structures* for the molecule. The following possible structures all have the formula C_3H_6O.

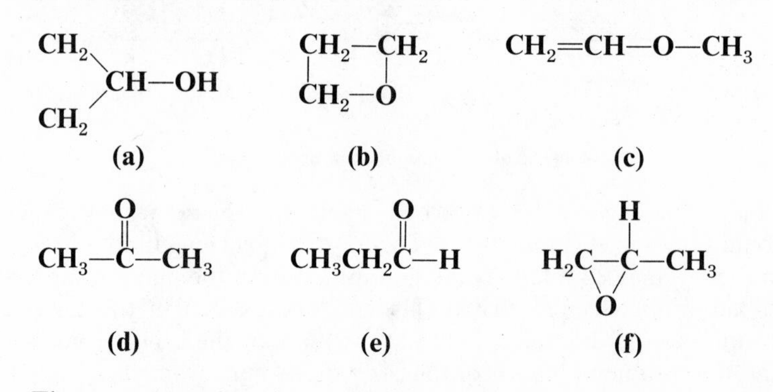

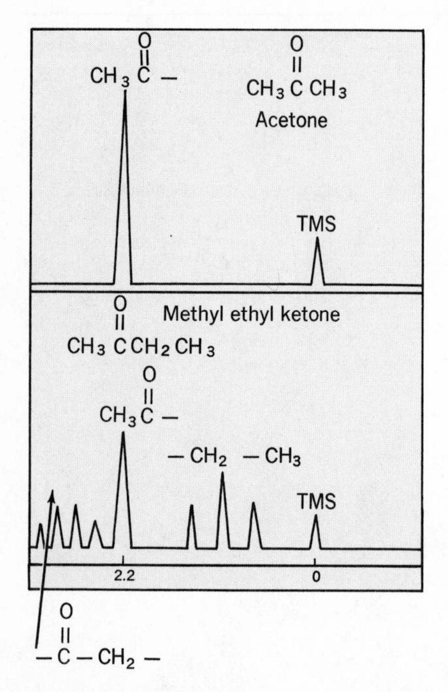

20-31 Nuclear magnetic resonance spectra for acetone and methyl ethyl ketone.

The next procedure is to obtain an *infrared spectrogram* for our unknown. The IR spectrogram indicates the presence or absence of a carbonyl group and, thus, narrows the list of possible compounds. The infrared spectrogram shown in Fig. 20-28 reveals the presence of the carbonyl group. This tells us that our molecule must have the structure represented by either (d) or (e). To decide which of these structures is correct, we next obtain a *nuclear magnetic resonance spectrogram* (Fig. 20-31). The small peak at 0, labeled TMS, is the internal reference standard. The letters TMS stand for tetramethylsilane $[(CH_3)_4Si]$, a substance in which all hydrogen atoms are equivalent. The units shown along the base of the graph are related to the frequency at which the groups indicated at the top of the peaks absorb energy relative to TMS. This spectrogram contains only one peak. This indicates that all six protons are equivalent. Hence, our compound is $CH_3\overset{\overset{\displaystyle O}{\|}}{C}CH_3$. An nmr spectrogram for CH_3CH_2CHO would contain more than one single peak since not all protons in propanal are equivalent.

To confirm our analysis, we *prepare a derivative* of the compound and *check the experimentally determined melting point of the derivative with the melting point* reported in the literature. For example, we might react the compound we have identified as acetone with 2,4-dinitrophenylhydrazine. The reaction is

$$\text{2,4-}(NO_2)_2C_6H_3NHNH_2 + CH_3\overset{\overset{\displaystyle O}{\|}}{C}CH_3 \longrightarrow \text{2,4-}(NO_2)_2C_6H_3NH{-}N{=}C(CH_3)_2$$

2,4-dinitrophenylhydrazine → **2,4-dinitrophenylhydrazone derivative of acetone**

The literature gives the melting point of the 2,4-dinitrophenylhydrazone derivative of acetone as 126°C. If the melting point of our experimentally prepared derivative checks closely with the reported value, we can assume with confidence that our unknown compound is acetone (CH_3COCH_3).

Special Feature: Biochemistry

DNA, RNA, AND PROTEIN SYNTHESIS

DNA (deoxyribonucleic acid) and RNA (ribonucleic acid), known as nucleic acids, play key roles in the manufacture of proteins in living cells. In addition, DNA can undergo self-replication and is the molecule which carries the genes responsible for transmitting hereditary or genetic information. We shall first examine the theory of protein synthesis and then briefly consider the self-replication of DNA. It should be emphasized that molecular biology is a field of intensive research so that theories are subject to constant modification. The following discussion should be considered more speculative than factual.

DNA is a polymer of phosphoric acid, deoxyribose, and organic nitrogen bases. A unit of the polymeric chain, known as a nucleo-

JAMES D. WATSON (1928–) **FRANCIS H. C. CRICK** (1916–)

Francis H. C. Crick, an English biochemist, and James D. Watson, an American biochemist, collaborated to construct one of the most complex scientific models ever made.

Crick was educated at University College of London and won a Ph.D. in physics at Cambridge University. During World War II he worked on radar research and other problems in physics. Later he worked in the field of X-ray diffraction as a tool for studying molecular structure. He was at this time engaged in research at Cambridge University.

Watson was a child prodigy who entered the University of Chicago at age 15 and was graduated in 1947. He completed his Ph.D. at the University of Indiana in 1950 in biochemistry. He spent a year of post-doctoral study at the University of Copenhagen and then went to Cambridge where he met Crick.

The two men were very much interested in the problem of the structure of the DNA (deoxyribonucleic acid) molecule. Biochemists knew that nucleic acids rather than ordinary proteins were responsible for transmitting inherited characteristics and that DNA was a key chemical in the mystery of the chromosome. The orthodox chemical analysis of this acid had been worked out by Alexander Todd, at Manchester. For this work Todd received the 1957 Nobel Prize. A far more difficult problem was to piece together the many components of this highly complex compound and determine its structure. Using X-ray diffraction techniques and paper chromatography, Crick and Watson proposed a double helix (a double coiled spring) made up of a "sugar-phosphate backbone with nitrogenous bases extending in toward the center of the helix from each of the two backbones."

It was known that DNA must be capable of self-replication (producing other species like itself). Crick and Watson showed that the two strands of the double helix could unwind and each serve as a model for its complement. This idea at first seemed strange to many scientists but research confirmed the model until it was accepted. For their work, they shared the 1962 Nobel Prize in Medicine.

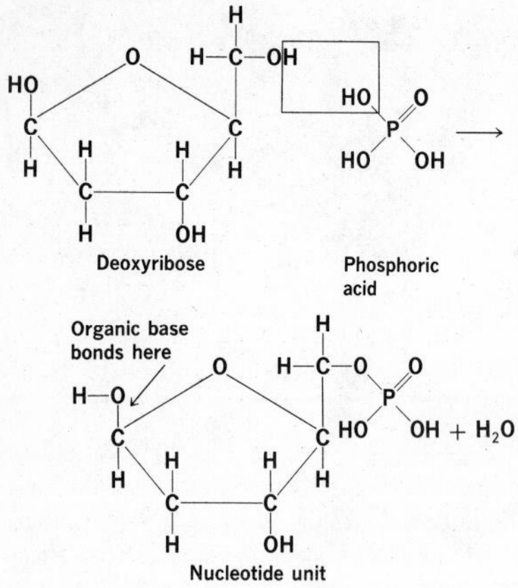

Deoxyribose

Phosphoric acid

Organic base bonds here

Nucleotide unit

tide unit is shown on this page. The DNA molecule has a backbone of *alternating phosphate* and *deoxyribose* to which an *organic base* is attached (Fig. 20-32). The organic bases are derivatives of *pyrimidine* and *purine.* Four bases found in nucleic acids are derived from pyrimidine whose formula is

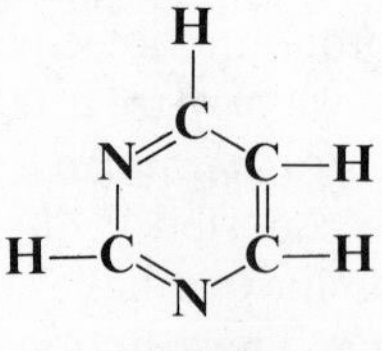

pyrimidine

Two of the derivatives of pyrimidine are *thymine* (T) and *cytosine* (C) whose formulas are represented more simply as

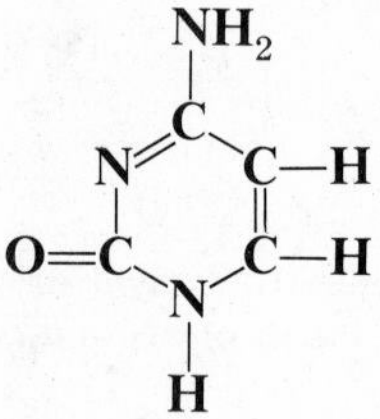

cystosine

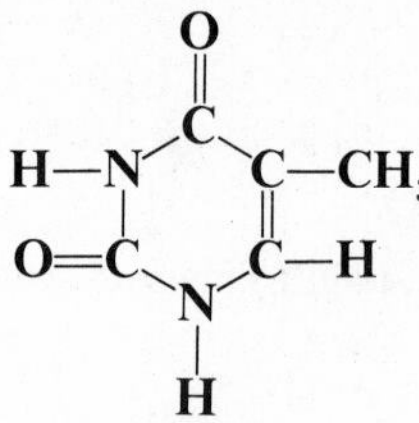

thymine

Two other bases found in nucleic acids are derived from purine whose formula is

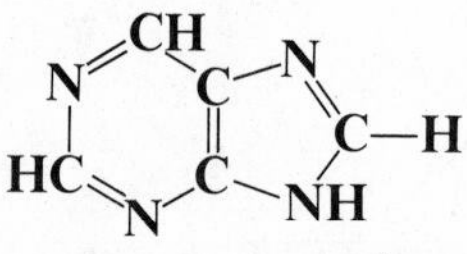

purine

The two organic bases which are derived from purine are *guanine* (G) and *adenine* (A). The formula for these compounds are

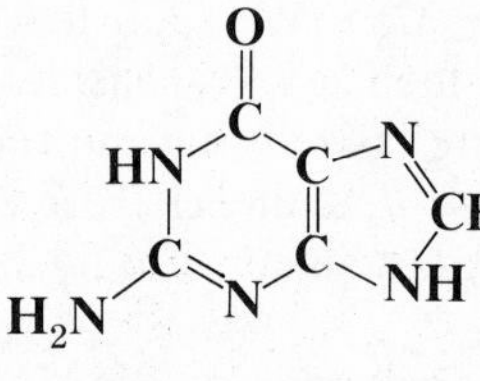

guanine

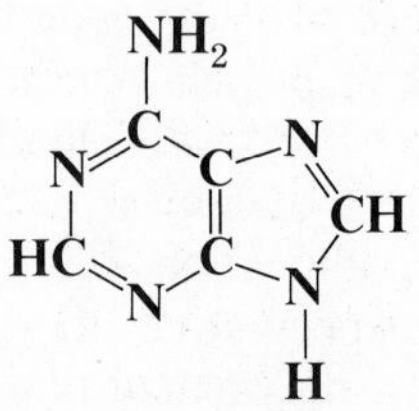

adenine

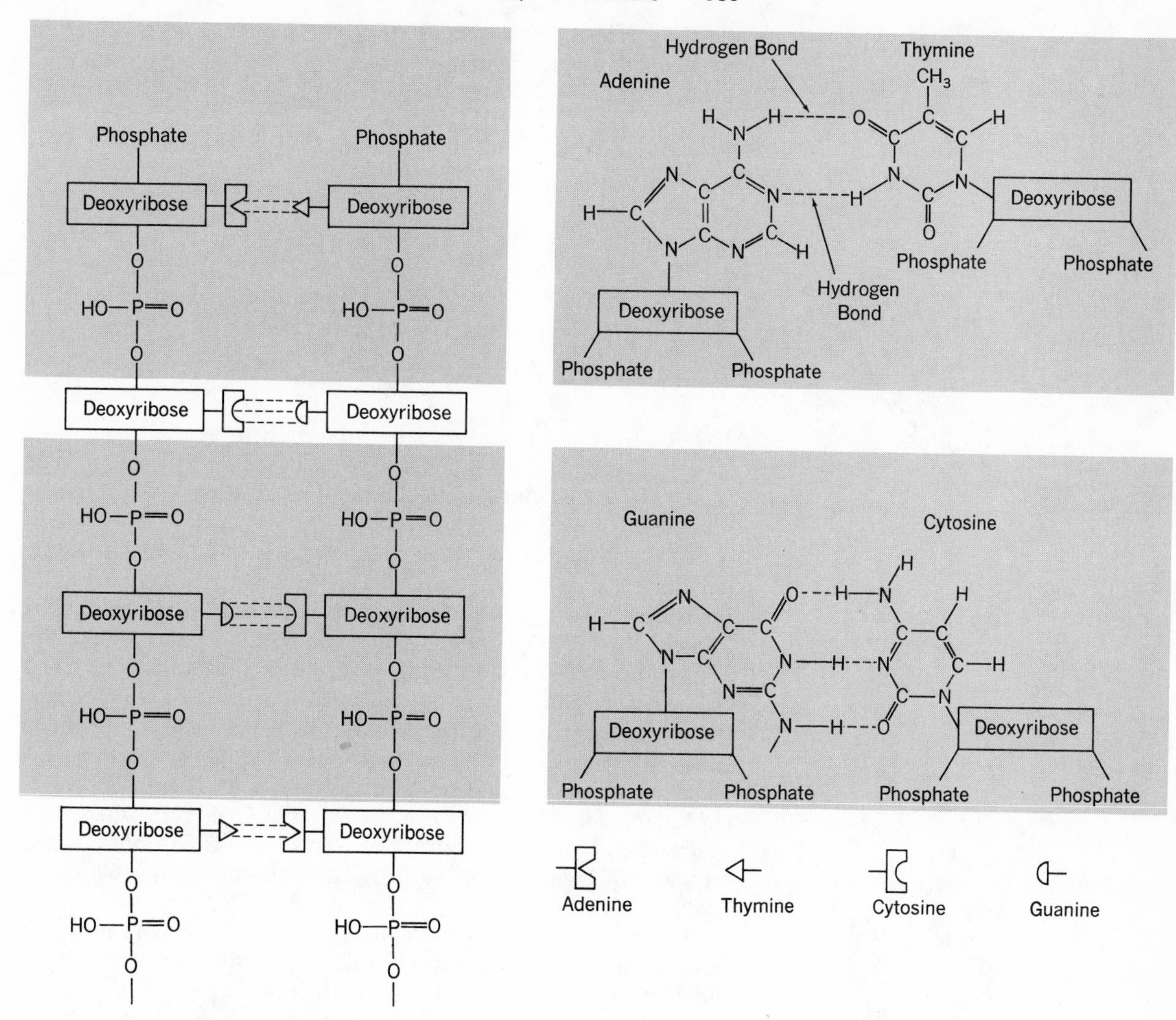

20-32 Two-dimension representation of a section of DNA showing the two hydrogen bonds between adenine and thymine, and the three hydrogen bonds between guanine and cytosine. The schematic diagram on the left shows the phosphate-deoxyribose backbone of DNA.

The DNA which exists in the nucleus of the cell is thought to be a double helix in which two molecules of DNA are coiled about each other (Fig. 20-33). The two strands of DNA are held together by hydrogen bonding between the bases. For maximum stability, a thymine (T) must pair with an adenine (A), and a guanine (G) must be paired with a cytosine (C).

A series of events occurring in both the nucleus and the cytoplasm result in protein synthesis. The *DNA* transfers its message by the formation of a complementary strand of *RNA*. *RNA* is similar in structure to *DNA* with the exception that *RNA* contains *ribose sugar* and *uracil* (U) instead of deoxyribose and thymine. In addition, *RNA* remains as a single strand. The *DNA* synthesizes an *RNA* strand which is complementary to the base sequence of the *DNA*. This complementation of base sequence is illustrated in a schematic diagram below.

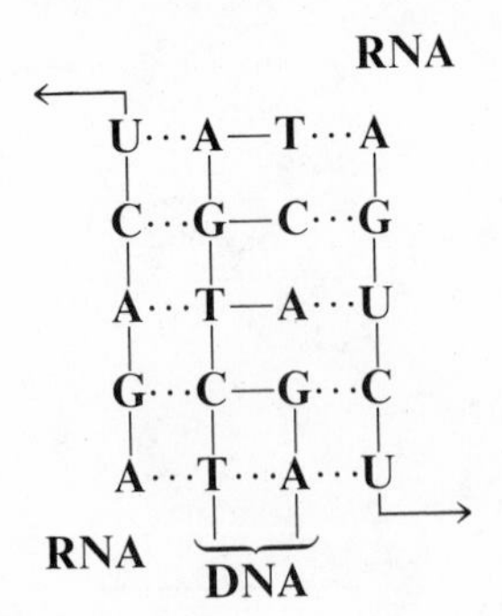

The *RNA* formed in the nucleus is referred to as *messenger-RNA* (*m-RNA*) since it carries the amino acid sequence for a specific protein from the nucleus to the cytoplasm where the protein is produced.

Protein is synthesized when the message of the m-*RNA* is translated into a sequence of amino acids. The substance that translates the code of the *m-RNA* is another *RNA* present in the cell called transfer *RNA* (*t-RNA*). When a particular amino acid is called for, a specific *t-RNA* attaches to a molecule of that amino acid and transports it to the surface of the *m-RNA*. The factor which determines the order of amino acids in a specific protein is thought to be the sequence of organic bases of the *m-RNA*. Every three organic bases appears to call for one specific amino acid. There is a specific *t-RNA* for each triplet of organic bases and in turn, there is a specific *t-RNA* for each amino acid. Thus, the *t-RNA* translates the sequence of organic bases in the *m-RNA* into a sequence of amino acids.

The amino acids then form peptide bonds producing polypeptides of high molecular mass. As this polypeptide is removed from the *m-RNA* it assumes its α-helical structure and is called protein.

The message of the *DNA* is repeated through cell division carrying the information from generation to generation. The characteristics which repeat from generation to generation to generation are carried by the *DNA* in a process called replication. A double strand of *DNA* separates within the nucleus of the cell. Each of the single strands picks up phosphates, deoxyribose sugars, and organic bases from the surrounding medium and builds a complementary strand. Thus, in the nucleus of a single cell, there are two doubly stranded *DNA* molecules with the same code or message as the original *DNA*. The cell then divides, leaving a *DNA* molecule in each of the newly formed nuclei.

A new generation begins when the ovum of a female and the sperm of a male unite and form a single cell. The organism grows by duplication of the *DNA* strands present in that original cell which contained the characteristics of each parent. This new orga-

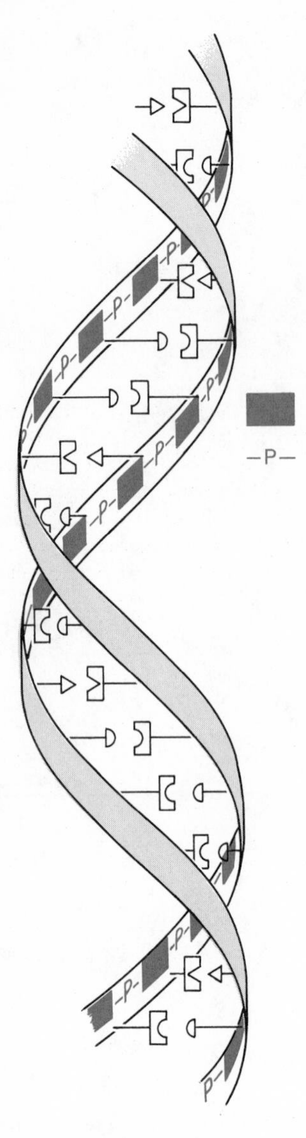

20-33 Three-dimensional representation of a section of DNA illustrating the double helix and the pairing of specific organic bases.

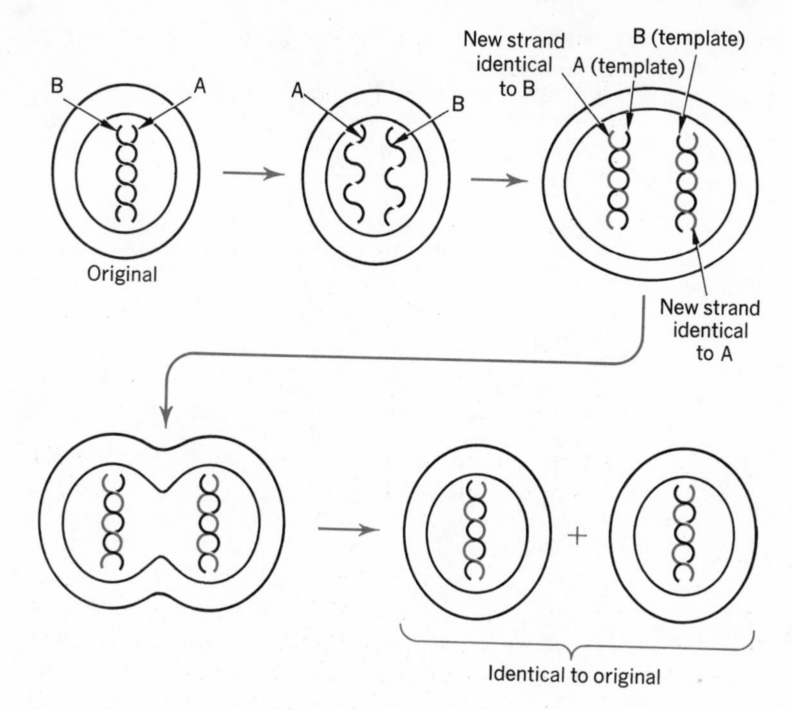

20-34 In self-replication, it is believed that DNA separates into two strands, each strand serving as the template for the synthesis of a new strand. Each new polynucleotide strand has a base sequence identical to that of the strand to which its template was bonded in the original DNA molecule. For example, adenine on the template would have been paired originally with thymine and could join another thymine by hydrogen bonding on the new strand. Thus two identical molecules are produced from the original.

nism has *DNA* which repeats the *DNA* of his parents. Therefore, the proteins and, hence, the features of the new organism have the characteristics of the parent organism.

LOOKING AHEAD

In the first twenty chapters of this text we have encountered and studied a vast number of chemical reactions, most of which involve only the outermost electrons of atoms and do not affect the nuclei. In our study of atomic structure, we cited experiments which revealed the general composition of the atomic nucleus. We did not consider, however, reactions in which the nuclei of atoms undergo a change. We are now ready to investigate in more detail, the composition, structure, nature, and reactions of atomic nuclei. We shall be particularly concerned with unstable or radioactive nuclei. A discussion of the origin, characteristics, and uses of the different forms of radiation constitute the major portion of the last chapter in this text. You will find the events and discoveries which led to man's harnessing atomic energy an exciting story.

QUESTIONS AND PROBLEMS

1. Define organic chemistry.

2. Explain how carbon can be tetravalent if the carbon atom in the ground state has a $2s^2$, $2p_x$, $2p_y$ electron configuration.

3. Give reasons for the fact that the rate of reaction of organic compounds is slower than that of inorganic compounds?

4. Use Table 20-2 to show that the melting points of similar organic compounds are proportional to their molecular masses.

5. Using tetrahedrons as models of carbon atoms, criticize the term "straight-chain" hydrocarbon.

6. (a) Name four classes of hydrocarbons. (b) Give a general description of each class.

7. (a) What is a structural isomer? (b) Which of these pairs are structural isomers?

```
        H  H                     Cl H
        |  |                     |  |
1. Cl—C—C—Cl        and     H—C—C—H,
        |  |                     |  |
        H  H                     H  Cl

       Br                        Br Br
       |                         |  |
2. H—C—C—H          and     H—C=C—H,
          |
          Br

       H  Cl                     Cl Cl
       |  |                      |  |
3. H—C—C—Cl         and     H—C—C—H,
       |  |                      |  |
       H  H                      H  H

          H                         C
          |                         |
       H—C—H                     H—C—H
       H  |  H                   H  |
       |  |  |                   |  |
4. H—C—C—C—H        and     H—C—C—H,
       |  |  |                   |  |
       H  H  H                   |  H
                                 H—C—H
                                   |
                                   H

          H
          |
       H  O  H                   H  H  H
       |  |  |                   |  |  |
5. H—C—C—C—H        and     H—C—C—C—OH,
       |     |                   |  |  |
       H     H                   H  H  H
```

8. (a) Write the general formulas for the alkane series, the alkene series, and the alkyne series. (b) Write the name and structural formula for the first member of each group.

9. Write the structural formulas and names of the first five members of the alkane series.

10. (a) What is a functional group? (b) Name three functional groups.

11. (a) How many isomers can there be for hexane? (b) Write the structural formula for each. (c) Give the systematic name of each.

12. What is a saturated hydrocarbon?

13. Propane can be burned, forming carbon, carbon monoxide, or carbon dioxide, in addition to water. Write an equation for each.

14. (a) What is the source of most hydrocarbons? (b) How are they separated from each other?

15. (a) What is cracking? (b) Why is it economically important?

16. Explain the presence of water in the exhaust pipe of an automobile which has just started up.

17. Draw the structural formulas of (a) 2-butene, (b) 1,3-pentadiene, (c) 3,4-dimethyl-1-hexene, (d) 1,5-hexadiene, (e) 3-methyl-2-chloro-1-butene.

18. (a) What is an addition reaction? (b) Use structural formulas to show how pentene can be changed to pentane by reaction with hydrogen. (c) If propene reacts with chlorine, what are some of the compounds formed?

19. Chemically, how would you distinguish hexane from hexene?

20. (a) Use the concept of π- and σ-bonding to explain why alkenes are more reactive than alkanes. (b) Why are the alkynes more reactive than the alkenes?

21. Show that 1,2-dichloroethene can exist in two isomeric forms while 1,2-dichloroethane does not.

22. Which has a higher boiling point, cis or trans-dichloroethene? Explain.

23. (a) What are polymers? (b) Explain that polyethylene is a solid while ethylene is a gas.

24. Use the concept of σ- and π-bonding orbitals to explain the difference in reactivity between ethane, ethene, and ethyne.

25. How does the structure of aromatic compounds differ from the structures of the members of the alkane and alkene series?

26. (a) What is meant by a resonance hybrid model? (b) What reasons can be given for the validity of a resonance model for the structure of benzene?

27. Draw the structural formulas for (a) ethanol, (b) ethanal, (c) 2-propanone, (d) 1,3,5-trimethylbenzene, (e) 1-methyl-3-nitrobenzene, (f) 1,3-dichlorobenzene, (g) 2,4,6-trinitro-methylbenzene, (h) butanoic acid, (i) 4-aminobenzenecarboxylic acid.

28. Identify these compounds. Give their systematic names and other names if they exist.

(a) $\begin{array}{c} Cl \\ | \\ Cl-C-Cl \\ | \\ H \end{array}$, (b) $\begin{array}{c} Cl \\ | \\ Cl-C-Cl \\ | \\ Cl \end{array}$, (c) $\begin{array}{c} Cl \\ | \\ F-C-F \\ | \\ Cl \end{array}$,

(d) $\begin{array}{c} \;\;H\;\;H \\ \;\;|\;\;\;| \\ H-C-C-OH \\ \;\;|\;\;\;| \\ \;\;H\;\;H \end{array}$, (e) $\begin{array}{c} H \\ | \\ H-C-OH \\ | \\ H \end{array}$,

(f) $\begin{array}{c} \quad H \\ \quad | \\ H\;\;O\;\;H \\ |\;\;\;|\;\;\;| \\ H-C-C-C-H \\ |\;\;\;|\;\;\;| \\ H\;\;H\;\;H \end{array}$, (g) $\begin{array}{c} H\;\;H\;\;H \\ |\;\;\;|\;\;\;| \\ O\;\;O\;\;O \\ |\;\;\;|\;\;\;| \\ H-C-C-C-H \\ |\;\;\;|\;\;\;| \\ H\;\;H\;\;H \end{array}$,

(h) $\begin{array}{c} H\;\;H\quad\;\;H\;\;H \\ |\;\;\;|\quad\;\;\;|\;\;\;| \\ H-C-C-O-C-C-H \\ |\;\;\;|\quad\;\;\;|\;\;\;| \\ H\;\;H\quad\;\;H\;\;H \end{array}$, (i) C_6H_5-O-H (benzene ring bearing O—H),

(j) $\begin{array}{c} H-C=O \\ | \\ H \end{array}$, (k) $\begin{array}{l} \quad\;\; H \\ \quad\;\; | \\ H-C-(OOC-C_{17}H_{35}), \\ \quad\;\; | \\ H-C-(OOC-C_{17}H_{35}) \\ \quad\;\; | \\ H-C-(OOC-C_{17}H_{35}) \\ \quad\;\; | \\ \quad\;\; H \end{array}$

29. What are the products of the mild oxidation of 1-propanol? of 2-propanol?

30. Butane can form primary, secondary, and tertiary alcohols. (a) Write the structural formulas for each of these compounds. (b) How do the three alcohols compare in boiling points and in viscosity? Explain. (c) What advantage would the tertiary alcohol have over the primary alcohol when used as an automobile antifreeze?

31. Show that this sequence of compounds represents an oxidation process.

$$C_2H_6 \longrightarrow C_2H_5OH \longrightarrow CH_3CHO \longrightarrow CH_3COOH \longrightarrow CH_3OH \longrightarrow CO_2$$

32. How would you distinguish between a saturated and an unsaturated acid, both having the same number of carbon atoms in the chain, by (a) physical measurements, (b) chemical properties?

33. (a) Define an ester. (b) Name the acids and alcohols which react and form banana oil, wintergreen oil, and glyceryl stearate.

34. (a) What is a soap? (b) Describe how a soap can be prepared from fat

35. How do soaps differ from synthetic detergents?

36. Explain the cleansing action of detergents.

37. (a) Draw the electron-dot structure of methyl amine. (b) Show that methyl amine can be considered a Lewis base. (c) Explain why methyl amine is soluble in water while methane is slightly soluble.

38. (a) What is an amino acid? (b) Show how amino acids can act as proton donors (acids) or proton acceptors (bases).

39. Use structural formulas to show how amino acids form a peptide bond. Write an equation for the reaction of aminopropionic acid with aminoacetic acid forming a dipeptide.

40. Write an equation which sums up the process of photosynthesis.

41. How do these sugars differ: monosaccharides, disaccharides, polysaccharides, pentoses, hexoses? Give an example of each.

42. Using structural formulas, show that a simple sugar can be a polyhydroxyaldehyde or a polyhydroxyketone.

43. How is the hydrolysis of polysaccharides accomplished in (a) the laboratory, (b) in organisms?

44. Explain why disaccharides and polysaccharides must be hydrolyzed before being used by the body.

45. Both starch and cellulose are polysaccharides. The human body can hydrolyze starch but not cellulose. Explain.

46. (a) Using structural formulas, write an equation to show the formation of sucrose from glucose and fructose. (b) Identify the dehydration and hydrolysis processes.

47. Using the chart on p. 664, identify these amino acids from their chemical names, (a) aminoacetic acid, (b) 2-aminopropionic acid, (c) 2-amino-3-methylbutanoic acid.

48. Use structural formulas and draw (a) a single peptide bond, (b) a polypeptide.

49. (a) What is meant by "denaturation" of a protein? (b) Name three denaturation agents.

50. Show by means of formulas that the digestion of polysaccharides and of proteins is essentially the same chemical process.

51. Distinguish between a condensation polymer and an addition polymer.

52. What reason is there to consider the structure of nylon related to that of proteins?

53. (a) Explain, on the molecular level, the elasticity of rubber. (b) How does vulcanizing harden rubber according to the model used in (a)?

54. (a) Describe the principle upon which molecular spectroscopy works. (b) What energies are involved in molecular spectroscopy?

55. Explain that organic, covalently bonded compounds are likely to have lower melting points than ionically bonded compounds.

56. Analysis shows a hydrocarbon to be composed of 80% C and 20% H. (a) What is its simplest formula? (b) Its density is 1.34 g/ℓ at STP. What is its molecular mass? (c) What is its molecular formula?

57. At STP, 11.2 liters of a hydrocarbon burn and form 22 g of CO_2 and 18.0 g of water. What is the probable formula of the hydrocarbon?

58. When a hydrocarbon is burned, the carbon dioxide weighs $1\frac{5}{6}$ times as much as the water formed. What is the formula of the hydrocarbon?

59. (a) Calculate the enthalpy (heat) of combustion of each of these hydrocarbons. (b) Which hydrocarbon gives off the most heat per gram when burned? Assume $H_2O(l)$ is one product.

Compound	ΔH_f° *(kcal/mole)*
ethane	+20.2
ethene	12.5
ethyne	54.2
propane	−24.8
n-butane	−29.8
$CO_2(g)$	−94.0
$H_2O(g)$	−57.8

60. When 32.0 ml of benzene is reacted with 79.9 g of liquid bromine, 30.0 g of bromobenzene (C_6H_5Br) is formed. (a) What is the theoretical yield of bromobenzene? (b) What is the percentage yield? The density of benzene is 0.879 g/ml. The equation for the reaction is

$$C_6H_6(l) + Br_2(l) \longrightarrow C_6H_5Br(l) + HBr(g)$$

61. When 3.5 g of ethyl alcohol is reacted with $K_2Cr_2O_7$ in sulfuric acid solution, 1.3 g of acetaldehyde is obtained. What is the percentage yield? The skeleton equation for the reaction is

$$C_2H_5OH(l) + K_2Cr_2O_7(aq) + H_2SO_4(aq) \longrightarrow C_2H_4O(l) + Cr_2(SO_4)_3(aq) + H_2O(l) + K_2SO_4(aq)$$

Ans. 39%.

62. Bromotoluene ($CH_3C_6H_4Br$) may be prepared by reacting toluene diazonium sulfate ($CH_3C_6H_4N_2HSO_4$) with a mixture of copper(I) bromide and hydrobromic acid. The equation for the reaction is

$$CH_3C_6H_4N_2HSO + Cu_2Br_2 + HBr \longrightarrow CH_3C_6H_4Br + N_2 + Cu_2Br_2 + H_2SO_4$$

How many grams of bromotoluene are formed by reacting 0.400 moles of toluene diazonium sulfate if the yield is 51.5%?

63. The nitrogen in an amine may be converted to NH_3 which may be converted into nitrogen gas by the reaction

$$2NH_3 + 3OBr^- \longrightarrow N_2 + 3H_2O + 3Br^-$$

If a 0.600-g sample of an amine yielded 25.0 ml of N_2 at STP, what is the percentage of nitrogen in the amine?

21

Nuclear Chemistry

Preview

"Although the achievements of science may, indeed, throw us back into barbarism, the abandonment of our search for knowledge and material betterment would only make vegetables of us."

—Vannevar Bush (1890–   )

For thousands of years man has depended on chemical reactions as a source of heat energy. Unfortunately, there is a limit to the amount of coal, oil, and gas to be found in the earth. During the latter part of the 19th and the first part of the 20th centuries, man has exhausted much of the fossilized fuel reserves which were formed during the course of billions of years. At the present time, these resources are rapidly diminishing, while our need for heat energy is greatly increasing. We are using these sources of energy at 10 times the rate of 100 years ago. This expenditure of our fossilized fuel supplies is analogous to a boy's spending, in a few days, a fortune which his father has amassed in a lifetime of earning.

Fortunately, we now do not have to depend entirely upon the burning of conventional fuels to supply all the energy required by our modern civilization. In 1942, Enrico Fermi and his co-workers made available a new source of energy when they unlocked some of the energy stored in the nuclei of atoms and ushered in the nuclear age.

Nuclear, as well as chemical reactions, are now being used as sources of heat energy throughout the world. In addition to heat energy, nuclear reactions also liberate energy in the form of radioactive rays. The radioisotopes which emit these rays are among the most important tools used by scientists.

In this chapter, we shall investigate the structure and properties of nuclei and then consider different kinds of nuclear reactions. We shall conclude by illustrating the applications of radioisotopes and discussing the hazards of radiation.

THE COMPOSITION AND STRUCTURE OF THE NUCLEUS

21-1 Mass-Energy Equivalence The German chemist, Landolt, late in the 19th century, performed a series of very precise experiments in order to determine whether any mass is lost during a chemical reaction. He repeated several experiments in sealed glass

ALBERT EINSTEIN
1879–1955

To most people, the name Einstein is synonymous with genius. He is probably the most famous scientist of the 20th century. Einstein was primarily a theoretician, not an experimenter. That is, he developed theories and suggested methods for testing them but left the actual experimental work to others.

Einstein is best known for his *theory of relativity* which he developed in 1905 while working in a patent office in Berne, Switzerland. This highly abstract theory relating time, space, mass, and motion played a major role in revolutionizing 20th century physics.

His famous equation, $E = mc^2$, was part of his special theory of relativity. Interpretation of this equation suggests that matter is a form of energy. It is interesting to note that Einstein received the Nobel Prize in 1921 for his discovery of the photoelectric law, not for his theory of relativity.

While on a lecture tour in 1933, he was stripped of his property, position, and citizenship in Nazi Germany. He accepted a position as Professor of Mathematics at The Institute for Advanced Studies at Princeton in the same year. Einstein had little interest in money. When he came to Princeton, he requested a salary of $3,000 per year. Princeton countered with an offer of $15,000. He once used a pay check for a book mark, then lost the book.

In addition to science, Einstein enjoyed playing the violin and was intensely interested in world peace. His famous 1939 letter to President Franklin Roosevelt warned of the possible preparation of nuclear weapons by Nazi Germany. This letter was instrumental in initiating the project which led to the development of nuclear weapons. His last scientific work was a 24-page paper on the unified field theory. This theory was an attempt to relate gravity and magnetism. As yet this theory has not been tested.

tubes, carefully weighing and reweighing to detect any gain or loss in mass. As a result of his experiments, he concluded that *mass is conserved in a chemical reaction.* By the end of the 19th century, chemists and physicists had agreed that both ***matter and energy are conserved in all physical and chemical changes.*** At this time there were two separate physical conservation laws, the ***Law of Conservation of Mass*** and the ***Law of Conservation of Energy.*** The work of the great physicist, Albert Einstein, led to the unification of the conservation laws into a single statement. His famous equation,

$$\mathbf{E = mc^2} \qquad 21\text{-}1$$

where m is the mass in g, E is the energy in ergs, and c is the velocity of light in cm/sec, shows that mass has an energy equivalent. The value of c is 3×10^{10} cm/sec. Thus c^2 is a very large number. This means that *a very small mass is equivalent to a large amount of energy.* Equation 21-1 implies that when a system gains energy, it also gains mass. It has been shown experimentally that the faster a particle moves, the greater its mass becomes. The mass of a stationary particle is called its ***rest mass.*** The mass of a particle in motion is called its ***relativistic mass.*** Nuclear scientists working with high-speed particles must consider relativistic masses. In our study of chemical reactions we were concerned only with the rest mass of substances.

Einstein used the term ***mass-energy*** to illustrate their interchangeability. The ***Law of Conservation of Mass-Energy*** states that *mass and energy can be transformed but that mass-energy cannot be created or destroyed.* Einstein visualized mass as concentrated energy and energy as deconcentrated mass. The equivalency of mass and energy was dramatized with the development of ***nuclear energy*** during World War II. In a nuclear explosion or reaction, a small but measurable amount of mass is transformed into an enormous amount of energy. This type of reaction, in which significant amounts of matter are transformed into energy is known as a ***nuclear reaction.*** The energy is known as ***nuclear*** or ***mass energy.*** Einstein's equation applies to all reactions, chemical and nuclear, but in the case of a chemical reaction, the decrease in mass is so small that it is negligible. For all chemical reactions, we used the laws of conservation of mass and energy separately.

21-2 Binding Energy of the Nucleus Nuclear and chemical reactions differ in several ways. Chemical reactions involve only the outermost electrons of atoms whereas nuclear reactions involve changes in the nuclei of atoms. Another significant difference between the two types of reactions is in the amount of energy liberated. *Nuclear reactions involve energies a million times greater than those of chemical reactions.* In order to understand the nature of nuclear reactions, let us consider the structure of nuclei.

The fundamental particles in the atomic nucleus are protons and neutrons. The number of protons represents the atomic number and is designated by Z. The sum of the neutrons and protons is called the mass number and is designated by A. All nuclei with the same atomic number (Z) are isotopes of the same element.

The nucleus of any atom may be identified by specifying the atomic number and the mass number. A specific nucleus is usually represented by writing the *atomic number* as a *subscript* at the *bottom left of the atomic symbol* and the *mass number* as a *superscript* at the *upper left.* For example, the nucleus of the most common isotope of oxygen is designated as $^{16}_{8}O$.

Earlier we indicated that the use of mass spectrometers enables scientists to determine isotopic masses with a high degree of precision. Using precisely determined masses, we can show that the sum of the masses of the individual particles in the nucleus is not equal to the actual mass of the nucleus. For example, a precise calculation shows that mass is transformed when protons and neutrons combine and form a helium nucleus.

amu **refers to atomic mass unit.**

Mass of 2 free neutrons 2×1.00867 = 2.01734 amu
Mass of 2 free protons 2×1.00728 = 2.01456 amu

Sum of the masses of 2 free neutrons and 2 free protons = 4.03190 amu
Mass of 2 protons and 2 neutrons merged in a helium nucleus = 4.00150 amu
Quantity of mass transformed during merging = 0.03040 amu

The mass transformed as a result of the merging is sometimes called the *mass defect.* Using Einstein's mass-energy equivalence law, $\Delta E = \Delta mc^2$, and the Law of Conservation of Mass-Energy, we can calculate the difference in the masses in terms of an energy equivalent known as the *binding energy* of the nucleus. To do this we must first express the amu in terms of an energy equivalent. By definition, one amu is exactly $\frac{1}{12}$ the mass of a single carbon-12 atom. One mole of carbon-12 atoms has a mass of exactly 12.00 g. Therefore, one atom of carbon-12 has a mass of

$$\frac{12.00 \text{ g}}{\text{mole}} \times \frac{1 \text{ mole}}{6.02 \times 10^{23} \text{ atoms}}$$

and $\frac{1}{12}$ of a carbon-12 atom has a mass of

$$\frac{1}{12} \times \frac{12.00 \text{ g}}{1 \text{ mole}} \times \frac{1 \text{ mole}}{6.02 \times 10^{23} \text{ atoms}} = 1.66 \times 10^{-24} \text{ g/atom}$$

Substituting this value in $\Delta E = \Delta mc^2$, we obtain

$$\Delta E = 1.66 \times 10^{-24} \text{ g} \times 9.00 \times 10^{20} \text{ cm}^2/\text{sec}^2$$
$$= 1.49 \times 10^{-3} \text{ ergs}$$

Converting ergs to electron volts yields

$$1.49 \times 10^{-3} \text{ ergs} \times \frac{1 \text{ ev}}{1.6 \times 10^{-12} \text{ ergs}} = 9.31 \times 10^{8} \text{ ev} = 931 \text{ mev}$$

This means that when a system loses one amu of mass, it also loses 931 million electron volts (mev) of energy. Thus, the binding energy of the helium nucleus is

$$0.03040 \text{ amu} \times \frac{931 \text{ mev}}{\text{amu}} = 28.3 \text{ mev}$$

The binding energy per nuclear particle (nucleon) is a measure of the stability of the nucleus. Binding energy is the energy required to separate the nucleus into individual particles, or conversely, the energy released when a specific nucleus is formed. The greater the binding energy per nucleon, the more stable is the nucleus. The binding energy per nucleon can be calculated by dividing the total binding energy by the total number of nucleons. We have shown that the total binding energy of a helium nucleus is 28.3 mev. The total number of nucleons in a helium nucleus is four. Thus, the binding energy per nucleon for helium is

$$\frac{28.3 \text{ mev}}{4 \text{ nucleons}} = 7.08 \text{ mev per nucleon}$$

By plotting the binding energy per nucleon against the mass number, we obtain the graph shown in Fig. 21-1. This shows the relative stabilities of various nuclei.

Iron, the most stable nucleus, has a mass number of 56. That is, more energy is required to separate two nucleons in the nucleus of an iron atom than for any other atom. The same point is expressed by saying that the formation of an iron nucleus from component particles liberates more energy than the formation of any other nucleus. The graph shows that the binding energy per nu-

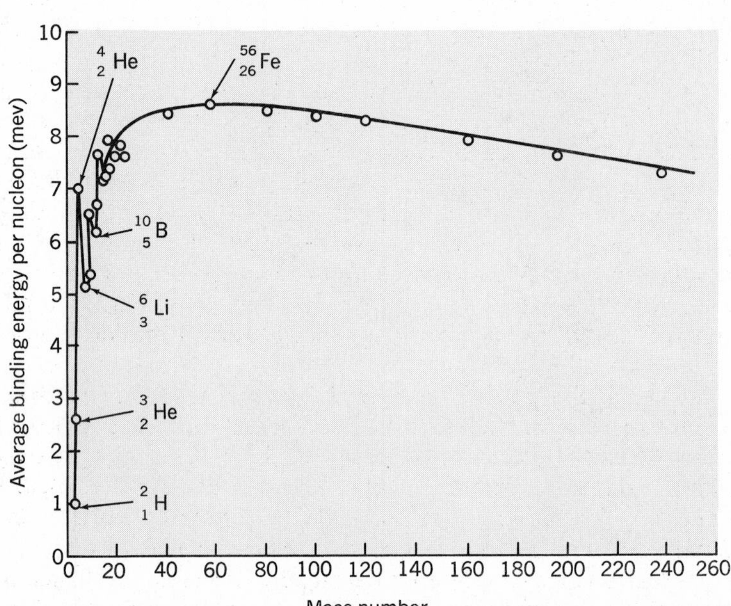

21-1 Binding-energy curve.

cleon in the nuclei of very light elements is much less than of iron and its neighboring elements. Thus, a large amount of energy is released when very light nuclei merge and form nuclei having mass numbers around 56. Nuclear reactions in which small nuclei merge and form more massive nuclei are called ***nuclear fusion*** or ***thermonuclear reactions.*** This type of reaction is responsible for the energy evolved by the sun. Scientists have succeeded in initiating uncontrolled fusion reactions in hydrogen bombs. They have not yet been able to control these reactions so that they can be put to use.

The graph also shows that the nuclei of very heavy elements have binding energies that are somewhat less than those of the intermediate elements. The difference is not as great as in the case of the very light nuclei, thus indicating that less energy will be released when very heavy nuclei split into smaller fragments. This type of nuclear reaction is known as ***nuclear fission.*** Fission reactions were first developed for use in atomic weapons. The energy evolved from fission reactions is now used to produce electricity and to perform other valuable functions.

It can be seen in Fig. 21-1 that $^{4}_{2}He$, $^{12}_{6}C$, and $^{16}_{8}O$ do not lie right on the curve but have high binding energies per nucleon. These nuclei are more stable than their neighbors. The observation that the number of nucleons in relatively stable nuclei are divisible by four suggests that *helium nuclei,* which have 4 nucleons (2 protons and 2 neutrons), may be found as *units in other nuclei.*

One model of nuclear structure draws an analogy to the electron structure of atoms. This model visualizes energy levels in the nucleus analogous to electron shells in an atom. The *nuclear shell model* postulates that 2 protons and 2 neutrons are needed to complete the lowest energy levels in the nucleus. These nucleons are believed to form units of helium nuclei which may be emitted if the nucleus becomes too unstable.

21-3 Nuclear Forces Nuclear forces are powerful, short-range forces of attraction which are responsible for the stability of the nucleus. Because of the close packing of positive and neutral particles, it might appear that forces of repulsion would cause all nuclei to be unstable and, therefore, to disintegrate spontaneously. It is true that any proton in a nucleus repels any other proton with a strong, long-range, coulombic force of repulsion. This force tends to decrease the stability of a nucleus and helps explain the instability of the heavier elements whose nuclei have large numbers of protons.

Opposing repulsion are the powerful short-range forces of attraction between proton and neutron, neutron and neutron, and proton and proton. These are known as *nuclear forces.* They operate only within distances of approximately 1×10^{-13} cm, the diameter of a nucleus. Nuclear forces are not related to electrostatic or gravitational forces and are not described by inverse square law.

Nuclear forces are not well understood. They are a fundamental property of the nucleons. These forces have been described as exchange forces associated with the interaction of matter waves or, in terms of particle language, with the interconversion of protons and neutrons. The conversion of protons to neutrons and vice versa is believed to involve the emission and absorption of intermediate sub-atomic particles called mesons. Mesons and other sub-atomic particles have been observed in radiation-laboratory studies. The origin, role, and organization of these particles constitutes one of the most formidable problems being investigated by nuclear scientists.

21-4 Relationship Between Nuclear Stability and the Neutron/Proton Ratio It is reasonable to suspect that a coulombic repulsion caused by the larger number of protons in the nuclei of heavier atoms might cause them to be unstable. The problem of explaining nuclear stability is a difficult one. It has been suggested that the forces involved in nuclear stability are related to proton repulsion and to the interconversion of neutrons and protons. Therefore a starting place is to see if there is some sort of rela-

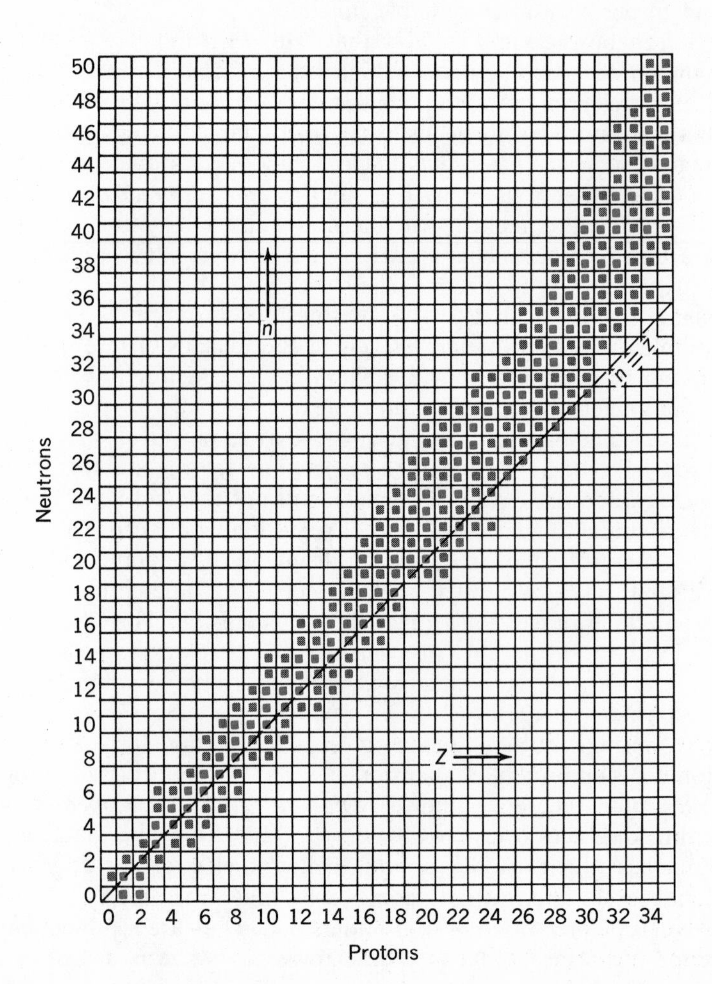

21-2 Distribution of protons and neutrons in nuclei. The colored squares show the region of stable nuclei.

tionship between the numbers of nucleons and the experimentally observed stabilities of nuclei. Some information related to the problem is obtained by investigating the neutron/proton ratio of a large number of isotopes. By plotting the number of protons against the number of neutrons in nuclei, a belt-like graph is obtained, as shown in Fig. 21-2. Because it is not practical to produce such a graph for all the elements in a limited area such as one page of this book, this graph shows only elements with atomic numbers 1 through 35.

Close examination of such a complete graph would reveal a few significant facts (a) stable nuclei, up to and including those in which $Z = 20$, have the same number of protons and neutrons, (b) the number of neutrons is greater than the number of protons for atoms whose atomic number falls between 20 and 83, (c) the nuclei of atoms with atomic numbers greater than 83 are classed as unstable, (d) atoms with even atomic numbers have more stable isotopes than those with an odd number of protons, and (e) many more stable isotopes have an even number of neutrons than an odd number. Detailed investigation reveals that out of hundreds of nuclei there are fewer than 10 stable nuclei which have both an odd number of protons and an odd number of neutrons. This implies that odd numbers of protons and neutrons produce instability. It further suggests that *pairing of neutrons and of protons may yield more stable nuclei.*

21-5 The Nuclear Shell Model By noting relationships between the number and kind of nucleons and the number of stable isotopes, scientists have developed the *nuclear shell model* which assigns certain numbers of nucleons to different energy levels in the nucleus. A detailed analysis of the data suggests that there may be a periodicity in the characteristics of nuclei, related to the nucleon population of the nuclear "shells." There appears to be an unusual nuclear stability associated with the presence of 2, 8, 20, 28, 50, 52, 82, and 126 nucleons (protons plus neutrons). These so-called "magic numbers" seem to be analogous to the noble-gas electron populations of 2, 10, 18, 36, 54, and 86 electrons.

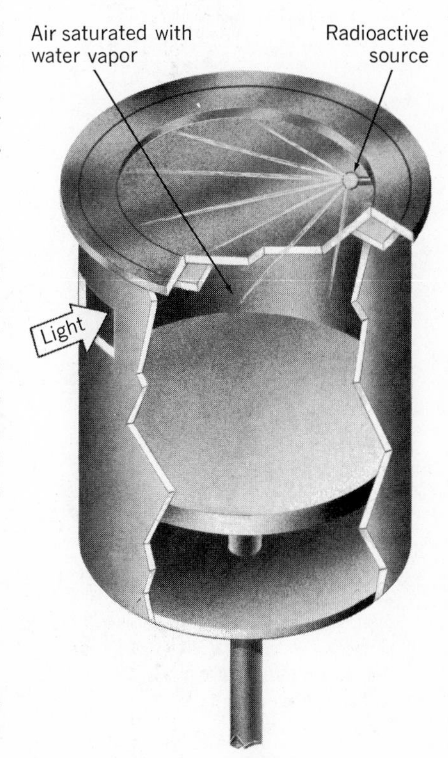

21-3 A cloud chamber is used to study and identify the fragments of disintegrating atoms. When the chamber containing moist air is suddenly expanded, the radioactive particles leave a path of ionized particles as they interact with molecules in the air. Droplets of water condense around the ionized particles, leaving a visible path.

RADIOACTIVITY

Any nucleus whose neutron-proton ratio falls outside the colored region in Fig. 21-2 emits some form of radiation and is said to be radioactive. *Radiation represents a decrease in energy for a nucleus and results in an increase in its stability.* The three types of nuclear radiation which we shall consider are *alpha-particle emission, beta-particle emission, and gamma-ray emission.* These and other particles may be detected by such devices as cloud chambers, photographic film, or bubble chambers. Let us examine each of these types of radiation in some detail.

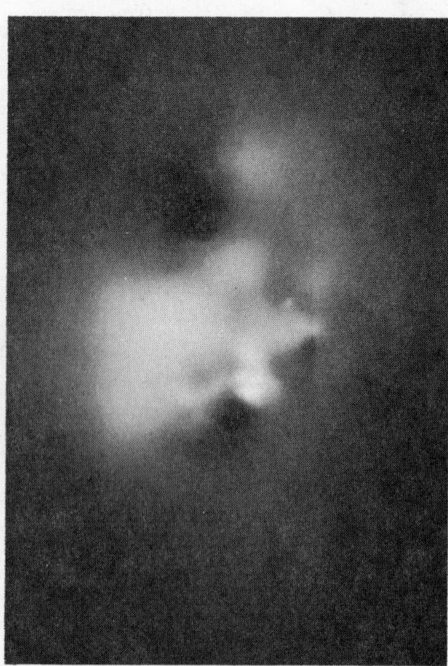

21-4 Photographic film is an excellent detector of radioactive rays. The photographic film was protected from ordinary light by black paper wrapping (left). The radioactive rays from the sample of ore penetrated the paper and exposed the film (right). (Fundamental Photographs).

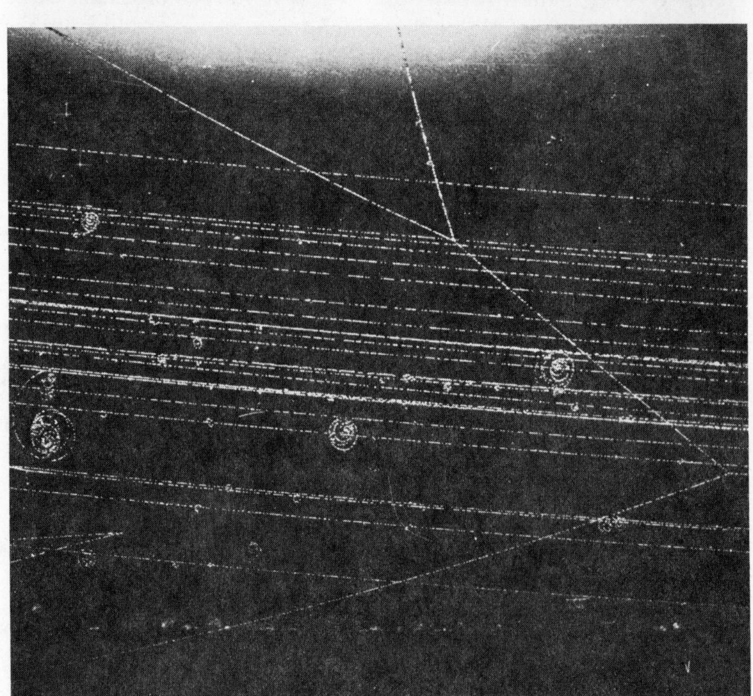

21-5 Tracks made by high-energy protons and other nuclear particles in a liquid-hydrogen bubble chamber. The collision at the upper left produces two charged particles, one of which is involved in another collision at the lower center. (Brookhaven National Laboratory)

21-6 Alpha-Particle Emission and Characteristics Loss of an alpha particle lowers the atomic number of the parent nucleus by two units and the mass number by four units. We have suggested that helium nuclei exist as units in other larger nuclei. A helium nucleus (^4_2He) has a charge of 2+ and a mass number of 4. A heavy nucleus with an atomic number greater than 82 could change its neutron/proton ratio and also decrease in both charge and mass, and thus reach a stable condition by emitting a helium nucleus. The emission of a helium nucleus by a radioactive nucleus is known as alpha-particle decay. This causes the parent nucleus to change into a daughter nucleus having an atomic number of 2 smaller and a mass number 4 smaller than the parent nucleus.

Nuclear reactions are represented by nuclear equations. In a nuclear reaction, the total number of nuclear particles and electric charges is conserved. This nuclear equation represents the alpha decay of radium, forming radon.

$$^{226}_{88}\mathbf{Ra} \longrightarrow {}^{4}_{2}\mathbf{He} + {}^{222}_{86}\mathbf{Rn} \qquad 21\text{-}2$$

Notice that both the superscripts and subscripts in Equation 21-2 balance.

FOLLOW-UP PROBLEM

Uranium 238 decays by alpha-particle emission. Write an equation for this decay reaction.

Alpha particles have high ionizing ability and low penetrating power. Alpha particles are relatively massive radioactive particles that travel at about one-tenth the speed of light. A moving, massive alpha particle easily dislodges electrons from atoms and molecules and produces what is known as ion pairs. An ion pair consists of the electron which is dislodged (the negative half of the ion pair) and the charged particle which remains after an electron has been removed (positive half). It has been estimated that the average alpha particle produces more than 100 000 ion pairs during its existence. The process of ionization robs the alpha particle of energy so that it slows down and finally captures a pair of electrons, thereby becoming a neutral helium atom.

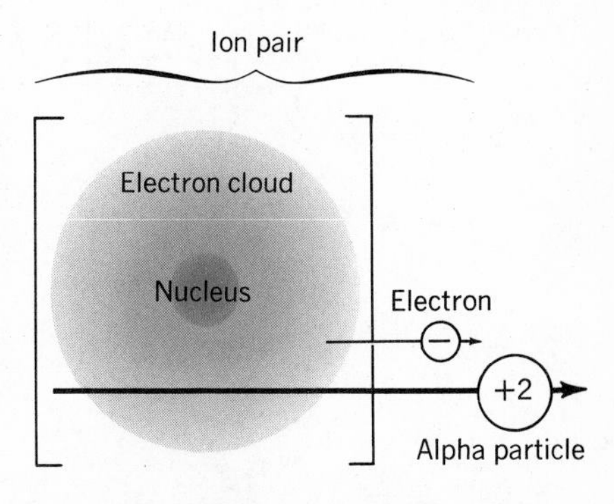

21-6 When a positively charged alpha particle passes near an atom, it attracts an electron from the atom, producing an ion pair.

Because of its relatively large mass and slow speed, an *alpha particle has relatively little penetrating power* compared to other types of radiation. In air, its range may be merely a few centimeters. A sheet of paper or the presumed dead layer of the skin stops it. This means that it does not present any appreciable external hazard. On the other hand, because of its high ionizing ability, alpha radiation presents a serious internal hazard. When working with alpha emitters, special care should be exercised to avoid breathing or ingesting alpha particles.

21-7 Gamma Radiation The emission of an alpha particle leaves the nucleus in a lower energy state. The difference between the energy state of the parent and the daughter nucleus is equal to the energy of the alpha particle. The amount of energy depends on the specific nucleus involved. The daughter nucleus is usually left in an excited state. It can return to a stable ground state by giving off a photon of electromagnetic energy. The energy of the photon is in the gamma-ray region of the electromagnetic spectrum. The rays which emanate from a large number of nuclei undergoing such transformation are known as *gamma radiation.*

Whenever a nucleus is left in an excited state by particle emission or bombardment, gamma rays (γ) are emitted as the nucleus goes from an excited to a ground state. The energy of the gamma ray is equal to the difference in energy between the initial excited state and the final state of the nucleus.

Gamma radiation has a shorter wavelength and is much more energetic than visible light. Since it has *no rest mass* and *no charge,* a gamma ray causes no change in the mass number or the atomic number of a nucleus. *It has less ionizing ability, but far greater penetrating power than an alpha particle.* When a gamma-ray photon passes through matter, it may interact with matter by dislodging electrons from the atoms or molecules. The ejected electrons generally have a considerable amount of energy and, in turn, produce additional ionization. Only those photons which are absorbed by the body produce harmful effects. A certain number of the photons pass through the body without dislodging electrons. Sheets of lead or thick concrete walls are effective absorbers of gamma rays.

21-8 Beta-Particle Emission and Characteristics Nuclei that lie outside the region of stability in Fig. 21-2 can change their neutron-to-proton ratios and become more stable by beta-particle emission. *Beta particles* (β) are high-energy electrons. If the electrons carry a positive charge, they are called *positrons* and are symbolized as ${}^{0}_{+1}e$.

An electron is not one of the fundamental particles of a nucleus. This means that a *beta particle results from the transformation of a neutron into a proton and an electron.* Experiments show that beta particles have a complete range of energies. This means we might expect a corresponding range of energies for gamma rays emitted as the nucleus returns to a stable state. Because a range of gamma radiation is not observed, the energy of a beta particle does not account for the difference in energy between the parent and daughter nuclei. To account for the missing energy, scientists postulated the existence of another particle. This particle, called a *neutrino,* is emitted along with a beta particle when a neutron is transformed into a proton. A neutrino usually is not shown in a nuclear equation.

The process of *beta emission* takes place when a nucleus has an unusually high ratio of neutrons to protons. Any nucleus which emits a negative beta particle becomes a new nucleus with the same mass but with an *atomic number one greater than the original.* This reduces the neutron-to-proton ratio, and the nucleus falls into the region of stability. The beta decay of phosphorus-32 is an example of this process.

$$ {}^{32}_{15}\mathrm{P} \longrightarrow {}^{32}_{16}\mathrm{S} + {}^{0}_{-1}\mathrm{e} $$

When the ratio of neutrons to protons is too low, a nucleus may increase the number of neutrons by transforming a proton into a neutron. This can take place in two different ways.

1. The nucleus can capture an orbital electron from the first quantum level (sometimes called the K shell). When this happens, an electron from the second major energy level (sometimes called the L shell) gives off energy in the form of an X ray as it drops into the lower vacancy. This process, known as *K-electron capture,* reduces the nuclear charge by one unit.
2. If the excess energy of the excited state has the mass equivalent of two beta particles, then the energy can be transformed into two particles, a beta particle and a positron. The negative beta transforms the proton into a neutron, and the positron is emitted. Thus the number of neutrons is increased.

$$^{1}_{1}\mathbf{H} + \mathbf{energy} \longrightarrow {}^{1}_{0}\mathbf{n} + {}^{0}_{+1}\mathbf{e} + \mathbf{excess\ energy}$$

Any excess energy may be carried off by neutrons and gamma radiation.

Unlike an alpha particle, a beta particle has a very small mass. Thus its ionizing power is about 0.01 that of the alpha particle. Also, because of its small size and high speed, a beta particle has a penetrating power of approximately 100 times that of an alpha particle and, consequently, presents more of an external radiation hazard. As it moves along, a beta particle loses energy whenever it ionizes atoms or molecules in its path. At the end of its range, it combines with some positive particle. The range of a beta particle varies with its energy, but in air the range is not more than a few meters. It can usually be stopped by thin sheets of aluminum.

21-9 Rate of Decay The rate at which different substances "decay" is a measure of their relative stabilities. Decay rates can be measured by *Geiger counters* or other counting devices. Measurements show that every radioactive substance decays at a characteristic rate which may be expressed in terms of a property called

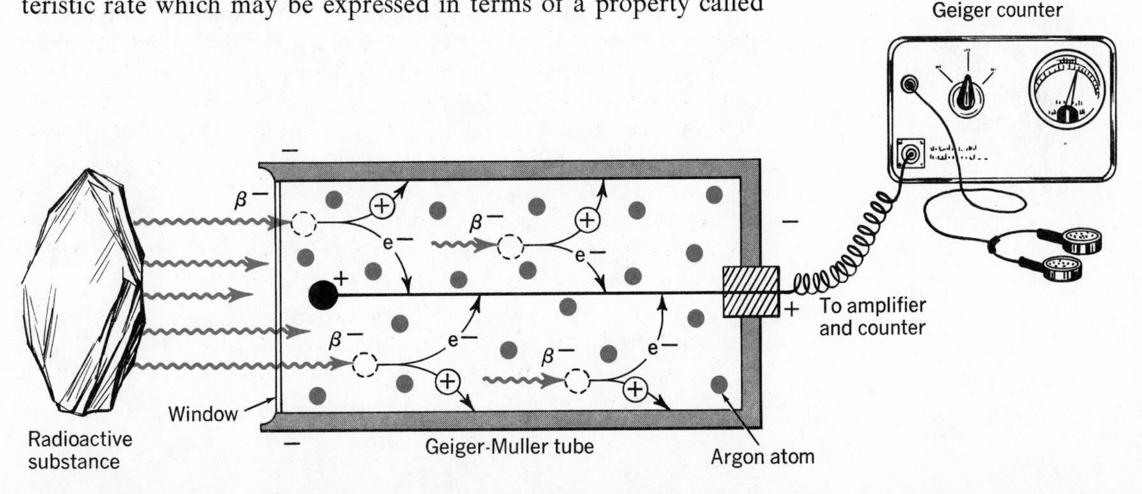

21-7 A Geiger counter. When ionizing radiation enters a Geiger-Müller tube, ion pairs are formed from the gas (argon) in the tube. The positive half of the ion pair is attracted to the negative plate and the negative half (electron) is attracted to the central wire. Each ion-pair collected is registered as an audible click.

its half-life. The *half-life* of an isotope is the time it takes for one-half of the atoms in a given sample to decay. The shorter the half-life, the more unstable the nuclei are. Uranium, with a half-life of 4.5 billion years, is relatively stable compared to an isotope of astatine which has a half-life of only 0.02 seconds. This means that 1 g of U-238 disintegrates at a rate such that a 1-g sample will be changed to 0.5 g of U-238 in 4.5 billion years. On the other hand, the rate of disintegration of astatine-217 is so rapid that it would take only 0.02 seconds for a 1-g sample to be reduced to a 0.5-g sample. Scientists do not work with 1-g samples of astatine-217.

The number of disintegrations per second or minute undergone by a radioactive source is a measure of its activity. One gram of radium emits about 37 billion (3.7×10^{10}) particles per second. This is taken as the basic unit of radioactivity and is called a ***curie.*** A more commonly used unit is the millicurie (37 million disintegrations per second). A radioactive source which is emitting 2×37 billion disintegrations per second is emitting particles at about twice the rate of a gram of pure radium and is said to be a 2-curie source. A high-curie source is analogous to a high-wattage light bulb. A 10-watt bulb is a more intense source of light than a 25-watt bulb, and emits more photons per second than the small bulb.

You might think that a gram of pure radium (1 curie) which emits 37 billion particles per second, would completely disappear in a few minutes or hours. A gram of radium, amazingly, contains over 2.5 thousand billion billion (2.5×10^{21}) atoms. The rate of disintegration slowly decreases until after approximately 1600 years (about one half-life) there would be only a one-half curie source left. The rate of decay after 1600 years would be about 18.5 billion disintegrations per second. Each disintegration results in a *radium nucleus changing into a radon nucleus and an alpha particle* (*helium nucleus*). The disintegration is represented by the nuclear equation

$$^{226}_{88}\mathbf{Ra} \longrightarrow {}^{222}_{86}\mathbf{Rn} + {}^{4}_{2}\mathbf{He}$$

The radon gas which is produced is also radioactive and decays through a series of steps to an isotope of lead ($^{206}_{82}Pb$) which is stable. A number of common isotopes with their half-lives and modes of decay are shown in Table 21-1.

TABLE 21-1
HALF-LIFE AND MODE OF DECAY FOR SELECTED ISOTOPES

Symbol	*Half-life*	*Mode of Decay*
$^{1}_{0}n$	12 min	β^-
H-3 (T-3)	12.3 yr	β^-
C-14	5770 yr	β^-
Na-24	15 hr	β^-
P-32	14.3 days	β^-
Cl-36	3×10^5 yr	β^-
K-40	1.3×10^9 yr	β^+, K electron capture
Fe-55	2.7 yr	K
Co-60	5.27 yr	β^-
Sr-90	28 yr	β^-
I-131	8.05 days	β^-
Pb-210	21 yr	β^-
Pb-214	26.8 min	β^-
Bi-210	5.0 days	β^-
Bi-214	19.7 min	β^-
Po-210	138.4 days	α
Po-214	164 microsec	α
Po-218	3.05 min	α
Rn-222	3.82 days	α
Ra-226	1622 yr	α
Th-234	24.1 days	β^-
Pa-234	6.66 hr	β^-
U-234	2.48×10^5 yr	α
U-236	2.39×10^7 yr	α
U-238	4.51×10^9 yr	α
U-239	23.5 min	β^-

21-10 The Rate Law for Radioactive Decay According to reaction rate principles (Chapter 12), radioactive decay is a first-order process. The rate law for this process is

$$\mathbf{2.3 \log \frac{C}{C_o} = -kt} \qquad \text{21-3}$$

If we let N_o = the number of nuclei in the sample at time t_o and N equal the number of nuclei at time t, then Equation 21-3 becomes

$$\mathbf{2.3 \log \frac{N}{N_o} = -kt} \qquad \text{21-4}$$

Solving mathematically, we can derive an expression for the half-life of a reactant. This expression

$$t_{1/2} = \frac{0.693}{k} \qquad 21\text{-}5$$

may be applied to the disintegration of a radioactive substance.

Example 21-1
(a) Calculate the rate constant, k, for the decay of Sr-90.
(b) What fraction of Sr-90 remains after 5.0 years?

Solution
(a) Obtain the half-life of Sr-90 from Table 21-1 and substitute in Equation 21-5. $t_{1/2} = 28$ yr

$$k = \frac{0.693}{28 \text{ yr}} = 2.5 \times 10^{-2} \text{ (yr)}^{-1}$$

(b) Substitute $k = 2.5 \times 10^{-2}$ yr^{-1} and $t = 5.0$ years in Equation 21-4 and solve for N/N_o.

$$2.3 \log \frac{N}{N_o} = -(2.5 \times 10^{-2} \text{ yr}^{-1})(5.0 \text{ yr})$$

$$2.3 \log \frac{N_o}{N} = 1.25 \times 10^{-1}$$

$$\log \frac{N_o}{N} = \frac{1.25 \times 10^{-1}}{2.3} = 5.4 \times 10^{-2}$$

$$\frac{N_o}{N} = 10^{0.054}$$

$$= 1.13$$

$$\text{fraction remaining} = \frac{N}{N_o} = \frac{1}{1.13}$$

$$= 0.88$$

$$= 88\%$$

FOLLOW-UP PROBLEM

(a) Calculate the rate constant for the decay of Co-60.
(b) What fraction has decayed after 4.0 years? **Ans. (a) 0.131 yr^{-1}, (b) 41%.**

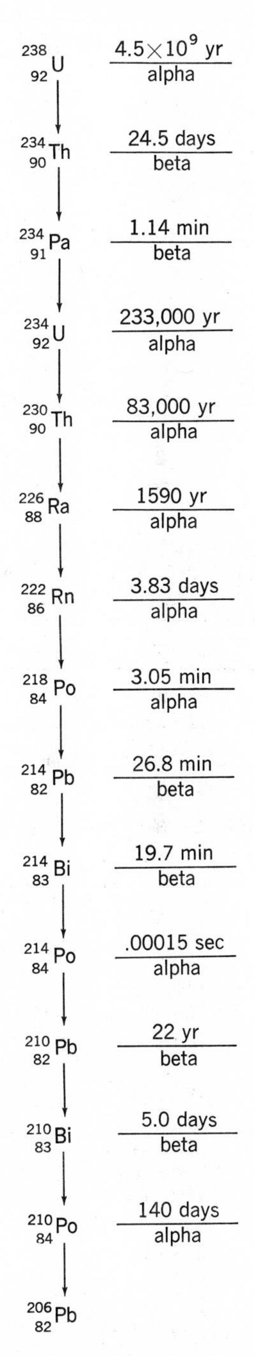

21-8 The uranium disintegration series. Half-lives of the products are given.

21-11 Age Dating Radioactivity can be used to determine the ages of certain minerals and other materials which have existed since prehistoric times. Let us examine the principles underlying this important process. Uranium-238 which has the extremely long half-life of 4.5 billion years, is present in many minerals. All

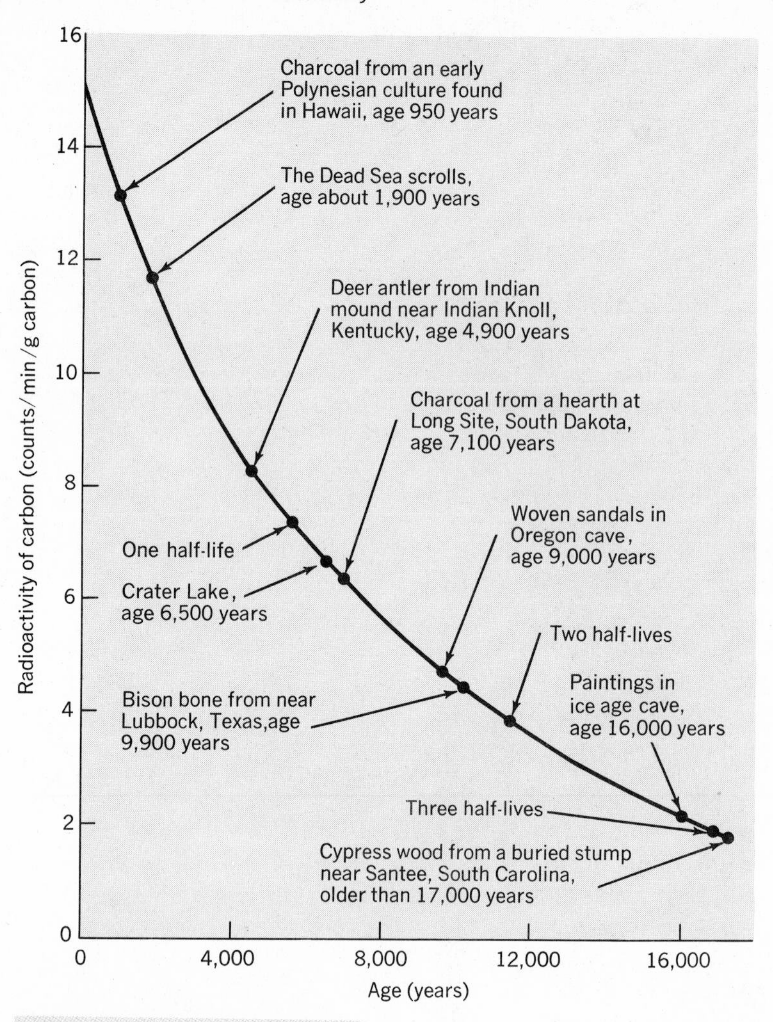

21-9 Half-life curve for ^{14}C. The ^{14}C activity of a preserved object may be related to its age by referring to this graph.

21-10 "Sure hope there's enough radioactive carbon in this paint to last 16 000 years."

naturally-occurring uranium-238 loses an alpha particle and becomes thorium ($^{234}_{90}Th$). The thorium loses a beta particle and becomes protactinium ($^{234}_{90}Pa$). These transmutations represent the first steps in a series of more than a dozen disintegrations which finally end when the stable, nonradioactive isotope, lead-206, is formed (Fig. 21-8). Scientists believe that this is the only way lead-206 is formed in nature. This means that every sample of uranium-238 mineral includes some lead-206. Because the rate of disintegration for each step is known, it is possible to calculate the age of a given uranium mineral by determining the proportion of uranium-238 atoms to lead-206 atoms. Through the use of this method, scientists have calculated that some minerals are about 3 billion years old.

Uranium-238, with its extremely long half-life, is valuable mainly to geologists and to those who are interested in long time intervals, such as the age of the earth. Measurements of the radioactivity in rocks and meteorites have shown that the earth is at least 4.5 billion years old.

The science of radioactive age-dating has proved valuable to historians and archeologists as well as to geologists.

We have all been intrigued by the riddle of the age of prehistoric objects, buried cities, and old documents; the dates at which certain catastrophic events took place; the first appearance of modern man; and many other fascinating problems related to the past.

Radioactive carbon acts as an atomic clock and enables scientists to determine the ages of prehistoric objects and the dates at which prehistoric events occurred. The process is known as radiocarbon dating. Radiocarbon dating has proved useful in many situations. It was by using this process that the Dead Sea Scrolls were shown to be over 1900 years old and, therefore, authentic. The explosion that made Crater Lake in Oregon was calculated through radiocarbon dating to have taken place only 6500 years ago, much more recently than was formerly believed. The same method was used to prove that an Egyptian sarcophagus in one of the reputable museums was only 100 years old and was, therefore, a hoax which had fooled even the experts.

Perhaps the most significant dates obtained by radiocarbon dating are those which reveal the date of the last ice age and the subsequent history of man in the United States. Scientists had previously believed that the glaciers last invaded what is now the United States about 23 000 years ago. Radiocarbon tests of the spruce wood and peat buried under the glaciers' trail show, however, that a spruce forest was killed and buried 9000 years ago.

It was once believed that the Indians had been in North America only a few hundred years before Columbus. The carbon-14 content of woven prehistoric sandals found in an Oregon cave shows that they are over 9000 years old. Other radiocarbon evidence indicates that men were hunting animals with stone-tipped spears in the United States as early as 10 000 B.C., that the temples of the Mayas

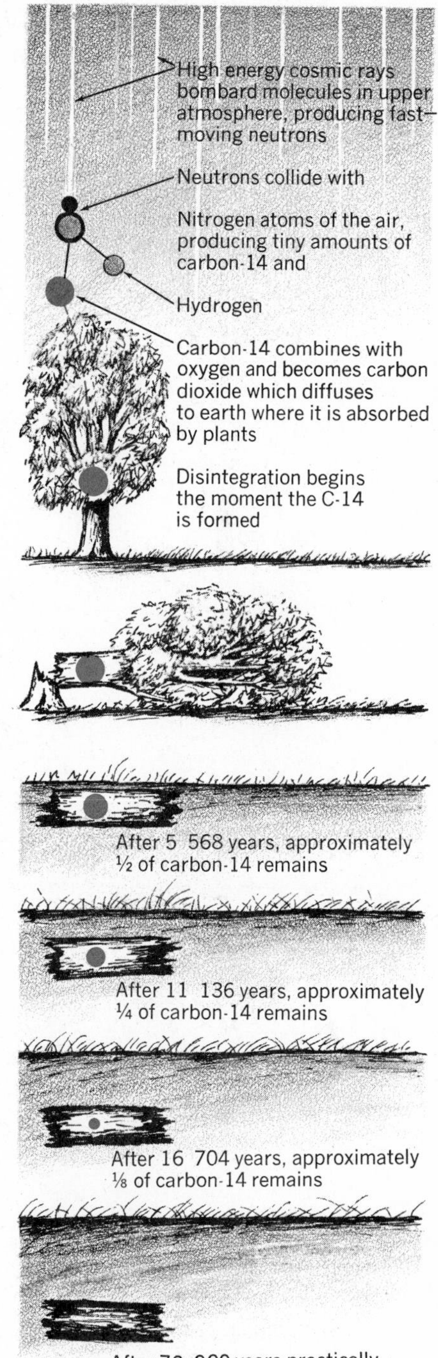

21-11 Formation, absorption, and disintegration of carbon-14.

were constructed about 300 B.C. and that some of the paintings on the walls of ice age caves are about 16 000 years old.

Dr. Willard F. Libby (1908–), a professor at the University of California at Los Angeles, won a Nobel prize for developing the technique of radiocarbon dating. Let us examine the principles underlying the process.

The sequence of events which makes radiocarbon dating possible begins when *high-energy cosmic rays,* which are high-velocity atomic nuclei from outer space, bombard the upper atmosphere producing large numbers of neutrons. These neutrons collide with the nitrogen in the air, changing some of it into carbon-14, as shown in the equation

$$^{14}_{7}\mathrm{N} + ^{1}_{0}\mathbf{n} \longrightarrow ^{14}_{6}\mathrm{C} + ^{1}_{1}\mathrm{H}$$

This carbon-14 is radioactive, with a half-life of approximately 5568 years. It combines with oxygen in the upper atmosphere and forms carbon dioxide. The CO_2 diffuses down to the earth and is absorbed by plants. Radiating carbon-14 atoms become fixed into the plant protoplasm and may be transferred from there to animals.

A gram of carbon from living plant or animal tissue radiates about 15 beta particles per minute. If the plant or animal dies, the intake of carbon-14 stops. By comparing the radioactivity of carbon from present-day plants and animals with that of ancient relics, scientists can very accurately determine their age. If the number of beta particles emitted from a gram of carbon from an ancient Indian campfire is one-fourth that of carbon from a recent plant or animal, then two half-life periods have passed, so the campfire would have burned about 11 136 years ago. Age-dating with carbon-14 is limited in its scope. That is, after several half-lives have elapsed, the activity is so low that precision is lost.

Radioactive age-dating is an important application of nuclear energy which has enabled scientists to discover the answers to many questions related to events which took place during prehistoric times. In the absence of written records, this process provides a means of obtaining important information about the past.

FOLLOW-UP PROBLEM

An object found in a cave has a C-14 disintegration rate of 1.7 disintegrations per minute per g of carbon. How old is the object?

Ans. Approximately 16 800 years.

Note that this decay rate is too slow to give a reliable answer.

21-12 Accelerators and Reactors Radioactivity is the simplest type of nuclear reaction and numerous examples of this reaction have been given in preceding sections. Natural radioactive substances are found in the earth's crust, whereas artificial radioactive substances are produced in nuclear reactors or by scientists using *particle accelerators* such as *cyclotrons, synchrotrons,* and *linear accelerators.* The particles involved in nuclear reactions are listed in Table 21-2.

In cyclotrons and other accelerators, alpha particles, protons, and other atomic "bullets" are given high kinetic energies and are used

TABLE 21-2
PARTICLES INVOLVED IN NUCLEAR REACTIONS

Particle	Charge (atomic number)	Mass Number	Symbol
Alpha particle	2+	4	${}^{4}_{2}He$
Beta particle	1−	0	${}^{0}_{-1}e$
Positron	1+	0	${}^{0}_{+1}e$
Gamma ray	0	0	γ
Neutron	0	1	${}^{1}_{0}n$
Proton	1+	1	${}^{1}_{1}H$
Deuteron	1+	2	${}^{2}_{1}H$

21-12 Particle accelerators such as this cosmotron are used in basic research devoted to determining the fundamental nature and structure of matter. (Brookhaven National Laboratory)

to bombard the nuclei of atoms. Various products may be obtained depending on the energies and particles involved. For example, when phosphorus is bombarded with alpha particles, these reactions may occur

$$^{31}_{15}\mathbf{P} + ^{4}_{2}\mathbf{He} \longrightarrow ^{34}_{17}\mathbf{Cl} + ^{1}_{0}\mathbf{n}$$
$$^{31}_{15}\mathbf{P} + ^{4}_{2}\mathbf{He} \longrightarrow ^{34}_{16}\mathbf{Cl} + ^{1}_{1}\mathbf{H}$$
$$^{31}_{15}\mathbf{P} + ^{4}_{2}\mathbf{He} \longrightarrow ^{33}_{16}\mathbf{S} + ^{1}_{1}\mathbf{H} + ^{1}_{0}\mathbf{n}$$

Some of the products may be radioactive, others not.

21-13 Air view of the Stanford University linear accelerator which is the largest and longest (two miles) in the world. (Courtesy of the Stanford Linear Accelerator Center)

ENERGY FROM THE NUCLEUS

21-13 Nuclear Fission As defined earlier, nuclear fission is a neutron reaction in which a heavy nucleus splits into fragments. The discovery and control of nuclear fission is one of the significant events in the history of science. Space limitation permits only a brief version of the story.

The story of modern man's effort to release and control the energy of the nucleus begins with the discovery of the neutron by James Chadwick in 1932. Soon after Chadwick's discovery, scientists recognized that this neutral particle would make an excellent projectile to bombard the nuclei of other atoms. *Because the neutron carries no electric charge, it is not repelled by positively charged nuclei.*

One of the first persons to recognize the value of the neutron as an atomic bullet was a young Italian physicist, Enrico Fermi (1901–1954). He and his co-worker bombarded practically every known element with neutrons and succeeded in producing a large number of radioactive isotopes. When Fermi bombarded uranium with neutrons and obtained radioisotopes, he incorrectly assumed that the uranium was absorbing a neutron and changing into elements with atomic numbers 93 and 94. His assumption was a natural one since it was known that elements with a high neutron-to-proton ratio emitted beta particles and formed elements with higher atomic numbers. The quantity produced was so small, however, that there was no chemical method available to detect and prove the existence of a new element. Although he was unaware of it at the time, Fermi was probably the first man to achieve nuclear fission, the process which, a few years later, provided a new source of energy and ushered in the Nuclear Age.

Three German scientists, Otto Hahn, (1879–1968), Fritz Strassman (1902–), and Lise Meitner (1878–1969) followed Fermi's work with great interest and decided to repeat his experiments. They finished their experiments just prior to Christmas in 1938 and published their results in *Naturwissenshaften,* a German scientific

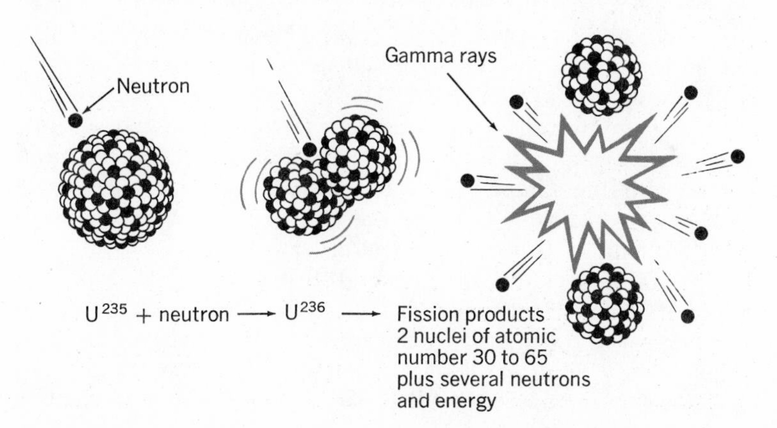

21-14 A U-235 nucleus absorbs a neutron and forms a U-236 nucleus which is highly unstable. This immediately splits into two fragments of roughly equal size and produces several neutrons and yields a relatively large amount of energy in the process.

journal, in January, 1939. They reported that the products of the bombardment of uranium with neutrons were not transuranium elements (those with atomic numbers higher than 92). Rather, they said that the uranium nuclei had split (fissioned), thus forming lighter nuclei. One of the products which Hahn identified was barium, a middle-sized atom. He even doubted the results of his own experiments which, he said, "were at variance with all previous experiences in nuclear physics." Lise Meitner, a refugee from Germany, gave the experiment its proper interpretation. She suggested that perhaps when uranium absorbs a neutron, it splits into two roughly equal fragments. She and her nephew, Otto Frisch, termed the process fission and made a correct guess that the product other than barium was krypton, another middle-sized atom.

These experiments were quickly reproduced and verified by scientists in many countries, who also discovered that uranium fission was accompanied by the release of neutrons as well as large amounts of energy. Scientists were quick to realize the enormous possibilities of this process. With the war clouds gathering in Europe, this new type of nuclear reaction resulted in a great deal of speculation and experimentation by the nuclear physicists. A popular description of scientific effort at the time was, "everyone was working hard, thinking hard, and trying hard to look nonchalant."

For the first time since the discovery of atomic energy by Antoine Henri Becquerel (1852–1908) in 1895, scientists realized the possibility of obtaining and using large amounts of energy from the nucleus. The idea that had excited so many of the scientists was the possibility of producing a self-sustaining chain reaction which would liberate a tremendous amount of energy. Scientists knew that when a neutron is absorbed by a uranium atom the uranium nucleus splits into fragments and releases energy and more neutrons. These neutrons might then split other uranium atoms and again release more energy and more neutrons. This process is known as a ***chain reaction.***

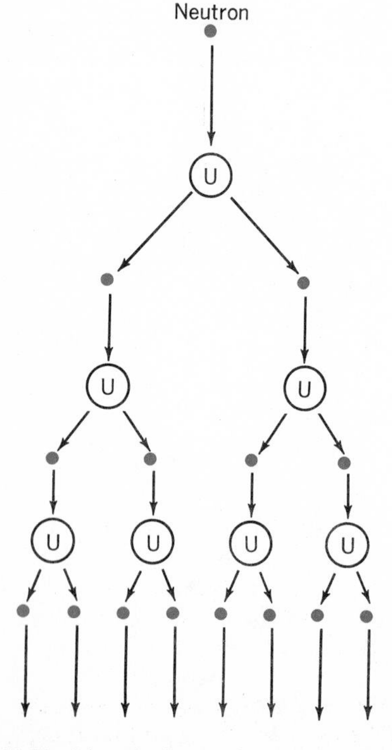

21-15 Schematic diagram of a chain reaction.

You can imagine the tremendous energy released when over a billion billion billion (10^{27}) atoms fission in 50 millionths of a second. Some scientists suggested that the release of this energy might lead to the construction of new and powerful bombs. There was much concern that Hitler might use this source of energy in his attempt at world conquest. The fact that the Germans had forbidden the export of uranium from Czechoslovakia intensified their fears. When Neils Bohr returned to his native Denmark from a visit with the famous German physicist, Werner Heisenberg (1901–), in October, 1941, he was convinced that the Germans were far along in their development of atomic energy. He returned to England and later, with several British scientists, came to America. These scientists urged the United States government to begin intensive research on the problem of obtaining energy from uranium. As a result of a letter written by Albert Einstein to President Franklin D. Roosevelt (1882–1945), a highly secret project was organized in August, 1942, under the code name DSM (Development of Substitute Materials) or as it later came to be called, the *Manhattan Project.* The scientists who began work on the project thought they were in a secret deadly race with the German physicists for the development of an atomic bomb.

Enrico Fermi, who had come to America from Italy, was sent to the University of Chicago to conduct research on the fission project. Fermi knew from previous experiments that U-235 would fission in the presence of slow neutrons. His first problem was to obtain enough U-235 atoms to fission and enough slow neutrons to keep the reaction going. Fermi expressed the problem, "The fundamental point in fabricating a chain-reacting machine is, of course, to see to it that each fission produces a certain number of neutrons and some of these neutrons will again produce fission. If the original fission causes more than one subsequent fission then, of course, the reaction goes. If an original fission causes less than one subsequent fission, then the reaction does not go."

Fermi suggested that this process might be set into action by piling up enough U-235 to obtain a chain-reacting structure. The first problem was to obtain the pure isotope of U-235. Since this was the atom which was to fission, it had to be available in extremely pure form. Because uranium-238 has a voracious "appetite" for slow neutrons, which it absorbs without fissioning, any U-238 present could reduce the number of neutrons to a point where a chain reaction might not take place. The *gaseous diffusion* method described below turned out to be the most efficient method of separating small amounts of U-235 from large amounts of U-238.

21-16 Uranium-235 is separated from Uranium-238 in this gaseous diffusion plant at Oak Ridge, Tennessee. (Courtesy A.E.C.)

ENRICO FERMI
1901–1954

Immediately after the discovery of the neutron by Sir James Chadwick of England in 1931, scientists throughout the world began experimenting with this new atomic particle. One of the most enthusiastic experimenters with neutrons was Enrico Fermi of Italy. In 1934, his experiments led him to suggest that, when a massive nucleus such as uranium absorbed a neutron, it subsequently lost one or more electrons (beta particles) and became a new element with an atomic number greater than 92. His research on radioactivity won him a Nobel prize in physics in 1938. After going to Sweden to accept the Nobel prize he never returned to fascist Italy. He came to the United States the following year where he became a professor of physics at Columbia University. In 1942, he went to the University of Chicago where he built and operated the first nuclear reactor. The successful operation of this reactor signalled the beginning of the nuclear age.

Fermi's research led him to the discovery of element 93, now called neptunium, and he postulated the existence of an atomic particle called the neutrino. The neutrino, which he proposed in order to account for conservation of spin during radioactive emissions, has since been identified.

The story of Fermi's life and work is told in the book, *Atoms in the Family*, written by his widow, Laura Fermi.

The first step in the separation process is to combine the uranium with the element fluorine and form a compound, uranium hexafluoride (UF_6). When heated, UF_6 vaporizes. The vapor is then forced through a series of porous barriers. The molecules of UF_6 containing U-235 atoms are three atomic mass units lighter than those UF_6 containing U-238 atoms. Thus, the UF_6 containing the U-235 isotope passes through the barriers a little more rapidly than the UF_6 containing the U-238 isotope. To provide enough U-235 to produce an atom bomb, a very large diffusion separation plant was constructed at Oak Ridge, Tennessee. This plant was so large that its operation consumed more electric power than the entire country of France during the war years. The plant had to be wired with silver, borrowed from Fort Knox, because there was not enough copper available.

Before Fermi and his associates could begin constructing their atomic pile (furnace) they needed, in addition to the pure uranium-235, a large quantity of graphite to act as a moderator. A *moderator* is used to slow down fast neutrons. It consists of rather densely packed, small atoms which act as nuclear shock absorbers. Each time a neutron collides with a moderator atom, it loses energy and slows down until it is traveling slowly enough to be captured by a U-235 nucleus. The moderator also serves as a reflector, thus conserving neutrons which might otherwise be lost. The material used as a moderator must be pure because impurities can absorb neutrons and prevent a chain reaction.

By 1942, Fermi had obtained six tons of uranium oxide enriched with U-235 and enough pure graphite to begin constructing the pile. The uranium oxide and graphite where piled in alternate layers in the shape of a great door knob eight feet high. Fermi remarked that handling the enormous quantity of graphite made the scientists look like coal miners by the end of the day. They finally hired a dozen husky students from the University of Chicago to complete the pile.

Since there was little government financing of the project at this point, one of Fermi's associates, Leo Szlizard (1896–1964), through a personal loan, enabled Fermi's group to purchase an aspirin-sized piece of radium for preparing a radium-beryllium neutron source needed to initiate the chain reaction. *Cadmium control* rods were inserted in the pile at various points. The element cadmium is an excellent absorber of neutrons, thus the rods when pushed into the reactor, acted as a neutron screen, allowing only a few neutrons to pass through. The number of free neutrons was controlled by inserting or removing the rods.

Work on the pile was completed by December 2, 1942. Final preparations were made for the experiment which would verify, or prove false, Fermi's prediction regarding a chain reaction. Neutron counting and recording equipment was placed in and around the pile. The control rods were removed from the pile and the radium-beryllium neutron source was placed into position for "kindling" the reaction. The critical moment had arrived.

The following telephone conversation between Dr. Arthur Compton, (1892–1962), the Manhattan Project leader in Chicago, and his friend, Dr. James Conant (1893–), on the East Coast, is a clue to the result of the experiment. "Hello," said Compton, "the Italian navigator has landed in the New World." Conant's reply was, "How were the natives?" Compton answered, "Very friendly." This dialogue was Dr. Compton's secret way of telling Conant that Fermi had succeeded in establishing a nuclear chain reaction.

The pile, at first, put out only one-half watt of energy, not enough to light even a flashlight bulb. Two weeks later, however, it was operating at about enough energy to light two 100-watt light bulbs. This event was really the beginning of the nuclear age. It was the first time man had "set a nuclear fire," or released energy which was not derived from the sun. As you know, atomic energy was first used in constructing atomic bombs.

Nuclear reactions occurring in the cores of nuclear reactors are a practical source of energy for our modern civilization. After World War II had ended, many people speculated on the possibility of atomic power for peaceful uses. Glowing predictions were made that this source of energy could be used to melt the snow as it falls, or to warm the water in the Arctic Ocean and convert the Arctic shores to tropical resorts. Others predicted that automobiles and trains could be operated for years on a pellet-sized piece of uranium fuel. Most of the predictions have turned out to be wishful thinking, at least for the present.

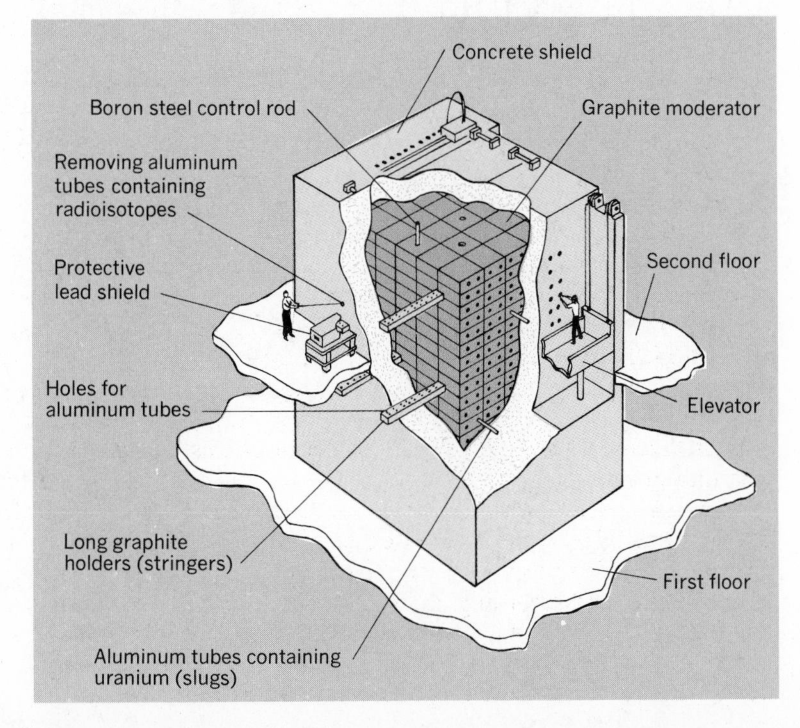

21-17 A cutaway view of the reactor at Oak Ridge, Tennessee. This reactor is used for making radioisotopes.

Today nuclear power reactors are still in their infancy. The energy in a nuclear reactor is obtained from the nuclei of fissionable isotopes, such as U-235, U-233, or plutonium-239, which is made from the more plentiful U-238. The reaction chamber of a nuclear reactor contains a neutron source and stainless steel cartridges filled with uranium or plutonium. These cartridges are embedded in or surrounded by some material such as heavy water or graphite, which slows down the neutrons so that they will be more easily absorbed by the fissionable metals. When the neutrons are absorbed, the fission of heavy nuclei produces fast neutrons as well as energy. When fast neutrons are slowed down and absorbed by other heavy nuclei, additional neutrons are released and a chain reaction is set up. Control rods are placed throughout the core of the reactor. The tempo of the chain reaction is controlled by inserting or removing the control rods. Heat from the core of the reactor is transferred by a heat-exchange system and used to convert water to steam. Steam is used to spin turbines coupled to generators which produce electric energy. In some parts of the United States 30 percent of the electric energy comes from nuclear reactions.

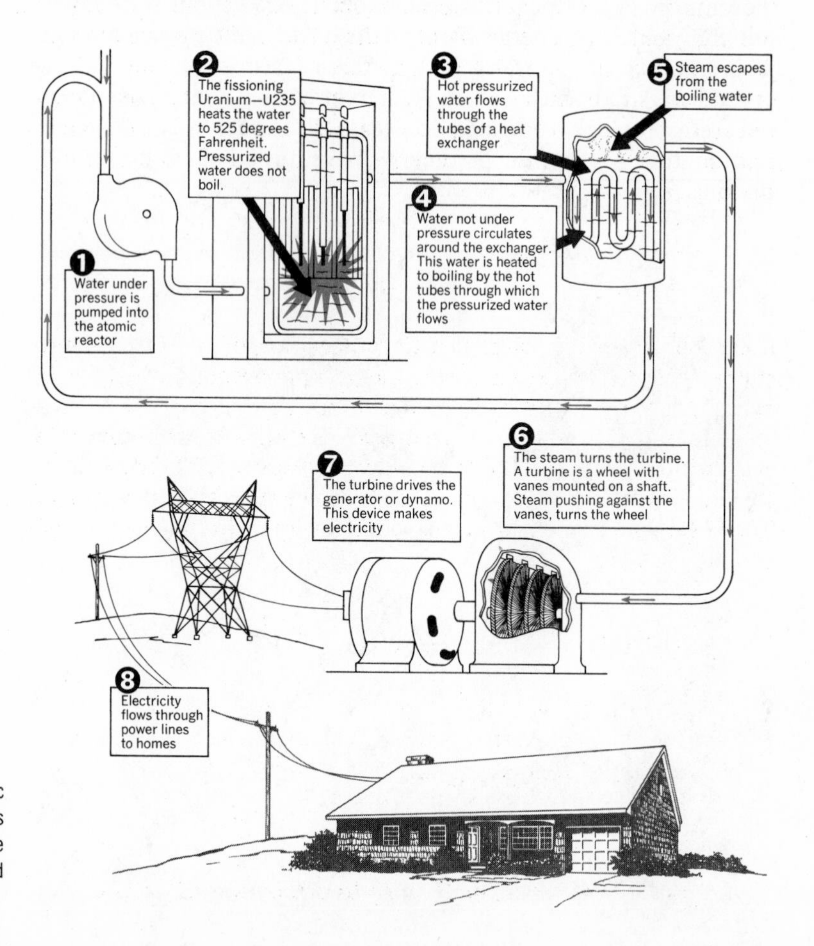

21-18 Schematic diagram showing how atomic energy is used to produce electric energy. Ten tons of slightly enriched U-235 can produce the same amount of electric energy as would be generated by burning 300 000 tons of coal.

21-14 Nuclear Fusion or Thermonuclear Reactions. Even before the discovery of radioactivity, man marveled at the seemingly limitless source of energy from the sun. It was known that this supply of energy could not be produced by ordinary burning, or the sun would have long ago reduced itself to cold ashes. Sir Arthur Eddington (1882–1944) first suggested in 1920 that the sun might produce all of its energy by fusing light nuclei into more complex ones. With the help of the Einstein equation $E = mc^2$, he reasoned that this process would produce a million times more heat per atom than any ordinary burning process.

In 1930, Dr. Hans A. Bethe (1906–), an American scientist from Cornell University, first gave a clear explanation of how this fusion process takes place in the sun. He developed the theory that the hydrogen in the interior of the sun, under the influence of the sun's strong gravitational field, is squeezed together into a paste which is many times denser than lead. Under this great pressure and at a temperature of several million degrees Celsius he postulated that a net of four protons are forced together, forming a single helium nucleus as a result of a series of nuclear processes called the carbon cycle (see chapter-end Question 20). The fusion of protons yields energy equivalent to the difference between the sum of the masses of the individual particles and that of the nucleus or nuclei and particles they form. Let us calculate the energy released when four hydrogen atoms are fused and form helium and compare this with the energy evolved during an energetic chemical reaction.

The net equation for the nuclear fusion reaction is

$$4\,{}^{1}_{1}\mathrm{H} \longrightarrow {}^{4}_{2}\mathrm{He} + 2\,{}^{\;\;0}_{+1}\mathrm{e}$$

It can be shown that there is a transformation of 0.0282 amu into energy when 4 hydrogen atoms merge and become one mole of helium, or 4.03188 g of hydrogen have been converted to 4.00259 g of helium and 0.00110 g of positrons. The mass transformed into energy is equivalent to 4.03188 g minus 4.00369 or 0.0282 g/2 moles of H_2. One mole of H_2 completely converted shows a loss of 0.0141 g. Substitution of these data in the Einstein equation, $E = mc^2$, yields

$$\begin{aligned}
E &= 0.0141\,\frac{\mathrm{g}}{\mathrm{mole}} \times \left(3.00 \times 10^{10}\,\frac{\mathrm{cm}}{\mathrm{sec}}\right)^2 \\
&= 12.7 \times 10^{18}\,\frac{\mathrm{g\ cm^2}}{\mathrm{sec^2\ mole}} \\
&= 12.7 \times 10^{18}\ \mathrm{ergs/mole} \\
&= 12.7 \times 10^{18}\,\frac{\mathrm{ergs}}{\mathrm{mole}} \times 2.4 \times 10^{-11}\,\frac{\mathrm{kcal}}{\mathrm{erg}} \\
&= 3.0 \times 10^{8}\ \mathrm{kcal/mole\ of\ hydrogen\ converted\ to\ He.}
\end{aligned}$$

Comparison of this value with the 60 kcal of energy liberated by the chemical reaction

$$H_2 + \frac{1}{2}O_2 \longrightarrow H_2O$$

reveals that a nuclear fusion reaction is millions of times as energetic as a chemical reaction. Calculations will show that in order to produce its existing energy output, the sun is converting matter into energy at the staggering rate of four million (4×10^6) tons per second.

The first fusion reactions carried out by scientists were those in the hydrogen bomb. To attain the high temperatures required for fusion, an "ordinary" fission-type atomic bomb supplied the "activation" energy. Many problems must be solved before fusion power becomes a practical source of energy for peacetime uses. In brief, some way must be found to contain and concentrate enough light nuclei at temperatures of several million degrees. Once controlled, the unlimited supply of hydrogen in the oceans of the world promises man an almost endless source of energy.

RADIOISOTOPES

21-15 Characteristics and Applications of Radioisotopes Some of the most important applications of nuclear energy involve the use of *radioisotopes* as *tracers.* The atoms of almost any element can be made radioactive by subjecting them to neutron bombardment in a nuclear reactor. Radioisotopes are chemically identical to their nonradioactive counterparts, but in reactions, their paths can be followed and their fates determined.

When injected into plants, animals, or industrial liquids, radioactive chemicals can be traced, using radiation detectors. Medical research workers have found that radioisotopes offer an extremely sensitive tool for exploring the inner working of the human body. Experiments are sometimes performed by students as an introduction to tracer techniques. The object of one part of such an experiment is to investigate the uptake of radioactive phosphorus by a tomato or tobacco plant.

21-19 A dahlia bloom exposed to gamma radiation has developed a somatic mutation. Note that one-half the petals of the normally red flower are white after exposure to cobalt-60. (Brookhaven National Laboratory)

The great usefulness of radioisotopes stems from the fact that these atoms can be detected in *low concentrations* or *trace quantities.* Using radiation detectors, as little as a billionth of a billionth of a gram (10^{-18} g) of some isotopes can be detected. In fact, one hundred thousandths of a gram (10^{-5} g) of carbon-14 can be accurately measured when spread through the tissues of 20 000 guinea pigs. Even with the powerful magnification of the electron microscope, you cannot see such small quantities of matter.

Radioisotopes are used to help unravel fundamental processes of growth and solve many problems in medicine, agriculture, and industry. Although agriculture is one of man's oldest occupations, the application of scientific principles to aid in its progress is only beginning. Many of the complex and difficult problems in agriculture, like those in medicine, have to do with the fundamental processes of growth. What minerals and organic nutrients do plants need? How do plant roots pick them up, and how are they used? What is the mechanism of photosynthesis, the little-understood process of nature that accounts for all the world's food and most of its fuel?

Radioisotopes are helping to provide some of the answers to these questions. At agricultural research farms, radioactive fertilizers are used to measure how efficiently the fertilizer is used by growing plants. You know that plants grow well on soil which is fertile, and that an important agricultural problem is the replenishment of depleted and overworked soils with fertilizers. Phosphorus-32 has been used more than any other radioisotope in studies on fertilizers. The movement of phosphorus in the soil has helped determine the best way to apply commercial fertilizers.

The most fundamental problem in agricultural studies is the riddle of photosynthesis. How does the plant combine carbon dioxide and water, in the presence of sunlight, and produce sugar and starches? Although it may appear to be a very simple chemical process, scientists have not yet reproduced the process in the laboratory. However, investigators are achieving a more detailed understanding of photosynthesis through the use of radioisotopes. In a typical experiment, carbon dioxide is tagged with carbon-14, fed to plants, and traced through a complicated chain of intermediate products in order to help unravel the mechanism of the reaction.

An economic benefit of radiation is currently saving livestock growers in the Southeast millions of dollars annually. All livestock in this area have long been plagued by screw-worms which are the larvae of the blow fly. These pests cause their damage by feeding on flesh of warm-blooded animals. One characteristic of radiation is that it can inhibit fertility in living organisms. This is one of the hazards of radiation to man. In the case of destructive pests such as screw-worms, however, this hazard becomes a useful benefit. An eradication project to eliminate the blow fly started in January, 1958. Millions of male blow flies were reared under controlled conditions, sterilized by exposure to cobalt-60, and released from planes. The sterilized males mated with females who then produced infertile eggs. Since the female blow fly mates only once during her life, only a few years were required to eliminate the pest.

Industry also makes widespread use of radioisotopes. More than 2000 companies in the United States are licensed to use radioisotopes in their research facilities and industrial processes. The isotopes are used for detecting internal flaws in weldings, as thickness gauges, and in studies related to the wear of engines, tires, and tools. These are but a few of the thousands of uses that could be cited to show the importance of radioisotopes as a scientific tool.

RADIATION EXPOSURE

Sunburn is actually a form of "radiation sickness" where only the surface is involved. In some cases of overexposure to X rays, the resulting damage appears to be quite like a sunburn, except that the damage is all the way through the body rather than only on the surface.

21-16 Sources and Types of Radiation Exposure Although you may not have the opportunity to perform experiments using radioisotopes or make a career of working in a nuclear-science laboratory, it is still just as important to know about the hazards of radiation as, for example, the dangers of electricity.

You may find it difficult to believe that, of most of the environmental hazards with which you live, radiation is by far the best understood. Current studies of smog, the use of insecticides and food additives, and on other problems, have resulted in only limited understanding of their effects on people.

One of the reasons for our extensive knowledge about radiation stems from the fact that it has been observed and studied since as early as 1896. In fact, you can read about one of the first reports of external radiation damage in Marie Curie's (1867–1934) biography of her husband, where she describes how he voluntarily placed his arm near a radium source for about ten hours. This resulted in a lesion resembling a burn which developed progressively and, after 20 days, formed a scab which required several months to heal.

An early example of internal radiation damage, that is, damage caused by the ingestion of radioactive materials into the body, is the case of the radium-dial painters. Beginning in 1922 and continuing to 1931, a group of workmen used camel-hair brushes to apply material containing radium to luminous watch dials and other articles. Because they frequently moistened their brushes by mouth, small amounts of radium compounds were consequently swallowed. Because radium tends to concentrate in the bone marrow, the tragic result was great internal damage.

It has also been known for some time that inhalation of a radioactive substance can cause cancer of the lungs. Before World War II, in the famous uranium mines of Schneeberg and Joachimsthal, the radon concentration in the air was about 30 times the maximum tolerable concentration. Lung cancer resulted for about 50 percent of the miners. The average latent period (between exposure and appearance of symptoms) was 17 years.

We are all exposed to natural radiation from outer space as well as from the earth's crust. Within the make-up of your own body there are radioactive elements such as potassium, radium and radioactive carbon. As opposed to man-made radiation, exposure to natural radiation has been inescapable and man has lived with it from the beginning of time. There are many sources of ionizing radiation that you encounter during your lifetime. These include cosmic rays, minerals, radioactive atoms in the body, "radium-dial" watches, chest and dental X rays, fluoroscopic examinations, and fallout from nuclear tests. As a citizen of the nuclear-space age, you should know what biological effects may result from radiation exposure and be aware of the tolerable limits of exposure. Also, you should be acquainted with methods of minimizing exposure,

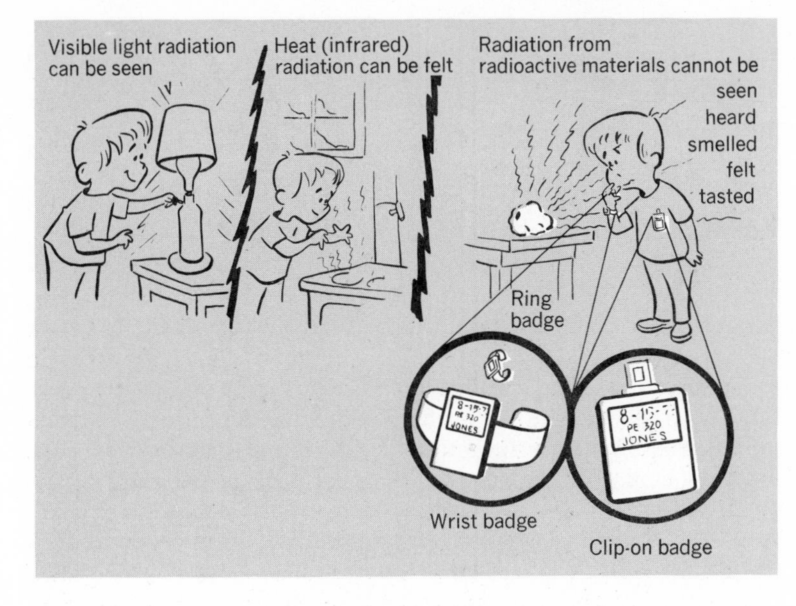

21-20 Because high-energy radiation cannot be sensed, atomic-energy workers and X-ray technicians wear film badges to detect and measure radiation. The sensitized films are wrapped in opaque paper, to keep out light, and are set in frames, so that some of the film is covered by a thin metal shield, while the remainder is behind a "window." Most of the beta radiations are absorbed by the metal, but the gamma rays penetrate and affect the photographic film. The film is developed and from the blackening observed in the different parts of the film, the extent of exposure to various radiations can be accurately estimated.

and know how exposure to radiation compares with other hazards encountered during your lifetime.

21-17 Methods of Minimizing Radiation Exposure Many demonstrations show that without instruments, man cannot detect the presence of radioactivity. It is unwise to expose yourself to material you suspect may be radioactive. Such material should be approached only with proper instruments. Various methods can be used to minimize radiation exposure. Some principal ones are

1. Maintain maximum *distance* from the source. The radiation exposure is inversely proportional to the square of the distance from the source. Thus, the dosage received at three meters is only one-ninth of that received at one meter. Distance is the best protection from a radioactive source.

2. Minimize the *time of exposure.* Discounting the radiation you receive from natural sources, the dose you may receive from a single artificial source is proportional to the time you spend near the source. Of course, the intensity of the source must be considered. The body can adjust to a great amount of exposure if it is spread out over a long period of time. The situation is analogous to a person sunbathing for a single ten-hour period as opposed to twenty half-hour periods on separate days. For most people, ten consecutive hours of sunbathing would result in a serious sunburn. People who work with radioactive materials keep an accurate record of their exposure and limit it.

3. Maximize the amount of *shielding* material between yourself and the source. The nature of the absorbing material is not critical. Because radioactive rays interact with electrons, the best shielding materials are those whose atoms furnish a large supply of electrons. Cost and convenience are usually the factors determining the kind

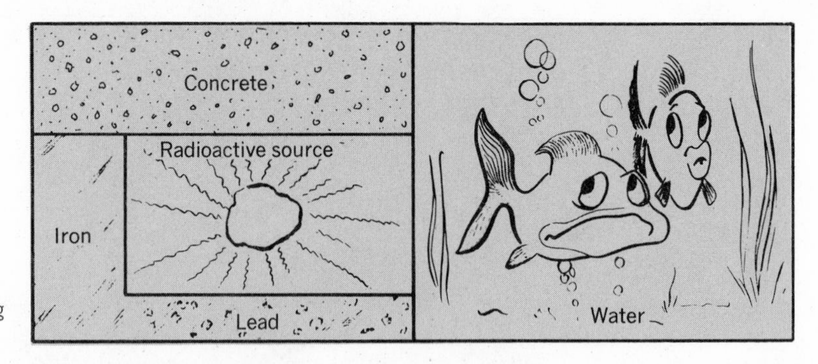

21-21 Relative efficiency of various shielding materials.

of material used. In controlled laboratory experiments, lead is often used as a shield because it offers a tightly packed, dense source of electrons. When testing nuclear explosives, scientists often use earth, concrete, or ordinary building materials as shields.

21-18 Radiation Dosage As mentioned earlier, electrons may be dislodged from atoms and ion-pairs may be formed when radiation passes through matter. A convenient unit of radiation dosage which is called the ***roentgen*** is defined as the amount of radiation that forms 2×10^9 (two billion) ion-pairs upon passing through one cm^3 of dry air. As of this date the upper permissible dosage set by the United States Atomic Energy Commission for its workers is 15 roentgens per year. Most scientists believe that a dose of 500 roentgens would be fatal to about one-half of the population. Although radiation in large doses is harmful to all living organisms the degree of harm varies with the individual.

Let us see how the maximum permissible dosage compares to the dosage you may receive from environmental sources. Just as you are almost constantly exposed to colds, you are likewise exposed to background radiation. Although a rock may not sneeze on you, it does emit radiation as does the soil beneath your feet. In fact, you receive about 0.1 roentgen or more per year from this source by just being a citizen, living in the United States. Background radiation varies, depending on your location. For example, if you live in a brick house rather than a wooden one, you might receive 0.04 roentgens per year instead of 0.01 roentgen. In changing your elevation from sea level to a mile-high altitude, increases the radiation dosage about 50 percent. There are many other unexpected sources of radiation. For instance, an airplane instrument panel with about 100 luminous dials may give the pilot about 1300 milliroentgens per year. Also, the average person receives about 2 milliroentgens per year from other people, particularly in a dense crowd. X rays account for about 0.1 roentgen per year.

Maximum permissible concentration is defined as a dose of ionizing radiation that, in the light of present knowledge, is not expected ever to cause appreciable bodily injury to a person.

21-19 Biological Effects of Radiation Exposure The biological effects of radiation may be divided into two categories, (a) hazards of radiation to the individual, and (2) hazards of radiation to the race. The effects of radiation exposure to the individual, broadly

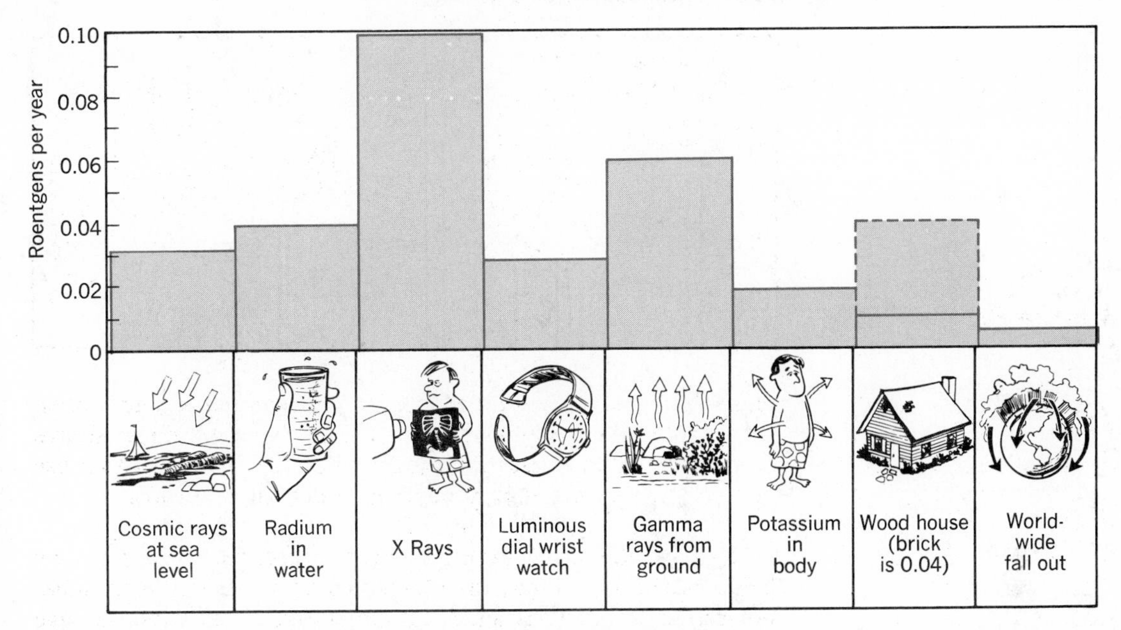

21-22 Average radiation dosage per year, obtained from selected environmental sources.

speaking, fall under two categories, short-term (early) effects and long-term (late or delayed) effects. The early effects appear within weeks to months after exposure whereas the late effects take years to decades to appear. For prolonged exposure at low dosage there will be no early effects. The nature and magnitude of delayed injury depends on many factors, such as the type of radiation, the quantity and duration of dosage, the age and general condition of the individual, and whether the exposure is for the whole body or parts of it. Short-term effects are radiation sickness, severe beta radiation burns, and nausea which may be followed later by loss of hair, and lesions. In the case of acute exposure there may be late effects in addition to early effects. The most hazardous of the late effects appear to be cancer, leukemia, and general shortening of life.

The process by which radiation inflicts damage is still largely and basically obscure. Its study is one of the most fundamental and fascinating of current scientific problems. The first step in a series of complex events is the ionization of the molecules of which cells are composed. As a result of ionization, the molecules undergo disruption. If the dosage is large enough, the cell may die in the process of division.

A small nonlethal dose may cause the cell to undergo a change (mutation). Mutations in body cells (somatic mutations) will probably affect the individual but if they take place in a germ cell (genetic mutation) the new characteristic will probably appear in later generations. The majority of mutations are detrimental.

Hazards of radiation involve genetic effects which show up in future generations. Genetic effects are produced only if radiation strikes the reproductive organs. It is generally agreed by geneticists that any amount of radiation exposure is a genetic risk, and that

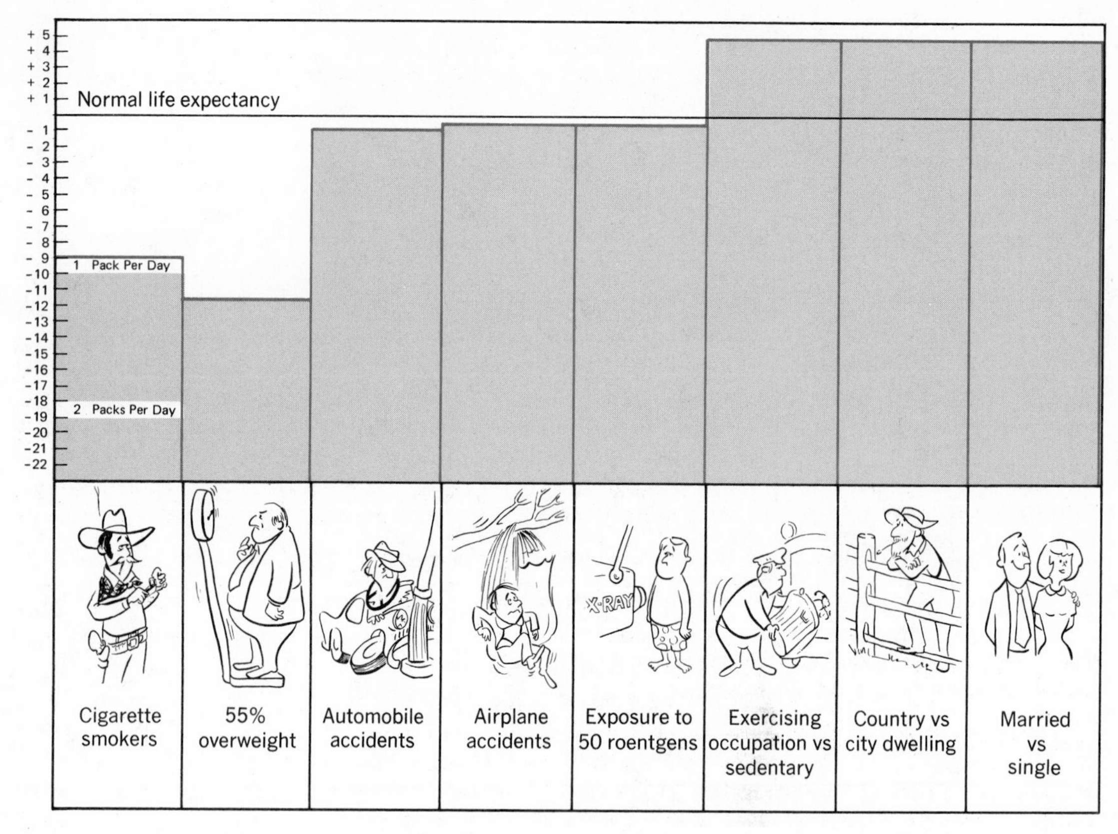

21-23 Effects of various agents and factors on life span.

genetic effects are cumulative and apparently depend only upon the total dosage received by the germ cells. Smaller dosages, of course, lower the chance of mutation, but even a minor mutational damage is a hazard.

It has been shown that another effect of radiation on living organisms is to shorten the life span of both the exposed individual and the unexposed offspring. The small amount of natural background radiation which an individual receives has a negligible effect on the life span. A person who makes a career of working in the radiation industry can, in some instances, be exposed to a maximum of 50 roentgens in ten years. In terms of shortening one's life span, this amount of radiation presents the same hazard as riding in an automobile. Figure 21-23 compares radiation hazard with other hazards in shortening life span. Note that statistically, smoking one pack of cigarettes a day seems to cut life expectancy by about 9 years, equivalent to 15 minutes per cigarette.

We may conclude that natural environmental high-energy radiation does not present any serious health hazard to the individual but we must remember that all high-energy radiation may produce genetic effects which are cumulative to both man and his offspring. Thus, in order to minimize damage to human genetic stock, adequate precautions should be taken wherever and whenever nuclear reactions and high-energy radiation are involved in man's activities.

QUESTIONS

1. Contrast the source and magnitude of energies liberated in chemical reactions and nuclear transformations.

2. Write a short paragraph to explain the meaning of the equation $\Delta E = \Delta mc^2$.

3. (a) What is the meaning of "binding energy"? (b) Which elements in the Periodic Table have the greatest binding energy per nucleon? (c) Which elements have the most stable nuclei?

4. What are the characteristics of particles emitted by radioactive nuclei: (a) α particle, (b) β particle, (c) γ ray, (d) β^+ ray?

5. Explain the fact that the coulombic force of repulsion between two protons in an atomic nucleus does not disrupt the nucleus.

6. (a) How is nuclear stability related to the neutron-proton ratio? (b) Give evidence.

7. (a) What is meant by radioactivity? (b) How is it detected? (c) What particles are emitted during radioactive decay? (d) How do you distinguish between the particles in (c)?

8. Why are gamma rays more likely to be dangerous than alpha particles?

9. Complete this table.

Structure of the Nucleus

Symbol	p	n	Symbol	p	n
${}^{1}_{1}H$			${}^{17}_{8}O$		
${}^{2}_{1}H$	1	1	${}^{39}_{19}K$	19	20
${}^{4}_{2}He$			${}^{40}_{19}K$		
${}^{5}_{2}He$			${}^{235}_{92}U$		
${}^{6}_{3}Li$	3	3	${}^{239}_{92}U$	92	147
${}^{7}_{3}Li$			${}^{239}_{93}Np$		
${}^{16}_{8}O$			${}^{239}_{94}Pu$		

10. Identify the particle resulting from these natural decay processes.

(a) ${}^{238}_{92}U \longrightarrow ? + {}^{4}_{2}He$

(b) ${}^{214}_{83}Bi \longrightarrow ? + {}^{0}_{-1}e$

(c) ${}^{239}_{93}Np \longrightarrow ? + {}^{0}_{+1}e$

(d) ${}^{226}_{88}Ra \longrightarrow {}^{222}_{86}Rn + ?$

(e) ${}^{239}_{92}U \longrightarrow {}^{239}_{93}Np + ?$

(f) ${}^{234}_{91}Pa \longrightarrow {}^{234}_{92}U + ?$

Ans. (b) ${}^{214}_{84}$Po.

11. Explain why a neutrino is required in beta-particle emission.

12. (a) Define half-life. (b) Explain why radioactive elements like strontium-90 are not likely to occur naturally.

13. (a) What is a curie? (b) What is a microcurie? (c) How does this unit help an investigator know how much of the substance he has?

14. Uranium-228 has a half-life of 10 minutes and undergoes alpha (α) emission when it decays. (a) Write the nuclear equation for the decay of U-228. (b) If you had a one-g sample, how much U-228 would you have after 10 minutes, 20 minutes, one hour, one day? (c) Write a generalized formula for the answer to part (b) of this question.

15. Explain how it is possible to determine the age of uranium-bearing rocks.

16. If radioactive wastes must be stored for seven half-lives before disposal (a) how long must P-32 be held? (See half-life table.) (b) What fraction of the P-32 originally set aside remains after this time?

17. Why are neutrons useful as "atomic bullets?"

18. Using Graham's Law (Index), calculate the ratio of the rates at which ${}^{235}UF_6$ diffuses compared with ${}^{238}UF_6$.

19. If S-32 is exposed to a neutron field, it captures a neutron and disintegrates, forming radioactive P-32. Write a nuclear equation for the reaction.

20. The carbon-nitrogen cycle has been used to explain the source of the sun's energy. It is

$$
\begin{aligned}
{}^{12}_{6}C + {}^{1}_{1}H &\longrightarrow {}^{13}_{7}N + \gamma \\
{}^{13}_{7}N &\longrightarrow {}^{13}_{6}C + {}^{0}_{1}e \\
{}^{13}_{6}C + {}^{1}_{1}H &\longrightarrow {}^{14}_{7}N + \gamma \\
{}^{14}_{7}N + {}^{1}_{1}H &\longrightarrow {}^{15}_{8}O + \gamma \\
{}^{15}_{8}O &\longrightarrow {}^{15}_{7}N + {}^{0}_{1}e \\
{}^{15}_{7}N + {}^{1}_{1}H &\longrightarrow {}^{12}_{6}C + {}^{4}_{2}He
\end{aligned}
$$

Show that this sequence is equivalent to

$$4{}^{1}_{1}H \longrightarrow {}^{4}_{2}He + 2{}^{0}_{+1}e$$

21. (a) Describe three uses of radioactive isotopes in research or industry. (b) What precautions are necessary when working with these materials?

Appendix 1

Mathematical Operations and Concepts Used in Chemistry

Appendix 1

Mathematical Operations and Concepts Used in Chemistry

A. Scientific Notation

Scientific work frequently involves use of large and small numbers. Multiplying and dividing such numbers as 0.000 000 065 3 and 605 000 000 000 would be extremely tedious operations unless these numbers were expressed as powers of 10. A convenient method of expressing these large and small numbers in exponential form is called *scientific notation.* The large or small number in this system is expressed as a number between 1 and 10 with some power of 10 used to indicate the placement of the decimal point. Multiplying by a positive power of 10 indicates that the decimal point must be moved to the right.

$$1.48 \times 10^6 = 1\,480\,000$$
$$1.23 \times 10^2 = 123$$
$$7.66 \times 10^9 = 7\,660\,000\,000$$

Multiplying by a negative power of 10 indicates that the decimal point must be moved to the left.

$$1.66 \times 10^{-4} = 1.66 \times \frac{1}{10^4}$$
$$= 1.66 \times 0.0001$$
$$= 0.000\,166$$

1. Multiplication: When multiplying numbers expressed in scientific notation the exponents of 10 are added.

$$(1.40 \times 10^4)(2.00 \times 10^6)$$
$$= 2.80 \times 10^4 \times 10^6 = 2.80 \times 10^{10}$$
$$(8.64 \times 10^4)(6.42 \times 10^{-6})$$
$$= 8.64 \times 6.42 \times 10^4 \times 10^{-6}$$
$$= 55.4 \times 10^{4+(-6)}$$
$$= 55.4 \times 10^{-2}$$
$$= 5.54 \times 10^{-1}$$

2. Division: When dividing numbers expressed in scientific notation the exponents of 10 are subtracted.

$$\frac{1.44 \times 10^7}{1.44 \times 10^3} = 1.00 \times 10^4$$
$$\frac{6.88 \times 10^6}{2.00 \times 10^7} = 3.44 \times 10^{-1}$$
$$\frac{4.50 \times 10^{-6}}{6.66 \times 10^{-7}} = 0.676 \times 10^{-6-(-7)}$$
$$= 0.675 \times 10^{-6+7}$$
$$= 0.675 \times 10^1$$
$$= 6.75$$

3. Addition or subtraction: Numbers expressed as powers of 10 cannot be added or subtracted unless the powers of ten are made equal.

$$1.55 \times 10^4 + 1.643 \times 10^5$$
$$= 0.155 \times 10^5 + 1.643 \times 10^5$$
$$= 1.798 \times 10^5$$
$$2.77 \times 10^3 - 8.6 \times 10^2$$
$$= 2.77 \times 10^3 - 0.86 \times 10^3$$
$$= 1.9 \times 10^3$$

The rule of division of exponents tells us that any number raised to the zero power is equal to 1.

$$\frac{5^2}{5^2} = 5^{2-2} = 5^0 = 1$$
$$\frac{X^m}{X^m} = X^{m-m} = X^0 = 1$$
$$\frac{8 \times 10^3}{4 \times 10^3} = 2 \times 10^0 = 2$$
$$\frac{7.6}{4.0 \times 10^{-3}} = \frac{7.6 \times 10^0}{4.0 \times 10^{-3}} = 1.9 \times 10^{0-(-3)}$$
$$= 1.9 \times 10^3$$

4. Extracting roots of numbers expressed in scientific notation: The exponent is divided by 2 in taking the square root and by 3 in taking the cube root.

$$\sqrt{1.6 \times 10^7} = \sqrt{16 \times 10^6}$$
$$= \sqrt{16} \times \sqrt{10^6} = 4.0 \times 10^3$$

$$\sqrt[3]{2.7 \times 10^{10}} = \sqrt[3]{27 \times 10^9}$$
$$= \sqrt[3]{27} \times \sqrt[3]{10^9}$$
$$= 3.0 \times 10^3$$

B. Derived Values

In many parts of scientific work, measurements obtained to a given degree of certainty are used to derive other data. For example, to determine the area of a rectangle, the length and width are measured to some degree of certainty. Their product is the area. The area represents a derived value since it was not determined by a single measurement. Uncertainty in measuring the length and width is compounded when determining the area. If the length were measured as 25.1 cm and the width as 12.4 cm, the uncertainty in each measurement would be ± 0.1 cm. Thus, the area lies between 25.0 cm $\times$ 12.3 cm and 25.2 cm $\times$ 12.5 cm. We shall follow the general rule that derived values can be no more certain than the original measured data. Derived values should be expressed with the same number of significant figures as the original measurements. The area of the above rectangle would be properly expressed to three significant figures. This is the same number of significant figures contained in the measurements from which the area was derived. Multiplying 24.1 cm $\times$ 12.4 cm, we obtain 311.24 cm^2. This should be rounded off to three significant figures, so our answer reads 311 cm^2.

Points to Remember Regarding Significant Figures

1. In addition and subtraction, a figure in the answer is significant only if each number in the problem contributes a significant figure at that particular decimal level.

Example:

```
25.632
 1.48        The last figure in
 2.766       these two numbers
------       are not significant
29.87
```

2. In multiplication and division, the answer can be expressed to only the same number of significant figures as that measurement containing the fewest number of significant figures.

(a) $\dfrac{12.05}{3.1} = 3.9$

(b) $6.42 \times 10^{-3} \times 4.0111 \times 10^{-5}$
$= 25.8 \times 10^{-8}$

3. When a mathematical operation requires a number of steps, determine the number of significant figures the answer should have and then round off each figure in the problem to one more significant figure. After solving the problem, round off the answer to the proper number of significant figures.

(a) Evaluate

$$21 \times \frac{763.4}{821} \times \frac{273}{381.6}$$

(b) Round off all parts of the problem to three significant figures.

$$21 \times \frac{763}{821} \times \frac{273}{382}$$

(c) Carry out the indicated operations and obtain the answer.

13.9

(d) Round off to two significant figures.

14

4. Pure mathematical numbers, which do not represent measurements, have an infinite number of significant figures. In contrast, if we count 12 beakers, the number 12 does not represent an approximation but means exactly 12.

5. Rounding off numbers.

(a) When the last digit of a figure is greater than 5, increase the last remaining digit by 1. 20.147 to 4 significant figures = 20.15

(b) When the last digit is less than 5, it can be dropped, leaving the last remaining digit unchanged. 12.33 to 3 significant figures = 12.3

(c) When the last digit is 5, the number

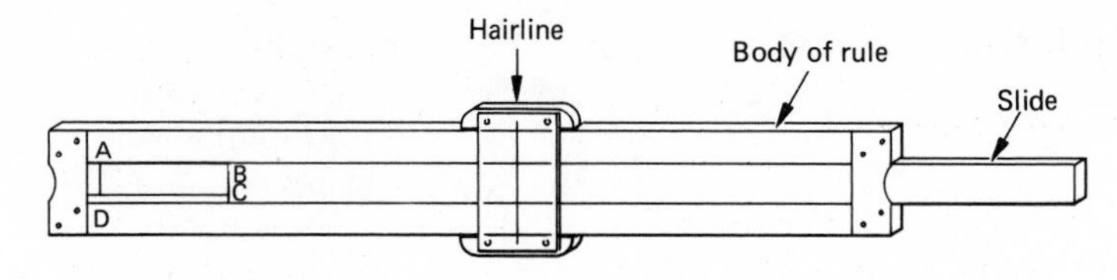

Fig. 1

is rounded off so that the last remaining digit is even.

21.5 to 2 significant figures = 22

26.25 to 3 significant figures = 26.2

6. Zeros are significant only if they are not being used exclusively to locate the decimal point.

Example 1: 0.0032 mm contains only 2 significant figures. The zeros are needed only to locate the decimal point and are, therefore, not significant.

Example 2: 302 ml contains 3 significant figures. The zero in this case indicates the number of tens.

Example 3: 25.0000 cm contains 6 significant figures. The zeros are all significant since they are not needed to locate the decimal point. The last zero in this case indicates an estimate made to 0.0001 cm.

Example 4: 6000 mm contains only 1 significant figure as written. This measurement most likely could have been made with greater certainty than 1 significant figure. The person making the measurement could indicate the number of significant figures by expressing the number in scientific notation.

6×10^3	1 significant figure
6.0×10^3	2 significant figures
6.00×10^3	3 significant figures
6.000×10^3	4 significant figures

C. The Slide Rule

1. Introduction

The slide rule was developed as an outgrowth of the study of logarithms. Although it is a mathematical tool, it is often used by chemists and engineers. The slide rule is a useful device for carrying out very close approximations of multiplication, division, and finding roots, as well as various combinations of these processes. Most data that the chemist deals with involve measurements which, themselves, are limited in precision. The obtainable accuracy of an ordinary slide rule is about 1 part in 1000 or one-tenth of one percent. This accuracy probably exceeds or equals that of the instruments used in obtaining the data, and is, therefore, more than sufficient.

2. Parts of the Slide Rule

The A, D, K, and L scales are on the body. The B and C scales are on the slide. The A and B scales are identical as are the C and D scales. The C and D scales begin and end with a 1. You must keep in mind that 1 can represent 1, 10, 100, or 1000, or even 1 000 000. The operator of the slide rule must supply the decimal point. The 1 at the left of the movable C scale is called the left index. The 1 at the right of the C scale is called the right index.

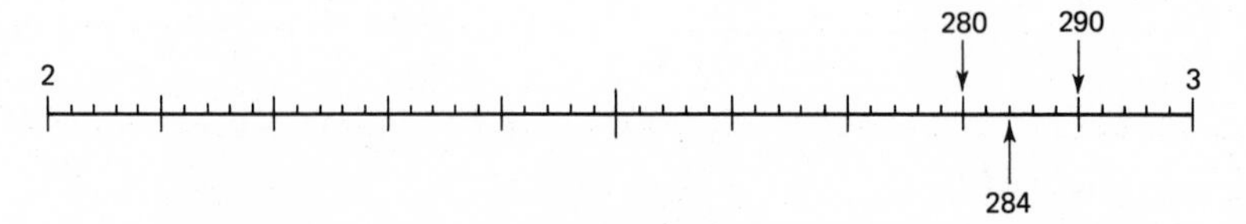

Fig. 2

3. Reading the Scales

The C and D scales are separated into 9 major divisions. These divisions are set off by the numbers 1, 2, 3, 4, 5, 6, 7, 8, 9, and 1, which are called primary marks. Each of these major divisions is further divided into 10 subdivisions but not all in the same way. Between 1 and 2 the subdivisions are numbered and each subdivision is broken into 10 parts. Between 2 and 4, the subdivisions are broken into 5 parts. For the remainder of the scale the subdivisions are broken into only two parts. Let us locate the number 284 on the D scale. You can see that there are 5 divisions between 280 and 290. Note that the hairline is placed $\frac{4}{10}$ or $\frac{2}{5}$ of the way between 280 and 290. We must keep in mind that our location of 284 has no bearing on the decimal point. This could be 0.0284, 2.84, or 28 400, etc. Locate these numbers with the hairline on the D scale. After you have located the number on your slide rule, compare your answer with Fig. 3.

(a) 181,	(f) 694,
(b) 260,	(g) 102,
(c) 337,	(h) 120,
(d) 405,	(i) 1055,
(e) 521,	(j) 794,

Fig. 3 shows

(a) 181 as $\frac{1}{10}$ of the way between 180 and 190.

(b) 260 is $\frac{6}{10}$ of the way between 200 and 300.

(c) 337 is midway between 336 and 338. The subdivisions in this portion of the scale between 330 and 340 are in $\frac{1}{5}$ths, each mark being $\frac{1}{5}$ or $\frac{2}{10}$ths. 337 is, therefore, $\frac{7}{10}$ or $3\frac{1}{2}/5$ of the way between 330 and 340.

(d) 405 is $\frac{1}{2}$ way between 400 and 410.

(e) 521 is located $\frac{1}{10}$ of the way between 520 and 530.

Note that you must estimate the last digit since there are only 2 divisions between 520 and 530. Notice that this portion of the scale is different than that between 1 and 2. There is a subdivision mark to locate the 181 as in part (a) of the exercise.

(f) 694 is $\frac{4}{10}$ of the way between 690 and 700.

(g) 102 is $\frac{2}{10}$ of the way between 100 and 110. The location of 102 is frequently confused by beginners with the location of 120.

(h) 120 is located $\frac{2}{10}$ of the way between 1 and 2.

(i) 1055 is located half-way between 1050 and 1060. A fourth significant figure can be found on this portion of the scale. The same accuracy is not possible in locating 9055.

(j) 794 is $\frac{4}{10}$ of the way between 790 and 800.

The reading of scales is by far the most important step to master in learning to use a slide rule. The mechanics of carrying out the operations of multiplication, division, and finding roots is simple if one is able to locate numbers with ease.

4. Multiplication

Multiplication is carried out with the C and D scales. Example: Multiply 2 by 3.

(a) Place the index (either end of movable C scale) over the first factor on the D scale.

(b) Place hairline over the second factor on the C scale.

(c) Read answer under hairline on D scale. See Fig. 4.

Notice that the *left* index of the C scale is placed over the 3 of the C scale and the answer is read under the hairline on the D scale. If

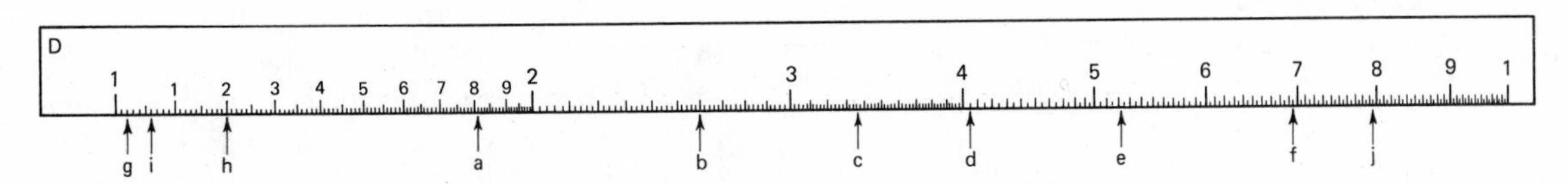

Fig. 3

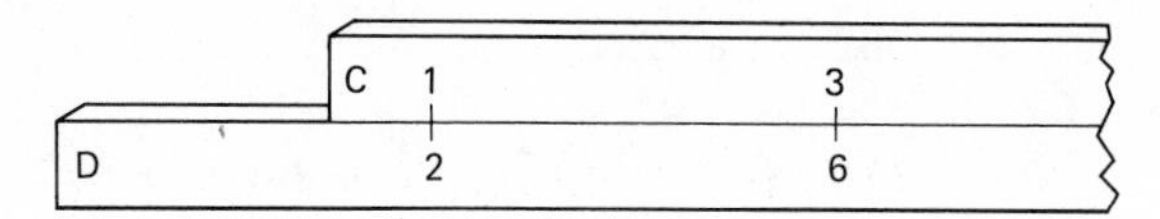

Fig. 4

the right index of the C scale had been used, you can see that the 3 of the C scale would have been off the D scale and it would have been impossible to find the answer. If this happens, simply move the C scale so the left index is over the 2 of the D scale. *Either index of the C scale may be used.* If it happens that you try to use the left index in multiplying and your answer falls off the D scale, then simply switch to the right index. In multiplying 2 × 9 you will find that only the right index of the C scale can be used. In multiplying 2 × 3 only the left index of the C scale can be used. The decimal point is best determined by making a rough calculation.

Follow-up Problems	Answers
(a) 4.5 × 1.5	(a) 6.75
(b) 1.75 × 5.5	(b) 9.63
(c) 4.33 × 11.5	(c) 49.8
(d) 0.0020 5 × 408	(d) 0.836
(e) 3.05 × 5.17	(e) 15.8
(f) 5.56 × 634	(f) 3530
(g) 743 × 0.0567	(g) 42.1
(h) 0.0495 × 0.0267	(h) 0.001 32
(i) 42.3 × 32.7	(i) 1380
(j) 5.78 × 6.35	(j) 36.7
(k) 0.0634 × 53 600	(k) 3400

5. Division

The process of division is also carried out using the C and D scales.

Example: Divide 6 by 2

(a) Express the division problem as a fraction and identify the numerator and denominator. 6 ÷ 2 may be expressed as $\frac{6}{2}$.

(b) Place the hairline over the numerator on the D scale.

(c) Place the denominator of the movable C scale over the numerator.

(d) Read the answer under the index on the D scale.

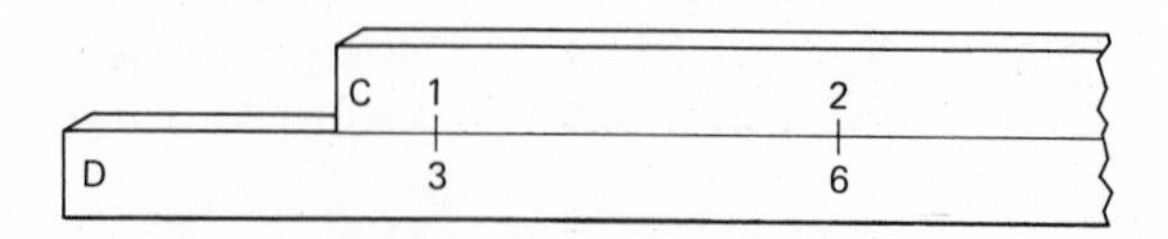

Fig. 5

Follow-up Problems	Answers
(a) 87.5 ÷ 37.7	(a) 2.32
(b) 3.75 ÷ 0.0227	(b) 165.2
(c) 0.685 ÷ 8.93	(c) 0.0767
(d) 0.003 77 ÷ 5.29	(d) 0.000 713
(e) 871 ÷ 0.468	(e) 1861
(f) 3.14 ÷ 2.72	(f) 1.15
(g) 3.42 ÷ 81.7	(g) 0.0419
(h) 0.004 65 ÷ 7.36	(h) 0.000 063 2
(i) 946 ÷ 0.0677	(i) 13.970
(j) 0.007 55 ÷ 0.338	(j) 0.0223
(k) 0.0948 ÷ 7.23	(k) 0.0131
(l) 149 ÷ 63.3	(l) 2.35

6. Combined multiplication and division

The slide rule is especially useful for solving problems of combined multiplication and division. These problems are often encountered in chemistry and are very time-consuming to solve arithmetically.

Example: Find the value of $\frac{2 \times 6}{3}$

(a) Place the hairline on 2 of the D scale.

(b) Move the 3 of the C scale under the hairline so that it is directly over the 2 of the D scale.

(c) Move the hairline to the 6 on the C scale.

(d) Read the answer under the hairline on the D scale. You will notice that in the above steps we have first divided the 2 by 3 and then multiplied the result by 6. An alternative method would have been to multiply 2 by 6, then divide by 3.

Example: Find the value of $\frac{16 \times 3 \times 12}{4 \times 6}$

(a) Place the hairline over the 16 on the D scale.

(b) Bring the 4 of the C scale under the hairline.

(c) Place the hairline over 3 of the C scale.

(d) Bring the 6 of C scale under the hairline.

(e) Note: at this point the next logical step is to move the hairline to 12 of C scale. You can see that the 12 of the C scale is off the D scale so it is not possible to perform this step. Therefore move the hairline to the right index of C scale. Then bring the left index under the hairline. Now move the hairline to the 12 of the C scale.

(f) Read the answer, 24, under the hairline on D scale. If there are more terms in the denominator of the problem than in the numerator, it may help to place the factor of "1" in the numerator until there is one more term in the numerator than in the denominator.

$$\frac{4 \times 3}{8 \times 6 \times 5 \times 2} = \frac{4 \times 3 \times 1 \times 1 \times 1}{8 \times 6 \times 5 \times 2}.$$

Then, alternately divide and multiply until the final answer is obtained.

One method of locating the position of the decimal point in an answer to a problem solved by using a slide rule involves approximating the magnitude of the answer. The simplest way to estimate the magnitude of an answer is to express all numbers as one digit numbers multiplied by ten raised to the proper power.

Example: Use the slide rule to solve the following problem and then locate the decimal point by approximating the magnitude of the answer.

Problem:

$$(1.87) \times \frac{(293)}{(273)} \times \frac{(760)}{(776)} \times (22.4) = 440$$

Approximate answer:

$$2 \times \frac{(3 \times 10^2)}{(3 \times 10^2)} \times \frac{(8 \times 10^2)}{(8 \times 10^2)} \times 2 \times 10^1 = 40$$

Correct answer: 44.0

Follow-up Problems	**Answers**
1. $\frac{12 \times 6.0}{4.0}$	1. 18
2. $\frac{1250 \times .0681}{12.1}$	2. 7.04
3. $\frac{18}{12 \times 3.0}$	3. 0.50
4. $12 \times 16 \times 14$	4. 2690
5. $\frac{47.3 \times 3.14 \times 18.0}{12.0}$	5. 223
6. $\frac{22.4 \times 760 \times 441}{12.0}$	6. 626 000
7. $\frac{885 \times 721 \times 1094}{780 \times 527}$	7. 17 00
8. $\frac{12.8 \times 1071 \times 672}{883 \times 11.4}$	8. 915
9. $\frac{3.74 \times 6.85 \times 7.75 \times 327}{78.8 \times 654 \times 3.82}$	9. 0.300
10. $\frac{25 \times 36 \times 116}{1.5 \times 2.0 \times 48 \times 220}$	10. 3.3

7. Square Root

In extracting roots we always keep the slide in place and simply move the hairline over a number on either the A scale or the D scale. You will notice that the A scale is in two parts which we will refer to as A left or A right. Each portion of the A scale is a miniature D scale only one-half as large. See Fig. 6.

Before extracting square roots, express in exponential notation the number whose square root is to be taken as a number between 1 and 100 times 10 raised to an *even* power.

(a) If the number has an odd number of digits, place the hairline over the number on A left. Read answer under the hairline on the D scale.

(b) If the number has an even number of digits, place the hairline over the number on

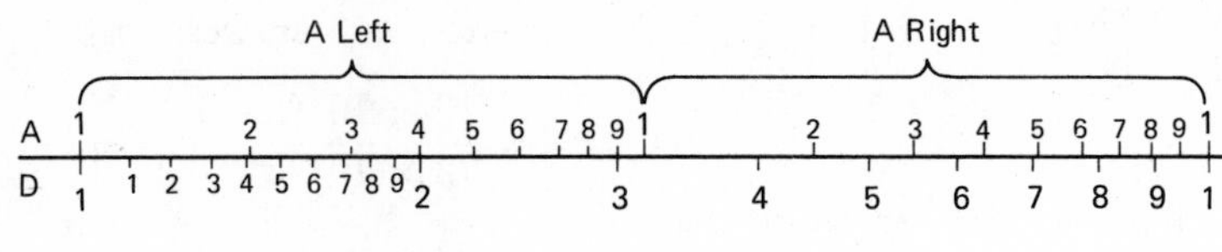

Fig. 6

A right. Read answer under the hairline on the D scale.

Example 1: Find the square root of 4

(a) Since 4 contains an odd number of digits, place the hairline over the 4 on A left.

(b) Read the answer on D scale under the hairline. You can see that 400 (4×10^2) also contains an odd number of digits and A left would also be used. The square root of 4 is 2 and the square root of 400 is $\sqrt{4} \times \sqrt{10^2} = 2 \times 10 = 20$. As in multiplication and division, the operator must perform a rough calculation to supply the decimal point.

Example 2: Find the square root of 40

(a) Since 40 contains an even number of digits, place the hairline over the 40 on A right.

(b) Read the answer under the hairline on the D scale. The answer is 6.32. Had you been looking for the square root of 4000, you would also have used A right and the answer would be 63.2.

Example 3: Find the square root of 0.09

(a) Converting 0.09 to exponential notation we obtain 9×10^{-2} so we use A left. The answer is $\sqrt{9} \times \sqrt{10^{-2}} = 3 \times 10^{-1} = 0.3$

Follow-up Problems (square roots)

Find the value of	**Answers**
1. $\sqrt{25}$	1. 5
2. $\sqrt{32}$	2. 5.65
3. $\sqrt{75}$	3. 8.66
4. $\sqrt{5720}$	4. 75.6
5. $\sqrt{2.06}$	5. 1.43
6. $\sqrt{0.335}$	6. 0.58
7. $\sqrt{0.0049}$	7. 0.07
8. $\sqrt{0.049}$	8. 0.221
9. $\sqrt{144}$	9. 12
10. $\sqrt{14.4}$	10. 3.8

8. Finding Logarithms

Most slide rules have an L scale which can be used in conjunction with the D scale to find a common logarithm of a number. When the hairline is placed over any number on the D scale, it is also over the common logarithm of that number on the L scale. We may think of a logarithm of a number as an exponent to the base 10. Since $10^0 = 1$ and $10^1 = 10$, we can see that logarithms (exponents) of numbers between 1 and 10 lie between 0 and 1. Thus, 0 on the L scale corresponds to 1 on the D scale and 1 on the L scale corresponds to 10 on the D scale. Figure 7 shows the relationship of the two scales.

Example 1: Find the logarithm of 2.4

(a) Place the hairline over 2.4 on the D scale.

(b) Read the answer, 0.38, on the L scale.

Example 2: Find the logarithm of 50

(a) Place the hairline over 5 of the D scale.

(b) Read the answer, 0.699, on the L scale.

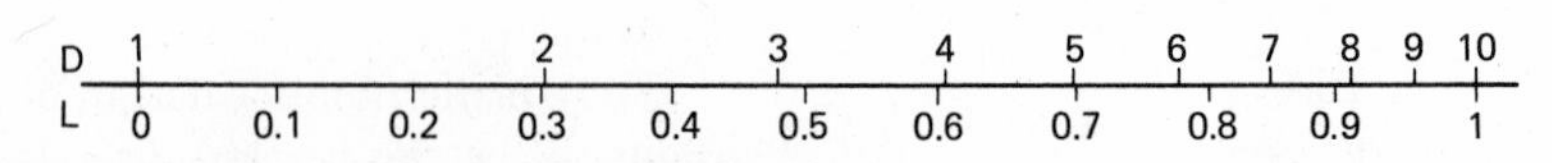

Fig. 7

Note that 0.699 is actually the log of 5. We are finding the log of 50.

$$50 = 10 \times 5$$
$$10^1 = 10 \text{ and } 10^{0.699} = 5$$

The rule for multiplying exponential numbers expressed to the same base is to add exponents.

$$50 = 10 \times 5 = 10^1 \times 10^{0.699} = 10^{1.699}$$
The logarithm of 50 = 1.699

Example 3: Find the logarithm of 0.003 47

(a) Express 0.003 47 as 3.47×10^{-3}.
(b) Place the hairline over 3.47 of the D scale.
(c) Read the answer, 0.54 on the L scale. The log of 10^{-3} is -3; therefore, the log of 0.003 47 = $-3 + 0.54$ or -2.46.

Example 4: Find the number whose logarithm is 2.88

(a) The number is equal to $10^{2.88}$ or $10^{0.88} \times 10^2$.
(b) Place the hairline over 0.88 on the L scale.
(c) Read the number under the hairline on the D scale. This number is 7.59. $10^{0.88} = 7.59$.
(d) The number whose logarithm is 2.88 is, therefore, 7.59×10^2 or 759.

Example 5: Find the number whose logarithm is -4.66.

(a) The number is equal to $10^{-4.66}$ which may be expressed as $10^{-5} \times 10^{0.34}$.
(b) Place the hairline over 0.34 on the L scale.
(c) Read the number, 2.18, on the D scale under the hairline. $10^{0.34} = 2.18$.
(d) The number whose logarithm is -4.66 is, therefore, 2.18×10^{-5} or 0.000 021 8.

Follow-up Problems

Find the logarithm of these numbers.

	Answers
1. 38.6	1. 1.586
2. 3.45	2. 0.538
3. 383	3. 2.583
4. 5710	4. 3.756
5. 0.654	5. -0.185
6. 0.0025	6. -2.602
7. 0.000 39	7. -3.409

Table 1 on page 734 shows the values of numbers on the D scale to 3 significant figures corresponding to the primary divisions of the L scale.

Find the number corresponding to these logarithms. In other words, what is the value of 10 raised to each of these powers?

1. 3.44,
2. 1.88,
3. 0.58,
4. -1.44,
5. -8.22,
6. 2750,
7. 76,
8. 3.8,
9. 0.0363,
10. 6.02×10^{-9}.

D. Quadratic Equations

A quadratic equation is any equation with a variable raised to an exponent of two. The equation $x^2 = 4$ is a quadratic expression and, like all quadratics, has two solutions for the variable x. Either $+2$ or -2 satisfies the value for x.

All quadratic equations can be expressed in the standard form, $ax^2 + bx + c = 0$. The general solution to all quadratic equations in this form is

$$x = \frac{-b \pm \sqrt{b^2 - 4ac}}{2a}$$

For the expression $x^2 = 4$, $a = 1$, $b = 0$, and $c = -4$.

In the expression $x^2 + 0.3x - 0.5 = 0$, $a = 1$, $b = 0.3$, and $c = -0.5$. In dealing with any physical reality, one of the possible roots may turn out to be meaningless, in which case it is simply discarded. For instance, if the term x in a given expression represents concentration, and one root of the equation turns out to be negative, you may simply disregard this root since it can have no real significance. It is impossible to have a negative concentration.

TABLE 1

The exponents (logarithms) to the base 10 are found on the L scale.		These correspond to the numbers between 1 and 10 which are found on the D scale.
10^0	=	1.00
$10^{0.1}$	=	1.26
$10^{0.2}$	=	1.58
$10^{0.3}$	=	1.99
$10^{0.4}$	=	2.51
$10^{0.5}$	=	3.16
$10^{0.6}$	=	3.99
$10^{0.7}$	=	5.01
$10^{0.8}$	=	6.30
$10^{0.9}$	=	7.94
10^1	=	10.0

An example of the use of the quadratic equation in connection with a chemistry problem is shown.

When ethyl alcohol and acetic acid react and form an ester, an equilibrium becomes established:

$$\underset{\text{alcohol}}{C_2H_5OH(l)} + \underset{\text{acid}}{CH_3COOH(l)} \rightleftharpoons \underset{\text{ester}}{CH_3COOC_2H_5(l)} + \underset{\text{water}}{H_2O(l)}$$

At room temperature the equilibrium constant for this reaction has a value of 4.0. Let us calculate the number of moles of ester produced when 3 moles of alcohol is reacted with one mole of acid in a one-liter container.

Solution

Let x = the number of moles of ester formed. Then

$$\underset{3-x}{C_2H_5OH(l)} + \underset{1-x}{CH_3COOH(l)} \rightleftharpoons \underset{x}{CH_3COOC_2H_5(l)} + \underset{x}{H_2O(l)}$$

$$K_e = \frac{[x][x]}{[3-x][1-x]} = 4$$

$$x^2 = 4(3 - 4x + x^2)$$

$$3x^2 - 16x + 12 = 0$$

Substitute in the quadratic equation

$$x = \frac{-b \pm \sqrt{b^2 - 4ac}}{2a}$$

$$= 16 \pm \frac{\sqrt{(16)^2 - 4(3)(12)}}{2.3}$$

$$= \frac{16 + 10.6}{6}$$

$$= 4.4 \text{ or } 0.9 \text{ moles of ester}$$

In order to produce 4.4 moles of ester, we would have needed at least 4.4 moles of each reactant. Since this amount of each reactant was not available, we must assume that this root is impossible. We shall then assume that the other root, 0.9 moles of ester, is the correct answer. Substitution of 0.9 moles/liter in the original equilibrium expression produces an equilibrium constant equal to 4.

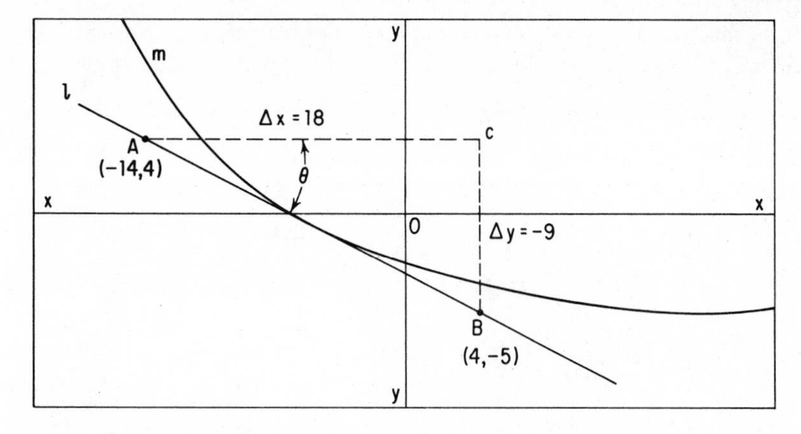

Fig. A The slope of the line, l, is given by the relationship $\frac{\Delta y}{\Delta x}$

$$\frac{\Delta y}{\Delta x} = \frac{-5 - (4)}{4 - (-14)} = \frac{-9}{18} = -\frac{1}{2}$$

E. The Slope of a Line

The slope of a line is given by the ratio of the change in x and y coordinates from one point to another on the line. We indicate the change in the y coordinate as Δy and the change in the x coordinate as Δx. In Fig. A above, points a and b are two points on line 1. The slope of this line is given as $\Delta y/\Delta x$ which in this case has a value of $-\frac{1}{2}$. The sign on the slope gives an indication of its direction of inclination. The slope of l_1 in Fig. B is negative while that of l_2 is positive.

Frequently it is desirable to find the slope of a curved line such as m in Fig. A. Obviously, the curved line does not have a constant slope at all points. Therefore, we describe its slope at a given point. The slope at a given point on a curve can be determined by drawing a line tangent to the curve at that point, and then determining the slope of the tangent. In Fig. A the slope of the curved line m at the point of tangency with line l is $-\frac{1}{2}$, the same as the slope of line l.

The slope of a line is also defined as the value of the trigonometric function called the tangent (tan). The tan of the angle theta, θ, in Fig. A, is defined as the opposite side of angle θ, cb, divided by the adjacent side of angle θ, ac. In this coordinate system the value of cb is -9 and the value of ac is 18. Thus, the tan $\theta = -\frac{1}{2}$.

F. Logarithms

A logarithm is an exponent. Logarithms have been developed as a useful tool for computations. The base number for common logarithms is 10.

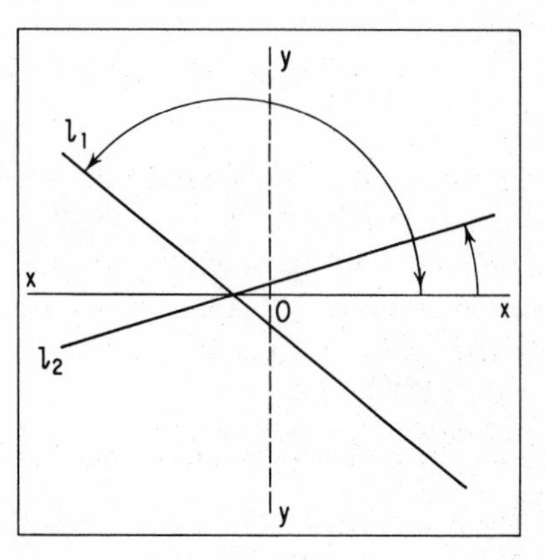

Fig. B If the inclination is greater than 90°, as in l_1, the slope is negative. If the inclination is less than 90°, as in l_2, the slope is positive.

$a^x = n$ (a is the base and x the exponent)

$10^3 = 1000$, therefore $\log_{10} 1000 = 3$

$10^2 = 100$, therefore, $\log_{10} 100 = 2$

$10^1 = 10$, therefore $\log_{10} 10 = 1$

$10^0 = 1$, therefore $\log_{10} 1 = 0$

$10^{-1} = \frac{1}{10^1} = 0.1$, therefore $\log_{10} 0.1 = -1$

$10^{-2} = \frac{1}{10^2}$

$= 0.01$, therefore $\log_{10} 0.01 = -2$

The logarithms of such numbers as 0.01, 0.1, 1, 10, 100 can be easily found by inspection. Logs that cannot be expressed as simple powers of 10 can be found on the log scale of slide rules or in tables (see Table 3 in Appendix 3). The table of logs includes four-place logs for all numbers between 1 and 100. Notice that the $\log_{10} 1 = 0$ and the $\log_{10} 10 = 1$. Therefore, the logs of all numbers between 1 and 10 lie between 0 and 1.

$$10^0 = 1; \log_{10} 1 = 0$$

The logs of all numbers between one and ten lie between 0 and 1.

$$10^1 = 10; \log_{10} 10 = 1$$

To find the logarithm of a number such as 6400, first express the number in scientific notation. Assume unless otherwise specified, that a log is taken to the base 10.

$$\log 6400 = \log (6.4 \times 10^3)$$

Since logarithms are exponents, the logs of products are added and the logs of quotients are subtracted.

$$\log ab = \log a + \log b$$

$$\log \frac{a}{b} = \log a - \log b$$

$$\log (6.4 \times 10^3) = \log 6.4 + \log 10^3$$

Since 6.4 lies between 1 and 10^1, the log 6.4 lies between zero and one. The log 6.4 is found in Appendix 3.

$$\begin{aligned} \log 6.4 &= 0.806 \text{ and } \log 10^3 = 3 \\ &= 0.806 + 3 = 3.806 \end{aligned}$$

Example 1: Find the log of 0.000 035 2

Solution

$$\begin{aligned} \log 0.000\,035\,2 &= \log (3.52 \times 10^{-5}) \\ &= \log 3.52 + \log 10^{-5} \\ &= 0.546 + (-5) \\ &= -4.454 \end{aligned}$$

One of the most common uses of logs in the field of chemistry is in connection with pH. The pH of a solution is defined as $-\log [H^+]$, where $[H^+]$ is the hydrogen-ion concentration expressed in moles per liter.

Example 2: Find the pH of a solution whose $[H^+]$ is 0.000 632.

Solution

$$\begin{aligned} \log (0.000\,632) &= \log (6.32 \times 10^{-4}) \\ &= \log 6.32 + \log 10^{-4} \\ &= 0.801 + (-4) \\ &= -3.199 \\ pH &= +3.12 \end{aligned}$$

Example 3: Find the $[H^+]$ for a solution whose pH = 6.6

Solution

$$\begin{aligned} pH &= 6.6 \\ [H^+] &= 10^{-6.6} \\ [H^+] &= 10^{-7} \times 10^{+0.4} \end{aligned}$$

To convert $10^{+0.4}$ to a nonexponential number you must find the antilog of 0.4. The number 0.4 above is an exponent and, therefore, a logarithm. The antilog is found by looking up the log in the body of the table, then finding the number which corresponds to it. The antilog of $\log_{10} 0.4 = 2.5$.

$$[H^+] = 10^{-7} \times 2.5 \text{ or } 2.5 \times 10^{-7}$$

Appendix 2

The Metric System

Appendix 2

The Metric System

The *metric system* is used in most countries of the world for everyday use as well as for scientific work. This system was first adopted by the French National Assembly in 1790, and soon after most other European countries also adopted it. In America, the English system is still in popular use although it lacks uniformity and relies on poor standards. The English system employs such units as the ounce, pound, ton, foot, inch, yard, quart, mile, and bushel. England has recently adopted a systematic plan for completely discarding English units and changing entirely to the metric system. The major advantage of this system is that it employs only a few basic units with fractional parts or multiples of the basic unit based upon the decimal system.

Once a basic unit, such as the meter, is defined in the metric system, all measurements of length are based on this unit. Prefixes are used to denote fractional parts or multiples of the basic unit.

The standard unit of mass is the gram. The kilogram (kg) was originally defined as the mass of 1000 cubic centimeters (cm^3) of water at 4°C. For practical purposes we will still use this definition. Accordingly, the gram (g) may be defined as the mass of 1 cm^3 of water at 4°C. The kilogram is now defined as the mass of a standard platinum-iridium cylinder at the International Bureau of Weights and Measures near Paris. The gram, which is the basic unit, is technically defined as 1/1000 of the mass of the standard kilogram. In working with small quantities, the milligram (mg) and microgram (μg) are also used.

The meter is the standard unit of length in the metric system. A fraction of a certain quadrant of the earth was arbitrarily chosen to serve as the standard of length in the metric system. Later the meter was more precisely defined as the distance between two marks on a platinum-iridium bar at the International Bureau of Weights and Measures near Paris. The meter was adopted as a standard of length by the United States in 1875 and for a long time served as a satisfactory standard. The requirements for modern technology are such that this type of standard is no longer adequate. An error of only a few millionths of a centimeter in any one part of the guidance system of a space vehicle could cause it to miss its target completely. In 1960, at an international conference on weights and measures, it was determined that the meter should be redefined as 1,650,763.73 times the wavelength (λ) of the orange-red spectral line of light emitted by krypton-86. Thus, the meter is now standardized to the atom rather than to one of the earth's quadrants. This standard is reproducible anywhere on earth and is indestructible. This refinement in the definition of a meter will probably be adequate for many years but further research in the field of subnuclear particles may require us to redefine the meter again. The meter, centimeter, and millimeter are much too large for conveniently expressing dimensions on the molecular or atomic scale. For this purpose the micron, the millimicron, and the Ångstrom are more useful. Fragments of large molecules such as proteins or viruses as seen through an electron microscope are measured in terms of millimicrons. The wavelengths of visible light, X rays, and infrared energy are measured in Ångstrom units (Å). The atomic diameters of atoms range from one to about five Ångstrom units.

Appendix 3

Useful Tables

Appendix 3

Useful Tables

TABLE 1 IMPORTANT CONSTANTS

Constant	Symbol	Value
Charge on the electron	e	4.8029×10^{-10} esu
		1.6021×10^{-19} coul
Electron rest mass	M_e	9.1091×10^{-28} g
		0.000 548 597 amu
Proton rest mass	M_p	1.6725×10^{-24} g
		1.007 276 63 amu
Neutron rest mass	M_n	1.6748×10^{-24} g
		1.008 665 4 amu
Planck's constant	h	6.6252×10^{-27} erg sec
Boltzmann's constant	k	1.3805×10^{-16} erg °K^{-1} molecule^{-1}
Speed of light	c	$2.997\,930 \times 10^{10}$ cm sec^{-1}
Avogadro's number	N	6.0229×10^{23} molecules mole^{-1}
Gas constant	R	0.082 054 ℓ atm °K^{-1} mole^{-1}
		8.3143 j °K^{-1} mole^{-1}
		1.9872 cal °K^{-1} mole^{-1}
Faraday constant	F	9.6486×10^{4} coul eq^{-1}

TABLE 2 ELECTRONIC ARRANGEMENT OF THE ELEMENTS

Energy Levels		1	2		3			4				5				6				7
Orbitals		1*s*	2*s*	2*p*	3*s*	3*p*	3*d*	4*s*	4*p*	4*d*	4*f*	5*s*	5*p*	5*d*	5*f*	6*s*	6*p*	6*d*	6*f*	7*s*
1	Hydrogen	1																		
2	Helium	2																		
3	Lithium	2	1																	
4	Beryllium	2	2																	
5	Boron	2	2	1																
6	Carbon	2	2	2																
7	Nitrogen	2	2	3																
8	Oxygen	2	2	4																
9	Fluorine	2	2	5																
10	Neon	2	2	6																
11	Sodium	2	2	6	1															
12	Magnesium	2	2	6	2															
13	Aluminum	2	2	6	2	1														
14	Silicon	2	2	6	2	2														
15	Phosphorus	2	2	6	2	3														
16	Sulfur	2	2	6	2	4														
17	Chlorine	2	2	6	2	5														
18	Argon	2	2	6	2	6														
19	Potassium	2	2	6	2	6		1												
20	Calcium	2	2	6	2	6		2												
21	Scandium	2	2	6	2	6	1	2												
22	Titanium	2	2	6	2	6	2	2												
23	Vanadium	2	2	6	2	6	3	2												
24	Chromium	2	2	6	2	6	5	1												
25	Manganese	2	2	6	2	6	5	2												
26	Iron	2	2	6	2	6	6	2												
27	Cobalt	2	2	6	2	6	7	2												
28	Nickel	2	2	6	2	6	8	2												
29	Copper	2	2	6	2	6	10	1												
30	Zinc	2	2	6	2	6	10	2												
31	Gallium	2	2	6	2	6	10	2	1											
32	Germanium	2	2	6	2	6	10	2	2											
33	Arsenic	2	2	6	2	6	10	2	3											
34	Selenium	2	2	6	2	6	10	2	4											
35	Bromine	2	2	6	2	6	10	2	5											
36	Krypton	2	2	6	2	6	10	2	6											
37	Rubidium	2	2	6	2	6	10	2	6			1								
38	Strontium	2	2	6	2	6	10	2	6			2								
39	Yttrium	2	2	6	2	6	10	2	6	1		2								
40	Zirconium	2	2	6	2	6	10	2	6	2		2								
41	Niobium	2	2	6	2	6	10	2	6	4		1								
42	Molybdenum	2	2	6	2	6	10	2	6	5		1								
43	Technetium	2	2	6	2	6	10	2	6	6		1								
44	Ruthenium	2	2	6	2	6	10	2	6	7		1								
45	Rhodium	2	2	6	2	6	10	2	6	8		1								
46	Palladium	2	2	6	2	6	10	2	6	10										
47	Silver	2	2	6	2	6	10	2	6	10		1								
48	Cadmium	2	2	6	2	6	10	2	6	10		2								
49	Indium	2	2	6	2	6	10	2	6	10		2	1							
50	Tin	2	2	6	2	6	10	2	6	10		2	2							
51	Antimony	2	2	6	2	6	10	2	6	10		2	3							
52	Tellurium	2	2	6	2	6	10	2	6	10		2	4							

TABLE 2 ELECTRONIC ARRANGEMENT OF THE ELEMENTS (CONT'D)

Energy Levels		1	2		3			4				5				6				7
Orbitals		1*s*	2*s*	2*p*	3*s*	3*p*	3*d*	4*s*	4*p*	4*d*	4*f*	5*s*	5*p*	5*d*	5*f*	6*s*	6*p*	6*d*	6*f*	7*s*
53	Iodine	2	2	6	2	6	10	2	6	10		2	5							
54	Xenon	2	2	6	2	6	10	2	6	10		2	6							
55	Cesium	2	2	6	2	6	10	2	6	10		2	6			1				
56	Barium	2	2	6	2	6	10	2	6	10		2	6			2				
57	Lanthanum	2	2	6	2	6	10	2	6	10		2	6	1		2				
58	Cerium	2	2	6	2	6	10	2	6	10	2	2	6			2				
59	Praseodymium	2	2	6	2	6	10	2	6	10	3	2	6			2				
60	Neodymium	2	2	6	2	6	10	2	6	10	4	2	6			2				
61	Promethium	2	2	6	2	6	10	2	6	10	5	2	6			2				
62	Samarium	2	2	6	2	6	10	2	6	10	6	2	6			2				
63	Europium	2	2	6	2	6	10	2	6	10	7	2	6			2				
64	Gadolinium	2	2	6	2	6	10	2	6	10	7	2	6	1		2				
65	Terbium	2	2	6	2	6	10	2	6	10	9	2	6			2				
66	Dysprosium	2	2	6	2	6	10	2	6	10	10	2	6			2				
67	Holmium	2	2	6	2	6	10	2	6	10	11	2	6			2				
68	Erbium	2	2	6	2	6	10	2	6	10	12	2	6			2				
69	Thulium	2	2	6	2	6	10	2	6	10	13	2	6			2				
70	Ytterbium	2	2	6	2	6	10	2	6	10	14	2	6			2				
71	Lutetium	2	2	6	2	6	10	2	6	10	14	2	6	1		2				
72	Hafnium	2	2	6	2	6	10	2	6	10	14	2	6	2		2				
73	Tantalum	2	2	6	2	6	10	2	6	10	14	2	6	3		2				
74	Tungsten	2	2	6	2	6	10	2	6	10	14	2	6	4		2				
75	Rhenium	2	2	6	2	6	10	2	6	10	14	2	6	5		2				
76	Osmium	2	2	6	2	6	10	2	6	10	14	2	6	6		2				
77	Iridium	2	2	6	2	6	10	2	6	10	14	2	6	9		0				
78	Platinum	2	2	6	2	6	10	2	6	10	14	2	6	9		1				
79	Gold	2	2	6	2	6	10	2	6	10	14	2	6	10		1				
80	Mercury	2	2	6	2	6	10	2	6	10	14	2	6	10		2				
81	Thallium	2	2	6	2	6	10	2	6	10	14	2	6	10		2	1			
82	Lead	2	2	6	2	6	10	2	6	10	14	2	6	10		2	2			
83	Bismuth	2	2	6	2	6	10	2	6	10	14	2	6	10		2	3			
84	Polonium	2	2	6	2	6	10	2	6	10	14	2	6	10		2	4			
85	Astatine	2	2	6	2	6	10	2	6	10	14	2	6	10		2	5			
86	Radon	2	2	6	2	6	10	2	6	10	14	2	6	10		2	6			
87	Francium	2	2	6	2	6	10	2	6	10	14	2	6	10		2	6			1
88	Radium	2	2	6	2	6	10	2	6	10	14	2	6	10		2	6			2
89	Actinium	2	2	6	2	6	10	2	6	10	14	2	6	10		2	6	1		2
90	Thorium	2	2	6	2	6	10	2	6	10	14	2	6	10		2	6	2		2
91	Protactinium	2	2	6	2	6	10	2	6	10	14	2	6	10	2	2	6	1		2
92	Uranium	2	2	6	2	6	10	2	6	10	14	2	6	10	3	2	6	1		2
93	Neptunium	2	2	6	2	6	10	2	6	10	14	2	6	10	4	2	6	1		2
94	Plutonium	2	2	6	2	6	10	2	6	10	14	2	6	10	6	2	6			2
95	Americium	2	2	6	2	6	10	2	6	10	14	2	6	10	7	2	6			2
96	Curium	2	2	6	2	6	10	2	6	10	14	2	6	10	7	2	6	1		2
97	Berkelium	2	2	6	2	6	10	2	6	10	14	2	6	10	8	2	6	1		2
98	Californium	2	2	6	2	6	10	2	6	10	14	2	6	10	10	2	6			2
99	Einsteinium	2	2	6	2	6	10	2	6	10	14	2	6	10	11	2	6			2
100	Fermium	2	2	6	2	6	10	2	6	10	14	2	6	10	12	2	6			2
101	Mendelevium	2	2	6	2	6	10	2	6	10	14	2	6	10	13	2	6			2
102	(Nobelium)	2	2	6	2	6	10	2	6	10	14	2	6	10	13	2	6	1		2
103	Lawrencium	2	2	6	2	6	10	2	6	10	14	2	6	10	14	2	6	1		2

TABLE 3 FOUR-PLACE LOGARITHMS OF NUMBERS

n	0	1	2	3	4	5	6	7	8	9
10	0000	0043	0086	0128	0170	0212	0253	0294	0334	0374
11	0414	0453	0492	0531	0569	0607	0645	0682	0719	0755
12	0792	0828	0864	0899	0934	0969	1004	1038	1072	1106
13	1139	1173	1206	1239	1271	1303	1335	1367	1399	1430
14	1461	1492	1523	1553	1584	1614	1644	1673	1703	1732
15	1761	1790	1818	1847	1875	1903	1931	1959	1987	2014
16	2041	2068	2095	2122	2148	2175	2201	2227	2253	2279
17	2304	2330	2355	2380	2405	2430	2455	2480	2504	2529
18	2553	2577	2601	2625	2648	2672	2695	2718	2742	2765
19	2788	2810	2833	2856	2878	2900	2923	2945	2967	2989
20	3010	3032	3054	3075	3096	3118	3139	3160	3181	3201
21	3222	3243	3263	3284	3304	3324	3345	3365	3385	3404
22	3424	3444	3464	3483	3502	3522	3541	3560	3579	3598
23	3617	3636	3655	3674	3692	3711	3729	3747	3766	3784
24	3802	3820	3838	3856	3874	3892	3909	3927	3945	3962
25	3979	3997	4014	4031	4048	4065	4082	4099	4116	4133
26	4150	4166	4183	4200	4216	4232	4249	4265	4281	4298
27	4314	4330	4346	4362	4378	4393	4409	4425	4440	4456
28	4472	4487	4502	4518	4533	4548	4564	4579	4594	4609
29	4624	4639	4654	4669	4683	4698	4713	4728	4742	4757
30	4771	4786	4800	4814	4829	4843	4857	4871	4886	4900
31	4914	4928	4942	4955	4969	4983	4997	5011	5024	5038
32	5051	5065	5079	5092	5105	5119	5132	5145	5159	5172
33	5185	5198	5211	5224	5237	5250	5263	5276	5289	5302
34	5315	5328	5340	5353	5366	5378	5391	5403	5416	5428
35	5441	5453	5465	5478	5490	5502	5514	5527	5539	5551
36	5563	5575	5587	5599	5611	5623	5635	5647	5658	5670
37	5682	5694	5705	5717	5729	5740	5752	5763	5775	5786
38	5798	5809	5821	5832	5843	5855	5866	5877	5888	5899
39	5911	5922	5933	5944	5955	5966	5977	5988	5999	6010
40	6021	6031	6042	6053	6064	6075	6085	6096	6107	6117
41	6128	6138	6149	6160	6170	6180	6191	6201	6212	6222
42	6232	6243	6253	6263	6274	6284	6294	6304	6314	6325
43	6335	6345	6355	6365	6375	6385	6395	6405	6415	6425
44	6435	6444	6454	6464	6474	6484	6493	6503	6513	6522
45	6532	6542	6551	6561	6571	6580	6590	6599	6609	6618
46	6628	6637	6646	6656	6665	6675	6684	6693	6702	6712
47	6721	6730	6739	6749	6758	6767	6776	6785	6794	6803
48	6812	6821	6830	6839	6848	6857	6866	6875	6884	6893
49	6902	6911	6920	6928	6937	6946	6955	6964	6972	6981
50	6990	6998	7007	7016	7024	7033	7042	7050	7059	7067
51	7076	7084	7093	7101	7110	7118	7126	7135	7143	7152
52	7160	7168	7177	7185	7193	7202	7210	7218	7226	7235
53	7243	7251	7259	7267	7275	7284	7292	7300	7308	7316
54	7324	7332	7340	7348	7356	7364	7372	7380	7388	7396

TABLE 3 FOUR-PLACE LOGARITHMS OF NUMBERS (CONT'D)

n	0	1	2	3	4	5	6	7	8	9
55	7404	7412	7419	7427	7435	7443	7451	7459	7466	7474
56	7482	7490	7497	7505	7513	7520	7528	7536	7543	7551
57	7559	7566	7574	7582	7589	7597	7604	7612	7619	7627
58	7634	7642	7649	7657	7664	7672	7679	7686	7694	7701
59	7709	7716	7723	7731	7738	7745	7752	7760	7767	7774
60	7782	7789	7796	7803	7810	7818	7825	7832	7839	7846
61	7853	7860	7868	7875	7882	7889	7896	7903	7910	7917
62	7924	7931	7938	7945	7952	7959	7966	7973	7980	7987
63	7993	8000	8007	8014	8021	8028	8035	8041	8048	8055
64	8062	8069	8075	8082	8089	8096	8102	8109	8116	8122
65	8129	8136	8142	8149	8156	8162	8169	8176	8182	8189
66	8195	8202	8209	8215	8222	8228	8235	8241	8248	8254
67	8261	8267	8274	8280	8287	8293	8299	8306	8312	8319
68	8325	8331	8338	8344	8351	8357	8363	8370	8376	8382
69	8388	8395	8401	8407	8414	8420	8426	8432	8439	8445
70	8451	8457	8463	8470	8476	8482	8488	8494	8500	8506
71	8513	8519	8525	8531	8537	8543	8549	8555	8561	8567
72	8573	8579	8585	8591	8597	8603	8609	8615	8621	8627
73	8633	8639	8645	8651	8657	8663	8669	8675	8681	8686
74	8692	8698	8704	8710	8716	8722	8727	8733	8739	8745
75	8751	8756	8762	8768	8774	8779	8785	8791	8797	8802
76	8808	8814	8820	8825	8831	8837	8842	8848	8854	8859
77	8865	8871	8876	8882	8887	8893	8899	8904	8910	8915
78	8921	8927	8932	8938	8943	8949	8954	8960	8965	8971
79	8976	8982	8987	8993	8998	9004	9009	9015	9020	9025
80	9031	9036	9042	9047	9053	9058	9063	9069	9074	9079
81	9085	9090	9096	9101	9106	9112	9117	9122	9128	9133
82	9138	9143	9149	9154	9159	9165	9170	9175	9180	9186
83	9191	9196	9201	9206	9212	9217	9222	9227	9232	9238
84	9243	9248	9253	9258	9263	9269	9274	9279	9284	9289
85	9294	9299	9304	9309	9315	9320	9325	9330	9335	9340
86	9345	9350	9355	9360	9365	9370	9375	9380	9385	9390
87	9395	9400	9405	9410	9415	9420	9425	9430	9435	9440
88	9445	9450	9455	9460	9465	9469	9474	9479	9484	9489
89	9494	9499	9504	9509	9513	9518	9523	9528	9533	9538
90	9542	9547	9552	9557	9562	9566	9571	9576	9581	9586
91	9590	9595	9600	9605	9609	9614	9619	9624	9628	9633
92	9638	9643	9647	9652	9657	9661	9666	9671	9675	9680
93	9685	9689	9694	9699	9703	9708	9713	9717	9722	9727
94	9731	9736	9741	9745	9750	9754	9759	9763	9768	9773
95	9777	9782	9786	9791	9795	9800	9805	9809	9814	9818
96	9823	9827	9832	9836	9841	9845	9850	9854	9859	9863
97	9868	9872	9877	9881	9886	9890	9894	9899	9903	9908
98	9912	9917	9921	9926	9930	9934	9939	9943	9948	9952
99	9956	9961	9965	9969	9974	9978	9983	9987	9991	9996

Index

Index

ATOMIC NUMBERS AND MASSES

Name of element	Symbol	Atomic number	Atomic weight	Name of element	Symbol	Atomic number	Atomic weight
Actinium	Ac	89	[227]	Mercury	Hg	80	200.59
Aluminum	Al	13	26.9815	Molybdenum	Mo	42	95.94
Americium	Am	95	[243]	Neodymium	Nd	60	144.24
Antimony	Sb	51	121.75	Neon	Ne	10	20.183
Argon	Ar	18	39.948	Neptunium	Np	93	[237]
Arsenic	As	33	74.9216	Nickel	Ni	28	58.71
Astatine	At	85	[210]	Niobium	Nb	41	92.906
Barium	Ba	56	137.34	Nitrogen	N	7	14.0067
Berkelium	Bk	97	[249*]	(Nobelium)	(No)	102	[254]
Beryllium	Be	4	9.0122	Osmium	Os	76	190.2
Bismuth	Bi	83	208.980	Oxygen	O	8	15.9994
Boron	B	5	10.811	Palladium	Pd	46	106.4
Bromine	Br	35	79.909	Phosphorus	P	15	30.9738
Cadmium	Cd	48	112.40	Platinum	Pt	78	195.09
Calcium	Ca	20	40.08	Plutonium	Pu	94	[242]
Californium	Cf	98	[251*]	Polonium	Po	84	[210*]
Carbon	C	6	12.01115	Potassium	K	19	39.102
Cerium	Ce	58	140.12	Praseodymium	Pr	59	140.907
Cesium	Cs	55	132.905	Promethium	Pm	61	[147*]
Chlorine	Cl	17	35.453	Protactinium	Pa	91	[231]
Chromium	Cr	24	51.996	Radium	Ra	88	[226]
Cobalt	Co	27	58.9332	Radon	Rn	86	[222]
Copper	Cu	29	63.54	Rhenium	Re	75	186.2
Curium	Cm	96	[247]	Rhodium	Rh	45	102.905
Dysprosium	Dy	66	162.50	Rubidium	Rb	37	85.47
Einsteinium	Es	99	[254]	Ruthenium	Ru	44	101.07
Erbium	Er	68	167.26	Samarium	Sm	62	150.35
Europium	Eu	63	151.96	Scandium	Sc	21	44.956
Fermium	Fm	100	[253]	Selenium	Se	34	78.96
Fluorine	F	9	18.9984	Silicon	Si	14	28.086
Francium	Fr	87	[223]	Silver	Ag	47	107.870
Gadolinium	Gd	64	157.25	Sodium	Na	11	22.9898
Gallium	Ga	31	69.72	Strontium	Sr	38	87.62
Germanium	Ge	32	72.59	Sulfur	S	16	32.064
Gold	Au	79	196.967	Tantalum	Ta	73	180.948
Hafnium	Hf	72	178.49	Technetium	Tc	43	[99*]
Helium	He	2	4.0026	Tellurium	Te	52	127.60
Holmium	Ho	67	164.930	Terbium	Tb	65	158.924
Hydrogen	H	1	1.00797	Thallium	Tl	81	204.37
Indium	In	49	114.82	Thorium	Th	90	232.038
Iodine	I	53	126.9044	Thulium	Tm	69	168.934
Iridium	Ir	77	192.2	Tin	Sn	50	118.69
Iron	Fe	26	55.847	Titanium	Ti	22	47.90
Krypton	Kr	36	83.80	Tungsten	W	74	183.85
Lanthanum	La	57	138.91	Uranium	U	92	238.03
Lawrencium	Lw	103	[257]	Vanadium	V	23	50.942
Lead	Pb	82	207.19	Xenon	Xe	54	131.30
Lithium	Li	3	6.939	Ytterbium	Yb	70	173.04
Lutetium	Lu	71	174.97	Yttrium	Y	39	88.905
Magnesium	Mg	12	24.312	Zinc	Zn	30	65.37
Manganese	Mn	25	54.9380	Zirconium	Zr	40	91.22
Mendelevium	Md	101	[256]				

A value given in brackets denotes the mass number of the isotope of longest known half-life, or for those marked with an asterisk, a better known one. The atomic masses of most of these elements are believed to have no error greater than ± 0.5 of the last digit given.